Introduction to animal parasitology

Introduction to animal parasitology

THIRD EDITION

J. D. SMYTH
Emeritus Professor of Parasitology, University of London

With a chapter on Immunoparasitology by PROFESSOR D. WAKELIN
University of Nottingham

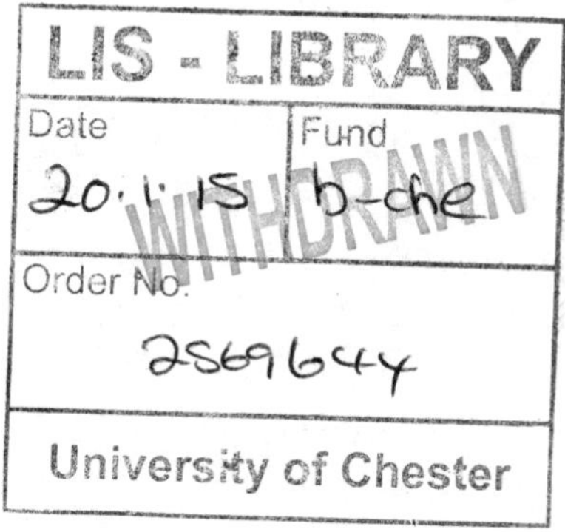
LIS - LIBRARY
Date 20.1.15 Fund b-che
WITHDRAWN
Order No. 2569644
University of Chester

CAMBRIDGE
UNIVERSITY PRESS

Published by the Press Syndicate of the University of Cambridge
The Pitt Building, Trumpington Street, Cambridge CB2 1RP
40 West 20th Street, New York, NY 10011-4211, USA
10 Stamford Road, Oakleigh, Melbourne 3166, Australia

© J. D. Smyth 1962, 1976
Third edition © Cambridge University Press 1994

First published in 1962 by Hodder and Stoughton Educational
Second edition first published in 1976 by Hodder and Stoughton Educational
Third edition published by Cambridge University Press 1994

A catalogue record for this book is available from the British Library

Library of Congress cataloguing in publication data

Smyth, J. D. (James Desmond), 1917–
Introduction to animal parasitology / by J. D. Smyth ; with a chapter on immunoparasitology by D. Wakelin. – 3rd ed.
p. cm.
Includes bibliographical references and indexes.
ISBN 0 521 41770 8 (hardback). – ISBN 0 521 42811 4 (pbk.)
1. Parasitology. I. Wakelin, Derek. II. Title.
QL757.S6 1994
591.52′49 – dc20 93-33961 CIP

ISBN 0 521 41770 8 hardback
ISBN 0 521 42811 4 paperback

Transferred to digital printing 2001

RO

This book is affectionately dedicated to my students, whose interest and curiosity largely stimulated its production

Contents

Preface to third edition

Since the Second Edition of this text was published, the spectacular advances which have been made in the fields of ultrastructure, speciation, biochemistry, molecular biology, genetics, epidemiology, neurobiology and immunology have made a major impact on all aspects of parasitology. For example, the concept of species has been revolutionised by the application of isoelectric focusing and antigenic analysis and many well-known species have been shown not to be simple, rigid species but to consist of a 'species complex' showing intraspecific variations in their metabolism or antigenic make-up: differences not revealed by simple morphological descriptions.

The text has been largely rewritten to incorporate these new concepts and extensive reference lists, covering the world literature (many in 1991–2) are provided, as in previous editions. In addition to numerous new tables, some 70 new text figures have been added and many of the older figures in the earlier edition have been updated and redrawn.

The general format of the Second Edition has been retained; as before, special attention has been paid to those organisms which are widely used for experimental models (especially for human parasites) but are often given little attention in parasitological texts. Most of the important parasites of man are also considered but no attempt is made to cover the whole field of human or animal parasitology. Morphological descriptions in general have been kept brief, systematics treated broadly and pathological effects only dealt with when they are related to some particular point of physiological interest. Parasitic Mollusca or Crustacea have been omitted, as such forms are not sufficiently common to fall within the scope of this book. Thus the treatment is confined to those groups – protozoa, platyhelminths, nematodes and acanthocephalans – which have been successful in invading the tissues or body fluids of animals. Arthropod vectors and so-called ectoparasites, such as fleas or lice, are also not covered here, since the morphology and biology of these groups have been adequately treated in a number of excellent textbooks on medical or veterinary entomology.

Special attention has been paid to presenting diagrams showing the life cycles of parasites in detail, relating these, where possible, to the properties of the environment. The preparation and presentation of such diagrams often helps to focus attention on some overlooked aspects of the life cycle and, in particular, draws attention to gaps in our knowledge about that particular parasite. As well as being the causative organisms of major human and animal diseases, parasites often serve as elegant models for the study of fundamental biological phenomena such as enzyme dynamics, membrane transport, and differentiation (especially sexual differentiation), common to many other biological fields; attention is especially drawn to those parasites which are valuable in this respect.

Acknowledgements

I am grateful to Professor D. Wakelin for agreeing to revise the chapter (Chapter 32) on the difficult and rapidly expanding field of Immunoparasitology.

I acknowledge, with gratitude, much help received from numerous colleagues in different countries, in commenting on various aspects of this text or for providing literature, illustrative material or assistance in other ways; especially Dr J. P. Dubey, Dr I. Fairweather, Professor D. W. Halton, Dr D. R. McManus, Professor R. C. A. Thompson, Professor R. C. Tinsley and Professor K. Vickerman, FRS.

My special thanks are due to my wife, Mim, who has given tireless help and support at every stage in the production of this book. I am also indebted to numerous authors, editors and publishers for permission to use material from their publications; appropriate acknowledgement is made in the text. I am especially grateful to Lynn Davy for her skilful subediting.

1

Parasitism: what is a parasite?

1.1 Animal associations

The majority of animals live independently in their natural habitats, seeking their own food materials and utilising free water and oxygen for their metabolic processes. Between some animals, however, a variety of patterns of association have developed, and these may be broadly divided into two groups: homogenetic associations – those between individuals of the same genotype; and heterogenetic associations – those between individuals of different genotype.

Individuals of the same species may form loosely united communities, such as herds of cattle or flocks of sheep, while others, such as some species of ants or bees, may form elaborately organised communities in which individual members often exhibit considerable division of labour or specialisation.

Heterogenetic associations are in general much more complex and a number of terms have been developed to describe them. Like many terms used in biology, these are essentially operational words which are definable only within broad limits and not in absolute terms. They do, nevertheless, serve a useful role in enabling us to file data into convenient, though not water-tight, compartments. Terms such as *commensalism*, *phoresis*, *symbiosis*, *mutualism* and *parasitism* have been widely used for various types of heterogenetic association, and their definition – which has always been controversial – has been much discussed by early workers such as Baer (1952), Caullery (1952) and Read (1970) and more recently by modern workers such as Kreier & Baker (1987), MacInnis (1976), Trager (1986), Schmidt & Roberts (1989), Mehlhorn (1988) and Noble *et al.* (1989). These terms were developed in a period when little data on the possible physiological and/or mathematical basis of such an association were available. Within recent years, the situation has changed somewhat, and although information is still meagre, the general increase in knowledge of animal physiology, biochemistry and population dynamics enables these associations to be considered on a broader basis. In particular, it is increasingly recognised that Parasitology is essentially a branch of Ecology and the phenomenon of parasitism can be considered as an ecological relationship between two *populations* of different species; much attention has been paid to the quantitative aspects of this relationship (Anderson, 1982, 1987; Anderson & May, 1985; Cox, 1982; Crofton, 1971*a,b*; Esch *et al.*, 1989; Kennedy, 1975, 1976, 1983; Rollinson & Anderson, 1985; Schmid & Robinson, 1972).

An important conclusion which has emerged from such studies is that – instead of parasites being randomly distributed within the host population (as might be expected) – they tend to be overdispersed, i.e. a few hosts harbour large numbers of parasites and many hosts harbour only a few. That this type of frequency distribution is a major characteristic of parasitism was first postulated by Crofton (1971*a,b*) and has since been confirmed by numerous studies on many host–parasite systems, especially those involving helminths.

A negative binomial has proved to be a good empirical description of this pattern of infection and an example of this is shown in Fig. 1.1. This shows the distribution of the metacercaria of the trematode *Diplostomum spathaceum*, which commonly occurs in the eyes of fish throughout the world (p. 259, Fig. 17.6). The lenses of a few fish are found to be infected with enormous numbers of larvae (over 400 in one fish) but most only contain relatively few, and its distribution closely follows a negative binomial pattern (Pennycuick, 1971).

In some host–parasite systems, such as the plerocercoid of the cestode *Schistocephalus solidus*, also in the stickleback (p. 315, Figs 22.9, 22.15), dispersion is reduced because some fish may die and the larvae are

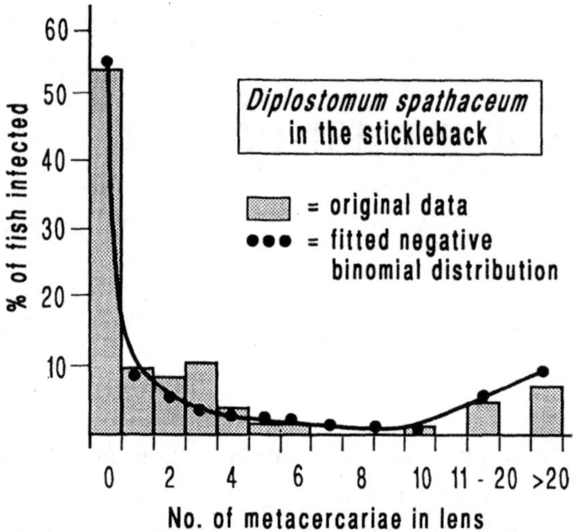

Fig. 1.1
Frequency distribution of the metacercariae of the trematode *Diplostomum spathaceum* (p. 259) in the eye (lens) of the 3-spined stickleback, *Gasterosteus aculeatus*, compared with the negative binomial expected frequency. (Modified from Pennycuick, 1971.)

generally distributed as a log normal (Fig. 22.13) rather than a negative binomial pattern.

Further quantitative aspects of the frequency distribution of parasites are discussed in detail in the references quoted above and are not considered further here.

The relativity of any definition of parasitism must also be emphasised, for this will depend on the relative weight placed on certain aspects by a particular observer. Thus different workers have emphasised factors such as (*a*) the intimacy of the association, (*b*) its pathogenic effect, (*c*) its metabolic or physiological dependence, (*d*) whether or not the host 'recognises' the parasite as 'foreign', (*e*) the ability of the parasite to 'recognise' the host site as being a suitable ecological niche. These factors are not, of course, independent, and complex interactions between them may occur in any particular host–parasite situation.

If the *degree* of association is considered first, most workers would agree that the terms *commensalism* and *phoresis* represent only loose associations whereas the terms *symbiosis*, *mutualism* and *parasitism* represent intimate associations in which the metabolism of one species is dependent to some degree on permanent association with an individual of another species. The concept developed here is that it is the metabolic dependence of one species on another which separates these intimate kinds of association very markedly from those of the looser kind. As explained further on p. 6, mutualism and symbiosis are here considered as special cases of parasitism in which mutual metabolic dependence occurs.

Although this text is concerned with the phenomenon of parasitism in particular, it is worth while to consider briefly the other types of heterogenetic associations in order to place parasitism in its true perspective.

○○○ 1.2 Commensalism

The term literally means 'eating at the same table', and there are a number of often-quoted classical examples of this type of loose association between animals of different species. One of the best known is that between certain species of hermit crabs and sea-anemones, in which the anemone lives on the shell sheltering a hermit crab. The sea-anemone benefits directly, having access to the remains of food caught and scattered by the crab, whereas the crab benefits by the presence of the sea-anemone which assists in warding off undesirable predators. In many cases, although this type of association is beneficial to one or both organisms, it is not usually obligatory for their existence. An exception is the association between the hermit crab *Eupagurus prideauxi* and the anemone *Adamsia palliata*, in which neither of the partners is able to survive alone. The crab crawls into a shell which is too small for itself and uses the pedal disc of the anemone as cover for the unprotected portion of its body.

Commensalism may thus be considered a type of loose association in which two animals of different species live together *without either being metabolically dependent on the other*, although one or both organisms may receive some benefit from the association. It is important to stress the absence of *metabolic* dependence in this type of association, for it is the absence of this feature, in particular, which in the definition considered further below (p. 3) separates a commensal sharply from a parasite.

○○○ 1.3 Phoresis

There appears to be no difference of opinion on this term, which is used for a particular type of association in which one organism merely provides shelter, support or transport for another organism of a different species.

The classical example is that of fishes belonging to the genus *Fierasfier*, which live within the respiratory trees of holothurians, or occasionally in starfish. These fish are relatively helpless and are readily attacked and devoured by other species. The holothurians appear to be undisturbed by the presence of the fish.

○○○ 1.4 Parasitism

In the type of association which forms the subject matter of this book, contact between the individuals of two different species differs markedly from that already described in that it is intimate and continuous. Many parasites have free-living stages in their life cycles and only during the periods when they make contact with their hosts can they actually be considered to lead a parasitic existence.

○ 1.4.1 General considerations

Of all the types of animal associations, perhaps the term parasitism has been the most difficult to describe. This appears to be largely due to the failure to recognise that the term has only a relative meaning. It has also been complicated by the insistence of many authors that a parasite must *necessarily* be harmful to its host: Crofton (1971*a*) proposes further that the term parasite should be restricted to organisms which are *potentially* capable of *killing* their host. In the writer's opinion, emphasis on the harmful effects of a parasitic association, more than any other single factor, has somewhat bedevilled a broad, biological approach to considerations of the phenomenon. It must be recognised, however, that any worker is entitled to restrict his usage of any term in any way, provided that he clearly defines his understanding of the term.

In an earlier edition of this text, while recognising the relative basis of terminology in general, the writer stressed the physiological and metabolic relationship between host and parasites (Smyth, 1962). At that time, little was known concerning the interaction of host and parasite metabolisms, but since then this field of parasitology has expanded considerably and much more information is now available. This view has been further developed by a number of authors (see Lincicome, 1963; Read, 1970) but is only one of the various aspects of the whole complex phenomenon of parasitism. Intricate physiological or immunological interactions between host and parasite are also involved, especially during the infective and establishment processes (Rogers, 1962; Sprent, 1963). Many of these aspects are discussed when the individual life cycles of different species are being considered; immunological reactions are dealt with in Chapters 32 and 33.

○ 1.4.2 Metabolic dependence

A parasite is thus considered here to be an organism which is not only in continuous, intimate association with another organism, the host (normally of a different species, but see below), but is also metabolically dependent, directly or indirectly, on it to some degree. The relative nature of this association is stressed and it is possible to draw up a list of parasite species which show an increasing degree of metabolic dependence on their hosts. At one end of this hypothetical scale (Fig. 1.2) is zero dependence, i.e. a free-living organism; at the other end is 100 per cent dependence or total parasitism. In between these two extremes lies a range of organisms which satisfy their metabolic requirements to a varying extent at the expense of the host.

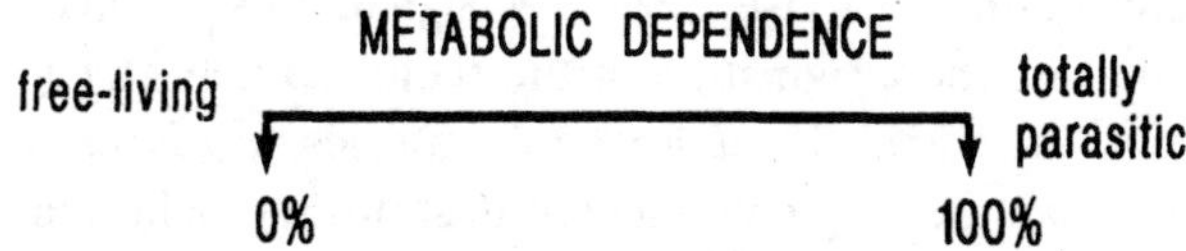

Fig. 1.2
Diagram illustrating the relative concept of parasitism based on the degree of metabolic dependence of a parasite on its host. A free-living organism shows zero dependence, whereas a blood-dwelling protozoon (e.g. *Plasmodium*) is 100 per cent dependent. All degrees between these two extremes are encountered.

MacInnis (1976) neatly proposes a similar fundamental concept of parasitism by defining it as an association in which '*one partner, the parasite, of a pair of interacting species, is dependent upon a minimum of one gene or its products from the other interacting species, defined as the host, for its survival*'.

We must make clear here what we mean by 'metabolic dependence'. Although at first sight it would appear that most parasites are only dependent on their hosts for food materials, a closer examination shows that the situation is more complex than this. Examples can be given of parasites which are dependent on their hosts for one or more of the following: (*a*) developmental stimuli; (*b*) nutritional materials; (*c*) digestive enzymes; (*d*) control of maturation and (more rarely) mitosis.

Each of these can now be considered in some further detail.

(*a*) **Developmental stimuli.** Stimuli of this nature may be defined as those which stimulate or 'trigger' the parasite to enter into its next phase of development. Thus, parasites have evolved devices whereby they can automatically 'recognise' that they are in the particular host environment suitable for establishment and subsequent development. This is achieved by responding to a particular parameter, or combination of parameters, of that environment. A wide variety of factors can be utilised by parasites, the only stipulation being that each host environment contains a unique combination of parameters; this is essential to ensure that the recognition and trigger for development only occurs at the appropriate site. Thus many parasites in an encysted state (e.g. the metacercarial cysts of trematodes p. 195, the oocysts of coccidia p. 95) or in a suspended state (e.g. the infective larvae of many nematodes, such as *Haemonchus contortus*, p. 416), utilise mainly the high $p\mathrm{CO_2}$ of the intestine as a 'signal' for excystment or ensheathment; moreover, this process is often enhanced by other factors such as a low E_h or the presence of bile. In contrast, many cestodes utilise bile as a signal; this normally stimulates the evagination of the scolex so that attachment can occur. In at least one species, *Echinococcus granulosus*, the evaginated scolex must further make contact with a suitable surface, i.e. the gut mucosa or a suitable surface *in vitro* (p. 515). Again several cestodes of birds (body temperature, 40 °C) with larval stages in fish (body temperature, 10–15 °C) utilise the sudden change in temperature from fish to bird host as the stimulus to trigger the plerocercoid larva to undergo maturation (p. 509) in the bird host.

How these triggers act in physiological or biochemical terms has been one of the major problems in parasitology and is discussed when the developmental biology of the various groups is considered.

(*b*) **Nutritional dependence.** It is self-evident that this is undoubtedly the commonest form of metabolic dependence. A range of levels occurs and parasites may be dependent for nutritional supplies on (*a*) the food of the host either before or after digestion; (*b*) the tissues of the host; or (*c*) secretions of the host. In most parasites, all the nutritional requirements are satisfied by the host but some parasites may obtain additional materials from extraneous sources. Consider, for example, the case of a monogenetic trematode (p. 169) parasitic on the gills of a fish. This parasite feeds on the blood vessels of the gills, but is able to digest the engorged blood by means of its own enzymes and utilise the breakdown products obtained. Owing to the close contact with sea water, however, it almost certainly has access to extraneous oxygen by diffusion. Thus, in the synthesis of the parasite tissue, some of the oxygen molecules utilised for synthetic or energy purposes will have been derived from non-host sources. Nutritional dependence on the host can thus not be said to be complete.

In contrast, species of blood flukes, e.g. *Schistosoma haematobium* (p. 237) in the mesenteric veins of man are 100 per cent dependent on the host for all nutritional supplies including oxygen. Thus, in the synthesis of parasite tissue the percentage of atoms of host origin which are metabolically involved could be used as a means of estimating the nutritional dependence on the host.

(*c*) **Digestive enzymes.** Dependence of a parasite on the host for its food materials is of little value unless it can utilise the food thus obtained. There is increasing evidence that the majority of trematodes and nematodes possess the digestive enzymes necessary to hydrolyse complex molecules. Many parasites, like the monogenetic trematode quoted above, can digest blood and tissue. The larva of *Diplostomum phoxini* (p. 504), for example, can digest an egg-albumen mixture *in vitro* and utilise the breakdown products for the processes of growth and differentiation. Such organisms are therefore not dependent on the host for digestive enzymes. On the other hand, adult tapeworms can probably only utilise molecules small enough to be taken in through their tegument (p. 284) and are therefore largely dependent on their host's ability to break down carbohydrates, fats and proteins enzymatically. There is some evidence, however, that larval cestodes (e.g. cysticerci) can take in large molecules by pinocytosis (p. 284) but the situation in adult worms is equivocal (p. 284). A process of *membrane* (= contact) digestion may also occur in cestodes (Fig. 24.1). Some parasites have lost the enzyme systems involved in certain fundamental processes. For example, cestodes have lost their capacity for *de novo* synthesis of lipids and have become entirely dependent on the host for the provision of these (Frayha & Smyth, 1983). However, it is important to note that the host may also utilise metabolic products from the parasite, so the process of exchange may sometimes be a two-way one (Lincicome, 1963). Striking examples of this are the plerocercoids (spargana) of the cestodes *Spirometra mansonoides* and *S. erinacei* (p. 310) which release a

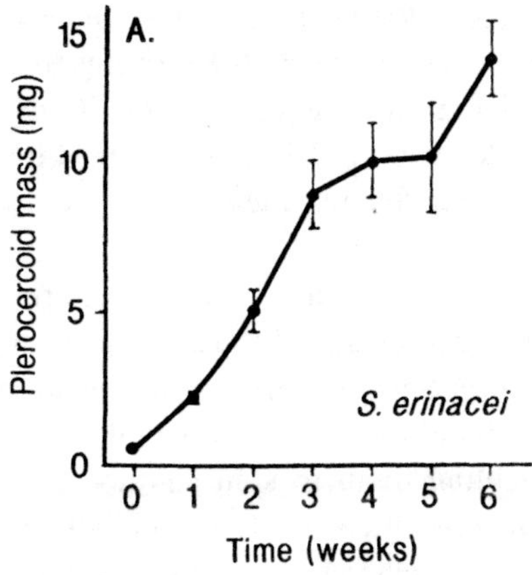

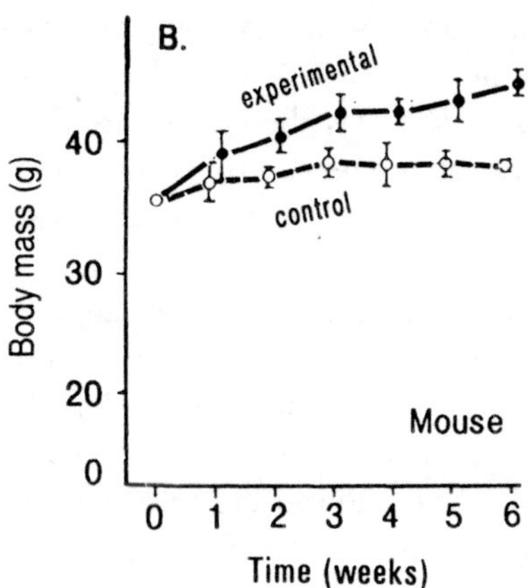

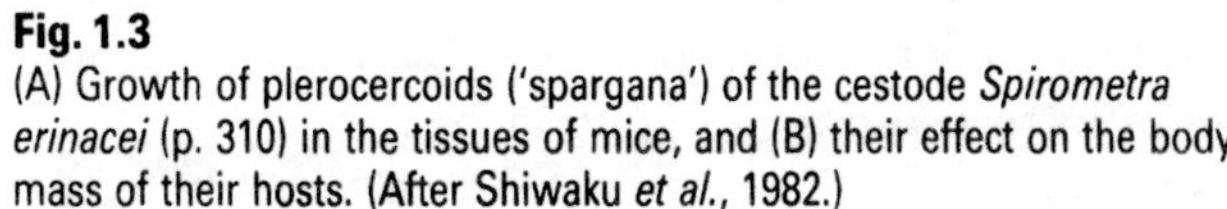
Fig. 1.3
(A) Growth of plerocercoids ('spargana') of the cestode *Spirometra erinacei* (p. 310) in the tissues of mice, and (B) their effect on the body mass of their hosts. (After Shiwaku *et al.*, 1982.)

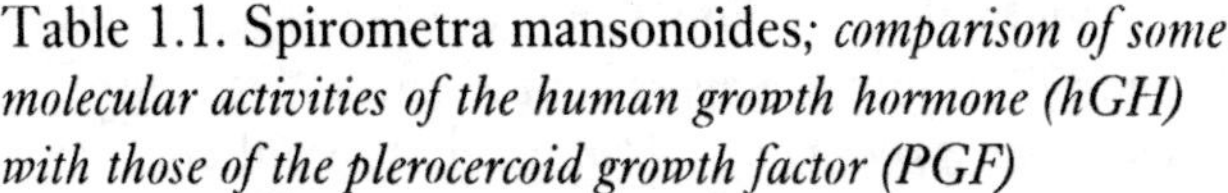
Table 1.1. Spirometra mansonoides; *comparison of some molecular activities of the human growth hormone (hGH) with those of the plerocercoid growth factor (PGF)*

	hGh	PGFF
Relative molecular mass	22 000	22 000
Isoelectric point	4.9	4.7
Reacts with hGH-specific monoclonal antibody (%)	100	61

Source: Data modified from Phares (1987).

substance (the *plerocercoid growth factor*) which induces increased growth weight (Fig. 1.3) in infected mice (Mueller, 1974; Odening, 1979; Shiwaku *et al.*, 1982). This factor has been shown to have physiological and biochemical characteristics similar to the human growth factor (Table 1.1) and Phares (1987) has put forward the intriguing hypothesis that the plerocercoid growth factor is produced by a human growth factor gene which has been 'stolen' by the parasite from the human host, a situation which has been referred to as 'genetic theft'!

(*d*) **Control of maturation.** A number of species have been shown to be dependent on the host for the control of their maturation processes. Such examples represent a remarkable stage in the evolution of parasitism, whereby an endocrine system developed to stimulate metabolic processes in one animal is utilised by another of a different species. This results in the synchronisation of the reproductive phases of host and parasite, an effect which has considerable survival value for the parasite. For example, the blood protozoon *Leucocytozoon* (p. 138) undergoes a multiplication phase resulting in a rapid increase in the number of its gamete-producing cells (gametocytes) only during the breeding season of its duck host, a time which corresponds to the natural occurrence of its vectors (blackflies).

Other parasites whose life cycles are influenced by the hormonal activity of the host are certain parasites of the frog – the protozoa *Opalina ranarum* (p. 146) and *Nyctotherus cordiformis* (p. 152), and the monogenean *Polystoma integerrimum* (p. 163). The life cycles of these organisms are beautifully synchronised with that of the amphibian host and involve release of cysts or eggs at a time when the frog is breeding in water, i.e. when potential tadpole hosts are assured. It has been shown experimentally that maturation of the parasites is related to the levels of sex hormones in the host and may be induced experimentally by injecting suitable hormones into the host (Mofty & Smyth, 1964).

As astonishing example of host–parasite synchronisation also occurs in a monogenetic trematode, *Pseudodiplorchis americanus*, parasitic in the spadefoot toad, *Schaphiopus couchii*, in the USA (Tinsley, 1982–1990; Tinsley & Jackson, 1986, 1988). This toad is entirely desert-adapted and enters the water to spawn only for a few hours in 1–3 nights during the summer rains (Figs 11.8–11.11). The uteri of the parasites contain over 200 fully developed larvae, which are encapsulated. When the toads enter the water the larvae are released within seconds of their deposition, the escaping oncomiracidia invading the nostrils of adult toads. Unlike *Polystoma*, however, the trigger for the egg release does not appear to be hormonal and its nature, at present, is unknown. The life cycle of *Pseudodiplorchis* is discussed further on p. 167.

A further remarkable example of synchronisation of host and parasite metabolic activities has been demonstrated by Wikgren *et al.* (1970) who showed that the mitotic activity pattern of plerocercoids of the cestode *Diphyllobothrium dendriticum* followed the same circadian rhythm as the comparable cycle of the host. Many other examples of circadian rhythms in parasites are known (Hawking, 1975).

From the above examples, it is clear that the physiological or metabolic dependence of a parasite on its host is a most complex matter involving much more than just nutritional factors and that each host–parasite system must be analysed individually before the full extent of this dependence can be evaluated.

○ 1.4.3 Mutualism and symbiosis

An association in which both associates benefit has long been referred to as *mutualism* by some authors and *symbiosis* by others. The literature on the definition of these words is confused. Mutualism is derived from the Latin *mutuus* (= exchanged), whereas symbiosis comes from the Greek *symbioun* (= to live together). The term *symbiosis* could thus broadly be used to include all the different kinds of relationship that exist in nature. By usage, however, it has come to be restricted to associations of a special kind in which the participating species are dependent on each other for existence. In cases of mutualism, on the other hand, the association is not obligatory for existence.

On the metabolic view put forward above, both mutualism and symbiosis are merely recognised as special cases of parasitism in which some metabolic by-products of the parasite are of value to the host. There are several well-known examples of the phenomenon. The association between wood-eating termites and hyperflagellates in their intestine is of the symbiotic type of parasitism. The termites are entirely dependent on the flagellates for certain nutritional requirements, notably the supply of nitrogen and carbohydrates obtained by the breakdown of wood. The flagellates are similarly dependent on the host for nutriment and the physical environment in which they live. The dependence of the host on the flagellates may be readily demonstrated by raising the termites to a temperature which is lethal to the protozoans, thus defaunating them. Under such conditions, the termites fail to survive, as they lack the enzyme systems necessary to digest a wood diet.

The association between intestinal ciliates and their ruminant hosts (discussed on p. 154) is of a similar nature. Several genera of ruminant-dwelling protozoans produce the enzymes cellulase and cellobiase, thus enabling them to split cellulose and utilise the breakdown products for their metabolism. The rate of fission of these ciliates is extremely rapid, and they soon die. On disintegration, they provide the host with about one-fifth of its total nitrogen requirements.

An example of an association which could be considered to be mutualistic is that between the coelenterate *Hydra viridis* and the alga *Zoochlorella* which lives within its endodermal cells. The alga produces oxygen which *Hydra* utilises, and *Zoochlorella* makes use of the nitrogenous waste products of *Hydra* for its synthetic processes. It is possible that a mutualistic association exists in many cases of parasitism, but sufficient physiological studies have not been made to reveal their existence.

○ 1.4.4 Types of parasite

Once the relative connotation of the term 'parasite' is accepted, parasites can be classified in other ways according to their life cycles, position in or on the host, or other features. It is common practice, for example, to speak of *ectoparasites* and *endoparasites*. *Ectoparasites* are organisms (e.g. fleas, lice, ticks) that live on the outside of their hosts, usually attached to the skin, feathers, hair, gills, etc.; such forms can never lead a completely parasitic existence, but utilise oxygen from outside the host. Many maintain only periodic contacts with their hosts and, according to the definition given earlier, cannot be considered parasites but essentially special kinds of predators. *Endoparasites* are parasites living within their hosts, in the gut, body cavity, lungs or other tissues; such forms nearly always live a completely parasitic existence. Certain parasites fall into both these groupings. The itch mite (*Sarcoptes scabei*), for example, burrows in tunnels in the skin and could satisfy the criteria of either an ectoparasite or an endoparasite. Again then, these terms cannot always be accurately defined, but they are convenient general terms. Parasitologists also speak of *facultative* parasites, organisms

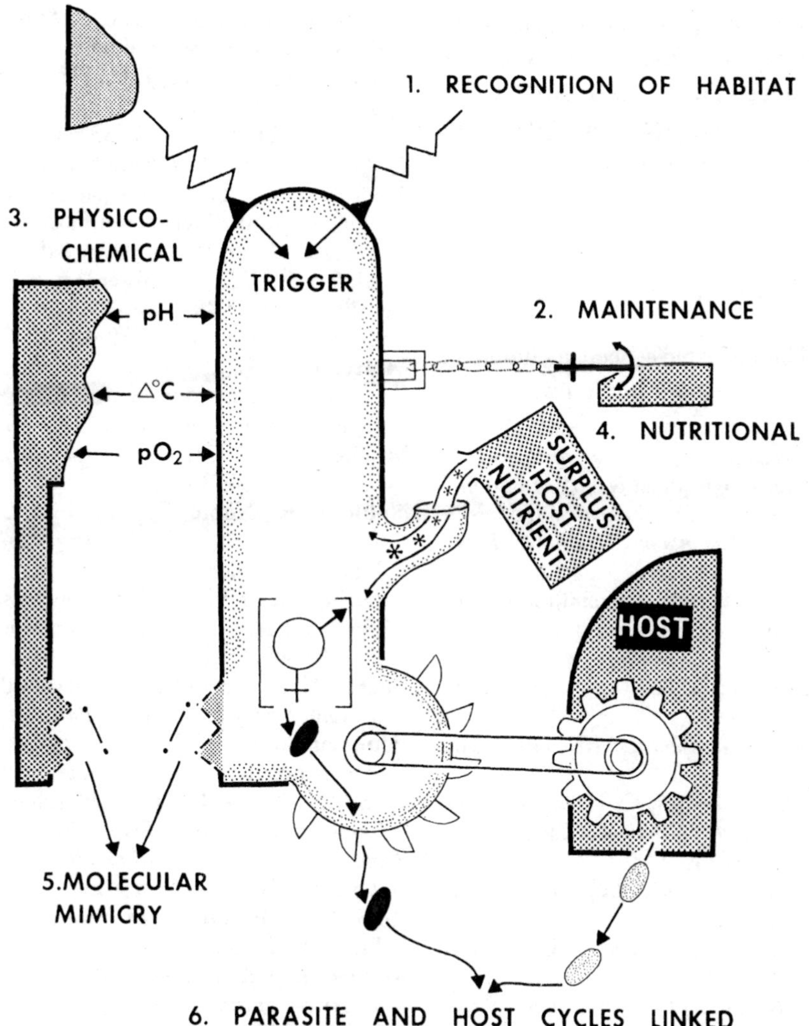

Fig. 1.4
A cartoon-like representation of an 'ideal parasite' which: (1) 'recognises' a host site suitable for establishment; (2) maintains its position there; (3) is adapted to the physico-chemical conditions of the host; (4) utilises host nutrient in a manner compatible with host survival; (5) presents a surface with a molecular configuration such that the host immune response is absent or minimised (see p. **000**); (6) has a life cycle synchronised with that of the host. (After Smyth, 1972.)

which can live either a parasitic or non-parasitic existence, and *obligate* parasites, which are obliged to live a parasitic existence and are incapable of surviving outside the host environment. In the latter definition the emphasis must be on a 'naturally occurring' environment, for many 'obligate' parasites can now be cultured in artificial environments of a complex nature (Chapters 33, 34).

To invade the body of another species of animal, and to live and multiply in or on it, could not have been achieved without considerable morphological, physiological, biochemical and immunological adaptations by the parasite. Some of these adaptations have been discussed here and are covered more fully in the text. It is not intended, however, to discuss the morphological adaptations of parasites to their way of life. These have been discussed in detail by numerous authors, especially Baer (1952) and Caullery (1952), and a considerable literature exists on this aspect of parasitism.

Much more information is now also available on the biochemical and physiological adaptations of parasites, especially protozoa (Coombs & North, 1991; Englund & Sher, 1988; Gutteridge & Coombs, 1977; Kreier & Baker, 1987) and helminths (Barrett, 1981; Bennett, Behm & Bryant, 1989; Trager, 1986; Smyth & Halton, 1983; Bryant & Behm, 1989; Smyth & McManus, 1989).

A coherent picture of the processes underlying parasitism will only emerge when all aspects of the phenomena are fully understood. Fig. 1.4 is a cartoon-like representation of the 'ideal parasite' which serves to underline some of the problems involved.

○○○ References

(References of general interest have been included in this list, in addition to those quoted in the text.)

Anderson, R. M. 1982. *Population Dynamics of Infectious Diseases: Theory and Applications*. Chapman & Hall, London.

Anderson, R. M. 1987. The role of mathematical models in helminth population biology. *International Journal for Parasitology*, **17**: 519–29.

Anderson, R. M. & May, R. M. 1985. Helminth infections of human: population dynamics and control. *Advances in Parasitology*, **24**: 1–101.

Baer, J. G. 1952. *Ecology of Animal Parasites*. University of Illinois Press, Urbana.

Barrett, J. 1981. *Biochemistry of Parasitic Helminths*. The Macmillan Press, London.

Bennett, E.-M., Behm, C. & Bryant, C. 1989. *The Comparative Biochemistry of Parasitic Helminths*. Chapman & Hall, London. ISBN 0 412 32730 9.

Bryant, C. & Behm, C. 1989. *Biochemical Adaptation in Parasites*. Chapman & Hall, London.

Caullery, M. 1952. *Parasitism and Symbiosis*. Sidgwick & Jackson, London.

Coombs, G. H. & North, N. J. (eds) 1991. *Biochemical Protozoology*. Taylor & Francis, London.

Cox, F. E. G. 1982. *Modern Parasitology*. Blackwell Scientific Publications, Oxford.

Crofton, H. D. 1971*a*. A quantitative approach to parasitism. *Parasitology*, **62**: 179–93.

Crofton, H. D. 1971*b*. A model of host-parasite relationships. *Parasitology*, **63**: 343–64.

Englund, P. T. & Sher, A. 1988. *The Biology of Parasitism: a Molecular and Immunological Approach*. Alan R. Liss, New York.

Esch G. W., Bush, A. O. & Aho, J. M. 1989. *Parasite Communities*. Chapman & Hall, London. ISBN 0 412 33540 9.

Frayha, G. & Smyth, J. D. 1983. Lipid metabolism in parasitic helminths. *Advances in Parasitology*, **22**: 309–87.

Gutteridge, W. E. A. & Coombs, G. H. 1977. *Biochemistry of Parasitic Protozoa*. The Macmillan Press, London.

Hawking, F. 1975. Circadian and other rhythms of parasites. *Advances in Parasitology*, **13**: 123–82.

Kennedy, C. R. 1975. *Ecological Animal Parasitology*. Blackwell Scientific Publications, Oxford.

Kennedy, C. R. (ed.) 1976. *Ecological Aspects of Parasitology*. North-Holland, Amsterdam.

Kennedy, C. R. 1983. General ecology. In: *Biology of the Eucestoda*, Vol. 1 (ed. C. Arme & P. W. Pappas), pp. 27–80. Academic Press, London.

Kreier, J. P. & Baker, J. R. 1987. *Parasitic Protozoa*. Allen & Unwin, Massachusetts.

Lincicome, D. R. 1963. Chemical basis of parasitism. *Annals of the New York Academy of Sciences*, **113**: 36–380.

MacInnis, A. J. 1976. How parasites find their hosts: some thoughts on the inception of host–parasite integration. In: *Ecological Aspects of Parasitology* (ed. C. R. Kennedy), pp. 3–20. North-Holland, Amsterdam.

Mehlhorn, H. 1988. *Parasitology in Focus*. Springer-Verlag, Berlin. ISBN 3 540 17838 4.

Mofty, M. M. El & Smyth, J. D. 1964. Endocrine control of encystation in *Opalina ranarum* parasitic in *Rana temporaria*. *Experimental Parasitology*, **15**: 185–99.

Mueller, J. F. 1974. The biology of *Spirometra*. *Journal of Parasitology*, **60**: 3–14.

Noble, E. R., Noble, G. A., Schad, G. & MacInnis, A. J. 1989. *Parasitology. The Biology of Animal Parasites*. Lea & Fibiger, Philadelphia.

Odening, K. 1979. Zum Erforschungsstand des 'Sparganum Growth Factor' von *Spirometra*. *Angewandte Parasitologie*, **20**: 185–92.

Phares, C. K. 1987. Plerocercoid growth factor: a homologue of human growth hormone. *Parasitology Today*, **3**: 346–9.

Pennycuick, L. 1971. Frequency distribution of parasites in a population of three-spined sticklebacks, *Gasterosteus aculeatus* L., with particular reference to the negative binomial distribution. *Parasitology*, **63**: 389–406.

Read, C. P. 1970. *Parasitism and Symbiology. An Introductory Text*. Ronair Press, New York.

Rogers, W. P. 1962. *The Nature of Parasitism*. Academic Press, New York.

Rollinson, D. & Anderson, R. M. 1985. *Ecology and Genetics of Host–Parasite Interactions*. Academic Press, London. ISBN 0 125 93690 7.

Schmid, W. D. & Robinson, E. J. 1972. The pattern of a host–parasite distribution. *Journal of Parasitology*, **58**: 907–10.

Schmidt, G. D. & Roberts, L. S. 1989. *Foundations of Parasitology*. 4th Edition. Times Mirror/Mosby College Publishing, St Louis.

Shiwaku, K., Hirai, K., & Torii, M. 1982. Growth-promoting effect of *Spirometra erinacei* (Rudolph, 1819) plerocercoids in mature mice: relationships between number of infected plerocercoids and growth-promoting effect. *Japanese Journal of Parasitology*, **31**: 353–60.

Smyth, J. D. 1962. *An Introduction to Animal Parasitology*. 1st Edition. English Universities Press, London.

Smyth, J. D. 1972. *Parasites as Models in Cellular Differentiation*. Inaugural Lectures, Vol. 9, pp. 55–66. Imperial College, University of London.

Smyth, J. D. & Halton, D. W. 1983. *The Physiology of Trematodes*. 2nd Edition. Cambridge University Press.

Smyth, J. D. & McManus, D. P. 1989. *The Physiology and Biochemistry of Cestodes*. Cambridge University Press.

Sprent, J. F. 1963. *Parasitism*. University of Queensland Press, Brisbane.

Tinsley, R. C. 1982. The reproductive strategy of a polystomatid monogenean in a desert environment. *Parasitology*, **85**: XV.

Tinsley, R. C. 1988. The effects of host sex on transmission success. *Parasitology Today*, **5**: 190–5.

Tinsley, R. C. 1990. Host behaviour and opportunism in parasite life cycles. In: *Parasitism and Host Behaviour* (ed. C. J. Barnard & J. M. Behnke), pp. 158–92. Taylor & Francis, London.

Tinsley, R. C. & Earle, C. M. 1983. Invasion of vertebrate lungs by the polystomatid monogeneans *Pseudodiplorchis americanus* and *Neodiplorchis scaphiopodis*. *Parasitology*, **86**: 501–17.

Tinsley, R. C. & Jackson, H. C. 1986. Intestinal migration in the life-cycle of *Pseudodiplorchis americanus* (Monogenea). *Parasitology*, **93**: 451–69.

Tinsley, R. C. & Jackson, H. C. 1988. Pulsed transmission of *Pseudodoplorchis americanus* between desert hosts (*Scaphiiophus couchii*). *Parasitology*, **97**: 437–53.

Trager, W. 1986. *The Biology of Animal Parasitism*. Plenum Press, New York.

Wikgren, B.-J. P., Knuts, G. M. & Gustafsson, M. K. S. 1970. Circadian rhythm of mitotic activity in the adult gull-tapeworm *Diphyllobothrium dendriticum* (Cestoda). *Zeitschrift für Parasitenkunde*, **34**: 242–50.

2

Niches, habitats and environments

2.1 Niches

Definition. As already emphasised, Parasitology is essentially a branch of Ecology in which the habitat and environment of an organism (the parasite) is provided by another organism (the host). Hence many of the principles which apply to free-living organisms can also be applied to parasites. One of the most important of these – and perhaps one of the most neglected – is the concept of the *niche.*

All living organisms can be said to occupy a biological *niche*, but few terms in biology have been so misunderstood and yet it is a term of great significance to parasitologists. It essentially refers to an (abstract) 'space' in the biotic environment in which life is possible. Some workers confuse the term with 'habitat', the latter being only the environmental component of the niche. The habitat is sometimes referred to as an 'environmental niche' of which the alimentary canal or the gills of fishes could be considered common examples. It is possible, for example, for two different species of parasites to occupy the same *habitat* (e.g. the intestine) while occupying different *niches*, because one utilises glucose as a nutritional substance, whereas the other utilises fructose; they thus occupy different (nutritional) niches.

These considerations have given rise to the concept of a *fundamental niche* as *'That unique combination of environmental factors, biotic and abiotic, which are capable of supporting life'*. Since there are an unlimited number of both biotic and abiotic factors to be considered, the (abstract) concept of a fundamental or ecological niche is sometimes referred to as an *n-dimensional hypervolume.*

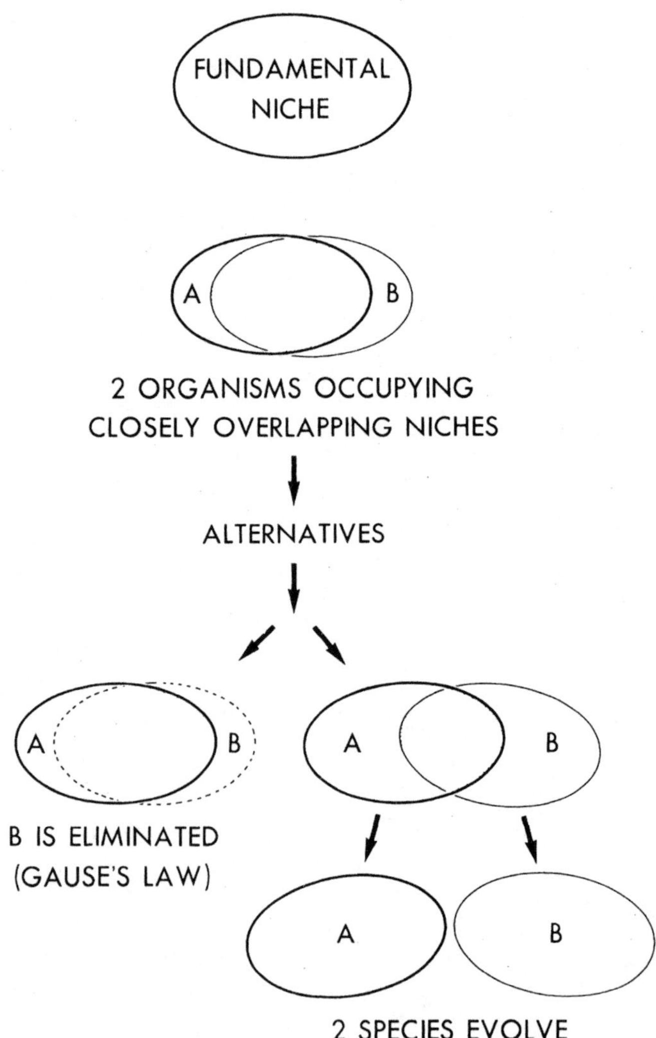

Fig. 2.1
Gause's Law. If two species attempt to occupy the same niche, for any length of time, either one will replace the other, or each species will evolve in such a way that their niches no longer overlap. (After Smyth, 1987.)

The question can be asked, what happens when two populations of the same species attempt to occupy the same niche? This problem was first considered by the Russian biologist Gause (1934), who postulated – what is now known as Gause's Law (Hypothesis) – that 'No two forms can occupy that same ecological niche for an indefinite period, eventually one form will replace the other'. This hypothesis was restated by Hardin (1960) as the 'Competitive Exclusion Principle' which simply states: 'Complete competitors cannot coexist'.

Gause's Hypothesis is illustrated in Fig. 2.1; it shows that if two populations of the same species (differing in minor characteristics) attempt to occupy the same niche, eventually either (*a*) one will replace the other or (*b*) their characteristics will evolve in such a way that their niches no longer overlap, i.e. a new species is formed. It has been speculated that the sheep and horse 'strains' of the hydatid organism, *Echinococcus granulosus*, have evolved in this way. Both strains can inhabit the dog gut, but their metabolisms (Table 24.3) have been shown to be substantially different (McManus & Smyth, 1978; Smyth, 1987).

○○○ 2.2 Habitats: general comments

One feature of the evolution of metazoan organisms is the increasing complexity of their alimentary, respiratory and circulatory systems. The development of such systems was, of course, advantageous to these evolving organisms and it resulted in a more efficient metabolism, but it was not without some inherent disadvantages. Each new organ system evolved – especially those containing cavities or surfaces – presented a habitat for potential parasites. The favoured habitats in vertebrates are (*a*) the alimentary canal and its associated glands; (*b*) the blood stream; (*c*) the respiratory system; and (*d*) the coelom, in that order. The nervous system and its derivations, the excretory system and the reproductive system have also been invaded, but less commonly so than the larger tubular systems. In the invertebrates, possible habitats are fewer, but most tissues have been invaded, especially by vector stages of protozoa or larval stages of helminths.

This chapter is concerned with the nature of the environmental conditions provided by some of these parasite habitats, only the more important being considered.

In organisms with complex heteroxenous life cycles (i.e. those involving more than one host), the question of the part played by environmental conditions in controlling the development of the parasite is one of considerable interest. The physico-chemical conditions of a habitat – pH, $p\text{CO}_2$, oxygen tension, oxidation–reduction potential, temperature, viscosity, osmotic pressure – are of major importance, as are the nature, quantity and availability of food materials. For example, egg-production in certain trematodes appears to be limited by the degree to which the environmental food supplies can satisfy the enormous synthetic demands of the egg-producing stages. The ecological pattern here is thus not very different from that of free-living forms.

Many parasites pass through a number of habitats, differing widely in their properties. *Cryptocotyle lingua*, an intestinal trematode of a bird (p. 228), for example, passes through water, mollusc tissue (first intermediate host) and fish tissue (second intermediate host) before reaching its final location in the bird intestine. The environmental conditions and food supply in these habitats may vary within wide limits.

Considerations of this kind have led to the concept of the physiological life cycle, which is concerned with changes in the physico-chemical characteristics of the habitat at each stage along with associated metabolic changes in the parasite; this is discussed further on p. 19. When this type of information is integrated with the morphological considerations of the life cycle, the combined result is a more informative and interesting overall picture of the biology and behaviour of the organism; where possible, life cycles have been treated on this basis throughout this book.

○○○ 2.3 The vertebrate alimentary canal

○ 2.3.1 General properties

The vertebrate gut may be regarded as one of the most hazardous habitats for a would-be parasite. It is dark; it is undergoing regular physiological changes relative to the feeding habits of the host; it contains a battery of protein-, fat- and carbohydrate-splitting enzymes; its pH may range from 1.5 to 8.4; it is almost oxygen-free, and it often undergoes violent movements. Each of the features, to a greater or lesser degree, presents a problem to an organism attempting to live and reproduce in it.

The physical and chemical properties of the gut are only dealt with briefly here. Most of the information is based on the vertebrate alimentary canal, the morphology and physiology of which has been reviewed in considerable detail by Stevens (1988). Various other aspects of intestinal physiology have been reviewed by Befus & Podesta (1976), Chadwick & Phillips (1982), Crompton (1970, 1973), Johnston (1981), Lewis

(1982), Madge (1975), Mettrick (1970), Morton (1979), Sanford (1982), Smyth & McManus (1989), Smyth & Halton (1983), Stevens (1988), Wrong *et al.* (1981) and Pounder *et al.* (1989). The alimentary canal of invertebrates has been poorly studied, the most comprehensive information being available for insects (Richards & Davies, 1977).

○ 2.3.2 General environmental conditions in different regions of the alimentary canal

The mouth, oesophagus and stomach are not common habitats for parasites. In fishes and amphibians, trematodes and copepods may occur in the mouth. In man, the flagellate *Trichomonas buccalis* occurs in the same site. The stomach of cattle may contain the nematode *Haemonchus contortus* and the larvae of bot flies of the genus *Gastrophilus*. By far the most popular sites for intestinal parasites are the duodenum, the ileum, the caecum and the large intestine. It is worthwhile considering in a little more detail the special structural details of these regions, especially as they affect the biology of parasites. The topography of the alimentary canal – especially with respect to size of villi and width and depth of crypts – undoubtedly plays an important role in determining host specificity of certain species. In rays, for example, it has been shown that the scoleces of various species of the cestode genus *Echeneibothrium* are closely adapted to the depths of the villi, crypts or reticulations of the mucosa (Williams, 1960).

Duodenum. This region is characterised by possessing special glands, Brunner's glands, which in many species are continuous with the pyloric glands. They are composed of ramifying tubules which empty by many ducts into the crypts of Lieberkühn in the overlying mucosa. These crypts make ideally sheltered environments for some protozoa. The nutritional level of the duodenal contents is substantially added to by shed mucosal cells (enterocytes), which, it is now recognised, are lost and replaced at a remarkable rate. The 'turnover time' of such cells in man is 2–6 days and in the rat less than 2 days.

It is difficult to visualise the exact physical conditions within the duodenum. Food is passed from the stomach in a partly digested condition having been attacked by salivary amylase and stomach pepsin. In the upper duodenum, as already noted, the pH rises sharply – probably rarely to alkalinity – and over the food are poured bile and pancreatic juice with their batteries of enzymes (Table 2.1). Protein breakdown continues, and proteases, peptones, polypeptides and some amino acids are released; monosaccharides result from further breakdown of carbohydrates, and glycerol and fatty acids from the hydrolysis of fats.

This region, then, is rich in highly nutrient food materials, although the amount of monosaccharides available at any instant is probably small, for polysaccharides are only slowly broken down, and glucose especially is actively absorbed. Radiological studies have shown that physically, like the stomach, the anterior

Table 2.1. *Digestive enzymes in the alimentary canal of man*

T = present in tissue, S = present in secretion.

Enzyme	Salivary glands	Oesophagus	Stomach	Pancreas	Small intestine	Caecum and colon
Amylase	S	—	—	S	T, S	—
Enterokinase	—	—	—	—	S	—
Erepsin	S	—	—	T, S	T, S	—
Sucrase	—	—	—	—	S	—
Lipase	T, S	T	T, S	T, S	S	—
Maltase	—	—	—	S	—	—
Pepsin	—	—	T, S	—	—	—
Phosphatase	S	—	T	—	T	T
Rennin (chymosin)	—	—	S	—	—	—
Trypsin	—	—	—	T	—	—
Urease	—	—	T	—	—	—

Source: Data from *Handbook of Biological Data*, 1956.

duodenum is a region of comparative calm, whereas further down the duodenum the food boli are torn and shredded, as one worker puts it, in a manner 'comparable with a mincing machine'. This process is not to be confused with peristalsis, which is one in which gentle waves of contraction pass over the stomach, down the pylorus and along the small intestine.

Ileum. Further down the small intestine, the availability of food material for intestinal parasites decreases owing to the absorption of amino acids and carbohydrates through the portal radicles, and of fat by the lymphatics. There is another possible source of nutriment available and this is the cells of the intestine itself. As in the duodenum, this region is one of considerable wear and tear, so that a constant sloughing off of cells takes place. These desquamated cells are rich in enzymes and probably contribute substantially to the enzymes available in the mixed intestinal juice (the *succus entericus*).

Colon and caecum. The caecum is an especially favoured site for parasites, such as entamoebae and scavenging bacteria. These are regions of comparative inactivity. The mucous membrane contains crypts, but no villi. Much mucus may be secreted. The chief function of the caecum and colon is to absorb water, at least in the Carnivora, for which most information is available. In the Herbivora, these regions are the site of cellulose digestion, a process aided by bacterial action.

Viscosity. The mucosa of the vertebrate gut is covered with a film of mucus, even in the fasting animal, and its character varies somewhat in the different regions. In general, it has a viscid surface suitable for the adhesion of many types of parasites. In autopsy examination of the gut, a tapeworm, for example, is usually found flattened and stretched with its surface adhering closely to the mucosa. This close adhesion to the intestinal mucosa is probably essential to cestodes for absorption of food materials and diffusion of waste materials, as well as being necessary to compress the strobila sufficiently to enable the cirrus in each proglottis to be bent into an adjacent proglottid, thereby permitting insemination and fertilisation to take place (see p. 510).

○ 2.3.3 Intestinal physiology

It is beyond the scope of this text to give more than an outline account of intestinal physiology; for specific details reference should be made to the reviews already quoted, that of Stevens (1988) being particularly informative (2.3.1).

The inter-relationships between the various parameters concerned in the physiological conditions pertaining in the intestine at any one time, as determined by Mettrick (1973*a*) for helminths, are shown in Fig. 2.2. It is at once evident that an extremely complex situation exists involving the quality and quantity of the ingested food, the hormonal activities of the host, the microorganisms present in the gut and numerous related factors. Some parameters of especial importance to parasites are considered below.

Intestinal function. Concepts of vertebrate intestinal function have developed and changed continuously within recent years. The main features, related to the intestine as a habitat for parasites, have been reviewed by Smyth & Halton (1983) and Smyth & McManus (1989) and may be summarised as follows:

(*a*) It is now recognised that dipeptides and tripeptides and probably even larger protein molecules, as well as amino acids, are absorbed by the mucosa.
(*b*) Carbohydrates are digested first to disaccharides in the lumen and then hydrolysed to monosaccharides on (or in) the mucosal surface.
(*c*) Carrier molecules are involved in the movement of many sugars and amino acids across the mucosal membrane. For example, transport of glucose appears to be coupled to an inward Na^+ gradient through a common carrier, forming a ternary complex (Na^+–carrier–solute).
(*d*) A highly organised, filamentous coating, the *glycocalyx*, covers the mucosal membrane; it is rich in ionised sugar moieties and has many functional correlates. These include providing binding sites for ions important to transport phenomena and digestive enzyme function.
(*e*) Within the interstices of the glycocalyx there exists a hydrodynamic layer of *unstirred water*. Opinions differ regarding the effect on the intestinal functions (Podesta, 1980; Smyth & McManus, 1989) but it is believed that it may affect permeability coefficients and, in the case of steroids and fatty acids, may play

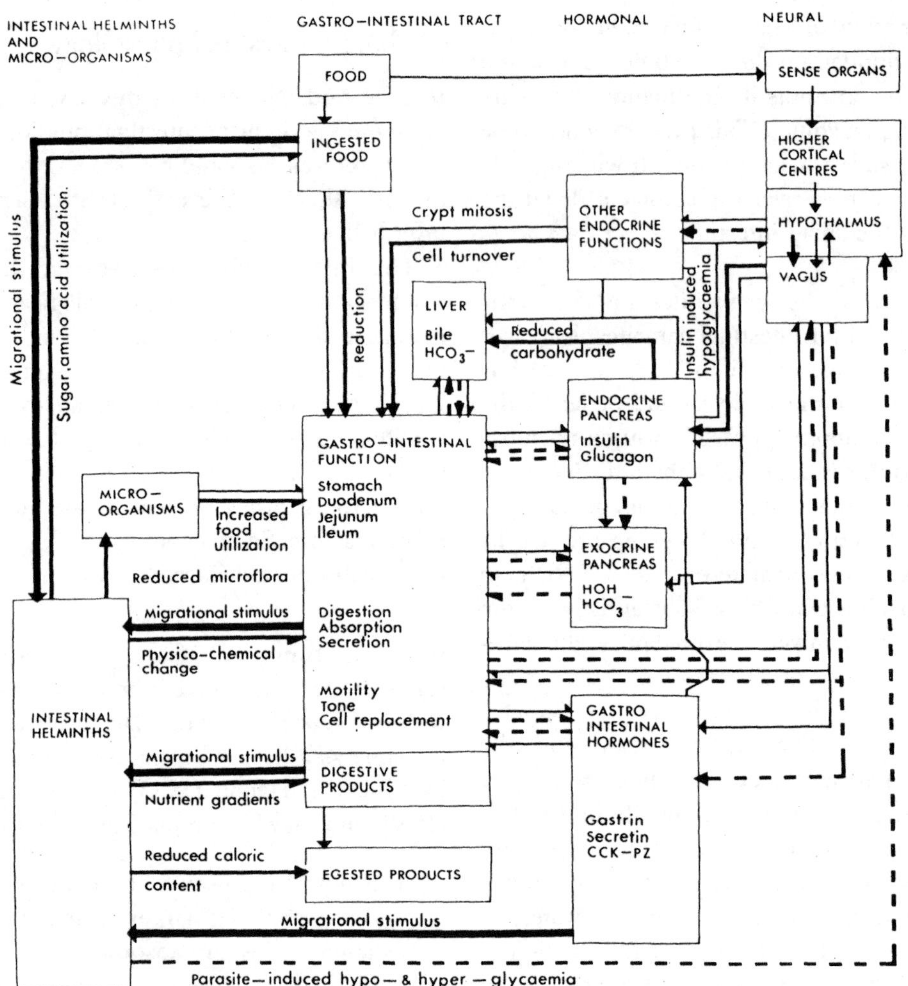

Fig. 2.2
Interactions between the various neural-hormonal-gastro-intestinal factors controlling the conditions in the rat intestine which could influence the growth and migration of the cestode *Hymenolepis diminuta*. (After Mettrick, 1973a.)

an important role in limiting the intestinal absorption rate.

○ 2.3.4 Physico-chemical characteristics

Although a substantial amount of information is available on the vertebrate gut, it must be born in mind that the presence of parasites themselves may substantially alter the levels of the various parameters. Surprisingly little work has been carried out on the effect of parasitism on physico-chemical characteristics of the host. The most comprehensive study appears to be that of Mettrick (1971*a,b,c*, 1973*a,b*) on the rat gut infected with the cestode *Hymenolepis diminuta*. He has shown that the presence of this parasite substantially alters the pH, pCO_2 and pO_2 of the gut (Figs 2.2, 2.3).

pH. The hydrogen-ion concentration alters sharply from region to region. In mammals, on which most work has been carried out, the pH of the mouth is usually about 6.7 but may show a range of 5.6–7.6. In the stomach, strongly acid conditions prevail but the degree of acidity varies remarkably with the condition of the animal. Typical figures for the pH of gastric juice are: man, 1.49–8.38; cattle, 2.0–4.1; sheep, 1.05–3.6; mouse, 3.26–6.24. In the duodenum, the pH in most mammals is just on the acid side, about 6.7, but shows a range of 5.1–7.8. That the duodenum is usually an acid habitat is not generally appreciated; it is commonly held to be alkaline or even strongly alkaline. This misconception is based on the values for the succus entericus, bile and pancreatic juice, which, although showing a range of figures, tend to be weakly alkaline.

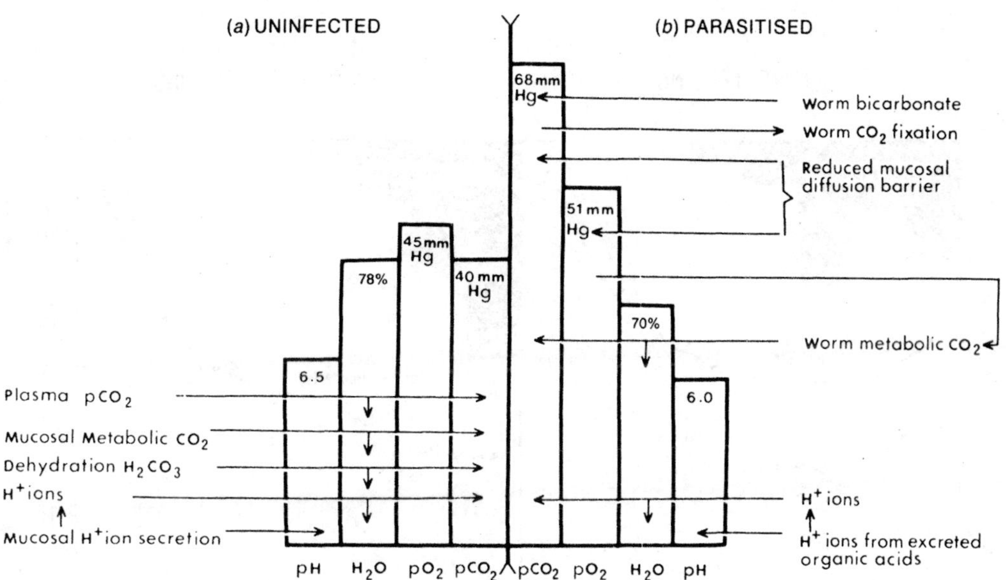

Fig. 2.3
Comparison between the physico-chemical characteristics of the intestine in (*a*) uninfected rats, and (*b*) rats infected with the cestode *Hymenolepis diminuta*. (After Mettrick, 1973*a*.)

The gastric contents shed from the stomach into the duodenum are strongly acid, so that, after feeding, a drop in pH could be expected. In the rat, a fall in pH continues for some four hours after feeding and then rises again (Mettrick, 1971*b*).

In the rat, the presence of the cestode *Hymenolepis diminuta* has a marked lowering effect on the pH (Figs 2.3, 2.4) a result consistent with the fact that these worms excrete fatty acids (Mettrick, 1971*a*). The pH of the large intestine in mammals has a range of about 5.2–7.9 (Wrong *et al.*, 1981).

When considering the pH of the vertebrate gut, it must be recalled that many parasites are partly embedded in the mucosal wall, between the villi or in the crypts of Lieberkühn, and the pH of these sites is not known. On theoretical grounds the pH is more likely to approach the pH of tissue sites than that of the lumen and is probably close to that of blood (approx. 7.4). In birds, the pH of the duodenum appears to be generally substantially lower than that of mammals, and figures as low as 3.0–4.0 have been reported.

Oxidation–reduction potential. This characteristic, which may have special importance for the electron transport of parasites, has not been extensively studied. In the rat, the E_h in the stomach has been given as +150 mV and that of the small intestine about −100 mV (Hopkins, 1967). That this negative potential is due to the presence of microflora is seen from the fact that in germ-free rats the E_h of the large intestine is +30 to +200 mV but becomes negative after contamination with a variety of different coliformes (Wrong *et al.*, 1981). 'Reducing conditions' i.e. a negative E_h, serve as a signal to trigger many hatching stages in parasite life cycles, such as the hatching of coccidian oocysts (Fig. 6.8), trematode cysts (p. 209) and nematode eggs (p. 382).

Oxygen. Quantification of intestinal gas has proved extremely difficult and the data are 'widely divergent' (Code, 1968). The oxygen tension is of special interest in relation to the possible aerobic or anaerobic metabolism of intestinal parasites. Early work on the rat (Rogers, 1949) found that the p_{O_2} on the mucosal surface was three times higher than in the lumen. The same order of result was obtained in the duck (range 0.5–25 mmHg) by Crompton *et al.* (1965). In contrast, using a tonometric method, it was concluded that the lumen p_{O_2} in dogs was a true reflection of the mucosal p_{O_2} (Hamilton *et al.*, 1968) (Fig. 2.5). These somewhat conflicting results may reflect the difficulties of making measurements of this kind. Although some oxygen from swallowed air is found in the oesophagus and stomach, the oxygen tension in the intestine is generally low. Data

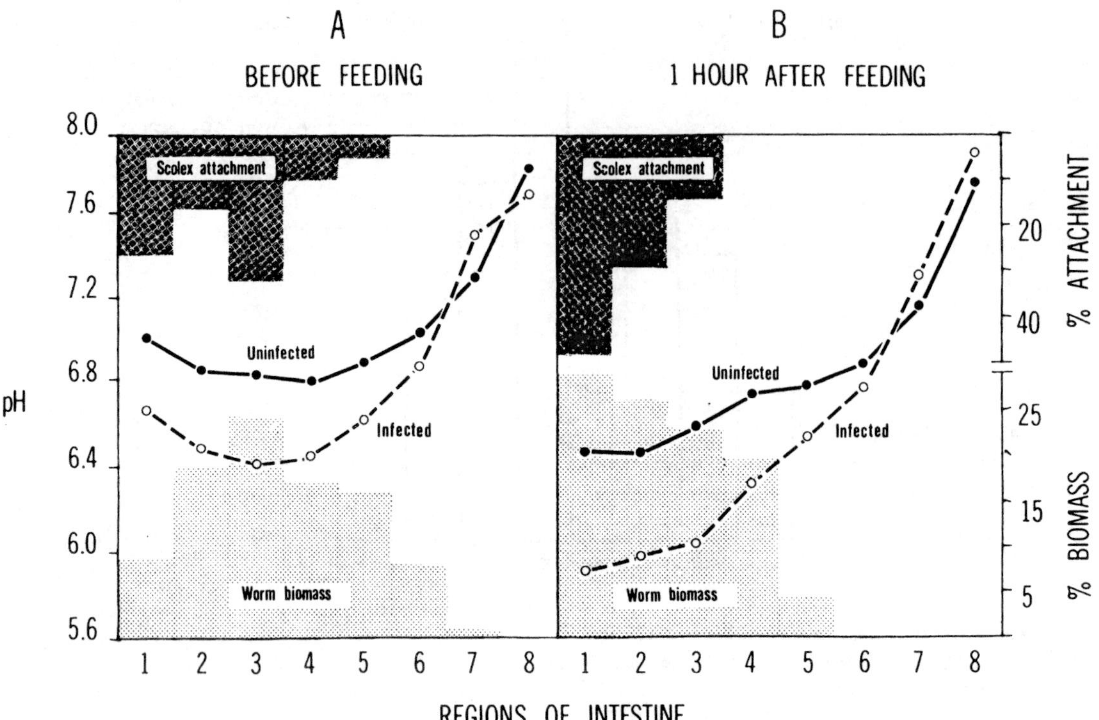

Fig. 2.4
Changes in pH after feeding in uninfected rats and in rats infected with the cestode *Hymenolepis diminuta*. Note how the percentage distribution of the scolex attachment sites and the total worm biomass shifts forwards after feeding. (After Mettrick, 1971*b*.)

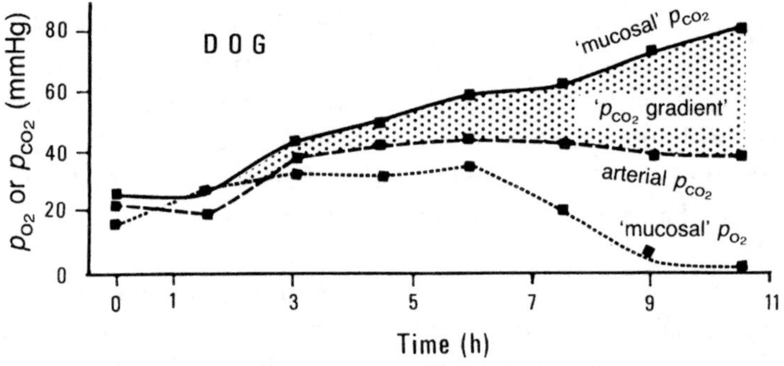

Fig. 2.5
The change of mucosal p_{O_2} and of mucosal blood p_{CO_2} gradient throughout a period of general anaesthesia in a dog. (After Hamilton, J. D., Dawson, A. M. & Webb, J. P. W. 1968. Observations on the small gut 'mucosal' pO_2 in anaesthetized dogs. *Gastroenterology*, **55:** 52–60. © American College of Gastroenterology.)

for some hosts are given in Table 2.2 but the technical difficulties of measurement are great and some early estimations may need re-examination by more reliable techniques.

On theoretical grounds, it might be expected that the liquid in contact with the mucosa of the gut would contain a higher $p\text{O}_2$ than the intestinal lumen.

It should be stressed that the $p\text{O}_2$ may be greatly affected by the nutritional state of the host. Thus in the dog, the level of oxygen falls from 4–6 per cent in a fasting dog to zero after a meat diet (Code, 1968). The oxygen available to a particular parasite in a particular site may also be affected by the presence of microflora in the immediate vicinity of the parasite.

Conversely, parasites in turn may influence the microbial fauna of the gut. Mettrick (1971*c*) has shown

Table 2.2. *Oxygen tensions in some habitats of parasites*

Figures are given in mmHg.

Habitat	Host species	O_2 tension
Skin	man	50–100
Subcutaneous tissue	man, rat, pig	20–43
Arterial blood	man, dog, fish	70–100
Venous blood (heart)	man, horse, duck	37–40
Venous blood (portal vein)	dog, cat	49–66
Peritoneal cavity	rabbit, rat, cat	28–40
Pleural cavity	man, monkey	12–39
Urine	man	14–60
Bile	cattle, sheep, dog	0–30
Abomasum (near mucosa)	sheep	4–13
Rumen (gases)	cattle, sheep, goat	0–2
Stomach (gases)	man	0–70
Small intestine (near mucosa)	sheep, rat	4–30
Small intestine (gases)	horse, cattle	0–6
Small intestine (mucosal)	dog	1–57
Small intestine (gases)	pig	8–65
Large intestine (gases)	horse, cattle, rabbit	0–5
Small intestine	duck	0.5–25

Source: Data from von Brand (1952) and Crompton (1970).

that the presence of *H. diminuta* in the rat gut reduced the number of microorganisms present to less than half that in the uninfected animal; the percentage of common aerobes present in the population was also greatly reduced.

Other gases. Of the other gases found in the vertebrate intestine, carbon dioxide is important to a parasite because this gas acts as a valuable source of carbon atoms. Carbon dioxide fixation occurs in many protozoan and helminth parasites and plays an important part in their metabolism. The p_{CO_2} also plays a major role in the processes of excystment of protozoan and trematode cysts (p. 209) and in the hatching of nematode eggs (p. 381). There seems no doubt also, that the high p_{CO_2} of the gut is one of the parameters which is readily detected by parasites, which can thus 'recognise' when they are in an intestinal environment.

In the dog the p_{CO_2} of the lumen is believed to reflect the p_{CO_2} of the mucosal cells (enterocytes). In 18 dogs examined, the p_{CO_2} gradient between mucosal and arterial blood varied from 4–44 mmHg to 12–130 mmHg (Fig. 2.5) during an 11 h period of anaesthesia (Hamilton *et al.*, 1968).

Bile. Bile is concerned with a number of phenomena of considerable importance to parasites, such as excystment of protozoan or trematode cysts, the hatching of helminth eggs and the evagination of cestode scoleces (p. 4). Most of these are essentially 'trigger mechanisms' of one kind or another; the surface-active properties of bile probably play a major role in these activities. This 'trigggering' action of bile assures that the parasite is released at the appropriate site in the alimentary canal.

The composition of bile varies substantially within different vertebrate groups and, in some cases at least, this difference may play a part in determining host specificity (Smyth & Haslewood, 1963; Smyth, 1962). Cholic acid is the commonest acid found in bile, but other common acids are deoxycholic, chenocholic, hyocholic and lithocholic. Alcohols (C_{27} or C_{28}) occur in the biles of amphibia and fish. Bile alcohols and acids normally occur in conjugated forms, acids conjugating with the amino acids taurine ($NH_2CH_2CH_2SO_3H$) and glycine (NH_2CH_2COOH).

Carnivore biles contain largely salts of cholic acids, whereas, in contrast, those of herbivores contain high levels of salts of deoxycholic acid. Since the cestode *Echinococcus granulosus* is very rapidly lysed by sodium deoxycholate but unaffected by cholate it is clear that this species could not establish itself in a herbivore gut (Smyth, 1962). Many other factors, of course, are also likely to play an important role in determining the host specificity.

Some properties of biles from different animals are shown in Table 2.3. The biochemistry of bile salts has been reviewed in detail by Haslewood (1967) and Heaton (1972).

○○○ 2.4 Blood

○ 2.4.1 General constituents

Blood and blood cells as environments for parasites have been reviewed by Smithers & Worms (1976) and Cox (1976), respectively. The constituents of human blood have been reviewed, in detail, by Lewis (1982) and Altman & Dittmer (1974). Like the alimentary canal, the quantity of soluble food materials in the blood stream will vary with the feeding habits of the host. As a source of nutriment, its value must be considered in relation to the morphology or physiology of the organism utilising it. To a cestode, which lacks a gut and is dependent on small molecules of absorbable dimensions, such

Table 2.3. *Some quantitative data on the major constituents of bile*

Analyses in mg per 100 ml.

	Total lipids	Bilirubin	Cholesterol	Fatty acids	Total bile acids
Ox	100–160	—	37	370	7200
Dog	—	92–170	80–100	1600–5000	7900–15 000
Guinea pig	140	—	—	—	780
Rabbit	—	87–131	10–120	—	1100–2600
Rat*	—	8–9	12.7	—	—
Pig	—	32–62	130–180	820–2000	8500–12 000
Man	—	1000	630	970	5180

*No gall bladder.
Source: Data from Dittmer (1961).

as amino acids or glucose (p. 355), it is a relatively poor medium compared with that provided by the duodenum. On the other hand, to a trematode such as *Schistosoma mansoni* (p. 265), which possesses both a gut and a well-developed digestive enzyme system, the quantity of protein in the plasma and blood cells represents a diet of potentially high nutritional value. In mammalian blood, apart from protein and the usual inorganic constituents, the substances likely to be of physiological importance to a parasite are fat, in the form of neutral fats (triglycerides), lecithin and cholesterol, amino acids and carbohydrates. Although the greater part of the amino acids are deaminated by the liver, some pass into the systemic circulation from which they are normally taken up by the tissues to repair protein wear and tear. The resting concentration of amino acids in blood (man) is 3–5 mg per 100 ml, but may rise to 10 mg per 100 ml after a protein meal. Glucose in the blood of a fasting animal may be 50–100 mg per 100 ml but may rise rapidly after feeding to 125 mg per 100 ml, to fall again as it diffuses into the tissues for energy purposes. The concentrations of nutrients in the blood may be influenced by the presence of parasites. Only a limited number of parasites are capable of direct utilisation of blood protein. The trematode *Schistosoma* (p. 265) and the malarial organism (*Plasmodium* sp.), which metabolises haemoglobin within an erythrocyte, are examples.

The gaseous content of the blood varies with the species and the source (i.e. whether arterial or venous; see Table 2.2). In man, the difference between the oxygen tension in arterial and venous blood shows a utilisation of 27 per cent, whereas in animals with a high metabolic rate (birds) it may be as high as 60 per cent. The carbon dioxide content is remarkably constant at about 40 mmHg, and the pH varies only slightly on each side of neutrality with an average of 7.4 in most warm-blooded animals. The relative viscosity is about 1.8.

In addition to providing metabolic materials such as oxygen, carbon dioxide and nutrients to parasites, blood contains globulins which are concerned with immune reactions, especially the production of antibodies; the role of these in immune phenomena is discussed in Chapter 32.

Most of the above information is based on human blood. Information on the physico-chemical characteristics of the body fluids of other organisms, both vertebrate and invertebrate, has been reviewed by Altman & Dittmer (1974) and Chapman (1980).

○ 2.4.2 Blood cells

For a succinct account of blood cells as habitats for parasites, see Cox (1976). Two groups of cells are found, red blood cells and white cells (leucocytes). In mammals, three types of white cells occur, lymphocytes, monocytes and polymorphonuclear granulocytes. A fourth type – thrombocytes – occurs in birds and reptiles.

○○○ 2.5 Tissues and other habitats

Muscle. Muscle is not generally a favoured site for parasites. This is probably related to the nature of its function, which results in an environment which is liable to change suddenly. Despommier (1976) has given a useful analysis of muscle as an environment for parasites. The most studied parasites of skeletal muscle are probably the protozoan *Sarcocystis* (p. 104) and the encysted stage of the nematode *Trichinella spiralis* (p. 390). In addition to containing high levels of ATP, muscle cells contain unusual components such as myoglobin, myosin, actin, creatine and creatine phosphate. The presence of some substances (e.g. lactic acid) may fluctuate wildly depending on the state of contraction of the muscle.

Liver. Protozoans (e.g. coccidians) and several helminths (especially trematodes and larval cestodes) are the main parasites found in the liver. This organ does not provide a very stable environment for its chemical composition may fluctuate widely depending on whether the diet is a balanced one, or one rich in proteins, fats or carbohydrates. Ranges for extremes are: fat, 1.6–52 per cent; glycogen, 0.07–11 per cent; water 35–73 per cent of liver mass. It is also rich in accessory growth factors, such as iron or vitamins, and plasma proteins. Its arterial and venous supply is excellent so that as a tissue habitat it is probably aerobic. From the abundance of parasites in or near the liver, it may be assumed that as an environment it is particularly favourable, and that many of the stored materials are available for uptake by the parasites. Its pH is in the region of 7.0 with a tendency towards acidity. A major function of the liver is to produce bile, which is not only involved in the absorption of lipid by the gut but also serves as an important 'trigger' in many parasite life cycles. For a comprehensive account of the liver and other digestive organs as parasite habitats, see Cheng (1976).

Body cavity (peritoneal cavity). Many larval helminths occur in the body cavity of mammals, especially rodents, and other vertebrates, especially fish and amphibia. The physico-chemical properties and the cellular constituents have been reviewed by Howell (1976).

Cerebrospinal fluid. The chemical composition of the cerebrospinal fluid somewhat resembles that of lymph. It is apparently a true secretion and not just a dialysate of blood plasma. It differs from other body fluids in the small protein content, 28 mg per 100 ml or approximately 0.3–0.5 per cent that of blood plasma. There is a correspondingly small antibody content, with the result that parasites may be able to persist in the central nervous system after antibodies have eliminated them from other parts of the body.

○○○ 2.6 Invertebrate habitats

Although invertebrates are parasitised by protozoa and adult worms, their main interest to the parasitologist lies in the fact that they serve as intermediate hosts to all the major parasitic groups: Protozoa, Trematoda, Cestoda, Nematoda and Acanthocephala.

The most favoured sites are probably the haemocoele, the coelom, the digestive gland or the muscles. To discuss the nature and composition of all these habitats in the arthropods and molluscs – the groups which act chiefly as intermediate hosts – is beyond the scope of this book.

○○○ 2.7 Importance of nutritional levels of environment in parasite life cycles

One of the features of parasitic organisms is that they may be exposed to a number of different environments during their life cycle. As many as four hosts may be involved (e.g. some trematodes) as well as several free-living phases (e.g. miracidia, cercariae) in water. Changes of environment may involve major changes in the physico-chemical conditions such as pH, E_h, pCO_2, pO_2, temperature or osmotic pressure. On theoretical grounds, the metabolism of a parasite will be adapted to each change of environment; such metabolic changes are likely to be accompanied by correlated biochemical and ultrastructural changes.

An example of such a switch is seen in the haemoflagellates *Trypanosoma rhodesiense* (p. 69) or *Leishmania mexicana* (Fig. 5.20) in which a change from the nutritionally rich, oxygen-rich, mammalian blood stream to the nutritionally poor, oxygen-poor, insect gut is accompanied by synthesis of mitochondria and associated biochemical changes. Similar changes probably occur in many helminth life cycles (e.g. *Fasciola hepatica* (Fig. 18.2), *Ascaris suum* (Fig. 30.2)).

The changes in the characteristics of environments are utilised by parasites to 'trigger' the appropriate development of the next stage in the life cycle. Thus, as emphasised earlier, a parasite may respond to (say) the p_{CO_2} in the gut or the presence of bile and in this way is able to 'recognise' that it is an environment suitable for establishment and development.

It is well recognised that the nutritional demands of an organism may vary with different phases in its life cycle. Thus, the demands of a cysticercus stage of a cestode, undergoing growth in size with little differentiation, are likely to be different from those of an adult cestode producing thousands of eggs daily. Two extreme levels of nutrition can be recognised, a 'survival' level at which cells or tissues metabolise without undergoing growth in size or differentiation, and a 'differentiation' level at which cells can undergo growth and differentiation into more complex structures such as eggs or sperm.

In natural life cycles, then, it is probable that nutritional barriers can similarly operate. This implies that the degree of differentiation, the rate of development, or the reproductive capacity of a parasite will be dependent, among other factors, on the nutritional value of the particular environment, especially as regards the availability of protein. In general, the metabolic demands of parasites of homoiotherms will be higher than those of poikilotherms. Marked differences exist, however, between species. For example, the egg production of the trematode *Schistosoma japonicum*, at 500 eggs daily, is one hundred times that of the related species *S. mansoni*, although both inhabit the portal system of mammals.

This is a somewhat over-simplified view of the factors underlying developmental patterns in helminth life cycles. It does not imply that nutritional factors are the only external factors regulating differentiation and maturation of parasites in their hosts. As indicated above, in many cases special stimuli are required by the parasite before it is metabolically or morphologically capable of maturation, and unless these stimuli have been applied, the nutritional factors are unable to operate. Thus, maturation of several protozoa (p. 146) and a monogenean (p. 163) has been shown to be dependent on the endocrine behaviour of the host, and some nematode larvae only exsheath (i.e. become potentially capable of reaching maturity) in response to physicochemical factors in a particular site in their host (p. 381).

It is also important to note that variation in the diet of the host – and therefore variation in the normal environment of a parasite – can have profound effects not just on growth and maturation but on characteristics such as virulence. For example, growth of the rodent malaria *Plasmodium berghei* and of *Toxoplasma gondii* is substantially inhibited by maintaining the host on a milk diet (which is deficient in *p*-aminobenzoic acid).

○○○ References

(References of general interest have been included in this list, in addition to those quoted in the text.)

Altman, P. L. & Dittmer, D. S. (eds) 1974. *Biological Handbooks: Blood and other Body Fluids.* (*Biology Data Book*, Vol. III, 2nd Edition.) Federation of American Societies for Experimental Biology, Bethesda, Maryland, USA.

Befus, A. D. & Podesta, R. B. 1976. Intestine. In: *Ecological Aspects of Parasitology* (ed. C. R. Kennedy), pp. 303–25. North-Holland, Amsterdam.

Chadwick, V. S. & Phillips, S. F. 1982. *Small Intestine.* Butterworth Scientific, London.

Chapman, G. 1980. *The Body Fluids and their Functions.* 2nd Edition. (*Studies in Biology*, No. 8.) Edward Arnold, London.

Cheng, T. C. 1976. Liver and other digestive organs. In: *Ecological Aspects of Parasitology* (ed. C. R. Kennedy), pp. 287–302. North-Holland, Amsterdam.

Code, C. F. (ed.) 1968. Alimentary Canal. *Handbook of Physiology*, Vol. 5. American Physiological Society, Washington.

Cox, F. E. G. 1976. Blood cells. In: *Ecological Aspects of Parasitology* (ed. C. R. Kennedy), pp. 387–407. North-Holland, Amsterdam.

Crompton, D. W. T. 1970. *An Ecological Approach to Acanthocephalan Physiology.* Cambridge University Press.

Crompton, D. W. T. 1973. The sites occupied by some parasitic helminths in the alimentary canal of vertebrates. *Biological Reviews*, **48:** 27–83.

Crompton, D. W. T., Shrimpton, D. H. & Silver, I. A. 1965. Measurement of the oxygen tension in the lumen of the small intestine of the domestic duck. *Journal of Experimental Biology*, **43:** 473–8.

Despommier, D. 1976. Musculature. In: *Ecological Aspects of Parasitology* (ed. C. R. Kennedy), pp. 269–85. North-Holland, Amsterdam.

Dittmer, D. S. (ed.) 1961. *Biological Handbooks: Blood and other Body Fluids.* 1st Edition. Federation of American Societies for Experimental Biology. Bethesda, Maryland, USA.

Gause, G. F. 1934. *The Struggle for Existence.* Williams & Wilkins, Baltimore.

Hamilton, J. D., Dawson, A. M. & Webb, J. P. W. 1968. Observations on the small gut 'mucosal' pO_2 and pCO_2 in anaesthetised dogs. *Gastroenterology*, **55:** 52–60.

Hardin, G. 1960. The competitive exclusion principle. *Science*, **131:** 1292–7.

Haslewood, G. A. D. 1967. *Bile Salts.* Methuen, London.

Heaton, K. W. 1972. *Bile Salts in Health & Disease.* Churchill Livingstone, London.

Hewitt, E. A. & Schelkopf, R. L. 1955. pH values and enzymatic activity of the digestive tract of a chicken. *American Journal of Veterinary Research*, **16**: 576–9.

Hopkins, C. A. 1967. The *in vitro* cultivation of cestodes with particular references to *Hymenolepis nana. Symposium of the British Society for Parasitology*, **5**: 27–47.

Howell, M. J. 1976. The peritoneal cavity of vertebrates. In: *Ecological Aspects of Parasitology* (ed. C. R. Kennedy), pp. 246–68. North-Holland, Amsterdam.

Johnston, L. R. 1981. *Physiology of the Gastrointestinal Tract* (2 vols). Raven Press, New York.

Kennedy, C. R. (ed.) 1976. *Ecological Aspects of Parasitology*. North-Holland, Amsterdam.

Lewis, S. M. 1982. The constituents of normal blood and bone marrow. In: *Blood and its Disorders* (ed. R. M. Hardisty & D. J. Weatherall), 2nd Edition, pp. 3–56. Blackwell Scientific Publications, Oxford.

McManus, D. P. & Smyth, J. D. 1978. Differences in the chemical composition and carbohydrate metabolism of *Echinococcus granulosus* (horse and sheep strains) and *E. multilocularis. Parasitology*, **77**: 103–9.

Madge, D. S. 1975. *The Mammalian Alimentary Canal: a functional Approach*. Edward Arnold, London.

Mettrick, D. F. 1970. Protein nitrogen, amino acid and carbohydrate gradients in the rat intestine. *Comparative Physiology and Biochemistry*, **37**: 517–41.

Mettrick, D. F. 1971*a*. *Hymenolepis diminuta*: pH changes in the rat intestinal contents and worm migration. *Experimental Parasitology*, **29**: 386–401.

Mettrick, D. F. 1971*b*. Effect of host dietary constituents on intestinal pH and the migratory behavior of the rat tapeworm, *Hymenolepis diminuta. Canadian Journal of Zoology*, **49**: 1513–25.

Mettrick, D. F. 1971*c*. *Hymenolepis diminuta*: the microbial fauna, nutritional gradients, and physico-chemical characteristics of the small intestine of uninfected and parasitized rats. *Canadian Journal of Physiology and Pharmacology*, **49**: 972–84.

Mettrick, D. F. 1973*a*. Competition for ingested nutrients between the tapeworm *Hymenolepis diminuta* and the rat host. *Canadian Journal of Public Health, Monograph Suppl.*, **64**: 70–82.

Mettrick, D. F. 1973*b*. The intestine as an environment for *Hymenolepis diminuta*. In: *Biology of the Tapeworm Hymenolepis diminuta* (ed. H. P. Arai), pp. 281–356. Academic Press, New York.

Morton, J. 1979. *Guts*. (*Studies in Biology* no. 7.) Edward Arnold, London.

Podesta, R. B. 1980. Concepts of membrane biology in *Hymenolepis diminuta*. In: *Biology of the Tapeworm Hymenolepis diminuta* (ed. H. P. Arai), pp. 505–9. Academic Press, New York.

Pounder, R. E., Allison, M. C. & Dhillon, A. P. 1989. *A Colour Atlas of the Digestive System*. Wolfe Medical Publications, London.

Richards, O. W. & Davies, R. G. 1977. *Imm's General Textbook of Entomology*, 10th Edition, Vol. 1: *Structure, Physiology and Development*. Chapman & Hall, London.

Rogers, W. P. 1949. On the relative importance of aerobic metabolism in small nematodes in the alimentary canal. I. Oxygen tensions in the normal environment of the parasites. *Australian Journal of Scientific Research. Series B. Biological Sciences*, **2B**: 157–65.

Sanford, P. A. 1982. *Digestive System Physiology*. Edward Arnold, London.

Smithers, P. A. & Worms, M. J. 1976. Blood fluids – helminths. In: *Ecological Aspects of Parasitology* (ed. C. R. Kennedy), pp. 349–69. North-Holland, Amsterdam.

Smyth, J. D. 1962. Lysis of *Echinococcus granulosus* by surface-active agents in bile and the role of this phenomenon in determining host specificity in helminths. *Proceedings of the Royal Society of London*, **B156**: 553–72.

Smyth, J. D. 1987. Changing concepts in the microecology, macroecology and epidemiology of hydatid disease. In: *Helminth Zoonoses* (ed. S. Geerts, V. Kumar & J. Brandt), pp. 1–11. Martinus Nijhoff, Dordrecht, The Netherlands.

Smyth, J. D. & Halton, D. W. 1983. *The Physiology of Trematodes*. Cambridge University Press.

Smyth, J. D. & Haslewood, G. A. D. 1963. The biochemistry of bile as a factor in determining host specificity in helminths. *Annals of the New York Academy of Sciences*, **113**: 234–60.

Smyth, J. D. & McManus, D. P. 1989. *The Physiology and Biochemistry of Cestodes*. Cambridge University Press.

Stevens, C. E. 1988. *Comparative Physiology of the Vertebrate Digestive System*. Cambridge University Press.

Von Brand, Th. 1952. *Chemical Physiology of Endoparasitic Animals*. Academic Press, New York.

Von Brand, Th. 1966. *Biochemistry of Parasites*. Academic Press, New York.

Williams, H. H. 1960. The intestine in members of the genus *Raja* and host specificity in the Tetraphyllidea. *Nature*, **188**: 514–16.

Wrong, O. M., Edmonds, C. J. & Chadwick, V. S. 1981. *The Large Intestine: its role in Mammalian Nutrition and Homeostasis*. MTP Press, Lancaster, England.

3

Protozoa: the amoebae*

Amoebae occur in almost every natural habitat capable of supporting life: water, soil, decaying plant material and sewage. Organisms with such a wide distribution can be readily ingested accidentally by potential hosts, with drinking water or food.

It is not difficult to visualise how, by suitable adaptations, physiological as well as morphological, some species became at first temporary, and finally permanent parasites of invertebrates and vertebrates, including man. Indeed, one of the most astonishing findings of recent years has been that several genera of amoebae – previously considered to be entirely free-living in soil and fresh water – are capable of invading the tissues of man and causing serious diseases often resulting in death; such amoebae are referred to as *amphizoic amoebae*. Particularly important genera are *Naegleria* and *Acanthamoeba*, which are considered later in this chapter.

Since many other genera (*Hartmanella*, *Vahlklamfia*, *Paramoeba*, *Vexillifera* and *Nuclearia*) have been found to infect invertebrates and vertebrates other than man, it is possible that some of these may still be involved in human and animal infections as yet unidentified.

Amoebae are essentially scavengers; they feed mainly on bacteria or detritus, being capable of ingesting particles as large as starch grains (about 2–5 μm). The parasitic species (e.g. *Entamoeba histolytica*) are capable (under certain conditions) of becoming carnivorous and invading host tissues. The use of pseudopodia by amoebae generally imposes a peculiar limitation on their habitats, i.e. the latter must normally present a surface on which organisms can move and capture and ingest food. This limitation is probably one of the most important controlling the distribution of amoebae in the body. A habitat such as the blood stream, which is suitable for other classes of parasitic protozoa, such as haemoflagellates (p. 57), would thus be unlikely to present suitable conditions for the long-term maintenance of a potentially parasitic amoeba.

In the body of the host, amoebae thus usually confine themselves to habitats which present suitable surfaces on which they may graze and which generally are rich in bacteria and decaying food material. In vertebrates, these conditions are provided by the mouth, the large intestine, and the caecum. Physical and chemical conditions of the intestinal environment, i.e. pH, oxygen tension, $p\mathrm{CO_2}$, osmotic pressure, E_h, bile concentration and composition (p. 17), etc., must also play some part in limiting their distribution.

○○○ 3.1 Type example: *Entamoeba muris*

Definitive hosts: mouse, rat, hamster; various other rodent species.
Location: caecum and large intestine.
Transmission: cysts.

For teaching and demonstration purposes, the human pathogen *Entamoeba histolytica* may not be easily available, but two non-pathogenic species, common in laboratory animals, make excellent teaching material: *Entamoeba muris* (Fig. 3.1) in mice, and *E. ranarum* (Fig. 3.7) in amphibia (Smyth & Smyth, 1980).

Occurrence. *E. muris* is common in mice, 43 per cent prevalence being reported in one commercial breeding population (Casebolt *et al.*, 1988).

General morphology of trophozoite. The most complete account available is that of Neal (1950). The

*The classification of the Protozoa is given in the Appendix (p. 42); amoebae belong to the Subphylum Sarcodina.

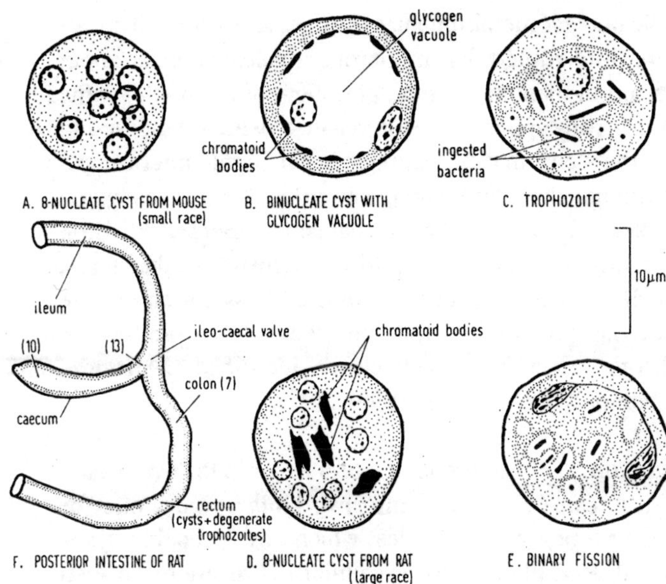

Fig. 3.1
A–E. Morphology of *Entamoeba muris*. F. Its distribution in thirteen autopsies; the numbers in brackets refer to the number of rats with infections at each site. (After Neal, 1950.)

general form of the active trophozoite is irregular (Fig. 3.1) and about 12–30 μm in length. In rats and mice there are two 'races', a 'small' race in mice, and a 'large' race in rats: numerous other 'races' or strains occur in other rodents. The ectoplasm is usually readily distinguished from the endoplasm. The pseudopodia are ectoplasmic, that is, they are extensions of the ectoplasm without corresponding endoplasmic streaming, such as occurs in most free-living amoebae. In fixed preparations, the ectoplasm is homogeneous and finely granular, and the endoplasm coarsely granular with food vacuoles and nucleus. The nucleus is characterised by a peripheral layer of chromatin in the form of a beaded ring (Fig. 3.1) and an eccentric karyosome. The peripheral concentration of chromatin is characteristic of parasitic species of *Entamoeba*, and this feature may be used to distinguish it from other genera. The following stages may be recognised:

(*a*) trophozoites: the mature adults, large and active.
(*b*) precystic stages: small amoeboid stages prior to encystment.
(*c*) cysts: encysted stages.
(*d*) metacystic stages: excysting stages with many nuclei giving rise to (*a*).

Except for *Entamoeba gingivalis*, which does not form cysts, departures from this pattern of development are unusual among parasitic amoebae.

Nutrition. *E. muris* is a non-pathogenic species. It is essentially a scavenger, feeding on whatever materials are at hand, and its food vacuoles normally contain bacteria or intestinal protozoans.

Reproduction. This is by binary fission, the nucleus dividing some time before the cytoplasm. The dividing nuclei are characterised by the appearance of chromatin granules with the loss of peripheral beading; whether these granules represent 'chromosomes' is open to question.

Encystment. In *E. muris*, the conditions under which encystment takes place are not fully understood. The precystic forms differ from the trophic forms in (*a*) their smaller size and (*b*) the fact that inclusions are not found, which results in the demarcation between ectoplasm and endoplasm being less clear. The organisms become rounded and secret a cyst wall. Ultrastructure studies on the cyst wall have shown that it is composed mainly of aggregates of filaments adjacent to the plasma membrane (Ravdin, 1988). The chemical nature of the wall is not completely known but a chitin-like component has been observed in *E. histolytica* and *E. invadens* cysts. A characteristic of the newly formed cyst is the appearance of a *glycogen vacuole* (Fig. 3.1), a large vacuole which reaches its maximum size at the binucleate stage, and which almost fills the cyst.

These vacuoles are characteristic of most parasitic amoebae in the cystic stage, and presumably serve as food reserves. The glycogen vacuole is readily revealed in iodine-stained cysts.

A cyst usually contains, in addition, refractive rod-like structures called *chromatoid bodies* or *chromidial bars*, which may be of any shape, but are usually bar-shaped with irregular, splintered or, less frequently, rounded ends; these structures have been shown to consist of RNA and probably serve as ribosome reserves for the metacystic stages (p. 27); these may be absent from mature cysts.

The cyst size varies somewhat. Typical measurements are: small race from mice, 9–19 μm (Fig. 3.1); large race from rats, 12–22 μm (Fig. 3.1).

Life cycle. Cysts are ingested with food or drinking water, or reach the mouth accidentally, and hatch in the duodenum. After excystment in the intestine, the amoebae ultimately become established in the caecum for which they have a marked predilection; rarely do they occur in quantity higher up the alimentary canal than the ileo-caecal valve, and only in reduced numbers in the colon (Fig. 3.1).

Alternative hosts. Although rats, mice and hamsters may be infected with strains of *E. muris*, only the rat strain will develop in all three hosts.

○○○ 3.2 Ultrastructure

○ *Entamoeba* spp. (Fig. 3.2)

Although the ultrastructure of *E. muris* has not been examined, it is possible, from studies of other species (see below) such as *E. histolytica* (Martínez-Palomo, 1982, 1986; Lushbaugh & Miller, 1988; Yong *et al.*, 1985) and other species, to build up a generalised picture of the fine structure of a 'typical' amoeba (Fig. 3.2).

Plasmalemma (Plasma membrane). The boundary membrane has the characteristic 'unit' structure, as in other protozoa, and is covered by a 'fuzzy' layer, probably of mucopolysaccharide. Acid phosphatase activity has been demonstrated in this membrane (Rondanelli *et al.*, 1977).

Nucleus. The structure, ultrastructure and cytochemistry of the nucleus have been extensively studied. The nuclear membrane has the characteristic unit structure interrupted by numerous nuclear pores, about 65 nm in diameter. At the periphery of the nucleus, the chromatin appears as aggregates, which correspond to the 'chromatin' aggregates of light microscopy. Although the DNA composition has been investigated in several species, the mechanism of nuclear division remains a mystery, in spite of intensive study. Thus, nuclear division proceeds without dissolution of the nuclear membrane and no centrioles, spindles or chromosomes have been clearly identified (Martínez-Palomo, 1982; Ravdin, 1988).

Vacuoles. Numerous vacuoles occur in the cytoplasm. These can be grouped into two kinds: pinocytotic (or endocytotic) and vacuoles which are probably lysosomal. Red blood cells or cellular debris are often seen in pinocytotic food vacuoles, especially in pathogenetic strains. Clearly, after ingestion, the degradation of the mammalian cell components implies the existence of a lysosomal system in the amoeba and there is evidence that this may be different from the mammalian cell model (Schlesinger, 1988).

Particulate and vesicular components. The ground cytoplasm contains ribosomes (20–30 nm) and glycogen particles (40–70 nm); the latter presumably become aggregated and surrounded by a membrane to form a glycogen vacuole (Fig. 3.1). The cytoplasm is also finely alveolated with scattered circular vesicles which probably represent a poorly developed endoplasmic reticulum.

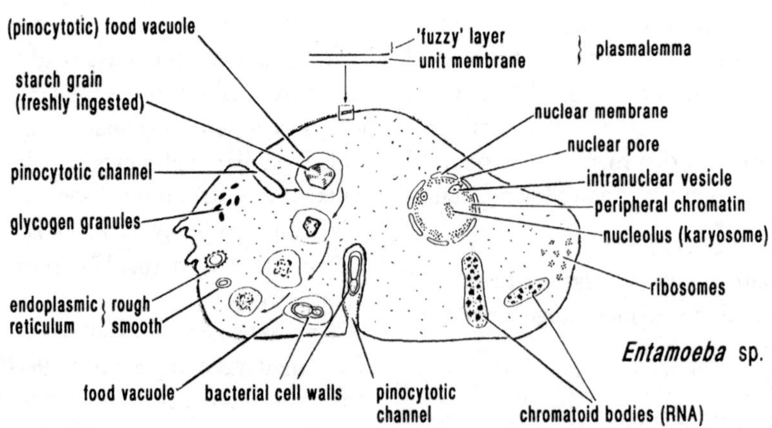

Fig. 3.2
Ultrastructure of *Entamoeba* spp. Parasitic amoebae lack mitochondria and hydrogenosomes. (Original.)

Table 3.1. *The best-described species of* Entamoeba

Species	Host	No. of nuclei in cysts
Entamoeba barretti	*Chelydra serpentina* (turtle)	8
Entamoeba cobayae	laboratory guinea pigs	8
Entamoeba coli	man	8
Entamoeba gallinarum	domestic fowls	8
Entamoeba marmotae	*Marmota monax*	8
Entamoeba muris	various rodents	8
Entamoeba aulastomi	leech	4
*Entamoeba dispar**	man	4
Entamoeba hartmanni	man	4
Entamoeba histolytica	man	4
Entamoeba invadens	various reptiles (esp. snakes)	4
Entamoeba knowlesi	tortoises	4
Entamoeba moshkovskii	sewage	4
*Entamoeba nuttalli**	monkeys	4
Entamoeba philippinensis	*Panesthia javanica* (roach)	4
Entamoeba pyrrhogaster	salamanders	4
Entamoeba ranarum	amphibians (esp. frogs)	4
Entamoeba terrapinae	turtles	4
Entamoeba bovis	cattle	1
Entamoeba chattoni	monkeys	1
Entamoeba debliecki	goats, pigs	1
Entamoeba polecki	man, pigs	1
Entamoeba suis	pigs	1
*Entamoeba caudata**	dog	unknown
Entamoeba gingivalis	man	?†

*Probably synonymous with *E. histolytica*.
†See text.
Source: Data from Neal (1966).

Golgi apparatus. This is absent in all species so far investigated (contrast the soil amoeba *Acanthamoeba*, p. 36).

Mitochondria. These are absent in *E. invadens*, *E. gingivalis* and *E. histolytica*, a feature presumably related to their anaerobic metabolism (p. 38). Numerous mitochondria are found in *Acanthamoeba* (p. 36).

Writing on amitochondrial protozoa in general, Müller (1988) speculated '. . . that the absence of mitochondria might be a primary feature that has been conserved from times when life was still anaerobic and mitochondria were unknown.' He points out that two major aspects of advanced energy metabolism are missing, namely electron transport-linked phosphorylation and utilisation of O_2 as a terminal acceptor for the electrons involved in this progress.

Chromatoid bodies. These are crystalloid structures, composed of RNA, dealt with further on p. 40.

○○○ 3.3 Entamoebae of man and other animals

Species of *Entamoeba* have been found in a number of invertebrate and vertebrate hosts (Table 3.1).

Five species of the genus *Entamoeba* infect man: *E. histolytica*, a species which is sometimes pathogenic in the caecum and colon; *E. hartmanni*, a harmless species closely related to *E. histolytica* and not regarded as a separate species by some workers; *E. coli*, a harmless species; *E. gingivalis*, in the mouth; and *E. polecki*, a species rare in man and probably normally occurring in pigs. Morphological and physiological characteristics of some of these species are given in Table 3.2. The relationship between these species is complicated by the discovery of numerous 'strains' of some species.

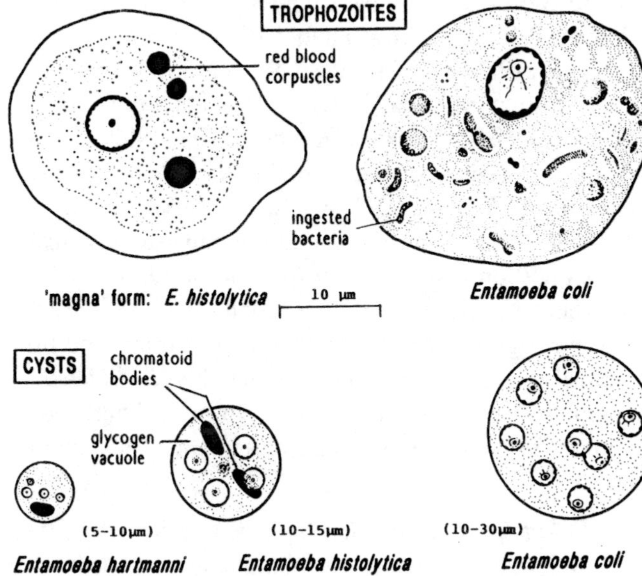

Fig. 3.3
Comparison of *Entamoeba histolytica*, *E. coli* and *E. hartmanni*, parasitic in man. (Adapted from Hoare, 1959.)

○ 3.3.1 *Entamoeba histolytica*

Distribution. Infections of *Entamoeba histolytica* commonly result in *amoebiasis*, the term 'histolytica' literally meaning 'tissue-dissolving', referring to the potential carnivorous habits of this organism. The qualification *potentially* is stressed, for, as discussed below, a high percentage of individuals infected with entamoebae show no symptoms of disease. Where clinical symptoms result, the disease is referred to as *invasive amoebiasis*; the non-invasive infection is sometimes called *lumenal amoebiasis*.

In the developing world, amoebiasis causes some 450 million infections *per annum*, about 50 million incidents and about 100 000 deaths (Ravdin, 1988). Invasive amoebiasis is prevalent in certain areas of the world including West and South East Africa, China, the whole of SE Asia, Mexico and the western portion of South America, and the Indian subcontinent. Possibly only about 10 per cent of infections result in invasive amoebiasis; the remainder of those infected appear to remain asymptomatic carriers who pass cysts in their stools (Mirelman, 1987, 1988; Sargeaunt, 1987).

Basic biology. The basic biology of *E. histolytica* is described in detail in a number of books and reviews (Martínez-Palomo, 1982, 1986; Ravdin, 1988; Kretschmer, 1990; Sepulveda, 1982) and has been the subject of several WHO Working Groups (WHO, 1980, 1984). A major question has been whether there is one or two species of '*E. histolytica*' or whether these exist as a number of intra-specific variants (strains or isolates). This question is discussed further on p. 29.

Where size is used as a criterion, trophozoites fall into three groups and the cysts into two groups and the controversy has centred on whether this '*E. histolytica* complex' represents one or more species. There now appears to be a large measure of agreement that there occur:

(*a*) *E. histolytica* 'magna' (?) trophozoites, diameter 20–40 μm; these do not form cysts and may be haematophagous (blood-ingesting).
(*b*) *E. histolytica* 'minuta' (?) trophozoites, diameter 7.0–15.9 μm, which form cysts, diameter 10–15 μm, feed on bacteria and detritus, and are never haematophagous.
(*c*) *E. hartmanni* trophozoites, diameter 5–11 μm; these form cysts of diameter 5–10 μm and which similarly are not haematophagous.

There is thus general acceptance (Elsdon-Dew, 1968) that two species, i.e. *E. histolytica* and *E. hartmanni*, exist; the dividing point in cyst diameter occurs at approximately 10 μm (Fig. 3.3). Differences between the nuclei of the two species have also been reported (Neal, 1966). In *E. histolytica* the peripheral chromatin is arranged in the form of small granules, while in *E. hartmanni* the chromatin occurs in larger peripheral masses. The isoenzyme pattern of *E. hartmanni* has also been shown to be markedly different from that of *E. histolytica* (Sargeaunt & Williams, 1979).

Morphology. The general morphology of *E. histolytica* and *E. hartmanni* closely resembles that of *E. muris*, except for a few points (Table 3.2). In the living trophozoite, the nucleus is almost invisible by light microscopy but can be seen by phase-contrast microscopy, whereas in *E. muris* the nucleus is clearly visible by the light microscope.

Although the morphology of *E. histolytica* is shown in Fig. 3.3 and described in Table 3.2, it is generally recognised that 'considerable variations in the structural features occur, especially in cultured forms (Albach & Booden, 1978). For a detailed, illustrated account of the morphology of *E. histolytica* and other intestinal amoebae, see Sargeaunt & Williams (1982). As already indicated, complementary methods, such as isoenzyme, protein or antigen analyses are now widely used for species or strain identification.

The movement of the pseudopodia in *E. histolytica*, as seen in a fresh film on a warm stage, is striking, the single ectoplasmic pseudopodium often (but not always) moving out in a manner usually described as 'explosive'. The precystic forms may lay down in their cytoplasm one or two chromatoid bodies, which are bar-like with rounded ends as opposed to the irregular form in *E. muris*. In encysted forms, the nucleus divides twice to give four nuclei (Fig. 3.3). Glycogen vacuoles are formed in the young cysts.

Life cycle (Fig. 3.4). After ingestion by man, and excystment in the small intestine, each tetranucleate amoeba produces eight uninucleate amoebae by a process of division. The amoebae may either invade the

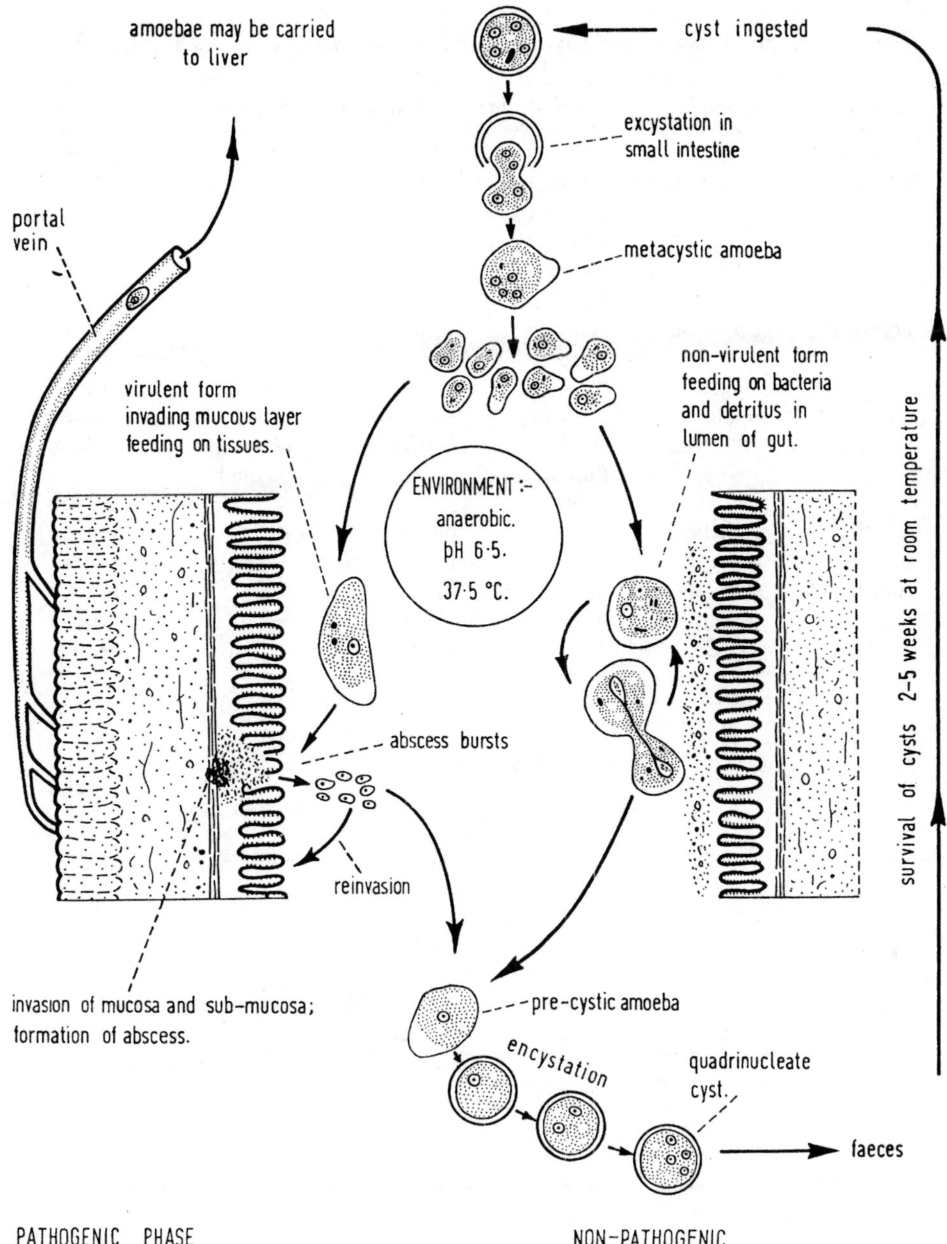

Fig. 3.4
Life cycle of *Entamoeba histolytica* in man. (Original.)

mucous membrane of the colon and caecum and multiply, or may remain in the intestinal lumen.

Pathogenicity. *E. histolytica* has long been regarded as an obligatory tissue parasite but the evidence now suggests that probably in the majority of persons it is harmless, the host serving as a 'symptomless carrier'. This type of infection is referred to as 'lumenal amoebiasis' (Elsdon-Dew, 1968). On the other hand, under conditions not understood, trophozoites (which become 'magna' trophozoites) may invade the tissues and feed on blood, forming abscesses in the mucosa and submucosa. Some may be carried by the portal blood stream to the liver.

Table 3.2. *Comparison of some morphological and physiological characteristics of species of* Entamoeba

	E. muris	*E. coli*	*E. histolytica*	*E. hartmanni*	*E. gingivalis*	*E. invadens*	*E. moskovskii*
Host	Mouse	Man	Man	Man	Man	Reptiles	Sewage
Trophozoite size	12–30 μm	20–30 μm	20–40 μm ('magna') 7.0–15.9 μm ('minuta')	5–11 μm	10–20 μm	9–38 μm	9–29 μm
Karyosome position	eccentric	eccentric	central; varies	central, varies	central	eccentric	central
Pseudopodium	sluggish, blunt	sluggish, blunt	blade-like, explosive	blade-like, explosive	broad, active	blade-like, explosive	*limax* type
Cyst size	9–22 μm	10–30 μm	10–15 μm ('minuta' only)	5–10 μm	cysts not formed	11–20 μm	7–16 μm
Chromatoid bodies in cysts	bar-like, irregular or splinter-like	sometimes present, splinter-like	often present, bar-like	often present, bar-like	—	often present, bar-like	often present, bar-like
Nuclei in cysts	usually 8	usually 8	usually 4	usually 4	—	usually 4	usually
Optimum temperature for development	37 °C	37 °C	37 °C	37 °C	37 °C(?)	24–30 °C	24 °C
Survival temperature range	?	?	10*–41 °C	?	?	16–35 °C	17–37 °C

*'Laredo' strain.

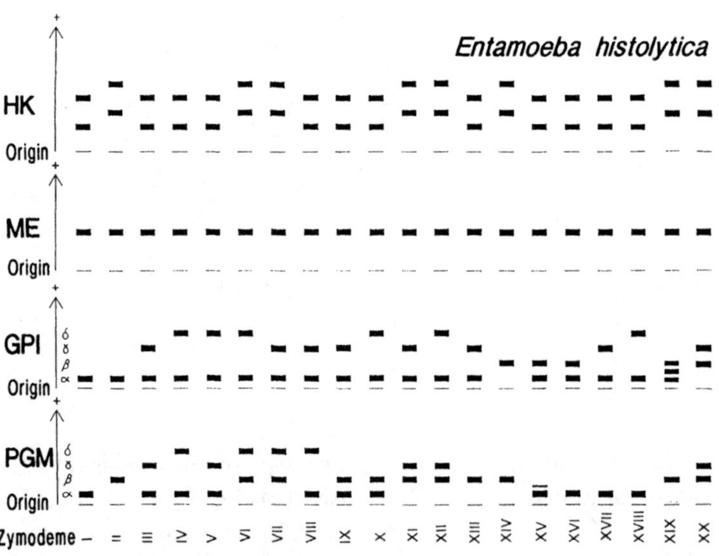

Fig. 3.5
Entamoeba histolytica: zymodemes identified using the following enzymes: **HK**, hexokinase; **ME**, L-malate: NADP$^+$ oxidoreductase (oxaloacetate decarboxylating); **GPI**, glucose phosphate isomerase; **PGM**, phosphoglucomutase. A *zymodeme* is a population of a species differing from similar populations in the electrophoretic mobility of certain enzymes. The markers for pathogenicity in amoebae are the absence of the α band together with the presence of the β band in **PGM**. Advanced bands in **HK** confirm the **PGM** results. The only exception is Zymodeme XIII, which lacks advanced bands. (After Sargeaunt, 1985, 1988.)

It must be stressed therefore that occurrence of *cysts* in faeces is evidence *only* of lumenal amoebiasis, and that the only unequivocal microscopic evidence of the invasive (pathogenic) condition is the finding of haematophagous trophozoites, as evidenced by the presence in them of red blood cells.

There is considerable controversy over the conditions which cause invasive amoebiasis (Mirelman, 1987; Martínez-Palomo, 1987; Sargeaunt, 1987) and the factors which result in lumenal trophozoites becoming transformed into invasive forms remain unknown. There is, however, unequivocal evidence of the existence of numerous strains of *E. histolytica* (and other species of amoeba). Those strains which are pathogenic and those which are not can now be identified with some degree of certainty. That strains existed has long been recognised from antigen studies but it was not until isoenzyme analysis (Fig. 3.5) became available that it was possible to attempt to separate the pathogenic and non-pathogenic strains (Sargeaunt & Williams, 1978, 1979).

The most successful studies have utilised thin-layer starch-gel electrophoresis to study isoenzyme patterns of the following enzymes: glucose phosphate isomerase (GPI; EC: 5319), phosphoglucomutase (PGM; EC: 2751), L-malate:NADP$^+$ oxidoreductase (oxaloacetate decarboxylating) (ME; EC: 11140) and hexokinase (HK; EC: 2711). A population of amoebae (or other organisms) with a unique combination of isoenzyme patterns (as expressed by electrophoretic mobility), is now referred to as a *zymodeme*. To date, 20 zymodemes of *E. histolytica* have been described from various parts of the world (Fig. 3.5) (Sargeaunt, 1988). Seven zymodemes have been obtained only from patients suffering from invasive amoebiasis, and 12 from asymptomatic 'cyst passers'; the remaining one was a result of genetic engineering. Most zymodemes have a global distribution with some local variations.

The general pattern emerging from these isoenzyme studies appears to be as follows (Sargeaunt, 1988): *E. histolytica* cultured from cases of invasive amoebiasis show the presence of a beta band in the absence of an alpha band in PGM. This is confirmed by fast-running bands in HK. An exception to this rule is zymodeme XIII, which does not have advanced bands in HK.

E. histolytica cultured from asymptomatic subjects who pass cysts 12.5 μm (average) in diameter, which have 1–4 vesicular nuclei and contain smooth-edged chromatoids, show an alpha band in PGM with slow-running bands in HK.

However, the clinical picture is not as straightforward as the above findings suggest, since (*a*) healthy subjects (i.e. those with non-pathogenic zymodemes) may nevertheless give antibody-positive serological reactions; and (*b*) some subjects infected with pathogenic zymodemes show no clinical disease although giving strong positive serological reactions (Jackson *et al.*, 1985) i.e. they are symptomless carriers. This evidence suggests that pathogenic zymodemes are in constant contact with the host tissues even in symptomless carriers.

A detailed account of the pathogenesis of invasive amoebiasis is given by Martínez-Palomo (1987). The view that pathogenic and non-pathogenic isolates of *E. histolytica* exist and can be detected by their isoenzyme profiles, as described above, has been challenged by the work of Mirelman (1987) who presented evidence that the isoenzyme pattern can be modified according to the growth medium conditions and the bacterial associates. However, the application of molecular biological techniques has shown that endonuclease digestion of amplified DNA yields fragments that are characteristic for pathogenic and non-pathogenic forms (Tannich & Burchard, 1991). The differentiation of 48 isolates into pathogenic and non-pathogenic strains by using this method corresponded to the clinical status of infected individuals and the classification obtained by isoenzyme analysis, a resulting supporting the view that pathogenic and non-pathogenic strains of *E. histolytica* constitute distinct subspecies.

○ 3.3.2 Other species of *Entamoeba*

Entamoeba coli (Table 3.2)

This is a non-pathogenic species whose distribution is world-wide, occurring in some 30 per cent of the world's population. It is found in the large intestine and distinguished from *E. histolytica* by a number of features, chief of which are: (*a*) it has slower movement; (*b*) its pseudopodium is mainly coarser and not clear like that of *E. histolytica*; (*c*) its nucleus is coarser and with an eccentric karyosome (Fig. 3.3); (*d*) it has a larger number of food vacuoles.

The precystic stages are difficult to distinguish from those of *E. histolytica*, critical observation of the nucleus being necessary. The cysts, however, are easily differentiated from those of *E. histolytica* by having eight nuclei and the chromatoid bodies, which, when present, have splintered, but never rounded ends (Fig. 3.3).

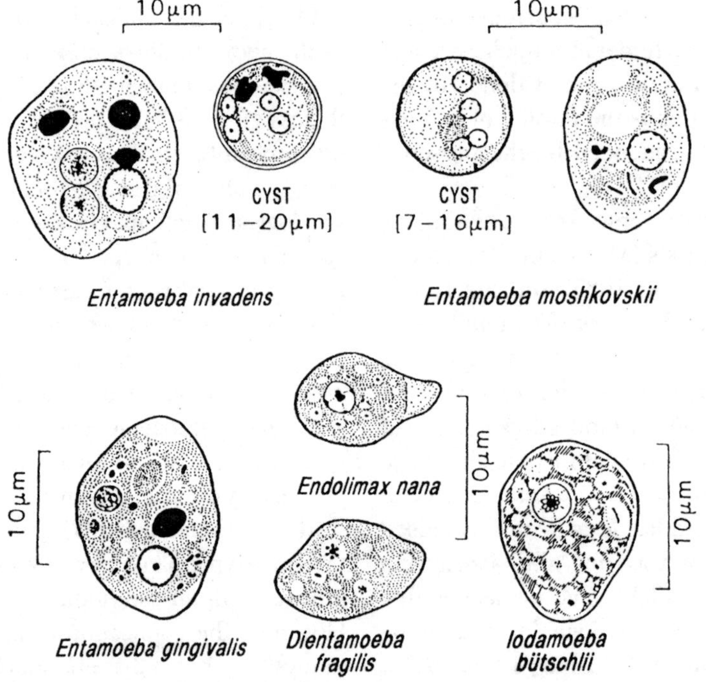

Fig. 3.6
Miscellaneous parasitic entamoebae. *E. gingivalis* is a non-pathogen (?) from the human mouth; *E. invadens*, a pathogen from reptiles; *E. moshkovskii*, a free-living species from sewage. The remaining species are relatively uncommon intestinal parasites of man. (Adapted from various authors.)

Table 3.3. *The frequency of* Entamoeba gingivalis *in histologic sections of 16 patients with moderate or advanced periodontitis*

Clinical classification	Present	Absent	Per cent present
Moderate periodontitis	1	5	17
Advanced periodontitis	9	1	90

Source: Data from Gottlieb & Miller (1971).

E. coli is entirely a scavenger, feeding on bacteria and detritus in the large intestine. It is not capable of eroding the intestinal mucosa, but, as an indiscriminate feeder, may phagocytose blood cells if these are available.

Entamoeba gingivalis (Table 3.2)

A common parasite of the human mouth; related species occur in the mouths of dogs, horses and donkeys. The spaces between the teeth and the soft pits of the gums offer ideal surfaces for amoebae, because in these sites bacteria and detritus abound. *E. gingivalis* resembles *E. histolytica*; it has a crystal-clear ectoplasm and moves actively (Fig. 3.6). Its food vacuoles are usually numerous and contain bacteria, leucocytes and occasionally red cells. The pathogenicity of this species has long been a matter of dispute and has never been established. In one study (Gottlieb & Miller, 1971) on sections of extracted teeth, the amoebae were never found within living gingival tissue but only within the plaque and cellular debris adjacent to the subcellular epithelium. Although amoebae were found more ferquently in patients with advanced periodontitis than in moderate infections (Table 3.3), in the absence of proof of tissue invasion, the question of pathogenicity must remain unanswered.

Entamoeba polecki

This is a species which normally occurs in pigs. It forms uninucleate cysts and shows some morphological differences from the cyst of *E. histolytica*. It is generally considered to be a rare parasite of man but may be more common than supposed. Thus, in a survey of 184 children in Papua New Guinea 19 per cent of the children were infected, rising to 43 per cent in one community

(Desowitz & Barnish, 1986). These authors suggest that these high figures indicate that this species might be transmitted from man to man.

Entamoeba moshkovskii (Table 3.2)

This species is not an animal parasite, but it may conveniently be discussed with the other forms. It is a free-living species, which was first discovered in a sewage-disposal plant in Moscow and has since been shown to be of world-wide occurrence (Neal, 1966). A large number of strains have been identified. Morphologically (Fig. 3.6) it closely resembles *E. histolytica* throughout its life cycle, including encystment and metacystic development, differing from it only in details (Neal, 1953). Yet all experiments to establish this species in rats or amphibians, animals which might act as hosts in sewage beds, have failed, and there seems no doubt that it is a true free-living form. This is emphasised by its optimum temperature of cultivation, which is 24 °C. At 37 °C, the optimum temperature for *E. histolytica*, *E. coli* and *E. muris*, the organism grows well but does not encyst. It is possible that *E. moshkovskii* may be related to the 'Laredo' strain of *E. histolytica*, which can grow at temperatures down to 10 °C (p. 28).

Entamoeba invadens

This species is increasingly being used as an experimental model for the study of amoebiasis. This is partly on account of its resemblance to *E. histolytica* and partly because it causes invasive amoebiasis in reptiles. It is probably relatively harmless in turtles (Meerovitch, 1961), which may have been the original hosts, but is highly pathogenic to snakes, which may die within two weeks (Neal, 1966). Its life cycle has been described by Geiman & Ratcliffe (1936) and its morphology by Kreier (1978).

Entamoeba ranarum (Fig. 3.7)

The morphology of this species is indistinguishable from that of *E. histolytica*. It is commoner in tadpoles than adult frogs. The detailed life cycle is described by Sanders (1931).

○○○ 3.4 Other intestinal amoebae

Endolimax nana

A small intestinal amoeba of man, measuring some 6–15 μm in diameter. It has a wide distribution, occurring in 15–30 per cent of the world's population. The nucleus is characterised by the absence of chromatin granules (Fig. 3.6) and the presence of a large karyosome, usually irregular in shape and sometimes divided into several parts. The trophozoite is sluggish in movement, like *Entamoeba coli*, and is similarly a harmless parasite. *Endolimax reynoldsi* from the lizard and *E. blattae* from the cockroach are closely related forms.

Iodamoeba bütschlii

A small amoeba, very common in monkeys and pigs, up to 50 per cent of the latter being infected in some areas. It also occurs in man, where its incidence in most countries is usually less than 10 per cent. The trophozoites

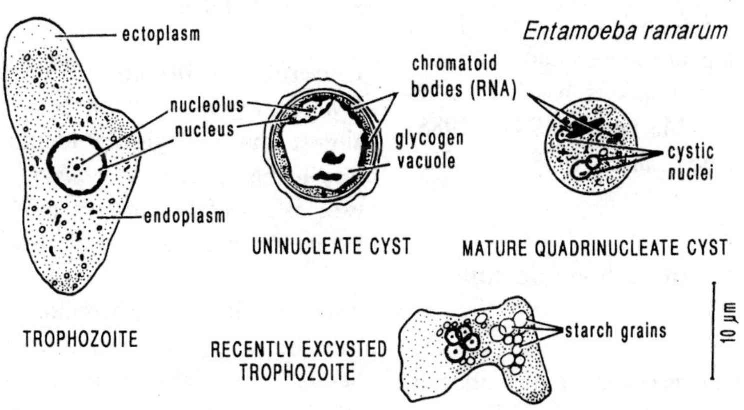

Fig. 3.7
Morphology of *Entamoeba ranarum*. (From Smyth & Smyth, 1980; after Sanders, 1931.)

(Fig. 3.6) are larger than *E. nana* but smaller than *Entamoeba histolytica*. Cysts usually have one, but sometimes two or three nuclei, and a large, sharply defined, glycogen vacuole.

Dientamoeba fragilis

This is a rather rare amoeba of man (Fig. 3.6). In an examination of 20 000 stools, the average incidence in the United States was 2.4 per cent (Kean & Malloch, 1966). The trophozoite diameter range is 4–20 μm but commonly 8–11 μm. It is normally binucleate, but uninucleate and quadrinucleate forms are found. The pathogenic role of this parasite has not been proved; like *Entamoeba*, it appears to induce no symptoms in some patients and abdominal pains in others. In a survey of 22 day-centres in Toronto, a prevalence of 7.8 per cent was reported, the highest being in children 7–10 years old (Keystone *et al.*, 1984).

○○○ 3.5 Tissue-invading soil amoebae

○ 3.5.1 General account

One of the most important discoveries within the past 25 years has been that certain species of free-living ('soil') amoebae are capable of invading mammalian tissues, causing severe pathogenic conditions and frequently death. These organisms occur as free-living forms in water, including mineral water (Table 3.4), moist soil, sewage or other decaying matter. They feed on bacteria and form exceptionally durable cysts. Taxonomically they belong to a rather ill-defined group of free-living amoebae often referred to as *limax* (Latin: slug) on account of their slug-like movements. The general biology of the pathogenic species has been comprehensively reviewed by Martínez (1983, 1985), Rondanelli (1987); Marciano-Ĉabral (1988) and Warhurst (1985).

○ 3.5.2 The discovery of pathogenic soil amoebae

History. The history of the discovery of the pathogenicity of this species makes enthralling reading (Ferrante, 1986). Free-living hartmannellid amoebae were reported from time to time in tissue cultures of mammalian cells and it was generally assumed that these represented airborne contaminants. Amoebae were also sometimes isolated from swab cultures of children's mouths. In 1958, Culbertson *et al.* showed that hartmannellid amoebae when injected into monkeys or mice caused fatal lesions of the brain and spinal cord; later work (Culbertson, 1961) confirmed these results. It soon became apparent that free-living amoebae could infect man and the first report of a human case came from Australia where four fatal cases of amoebic meningoencephalitis were described. Other reports soon followed and infections have now been reported from all over the world.

○ 3.5.3 Species involved

Species which have an *exozoic* (i.e. non-parasitic) phase which is capable of transforming into an *enzoic* (i.e. parasitic) phase have been called *amphizoic* amoebae (Rondanelli, 1987). These amphizoic species belong to the genus *Naegleria*, within the family Vahlkamfidae, and to the genus *Acanthamoeba*, within the family Hartmanellidae. A feature of the Vahlkamfidae is the ability to transform, under certain conditions, from an amoeboid to a flagellate phase: the so-called *amoebo-flagellate transformation* (Figs. 3.8, 3.9).

This ability enables *Naegleria* to be distinguished from *Acanthamoeba*. Although it was originally thought that only species of *Naegleria* were pathogenic, it is now established that species of *Acanthamoeba* are also infective to man.

○ 3.5.4 Type examples: *Naegleria gruberi* and *N. fowleri*

General morphology. Although the species of *Naegleria* pathogenic to man is *N. fowleri*, this species is almost morphologically identical with the common, free-living species *N. gruberi*, which can therefore be used as a useful laboratory model (Fig. 3.8). The minor points of difference are discussed below.

Trophozoite. The trophozoite exhibits a characteristic 'limax' type of motion marked by the formation of hemispherical hyaline bulges, often explosively, and usually in such a manner as to produce a sinuous path of movement (Page, 1967*a*). The anterior end of the trophozoite is formed by a single pseudopod of clear hyaline ectoplasm, usually blade-like. The posterior end

Table 3.4. *Amoebae found in culture samples of three brands of mineral water in Mexico*

		Culture media						
		Chalkley						
			Hay		Rice		Willmer	
Brand	Name and stage in life cycle of organism	Wheat 28°C	RT[a]	28°C	RT[a]	28°C	RT[a]	28°C
A	*Naegleria gruberi* (trophic and cystic stages)				+		+	
B	*N. gruberi* (trophic and cystic stages)	+	+		+++	+	++	+
B	*Acanthamoeba astronyxis* (cystic stage)		+			+		
B	*Vahlkampfia vahlkampfi* (trophic stage)	+			++	++		
B	*Bodomorpha minima* (trophic stage)					+	+	
C	*N. gruberi* (trophic and cystic stages)			+	+		+	+

[a] Room temperature (15–25 °C).
Source: Data from Rivera *et al.* (1981).

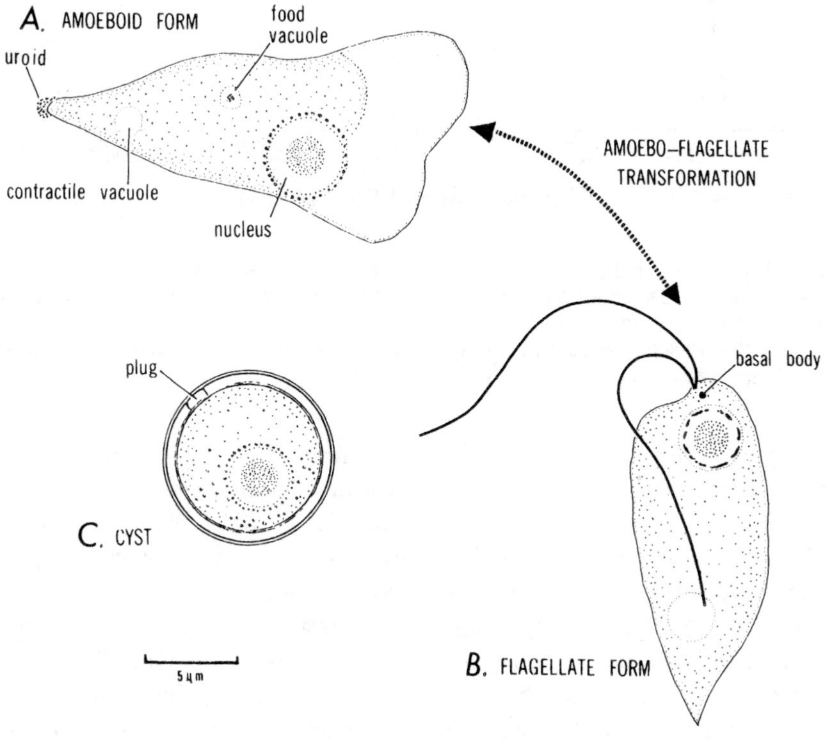

Fig. 3.8
Naegleria gruberi: a soil amoeba, related species of which are parasitic in man. Under certain physiological conditions, the amoeboid form undergoes transformation into a flagellate phase; see Fig. 3.9. (Modified from Page, 1967*a*.)

seems to persist during locomotion as a relatively stable structure, about 2 μm in diameter, referred to as the 'uroid' or 'uroid process'; 1–4 fine filaments may extend from it, sometimes trailing entangled debris (Carter, 1970). The mean size of the trophozoite is 22 μm long (range 15–30 μm) and 7 μm wide (range 6–9 μm). There is usually only a single nucleus but up to four may occasionally occur. There are 1–6 contractile vacuoles, usually 3–4. The trophozoites of *N. fowleri* exhibit remarkable 'feeding cups' or *amoebostomes*', which are cytoplasmic extensions of the surface. The number and size of these vary with the species and strain of *Naegleria*

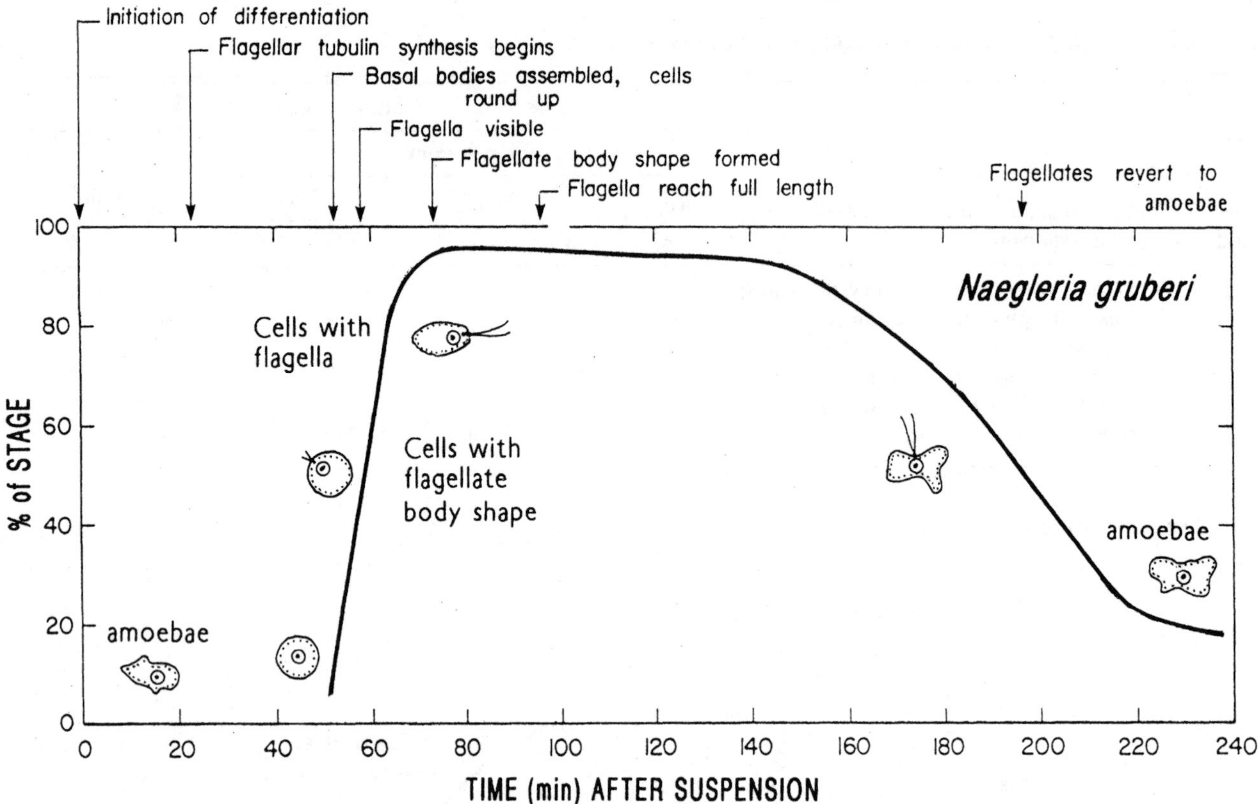

Fig. 3.9
Naegleria gruberi, showing the stages of differentiation during the amoebo-flagellate transformation. The amoebae were initially grown at 33 °C on NM agar in association with the bacterium *Klebsiella pneumoniae*. At zero time, the amoebae were suspended in T buffer and washed relatively free of bacteria and resuspended in T buffer at a concentration of 10^7 organisms ml^{-1} at 25 °C. Samples were taken at intervals, stained with Lugol's iodine and counted; scatter points are omitted here. (Modified from Fulton, 1977. Reproduced with permission from *Annual Reviews of Microbiology*, **31**, 1977.)

and serve to ingest bacteria, yeast cells and cellular debris (Marciano-Ĉabral, 1988).

The amoebo-flagellate transformation (Figs 3.8, 3.9). In the soil, the amoebae normally exist as trophozoites, but transformation to the flagellate stage takes place with the dilution of the soil substrate by rain water. This process can readily be carried out in the laboratory by transferring the trophozoites to distilled water (Page, 1967*a*). Differentiation into the flagellate stage – which is non-feeding and non-dividing – involves change in cell shape and synthesis of the organelles of the flagellar apparatus, i.e. basal bodies, a rootlet and flagella. The complete process of the amoebo-flagellate transformation has been reviewed in detail by Fulton (1977).

The cyst. If trophozoites are grown on agar medium (Chapter 33), cysts appear within about 48 h. In nature, many biotic or abiotic factors are thought to induce cyst formation; these include food deprivation, desiccation, crowding, accumulation of waste products, anaerobic conditions, pH changes, salt concentration and exposure to the metabolic products of bacteria (Marciano-Ĉabral, 1988).

Cysts are readily identified as highly refractive, circular or oval bodies (about 15–18 μm in diameter) which when clumped in clusters of 3–12 become somewhat polygonal. As it is the trophozoite stage which is infective to man (see below), unlike *Entamoeba histolytica*, the cysts of *Naegleria fowleri* are not infective. However, the cysts may play a major role in the epidemiology of the disease in that their resistance to unfavourable conditions allows the cysts to be dispersed widely, although the cysts of *N. fowleri* do not appear to be as resistant as those of *N. gruberi* (Mehlotra, 1987). The cysts possess one or more pores, through which the trophozoite escapes during excystment.

Electron microscopy has demonstrated differences in

the cyst wall of *N. gruberi* and *N. fowleri* (Marciano-Ĉabral, 1988; Schuster, 1975). The cyst of *N. gruberi* is encompassed within a double-walled structure about 20 µm thick, the outer layer of which is absent within the cyst of *N. fowleri*.

Cell division (Fig. 3.10). Nuclear division in *Naegleria* spp. is unusual in that the nuclear membrane remains intact throughout cell division and a nucleolus (karyosome) persists. This process is referred to as *promitosis*. The nucleolus is exceptionally large and occupies about two-thirds of the endonuclear space. The mode of division is quite different from that of *Acanthamoeba* spp. (Fig. 3.10), which follows the normal pattern (*metamitosis*) found in metazoan cells. The process can be studied by the use of the simple method developed by Singh (1950).

In the early stages of promitosis, enlargement of the nucleus and condensation of the chromosomal material occurs, although distinct chromosomal elements are not recognisable and the chromosome number remains unknown in *Naegleria* spp. The nucleus elongates and the nucleolus becomes dumb-bell shaped. Spindle microtubules can be seen in the nucleus; the 'chromosomes' orientate along the equatorial plate and later separate. In late prophase or early metaphase the nucleolus stretches further and divides into two 'polar masses' with the equatorial plate in between (Fig. 3.10). Anaphase is characterised by the appearance of a large 'interzonal' body formed by the conglomeration of nuclear material (Rondanelli, 1987); centrioles have not been observed. The elongated nucleus finally separates into daughter nuclei.

Ultrastructure. This has been reviewed in detail by Rondanelli (1987). In contrast to *Entamoeba histolytica*, the trophozoite of *Naegleria* possesses mitochondria.

Mode of infection. Infection is brought about by exposure to water in swimming pools, puddles, moist soil, sewage etc., the amoebae entering the nasal cavity by inhalation or aspiration of water containing trophozoites. The organisms invade the neuroepithelium and rapidly process along the nerve tracts to invade the brain.

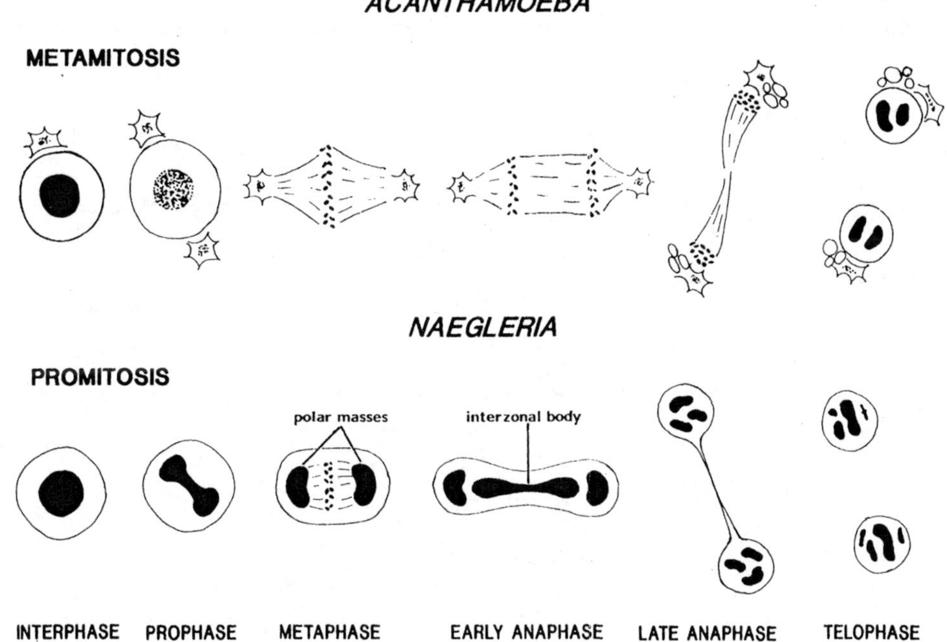

Fig. 3.10
Comparison of cell division in *Naegleria* sp. and *Acanthamoeba* sp. (Modified from Rondanelli, 1987.)

Pathogenesis. The disease caused by *N. fowleri* is known as *Primary Amoebic Meningoencephalitis* (*PAM*) or *Naegleriamoebiasis*, and its pathology has been extensively reviewed (Carter, 1970; Martínez, 1983; Rondanelli, 1987; Singh, 1975; Warhurst, 1985). It resembles acute purulent bacterial meningitis and its early stages cannot be readily distinguished from it. One sign is that PAM usually occurs in children and young adults previously in good health. Numerous other clinical effects have been recorded. The incubation period is generally only 1–2 days but may be as long as 2 weeks; the clinical course is rapid and death may occur in 5–6 days.

○ 3.5.5 Other *Naegleria* species

Other species of *Naegleria*, which may or may not prove to be pathogenic, are *N. australiensis*, *N. jardini* and *N. lovaniensis* (Rondanelli, 1987).

○ 3.5.6 Type example: ***Acanthamoeba castellanii*** (Fig. 3.11)

General comment. It was formerly thought that only amoebae of the genus *Naegleria* were pathogenic, but it is now well established that species of *Acanthamoeba* are also invasive and can cause chronic encephalitis, skin ulcers, keratinitis, etc. The general biology of *Acanthamoeba* spp. and their pathogenicity has been reviewed by Martínez (1983, 1985) and Rondanelli (1987).

Occurrence. *Acanthamoeba* spp. are common in soil and water and also occur as 'normal' flora in man; they are probably common in healthy people. They have been isolated from fresh water, thermal factory discharges, air, air conditioning systems and even bottled water (Table 3.4) and kidney dialysis machines (Rondanelli, 1987).

Morphology (Fig. 3.11). The trophozoite is somewhat larger than that of *Naegleria* but, in contrast, it shows a polydirectional mode of locomotion. The pseudopodia are extremely variable in form, being either lobose or filose or sometimes both. One characteristic is that the needle-like projections or *acanthopodia* (Fig. 3.11) which extend from the body. The cytoplasm is largely comparable to that of *Naegleria*. The ultrastructure has been studied in detail by early workers, such as Vickerman (1962) and more recently reviewed by Rondanelli (1987); these studies have revealed the presence of very prominent mitochondria.

Cyst (Fig. 3.11). Cyst formation is readily induced *in vitro* by allowing slants of monoxenic cultures to become dehydrated; as cysts can be washed off with water, axenic cultures are easily started (Neff, 1957). The cyst diameter varies from 15 to 28 μm and the wall is double-

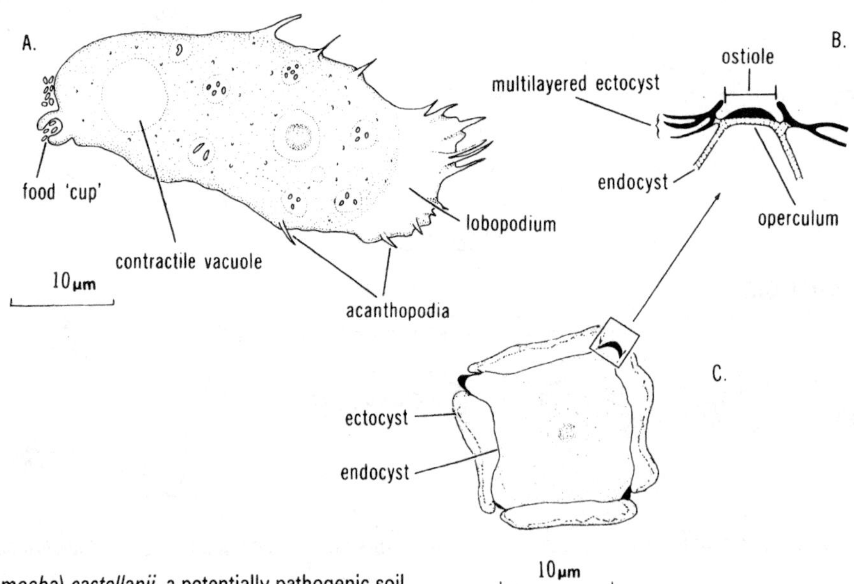

Fig. 3.11
Hartmannella (*Acanthamoeba*) *castellanii*, a potentially pathogenic soil amoeba. A, trophozoite; B, ostiole (enlarged); C, cyst. (After Page, 1967*b*; Volkonsky, 1931.)

layered, the ectocyst being multilayered (Fig. 3.11C). The endocyst shows a typical polygonal arrangement that distinguishes *Acanthamoeba* cysts from those of *Naegleria* and other free-living species. In older cysts, protuberances appear, each of which marks the position of an *ostiole* sealed by a cap or *operculum*, as many as eight being present. Excystment takes place readily in an *in vitro* system rich in microorganisms, the amoebae escaping through an ostiole. The $p\text{CO}_2$ may act as a trigger, as in *Naegleria* (Averner & Fulton, 1966) and coccidian cysts.

Cell division (Fig. 3.10). In contrast to *Naegleria*, which divides by promitosis, cell division in *Acanthamoeba* follows the typical pattern in the Metazoa, now referred to as *metamitosis*.

Pathology. A number of species of *Acanthamoeba* appear to be involved in human infections, but with a few exceptions (Rondanelli, 1987) they have not been cultured from human cases, which has made exact identification difficult. Species which are probably implicated are *A. castellanii*, *A. culbertsoni*, *A. palestinensis* and *A. astronyxis*. Invasion of the body can occur in a number of ways. Like *Naegleria fowleri*, invasion of the central nervous system (CNS) can occur via the olfactory epithelium, resulting in Granulomatous Amoebic Encephalitis (GAE). The CNS can also be invaded (via the blood stream) from a primary focus elsewhere in the body. Other points of entry are the lower respiratory tract, ulcerations in the skin, or open wounds or cracks in any mucosal or serosal surface. A major site of infection is the eye, which appears to be infected directly, resulting in corneal ulceration or keratinitis. Ill-fitting contact lenses may cause damage to the corneal epithelium, which may be infected secondarily by 'opportunistic' amoebae. Infections produced by *Acanthamoeba* and *Naegleria* have been compared by Martínez (1980).

***In vitro* culture.** *A. castellanii* makes an excellent laboratory model for teaching and research, in that it can be readily cultured *in vitro* (Neff, 1957). Not only can it feed on a range of common bacteria, such as *Escherichia coli* or *Aerobacter aerogenes*, but it grows equally well on dead (autoclaved) baker's yeast (*Saccharomyces cerevisiae*) over a non-nutrient agar base; the latter provides a very simple culture system.

○○○ 3.6 Physiology of intestinal amoebae

○ 3.6.1 Physico-chemical considerations

Introduction. Almost all common species of parasitic amoebae can be cultured successfully *in vitro* in a simple diphasic medium with added starch (Fig. 33.1) and bacteria, which facilitate growth, normally being present. However, most species can also be grown axenically in liquid media (Table 33.2).

Species from mammalian hosts, established in culture, are normally maintained at body temperature (approx. 37 °C). However, the so-called Laredo-type strain of *Entamoeba histolytica* can grow at a temperature as low as 10 °C (Richards *et al.*, 1966). The reptilian parasite *E. invadens* can be grown within the range 25–30 °C with a lower critical temperature of 12.5 °C (Lachance, 1963).

Biological oxygen relationship. *In vivo* and *in vitro*, intestinal amoebae grow in sites which are probably virtually anaerobic. In cotton-wool-stopped tubes (Fig. 33.1) growth only takes place at the bottom of the tube where conditions are anaerobic. The reducing activity

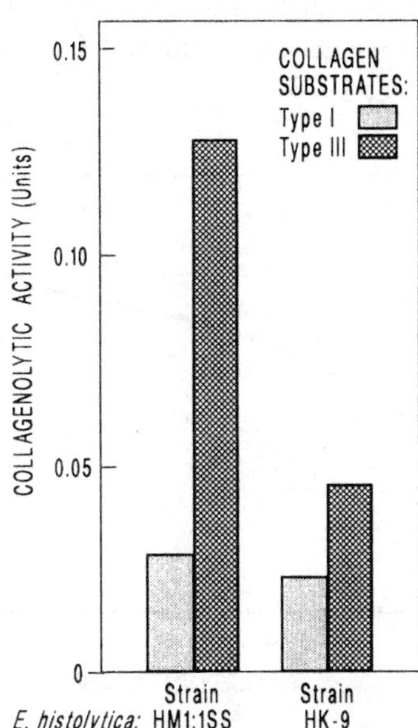

Fig. 3.12
Collagenolytic activity of two strains of *Entamoeba histolytica* on two different types of collagen. (Modified from Rojkind *et al.*, 1988.) In: *Amoebiasis: Human Infection by Entamoeba histolytica* (ed. J. I. Ravdin). Churchill-Livingstone, New York, 1988.)

of a culture medium is clearly related to the oxygen tension. Thus anaerobic organisms generally proliferate only at low oxidation–reduction potentials (E_h) and the addition of reducing agents, such as cysteine (Fig. 33.3), is often used in culture media. In contrast, the soil amoeba *Acanthamoeba* (*Hartmannella*) *castellanii* (Fig. 3.11) is a strict aerobe.

Osmotic pressure. *E. histolytica* remains viable over a wide range of osmotic pressure ($\Delta = 0.17$ °C to 0.95 °C); similar results have been reported for other intestinal amoebae.

○ 3.6.2 General metabolism

The metabolism of *Entamoeba* spp. is of special interest as it lacks both mitochondria and hydrogenosomes. Various aspects have been reviewed by Avron & Chayen (1988), Bakker-Grunwald (1991), Bryant & Behm (1989), McLaughlin & Aley (1985), Müller (1988), Reeves (1984) and Weinbach (1988).

Chemical composition. Analysis of intestinal amoebae shows no unusual features. The trophozoites of *E. histolytica* have a dry mass (in grams) 11–16 per cent wet cell volume (in millilitres) and a chemical composition of approximately 20–40 per cent carbohydrate (glycogen), 40–50 per cent protein, 5–6 per cent lipid, 7 per cent pentose nucleic acid and 6–8 per cent ash (Reeves & Ward, 1965).

Carbohydrate metabolism. Like most other intestinal parasites, *Entamoeba histolytica* utilises carbohydrate as its main energy source. Glucose is taken up largely by a carrier-mediated, saturable, equilibrative system in the plasma membrane (Fig. 3.13), but some is also taken up by a pinocytotic process although at physiological glucose concentrations (up to 5 mM) the latter process contributes less than 1 per cent of the total glucose uptake (Bakker-Grunwald, 1991; Serrano & Reeves, 1974, 1975).

All evidence suggests that the organism obtains its energy by glycolytic phosphorylation via the Embden–Meyerhof pathway. However, because the parasite lacks lactate dehydrogenase, lactate is not an end product. Pyruvate is converted to ethanol and CO_2, which are the main products of anaerobic metabolism (Table 3.5). D-galactose can substitute for glucose and will support growth in axenic culture; when [^{14}C]galactose is used the label appears in glycogen and also in the ethanol and acetate end-products but only a trace in the CO_2. Although trophozoites contain abundant glycogen, the pathways of its synthesis have not been determined. Several key enzymes concerned with glycogen synthesis have been found, but the chief enzyme concerned, glycogen synthetase, is lacking (Weinbach, 1988).

Intermediary metabolism. Since *E. histolytica* lacks mitochondria it is not surprising that it also lacks a functional Krebs cycle. A unique feature of the glycolysis is that the general reactions beyond phosphoenolpyruvate

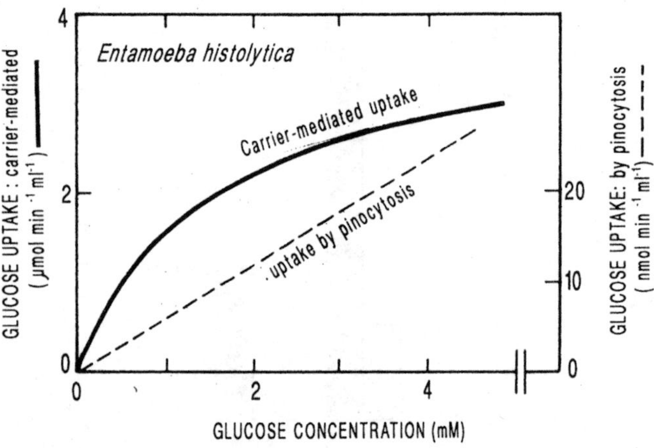

Fig. 3.13
Entamoeba histolytica: uptake of glucose by carrier-mediated and pinocytotic mechanisms. At the above concentrations, pinocytotic uptake contributes less than 1 per cent of the total (carrier-mediated plus pinocytotic) uptake. (After Bakker-Grunwald, 1991; based on data from Serrano & Reeves, 1974, 1975.)

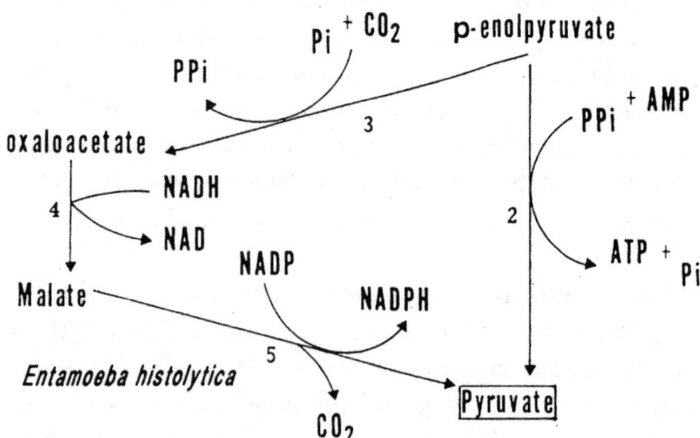

Fig. 3.14
Entamoeba histolytica: pathways for pyruvate formation from phosphoenolpyruvate (PEP). Numbers indicate enzymes: 2 = pyruvate phosphate dikinase; 3 = phoshoenolpyruvate carboxytransphorylase; 4 = malic dehydrogenase; 5 = malic enzyme. (After McLaughlin & Aley, 1985.)

Table 3.5. *Major metabolic end products of some anaerobic parasitic protozoa*

Subphylum	Species	Host and habitat	Hydrogenosome	Major end products: Anaerobic	Major end products: Aerobic
Rhizopoda	*Entamoeba histolytica*	Man: large intestine	Absent	Ethanol, CO_2	Ethanol, acetate, CO_2
	Entamoeba invadens	Reptile: intestine	Absent	No data	No data
Mastigophora	*Giardia lamblia*	Man: small intestine	Absent	Ethanol, acetate, CO_2	Ethanol, acetate, CO_2
	Trichomonas vaginalis	Man: urinogenital tract	Present	Glycerol, lactate, acetate, H_2O, CO_2	Glycerol, lactate, acetate, CO_2
	Tritrichomonas foetus	Cattle: urinogenital tract	Present	Glycerol, succinate, acetate, H_2, CO_2	Glycerol, succinate, acetate, CO_2

Source: Data from Müller (1988).

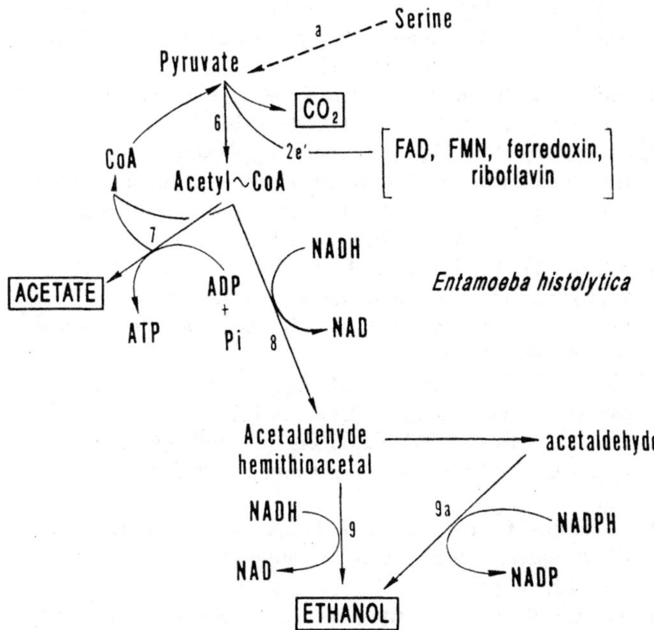

Fig. 3.15
Entamoeba histolytica: pathways for pyruvate degeneration. Numbers indicate enzymes: 6 = pyruvate : ferrodoxin oxidoreductase; 7 = acetate thiokinase; 8 = acetaldehyde dehydrogenase; 9 = NADH-linked alcohol dehydrogenase; 9a = NADPH-dependent alcohol dehydrogenase; a = serine dehydratase. (After McLaughlin & Aley, 1985.)

(PEP) are catalysed by a PP_i-dependent enzyme, pyruvate phosphate di-kinase (P_i, inorganic phosphate; PP_i, inorganic pyrophosphate):

$$PEP + AMP + PP_i \rightarrow pyruvate + ATP + P_i$$

However, the reverse reaction predominates, resulting in the formation of pyruvate with a net yield of 1 mol ATP. An alternative route for the formation of pyruvate from PEP has been postulated (Fig. 3.14) involving a unique enzyme, phosphoenolpyruvate carboxyltransphosphorylase. The ultimate fate of the pyruvate depends on the prevailing degree of anaerobiosis. Under aerobic conditions, both ethanol *and* acetate are formed, as well as CO_2, whereas under anaerobic conditions only ethanol and CO_2 are produced (Table 3.5). The pathway to ethanol is shown in Fig. 3.15.

Electron transport. This is very poorly understood. Although no cytochromes or haemin have been found, there is clear evidence of a Fe–S protein, ferredoxin being implicated and this has been isolated from *E. histolytica.* The enzyme pyruvate:ferredoxin oxidoreductase is functional in *E. histolytica*, as shown by the production of ethanol from acetyl-CoA (Fig. 3.15) but the mechanism of reoxidation of ferredoxin, which is reduced in the process, remains unknown (Müller, 1988). Flavins have also been found in quantity.

Protein metabolism. The protein metabolism of amoebae has been little studied and there is almost no information of protein catabolism. As might be expected in a tissue-invading parasite, a number of enzymes showing proteolytic activity – proteases and collagenases – have been detected. The chief of these is a cysteine proteinase, *amoebapain*, with a maximum activity at pH

4.5 (Scholze, 1991); this enzyme may be identical with a similar enzyme, *histolysin*, also isolated from this protozoan (Luaces & Barrett, 1988). A collagenase(s) also occurs and the evidence suggests that this is either transient or permanently localised on the plasma membrane of the trophozoite (Gadasi & Kobiler, 1983) as it is not found free in culture media. Strain differences in the proteolytic activities of the collagenases (Fig. 3.12) have been demonstrated and this may partly account for differences in pathogenicity between different strains.

Lipid metabolism. Very little information on this area of the metabolism is available. All the lipids in *E. histolytica* appear to be derived from the medium, but the mechanism of their uptake is unknown; lipids do not appear to serve as a source of energy (Weinbach, 1988).

Nucleic acids. There are a number of unusual features regarding the nucleic acids of *E. histolytica*, many of which have been reviewed by Albach (1989) and Weinbach (1988). Thus, the 'normal' *E. histolytica* has five times as much DNA as the Loredo-strain, and *E. moshkovskii* has approximately ten times as much (Gelderman *et al.*, 1971). It has been suggested that polyploidy (perhaps up to 14*n*!) may occur in some strains (Beyers, 1986). The distribution and synthesis of RNA – and to a lesser extent DNA – in the nucleus of *E. histolytica* appears to be contrary to what is normally expected to be present in eurokaryotes: (*a*) most RNA (ribosomal?) is synthesised or accumulates in the peripheral chromatin; (*b*) the DNA is so evenly dispersed in the nucleus that it cannot be visualised by standard procedures; (*c*) it has been speculated that the DNA may accumulate around the endosomal area (not a nucleolus) prior to nuclear division (Albach, 1989).

The nucleic acid metabolism of *E. histolytica* resembles that of other anaerobic intestinal protozoa such as *Trichomonas vaginalis* and *Giardia lamblia* (Chapter 4) in being deficient in *de novo* synthesis of purines and (probably) pyrimidines. In these flagellates, the deficiency is overcome by the development of a 'salvage' pathway which in *T. vaginalis* depends on the salvage of adenosine and guanosine. *E. histolytica* appears to follow a similar pathway (Lo & Wang, 1985) details of which are elaborated further in Chapter 4. No information on the pyrimidine salvage pathway for *E. histolytica* appears to be available, but it is also likely to resemble that of *Trichomonas* and *Giardia*.

○○○ References

(References of general interest have been included in this list, in addition to those quoted in the text.)

Albach, R. A. 1989. Nucleic acids of *Entamoeba histolytica*. *Journal of Protozoology*, **36**: 197–205.

Albach, R. A. & Booden, T. 1978. Amoebae. In: *Parasitic Protozoa*, Vol. II (ed. J. P. Krier), pp. 455–505. Academic Press, London.

Averner, M. & Fulton, C. 1966. Carbon dioxide: signal for excystment of *Naegleria gruberi*. *Journal of General Microbiology*, **42**: 245–55.

Avron, B. & Chayen, A. 1988. Biochemistry of *Entamoeba histolytica*: a review. *Cell Biochemistry and Function*, **6**: 71–86.

Bakker-Grunwald, T. 1991. Transport and compartmentation in *Entamoeba histolytica*. In: *Biochemical Protozoology* (ed. G. Coombs & M. North), pp. 367–76. Taylor & Francis, London. ISBN 0 748 0001 X.

Beyers, T. J. 1986. Molecular biology of DNA in *Acanthamoeba*, *Amoeba*, *Entamoeba*, and *Naegleria*. *International Review of Cytology*, **99**: 311–41.

Bryant, C. & Behm, C. 1989. *Biochemical Adaptation in Parasites*. Chapman & Hall, New York. ISBN 0 412 32530 6.

Carter, R. F. 1970. Description of a *Naegleria* species isolated from two cases of primary amoebic meningoencephalitis and of the experimental pathological changes induced by it. *Journal of Pathology*, **100**: 217–44.

Casebolt, D. B., Lindsey, J. R. & Cassell, G. H. 1988. Prevalence rates of infectious agents among commercial breeding populations of rats and mice. *Laboratory Animal Science*, **38**: 327–9.

Coombs, G. H. & North, M. J. 1991. *Biochemical Protozoology*. Taylor & Francis, London. ISBN 0 748 4001 X.

Culbertson, C. G. 1961. Pathogenetic acanthamoeba (*Hartmannella*). *American Journal of Clinical Pathology*, **35**: 195–202.

Culbertson, C. G., Smith, J. W. & Minner, J. R. 1958. *Acanthamoeba*: observations on animal pathogenicity. *Science*, **127**: 1506.

Desowitz, R. S. & Barnish, G. 1986. *Entamoeba polecki* and other intestinal protozoa in Papua New Guinea highland children. *Annals of Tropical Medicine and Parasitology*, **80**: 399–402.

Elsdon-Dew, R. 1968. The epidemiology of amoebiasis. *Advances in Parasitology*, **6**: 1–62.

Ferrante, A. 1986. Discovery and control of primary amoebic meningoencephalitis (PAM). *Parasitology Today*, **2**: S10.

Fulton, C. 1977. Cell differentiation in *Naegleria gruberi*. *Annual Review of Microbiology*, **31**: 597–629.

Gadasi, I. & Kobiler, D. 1983. *Entamoeba histolytica*: correlation between virulence and content of proteolytic enzymes. *Experimental Parasitology*, **55**: 105–10.

Geiman, Q. S. & Ratcliffe, H. L. 1936. Morphology and

life-cycle of an amoeba producing amoebiasis in reptiles. *Parasitology*, **28**: 208–28.

Gelderman, A. H., Keister, D. B., Bartgis, I. L. & Diamond, L. S. 1971. Characterization of the deoxyribonucleic acid of representative strains of *Entamoeba histolytica*, *E. histolytica*-like amebae, and *E. moshkovskii*. *Journal of Protozoology*, **57**: 906–11.

Gottlieb, D. S. & Miller, L. H. 1971. *Entamoeba gingivalis* in periodontal disease. *Journal of Periodontology*, **42**: 412–15.

Hoare, C. A. 1959. *Handbook of Medical Protozoology*. Balliere, Tindall & Cox, London.

Hyde, J. E. 1990. *Molecular Parasitology*. Open University Press, Buckingham, UK. ISBN 0 471 93227 2.

Jackson, T. F. H., Gathiram, V. & Simjee, A. E. 1985. Serological studies of antibody responses to the zymodemes of *Entamoeba histolytica*. *The Lancet*, *i* (No. 8431), 716–17.

Kean, B. H. & Malloch, C. L. 1966. The neglected ameba: *Dientamoeba fragilis*. *American Journal of Digestive Diseases*, **11**: 735–46.

Keller, O., Orland, F. J. & Baird, G. R. 1967. Ultrastructure of *Entamoeba gingivalis*. *Journal of Dental Research*, **46**: 1010–18.

Keystone, J. S., Yang, J., Grisdale, D., Harrington, M., Pillon, L. & Andreychuk, R. 1984. Intestinal parasites in metropolitan Toronto day-care centres. *Canadian Medical Association Journal*, **131**: 733–5.

Kreier, J. P. 1978. *Parasitic Protozoa*, Vol. II. Academic Press, New York.

Kretschmer, R. R. (ed.) 1990. *Amebiasis: infection and disease by Entamoeba histolytica*. CRC Press, Boca Raton, Florida. ISBN 0 849 35342 4.

Lachance, P. J. 1963. Experimental studies on *Entamoeba*. 1. Temperature: a microclimatic factor in the cultivation of three species of *Entamoeba*. *Canadian Journal of Zoology*, **41**: 1079–93.

Levine, N. D. (Chairman) *et al.* 1980. A newly revised classification of the Protozoa. (The Committee on Systematics and Evolution of the Society of Protozoologists.) *Journal of Protozoology*, **27**: 37–58.

Lo, H.-S. & Wang, C. C. 1985. Purine salvage in *Entamoeba histolytica*. *Journal of Parasitology*, **71**: 662–9.

Luaces, A. L. & Barrett, A. J. 1988. Affinity purification and biochemical characterization of histolysin, the major cysteine proteinase in *Entamoeba histolytica*. *Biochemical Journal*, **25**: 903–9.

Lushbaugh, W. B. & Miller, J. H. 1988. The morphology of *Entamoeba histolytica*. In: *Amebiasis: Human Infection by Entamoeba histolytica* (ed. J. Ravdin), pp. 41–68. John Wiley and Sons, New York. ISBN 0 471 82817 3.

McLaughlin, J. & Aley, S. 1985. The biochemistry and functional morphology of the *Entamoeba*. *Journal of Protozoology*, **32**: 221–40.

Marciano-Ĉabral, F. 1988. Biology of *Naegleria* spp. *Microbiological Reviews*, **52**: 114–33.

Martínez, A. J. 1980. Is Acanthamoebic encephalitis an opportunistic infection? *Neurology* (N.Y.), **30**: 567–74.

Martínez, A. J. 1983. Free-living amoebae: pathogenic aspects; a review. *Protozoological Abstracts*, **7**: 293–306.

Martínez, A. J. 1985. *Free-living Amebas: Natural History, Prevention, Diagnosis, Pathology and Treatment of Disease*. CRC Press, Boca Raton, Florida.

Martínez-Palomo, A. 1982. *The Biology of Entamoeba histolytica*. Research Studies Press, Wiley & Sons, Chichester (UK) and New York.

Martínez-Palomo, A. 1986. *Amebiasis*. Elsevier, Amsterdam, ISBN 0 444 80728 4.

Martínez-Palomo, A. 1987. The pathogenesis of amoebiasis. *Parasitology Today*, **3**: 111–17.

Meerovitch, E. 1961. Infectivity and pathogenicity of polyxenic and monoxenic *Entamoeba invadens* to snakes kept at normal and high temperatures and the natural history of reptile amoebiasis. *Journal of Parasitology*, **47**: 791–4.

Mehlotra, R. K. 1987. Infective stages of free-living amoebae. *Parasitology Today*, **3**: 185.

Mirelman, D. 1987. Effect of culture conditions and bacterial associations on the zymodemes of *Entamoeba histolytica*. *Parasitology Today*, **3**: 37–40.

Mirelman, D. 1988. Ameba–bacterial relationships in amebiasis. In: *Amebiasis: Human Infection with Entamoeba histolytica* (ed. J. Ravdin), pp. 351–69. John Wiley and Sons, New York. ISBN 0 471 82817 3.

Müller, M. 1988. Energy metabolism of protozoa without mitochondria. *Annual Review of Microbiology*, **42**: 465–88.

Neal, R. A. 1950. An experimental study of *Entamoeba muris* (Grassi, 1879); its morphology, affinities and host–parasite relationship. *Parasitology*, **40**: 343–65.

Neal, R. A. 1953. Studies on the morphology and biology of *Entamoeba moshkovskii* Tshalaia 1941. *Parasitology*, **43**: 253–68.

Neal, R. A. 1966. Experimental studies on *Entamoeba* with reference to speciation. *Advances in Parasitology*, **4**: 1–51.

Neff, R. J. 1957. Purification, axenic cultivation and description of a soil amoeba, *Acanthamoeba* sp. *Journal of Protozoology*, **4**: 176–82.

Page, F. C. 1967*a*. Taxonomic criteria for limax amoeba with descriptions of 3 new species of *Hartmanella* and 3 of *Vahlkampfia*. *Journal of Protozoology*, **14**: 499–521.

Page, F. C. 1967*b*. Re-definition of the genus *Acanthamoeba* with descriptions of three species. *Journal of Protozoology*, **14**: 709–24.

Pernin, P. 1984. Isoenzyme patterns of pathogenic and nonpathogenic thermophilic *Naegleria* strains by isoelectric focusing. *International Journal for Parasitology*, **14**: 459–65.

Pernin, P. & Grelaud, G. 1989. Application of isoenzymatic typing to the identification of nonaxenic strains of *Naegleria* (Protozoa, Rhizopoda). *Parasitology Research*, **75**: 595–8.

Ravdin, J. (ed.) 1988. *Amebiasis: Human Infection by Entamoeba histolytica*. John Wiley and Sons, New York. ISBN 0 471 82817 3.

Reeves, R. E. 1984. Metabolism of *Entamoeba histolytica*. *Advances in Parasitology*, **23**: 105–42.

Reeves, R. E. & Ward, A. B. 1965. Large lot cultivation of *Entamoeba histolytica*. *Journal of Parasitology*, **51**: 321–4.

Richards, C. S., Goldman, M. & Cannon, L. T. 1966. Cultivation of *Entamoeba histolytica* and *Entamoeba*-like strains at reduced temperature and behavior of the amebae in diluted media. *American Journal of Tropical Medicine and Hygiene*, **15**: 648–55.

Rivera, F., Galvan, M., Robles, E., Leal, P., Gonzalez, L. & Lacy, A. M. 1981. Bottled mineral waters polluted by Protozoa in Mexico. *Journal of Protozoology*, **28**: 54–6.

Rojkind, M., Rosales-Encina, J. L. & Muñoz, M. L. 1988. The collagenase of *Entamoeba histolytica*. In *Amoebiasis: Human Infection by Entamoeba histolytica* (ed. J. I. Ravdin), pp. 263–72. John Wiley and Sons, New York. ISBN 0 471 82871 3.

Rondanelli, E. G. (ed.) 1987. *Amphizoic Amoebae Human Pathology*. (Infectious Diseases Color Atlas Monographs No. 1.) Piccin Nuova Libraria, Padua, Italy.

Rondanelli, E. G., Carosi, G., Filice, C., Carnevale, G., Scaglia, M. & Barbarini, G. 1977. Ultrastruttura di *Entamoeba histolytica*. *Giornale di Malatte Infettive e Passitarie*, **29**: 592–606.

Sanders, E. P. 1931. The life cycle of *Entamoeba ranarum* Grassi, 1987. *Archiv für Protistenkünde*, **74**: 365–71.

Sargeaunt, P. G. 1985. Zymodemes expressing possible genetic exchange in *Entamoeba histolytica*. *Transactions of the Royal Society of Tropical Medicine and Hygiene*, **79**: 86–9.

Sargeaunt, P. G. 1987. The reliability of *Entamoeba histolytica* zymodemes in clinical diagnosis. *Parasitology Today*, **3**: 40–3.

Sargeaunt, P. G. 1988. Zymodemes of *Entamoeba histolytica*. In: *Amebiasis: Human Infection by Entamoeba histolytica* (ed. J. I. Ravdin), pp. 370–87. John Wiley and Sons, New York. ISBN 0 471 82871 3.

Sargeaunt, P. G. & Williams, J. E. 1978. The differentiation of invasive and non-invasive *Entamoeba histolytica* by isoenzyme electrophoresis. *Transactions of the Royal Society of Tropical Medicine and Hygiene*, **72**: 519–21.

Sargeaunt, P. G. & Williams, J. E. 1979. Electrophoretic isoenzyme patterns of the pathogenic and non-pathogenic intestinal amoebae of man. *Transactions of the Royal Society of Tropical Medicine and Hygiene*, **73**: 225–7.

Sargeaunt, P. G. & Williams, J. E. 1982. The morphology in culture of the intestinal amoebae of man. *Transactions of the Royal Society of Tropical Medicine and Hygiene*, **76**: 465–72.

Schlesinger, P. 1988. Lysosomes and *Entamoeba histolytica*. In: *Amebiasis: Human Infection by Entamoeba histolytica* (ed. J. I. Ravdin), pp. 297–313. John Wiley and Sons, New York. ISBN 0 471 82817 3.

Scholze, H. 1991. Amoebapain, the major proteinase of pathogenetic *Entamoeba histolytica*. In: *Biochemical Protozoology* (ed. G. Coombs & M. North), pp. 251–6. Taylor & Francis, London. ISBN 0 748 4001 X.

Schuster, F. L. 1975. Ultrastructure of cysts of *Naegleria* spp.: a comparative study. *Journal of Protozoology*, **22**: 352–9.

Sepulveda, A. B. 1982. Progress in amebiasis. *Scandinavian Journal of Gastroenterology*, **17** (Suppl. 77): 153–64.

Serrano, R. & Reeves, R. E. 1974. Glucose transport in *Entamoeba histolytica*. *Biochemical Journal*, **144**: 43–8.

Serrano, R. & Reeves, R. E. 1975. Physiological significance of glucose transport in *Entamoeba histolytica*. *Experimental Parasitology*, **41**: 370–84.

Singh, B. N. 1950. A culture method for growing small free-living amoebae for the study of their nuclear division. *Nature*, **165**: 65–6.

Singh, B. N. 1975. *Pathogenic and Non-pathogenic Amoebae*. Macmillan Press, London.

Smyth, J. D. & Smyth, M. M. 1980. *Frogs as Host–Parasite Systems. I.* Macmillan Press, London. ISBN 0 333 23565 7.

Tannich, E. & Burchard, G. D. 1991. Differentiation of pathogenetic from nonpathogenetic *Entamoeba histolytica* by restriction fragment analysis of a single gene amplified *in vitro*. *Journal of Clinical Microbiology*, **29**: 250–5.

Vickerman, K. 1962. Patterns of cellular organisation in *Limax* amoebae. *Experimental Cell Research*, **26**: 497–519.

Volkonsky, M. 1931. *Hartmannella castellanii* Douglas et classification des Hartmannelles (*Hartmannellinae*) nov. subfam., *Acanthamoeba* nov. gen., *Glaeseria* nov. gen. *Archives de Zoologie expérimental et générale*, **72**: 317–39.

Warhurst, D. C. 1985. Pathogenic free-living amoebae. *Parasitology Today*, **1**: 24–8.

Weinbach, E. C. 1988. Metabolism of *Entamoeba histolytica*. In: *Amoebiasis: Human Infection by Entamoeba histolytica* (ed. J. I. Ravdin), pp. 69–78. John Wiley and Sons, New York. ISBN 0 471 82817 3.

WHO. 1980. Parasite-related diarrhoeas. Report of a subgroup of the scientific working group on epidemiology and etiology. WHO Ref: CDD/PAR/80.1.

WHO. 1984. Informal meeting on strategies for control of amoebiasis. WHO Ref: CDD/PAR/84.2.

Yong, T. S., Chung, P. R. & Lee, K. T. 1985. [Electron-microscopic studies on fine structure and enzyme activity in the axenic and conventional strains of *Entamoeba histolytica*.] (In Korean.) *Korean Journal of Parasitology*, **23**: 269–84.

○○○ Appendix: Protozoa: abbreviated classification

A revised classification of the Protozoa was prepared by the Society of Protozoologists, the full text of which is published in the *Journal of Protozoology*, **27**: 37–58 (1980). Extracts from this classification are given below, in an abbreviated form, to cover the most studied groups of parasitic protozoa. For details of the terminology, the original paper should be consulted.

Phylum I. SARCOMASTIGOPHORA: simple type of nucleus; sexuality, when present, essentially syngamy; flagella, pseudopodia or both types of locomotor organelles.

Subphylum I. MASTIGOPHORA ('FLAGELLATA'): one or more flagella typically present in trophozoites; asexual reproduction basically by intrakinetal (symmetrogenic) binary fission; sexual reproduction known in some groups.

Class 2. Zoomastigophorea: chloroplasts absent; one to many flagella; amoeboid forms, with or without flagella; sexuality known in a few groups; a polyphyletic group.

Order 2. Kinetoplastida: one or two flagella arising from a depression; flagella typically with paraxial rod in addition to axoneme; single mitochondria (sometimes non-functional) extending length of body: Feulgen-positive (DNA-containing) kinetoplast.

Suborder 2. Trypanosomatina

e.g. *Trypanosoma*, *Leishmania*

Order 4. Retortamonadida: two to four flagella, one turned posteriorly; mitochondria and Golgi apparatus absent.

e.g. *Chilomastix*, *Retortamonas*

Order 5. Diplomonadida: one or two karyomastigonts; in general with two karyomastigonts with two-fold rotational symmetry or, in one genus, primary mirror symmetry.

Suborder 2. Diplomonadina

e.g. *Giardia*

Order 7. Trichomonadida: Typical karyomastigonts with 4–6 flagella; in mastigonts of typical genera, one flagellum current, free or with proximal or entire length adherent to body surface; undulating membrane, if present, associated with adherent segment of recurrent flagella; pelta and noncontractile axostyle in each mastigont, except for one genus; hydrogenosomes present; true cysts infrequent, known in very few species.

e.g. *Trichomonas*, *Dientamoeba*.

Subphyllum II. OPALINATA: numerous cilia in oblique rows over entire body surface; cytostome absent; nuclear division acentric; binary fission generally interkinetal (symmetrogenetic); all parasitic.

e.g. *Opalina*.

Subphyllum III. SARCODINA ('AMOEBAE'): pseudopodia or locomotive protoplasmic flow without discrete pseudopia; flagella, when present, usually restricted to developmental or other temporary stage.

Superclass 1. RHIZOPODA. Locomotion by lobopodia, filopodia, or reticulopodia, or by protoplasmic flow without production of discrete pseudopodia.

Class 1. Lobosea. Pseudopodia lobose or more or less filiform but produced from broader hyaline lobe; usually uninucleate.

e.g. *Entamoeba*, *Acanthamoeba*, *Naegleria*.

Phylum III. APICOMPLEXA: Apical complex, generally consisting of polar ring(s), rhoptries, micronemes; conoid and subpellicular microtubules present at some stage; cilia absent; sexuality by syngamy.

Class 2. Sporozoea. Conoid, if present, forming complete cone; reproduction generally asexual and sexual; oocysts with infective sporozoites; flagella, if present, only in microgametes.

Subclass 1. Gregarinia. Mature gamonts large, extracellular; mucron or epimerite in mature organism; generally syzygy of gamonts; gametes usually isogamous; zygotes forming oocysts within gametocytes; life cycle characteristically consisting of gametogony and sporogony; in digestive tract or body cavity of invertebrates.

e.g. *Monocystis*.

Subclass 2. Coccidia. Gamonts ordinarily present, typically intracellular, without mucron or epimerite; syzygy generally absent; gametes anisogamous; life cycle characteristically consisting of merogony, gametogony and sporogony; most species in vertebrates.

e.g. *Isospora*, *Toxoplasma*, *Plasmodium*.

Subclass 3. Piroplasmia. Piriform, round, rod-shaped or amoeboid; conoid absent; parasitic in erythrocytes or other circulating or fixed cells; ticks are vectors in known species.

e.g. *Babesia*, *Theileria*.

Phylum IV. MICROSPORA. Not considered here.

Phylum V. ASCETOSPORA. Not considered here.

Phylum VI. MYXOZOA. Not considered here.

Phylum VII. CILIOPHORA. Simple cilia or compound ciliary organelles typical in at least one stage of the life cycle; with rare exceptions, two types of nuclei; sexuality involving conjugation, autogamy and cytogamy.

e.g. *Balantidium*, *Nyctotherus*.

○ 4 ○

Flagellates: intestinal and related forms

Flagellates possess one marked advantage over their amoeboid relatives: they can swim. The possession of flagella enables these protozoans to move actively in environments unsuitable for amoebae. The dependence on surfaces, so marked in amoebae, is not a behavioural feature and flagellates have invaded not only the alimentary canal and tissue sites but also essentially liquid environments such as the blood stream, lymph vessels and cerebrospinal canal. Many species possess considerable ability to survive in a number of environments since they can change from a flagellated free-swimming to a non-flagellated tissue stage and *vice versa*.

A further adaptation for life in a liquid habitat is seen in the trypanosomes, in which the body has a streamlined torpedo shape. Locomotion is carried out by one or more flagella; many species also possess a so-called *undulating membrane* (see p. 47) sometimes bordered by a flagellum (Fig. 4.1). Intestinal species usually have three to five flagella, with the exception of the hyperflagellates from the intestine of white ants, which have many more. Some flagellates are additionally strengthened by the possession of a stiff rod or *axostyle* (Figs 4.1, 4.2) extending the length of the body.

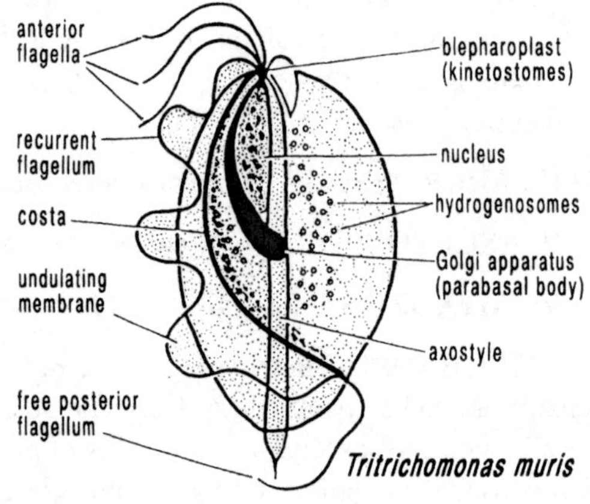

Fig. 4.1
Tritrichomonas muris from the caecum of the rat; see also Fig. 4.2. (Modified from Hegner *et al.*, 1938.)

The mechanisms of flagellar movement are complex and beyond the scope of this book; for a general review of this topic see Jahn & Bovee (1967).

○○○ 4.1 Classification (Levine *et al.*, 1980)

The classification of protozoa given here is based on the revision of Levine *et al.* (1980). Only the majority of parasitic groups of flagellates, as detailed in Chapter 3 (Appendix) are dealt with here:

Phylum I. SARCOMASTIGOPHORA
Subphylum 1. Mastigophora (Flagellata)
  Class 2. Zoomastigophorea
    Order 2. Kinetoplastida: most species parasitic
    Order 4. Retortamonidida: all species parasitic
    Order 5. Diplomonadida: most species parasitic
    Order 7. Trichomonadida: most species parasitic
Subphyllum II. Opalinata. (The proper taxonomic allocation of this group is in dispute and for convenience is considered with the Ciliophora in Chapter 10.)

This is the strictly zoological classification, but it is usual to speak in a general way about ‘intestinal flagellates’, which occupy the alimentary canal, and ‘haemoflagellates’, which occur in the blood, lymph or tissues of the vertebrate host. Grouped with the ‘intestinal’ forms are flagellates which invade other tubular systems, such as the genital tract, and it is convenient to consider them together. Examples of Order 7 are dealt with before the other Orders.

Table 4.1. *Common intestinal flagellates of man and laboratory animals*

Species	Host	Habitat
Chilomastix mesnili	man	large intestine
Chilomastix bettencourti	rat, mouse	large intestine
Chilomastix cuniculi	rabbit	large intestine
Chilomastix gallinarum	chicken	caeca
Pentatrichomonas hominis	man	large intestine
Trichomonas tenax	man	mouth
Trichomonas gallinae	pigeon	oesophagus, crop
Trichomonas vaginalis	man	genital tract
Tritrichomonas muris	rat	caecum
Tritrichomonas foetus (= *suis?*)	cattle	genital tract
Tritrichomonas augusta	frog	rectum
Histomonas meleagridis	turkey	caeca
Giardia lamblia	man	duodenum
Giardia muris	mouse	duodenum
Giardia caviae	guinea pig	duodenum
Giardia duodenalis	rabbit, rat	duodenum
Giardia agilis	tadpole	duodenum

LIBRARY, UNIVERSITY OF CHESTER

○○○ 4.2 Trichomonads

○ 4.2.1 General account

Occurrence. 'Trichomonad' is a general term used for members of the Order Trichomonadida, which includes a large number of species parasitising the tubular organs of a wide range of hosts (see Table 4.1). Species readily obtained for laboratory study are *Tritrichomonas muris* in laboratory mice and rats and *Trichomonas gallinae* from the crop of pigeons.

Systematics. The systematics of trichomonads is based largely on the number and arrangement of the anterior flagella. Whereas previously most of the commonly studied forms were placed in the genus *Trichomonas*, more detailed study has led to these species being placed into three genera: *Trichomonas*, *Pentatrichomonas* and *Tritrichomonas*. For discussion on the systematics of these organisms see Jírovec & Petrů (1968); Honigberg (1978*a,b*). Studies with isoenzyme analysis have revealed that extensive levels of biochemical polymorphism and genetic divergence occur and numerous strains of most species exist (Nadler & Honigberg, 1988).

Locomotion. Trichomonads swim with a characteristic wobbly movement, quite unmistakable once observed.

Morphology: *light microscopy*. Only a general description of a typical trichomonad (based on *Tritrichomonas muris*, Fig. 4.1) is given here: a further account at the ultrastructural level is given below. A trichomonad has an ovoid body with a single ovoid nucleus. Its most striking feature, visible in warm, fresh preparations, is the presence of 3–5 free anterior flagella and one (termed the *recurrent*) flagellum running posteriorly. The pellicle is extended out on one side (considered to be dorsal) to form a frill-like membrane, the undulating membrane, which appears to bear (see p. 47) the recurrent flagellum attached to its outer margin. The movements of this marginal flagellum are translated to the membrane, which thus acts as a supplementary locomotory organ. The undulating membrane has a deeply staining basal rod, the *costa*, along the line of its attachment to the body. The body is supported by a stiff rod or *axostyle*, which passes through its long axis and protrudes posteriorly like a tail. Most species have a sausage-shaped *parabasal* body; this represents the Golgi apparatus of trichomonads. In some species there is a well-developed cytostome but this is vestigial or lacking in others.

Ultrastructure (Fig. 4.2). The basic ultrastructure of trichomonads has been extensively studied by TEM (transmission electron microscopy) and SEM (scanning electron microscopy) (Honigberg, 1989). Early work has been reviewed by Honigberg (1978*a,b*). Some later specialised studies are those of Honigberg *et al.* (1971), Warton & Honigberg (1979), Juliano *et al.* (1986), Benchimol & De Souza (1983, 1987) and Benchimol *et al.* (1986).

The following is a brief, *generalised* account of the ultrastructure of a trichomonad, based largely on *T. muris*; differences between species are, however, not great. Many of the details of the ultrastructure are clear from Fig. 4.2.

The mastigont system. A set of flagella, together with its associated organelles, is known as a *mastigont.* The flagella, which have the 'typical' structure found in other flagellates, i.e. nine pairs of peripheral and one pair of central fibrils, pass into cylindrical *basal bodies* or *kinetosomes* which have essentially the same 9-fibril structure as the flagella. The region containing the kinetosomes makes up the *blepharoplast* region of light microscopy (Fig. 4.1).

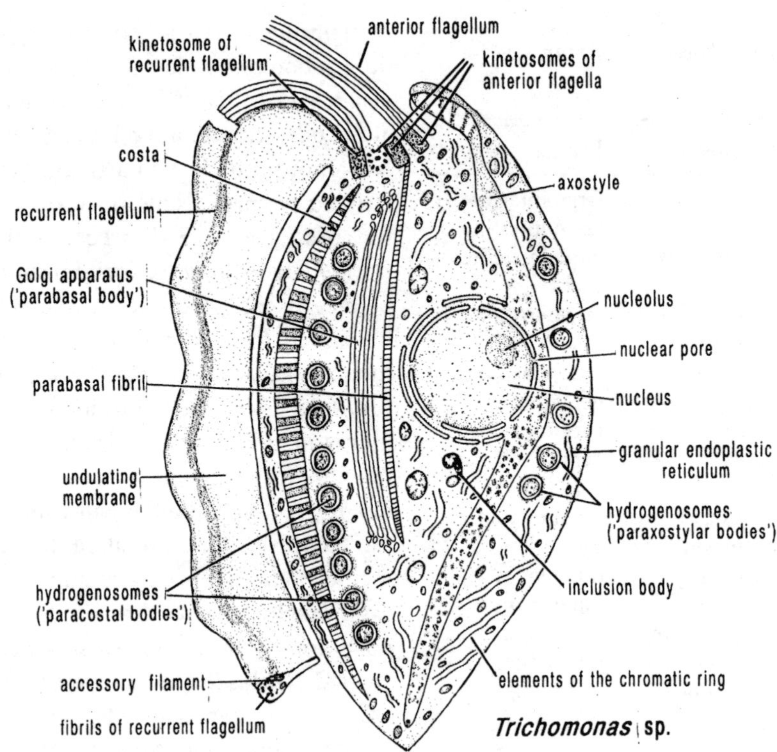

Fig. 4.2
Ultrastructure of a trichomonad. (Modified from Pitelka, 1963, from E. Anderson, unpublished.)

Axostyle. The axostyle is essentially tubular, the anterior part or *capitulum* being cup-like and extending into the kinetosome complex without actually making contact with it. The axostyle has a fibril substructure, the number of rows of fibrils depending on the species. Associated with the axostyle are the hydrogenosomes, which have a specific role in metabolism (see below).

The parabasal complex (Golgi apparatus). This consists of the *parabasal body* and *parabasal filament*. The parabasal body consists of stacks of flattened, pancake-like cisternae, surrounded by numerous spherical vesicles, and undoubtedly is homologous with the Golgi apparatus of higher organisms. The parabasal filament, which appears to originate in the dense granules of the blepharoplast region, consists of a cross-banded structure like the costa (see below), but is much thinner.

The costa. The costa originates in the blepharoplast region and is closely associated with the kinetosome of the recurrent flagellum. A striking ultrastructural feature of the costa is the regular cross-striations, which have a periodicity of about 42 nm. The costa also has fine longitudinal filaments, arranged in a herringbone pattern, a substructure which would explain its presumed function as a strong but flexible supporting structure to the undulating membrane and its associated flagellum.

Hydrogenosomes. It is now recognised that mitochondria as such are lacking in trichomonads. In their place are found spherical organelles termed *hydrogenosomes*, approximately 0.51 μm in diameter, with a double membrane surrounding a granular matrix (Benchimol & De Souza, 1983; Lindmark *et al.*, 1975). These are the paraxostylar granules of light microscopy and are particularly numerous in *Trichomonas vaginalis*, where their arrangement around the axostyle constitutes a consistent diagnostic arrangement (Honigberg *et al.*, 1983). Although these bodies share with mitochondria the possession of circular DNA, a high concentration of cardiolipin and a cyanide-insensitive superoxide dismutase, they function in a completely different way, in that protons, rather than molecular oxygen, act as terminal electron acceptors (Mehlhorn, 1988). Their biochemical role is further discussed on p. 53.

Undulating membrane and recurrent flagellum. The undulating membrane is formed by the pellicle curving over and enclosing the recurrent flagellum. The flagellum lies in a shallow 'recess' in the membrane but is not actually connected with it as appears to be the case under light microscopy.

Miscellaneous inclusions. The endoplasmic reticulum is found frequently as a corona around the nucleus. Ribosomes, and lipid and glycogen granules, are common in the cytoplasm, together with vacuoles of varying size; some of the latter may represent food vacuoles.

Feeding. Many species (e.g. *T. vaginalis*) lack a cytostome but solid food can apparently be taken in by the posterior region. Thus, *Tritichomonas foetus* has been shown to ingest horseradish peroxidase, ferritin and 0.08 μm latex beads by pinocytosis (Benchimol *et al.*, 1986). The pinocytotic vesicles fuse with each other and with lysosomes, forming large vacuoles.

Reproduction. This is effected by binary longitudinal fission.

Habitat. Trichomonads occur in a great variety of environments, a feature possibly related to the possession of several flagella, and also to the fact that, unlike the amoebae, whose nutrition is predominantly holozoic, their nutrition is both holozoic and saprozoic. Although the majority occur in the caecum and large intestine, the mouth, throat, oesophagus, crop, vagina, prostate and uterus are favourable sites for some species.

Transmission. There is no evidence that any species of *Trichomonas* is capable of forming cysts, although some species (e.g. *Tritrichomonas muris*) become rounded and form 'pseudocysts' which appear to play a prominent role in transmission. Thus, it has been shown that pregnant and lactating hamsters infected with *T. muris* shed 'pseudocysts' in their faeces (Mattern & Daniel, 1980). The pseudocysts do not have a cyst wall but appear to have internalised their locomotor organelles (flagella and undulating membrane). Newborn hamsters, living in an environment heavily contaminated with the mother's faeces, all become infected by the 7th day, presumably by ingesting the pseudocysts. *In vitro*, pseudocysts readily 'excyst' in a suitable liquid medium; the locomotor organs become externalised and there is full return to motility.

○ 4.2.2 Particular genera of trichomonads

The following are the characteristics of the major genera (Jírovec & Petrů, 1968):

1. Genus *Trichomonas*

Four anterior flagella, the fifth (recurrent) flagellum not trailing beyond the undulating membrane; undulating membrane shorter than the body; axostyle thin; parabasal body rod-shaped or with lateral bifurcation, e.g. *T. vaginalis*, *T. tenax*, *T. gallinae*.

2. Genus *Pentatrichomonas*

Five anterior flagella: four in a group and one flagellum beating in an independent rhythm; recurrent flagellum extending beyond undulating membrane to form a long trailing flagellum; undulating membrane extending to termination of body; costa well developed; parabasal body composed of one or more granules, surrounded by an elliptical or spherical zone, e.g. *Pentatrichomonas hominis*.

3. Genus *Tritrichosomonas*

Three anterior flagella; recurrent flagellum as in *Pentatrichomonas*; axostyle thick, sharply pointed at posterior end, surrounded by one or more periaxostylic rings at point of projection from body; parabasal body rod-shaped, often very elongated, e.g. *Tritrichomonas muris*, *T. suis*.

○ 4.2.3 Particular species

Although most species of trichomonads are harmless, several have pathogenic effects in man and domestic animals. Those of proven medical importance in man are *Trichomonas vaginalis* and *T. tenax*, and those of veterinary importance are *Tritrichomonas foetus* (in cattle) and *Trichomonas gallinae* (in birds). The systematics, biology, epidemiology and pathogenicity of trichomonads have been the subject of several International Symposia (Symposium, 1981, 1986) and major reviews (Jírovec & Petrů, 1968; Honigberg, 1978*a*, 1989).

Trichomonas vaginalis

This species is a relatively common pathogenic species of the female and male urinogenital tract. The details of its morphology are shown in Fig. 4.3 and Table 4.2. A detailed account of its biology and morphology has been given by Honigberg (1989). The organism causes *trichomoniasis*, resulting in vaginitis in women and urethritis in men; 200 million people annually are probably infected (Quinn & Holmes, 1984).

Table 4.2. *Morphological and physiological data on the three trichomonad species parasitic in man*

	Trichomonas vaginalis	*Trichomonas tenax*	*Pentatrichomonas hominis*
Body size (mean)	4–32 μm × 2.4–14.4 μm 10 μm × 7 μm	4.2–12.8 μm × 2.1–14.7 μm 7.4 μm × 5.3 μm	7–15 μm × 4–10 μm ?
Body shape	oval	piriform	piriform
Length : width ratio	1.4 : 1	1.4 : 1	?
Nucleus size	2.4–6.4 μm × 1.2–3.2 μm	1.5–3.3 μm × 1–2.5 μm	?
Chromosome no.	5	3	5
No. of flagella	4	4	5
Hydrogenosomes	present	paracostal only	present?
Golgi apparatus (parabasal body)	mostly V-shaped	rod-shaped	disc-shaped
Optimum pH	5.8–6.0	7.0–7.5	7.0
Optimum temp.	37 °C	31–32 °C	30–37 °C
Axenic *in vitro* culture	easy	very difficult	difficult

Source: Data from Jírovec & Petrů (1968) and Honigberg (1989).

As is the case with some other protozoan parasites (e.g. *Entamoeba histolytica*) the conditions under which *T. vaginalis* induces pathogenic symptoms in humans are not known. In many cases, the organism appears to be harmless; in others it produces severe symptoms. In the male 50–70 per cent of infected men appear to be asymptomatic and the infection may be self-limited. In some men, however, not only the urethra, but all the genital organs and glands may be infected. Similarly, in the female, although the vagina is most commonly infected, the organism may spread to all parts of the urinogenital tract. The clinical manifestations range from asymptomatic carriage to severe vaginitis. There is usually intense inflammation with itching and copious discharge (leucorrhea).

Transmission is by sexual intercourse and the disease has now come to be regarded as a true (and serious) venereal disease. Infection via damp towels or clothing cannot, however, be ruled out. As with other protozoa, a number of strains, some of which are more pathogenic than others, have now been recognised. Treatment by the drug metronidazole is very effective, although resistant strains have now appeared.

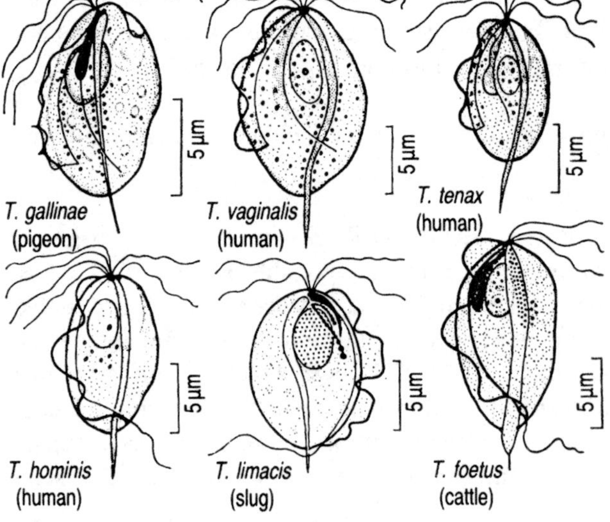

Fig. 4.3
Trichomonads from man and other animals. (After various authors.)

Trichomonas tenax (Synonym: *T. buccalis*) (Fig. 4.3)

This species occurs in the human mouth and has an incidence of 11–25 per cent. Its presence tends to be associated with caries, pyorrhoea or other infections of the gums. Its role as a pathogen is unproven although it has been associated with a fatal case of pulmonary trichomoniasis (Hersh,1985). Its basic biology has been reviewed by Honigberg (1978*b*, 1989). It is generally smaller than *T. vaginalis* (Table 4.2).

Trichomonas gallinae (Synonym: *T. columbae*) (Fig. 4.3)

A species common to a number of birds, including turkeys and chickens, but the common or domestic pigeon is probably the natural host. It occurs in the pre-gastric alimentary canal – the mouth, pharyngeal region, oesophagus and crop – but never in the post-gastric regions. From these sites it may invade other tissues, but various strains seem to favour different localities; severe infections may prove fatal. Its biology and pathogenicity has been reviewed by Honigberg (1978*a*).

In the pigeon, infection approaches 100 per cent, for the mode of transmission from parent to young is exceptionally efficient. The pigeon's 'milk' in the crop is a favourite habitat for the organism, and a drop is usually teeming with active trophozoites. Within minutes of hatching, offspring may become infected by transference of 'milk' from the parent bird. The epidemiological picture is reflected in the high prevalence in bird colonies. A recent survey reports the following: feral town pigeons, 54 per cent and carrier pigeons, 61 per cent, but wild pigeons, only 12 per cent (Gregurić *et al.*, 1986).

Pentatrichomonas hominis (Synonym: *Trichomonas hominis*) (Fig. 4.3)

The commonest intestinal flagellate of man; it appears to be non-pathogenic. Its morphology is compared with *T. vaginalis* and *T. tenax* in Table 4.2. It is most readily distinguished from these forms by the fact that the undulating membrane runs the whole length of the body, with the recurrent flagellum trailing posteriorly. Although the characteristic (and diagnostic) number of flagella is five, there is some deviation from this number. It has a wide host range and flagellates indistinguishable from it have been found in non-human primates, cats, dogs and various rodents (Honigberg, 1978*b*). Its isoenzymes have been studied by Gradus & Matthews (1985).

Tritrichomonas muris (Fig. 4.1)

This species occurs in the caecum of rodents and is usually readily obtainable. Like *P. hominis*, the recurrent flagellum trails posteriorly. It appears to be a non-pathogenetic species. Unlike *T. vaginalis*, it forms 'pseudocysts' (p. 47), which play a role in transmission.

Tritrichomonas foetus (Fig. 4.3)

This species occurs in the genital tract of cattle. Its structure and general biology is reviewed by Honigberg (1978*a*) and its pathogenesis, control, resistance, diagnosis and treatment by Fitzgerald (1986). In cattle, it is transmitted venereally from infected bulls to heifers, in which it attacks the vagina and uterus, causing abortion and other pathogenic disturbances. The females are self-curing, but the males are apparently infected for life. A number of strains have been recognised.

○○○ 4.3 Intestinal flagellates other than trichomonads

○ 4.3.1 Minor species

Embadomonas intestinalis (Fig. 4.4)

A rare human parasite with an ovoid body, 5–6 μm in length, with two flagella, producing pear-shaped cysts.

Enteromonas hominis (Fig. 4.4)

Another rare form in man with an ovoid body, 4–10 μm in length, with two flagella. Produces elongate, oval cysts.

Chilomastix mesnili (Fig. 4.4)

The largest intestinal flagellate of man, 6–20 μm in length. Characterised especially by the large cytostome in the form of a slightly spiral groove whose lateral margins are supported by two filaments. There are three anterior flagella and a fourth lies in the oral groove. It produces pear-shaped cysts resembling those of *Embadomonas*, with a characteristically coiled darkly stained filament.

Chilomastix occurs in the human large intestine; incidence is about 1–10 per cent. Numerous species have been reported from other animals. The species in quail appears to be highly pathogenic to young birds (Davis *et al.*, 1964).

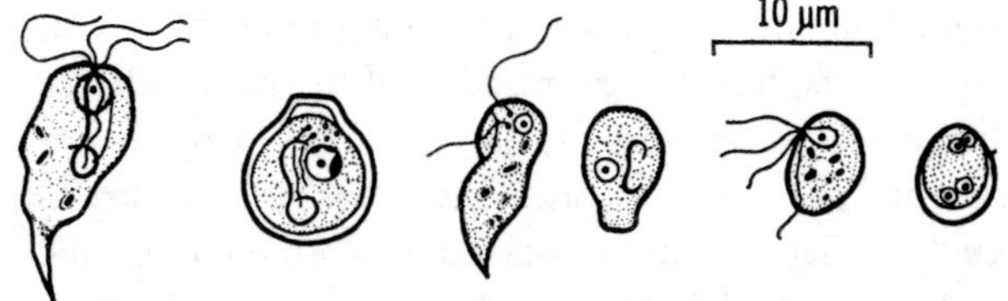

Fig. 4.4
Trophozoites (left) and cysts (right) of the smaller and rarer flagellates of man. (After Hoare, 1949, and various authors.)

○ 4.3.2 Genus *Giardia*

General account. A flagellate of this genus is quite unlike any of the preceding species in shape or habits. It has been described in front view as looking like a 'tennis racquet without a handle' and it has a comical, face-like appearance (Fig. 4.5). The body is a 'tear-drop shaped' with a convex dorsal surface and a concave ventral one. The latter possesses two depressions, sometimes termed *adhesive discs* (= suckers) which make contact with the intestinal cells of the host (Fig. 4.5). A single or double *median body*, unique to this genus, is found just below the adhesive discs. *Giardia* exhibits

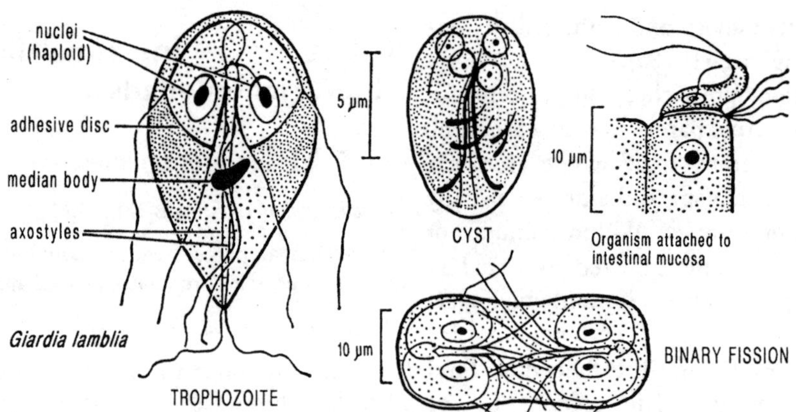

Fig. 4.5
Giardia lamblia (syn. *G. duodenalis*) from the human duodenum. (After various authors.)

perfect bilateral symmetry, and there are double sets of nuclei, flagella and kinetosomes (Fig. 4.5). Each nucleus has been shown to contain a haploid number of chromosomes (Kabnick & Peattie, 1990).

Ultrastructural studies, reviewed by Honigberg (1978*b*) and Owen (1980), show that the so-called adhesive disc is in fact a structure composed of supportive, rather than contractile, elements which remain a fixed shape; the term 'striated' disc has been suggested. The median body superficially resembles the axostyle of trichomonads, but the organisation of the microtubules which compose it, and the fact that the body is not always present, make it a distinctive structure. No structures identifiable as mitochondria, smooth endoplasmic reticulum or Golgi complex have been identified in *Giardia*. The central flagella are 'ribbon-like', and thus morphologically adapted to their suggested pumping function (see below). Studies with ferritin have demonstrated that particulate material can be ingested by pinocytosis.

Encystment. Species of *Giardia* produce characteristic oval cysts with thick walls (Fig. 4.5), the remains of the disintegrated flagella forming a central 'streak' visible in iodine or MIF (merthiolate-iodine-formaldehyde) preparations.

Habitat and nutrition. Species of *Giardia* are confined in their distribution to the small intestine, particularly the duodenum, occasionally invading the bile ducts. In severe infections they may carpet large areas of the mucosa. Nutritionally the duodenum is the richest habitat in the alimentary canal and, as mentioned above, the organism makes very close contact with the mucosa. Speeded up micro-cinematography (Soloviev, 1966) reveals that the central pair of flagella (which are flattened) work as a pump removing fluid from the sucker disc. It has been suggested that they serve for pumping out nutrient substances from the microvilli of the intestinal mucosa.

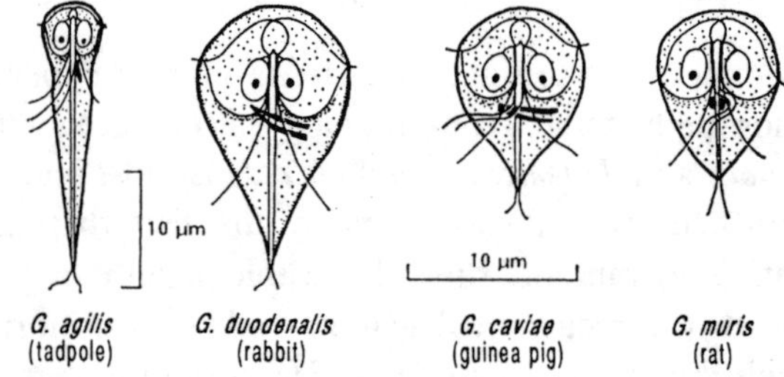

Fig. 4.6
Giardia species commonly present in laboratory animals. (After Hegner *et al.*, 1938.)

G. muris (Fig. 4.6)

This species is common in rats, mice and other rodents; the rat and rabbit forms are considered by some to be 'races' of a distinct species, *G. duodenalis*. The morphology has been described in detail by Iwańczuk (1968). Trophozoites occur all along the intestine from the pylorus to the caecum (Fig. 4.7) but the highest percentage occurs 100–150 mm from the pyloric sphincter. Encystment takes place chiefly in the region 220–270 mm from the sphincter.

G. muris is considered to be a good model for the study of human giardiasis (Majewska, 1985).

Giardia lamblia (*G. duodenalis* = *G. intestinalis*); giardiasis

The species which infects man, *G. lamblia*, has a worldwide distribution with an incidence of 1–30 per cent. In the USA, it is now considered to be the most common intestinal parasite of man and the leading cause of diarrhoea due to protozoan infection in humans. The resulting infection is usually referred to as *giardiasis*, but the terms *lambliasis* or *flagellate diarrhoea* are sometimes used. Children seem especially susceptible and mass infections occasionally break out in kindergartens or day-care centres (Woo & Paterson, 1986). It is also the most frequently reported intestinal parasite in Britain (Knight & Wright, 1978).

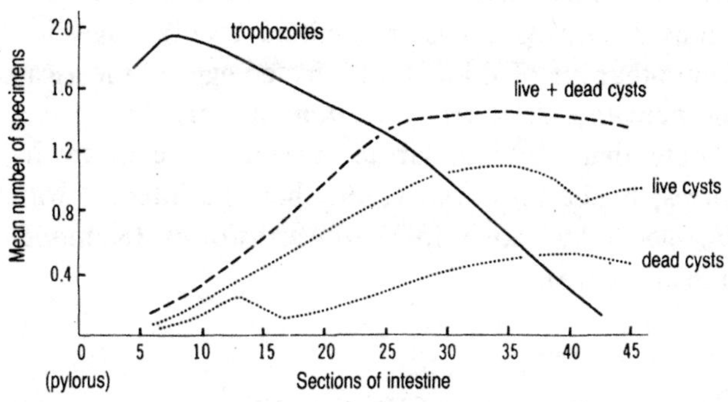

Fig. 4.7
Distribution of trophozoites and cysts of *Giardia muris* along the intestine of mice. (After Iwańczuk, 1968.)

There is a vast literature on *Giardia* and giardiasis, covering biology, host specificity, pathogenicity, transmission, epidemiology and immunology. Some valuable reviews are the following: Ackers (1980), Boreham *et al.* (1990), Den Hollander *et al.* (1988), Erlandsen & Meyer (1984), Farthing (1989), Kasprzak & Pawlowski (1989), Knight & Wright (1978), Majewska & Kasprzak (1986), Meyer (1985, 1990), Meyer & Radulescu (1979), Symposium (1980) and Thompson *et al.* (1990).

Transmission. Cysts (Fig. 4.5) are passed in the faeces. Transmission can be described as faecal–oral, i.e. *Giardia* cysts passed in the faeces of one person result in a new infection when swallowed by another person. According to Kasprzak & Pawlowski (1989)

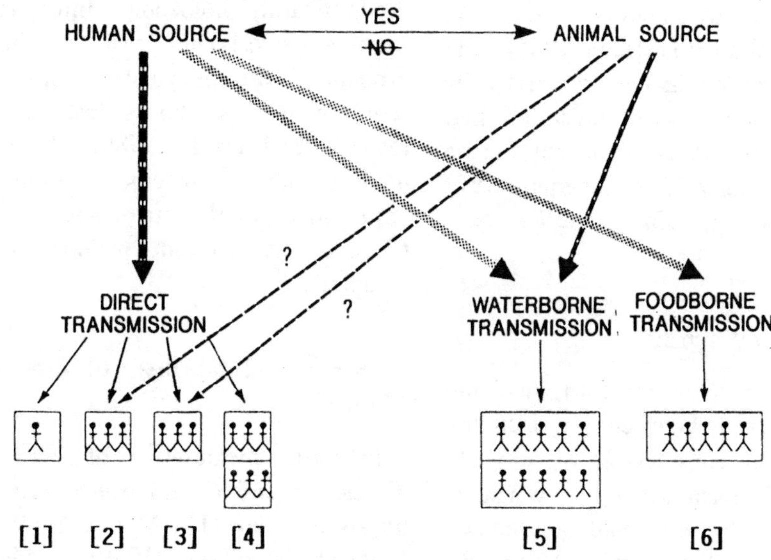

Fig. 4.8
Possible pathways of transmission of giardiasis within or to a human population. (After Kasprzak & Pawlowski, 1989.)

'. . . this transmission can be very short as in the cases of sexual partners, family members, institutional inmates or groups of professional workers, relatively short as in the food-borne outbreaks, and relatively long in water-borne epidemics' (Fig. 4.8). Water-borne infections appear to be a major cause of many epidemics (Meyer, 1988).

Human giardiasis: is it a zoonosis*?

A major controversy in the epidemiology of giardiasis is whether or not it is a zoonosis transmitted to man from animals such as cats and dogs. Animals considered as possible sources of infection include rodents (rats, voles, beavers and muskrats), dogs and cats, and hoofed animals; beavers appear to be the animals most commonly infected (Bemrick & Erlandsen, 1988; Eckert *et al.*, 1989; Thompson *et al.*, 1988, 1990).

Moreover, it has been shown that *Giardia* cysts from a human source can be infective to numerous animals: dogs, guinea pigs, raccoons and sheep (Faubert, 1988). Evidence from experiments using germ-free animals, however, suggests that cats and dogs, although susceptible to their own *Giardia* species (*G. simoni* and *G. canis*, respectively), are not infected by *G. lamblia* and do not appear to be responsible for transmission of infection in the human population (Woo & Paterson, 1986). The available evidence has been succinctly summarised by Kasprzak & Pawlowski (1989) who concluded that although zoonotic giardiasis may occur, it may not be common and that '. . . in reality the major methods of transmission of giardiasis remain basically host-specific'.

○ 4.3.3 *Histomonas meleagridis*

A small flagellate parasite of the turkey, fowl, quail and ruffed grouse. Its size range is between 8 and 20 μm, with its appearance showing considerable variation. In general, it has a rounded form which, on warming, shows sluggish activity with amoeboid tendencies. There is usually only a single flagellum but large individuals with two to four flagella sometimes occur. There is a single nucleus and kinetosome. Ultrastructure studies (Lee, 1969) have shown that this organism basically resembles a trichomonad; mitochondria are lacking, although cytoplasmic sacs about the size of mitochondria have been found. The organism apparently sheds its flagella and becomes amoeboid before or as it invades the caecal tissues, the invasive stage feeding by pinocytosis.

H. meleagridis is parasitic in the caecum, a site from which it migrates causing characteristic lesions associated with the disease *histomoniasis*, often referred to as 'blackhead'. Its biology and pathogenicity have been reviewed by Honigberg (1987*a*,*b*). Under natural conditions *H. meleagridis* is transmitted by oral ingestion (from other infected birds) of infected eggs of the ascaroid nematode *Heterakis gallinarum*. It is also interesting to note that *Heterakis* larvae remain viable in earthworms; hence birds can readily become infected with *Histomonas* by eating infected earthworms (Kemp & Franson, 1975).

*A *zoonosis* is defined (WHO, 1979) as a disease or infection naturally transmitted between *vertebrate* animals and man.

○○○ 4.4 Physiology and biochemistry of intestinal flagellates

○ 4.4.1 *In vitro* cultivation

In vitro, the nutrition of these flagellates appears to be predominantly holozooic. Intestinal forms feed on detritus, bacteria and yeasts; buccal and vaginal forms probably feed on leucocytes. Some of the basic growth requirements as shown from *in vitro* studies are reviewed by Linstead (1989). The use of antibiotics has made possible the axenic isolation and subsequent *in vitro* culture of the major species of *Trichomonas* and *Giardia*; the relevant techniques are reviewed in Chapter 33.

○ 4.4.2 Metabolism of *Trichomonas* and *Giardia*

The basic metabolism of *Trichomonas*, based on *T. vaginalis* and *Tritrichomonas foetus*, has been comprehensively reviewed by Müller (1989) and that of *Giardia lamblia* by Jarroll *et al.* (1989) and Meyer (1990). Various specific aspects of the metabolism of these and related species have been reviewed by Coombs & North (1991), Müller (1988, 1989), Lindmark *et al.* (1989) and Fairlamb (1989).

Like *Entamoeba histolytica* (p. 38), species of *Trichomonas* and *Giardia* lack mitochondria although *Trichomonas* contains membrane-bound organelles, known as *hydrogenosomes* (see later), which constitute a separate

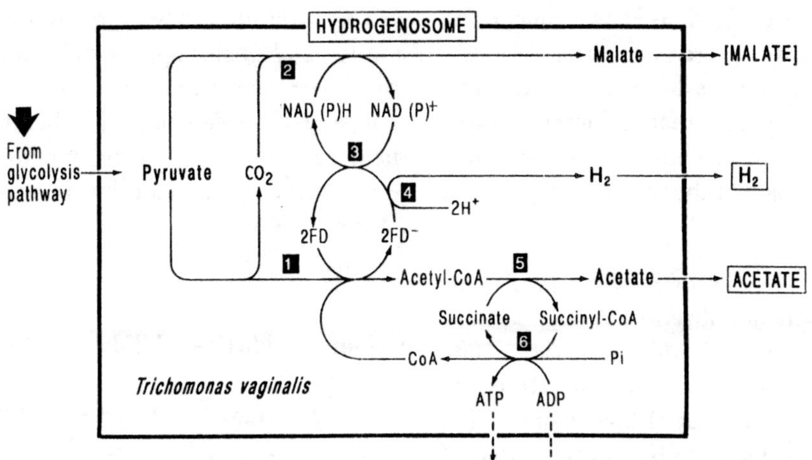

Fig. 4.9
Map of metabolism within the hydrogenosomes of *Trichomonas vaginalis* and *Tritrichomonas foetus*. [1] Pyruvate ferrodoxin oxidoreductase, [2] malate dehydrogenase (decarboxylating, NAD), [3] NAD : ferredoxin oxidoreductase, [4] H_2 : ferredoxin oxidoreductase, [5] acetate : succinate CoA-transferase, [6] succinate thiokinase. Hydrogen and acetate are the major end products and malate a minor one; see also Table 3.5. (Modified from Steinbüchel & Müller, 1986.)

compartment of energy metabolism (Fig. 4.9), as mitochondria do in most other protozoa. The '*paracostal bodies*' and the '*paraxostylar bodies*' described by light microscopists (Fig. 4.2) have now been identified as hydrogenosomes. *Giardia* lacks both hydrogenosomes and mitochondria but contains lysosome-like organelles which accumulate ferritin and stain positively for acid phosphatase and aryl sulphatase (Jarroll *et al.*, 1989). All the enzyme activities of *Giardia* thus take place in the cytosol.

Hydrogenosomes. These organelles are so-called because they produce molecular hydrogen as a metabolic product. They are predominately spherical structures, 0.5–1 µm in diameter, and are surrounded by an envelope consisting of two closely opposed membranes; unlike mitochondria, the inner membrane does not form cristae. Like mitochondria, however, hydrogenosomes constitute a separate compartment of energy metabolism which results in the eventual conversion of pyruvate to acetate, malate, CO_2 and H_2 (Fig. 4.9). In intact *T. vaginalis*, hydrogenosomes function under both anaerobic and aerobic conditions, as reflected by the production of CO_2 (Müller, 1989). The carbon flow through the hydrogenosomes is not affected by the presence or absence of O_2, even though the electrons through the pyruvate oxidation have a different fate. Under anaerobic conditions, protons serve as the terminal acceptors and under aerobic conditions O_2 serves. Other possible activities of the hydrogenosomes are reviewed by Lindmark *et al.* (1989).

Carbohydrate metabolism. Like most anaerobic protozoa, the main source of energy are carbohydrates and their metabolism is fermentative. Glucose is phosphorylated by hexokinase or glycogen by glycogen hydrogen phosphorylase and enters the glycolysis pathway and by steps is converted to equimolar amounts of dihydroxyacetone phosphate and glyceraldehyde 3-phosphate. The latter is then metabolised, via the classical Embden–Meyerhof pathway, to PEP and finally to pyruvate. In *T. vaginalis*, various intermediates of the glycolysis give rise to glycerol, lactate, CO_2 and H_2 as endproducts (Table 3.5). *Tritrichomonas foetus*, however, produces succinate as its chief end product. Glycerol is produced by the reduction of dihydroxyacetone phosphate to *sn*-glycerol 3-phosphate, which is further hydrolysed to glycerol and P_i. Within the hydrogenosome, pyruvate is oxidatively decarboxylated, following the pathways shown in Fig. 4.9 with the formation mainly of H_2 and acetate (and to a lesser extent malate) as end products.

In the cytosol, pyruvate can further give rise to lactate via lactate dehydrogenase. In both *Trichomonas vaginalis* and *G. lamblia* the oxidative decarboxylation of pyruvate, leading to the production of acetyl-CoA, is catalysed by an unusual reversible, iron–sulphur enzyme, *pyruvate:ferredoxin oxidoreductase*, and not by the irreversible pyruvate dehydrogenase as in most organisms; it also uses another iron–sulphur protein, ferrodoxin. In *Giardia*, the oxidative decarboxylation of pyruvate takes place in the cytosol, as hydrogenosomes are lacking. In addition to acetate, ethanol is produced by a pathway similar to that utilised by *Entamoeba histolytica* (Fig. 3.15).

Amino acid metabolism. Little is known concerning this area of metabolism and there is no available information of the essential or unessential nature of individual amino acids. It is known that aminotransferases are present and pyruvate, α-ketoglutarate, oxaloacetate and phenylpyruvate can accept the amino group from a number of amino acids.

Lipid metabolism. *T. vaginalis* requires fatty acids and cholesterol as essential nutrients although *de novo* biosynthesis of these substances does not occur. The situation appears to be the same in *G. lamblia* as the phospholipid, fatty acid and neutral lipid composition of trophozoites has been found to be similar to that of the medium in which they were grown (Jarrol *et al.*, 1989).

Nucleic acid metabolism. This has been reviewed by Wang (1989) for *T. vaginalis* and by Jarroll *et al.* (1989) for *G. lamblia*. In *Trichomonas*, there is no detectable *de novo* synthesis of purine nucleotide or pyrimidine rings and no activity of the enzyme dihydrofolate reductase or purine or pyrimidine phosphoribosyltransferase. This represents a deficiency in nucleic acid metabolism greater than that reported in any other organism. Lack of *de novo* purine or pyrimidine nucleotide synthesis also occurs in *G. lamblia* (Jarroll *et al.*, 1989) and *Entamoeba histolytica* (p. 40)).

All these species are in close contact with host cells of the intestinal or genital tract, which – being subject to inflammatory attack – are likely to have a rapid turnover. The parasites are thus in an environment rich in nucleic acids which could be effectively utilised in view of their active phagocytic and lysomal activities. Results have shown that all these species have simplified their nucleic acid metabolism and survive by 'salvaging' preformed nucleosides and bases.

In *T. vaginalis*, the only major purine salvage enzymes are adenosine kinase, guanosine kinase, adenosine phosphorylase, and purine nucleoside phosphorylase. This points to adenosine nucleoside and guanosine as the two intermediate precursors of adenine and guanine through the action of adenine and guanine kinase; the major purine salvage pathways thus being (Wang, 1989):

Adenine ↔ Adenosine → AMP → ADP → ATP
Guanine ↔ Guanosine → GMP → GDP → GTP.

A similar salvage pathway appears to operate in *Giardia lamblia* (Aldritt *et al.*, 1985; Jarroll *et al.*, 1989).

The pyrimidine salvage pathways in both *T. vaginalis* and *G. lamblia* also appear to be similar. These involve the conversion of exogenous uracil and cytidine to uridine, and of cytidine to uridine before incorporation into the nucleotide pool. The major pyrimidine salvage pathway in these species can thus be summarised as follows (Wang, 1989):

Uracil
↕
Uridine ↔ UMP → UDP → UTP
↕
Cytidine ↔ CMP → CDP → CTP.

G. lamblia also relies on salvage of exogenous thymidine for ribosylthymine monophosphate (TMP) synthesis.

○○○ References

(References of general interest have been included in this list, in addition to those quoted in the text.)

Ackers, J. P. 1980. Giardiasis: basic parasitology. *Transactions of the Royal Society of Tropical Medicine & Hygiene*, **74**: 427–9.

Aldritt, S. M., Tien, P. & Wang, C. C. 1985. Pyrimidine salvage in *Giardia lamblia*. *Journal of Experimental Medicine*, **161**: 437–45.

Bemrick, W. J. & Erlandsen, S. L. 1988. Giardiasis – is it really a zoonosis? *Parasitology Today*, **4**: 69–71.

Benchimol, M., Dacunha, E., Silva, N. L., Elias, C. A. & De Souza, W. 1986. *Tritrichomonas foetus*: ultrastructure and cytochemistry of endocytosis. *Experimental Parasitology*, **62**: 405–15.

Benchimol, M. & De Souza, W. 1983. Fine structure and cytochemistry of the hydrogenosome of *Tritrichomonas foetus*. *Journal of Protozoology*, **30**: 422–5.

Benchimol, M. & De Souza, W. 1987. Structural analysis of *Tritrichomonas foetus*. *Journal of Submicroscopic Cytology*, **19**: 139–47.

Boreham, P. F. L., Upcroft, J. A. & Upcroft, P. 1990. Changing approaches to the study of *Giardia* epidemiology: 1681–2000. *International Journal for Parasitology*, **20**: 479–87.

Coombs, G. & North, M. 1991. *Biochemical Protozoology*. Taylor & Francis, London. ISBN 0 748 4001 X.

Davis, D. E., Schwartz, L. D. & Jordan, H. E. 1964. A case report: *Chilomastix* sp. infection in a pen-raised quail. *Avian Disease*, **8**: 465–70.

Den Hollander, N., Riley, D. & Befus, D. 1988. Immunology of giardiasis. *Parasitology Today*, **4**: 124–31.

Eckert, J., Thompson, R. C. A. & Lymbery, A. J. 1989. Further contribution to the giardiasis debate. *Parasitology Today*, **5**: 161.

Erlandsen, S. L. & Meyer, E. A. (eds) 1984. *Giardia and Giardiasis*. Plenum Press, New York.

Fairlamb, A. H. 1989. Novel biochemical pathways in parasitic protozoa. *Parasitology*, **99S**: S93–112.

Farthing, M. J. G. 1989. Immune responses in human giardiasis. *Saudi Medical Journal*, **10**: 1–8.

Faubert, G. M. 1988. Evidence that giardiasis is a zoonosis. *Parasitology Today*, **4**: 66–8.

Fitzgerald, P. R. 1986. Bovine trichomoniasis. *Veterinary Clinics of North America, Food Animal Practice*, **2**: 277–82.

Gradus, M. S. & Matthews, H. M. 1985. Electrophoretic analysis of soluble proteins and esterase, superoxide dismutase and acid phosphate isoenzymes of members of the protozoan family *Trichomonadidae*. *Comparative Biochemistry and Physiology, B*, **81**: 229–33.

Gregurić, J., Jerčić, J., Mužinć, J., Szeleszcuk, P. & Pecaric, A. 1986. (Effect of ecological environment on the prevalence of *Trichomonas gallinae* in pigeons.) (In Yugoslav.) *Veterinarski Glasnik*, **40**: 657–60.

Hegner, R., Root, F. M., Augustine, D. L. & Huff, C. G. 1938. *Parasitology with special reference to man and domestic animals*. Appleton-Century Crofts, New York.

Hersh, S. M. 1985. Pulmonary trichomoniasis and *Trichomonas tenax*. *Journal of Medical Microbiology*, **20**: 1–10.

Hoare, C. A. 1949. *Handbook of Medical Protozoology*. Ballière, Tindal & Cox, London.

Honigberg, B. M. 1978*a*. Trichomonads of veterinary importance. In: *Parasitic Protozoa* (ed. J. P. Kreier), pp. 163–273. Academic Press, New York.

Honigberg, B. M. 1978*b*. Trichomonads of importance to human medicine. In: *Parasitic Protozoa* (ed. J. P. Kreier), pp. 275–454. Academic Press, New York.

Honigberg, B. M. (ed.) 1989. *Trichomonads parasitic in humans*. Springer-Verlag, New York. ISBN 3 540 96903 9.

Honigberg, B. M., Mattern, C. F. T. & Daniel, W. A. 1971. Fine structure of the mastigont system in *Tritrichomonas foetus* (Riedmuller). *Journal of Protozoology*, **18**: 183–98.

Honigberg, B. M., Mattern, C. F. T. & Warton, A. 1983. Fine structure of *Trichomonas vaginalis* in relation to that of other trichomonads. *Wiadomosći Parazytologiczne*, **29**: 5–7.

Hyde, J. E. 1990. *Molecular Parasitology*. Open University Press, Buckingham, UK. ISBN 0 471 93227 2.

Iwańczuk, I. 1968. Studies on the distribution of *Lamblia muris* in the alimentary tract of white mice. *Acta Parasitologica Polonica*, **15**: 367–73.

Jahn, T. L. & Bovee, E. C. 1967. Motile behaviour in Protozoa. In: *Research in Protozoology* (ed. T.-T. Chen), Vol. 1, pp. 41–200. Pergamon Press, Oxford.

Jarroll, E. L., Manning, P., Berrada, A., Hare, D. & Lindmark, D. G. 1989. Biochemistry and metabolism of *Giardia*. *Journal of Protozoology*, **36**: 190–2.

Jírovec, O. & Petrů, M. 1968. *Trichomonas vaginalis* and trichomoniasis. *Advances in Parasitology*, **6**: 117–88.

Juliano, C., Rubino, S., Zicconi, D. & Cappuccinelli, P. 1986. An immunofluorescent study of the microtubule organization in *Trichomonas vaginalis*. *Journal of Protozoology*, **33**: 56–9.

Kabnick, K. S. & Peattie, D. A. 1990. *In situ* analyses reveal that the two nuclei of *Giardia lamblia* are equivalent. *Journal of Cell Science*, **95**: 353–60.

Kasprzak, W. & Pawlowski, Z. 1989. Zoonotic aspects of giardiasis: a review. *Veterinary Parasitology*, **32**: 101–8.

Kemp, R. L. & Franson, J. C. 1975. Transmission of *Histomonas melagridis* to domestic fowl by means of earthworms recovered from pheasant yard soil. *Avian Diseases*, **19**: 741–4.

Knight, R. & Wright, S. G. 1978. Intestinal protozoa. Progress report. *Gut*, **19**: 241–8.

Kreier, J. P. (ed.) 1978. *Parasitic Protozoa*, Vol. 2. Academic Press, New York.

Lee, D. L. 1969. The structure and development of *Histomonas meleagridis* (Mastigamoebidae: Protozoa) in the female reproductive tract of its intermediate host, *Heterakis gallinarus* (Nematoda). *Parasitology*, **59**: 877–84.

Levine, N. D. (Chairman) *et al.* 1980. A newly revised classification of the Protozoa. (The Committee on Systematics and Evolution of the Society of Protozoologists.) *Journal of Protozoology*, **27**: 37–58.

Lindmark, D. G., Eckenrode, B. L., Halberg, L. A. & Dinbergs, I. D. 1989. Carbohydrate and hydrogenosomal metabolism of *Tritrichomonas foetus* and *Trichomonas vaginalis*. *Journal of Protozoology*, **36**: 214–16.

Lindmark, D. G., Muller, M. & Shio, H. 1975. Hydrogenases in *Trichomonas vaginalis*. *Journal of Parasitology*, **61**: 552–4.

Linstead, D. 1989. Cultivation. In: *Trichomonads Parasitic in Humans* (ed. B. M. Honigberg), pp. 91–111. Springer-Verlag, New York. ISBN 0 387 96903 9.

Majewska, A. C. 1985. (*Giardia muris* in rat as a model of giardiasis.) (In Polish, English summary.) *Wiadomosći Parazytologiczne*, **32**: 27–49.

Majewska, A. C. & Kasprzak, W. 1986. (Host specificity of species of *Giardia*.) (In Polish, English summary.) *Wiadomosći Parazytologiczne*, **32**: 141–56.

Mattern, C. F. T. & Daniel, W. A. 1980. *Tritrichomonas muris* in the hamster: pseudocysts and infection in the newborn. *Journal of Protozoology*, **27**: 435–9.

Mehlhorn, H. 1988. *Parasitology in Focus*. Springer-Verlag, Berlin.

Meyer, E. A. 1985. The epidemiology of giardiasis. *Parasitology Today*, **4**: 101–5.

Meyer, E. A. 1988. Waterborne *Giardia* and *Cryptosporidium*. *Parasitology Today*, **4**: 200. ISBN 0 444 81258 X.

Meyer, E. A. (ed.) 1990. *Giardiasis*. Elsevier Science Publishers, The Netherlands.

Meyer, E. A. & Radulescu, S. 1979. *Giardia* and giardiasis. *Advances in Parasitology*, **17**: 1–47.

Müller, M. 1988. Energy metabolism of protozoa without mitochondria. *Annual Review of Microbiology*, **42**: 465–88.

Müller, M. 1989. Biochemistry of *Trichomonas vaginalis*. In: *Trichomonads Parasitic in Humans* (ed. B. M. Honigberg), pp. 53–83. Springer-Verlag, New York. ISBN 0 387 96903 9.

Nadler, S. A. & Honigberg, B. M. 1988. Genetic differentiation and biochemical polymorphism among trichomonads. *Journal of Parasitology*, **74**: 797–804.

Owen, R. L. 1980. Ultrastructural basis of *Giardia* function. *Transactions of the Royal Society of Tropical Medicine and Hygiene*, **74**: 429–33.

Pitelka, D. R. 1963. *Electron-Microscopic Structure of the Protozoa*. Pergamon Press, Oxford.

Quinn, T. C. & Holmes, K. K. 1984. Trichomoniasis. In: *Tropical and Geographic Medicine* (ed. K. S. Warren & A. F. Mahmould), pp. 335–41. McGraw-Hill Book Company, New York.

Soloviev, M. M. 1966. (The biological design of the central flagella in *Giardia*.) (In Russian, English summary.) *Meditisinskaya Parasitologiya i Parasitarnye Bolezni, Moscow*, **35**: 91–3.

Steinbüchel, A. & Müller, M. 1986. Anaerobic pyruvate metabolism of *Tritrichomonas foetus* and *Trichomonas vaginalis*. *Molecular and Biochemical Parasitology*, **20**: 57–65.

Symposium. 1980. Symposium on Giardiasis. *Transactions of the Royal Society of Tropical Medicine and Hygiene*, **74**: 427–48.

Symposium. 1981. International Symposium on Trichomonadosis. Bialystok, 13–15 July, 1981. *Wiadomosći Parazytologízne*, **29**: 1–236.

Symposium. 1986. International Symposium on Trichomonads and Trichomoniasis. Prague, 3–7 July, 1985. *Acta Universitatis Carolinae, Biologia*, Vol. 30.

Thompson, R. C. A., Lymbery, A. J. & Meloni, B. P. 1988. Giardiasis a zoonosis in Australia? *Parasitology Today*, **4**: 201.

Thompson, R. C. A., Lymbery, A. J. & Meloni, B. P. 1990. Genetic variation in *Giardia* Kunstler, 1882: taxonomic and epidemiological significance. *Protozoological Abstracts*, **14**: 1–28.

Tillotson, K. D., Buret, A. & Olson, M. E. 1991. Axenic isolation of viable *Giardia muris* trophozoites. *Journal of Parasitology*, **77**: 505–8.

Wang, C. C. 1989. Nucleic acid metabolism in *Trichomonas vaginalis*. In: *Trichomonads Parasitic in Humans* (ed. B. M. Honigberg), pp. 84–90. Springer-Verlag, New York.

Wang, C. C. & Aldritt, S. 1983. Salvage networks in *Giardia lamblia*. *Journal of Experimental Medicine*, **158**: 1703–7.

Warton, A. & Honigberg, B. M. 1979. Structure of trichomonads as revealed by scanning electron microscopy. *Journal of Protozoology*, **26**: 56–62.

WHO. 1979. Parasitic Zoonoses. *Technical Report Series 637*.

Woo, P. T. K. & Paterson, W. B. 1986. *Giardia lamblia* in children in day-care centres in southern Ontario, Canada, and susceptibility of animals to *G. lamblia*. *Transactions of the Royal Society of Tropical Medicine and Hygiene*, **80**: 56–9.

5

Haemoflagellates

○○○ 5.1 Haemoflagellates as biological models

Haemoflagellates are best known for causing serious diseases in man (e.g. sleeping sickness) and domestic animals (e.g. 'surra' in horses) in the tropics. However, although the pathogenic species have received most attention for the diseases they cause, these organisms have also excited the interest and attention of molecular biologists, immunologists and geneticists on account of a number of unusual features which make them interesting models for the study of various fundamental biological phenomena, especially in relation to:

(*a*) the form of the mitochondrion and its behaviour during the different stages of the life cycle (see Fig. 5.12);
(*b*) the ability of the surface coat of trypanosomes to undergo continuous antigenetic variation, which results in the avoidance of the host's immune response (see section 5.2).
(*c*) the possibility that a sexual process is involved in the life cycle (see section 5.5.5).

Various aspects of these are dealt with later.

The haemoflagellates are not confined in their distribution to the blood stream, but many species have adapted themselves to an intracellular existence and have invaded various tissues, especially those of the reticulo-endothelial (lymphoid-macrophage) system. This system plays an important role in the defence mechanism of the body; its essential components are the macrophages which normally ingest and immobilise invaders of the blood stream and can similarly treat certain haemoflagellates. Many species, however, are resistant to the lytic action of these cells, and one particular group, the leishmanias, have made them their exclusive habitat. Thus, these remarkable organisms survive and reproduce in the very cells the body utilises to ingest invaders; their ability to survive in cells of the reticulo-endothelial system is shared with the exo-erythrocytic stages of malarial organisms.

Some species of organisms discussed under the heading of 'haemoflagellates' are intestinal parasites of invertebrates, but their relationships with the blood-dwelling forms are so close that it would be creating an artificial distinction to treat them separately, and hence they are considered here also.

○○○ 5.2 General account

Occurrence. Although the species most frequently studied are those which occur in mammals, particularly man and domestic animals, haemoflagellates occur in most vertebrates and many invertebrates, even in such diverse forms as the siphonophore of coelenterates. In fact, trypanosomes were first recognised in the blood of fish and amphibia (Hegner *et al.*, 1938). It is significant that, whereas those parasitic in vertebrates generally require an intermediate host (usually a blood-sucking insect but occasionally a leech (e.g. *Cryptobia vaginalis*, p. 64)), those parasitic in invertebrates can undergo their entire life cycle within the same host. This strongly suggests that the whole group was originally parasitic in the alimentary canal of invertebrates.

○ 5.2.1 Morphology: light microscopy

The haemoflagellates have a number of basic morphological stages which were previously termed the *leptomonad*, *crithidial*, *trypanosome* and *leishmanial* stages (but now renamed: see p. 61). Each of these, under certain conditions, can be transformed into another stage. This happens in many life cycles or during *in vitro* culture.

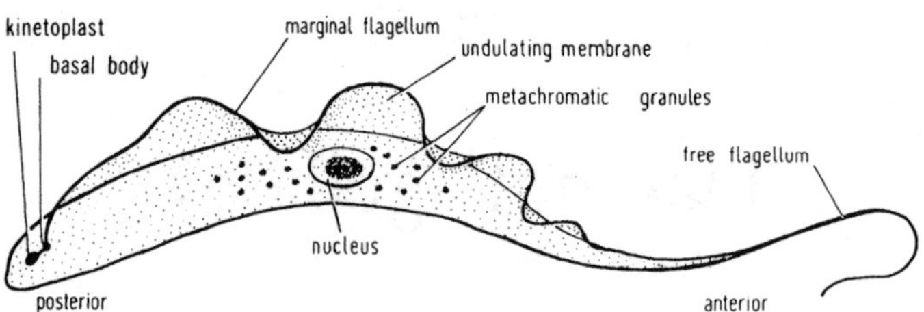

Fig. 5.1
A typical trypanosome; based on *T. gambiense.* (After Chandler, 1955.)

Until recently, the physiological significance of these transformations was not understood. However, electron microscopy and cytochemistry have now thrown light on these questions.

A typical 'trypanosome' (Fig. 5.1) is an elongate organism, 15–30 μm in length (including flagellum) and 1.5–3 μm in width, with a single nucleus containing a large central nucleolus (karyosome). Movement is effected by a single flagellum which arises at the posterior end of the body and runs along the free edge of an 'undulating membrane' (but see p. 60) continuing anteriorly as a free flagellum. Locomotion occurs with the flagellum leading; both conducting and contractile elements are probably present in the flagellum. The flagellum originates in a complex comprising a *basal body* and a strongly basophilic *kinetoplast* (Fig. 5.1) but the terminology used by different authors for this region is somewhat ambiguous and is dealt with further below. At the light microscope level, the kinetoplast is unusual in giving positive reactions for *both* DNA (i.e. it stains with Feulgen) *and* mitochondria (i.e. it stains intravitally with Janus B. Green). This unusual reaction is now known to be due to the fact that the kinetoplast is a filamentous body which lies within a single mitochondrion (see section 5.2.2), a situation revealed by electron microscopic and cytochemical studies.

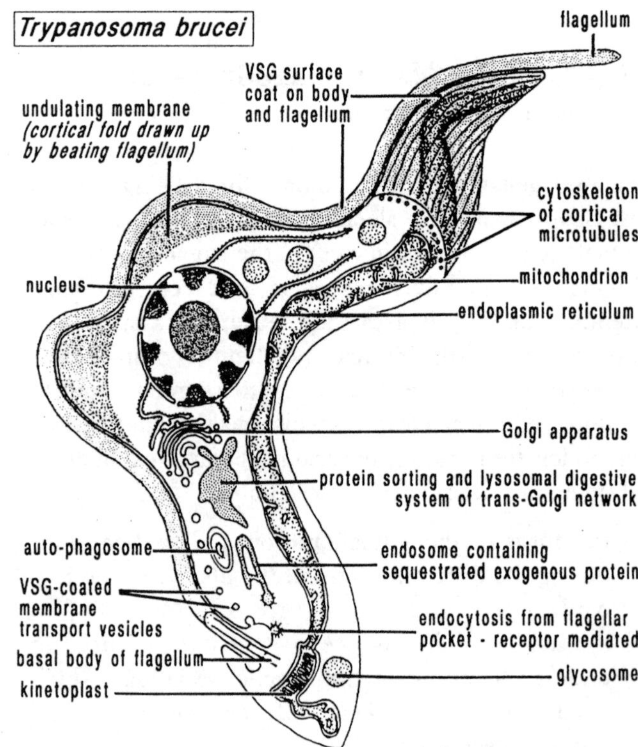

Fig. 5.2
Diagrammatic representation of the principal structures in a trypanosome (blood stream intermediate form of *Trypanosoma brucei*) as revealed by electron microscopy; see p. 59. VSG = variant surface glycoprotein. (After Vickerman *et al.* 1992.)

○ 5.2.2 Ultrastructure

General account. The ultrastructure of a 'typical' trypanosome of vertebrates is shown in Fig. 5.2; for species studied see p. 66. For excellent comprehensive reviews, see Rudzinska & Vickerman (1968), Vickerman (1976) and Vickerman & Preston (1976). Most of the fine structure is evident from Fig. 5.2 but attention is drawn to the following significant features.

Kinetoplast. This body is also known as the *kinetonucleus* and was earlier often referred to as a *parabasal body*. Since the latter is now known to represent the Golgi apparatus in flagellates, the use of this term should be avoided. The kinetoplast is a filamentous body which contains kinetoplast DNA (k-DNA) and has been shown to lie within a single mitochondrion which runs the whole length of the body. It is surrounded by a capsule continuous with that of the mitochondrion. The kinetoplast (Figs 5.2 and 5.3) is commonly disc-shaped or doughnut-shaped or may be an elongated plate. The filaments of the kinetoplast, which are oriented parallel to the long axis of the cell, appear 'ladder-like' and are

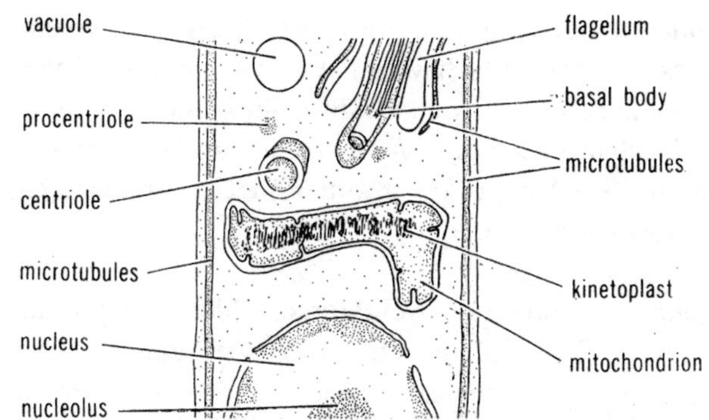

Fig. 5.3
Relationship between organelles in the epimastigote form of *Trypanosoma lewisi*. (After Burton & Dusanic, 1968.)

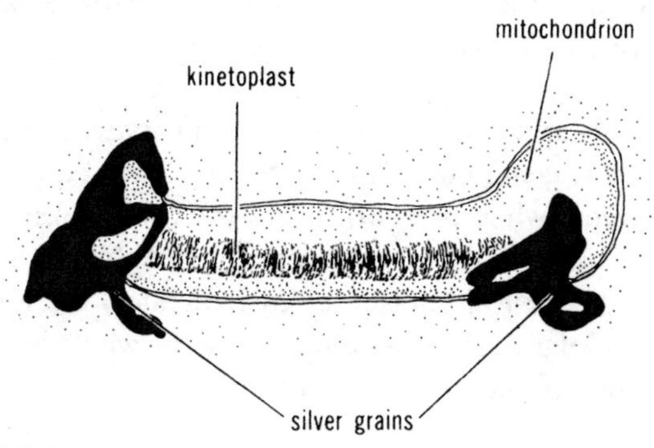

Fig. 5.4
Longitudinal profile of kinetoplast of epimastigote of *Trypanosoma lewisi* after 7 h culture in [^{3}H]thymidine. The silver grains represent the site of the DNA synthesis. (Adapted from Burton & Dusanic, 1968.)

10–16 nm in diameter with walls 4–6 nm thick. The presence of DNA in the kinetoplast is further confirmed by the fact that *both* the kinetoplast *and* the nucleus incorporate radiothymidine as shown in Fig. 5.4 (Burton & Dusanic, 1968). According to Vickerman & Preston (1976), this suggests that unlike mitochondrial DNA (mt-DNA) of other cells, the synthesis of kinetoplast DNA (k-DNA) is temporarily linked to nuclear DNA synthesis; in general, the replication (S) periods of nuclear and k-DNA are synchronous. Since some variations in phasing may occur, it is unlikely that there is a common initiator of DNA synthesis in nucleus and kinetoplast.

It has been estimated that in a single mitochondrion, there is a typical mitochondrial genome of about 20 kb present in about 100 copies. In addition, there are several thousand mini-circles of about 1 kb in about 100 copies which are locked together to form the kinetoplast. 'The putative RNAs are the smallest molecules known and should prove to be excellent tools for the study of ribosomes in general' (Barry, 1986*b*).

Dyskinetoplastry. Some species of trypanosomes (e.g. *T. equinum*) appear to have lost their kinetoplast and are referred to as *dyskinetoplastic* variants. Treatment with certain drugs (e.g. acriflavin, tryptaflavin) can also induce dyskinetoplastry. The kinetoplast and mitochondrion is actually present but its envelope persists only as a 'ghost' and DNA is lacking. The significance of this phenomenon in flagellates, in general, has been summarised by Vickerman & Preston (1976).

Mitochondrion. It is difficult to generalise on the form of the mitochondrion as this varies in different groups of trypanosomes and may vary in the stage of life cycle examined. Thus in bloodstream stages of *Trypanosoma lewisi*, *T. cruzi*, *T. vivax* and *T. congolense*, the mitochondria take the form of a single or branched tube, running most of the length of the body, with characteristic cristae. In trypanosomes of the *Brucei-Evansi* Group (p. 68), on the other hand (e.g. *T. brucei rhodesiense*) the mitochondrion in the blood stream phase lacks cristae and is inactive (as shown by cytochemical and biochemical tests) but becomes active in the insect vector. The biochemical implications of this are discussed later (p. 82). In those species or stages in which it is active, the mitochondrion can be beautifully demonstrated cytochemically by testing for the enzyme nicotinamide adenine dinucleotide (NAD) diaphorase using the tetrazolium salts NBT or MTT (Vickerman, 1965).

Nucleus. This shows no special features.

Surface coat. Many haemoflagellates have, in addition to a plasma membrane of the usual trilaminar structure, an outside coat covering this membrane (Fig. 5.2). This consists of about 10^7 identical glycoprotein molecules known as *variant surface glycoproteins* (VSGs). In each trypanosome only one VSG is expressed at a time and for each VSG there is a separate complete gene (termed the basic copy). Several VSG genes have been sequenced, mainly from RNA. The VSGs apparently protect the plasma membrane from the host immune attack (see *Antigenic variation*) by masking sites in the plasma membrane which are capable of activating complement (see Chapter 32). In all trypanosomes the glycoprotein coat is lost after entering the vector and commencing cyclical development or during *in vitro* culture.

The flagellar apparatus (the mastigont system). This is essentially a group of organelles associated with

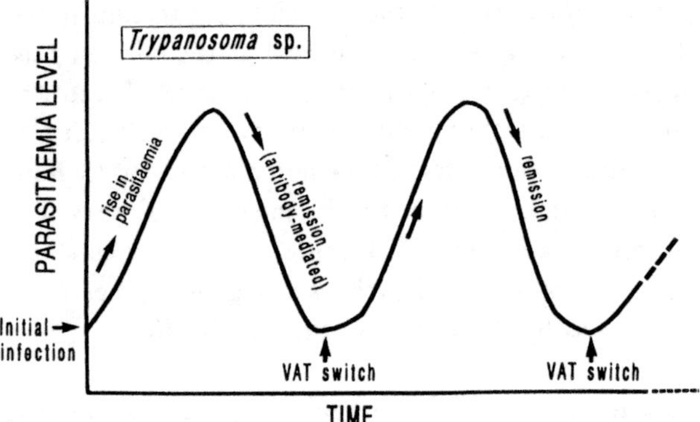

Fig. 5.5
Generalised diagram illustrating how the parasitaemia in trypanosome infections rises and falls as one metacyclic VAT (variable antigen type) is eliminated by immunity and replaced by another type with a different VSG (variant surface glycoprotein). See also Figs 5.6 and 32.3. (Original.)

the basal body of the flagellum. The flagellum originates in a cylindrical *basal body*, also known as a *basal granule*, *blepharoplast* or *kinetosome* (as in trichomonads). In order to avoid confusion with the kinetoplast, the descriptive term *basal body* is used here. The basal body has the classical centriole construction. Also running along the axial fibre bundle of the flagellum is an *intraflagellar body* (not shown in Fig. 5.2). For a detailed account of the mastigont system see Vickerman & Preston (1976).

'Undulating membrane'. The intraflagellar body may have a supporting or contractile function and its presence is believed to be responsible for the appearance of an 'undulating membrane' under light microscopy. Ultrastructure studies indicate that the undulating membrane is not a permanently differentiated structure but a series of transient crescents drawn up by the beating of the flagellum. In other words it would appear that the 'undulating membrane' is a light microscope optical illusion and has no reality, at least in the species studied. Since, however, this feature may be 'seen' under light microscopy, the term has been retained for diagnostic purposes.

Reservoir, cytostome and contractile vacuole. In some forms, e.g. *Trypanosoma mega*, but apparently not in the blood stream salivarian trypanosomes, the pellicle invaginates latero-posteriorly to form a membrane-bound space, termed by some workers the 'reservoir', from which leads, in some species, a cytostome (Vickerman & Preston, 1976). A contractile vacuole associated with the reservoir has also been described and may be a common but somewhat overlooked feature of trypanosomes. Experiments have shown that ferritin is taken up by the reservoir by endocytosis into vacuoles around the flagellar pocket of *T. brucei rhodesiense* and it may be that the flagellar pocket in other species similarly has an endocytotic function.

Other cytoplasmic inclusions. The endoplasmic reticulum (Fig. 5.2) is represented by tubules which emerge from the nuclear envelope and ramify through the cytoplasm. Ribosomes occur as granules abundantly distributed in the cytoplasm. In addition, the cytoplasm contains numerous granules of unknown function, and various vacuoles, some of which may be endocytotic (pinocytotic) vacuoles and lysosomes. Various acid hydrolases have been demonstrated in several species.

The Golgi apparatus. The Golgi apparatus of kinetoplastids is generally located anteriorly apposed to the cytopharynx (when present). It has the usual eukaryote structure with stacks of cisternae. Remarkably, in *Trypanophis grobbeni* (a parasite of coelenterates) there are multiple Golgi stacks extending almost the whole length of the body (Cachon *et al.*, 1972).

Species studied. The basic ultrastructure and/or the cytochemistry of the Trypanosomatidae has been extensively examined by early workers: see 2nd Edition of this book.

○ 5.2.3 Antigenic variation

It has been known for many years that some species of trypanosomes have considerable capacity for antigenic variation, the molecular basis of which was discussed above (see *Surface coat*). Thus, in well-established infections in vertebrate hosts, the surface antigens change (Fig. 5.5) every few days, with the result that specific antibodies produced by the host are no longer effective owing to the appearance of new variant surface antigens (VSGs). The number of trypanosomes in the host blood thus fluctuates from time to time, each peak being replaced by trypanosomes of variant antigen type (VAT). The range of VATs of a species is known as its *repertoire*. The extent of the repertoire has been examined for some species and comparisons made of *stocks* from different geographical regions, a *stock* being a population derived by serial passage *in vivo* or *in vitro* from a primary isolation. In one stock of the horse trypanosome *T. equiperdum* (p. 67), for example, there are at least

Trypanosoma b. rhodesiense

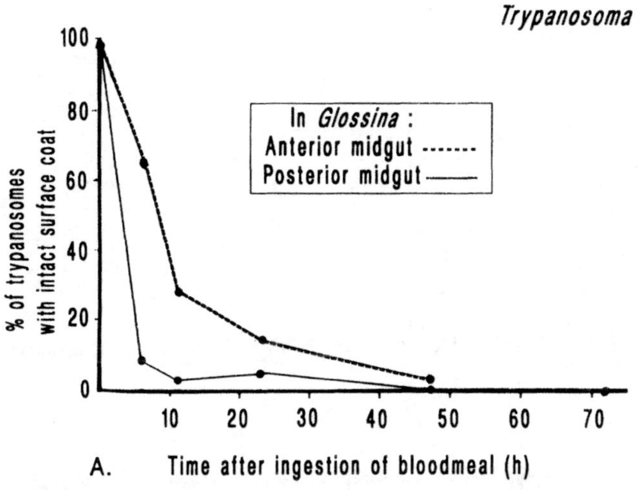

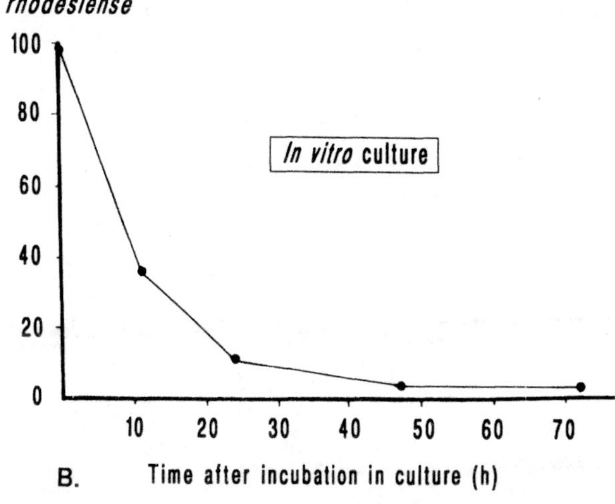

Fig. 5.6
Rate of loss of the surface coat from the blood stream forms of a pleomorphic line of *Trypanosoma b. rhodesiense* expressing a single variable antigen during transformation to procyclic forms (A) in the gut of the tsetse fly *Glossina morsitans*, and (B) during *in vitro* culture. Loss of variable antigen occurred at similar rates in the anterior midgut region of the tsetse fly and in culture *in vitro*, but 2–3 times faster in the posterior midgut region of the fly. (Modified from Turner *et al.*, 1988.)

101 different VATs (Barry, 1986*a*). Populations of trypanosomes which have the same repertoire are said to have the same *serodeme*.

○○○ 5.3 Morphological stages of haemoflagellates

Early workers referred to the various stages in the life cycle of haemoflagellates by names (e.g. leptomonad) derived from those genera (e.g. *Leptomonas*) in which the corresponding stages are the most characteristic forms.

These stages were known as:

(i) **leptomonad:** based on the flagellated stage of *Leptomonas* (Fig. 5.7)
(ii) **crithidial:** forms with short undulating membrane attributed to *Crithidia* (Fig. 5.7)
(iii) **leishmanial** (Leishman–Donovan bodies): based on rounded bodies of *Leishmania* (Fig. 5.7)
(iv) **trypanosome, trypanomorphic:** based on blood and metacyclic forms (p. 70) of *Trypanosoma* (Fig. 5.7) and for stages of *Herpetomonas* which bear a superficial resemblance to trypanosomes (Fig. 5.7)
(v) **herpetomonad:** used specifically for flagellates of *Herpetomonas*.

The use of these terms has been somewhat misleading, especially with respect to their association with the names of genera.

A new, more rational system, based on the positioning of the flagellum (Greek *mastigote* = a whip) which avoids these difficulties was put forward (Hoare & Wallace, 1966) and is now widely used as follows:

(i) **amastigote** (Fig. 5.7) (for former 'leishmanial' stage): rounded forms devoid of external flagellum; as in genus *Leishmania* and others.
(ii) **promastigote** (Fig. 5.7) (for former 'leptomonad' stage): forms with antenuclear kinetoplast; flagellum arising near it and emerging from anterior end of body; as in genus *Leptomonas* and others.
(iii) **opisthomastigote** (Fig. 5.7) (for former 'trypanomorphic' stages): forms with postnuclear kinetoplast; flagellum arising near it passing through body and emerging anteriorly. As in genus *Herpetomonas* only.
(iv) **epimastigote** (Fig. 5.7) (for former 'crithidial' stage): juxtanuclear kinetoplast with flagellum arising near it and emerging from side of body to run along short 'undulating membrane'. As in genus *Blastocrithidia* and in certain species of *Trypanosoma* (p. 66).
(v) **trypomastigote** (Fig. 5.7) (for former trypanosome stage): the 'true' trypanosome type; postnuclear kinetoplast; flagellum arising near it to run along a long 'undulating membrane'. As in the genus *Trypanosoma*. Note that some trypanosomes are also of the epimastigote type.

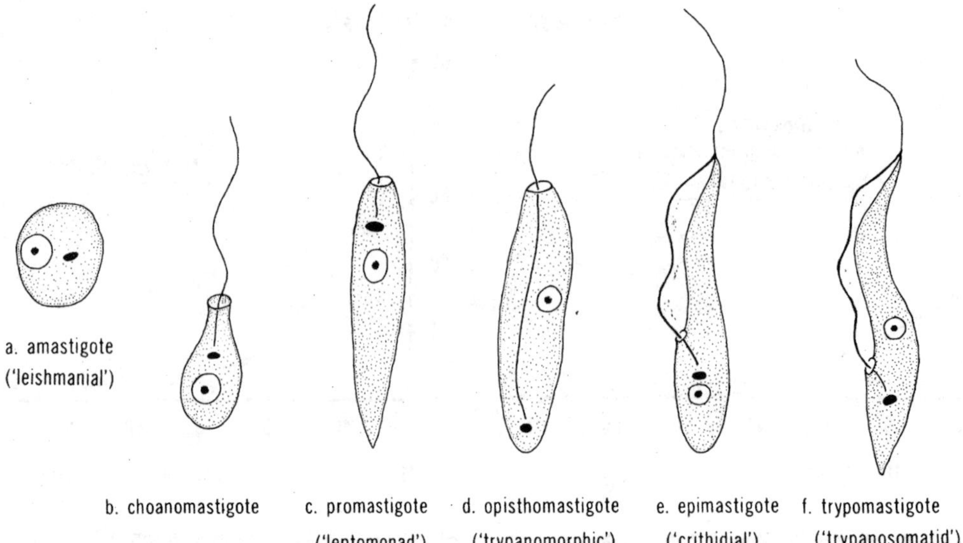

Fig. 5.7
Developmental stages of trypanosomatid flagellates: relationships between old and new terminologies. (Based on Hoare & Wallace, 1966.)

(vi) **choanomastigote** (Fig. 5.7): the peculiar 'barley corn' form, typical of the genus *Crithidia.* Kinetoplast usually antenuclear, flagellum arising from wide funnel-shaped reservoir and emerging at anterior end of body.

As terms describing the morphological changes during the life cycle – such as *leishmanial–leptomonad* transformation – are deeply entrenched in the literature, it is difficult to abandon them completely. In order to avoid ambiguity therefore, both terminologies are occasionally used here, where appropriate.

Only a limited number of the above stages occur in vertebrates, but any or all may occur in the alimentary canal of invertebrates.

In vitro studies have contributed much to our knowledge of the differentiation of the various stages of haemoflagellates, as it is frequently possible to obtain several stages of one species in a single culture. For example, when *T. cruzi* is cultured in tissue cultures, epimastigote, amastigote and trypomastigote stages develop. The morphological transformation from one stage to another is often accompanied by marked biochemical, nutritional, physiological or antigenic changes which have been widely studied. This question is dealt with further below (p. 69).

○○○ 5.4 Classification: Family Trypanosomatidae

The Family belongs to the Order Kinetoplastida, Suborder Trypanosomatina (see Appendix, p. 42). Making use of the terminology described above, the revised classification of the Trypanosomatidae is given below; the following division of genera is based on Vickerman (1976). [The little-known genera *Endotrypanum* and *Rhynchoidomonas* and the genus *Phytomonas* (which parasitises plants) are not dealt with here.]

1. Genus *Leptomonas*

Exclusively parasites of invertebrates: protozoa, nematodes, molluscs and insects. The most easily obtainable form is probably *Leptomonas ctenocephali* in the intestine and malpighian tubules of the dog flea, *Ctenocephalides canis*; promastigote and amastigote stages occur in the life cycle (Fig. 5.8).

2. Genus *Herpetomonas*

Exclusively parasites of invertebrates. Examples are *Herpetomonas muscarum* and *H. culicis* from the gut of Diptera. The ultrastructure of several stages has been examined (Clark & Wallace, 1960); promastigote, opisthomastigote and amastigote stages occur in the life cycle (Fig. 5.8).

3. Genus *Blastocrithidia* (syn. *Crithidia pro parte*)

Exclusively parasites of insects, especially phytophagous bugs or bugs feeding on insects. 'Water striders' are commonly

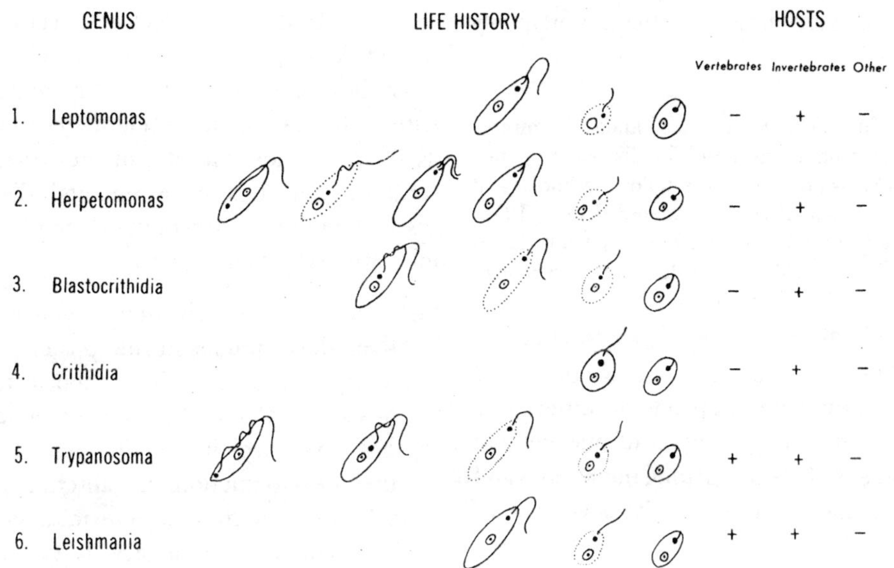

Fig. 5.8
Characteristics of the genera of the family Trypanosomatidae. The forms figured in dotted lines have never been demonstrated. (After McGhee, 1968.)

infected with these forms, a representative species being *Blastocrithidia gerridis* which has been reported from numerous species of *Gerris*. For life cycle stages see Fig. 5.8.

4. Genus *Crithidia* (syn. *Strigomonas*)

Exclusively parasites of insects; often clustered together or attached like the pile of a carpet to the gut wall. Only epimastigote and amastigote stages in life cycle; Fig. 5.8. Best-known species probably *Crithidia fasciculata*, a parasite of mosquitoes such as *Anopheles maculipennis* or *Culex pipiens*.

All the above four genera, which will not be considered further here, are *monogenic* parasites, i.e. they occur in only one host. In all these too, the amastigote forms represent the infective stage and are passed in the faeces; infection is by ingestion.

The following genera which contain a number of important species of digenetic parasites (i.e. with a two-host life cycle) are dealt with in detail below:

5. Genus *Trypanosoma*

Blood or tissue parasites of vertebrates with invertebrate intermediate hosts; contain trypomastigote, promastigote, amastigote and epimastigote stages in the life cycle (Fig. 5.8).

6. Genus *Leishmania*

Tissue and blood parasites of vertebrates with invertebrates acting as intermediate hosts; only amastigote and promastigote stages occur in life cycle (Fig. 5.8).

○○○ 5.5 Genus *Trypanosoma*

○ 5.5.1 General account

Trypanosomes occur in the blood and some tissues of the majority of vertebrate animals, both warm- and cold-blooded. The life cycle involves an intermediate host, usually an insect, occasionally a leech, and in one species transmission is venereal. It is probably true to say that the great majority of trypanosomes are non-pathogenic and live at peace with their hosts, some of which can withstand heavy infections without developing any apparent symptoms.

Many of the best-known species are, however, pathogenic, and the morphology and physiology of these have been most extensively studied. The disease produced by such forms is termed 'trypanosomiasis'. One curious feature of some pathogenic species is that they may also be parasitic in other animals in which they are apparently harmless. In these natural hosts, the tissues have become physiologically adapted to the presence of the parasites through long periods of association, and such animals act as 'reservoir' hosts for these potentially pathogenic species.

Literature. A vast literature exists on the biology, immunology and pathology of trypanosomes and the disease, trypanosomiasis. Much of this is covered in the following books and reviews: Baker, 1982; Ciba, 1974; Lumsden & Evans, 1976; Mehlhorn, 1988; Molyneux & Ashford, 1983; Newton, 1985; Seed *et al.*, 1980; Taylor, 1986; Tizard, 1985; Vickerman, 1976, 1985; Vickerman & Preston, 1976.

Morphology. This has already been discussed (p. 58). Species determination depends on differences in size and shape of the body, position of nucleus and degree of development of the undulating membrane and flagellum; antigenic and isoenzyme analyses are also of value.

Considerable differences may occur between 'strains', often developed after passage through laboratory hosts, a process which may result in antigenic or other changes (Lumsden, 1965). Preservation of a particular 'strain' has been much helped by the discovery that trypanosomes may be stored in a viable condition at low temperatures (−80 °C). The term 'stabilate' is used for organisms thus preserved in a viable condition on a unique occasion.

Life cycle. Transmission from one vertebrate host to another is carried out by blood-sucking invertebrates, usually an insect, but with amphibian trypanosomes (e.g. *T. inopinatum*) the vector is a leech. Shortly after infected blood is drawn into the insect gut, the trypomastigote forms become transformed into epimastigote forms; the latter give rise (directly or by fission) to further trypanosomal forms, usually referred to as *metacyclic* trypanosomes. The mode of transmission, just outlined, is known as *cyclical* to distinguish it from *mechanical transmission*, a process in which trypanosomes survive, for a short time, probably only minutes, on and about the mouth parts of an insect and are inoculated into a new host when the insect bites again, without undergoing any developmental cycle. The evidence that mechanical transmission actually takes place, however, appears to be tenuous.

It is important to distinguish between these two types of transmission. If trypanosomes are transmitted by the mechanical method, they must be transmitted by bites which rapidly succeed one another, while those transmitted by the cyclical method cannot be transmitted until sufficient time has elapsed to enable them to reach an infective stage by a particular developmental sequence in the insect vector, a cycle usually requiring 15–35 days.

Cyclical development may culminate in two sites within the insect, the hindgut (technically referred to as the 'posterior station'), or the foregut (the 'anterior station'), and this environmental discrimination has resulted in the division of the Family Trypanosomatidae into two sections as follows:

(*a*) **Stercoraria**, which contains genera which complete their development in the 'posterior station', that is to say that infective forms appear in the faeces and hosts are infected by the contaminative route.
(*b*) **Salivaria**, which contains genera which complete their development in the 'anterior station' (i.e. in the salivary system) and transmission takes place by inoculation of the metacyclic stage.

These sections are further considered on p. 67.

Terminology. In the classification of trypanosomes, the character of the flagellum often serves as a useful criterion and species are said to be *monomorphic* or *polymorphic.*

Monomorphic:

If *all* the individual trypanosomes of the species possess a free flagellum, or alternatively if *none* of the individuals possesses a free flagellum.

Polymorphic:

If some of the individuals of the species possess a free flagellum while others do not.

○ 5.5.2 Type example: *Trypanosoma lewisi*

definitive host: rat
location: blood stream
transmission: contaminative, via *Nosopsyllus fasciatus*, the rat flea

Occurrence. The species is common in *wild* rats of the genus *Rattus* in all parts of the world. It is conveniently studied in the laboratory in the albino rat. It is a non-pathogenic species, often occurring in concentrations of 20 000–60 000 trypanosomes per cubic millimetre.

Life cycle (Fig. 5.9). During the early days of an infection, rapid reproduction – mainly by multiple fission – takes place. This gradually ceases, and on about the tenth day there is a 'crisis' when most of the para-

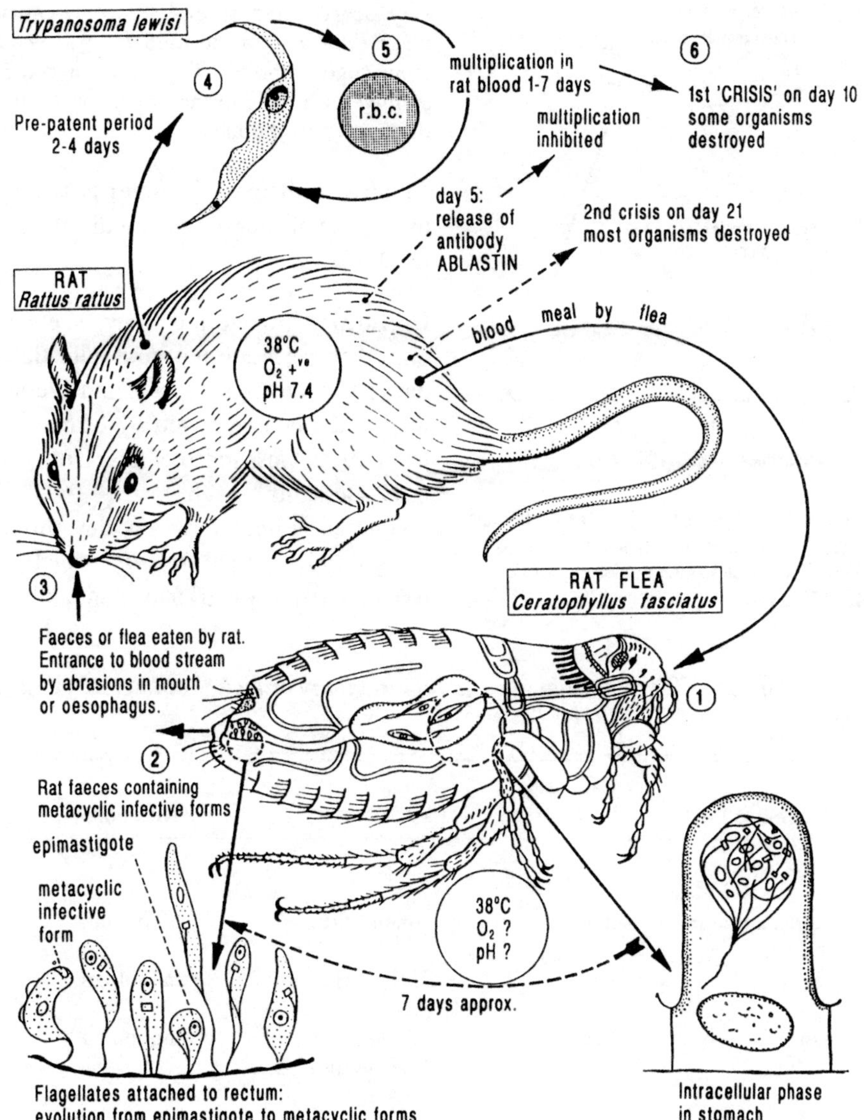

Fig. 5.9
Life cycle of *Trypanosoma lewisi* in the rat; see also Fig. 5.10. (Partly after Hegner *et al.*, 1938.)

sites, which are now mainly long forms (the so-called 'adults'), are destroyed; this is termed the *first trypanocidal crisis*. The remaining trypanosomes continue to live for several weeks until, at about 21 days, a second 'crisis' occurs when all (or practically all) the organisms are destroyed; this is known as the *second* (*terminal*) *crisis* (Fig. 5.10). Immunity is thus rapidly established, the basis of which is discussed further below and in Chapter 32. The laboratory maintenance of this organism therefore requires repeated passage through fresh rats at regular intervals; intraperitoneal inoculation is the most suitable and convenient method.

Transmission is of the contaminative type and is carried out by the ingestion of flea faeces or an entire flea; both of these may occur when a rat licks its fur. The common rat flea of Europe and N. America, *Nosopsyllus fasciatus*, is the usual host, but transmission can also be affected by the human flea, *Pulex irritans*, the Indian rat flea, *Xenopsylla cheopis*, or the dog flea, *Ctenocephalides canis*.

After a flea takes a blood meal, the trypanosomes multiply in the midgut, finally invading the cells of the gut wall. Within these cells, rapid multiplication takes place and daughter trypanosomes are released to infect other cells. This process

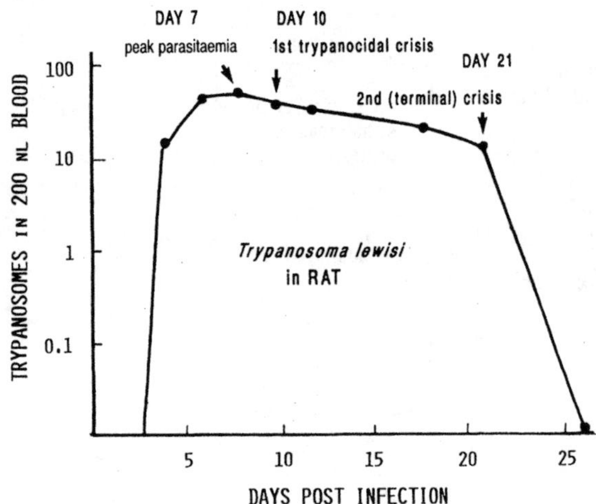

Fig. 5.10
Trypanosoma lewisi: the rise and fall of parasitaemia in the rat (female Sprague–Dawley); initial inoculum 10^7 trypanosomes. (Modified from Giannini & D'Alesandro, 1984. Reproduced with permission from *Infection and Immunity*, **43:** 619, 1984.)

may be repeated several times. Eventually, the organisms make their way to the hindgut and rectum and give rise to epimastigote forms. The latter become transformed into stumpy metacyclic forms which constitute the infective stages discharged in the faeces.

***In vitro* culture.** *T. lewisi* is readily cultured in any of the usual blood agar media used for trypanosomes (p. 496).

Immunity (for terminology, see Chapter 32). The remarkable cessation of multiplication after about 7–10 days has been the subject of considerable research. Early work has been reviewed by Ormerod (1963) and later work by D'Alessandro & Clarkson (1980) and Giannini & D'Alessandro (1984).

The early held view was that this was a response to an antibody termed *ablastin*, which – while not being trypanocidal – was trypanostatic. The nature of ablastin

Table 5.1. *Characteristics of species of* Trypanosoma *of medical, veterinary or biological interest, parasitic in mammals; subgeneric names omitted*

Species	Definitive hosts	Laboratory hosts	Vectors	Distribution	Pathology
Section STERCORARIA					
T. theileri	cattle, antelopes	none	tabanid flies	cosmopolitan	non-pathogenic (?)
T. melophagium	sheep	none	*Melophagus ovinus* (ked)	cosmopolitan	non-pathogenic
T. lewisi	rats	rats	*Ceratophyllus fasciatus* and other fleas	cosmopolitan	non-pathogenic
T. musculi (= *T. duttoni*)	mice	mice, rats*	fleas	cosmopolitan	non-pathogenic
T. nabiasi	rabbits	rabbits	*Spilopsyllus cuniculi*	cosmopolitan	non-pathogenic
T. rangeli	man, monkeys, dogs, opossums	rats, mice	triatomid bugs	Central and S America	non-pathogenic
T. cruzi	man, dogs, cats, armadillos, opossums, raccoons	rodents and other mammals	triatomid bugs	Americas	pathogenic: Chagas' disease
T. theodori	goats	?	*Lipoptena capreoli*	Palestine	non-pathogenic
Section SALIVARIA					
T. vivax	cattle, sheep, equines, goats, antelopes, dogs	rodents*	tsetse flies	Tropical Africa (also Mauritius, Antilles, S America)	pathogenic: Souma
T. uniforme	antelopes, cattle, sheep, goats	none	tsetse flies	Central and E Africa, Angola	pathogenic

University of Chester, Seaborne Library

Title: Animal behavior and wildlife conservation / edited by Marco Festa-Bianchet and Marco Apollonio.
ID: 36023408
Due: 11-05-18

Title: Animal personalities : behavior, physiology, and evolution / edited by Claudio Carere and Dario Maes
ID: 36188626
Due: 25-05-18

Title: Introduction to animal parasitology / J. D. Smyth.
ID: 36139863
Due: 25-05-18

Total items: 3
04/05/2018 15:22

Renew online at:
http://libcat.chester.ac.uk/patroninfo

Thank you for using Self Check

3

Table 5.1. *(cont.)*

Species	Definitive hosts	Laboratory hosts	Vectors	Distribution	Pathology
T. congolense	cattle, sheep, zebras, warthogs, equines	rodents	tsetse flies	Tropical Africa	pathogenic
T. dimorphon	cattle, sheep, equines, pigs	rodents	tsetse flies	Tropical Africa	pathogenic
T. simiae	warthogs, pigs, camels; possibly cattle, equines	monkeys, rabbits*	tsetse flies	Tropical Africa	pathogenic
T. suis	pigs	none	tsetse flies	Zaire, Tanzania	pathogenic
T. b. brucei	domestic mammals, antelopes	rodents	tsetse flies	Tropical Africa	pathogenic: Nagana
T. b. rhodesiense	man, antelopes	rodents	tsetse flies	E Africa	pathogenic: sleeping sickness
T. b. gambiense	man	rodents	tsetse flies	Tropical Africa	pathogenic: sleeping sickness
T. evansi	dog, cattle, equines	rodents	tabanid flies	cosmopolitan (tropical and subtropical)	pathogenic†
T. equinum	equines, dogs, cattle	rodents	tabanid flies, vampire bats	S America	pathogenic: Mal de Caderas
T. equiperdum	equines	rabbits, dogs	(venereal transmission)	S Europe, Asia, N Africa	pathogenic: dourine

*'Adapted' by injection of homologous serum.
†Surra (Old World); Derrendgadera, Murrina (New World).
Source: Data partly from Hoare (1966).

has been controversial for many years on account of its non-absorption from immune serum by living suspensions of trypanosomes. It has now been shown to be a specific immunoglobulin G (IgG) antibody (D'Alessandro & Clarkson, 1980; Giannini & D'Alessandro, 1984) which increases in amount on the surface of the trypanosome during the course of infection. This antibody IgG (SCIgG) is trypanostatic, apparently exerting its influence by interfering with active transport. The second crisis may be due to the fact that the specially incorporated SCIgG may be the target of a second trypanocidal antibody which is an M immunoglobulin (Giannini & D'Alessandro, 1984).

Development in heterologous hosts. *Trypanosoma lewisi* exhibits a well-defined host-specificity, developing optimally in species of the genus *Rattus*. The organism can become established in mice *provided* that the mice receive a supplement of rat (= homologous) serum, the minimal volume per mouse being about 0.05–0.01 ml^3 (Lincicome & Francis, 1961). This suggests that rat serum contains a factor essential for the development of *T. lewisi*. The factor is present to some extent in other rodents, in which transient infections can sometimes be established by inoculation of large numbers of blood stream forms.

○ 5.5.3 Classification of mammalian trypanosomes

As mentioned earlier, Hoare (1966) divided the genus *Trypanosoma* into two sections (Table 5.1): A. *Stercoraria*, B. *Salivaria*, on the basis of whether they develop in the hindgut ('posterior station') or salivary glands of the insect vector.

Section A **Stercoraria**

Free flagellum always present; kinetoplast large and not terminal; posterior end of body pointed; multiplication in mammalian host discontinuous, typically taking place in the epimastigote or amastigote stages; development in vector in posterior station and transmission contaminative (in *T. rangeli* also in anterior station, with inoculative transmission); except *T. cruzi*, typically non-pathogenic e.g. *T. lewisi*, *T. theileria*.

Section B **Salivaria**

Free flagellum present or absent; kinetoplast terminal or sub-terminal; posterior end of body usually blunt; multiplication in mammalian host continuous, taking place in trypomastigote stage; development in vector in anterior station (except in mechanical vectors, but see p. 64); transmission is inoculative (except in *T. equiperdum* where it is venereal); typically pathogenic, e.g. *T. vivax*, *T. congolense*, *T. simiae*, *T. suis*, *T. b. brucei*, *T. b. rhodesiense*, *T. b. gambiense*, *T. evansi*.

A list of some mammalian trypanosomes of medical and veterinary interest is given in Table 5.1. These have been reviewed by Lumsden & Evans (1976) and Stephen (1986).

○ 5.5.4 Trypanosomes infecting man: trypanosomiasis

There are two main diseases of man caused by species of *Trypanosoma*:

Sleeping sickness or African trypanosomiasis: caused by *T. brucei gambiense* and *T. brucei rhodesiense*; occurs in tropical Africa. Vectors: tsetse flies of the genus *Glossina*.

Chagas' disease or American trypanosomiasis: caused by *T. cruzi*; occurs in Central and South America. Vectors: bugs of the family Reduviidae.

T. brucei and sleeping sickness

General comment. The study of this species is complicated by the fact that it multiples in the blood of a range of animals – domestic animals and wild game as well as man – and is one of numerous species transmitted by tsetse flies; the disease is thus a zoonosis. Domestic animals (cattle, sheep, dogs, pigs and goats) and many species of wild game (Table 5.2) can act as reservoir hosts. The bushbuck (*Tragelaphus scriptus*) is a particularly favoured host of *Glossina*.

Only certain stocks of this trypanosome are able to infect man, but these are morphologically indistinguish-

Table 5.2. *Animal reservoir hosts of* T.b. rhodesiense *and* T.b. gambiense

1. Known or potential reservoir hosts of *T.b. rhodesiense*-like organisms	
Domestic animals	Game animals
Cattle	Bushbuck, *Tragelaphus scriptus*
Sheep	Giraffe, *Giraffa camelopardalis*
Dogs	Hartebeest, *Alcelaphus buselaphus*
Goats	Hippopotamus, *Hippopotamus amphibius*
	Reedbuck, *Redunca redunca*
	Waterbuck, *Kobus defassa*
	Warthog, *Phacochoerus aethiopicus*
	Hyena, *Crocuta crocuta**
	Lion, *Panthera leo**
2. Animals found infected with *T.b. gambiense*-like organisms which by available biochemical methods have been found to be indistinguishable from parasites isolated from man in the same area	
Domestic animals	Game animals
Pigs	Kob, *Kobus kob*
Dogs	Hartebeest, *Alcelaphus buselaphus major*
Cattle	
Sheep	

*These animals are rarely bitten by *Glossina*. Infection is acquired mainly through feeding on other infected animals.
Source: Data from WHO (1986).

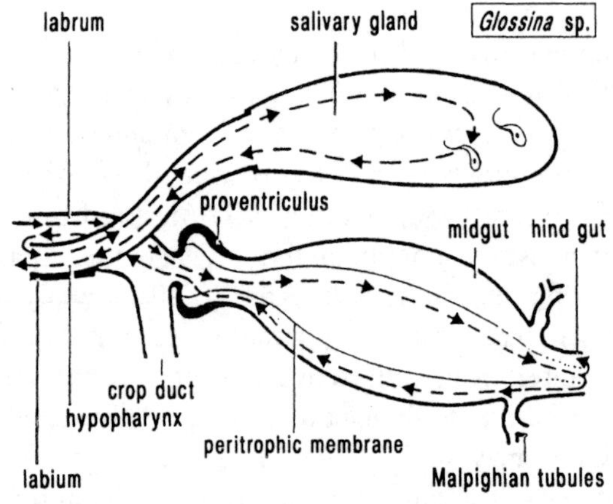

Fig. 5.11
Pathway presumed to be followed by trypanosomes taken into the tsetse fly *Glossina palpalis*. The organisms pass down the midgut to the open end of the peritrophic membrane and then back, outside this membrane, before reaching the salivary duct. Within the salivary gland, they transform to infective forms and pass down the mouthparts to be injected with saliva into the host. It has also been shown that some organisms may penetrate through the peritrophic membrane into the midgut cells. (After Ellis, 1986.)

able from the non-human-infective forms which are now assigned to the sub-species *T. brucei brucei*. The human-infecting form causes sleeping sickness; that which causes an acute form of the disease in East Africa is known as *T. brucei rhodesiense* and that which induces the chronic West African form of trypanosomiasis is designated *T. brucei gambiense*. All these forms are transmitted by the tsetse fly, which ingests the infective metacyclic forms in its saliva during feeding (see Fig. 5.11 and p. 72).

Polymorphism; pleomorphism. Both *T. b. gambiense* and *T. b. rhodesiense* can develop in laboratory animals, such as rats and monkeys, and their development has been extensively studied in these and other mammals. Development is characterised by the occurrence of three main types of blood forms, a phenomenon generally referred to as *polymorphism*; these are:

(*a*) **slender** forms (Fig. 5.12): long and thin, about 29 μm long, free flagellum;
(*b*) **stumpy** forms (Fig. 5.12): thick and short, average length 18 μm, typically no free flagellum, but a short one may be present.
(*c*) **intermediate** forms (Fig. 5.12): about 23 μm long with a moderately thick body and a free flagellum of medium length.

The use of the term *polymorphism* is an unfortunate one as this term arose in relation to *genetic polymorphism* (in moths); since the variation in the trypanosome form described above occurs within what is essentially a clone, the alternative terms *pleomorphism* or *clonal pleomorphism* have been proposed (Ormerod, 1967).

The distinction between 'stumpy' and 'intermediate' forms may not always be clear.

Differences between blood stream and fly gut forms. Ultrastructure studies have revealed marked cytological and metabolic differences between the blood stream and fly gut forms (Vickerman, 1962, 1985). In the slender blood stream forms, the mitochondrion associated with the kinetoplast is poorly formed: it consists of a double-membrane-walled tube extending the length of the body, with few cristae present (Fig. 5.12). In contrast, in the fly gut form (such as obtained in culture), the mitochondrion takes the form of an extensive tubular network with well-developed cristae; this transformation is believed to take place when the slender forms transform into the stumpy forms. These ultrastructural differences are reflected in differences in metabolism, summarised as follows:

A. 'Slender' blood stream forms

1. Break down glucose to pyruvic acid, which is then excreted.
2. Respiration not inhibited by cyanide, therefore cytochromes not involved.
3. Inhibitors of Krebs cycle enzymes do not affect glucose or oxygen consumption.
4. From 1–3, it is concluded that mitochondria are inactive.

B. 'Stumpy' blood stream forms; tsetse mid-gut forms

1. Respiration inhibited by cyanide.
2. Cytochromes present (spectroscopically).
3. Krebs cycle substrates (e.g. α-ketoglutaric acid) used.
4. Pyruvic acid not excreted.
5. From 1–4, it is concluded that mitochondria are now active.

That the mitochondria are active in B can be demonstrated by incubation in $NADH_2$ with a tetrazolium salt (Nitro MT), the presence of the active enzyme (NAD diaphorase) being demonstrated cytochemically by the appearance of a blue precipitate (shown stippled in Fig. 5.12) as described earlier (p. 59). It has been speculated (Vickerman, 1966) that in the blood stream form (in the slender forms only?) in which the mitochondrial 'ghost' (or *promitochondrion*) is non-functional, the organisms have a wasteful method of respiring, i.e. glucose is only broken down to pyruvic acid (which is discarded) and oxygen is used to oxidise (via an L-α-glycerophosphate cycle) the pyridine nucleotides reduced in glycolysis. This alternative terminal respiratory system is localised in dense (extramitochondrial) bodies, the glycosomes. On entering the tsetse midgut, a more efficient method of energy production apparently operates; the pyruvic acid (which was previously being excreted) is now used completely and the pyridine nucleotides, reduced in the course of glycolysis and the Krebs cycle, are now oxidised via a conventional cytochrome system with oxygen as the final acceptor. These systems are discussed further under *Physiology* (section 5.7).

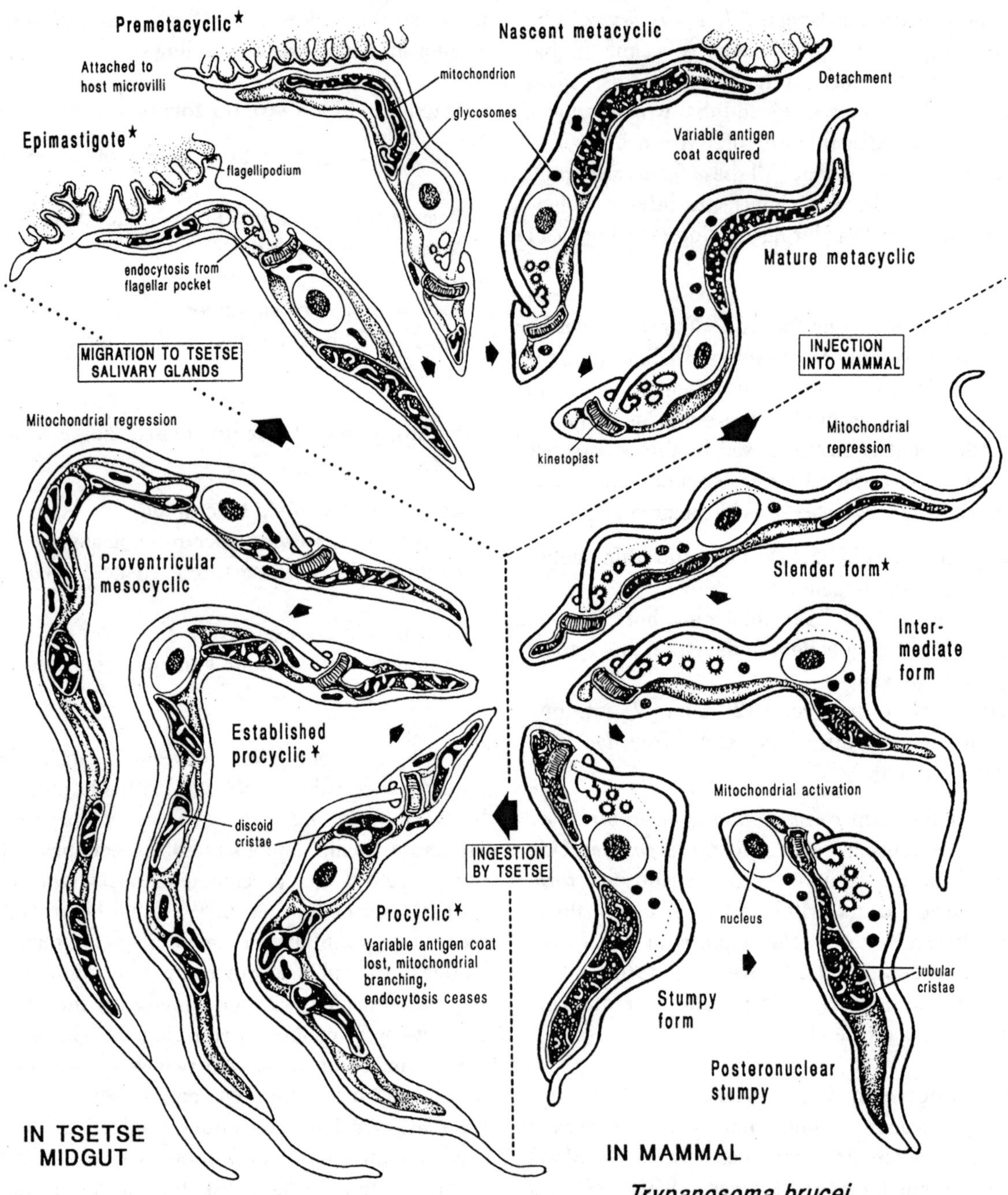

Fig. 5.12
Trypanosoma brucei: schematic representation of developmental cycle in a mammal and in the tsetse fly vector. Stages with the variable antigen coat are shown on the right, uncoated forms on the left. The tubular mitochondrion is depicted partly in section to show changes in its cristae. The posteronuclear stumpy form is included as an example of a form produced in some stocks, but plays no essential part in the cycle. Division occurs in asterisked stages. (After Vickerman, 1985.)

Development in mammalian host. When it bites, the tsetse fly deposits metacyclic trypanosomes in the dermal connective tissue, inducing a local inflammatory reaction which results in the development of a 'chancre' characterised by erythema, local heat, oedema, and tenderness; this is rare in *T. b. gambiense* sleeping sickness (WHO, 1986). From this region the trypanosomes migrate to the lymphatics and hence to the blood stream. From both these sites, after undergoing intense multiplication, they penetrate the blood and lymph capillaries, into the connective tissue and, after a month or so, cross the choroid plexus into the brain and cerebrospinal fluid. Multiplication is by binary fission with a doubling time of about 6 h (Vickerman, 1985). As already explained on p. 60, the level of parasitaemia fluctuates, owing to the remission as the prevailing antigen type is destroyed only to rise again as a new antigen type appears (Fig. 5.5). In the early stages of infection (Fig. 5.12) 'slender' forms predominate, eventually to give way to 'stumpy' forms via an 'intermediate' form stage. The invasion of the nervous system results in 'sleeping sickness', characterised by a general physical and mental depression and a desire to sleep. Useful reviews of the pathology are those of Poltera (1985) and WHO (1986).

There is some evidence that the life cycle may prove to be more complex than previously thought because of the finding that, in experimental infections of *T. b. brucei* in rats and mice, intracellular stages have been identified in the ependymal cells (which line the ventricular surface of the brain), damaging them and stripping them from the cerebral aqueduct (Abolarin *et al.*, 1986; Ormerod & Hussein, 1986). It has been speculated that these results are the result of mechanical destruction of serotonin-specific neurons secondary to the destruction of ependymal cells (Ormerod *et al.*, 1988). This result may be of some significance in accounting for the neuropsychiatric disturbance related to sleeping sickness.

The two species are normally said to differ *nosologically* (i.e. in the type of disease they produce – but see below). *T. b. gambiense* produces 'Gambian' sleeping sickness, which is a chronic human disease that may last for several years before ending in death. In contrast, the disease produced by *T. b. rhodesiense* – 'Rhodesian' sleeping sickness – is an acute disease i.e. it runs an acute course for about six months before terminating fatally. Intermediate strains are known (such as the Zambezi–Botswana strain) which, although virulent to laboratory animals, produces a disease intermediate between the Gambian and the Rhodesian types. In many areas it is becoming increasingly difficult to draw a hard and fast line between the diseases caused by *T. b. gambiense* and *T. b. rhodesiense*. The only difference between these species and *T. b. brucei* appears to be their ability to infect man.

○ 5.5.5 Sexuality in the *T. brucei* complex

It has long been assumed that trypanosomes of the *T. brucei* group multiply by asexual fission and that sexual processes are not involved. However, more recent studies on isoenzyme electrophoretic variation in different stocks of trypanosomes has provided strong evidence that genetic exchange can take place (Gibson *et al.*, 1980; Jenni *et al.*, 1986; Tait, 1983; Tait & Turner, 1990; Vickerman, 1986). This interpretation has been confirmed in the laboratory by demonstrating experimentally, for the first time, a cross between stocks of trypanosomes. The technique followed was to feed tsetse flies with two trypanosome stocks (mixed in a 1 : 1 ratio), hybrid clones being identified by having isoenzyme patterns different from those of their parents. One interesting result was a hybrid identical to one of the parents on the basis of nuclear gene markers, but possessing the kinetoplast DNA of the other parent! How the exchange of genetic material takes place is not known, but there is some evidence that fusion of trypanosomes may occur in the tsetse fly (Ellis, 1986).

Great difficulty has been encountered in determining the *ploidy* of trypanosomes, i.e. whether they are haploid, diploid or polyploid. In most organisms, ploidy is worked out from cytological observations on the chromosomes. However, in trypanosomes this is difficult as the chromosomes do not appear to condense into cytologically recognisable structures, although three types of chromosomes, 'mini', 'intermediate' and 'large', have been described (Fig. 5.13). There appear to be 13 large chromosomes.

Although, on the basis of their DNA content, it was formerly thought that metacyclic stages were haploid, it is now thought that trypanosomes are diploid throughout their entire life cycle but may contain variable amounts of DNA (Ash, 1990). Although the sexual process clearly involves meiosis and syngamy, the order of these processes is still unknown and much further work is needed on this problem. Three models of gene exchange have been proposed and these are clear from Fig. 5.14. However, these models remain controversial and the ploidy of the different stages in the life cycle remains unclear. Tait & Turner (1990) have provided

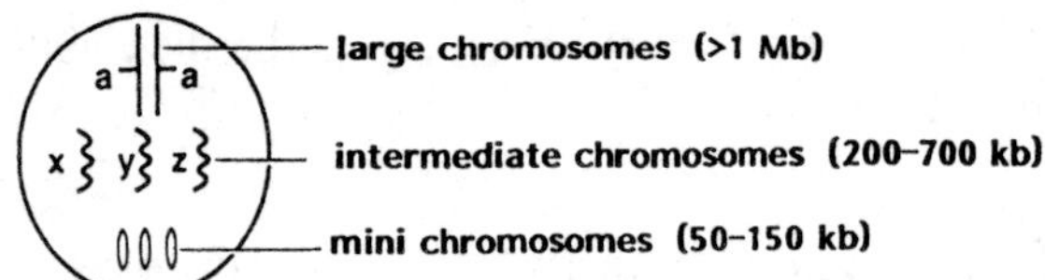

Fig. 5.13
Trypanosoma b. brucei: schematic representation to illustrate the three classes of chromosomes. There are believed to be approximately 100 minichromosomes, a variable number of intermediate chromosomes (between 1 and 8) and probably 13 large chromosomes. (Modified from Tait & Turner, 1990.)

a succinct analysis of the attributes of these various models.

In contrast to *T. brucei*, it should be noted that in *T. cruzi* (p. 73) there appears to be no (or very restricted) sexuality (Tibayrenc & Ayala, 1987).

Transmission. The vector of sleeping sickness is the tsetse fly, a name derived from the native name for flies of the genus *Glossina*. Although it has been found experimentally that *T. b. gambiense* and *T. b. rhodesiense* will develop in any species of *Glossina* (an exceptional situation for insect-borne parasites), under field conditions only flies of either the *G. palpalpis* or *G. morsitans* groups act as vectors. The biology and ecology of the tsetse fly have been extensively reviewed (Allsopp, 1984; Challier, 1982; Molyneux, 1982; Jordan, 1985; Schofield, 1985; WHO, 1986).

Cyclical development in *Glossina*. Ingested parasites multiply rapidly in the midintestine, for a period of 10–15 days, after which they begin to give rise to slender proventricular forms, which become concentrated in the proventricular region. These make their way into the salivary glands without penetration into the body cavity. Here they attach themselves to the walls by means of their flagella or lie free in the lumen, developing into epimastigote forms. These undergo further multiplication and give rise to small forms, with or without a short flagellum and somewhat resembling the stumpy blood forms; these are the infective metacyclic trypanosomes. Normal cyclical development takes 20–30 days at 24–29 °C. Many flies seem resistant to infection, and in nature only about 1–2 per 1000 wild flies are found to be infected.

Reservoir hosts. In addition to man, as already indicated, numerous other animals may act as reservoirs

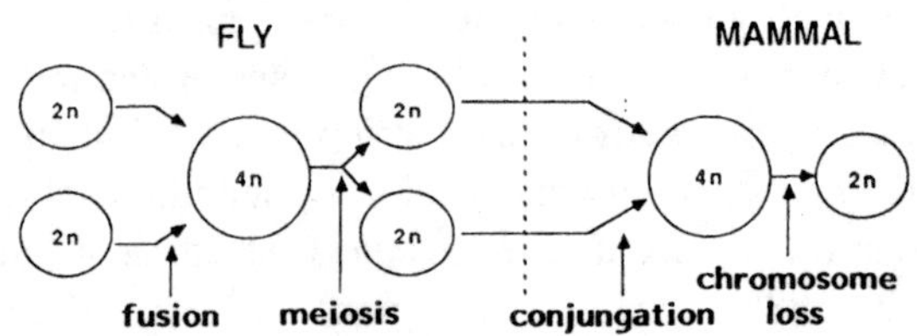

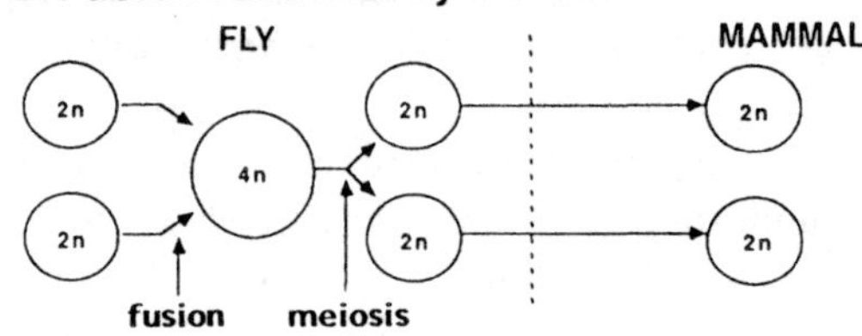

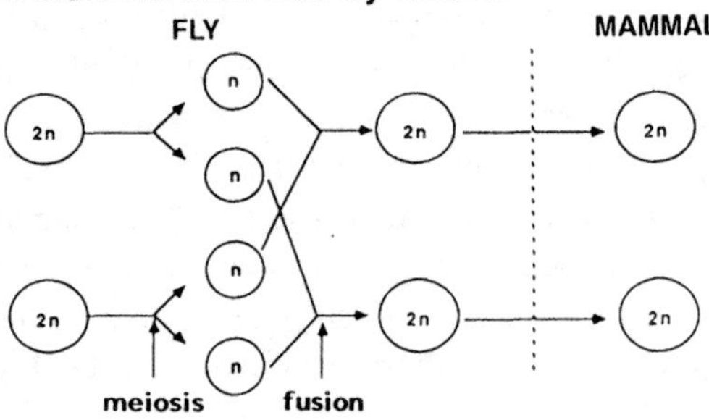

Fig. 5.14
Trypanosoma b. brucei: three possible models of the mechanism of gene exchange when two different stocks (mixed 1 : 1) of this trypanosome are fed to tsetse flies, the resultant hybrid clones being identified by having isoenzyme patterns different from those of their parents. *A*. A fusion model. *B*. Fusion followed by meiosis; this differs from Model *A* only in that the metacyclics do not change their ploidy on entry into the blood stream. *C*. Meiosis followed by fusion; it is uncertain whether selfing can occur between haploid products of the same cell. (After Tait & Turner, 1990.)

for the trypanosomes of sleeping sickness (Table 5.2). Infections may last up to four years without apparent loss of transmissibility to man. The importance of reservoir hosts, in relation to sleeping sickness, presents an exceedingly complex problem, for the incidence of biting of such hosts (i.e. cattle) may be influenced by the presence of other animals, such as reptiles, which may serve as a food source but which do not act as reservoir hosts.

The significance of human African trypanosomiasis as a world problem and the epidemiology of the disease has been reviewed by Ormerod, 1969; Gibson & Miles, 1985; Molyneaux & Ashford, 1983; WHO, 1986.

Prevention. This involves, as usual, breaking the life cycle of the parasite at some stage by (*a*) treating patients

Table 5.3. *Chagas' disease in Latin America*

Country	Total population (millions)	Per cent rural	Estimated cases (millions)	Main vector†
Argentina	26.393	21	2.640	*T. infestans*
Belize	0.145	ND*	0.003	*T. dimidiata*
Bolivia	4.647	77	1.858	*T. infestans*
Brazil	119.024	36	6.340	*T. infestans*
Chile	10.857	20	0.367	*T. infestans*
Colombia	26	29	1.217	*R. prolixus*
Costa Rica	2.110	59	0.130	*T. dimidiata*
Ecuador	6.521	58	0.180	*T. dimidiata*
El Salvador	4.300	60	0.322	*T. dimidiata*
Fr. Guiana	0.080	ND*	0.021	*R. pictipes?*
Guatemala	7.110	64	0.730	*T. dimidiata*
Guyana	0.835	73	0.208	*R. prolixus?*
Honduras	3.400	68	0.213	*R. prolixus*
Mexico	69.950	34	3.798	*T. barberi*
Nicaragua	2.400	52	0.114	*T. dimidiata*
Panama	1.630	49	0.226	*R. pallescens*
Paraguay	2.880	56	0.397	*T. infestans*
Peru	16.800	42	0.643	*T. infestans*
Suriname	0.352	ND*	0.147	*R. pictipes?*
Uruguay	2.886	17	0.278	*T. infestans*
Venezuela	13.913	30	4.865	*R. prolixus*
Total	322.233	36	24.697	

*ND: No data.
†Vectors: *T.* = *Triatoma*; *R.* = *Rhodnius*.
Source: Data from Schofield (1985).

with drugs, (*b*) exterminating the flies or (*c*) reducing contacts between them and the human population. The position of possible reservoir hosts must also be taken into account. Intelligent use of insecticides and determined attacks on tsetse fly breeding areas, as well as destruction of game animals, are generally indicated. Existing chemotherapy has been reviewed by Gutteridge (1985).

Trypanosoma cruzi

This species causes a disease known as Chagas' disease, named after Carlos Chagas who first described it in Brazil in 1909; it is also referred to as American trypanosomiasis. It occurs throughout South and Central America, especially Brazil, Argentina and Mexico. The WHO estimates that over 24 million people are infected, or are at least serologically positive for *T. cruzi* (Table 5.3). This represents nearly 8 per cent of the population of Latin America (Schofield, 1985). Until recently, it was thought that the prevalence in Mexico was relatively low but it is now recognised as a major public health problem in that country, previously undetected owing to poor awareness at both a local and an institutional level (Salazar Schettino *et al.*, 1988).

Work on this organism has been the subject of numerous reviews and books, of which the following are representative: Brener, 1980; Dvorak *et al.*, 1985; Gutteridge, 1985; Hudson & Britten, 1985; Marsden, 1984; Molyneux & Ashford, 1983; Nogueira & Coura, 1984; Schofield, 1985; Van Meirvenne & Le Ray, 1985; Vickerman, 1974; WHO, 1984; Zeledón, 1974.

Morphology. *T. cruzi* has a monomorphic form, about 20 µm in length, and characteristically curved (Fig. 5.15). The kinetoplast is large, considerably larger than in any of the species discussed already, and sometimes appears as a bulge at the posterior end. The flagellum is of medium length.

Ultrastructure. The ultrastructure of *T. cruzi* is generally similar to that already described (p. 58) with some differences, the most marked being the amplification of the kinetoplast in metacyclic and bloodstream trypomastigotes and the presence of a cytopharynx (Vickerman, 1974).

Surface antigens. Unlike *T. brucei*, whose bloodstream forms 'evade' the host's immune response by repeatedly changing a single glycoprotein on their surface (p. 60; Figs 5.5, 5.12), *T. cruzi* expresses several glycoproteins simultaneously, the surface antigens being highly polymorphic (Plata *et al.*, 1984).

Transmission. The vectors of *T. cruzi* are brightly coloured bugs of the family Reduviidae, subfamily Triatominae, all stages of which (larva, nymph and imago) are susceptible to infection. Transmission is facilitated by poor housing conditions and socio-behavioural patterns; for example, some communities believe these bugs can cure warts or have aphrodisiac powers if eaten (Salazar Schettino *et al.*, 1988). The vector colonises wall cracks and vegetal roofs of houses, coming out at night and feeding on exposed parts of a sleeper's body. Soon after a blood meal, a bug defecates and metacyclic forms from the hindgut released in the faeces may be rubbed into the skin, penetrating it; the eyes and lips are particularly affected. The main vectors are listed in Table 5.4. Another form of transmission, which is rapidly on the increase, is by blood transfusion; more rarely, intra-uterine transmission via a damaged placenta can occur (Zeledón, 1974). It has also been speculated that infected metacyclics may be transmitted via the anal gland secretions of opossums (Schofield, 1988).

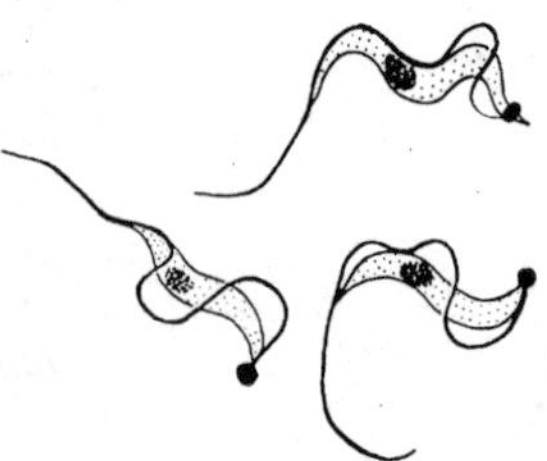

Fig. 5.15
Trypanosoma cruzi, from experimentally infected mice. (After Brumpt, 1949.)

Reservoir hosts. Chagas' disease is essentially a zoonosis (p. 52) and about 150 species of wild mammals from seven orders have been incriminated as reservoir hosts (Zeledón, 1974). The most important host reservoirs are *Dasypus novemcinctus* in S. America;

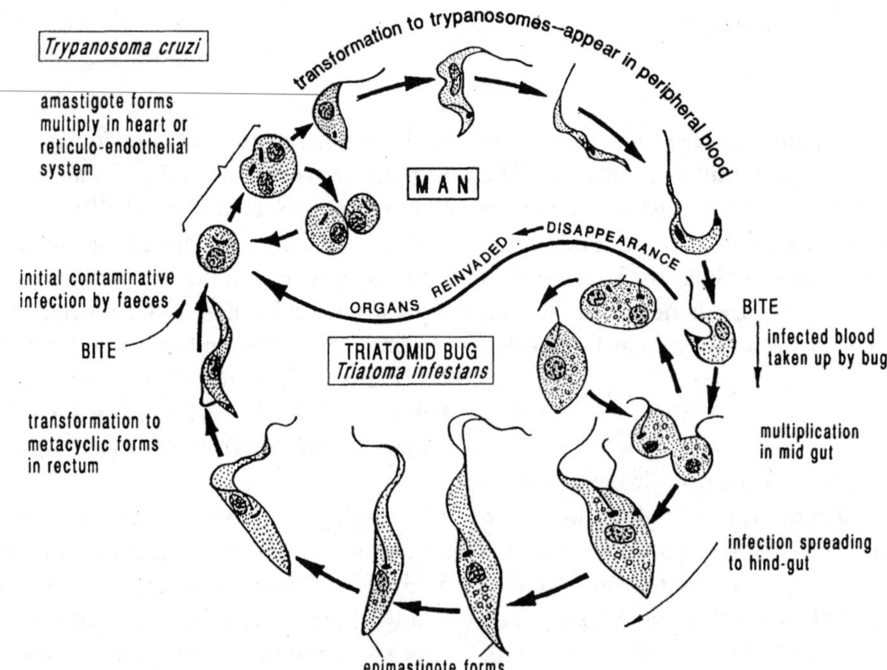

Fig. 5.16
Life cycle of *Trypanosoma cruzi* in man and in the triatomid bug *Triatoma infestans*; other arthropods, bed bugs, ticks and keds can also act as vectors. (After Brumpt, 1949.)

Table 5.4. *The five most important vectors of* Trypanosoma cruzi *in Latin America*

Triatoma infestans: Widespread throughout Argentina, Chile, Bolivia, Brazil, Paraguay, Uruguay and southern Peru. Largely domestic.
Triatoma brasiliensis: Main domestic vector in the arid caatinga of northern Brazil.
Triatoma dimidiata: Distributed throughout the humid areas from northern Peru into Mexico; also found in Venezuela. Often found in caves and armadillo burrows.
Panstrongylus megistus: Maintains sylvatic foci associated with ground-nesting mammals throughout the centre and east of Brazil, extending into eastern Paraguay and northern Argentina (Misiones).
Rhodnius prolixus: Important domestic vector in Venezuela, Colombia and parts of Central America (but not Panama). Sylvatic foci mainly in crowns of palm trees. Since the eggs of *R. prolixus* adhere to palm leaves, which are used for making roofs, this is a major route for house infestation.

Source: Data from Schofield (1985).

Didelphis spp. (opposums) in S. America and the USA; rats in Panama; and dogs, pigs and cats in S. America. An extensive account of animal reservoirs in Brazil and their role in the epidemiology of Chagas' disease has been given by Deane (1964).

Speciation. There is much variation in behaviour of laboratory strains of *T. cruzi*, a fact which suggested that the species is a heterogeneous complex of organisms. Isoenzyme profiles from numerous countries have confirmed that three principal 'strain' groups or zymodemes (p. 52) exist and there is also heterogeneity within the zymodemes (Gibson & Miles, 1985). Variation in kinetoplast DNA (kDNA) buoyant density has also been reported between different strains of *T. cruzi* isolates.

Human pathology. Infections occur most frequently in infants and children; in adults there seems to be a high natural resistance, but this may be greatly weakened by malnutrition. When inoculation of infective metacyclics occurs, *acute* Chagas' disease results. Swellings and inflammation take place. The initial swelling – which is related to the metacyclic organisms' being phagocytosed by macrophages and other local cells – is known as the primary 'chagoma'. If the ocular mucosa is inoculated, a characteristic unilateral conjunctivitis and oedema of the eyelids may occur, this is known as

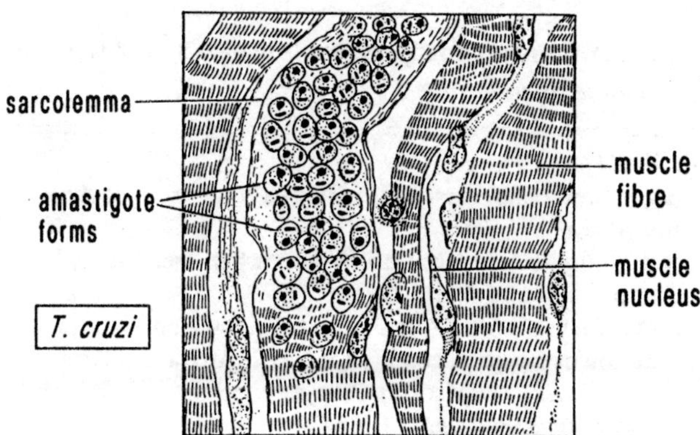

Fig. 5.17
Amastigote forms of *Trypanosoma cruzi* in the muscle fibres of a rat. (After Brumpt, 1949.)

'Romana's sign'. Trypanosomes show a marked predilection for the heart, in which they penetrate the myocardial fibres, lose their flagella and assume the amastigote form. Binary division follows and the parasite multiplies for a period of about 5 days, a cyst-like cavity (the *pseudocyst*, Fig. 5.17) being formed within the invaded tissue. A proportion of the organisms become transformed to the trypomastigote stage, perforate the pseudocyst membrane and escape into the blood stream to invade other tissues. This results in *chronic* Chagas' disease involving heart failure and breathing difficulties. For succinct accounts of the pathology of the disease, see Nogueira & Coura (1984) and Köberle (1968).

Trypanosoma rangeli

This species is worth mentioning briefly as it occurs in man, in whom it does not reproduce and is not pathogenic. In nature, it has a wide host range and has been experimentally transmitted to numerous species of laboratory animals. Its medical importance is due to the fact that it confuses the diagnosis of Chagas' disease by occurring in the same vector species and having more or less the same distribution. Its biology has been reviewed in detail by D'Alessandro (1976).

○○○ 5.6 Genus *Leishmania*

○ 5.6.1 General account

The flagellates of the genus *Leishmania* are mainly parasites of man and other mammals (especially dogs and rodents). They cause diseases collectively known as

Table 5.5. Leishmania *species and subspecies, and forms of human leishmaniasis*

Old World forms
L. major 'wet' cutaneous: widespread in rural areas of Asia and Africa
L. tropica 'dry' cutaneous: uncommon; urban areas of Europe, Asia and North Africa
L. aethiopica cutaneous, often diffuse: Ethiopia and Kenya
L. donovani donovani visceral (Kala Azar): Africa and Asia
L. donovani archibaldi visceral: Kenya
L. donovani sinensis visceral: China
L. donovani infantum infantile visceral: Mediterranean region
New World forms
L. donovani chagasi visceral: South America
L. braziliensis braziliensis mucocutaneous (Espundia): South America, especially Brazil
L. braziliensis guyanensis cutaneous: South America
L. braziliensis panamensis mucocutaneous: South and Central America
L. mexicana mexicana } cutaneous: South and Central America
L. mexicana amazonensis } cutaneous: South and Central America
L. mexicana pifanoi } cutaneous: South and Central America
L. peruviana cutaneous: South America, mainly Andean region

Source: Data from Barker (1987).

leishmaniases, which can be grouped under three headings:

(*a*) **Cutaneous** leishmaniasis;
(*b*) **Mucocutaneous** leishmaniasis (or *espundia*);
(*c*) **Visceral** leishmaniasis or Kala-azar (black disease); also known as 'dum-dum fever' or 'ponos'.

These are serious debilitating and disfiguring diseases, which occur throughout the Old and New Worlds (Table 5.5), their pathology is briefly considered later.

The organisms are exceptional in living in the cells of the reticulo-endothelial system: cells which normally phagocytose most protozoan parasites. The process of infection of macrophages by cells of *L. donovani* has been followed *in vitro* (Miller & Twohy, 1967). Promastigote stages become attached to a macrophage by the tip of their flagella and become surrounded and engulfed by pseudopodia. At first, the parasite is surrounded by a vacuole, which later disappears. Under favourable conditions, transformation to amastigote forms takes 1–4 hours and a population increase of parasites can be observed at 48 hours. These organisms are of special

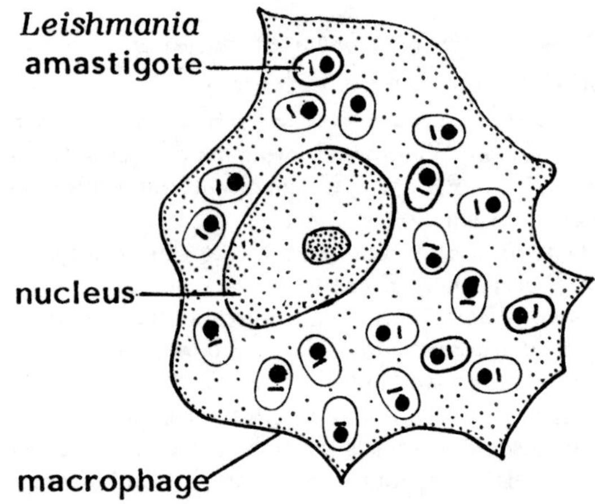

Fig. 5.18
Amastigotes of *Leishmania* sp. in a macrophage of a vertebrate host.

interest to biologists and epidemiologists on account of their widespread distribution in other animals. Leishmaniasis as a zoonosis is discussed later (p. 79).

An extensive literature exists on these diseases. The Technical Report of WHO (1984) is especially recommended as providing a comprehensive, yet succinct and readable summary. Other major reviews or books are those of Adler (1964), Ciba (1974), Chang & Bray (1985), Hart (1989), Hommel (1978), Lainson (1982, 1985), Molyneux & Ashford (1983) and Peters & Killick-Kendrick (1987).

○ 5.6.2 Morphology

(*a*) **Light microscopy.** In man, the parasites occur in the amstigote form only (Fig. 5.18) but in the insect vector the promastigote form is assumed. The amastigote forms are small, ovoid or round bodies (often called Leishman–Donovan bodies) about 2–5 μm in diameter. Under light microscopy, the cells are seen to contain a central nucleus, a rod-shaped kinetoplast and (sometimes) a basal body (from which the flagellum arises.

(*b*) **Ultrastructure and the kinetoplast.** Ultrastructure studies reveal a fine structure comparable to that of the trypomastigote stage (p. 58). The kinetoplast is shown to be a specialised portion of a highly extended, convoluted mitochondrion containing DNA (Rudzinska *et al.*, 1964), which shows some physical properties different from nuclear DNA. Recent work has been reviewed by Molyneux & Killick-Kendrick (1987). In general, the leishmanias show simpler mitochondrial cycles than the salivarian trypanosomes.

Habitat and life cycle. Leishmanias are unusual in living entirely as amastigotes within the cells of the reticulo-endothelial system to which they appear to have become perfectly adapted, since the proteolytic enzymes which attack other invaders of the bloodstream do not destroy them. The life cycle is shown in Fig. 5.19. In man, despite the wide range of conditions seen in leishmanial infections, all clinical and geographical strains of the disease share a common histological feature, i.e. the early accumulation of mononuclear phagocytic cells (or hypoplasia) in the tissues invaded. The species invading the skin induce an initial histiocytoma in the skin, while the viscera-invading species induce hyperplasia of the reticulo-endothelial cells of the organs invaded (WHO, 1984).

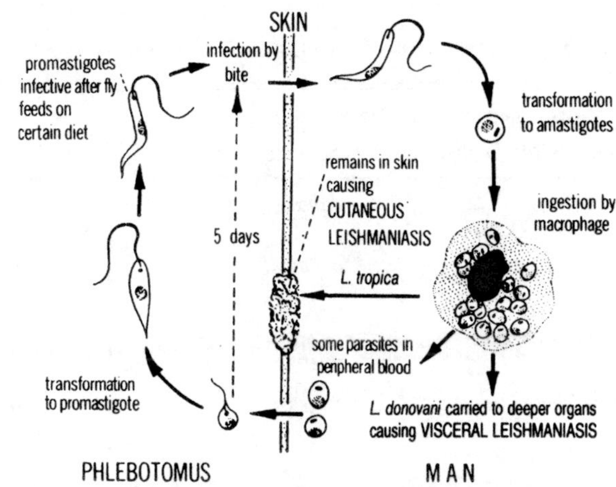

Fig. 5.19
Life cycle of *Leishmania donovani* and *L. tropica* in man and vector, *Phlebotomus argentipes*. (Partly after Hoare, 1949.)

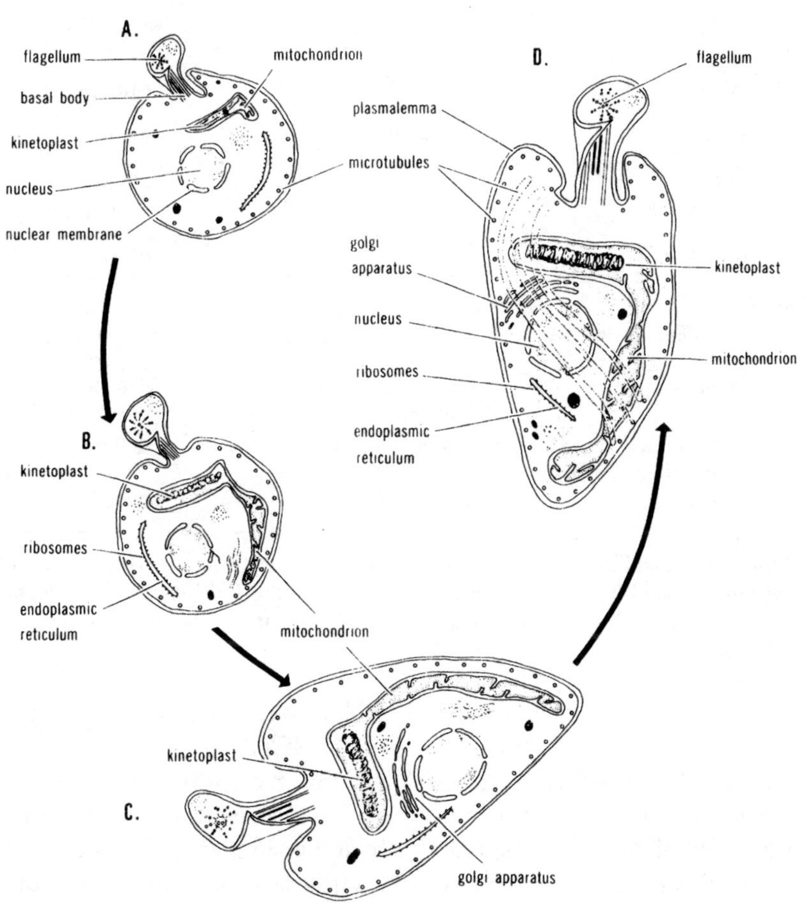

Fig. 5.20
Amastigote–promastigote transformation in *Leishmania mexicana* during *in vitro* culture for 0–21 h. A. Intracellular stage; B. cultured 1 h; C. cultured 6 h; D. cultured 21 h. (Modified from Creemers & Jardin, 1967.)

Table 5.6. *Proven vectors* of human leishmaniasis*

Old World: genus *Phlebotomus*			
Subgenus of *Phlebotomus*	Species	Parasite transmitted and main areas of endemic foci	
Phlebotomus	*duboscqi*	*L. major*	Senegal and parts of West Africa
	papatasi		South-West Asia, North Africa, and former USSR
	salehi		India (Rajasthan)
Paraphlebotomus	*sergenti*	*L. tropica*	Afghanistan, India, Islamic Republic of Iran, Iraq, Pakistan, and former USSR
Larroussius	*ariasi*	*L. donovani*	France
	longicuspis		North Africa
	major syriacus		South-West Asia
	orientalis		Sudan
	perniciosus		Italy and North Africa
	smirnovi		former USSR
	tobbi		Cyprus
	longipes	*L. donovani*	Ethiopia
	pedifer		Ethiopia and Kenya
Synphlebotomus	*martini*	*L. donovani*	Kenya
Adlerius	*chinensis*	*L. donovani*	China
	longiductus		former USSR
Euphlebotomus	*argentipes*	*L. donovani*	India
New World: genus *Lutzomyia*			
Subgenus of *Lutzomyia*	Species	Parasite transmitted and main areas of endemic foci	
Nyssomyia	*flaviscutellata*	*L.m. amazonensis*	Amazon basin in Brazil
	olmeca	*L.m. mexicana*	South Mexico, Central America
	trapidoi	*L.b. panamensis*	Panama
	umbratilis	*L.b. guyanensis*	North Brazil, French Guiana
Psychodopygus	*wellcomei*	*L.b. braziliensis*	Brazil

*Proven vector: anthropophilic and must bite animal reservoirs: found naturally infected with a parasite indistinguishable from that found in man and reservoirs. Other sandflies including species in the *Phlebotomus (Paraphlebotomus) caucasicus* group and *Lutzomyia longipalpis* are suspected of being vectors, but final incrimination awaits further study.
Source: Data from WHO (1984).

○ 5.6.3 Life cycle

Transmission. All forms of leishmaniasis of man are transmitted by the bite of female sandflies of the subfamily Phlebotominae, which contains about 600 species and subspecies; some 70 of these are proven or suspected vectors of leishmaniasis. The vectors in the Old World belong to the genus *Phlebotomus* and in the New World to the genus *Lutzomyia* (Table 5.6). Infection may occasionally occur by infected flies being accidentally inhaled or by infected flies being squashed on the skin (WHO, 1984).

Within the gut of the vector, transformation into the highly active, elongated promastigote forms commences about 24–36 h after the infective feed. From this time onwards rapid multiplication takes place but the speed of development varies with species. The pharynx becomes packed with active flagellates and the proboscis is invaded, at which time the sandfly becomes infective when it next bites a man or another host.

Reservoir hosts. The distribution and transmission is complicated by the fact that the disease is a zoonosis and there are a wide range of reservoir hosts. The chief kinds of leishmaniasis are generally associated with a particular type of animal reservoir. Thus, visceral leishmaniasis is associated with wild Canidae, such as jackals and dogs. The remaining types of leishmaniasis are associated with wild rodents and, again, dogs. The known reservoir hosts are reviewed by WHO (1984). It should be noted that the parasite found in lizards, and once thought to be *Leishmania*, has been shown to be a species of *Trypanosoma* (Chance, 1985).

○ 5.6.4 Leishmaniasis: the disease

Consideration of the detailed pathology of the disease is beyond the scope of this book, but the main features are summarised briefly below. The geographical distributions of some of the various species and subspecies are summarised in Table 5.5.

Cutaneous leishmaniasis. This can be classified by geographical occurrence, under two headings:

(*a*) *Old World cutaneous leishmaniasis*; also known as 'bouton d'Orient', 'oriental sore', 'Baghdad boil', 'Delhi sore' and 'Aleppo button'. This is normally caused by three species: *L. tropica*, *L. major* and *L. aethiopica*. This infection produces a skin ulcer which leaves an unsightly scar on healing.

(*b*) *New World cutaneous leishmaniasis*; this is more severe and chronic than the Old World form. Diffuse cutaneous leishmaniasis causes widespread thickening of the skin, which does not heal spontaneously and tends to relapse after treatment. Caused by numerous species and sub-species: *L. braziliensis braziliensis*, *L. b. panamensis*, *L. b. guyanensis*, *L. mexicana mexicana*, *L. m. amazonensis*, *L. m. venezuelensis*, *L. m. garnhami*, *L. peruviana*, *L. m. pifanoi*.

Monocutaneous leishmaniasis ('espundia'). This initially produces lesions like those of cutaneous leishmaniasis, but later – often after 30 years – metastastic spread may occur into the oronasal and pharyngeal mucosa, resulting in highly disfiguring tissue damage ('tapir nose').

Visceral leishmaniasis ('Kala-azar'). This is caused by *L. donovani* and its subspecies, and may be classified as *endemic*, *sporadic* or *epidemic*.

(*a*) *Endemic visceral leishmaniasis*

This especially affects children (1–4 years) in the Mediterranean area, South-West Asia, China and Latin America. In East Africa and India, the peak age is 5–9 years or older. The common symptoms are fever, malaise, weight loss, anorexia accompanied by anaemia, skin darkening, and enlargement of the spleen.

(*b*) *Sporadic visceral leishmaniasis*

Affects non-indigenous people entering an epidemic area. Characterised by a markedly sudden onset of fever, 3 weeks to 2 years after exposure.

(*c*) *Epidemic visceral leishmaniasis*

All ages are susceptible except those old enough to have been affected during a previous epidemic.

Taxonomy. There are very few morphological differences between the various members of the genus *Leishmania*; the classification which has developed is based initially on the clinical disease produced. This is supported by a variety of biological, geographical and epidemiological characteristics. A generally accepted list of named species and subspecies is given in Table 5.5. The application of modern techniques of biochemistry, molecular biology and immunology have recently transformed the situation and much more precise typing of isolates from different areas is now possible. By far the most widely used method for the characterisation of *Leishmania* and other parasitic protozoa (e.g. *Entamoeba histolytica*, see Fig. 3.5) has been the analysis of isoenzymes by electrophoresis. This has the advantage of being able to utilise small quantities of material from host, vector and reservoir hosts.

For example, Le Blancq *et al.* (1986) and Le Blancq & Peters (1986) undertook a major study of the isoenzymes of *Leishmania major* and *L. tropica* which enabled them to determine the geographical and hostal distribution of the zymodemes (populations defined by enzyme profiles) of these two species. They examined a collection of 135 stocks of *Leishmania major* from man, reservoir hosts and sandflies using thin-layer starch-gel electrophoresis of 13 enzymes. Polymorphism in four enzymes (phosphogluconate dehydrogenase, glucose-phosphate isomerase, proline iminopeptidase and esterase) revealed the existence of six zymodemes (Fig. 5.21) in the collection. Moreover, specimens from sandflies and several species of burrowing rodents could not be distinguished from those found in man in the same areas,

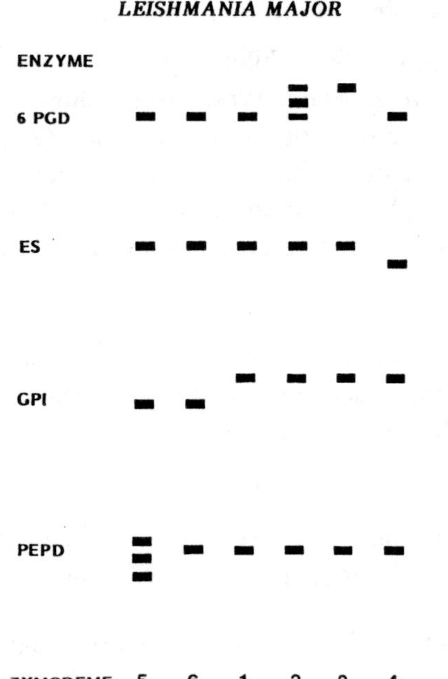

Fig. 5.21
Zymodemes identified in *Leishmania major* by polymorphism in four enzymes: **6 PGD** (phosphogluconate dehydrogenase); **ES** (esterase); **GPI** (glucosephosphate isomerase; **PEPD** (proline iminopeptidase). Homogeneity in this species was demonstrated by identical electrophoretic mobilities in nine enzymes. (After Le Blancq *et al.*, 1986.)

Table 5.7. *Modern techniques for idenfication of Leishmania*

Biological characters
Development in sandflies
Virulence of clones in rodents[a]
Immunological characters
Noguchi–Adler test
Excreted factor serotyping[a]
Monoclonal antibodies[b]
In vivo cross-immunity tests
Biochemistrry
DNA sequence analysis
DNA (nuclear) buoyant density
DNA–RNA hybridisation
Restriction-endonuclease fragment analysis[b]
Isoenzyme characterisation[c]
Cell membrane structure (e.g. lectins)
Fatty acid analysis
Radiorespirometry[b]

[a]Method demonstrates infra-specific differences, but of limited use at the species level.
[b]Methods that are limited to a few laboratories: preliminary results suggest they are potentially useful for identification at the species or infra-specific level.
[c]The most commonly used method at the moment to identify strains at the species or infra-specific level.
Source: Data from WHO (1984).

thus enabling them to confirm the vectors and the reservoir hosts.

In a related study (Le Blancq & Peters, 1986) on *L. tropica*, 27 stocks from widely separated geographical areas were examined and 18 different enzyme profiles were revealed – hence 18 zymodemes – indicating a considerable degree of intra-specific variation. Only one enzyme band was found in common with the enzyme profile of *L. major*, thus confirming the status of *L. tropica* as a separate species.

Numerous other similar studies have been carried out using isoenzyme analysis; many of these are reviewed by the above authors. Other modern techniques such as those shown in Table 5.7 have also been used to study intraspecific differences. Some relevant recent studies are reported by Barker (1987), Blackwell (1986), Chance (1985), Hart (1989) and Peters & Killick-Kendrick (1987).

○○○ 5.7 Physiology and biochemistry of haemoflagellates

○ 5.7.1 Biochemistry of *Trypanosoma*

General comments. Because of the importance of species of *Trypanosoma* and *Leishmania* as pathogens of major tropical diseases, their physiology and biochemistry has been extensively investigated. Moreover, the organisms are also of special interest to biologists and biochemists on account of the metabolic switches which occur in the blood stream and vector stages, as indicated in Fig. 5.12. The vast amount of research, which is only briefly summarised here, has been the subject of a number of valuable reviews and books of which the following are representative: Bowman (1974), Blum (1991), Coombs & North (1991), Fairlamb (1981, 1989), Fairlamb & Opperdoes (1986), Kilgour (1980), Marr (1980), Opperdoes (1987, 1991), Opperdoes *et al.* (1990), Klein & Miller (1981) and Turrens (1991).

In addition to the changing activity of the mitochondrion during the life cycle, as described earlier (Fig. 5.12), members of the Kinetoplastida possess two other unique features: (*a*) the presence of a peculiar organelle, the *glycosome*, and (*b*) the presence of *trypanothione*, a novel thiol-containing substance, first identified by Fairlamb & Cerami (1985), which plays an essential role in protecting against oxidative stress. These unusual features are briefly discussed below.

(*a*) *The glycosome*

The morphology and biochemical activities of this organelle have been reviewed by Opperdoes (1987, 1991). Early reports indicated that the glycolytic enzymes in the Trypanosomatidae might be particle-bound. This was confirmed by the association of specific enzymes with a microbody organelle which was eventually named the *glycosome*. It plays a dominant role in trypanosome metabolism (Fig. 5.22) and, to date, nine enzymes have been identified in it. These include all the early enzymes of the glycolytic pathway from hexokinase to phosphoglycerate kinase, together with two enzymes of the glycerol pathway (glycerol kinase and glycerol-3-phosphate dehydrogenase) and phosphoenolpyruvate kinase, an enzyme involved in CO_2 fixation; malate dehydrogenase may also be present. Three of these enzymes may have isoenzymes in the cytosol. About 240 glycosomes appear to be present in a blood stream form. Other functions of the glycosome include CO_2 fixation, pyrimidine biosynthesis, β-oxidation of fatty acids, biosynthesis of lipids and purine salvage (Opperdoes, 1991).

(*b*) *Trypanothione*

In most aerobic organisms, glutathione, (γ-glutamylcysteinylglycine, GSH) plays an essential role as a scavenger of free radicals and serves to maintain the correct intracellular redox potential for enzymatic activity; it is also involved in some other metabolic activities (Smith *et al.*, 1991). On account of the high activity of glutathione reductase (GR), less than 0.5 per cent of the cellular glutathione occurs as glutathione disulphide (GSSG). In trypanosomatids – in contrast to mammalian cells – most of the GSH occurs as the polyamine N^1,N^2-bis(glutathionyl)spermine now known as *trypanothione* (Fairlamb & Cerami, 1985). Furthermore, the classical glutathione reductase activity of mammalian cells is lacking in trypanosomatids; instead, they contain a unique disulphide oxidoreductase, trypanothione reductase (TR). Thus, trypanothione and trypanothione reductase appear to play a major role in the maintenance of the correct intracellular thiol redox potential in trypanosomatids and their uniqueness makes them a suitable target for drug design (Murgolo *et al.*, 1990; Smith *et al.*, 1991).

Carbohydrate metabolism

Blood stream forms. The blood stream trypomastigote forms of *T. brucei* possess no energy stores, are unable to oxidise amino acids or fatty acids and are thus entirely dependent on an exogenous supply of carbohydrate (glucose, fructose, mannose or glycerol) for their energy production (Fairlamb & Opperdoes, 1986). Short-stumpy trypomastigotes can also respire d-oxoglutarate at a rate about half that of glucose. Aerobic glycolysis proceeds through the Embden–Meyerhof pathway, all the enzymes of which have been detected in the glycosomes, the mitochondrion and the cytosol. The distribution of the enzymes between the various components of the cells are shown in Fig. 5.22. The glycolysis process differs from the normal, vertebrate pattern in that the end product is almost exclusively (98 per cent) pyruvate, with traces of CO_2 and glycerol (Table 5.8), the process supplying all the ATP synthesised by the organisms, and proceeds at a rate faster than observed in any other organism (Opperdoes *et al.*, 1990). Pyruvate is not metabolised further by long-slender forms as they lack the enzymes pyruvate decarboxylase and lactate dehydrogenase (Fairlamb & Opperdoes, 1986). However, some pyruvate is further metabolised to acetate, CO_2 and succinate in the short-stumpy forms.

In *T. brucei*, molecular oxygen serves as the terminal electron acceptor, resulting in the formation of water. A glycerol 3-phosphate shuttle transports the reducing equivalents from NADH in the cytosol into the mitochondrion, where, via a glycerophosphate oxidase system, it passes to molecular oxygen. This enzyme appears to be unique to salivarian trypanosomes.

Under anaerobic conditions, long-slender blood stream forms continue to utilise glucose, as under aerobic conditions, but, because the glycerophosphate is inoperative, the glucose is metabolised into equimolar amounts of pyruvate and glycerol.

Procyclic (vector) forms. As shown earlier (Fig. 5.12) the procyclic vector forms undergo a metabolic switch,

Table 5.8. *End products of aerobic glucose metabolism in trypanosomes of the brucei subgroup*

Morphological type	Product[a]					
	Pyruvate	Glycerol	Acetate	Lactate	Succinate	CO_2
Mammal						
Long-slender trypomastigote	++++	±[b]	–	–	–	±
Short-stumpy trypomastigote	++++	±[b]	++	–	±	+
Insect						
Procyclic trypomastigote	–	–	+	0	+	+++
Epimastigote	0	0	0	0	0	0
Metacyclic trypomastigote	0	0	0	0	0	0

[a]The amount of carbon recovered in each product is expressed as a percentage of the carbon of glucose utilised, where ++++ represents 51%; +++, 26–50%; ++, 6–25%; +, 1–5%; ±, 1%; 0, no available data.
[b]Variable amounts of glycerol have been reported by various authors, which is probably due to partial anaerobiosis.
Source: Data from Fairlamb & Opperdoes (1986).

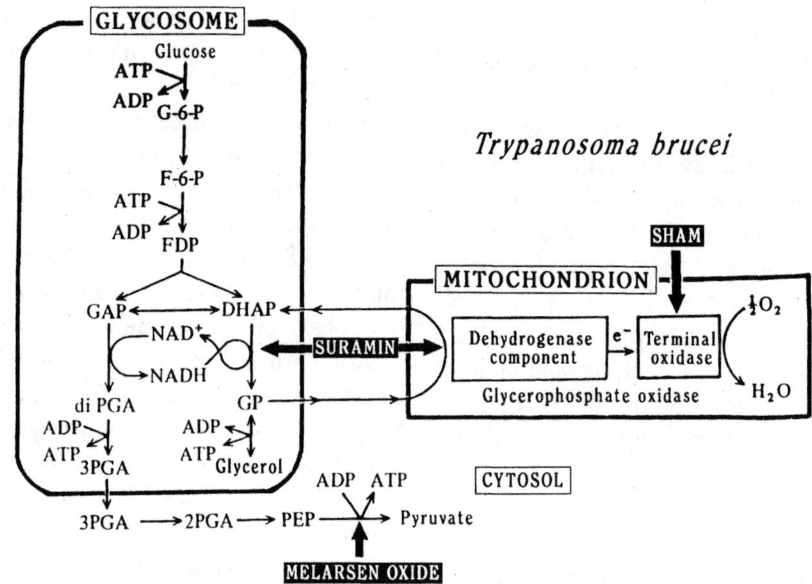

Fig. 5.22
T. brucei: aerobic and anaerobic glycolysis and principal sites of *in vivo* inhibition by the trypanocidal drugs *melarsen*, *suramin* and *SHAM* (= salicylhydroxoamid acid). Abbreviations: **G-6-P**, glucose-6-phosphate; **F-6-P**, fructose-6-phosphate; **FDP**, fructose-1,6-diphosphate; **GAP**, glyceraldehyde-3-phosphate; **DHAP**, dihydroxyacetone phosphate; **GP**, *sn*-glycerol-3-phosphate; **di PGA**, 1,3-diphosphoglycerate; **3PGA**, 3-phosphoglycerate; **2PGA**, 2-phosphoglycerate; **PEP**, phosphoenolpyruvate. (After Fairlamb, 1981.)

with the development of a fully developed mitochondrion (kinetoplast) with cristae and most of the Krebs cycle enzymes, although citrate synthase appears to be missing (Fairlamb & Opperdoes, 1986). In the cultured procyclic forms, glucose is consumed at a slower rate than in the blood stream forms and the pyruvate is oxidised further, the main end product being CO_2 together with acetate and citrate (Table 5.8). However, glucose is scarce in the insect vector and is probably only a very minor source of energy. The enzyme NADH–fumarate reductase, which can reverse the Krebs cycle by reducing fumarate to succinate, is present, a situation that also occurs in parasitic helminths (Chapter 24). Succinate is also an end product of proline catabolism in the midgut stages of *T. brucei*, which closely resemble procyclic culture forms. Utilisation of proline correlates well with the environment provided by the tsetse fly as the haemolymph contains excessively high levels of proline, although deficient in carbohydrate. In tsetse flies, therefore, trypanosomes depend largely for their subsistence on proline; by developing NADH–fumarate reductase, they are able to produce succinate which may be used as the main respiratory substrate or as an electron sink to be excreted outside the cell under anaerobic conditions (Turrens, 1991).

In general, other *Trypanosoma* species utilise substrates and metabolic pathways similar to those of *T. brucei*.

General metabolism

Terminal transport system. Blood stream forms of *T. brucei* and *T. b. rhodesiense* lack cytochromes and their respiration is thus not inhibited by cyanide, azide or antimycin A. In contrast, the rodent parasite *T. lewisi* and *T. cruzi* have cytochrome systems and hence their respiration is inhibited by cyanide. In *T. brucei* and *T. b. rhodesiense*, the NADH resulting from aerobic glycolysis is reoxidised via the 'glycerophosphate shuttle' (Fig. 5.22) comprising the NAD$^+$-dependent *sn*-glycerol-3-phosphate dehydrogenase in the glycosome and the *sn*-glycerol-3-phosphate oxidase in the mitochondrion, oxygen acting as the hydrogen acceptor with water as the end product (Fairlamb, 1981).

Protein metabolism. Relatively few studies on amino acid synthesis in haemoflagellates have been carried out. Blood stream stages of *T. b. rhodesiense* can synthesise alanine, glycine, serine and aspartic and glutamic acids, but, in general, little incorporation of carbon of carbohydrate origin occurs, consistent with the separation of energy metabolism from synthetic pathways. In *T. cruzi*, however, some radioactivity from labelled glucose appears in alanine, aspartic acid and glutamic acid (Von Brand, 1966).

Lipid metabolism. Although not a great deal of information is available on the lipid metabolism of haemoflagellates, this area is increasingly being examined as providing targets for developments in chemotherapy and chemotaxonomy. Lipids present in trypanosomatids are of a very diverse nature and include triacylglycerols, phospholipids, fatty acids, glycolipids, alkoxydiacylglycerols and sterols. Of these, phospholipids constitute the major proportion, accounting for 73–77 per cent of the total lipids. Among the glycolipids, glycosylphosphatidylinositols are of special significance in trypanosomatids in that they link readily to the protein molecules which form the variant surface glycoproteins (VSGs; Fig. 5.2). Recent work in this area has been reviewed in detail by Haughan & Goad (1991).

○ 5.7.2 Biochemistry of *Leishmania* and *T. cruzi*

These species generally follow the early stages of glycolysis as in *T. brucei*, but the pyruvate is metabolised further. In *Leishmania major*, the products of glucose catabolism have been identified as succinate, glycerol, D-lactate, pyruvate and alanine. CO_2 is also produced in the presence of oxygen. The nature and quantities

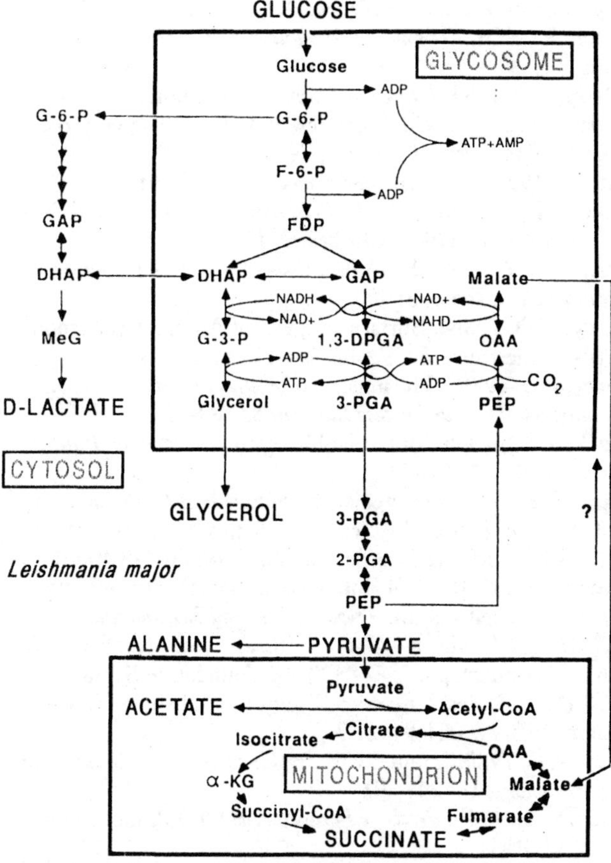

Fig. 5.23
Leishmania major promastigotes: probable pathways of glucose catabolism; but some pathways have yet to be confirmed. Abbreviations: **MEG**, methylgloxal; **G-3-P**, glucose-3-phosphate; **PEP**, phosphoenolpyruvate; **α-KG**, α-ketoglutarate; **OAA**, oxaloacetate; **1 : 3-DPA**, 1 : 3-diphosphoglycerate. For other abbreviations, see Fig. 5.22. (Modified from Blum, 1991.)

of the end products, however, are greatly affected by the osmolarity and the p_{O_2} and the p_{CO_2}. A partly confirmed scheme of the possible pathways is shown in Fig. 5.23. Valuable reviews on various aspects of the intermediate metabolism of *Leishmania* spp. and *T. cruzi* are those of Blum (1991), Darling *et al.* (1989), Fairlamb (1981, 1989) and Marr (1980). The lipid biochemistry has been reviewed by Haughan & Goad (1991).

○○○ References

(References of general interest have been included in this list, in addition to those quoted in the text.)

Abolarin, M. O., Stamford, S. A. & Ormerod, W. E. 1986. Interaction between *Trypanosoma brucei* and the ependymal cell of the choroid plexus. *Transactions of the Royal Society of Tropical Medicine & Hygiene*, **80**: 618–25.

Adler, S. 1964. *Leishmania. Advances in Parasitology*, **3**: 35–96.

Allsopp, R. 1984. Control of tsetse flies (Diptera: Glossinidae) using insecticides: a review and future prospects. *Bulletin of Entomological Research*, **74**: 1–23.

Ash, C. 1990. Editorial. *Parasitology Today*, **6**: 61.

Baker, J. R. (ed.) 1982. *Perspectives in Trypanosomiasis Research*. John Wiley, Chichester, UK.

Barker, D. C. 1987. DNA diagnosis of human leishmaniasis. *Parasitology Today*, **3**: 177–84.

Barry, J. D. 1986*a*. Surface antigens of African trypansomiasis in the tsetse fly. *Parasitology Today*, **2**: 143–5.

Barry, J. D. 1986*b*. The molecular biology of African trypanosomiasis. *Tropical Disease Bulletin*, **83**: R1–25.

Blackwell, J. 1986. Molecular biology of *Leishmania. Parasitology Today*, **2**: 45–53.

Blum, J. J. 1991. Intermediate metabolism of *Leishmania*. In: *Biochemical Protozoology* (ed. G. Coombs & M. North), pp. 123–33. Taylor & Francis, London. ISBN 0 748 40001 X.

Bowman, I. B. R. 1974. Intermediary metabolism of pathogenetic flagellates. In: *Trypanosomes and Leishmaniasis with Special Reference to Chagas' Disease* (ed. K. Elliot & M. O'Connor), pp. 255–84. (Ciba Foundation Symposium 20 (New Series).) Elsevier Scientific Publishing, Amsterdam. ISBN 9 021 94021 3.

Brener, Z. 1980. Immunity to *Trypanosoma cruzi. Advances in Parasitology*, **18**: 247–92.

Brumpt, E. 1949. *Précis de Parasitologie*. 6th Edition. Masson et Cie, Paris.

Bryant, C. & Behm, C. 1989. *Biochemical Adaptation in Parasites*. Chapman & Hall, New York. ISBN 0 412 32530 6.

Burton, P. R. & Dusanic, D. G. 1968. Fine structure and replication of the kinetoplast of *Trypanosoma lewisi. Journal of Cell Biology*, **39**: 318–31.

Cachon, J., Cachon, M. & Charnier, M. 1972. Ultrastructure du bodonidé *Trypanophis grobbeni* Poche, parasite des siphonophores. *Protistologica*, **8**: 223–36.

Challier, A. 1982. The ecology of tsetse (*Glossina* spp.) (Diptera, Glossinidae): a review. *Insect Science Applications*, **3**: 97–143.

Chance, M. C. 1985. The biochemical and immunological taxonomy of *Leishmania*. In: *Human Parasitic Diseases* (ed. K. P. Chang & R. S. Bray), Vol. 1, pp. 93–110. Elsevier, New York.

Chang, K. P. & Bray, R. S. (eds) 1985. *Human Parasitic Diseases*. Vol. 1. *Leishmaniasis*. Elsevier, New York.

Ciba. 1974. *Trypanosomiasis and Leishmaniasis with special reference to Chagas' disease.* (Ciba Foundation Symposium, No. 20.) Elsevier Scientific Publishers, Amsterdam. ISBN 9 021 94021 3.

Clark, T. B. & Wallace, F. G. 1960. A comparative study of the kinetoplast ultrastructure in the Trypanosomatidae. *Journal of Protozoology*, **7**: 115–24.

Coombs, G. & North, M. 1991. *Biochemical Parasitology*. Taylor & Francis, London. ISBN 0 748 40001 X.

Creemers, J. & Jardin, J. M. 1967. Étude de l'ultrastructure et de la biologie de *Leishmania mexicana* Biagi, 1953. I. Les modifications qui surviennent lors de la transformation *leishmania-leptomonas. Bulletin de la Société de Pathologie Exotique*, **60**: 53–8.

D'Alessandro, A. 1976. Biology of *Trypanosoma* (*Herpetosoma*) *rangeli* Tejera, 1920. In: *Biology of the Kinetoplastida*, Vol. 1 (ed. W. H. R. Lumsden & D. A. Evans), pp. 328–403. Academic Press, London. ISBN 0 124 60201 0.

D'Alessandro, P. A. & Clarkson, A. B. 1980. *Trypanosoma lewisi*: avidity and absorbability of ablastin, the rat antibody inhibiting parasite reproduction. *Experimental Parasitology*, **50**: 384–96.

Darling, T. N., Davis, D. G., London, R. E. & Blum, J. J. 1989. Metabolic interactions between glucose, glycerol, alanine, and acetate in *Leishmania braziliensis panamensis* promastigotes. *Journal of Protozoology*, **36**: 217–25.

Deane, L. M. 1964. Animal reservoirs of *Trypanosoma cruzi* in Brazil. *Revista Brasileira de de Malariologia e Doenças Tropicais*, **16**: 27–48.

Dvorak, J. A., Gibson, C. C. & Maelkelt, A. 1985. *Chagas' Disease. A Medlars-based, computer-processed bibliography*. US Dept of Health and Human Services, National Institutes of Health, Bethesda, Maryland, USA.

Ellis, D. S. 1986. Sex and the single trypanosome. *Parasitology Today*, **2**: 184–5.

Fairlamb, A. H. 1981. Workshop report. In Klein & Miller (1981), pp. 1–3.

Fairlamb, A. H. 1989. Novel biochemical pathways in parasitic protozoa. *Parasitology*, **99**S: S93–112.

Fairlamb, A. H. & Cerami, A. 1985. Identification of a novel, thiol-containing co-factor essential for glutathione reductase enzyme activity in trypanosomatids. *Molecular and Biochemical Parasitology*, **14**: 187–98.

Fairlamb, A. H. & Cerami, A. 1992. Metabolism and functions of trypanothione in the Kinetoplastida. *Annual Reviews of Microbiology*, **46**: 695–729.

Fairlamb, A. H. & Opperdoes, F. R. 1986. Carbohydrate metabolism in African trypanosomes, with special reference to the glycosome. In: *Carbohydrate Metabolism in Cultured Cells* (ed. M. J. Morgan), pp. 183–224. Plenum Publishing, New York.

Giannini, S. H. & D'Alessandro, P. A. 1984. Isolation of protective antigens from *Trypanosoma lewisi* by using tryptostatic (ablastic) immunoglobulin G from the surface coat. *Infection and Immunity*, **43**: 617–21.

Gibson, W. C., Marshall, T. F. de C. & Godfrey, D. G. 1980. Numerical analysis of enzyme polymorphism: a new approach to the epidemiology and taxonomy of trypanosomes of the subgenus *Trypanozoon. Advances in Parasitology*, **18**: 175–246.

Gibson, W. C. & Miles, M. A. 1985. Application of new technologies to epidemiology. *British Medical Bulletin*, **41**: 115–21.

Gutteridge, W. E. 1985. Existing chemotherapy and its limitations. *British Medical Bulletin*, **41**: 162–8.

Hart, D. T. (ed.) 1989. *Leishmaniasis: the current Status and new Strategies for Control.* (Proceedings of a NATO Advanced Study Institute on Leishaniasis. The First Centenary (1885–1988) New Strategies of Control. Held September 1987, Zakinthos, Greece.) Plenum Press, New York.

Haughan, P. A. & Goad, L. J. 1991. Lipid biochemistry of trypanosomatids. In: *Biochemical Parasitology* (ed. G. Coombs & M. North), pp. 312–28. Taylor & Francis, London. ISBN 0 748 40001 X.

Hegner, R., Root, F. M., Augustine, D. L. & Huff, C. G. 1938. *Parasitology – with Special Reference to Man and Domesticated Animals.* Appleton-Century-Crofts, New York.

Hoare, C. A. 1949. *Handbook of Medical Protozoology*. Baillière, Tindall & Cox, London.

Hoare, C. A. 1966. The classification of mammalian trypanosomes. *Ergebnisse der Mikrobiologie, Immunitätsforschung und experimentellen Therapie*, **39**: 43–57.

Hoare, C. A. & Wallace, F. G. 1966. Developmental stages of trypanosomatid flagellates, a new terminology. *Nature*, **212**: 1385–6.

Hommel, M. 1978. The genus *Leishmania*: biology of the parasites and clinical aspects. *Bulletin de l'Institut Pasteur*, **75**: 5–102.

Hudson, L. & Britten, V. 1985. Immune responses to South American trypanosomiasis and its relatinoship to Chagas' disease. *British Medical Bulletin*, **42**: 175–80.

Jenni, L., Marti, S., Schweizer, J., Betschart, B., Le Page, R. W. F., Wells, J. M., Tait, A., Paindavoine, P., Pays, E. & Steinert, M. 1986. Hybrid formation between African trypanosomes during cyclical transmission. *Nature*, **322**: 173–4.

Jordan, A. M. 1985. The vectors of African trypanosomiasis: research towards non-insecticidal methods of control. *British Medical Bulletin*, **41**: 181–6.

Kilgour, V. 1980. *Trypanosoma*: intricacies of biochemistry, morphology and environment. *International Journal of Biochemistry*, **12**: 325–32.

Klein, R. A. & Miller, P. G. G. 1981. Alternate metabolic pathways in protozoan energy metabolism. (Report of Workshop, EMOP, Cambridge, 1980.) *Parasitology*, **82**: 1–30.

Köberle, F. 1968. Chagas' disease and Chagas' syndromes: the pathology of American trypanosomiasis. *Advances in Parasitology*, **6**: 63–116.

Lainson, R. 1982. Leishmaniasis. In: *Parasitic Zoonoses* (ed. L. Jacobs & P. Arambulo), pp. 41–103. CRC Press, Boca Raton, Florida.

Lainson, R. 1985. Our present knowledge of the ecology and control of leishmaniasis in the Amazon region of Brazil. *Revista da Sociedade Brasileira de Medicina Tropical*, **18**: 47–56.

Le Blancq, S. M., Schnur, L. F. & Peters, W. 1986. *Leishmania* in the Old World: 1. The geographical and hostal distribution of *L. major* zymodemes. *Transactions of the Royal Society of Tropical Medicine and Hygiene*, **80**: 99–112.

Le Blancq, S. M. & Peters, W. 1986. *Leishmania* in the Old World: 2. Heterogeneity among *L. tropica* zymodemes. *Transactions of the Royal Society of Tropical Medicine and Hygiene*, **80**: 113–19.

Lincicome, D. R. & Francis, E. H. 1961. Quantitative studies on heterologous sera inducing development of *Trypanopsoma lewisi* in mice. *Experimental Parasitology*, **11**: 68–76.

Lumsden, W. H. R. 1965. Biological aspects of trypanosomiasis research. *Advances in Parasitology*, **3**: 1–57.

Lumsden, W. H. & Evans, D. 1976. *Biology of the Kinetoplastida.* Academic Press, London. ISBN 0 124 60201 0.

McGhee, R. B. 1968. Development and reproduction (vertebrate and arthropod host). In: *Infectious Blood Diseases of Man and Animals* (ed. D. Weinman & M. Ristic), vol. 1, pp. 307–41. Academic Press, New York.

Marr, J. J. 1980. Carbohydrate metabolism in *Leishmania.* In: *Biochemistry and Physiology of Protozoa* (ed. M. Levandowsky and S. H. Hunter), 2nd Edition, Vol. 3, pp. 313–40. Academic Press, New York.

Marsden, P. D. 1984. Chagas' disease: Clinical aspects. *Recent Advances in Tropical Medicine*, **1**: 63–87.

Mehlhorn, H. (ed.) 1988. *Parasitology in Focus*. Springer-Verlag, Berlin. ISBN 3 540 17838 4.

Miles, M. A. & Cibulskis, R. E. 1986. Zymodeme classification of *Trypanosoma cruzi. Parasitology Today*, **2**: 94–7.

Miller, H. C. & Twohy, D. W. 1967. Infection of macrophages in culture by leptomonads of *Leishmania donovani. Journal of Protozoology*, **14**: 781–9.

Molyneux, D. H. 1982. Trends in research on tsetse biology. In: *Perspectives in Trypanosomiasis Research. Proceedings of the 21st Trypanosomiasis Seminar.* London, 1981, (ed. J. R. Baker), pp. 85–102. Research Studies Press, Chichester, UK.

Molyneux, D. H. & Ashford, R. W. 1983. *The Biology of Trypanosoma and Leishmania, Parasites of Man and Domestic Animals*. Taylor & Francis, London.

Molyneux, H. H. & Killick-Kendrick, R. 1987. Morphology, ultrastructure and life-cycles. In: *The Leish-*

maniases in Biology and Medicine (ed. W. Peters & R. Killick-Kendrick), Vol. 1, pp. 121–76. Academic Press, London.

Murgolo, N. J., Cerami, A. & Henderson, G. B. 1990. Trypanothione. In: *Parasites: Molecular Biology, Drug and Vaccine Design* (ed. N. Agabian & A. Cerami), pp. 263–77. Wiley-Liss, New York. ISBN 0 471 56764 7.

Newton, B. A. (ed.) 1985. Trypanosomiasis. *British Medical Bulletin*, **41**: 103–99.

Nogueira, N. & Coura, J. R. 1984. American Trypanosomiasis (Chagas' disease). In: *Tropical and Geographical Medicine* (ed. K. S. Warren & A. A. F. Mahmoud), pp. 253–69. McGraw-Hill, New York.

Opperdoes, F. R. 1987. Compartmentation of carbohydrate metabolism in trypanosomes. *Annual Reviews of Microbiology*, **41**: 127–51.

Opperdoes, F. R. 1991. Glycosomes. In: *Biochemical Protozoology* (ed. G. Coombs & M. North), pp. 134–44. Taylor & Francis, London. ISBN 0 748 40001 X.

Opperdoes, F. R., Wierenga, R. K., Noble, M. E. M., Hol, W. G. J., Willson, M., Kuntz, D. A., Callens, M. & Perié, J. 1990. Unique properties of glycosomal enzymes. In: *Parasites: Molecular Biology, Drug and Vaccine Design* (ed. N. Agabian & A. Cerami), pp. 233–46. Wiley-Liss, New York. ISBN 0 471 56764 7.

Ormerod, W. E. 1963. The initial stages of infection with *Trypanosoma lewisi*; control of parasitemia by the host. In: *Immunity to Protozoa; a Symposium of the British Society for Immunology* (ed. P. C. C. Garnham *et al.*), pp. 213–27. Blackwell, Oxford.

Ormerod, W. E. 1967. Taxonomy of the sleeping sickness trypanosomes. *Journal of Parasitology*, **53**: 313–27.

Ormerod, W. E. 1969. Human and animal trypanosomiasis as world health problems. *Pharmacology and Therapeutics*, **6**: 1–40.

Ormerod, W. E., Bacon, S. J. & Smith, A. D. 1988. Degeneration of serotonin-specific neurons in the brain in experimental *Trypanosoma brucei* infection. *Bulletin de la Société de Pathologie Exotique*, **81**: 480–1.

Ormerod, W. E. & Hussein, M. S.-A. 1986. The ventricular ependyma of mice infected with *Trypanosoma brucei*. *Transactions of the Royal Society of Tropical Medicine and Hygiene*, **80**: 626–33.

Peters, W. & Killick-Kendrick, R. (eds) 1987. *The Leishmaniases in Biology and Medicine.* (2 vols.) Academic Press, London.

Plata, F., Pons, F. G. & Eisen, H. 1984. Antigenic polymorphism of *Trypanosoma cruzi*: clonal analysis of tryptomastigote surface antigens. *European Journal of Immunology*, **14**: 392–9.

Poltera, A. A. 1985. Pathology of human African trypanosomiasis with reference to experimental African trypanosomiasis and infections of the central nervous system. *British Medical Bulletin*, **41**: 169–74.

Rudzinska, M. A., D'Alessandro, P. A. & Trager, W. 1964. The fine structure of *Leishmania donovani* and the role of the kinetoplast in the leishmanial-leptomonad transformation. *Journal of Protozoology*, **11**: 166–90.

Rudzinska, M. A. & Vickerman, K. 1968. The fine structure. In: *Infectious Blood Diseases of Man and Animals: Diseases caused by Protista* (ed. D. Weinman & M. Ristic), Vol. 1, pp. 217–306. Academic Press, New York.

Salazar Schettino, P. M., de Haro Arteaga, I. & Uribarren Berrueta, T. 1988. Chagas' disease in Mexico. *Parasitology Today*, **4**: 348–51.

Schofield, C. J. 1985. Control of Chagas' disease vectors. *British Medical Bulletin*, **41**: 187–94.

Schofield, C. J. 1988. Complement mediated evolution? April reflections on parasitic protozoa. *Parasitology Today*, **4**: 89–90.

Seed, J. R., Bogucki, M. S. & Merrit, S. C. 1980. Interactions between immunoglobulins and the trypanosome cell surface. In: *Cellular Interactions in Symbiosis and Parasitism* (ed. C. B. Cook, W. Pappas & E. D. Rudolph), pp. 131–43. Ohio University Press, Columbia.

Smith, K., Mills, A., Thornton, J. M. & Fairlamb, A. H. 1991. Trypanothione metabolism as a target for drug design: molecular modelling of trypanothione reductase. In: *Biochemical Protozoology* (ed. G. Coombs & M. North), pp. 482–92. Taylor & Francis, London. ISBN 0 748 40001 X.

Stephen, L. E. 1986. *Trypanosomiasis – a Veterinary Perspective.* Pergamon Press, Oxford.

Tait, A. 1983. Sexual processes in the kinetoplastida. *Parasitology*, **86**: 29–57.

Tait, A. & Turner, C. M. R. 1990. Genetic exchange in *Trypanosoma brucei*. *Parasitology Today*, **6**: 70–5.

Taylor, A. E. R. 1986. Trypanosomiasis. (A review of recent abstracts, 1983–1984.) *Tropical Disease Bulletin*, **83**: R1–60.

Tibayrenc, M. & Ayala, F. J. 1987. *Trypanosoma cruzi* populations: more clonal than sexual. *Parasitology Today*, **3**: 189–90.

Tizard, I. 1985. *Immunology and Pathogenesis of Trypanosomiasis.* CRC Press, Boca Raton, Florida.

Turner, M. J. 1985. VSGs on the surface of trypanosomes. *British Medical Bulletin*, **41**: 137–43.

Turner, R. M. B., Barry, J. D. & Vickerman, K. 1988. Loss of variable antigen during transformation of *Trypanosoma brucei rhodesiense* from bloodstream to procyclic forms. *Parasitology Research*, **74**: 507-11.

Turrens, J. F. 1991. Mitochondrial metabolism of African trypanosomes. In: *Biochemical Protozoology* (ed. G. H. Coombs & M. J. North), pp. 145–53. Taylor & Francis, London. ISBN 0 748 40001 X.

Van Meirvenne, N. & Le Ray, D. 1985. Diagnosis of African and American trypanosomiasis. *British Medical Bulletin*, **41**: 156–61.

Vickerman, K. 1962. The mechanism of cyclical development in trypanosomes of the *Trypanosoma brucei* sub-group: an hypothesis based on ultrastructural observations. *Transactions of the Royal Society of Tropical Medicine and Hygiene*, **56**: 487–95.

Vickerman, K. 1965. Polymorphism and mitochondrial activity in sleeping sickness trypanosomes. *Nature*, **208**: 762–6.

Vickerman, K. 1966. Genetic systems in unicellular animals. *Science Progress*, **54**: 13–26.

Vickerman, K. 1969. On the surface coat and flagellar adhesion in trypanosomes. *Journal of Cell Science*, **5**: 163–93.

Vickerman, K. 1974. The ultrastructure of pathogenic flagellates. *Ciba Foundation Symposium*, **20**: 171–98.

Vickerman, K. 1976. The diversity of the kinetoplastid flagellates. In: *Biology of the Kinetoplastida* (ed. W. H. R. Lumsden & D. Evans), pp. 1–34. Academic Press, London.

Vickerman, K. 1985. Developmental cycle and biology of pathogenic trypanosomes. *British Medical Bulletin*, **41**: 105–14.

Vickerman, K. 1986. Clandestine sex in trypanosomes. *Nature*, **322**: 113–14.

Vickerman, K., Myler, P. J. & Stuart, K. O. 1992. African trypanosomiasis. In: *Immunology and Biology of Parasitic Infections* (ed. K. S. Warren), 3rd Edition, pp. 179–212. Blackwell Scientific Publishers, Oxford.

Vickerman, K. & Preston, T. M. 1976. Comparative cell biology of the kinetoplastid flagellates. In: *Biology of the Kinetoplastida* (ed. W. H. R. Lumbsden & D. Evans), pp. 35–130. Academic Press, London.

Vickerman, K. & Tetley, L. 1977. Biology and ultrastructure of trypanosomes in relation to pathogenesis. In: *Pathogenicity of Trypanosomes* (ed. G. Losos & L. Chouinard), pp. 23–31. International Developmental Research Centre, Ottawa.

Vickerman, K., Tetley, L., Hendry, K. A. K. & Turner, C. M. R. 1988. Biology of African trypanosomes in the tsetse fly. *Biology of the Cell*, **643**: 109–19.

Von Brand, Th. 1966. The physiology of *Leishmania. Revista de Biologica Tropical*, **14**: 13–25.

WHO 1984. *The leishmaniases*: Report of a WHO expert Committee. *Technical Report Series* No. 701. 140 pp.

WHO 1986. *Epidemiology and control of African trypanosomiasis*. Report of a WHO expert Committee. *Technical Report Series* No. 739. 127 pp.

WHO 1987. *Tropical Disease Research: a Global Partnership*. Eighth programme report of UNDP/World Bank/WHO Special Programme for research and training in tropical diseases. No. 7. *The Leishmaniases* (ed. J. Maurice & A. M. Pearce). WHO, Geneva.

Zeledón, R. 1974. Epidemiology, modes of transmission and reservoir hosts of Chagas' disease. *Ciba Foundation Symposium*, **20**: 51–85.

6

Sporozoea: gregarines and coccidia

Under the most recently approved classification (Levine *et al.*, 1980; see Appendix, p. 42), those parasitic protozoa originally grouped together under the long-established term 'Sporozoa' have now been assigned to four phyla: Apicomplexa, Microspora, Myxozoa and Ascetospora. A detailed classification of these phyla is given in the Appendix (p. 42).

This chapter is concerned only with the Subclasses Gregarinia and Coccidia of the class Sporozoea of the Phylum Apicomplexa, which include the forms commonly referred to as 'gregarines' and 'coccidia'.

Unlike the groups of Protozoa already studied, the Sporozoea lack organs of locomotion. The majority are intracellular parasites depending for nourishment on soluble cellular materials and are unable to ingest particulate food or bacteria. In vertebrates, they occur mainly in the bloodstream, the reticulo-endothelial system and the mucous lining of the intestine; in invertebrates, they are found chiefly in the digestive or excretory systems. Members of one subclass, the Gregarinia, however, are extracellular parasites, and these either have organs of attachment or can move by a mysterious (p. 93) amoeboid-like movement. For transmission from host to host, sporozoeans depend either on the production of highly resistant spores (which give the group its name) or make use of vectors, usually arthropods.

6.1 **Classification** (based on Levine *et al.*, 1980)

(Note: much of the revised classification is based on characters which can only be seen by electron microscopy.)

Phylum **APICOMPLEXA**

Apical complex (see Fig. 6.9), generally consisting of polar ring(s), rhoptries, micronemes; conoid and subpellicular microtubules present at some stage.

Class 1 **Perkinsea**

Contains only two species of the genus *Perkinsus*, parasites of oysters and abalones. Not considered here.

Class 2 **Sporozoea**

Spores typically present; conoid, if present, forming complete cone; oocysts with one or many sporozoites; nucleus single; cilia and flagella present only in microgametes of some groups; sexuality, when present, syngamy; all species parasitic.

Subclass 1 **Gregarinia.** Mature trophozoites extracellular, large; parasites of gut and body cavities of invertebrates; e.g. *Gregarina polymorpha*.

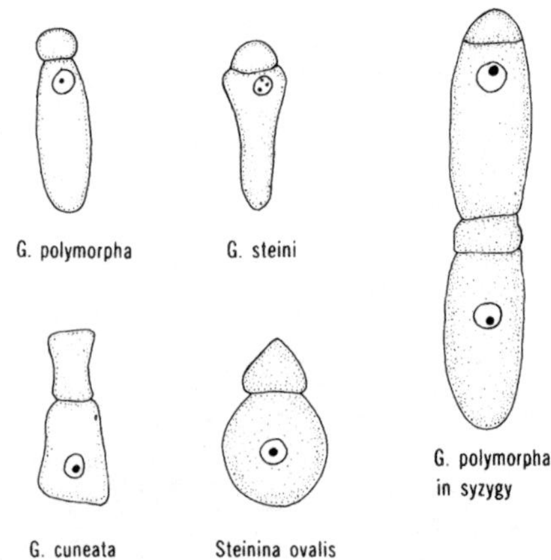

Fig. 6.1
Gregarines from gut of larval *Tenebrio molitor*. (After Devauchelle & Oger, 1968.)

Subclass 2 **Coccidia.** Mature trophozoites small, typically intracellular; syzygy (p. 92) generally absent; most species in vertebrates; e.g. *Eimeria*, *Cryptosporidium*, *Toxoplasma*, *Plasmodium* (Chapter 7).

Subclass 3 **Piroplasmia.** Piriform, round, rod-shaped or amoeboid; asexual and (probably) sexual reproduction; parasitic in erythrocytes and other circulating and fixed cells; e.g. *Babesia*, *Theileria* (Chapter 9).

○○○ 6.2 Subclass 1 Gregarinia

○ 6.2.1 General account

Contains the majority of the so-called 'gregarines', many of which are common parasites of arthropods and annelids. The best-known forms are undoubtedly species of *Monocystis* in the seminal vesicles of the earthworm *Lumbricus terrestris* (often studied in elementary biology classes) and other earthworms, e.g. *Allobophora* spp., found throughout the world. Cox (1968) and Duhlinska (1977) have produced useful, illustrated lists of *Monocystis* species found in Europe; many of these undoubtedly have a cosmopolitan distribution. Another common genus is *Gregarina*, species of which are very common in the gut of various insects.

Other species usually readily obtainable and especially useful for teaching purposes are:

Lankesteria culicis: in mosquitoes of the genus *Aedes*, which occurs in India, South America, Africa and the USA. A prevalence of 95 per cent has been reported for this species (Walsh & Callaway, 1969) and also for a related species, *L. barretti* (Vávra, 1969).

Gregarina foricula: in the earwig *Foricula auriculia*, in which a prevalence of 50–100 per cent has been reported in the UK (Ball *et al.*, 1985).

Gregarina blattarum: in cockroaches (*Blatta orientalis*, *Blatella germanica*, *Periplaneta americana*) throughout the world.

Gregarina blaberae: in the cockroach *Blaberus craniifer*.

Gregarina garnhami: in the locust *Schistocerca gregaria* (see Fig. 6.5).

○ 6.2.2 Morphology: light microscopy

The general morphology of a gregarine at a light microscope level is too well known for comment. In some species, e.g. the common *Monocystis agilis* – one of the *acephaline* gregarines – the body is a single, undivided unit; in others, the cephaline gregarines, it is clearly partitioned into a smaller, anterior *protomerite* and a larger, posterior *deutomerite* containing the nucleus (Fig. 6.5). In some, the protomerite is drawn out into a region specialised for attachment to the host tissue, known as the *hold-fast organelle* or *epimerite* (Fig. 6.2).

The epimerite is clearly an organ of great physiological importance to the organism; in many species (e.g. *G. garnhami*) it is simple (Fig. 6.5) but in others (e.g. *Polyrhabdina spionis*, *Lecythion thalassemae* and *Nina gracilis*) it is expanded, sometimes into filaments, the whole bearing a striking resemblance to a cestode scolex (Fig. 6.2). It is probable that, like the cestode scolex (p. 349), the epimerite has a dual function, serving as an organ of both attachment and absorption, food materials being absorbed from the host cell. It is especially interesting to note that when the *cephalont* (i.e. the three-segmented gregarine) is fully developed (stage 5, Fig. 6.5), the protomerite and the deutomerite develop projections on their outer covering reminiscent of the microvilli of vertebrate intestinal cells or the microtriches of the cestode tegument (Harry, 1965). It seems reasonable to conclude that these projections are concerned with absorption and that the organism takes in nutrients from the host cell via the placenta-like epimerite and from the gut contents via the deutomerite.

○ 6.2.3 Morphology: ultrastructure

Early work on the ultrastructure has been reviewed by Kreier (1977), Pitelka (1963) and Smyth (1976). The general structure as revealed in EM studies is shown in Fig. 6.3. These early studies revealed that the outer covering of the 'adult' cephalont represented a highly differentiated cell surface showing large longitudinal folds with subpellicular microtubules, as in species of Selendiidae (Schrével, 1971); numerous narrow longitudinal folds, as in Eugregarina (Vivier *et al.*, 1970); or hair-like structures, as in *Diplauxis* (Vivier & Petitprez, 1968). A remarkable and unique feature of all these cell surfaces is the presence of two cytomembranes, which lie under the plasma membranes (Fig. 6.4), resulting in what is known as a *three-cortical-membrane* pattern. This unusual membrane has been extensively examined by conventional electron microscopy and cytochemical and biochemical techniques in a number of other coccidia; representative species studied include *Aggregata eberthi* (Porchet *et al.*, 1982) and *Gregarina blaberae* (Philippe & Schrével, 1982; Schrével *et al.*, 1983).

The cell surface of *Gregarina blaberae* has been further examined by the use of freeze–fracture techniques

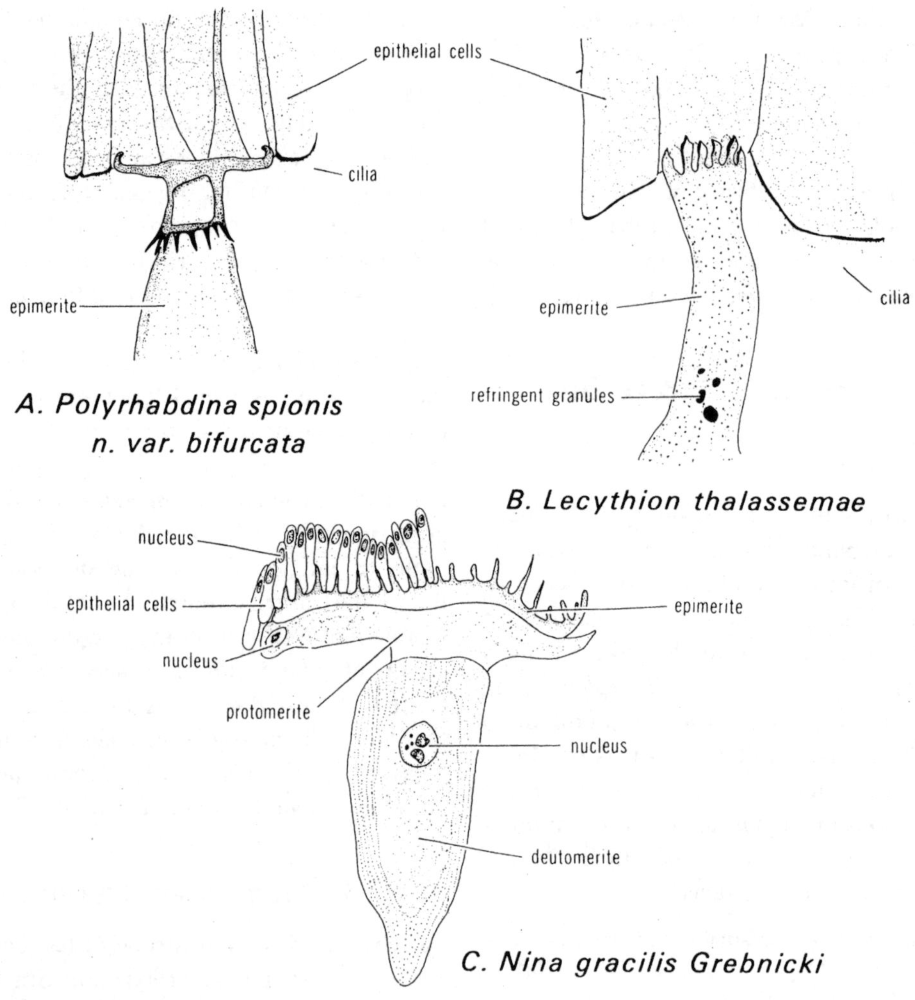

Fig. 6.2
Method of attachment of the epimerite of A. *Polyrhabdina spionis* from *Scolelepis fuliginosa*; B. *Lecythion thalassemae* from *Thalassema neptuni*; C. *Nina gracilis* from *Scolopendra subspinipes*. (After MacKinnon & Ray, 1931 and Goodrich, 1938.)

(Schrével *et al.*, 1983), which have confirmed the presence of two cytomembranes lying uniformly under the plasma membrane. From an examination of the density and distribution of the intramembraneous particles, it was concluded that the two cytomembranes were in topological continuity, forming either side of a flattened vesicle or cisterna. The cell surface exhibited numerous longitudinal folds with three types of cortical-associated structures, namely 12 nm filaments, an internal lamina and structures described as 'ribbed dense structures' (Fig. 6.4).

These cortical membrane-associated structures are also characteristic of many other species which possess longitudinal folds; the latter are only found in gregarines that exhibit gliding movement and are undoubtedly involved in some way in this unusual form of locomotion. This phenomenon is discussed further on p. 93.

○ 6.2.4 Life cycle: *Gregarina garnhami*

The life cycles of gregarines present problems of special interest, for many are correlated with both the life cycle and the external environment of the host. This has been demonstrated especially in *G. garnhami*, parasitic in the midintestine and caecae of the desert locust *Schistocerca gregaria* (Canning, 1956). After ingestion, hatching of the spores takes place and the released sporozoites penetrate the host cell (Stage 1, Fig. 6.5). The intracellular region grows rapidly at first; when it reaches a length of about 13–16 μm, the growth of the anterior

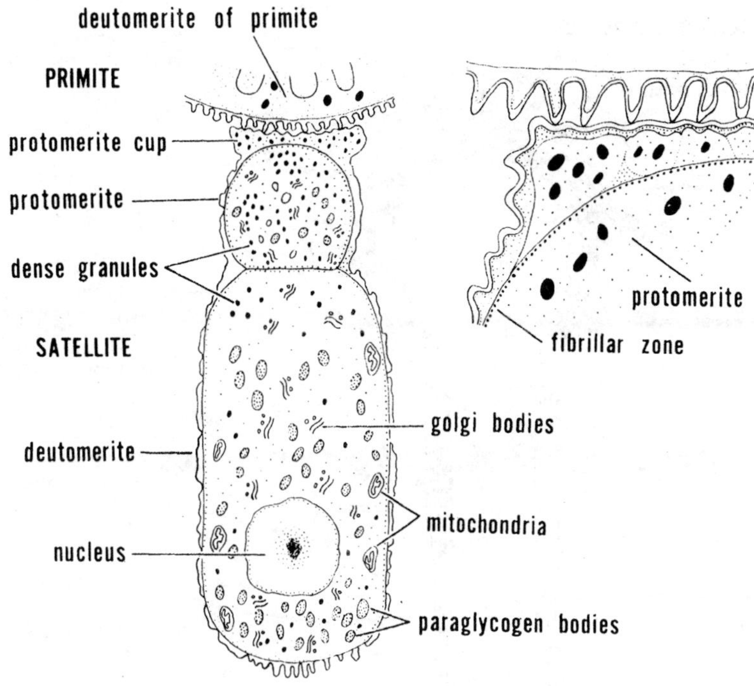

Fig. 6.3
The ultrastructure of *Gregarina polymorpha* from *Tenebrio molitor*, during syzygy. The inset gives in detail the junction between primite and satellite. (After Devauchelle, 1968.)

region slows down and that of the posterior region increases, the organism eventually becoming divided into the *protomerite* and the *deutomerite* (Harry, 1965).

When the gregarines become separated from the host cell, they are able to form a sexual association termed *syzygy* (see p. 92), in which cephalonts form a permanent end-to-end association; they often move around in this joined condition. The trophozoites then encyst and form a thick gelatinous cyst wall around themselves, the whole forming an *association cyst* (Fig. 6.5). In each organism, the protomerite and deutomerite merge and become *gametocytes*, which bud off gametes. Each gamete conjugates with a gamete from the other organism and zygotes are formed. A zygote forms a resistant spore coat and its contents divide to form sporozoites within a *sporocyst*. These are shed in the host's faeces. When placed in a humid chamber, spore tubes (sporoducts) develop within 48 hours (Fig. 6.5) and spores are released; this is clearly, a mechanism which assures that spores will be released only in a moist environment.

G. garnhami completes its life cycle – which is also beautifully synchronised with that of its host – in 11–13 days at 25 °C (Corbel, 1964). After an initial infection, a few cysts appear at 9 days but the bulk appear at the end of the fifth larval stage, i.e. in the 4 days preceding the imaginal moult (day 36–40). This sudden appearance of the cysts is related to the fact that, during this period of development, the gut epithelium is shed and the gregarines are released into the body cavity, where syzygy can take place, with subsequent sporocyst formation.

The synchronisation of host and gregarine life cycles is further demonstrated by the fact that if moulting of the host is retarded by removal of the ventral gland, no massive emission of cysts takes place until moulting occurs (Corbel, 1964). If the final moult is suppressed completely, no formation of cysts occurs.

Host–parasite synchronisation. There are numerous other examples of the sexual process in gregarines being synchronised to the metamorphosis and/or sexual development of the host. In polychaetes, some representative examples of this phenomenon are *Gonospora varia* in *Audouinia* (*Cirratulus*) *tentaculata* and *Gonospora arenicolae* in *Arenicola ecaudata*; other examples are reviewed by Malavasi *et al.* (1976). In insects, an interesting

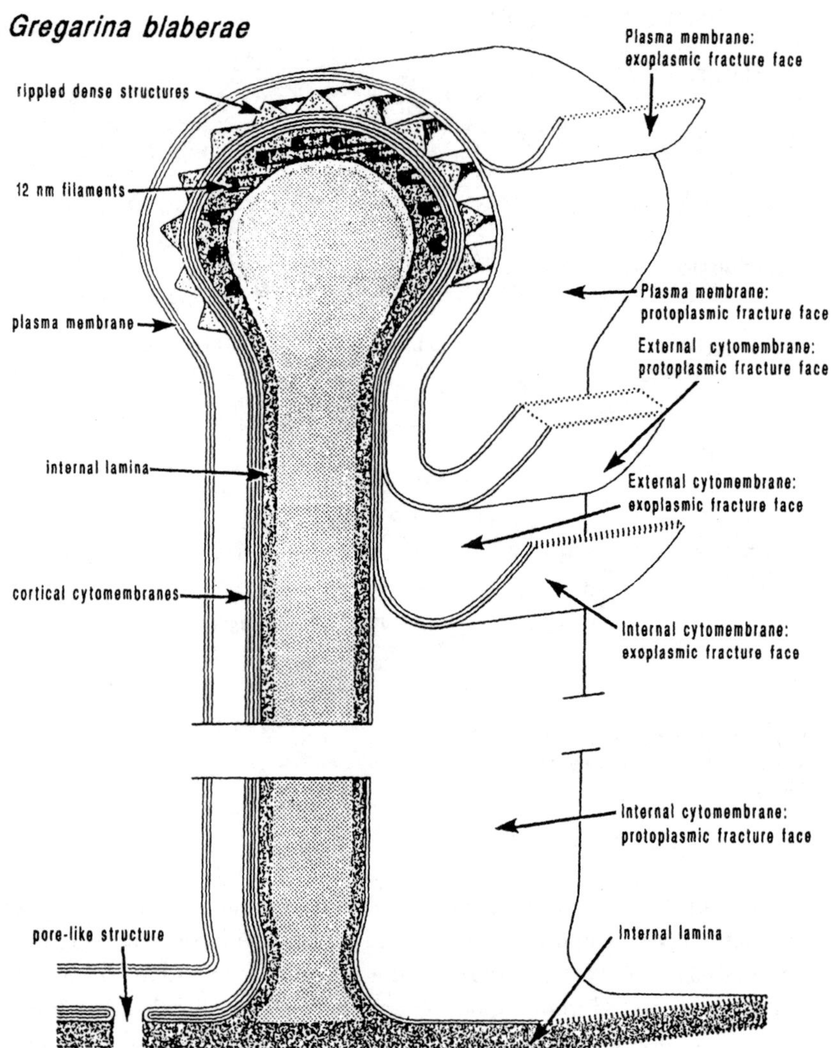

Fig. 6.4
Gregarina blaberae: diagram of the structure of a surface fold as revealed by freeze–fracture. (After Schrével *et al.*, 1983. Reproduced with permission of the Company of Biologists from the *Journal of Cell Science*, **61**: 619, 1983.)

example is *Schneideria schneiderae* in the intestinal caeca of the larva of the fly *Trichosia pubescens* (Malavasi *et al.*, 1976). The life cycles of the two organisms are synchronised so that the protozoa are at the sporozoite stage when the adult flies hatch. The adult flies disseminate the sporozoites, which are eaten by the larvae of the next generations. That the host's hormones are involved in the host–parasite synchronisation process was demonstrated by the fact that injections of ecdysterone caused the trophozoites to leave the host cells and begin sexual pairing (syzygy) soon after. Ligaturing the neurosecretory centres of the fly confirmed the influence of the host's hormone balance. Thus, in a normal pupa, the frequency of paired trophozoites in the intestine was 96.4 per cent, whereas in ligatured controls it was only 10 per cent (Malavasi *et al.*, 1976). It is not known whether the effect of the hormone was directly on the gregarine – causing the trophozoite to leave the host – or indirectly, causing the host cell to expel the gregarine.

The phenomenon has also been studied in the cockroach *Blaberus craniifer* (Tronchin *et al.*, 1986). The production of spores is markedly dependent on the temperature at which the host is maintained, with an optimum of 25 °C (Fig. 6.6). However, in this case, a direct control of the sexual phase of the gregarine by ecdysone could not be established.

It should be noted that the dependence of the differ-

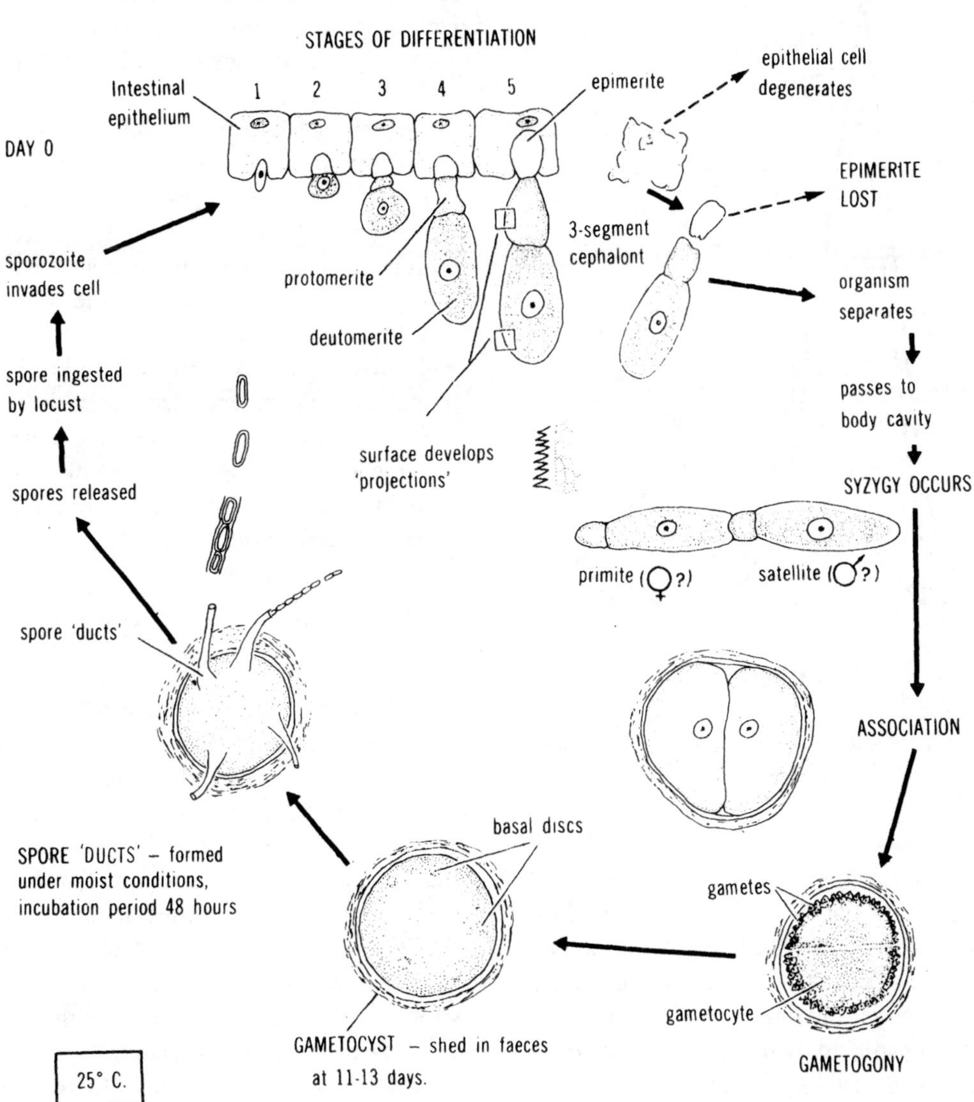

Fig. 6.5
Life cycle of the gregarine *Gregarina garnhami* in the desert locust, *Schistocerca gregaria*; the use of the symbols ♀ and ♂ are notional only; 'spores' are sometimes referred to as 'oocysts' or 'sporocysts'. (Partly based on Harry, 1965.)

entiation of the sexual phase of protozoan parasites on the hormone balance of the host is well established in other protozoan groups, for example in the ciliates, *Nyctotherus* (p. 152) and *Opalina* (p. 146) and in the flagellates of the woodroach, *Cryptocercus punctulatus* (Cleveland *et al.*, 1960).

Gregarine movement. The trophozoites of gregarines exhibit an eerie gliding movement, the basis of which has long puzzled cell biologists. Various theories have been put forward (reviewed by King, 1988; Mackenzie, 1980; Schrével *et al.*, 1983). The reported gliding rate for *Gregarina* trophozoites is 1–10 μm s^{-1} (King, 1988). A similar movement has been described for the trophozoites of *Plasmodium* and *Eimeria*. Since the movement is accompanied by a visible mucus trail, early workers believed that a kind of mucus jet-propulsion operated, but it is now thought that the mucus probably acts as a lubricant. The characteristic folds, described above (Fig. 6.4), and involving contractile proteins or antagonistic systems, were also proposed as being implicated. Although early workers were unable

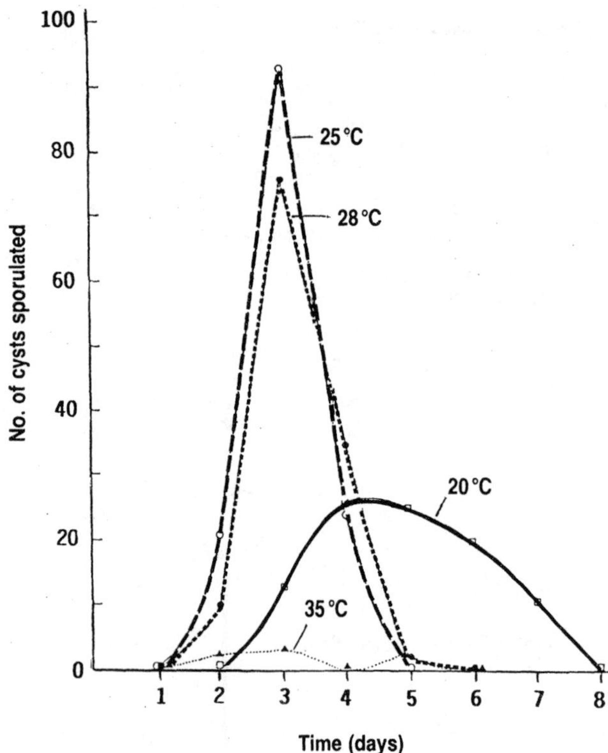

Fig. 6.6
Effect of temperature on the sporulation of sporocysts in the gregarine *Gregarina blaberae* (from a cockroach, *Blaberus craniifer*). (After Tronchin *et al.*, 1986.)

to detect protein with a similar electrophoretic mobility to that of muscle actin (Mackenzie, 1980), a later hypothesis proposes a complex model involving a 'linear motor' powered by actin–myosin interactions, the gliding movement being studied by the movement of beads over the surface (King, 1988). The proposed system is shown in Fig. 6.7 and the mechanism elaborated in the legend.

○○○ 6.3 Subclass 2 Coccidia

Coccidians attack a wide range of invertebrate and vertebrate hosts. The majority are parasites of the alimentary tract and its associated glands. Asexual reproduction takes place by merogony (schizogony) and sexual reproduction by anisogamy with the production of highly resistant spores. Many species are pathogenic. Knowledge of the Coccidia has been reviewed by Hammond & Long (1973), Kreier (1977) and Levine (1988).

Order 1. Agamococcidiida. Rare, not dealt with here.
Order 2. Protococcidiida. Rare, not dealt with here.
Order 3. Eucoccidiida. Merogony (= schizogony) present; asexual and sexual phases in life cycle.

Suborder 1. Adeleina. Gametes associated in syzygy; few microgametes; sporozoites enclosed in envelope. E.g. *Adelea, Haemogregarina, Sarcocystis.*

Suborder 2. Eimeriina. Gametes develop independently; many microgametes; sporozoites enclosed in sporocyst; zygote non-motile. E.g. *Eimeria, Isospora, Toxoplasma.*

Suborder 3. Haemosporina (Chapters 7 and 8). Gametes develop independently; moderate number of microgametes; sporozoites naked; zygote motile in some forms; merogony in vertebrate and sporogony in invertebrate host. E.g. *Plasmodium, Haemoproteus, Leucocytozoon.*

○ 6.3.1 Life cycle of a typical coccidian (Fig. 6.8)

A typical life cycle consists essentially of several asexual generations (merogony) followed by a sexual generation, generally ending in the development of an encysted stage or *oocyst,* which is passed in the faeces.

In a typical cycle (Fig. 6.8), a sporozoite enters an epithelial cell and becomes rounded to form a trophozoite. This gradually increases in size and undergoes merogony in the course of which the nucleus of the organism (now *meront* = *schizont*) divides by rapid succession of mitoses, so rapid in fact that a new spindle may be formed before preceding ones are completed, thus creating an impression of multiple nuclear division. When merogony is completed, *merozoites* are released; these enter fresh cells and the cycle is repeated. The asexual cycle may continue for several generations, but eventually, under the influence of a stimulus, possibly associated with antibody formation by the host, some of the merozoites become differentiated into male and female forms which initiate the sexual cycle of development.

The sexual forms of the merozoites enter cells, become rounded, and differentiate into microgametocytes and macrogametocytes. The chromosomes of Coccidia have not been much studied; in *Eimeria tenella,* according to Canning & Anwar (1968), the diploid number is ten. These authors also give an account of meiosis in this species. The microgametocytes grow,

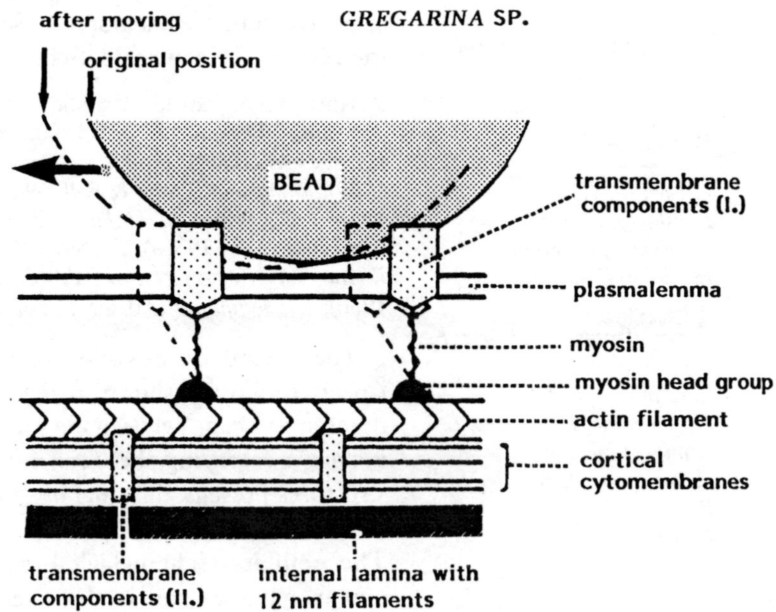

Fig. 6.7
Model of the 'gliding movement' in gregarines, as proposed by King (1988). It utilises the linear motion of polystyrene beads over the surface powered by actin–myosin interactions. Transmembrane components (I.) span the plasma membrane and act as the link between the external ligand (e.g. a bead) and the cytoplasm where interactions with myosin molecules might occur. The model proposes that the myosin head group(s) containing the ATP binding site and the actin binding site make contact with actin filaments hypothesised to be associated with the outer face of the cortical cytomembranes. It is possible that a second set of transmembrane components (II.) link the actin filaments to the internal (12 nm) filaments, thereby providing a more 'robust' trackway. In the muscle crossbridge cycle, it is thought that the power stroke is developed by the myosin head undergoing a conformational charge on release of bound ADP. The above model shows (dotted line) the transmembrane components (I.) and the bead moving to the left. On addition of ATP the myosin head would be released from the actin filament and another cycle would begin. (Modified from King, 1988.) See also Fig. 6.4.

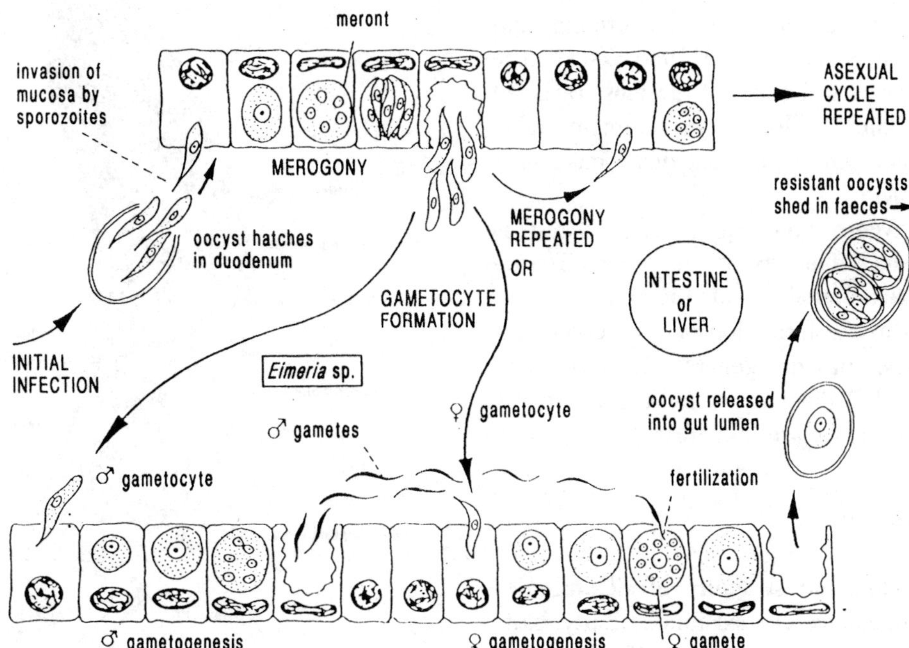

Fig. 6.8
Life cycle of a coccidian, based on *Eimeria* sp. (Based partly on Hoare, 1949.)

Table 6.1. *Some common species of* Eimeria *in mammals and birds*

Host	*Eimeria* species
Cattle, zebu, water-buffalo	*E. bovis, E. zurnii*
Sheep, goats	*E. ashsata, E. arloingi*
Pigs	*E. debliecki, E. scabra*
Rabbits	*E. stiedai, E. irresidua, E. magna*
Mouse	*E. vermiformis, E. contorta, E. falciformis*
Dog	*E. canis*
Chickens	*E. acervulina, E. maxima, E. mitis, E. tenella*
Turkeys	*E. adenoeides, E. meleagrimitis*
Ducks	*E. bucephalae*
Geese	*E. truncata*

undergo nuclear division and produce a number of comma-like microgametes which escape into the gut lumen. Meanwhile, each macrogamete has increased in size, and its nucleus has undergone reduction division. The cytoplasm of the macrogamete is characterised by possessing refractive globules around its edges. A microgamete penetrates a macrogamete and forms a zygote around which is developed a highly resistant spore case produced by the fusion of the refractive granules. Part of this spore case may contain a 'tanned protein' and chitin (Horton-Smith & Long, 1963) but its exact nature is unknown; it appears to be one of the most resistant organic materials known. Oocysts may remain viable even after treatment with 5 per cent dichromate or 1 per cent chromic acid.

The encysted zygote, now known as an *oocyst*, is passed with the faces. Oocysts of the genera *Eimeria* and *Isospora* require a period outside the body to *sporulate*, which results in the formation of sporocysts containing sporozoites. In the genera *Sarcocystis* and *Cryptosporidium* (p. 98) the oocysts are already sporulated when laid and hence are immediately infective.

○ 6.3.2 Morphology

Ultrastructure. The ultrastructure of all stages of coccidia has been extensively examined. Early basic studies (pre-1970) have been reviewed in Pitelka (1963) and in the second edition of this text (Smyth, 1976). Later work has been summarised in the comprehensively illustrated monograph of Scholtyseck (1979) and the review of Chobotar & Scholtyseck (1982).

More recent individual studies include those on: *Eimeria acervulina* (Sénaud *et al.*, 1980; Pittilo & Ball, 1984); *E. bakuensis* (Ball & Pittilo, 1988); *E. bateri* (Martínez *et al.*, 1985); *E. maxima* (Pittilo & Ball, 1985); *E. nieschulzi* (Dubremetz *et al.*, 1989); *E. stiedai* (Ball *et al.*, 1988); *E. stigmosa* (Gajadhar *et al.*, 1986*a*); *E. tenella* (Beesley & Latter, 1982); *E. truncata* (Gajadhar *et al.*, 1986*b*; Gajadhar & Stockdale, 1986); *E. vermiformis* (Adams & Todd, 1983).

The general features of the fine structure of a coccidian merozoite are shown in Fig. 6.9, but some minor differences occur between the various genera. Very little is known regarding the exact function of most of the structures present, the chief of which are the following.

The pellicle. The pellicle basically consists of three unit membranes and is characteristic of all coccidian sporozoites and merozoites so far studied.

Polar rings. Polar rings appear as osmiophilic thickenings of the inner pellicular complex and serve to anchor the microtubules. Two polar rings are present in most coccidia.

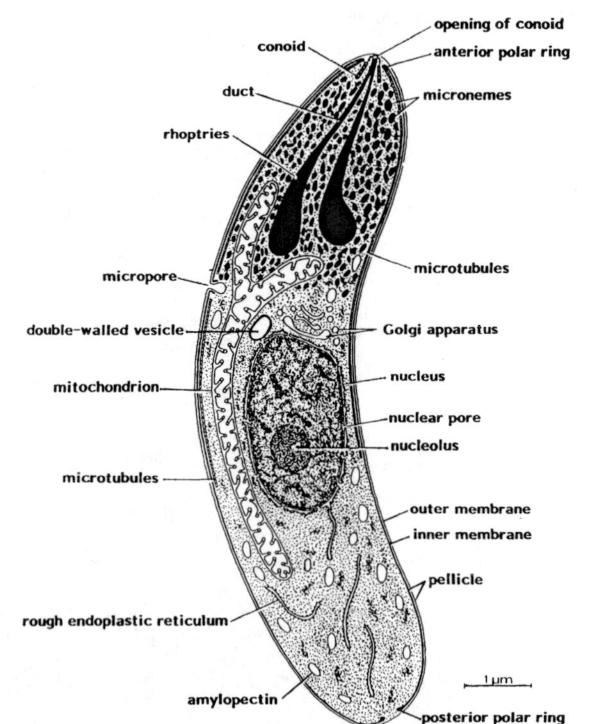

Fig. 6.9
Coccidian merozoite: diagrammatic representation of the ultrastructure as seen in longitudinal section. (After Scholtyseck, 1979.)

Microtubules. In sporozoites and merozoites, microtubules make up the conoid and occur as subpellicular organelles. In the cyst-forming coccidia (*Frenkelia*, *Besnoita*, *Sarcocystis*, *Toxoplasma* and *Hammondia*; see later), 22 microtubules are consistently present; in *Eimeria* there are usually 24–32.

Conoid. The conoid occurs in all coccidia and gregarines but not in haemosporidians and piroplasmas. It consists of a truncated hollow cone of spirally arranged fibrillar structures. Its function is unknown but it is thought that it probably functions as an organelle of penetration of host cells; it may be capable of protrusion.

Rhoptries. These are electron-dense, osmiophilic structures located at the anterior region of merozoites and sporozoites. Although the name rhoptries (singular *rhoptry*) is now in regular use, names such as 'taxonemes', 'club-shaped organelles' or 'paired organelles' were used in the early literature. The number of rhoptries varies from 2 to more than 20. Their function is unknown but they may secrete a proteolytic enzyme to assist in the (presumed) mechanical function of the conoid (Scholtyseck, 1979).

Micronemes. These are small, very dense, convoluted, osmiophilic structures located in the anterior region. Their function is unknown.

The micropore. This is an organelle formed by an evagination of the pellicle. In transverse section it consists of two concentric rings, which correspond to two lineal elements of the micropore wall as seen in longitudinal section. Up to nine micropores have been reported in the microgametes of *Eimeria auburnensis*. Although not proven, accumulating evidence suggests that the micropore functions as a cytostome with the formation of food vacuoles (Chobotar & Scholtyseck, 1982).

Other organelles. There is a single *mitochondrion*, which is often branched or Y-shaped. A *Golgi apparatus* is represented by lamellar and vesicular structures anterior to the nucleus. *Rough endoplastic reticulum* is present as parallel lines of ribosomes. *Vesicles* of various sizes and contents are present in the cytoplasm. Numerous ovoid bodies containing *amylopectin* are widely distributed in the lower half of the organism.

○ 6.3.3 Mammalian and avian coccidia

Suborder *Eimeriina*

The best-known genera are: *Eimeria*, *Isospora*, *Cryptosporidium*, *Toxoplasma*, *Sarcocystis* and *Besnoita*. These are dealt with briefly below.

Genus Eimeria. Oocysts with four sporocysts, each with two sporozoites. Species of *Eimeria* follow the 'typical' life cycle described in Fig. 6.8; they are parasites of both invertebrates and vertebrates, especially birds and mammals. There are a very large number of species, many of which are of veterinary importance; the best-known are listed in Table 6.1. An early catalogue of *Eimeria* species is that of Pellérdy (1963–8); more recent accounts of speciation are those in Kreier (1977), Levine & Ivans (1986) and Levine (1988).

E. stiedae (rabbit) and *E. tenella* (chick) are likely to provide the most readily available material for study. Both are pathogenic and these and other species cause great economic loss to breeders of rabbits and poultry; but most other domestic animals are also subject to coccidiosis (Table 6.1). The problems of poultry coccidiosis and its control have been reviewed by Long & Jeffers (1986).

Genus Isospora. The status of this genus is uncertain, although, until recently, all coccidian protozoa with sporulated oocysts containing two sporocysts, each with four sporozoites, were assigned to the genus *Isospora* and were assumed to have a *monoxenous* (i.e. a one-host) life cycle. However, it was later found that some species (compare *Toxoplasma*, p. 99) had a *heteroxenous* life cycle (i.e. with an asexual stage in one host and a sexual stage in another). For example, it was found that *Isospora bigemina* was really a synonym for *Sarcocystis miescheriana* (see p. 104). Similarly, *Isospora hominis*, with sexual stages in man, was found to be only a stage of *Sarcocystis cruzi* with cysts in the muscle of bovines (Kreier & Baker, 1987). Numerous other species of *Isospora* have also proved to be species of *Sarcocystis* and the whole taxonomic situation is very confused. Levine (1987) gives a valuable review of the problems involved. *Isospora belli*, in the intestine of man, appears to be a 'true' *Isospora* species and can cause severe disease with fever, persistent diarrhoea and occasionally death, especially in AIDS patients (Forthal & Guest, 1984). A species in animals, *I. suis*, characteristically found in the gut of

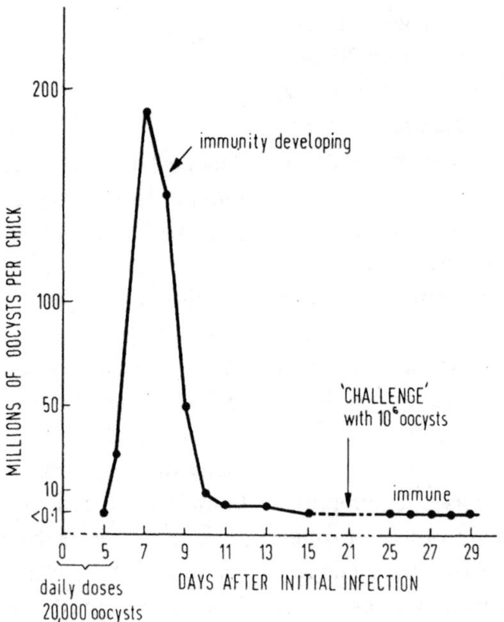

Fig. 6.10
Development of immunity by chickens against *Eimeria acervulina*. Initial parasitaemia was produced by daily doses of 20 000 oocysts and the immune reaction became apparent on the eighth day. A 'challenge' of 10^6 oocysts on the twenty-first day demonstrated the effectiveness of the immunity. (After Cuckler & Malanga, 1956.)

suckling pigs, is also markedly pathogenic; it causes villous atrophy and necrotic enteritis.

Immunity. There is an extensive literature on immunity to coccidiosis, especially in poultry and turkeys; this is discussed in Chapter 32. Chickens may become partly or wholly resistant to coccidial infections (Fig. 6.10).

Genus Cryptosporidium. Oocysts *without sporocysts*, but with four naked sporozoites. The organisms are only 2–6 μm in diameter and – in contrast to the 'typical' coccidian (Fig. 6.8) – live not within the host cell proper but just under the surface membrane or in the brush border (Fig. 6.11). *Cryptosporidium* spp. have long been known to infect domestic animals, especially cattle, resulting in *cryptosporidiosis*, the chief clinical condition of which is diarrhoea. Since about 1975, however, it has been identified as an opportunistic parasite of man – especially those with the acquired immunodeficiency disease AIDS – and it is now recognised as a serious health problem throughout the world. Among AIDS patients the disease has been reported to have a prevalence of 3–4 per cent in the USA and over 50 per cent in Africa and Haiti (Smith & Rose, 1990).

Succinct accounts of the life cycle, epidemiology and pathogenesis of cryptosporidiosis are those of Canning (1990), Cook (1987), Fayer & Ungar (1986), Smith & Rose (1990) and Tzipori (1985). A comprehensive bibliography covering all aspects of the disease is also available (Cook, 1987).

Some authors (Smith & Rose, 1990) recognise only four species of *Cryptosporidium*, i.e. *C. parvum* and *C. muris* in mammals, and *C. baileyi* and *C. melagridis* which infect birds. In contrast, Meyer (1988) lists 20 named species, but the speciation is constantly under review; *C. parvum* appears to be the only species infecting man.

Life cycle. As Fig. 6.11 shows, the life cycle of *Cryptosporidium* spp. closely follows that of the 'typical' coccidean shown in Fig. 6.8, except in the following respects: (*a*) the oocyst is without sporocysts and contains only four naked sporozoites; (*b*) the oocyst is sporulated when excreted (contrast *Eimeria* and *Isospora*) and is therefore immediately infective; and (*c*) when the organisms invade the intestinal epithelium of the host, they develop within the cell membrane and are said to be 'intracellular but extracytoplasmic' (Canning, 1990).

Only some 80 per cent of oocysts are thick-walled and resistant, and these are responsible for transmission between hosts. The remaining 20 per cent or so are

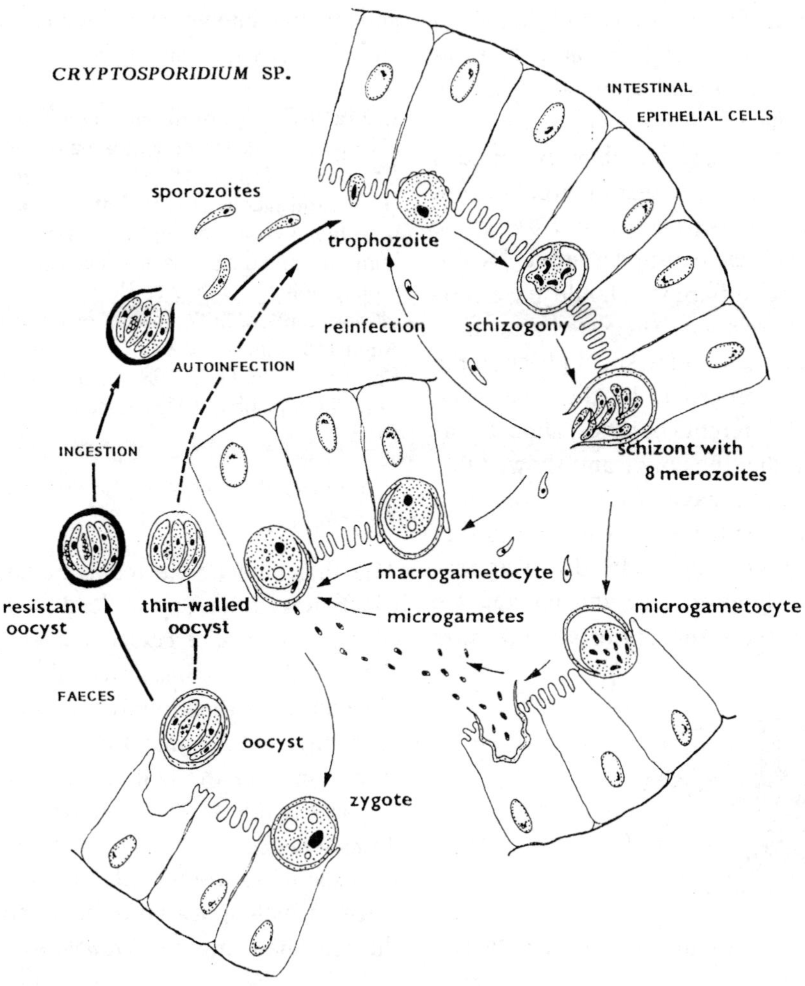

Fig. 6.11
Life cycle of *Cryptosporidium* sp. Note that in contrast to the 'typical' coccidium life cycle shown in Fig. 6.8: (a) this organism only invades the cell membrane of the intestinal cells and not the cytoplasm; (b) the oocysts are sporulated when laid and contain four sporozoites (without sporocysts) and hence are immediately infective; (c) 'thin-walled' oocysts are capable of hatching in the original host, thus causing autoinfection. (Modified from Smith & Rose, 1990.)

thin-walled, with a single membrane; these hatch *in the same host* and are responsible for *autoinfection*. Oocysts from humans are also infective to domestic animals, such as cattle or sheep, which thus act as important reservoirs of human infection. Infected calves and lambs are reported as excreting up to 10^{10} oocysts daily (Smith & Rose, 1990). The oocysts are remarkably insensitive to chlorination and as a result much human cryptosporidiosis must be regarded as being waterborne via drinking water. The deficiencies in our present water treatment systems in controlling the disease are reviewed in detail by Smith & Rose (1990).

○○○ 6.4 Tissue cyst-forming coccidia

○ 6.4.1 *Toxoplasma gondii*

This remarkable species was known as a potential parasite of man for many years but its true nature as a coccidian was only discovered some 20 years ago (1969–70). It was known to occur in all warm-blooded animals – mammals and birds – but its nature and means of transmission remained a mystery.

The two stages known were the *trophozoite* (Figs 6.12, 6.13) and the *cyst*, also known as a pseudocyst (see below). The trophozoites were intracellular parasites which could invade almost any nucleated cell within the host. The cysts had well-formed walls and contained organisms which multiplied within the cysts, as many as 1000–3000 organisms being found within a single cyst. The trophozoite and the cystic form appeared to be almost identical in morphology, but they exhibited a striking physiological difference in that the trophozoites were rapidly killed in acid pepsin whereas the cystic forms could survive 1 h of acid pepsin digestion (Jacobs, 1967). This physiological difference should have provided a clue to the life cycle as it suggested that there was a further development of an intestinal phase in the life cycle. This, in fact, proved to be the case when workers in a number of different countries, almost simultaneously, pieced together the puzzle and showed that *T. gondii* developed a *sexual* stage in the intestine of a cat and was, in fact, a coccidian closely related to (but not identical to) an *Isospora* species. This discovery revolutionised this area of protozoology and opened the door to the solution of the taxonomy of several other genera now known as *Sarcocystis*, *Atoxoplasma*, *Besnoita* and *Frankelia* (see later).

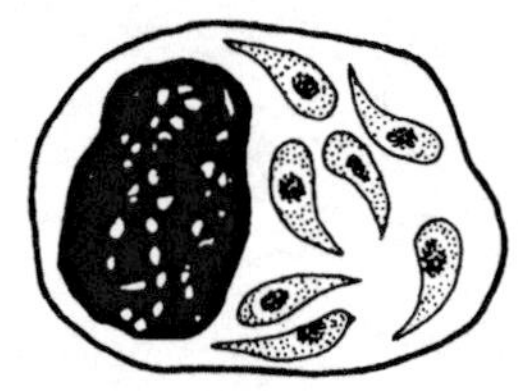

Fig. 6.12
Toxoplasma gondii, trophozoites in macrophage. (After Hoare, 1949.)

Literature. A vast literature on *Toxoplasma gondii* and toxoplasmosis exists, running into many thousands of publications. Early work (up to 1973–5) has been reviewed by Dubey (1977) and Jacobs (1973). Later work, covering all aspects of the biology, ultrastructure, epidemiology, pathogenesis, immunity and diagnosis has been reviewed in a comprehensive monograph by Dubey & Beattie (1988). Selective examples of more limited, but valuable, reviews are those of: Apt, 1985; August & Chase, 1987; Chowdhury, 1986; Dubey, 1986; Frenkel, 1988; Hall, 1986; Holliman, 1988; Hughes, 1985; Ito, 1985; Jacobs, 1973; Lysenko & Asatova, 1986; Wang, 1982.

Life cycle: discovery of the coccidian nature of *Toxoplasma*

The first clue to the main method of transmission – which was ultimately to lead to the discovery of the coccidian nature of *Toxoplasma* – came when Hutchison (1965) fed *Toxoplasma*-infected mice to cats, the faeces of which were then found to be infective to mice. It was originally thought that the protozoan was transmitted via the eggs of the nematode *Toxocara cati*, but shortly afterwards this worker and several others showed that *Toxoplasma* could be transmitted by the faeces of *worm-free* cats; these workers identified a *new cystic form* in the faeces, which appeared to be an oocyst, thus indicating the coccidian nature of *Toxoplasma* (Work & Hutchison,

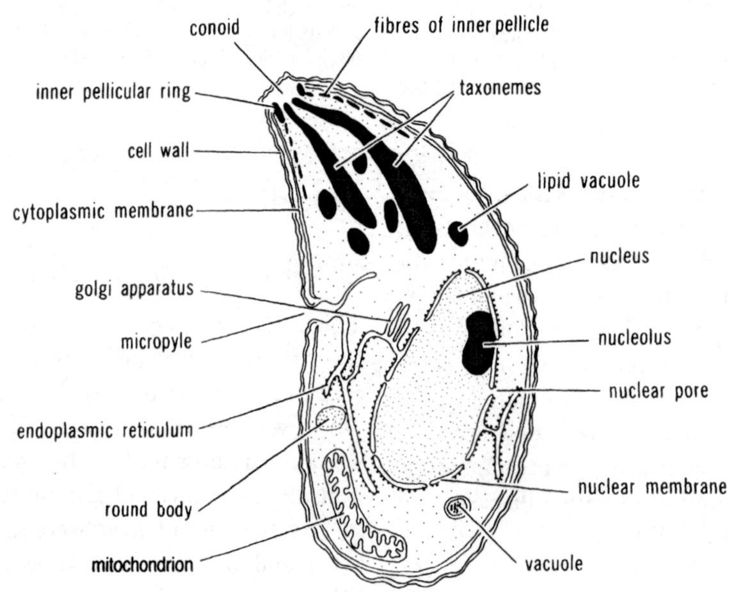

Fig. 6.13
Toxoplasma gondii ultrastructure. Original, but based on the photographs of Sheffield & Melton, 1968.)

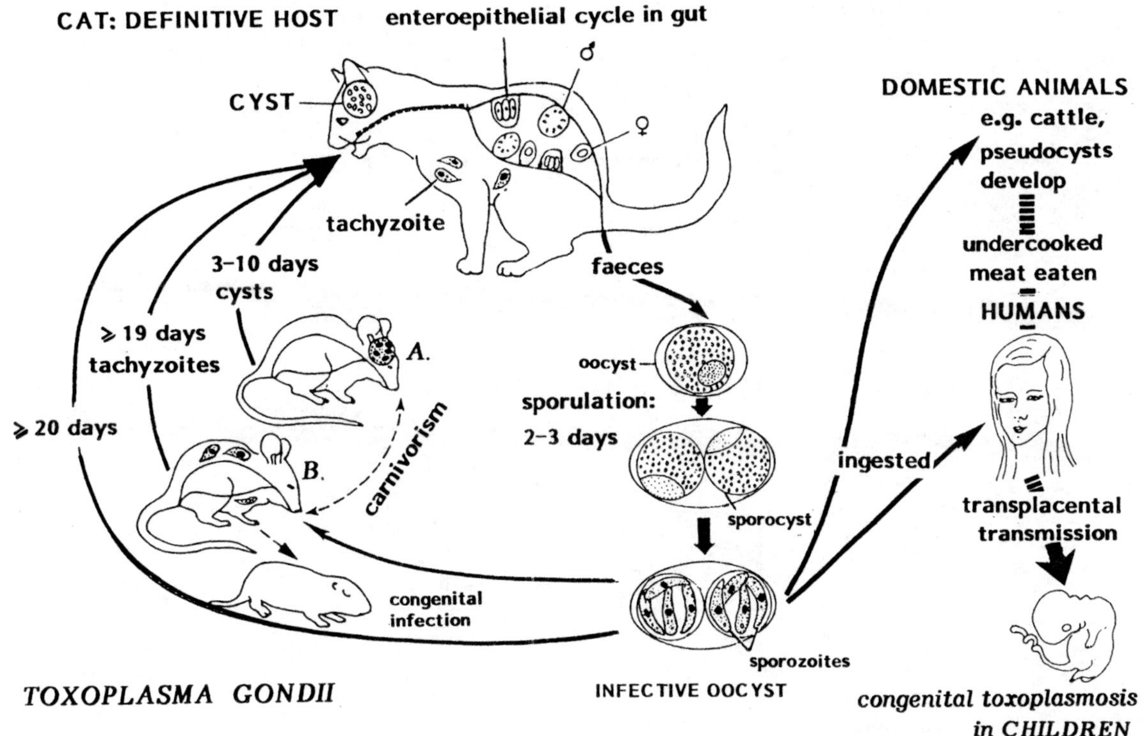

Fig. 6.14
Life cycle of *Toxoplasma gondii*. Ingestion of sporulated oocysts of a mouse with cysts (Mouse **A**) leads to the development of the enteroepithelial (sexual) cycle in the cat intestine. The pre-patent period up to the shedding of the oocysts varies with the stage of *T. gondii* ingested: only 3–10 days after ingesting cysts (Mouse **A**), but about 19–20 days or longer after direct infection with oocysts or ingestion of a mouse (Mouse **B**) containing only tachyzoites. Women most at risk of delivering an infected infant are those who acquire the infection just prior to gestation. (Modified from Dubey, 1977 and Smyth, 1976.)

1969*a*,*b*; Frenkel, 1970; Frenkel *et al.*, 1969; Sheffield & Melton, 1968; Overdule, 1970). On feeding these oocysts to cats, sexual stages were found in the epithelial cells of the cat intestine, thus indicating that *Toxoplasma gondii* was an *Isospora*-like coccidian which had merogonic (schizogonic) and gametogonic stages (Fig. 6.14) only in cats, but which had a multiplicative, vegetative state in many other paratenic hosts. The means of transmission became clear: cats became infected by ingesting oocysts or cysts in tissues of paratenic hosts such as mice, or transplacentally. Man could clearly become infected by ingesting oocysts or consuming the cystic forms in undercooked meat, or transplacentally.

Intestinal development in cats. After the ingestion of the infective stages (oocysts or cysts, see Fig. 6.14) the intestinal epithelium is invaded and there follows several series of *asexual generations* during which the parasites divide by an unusual process known as *endodyogeny*.

Endodyogeny was first described by Goldman *et al.* (1958), and is essentially a budding process in which two daughter cells are formed inside the original ('mother') cell, which is then consumed by the developing daughter cells. This process is shown diagrammatically in Fig. 6.15. The first sign of endodyogeny is the appearance of two areas of a 'new' internal membrane. This represents the external membranes of the new daughter cells. Each of these membranes grows within the mother cell, taking a piece of nucleus and various cell organelles. A process of *endopolygeny* (division into several organisms simultaneously by internal budding) and *ectomerogeny* (similar to endopolygeny except that the primordia for the individuals arise at the surface rather than internally) can also occur (Canning, 1990).

As in the 'typical' coccidian sexual cycle (Fig. 6.8), asexual multiplication is followed by macro- and microgametogenesis and fertilisation takes place, the zygotes developing into a thick-walled oocyst. The latter is not infective when laid and requires 2–3 days' incubation – under appropriate environmental conditions – to sporulate, resulting in an infective oocyst containing two sporocysts each with four sporozoites.

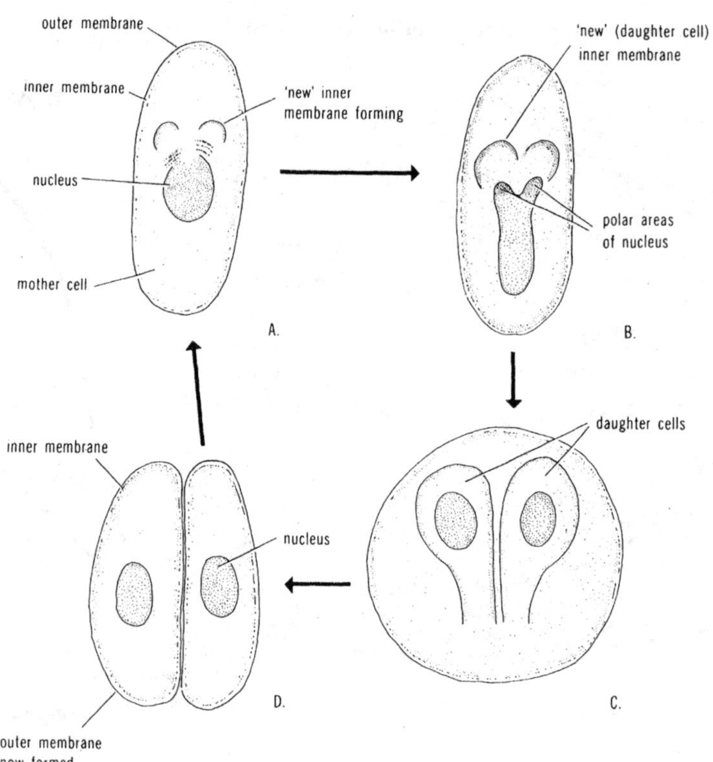

Fig. 6.15
Division of *Toxoplasma gondii* by *endodyogeny*, an unusual process in which two daughter cells are produced inside a mother cell which is then consumed by the offspring prior to their separation. Very diagrammatic; other cytological organelles omitted. (Original but based on electron micrographs of Sheffield & Melton, 1968.)

Extra-intestinal development in cats. Some of the organisms formed during the asexual intestinal development pass into other tissues and organs. The mesenteric lymph nodes are infected first, followed by the liver, lungs and other tissues. By the sixth day the trophozoites – now known as *zoites* – may be found in many tissues and persist for about 1–2 weeks after infection. By the fourteenth day, *Toxoplasma* lesions begin to subside, but after acute infections cysts commonly persist in the brain and muscles of cats, as in other animals (Dubey & Frenkel, 1972).

Intermediate (paratenic) hosts. Over 300 species of mammal and about 30 species of bird have been identified as paratenic hosts. *Toxoplasma* can only complete its sexual development in the cat. In the 'wrong' host, when oocysts are ingested the sporozoites hatch, cross the gut wall and invade macrophages and almost any other cell types (except erythrocytes) and – before immunity develops – divide rapidly by endodyogeny until the cell is full of them. The accumulation of organisms, bounded only by the plasmalemma of the host cell, is known as a *pseudocyst* to distinguish it from a true 'cyst', a term which is only used in parasitology for a protective membrane surrounding an organism in which at least part has been produced by the parasite.

Within a pseudocyst, the zoites divide rapidly and are known as *tachyzoites* (Greek *tachos* = speed). The host cell eventually bursts and releases the tachyzoites, which invade other cells. When this 'proliferative' phase is slowed down by the influence of developing host immunity and finally ends, the accumulations of zoites form 'true' cysts and the contained zoites are known as *bradyzoites* as they develop slowly (Greek *brady* = slow). The tissue cysts are found in all parts of the body, especially the central nervous system, lungs and muscles. These cysts, which may measure up to 60 μm in diameter, may survive for years, perhaps for the lifetime of the host (Canning, 1990). The danger in man is that these latent infections may become active again in individuals with an acquired immunodeficiency syndrome (AIDS) or other immune-compromised condition (see *Pathology*, below).

The prepatent period in the cat, i.e. the time from

Table 6.2. Toxoplasma gondii*: prevalence and distribution of indirect fluorescent antibody (IFA) titres for* T. gondii *in cats in Beirut*

Based on antibody titres >1:64, the prevalence was 79.7 per cent for stray cats and 72.6 per cent for owned cats; the overall prevalence was 78.1 per cent.

IFA titres	Stray cats		Owned cats		Total	
	number	%	number	%	Number	%
<1:64	51	20.3	20	27.4	71	21.9
1:64	7	2.8	5	6.8	12	3.7
1:128	3	1.2	2	2.7	5	1.5
1:256	9	3.6	7	9.6	16	5.0
1:512	14	5.6	3	4.1	17	5.3
1:1024	38	15.1	11	15.1	49	15.1
≥1:2048	129	51.4	25	34.3	154	47.5
Total	251	100.0	73	100.0	324	100.0

Source: Data from Deeb *et al.* (1985).

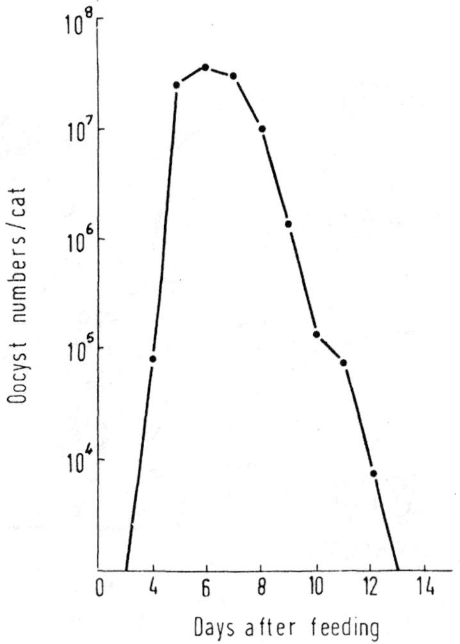

Fig. 6.16
Toxoplasma gondii: development of immunity in cats. Average oocyst numbers per day per cat from two weanling kittens fed brain and carcase of one mouse chronically infected with *Toxoplasma*. (After Dubey & Frenkel, 1972.)

ingestion of infective material to oocyst excretion in the faeces, varies with the stage ingested. Thus it is more than 20 days after ingestion of oocysts (Fig. 6.14), more than 19 days after ingestion of a mouse infected with tachyzoites (mouse B, Fig. 6.14), or 3–10 days after ingestion of a mouse with cysts (mouse A, Fig. 6.14). However, immunity is rapidly developed in the cat (Fig. 6.16).

Human infection. Man normally acquires infection either by direct ingestion of oocysts from a cat or by eating raw or undercooked meat. Cooks or butchers or those who handle raw meat are particularly at risk. Man and animals can be infected by the congenital route; acute infection in children can cause severe clinical symptoms (see *Pathology*, below). Recorded infections of *toxoplasmosis*, as measured by serological surveys, in various countries are astonishingly high. In the USA, the prevalence varies from 1 to 30 per cent depending on locality. In the UK, the prevalences of anti-*Toxoplasma* antibodies range from less than 0.1 per cent in one-year-old children to 45 per cent in those over 50 years old (Fleck, 1969; Canning, 1990). In France, where there is a preference for undercooked meat, figures of 80–90 per cent for adults have been reported. A prevalence of 14.5 per cent has been reported in cats in the USA (Childs & Seegar, 1987) and 78 per cent in Beirut (Table 6.2) (Deeb *et al.*, 1985).

Pathology: in adults. In man, toxoplasmosis in adults is almost symptomless, but may result in fever and swelling. An important factor in the pathogenicity, however, is the immune status of the individual: when the status of an individual already harbouring cysts is compromised, as in patients with Hodgkin's disease or AIDS, or in recipients of transplanted tissues or chemotherapy, disseminated toxoplasmosis may occur. This may lead to ocular toxoplasmosis and ultimately to fatal CNS disorders such as encephalitis (Hughes, 1985; Holliman, 1988). Frenkel (1988) has reviewed the physiopathogenicity of this disease.

Congenital toxoplasmosis. Congenital toxoplasmosis can occur when a pregnant woman acquires toxoplasmosis for the first time, although the mother herself may be asymptomatic. Transmission can take place via the placenta and the risk of infection to the foetus has been quoted as 45 per cent. Of the infants born, about 60 per cent will be subclinical, but of the remainder 30 per cent will either suffer symptoms ranging from hydrocephalus to intracerebral calcification, choriodoretinitis and mental subnormality or will die (some 9 per cent) (Hughes, 1985; Frenkel, 1988). The reported prevalence of congenital toxoplasmosis paints a disturbing picture (Hughes, 1985; Apt, 1985). In the USA, it has been estimated that about 3000 infants are born annually with congenital infection. In France and Austria, the incidence of congenital toxoplasmosis may exceed 3–4 per 1000 births. In the UK, 91 cases were reported between 1975 and 1980.

○ 6.4.2 Genus *Sarcocystis*

For over a century, species of *Sarcocystis* have been known from their cysts in the muscles of mammals, birds and reptiles. These cysts (Fig. 6.17) were known as *Miescher's tubules* after their discoverer and later became known as *sarcocysts*, a term still widely used. In modern terminology they are now termed *zoitocysts*. The discovery of the coccidian nature of *Toxoplasma* (p. 100) prompted workers to carry out *in vitro* and *in vivo* experiments with *Sarcocystis* and it was found also to be a coccidian. It was shown that all species of *Sarcocystis*

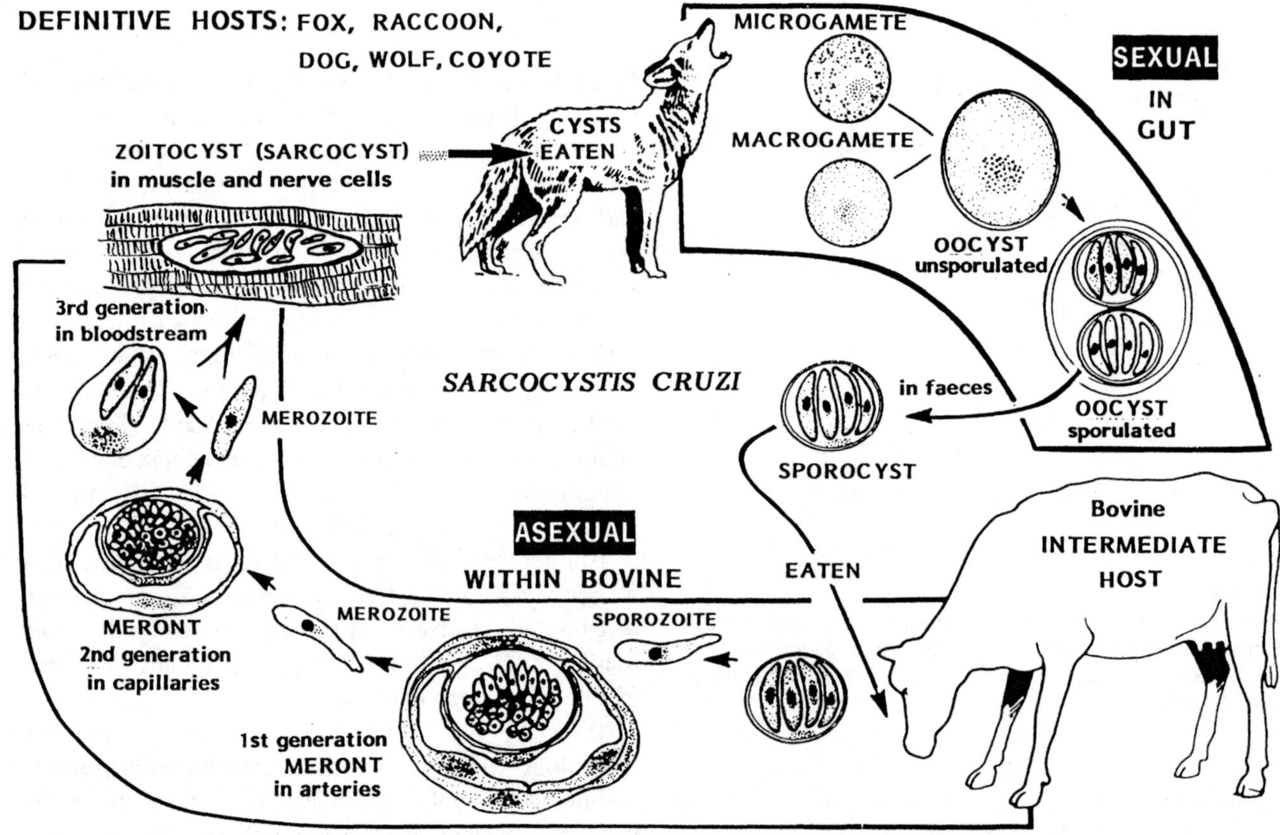

Fig. 6.17
Life cycle of *Sarcocystis cruzi*. (Modified and reproduced with permission from *Sarcocystosis of Animals and Man* by J. P. Dubey, C. A. Speer & R. Fayer, p. 10. © CRC Press, Boca Raton, Florida, 1989.)

have obligatory two-host (i.e. heteroxenous) life cycles, with herbivores (both wild and domestic animals) primarily acting as intermediate hosts and carnivores as the definitive hosts, in which the sexual stages developed.

Literature. There is an extensive literature on *Sarcocystis*, its taxonomy, ultrastructure, general biology, epidemiology and pathology. These areas are covered in depth in a comprehensive monograph of Dubey *et al.* (1989). Other useful studies or reviews are those of: Dubey, 1977; Gardiner *et al.*, 1988; Herbert & Smith, 1987; Mehlhorn & Heydorn, 1973; Levine & Ivans, 1986.

Life cycle. The life cycle is shown in Fig. 6.17. The definitive host (including man) becomes infected by eating flesh containing zoitocysts (Fig. 6.18). On digestion, bradyzoites are released and develop directly into gametes (i.e. without undergoing meregony; contrast *Toxoplasma*). The resulting zygotes develop into oocysts which sporulate within the gut (again, contrast *Toxoplasma*) and contain two sporocysts, each with four sporozoites. The oocysts are thin-walled and usually burst, resulting in fully sporulated sporocysts being shed in the faeces.

Species in domestic animals. *Sarcocystis* zoitocysts are common in farm animals – cattle, sheep, pigs, horses, buffalo, camels – and in game animals. Although it is generally held that the presence of the cysts is innocuous, recent work has shown that some species can be highly pathogenic to their hosts. Moreover, economic losses apparently due to sarcocystosis (sarcosporidiosis) have been reported from Australia, Canada, Germany, the UK and the USA. The main economic effects are related to abortion and reduced live mass again (Herbert & Smith, 1987).

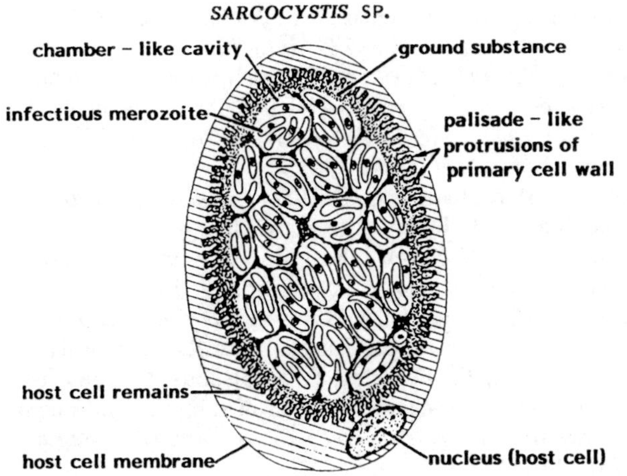

Fig. 6.18
Infective zoitocyst ('sarcocyst') of *Sarcocystis* sp. within a muscle fibre; 2 months development. (After Mehlhorn & Heydorn, 1973.)

Species in man. Man is the definitive host for two species of *Sarcocystis*, *S. hominis* and *S. suihominis*, with sexual stages in the intestine, and man can also act as accidental intermediate host for several (unidentified) species of animal *Sarcocystis*, resulting in zoitocysts in the muscles.

Intestinal sarcocystosis. Man acquires *S. hominis* by eating uncooked beef containing zoitocysts. Volunteer infections in man have shown that *S. hominis* is only mildly pathogenic in humans, causing nausea, stomach pains and diarrhoea. Sporocysts begin to be passed in the faeces after 14–18 days (Dubey *et al.*, 1989). *S. suihominis* is acquired by eating undercooked, infected pork. It is more pathogenic than *S. hominis*, causing vomiting, nausea, stomach pains, diarrhoea and dyspnoea within 24 h. Sporocysts are shed after 11–13 days.

Muscular sarcocystosis. Zoitocysts of unknown species of *Sarcocystis* have been reported in the striated muscles of man, only 40 cases being reported as of 1979. This surprisingly low prevalence is probably related to the fact that these reports were only incidental findings, as, until recently, no systematic survey of human muscles for sarcosystosis has been made. A survey in Denmark by Greve (1985) on 112 specimens of human muscles using the trichinoscopic technique, as used for examining pork for *Trichinella spiralis* cysts (p. 394) (by viewing tissue pressed between glass plates) detected four cases of *Sarcocystis* infection. As this figure corresponds to 3.6 per cent and is comparable to the prevalence found in sows and boars, it seems to indicate that humans are as likely to be infected as pigs. This suggests that human sarcocystosis may be much more widespread than previously thought and has been overlooked as a zoonosis. It is also pointed out that as sarcocystosis in animals can cause abortion, it is possible that a similar situation may occur in humans (Greve, 1985).

Genus Frenkelia. These parasites are morphologically similar to *Sarcocystis* and occur as small cysts in the brains of muskrats and voles. The buzzard, *Buteo buteo*, is the definitive host. *Frenkelia* is not transmissible by inoculation of infected tissues from one intermediate host to another, but congenital infection can occur (Dubey, 1977).

Genus Besnoitia. This is a small group of *Toxoplasma*-like organisms which parasitise mainly tissue cells and produce characteristic cysts with very thick walls. The intermediate hosts are cattle, deer, hares, rodents and lizards but not man. They can cause considerable economic loss in cattle (Dubey, 1977). The definitive hosts are various domestic and wild felids.

○○○ References

(References of general interest have been included in this list, in addition to those quoted in the text.)

Adams, J. H. & Todd, K. S. Jr 1983. Transmission electron microscopy of intracellular sporozoites of *Eimeria vermiformis* (Apicomplexa, Eucoccidia) in the mouse. *Journal of Protozoology*, **30**: 114–18.

Apt, W. 1985. Toxoplasmosis in developing countries, *Parasitology Today*, **1**: 44–6.

August, J. R. & Chase, T. M. 1987. Toxoplasmosis. *Veterinary Clinics of North America. Small American Practice*, **55**: 55–71.

Ball, S. J., Burgoyne, J. E. & Burgoyne, R. D. 1985. Cephaline gregarines of the earwig, *Foricula auricularia*, in England. *Journal of Natural History*, **20**: 453–8.

Ball, S. J., Hutchinson, W. & Pittilo, R. M. 1988. Ultrastructure of microgametogenesis of *Eimeria stiedae* in rabbits (*Oryctolagus cuniculus*). *Acta Veterinaria Hungaricae*, **36**: 229–32.

Ball, S. J. & Pittilo, R. M. 1988. Ultrastructural observations on microgametogenesis in *Eimeria bakuensis* (Syn. *E. ovina*) of sheep. *Parasitology Research*, **74**: 431–5.

Beesley, J. E. & Latter, V. S. 1982. The sporulation of *Eimeria tenella* as revealed by a novel preparative method. *Zeitschrift für Parasitenkunde*, **67**: 255–60.

Canning, E. U. 1956. A new eugregarine of locusts, *Gregarina garnhami* n.sp., parasitic in *Schistocerca gregaria* Forsk. *Journal of Protozoology*, **3**: 50–62.

Canning, E. U. 1964. Observations on the life history of *Mattesia trogodermae* sp.n., a schizogregarine parasite of the fat body of the Khapra beetle, *Trogoderma granarium*. *Journal of Insect Pathology*, **6**: 305–17.

Canning, E. U. 1990. Protozoan infections. *Transactions of the Royal Society of Tropical Medicine and Hygiene*, **84** (suppl. 1): 19–24.

Canning, E. U. & Anwar, M. 1968. Studies on meoitic division in coccidial and malarial parasites. *Journal of Protozoology*, **15**: 290–8.

Childs, J. E. & Seegar, W. S. 1987. Epidemiological observations on infection with *Toxoplasma gondii* in three species of urban mammals from Baltimore, Maryland, USA. *International Journal of Zoonoses*, **13**: 249–61.

Chobotar, B. & Scholtyseck, E. 1982. Ultrastructure. In: *The Biology of the Coccidia* (ed. P. L. Long), pp. 101–65. Edward Arnold, London.

Chowdhury, M. N. H. 1986. Toxoplasmosis: a review. *Journal of Medicine*, **17**: 373–96.

Cleveland, L. R., Burke, A. W. Jr. & Karlson, P. 1960. Ecdysone induced modifications in the sexual cycles of the protozoa of *Cryptocercus*. *Journal of Protozoology*, **7**: 229–39.

Cook, G. C. 1987. *Cryptosporidium sp. and other Intestinal Coccidia: a Bibliography*. Bureau of Hygiene and Tropical Diseases, London.

Corbel, J. C. 1964. Infestions expérimentales de *Locusta migratoria* L. (Insecte, Orthoptère) pour *Gregarina garnhami* Canning (Sporozoaire, Grégarinomorphe): relation entre le cycle de l'hôte et celui du parasite. *Comptes Rendus de Séances de l'Académie des Sciences*, **259D**: 207–10.

Cox, F. E. G. 1968. Parasites of British earthworms. *Journal of Biological Education*, **2**: 151–64.

Cuckler, A. C. & Malanga, C. M. 1956. The effect of nicarbazin on the development of immunity to avian coccidia. *Journal of Parasitology*, **42**: 593–605.

Deeb, B. J., Sufan, M. M. & Digiacomo, R. F. 1985. *Toxoplasma gondii* infection of cats in Beirut, Lebanon. *Journal of Tropical Medicine and Hygiene*, **88**: 301–6.

Devauchelle, G. 1968. Etude de l'ultrastructure des gregarines de *Gregarina polymorpha* (*Hamm.*) en Syzgie. *Journal of Protozoology*, **15**: 629–36.

Devauchelle, G. & Oger, C. 1968. Analysis biométrique des grégarines de *Tenebrio molitor*. *Protistologica*, **4**: 37–52.

Dubey, J. P. 1977. *Toxoplasma*, *Hammondi*, *Besnoitia*, *Sarcocystis*, and other tissue cyst-forming coccidia of man and animals. In: *Parasitic Protozoa* (ed. J. P. Kreier), Vol. III, pp. 101–237. Academic Press, London.

Dubey, J. P. 1986. Toxoplasmosis. *Journal of American Veterinary Medical Association*, **189**: 166–70.

Dubey, J. P. 1993. *Toxoplasma*, *Neospora*, *Sarcocystis* and other tissue cyst-forming coccidia of humans and animals. In: *Parasitic Protozoa* (ed. J. P. Kreier), Vol. 6, pp. 1–137. Academic Press, New York.

Dubey, J. P. & Beattie, C. P. 1988. *Toxoplasmosis of Animals and Man*. CRC Press, Boca Raton, Florida.

Dubey, J. P. & Frenkel, J. K. 1972. Cyst induced toxoplasmosis in cats. *Journal of Protozoology*, **19**: 155–77.

Dubey, J. P., Speer, C. A. & Fayer, R. 1989. *Sarcocystosis of Animals and Man*. CRC Press, Boca Raton, Florida.

Dubremetz, J. F., Ferreira, E. & Dissous, C. 1989. Isolation and partial characterization of rhoptries and micronemes from *Eimeria nieschulzi* zoites. *Parasitology Research*, **75**: 449–54.

Duhlinska, D. 1977. Protozoan parasites of lumbricid earthworms from Bulgaria. *Acta Zoologica Bulgarica*, **6**: 151–64.

Fayer, R. & Ungar, B. L. P. 1986. *Cryptosporidium* spp. and cryptosporidiosis. *Microbiological Reviews*, **50**: 458–83.

Fleck, D. G. 1969. Toxoplasmosis. *Public Health*, **83**: 131–5.

Forthal, D. N. & Guest, S. S. 1984. *Isospora belli* enteritis in three homosexual men. *American Journal of Tropical Medicine and Hygiene*, **33**: 1060–4.

Frenkel, J. K. 1970. Pursuing *Toxoplasma*. *Journal of Infectious Diseases*, **122**: 553–9.

Frenkel, J. K. 1988. Pathophysiology of toxoplasmosis. *Parasitology Today*, **4**: 273–8.

Frenkel, J. K., Dubey, J. P. & Miller, N. L. 1969. *Toxoplasma gondii*: fecal forms separated from eggs of the nematode *Toxocara cati*. *Science*, **164**: 432–3.

Gajadhar, A. A., Rainnie, D. J. & Cawthorn, R. J. 1986*a*. Description of the goose coccidian *Eimeria stigmosa* (Klimeš, 1963), with evidence of intramuscular development. *Journal of Parasitology*, **72**: 588–94.

Gajadhar, A. A. & Stockdale, P. H. G. 1986. Ultrastructural studies on microgametogenesis and macrogametogenesis of *Eimeria truncata* in the lesser snow goose. *Journal of Protozoology*, **33**: 345–51.

Gajadhar, A. A., Stockdale, P. H. G. & Cawthorn, R. J. 1986*b*. Ultrastructural studies on the zygote and oocyst formation of *Eimeria truncata* of the lesser snow goose. *Journal of Protozoology*, **33**: 341–4.

Gardiner, C. H., Fayer, R. & Dubey, J. P. 1988. *An Atlas of Protozoan Parasites in Animal Tissues*. US Government Publication.

Goldman, M., Carver, R. K. & Sulzer, A. J. 1958. Reproduction of *Toxoplasma gondii* by internal budding. *Journal of Parasitology*, **44**: 161–71.

Greve, E. 1985. Sarcosporidiosis – an overlooked zoonosis: man as an intermediate and final host. *Danish Medical Bulletin*, **32**: 228–30.

Hall, S. M. 1986. Toxoplasmosis. *Journal of Small Animal Practice*, **27**: 705–16.

Hammond, D. M. & Long, P. L. 1973. *The Coccidia*. Butterworth, London.

Harry, O. G. 1965. Studies on the early development of the eugregarine *Gregarina garnhami*. *Journal of Protozoology*, **12**: 296–305.

Herbert, I. V. & Smith, T. S. 1987. Sarcocystosis. *Parasitology Today*, **3**: 17–21.

Hoare, C. A. 1949. *Handbook of Medical Protozoology*. Ballière, Tindall & Cox, London.

Holliman, R. E. 1988. Toxoplasmosis and the acquired immune deficiency syndrome. *Journal of Infection*, **16**: 121–8.

Horton-Smith, C. & Long, P. L. 1963. Coccidia and coccidiosis in the domestic fowl and turkey. *Advances in Parasitology*, **1**: 67–107.

Hughes, H. P. A. 1985. Toxoplasmosis – a neglected disease. *Parasitology Today*, **1**: 41–4.

Hutchison, W. M. 1965. Experimental transmission of *Toxoplasma gondii*. *Nature*, **206**: 961–2.

Ito, S. 1985. (Life cycles of isoporan coccidia.) *Progress in Veterinary Science*, pp. 71–105 (no vol. no. given). (In Japanese.)

Jacobs, L. 1967. *Toxoplasma* and toxoplasmosis. *Advances in Parasitology*, **5**: 1–45.

Jacobs, L. 1973. New knowledge of *Toxoplasma* and toxoplasmosis. *Advances in Parasitology*, **11**: 631–69.

King, C. A. 1988. Cell motility of sporozoan protozoa. *Parasitology Today*, **4**: 315–19.

Kreier, J. P. 1977. *Parasitic Protozoa*, Vol. III. Academic Press, New York.

Kreier, J. P. & Baker, J. R. 1987. *Parasitic Protozoa*. Allen & Unwin, London.

Levine, N. D. 1987. Whatever became of *Isospora bigemina*? *Parasitology Today*, **3**: 101–3.

Levine, N. D. 1988. *The Protozoa Phylum Apicomplexa*. 2 Vols. CRC Press, Boca Rato, Florida.

Levine, N. & Ivans, V. 1986. *The Coccidian Parasites (Protozoa, Apicomplexa) of Artiodactyla*. University of Illinois Press, Urbana.

Levine, N. D. (Chairman) *et al.* 1980. A newly revised classification of the protozoa. (The Committee on Systematics and Evolution of the Society of Protozoologists.) *Journal of Protozoology*, **27**: 37–58.

Long, P. L. & Jeffers, T. K. 1986. Control of chicken coccidiosis. *Parasitology Today*, **2**: 236–40.

Lysenko, A. Ya. & Asatova, M. M. 1986. (Toxoplasmosis: facts and suppositions. Communication 3. Congenital toxoplasmosis.) *Meditsinskaya Parazitologiya i Parazitarnye Bolezni*, **6**: 65–76. (In Russian.)

Mackenzie, C. 1980. Gliding in *Gregarina garnhami* – movement without actin. *Cell Biology International Reports*, **4**: 769.

Mackinnon, D. L. & Ray, H. N. 1931. Observations on dicystid gregarines from marine worms. *Quarterly Journal of Microscopical Science*, **74**: 439–66.

Malavasi, A., Da Cunha, A. B., Morgante, J. S. & Marques, J. 1976. Relationships between the gregarine *Schneideria schneiderae* and its host *Trichosia pubescens* (Diptera, Sciaridae). *Journal of Invertebrate Pathology*, **28**: 363–71.

Martínez Gómez, F., Hernández Rodríguez, S. & Naváprete, I. 1985. (Electron microscopic study of the life cycle of *Eimeria bateri* in *Coturnix coturnix japonica*.) *Revista Ibérica de Parasitología*, **45**: 1–7. (In Spanish.)

Mehlhorn, H. & Heydorn, A. O. 1973. The Sarcosporidia (Protozoa, Sporozoa): life cycle and fine structure. *Advances in Parasitology*, **16**: 43–92.

Meyer, E. A. 1988. Waterborne *Giardia* and *Cryptosporidium*. *Parasitology Today*, **2**: 200.

Overdule, J. P. 1970. The identity of *Toxoplasma* Nicolle & Manceaux, 1908 with *Isospora* Schneider, 1881. *Proceedings of the Koninklijke Nederlandse Akademie van Wetenschappen*, **73**: 128–51.

Pellérdy, L. 1963–8. *Catalogue of Eimeriidea (Protozoa, Sporozoa). Supplement (1968)*. Akadémai Kiado, Budapest.

Philippe, M. & Schrével, J. 1982. The three cortical membranes of the gregarines (parasitic protozoa). Characterisation of the membrane proteins of *Gregarina blaberae. Biochemical Journal*, **201**: 455–64.

Pitelka, D. R. 1963. *Electron-Microscope Structure of Protozoa.* Pergamon Press, Oxford.

Pittilo, R. M. & Ball, S. J. 1984. Electron microscopy of *Eimeria acervulina* macrogametogeny and oocyst wall formation. *Parasitology*, **89**: 1–7.

Pittilo, R. M. & Ball, S. J. 1985. Ultrastructural observations on the sporogony of *Eimeria maxima. International Journal for Parasitology*, **15**: 617–20.

Porchet, E., Torpier, G. & Richard, A. 1982. Etude après cyofracture du mérozoite de la coccidie *Aggregata eberthi. Zentralblatt für Bakteriologie, Parasitenkunde, Infektionskrankheiten und Hygiene*, Abt. I, **66**: 257–71.

Rommel, M., Heydorn, A. O. & Gruber, F. 1972. Beiträge zum Lebenszyklus der Sarkosporidien, I. Die Sporocyste von *S. tenella* in den Fäzes der Katze. *Berlin-Münchner tierärztliche Wochenschrift*, **85**: 101–5.

Scholtyseck, E. 1979. *Fine Structure of Parasitic Protozoa.* Springer-Verlag, Heidelberg.

Schrével, J. 1971. Contribution à l'étude des Selendiidae, parasites d'Annélides Polychetès. II. Ultrastructure de quelques trophozoites. *Protistologica*, **7**: 101–30.

Schrével, J., Caigneux, E., Gros, D. & Philippe, M. 1983. The three cortical membranes of the gregarines. I. Ultrastructural organization of *Gregarina blaberae. Journal of Cell Science*, **61**: 151–74.

Sénaud, J., Augustin, H. & Doens-Juteau, O. 1980. Observations ultrastructurales sur le développement sexué de la coccidie *Eimeria acervulina* (Tyzzer, 1929) dans l'epithelium intestinal du poulet: La microgamètogénèse et la macrogamètogénèse. *Protistologica*, **16**: 241–57.

Sheffield, H. G. & Melton, M. L. 1968. The fine structure and reproduction of *Toxoplasma gondii. Journal of Parasitology*, **54**: 209–26.

Smith, H. V. & Rose, J. B. 1990. Waterborne cryptosporidiosis. *Parasitology Today*, **6**: 8–12.

Smyth, J. D. 1976. *An Introduction to Animal Parasitology.* 2nd Edition. Edward Arnold, London.

Tronchin, G., Philippe, M., Mocquard, J. P. & Schrével, J. 1986. Cycle biologique de *Gregarina blaberae*: description, chronologie, étude de la croissance, influence du cycle de l'hôte, *Blaberus craniifer. Protistologica*, **22**: 127–42.

Tzipori, S. 1985. *Cryptosporidium*: notes on epidemiology and pathogenesis. *Parasitology Today*, **1**: 159–65.

Vávra, J. 1969. *Lankesteria barretti* n.sp. (Eugregarinida, Diplocystidae), a parasite of the mosquito *Aedes triseriatus* (say) and a review of the genus *Lankestria* Mingazzini. *Journal of Protozoology*, **16**: 546–70.

Vivier, E. & Petitprez, A. 1968. Les ultrastructures superficielles et leur évolution au niveau de la jonction chez les couples de *Diplauxis hatti*, grégarine de *Perinereis cultifera. Comptes Rendus Hebdomadaires des Séances de l'Académie Agriculture de France*, **266**: 491–3.

Vivier, E., Devaucheille, G., Petitprez, A., Porchet-Hennere, E., Prensier, G., Schrével, J. & Vinckier, D. 1970. Observation de cytologie comparée chez les sporozoaires. I. Les structures superficielles chez les formes végétatives. *Protistologie*, **6**: 127–50.

Walsh, R. D. & Callway, C. S. 1969. The fine structure of the gregarine *Lankesteria culicis*, parasitic in the yellow-fever mosquito, *Aedes aegypti. Journal of Protozoology*, **16**: 536–45.

Wang, C. C. 1982. Biochemistry and Physiology of Coccidia. In: *The Biology of Coccidia* (ed. P. L. Long), pp. 167–228. Edward Arnold, London.

Work, K. & Hutchison, W. M. 1969*a*. A new cystic form of *Toxoplasma gondii. Acta Pathologica et Microbiologica Scandinavica*, **77**: 191–2.

Work, K. & Hutchison, W. M. 1969*b*. The new cyst of *Toxoplasma gondii. Acta Pathologica et Microbiologica Scandinavica*, **77**: 414–24.

7

Sporozoea: Haemosporina; malaria; basic biology

○○○ 7.1 General considerations

As an environment, the vertebrate red blood cell has certain advantages to offer. It is thin-walled and in constant motion, with the result that absorption of food materials and elimination of waste products of metabolism are relatively easily accomplished. In addition, it contains rich supplies of protein and oxygen. These very features which make red blood cells efficient metabolising units thus similarly equip them to serve as admirable habitats for parasites.

The Haemosporina are the only protozoan parasites capable of 'invading' the red blood corpuscles of vertebrates; many species, if not all, have multiplicative phases in the reticulo-endothelial system. Although the word 'invading' is commonly used in relation to the infection of red blood cells by the malarial organism, it is now known that the organism does not, in fact, penetrate the cell membrane but enters by endocytosis and is enclosed in a parasitophorous membrane; see p. 111 and Fig. 7.3. In spite of the importance of some of the species (i.e. those causing human malaria), and the vast amount of research that has been done upon them, some phases of the life cycles are still either imperfectly known or entirely unknown.

All Haemosporina undergo the same general type of developmental cycle, which involves:

(*a*) initial or continual schizogony in the vertebrate host with initiation of gametogeny;
(*b*) formation of gametes in the arthropod host and subsequent fertilisation and formation of a zygote.
(*c*) formation of sporozoites from the zygote by repeated nuclear divisions followed by cytoplasmic divisions.

Since the transfer between vertebrate and invertebrate host is made by withdrawal or injection of parasites during the bloodsucking act, there are no resistant stages exposed to the hazards of the outside world; thus the production of protective spore cases, such as occur in coccidians, is unnecessary for survival.

○○○ 7.2 Subclass Coccidia: suborder Haemosporina

Family 1 Plasmodiidae

Genus *Plasmodium:* parasites of mammals, birds and lizards. Exo-erythrocytic schizogony; sexual reproduction in blood-sucking insects.

Family 2 Haemoproteidae

Schizogony in endothelial cells of vertebrates, no erythrocytic schizogony; gametocytes in erythrocytes or lymphocytes.

Genus *Leucocytozoon*: parasites of birds; schizogony in visceral and endothelial cells of vertebrates; gametocytes in white blood cells.

Genus *Haemoproteus*: parasites of birds and reptiles; gametocytes in erythrocytes, with pigment granules, halter-shaped when fully formed.

Family 3 Babesiidae

Minute non-pigmented parasites of the erythrocytes of mammals.

Genus *Babesia*: in erythrocytes of cattle, pear-shaped, arranged in couples; sexual reproduction in female ticks.

Of uncertain systematic position

Genus *Theileria*: minute intracellular parasites in erythrocytes and endothelial cells of mammals.

○○○ 7.3 Genus *Plasmodium*: the malarial organism

Parasites of the genus *Plasmodium* are responsible for the disease 'malaria' in both animals and man. Although the species attacking man have been most extensively studied, considerable use has been made of species in laboratory animals, especially *P. knowlesi* and *P. cynomolgi* in monkeys, *P. relictum*, *P. cathemerium* and *P. gallinaceum* in birds, and *P. berghei*, *P. yoelii* and *P. chabaudi* in rodents. Species also occur in other mammals such as squirrels and bats and in amphibians and reptiles, but these have not been much studied. Although it has been known for many years that some species of monkey malaria (e.g. *P. cynomolgi*) are transmissible to man, it is only comparatively recently that it has been found that human malaria can be transmitted to several species of monkeys. The Colombian night monkey, *Aotus trivirgatus*, has proved to be an especially valuable laboratory model for malarial research (Bruce-Chwatt, 1980).

The genus *Plasmodium* contains the 'true' malarial parasites (in contrast to the haemoproteids, p. 138); the recognition that profound differences exist between the various species has led to the further subdivision into seven *subgenera*, which are as follows:

Plasmodium: parasitic in primates, with exo-erythrocytic schizogony in parenchymal cells of liver; erythrocytic schizonts large; gametocytes round.

Vinckeia: parasitic in non-primate mammals with exo-erythrocytic schizogony in parenchymal cells of liver; erythrocytic schizonts usually small; gametocytes round.

Laverania: parasitic in primates, with exo-erythrocytic schizogony in parenchymal cells of liver; gametocytes crescentic with perinuclear pigment.

Haemamoeba: parasitic in birds; exo-erythrocytic schizogony in reticulo-endothelial cells; erythrocytic schizonts large; gametocytes round.

Giovannolaia: parasitic in birds, exo-erythrocytic schizogony in reticulo-endothelial cells; erythrocytic schizonts large; gametocytes elongate.

Novyella: parasitic in birds; exo-erythrocytic schizogony in reticulo-endothelial cells; erythrocytic schizonts small; gametocytes elongate.

Huffia: parasitic in birds; exo-erythrocytic schizonts in the haemopoietic (i.e. blood-forming) system; gametocytes elongate.

Using the subgeneric designations, the correct zoological nomenclature for the four species infecting man is: *Plasmodium (Plasmodium) vivax*, *P. (P.) ovale*, *P. (P.) malariae* and *P. (Laverania) falciparum*. However, as is common practice in the literature, the subgeneric names have been omitted in this text.

Literature. A vast volume of literature exists on human and animal malaria. The definitive work covering most aspects of malariology is undoubtedly the two-volume treatise of Wernsdorfer & McGregor (1988). A selection of other recent books or reviews, summarising recent advances, are those of: Aikawa (1980, 1988), Aikawa & Seed (1980), Beale & Walliker (1988), Bruce-Chwatt (1980), Garnham (1988), Knell (1989), Kreier (1977), Meis & Verhave (1988), Sinden (1983*a*), Weber (1988). See also p. 126 and References to Chapter 8.

○ 7.3.1 History

Human malaria has been recognised since the earliest period of man's recorded history, and the occurrence of mosquitoes trapped in amber suggests its prevalence in pre-historic times. A variety of names was used to describe the disease: the shakes; march, Roman, jungle, intermittent fever; ague; chills. It was early thought (with good reason!) that there was an aetiological relationship between swamps and these fevers. Italians referred to the bad air in fever-producing areas as *mala aria*, written *mal'aria*, and some time in the middle ages the apostrophe was dropped, producing the term *malaria* as we now know it.

The work of Laveran, Manson, Ross and others showed the occurrence of the developmental cycle in the blood corpuscles and the transmission through mosquitoes, and by the early part of the present century it was generally believed that the broad outlines of the life cycle were fully known. When sporozoites were injected by a mosquito bite, they were thought to enter red cells directly and undergo schizogony; the actual penetration of a corpuscle was described in detail by Schaudinn. This simple account of the schizogonic cycle was accepted for nearly thirty years and published and republished in text-books of medicine and zoology. This account

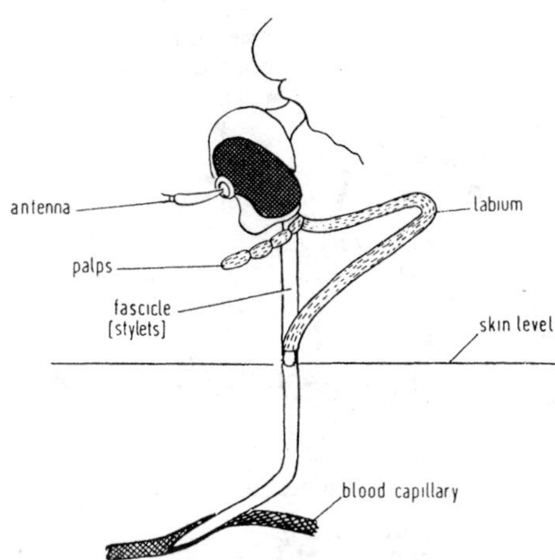

Fig. 7.1
Procedure adopted by *Aedes aegypti* on biting a frog's foot. Note how the fascicle is introduced directly into a capillary. (From Boyd, 1949, after Gordon & Lumsden, 1939.)

could not explain two facts: (*a*) that there was always a time lag of six to nine days after a mosquito bite, during which time the organisms could not be detected in the blood; and (*b*) that in certain types of malaria, relapses were more frequent than in others, as if there existed a hidden reservoir of infection somewhere in the body.

What Schaudinn actually saw must remain a mystery, for it was found, first in birds, then in monkeys, and later in man, that the sporozoites, on entering the blood, went not directly to the red blood cells, as formerly thought, but within half an hour were carried to the recitulo-endothelial system (usually the liver) where they underwent a schizogony cycle. After schizogony, the majority of merozoites re-entered the bloodstream and began the classical erythrochtic cycle. Some sporozoites, however, which enter liver cells, become dormant organisms (hypnozoites) in some species and may cause later relapses. The discovery of these exo-erythrocytic (EE) tissue forms provided a satisfactory explanation for the early disappearance of the sporozoites, the inability of freshly infected blood to be infective to another individual, and the occurrence of relapses; it also brought the life cycle of *Plasmodium* into line with that of *Leucocytozoon* and *Haemoproteus*.

○ 7.3.2 General features of the malarial life cycle

As far as it is known, all plasmodia of animals, including man, spend a part of their life in a vertebrate and part in a mosquito host. It has been suggested that certain species of plasmodia of reptiles and bats may have arthropod hosts other than mosquitoes but this has not been confirmed.

There are three phases of development in the life cycle of most species of plasmodia; all phases are not known for every species.

(*a*) **Exo-erythrocytic (EE) stages** in the tissues, usually the liver.
(*b*) **Erythrocytic schizogony** in the erythrocytes.
(*c*) **The sexual process**, beginning with the development of gametocytes in the vertebrate host and continuing with the differentiation of micro- and macrogametes, fertilisation and sporogony in the mosquito.

Asexual development in the vertebrate host

Exo-erythrocytic schizogony. A generalised diagram of the life cycle is shown in Fig. 7.2. When an infected mosquito bites man, numerous sporozoites (several hundreds) may be injected. Since a mosquito usually feels about with its proboscis until it strikes a small capillary (Fig. 7.1) the sporozoites are probably injected directly into the blood stream. Here they remain for about 30 min and then disappear from the blood stream. Many are destroyed by phagocytes but some enter the hepatocytes in the liver (via the Küpffer cells). Here they multiply rapidly by schizogony, a process referred to as *pre-erythrocytic schizogony*. When schizogony is completed, the merozoites are released from the schizont and invade the erythrocytes and the classical erythrocytic cycle begins.

In the human malarial species, it was originally thought that there were *two* successive pre-erythrocytic cycles, but it is now known that in *Plasmodium vivax* and *P. ovale*, some injected sporozoites may differentiate into stages termed *hypnozoites* (Greek *hypnos* = sleep; *zoon* = animal) which may remain dormant in the liver cells for some time only to undergo schizogony causing relapse of disease when the red cells are invaded; see Fig. 7.2.

Invasion of the erythrocyte (Fig. 7.3). It was formerly thought that the merozoites released from a schizont directly 'penetrated' the membrane of a red blood cell and then developed. Largely owing to the work of Aikawa and his colleagues (Aikawa, 1980; Aikawa & Seed, 1980) it has been shown that the merozoites enter erythrocytes by endocytosis. This key event in the malarial life cycle has been extensively examined, as the inhibition of this process can be regarded as a crucial goal in the development of a malarial vaccine. Recent reports on this process are those of Hermentin

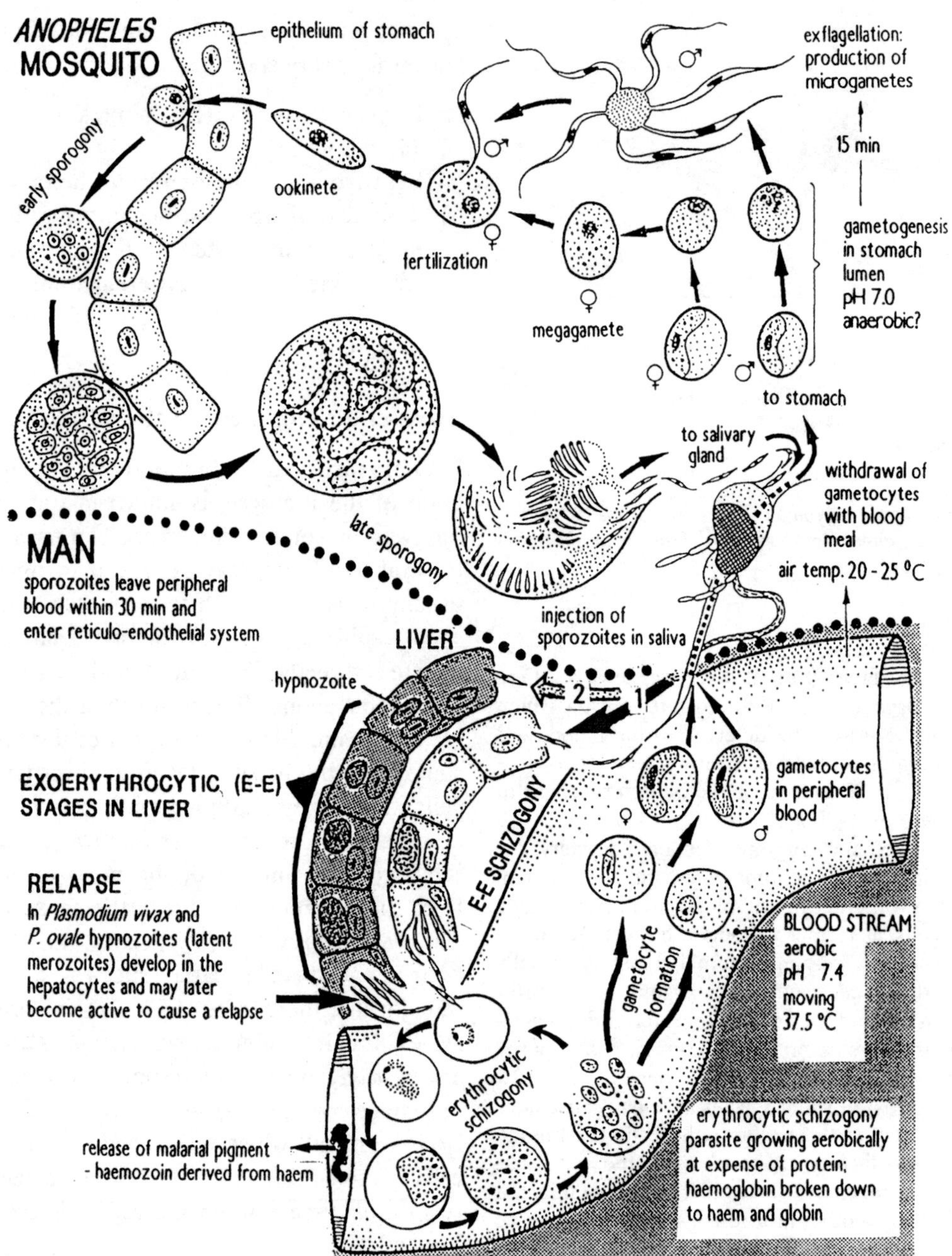

Fig. 7.2
Life cycle of *Plasmodium* sp. in man. 1. Indicates normal pathway of merozoites invading liver cells giving rise to (pre-erythrocytic) E-E schizogony. 2. Indicates additional pathway in some species, in which hypnozoites remain dormant until stimulated into activity resulting in relapses. (Original.)

(1987), Pasvol & Wilson (1989) and Tait & Sacks (1988) and a (simplified) possible sequence of events during invasion is shown in Fig. 7.3. The process involves recognition and attachment of the merozoite to the erythrocyte membrane and it is likely (but not proven) that a specific receptor for erythrocyte entry may be involved. If, on initial contact with an erythrocyte, the apical complex end of the parasite is not directed towards the blood cell surface, a re-orientation takes place. After re-orientation, some deformation of the erythrocyte membrane may occur and invagination and endocytosis follows. A distinct junction is formed

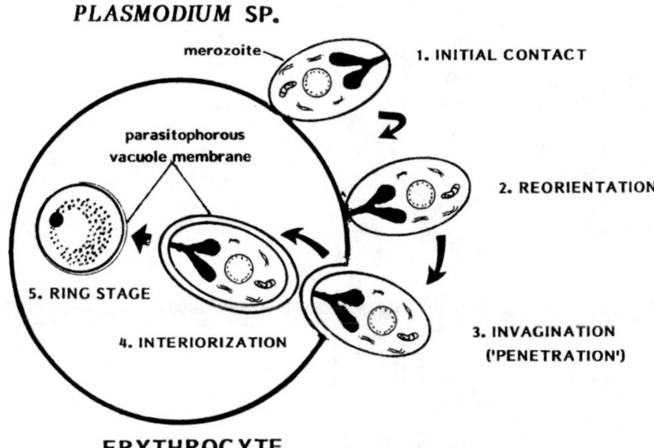

Fig. 7.3
Sequence of events during the invasion of an erythrocyte by a merozoite of *Plasmodium* sp. The erythrocyte membrane is not 'penetrated' but the parasite enters by endocytosis and becomes contained in a parasitophorous vacuole. (Partly after Hermentin, 1987.)

between the erythrocyte membrane and the merozoite surface during endocytosis and moves along the confronting membranes. When entry into the host cell is completed the merozoite is surrounded by a *parasitophorous* vacuole that has originated from the erythrocyte membrane; this grows with the developing parasite and is retained until the formation of the next generation of *Plasmodium* merozoites (Aikawa, 1980).

Erythrocytic schizogony (Fig. 7.2). Within an erythrocyte, the parasite is first seen microscopically as a minute speck of chromatin surrounded by scanty protoplasm. The uninucleate plasmodium gradually becomes ring-shaped and is known as a ring or trophozoite. It grows at the expense of the erythrocyte and assumes a form differing widely with the species but usually exhibiting active pseudopodia. Pigment granules of haemozoin appear early in the growth phase. As the chromatin begins to divide, the parasite is known as a *schizont*. The dividing chromatin tends to take up peripheral positions, and a small portion of cytoplasm gathers around each. The mature schizont is often known as a *segmenter*. The infected erythrocyte ruptures and releases a number of merozoites which attack new corpuscles, thus repeating the cycle of erythrocytic schizogony. The infection about this time enters the phase of *patent parasitaemia* with parasites detectable in blood smears.

In some species, merozoites show a distinct predilection for erythrocytes of a certain age. Many avian species attack almost exclusively the young erythrocytes. In the human plasmodia, the merozoites of *P. vivax* attack young immature corpuscles (reticulocytes), those of *P. malariae* attack the older ones, and those of *P. falciparum* indiscriminately enter any available. This partly (but see p. 128) explains the virulence of *P. falciparum* infections in which 10 per cent or more of the erythrocytes may be attacked, whereas *P. vivax* seldom occurs in even 1 per cent of the corpuscles, and *P. malariae* seldom in more than 0.2 per cent. Schizogony may continue for many months or even years; some cases of *P. malariae* have been recorded in which the infection seems to have persisted for 30–40 years.

All schizonts release their contained merozoites within a period of several hours together with a pigment and other waste products, and there is a sudden paroxysm of fever in the host, characterised by a marked rise in temperature. This periodicity in reproduction is one of the most remarkable features of malarial organisms, human and avian species being particularly regular. The length of the cycle is usually some multiple of 24 hours. In *P. vivax*, *P. ovale* and *P. falciparum* it is 48 hours and in *P. malariae* it is 72 hours, but in some monkey species (e.g. *P. knowlesi*) and some avian species (e.g. *P. cathemerium*) it is 24 hours.

The terms 'tertian' and 'quartan' as applied to human malarias are derived from the old Roman reckoning which counted the day on which something happened as the first day, the second day following being the third (tertian) and the third following day the fourth (quartan). In the early stages of *vivax* and *falciparum* infections, paroxysms of fever commonly occur daily (quotidian), an effect probably due to the fact that merozoites are not all released into the bloodstream at the same time.

Formation of gametocytes (gametocytogenesis). Some merozoites on entering red cells become sexual gametocytes, instead of asexual schizonts. Male gametocytes are termed *microgametocytes*, and female *macrogametocytes*. The stimulus that induces a merozoite to undergo gametocytogenesis, instead of schizogony, is largely unknown. It is generally held to be related to 'stress' due to rising asexual parasitaemia, the effects of drug suppression, nutrient depletion and/or rising immunity to the asexual stages; reliable supporting evidence is, however, lacking (Sinden, 1983*b*). The process of invasion of an erythrocyte, destined to become a gametocyte, appears to be the same as that of a merozoite destined for asexual development (Fig. 7.3). Thus the gametocyte also becomes surrounded by two

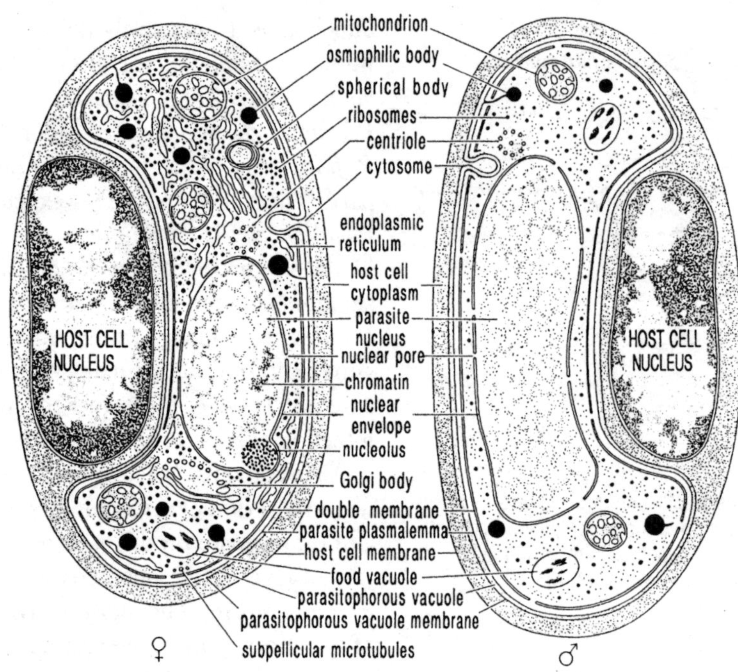

Fig. 7.4
Generalised diagram of the ultrastructure of gametocytes of the Haemosporina. (After Carter & Graves, 1988; reproduced with permission from *Malaria. Principles and Practice of Malariology* (ed. W. H. Wernsdorfer & I. McGregor), vol. 1, p. 367. Churchill-Livingstone, Edinburgh and London, 1988.)

continuous membranes, the host cell membrane and the parasitophorous vacuole membrane, as shown in Fig. 7.4. Although a mitochondrion is shown in both male and female gametocytes, the microgamete itself lacks a mitochondrion (Sinden, 1983*a*).

Gametogenesis: formation of gametes. When gametocytes are ingested by a mosquito, the cells rapidly undergo gamete production, generally referred to as 'exflagellation', although, strictly speaking, use of this term should be restricted to one phase of microgametogenesis (see below). The process has been reviewed in considerable detail by Carter & Graves (1988) and Sinden (1983*a*,*b*). The process is readily carried out *in vitro* and a number of factors stimulate this process, e.g. presence of serum or biocarbonate ions. This does not necessarily indicate that the same factors effect this process *in vivo*.

The formation of female gametes – *macrogametogenesis* – involves little more than their escape from the host cell with relatively few structural changes. The formation of male microgametes – *microgametogenesis* – is, however, a somewhat explosive phenomenon and the '. . . ultrastructural changes are rapid and stunning. Within 15 seconds the cytoplasmic microtubule organizing centre becomes transformed into two orthogonal planar tetrads of kinetosomes upon which axonemes immediately condense' (Sinden, 1983*a*). A complex cytoplasmic reorganisation follows, involving rapid nuclear divisions which result in the production of eight motile haploid gametes. The free male gamete normally has a single axoneme, a single highly condensed nucleus and a single kinetosome. Sinden & Croll (1975) recognised three phases in microgametogenesis: (*a*) maturation, (*b*) exflagellation and (*c*) escape. These events are shown in Fig. 7.5. In *P. y. nigeriensis*, maturation takes about 7 min and exflagellation less than 1 min, but the escape of microgametes may continue for 30–40 min. Gametocytogenesis has also been described at the ultrastructural level in *P. falciparum* (Sinden, 1982).

After a microgamete fertilises a macrogamete, the resulting zygote develops into a mobile elongated ookinete. This penetrates the gut wall of the mosquito between the cells and develops as an oocyst between the epithelium and the basement membrane. The oocyst matures in 10–20 days depending on temperature, species and perhaps individual mosquitoes, growing to a body 50–60 μm in size. The chromatin divides repeatedly until there are hundreds of tiny nuclear masses, and

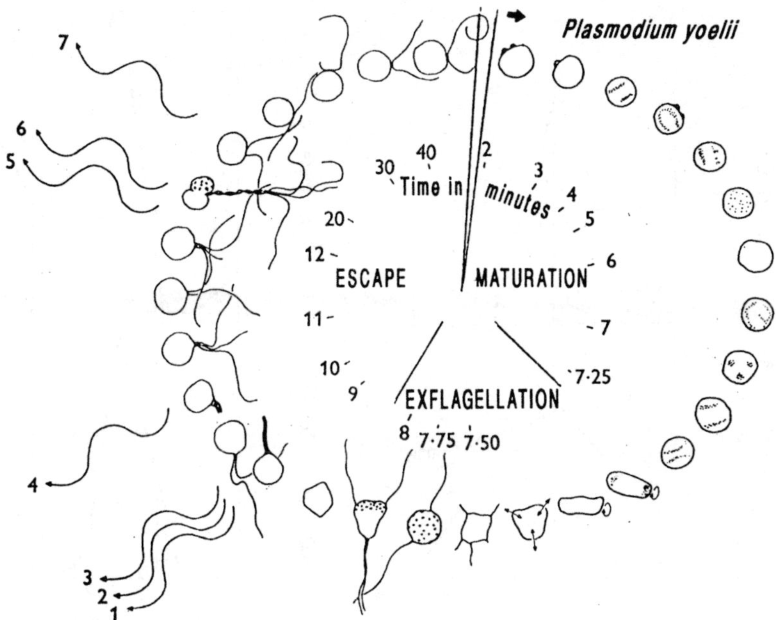

Fig. 7.5
Exflagellation in *Plasmodium yoelii* traced from a video tape recording. The time axis is not linear. (After Sinden & Croll, 1975.)

the cytoplasm follows suit, resulting in the production of enormous numbers of thread-like sporozoites, up to one thousand. When mature, an oocyst bursts and the released sporozoites make their way to the salivary glands where they become intracellular or extracellular organisms, or remain free in the ducts.

There is evidence that the morphological cycle of maturation of gametocytes corresponds to the period of infectivity to mosquitoes (Hawking *et al.*, 1968). Thus, the gametocytes of *P. knowlesi*, *P. cynomolgi* and *P. cathemerium* are apparently 'ripe' to infect mosquitoes only for a brief period (5–12 hours) normally during night-time (early morning with *P. cathemerium*) (Fig. 7.6). The development of 'ripe' gametocytes is related to a cycle of exflagellation and gametocyte formation (Fig. 7.2). This period of high infectivity of the gametocytes corresponds closely to the limited period during which vector mosquitoes normally bite. It is clear that such a synchronisation of gametocyte and mosquito cycles would have a marked selective advantage.

Although avian, rodent and simian species of malaria differ somewhat in certain details of morphology and life cycle, these differences are not very great and the general picture outlined above can be accepted as the basic pattern of the life cycle.

○○○ 7.4 Ultrastructure

The various genera in the Sporozoea share many ultrastructural and morphological features. In *Plasmodium* spp. some differences exist between mammalian-, reptilian- and avian-type species. The ultrastructure of a very large number of species has been examined. The extensive early work in this area has been documented in the second edition of this text (Smyth, 1976) and by Aikawa (1971). Relevant reviews of more recent work are those of Aikawa & Seed (1980), Aikawa (1988) and Scholtyseck (1979).

○ 7.4.1 Merozoite

General features. The general features of the fine structure of a merozoite are clear from Fig. 7.7. The organism is oval in shape and surrounded by a pellicular complex. The size varies slightly with species, but it is generally about 1.5 μm long and 1 μm in diameter. The pellicle consists of an outer and two closely aligned inner membranes. Beneath the inner membrane there is a row of subpellicular microtubules, which originate from the distal polar ring of the apical complex and pass posteriorly. A circular cytostome, which is involved in ingestion of host cytoplasm, occurs laterally. The nucleus is

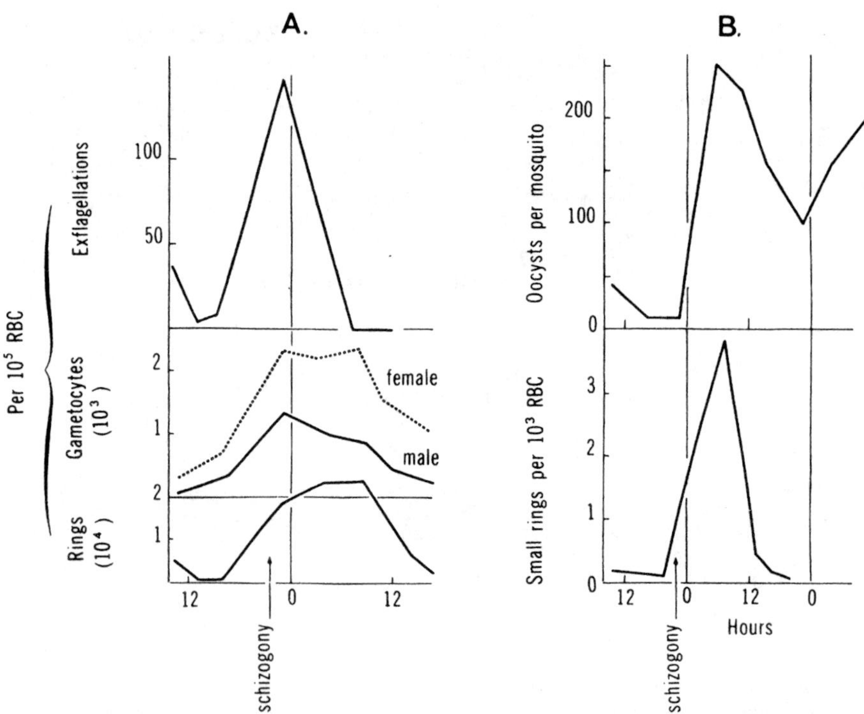

Fig. 7.6
A. Cycle of infectivity of gametocytes of *Plasmodium cathemerium* in canaries as assessed by the oocyst production in the mosquito *Culex fatigans*. B. Cycle of exflagellations compared with gametocyte production and schizogony. (After Hawking *et al.*, 1968.)

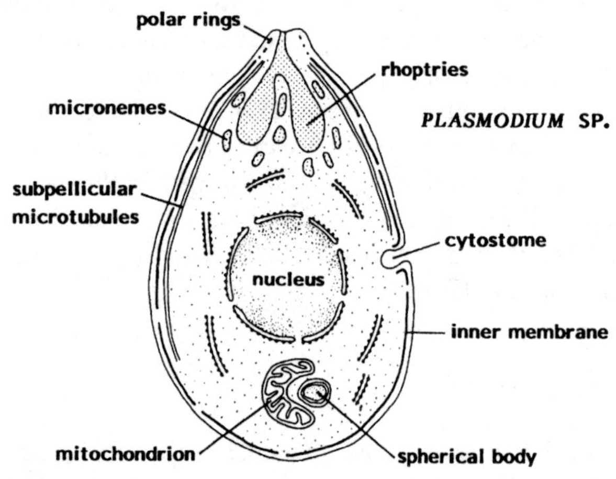

Fig. 7.7
Schematic diagram of the ultrastructure of an erythrocytic merozoite of *Plasmodium* sp. (Adapted from Aikawa & Seed, 1980.)

unusual in lacking a nucleolus; the significance of this is not clear. The cone-shaped apical end contains three polar rings, two electron-dense rhoptries and several micronemes. The rhoptries and micronemes probably play a part in the entry of the merozoite into the host cell (Aikawa & Seed, 1980).

Mitochondrion. The mitochondria of avian malarial parasites are cristate, whereas those of most mammalian parasites are not, or, if cristae are present, they are fewer in number (see p. 121). This has led to the widely held view that avian malarial parasites possess a functional citric acid cycle whereas those of mammals do not. This view has been questioned, as in most mitochondria the enzymes of this cycle are not part of the structure of the cristae and therefore the presence of cristae does not necessarily prove the presence of a functioning cycle (Homewood & Neame, 1980). This question is discussed further in section 7.4.2. A 'spherical body', measuring about 310 nm in diameter, is found closely associated with the mitochondrion. Its function is unknown; it may serve as an energy reserve.

○ 7.4.2 Trophozoite

On entry into an erythrocyte, a merozoite transforms into a trophozoite and becomes rounded, a process which appears to be caused by the rapid degradation of the inner membrane and microtubules of the pellicular complex. The exact process of ingestion of cytoplasm through the cytostome appears to vary in different species, but food vacuoles, in which digestion occurs, are formed. The process of digestion is characterised by the formation of particles of the malarial pigment haemozoin. A trophozoite contains several mitochondria and abundant ribosomes but the endoplasmic reticulum is scanty and the Golgi apparatus inconspicuous, being composed of small vesicles (Aikawa & Seed, 1980).

EE stages, oocysts. The detailed ultrastructure of these stages is beyond the scope of this book. For details see the reviews of Aikawa (1988) and Aikawa & Seed (1980).

○○○ 7.5 Genetics of *Plasmodium*

Various aspects of the genetics of *Plasmodium* have been reviewed by Beale & Walliker (1988), Walliker (1983*a*,*b*) and Sinden (1982, 1983*a*,*b*).

○ 7.5.1 Mendelian crossing experiments

Since genetic experiments on human patients involving the species of *Plasmodium* infecting man are not ethically possible, almost all the genetic studies *in vivo* have been carried out on rodent species. However, development of *in vitro* culture techniques for *P. falciparum* have also produced some valuable results (Walliker *et al.*, 1987); see below. The rodent species used have been mainly *Plasmodium yoelii* and *P. chabaudi* (p. 133). These occur naturally in Central Africa in the shiny thicket rat, *Thamnomys rutilans* (Kreier, 1977), but are readily maintained in laboratory rodents, especially mice.

Mendelian crossing experiments of the classical type are carried out by mixing clones from two different isolates (with known isoenzyme profiles) and inoculating them into mice. This allows the gametes formed from gametocytes to undergo cross-fertilisation in the mosquito; self-fertilisation of gametes from each parent line will, of course, also occur. The ookinetes which result from these experiments, which will thus consist of a mixture of hybrid and parental zygotes, are allowed to develop to oocysts; the sporocysts produced are used to infect more mice. The isoenzyme profiles of these new infections are then examined for parental and recombinant forms. The result of a typical experiment utilising the enzymes GPI (glucose phosphate isomerase), PGD (6-phosphogluconate dehydrogenase) and LDH (lactate dehydrogenase) is shown in Fig. 7.8. From this it can be seen that a substantial degree of

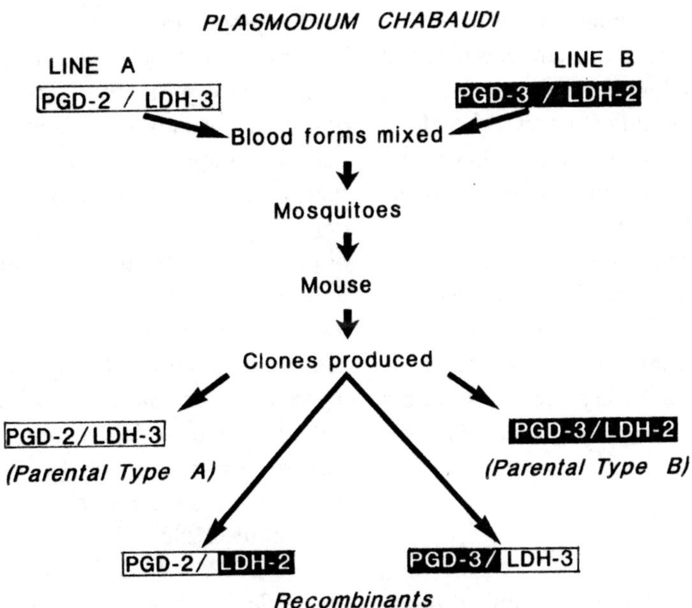

Fig. 7.8
A typical Mendelian-type experiment crossing two lines of *Plasmodium chabaudi*, differing in two enzymes, **PGD** (6-phosphogluconate dehydrogenase) and **LDH** (lactate dehydrogenase). After hybridisation, four clones of offspring are obtained: two parental types and two recombinants. (Based on Beale & Walliker, 1988; reproduced with permission from *Malaria. Principles and Practice of Malariology* (ed. W. H. Wernsdorfer & I. McGregor), vol. 1, p. 379. Churchill-Livingstone, Edinburgh and London, 1988.)

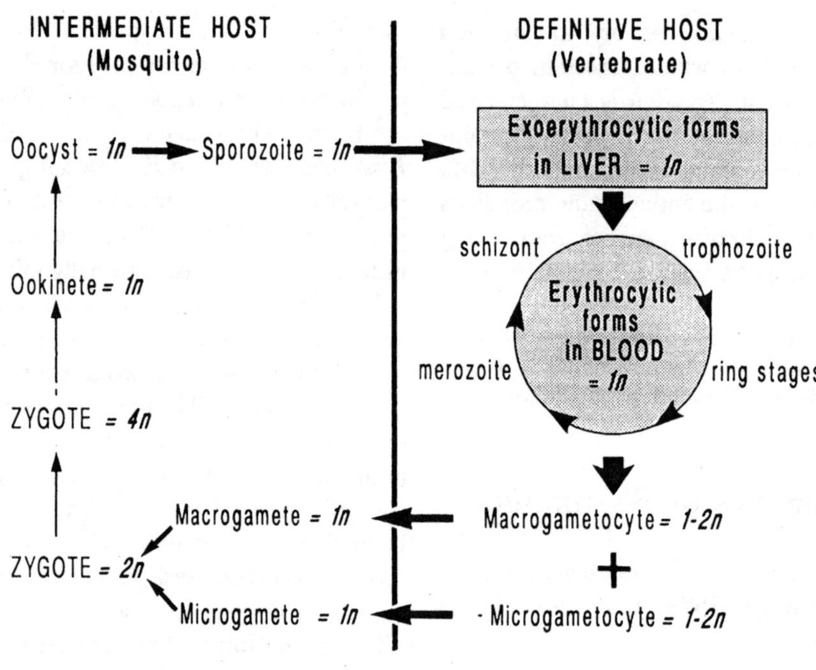

Fig. 7.9
Probable DNA content in *Plasmodium* sp. during the life cycle. (Modified from Weber, 1988.)

cross-fertilisation occurs, as witnessed by the appearance of recombinants.

Results from this type of experiment confirm the fact that recombination and segregation of the parental enzyme forms takes place before the emergence of the parasites into the blood, thus indicating that the blood forms are haploid (Fig. 7.9). This result supports the evidence from cytological studies that meiosis probably occurs in the ookinete (Sinden & Hartley, 1985). Utilising *in vitro* techniques, hybridisation between two *P. falciparum* lines has also been demonstrated (Walliker *et al.*, 1987). This involves making use of a mixture of cultured gametocytes, injecting the sporozoites produced into a chimpanzee, culturing the blood forms from the host and examining their enzyme profiles for recombinants.

○ 7.5.2 The *Plasmodium* genome

For many years almost no reliable data on the chromosomal organisation of *Plasmodium* was available. This was largely related to the fact that the amount of DNA in the nucleus is small; even electron microscope studies failed to detect condensed mitotic chromsomes. Later work on meiosis in cultured ookinetes (Sinden & Hartley, 1985) and application of recombinant DNA methodology and pulsed electrophoretic techniques have now greatly clarified the situation.

In early experiments with pulsed-field gel electrophoresis, chromosome-sized molecules up to at least 2000 kb in length were separated and it was concluded that the chromosome number must be at least 10 (Van der Ploeg *et al.*, 1985). Further extensions of this technique have shown unequivocally that the chromosome number is 14, the DNA all being in equimolar amounts; moreover, specific genes have now been identified on some chromosomes (Kemp *et al.*, 1987). That the karyotype of *P. falciparum* is 14 is further supported by ultrastructural serial sections and three-dimensional reconstruction in which 14 kinetochores were counted with mitotic spindles in young schizonts (Prensier & Slomianny, 1986).

The DNA content of the malarial parasite during its life cycle has been a matter of some controversy for many years. A current interpretation of the situation is shown in Fig. 7.9. Evidence from various sources (reviewed by Weber, 1988; Beale & Walliker, 1988) indicates that DNA synthesis begins about halfway through the asexual cycle with the nuclei being haploid (n). Microfluorimetric measurements indicate that the DNA contents in the micro- and macrogametocytes are intermediate between haploid (n) and diploid ($2n$) levels. In the mosquito, fertilisation of the haploid (or near-haploid) gametocyte results in a diploid zygote; fertilisation is followed by DNA replication, giving $4n$ nuclei. Two ('presumed') meiotic divisions follow, resulting in a haploid ookinete. Microfluorometric and other evidence indicate that the sporozoites and EE liver stages are also haploid (Weber, 1988).

Feeding process. Feeding takes place via the cytostome (Fig. 7.7), a depression in the plasma membrane surrounded by two dense rings. Both cytostomal feeding and feeding by pincytosis has been reported in some *Plasmodium* species. Food vacuoles are formed after cytostomal ingestion of host cytoplasm. The hydrolytic enzymes acid phosphatase, endoarylainidase and aminopeptidase – which are associated with lysosomes – have been demonstrated in the food vacuoles and presumably aid digestion of the haemoglobin, which serves as a major source of amino acids (see section 7.6.5).

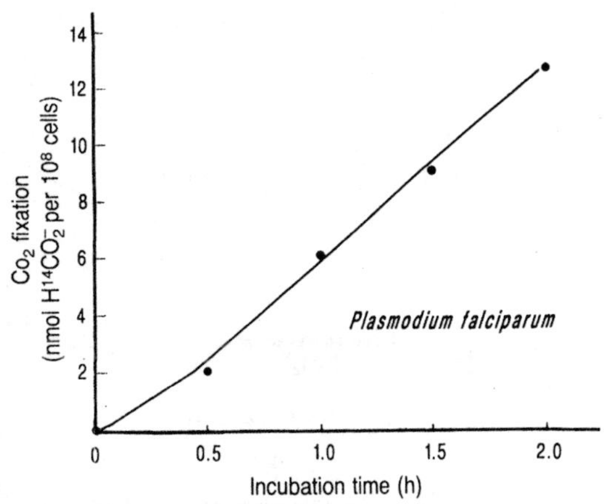

Fig. 7.10
CO_2 fixation in *Plasmodium falciparum*. An *in vitro* culture with 24 per cent parasitaemia in Medium RMPI containing 23.3 mM bicarbonate was incubated with [^{14}C]bicarbonate for the times indicated. (After Blum & Ginsburg, 1984.)

○○○ 7.6 Biochemistry and physiology

○ 7.6.1 General comments

The extensive literature in this field has been the subject of a number of valuable reviews. Early work has been summarised by Sherman (1979). Representative reviews of post-1980 research are those of Fairlamb (1989), Homewood & Neame (1980), Meis & Verhave (1988), Sherman (1984, 1991), Scheibel (1988), Scheibel & Sherman (1988), and Weber (1988).

When considering the biochemistry of *Plasmodium* it must be borne in mind that many studies have been carried out on parasitised whole blood, or suspensions containing other cells, and, in the case of avian malaria, erythrocytes containing nuclei. Results must therefore be interpreted with caution and the possible biochemical activity of the ('contaminating') host cells taken into account (Scheibel, 1988).

Our knowledge of the physiology and biochemistry of the malarial organism has also been greatly increased by the development of techniques for the *in vitro* culture (Fig. 7.10) of both intra-erythrocytic and EE stages (see Chapter 33). Modern molecular biological techniques have also expanded our understanding of the genetic basis of various biochemical aspects of the malarial life cycle (Weber, 1988).

○ 7.6.2 Carbohydrate metabolism

Glycolysis. As in trypanosomes (Chapter 5), intra-erythrocytic stages of *Plasmodium* lack carbohydrate reserves and consequently their primary source of energy is glucose from the blood stream. Although the p_{O_2} in the blood is high, *P. falciparum* does not oxidise glucose completely to CO_2 and H_2O. It has been shown that, *in vitro*, infected human red cells (10^9) consume 122 ± 34 nmol glucose per 24 h compared with 4.6 ± 1.5 μmol glucose per 24 h for uninfected cells. Glucose consumption was lower in ring forms than other stages and the schizont stages were shown to produce most of the lactate (Jensen *et al.*, 1983; Sherman, 1991). *In vitro*, very high concentrations of oxygen, such as 21 per cent O_2 (i.e. as in air) are deleterious to development but very low oxygen (0.5 per cent) could be tolerated; about 3 per cent O_2 appears to be a good concentration for growth in culture.

The lactate produced was found to have a marked inhibitory effect on growth *in vitro*; this emphasises the importance of replenishing the medium frequently in cultures.

All the enzymes of the Embden–Meyerhof pathway of glycolysis have now been identified (Fig. 7.11)

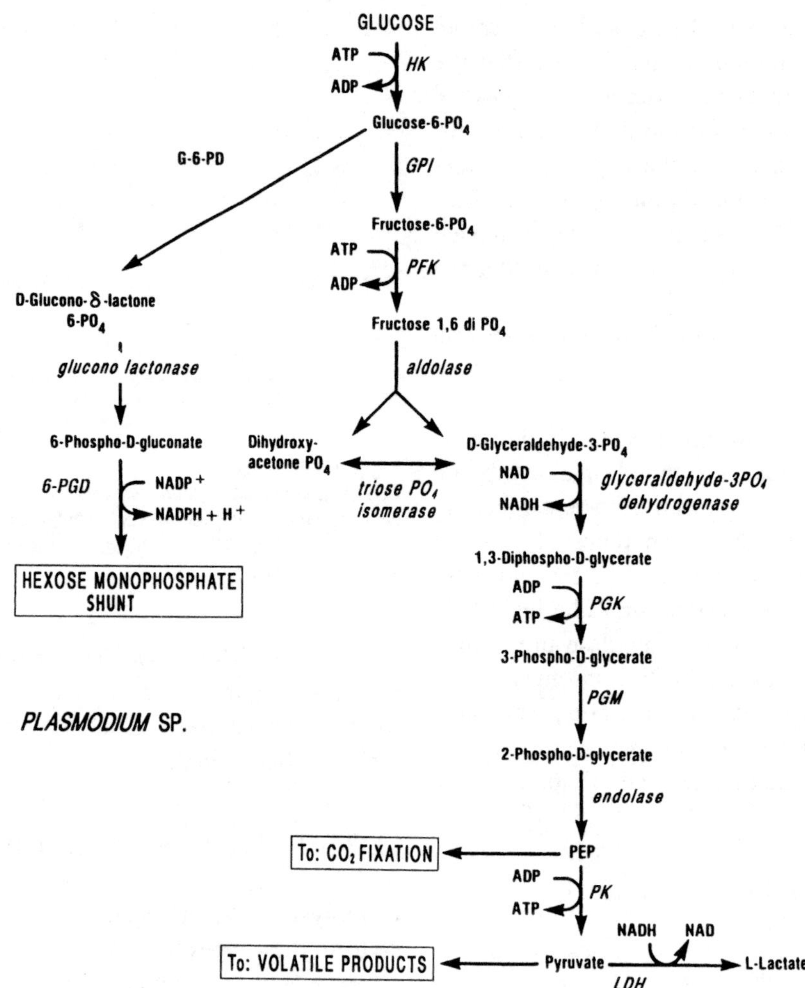

Fig. 7.11
Pathways of the glycolytic conversion of glucose in *Plasmodium* spp. Abbreviations (in italics) for enzymes, as follows: **HK**, hexokinase; **GPI**, glucose phosphate isomerase; **PFK**, phosphofructokinase; **PK**, pyruvate kinase; **LDH**, lactate dehydrogenase; **6-PGD**, 6-phosphogluconate dehydrogenase; **PGK**, phosphoglycerate kinase; **PGM**, phosphoglyceromutase. (Modified from Scheibel, 1988; reproduced with permission from *Malaria. Principles and Practice of Malariology* (ed. W. H. Wernsdorfer & I. McGregor), vol. 1, p. 182. Churchill-Livingstone, Edinburgh and London, 1988.)

although only a few have been characterised (Fairlamb, 1989). The non-glycolytic enzyme NADP-specific glutamic dehydrogenase, has also been identified, and since this is not present in uninfected cells, it can be used as a useful marker for the cytosol of *Plasmodium*. Pyruvate does not appear as an end product; most of it is converted to volatile products such as formate and acetate.

Pentose phosphate pathway. It was originally thought that the hexose monophosphate shunt was inoperative in the malarial parasite, because the first enzyme of the shunt, G6PD (glucose-6-phosphate dehydrogenase) was lacking. However, this enzyme has since been identified in *P. falciparum*, *P. knowlesi* and *P. berghei* (Sherman, 1991). Although the parasite G6PD has a higher affinity for glucose-6-phosphate than does the host enzyme, the contribution to the ribose pool of the parasite appears to be minor. It is of special interest to note that in certain parts of the world, a deficiency in G6PD in the human erythrocyte – which is an X-linked abnormality – occurs more commonly in highly malarious regions, suggesting that such a deficiency provides some selective advantage for human survival. This speculation is also supported by the results from *in vitro* experiments in which growth of plasmodia was retarded in G6PD-deficient cells (Sherman, 1991).

Carbon dioxide fixation. All the species of *Plasmodium*, so far studied, are capable of CO_2 fixation. Data are available for *P. falciparum*, *P. berghei*, *P. knowlesi* and *P. lophurae* (Blum & Ginsburg, 1984; Sherman, 1979, 1984). In the last three species, the end-products have been identified as alanine, asparate, glutamate, and citrate, with α-ketoglutarate and oxaloacetate as intermediate products. In *P. falciparum*, the rate of CO_2 fixation is essentially linear, as shown by incubation of infected cells with [^{14}C]bicarbonate; see Fig. 7.10.

Krebs cycle. The question of whether or not a functional Krebs cycle operates in malaria parasites has yet to be resolved. Scheibel (1988) summarised the complex situation by commenting that obtaining unequivocal evidence was 'fraught with pitfalls'! In experiments relating to that shown in Fig. 7.10, Blum & Ginsburg (1984) found that *P. falciparum*-infected erythrocytes, grown *in vitro*, did not release $^{14}CO_2$ when incubated in the presence of [1-^{14}C]glutamate, despite the presence of the enzyme glutamate dehydrogenase. They concluded that this implied the absence of α-ketoglutarate dehydrogenase activity and the lack of a functional Krebs cycle.

Mitochondrial metabolism. It has long been thought that the mitochondria of avian *Plasmodium* were cristate, whereas those from mammals were not. This led to the speculation (not confirmed; see above) that the avian species had a functional Krebs cycle, whereas those in mammals had not. It is now recognised that this is an oversimplification of a complex situation and that the mitochondria of some species in mammals are cristate but in others they are not. Adequate fixation has undoubtedly been a problem in the past, but this has now been overcome by the isolation of mitochondria, before fixation, by treating the parasites with saponin (0.15 per cent v/v) in serum-free RPMI 1640 at room temperatures (Fry & Beesley, 1991). Electron microscopy of saponin-released mitochondria has shown that those of *P. yoelii* (from rats) have a clear double membrane and easily recognisable cristae, whereas the mitochondria of *P. falciparum*, which also have a pronounced double membrane, show a variable internal structure in which 'only occasionally can well-defined cristae or whorls be distinguished' (Fry, 1991).

In spite of these apparent morphological differences, the mitochondria from both species exhibit comparable metabolic activities. Thus, when the respiratory activity of the isolated mitochondria was examined (Fry & Beesley, 1991), three substrates were found to be readily utilised – NADH (45 per cent), α-glycerophosphate (33 per cent) and succinate (15 per cent), with lower rates for proline, dihydroorotate and glutamate (*P. falciparum*). These results generally support the conclusion, indicated earlier, that a complete Krebs cycle is lacking.

Electron transport. The successful technique for obtaining isolated mitochondria has enabled these organelles to be examined spectroscopically for the first time (Fry & Beesley, 1991; Fry, 1991). The low temperature difference spectra of dithionite-reduced minus oxidised mitochondria of *P. yoelii* and *P. falciparum* have been examined, as shown in Fig. 7.12. The cytochrome

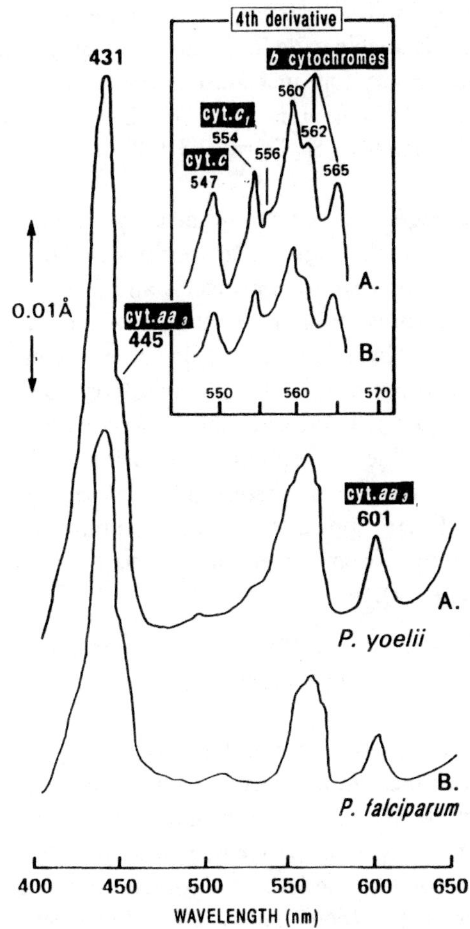

Fig. 7.12
Cytochromes in *Plasmodium* spp. Dithionite-reduced minus oxidised low temperature difference spectra of mitochondrial fractions of *Plasmodium yoelii* and *P. falciparum* showing the presence of a range of cytochromes in both species. (Modified from Fry, 1991.)

profiles of both species were virtually identical except for an apparently higher cytochrome : protein ratio in *P. yoelii* and an absorbance maximum at 556 nm (b cytochrome?) which appeared to be absent or unresolved in *P. falciparum*. Since all the a-, b- and c-type cytochromes appear to be present it can be concluded that *Plasmodium* has a 'classical' electron transport train. This view is further supported by the fact that the 'standard' mitochondrial transport inhibitors are effective. It is beyond the scope of this text to review all aspects of the mitochondrial metabolism; for further details the review of Fry (1991) should be consulted.

○ 7.6.3 Nucleic acid metabolism

Pyrimidine biosynthesis. Unlike some flagellate protozoa (e.g. *Trichomonas*; p. 54), *Plasmodium* is able to synthesise pyrimidines *de novo*, although the erythrocyte it parasitises is unable to do so. Hence a salvage mechanism is unnecessary for survival. The functional *de novo* pathway in *Plasmodium* involved in this synthesis is reviewed by Scheibel & Sherman (1988).

Purine salvage. In contrast to the above, a functional *de novo* biosynthetic pathway for purines has not been identified in any *Plasmodium* species and a purine salvage pathway must operate. A number of the enzymes involved have been identified, including adenosine kinase, adenosine deaminase, adenine phosphoribosyltransferase, purine nucleoside phosphorylase and hypoaxanthine-guanine phosphoribosyltransferase (Reyes *et al.*, 1982). The purine preferred by *P. falciparum* and *P. lophurae* (avian) appears to be hypoaxanthine, which is probably derived from the erythrocyte via the following (Sherman, 1991):

$$\text{ATP} \rightarrow \text{ADP} \rightarrow \text{AMP} \rightarrow \text{IMP} \rightarrow \text{inosine} \rightarrow \text{hypoaxanthine.}$$

Lipid metabolism. During the various stages in the life cycle of *Plasmodium* many membrane-containing organelles are formed – mitochondria, food vacuoles, endoplasmic reticulum, nucleus, cytostome, rhoptries, pelliculate complex and the parasitophorous vacuolar membrane – the biochemistry and ultrastructure of which have been reviewed by Sherman (1985). Biogenesis of these membranes involves a marked increase in the lipid content of the infected erythrocyte and, since the parasite has no lipid reserves and appears to be incapable of fabricating most lipids, these stages must rely on obtaining them by 'dynamic exchanges with the blood plasma an activity akin to that of the red cell' (Scheibel & Sherman, 1988). More detailed consideration of the lipid metabolism is to be found in the reviews of Sherman (1979–91) and Scheibel & Sherman, 1988).

○ 7.6.4 Protein metabolism

Amino acids required for the synthesis of proteins in *Plasmodium* are obtained from three sources; (*a*) biosynthesis from carbon sources, e.g. by CO_2 fixation; (*b*) uptake of free amino acids from the plasma or blood cells of the host; and (*c*) proteolysis of host haemoglobin.

CO_2 fixation. As mentioned earlier, it has long been recognised that *Plasmodium* spp., and blood cells infected with the parasite, fix CO_2 (Fig. 7.10) and synthesise amino acids such as alanine and aspartic and glutamic acids (Ting & Sherman, 1966; Sherman & Ting, 1968; Sherman, 1979). The free amino acid pool of infected blood cells is thus increased in these amino acids. In fact, very little of the amino acids formed by CO_2 fixation appears to be incorporated into the proteins of the parasite but may be involved in other activities (Scheibel & Sherman, 1988; Sherman, 1979).

Uptake mechanism of free amino acids. *Plasmodium*-infected erythrocytes can take up and incorporate free amino acids from the external environment at a higher rate than normal, uninfected erythrocytes. For example, *P. knowlesi*-infected (monkey) cells accumulate isolecucine, methionine, leucine, histidine and cysteine (Sherman, 1977*a*,*b*, 1979). It was also found that intraerythrocytic stages of *P. knowlesi* when grown *in vivo* or *in vitro* required extracellular sources of methionine and isoleucine, and it was speculated that this was due to the fact that primate haemoglobin is low in these acids. Further work, however, revealed that monkey erythrocytes contain adequate amounts of cystine and methionine to support parasite growth, the actual concentrations of these substances in the erythrocytes being respectively 2.4 and 6.0 times those found in the mature stages of the parasite (Sherman, 1979).

The mechanism of uptake of amino acids has only been examined for a few species. In uninfected duck erythrocytes, uptake of amino acids is saturable and appears to be a carrier-mediated process. In contrast, in duck erythrocytes infected with *P. lophurae*, the entry process is not saturable – except for glycine – and appears to be by diffusion. When isolated, erythrocyte-free, stages of *P. knowlesi* were examined it was found that only arginine, lysine, glutamatic acid, aspartic acid and cysteine appeared to be taken up by a carrier-mediated process. However, this result must be treated with caution, as the erythrocyte-free parasites were obtained by saponin treatment and there is some question that such treatment may have affected their permeability (Sherman, 1979). In support of the active transport hypothesis – an energy-requiring process – is the fact that erythrocyte-free parasites synthesise protein best when supplemented with glucose and red cell extract (Sherman, 1979).

Proteolysis of haemoglobin and malarial pigment formation. In addition to proteolytic activity being involved in the invasion of host cells during various stages of the life cycle (Schrével *et al.*, 1991), it also plays a major role in the degradation of haemoglobin as a source of amino acids, for the quantities of amino acids synthesised by the parasite or available from the erythrocyte are insufficient for the parasite's requirements. Rosenthal (1991) reviewing the proteinases in malarial parasites, summarises the evidence that *Plasmodium* degrades host haemoglobin into free amino acids, as follows:

(*a*) during the life cycle of the intraerythrocytic stages, the haemoglobin content of infected erythrocytes decreases by 25–75 per cent;
(*b*) the concentration of free amino acids is greater in infected than in uninfected erythrocytes;
(*c*) the composition of the amino acid pool of infected erythrocytes is similar to the amino acid composition of haemoglobin;
(*d*) when erythrocytes containing radiolabelled haemoglobin are infected with *Plasmodium*, the labelled amino acids appear in the proteins of the parasites.

It has long been recognised that the brown-black pigment (haemozoin) in infected humans was related to the destruction of haemoglobin by plasmodia. In *P. falciparum*, haemoglobin degradation occurs predominantly in the trophozoites and early schizonts. Erythrocyte cytoplasm, ingested by trophozoites, is transported within vesicles to the large central food vacuole. In the food vacuole it is broken down by various proteinases into *haem*, a major component of the malarial pigment, and *globin*, which is subsequently hydrolysed into its constituent amino acids (Rosenthal, 1991).

Molecular biology of the malarial parasite. Substantial progress has been made in our understanding of the genome of *Plasmodium* and of its genetics. Research in this field has been comprehensively reviewed by Weber (1988).

○○○ References

(References of general interest have been included in this list, in addition to those quoted in the text.)

Aikawa, M. 1971. *Plasmodium*: the fine structure of malarial parasites. *Experimental Parasitology*, **30**: 284–320.

Aikawa, M. 1980. Host cell invasion by malarial parasites. In: *Cellular Interactions in Symbiosis and Parasitism* (ed. C. B. Cook, P. W. Pappas & E. D. Rudolph), pp. 31–46. Ohio State University Press, Columbus, USA.

Aikawa, M. 1988. Fine structure of malaria parasites in the various stages of development. In: *Malaria* (ed. W. H. Wernsdorfer & I. McGregor), Vol. 1, pp. 97–129. Churchill-Livingstone, Edinburgh and London.

Aikawa, M. & Seed, T. M. 1980. Morphology of Plasmodia. In: *Malaria* (ed. J. P. Kreier), Vol. 1, pp. 285–344. Academic Press, New York.

Ashong, J. O., Blench, I. P. & Warhurst, D. C. 1989. The composition of haemozoin from *Plasmodium falciparum. Transactions of the Royal Society of Tropical Medicine and Hygiene*, **83**: 167–72.

Beale, G. H. & Walliker, D. 1988. Genetics of malaria parasites. In: *Malaria* (ed. W. H. Wernsdorfer & I. McGregor), Vol. 1, pp. 379–93. Churchill Livingstone, Edinburgh and London.

Blum, J. J. & Ginsburg, H. 1984. Absence of α-ketoglutarate dehydrogenase activity and presence of CO_2-fixing activity in *Plasmodium falciparum* grown *in vitro* in human erythrocytes. *Journal of Protozoology*, **31**: 167–9.

Boyd, M. F. 1949. *Malariology*. 2 Vols. Saunders & Co., Philadelphia.

Breuer, W. V. 1985. How the malaria parasite invades its host cell, the erythrocyte. *International Review of Cytology*, **96**: 191–238.

Bruce-Chwatt, L. 1980. *Essential Malariology*. 2nd Edition. William Heinemann Medical Books, London.

Carter, R. & Graves, P. M. 1988. Gametocytes. In: *Malaria* (ed. W. H. Wernsdorfer & I. McGregor), Vol. 1, pp. 253–305. Churchill-Livingstone, Edinburgh and London.

Cook, L., Grant, P. T. & Kermack, W. O. 1961. Proteolytic enzymes of the erythrocytic forms of rodent and simian species of malarial plasmodia. *Experimental Parasitology*, **11**: 372–9.

Cox, F. E. G. & Vickerman, K. 1966. Pinocytosis in *Plasmodium vinckei. Annals of Tropical Medicine and Parasitology*, **60**: 293–6.

Divo, A. A., Geary, T. G., Jensen, J. P. & Ginsburgh, H. 1985. The mitochondrion of *Plasmodium falciparum* visualized by rhodamine-123 fluorescence. *Journal of Protozoology*, **32**: 442–5.

Fairlamb, A. H. 1989. Novel biochemical pathways in parasitic protozoa. *Parasitology*, **99S**: S93–112.

Fletcher, A. & Maegrath, B. G. 1972. The metabolism of the malaria parasite and its host. *Advances in Parasitology*, **10**: 31–48.

Foote, S. J. & Kemp, D. J. 1989. Chromosomes of malarial parasites. *Trends in Genetics*, **5**: 337–42.

Fry, M. 1991. Mitochondria of *Plasmodium*. In: *Biochemical Protozoology* (ed. G. Coombs & M. North), pp. 154–67. Taylor & Francis, London.

Fry, M. & Beesley, J. E. 1991. Mitochondria of mammalian *Plasmodium* spp. *Parasitology*, **102**: 17–26.

Garnham, P. C. C. 1988. Malaria parasites of man: life cycles and morphology (excluding ultrastructure). In: *Malaria* (ed. W. H. Wernsdorfer & I. McGregor), Vol. 1, pp. 61–96. Churchill-Livingstone, Edinburgh.

Gero, A. M., Brown, G. V. & O'Sullivan, W. J. 1984. Pyrimidine *de novo* synthesis during the life cycle of the intraerythrocytic stage of *Plasmodium falciparum*. *Journal of Parasitology*, **70**: 536–41.

Hawking, F., Worms, M. J. & Gammage, K. 1968. 24- and 48-hour cycles of malaria parasites in the blood; their purpose, production and control. *Transactions of the Royal Society of Tropical Medicine and Hygiene*, **62**: 731–60.

Hermentin, P. 1987. Malaria invasion of human erythrocytes. *Parasitology Today*, **3**: 52–5.

Homewood, C. A. & Neame, K. D. 1980. Biochemistry of malarial parasites. In: *Malaria* (ed. J. P. Kreier), Vol. 1, pp. 345–405. Academic Press, New York.

Inselburgh, J. & Banyal, H. S. 1984. Synthesis of DNA during the asexual cycle of *Plasmodium falciparum* in culture. *Molecular and Biochemical Parasitology*, **10**: 79–87.

Jensen, M. D., Conley, M. & Helstowski, L. D. 1983. Culture of *Plasmodium falciparum*: the role of pH, glucose and lactate. *Journal of Parasitology*, **69**: 1060–7.

Kemp, D. J., Thompson, J. K., Walliker, D. & Corcoran, L. M. 1987. Molecular karyotype of *Plasmodium falciparum*: conserved linkage groups and expendible histidine-rich protein genes. *Proceedings of the National Academy of Sciences*, **84**: 7672–6.

Knell, A. J. 1989. *Malaria*. Oxford University Press.

Kreier, J. 1977. *Parasitic Protozoa*. Vol. III. Academic Press, New York.

Kreier, J. (ed.) 1980. *Malaria*. Vol. 1. Academic Press, New York.

Meis, J. F. G. M. & Verhave, J. P. 1988. Exoerythrocytic development in malarial parasites. *Advances in Parasitology*, **27**: 1–61.

Pasvol, G. & Wilson, R. J. M. 1989. Red cell deformability and invasion by malarial parasites. *Parasitology Today*, **5**: 218–21.

Perkins, M. E. 1989. Erythrocytic invasion by the malarial merozoite. *Experimental Parasitology*, **69**: 94–9.

Peters, W. & Richards, W. H. G. 1984. *Handbook of Experimental Pharmacology*, vol. 68/1. *Antimalarial drugs*. I. Springer-Verlag, Berlin. ISBN 3 540 12616 3.

Prensier, G. & Slomianny, Ch. 1986. The karyotype of *Plasmodium falciparum* determined by ultrastructural serial sectioning and 3D reconstruction. *Journal of Parasitology*, **72**: 731–6.

Reyes, P., Rathod, P., Sanchez, P., Mrema, J., Reickmann, K. & Heidrich, H. 1982. Enzymes of purine and pyrimidine metabolism from the human malaria parasite, *Plasmodium falciparum*. *Molecular and Biochemical Parasitology*, **5**: 275–90.

Rosenthal, P. J. 1991. Proteinases of malarial parasites. In: *Biochemical Protozoology* (eds. G. Coombs & M. North), pp. 257–69. Taylor & Francis, London.

Roth, E. F. Jr, Calvin, M. C., Max-Audit, I., Rosa, J. & Rosa, R. 1988. The enzymes of the glycolytic pathway in erythrocytes infected with *Plasmodium falciparum* malaria parasites. *Blood*, **72**: 1922–5.

Scheibel, L. 1988. Plasmodial metabolism and related organellar function during various stages of the life cycle: carbohydrates. In: *Malaria* (ed. W. Wernsdorfer & I. McGregor), Vol. 1, pp. 171–204. Churchill-Livingstone, Edinburgh and London.

Scheibel, L. & Sherman, I. W. 1988. Plasmodial metabolism and related organellar function during various stages of the life cycle: proteins, lipids, nucleic acids and vitamins. In: *Malaria* (ed. W. Wernsdorfer, W. & I. McGregor), Vol. 1, pp. 219–52. Churchill-Livingstone, Edinburgh and London.

Scholtyseck, E. 1979. *Fine structure of Parasitic Protozoa*. Springer-Verlag, Heidelberg.

Schrével, J., Barrault, C., Grellier, P., Mayer, R. & Monsigny, M. 1991. *Plasmodium* proteinases during the erythrocytic phase of infection. In: *Biochemical Protozoology* (ed. G. Coombs & M. North), pp. 270–80. Taylor & Francis, London.

Sherman, I. W. 1977*a*. Transport of amino acids and nucleic acid precursors in malarial parasites. *Bulletin of the World Health Organization*, **55**: 211–25.

Sherman, I. W. 1977*b*. Amino acid metabolism and protein synthesis and malarial parasites. *Bulletin of the World Health Organization*, **55**: 265–76.

Sherman, I. W. 1979. Biochemistry of *Plasmodium* (malarial parasites). *Microbiological Reviews*, **43**: 453–95.

Sherman, I. W. 1984. Metabolism. In: *Handbook of Experimental Pharmacology*, Vol. 68/1. *Antimalarial drugs* (ed. W. Peters & W. H. G. Richards), pp. 31–81. Springer-Verlag, Berlin.

Sherman, I. W. 1985. Membrane structure and function of malaria parasites and the infected erythrocyte. *Parasitology*, **91**: 609–45.

Sherman, I. W. 1991. The biochemistry of malaria: an overview. In: *Biochemical Protozoology* (ed. G. Coombs & M. North), pp. 6–34. Taylor & Francis, London.

Sherman, I. W. & Ting, I. P. 1968. Carbon dioxide fixation in malaria. II. *Plasmodium knowlesi* (monkey malaria). *Comparative Biochemistry and Physiology*, **24**: 639–42.

Sinden, R. E. 1982. Gametocytogenesis of *Plasmodium falciparum in vitro*: an electron microscope study. *Parasitology*, **84**: 1–11.

Sinden, R. E. 1983*a*. Sexual development of malarial parasites. *Advances in Parasitology*, **22**: 153–216.

Sinden, R. E. 1983*b*. The cell biology of sexual development in plasmodium. *Parasitology*, **86**: 7–28.

Sinden, R. E. & Croll, N. A. 1975. Cytology and kinetics of microgemetogenesis and fertilization in *Plasmodium yoelii nigeriensis*. *Parasitology*, **70**: 53–65.

Sinden, R. E. & Hartley, R. H. 1985. Identification of the meiotic division of malarial parasites. *Journal of Protozoology*, **32**: 742–4.

Smyth, J. D. 1976. *An Introduction to Animal Parasitology*. 2nd Edition. Edward Arnold, London.

Tait, A. & Sacks, D. L. 1988. The cell biology of parasite invasion and survival. *Parasitology Today*, **4**: 228–34.

Ting, I. P. & Sherman, I. W. 1966. Carbon dioxide fixation in malaria-I. Kinetic studies in *Plasmodium lophurae*. *Comparative Biochemistry and Physiology*, **19**: 855–69.

Trager, W. & Jensen, J. B. 1976. Human malaria parasites in continuous culture. *Science*, **193**: 673–5.

Van der Ploeg, L. H. T., Smits, M., Ponnudurai, T., Vermeulen, A., Meuwissen, J. H. E. Th. & Langsley, G. 1985. Chromosome-sized molecules of *Plasmodium falciparum*. *Science*, **22**: 658–61.

Walliker, D. 1983*a*. The genetic basis of diversity in malaria parasites. *Advances in Parasitology*, **22**: 217–59.

Walliker, D. 1983*b*. *The Contribution of Genetics to the Study of Parasitic Protozoa*. Research Studies Press (John Wiley & Sons), London.

Walliker, D., Carter, R., Quakyi, I. A., Wellems, T. E., McCutchan, T. F. & Szarfmann, A. 1987. Genetics of *Plasmodium falciparum*. In: *Molecular Strategies of Parasite Invasion. UCLA Symposia on Molecular & Cellular Biology* (ed. N. Aabian, H. Goodman & N. Nogueira), Vol. 42, pp. 259–67. Alan R. Liss, New York.

Weber, J. L. 1988. Molecular biology of malarial parasites. *Experimental Parasitology*, **66**: 143–70.

Wernsdorfer, W. H. & McGregor, I. (eds) 1988. *Malaria. Principles and Practice of Malariology*. 2 vols. Churchill-Livingstone, Edinburgh and London.

8

Sporozoea: malaria in man and the animal kingdom

8.1 Mammalian malaria

8.1.1 Human malaria

Importance. It is not always realised that malaria is one of the world's greatest killers, ranking in this respect with cancer and heart disease. It remains endemic in some 102 countries with more than half the world's population at risk. There are probably more than 100 *million* cases of the disease throughout the world, of which perhaps a million are fatal. In spite of control programmes in many countries, the situation has shown little improvement within the past 15 years and, partly due to world economic and/or political problems, there have been downtrends in some countries, notably India and China (WHO, 1989). The problem has grown worse in some rural areas which have undergone intensive economic development, especially parts of Asia and in the Amazon region of Latin America. The global situation is reflected in Fig. 8.1; the major progress was made between 1957 and 1967, but the number of the protected world population fell by 1977 and had only marginally improved by 1984.

It should also be emphasised that non-endemic countries are not immune to infection; in the United Kingdom, for example, in 1987, nearly 2000 cases of imported malaria were reported with 3 deaths.

An enormous literature on malaria in man and animals exists, the most comprehensive recent account being that of Wernsdorfer & McGregor (1988). Other relevant reviews on various aspects of malariology are those of Garnham (1984), Killick-Kendrick & Peters (1977), Kreier (1980), Nussenzweig & Nussenzweig (1990) and Warrell *et al.* (1990). Literature on drug resistance of *Plasmodium* spp. – which has become a major problem in malaria control – has been reviewed by Cowman & Foote (1990) and Peters (1970, 1990).

A major research effort has been directed towards developing a vaccine against the disease. In spite of early optimism, success in this has so far eluded workers (Nussenzweig & Nussenzweig, 1990) and remains a major unachieved (and perhaps unachievable?) goal. This topic is discussed further in Chapter 32.

It was formerly thought that malaria was one of the few diseases of which the life cycle of the causal parasite was completely known and the main barrier to its control and elimination appeared to be largely financial and physical. Yet

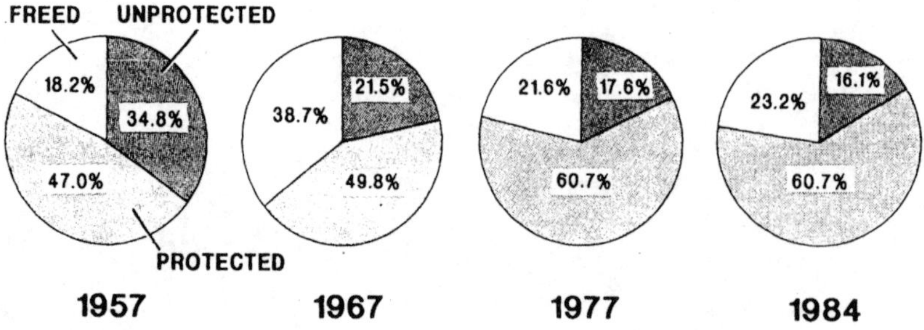

Fig. 8.1
Status of malaria control and eradication, 1957–84, showing the percentage of population *freed* from malaria, *protected* against malaria, and *unprotected* against malaria. (After Onari, 1988; reproduced with permission from *Malaria. Principles and Practices of Malariology* (ed. W. H. Wernsdorfer & I. McGregor), vol. 1, p. 1723. Churchill-Livingstone, Edinburgh and London, 1988.)

the disease has lived up to its reputation of continually confounding the parasitologist (see *History*, p. 110)! Two further problems have now arisen: (*a*) the appearance of drug-resistant strains and (*b*) the discovery that man may become infected by species of simian (monkey) malaria. These problems are dealt with later.

Excluding simian malaria, which is discussed later (p. 129), there are four species of *Plasmodium* that infect man and result in four kinds of malarial fever.

P. vivax: benign, simple or tertian malaria.
P. falciparum: aestivo-autumnal, malignant tertian, pernicious quotidian, subtertian or tropical malaria.
P. malariae: quartan ague, or quartan malaria.
P. ovale: ovale tertian malaria.

Of the above species, *P. vivax* shows the widest distribution, being prevalent throughout the tropics and many temperate regions. Vivax malaria is characterised by relapses: reappearances of symptoms after a latent period of up to 5 years, as is infection with *P. ovale*, which occurs chiefly in tropical Africa. As explained earlier (p. 112; Fig. 7.2), such relapses are due to the sudden activation of hypnozoites (sleeping merozoites) in liver cells. *P. falciparum* is most common in tropical and subtropical areas and causes the most dangerous, malignant form of malaria, which fortunately does not have relapses. *P. malariae* is widely distributed but is much less common than *P. vivax* or *P. falciparum*. Although falciparum malaria or malariae malaria do not show relapses, they are subjected to 'recrudescences': repeated manifestations of infection after a relatively short latent period between 3 months and 1 year.

Also, as mentioned (p. 127), human malaria can develop in certain monkey species.

The pattern of the life cycles of these species follows the general outline already given (Fig. 7.2) but there are important physiological differences which are often reflected in the nature of the diseases produced. Some of the morphological characteristics are summarised in Table 8.1 and illustrated in Fig. 8.2.

Many of these characteristics are sufficiently definite to be used as criteria for identification in blood films, but species differentiation may not be possible in the earliest stages. In *P. vivax* and *P. ovale* infections, erythrocytes become enlarged and paler in colour when the parasites have grown beyond the ring stage. In stained preparations, infected corpuscles show characteristic dots called Schüffner's dots (*P. vivax* and *P. ovale*), Ziemann's dots (*P. malariae*) or Maurer's dots (*P. falciparum*). The pigment granules vary in size and shape but their appearance and colour depend on a number of factors such as intensity of light, type of filter used, etc.

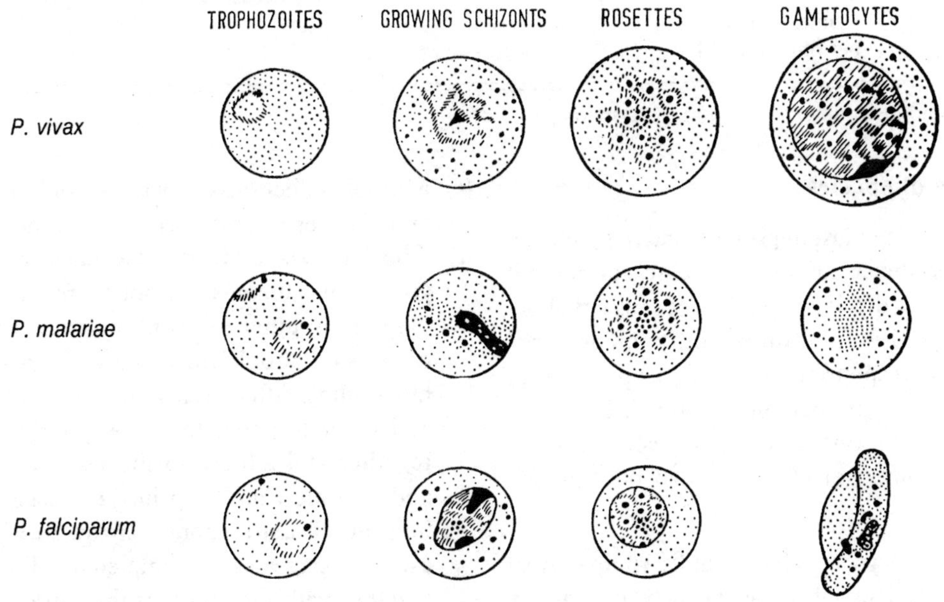

Fig. 8.2
Comparison of the various stages of the three common species of *Plasmodium* infecting man. (After various authors.)

Table 8.1. *Comparative characters of plasmodia of man*

Morphological descriptions based on stained thin films.

	P. falciparum	*P. vivax*	*P. malariae*	*P. ovale*
Incidence in world malarial infections	50 per cent	43 per cent	7 per cent	Rare
Erythrocytes attacked	Any indiscriminately	Reticulocytes only	Older erythrocytes only	?
Exo-erythrocytic forms	Do not persist	Persist as hypnozoites	Do not persist	Persist as hypnozoites
Schizogony cycle (h)	48	48	72	48
Rings	1/5–1/6 diam. r.b.c. Multiple infections common	1/3 diam. r.b.c. Often 2 rings or more per r.b.c.	1/3 diam. r.b.c. Double infections rare	As *P. malariae*
Late trophozoite	Medium, compact; band forms frequent; vacuole inconspicuous; rare in peripheral blood	Large, very amoeboid; prominent vacuole	Small, not amoeboid; often band-shaped; vacuole inconspicuous	Small, not amoeboid; vacuole inconspicuous
Mature schizont	Smaller than r.b.c. Rare in peripheral blood	Larger than r.b.c.	Smaller than r.b.c.	Larger than in *P. malariae*
Number of merozoites	12–24, usually 8–12	8–16, usually 12–15	6–12, usually 8	6–12, usually 8
Microgametocytes*	Crescents usually sausage-shaped	Spherical; compact	Similar to *P. vivax* but smaller and less numerous	As *P. malariae*
Macrogametocytes	Crescents often longer and more slender	Spherical; compact	Similar to *P. vivax* but small and less numerous	As *P. malariae*
Alterations in r.b.c.	Normal size; appear 'brassy'. Maurer's dots or 'clefts' may be visible; rare in peripheral blood	Enlarged and pale; Schüffner's dots present	May appear smaller; fine dots (Ziemann's dots) occasionally seen	r.b.c. oval; Schüffner's dots prominent and appear early

*Usually smaller and less numerous than macrogametocytes.

Characteristics of species

P. falciparum**.** The most important malarial parasite, the disease it produces running an acute course and often terminating fatally. It is a significant cause of abortion or stillbirth and even death of non-immune pregnant women. It is responsible for some 50 per cent of all the malarial cases throughout the world. Its distribution is restricted to warm and tropical countries.

The special features of its life cycle may be summarised as follows:

(*a*) It attacks erythrocytes of all ages indiscriminately so that a high density of parasites may be rapidly reached. In extreme cases up to 48 per cent of the red cells may be parasitised.

(*b*) Multiple infections (polyparasitism) resulting in several ring forms in a corpuscle are not uncommon.

(*c*) The later stages in the asexual cycle, that is, the growth to schizonts, do not occur in the peripheral blood as in other forms of malaria, except in severe cases, so that only rings and crescents are found in blood films. After twenty-four hours, the ring forms and older trophozoites show a tendency to clump together and adhere to the visceral capillary walls and become caught up in the vessels of the heart, intestine, brain or bone marrow in which the later asexual stages are completed. This behaviour, together with the fact that the subtertian malaria is more toxic, are the principal reasons why this type is so dangerous. The severe and complicated malaria which may develop from falciparum malaria has been reviewed in detail by Warrell *et al.* (1990).

(*d*) Sporulation is not so well synchronised as in other species so that fever paroxysms may be longer drawn out.

(*e*) EE forms do not persist in the tissues and hence relapses do not occur.

P. vivax. Causes the benign tertian form of malaria which is responsible for about 43 per cent of all cases in the world and has the widest geographical distribution. Although generally not life-threatening, it can cause severe, acute illness. Several points in its life cycle may be noted:

(*a*) The degree of infection is low, for only the young immature corpuscles (reticulocytes) are attacked; about 2 per cent of erythrocytes are parasitised.

(*b*) The periodicity of the asexual cycle is closely synchronised.

(*c*) Hypnozoites develop in the liver, so that relapses may occur (Fig. 7.2).

The morphological features of a sub-species *bastianellii* of the simian species *P. cynomolgi* (see p. 131) bear a striking resemblance to those of *P. vivax*; differentiation between these two is difficult.

P. malariae. A relatively rare parasite producing quartan malaria which is responsible for about seven per cent of the malaria in the world. Particular points of interest are:

(*a*) Infected erythrocytes are not larger than uninfected ones and sometimes even smaller.

(*b*) Mature erythrocytes are attacked and rarely reticulocytes, so that the density of parasites is very low; about 0.2 per cent of erythrocytes are parasitised.

(*c*) It is often difficult to distinguish between a large trophozoite and an immature gametocyte.

P. ovale. This is a species rarely encountered; it is confined essentially to the tropics and subtropics although reported from many continents. The type of fever it produces (ovale tertian) is milder than the benign tertian of *P. vivax*. Special points of interest are:

(*a*) It morphologically resembles *P. malariae* in most of its stages.

(*b*) The changes produced in the erythrocytes in general are similar to those produced by *P. vivax*, but Schüffner's dots appear considerably earlier (in the ring stage) and are coarser and more numerous.

(*c*) In the oocyst the pigment granules are (usually) characteristically arranged in two rows crossing each other at right angles.

(*d*) Hypnozoites develop in the liver so that relapses may occur (Fig. 7.2).

Vectors of human malaria. As far as is known, the plasmodia of man will develop only in female mosquitoes of the genus *Anopheles* and not in any other arthropod. Almost any species can be infected with plasmodia in the laboratory, but many species are poor vectors and not natural ones. For a mosquito to be an efficient vector, it must present certain characteristics: (*a*) it must be susceptible to infection and present physico-chemical and nutritional characteristics suitable for the development of plasmodia; (*b*) it must bite man in preference to animals; (*c*) it must not be shy of human habitation; (*d*) its span of life must be sufficiently long to permit sexual development of the plasmodium.

A comprehensive list of the *Anopheles* of human malaria and their distribution is available (White, 1989); see Table 8.2.

Control measures. It is beyond the scope of this text to discuss the many and varied problems of eradication methods based on the elimination of the mosquito vector. These problems are not only biological and technical but frequently involve social, economic and political questions. Recent reviews on these complex problems are those of Bruce-Chwatt (1980), Onori (1988) and Wernsdorfer & Wernsdorfer (1988).

It is not intended here to deal with morphological or physiological features of mosquitoes, or to provide taxonomic details. These are treated in works such as Garnham (1966), Gillett (1972), Gillies & De Meillon (1968) and Gillies & Coetzee (1987).

○ 8.1.2 Simian malaria: speciation

The chimpanzee, gorilla, orangutan and gibbon all contract various species of malaria, as do some monkey species. The best known of these are listed in Table 8.3. Some of these closely resemble those which occur in man. Thus *P. schwetzi* is the ape equivalent of *P. vivax*; *P. reichenowi* is the ape equivalent of *P. falciparum*, and *P. rodhaini* is indistinguishable from *P. malariae* and may

Table 8.2. *Vectors of malaria throughout the world*

The main species of vectors are printed in bold type; others are local or secondary vectors. Incidental vectors are omitted. Species no longer valid are put in parentheses. Names of subgenera: *A*, *Anopheles; C, Cellia; K, Kerteszia; N, Nyssorhynchus.*

Zones	Vectors	
1. North American	*(A.(A.)freeborni)* *(A.(A.)quadrimaculatus)*	*(A.(N.)albimanus)*
2. Central American	*A.(A.)aztecus* *A.(A.)punctimacula* **A.(N.)albimanus** *A.(N.)albitarsis*	**A.(N.)aquasalis** *A.(N.)argyritarsis* **A.(N.)darlingi**
3. South American	**A.(A.)pseudopunctipennis** **A.(A.)punctimacula** **A.(K.)bellator** **A.(K.)cruzii** **A.(N.)albimanus** **A.(N.)albitarsis** **A.(N.)aquasalis**	*A.(N.)argyritarsis* *A.(N.)braziliensis* **A.(N.)darlingi** **A.(N.)nuneztovari s.l.** *A.(N.)benarrochi* *A.(N.)triannulatus*
4. North Eurasian	**(A.(A.)atroparvus)** *(A.(A.)messeae)* *A.(A.)sacharovi*	**A.(A.)sinensis** *A.(C.)pattoni*
5. Mediterranean	**A.(A.)atroparvus** *A.(A.)claviger* **A.(A.)labranchiae** *A.(A.)messeae*	**A.(A.)sacharovi** *A.(C.)hispaniola* **A.(C.)superpictus**
6. Afro-Arabian *(Desert)*	*A.(C.)culicifacies* *A.(C.)fluviatilis* *A.(C.)hispaniola*	*A.(C.)multicolor* **A.(C.)pharoensis s.l.** **A.(C.)sergentii**
7. Afrotropical	**A.(C.)arabiensis** **A.(C.)funestus** **A.(C.)gambiae** **A.(C.)melas**	*A.(C.)merus* *A.(C.)moucheti* *A.(C.)nili* *A.(C.)pharoensis s.l.*
8. Indo-Iranian	*A.(A.) sacharovi* *A.(C.)annularis* **A.(C.)culicifacies s.l.** **A.(C.)fluviatilis**	*A.(C.)pulcherrimus* *A.(C.)stephensi* *A.(C.)superpictus* *A.(C.)tessellatus*
9. Indo-Chinese hills	*A.(A.)nigerrimus* *A.(C.)annularis* *A.(C.)culicifacies s.l.* **A.(C.)dirus s.l.**	**A.(C.)fluviatilis** *A.(C.)maculatus s.l.* **A.(C.)minimus**
10. Malaysian	**A.(A.)campestris** **A.(A.)donaldi** **A.(A.)letifer** **A.(A.)nigerrimus** *A.(A.)whartoni* **A.(C.)aconitus** **A.(C.)balabacensis** **A.(C.)dirus s.l.**	**A.(C.)leucosphyrus** **A.(C.)ludlowae** **A.(C.)maculatus s.l.** **A.(C.)minimus** *A.(C.)philippinensis* **A.(C.)subpictus s.l.** **A.(C.)sundaicus**
11. Chinese	*A.(A.)lesteri anthropophagus* **A.(A.)sinensis**	*A.(C.)pattoni*

Table 8.2. *(cont.)*

Zones	Vectors	
12. Australasian	*A.(A.)bancrofti* *A.(C.)annulipes s.l.* **A.(C.)farauti s.l.** *A.(C.)karwari*	*A.(C.)kolensis* *A.(C.)punctulatus* *A.(C.)subpictus s.l.*

Source: Data from White (1989).

Table 8.3. *Some species of simian malaria and their infectivity for man*

Species	Natural host	Morphology resembles	Infective to man
Oriental Region			
P. cynomolgi	monkeys of genus *Macaca*	*P. vivax*	Yes
P. c. bastianellii	monkeys of genus *Macaca*	*P. vivax*	Yes
P. c. ceylonensis	monkeys of genus *Macaca*	*P. vivax*	No
P. coatneyi	monkeys of genus *Macaca*	—	No
P. eylesi	white-handed gibbon	—	Yes
P. fieldi	monkeys of genus *Macaca*	—	No
P. fragile	monkeys of genus *Macaca*	—	No
P. knowlesi	monkeys of genus *Macaca*	—	Yes
P. inui	monkeys of genus *Macaca*	—	Yes
Neotropical Region			
P. brasilianum	numerous monkey species	*P. malariae*	Yes
P. simium	howler monkey	*P. vivax;* *P. ovale*	Yes
Ethiopian Region			
P. reichenowi	chimpanzee, gorilla	*P. falciparum*	No
P. rodhaini	chimpanzee	*P. malariae*	Yes
P. gonderi	drill, mangabey	*P. vivax*	No
P. schwetzi	chimpanzee, gorilla	*P. vivax*	Yes

Source: Data from Garnham (1966, 1967) and Coatney (1968).

be a synonym. Details of these species can be found in Bruce-Chwatt (1980), Kreier (1980) and Collins (1988).

As discussed in Chapter 7, some species of monkey, notably *Aotus trivirgatus*, have proved to be valuable experimental animal hosts for the human *P. vivax* and *P. falciparum*. The use of this primate species in research has been so successful that its continued existence in the wild became so threatened that its export from Latin America has been prohibited.

○ 8.1.3 Simian malaria: infectivity to man

As mentioned earlier, one of the surprising research developments has been the discovery that certain species of simian malaria are capable of infecting man. Although the blood forms of one species, *P. knowlesi*, were known to be transmissible to man, it was generally held that monkey malaria was not usually infective to man. This position changed dramatically in 1960 when the unexpected occurrence of a human infection with *P. cynomolgi bastianellii*, contracted in a laboratory, was reported.

A graphic account of the discovery of this infection has been given by Coatney (1968) who describes (in a lecture to the American Society of Tropical Medicine and Hygiene) how on 5 May 1960 he received a telephone call from Memphis, Tennessee and heard the late Dr Don Eyles say 'Bob, I have monkey malaria.' He goes on to say 'I was incredulous. Could it be true? It was, for Eyles' preliminary diagnosis was confirmed quickly. *Plasmodium cynomolgi* was the offending parasite.'

Since that unexpected finding, a number of species have been found to be transmissible to man. These are listed in Table 8.3. For detailed accounts of monkey malaria as it affects man see Coatney (1968), Kreier (1980), Collins (1988) and Garnham (1966, 1967).

Initially, experimental infections in man are usually light, and it is easy to see how such infections were missed in the past. Virulence may develop later and gametocytes may sometimes be produced. In some species (e.g. the *P. cynomolgi* forms) the symptoms which develop are quite out of proportion to the low parasitaemia, in their severity. Although many examples of the laboratory transmission of monkey malaria to man are now known, it is probable that only rarely in nature does it occur as a true zoonosis, the only examples at present being confined to Brazil and Malaya. Garnham (1967) concludes that 'monkey malaria is unlikely to interfere seriously with malaria eradication'.

○ 8.1.4 Rodent malaria: *P. berghei*

Discovery. In 1943, a Belgian antimalarial team began working on the mosquitoes of certain forest districts in Katanga, Congo. The mosquito *Anopheles dureni*, which has a localised distribution (on shady trees by the Kisanga River) was found to be frequently gorged with non-human blood, and to show a high sporozoite index. This engorged blood when tested gave positive reactions to anti-rat serum, so the blood of rodents occurring in the district was examined for parasites. This led to the discovery of a new species of malaria, *P. berghei*, in the blood of the thicket rat, *Grammomys surdaster* (Vincke & Lips, 1948). This was a finding of immense significance, for it has been found possible to transmit this species of malaria to a number of laboratory rodents with the result that laboratory research on malaria has been greatly facilitated.

General account. Since the discovery of *P. berghei* a number of other species of rodent malarias have been found; these are listed in Table 8.4.

Major reviews on the biology of rodent malaria are those of: Garnham (1966, 1980), Killick-Kendrick & Peters (1977), Carter & Diggs (1977), Kreier (1980) and Cox (1988).

Only a limited number of rodent forms have been used routinely for laboratory studies; isoenzyme analysis has revealed the existence of many subspecies and strains. One disadvantage with working with these species is that each has its own special characteristics, which may vary from host to host; this makes it difficult for individual workers in different laboratories to compare results. A further problem is that the infection may become profoundly altered after repeated blood passages; for example, some organisms may lose their ability to produce gametocytes.

Life cycle. All rodent species have similar life cycles. There is typically only one exo-erythrocytic phase in the liver, which takes 48–60 h to complete. A second schizogonic phase may sometimes occur in some species. The normal erythrocytic cycle occurs with a periodicity of 24 h, schizonts producing 6–18 merozoites. The gametocyte and sporogonic phases are similar to those described in man (Chapter 7).

Species fall into three groups: the *P. berghei–P. yoelii* group, the *P. vinckei* group and the *P. chabaudi* group; for details see Cox (1988). Only *P. berghei*, which is the most widely used laboratory rodent model, is described here.

Plasmodium berghei

Biocenose. The characteristic biocenose* of *P. berghei* is gallery forest of Upper Katanga between an altitude of 1000 and 17 000 metres with a population of various wild rodents (especially the semi-arboreal rat *Grammomys surdaster*) and the mosquito *Anopheles dureni*. It is interesting to note that this natural focus has an unusually low temperature – between 19 °C and 21 °C – and it is only when infected mosquitoes are kept at this temperature in the laboratory that they produce viable sporozoites; higher temperatures produce abnormal sporogony.

Vectors. The natural invertebrate host is *Anopheles dureni*, a species of mosquito difficult to maintain in the laboratory; as indicated above, a low temperature is essential to assure sporogony. Of the other possible mosquito hosts, only *A. stephensi* appears to be really satisfactory (Garnham, 1967); mosquitoes of the *maculipennis* complex, as a whole, appear to be unsuitable. Laboratory transmission can, of course, be readily performed by blood inoculation.

*Essentially a community of organisms, both plants and animals, living together in a particular habitat or *biotype*. The physical environment is not included but the biocenose and biotype form a natural unit, the *biocenose-biotype*; this is essentially the same as the *ecosystem* of some biologists.

Table 8.4. *Species of* Plasmodium *which occur naturally in rodents*

Species	Hosts	Vectors	Susceptibility of laboratory rodents
P. berghei	Thicket rats *(Grammomys surdaster)*	*Anopheles* spp.	+
P. yoelii	Thicket rats *(Thamnomys rutilans)*	*Anopheles* spp.	+
P. chabaudi	Thicket rats *(T. rutilans)*	*Anopheles* spp.	+
P. vinckei	Thicket rats *(T. rutilans)*	*Anopheles* spp.	+
P. aegyptensis	Nile rat	Unknown	?
P. atheruri	Brush-tailed porcupine	*Anopheles* spp.	+
P. anomaluri	African scaly-tailed flying squirrels	Unknown	?
P. landauae	African scaly-tailed flying squirrels	Unknown	–
P. pulmophilum	African scaly-tailed flying squirrels	Unknown	–
P. booliati	Asian flying squirrels	Unknown	?
P. watteni	Asian flying squirrels	Unknown	?

Source: Data from Cox (1988).

Laboratory hosts. A wide range of laboratory hosts have been tested for susceptibility to *P. berghei.* Young rats and hamsters are easily infected. Adult rats are refractory to infection. In mice and young rats, *P. berghei* is almost always fatal.

Life cycle. The life cycle follows the typical pattern (Fig. 7.2). Although there are some marked differences between the various subspecies, the following general points of interest may be observed in *P. berghei* infections (in mice):

(*a*) The exo-erythrocytic stages occur in the parenchymal cells of the liver and mature 48 hours after the introduction of sporozoites.
(*b*) The erythrocytic cycle takes 22–24 hours, and commences with uninucleate rings, containing a relatively large vacuole; the mature schizont contains 10–18 nuclei.
(*c*) The trophozoites have a predilection for reticulocytes, up to 100 per cent of which may be infected, whereas only 10 per cent of erythrocytes may be infected. In contrast, gametocytes develop in mature erythrocytes exclusively.
(*d*) Polyparasitism is not uncommon, the cytoplasm of the parasites merging into a confluent mass.

○○○ 8.2 Bird malaria

Most of the important milestones in understanding the malarial life cycle were discovered using bird malaria as experimental material. Ross's discovery of the mosquito transmission was the most prominent contribution but other substantial contributions include the introduction of various malarial drugs first tried out on birds, and more recently, the establishment of the extra-erythrocytic stages of the life cycle. However, the discovery of rodent malaria has resulted in this form of malaria replacing avian malaria as a laboratory model, the rodent host being easier to maintain in a laboratory. An early monograph of bird malaria is that of Hewitt (1940); more recent work is dealt with by Seed & Manwell (1977), Garnham (1980), and McGhee (1988).

Occurrence and geographical distribution. Plasmodia in birds is widespread throughout the world, except possibly in the far North; the incidence varies within the range 1–19 per cent with a mean of 5.8 per cent (estimated on some 7000 birds). The climatic barriers which limit the distribution of human malaria do not exist in bird malaria, and *the incidence of infection is as high in temperate regions as it is in the tropics.*

Passerine birds are more frequently infected than any other group, sparrows showing the highest incidence of infection. Domestic birds, e.g. pigeons, geese, ducks, turkeys, chickens and doves, are rarely infected.

Experimental hosts. One of the difficulties that beset early workers was the fact that they could never be sure that the laboratory hosts which they used, mainly sparrows, chaffinches, pigeons, linnets and larks, were free from parasites. The discovery of the canary as a suitable laboratory host for most species of avian malaria soon led to its adoption as the standard experimental host. Canaries (especially females) have the advantage of being cheap, easily maintained and virtually never

Table 8.5. *Some species of avian malaria and their natural and laboratory hosts*

Many 'strains' of most species are known.

Subgenera	Example	Natural host	Laboratory host	Suitable laboratory vector	Comments
Haemamoeba	*P. relictum*	house sparrow*	canaries	*Culex pipiens*	highly pathogenic
	P. cathemerium	house sparrow	canaries	*Culex pipiens*	pathogenic
	P. gallinaceum	jungle fowl	domestic fowl	*Aedes aegypti**	brain especially infected
Huffia	*P. elongatum*	house sparrow†	ducklings; canaries	*Culex pipiens*	highly pathogenic
Giovannolaia	*P. circumflexum*	redwing†	canaries	*Culiseta annulata*	little studied
	P. lophurae	fire-back pheasant	chicks; ducks	*Aedes aegypti*	pathogenic
Novyella	*P. juxtanucleare*	jungle fowl	domestic fowl	*Culex p. pallens?*	pathogenic

*Host range exceptionally wide.
†Host range fairly wide.

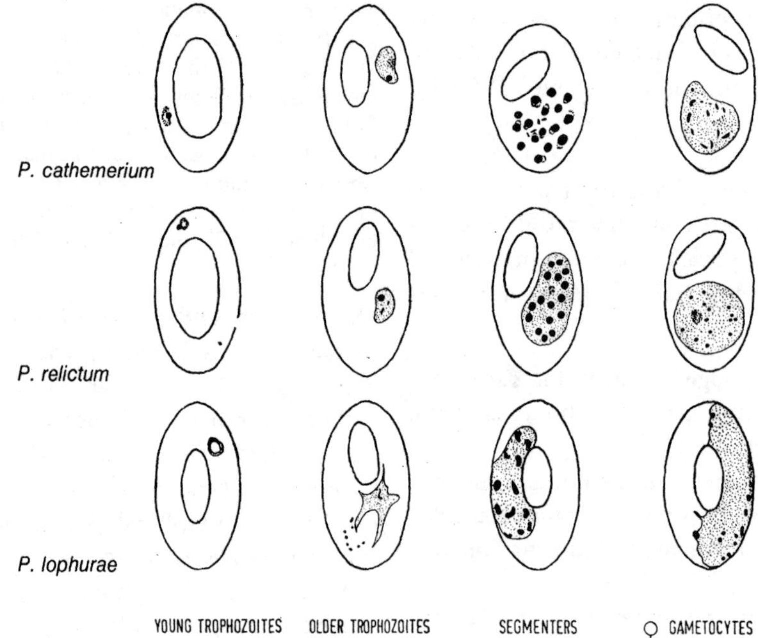

Fig. 8.3
Comparison of the various stages of three species of avian plasmodia. (After Hewitt, 1940.)

infected naturally with malarial parasites. Chicks have later replaced canaries; supplies of these may be cheaply obtained.

Species. There are numerous species of avian malaria known, of which only a small number have been used extensively in laboratory studies. *Plasmodium relictum* and *P. cathemerium* (Fig. 8.3) have been most widely used for experimental work and they also occur more commonly than any other species in nature. They are particularly suitable in being readily infective to canaries and transmissible by *Culex pipiens*, one of the few species of mosquitoes easily maintained in the laboratory. The small size of canaries does not permit the withdrawal of more than about 600–800 mg of blood, and since the total amount of blood only amounts to about 1000 mg the donor always dies. Where large amounts of parasitised blood are required, more suitable laboratory species are *P. gallinaceum* or *P. lophurae*. The former species was first imported into France by Brumpt in 1936 in domestic fowl and is readily maintained in chicks in the laboratory.

Life cycle

General account. Although the life cycle of avian malarias follows the general pattern of the mammalian parasites, some differences should be noted:

(*a*) A major difference is seen in the exo-erythrocytic (EE) stages of the cycle. Thus, human malaria has a non-repetitive phase which leads irreversibly to the blood infection and the erythrocytic cycle. In avian malaria, however, repetitive EE schizogony occurs, which is also replenished from erythrocytic merozoites. One can thus speak of an EE 'cycle' in avian malaria, whereas an EE 'phase' (i.e. a single event) accurately defines the situation in mammalian malaria (McGhee, 1988).

(*b*) The site of EE development also differs in mammalian and avian malarias, the bird forms developing in the lymphoid–macrophage system and not in the liver as in the mammalian species.

(*c*) Species of *Anopheles* are rarely utilised as vectors; the natural vectors are usually mosquitoes of the genera *Culex* or *Aedes*.

Accurate diagnosis of species requires information on numerous biological characteristics especially those concerned with the erythrocytic stages, EE stages, sporogonic stages and vertebrate host specificity. Detailed data for identification are beyond the scope of this book, but preliminary identification into subgenera can be made from the following key. For more precise identification the works of Garnham (1966, 1980), Seed & Manwell (1977), McGhee (1988) or Huff (1965) should be consulted.

Key to the subgenera of avian *Plasmodium* (Garnham, 1966)

1.	Round gametocytes	*Haemamoeba*
	Elongate gametocytes	2
2.	Schizogony in primitive blood-forming cells present	*Huffia*
	Schizogony in primitive blood-forming cells absent	3
3.	Erythrocytic schizonts large with plentiful cytoplasm	*Giovannolaia*
	Erythrocytic schizonts small with scanty cytoplasm	*Novyella*

The characteristics of some avian species which have been used for experimental work are listed in Table 8.5.

The prepatent period in avian malaria varies with the species, the method employed in inoculation (i.e. by mosquito bite, sporozoite injection or blood transfer) and particularly the size of the inoculum. With heavy inocula (2×10^7 organisms) the prepatent period may be as short as one day, with light injections (1×10^3) as long as nine days. The length ofthe patent period similarly varies with the nature of the infection, whether benign or acute, and covers the range 5–23 days. The subpatent periods in bird malarias may be as long as eight months. During this period, parasites are present in such small numbers that they cannot be demonstrated by the usual routine methods in the peripheral blood, and their presence can only be confirmed by subinoculation into other unparasitised animals.

○○○ 8.3 Amphibian malaria

Several cases of so-called amphibian 'malaria' have been reported (e.g. *P. bufonis* in *Bufo americanus*) but the validity of such identification as true malarial parasites still remains to be verified.

○○○ 8.4 Reptilian malaria

There are about 25 known species of malaria in reptiles; these have been reported from five continents. It occurs commonly in all the major families of lizards and occasionally in snakes, but not in crocodiles, turtles or tuataras. The life cycles and general biology are poorly known, the parasites being found in erythrocytes and sometimes in leucocytes. The mechanism of transmission remains unknown, the suspected vectors being mosquitoes, hemiptera and acarine ectoparasites. The systematics are also controversial. What is known of the biology and epidemiology of reptile malarias has been succinctly reviewed by Ayala (1977).

○○○ References

(References of general interest have been included in this list, in addition to those quoted in the text.)

Ayala, S. C. 1977. Plasmodia of reptiles. In: *Parasitic Protozoa* (ed. J. P. Kreier), Vol. III, pp. 267–309. Academic Press, New York.

Bruce-Chwatt, L. J. 1980. *Essential Malariology*. William Heinemann Medical Books, London.

Carter, R. & Diggs, C. L. 1977. Plasmodia of rodents. In: *Parasitic Protozoa* (ed. J. P. Kreier), Vol. III, pp. 359–465. Academic Press, New York.

Coatney, G. R. 1968. Simian malarias in man: facts, implications and predictions. *American Journal of Tropical Medicine and Hygiene*, **17**: 147–55.

Collins, W. E. 1988. Major animal models in malaria research: simian. In: Wernsdorfer, W. H. & McGregor, I. (eds) 1988 (q.v.), Vol. 2, pp. 1473–501.

Cowman, A. F. & Foote, S. J. 1990. Chemotherapy and drug resistance in malaria. *International Journal for Parasitology*, **20**: 503–13.

Cox, F. E. G. 1988. Major animal models in malaria research: rodent. In: Wernsdorfer, W. H. & McGregor, I. (eds) 1988 (q.v.), Vol. 2, pp. 1503–43.

Garnham, P. C. C. 1966. *Malaria Parasites and Other Haemosporidea*. Blackwell Scientific Publications, Oxford.

Garnham, P. C. C. 1967. Malaria in mammals excluding man. *Advances in Parasitology*, **5**: 139–204.

Garnham, P. C. C. 1980. Malaria in its various vertebrate hosts. In: *Malaria* (ed. J. P. Kreier), Vol. 1, pp. 95–144. Academic Press, New York.

Garnham, P. C. C. 1984. The present state of malaria research; an historical survey. *Experientia*, **40**: 1305–10.

Gillett, J. D. 1972. *Common African Mosquitoes and their Medical Importance*. William Heinemann Medical Books, London.

Gillies, M. T. & Coetzee, M. 1987. A supplement to the Anophelinae of Africa south of the Sahara. *Publications of the South African Institute for Medical Research*, No. 55. Johannesburg.

Gillies, M. T. & De Meillon, B. 1968. The Anophelinae of Africa south of the Sahara. *Publications of the South African Institute for Medical Research*, No. 54. Johannesburg.

Hewitt, R. 1940. *Bird Malaria*. John Hopkins Press, Baltimore.

Huff, C. G. 1965. Susceptibility of mosquitoes to avian malaria. *Experimental Parasitology*, **16**: 107–32.

Killick-Kendrick, R. & Peters, W. (eds) 1977. *Rodent Malaria*. Academic Press, London.

Kreier, J. P. (ed.) 1980. *Malaria*. Vol. 1. Academic Press, New York.

McGhee, R. B. 1988. Major animal models in malaria research: avian. In: Wernsdorfer, W. H. & McGregor, I. (eds) 1988 (q.v.), Vol. 2, pp. 1545–67.

Nussenzweig, V. & Nussenzweig, S. 1990. Sporozoite malaria vaccine: where do we stand? *Annales de Parasitologie Humaine et Comparée*, **65**, Suppl. I: 49–52.

Onori, E. 1988. Malaria control constraints. In: Wernsdorfer, W. H. & McGregor, I. (eds) 1988 (q.v.), Vol. 2, pp. 1721–39.

Peters, W. P. 1970. *Chemotherapy and Drug Resistance in Malaria*. Academic Press, London.

Peters, W. P. 1990. *Plasmodium*: resistance to antimalarial drugs. *Annales de Parasitologie Humaine et Comparée*, **65**, Suppl. I: 103–106.

Seed, T. M. & Manwell, R. D. 1977. Plasmodia of birds. In: *Parasitic Protozoa* (ed. J. P. Kreier), Vol. III, pp. 311–57. Academic Press, New York.

Targett, G. A. T. 1991. *Malaria: Waiting for the Vaccine*. John Wiley and Sons, London.

Vincke, I. H. & Lips, M. 1948. Un noveau plasmodium d'un rongeur sauvage du Congo, *Plasmodium berghei n.sp. Annales de la Société Belge de Medecine Tropical*, **28**: 97–104.

Warrell, D. A., Molyneux, M. E. & Beales, P. F. (eds) 1990. *Severe and Complicated Malaria*. 2nd Edition. (WHO Division of Control of Tropical Diseases.) *Transactions of the Royal Society of Tropical Medicine & Hygiene*, **84**, Suppl. 2: 1–65.

Wernsdorfer, W. H. & McGregor, I. (eds) 1988. *Malaria. Principles and Practice of Malariology*. 2 vols. Churchill-Livingstone, Edinburgh and London.

Wernsdorfer, W. H. & Trigg, P. I. 1988. Recent progress of Malaria research: chemotherapy. In: Wernsdorfer, W. H. & McGregor, I. (eds) 1988 (q.v.), Vol. 2, pp. 1569–674.

Wernsdorfer, G. & Wernsdorfer, W. H. 1988. Social and economic aspects of malaria. In Wernsdorfer, W. H. & McGregor, I. (eds) 1988 (q.v.), Vol. 2, pp. 1421–71.

White, G. B. 1989. Malaria. In: *Geographical distribution of arthropod-borne diseases and their principal vectors. WHO Report*: VBC/89.967, pp. 7–22. WHO, Geneva.

WHO. 1989. *Tropical Diseases. Progress in International Research, 1987–88*. (9th Programme Report of the UNDP/World Bank/WHO Special Programme for Research and Training in Tropical Diseases (TDR).) WHO, Geneva. ISBN 92 4 1561297.

9

Sporozoea: Haemosporina other than Plasmodia; Piroplasmia

9.1 Family Haemoproteidae

9.1.1 Genus *Leucocytozoon*

General account. Species of this genus are malaria-like parasites of birds which have no erythrocytic trophozoite stages but which undergo exo-erythrocytic schizogony in the parenchyma of a number of organs, especially the liver, spleen, heart and brain. The organism is unusual in forming two types of gametocytes (round and elongate) and in having a life cycle synchronised with that of the reproductive pattern of the host. Vectors are blackflies of the genus *Simulium*.

The most studied species is *Leucocytozoon simondi* (in ducks) on which the following account is based.

The development is characterised by the appearance of two kinds of schizonts, *hepatic* schizonts and *megaloschizonts*, and two kinds of gametocytes, *round* and *elongate*. These appear to be formed as follows (Desser, 1967):

(*a*) On injection of sporozoites by the vector, the first asexual cycle takes place exclusively in the parenchymal cells of the liver, *hepatic* schizonts being formed. The prepatent period is 5 days (Fig. 9.1).

(*b*) Progeny from the hepatic schizonts follow one of three courses:

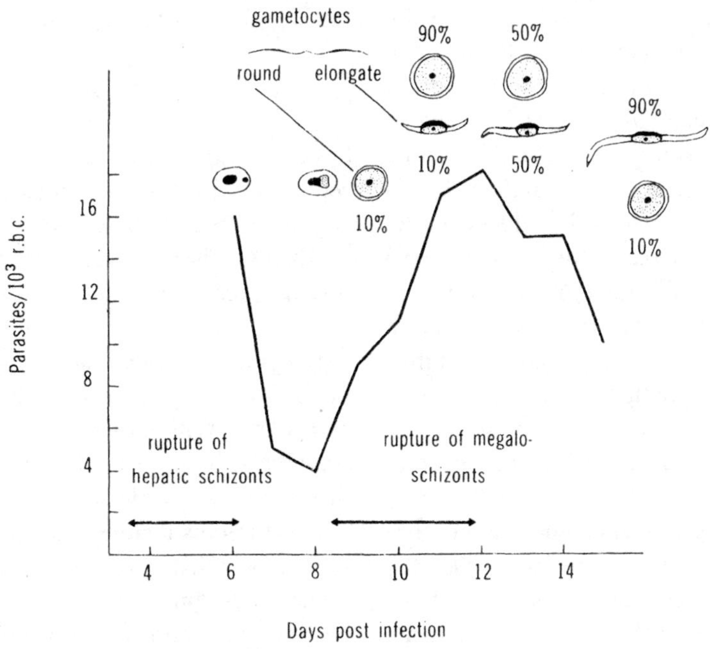

Fig. 9.1
Leucocytozoon simondi: level of parasitaemia in artificially infected ducklings and appearance of gametocytes in relation to the initial asexual cycle. (After Desser, 1967.)

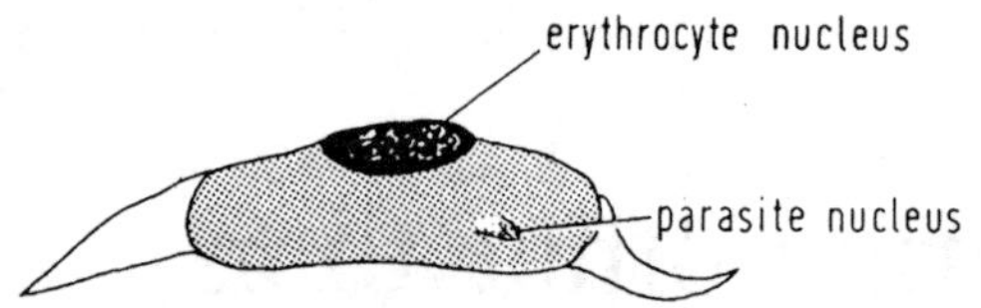

Fig. 9.2
Gametocyte of *Leucocytozoon neavei* in erythrocyte of guinea-fowl. (After Richardson, 1948.)

(i) some invade further parenchymal cells to initiate another hepatic cycle;
(ii) some penetrate blood cells (mainly reticulocytes and erythrocytes) to form *round* gametocytes;
(iii) some are phagocytosed by macrophages and grow into *megaloschizonts* throughout the body.

(*c*) Elongate gametocytes (Fig. 9.2) are formed only from the progeny of megaloschizonts. This developmental pattern is summarised in Figs 9.1 and 9.3.

In the insect vector, a sexual process takes place which results in a sporogony somewhat similar to that found in *Plasmodium*; sporozoites pass to the salivary gland and are injected with saliva during biting.

Synchronisation of host and parasite reproductive cycles. It has been shown that there is a striking relationship between the degree of parasitaemia with *L. simondi* and the vertebrate host physiology (Chernin, 1952). During autumn and winter months (October to January) parasitaemia occurs only at a low level. Almost coincidental with the onset of the sexual maturity of the bird, characterised especially by egg-laying in the female, there is a marked relapse, as manifested by a high percentage of positively infected blood smears. By subjecting female birds to increased hours of artificial light per day in autumn and winter, it is possible to precipitate egg-laying weeks or even months earlier with a concomitant shift in the time of the relapse. It is thus clear that the reproductive pattern of the parasite is related to the complexity of metabolic, hormonal and anatomical changes taking place during the onset of sexual activity.

The nature of the 'relapse phenomenon' is obscure but it has been speculated that in the late spring the action of some hormone or chemical substance softens the capsule of a 'relapse schizont' and allows the parasite to escape and complete its development (Desser *et al.*, 1968).

Table 9.1. *Prevalence of* Leucocytozoon smithi *and* Haematoproteus meleagridis *in male and female wild turkeys* in Florida during 1969–72*

Sex	No. of birds examined	Percentage infection			% negative
		H	L	H + L†	
Males	185	88	73	65	4
Females	299	82	72	61	8

*Excluding turkeys <1 month old.
†H, *H. meleagridis*; L, *L. smithi*.
Source: Data from Forrester *et al.* (1974).

Other species. Some 50 species of *Leucocytozoon* have been reported, one species of major economic importance being *L. smithi* (Table 9.1). The disease, leucocytozoonosis – often known as 'turkey malaria' or 'gnat' fever – caused by this species, results in mortalities as high as 75 per cent in young turkeys and greatly affects reproduction in older infected birds. A number of devastating outbreaks of the disease have occurred in the USA. The known blackfly vectors are *Simulium occidentale*, *S. jenningsi*, *S. slossonae*, *S. congareenarum* and *Prosimulium hirtipes*. The biology and control of this species has been reviewed by Long *et al.* (1987), Fallis & Desser (1977), Bennett (1987) and Mongardi (1988).

○ 9.1.2 Genus *Haemoproteus*

Species of this genus are blood parasites of birds and reptiles in which, like *Leucocytozoon*, no erythrocytic schizogony occurs. The best-known are *H. columbae*, which produces a disease, sometimes fatal, with the misleading title of 'pigeon-malaria', and *H. meleagridis* in turkeys.

Life cycle. *H. columbae* occurs in the endothelial cells of the blood vessels, especially those of the lungs, liver and spleen. Within these cells, the organisms divide into cytomeres. The host cell eventually ruptures, releasing the cytomeres inside which merozoites are developing. The cytomeres disintegrate and release the merozoites, some of which may repeat the schizogony in other endothelial cells, but others penetrate red blood cells and become male and female gametocytes. These gametocytes (Fig. 9.4) are shaped like a curved sausage and encircle the nucleus in a halter-fashion, and are some-

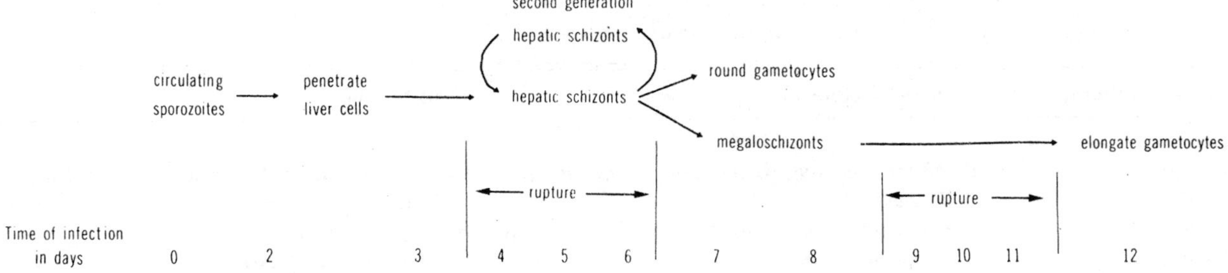

Fig. 9.3
Leucocytozoon simondi: inter-relationship between various stages of the parasite. (After Desser, 1967.)

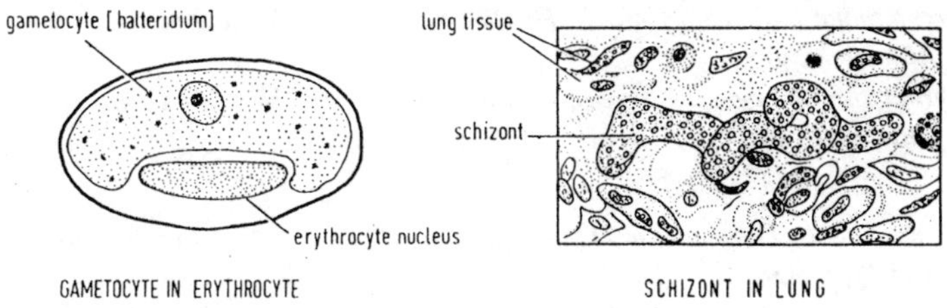

Fig. 9.4
Haemoproteus columbae: blood and tissue stages in the pigeon. (After Chandler, 1955.)

times referred to as *Halteridia*, as they were at one time called by the generic name of *Halteridium*.

H. meleagridis is a common parasite in wild turkeys, often occurring in birds infected with *Leucocytozoon smithi* (Table 9.1). It is only mildly pathogenetic, much less so than *L. smithi*. Its biology has been reviewed by Long *et al.*, 1987.

○○○ 9.2 Subclass Piroplasmia

Members of this subclass, commonly referred to as the 'piroplasms' on account of the pear-like shape of their merozoites, are small intra-erythrocytic parasites of mammals, particularly cattle and, more rarely, man.

○ 9.2.1 Family Babesiidae

Genus *Babesia*

Species of *Babesia* cause disease in domestic animals, sheep, cattle, goats, horses, pigs, dogs and cats (Table 9.2). Man can also be infected. Bovine babesiosis is economically an important disease with a cosmopolitan distribution and one which has a major impact on the livestock industries in many countries. It is known under a variety of names, the most common being 'redwater fever' or 'tick fever'. In the developing world, it is known under other names such as 'Texas fever', 'piroplasmosis', 'ranilla', 'splenic', 'La Tristeza', etc. (McCosker, 1981).

The two major species infecting cattle in North and South America, Australia and South Africa are *Babesia bovis* and *B. bigemina*. The vectors are one-host ticks, *Boophilus microplus*, *B. decoloratus* and *B. annulatus*. The development in the vector is unusual in that the parasites are passed to the eggs and hence to the larval stages which can thus become infective after the adult tick (which drops off the host) dies. This process is known as *transovarian* transmission.

The disease has been controlled in many countries by eradication of the tick vector, control of animal movements and vaccination, the latter being particularly effective in Australia, where up to 750 000 doses of vaccines are used each year in the State of Queensland. For a succinct review of biology and control of cattle babesiosis, see Young & Morzaria (1986).

Table 9.2. *Common species of* Babesia *in domestic animals and man*

Host and *Babesia* sp.	Principal vectors	Distribution
Bovine		
B. bigemina	*Boophilus annulatus, B. microplus, B. decoloratus, B. geigyi*	Central and South America, Southern Europe, Africa, Asia, Australia
B. bovis	*B. annulatus, B. microplus, B. geigyi, Ixodes ricinus*	Same as *B. bigemina*
B. major	*Haemaphysalis punctata*	Europe and North Africa
B. divergens	*I. ricinus*	Northern Europe
B. jakimovi	*I. ricinus*	Siberia, northern Russia
B. ovata	*Haemaphysalis longicornis*	Japan
B. occultans	*Hyalomma m. rufipes*	South Africa
Equine		
B. caballi	*Dermacentor (Anocentor) nitens, Hyalomma, Rhipicephalus* sp.	World-wide
B. equi	*Dermacentor (Anocentor) nitens, Hyalomma, Rhipicephalus* sp.	World-wide
Canine		
B. canis	*Rhipicephalus, Dermacentor, Haemaphysalis* (?) and *Hyalomma* (?)	Southern Europe, North America, Africa, Asia
B. gibsoni	*Rhipicephalus sanguineus, Haemaphysalis bispinosa*	Far East, Asia, but expanding its distribution to Europe and North America
Ovine/caprine		
B. motasi	*Haemaphysalis, Rhipicephalus*	Europe, Asia, Africa
B. ovis	*Rhipicephalus*	Southern Europe, Middle East, Africa
B. trautmanni	*Boophilus* (?), *Rhipicephalus* (?), *Dermacentor* (?), *Hyalomma* (?)	Southern Europe, former USSR, Africa
B. perroncitoi	Unknown	Southern Europe, Africa
Feline		
B. herpailuri	Unknown	Africa
B. felis	Unknown	Africa and southern Asia
Man		
B. microti	*I. dammini*	Northeastern states of US
B. divergens	*I. ricinus*	Northern Europe

Source: Data from Kuttler (1988).

Literature. Other major reviews and monographs on various aspects of babesiosis are those of Friedhoff (1981), McCosker (1981), Mehlhorn & Schein (1984), Piesman (1987), Raether (1988), Ristic (1979, 1988), Ristic & Kreier (1981), Ristic *et al.* (1984) and WHO (1987).

Morphology and life cycle. The morphology and life cycle has been reviewed by Mehlhorn & Schein (1984) and Ristic (1988); it is best known for *B. canis* in the dog (Fig. 9.5). It is only within recent years that the (so-called) sexual phase of the life cycle has been clarified. The host becomes infected by the injection of sporozoites from an infected tick. The sporozoites penetrate an erythrocyte and appear as single or paired pear-shaped bodies which reproduce by binary fission. The released merozoites then reinfect further erythrocytes. Unlike *Plasmodium* (Fig. 7.3), however, although a parasitophorous membrane is formed from the host membrane, it soon breaks down, the parasite becoming free within the erythrocyte. Destruction of the blood corpuscles may result in the development of haemoglobinuria, hence the name 'redwater fever'. Not all merozoites reinvade erythrocytes; some become oval-shaped and undergo no further development until taken up into the gut of a tick when it gorges blood. These stages are *gamonts*; whereas normal merozoites in erythrocytes become lysed in the gut, the gamonts develop in the gut cells as bizarre-shaped 'sexual' stages which possess at least one 'thorn' and a number of flagella-like protrusions; these bodies are now known as *ray-bodies* or *strahlenkörper* (Fig. 9.5). The function of the 'thorn' is unknown; its ultrastructure is summarised by Mehlhorn & Schein (1984). Two ray-bodies fuse to form a zygote; this becomes a *kinete*, which leaves the gut and enters cells of various organs, eventually reaching the oocytes in the ovary via the haemolymph. Thus the eggs laid by the adult are already infected (trans-ovarian infection) and hence the larvae which develop

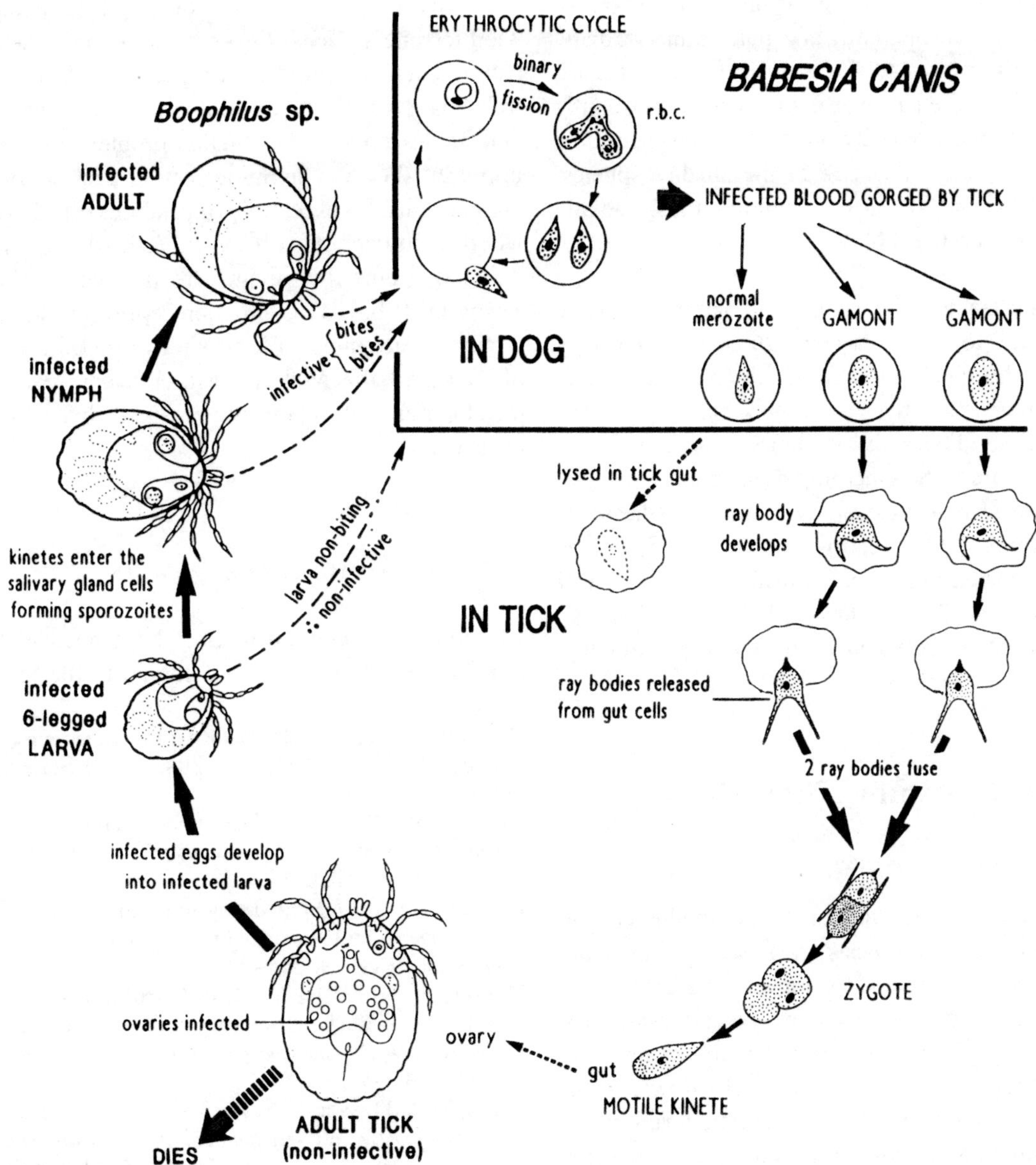

Fig. 9.5
Life cycle of *Babesia canis* in the dog. A dog becomes infected by the bite of an infected tick, the sporozoites being injected with the saliva. These penetrate erythrocytes and asexual reproduction by binary fission follows. This results in the formation of merozoites, most of which reinvade further erythrocytes; some merozoites form ovoid *gamonts*. When a tick bites an infected dog and engorges with blood, the normal merozoites are lysed in the tick gut, but the gamonts survive and differentiate into *ray bodies* (*strahlenkörper*) within the gut cells. When fully differentiated, these ray bodies leave the gut cells and fuse in pairs. The zygotes so formed give rise to motile kinetes, some of which invade oocytes; thus the eggs, when laid by the adult, are already infected. Within the developing larvae, some kinetes penetrate the salivary gland cells, where each kinete gives rise to many thousands of sporozoites. When the infected tick (or tick nymph) bites another dog, the sporozoites are injected in the saliva and the cycle starts again. (Based partly on Mehlhorn & Schein 1984; Smyth 1976.)

are also infected. Within the larvae some kinetes penetrate the cells of the salivary glands forming a large multinuclear *sporont* which finally gives rise to thousands of sporozoites. The six-legged larva which develops from the eggs is non-biting and is therefore not a vector, but the nymph and the adult, which are both infective and biting, serve as vectors. In species which utilise two- and three-host ticks, transmission is not only trans-ovarian but can also apparently take place trans-stadially (Riek, 1966). In trans-stadial transmission, parasites are ingested by either the larvae or nymph and subsequently transmitted as sporozoites by the succeeding nymph or adult. It also appears that two of the smaller species, *B. equi* and *B. microti*, are also transmitted trans-stadially (Young & Morzaria, 1986).

Human babesiosis. Babesiosis in man is relatively rare and is due to the cattle form, *B. bovis*, which may be fatal, and the species, *B. microti*, which is less pathogenic and, until recently, was largely confined to the United States. There is evidence, however, that *B. microti* appears to be emerging as an important tick-borne disease of man in temperate climates (Piesman, 1987), and WHO (1987) reports that since 1975, 7 cases have been reported in France, a further 7 in the rest of Europe and some 200 in the USA. *B. microti* is also known to infect rodents and insectivores; presumably, these act as reservoir hosts.

○○○ 9.3 Family Theileriidae

○ 9.3.1 Genus *Theileria*

Species of *Theileria* are the cause of 'theileriasis' in cattle. The best-known species are *T. parva*, *T. annulata* and *T. mutans*. The life cycle somewhat resembles that of *Babesia* except that the stages in the vertebrate host occur in the lymphocytes of the lymphatic system and in the endothelial cells. As in *Babesia*, *Theileria* species form ovoidal or spherical intra-erythrocytic forms which become gamonts and, when ingested by ticks, develop into either ray bodies (which give rise to microgametes) or round bodies (which are considered to be macrogametes). A microgamete and a macrogamete fuse and the kinete formed from the resultant zygote eventually migrates to the salivary gland cells. Here active asexual reproduction takes place and thousands of sporozoites are formed. The latter are injected on biting the host. The life cycle has been reviewed in detail by Mehlhorn & Schein (1984).

T. parva is the cause of the deadly East Coast Fever in cattle in southern Africa, the principal vector being the three-host tick *Rhipicephalus appendiculatus*. The epidemiology and dynamics of transmission have been reviewed by Gettinby & Byrom (1989).

T. annulata is the cause of tropical theileriosis or Mediterranean Coast Fever in cattle and is widely distributed across southern Europe, north Africa and central Asia. Its main tick vectors are *Hyalomma excavatum* and *H. detritum*. It is a major problem in Iran where imported (i.e. non-immune) cattle, such as *Bos taurus*, are particularly susceptible (mortalities 40–80 per cent), whereas indigenous cattle (*B. indicus*) are much less affected (mortality about 10 per cent) (Hashemi-Fesharki, 1988). This situation represents a major constraint on attempts to improve livestock in many parts of Asia and the Middle East. Active research on the development of a vaccine against theileriosis is in progress.

○○○ References

(References of general interest have been included in this list, in addition to those quoted in the text.)

Bennett, G. F. 1987. Haematozoa. In: *Comparative Bird Malaria* (ed. E. W. Burr), pp. 120–8. Iowa State University Press, Iowa.

Chernin, E. 1952. The relapse phenomenon in *Leucocytozoon simondi* infections of the domestic duck. *American Journal of Hygiene*, **56**: 101–18.

Desser, S. S. 1967. Schizogony and gametogony of *Leucocytozoon simondi* and associated reactions in avian host. *Journal of Protozoology*, **14**: 244–54.

Desser, S. S., Fallis, A. M. & Garnham, P. C. C. 1968. Relapses in ducks chronically infected with *Leucocytozoon simondi* and *Parahaemoproteus nettionis*. *Canadian Journal of Zoology*, **46**: 281–4.

Fallis, M. & Desser, S. S. 1977. On species of *Leucocytozoon*, *Haemoproteus*, and *Hepatocystis*. In: *Parasitic Protozoa* (ed. J. P. Kreier), Vol. III, pp. 239–66. Academic Press, New York.

Forrester, D. J., Hon, L. T., Williams, L. E. & Austin, D. H. 1974. Blood protozoa of wild turkeys in Florida. *Journal of Protozoology*, **21**: 494–7.

Friedhoff, K. T. 1981. Morphologic aspects of *Babesia* in the tick. In: *Babesia* (ed. M. Ristic & J. P. Kreier), pp. 143–69. Academic Press, New York.

Gettinby, G. & Byrom, W. 1989. The dynamics of East

Coast Fever: a modelling perspective for the integration of knowledge. *Parasitology Today*, **5**: 68–73.

Hashemi-Fesharki, R. 1988. Control of *Theileria annulata* in Iran. *Parasitology Today*, **4**: 36–42.

Kuttler, K. L. 1988. World-wide impact of babesiosis. In: *Babesiosis of Domestic Animals and Man* (ed. M. Ristic), pp. 1–21. CRC Press, Boca Raton, Florida.

Long, P. L., Current, W. L. & Noble, G. P. 1987. Parasites of the Christmas Turkey. *Parasitology Today*, **3**: 360–6.

McCosker, P. J. 1981. The global importance of babesiosis. In: *Babesiosis* (ed. M. Ristic & J. P. Kreier), pp. 1–124. Academic Press, New York.

Mehlhorn, H. & Schein, E. 1984. The piroplasms: life cycle and sexual stages. *Advances in Parasitology*, **23**: 37–103.

Mongardi, D. 1988. (Leucocytozoonis: a slow but spreading disease.) *Rivista di Avicoltura*, **57**: 55. (In Italian.)

Piesman, J. 1987. Emerging tickborne diseases in temperate climates. *Parasitology Today*, **3**: 197–9.

Raether, W. 1988. Chemotherapy and other measures of parasitic diseases in domestic animals and man. In: *Parasitology in Focus* (ed. H. Mehlhorn), pp. 739–866. Springer-Verlag, Berlin.

Riek, R. F. 1966. The development of *Babesia* spp. and *Theileria* spp. in ticks with special reference to those occurring in cattle. In: *Biology of Parasites* (ed. E. J. L. Soulsby), pp. 127–39. Academic Press, New York.

Ristic, M. 1979. *Babesiosis*. Academic Press, London.

Ristic, M. 1988. *Babesiosis of Domestic Animals and Man*. CRC Press, Boca Raton, Florida.

Ristic, M., Ambroise-Thomas, P. & Kreier, J. 1984. *Malaria and Babesiosis*. (New Prospects in Clinical Medicine, No. 7.) Nijhoff, Dordecht.

Ristic, M. & Kreier, J. P. (eds) 1981. *Babesiosis*. Academic Press, New York.

WHO. 1987. Parasitic diseases: human babesiosis. *Weekly Epidemiological Record*, **62** (No. 28): 226–7.

Young, A. S. & Morzaria, T. P. 1986. Biology of *Babesia*. *Parasitology Today*, **2**: 211–19.

10

Opalinata and Ciliophora

10.1 'Opalinids' and 'ciliates'

These groups constitute the two most highly organised groups of Protozoa and contain both free-living and parasitic species. Some species provide superb models for the study of fundamental cellular phenomena, such as asexual/sexual differentiation. Both groups were originally classified as 'ciliates' and included in the Subphylum Ciliophora. However, later classifications have recognised major differences between them and placed the opalinids with the flagellates in the Subphylum Opalinata of the Phylum Sarcomastigophora and the ciliates in a separate Phylum Ciliophora. This classification is based on that of Levine *et al.* (1980), which is not universally accepted, so the phylogeny of the opalinids remains a matter of considerable dispute (Patterson, 1985). Small & Lynn (1981) have also proposed a new classification of the Ciliophora 'based primarily on the concept of the structural conservatism of the cortical fibrillar structures'.

The morphology and classification of ciliates have been reviewed in detail in a valuable monograph by Corliss (1979).

10.2 Classification (After Levine *et al.*, 1980)

Subphylum Opalinata (see comment above)

Numerous cilia-like oragnelles in oblique rows cover the entire body surface; cytostome absent; two to many nuclei of one type; all parasitic; nuclear division acentric; binary fission generally interkinetal. Contains only the Order Opalinada; e.g. *Opalina*, *Zelleriella*.

Phylum Ciliophora. A fundamental characteristic of ciliates is the possession of cilia at some stage in their life history and the possession of dimorphic nuclei, which include a vegetative *macro*nucleus and a generative *micro*nucleus (Fig. 10.1). The classification of the Ciliophora is exceptionally complex; for full details see Levine *et al.* (1980). Only the following genera are considered here.

Class 1 Kinetofragminophorea
- Order Trichostomatida, e.g. *Balantidium*
- Order Entodiniomorpha, e.g. *Diplodinium*

Class 2 Oligohymenophorea
- Order Hymenostomatidae, e.g. *Ichthyophthirius*

Class 3 Polymenophorea
- Order Heterotrichida, e.g. *Nyctotherus*
- Order Hypotrichida, e.g. *Kerona*

10.3 Subphylum Opalinata

10.3.1 General account

Opalinids are multinucleate or binucleate (flagellated? ciliated?) protozoa inhabiting the rectum of amphibia, reptiles and fishes; the genera *Protoopalina* and *Zelleriella* are binucleate, and the genera *Cepedea* and *Opalina* are multinucleate (Fig. 10.1), the latter commonly possessing as many as 200 nuclei.

As mentioned earlier, the group were previously classified with the ciliates but may have more affinities with the flagellates, with which they share some ultrastructural features and with which they are currently classified (Levine *et al.*, 1980). These features include the fine structure of the kinetosomes, mitochondria, nuclear division and surface folds with their supporting ribons of microtriches. Patterson (1985) has proposed that opalinids should be classified with the (family) protermonads within a new Order Slopalinada, but this proposal has yet to be accepted.

The group has a number of features of great interest to cytologists, e.g. the division of nuclei is asynchronous (i.e. does not appear to divide simultaneously with the cytoplasm) and the differentiation of the sexual phase

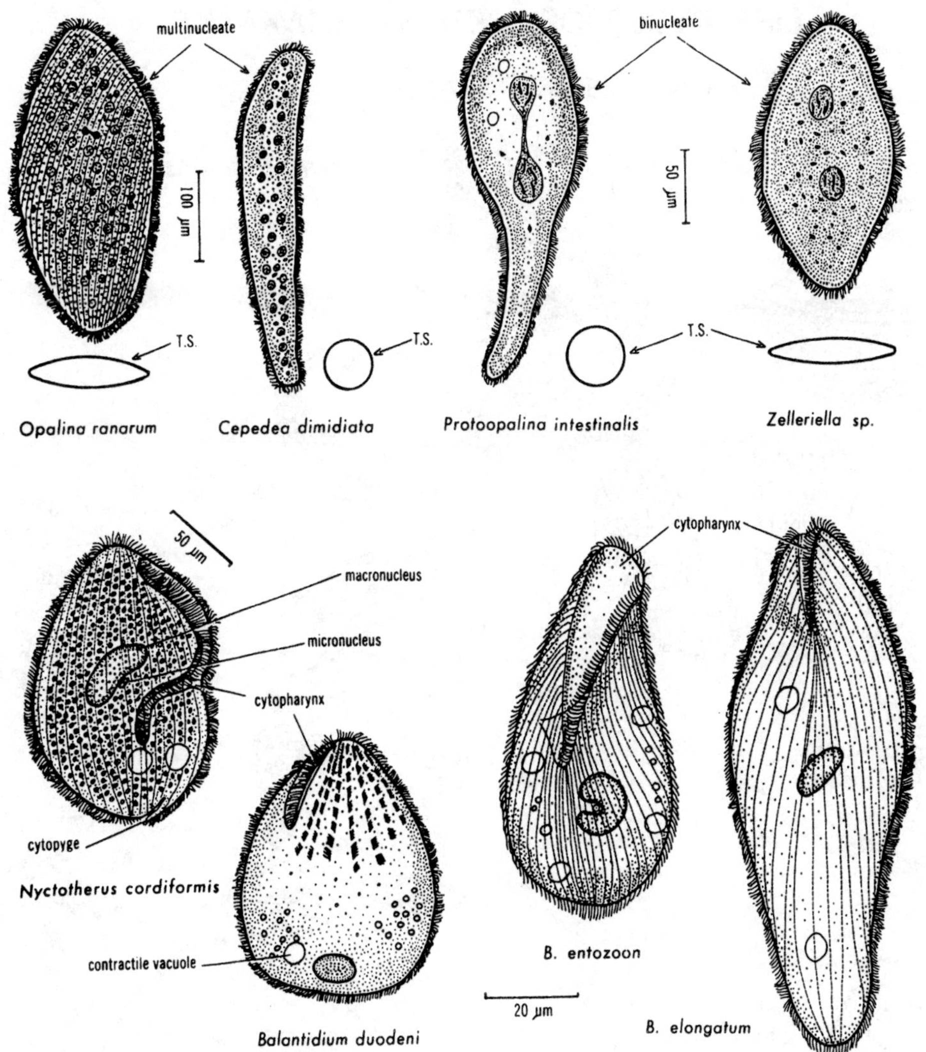

Fig. 10.1
Opalinids and ciliates from various *Rana* spp. (After various authors.)

of their life cycle is controlled by the hormonal activities of the host. The fact that they can readily be cultured *in vitro* (Yang, 1960; see Chapter 33) makes this group an especially valuable experimental model for teaching and research. The ultrastructure of opalinids has been much studied, especially with a view to determining their possible affinities with the flagellates or ciliates. Early work has been reviewed by Pitelka (1963) and more recent studies by Patterson (1985) and Wessenberg (1978). The general biology and life cycles of opalinids have been reviewed by Wessenberg (1961, 1978). A valuable list of the Opalinidae in Australian Anura has been published by Delvinquier (1987). The effect of abiotic factors on the prevalence of *Opalina* sp. and other rectal protozoa in the Indian frogs *Rana tigrina*, *R. limnocharis* and *R. breviceps* has been investigated by Rao *et al.* (1985). It was found that low rainfall and low oxygen concentrations favoured the increase of infection levels.

○ 10.3.2 *Opalina ranarum*

Life cycle (Fig. 10.2). The best-known species are *Opalina ranarum* (Fig. 10.1) in Europe and *O. obtrigonoidea* in North America from various species of frogs and toads. The body organisation is simple and there is no cytostome. The body, which is often irregular

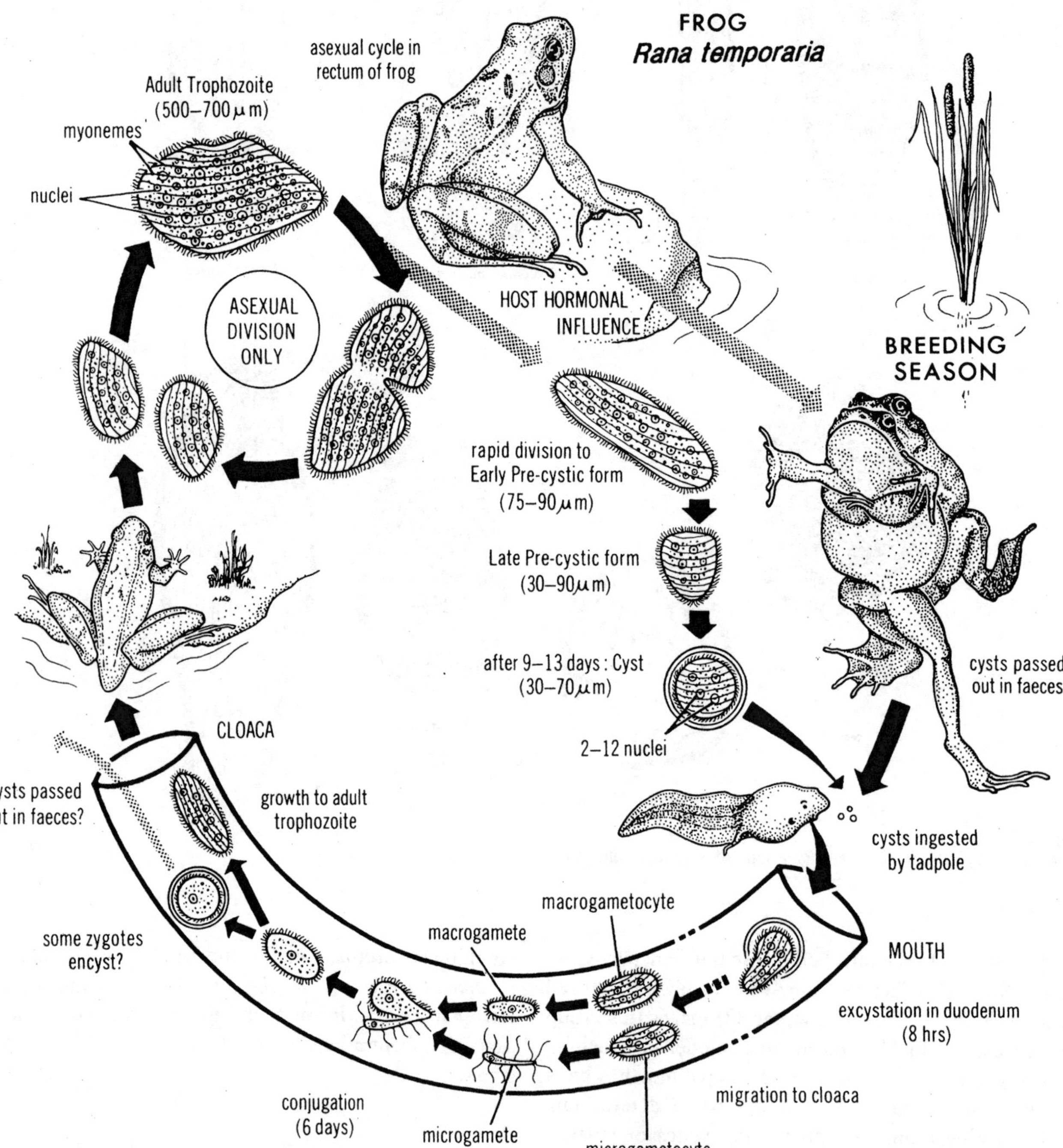

Fig. 10.2
Life cycle of *Opalina ranarum* from *Rana* spp. (From Smyth & Smyth, 1980.)

in outline, is flattened dorso-ventrally and the pellicle is relatively tough. The size and distribution of the 'cilia' are extremely uniform. There is a posterior 'excretory pore'. The life cycle is of particular physiological interest in that its sexual reproductive phase is synchronised with that of the host and is under its endocrinal control, a condition which represents a degree of metabolic dependence (p. 3) on the host more complete than in the majority of parasites. A few other parasites have a reproductive pattern similarly related to that of the host: the ciliate *Nyctotherus cordiformis* (p. 152) and other amphibian ciliates, the haemosporidean *Leucocytozoon simondi* (p. 138), various hypermastigina in insects and the monogenetic trematode *Polystoma integerrimum* (p. 163). Although the general pattern of the life cycle is well established, small points of detail regarding the later stages of gametogeny and post-conjugation behaviour are in dispute.

During the non-breeding terrestrial phase of the amphibian host, only 'adult' trophozoites or *trophonts* are found (Fig. 10.2); these divide occasionally by binary fission. As the breeding season approaches (February–March in Europe) a marked change takes place, and about fourteen days before the hosts enter the water to copulate, the fission rate accelerates and small precystic forms or *tomonts* appear. These tomonts encyst to give cysts with 2–12 nuclei, four being the average number. In copulating frogs immersed in water a few cysts may be found, appearing slightly earlier in females than in males, and in both sexes cyst formation is in advance of that in *Nyctotherus* (p. 152). After ovulation and copulation are completed, the percentage of cysts rises sharply (Fig. 10.3), but falls gradually later as they are passed out with the faeces. About three months after copulation cysts are no longer found. Cyst formation may be completely inhibited by enforced hibernation of potentially sexually mature frogs by maintaining them in a dry habitat (El Mofty, 1961).

Nucleic acid synthesis is of special interest in this organism. During winter and spring *no DNA synthesis takes place* but some RNA synthesis occurs in 20 per cent of nuclei and in the cytoplasm. During the summer, some DNA synthesis occurs; about 60 per cent of nuclei may be labelled with [^{14}C]-adenine but only 17 per cent of these (Fig. 10.4) show DNA synthesis (Nilova & Sukhanova, 1964).

Cysts are passed into the water and excyst if ingested by a tadpole, giving rise to small multinucleate forms (Fig. 10.2). By further divisions, these give rise to uni-

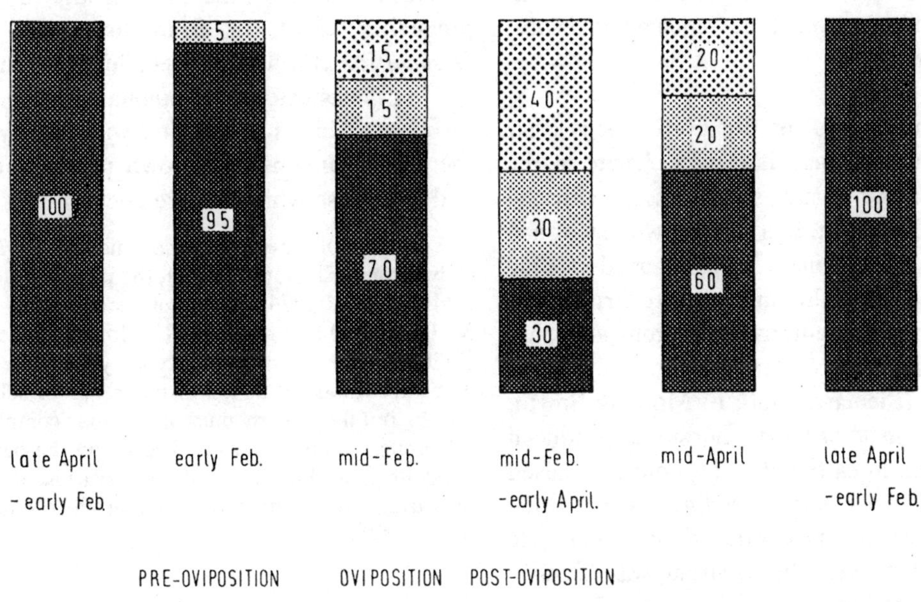

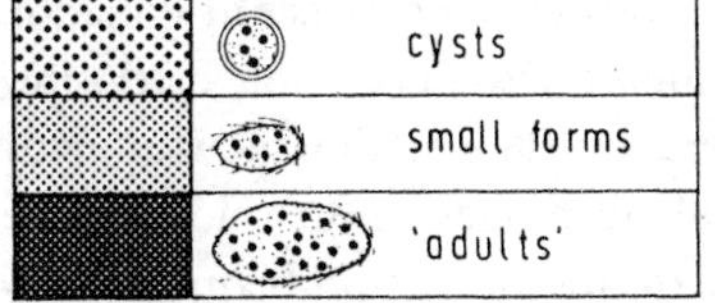

Fig. 10.3
Relation of cyst formation in *Opalina ranarum* to the breeding cycle of *Rana temporaria*. (After El Mofty & Smyth 1964.)

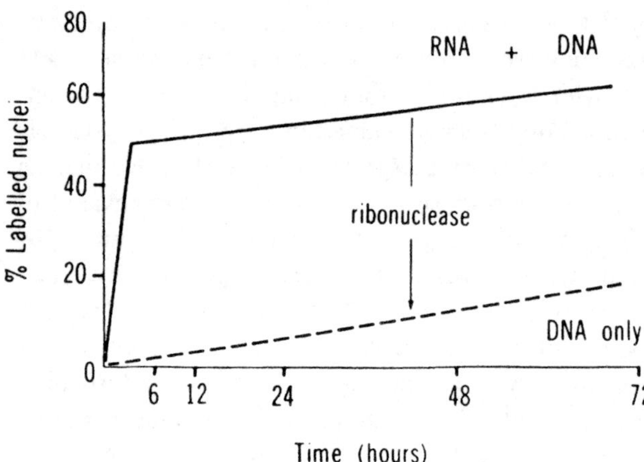

Fig. 10.4
Labelling of nuclei of *Opalina ranarum* (during summer) after incubation for various periods in ^{14}C-adenine (1 mC ml^{-1}). The continuous line shows the total number of labelled nuclei; the dotted line, the percentage nuclei after treatment with ribonuclease to remove the RNA. Note that during winter and spring, only RNA synthesis takes place; DNA is not formed. (Modified from Nilova & Sukhanova, 1964.)

nucleate heterogametes which conjugate to form a zygote. According to some observations, these zygotes divide to form trophonts. According to others, some zygotes encyst to form *cystozygotes*, which pass out with the faeces and aid in the distribution of the species when ingested by tadpoles; cystozygotes excyst to form uninucleate individuals which, by repeated divisions, give rise to the trophonts.

***In vivo* endocrine control of life cycle.** It is clear from the above account that there is a change in the growth pattern of the ciliate, synchronised with the breeding season of the host. It is well known that gonad maturation is under endocrinal control of gonadotrophic hormones secreted from the anterior pituitary. Injections of pregnancy urine, adrenaline or gonadotrophin into immature frogs induce cyst formation in *Opalina* within 9–13 days (Bieniarz, 1950; El Mofty & Smyth, 1964). Injection of male or female hormones produced similar results, even in castrated or hypophysectomised frogs, in the period prior to the breeding season (Table 10.1). Testosterone propionate induced the sexual cycle during any time of the year, but oestrone was effective only prior to the breeding season. It was speculated (El Mofty & Smyth, 1964) that the effect of gonadotrophic substances in inducing cyst formation in the protozoa could be explained by their action in stimulating the release of gonadal hormones. The role of adrenaline was not clarified. In another amphibian, *Bufo regularis*, an Egyptian toad, it has been shown that injections of adrenaline or adrenocorticotrophin (ACTH) stimulate the induction of sexual reproduction resulting in encystation in the *Opalina sudafricana* (El Mofty & Sadek, 1973); the effect was obtained in both normal and gonadectomised male and female toads. These authors speculated that adrenaline stimulates the release of ACTH from the anterior pituitary, which in turn induces the secretion of androgenic steroids from the inter-renal tissue. They suggested that it is the breakdown products of the androgenic steroids which cause encystation of the parasites. The positive results in hypophysectomised male and female toads may be explained by the supposition that the adrenaline may directly stimulate the interstitial tissues of the testes and ovaries to secrete androgens, the breakdown products of which induce encystation.

***In vitro* control of life cycle.** Opalinids may be cultured *in vitro* under either non-sterile or axenic conditions using a saline–serum–albumin medium similar to that used for intestinal protozoa (Chapter 33). None of the substances (hormones and adrenaline) listed in Table 10.1 induce cyst formation *in vitro* of either *O. ranarum* or *O. sudafricana* (El Mofty, 1961; El Mofty & Sadek, 1973). On the other hand, urine from human pregnancy cases, or from toads (in the case of *O. sudafricana*) which had been injected with adrenaline, ACTH or testosterone propionate, produced a positive response. This is in keeping with the hypothesis that the urine contains breakdown products of androgens and it is these which induce encystment.

A number of other substances, nicotine (El Mofty, 1973*a*), ecdysterone (El Mofty, 1973*b*), the plant hormone gibberellic acid (El Mofty, 1974*b*), numerous carcinogenic substances (El Mofty, 1974*a*) and toad bile (El Mofty & Sadek, 1975), have similarly been shown to induce encystation in *O. sudafricana.* The exact mode of action of this range of substances is not known, but they clearly must share some common physiological property. They may act directly on the parasite, or their breakdown products in the urine may cause the effect or they may trigger some host secretory product which reacts with the parasite.

○○○ 10.4 Phylum Ciliophora

The best-known parasitic ciliate species occur in the alimentary canal of vertebrates. The rumen and reticulum of ruminants maintains a large ciliate fauna, some

Table 10.1. *Response of* Opalina ranarum *to injections of various hormones into its amphibian host*

The onset of the 'sexual' reproductive cycle of the protozoon was signalled by the appearance of cysts in the rectum: +, cyst present; –, cyst absent; ○, no experiment carried out. Results of experiment just prior to the breeding season must be interpreted with caution, as during this period the endocrine balance is easily disturbed and control experiments (with saline) often give positive results.

	Pre-breeding season			Post-breeding season		
	Normal	Hypophysectomised	Gonadectomised	Normal	Hypophysectomised	Gonadectomised
Pregnancy urine	+	+	–	–	○	○
Chorionic gonadotrophin	+	+	–	–	○	○
Serum gonadotrophin	+	+	–	–	○	○
Progesterone	–	○	○	–	○	○
Oestrone	+	+	–	–	○	○
Testosterone propionate	+	+	+	+	○	○
Adrenaline	+	+	–	+	○	○

Source: Data from El Mofty (1961) and El Mofty & Smyth (1964).

thirty-nine different species being reported from this habitat. Many invertebrates also house parasitic ciliates and species have been found in the gastro-vascular cavity of medusae, the intestine of insects and annelids, the coelom and blood vessels of crustaceans, the liver of molluscs, the gonads of echinoderms and other such unusual habitats. Some are ectoparasitic. The majority of endoparasitic species are harmless, but several species are pathogenic. The group has never succeeded in becoming established in the vertebrate blood stream. Like amoebae, most ciliates are scavengers and feed on detritus; only one species, *Balantidium coli*, is a pathogen of man.

The general biology, morphology and life cycles of ciliates have been reviewed by Corliss (1979).

○ 10.4.1 *Ichthyophthirius multifiliis*

This ciliate is a well-known scourge of aquaria, causing ichthyophthiriasis, also known as 'ick' or 'white-spot', which is highly pathogenic to fish, killing them by interfering with their gill function and osmoregulation (Ventura & Papernal, 1985; Hines & Spira, 1974). The terminology describing the various stages of the life cycle is somewhat confusing. The trophozoite is known as a *trophont* when it is growing on a fish and a *tomont* when it leaves the fish and becomes a reproducing non-feeding stage. Some authors simply use the term trophozoite for both these stages. A tomont (Fig. 10.5) has an oval body with uniform ciliation and a longitudinally striated pellicle. There is an anterior cytostome with a short cytopharynx, a large horseshoe-shaped macronucleus and a small micronucleus. The ultrastructure has been studied by Ewing & Kocan (1986) and McCartney *et al.* (1985).

The macronucleus of *I. multifiliis* is of special interest in being highly polyploid. In the young free-swimming stages, the degree of polyploidy is $48n$ (i.e. $24 \times 2n$) but in the mature trophont it has been estimated to be $1200n$! (Uspenskaja & Ovchinnikova, 1966).

There are several hundred actively functioning contractile vacuoles, spaced evenly over the organism.

Life cycle (Fig. 10.5). *I. multifiliis* is essentially an endoparasite, because it settles *beneath* the epidermis where it develops in characteristic thin-walled pustules: the so-called 'white spots'. The trophonts do not reproduce on the fish, but when they reach a certain size, they leave the pustules as non-feeding tomonts, which fall to the bottom and encyst on plant surfaces or on the bottom of the tank. Within a cyst, a multiplication reminiscent of sporozoan schizogony takes place, the cytoplasm dividing into 100–1000 small ciliated cells, which metamorphose into elongated forms known as *tomites*. These break through the cyst wall and seek new hosts by actively swimming, rapidly penetrating the host epidermis where a pustule is formed around them, thus completing the life cycle. As could be predicted, the multiplicative phase is very dependent on temperature (Fig. 10.6).

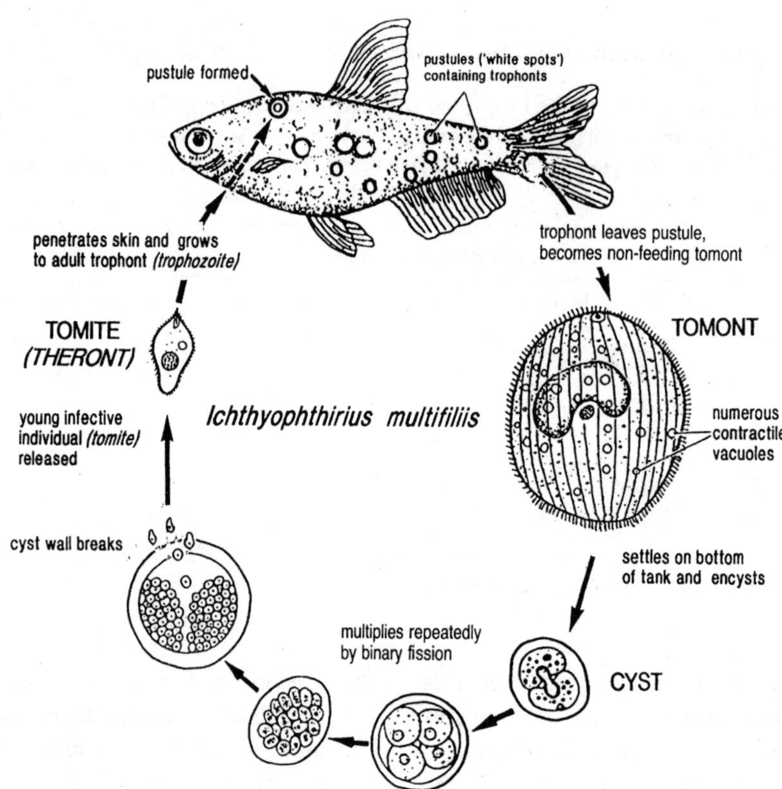

Fig. 10.5
Life cycle of *Ichthyophthirius multifiliis*. (Based on several authors.)

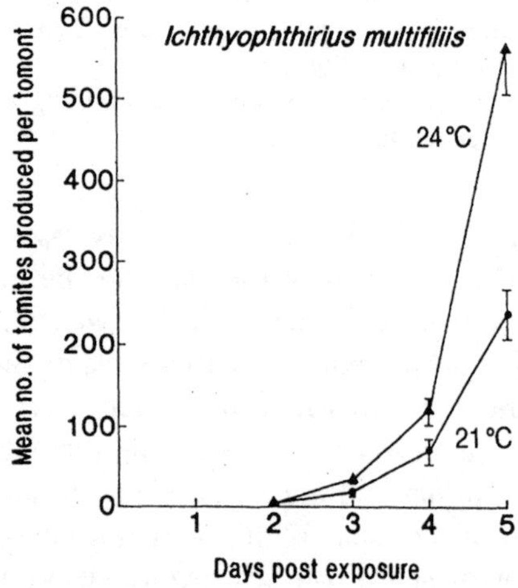

Fig. 10.6
Ichthyophthirius multifiliis: effect of fish tank temperature on tomite production. (After Ewing *et al.*, 1986.)

Population dynamics. Because it multiplies so rapidly and can be readily maintained in the laboratory in a closed tank system, the population dynamics of *I. multifiliis* are readily studied. It is an unusual model, because unlike most parasites, death of the host does not result in death of the parasite so that its reproductive capacity is not entirely destroyed. Relevant studies on this problem are those of McCallum (1982, 1985) and Ewing & Kocan (1986).

○ 10.4.2 *Balantidium coli* (Fig. 10.7)

This is the largest protozoan parasite of man and pigs, with a size variously reported in the range of 30–200 μm long, by 20–120 μm in breadth; the upper limits refer to specimens from pigs. The general biology and pathology has been reviewed by Juniper (1984).

The general outline is egg-shaped, bearing an anterior slit-like peristome opening into a cytostome. The whole body is covered with small fine cilia with an adoral zone around the peristome. There is a well-marked macronucleus, an inconspicuous micronucleus, and two contractile vacuoles. Food vacuoles are commonly present. Reproduction is by binary fission and conjugation.

The main animal reservoir of the species is probably the pig, but the organism also occurs in monkeys and rats. Guinea pigs serve as useful laboratory hosts. Thus in Venezuela, where human infections with *B. coli* are

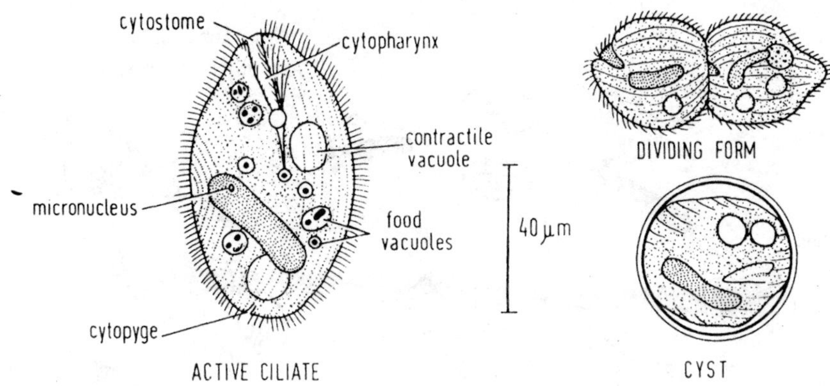

Fig. 10.7
Balantidium coli. (Adapted from Hoare, 1949.)

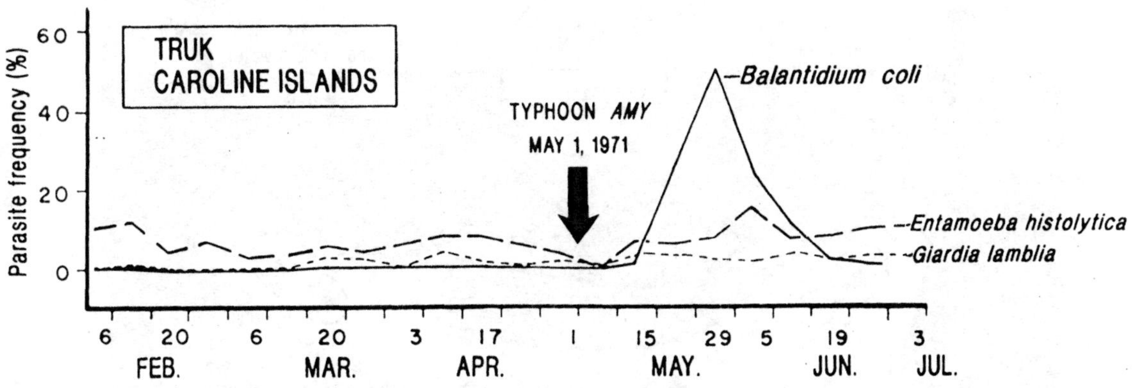

Fig. 10.8
Effect of typhoon AMY on the frequency of *Balantidium coli* in humans on the Island of Truk, in the Caroline Islands, where there is a large population of pigs, the reservoir host of *B. coli.* (Modified from Walzer *et al.*, 1973.)

common, 80 per cent of infections are asymptomatic, but in 20 per cent a syndrome identical to amoebic dysentery is seen (Wenger, 1967). Hence, like *E. histolytica* (p. 29), 'physiological' strains of *B. coli* almost certainly occur, some of which may produce a pathogenic condition in susceptible strains of certain host species. The ability to secrete hyaluronidase probably helps the organism to invade the mucosa; the mucosal lesions are very similar to those of amoebiasis. More severe infections can lead to perforation of the colon, hepatic abscesses and appendicitis. In very rare cases, death may result (Dorfman *et al.*, 1984).

Under certain undefined conditions, encystment takes place and cysts are passed in the faeces.

As mentioned, the pig is the major animal reservoir from human infection. Prevalences as high as 68 per cent in confined pigs have been reported (Martio & Hale, 1986); about 25 per cent of patients give a history of contact with pigs. Human to human transmission has also been reported from institutions.

One of the best documented epidemics of balantidiasis occurred in Truk in the Caroline Islands following a devastating typhoon. The inhabitants live in close association with pigs and the epidemic resulted from contamination of ground and surface water by pig faeces. As it happened, the protozoan infections of the inhabitants were being carefully monitored at that time and the sudden rise in infections with *Balantidium* is clearly seen in Fig. 10.8.

Other species of *Balantidium*

Numerous species occur in frogs and salamanders, e.g. *B. entozoon*, *B. rotundum*, *B. gracilis*. Other species available for laboratory study are *B. blattarum* in cockroaches and *B. pisciola* in fishes. *B. duodeni* and *B. elongatum* (Fig. 10.1) are relatively common in *Rana esculenta* (Smyth & Smyth, 1980).

○ 10.4.3 Genus *Nyctotherus*

Species of this genus are amongst the best-known of the protozoan parasites of frogs and toads. *Nyctotherus*

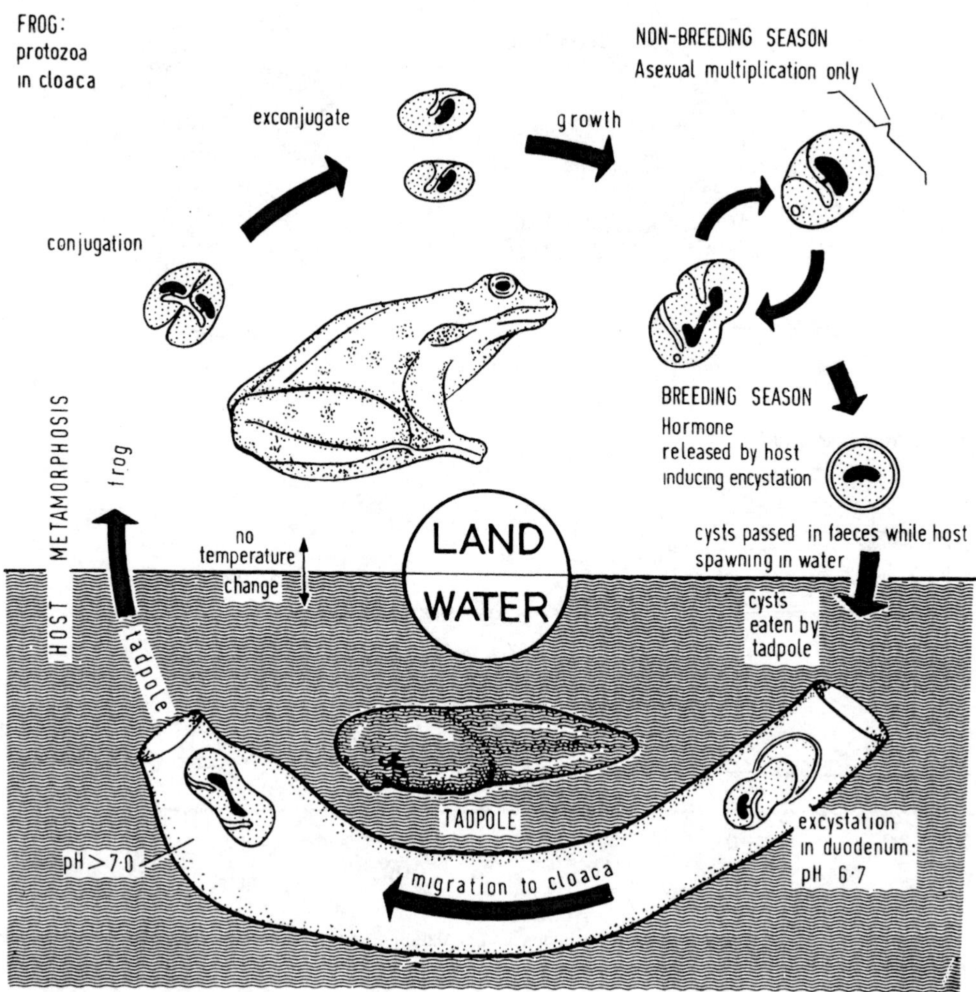

Fig. 10.9
Nyctotherus cordiformis: life cycle showing relation to breeding cycle of *Rana temporaria*. (Original.)

cordiformis, which has a cosmopolitan distribution, can be considered the type example; it is relatively common in *R. pipiens* and N. America and in *R. temporaria* in Europe.

N. cordiformis (Figs 10.1, 10.9). Occurs in the rectum of frogs and tadpoles. It is completely ciliated with a well-marked curved peristome leading to a cytostome, and a cytopharynx. There is a kidney-shaped macronucleus, a small micronucleus, a posterior contractile vacuole, and an excretory pore.

Life cycle (Fig. 10.9). The reproductive cycle of *Nyctotherus*, like that of *Opalina* (p. 146), is linked with the sexual cycle of the amphibian host. During the year, *Nyctotherus* divides occasionally by binary fission. In spring, parallel with the frog breeding season, a change from asexual to sexual multiplication takes place, and trophozoites divide with greater frequency, finally forming mononuclear precystic forms. These encyst and pass out into the water with the host faeces. The cysts are ingested by tadpoles developed from eggs laid in the same water, and the young individuals, the preconjugants, hatch out. These conjugate and undergo a nuclear process similar to the well-known process in *Paramoecium*. The conjugants separate and the exconjugants, which occur almost exclusively in recently metamorphosed frogs, undergo repeated binary fission as before (El Mofty, 1961; Bieniarz, 1950).

Preliminary experiments on frogs have given similar results to those already described for *Opalina ranarum* (p. 148). When pregnant human urine or male or female

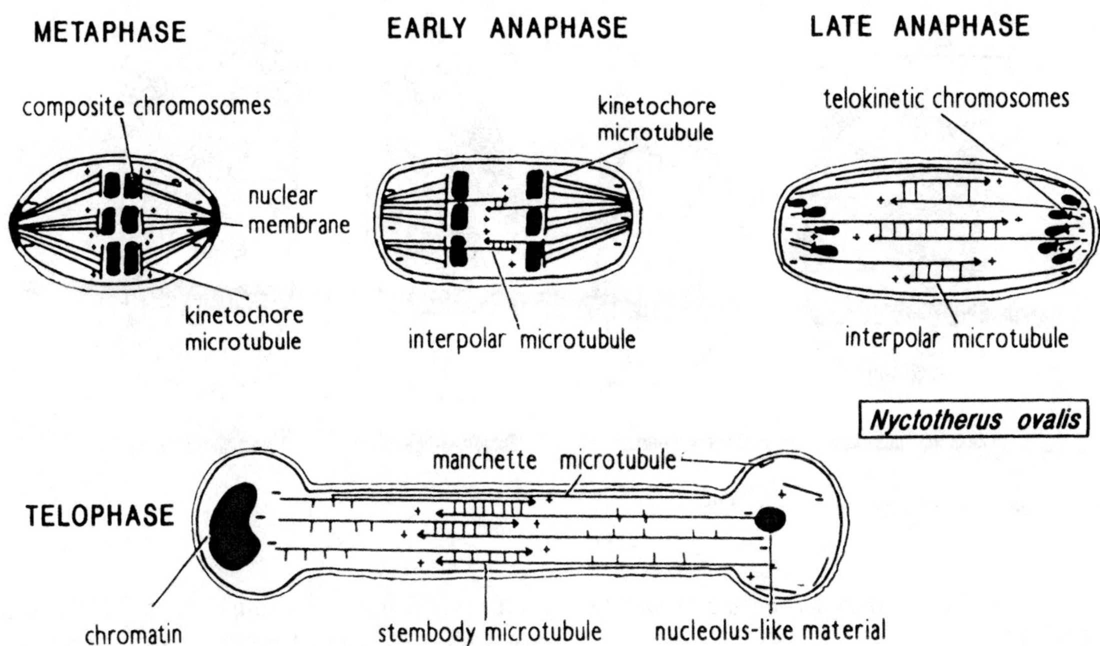

Fig. 10.10
Mitosis in the micronucleus of the ciliate *Nyctotherus ovalis*, from the cockroaches *Blaberus fuscus* and *Blaptica dubia*. Unusually, the chromosomes adhere to form 'composite chromosomes' before mitosis. These separate again ('telokinetic chromosomes') after mitosis in late anaphase/early telophase. The microtubule organisation during mitosis is also shown. (Modified from Eichenlaub-Ritter, 1982; reproduced with permission from *Microtubules in Microorganisms* (ed. P. Cappuccinelli & N. R. Norris), p. 370. Marcel Dekker, New York, 1992.)

hormones are injected into frogs under certain conditions, encystation of *Nyctotherus* takes place, suggesting that, as in *Opalina*, the process is related to the level of the gonadal hormones in the host.

Other species. There is a very large number of species in the genus *Nyctotherus*, not only in amphibia and reptiles, but also in invertebrates, especially arthropods. For example, 16 species of *Nyctotherus* have been recorded from myriapods (Albaret, 1970) and a common species in the cockroaches *Blaptica dupla* and *Blaberus fuscus* is *N. ovalis*.

N. ovalis has proved to be a valuable model for the study of mitosis in the micronucleus of ciliates at an ultrastructure level, especially in relation to microtubule development (Eichenlaub-Ritter, 1982; Eichenlaub-Ritter & Ruthmann, 1982*a,b*). Mitosis has proved difficult to study in the ciliate micronucleus although it has been well studied in the macronucleus. In *N. ovalis*, the fixed position of the nucleus has helped the ultrastructural examination of all stages of the process, and there is also an unusual arrangement of chromatin which leads to the appearance of holokinetic composite chromosomes; a peculiar kinetochore structure has also been revealed. During metaphase, the entire poleward-facing surface of the chromosomes becomes covered in trilaminar kinetochore (= centromere) plates. For details of this complex process and the role of the microtubules (Fig. 10.10), the original papers should be consulted.

Nyctotheroides puytoraci from the rectum of *Bufo regularis* has also proved to be an interesting experimental model in that, like *Opalina ranarum* (p. 148) in *Rana esculenta*, injections of carcinogenic substances induce the sexual process (i.e. encystation) (Sadek, 1979, 1980). Encystment has also been induced in *Neonyctotherus reticulatus* by injection of gonadotrophin (Affa'a, 1986).

○ 10.4.4 Order Entodiniomorphida

Symbiotic protozoa occur in quantity in ungulates and rodents; none occur in carnivorous animals. The most favoured sites are the rumen and reticulum of cattle, which may contain anything up to thirty-nine different species of ciliates. These belong mainly to the genera *Isotricha*, *Bütschlia*, *Entodinium* and *Diplodinium*. Their morphology ranges from relatively simple species such as *Bütschlia* to *Diplodinium* (Fig. 10.11), the most complicated protozoon known.

The general biology and physiology of the group have been discussed in a wonderfully comprehensive review

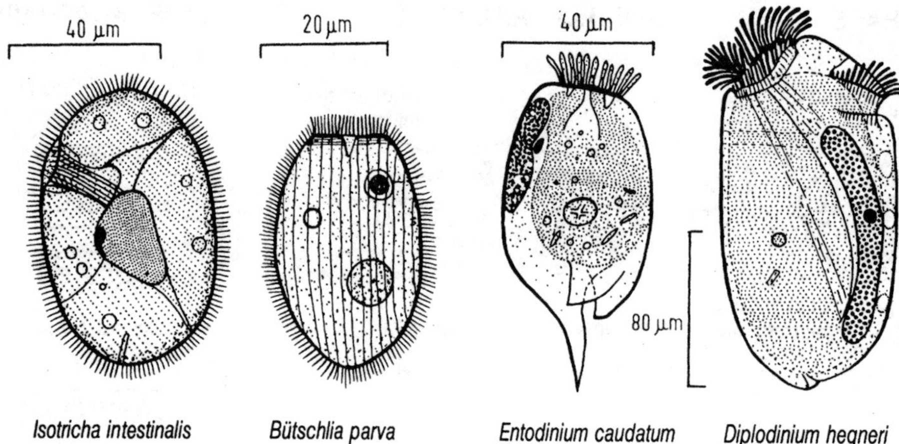

Fig. 10.11
Ciliates from the rumen of cattle. (After Hegner *et al.*, 1938.)

by Dehority (1986). Earlier work has been reviewed by Hungate (1978).

Hume & Warner (1980) consider that 'the first mammals were probably insectivore/carnivore types, with a simple digestive system in which little microbial activities occurred'. At some point in time it became advantageous (or necessary) for these early mammals to include plant material in their diet and digest it. Since the major structural carbohydrate in plants is cellulose, the enzyme cellulase was necessary to utilise this as a source of energy. As this enzyme is absent in all vertebrate digestive systems, the problem was resolved by developing a symbiotic relationship with cellulolytic microorganisms – protozoa and bacteria – accompanied by evolutionary changes in the anatomy of the herbivore digestive tract (Dehority, 1986).

It has been suggested that the ancestors of the present-day rumen ciliate protozoa were fresh-water ciliates which were swallowed when the animals drank from waterholes or lakes (Dogiel, 1947).

The Isotrichidae, *Isotricha* and *Dasytricha*, do not ingest plant material but can ferment glucose, fructose, raffinose and pectin. *Dasytricha* can also ferment maltose and cellobiose and *Isotricha* can ingest and utilise small vegetable starch grains. Most of the higher rumen ciliates utilise one or more of the plant polysaccharides, i.e. cellulose, pectin, starch and hemicellulose. Although early work suggested that the cellulase might have been of intracellular bacteria origin, it has now been shown unequivocally that the cellulase activity is of protozoan origin. The genera *Epidinium* and *Ophryoscolex*, however, do not appear to be cellulolytic.

The metabolism of these ciliates is intense and they can multiply very rapidly although their life period is only about twenty-four hours. Consequently, they are continually dying and disintegrating. This provides the host with a substantial proportion of its nitrogen and carbohydrate requirements. However, it must be emphasised that – contrary to the generally held view – the protozoa are *not* essential to the host and in their absence the fermentation processes are taken over by an increased bacterial population. However, defaunated animals do not grow as well as their faunated controls (Dehority, 1986).

○ 10.4.5 Order Hypotrichida

An example of this sub-order is the common species *Kerona pediculus*, which occurs on the fresh-water coelenterate *Hydra*, especially *H. fusca* and *H. vulgaris*. It is a true ectoparasite, feeding on the living cells of the *Hydra*, which it may ultimately destroy. It has almost the same structure as the typical free-living hypotrich, but is especially adapted for gliding over the surface of *Hydra*; its anterior and lateral edges are flexible.

○○○ References

(References of general interest have been included in this list, in addition to those quoted in the text.)

Affa'a, F. 1986. Induction par la gonadotrophine et l'enkystement et de la conjugasion chez les ciliés parasites de crapauds du Cameroon. *Annales des Sciences Naturelles Zoologie et Biologie Animale*, **8**: 1–4.

Albaret, J.-L. 1970. Observations sur les nyctotheres des genres *Nyctotherus* Leidy et *Metanyctotherus* N.Gen., ciliés hétérotriches parasites de Myriapides Africains. *Protistologica*, **6**: 225–39.

Bieniarz, J. 1950. Influence of vertebrate gonadotrophic hormones on the reproductive cycle of certain protozoa in frogs. *Nature*, **165**: 650–1.

Corliss, J. O. 1979. *The ciliated Protozoa. Characterization, Classification and Guide to the Literature.* 2nd Edition. Pergamon Press, Oxford.

Dehority, B. A. 1986. Protozoa of the digestive tract of herbivorous mammals. *Insect Science and its Applications*, **7**: 279–96.

Delvinquier, B. L. J. 1987. Opalinidae in Australian Anura. *Proceedings of the Royal Society of Queensland*, **98**: 93–122.

Dogiel, V. A. 1947. The phylogeny of the stomach-infusorians of ruminants in the light of palaeontological and parasitological data. *Quarterly Journal of Microscopical Science*, **88**: 337–43.

Dorfman, S., Rangel, O. & Bravo, L. G. 1984. Balantidiasis: report of a fatal case with appendicular and pulmonary involvement. *Transactions of the Royal Society of Tropical Medicine and Hygiene*, **78**: 833–4.

Eichenlaub-Ritter, U. 1982. Micronuclear mitosis in the ciliate *Nyctotherus ovalis* Leidy: an unusual chromosome arrangement and kinetochore structure and its implications for spindle structure and function. In: *Microtubules in Microorganisms* (ed. P. Cappuccinelli & N. R. Morris), pp. 352–75. Marcel Dekker, New York.

Eichenlaub-Ritter, U. & Ruthmann, A. 1982*a*. Holokinetic composite chromosomes with 'diffuse' kinetochores in the micronuclear mitosis of a heterotrichous ciliate. *Chromosoma*, **84**: 701–6.

Eichenlaub-Ritter, U. & Ruthmann, A. 1982*b*. Evidence for the 'classes' of microtubules in the interpolar space of the mitotic micronucleus of a ciliate and the participation of the nuclear envelope in conferring stability to microtubules. *Chromosoma*, **85**: 687–706.

El Mofty, M. M. 1961. The life cycle of *Opalina ranarum* and an investigation into the factors responsible for controlling its reproductive pattern. Ph.D. Thesis, Trinity College, University of Dublin, Ireland.

El Mofty, M. M. 1973*a*. Induction of sexual reproduction in *Opalina sudafricana* by injecting its host (*Bufo regularis*) with nicotine. *International Journal for Parasitology*, **3**: 265–6.

El Mofty, M. M. 1973*b*. Ecdysterone induced sexual reproduction in *Opalina sudafricana* parasitic in *Bufo regularis*. *International Journal for Parasitology*, **3**: 863–8.

El Mofty, M. M. 1974*a*. A new biological assay, utilizing parasitic protozoa, for screening carcinogenetic substances which induce bladder cancer in man and other mammals. *International Journal for Parasitology*, **4**: 47–54.

El Mofty, M. M. 1974*b*. Induction of sexual reproduction in *Opalina sudafricana* by injecting its host *Bufo regularis* with gibberellic acid. *International Journal for Parasitology*, **4**: 203–6.

El Mofty, M. M. & Sadek, I. A. 1973. The mechanism of action of adrenaline in the induction of sexual reproduction (encystation) in *Opalina sudafricana* parasitic in *Bufo regularis. International Journal for Parasitology*, **3**: 425–31.

El Mofty, M. M. & Sadek, I. A. 1975. The effect of fresh toad bile on the induction of encystation in *Opalina sudafricana* parasitic in *Bufo regularis. International Journal for Parasitology*, **5**: 219–24.

El Mofty, M. M. & Smyth, J. D. 1964. Endocrine control of encystation in *Opalina ranarum* in parasitic in *Rana temporaria. Experimental Parasitology*, **15**: 185–99.

Ewing, M. S. & Kocan, K. M. 1986. *Ichthyophthirius multifiliis* (Ciliophora) development in gill epithelium. *Journal of Protozoology*, **33**: 369–74.

Hegner, R., Root, F. M., Augustine, D. L. & Huff, C. G. 1938. *Parasitology*. Appleton-Century-Crofts, New York.

Hines, R. S. & Spira, D. T. 1974. Ichthyophthiriasis in the mirror carp, *Cyprinus carpio* (L). IV. Physiological dysfunction. *Journal of Fish Biology*, **6**: 365–71.

Hoare, C. A. 1949. *Handbook of Medical Protozoology*. Baillière, Tindall & Cox, London.

Hume, I. D. & Warner, A. C. I. 1980. Evolution of microbial digestion in mammals. In: *Digestion Physiology and Metabolism in Ruminants* (ed. Y. Ruckebusch & P. Thivend), pp. 665–84. MTP Press, Lancaster.

Hungate, R. E. 1978. The rumen protozoa. In: *Parasitic Protozoa* (ed. J. P. Kreier), Vol. II, pp. 655–95. Academic Press, New York.

Juniper, K. 1984. Balantidiasis. In: *Tropical and Geographical Medicine* (ed. K. S. Warren & A. F. Mammoud), pp. 326–8. McGraw-Hill, New York.

Levine, N. D. (Chairman). 1980. A newly revised classification of the Protozoa. *Journal of Protozoology*, **27**: 37–58.

McCallum, H. I. 1982. Infection dynamics of *Ichthyophthirius multifiliis. Parasitology*, **85**: 475–88.

McCallum, H. I. 1985. Population effects of parasite survival of host death: experimental studies on the interaction of *Ichthyophthirius multifiliis* and its fish host. *Parasitology*, **90**: 529–47.

McCartney, J. B., Fortner, G. W. & Hansen, M. F. 1985. Scanning electron microscope studies on the life cycle of *Ichthyophthirius multifiliis. Journal of Parasitology*, **71**: 218–26.

Martio, G. & Hale, O. M. 1986. Parasite transmission in confined hogs. *Veterinary Parasitology*, **19**: 301–4.

Nilova, V. K. & Sukhanova, K. M. 1964. Synthesis of nucleic acids in *Opalina ranarum* Ehrbg. *Nature*, **204**: 459–60.

Patterson, D. J. 1985. The fine structure of *Opalina ranarum* (Family Opalinidae); opalinid phylogony and classification. *Protistologica*, **21**: 413–28.

Pitelka, D. R. 1963. *Electron-microscopic Structure of Protozoa.* Pergamon Press, Oxford.

Rao, L. S., Jaleel, M. A., Hussain, M. & Khan, M. A. 1985. Influence of abiotic factors on the protozoan infection in amphibian hosts. *Proceedings of the Indian Academy of Parasitology*, **6**: 63–8.

Sadek, I. A. 1979. *Nyctotheroides puytoraci*: stimulation of encystment in toads, *Bufo regularis*, injected with rheumatoid arthritis patient's urine. *Experimental Parasitology*, **48**: 239–44.

Sadek, I. A. 1980. Inhibition by vitamin A of cyst formation in *Nyctotheroides puytoraci* induced by injecting its host, *Bufo* *regularis*, with 20-methylcholanthrene. *Journal of Protozoology*, **27**: 313–15.

Small, E. B. & Lynn, D. H. 1981. A new macrosystem for the Phylum Ciliophora Dolflein, 1901. *Biosystems*, **14**: 387–401.

Smyth, J. D. & Smyth, M. M. 1980. *Frogs as host–parasite systems. I. An Introduction to Parasitology through the Parasites of* Rana temporaria, R. esculenta *and* R. pipiens. The Macmillan Press, London.

Uspenskaja, A. V. & Ovchinnikova, L. P. 1966. (Quantitative changes of DNA and RNA during the life cycle of *Ichthyophthirius multifiliis*.) *Acta Protozoologica*, **4**: 127–41. (In Russian.)

Ventura, M. A. & Papernal, L. 1985. Histopathology of *Ichthyophthirius multifiliis* infections in fishes. *Journal of Fish Biology*, **27**: 185–203.

Walzer, P. D., Judson, F. N., Murphy, K. B., Healy, G. R., English, D. K. & Schultz, M. G. 1973. Balantidiasis outbreak in Truk. *American Journal of Tropical Medicine and Hygiene*, **22**: 33–41.

Wenger, F. 1967. Liver abscess caused by *Balantidium coli. Kasmera*, **2**: 433–41.

Wessenberg, H. 1961. Studies on the life cycles and morphogenesis of *Opalina*. *University of California Publications in Zoology*, **6**: 315–70.

Wessenberg, H. S. 1978. Opalinata. In: *Parasitic Protozoa* (ed. J. P. Kreier), Vol. 2, pp. 551–81. Academic Press, New York.

Yang, W. C. T. 1960. On the continuous culture of opalinids. *Journal of Parasitology*, **46**: 32.

11

Helminth parasites: Platyhelminthes: Monogenea

The term 'worms' is loosely applied to an assemblage of organisms with elongated bodies and a more or less creeping habit. The term has no exact zoological meaning but is retained as a convenient operational word for defining a kind of organism well recognised but precisely indefinable. The term *helminth* (derived from the Greek words *helmins* or *helminthos*), although literally also meaning 'worm', zoologically speaking has a more precise connotation and is nowadays restricted to members of the phyla Platyhelminthes, Nematoda and Acanthocephala. Although the first phylum includes the free-living turbellarians, the study of helminths – or helminthology – has come to be regarded as being confined to the study of parasitic forms. Helminths typically parasitise vertebrates, although invertebrates, especially arthropods and molluscs, act as intermediate hosts.

Helminths may be classified as follows:

Phylum Platyhelminthes

Body dorso-ventrally flattened, bilaterally symmetrical, without definite anus, body cavity lacking. Organs embedded in specialised connective tissue known as parenchyma. Usually hermaphrodite. Respiratory and circulatory systems absent, flame cells in excretory system.

Class 1 **Turbellaria**

Fresh-water, marine or terrestrial platyhelminths with ciliated epidermis. Almost entirely free-living.

Class 2 **Monogenea**

Ecto- or endoparasites, bearing as adhesive organs a large posterior disc or *haptor* (Fig. 11.1) and an anterior adhesive region (often absent or poorly developed); tegument syncytial; life cycle simple, no alternation of hosts.

Class 3 **Trematoda** (flukes)

Ecto- or endoparasitic; tegument syncytial and (to a limited degree) absorptive; alimentary canal and adhesive organs well developed; life cycles complex.

Class 4 **Cestoidea** (tapeworms)

Elongated endoparasites lacking an alimentary canal; tegument syncytial and (highly) absorptive; usually divided into segments; adhesive organs at anterior end; adult stages parasitic in alimentary canal of vertebrates; life cycle complicated with two or more hosts.

Phylum Nematoda (roundworms)

Unsegmented, cylindrical worms with a resistant cuticle. Sexes separate, gonads tubular and continuous with their ducts. Body cavity a pseudocoele (p. 385). Free-living and parasitic.

Phylum Nematomorpha (hairworms)

Unsegmented, long thread-like free-living worms occurring in soil or water. Sexes separate, adults free-living, larvae parasitic in arthropods (not considered here).

Phylum Acanthocephala (spiny-headed worms)

Unsegmented, cylindrical worms with an armed protrusible proboscis. Pseudocoele; alimentary canal lacking. Sexes separate, parasitic in the adult stage in the intestine of vertebrates.

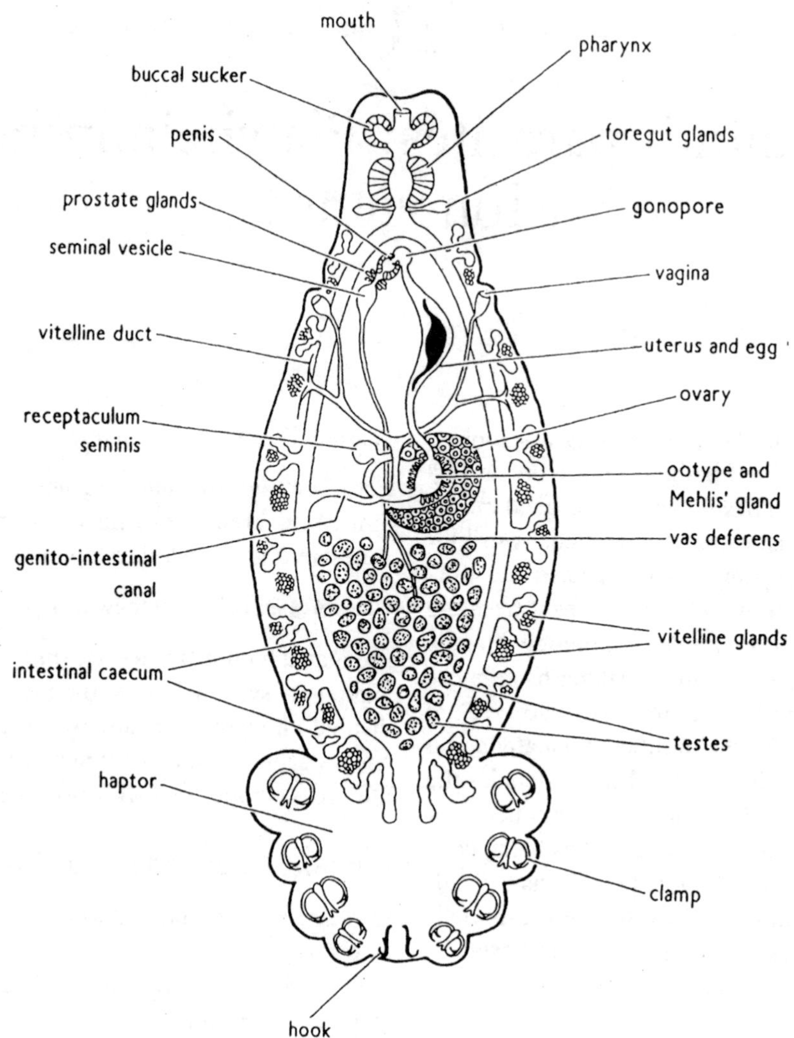

Fig. 11.1
A generalised diagram of monogenean anatomy. (After Smyth & Halton, 1983.)

○○○ 11.1 Trematoda: introduction

The term Trematoda is derived from the Greek *trematodes* (having holes), referring to the suckers which form a characteristic feature of the group. Trematodes occur in a wide range of host environments; the majority of species are endoparasitic but some are ectoparasitic. The larval stages may occur in invertebrate (especially mollusc) hosts and vertebrate hosts, but the majority of adult stages occur in or on vertebrates. Specimens range in length from 1 mm to 75 mm and are usually grey or creamy white in colour, but many acquire a characteristic coloration due to the assimilation of food materials from the host. The mouth and reproductive apertures are typically anterior. The digestive system is well developed and food material consists of intestinal debris, blood, mucus or other tissue exudants depending on the host environment. They have previously been divided into three groups: the *Monogenea* – typically external parasites of fish with direct life cycles (i.e. with no intermediate host); the *Aspidogastrea* – endoparasitic species with the entire ventral surface an adhesive organ; the *Digenea* – endoparasitic species with simpler adhesive organs and indirect life cycle (i.e. involving one or more intermediate hosts).

On the basis that the larva (*oncomiracidium*; Fig. 11.4) of the Monogenea resembles more the larva of gyrocotylideans, amphilinideans (p. 278), and cestodes, some workers, notably Bychowsky (1957), Llewellyn (1970, 1987) and Mackiewicz (1981, 1982) have removed this group from the Trematoda and placed them nearer the Gyrocotylidea (p. 279). Other workers (e.g. Dubois, 1970) believe that on structural grounds (ametamery, internal anatomy, digestive system, nervous system and egg-formation) the Monogenea and Digenea should be placed in the one class Trematoda. Detailed consideration of the affinities of these groups, which are unusually controversial (see section 11.2.2), however, is beyond the scope of this book, and the Monogenea are here placed in a separate class. For convenience, on account of their relatively simple life cycles, they are discussed before the Trematoda, which have complex life cycles.

Body covering. Although there are considerable morphological differences between these groups (see below) the outside covering is strikingly similar. Earlier workers referred to this as a *cuticle*, and the origin and homologies of this structure have long been a matter of dispute. The body covering in the Platyhelminthes is now referred to as the *tegument*. It is known to be cytoplasmic in nature and arises as a distal extension of fusiform cells which lie deep among the parenchymal cells. The tegument in these groups closely resembles that of cestodes and, in that group, is probably also absorptive and, possibly, secretory in nature (Smyth & Halton, 1983).

○○○ 11.2 Class Monogenea

○ 11.2.1 Type example: *Polystoma integerrimum*

This species is a parasite of the excretory bladder of frogs. Although common in Europe and N. America, it often has a markedly local distribution. Its anatomy and life cycle have been reinvestigated by Williams (1960–1), Kohlmann (1961), Streble (1964) and Combes (1968).

Host specificity. In Europe, the amphibian host is *Rana temporaria*, about 27 per cent of which may be infected with 1–8 worms (Kohlmann, 1961). Closely related species are *P. pelobatis* in *Pelobates cultripes*, and *P. gallieni* in *Hyla meridionalis*. Host specificity appears to be rigid, because experiments to transfer these species to tadpoles of the wrong host proved unsuccessful (Combes, 1966). In North America, the tree frogs *Hyla versicolor* (in New England) and *H. cinerea* (in Florida) act as hosts, up to 50 per cent of which may be infected. The North American polystome is considered to be a different species, *Polystoma nearticum* on account of minor morphological differences (Paul, 1938); see p. 163.

External features (Fig. 11.2). The most striking feature is the posterior adhesive organ or *haptor*, an apparatus consisting of six posterior suckers on a muscular disc. Two hooks lie slightly anterior to the most posterior pair of suckers. The anterior adhesive organ takes the form of an oral sucker surrounding the terminal mouth. The male and uterine openings form a common genital opening in the mid-ventral line. Two marginal swellings at the anterior end indicate the positions of the two vaginal openings.

Body wall. The body in monogeneans (Fig. 11.5) is covered by syncytial tegument (see section 11.2.2) beneath which are three muscle layers (outer circular, oblique and inner longitudinal). As in other platyhelminths, the internal spaces between the organs are filled with a mesenchymatous parenchyma made up of typical

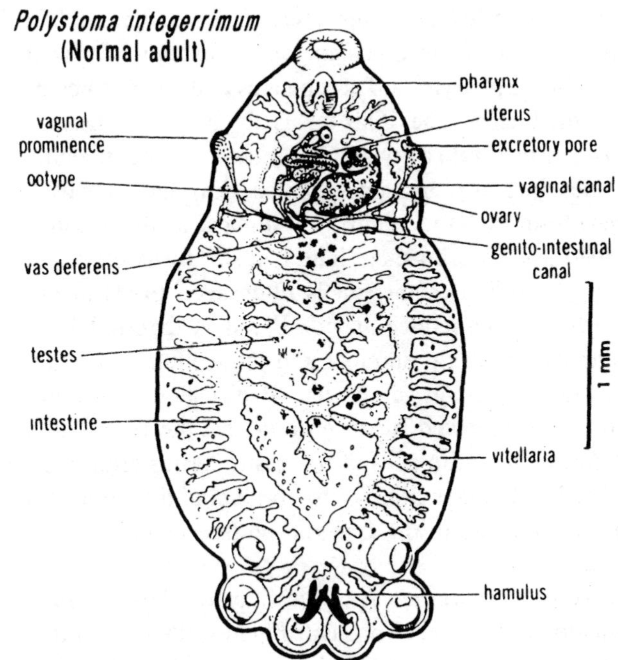

Fig. 11.2
Polystoma integerrimum: normal adult from bladder of frog. Genital organs become mature only during breeding season of host. (From an original drawing by Miss J. B. Williams.)

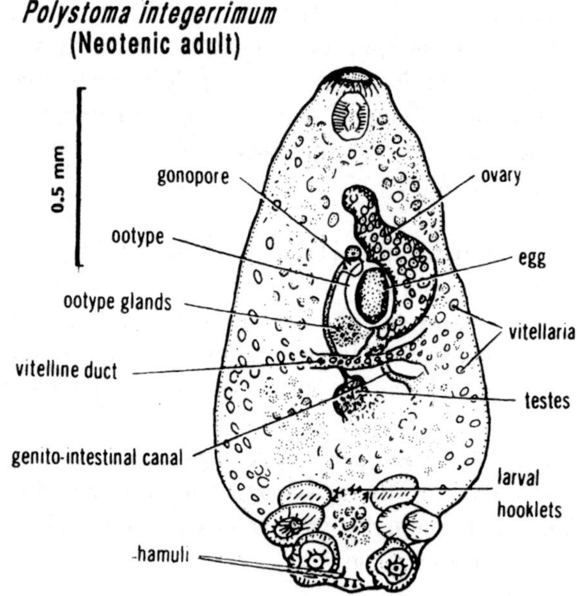

Fig. 11.3
Polystoma integerrimum: neotenic adult from external gills of a frog tadpole. (From an original drawing by Miss J. B. Williams.)

polygonal cells. The ultrastructure in this species has been studied by Bresciani (1973).

Digestive system. The mouth leads into a short muscular pharynx leading to the intestine. A peculiar median passage, the *bucco-intestinal canal*, extends forward beneath the pharynx and oesophagus and communicates with the buccal tube. This forms an alternative passage between the oral cavity and the intestine. The intestine divides into two branches, which bear lateral caeca; those from each branch may anastomose so that a widespread system is formed. The parasite feeds on blood and the outline of the gut is often seen to stand out clearly with contained blood. Blood is digested by a combination of intraluminar and intracellular digestion processes; haemolysis and partial digestion in the gut lumen are followed by absorption and the completion of digestion by the cells of the gastrodermis (Jennings, 1959, 1968). Haematin is a by-product of digestion and can be detected in the frog's urine (see p. 164).

Excretory system. A typical platyhelminth protonephridial system is present. Flame cells lead to fine canals which join to form paired excretory canals opening anteriorly and dorsally at about the same level of the pharynx.

Nervous system. This consists of a nerve ring surrounding the pharynx from which nerves extend anteriorly and posteriorly. The muscles of the haptor are supplied with fibres from the ventral nerves.

Reproductive system. Male: in general, this resembles that of rhabdocoele turbellarians. The testes are scattered with fine ducts connecting to a penis leading to the median genital pore.

Female: the oviduct arises from a single ovary. Shortly after its origin there opens into it (*a*) a *genito-intestinal canal* which leads to the intestine and (*b*) a short median vitelline duct. The latter bifurcates to give transverse vitelline ducts which run anteriorly and posteriorly. Into the anterior ducts are connected a pair of vaginae which open by small pores at marginal vaginal swellings. The oviduct passes forward to Mehlis' gland, and so to the uterus.

Thus, in *Polystoma*, the vaginae enter the vitelline ducts and pass into the oviduct, as in many Monogenea; hence the sperms enter the oviduct with the yolk. The occurrence of a genito-intestinal canal is not a general feature of Monogenea, but only occurs in the *Polyopisthocotylea*. It also occurs in some triclads, and may be a relic of the time when, in some primitive species (e.g. *Acoela*), the ova escaped from the parenchyma into the gut and out through the mouth. It has also been suggested that it could act as an escape arrangement for surplus yolk and ova.

The egg-shell in monogeneans is formed in the same way as in the digenetic trematodes and pseudophyllidean cestodes by the release of large semi-liquid globules from the so-called vitelline cells (Fig. 11.6). These run together and form an egg-shell (= capsule) which, as reported for many Digenea, appears to be made of sclerotin, a quinone-tanned protein. The process of protein stabilisation is, however, a complex one and there is evidence that alternative hardening systems may operate in some species (see p. 187). The process of egg-shell formation has a special significance in the Platyhelminthes, in general, and is discussed in detail later (p. 184; Table 13.1).

Life cycle (Fig. 11.4). The life cycle of *Polystoma integerrimum* is one of exceptional interest as being a helminth parasite whose maturation is apparently synchronised with the sexual maturation of its frog host (compare *Diplozoon nipponicum*, p. 172). In spring, when

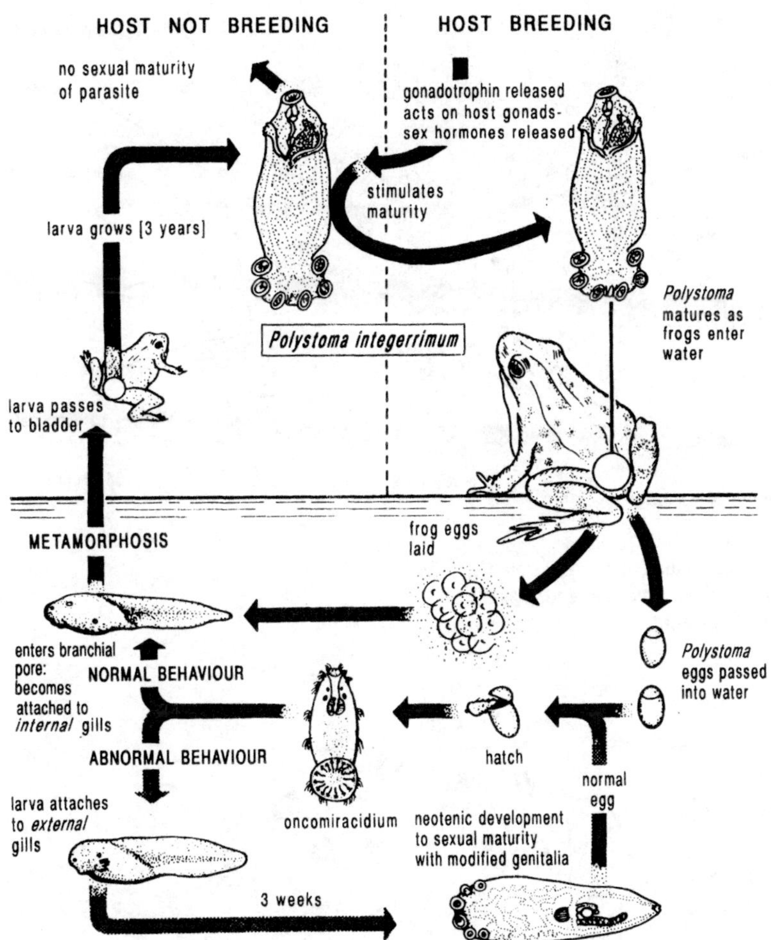

Fig. 11.4
Life cycle of *Polystoma integerrimum*, showing synchronisation of sexual maturity with breeding cycle of the amphibian host. (Original based on the work of Miretski, 1951.)

the frog is preparing to enter the water preparatory to copulation, the genitalia of the parasite begin to ripen. When frogs actually enter the water, maturation of *Polystoma* is completed and large quantities of eggs are released. These eggs hatch in about the same time as that required for the eggs produced by the host frogs to reach the *internal* gill tadpole stage. The hatched *larva* or *oncomiracidium* is barrel-shaped with a large posterior sucker bearing sixteen hooks, a primitive gut, and five incomplete bands of cilia. This larva makes its way into the branchial pore of the tadpole and attaches itself to the gills where it feeds on mucus and detritus.

It was formerly believed at metamorphosis that the juvenile *Polystoma* migrated from the gills to the bladder via the gut. However, failure to discover migrating forms in the tadpole gut led to the discovery that the juveniles migrate at night over the ventral surface of the tadpole to the cloaca and hence to the bladder – the whole process of migration lasting only 1 minute or so! (Coombes, 1967*a*). It may be noted that in another polystomid, *Protopolystoma xenopodis*, in the clawed toad *Xenopus laevis*, the free-swimming larvae enter the tadpole cloaca directly, passing to the bladder via the kidney (Thurston, 1964). Its biology has been reviewed by Macnae *et al.* (1973).

The cycle described above is probably the most commonly enacted one, but an alternative cycle can occur, if, by chance, a larva becomes attached to the *external* gills of a tadpole. In this case, rapid *neotenic* development takes place and a miniature sexually mature fluke is formed within about twenty days. The neotenic adult (Fig. 11.3) differs greatly from the normal adult matured

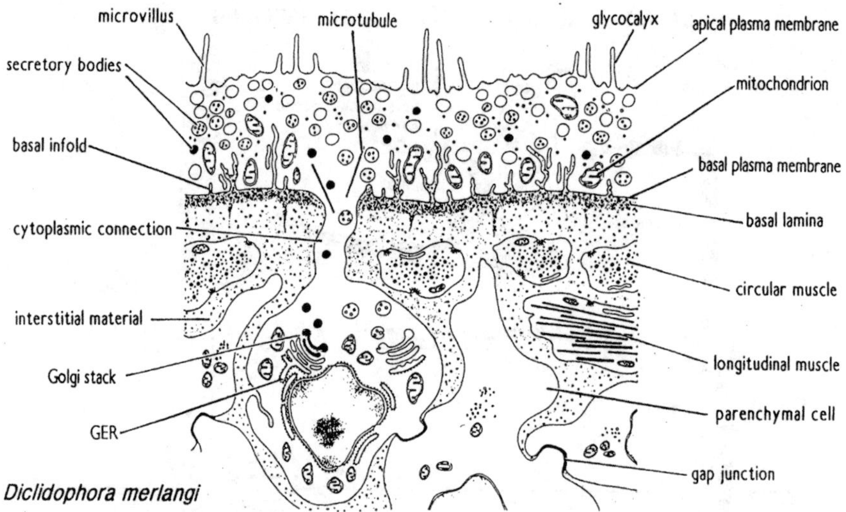

Fig. 11.5
The ultrastructural organisation of the tegument and associated structures of *Diclidophora merlangi*; GER = granular endoplasmic reticulum. (Based on Morris & Halton, 1971.)

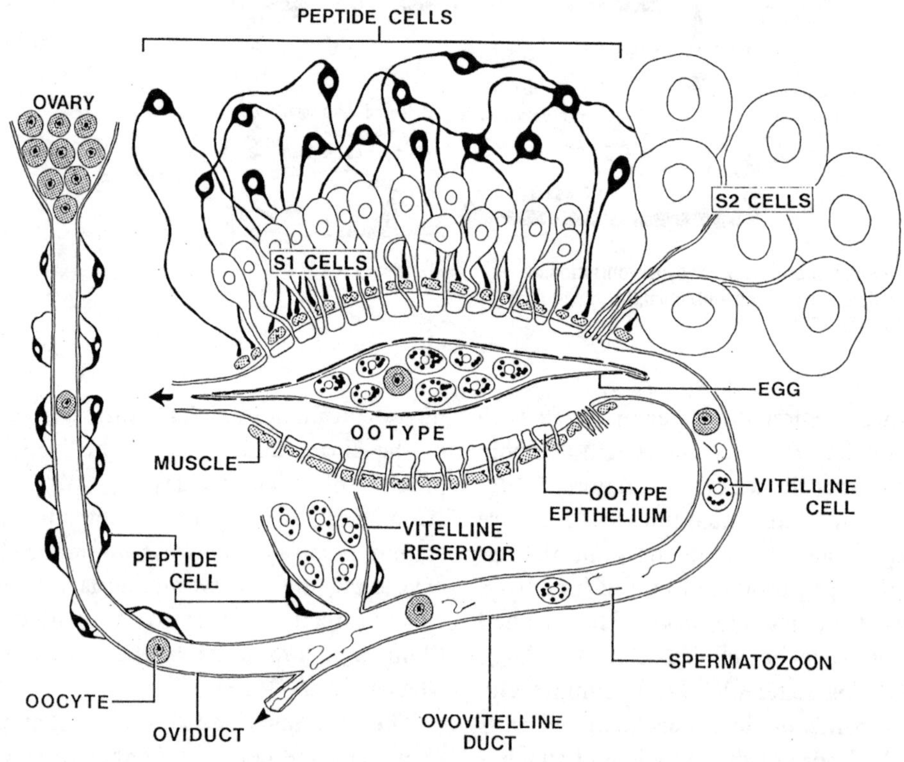

Fig. 11.6
Diclidophora merlangi: female reproductive system showing the general anatomy. The distribution of peptidergic nerve fibres was demonstrated by positive immunostaining for various peptides. The S1 and S2 cells represent the two types of Mehlis' gland cells identified. (After Maule *et al.*, 1990.)

Table 11.1. *Occurrence of Monogenea on marine fish examined at Plymouth, England, during 1953–5*

Host	Parasite	Per cent infection
Merlangius merlangus (whiting)	*Diclidophora merlangi*	8.7
Gadus luscus (pout, bib)	*Diclidophora luscae*	21.0
Merluccius merluccius (hake)	*Anthocotyle merlucci*	7.6
Trigla cuculus (red gurnard)	*Plectanocotyle gurnardi*	95.0
Trachurus trachurus (horse mackerel)	*Pseudaxine trachuri*	21.6
Trachurus trachurus (horse mackerel)	*Gastrocotyle trachuri*	62.2
Belone belone (gar fish)	*Axine belones*	77.8
Morone labrax (bass)	*Microcotyle labracis*	66.7
Scomber scombrus (mackerel)	*Octosoma scombri*	100.0

Source: Data from Llewellyn (1956).

in the bladder, having only a single testis, functionless copulatory organs, and rudimentary uterus, vaginae and vitellaria. Cross-fertilisation takes place although the definitive sperm is only a late spermatid. Notwithstanding these extraordinary morphological abnormalities, neotenic adults developing in this way give rise to fertile eggs; these hatch to produce larvae which undergo normal development.

A *direct* internal cycle has also been reported for *P. integerrimum* (rarely) and *P. pelobatis* (more commonly); forms in the bladder become ovoviviparous and the larvae produced take up immediate residence in the bladder (Combes, 1967*b*, 1968). The biology of other species of *Polystoma* has been reviewed by Tinsley (1973).

Similar differences between the egg hatching of other species of *Polystoma* and their hosts have also been noted and presumably the same phenomenon occurs in these species also.

Endocrine control of maturation. Experiments by Miretski (1951) have demonstrated that the maturity of the fluke is apparently controlled directly or indirectly by the hormonal activity of the frog. Thus when infected, immature frogs were injected with hypophysis extract, the polystomes matured within four to eight days and produced quantities of eggs for a period of about a week. This latter time corresponds approximately to the time frogs spend on spawning. This beautifully synchronised mechanism results in the eggs of the fluke being released only when the frogs enter water to breed, thus assuring that by the time they have passed through the embryo stage and hatched, abundant tadpoles will be available for re-infection. This correspondence of host and parasite reproduction is all the more remarkable because it involves not only oviposition, but, in addition, the parasite reaches sexual maturity only when the host does, after three years; their gonadal cycles also run in parallel, with regression and resorption in late spring, and growth and preparation in summer and autumn. The nature of the endocrinal control has not been established. Injection of hypophysis extract would result in increased levels of gonadotrophins and gonadal hormones, either of which could be responsible for the maturation effect.

An even more remarkable example of host–parasite reproduction synchrony is seen in *Pseudodiplochis americanus* in the North American toad *Scaphiopus couchii*, which spawns almost instantly only when it rains: see Figs 11.9, 11.10 and p. 167. It should be noted that protozoan parasites (e.g. *Opalina*, *Nyctotherus*) in the frog are similarly under the influence of the hormonal system of the host. These protozoa, moreover, may be induced also to enter their sexual phase of development by injection of substances other than hormones (p. 148) and it would be interesting to see if *Polystoma* similarly responds to the injection of the host with these substances.

Polystoma nearticum (Fig. 11.7). In the USA, this species, which closely resembles the European *P. integerrimum* in its morphology and life cycle, occurs commonly in the tree frogs *Hyla versicolor* and *H. cinerea*, with prevalences of 23–50 per cent being reported from the former host (Paul, 1938; Hausfater *et al.*, 1990). Like *P. integerrimum* it occurs in two forms, a 'bladder' form (Fig. 11.7A) in adult frogs and a branchial 'neotenic' form (Fig. 11.7B) on the gills of tadpoles. The branchial form matures within 22 days, but the bladder form probably only reaches maturity after three years when the frog becomes mature (Paul, 1938). When environmental conditions are suitable, the frogs descend from the trees and mate in the water, the parasites releasing eggs at the same time.

Table 11.2. *Occurrence of Monogenea on fresh-water fish in the river Roding, Essex, England*

Host	Parasite*	Average incidence (%)	Average intensity (parasites per fish)
Rutilus rutilus (roach)	*Dactylogyrus sphyrna*	15.4	2.0
	Dactylogyrus n.sp.	88.7	34.1
	Gyrodactylus n.sp. 1	1.6	3.0
	Diplozoon paradoxum	8.1	1.0
Leuciscus leuciscus (dace)	*Dactylogyrus cordus*	61.9	16.4
	Diplozoon paradoxum	11.9	1.0
	Gyrodactylus n.sp. 2	7.1	5.7
Gobio gobio (gudgeon)	*Dactylogyrus gobii*	4.3	1.8
	Gyrodactylus n.sp. 3	14.0	2.0
	Diplozoon paradoxum	60.2	3.6
Phoxinus phoxinus (minnow)	*Dactylogyrus phoxini*	7.0	2.3
	Diplozoon paradoxum	44.2	2.3
Leuciscus cephalus (chub)	*Dactylogyrus vistulae*	75.0	16.1
Gasterosteus aculeatus (stickleback)	*Gyrodactylus* n.sp. 4	50.0	9.0

*Many species unidentifiable and probably represent new species.
Source: Data by courtesy of Dr J. Shillcock.

Detection of *Polystoma* in frogs. Jennings (1956) has developed a technique for the detection of *Polystoma integerrimum* in frogs based on the presence of haematin in the urine of infected hosts. By gently squeezing a frog a little urine is collected into a 60 × 6 mm tube to which is added an alkaline solution of luminol and hydrogen peroxide. The presence of haematin, and hence *Polystoma*, is indicated by the development of an intense blue luminescence. In carrying out this test, faecal contamination must be avoided, as the peroxidases of the faecal protozoa act with the luminol–peroxide reagent to give a false positive reaction.

Table 11.3. *Host specificity of Monogenea of marine fishes*

No. of species on:				
1 species of fish	1 genus of fish	1 family of fish	1 order of fish	Total
340 (78%)	388 (89%)	420 (96%)	429 (98%)	435

Source: Data from Rohde (1979).

○ 11.2.2 General account

Early reference works on the Monogenea are those of Sproston (1946), Bychowsky (1957), Baer & Joyeux (1961), Hyman (1951) and Yamaguti (1963). The more recent literature has been listed in a valuable bibliography by Hargis & Thoney (1983). Reviews on the evolutionary status of the group – which has proved to be exceptionally controversial – are those of Lebedev (1986, 1988), Llewellyn (1987), Mackiewicz (1982) and Malmberg (1990). The monogeneans of fish have received special attention in many countries, some species lists being those for Australia (Rohde, 1978), Canada (Margolis & Kabata, 1984), Venezuela (Nasir & Fuentes Zambrano, 1983), N. America (Schell, 1985), the USA (Crane, 1972), Russia (USSR, 1985), United Kingdom (Llewellyn, 1956; Llewellyn *et al.*, 1984; see also Tables 11.1, 11.2); Worldwide (Chubb, 1977).

Distribution. Monogeneans are chiefly ectoparasites of fish, amphibia and reptiles, being found on the gills, the gill chamber, skin, and buccal cavity, but are also not uncommon in the uterus and body cavity; exceptionally they are found in the heart (*Amphibdella torpedinis*) and the excretory system (*Acolpenteron petruschewskyi*). Monogenea of fish are remarkably host-specific and are generally restricted to a single species, a single genus or a single family of host. Thus in one survey (Table 11.3), it was found that only 2 per cent of 435 species occur on more than one order of fish (Rohde, 1979).

Adhesive attitudes. The adhesive attitudes, form and operation of the attachment organs have been studied in considerable detail. As a general rule, gill forms attach with their posterior adhesive organs near the host gill and with the anterior end nearer the distal end of the primary lamellae (Llewellyn, 1956). This results in the worm lying with its mouth downstream to the current passing over the gill.

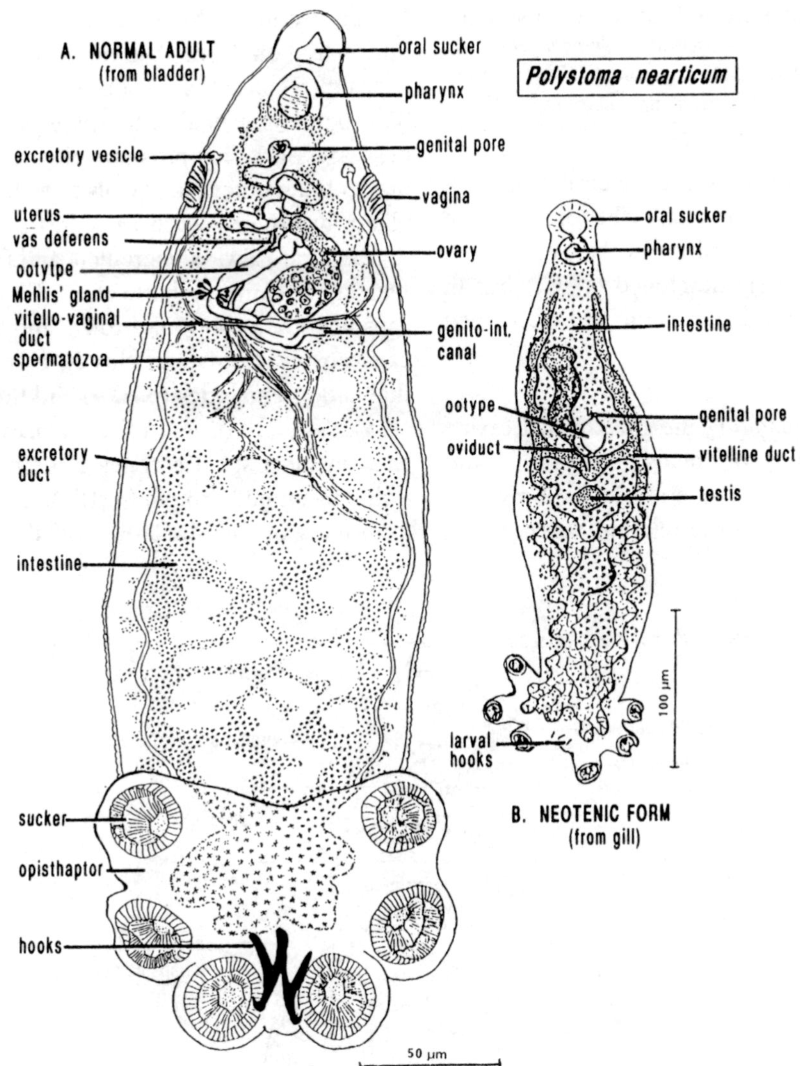

Fig. 11.7
Polystoma nearticum: a monogenean from the gray tree frog, *Hyla versicolor*, in the USA. The 'normal' form occurs in the bladder of adult frogs, the neotenic form on the gills of tadpoles. (After Paul, 1938.)

Worms also show a tendency to attach to the secondary gill lamellae and different species may show a marked predilection for a particular gill or gills. Thus, *Diclidophora merlangi* (Fig. 11.12) occurs most frequently on the first gill of the whiting, whereas *D. luscae* occurs on the second and third gill of the pout.

Body wall (Fig. 11.5). The ultrastructure of the body wall of monogeneans has been reviewed in detail by Smyth & Halton (1983). It consists of a syncytial protoplasmic layer or *tegument* essentially similar to digenean tegument. Externally, the surface plasma membrane is covered by a thin, amorphous coat or glycocalyx, which is PAS-positive and gives B metachromasia with toludine blue. The tegumental matrix contains secretory bodies produced in the inner nucleated regions which are passed to the outer layer, together with mitochondria and other inclusions.

Although unequivocal evidence is lacking, the function of the tegument appears to be secretory and absorptive. In most adult Monogenea the tegument is amplified by microvilli and surface pits may be present, as in the Digenea; other details are clear from Fig. 11.5.

Data on the ultrastructure are available for the following: *Entobdella soleae*, *Acanthocotyle elegans*, *Gyrodactylus* sp., *Amphibdella flavolineata*, *Diclidophora merlangi*, *Plectanocotyle gurnardi*, *Gyrodactylus eucaliae*, *Euzetrema knoepffleri*, *Diplectanum aequans* (Smyth & Halton, 1983).

Other details of the fine structure are clear from Fig. 11.5. Points of special interest are the presence of secretory bodies and mitochondria. In the outer tegument, the latter have poorly developed cristae, but in the muscles, the mitochondria contain well-developed cristae.

Nutrition. Species inhabiting the gills, buccal cavity, and bladder feed mainly on blood, epithelium and mucus; skin forms feed on epidermal cells. Smyth & Halton (1983) have reviewed the relevant literature.

Reproduction. In general, the male system shows little variation from that described in *Polystoma*. The testes may consist of a solid body or scattered capsules. The penis is armed with spines in some species and possesses so-called prostatic glands. The ootype region may contain other gland cells as well as those of Mehlis' gland, but their role in capsule formation is unknown. The eggs are always operculate and frequently bear long filaments at one or both ends (Fig. 11.12). The egg-capsules are yellow or brown and the mechanism of egg-capsule formation closely resembles that of the digenetic trematodes (p. 184) and the pseudophyllidean cestodes (p. 288). As in the majority of these, the egg-shell in most species appears to consist of sclerotin, a quinone-tanned protein (p. 184). The shell material has its origins in the vitellaria, and thus the latter can be

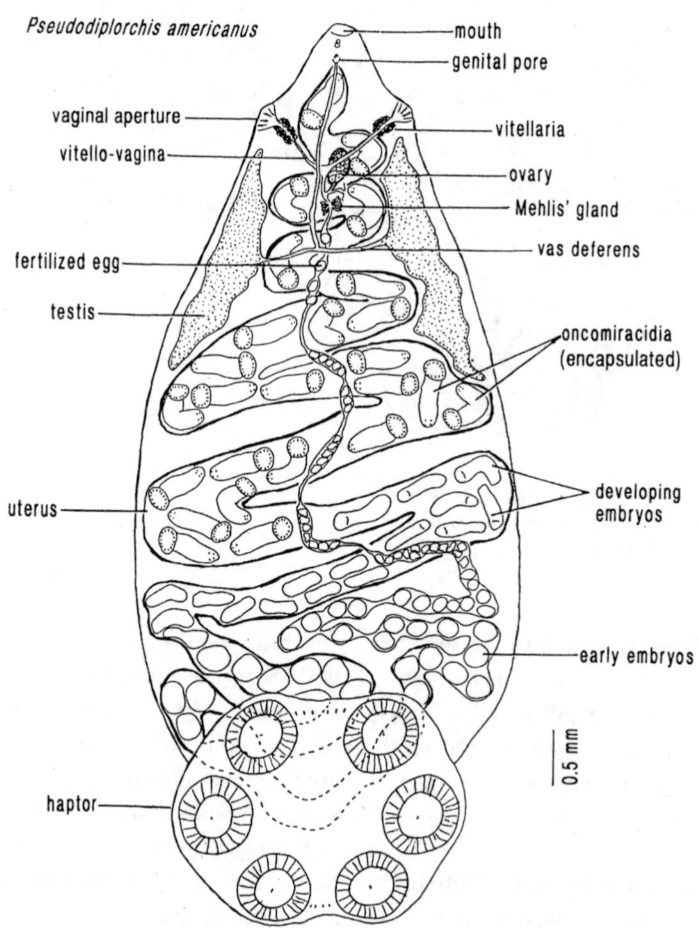

Fig. 11.8
Pseudodiplorchis americanus: a monogenean from the bladder of the American spadefoot toad, *Scaphiopus couchii*. The adult has a long uterus in which eggs develop during the period of host hibernation; this results in the eggs giving rise to large numbers of larvae (oncomiracida). These remain encapsulated in thin flexible capsules (not shown) until the host enters the water when they are released. The life cycle is shown in Figs 11.9 and 11.10. (Based on Tinsley, personal communication.)

brilliantly and specifically stained by histochemical reagents (such as Fast Red Salt B) for phenol or for the enzyme phenolase (Johri & Smyth, 1956; Smyth & Halton, 1983); see Table 13.1. The formation of the egg-shell in platyhelminths is a complex phenomenon and that in Monogenea is discussed when the Digenea are considered (p. 184). In the viviparous form *Gyrodactylus* (Fig. 11.11), vitellaria are lacking, the ovary being a germovitellarium (i.e. an ovary with specialised yolk-producing regions). Some species have developed remarkable reproductive strategies related to their ecology; see p. 168.

Physiology. Although a few figures for the chemical composition of monogeneans are available (Smyth & Halton, 1983), the biochemistry and general physiology of this group remain almost entirely unstudied. It is interesting to note that the hooks (sclerites) of larval monogeneans contain cystine sulphur, as in the eucestodes and cestodarians (Lyons, 1966), a finding which further strengthens the conclusion that cestodes and monogeneans are closely related.

Some experiments on the mechanism of host-finding by oncomiracidia have been carried out (Kearn, 1967). By using agar blocks which had been in contact with the fish surface, it has been shown that chemotaxis is involved, the attractant apparently being a specific substance secreted by the skin. More recent experiments on egg-hatching and host-finding have been reviewed by Kearn (1986*a*,*b*).

○ 11.2.3 Miscellaneous Monogenea

Pseudodiplorchis americanus (Fig. 11.8)

The adult of this species lives in the bladder of the spadefoot desert toad of North America, *Scaphiopus couchii*, which has a remarkable life cycle (Tinsley, 1989). It hibernates for 9–10 months of the year in deep burrows, but must enter the water to breed. However, it is only able to do this for 1–3 days in each year when heavy rainstorms (in June/July in Arizona) produce temporary pools in the desert (Fig. 11.9). As the toads are nocturnal, they only enter the water after dark (about 9 pm) and leave before dawn (4 am), thus defining a very narrow 'window' of opportunity of about 7 h when parasite transmission is possible (Fig. 11.10). This represents a restriction of parasite transmission probably greater than in any other helminth.

Although the adult worm appears to be ovoviviparous, the eggs are actually enclosed in an electron-dense shell, which is subsequently replaced by a capsule consisting of a flexible, multilaminate sac derived from the uterine wall. Remarkably, the capsule lining extends into fine cytoplasmic processes which make a placenta-like connection with the developing embryo. This unique system appears to be a mechanism for providing nutrients for the developing embryo (Cable & Tinsley, 1991). The oncomiracidia often hatch out of their capsules in the bladder, thus giving the appearance of being ovoviviparous. On contact with the water, the oncomiracidia are immediately infective to the spawning toads so that almost 'instantaneous' transmission takes place (Tinsley, 1989). The oncomiracida invade their new hosts via the nostrils and within a week migrate into the mouth and associated cavities and hence through the glottis to the lungs. By day 21 the parasites begin to return to the oral cavity and from day 28 onwards migrate to the urinary bladder. The biology of this remarkable parasite has been investigated in detail in a series of studies by Tinsley and his co-workers (Cable & Tinsley, 1991; Tinsley & Jackson, 1986, 1988; Tinsley, 1989, 1990).

Gyrodactylus sp. (Fig. 11.11)

This is a small ectoparasite of fresh-water and marine fish, often easily obtainable from aquarium fish. It is unusual in that it is not only viviparous but the larva forming in the uterus of the parent may contain a further embryo, which itself contains a mass of embryonic cells, the whole forming a kind of Chinese box puzzle of developing forms. Up to four generations may occur in one individual. This is probably due to polyembryony.

Calicotyle kroyeri (Fig. 11.13)

A not uncommon species occurring in the cloacal region of skates or rays. It is a squat form almost as broad as long. The haptor is sucker-like with seven radial septa and two hooks. The vaginae are paired and open ventrolaterally slightly behind the level of the genital aperture. Eggs are not numerous. The oncomiracidium has been described by Kearn (1970).

Polystomoides oris

A species closely related to *Polystoma integerrimum*; parasitic in the mouth cavity of the fresh-water painted turtle, *Chrysemys picta*. Its basic morphology is similar

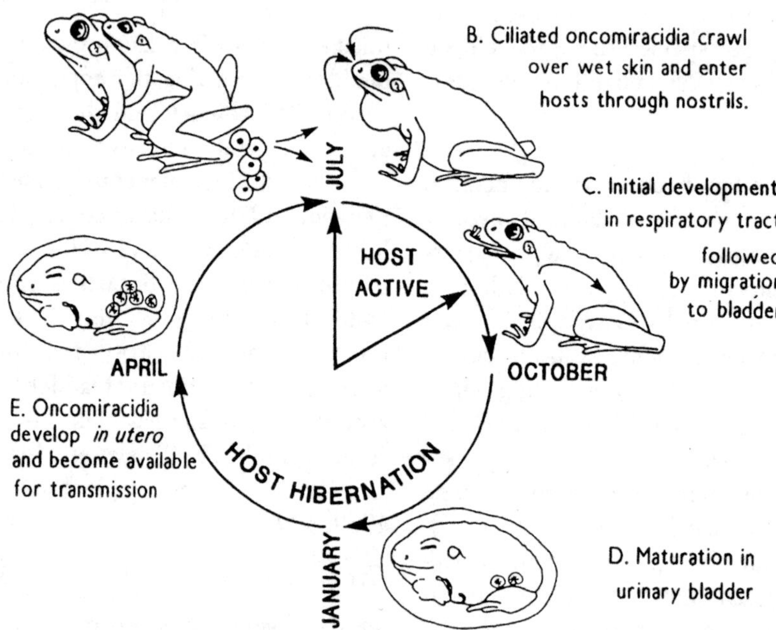

Fig. 11.9
The life cycle of the monogenean *Pseudodiplorchis americanus*, which is synchronised with the annual cycle of breeding activity of its host, the American spadefoot toad, *Scaphiopus couchii*. The parasite releases its (encapsulated) oncomiracidia only when a summer rainstorm forms pools in which the toads can spawn and the parasite larvae can escape. See also Fig. 11.10. (Modified from Tinsley, 1989.)

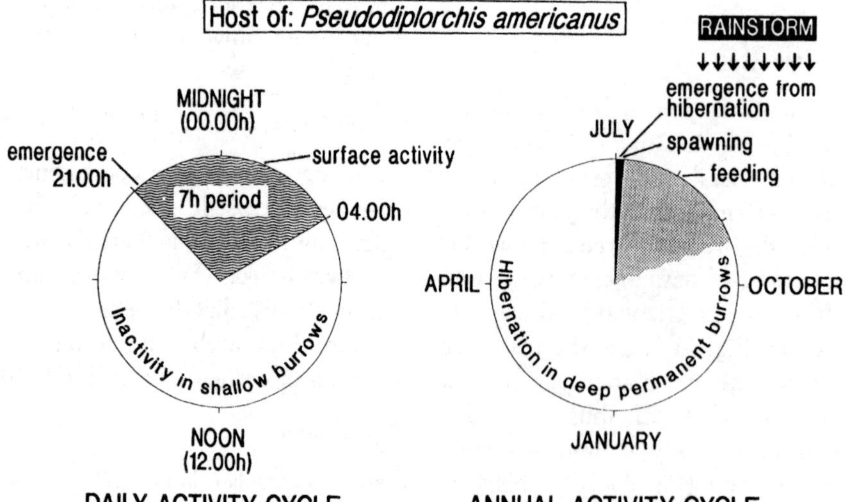

Fig. 11.10
Daily and annual patterns of activity of the American spadefoot, *Scaphiopus couchii*. The toad can spawn only during a short period (about 7 h) when pools are formed during a summer rainstorm. See also Fig. 11.9. (Modified from Tinsley, 1990.)

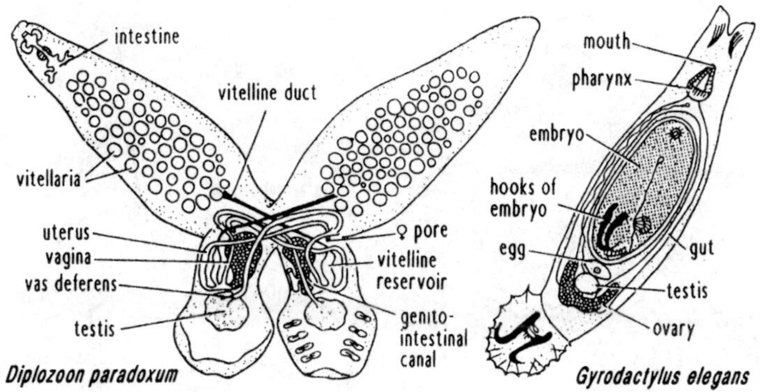

Fig. 11.11
Two aberrant monogenetic trematodes: *Diplozoon paradoxum*, from the gills of a minnow; *Gyrodactylus elegans*, from the skin of aquarium fish. (After Dawes, 1946.)

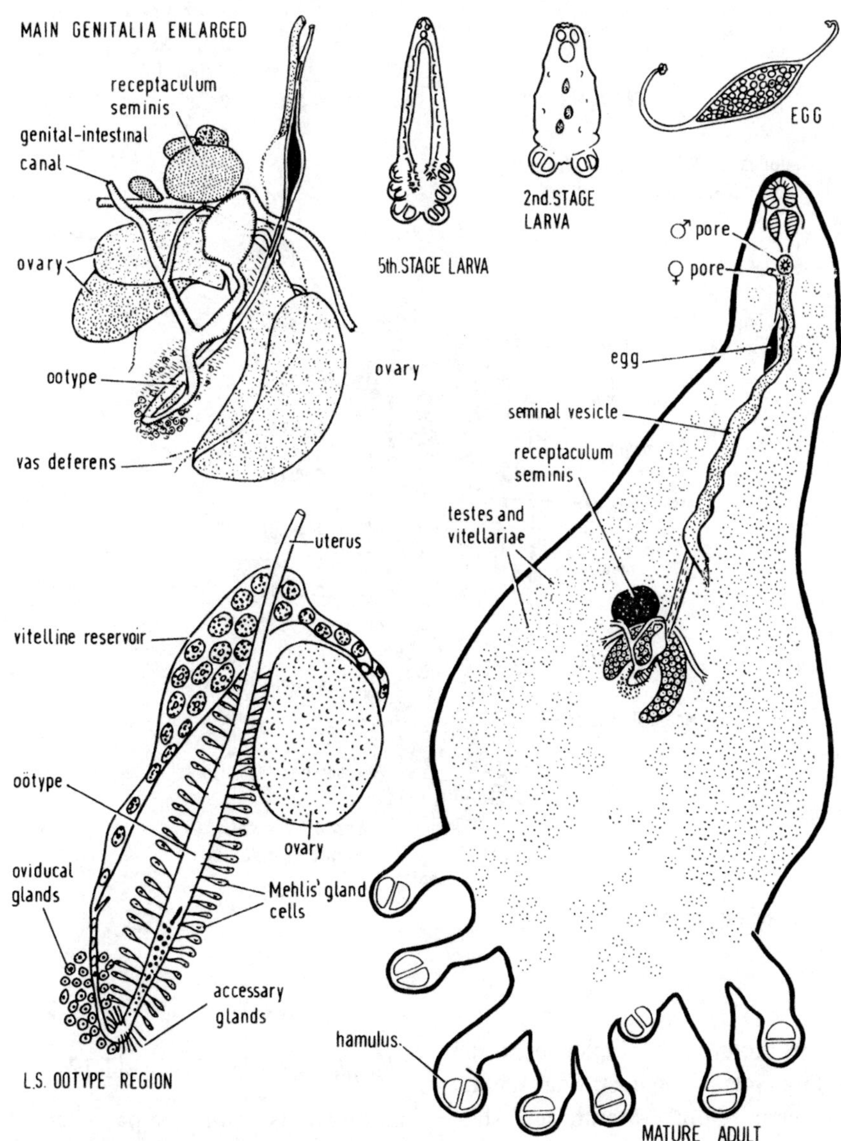

Fig. 11.12
Diclidophora merlangi: anatomy of adult and larvae. (From Frankland, 1955; B. Rennison, personal communication, 1953.)

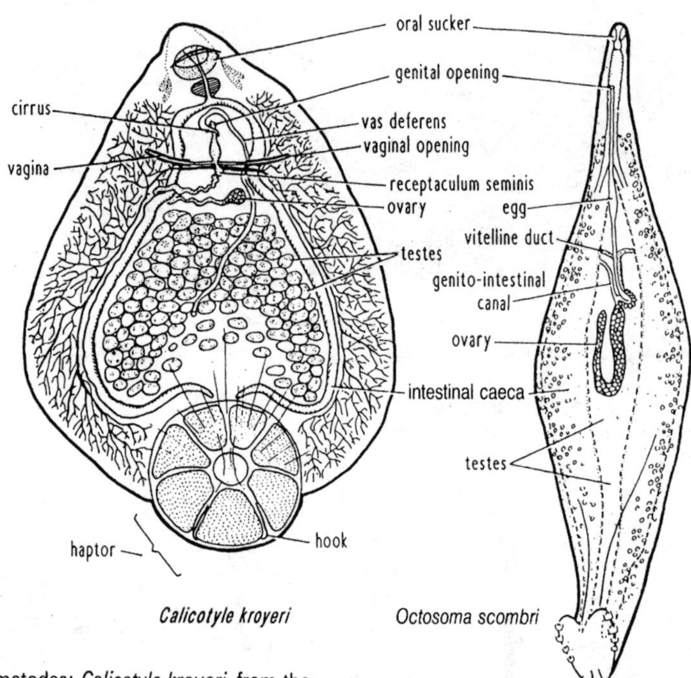

Fig. 11.13
Two common monogenetic trematodes: *Calicotyle kroyeri*, from the cloaca of skates and rays, and *Octosoma scombri*, from gills of mackerel. (After Goto, 1894.)

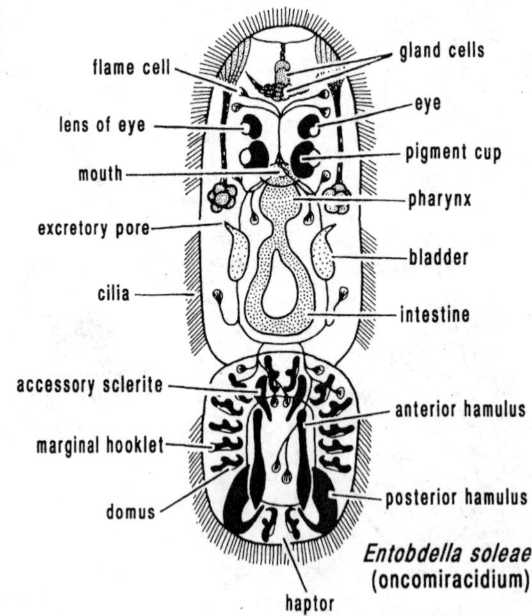

Fig. 11.14
Oncomiracidium of *Entobdella soleae*; ventral view. (After Kearn, 1963*c*.)

to that of *Polystoma* but its life cycle is simpler, bearing no relation to the breeding cycle of the host. Eggs hatch at laboratory temperatures in about a month, and the hatched larvae enter the mouth of other turtles and reach maturity in about a year.

Diclidophora merlangi (Fig. 11.12)

This species is a common parasite on the gills of whiting (*Merlangius merlangus*) and probably the most easily obtainable monogenetic trematode suitable for class material.

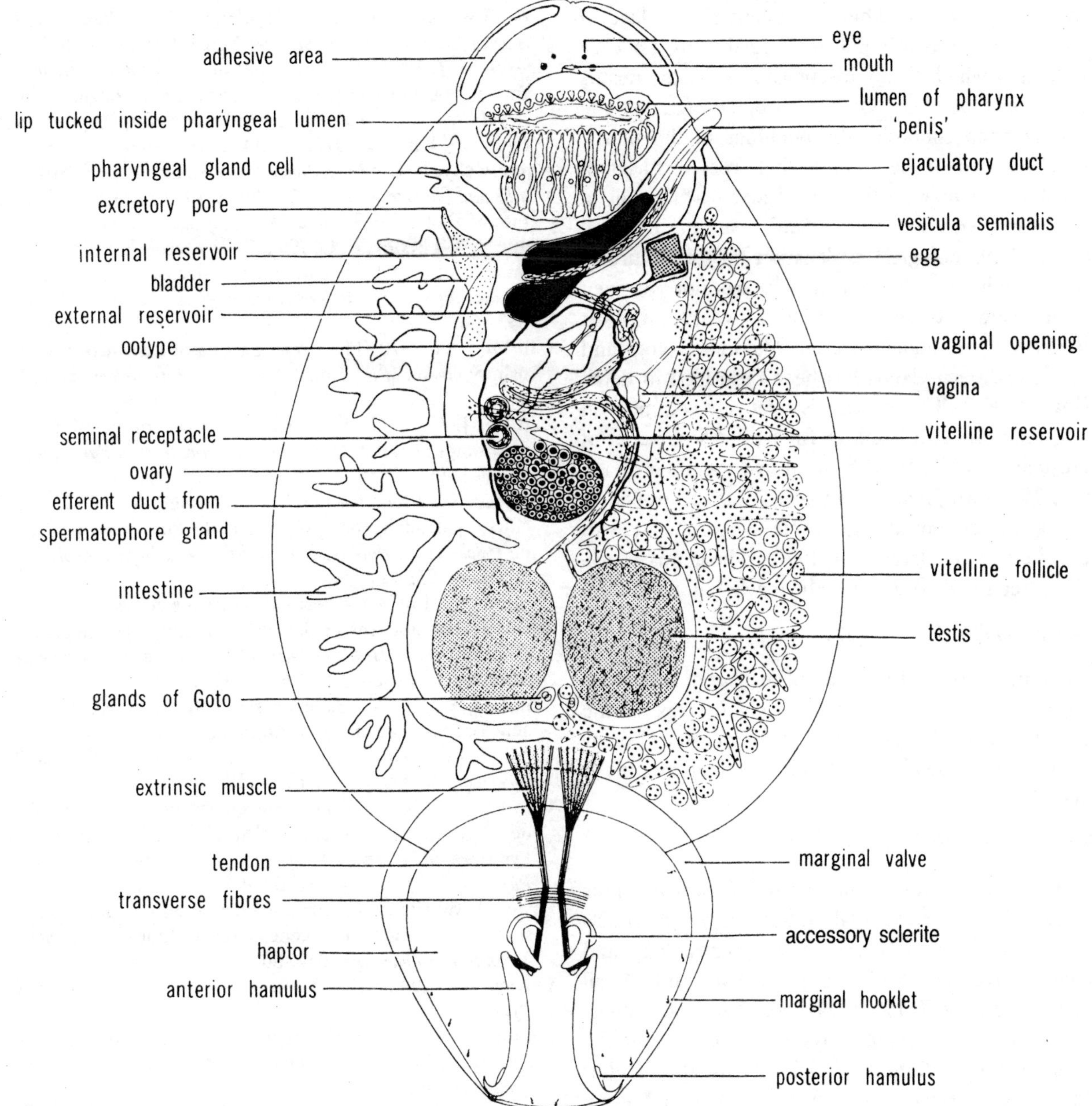

Fig. 11.15
General morphology of *Entobdella soleae*; ventral view. (After Kearn, 1971.)

Unfortunately, as its branched gut is usually filled with pigment, derived from its blood meals, the general morphology is difficult to follow in whole mounts. It lacks vaginae and has an elongated folded ovary. The haptor consists of four pairs of pedunculate clamp-like suckers each bearing a complex framework of sclerites. The mouth is not quite terminal and the gut branches extensively into the peduncles of the suckers. This species has proved to be an excellent laboratory model and its general biology, ultrastructure, physiology and neurobiology have been examined in a series of classical studies by Halton and his colleagues (reviewed in Smyth & Halton, 1983). See also Halton (1974–9), Halton *et al.* (1987) and Maule *et al.* (1989, 1990).

Diplozoon paradoxum (Fig. 11.11)

A remarkable species from the gills of fresh-water fish comprising two individuals which become completely fused to one another. The attachment takes place in the larval stage, a pair being held together by a sucker arrangement which finally disappears as the organisms become fused together. When mature, the vagina of one individual opens opposite the openings of the uterus and vas deferens of the other so that cross-fertilisation is possible and may be the rule. Larvae which fail to find a partner subsequently die and never attain sexual maturity. A physiological explanation for the failure of single individuals to become mature has not yet been produced and this species appears to present most interesting experimental material. The morphology and life cycle of *D. paradoxum* has been described in detail by Bovet (1967). This species is relatively common on bream, minnows and roach in England; 55 per cent of bream may be infected with up to five specimens each (Wiles, 1968). In Japan, *D. nipponicum* is of interest in having a sexual reproductive cycle synchronised with that of the fish host (Kamegai, 1970). The genus *Diplozoon* has been reviewed by Reichenbach-Klinke (1961).

Octosoma scombri (Fig. 11.13)

A common parasite of the gills of mackerel, infections of up to 100 per cent occurring in some areas off the coast of Great Britain. The general morphology resembles that of *Diclidophora*, but a vaginal opening is present.

Entobdella soleae (Fig. 11.15)

An ectoparasite of the common sole, *Solea solea*. This is a most useful species for laboratory study and aspects of its life cycle, larval development, physiology and behaviour have been worked out in considerable detail (Kearn, 1963*a*, 1967, 1971). The eggs hatch after incubation in sea water for 27 days at 13–17 °C and the ciliated oncomiracidium (Fig. 11.14) eventually attaches to the upper surface of the fish (which is a bottom feeder); it later migrates to the ventral surface where sexual maturity is reached.

○○○ References

(References of general interest have been included in this list in addition to those quoted in the text.)

Baer, J. & Joyeux, C. 1961. Platyhelminthes, Mésozaires, Acanthocéphales, Nemertiens. In: *Traité de Zoologie* (ed. P. Grassé), Vol. IV, pp. 561–692. Masson et Cie, Paris.

Bovet, J. 1967. Contribution à la morphologie et à la biologie de *Diplozoon paradoxum* V. Nordmann, 1832. *Bulletin de la Société Neuchâteloise des Sciences Naturelles*, **90:** 63–159.

Bresciani, J. 1973. The ultrastructure of the integument of the monogenean *Polystoma integerrimum* (Frolich, 1791). *Royal Veterinary and Agricultural University, Copenhagen, Denmark, Yearbook* 1973, pp. 14–27.

Bychowsky, B. E. 1957. *Monogenetic Trematodes, their Systematics and Phylogeny*. (English Edition, 1961.) American Institute of Biological Science, Washington, DC.

Cable, J. & Tinsley, R. C. 1991. Intra-uterine larval development of the polystomatid monogeneans, *Pseudodiplorchis americanus* and *Neodiplorchis scaphiopodis*. *Parasitology*, **103:** 253–66.

Chubb, J. C. 1970. The parasite fauna of British freshwater fish. *Symposia of the British Society for Parasitology*, **8:** 119–44.

Chubb, J. C. 1977. Seasonal occurrence of helminths in freshwater fish. Part I. Monogenea. *Advances in Parasitology*, **15:** 133–99.

Combes, C. 1966. Recherches expérimentales sur la spécificité parasitaire des polystomes de *Rana temporia* L. et de *Pelobates cultripes* (Cuv.). *Bulletin de la Société Zoologique de France*, **91:** 439–44.

Combes, C. 1967*a*. Corrélations entre les cycles sexuels des amphibiens anoures et des Polystomatidae (Monogenea). *Comptes Rendus Hebdomadaires des Séances de l'Académie des Sciences, Paris*, Series D, **264:** 1051–2.

Combes, C. 1967*b*. Biologie des Polystomatidae (Monogenea): Existence et démonstration experimental des possibilités de cycle interne directe. *Bulletin de la Société Zoologique de France*, **92:** 129–33.

Combes, C. 1968. Biologie, écologie cycles et biogéographie de Digenes et Monogenes d'Amphibiens dans l'est des Pyrénées. *Mémoires du Muséum National d'Histoire Naturelle, Paris, Series A, Zoology*, **50:** 1–195.

Crane, J. W. 1972. Systematics and new systematics and new species of marine Monogenea from California. *Wassmann Journal of Biology*, **30:** 109–66.

Dawes, B. 1946. *The Trematoda.* Cambridge University Press.

Dubois, G. 1970. Les Monogènes: class autonome ou sous-classe de Trématodas? *Annales de Parasitologie Humaine et Comparée*, **45:** 247–50.

Frankland, H. M. T. 1955. The life history of binomics of *Diclidophora denticulata. Parasitology*, **45:** 313–53.

Gussev, A. V. 1973/74. Freshwater Indian Monogenoidea. Principles of systematics analysis in the world faunas and their evolution. *Indian Journal of Helminthology*, **25/26:** 241 pp.

Halton, D. W. 1974. Hemoglobin absorption in the gut of a monogenean trematode, *Diclidophora merlangi. Journal of Parasitology*, **60:** 59–66.

Halton, D. W. 1975. Intracellular digestion and cellular defecation in a monogenean, *Diclidophora merlangi. Parasitology*, **76:** 29–37.

Halton, D. W. 1978. Trans-tegumental absorption of L-alanine and L-leucine by a monogenean, *Diclidophora merlangi. Parasitology*, **76:** 29–37.

Halton, D. W. 1979. The surface topography of a monogenean, *Diclidophora merlangi*, revealed by scanning electron microscopy. *Zeitschrift für Parasitenkunde*, **61**: 1–12.

Halton, D. W., Maule, A. G., Johnston, C. F. & Fairweather, I. 1987. Occurrence of cholinesterase and ciliated sensory structures in a fish gill fluke, *Diclidophore merlangi*. *Parasitological Research*, **74**: 151–4.

Hargis, W. J. & Thoney, D. A. 1983. *Bibliography of the Monogenea. Literature of the World, 1758–1982*. Special Scientific Report No. 112, College of William & Mary in Virginia, USA.

Hausfater, G., Gerhardt, H. C. & Klamp, G. M. 1990. Parasites and mate choice in gray tree frogs, *Hyla versicolor*. *American Zoologist*, **30**: 299–311.

Hyman, L. 1951. *The Invertebrates II. Platyhelminthes and Rhynchocoela*. McGraw-Hill, New York and London.

Jennings, J. B. 1956. A technique for the detection of *Polystoma integerrimum* in the common frog (*Rana temporaria*). *Journal of Helminthology*, **30**: 119–204.

Jennings, J. B. 1959. Studies on digestion in the monogenetic trematode *Polystoma integerrimum*. *Journal of Helminthology*, 33: 197–204.

Jennings, J. B. 1968. Nutrition and digestion. Chapter 2. In: *Chemical Zoology*, Vol. 2 (ed. M. Florkin & B. T. Scheer), pp. 303–26. Academic Press, New York.

Johri, L. N. & Smyth, J. D. 1956. A histochemical approach to the study of helminth morphology. *Parasitology*, **46**: 107–16.

Kamegai, S. 1970. On *Diplozoon nipponicum* Goto, 1891. Part III. The seasonal development of the reproductive organs of *Diplozoon nipponicum* parasitic on *Cyprinus carpio*. *Research Bulletin of the Meguro Parasitological Museum*, **3**: 21–5.

Kearn, G. C. 1963*a*. The life cycle of the monogenean *Entobdella soleae*, a skin parasite on the common sole. *Parasitology*, 53: 253–63.

Kearn, G. C. 1963*b*. The egg, oncomiracidium and larval development of *Entobdella soleae*, a monogenean skin parasite of the common sole. *Parasitology*, **53**: 435–47.

Kearn, G. C. 1967. Experiments on host-finding and host-specificity in the monogenean skin parasite *Entobdella soleae*. *Parasitology*, **57**: 585–605.

Kearn, G. C. 1970. The oncomiracidia of the monocotylid monogeneans *Dictyocotyle coeliaca* and *Calicotyle kroyeri*. *Parasitology*, **61**: 153–60.

Kearn, G. C. 1971. The physiology and behaviour of the monogenean skin parasite *Entobdella soleae* in relation to its host (*Solea solea*). In: *Ecology and Physiology of Parasites* (ed. A. M. Fallis), pp. 161–87. (Toronto University Press.) Adam Hilger, London.

Kearn, G. C. 1986*a*. Role of chemical substances from fish in hatching and host-finding in monogeneans. *Journal of Chemical Ecology*, **12**: 1651–8.

Kearn, G. C. 1986*b*. The eggs of monogeneans. *Advances in Parasitology*, **25**: 175–273.

Kohlmann, F. W. 1961. Untersuchungen zur Biologie, Anatomie und Histologie von *Polystoma integerrimum* Fröhlich. *Zeitschrift für Parasitenkunde*, **20**: 495–524.

Lebedev, B. I. 1986. (Monogenea: suborder Gastrocotylinea.) Nauka, Leningrad. (In Russian.)

Lebedev, B. I. 1988. Monogenea in the light of new evidence and their position among platyhelminths. *Angewandte Parasitologie*, **29**: 149–67.

Llewellyn, J. 1956. The host-specificity, micro-ecology, adhesive attitudes and comparative morphology of some trematode gill parasites. *Journal of Marine Biology*, **35**: 113–27.

Llewellyn, J. 1970. Taxonomy, genetics and evolution of parasites: Monogenea. *Journal of Parasitology*, **56** (Section 2, Part 3): 493–504.

Llewellyn, J. 1987. Phylogenetic inference from Platyhelminth life-cycle stages. *International Journal for Parasitology*, **17**: 281–9.

Llewellyn, J., Green, J. E. & Kearn, G. C. 1984. A check-list of monogenean (Platyhelminth) parasites of Plymouth hosts. *Journal of the Marine Biological Association of India*, **64**: 881–7.

Lyons, K. M. 1966. The chemical nature and evolutionary significance of monogenean attachment sclerites. *Parasitology*, **56**: 63–100.

Mackiewicz, J. S. 1981. Caryophyllidea (Cestoidea): Evolution and classification. *Advances in Parasitology*, **19**: 140–206.

Mackiewicz, J. S. 1982. Parasitic platyhelminth evolution and systematics: perspectives and advances since ICOPA IV 1978. In: *Parasites – Their World and Ours* (ed. D. F. Mettrick & S. S. Desser), pp. 178–88. Elsevier, Amsterdam.

Macnae, W., Rock, L. & Makowski, M. 1973. Platyhelminths from the South African Clawed Toad or Platanna (*Xenopis laevis*). *Journal of Helminthology*, **47**: 199–235.

Malmberg, G. 1990. On the ontogeny of the haptor and the evolution of the Monogenea. *Systematic Parasitology*, **17**: 1–65.

Margolis, L. & Kabata, Z. (eds) 1984. *Guide to the parasites of fishes of Canada*. Part 1. General Introduction (by Margolis, L. & Kabata, Z.); Monogenea (by Beverley-Burton, M.). Department of Fisheries and Oceans, Ottawa. ISBN 0 660 11742 8.

Maule, A. G., Halton, D. W., Allen, J. M. & Fairweather, I. 1989. Studies on the motility *in vitro* of an ectoparasitic monogenean, *Diclidophora merlangi*. *Parasitology*, **98**: 85–93.

Maule, A. G., Halton, D. W., Johnston, C. F., Shaw, C. & Fairweather, I. 1990. The serotonic, cholinergic and peptidergic components of the nervous system in the monogenean parasite, *Diclidophora merlangi*: a cytochemical study. *Parasitology*, **100**: 255–73.

Miretski, O. Y. 1951. Experiment on controlling the processes of vital activity of the helminth by influencing the condition of the host. *Doklady Obschee Sobranie Akademiya Nauk U.S.S.R.*, **78**: 613–15.

Morris, G. P. & Halton, D. W. 1971. Electron microscope studies of *Diclidophora merlangi* (Monogenea: Polyopisthocotylea). II. Ultrastructure of the tegument. *Journal of Parasitology*, **57**: 49–61.

Nasir, P. & Fuentes Zambrano, J. L. 1983. (Monogenetic

trematodes of Venezuela.) *Rivista di Parasitologia*, **44**: 335–80. (In Spanish.)

Paul, A. A. 1938. Life history studies on North American freshwater polystomes. *Journal of Parasitology*, **57**: 49–61.

Reichenbach-Klinke, H.-H. 1961. Die Gattung *Diplozoon* v. Nordmann zugleich Neubeschreibung einer Species und zweier Subspecies sowie Revision der Gattung. *Zeitschrift für Parasitenkunde*, **20**: 541–57.

Rohde, K. 1978. Latitudinal differences in host-specificity of marine Monogenea and Digenea. *Marine Biology*, **47**: 125–34.

Rohde, K. 1979. A critical evaluation of intrinsic and extrinsic factors responsible for niche restriction in parasites. *The American Naturalist*, **114**: 648–71.

Schell, S. C. 1985. *Handbook of Trematodes of North America north of Mexico*. University Press of Idaho, Idaho, USA.

Smyth, J. D. (ed.) 1990. *In Vitro Cultivation of Parasitic Helminths*. CRC Press, Boca Raton, Florida.

Smyth, J. D. & Halton, D. W. 1983. *The Physiology of Trematodes*. Cambridge University Press.

Sproston, N. G. 1946. A synopsis of monogenetic trematodes. *Transactions of the Zoological Society of London*, **25**: 1–118.

Streble, H. 1964. Der Sangwurm *Polystoma integerrimum*. Ein Parasit unserer Frösche. *Mikrokosmos*, **53**: 1–7.

Thurston, J. P. 1964. The morphology and life cycle of *Protopolystoma xenopi* Price (Bychowsky) in Uganda. *Parasitology*, **54**: 441–50.

Tinsley, R. C. 1973. Observations on Polystomatidae (Monogenoidea) from East Africa with a description of *Polystoma makereri* n.sp. *Zeitschrift für Parasitenkunde*, **42**: 251–63.

Tinsley, R. C. 1989. The effects of host sex on transmission success. *Parasitology Today*, **5**: 190–5.

Tinsley, R. C. 1990. Host behaviour and opportunism in parasite life cycles. In: *Parasitism and Host Behaviour* (ed. C. J. Barnard & J. M. Behnke), pp. 158–92. Taylor & Francis, London.

Tinsley, R. C. & Jackson, H. C. 1986. Intestinal migration in the life-cycle of *Pseudodiplorchis americanus* (Monogenea). *Parasitology*, **93**: 451–69.

Tinsley, R. C. & Jackson, H. C. 1988. Pulsed transmission of *Pseudodiplorchis americanus* between desert hosts (*Scaphiopus couchii*). *Parasitology*, **97**: 437–52.

USSR, Akademiya Nauk SSR Zoologicheskii Institut. 1985. *Keys to the parasites of freshwater fish of the USSR. Vol. 2. Parasitic Metazoa (Part 1)*. Nauka, Leningrad. (Helm. Abstract No. 54: 4942.)

Wiles, M. 1968. The occurrence of *Diplozoon paradoxum* Nordmann, 1832 (Trematoda: Monogenea) in certain waters of northern England and its distribution on the gills of certain Cyprinidae. *Parasitology*, **58**: 61–70.

Williams, J. B. 1960–1. The dimorphism of *Polystoma integerrimum* (Frölich) Rudolphi and its bearing on relationships within the Polystomatidae: Parts I–III. *Journal of Helminthology*, **34**: 151–92, 323–46; **35**: 181–202.

Yamaguti, S. 1963. *Systema Helminthum*. Vol. IV. *Monogenea and Aspidocotylea*. John Wiley & Sons, New York.

12

Trematoda: Aspidogastrea

The order contains only a single Family, Aspidogastridae, with nine genera. They are all parasites of poikilothermic animals: molluscs, elasmobranchs, teleosts and turtles and (more rarely) crustaceans. They are mostly endoparasitic but some may be ectoparasites (ectocommensals?) on molluscs. Although the morphology is well known, knowledge of the physiology and life cycles of the group is scanty.

Early monographic works on the group are those of Dollfus (1958) and Yamaguti (1963). More recently, the morphology, ultrastructure, development and general biology has been reviewed by Rohde (1972) and Smyth & Halton (1983).

The group, apparently closely related to the Digenea (Rohde, 1971), appears to be poorly adapted to the parasitic way of life and can survive for long periods in relatively simple media such as saline.

The most studied species appear to be *Aspidogaster conchicola*, *Multicotyle purvisi* and *Cotylaspis insignis* (Table 12.1); prevalence has been especially investigated in molluscs in the USA (Danford & Joy, 1984; Huehner, 1984; Hendrix & Short, 1965; Pauley & Becker, 1968).

External features. The most striking external feature is the enormous adhesive apparatus which occupies almost the entire ventral surface. This organ is made up of numerous compartments or alveoli, each of which can act as an efficient sucker (Fig. 12.1).

Tegument. The structure of the tegument closely follows that of the Digenea (p. 178) (Bailey & Tomkins, 1971; Halton & Lyness, 1971; Rohde, 1972).

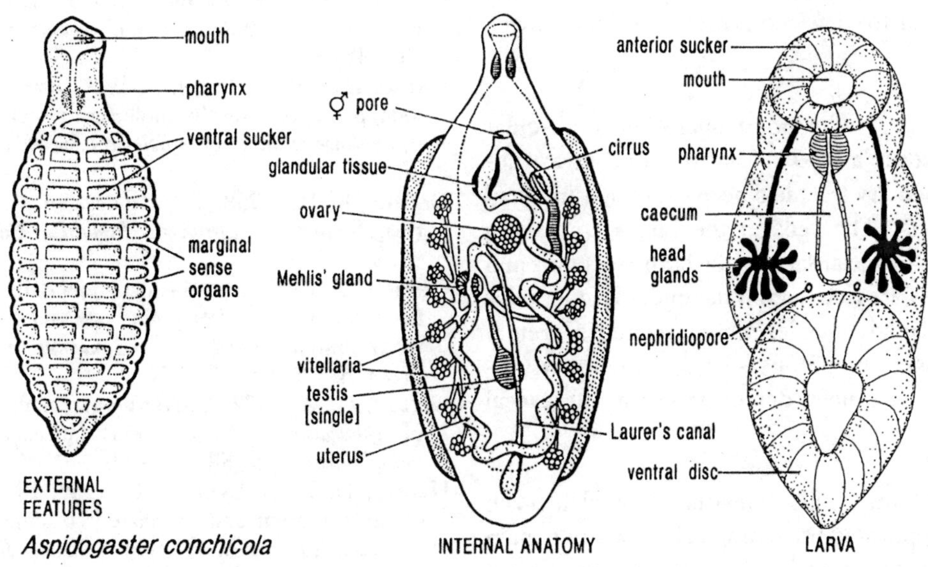

Fig. 12.1
Morphology of *Aspidogaster conchicola*. (Adapted from various authors.)

Table 12.1. *Some species of Aspidogastrea*

Species	Host	Habitat
Aspidogaster conchicola	fresh-water clams	pericardial cavity
	reptiles	intestine
	fish	intestine
	snails	visceral mass
Lophotaspis vallei	marine turtles	oesophagus, stomach
	snails	mantle cavity
	clams	pericardial cavity
Multicotyle purvisi	snails	pallial complex
	turtles	intestine
Stichocotyle nephropis	lobsters	wall of intestine
	rays	bile bladder
Macraspis elegans	ratfish	bile bladder
Cotylaspis insignis	fresh-water clams	kidney, mantle cavity
Cotylogasteroides occidentalis	fresh-water clams	gill cavity
	snails	mantle cavity
	fish	intestine
Rugogaster hydrolagi	ratfish	rectal gland

Source: Data from Mehlhorn (1988).

General internal features. As in other trematodes, the internal organs are supported by loose parenchyma. In the Aspidogastrea, however, a unique *septum*, not found in other trematodes, occurs. This divides the body into upper and lower compartments, the upper containing the alimentary canal, the terminal genital ducts and the vitellaria, and the lower the ovary, oviduct, ootype and testes.

Alimentary canal. The mouth opens into a muscular pharynx and leads to a sac-like caecum. The ultrastructure of the alimentary tract has been examined in detail by Halton (1972). The cells of the caecal epithelium exhibit structural specialisations for both secretion and absorption. They are rich in granular endoplasmic reticulum (GER), Golgi complexes and related secretory bodies and possess numerous surface lamellae. Lysosomal enzymes have been demonstrated in the caecal cells.

Excretory system. This consists of a flame-cell system with complicated branching. There are two main lateral excretory canals which open posteriorly by two funnel-like pores.

Reproductive system. The reproductive system has the typical trematode pattern arranged as shown in Fig. 12.1, but with an unusual feature in the elongate blind Laurer's canal. Both self-fertilisation and cross-fertilisation have been observed. The egg-shell material is produced by the vitelline cells, which give positive histochemical reactions for polyphenols, protein and phenolase (Gerzeli, 1968); this suggests that the egg-shell consists of sclerotin or tanned protein as in the majority of trematodes (see Table 13.1 and p. 184).

Life cycle. With the exception of *Aspidogaster conchicola* and *Multicotyle purvisi*, the life cycles of most species are poorly known. Most research has concentrated on the means of transmission. Huehner & Etges (1977) demonstrated unequivocally that the snail *Viviparus malleatus* became infected by ingesting embryonated, unhatched eggs of *A. conchicola*. They also concluded that the hatched juveniles of this species are incapable of acting as a transmission stage, although the infection of snails by hatched larvae of *M. purvisi* has been described (Rohde, 1972).

○○○ References

(References of general interest have been included in this list, in addition to those quoted in the text.)

Bailey, H. H. & Tomkins, S. J. 1971. Ultrastructure of the integument of *Aspidogaster conchicola. Journal of Parasitology*, **57**: 848–54.

Danford, D. W. & Joy, J. E. 1984. Aspidogastrid (Trematoda) parasites of bivalve molluscs in western West Virginia. *Proceedings of the Helminthological Society of Washington*, **51**: 301–4.

Dollfus, R. Ph. 1958. Trématodes. Sous-classe Aspidogastrea. *Annales de Parasitologie Humaine et Comparée*, **33**: 305–95.

Gerzeli, G. 1968. Aspetti istochemici della formazione degli involucri ovuari in *Aspidogaster conchicola* von Baer (Trematoda). *Instituto Lombardo – Academia di Scienze e Lettere (Rend. Sc.)* B, **102**: 263–76.

Halton, D. W. 1972. Ultrastructure of the alimentary tract of *Aspidogaster conchicola* (Trematoda: Aspidogastrea). *Journal of Parasitology*, **58**: 455–67.

Halton, D. W. & Lyness, R. A. W. 1971. Ultrastructure of the tegument and associated structures of *Aspidogaster conchicola* (Trematoda: Aspidogastrea). *Journal of Parasitology*, **57**: 1198–210.

Hendrix, S. S. & Short, R. B. 1965. Aspidogastrids from

northeastern Gulf of Mexico river drainages. *Journal of Parasitology*, **51**: 561–9.

Huehner, M. K. 1984. Aspidogastrid trematodes from freshwater mussels in Missouri with notes on the life cycle of *Cotylaspis insignis*. *Proceedings of the Helminthological Society of Washington*, **51**: 270–4.

Huehner, M. K. & Etges, F. J. 1977. The life cycle and development of *Aspidogaster conchicola* in the snails *Viviparus malleatus* and *Goniobasis livescens*. *Journal of Parasitology*, **63**: 669–74.

Mehlhorn, H. 1988. *Parasitology in Focus*. Springer-Verlag, Berlin.

Pauley, G. B. & Becker, C. D. 1968. *Aspidogaster conchicola* in mollusks of the Columbia river system with comments on the host's pathological response. *Journal of Parasitology*, **54**: 917–20.

Rohde, K. 1971. Phylogenetic origin of trematodes. In: *Perespektiven der Cercarienforschung* (ed. K. Odening). *Parasitologische Schriftenrheihe*, **21**: 17–21.

Rohde, K. 1972. The Aspidogastrea, especially *Multicotyle purvisi* Dawes, 1941. *Advances in Parasitology*, **10**: 77–151.

Smyth, J. D. & Halton, D. W. 1983. *The Physiology of Trematodes*. Cambridge University Press.

Yamaguti, S. 1963. *Systema Helminthum*. IV. *Monogenea and Aspidogastrea*. Interscience Publishing Co., New York and London.

13

Trematoda: Digenea

The great majority of digenetic trematodes are inhabitants of the vertebrate alimentary canal or its associated organs, especially the liver, bile duct, gall bladder, lungs, pancreatic duct, ureter and bladder. These are organs containing cavities rich in potential semi-solid food materials such as blood, bile, mucus and intestinal debris. The digenetic trematodes are readily distinguished microscopically from the Monogenea by their relatively simple external structure, in particular the absence of complicated adhesive organs; only simple suckers are present. They also differ markedly in having complex heteroxenous life cycles involving at least one intermediate host; from this feature the term *Digenea* is derived.

With two exceptions, in all life cycles the first intermediate host is a mollusc, usually a gastropod, occasionally a lamellibranch or a scaphopod but never a member of any other molluscan order. In the exceptional cases, the first intermediate host is an annelid. The larval phases are unusual in undergoing polyembryony so that enormous numbers of larvae may result from small initial infections. The biology of the group in general has been reviewed by Erasmas (1972) and the physiology by Smyth & Halton (1983).

13.1 General morphology

The morphology of a 'generalised' fluke is given in Fig. 13.1, to show the relationship between the various organs.

External features. Most species are flattened dorsoventrally but some have thick fleshy bodies and some are round in section. There are typically two suckers, an anterior *oral* sucker surrounding the mouth at the anterior end, and a *ventral* sucker, sometimes termed an *acetabulum*, on the ventral surface. In one group, the holostomes (= strigeids, p. 197) an additional large adhesive organ, the *holdfast* (p. 197) occurs behind the acetabulum. One remarkable species, *Transversotrema patialensis* (Fig. 13.15), ectoparasitic on tropical fish, possess a pair of eye spots (Whitfield & Wells, 1973).

Tegument (Fig. 13.2). The body covering of trematodes has long been referred to as a 'cuticle', the origin of which was obscure. Ultrastructural studies have revealed, however, that far from being an inactive, largely protective layer – as formerly thought – the body covering is, in fact, a metabolically active surface of considerable physiological importance to the parasite. Moreover, it is strikingly similar to the body covering of monogeneans (Fig. 11.5), aspidogastreans and cestodes (Fig. 20.2), with which it is clearly homologous (Smyth & Halton, 1983).

The word '*cuticle*' tends to have the biological connotation of a tough, resistant, inert structure; the term *tegument* was suggested as being a more appropriate term (Rothman, 1959) and this term has now been widely adopted. Its ultrastructure has been extensively studied and the relevant literature has been reviewed by Smyth & Halton (1983). Lumsden (1975) has also published a comprehensive account of the surface ultrastructure and cytochemistry of parasitic helminths in general.

The tegument is essentially a syncytial epithelium which comprises an outer, *anucleate* layer of cytoplasm connected by cytoplasmic strands ('connections' in Fig. 13.2) to nucleated portions of the cytoplasm known as *tegumental cell bodies*, but also referred to by different authors as *tegumental cells*, *cytons*, or *perikarya*, located in the parenchyma beneath the basal lamina and muscle layers. The entire tegumental system – except in the blood flukes (see below and Fig. 13.3) – is bounded

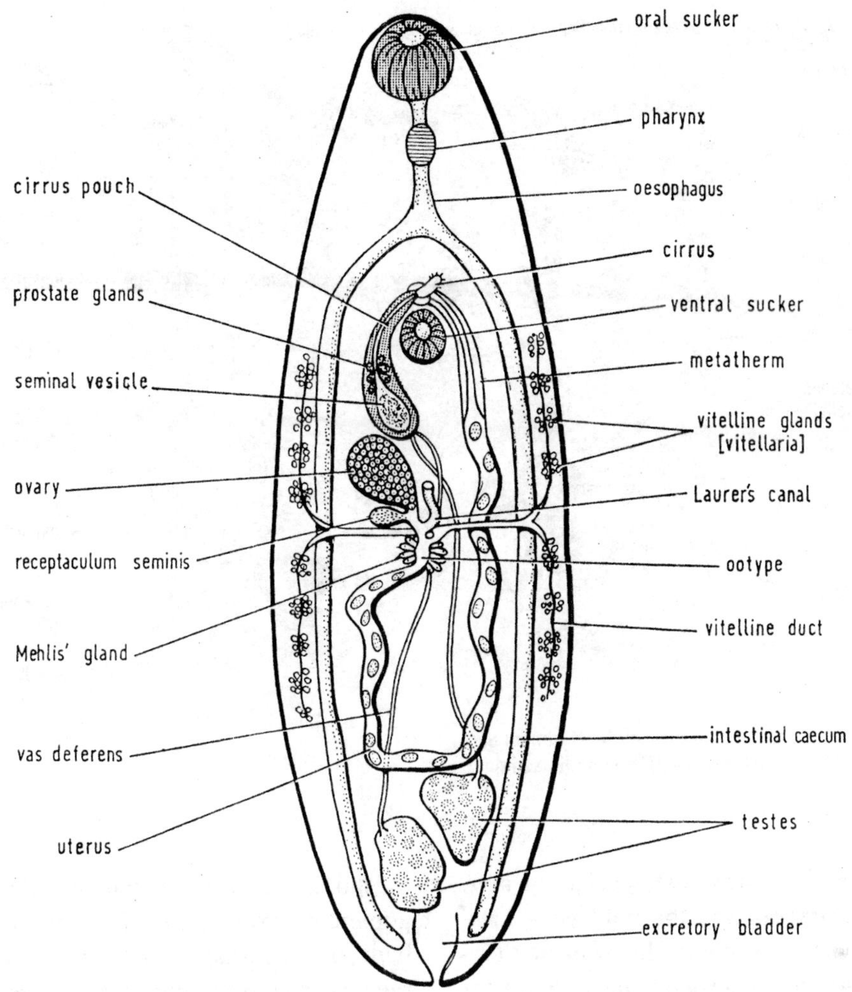

Fig. 13.1
A generalised diagram of trematode anatomy. (Modified from Cable, 1949.)

apically (externally) and basally by typical trilaminate plasma membranes. Its matrix contains the usual cytoplasmic inclusions: mitochondria, ribosomes, endoplasmic reticulum, Golgi complexes and various secretory bodies. In addition to its obvious protective role, the tegument has a multiplicity of functions: (*a*) the absorption of nutrients, (*b*) the synthesis and secretion of various materials, (*c*) excretion and osmoregulation and (*d*) a sensory role, as exemplified by the presence of numerous sense organs.

In common with many other cell surfaces, the apical plasma membrane possesses a carbohydrate-rich (PAS^{+ve}), polyanionic coating or *glycocalyx*, seen under electron microscopy as a fuzzy investment (Fig. 13.2). As an *integral* part of the tegument (i.e. not just an adherent extraneous layer) it undoubtedly plays an important role in the protective, absorptive and immunological properties of the tegument.

It should be noted that in blood flukes (e.g. *Schistosoma* spp.) although a glycocalyx is present in the cercaria it is lost during the cercarial–schistosomulum transformation (Fig. 13.3; p. 181). During this process, the original trilaminate surface membrane (and its associated glycocalyx) is *replaced* by a new heptalaminate membrane (Hockley & McLaren, 1973; McLaren & Hockley, 1976). The structure of the latter membrane appears to be an adaptation to life in the blood stream of its host where it undergoes rapid turnover.

Absorptive properties. Although trematodes possess a well-developed gut, they are also capable of taking in material through the tegument (see Chapter 18) and

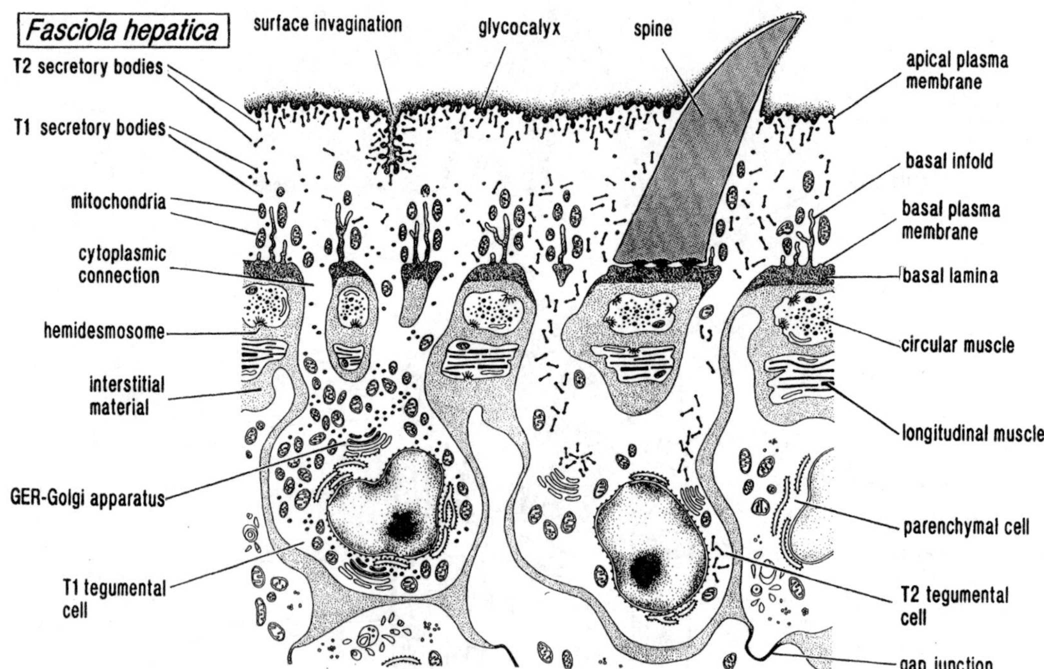

Fig. 13.2
Fasciola hepatica: the fine-structured organisation of the tegument and related structures. GER = granular endoplasmic reticulum. (Based on Threadgold, 1963, 1967.)

in some species (especially strigeids) specialised surface areas – such as lappets and the holdfast – have developed secretory enzyme systems which enable them to engage in extra-corporeal digestion of the tissues with which they make contact (Smyth & Halton, 1983).

Secretory properties. The synthetic function of the trematode tegument is indicated by the numerous membrane-bound inclusion bodies which originate in the tegumental cells and are transplanted to the outer syncytial region. These inclusions are usually disc-shaped with an electron-dense content. In *Fasciola*, two distinct types of tegumental cells, Type 1 and Type 2 (Fig. 13.2) produce two tegumental inclusive bodies. Type 1 cells produce dense vesicular inclusions (T 1 bodies) and the (less common) Type 2 cells form biconcave discoid inclusions with fairly electron-lucent contents (T 2 bodies).

Although tegumental secretory bodies are universal in trematodes their functional significance is still largely undetermined. Numerous enzymes have been demonstrated in the tegument which may eventually provide clues to their activities. For further details of the physiological activities of the tegument, see Chapter 18.

It should be stressed that the ultrastructure of the tegument may vary in the different functional regions of a body. This, perhaps, is best illustrated in the strigeids *Cyathocotyle bushiensis* and *Apatemon gracilis* in which differences have been demonstrated between the tegument of (*a*) the general body surface, (*b*) the lappets and associated glands and (*c*) the adhesive organ (Erasmus, 1969*a*,*b*) (see p. 266).

The body wall generally has three layers of muscles: an outer circular, a middle diagonal and an inner longitudinal. An extra layer of circular muscles is occasionally present, as are vertical muscles transversing the parenchyma. The muscle fibres are apparently unstriated.

Parenchyma. The tissue framework of the body is made up of *parenchyma*, a network of cells and fibres enclosing irregular spaces.

Nervous system. The anterior nervous system is relatively simple and consists of paired cerebral ganglia and a submuscular plexus in the form of longitudinal cords and transverse connectives. The cerebral ganglia are connected by a broad commissure. Three pairs of nerves pass anteriorly from the ganglia and three pairs

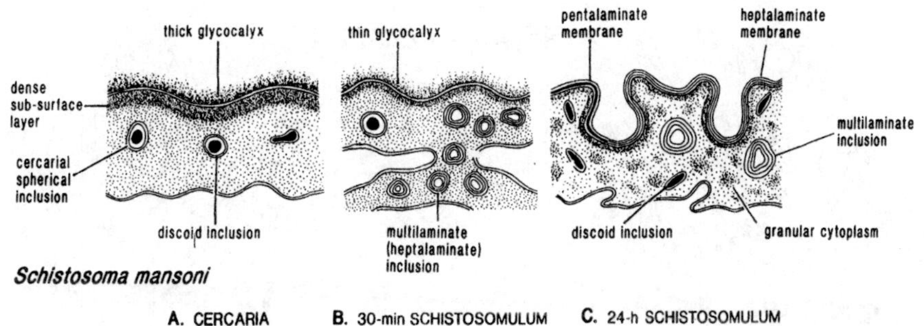

Fig. 13.3
Schistosoma mansoni: changes in the tegument morphology during the cercaria–schistosomulum transformation which occurs after skin penetration by the cercaria. (After Hockley & McLaren, 1973.)

pass posteriorly, and in the latter group the ventral pair is especially well developed. Various fibrils arise from these nerves to various parts of the body. Most of these fibrils are motor, the sensory elements being feebly developed. Typical of invertebrates, the axons are unmyelinated and possess axonemal synapses; the neuromuscular junctions are characterised by pre-synaptic vesicles. Presumed sense organs which terminate nerve processes in or below the tegument of adult trematodes are abundant, both uniciliated and non-ciliated types being found. Eye spots occur in some larval forms, particularly miracidia, and in one ectoparasite species (Fig. 13.15). 'Presumptive' neurosecretory cells have been identified (by staining methods only) in several trematodes, *Dicrocoelium lanceolatum*, *Fasciola hepatica*, *Opisthodiscus diplodiscoides*, *Leucochloridiomorpha constantiae*; Harris & Cheng (1972) and Smyth & Halton (1983) have reviewed the relevant literature. Recently, the application of peptide immunocytochemistry, using antisera to vertebrate regulatory peptides, has led to increasing interest in the investigation of neuropeptides in parasitic platyhelminths and a substantial literature exists. Recent work in this field has been succinctly reviewed by Halton *et al.* (1990) and McKay *et al.* (1990). For further consideration of this topic, see Chapter 18.

Digestive system. The alimentary canal is usually well developed with a prominent pharynx and oesophagus. In the gasterostomes (Figs 13.21 and 14.1) it is a simple blind sac, but in the majority of species it consists of branched caeca which occasionally branch further (e.g. *Fasciola*). The caeca may reach to the posterior end or only halfway down the body. Ultrastructural studies have revealed the presence of lamellar processes extending into the lumen of the caecum; the evidence indicates that the gastrodermis is both secretory and absorptive in nature (Fig. 13.4). In addition to ribosomes and mitochondria, the bulk of the cytoplasm is occupied by substantial quantities of GER, together with mitochondria and membranous inclusions. Secretion at the surface takes place by exocytosis and the protein material, where it is secreted, is chiefly enzymatic; numerous hydrolytic enzymes have been detected in the gastrodermis. The physiology of digestion is considered further in Chapter 18.

The food of trematodes generally consists of blood, mucus or tissue and well-developed enzyme systems are involved (see Chapter 18) (Jennings, 1968; Smyth & Halton, 1983).

Excretory system. This is essentially protonephridial, similar to that of turbellarians with the flame cell as the excretory unit. The arrangement of the flame cells, i.e. their number and manner of branching of their ducts, are of systematic importance and often of diagnostic value, especially in larvae (Rohde, 1990; Stunkard, 1946; La Rue, 1957). Although the arrangement differs in different families, it is similar in closely related species.

The basic flame-cell pattern is most readily studied in the cercarial stage where it is not masked by opaque tissues. As a cercaria develops to an adult fluke, the excretory system which is formed is an exact multiple of the simple cercarial type. Thus in the cercariae of the blood flukes of the genus *Schistosoma*, there are anterior and posterior pairs of nephridia on each side of the body (Fig. 13.5) and this arrangement is given the following formula:

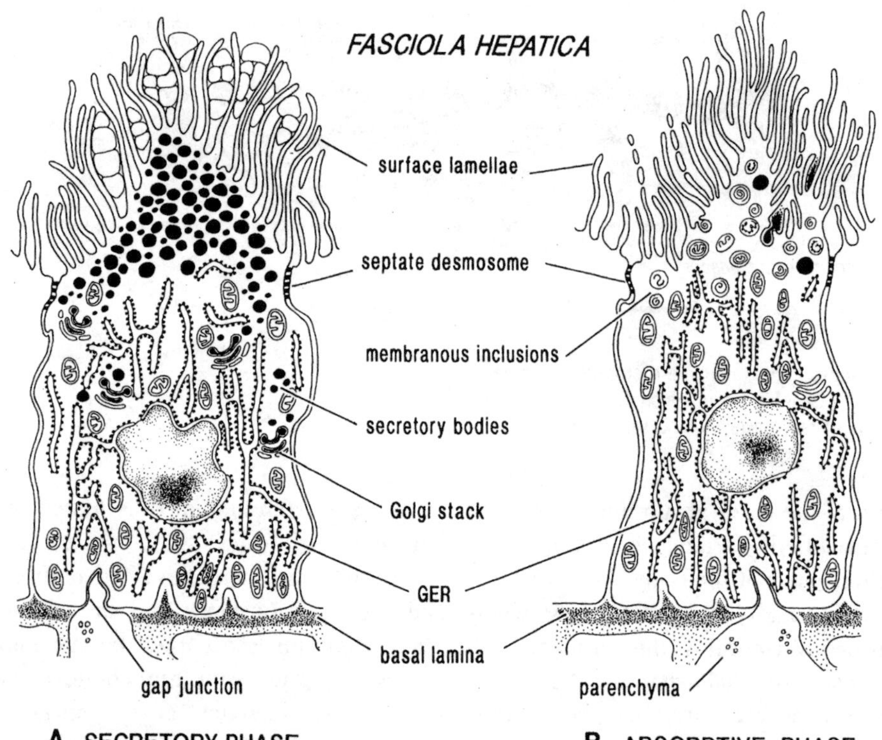

Fig. 13.4
Fasciola hepatica: ultrastructure of a gut cell during (A) the secretory phase and (B) the absorptive phase. (After Robinson & Threadgold, 1975.)

$$2(2 + 2)$$

but since the miracidium contains only two flame cells on either side, an anterior one and a posterior one, this $= 2(1 + 1)$. It is clear that each of these has given rise to another so that the system in the cercaria (which is fork-tailed) is more expressively written as

$$2[(1 + 1) + (1 + (1))] = 8$$

where each figure 1 represents a branch. Fork-tailed cercariae have flame cells in the tail stem and these are placed in parentheses in the formula. A good example of a fork-tailed cercaria is *Cercaria ocellata* from *Lymnea stagnalis* with a formula of

$$2[(1 + 1 + 1) + (1 + 1 + 1 + (1))] = 14$$

In the Dicrocoeliidae (Fig. 13.5) the miracidium has but a single flame cell on each side, formula $2[(1)]$, which in the sporocyst gives rise to two on each side $= 2[(1 + 1)]$, each of which gives rise to three flame cells in the cercaria

$$= 2[(1 + 1 + 1) + (1 + 1 + 1)] = 12$$

A more elaborate system is to use additional letters

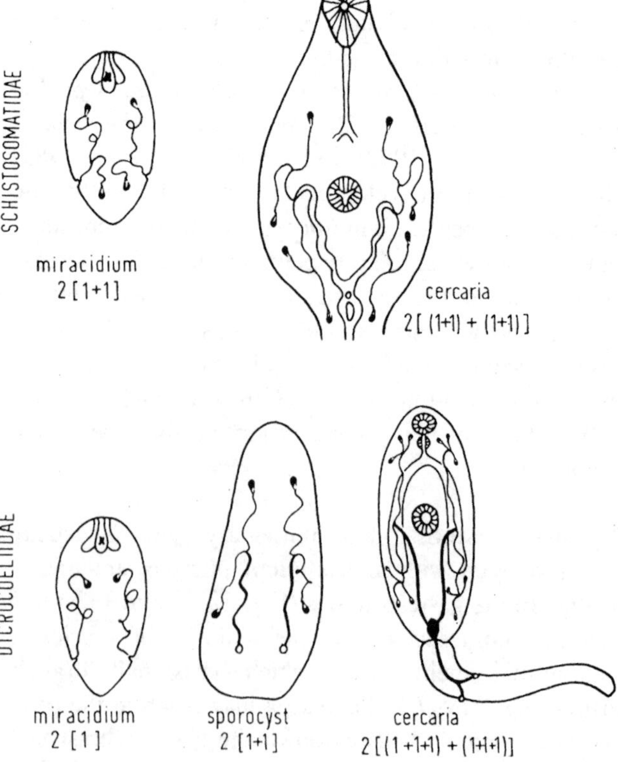

Fig. 13.5
Flame-cell patterns in larval trematodes. (After Dawes, 1946.)

or other symbols to indicate the branching pattern. Thus, cell No. 1 may give rise to 2 cells, 1a + 1b; if 1b divided further, the two cells formed would be termed $1b^{1} + 1b^{11}$; if b^{11} divided again, the cells would be $b^{11*} + b^{11**}$, and so on. An example of this formula is that of the cercaria of *Diplostomum phoxini* (p. 256) which has a formula

$$2[(1a + 1b + 2) + (3 + 4a + 4b^{1} + (4b^{11*} + 4b^{11**}))] = 16$$

as shown in Fig. 17.3.

There are normally two main lateral excretory canals which fuse to form an *excretory bladder* opening posteriorly by a pore. La Rue (1957) has utilised the form and development of the excretory bladder as a basis for dividing the Digenea into two superorders, the *Anepitheliocystidia* and the *Epitheliocystidia*. In the former the bladder is not epithelial but is merely the primitive bladder which is retained. In the latter, the primitive bladder is replaced by a thick-walled epithelial bladder whose cells are of mesodermal origin. Erasmus (1972) gives a detailed key based on La Rue's classification.

Reproductive system. Except for two families which are unisexual (the Schistosomatidae and the Didymozoidae), the digenetic trematodes are hermaphrodite. The reproductive system is extensively developed and although varying in detail, follows essentially the same pattern in most species.

Male

Protandry is the general rule in the Digenea. There are usually two testes, vasa efferentia, a vas deferens, seminal vesicle, a ductus ejaculatorius and a cirrus or penis enclosed in a sac or pouch. So-called *prostate* glands may be present. Minor variations from this general plan occur. Spermatogenesis follows the pattern typical of the platyhelminths, 64 spermatozoa arising in each sperm bundle. The flagella of spermatozoa follow the standard pattern of other flagella and consist typically of two central filaments with nine peripheral doublet filaments.

Some relevant studies on the ultrastructure of trematode spermatozoa are those on *Haematoloechus medioplexus* (Burton, 1972, 1973; Burton & Silveira, 1971), *Pharyngastomoides procyonis* (Grant *et al.*, 1976), *Schistosoma mansoni* (Kitajima *et al.*, 1976), *Paragonimus ohirra* and *Eustrema pancreaticum* (Fujino *et al.*, 1977).

Female

The female apparatus contains a single ovary with an

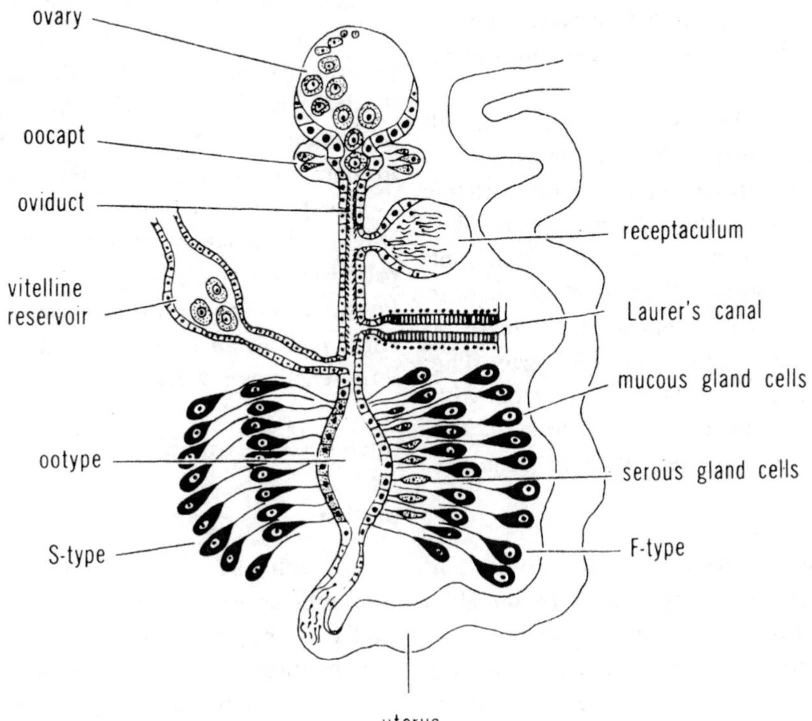

Fig. 13.6
Schematic representation of the egg-forming system (Oogenotop) of digenetic trematodes. (After Ebrahimzadeh, 1966.)

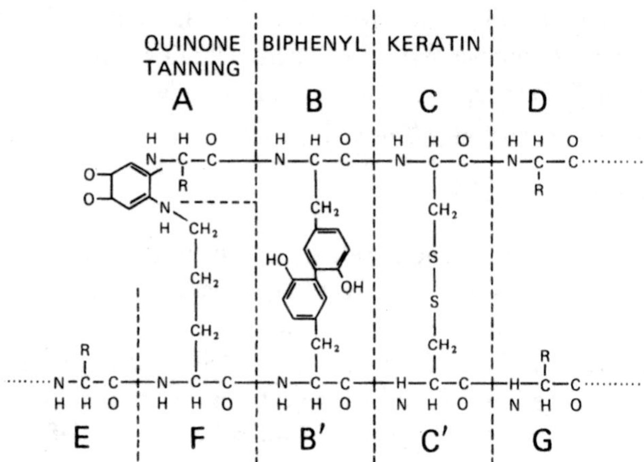

Fig. 13.7
Some of the covalent links which occur in structural proteins. **ABCD** represents one protein, **EFB′C′** another. The N-terminal amino acid **A** is linked to an *o*-quinone which is also linked to a lysyl residue **F** (= quinone tanning); **BB′** represents two tyrosyl residues coupled by a biphenyl linkage; **CC′** represents two cysteinyl residues coupled by a cystine linkage (as in keratin); **D, E** and **G** are not cross-linked. (After Brunet, 1967.)

oviduct, a receptaculum seminis, a pair of so-called 'vitelline glands' (= vitellaria) with ducts, a chamber where eggs are formed (the ootype), and a complex collection of gland cells collectively known as Mehlis' gland. The relationship between these structures is shown in Fig. 13.6. The whole egg-forming apparatus is referred to by some German authors (Gönnert, 1962; Ebrahimzadeh, 1966) as the 'Oogenotop'.

In addition, digenetic trematodes possess a canal termed Laurer's canal, which leads from the oviduct to the dorsal surface of the body and is homologous with the vagina of Monogenea and Cestoda. Many, and possibly all trematodes, possess a muscular chamber or ovicapt (= oocapt): an enlarged portion of the oviduct where it joins the ovary. This probably controls the release of ova and spaces out their descent down the uterus. The variations in the structure of the female genitalia, with particular reference to role in the formation of the shelled egg, have been examined in great detail by Ebrahimzadeh (1966).

Fertilisation. Little is known regarding fertilisation in trematodes. Self-fertilisation appears to be possible but may not be common as several flukes usually occur together. When copulation occurs, the cirrus of one fluke is thrust into the genital opening of another. Spermatozoa are stored in the receptaculum and released as required. The insertion of a cirrus into the opening of Laurer's canal has also been reported and this may represent an alternative insemination mechanism in some species.

○○○ 13.2 Egg formation

○ 13.2.1 General account

The formation of the egg in the Digenea and the Monogenea and especially its protective egg-shell or *capsule* presents a number of intriguing cytological and biochemical problems and has been the subject of numerous studies (Table 13.1). Early work, which was largely cytological and cytochemical, has been reviewed by Smyth (1966), Clegg & Smyth (1968) and Smyth & Clegg (1959). More recent work, which has involved ultrastructural, biochemical and molecular biological techniques, has been reviewed by Smyth & Halton (1983), Cordingley (1987), Waite & Rice-Ficht (1987, 1989, 1992) and Waite (1990).

○ 13.2.2 Structure of egg-shell

In various phyla, structural proteins may be rendered more resistant or stabilised by means of covalent cross linkages, which may take a number of forms (Fig. 13.7). In the majority of Digenea, most Monogenea, many pseudophyllidean cestodes, the Aspidogastrea and the Turbellaria, the egg-shell is believed to be made up of *sclerotin*, a 'tanned' protein similar to that which occurs in many invertebrates (e.g. the insect cuticle).

The process of 'quinone tanning' or *sclerotinisation* (Fig. 13.7) is accomplished by means of an *o*-quinone produced enzymically, via the enzyme *phenolase* (also known as polyphenoloxidase, catechol oxidase and tyrosinase), from an *o*-phenol in the presence of oxygen. The reactions involved are as follows:

phenol —phenolase→ quinone ↘
'tanned' protein = sclerotin
protein ——→ protein ↗

Since, with rare exceptions (e.g. *Syncoelium*), the egg-shell precursors (or *presclerotin*) have their origin in the vitellaria, the latter may be expected to give positive cytochemical reactions for proteins, polyphenols and phenolase. Some of these tests give brightly coloured end products and it is thus often possible to 'stain' the vitellaria and eggs in a highly selective manner, even in

Table 13.1. *Structural proteins in the egg-shell or vitellaria of Monogenea, Aspidogastrea and Digenea*

Symbols: +, positive; −, negative; ·, no information.

		Histochemical tests for:						
		Sclerotin			Keratin	Elastin	Others	
Family	Species	Phenol	Protein	Phenolase	-SS-bonds	Non-specific?	Di-tyrosine	References
Monogenea								
Capsalidae:	*Entobdella soleae*	+	·	+	·	·	·	Llewellyn (1965)
Diclidophoridae:	*Diclidophora luscae*	+	·	+	·	·	·	Llewellyn (1965)
	Diclidophora merlangi	+	+	+	·	·	·	Rennison (1953)
Discocotylidae:	*Octomacrum lanceatum*	+	+	+	·	·	·	Hathaway & Herlevich (1976)
Gastrocotylidae:	*Gastrocotyle trachuri*	·	·	+	·	·	·	Freeman & Llewellyn (1958)
Hexabothriidae:	*Pricea multae*	+	·	+	·	·	·	Ramalingam (1970*a*,*b*)
	Rajonchocotyle batis	·	·	+	·	·	·	Rigby & Marx (1962)
Monocotylidae:	*Calicotyle* sp.	+	·	·	·	·	·	Smyth (1966)
Polystomatidae:	*Polystoma integerrimum*	+	·	+	·	·	·	Köhlmann (1961)
	Polystomoides sp.	+	+	+	·	·	·	Fried & Stromberg (1971)
	Protopolystoma xenopodis	·	·	+	·	·	·	Thurston (1964)
Protomicrocotylidae:	*Protomicrocotyle* sp.	·	·	+	·	·	·	Ramalingam (1970*b*)
Aspidogastrea								
Aspidogastridae:	*Aspidogaster conchicola*	+	+	+	·	·	·	Gerzeli (1968)
Digenea								
Allocreadiidae:	*Macrolecithus papilliger*	+	+	+	·	·	·	Rees (1936)
Aporocotylidae:	*Orchispirium heterovitellatum*	−	+	−	−	+	·	Madhavi & Rao (1971)
Bucephalidae:	*Bucephaloides gracilescens*	+	·	−	·	·	·	Smyth & Clegg (1959)
Cyathocotylidae:	*Cyathocotyle bushiensis*	+	+	+	·	·	·	Erasmus (1972)
	Holostephanus lühei	+	+	+	·	·	·	Erasmus (1972)
Echinostomatidae:	*Artyfechinostomum mehrai*	+?	+?	+?	·	·	·	Madhavi (1971)
	Echinoparyphium recurvatum	+	+	+	·	·	·	Fried & Stromberg (1971)
	Echinostoma revolutum	+	+	+	·	·	·	Fried & Stromberg (1971)
	Isoparorchis hypselobagri	+	+	+	·	·	·	Srivasta & Gupta (1978)
Fasciolidae:	*Fasciola hepatica*	+*	+*	+*	+†	·	+†	Smyth & Clegg (1959)*, Ramalingam (1973)†
	Fasciola indica	+	+	+	·	·	·	Lal & Johri (1967)
Fellodistomatidae:	*Lintonium vibex*	+	+	+	·	·	·	Coil (1972)
	Proctoeces subtenuis	·	·	−	·	·	·	Freeman & Llewellyn (1958)
Gorgoderidae:	*Gorgoderina attenuata*	+	+	·	·	·	·	Nollen (1971)
	Gorgoderina sp.	+	+	−	·	·	·	Smyth & Clegg (1959)
Halipegidae:	*Halipegus eccentricus*	+	+	+	·	·	·	Guilford (1961)
Hemiuridae:	*Syncoelium spathulatum*	+	+	·	·	·	·	Coil & Kuntz (1963)
Heterophyidae:	*Cryptocotyle lingua*	+	·	+	·	·	·	Smyth & Clegg (1959)
Lecithodendriidae:	*Brandesia turgida*	+	+	·	·	·	·	Gerzeli (1968)
	Ganeo tigrinum	+	+	+	·	·	·	Kandhaswami (1980)
	Pleurogenes claviger	+	+	·	·	·	·	Gerzeli (1968)
Notocotylidae	*Ogmocotyle indica*	+	+	+	·	·	·	Coil (1966)
Opisthorchiidae:	*Clonorchis sinensis*	+	+	+	·	·	·	Ma (1963)

Table 13.1. *(cont.)*

Symbols: +, positive; −, negative; ·, no information.

		Histochemical tests for:						
		Sclerotin			Keratin	Elastin	Others	
Family	Species	Phenol	Protein	Phenolase	-SS-bonds	Non-specific?	Di-tyrosine	References
Paramphistomatidae:	*Carmyerius spatiosus*	·	+	·	+	·	·	Madhavi (1966)
	Carmyerius synethes	−	+	−	+	·	·	Eduardo (1976)
	Diplodiscus amphichrus	−	+	−	+?	·	·	Kanwar & Agrawal (1977)
	Diplodiscus mehrai	·	+	·	+	·	·	Madhavi (1968)
	Gastrodiscus secundus	·	+	·	+	·	·	Madhavi (1966)
	Gastrothylax crumenifer	−	+	−	+	·	·	Eduardo (1976)
	Megalodiscus temperatus	−	+	−	+	·	·	Nollen (1971)
	Paramphistomum cervi	·	+	·	+	·	·	Madhavi (1966)
Philophthalmidae:	*Philophthalmus megalurus*	·	+	+	·	·	·	Nollen (1971)
Plagiorchiidae:	*Dolichosaccus rastellus*	·	·	+	·	·	·	Smyth (1954)
	Glypthelmins sp.	+	+	+	·	·	·	Fried & Stromberg (1971)
	Haematoloechus medioplexus	+	+	+	·	·	·	Fried & Stromberg (1971)
	Haplometra cylindracea	+	+	+	·	·	·	Smyth (1954)
	Macrodera longicollis	+	+	·	·	·	·	Gerzeli (1968)
Schistosomatidae:	*Schistosoma japonicum*	+	+	+	·	·	·	Ho & Yang (1973)
	Schistosoma mansoni	+	+	+	·	·	·	Clegg & Smyth (1968), Piva & De Carneri (1961)
Strigeidae:	*Apatemon gracilis*	+	+	+	·	·	·	Erasmus (1972)
	Diplostomum phoxini	+	+	+	·	·	·	Bell & Smyth (1958)
	Diplostomum spathaceum	+	+	+	·	·	·	Erasmus (1972)

whole mount preparations; suitable histochemical methods are given by Smyth & Clegg (1959) and Johri & Smyth (1956). Waite & Rice-Ficht (1989) showed that the vitellaria of egg-containing tubules of *Fasciola* contained DOPA (3,4-dihydroxyphenyl-L-alanine) and would stain bright red with nitrate–molybdate.

Using largely histochemical methods, the egg-shell formation has been examined in a number of species (Table 13.1) and the process is shown diagrammatically in Fig. 13.8. In the majority of these the vitellaria give positive results for the three components of quinone-tanning systems – proteins, phenols and polyphenol oxidase – a result which would appear to confirm the formation of a sclerotin egg-shell.

The shell globules which coalesce to form the shell remain (cytochemically) positive for these three components. However, once the quinone is formed and it binds with the protein to form the egg-shell by the quinone-tanning process, the hardening egg-shell is not positive for either the phenol or the protein, although sufficient phenolase (polyphenol oxidase) remains to give weakly positive reactions (Fig. 13.8).

Quinone-tanning is a widespread phenomenon throughout the animal kingdom and the systems involved are characterised by the possession of DOPA-containing proteins. In many organisms, the quinone-tanning systems are more complex than the system described above. In *Fasciola hepatica*, the origin and fate of the DOPA in the egg-shell precursor protein, as envisaged by Waite & Rice-Ficht (1987), is shown in Fig. 13.9. Other workers have used molecular biological techniques combined with *in vitro* culture to identify and express an egg-shell protein gene for this species (Zurita *et al.*, 1987, 1989). The identification and local-

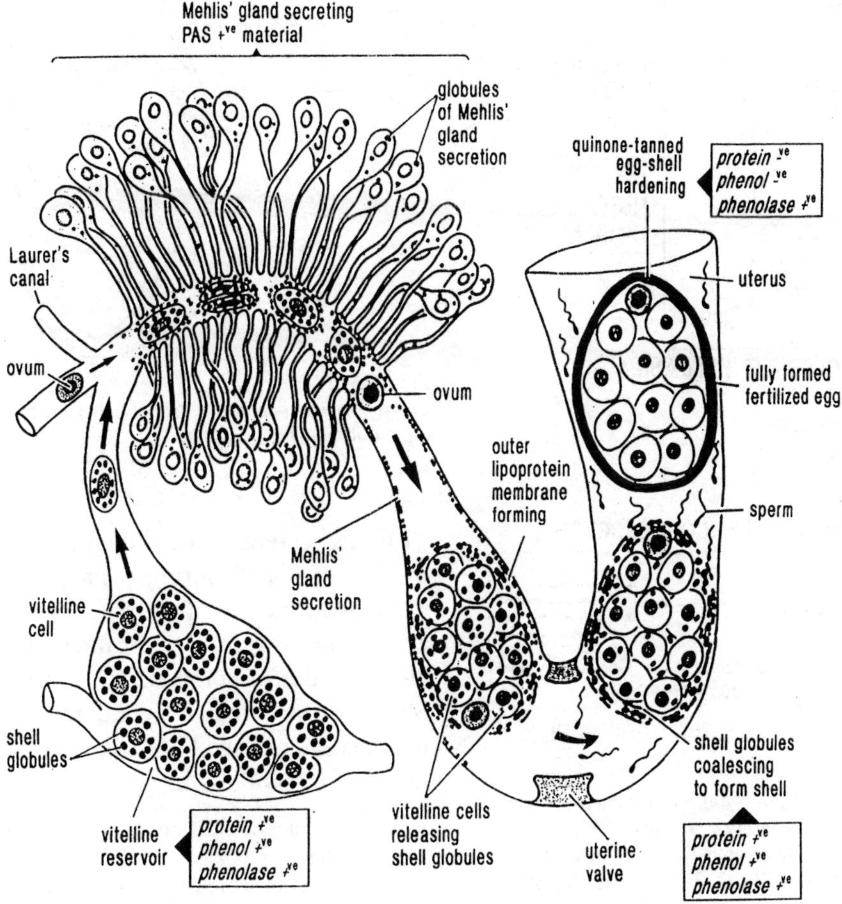

Fig. 13.8
Mechanism of egg-shell formation in a digenetic trematode; based on *Fasciola hepatica*. (After Smyth & Clegg, 1959.)

isation of products of a putative egg-shell precursor gene in the vitellaria of *Schistosoma mansoni* and *S. haematobium* have also been examined (Köster *et al.*, 1988; Bobek *et al.*, 1989).

In other species, however, e.g. *Gorgoderina* sp. (Table 13.1), although phenols and proteins are present in the vitellaria, phenolase is absent and presumably an alternative stabilisation system operates. This may be the di-tyrosine or –S–S– bonding systems shown in Fig. 13.7. In the Paramphistomidae (e.g. *Paramphistomum cervi*), the chief structural protein in the egg-shell appears to be keratin as evidenced by the presence of –S–S– bonds (Madhavi, 1966, 1968). Histochemical methods for keratin detection are, however, notoriously unreliable and these results should be accepted with caution (Nollen, 1971). It is evident from this brief account that the process of egg-shell stabilisation in the parasitic platyhelminths may prove to be a more complex process than formerly thought and several types of bonding systems may be involved in different species. Waite (1990) has succinctly reviewed the process of quinone-tanning in general and has provided a valuable survey of its occurrence throughout the animal kingdom.

○ 13.2.3 Role of Mehlis' gland

The part played by Mehlis' gland in the above process has long been a matter of dispute. Various functions suggested by different authors (reviewed by Smyth & Halton, 1983) include (*a*) lubrication for the passage of eggs, (*b*) activation of spermatozoa, (*c*) release of shell globules from the vitelline cells, (*d*) activation of the quinone-tanning process and (*e*) providing a membrane which serves as a template on which shell droplets

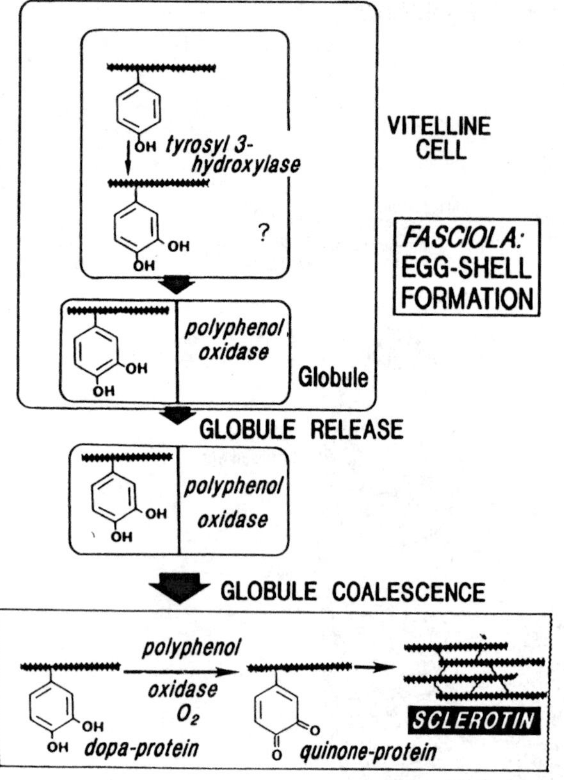

Fig. 13.9
Model illustrating the origin of DOPA (3,4-dihydroxyphenylalanine) in the egg-shell precursor protein of *Fasciola hepatica*. This amino acid could be formed by the action of a putative tyrosine 3-hydroxylase on a protoegg-shell precursor protein. The DOPA-containing egg-shell precursor is then stockpiled along with polyphenol oxidase in the biphasic vitelline globules. The vitelline globules are released from the vitelline cells and coalesce during egg-shell formation, during which polyphenol oxidase acts on the DOPA residues, converting them into cross-linking *o*-quinones. (After Waite & Rice-Ficht, 1987; reprinted with permission from *Biochemistry*, **26:** 7823. © 1987 American Chemical Society.)

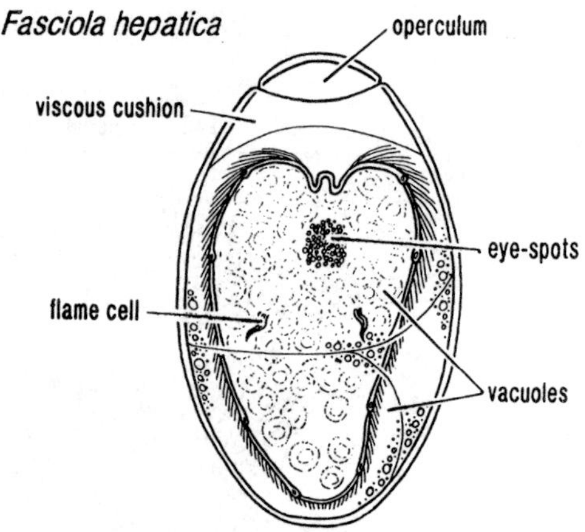

Fig. 13.10
Egg of *Fasciola hepatica*, showing fully developed miracidium formed after embryonation in water.

accumulate to form the egg-shell. There is little concrete evidence in support of or against any of these propositions because the evidence is based on static data from cytochemistry and electron microscopy.

However, the bulk of evidence suggests that the gland is involved in some way in laying down a membrane, possibly homologous with the 'peri-vitelline' material lining the shell (Clegg, 1965; Wilson, 1967) which could form a template on which the egg-shell is moulded. There is also evidence that a lipoprotein membrane is formed *outside* the shell (Fig. 13.8). Since Mehlis' gland appears to consist of at least two types of cells (variously labelled S, S_1, S_2, F cells by different authors) both in the Digenea (Fig. 18.6) and in the Monogenea (Fig. 11.6), its secretions may be involved in several activities of the egg-shell forming process. In most species examined, the Mehlis' gland cells are PAS-positive and diastase-fast, thus indicating secretions of mucopolysaccharides of some kind. As mentioned, a complex of four or five putative peptidergic nerve cells has been demonstrated (Fig. 11.6; Fig. 18.6), and these may be involved in coordinating the egg-forming mechanism.

An unusual modification of the above egg-forming mechanism has been reported in *Syncoelium spathulatum* (Coil & Kuntz, 1963) in which the phenolic and protein materials of the egg-shell are released from gland cells in the uterine wall.

○○○ 13.3 Trematode life cycles

Digenetic trematodes have complex life cycles, which, with rare exceptions, always involve a mollusc host. Six larval stages may occur in a cycle – a miracidium, sporocyst, redia, cercaria, mesocercaria (rare) and metacercaria – but the majority have four or five.

○ 13.3.1 Egg hatching, in water

Eggs reach the outside world via the three main body-waste materials: sputum, faeces or urine. In some cases, the eggs develop embryos while still in the body, but the majority remain undeveloped until suitable conditions for embryonation are reached in the outside world. Water or moisture is normally required for further development and eggs which fail to reach an aqueous environment desiccate rapidly. Unlike those of nematodes and some cestodes, trematode eggs cannot withstand desiccation. A high oxygen tension and a suitable temperature are also necessary for embryonation.

The hatching of eggs containing mature miracidia (Fig. 13.10) is controlled by a number of factors, the most important being light, temperature, salinity, and the gas phase (reviewed by Smyth & Halton, 1983).

Temperature. Fig. 13.11 shows that at 28 °C the eggs of *Schistosoma mansoni* hatch rapidly, but high (37 °C) or low (4 °C) temperatures almost completely inhibit the process, although hatching is restored after eggs at these temperatures are returned to 28 °C.

Light. Light plays a major role in stimulating the hatching processes in most operculate eggs. It has been suggested that on exposure to light a 'hatching substance' (enzyme?) is released which attacks the operculum-bonding material from the inside (Rowan, 1956, 1957). Giving some support to this view is the finding that the water in which *D. dendriticum* eggs are hatched induces hatching in *Fasciola* (Ractliffe, 1968).

An alternative hypothesis to explain hatching has been put forward by Wilson (1968) who speculates that the mechanism is as follows: (*a*) light stimulates the miracidium into activity; (*b*) this activity results in altering the permeability of the internal surface of the viscous cushion; (*c*) the change in permeability allows the fluid egg contents to reach the cushion, which becomes hydrated, the increased internal pressure finally rupturing the operculum.

Osmotic pressure. It is clearly to the advantage of the organism that embryonated eggs should not hatch prematurely while still within the host. The inhibitory effect of darkness and high body temperature has already been noted, but with eggs of certain species, osmotic pressure has an even more marked inhibitory effect within quite narrow limits, and this

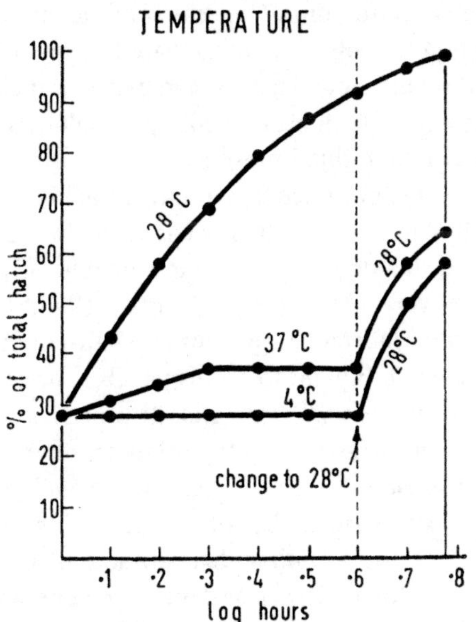

Fig. 13.11
Effect of temperature on hatching of eggs of *Schistosoma mansoni*. (After Standen, 1951.)

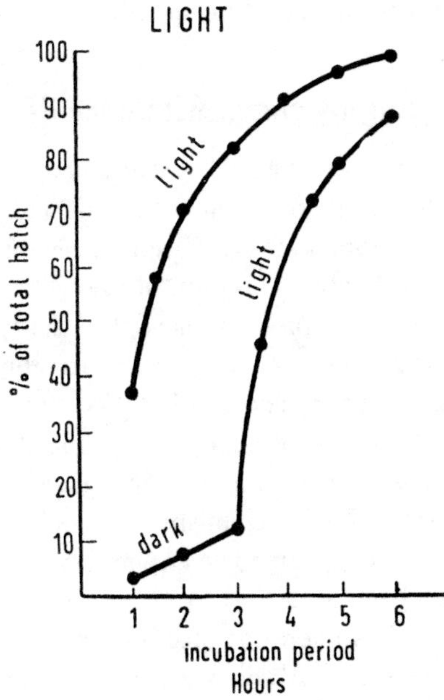

Fig. 13.12
Effect of light on hatching of eggs of *Schistosoma mansoni*. (After Standen, 1951.)

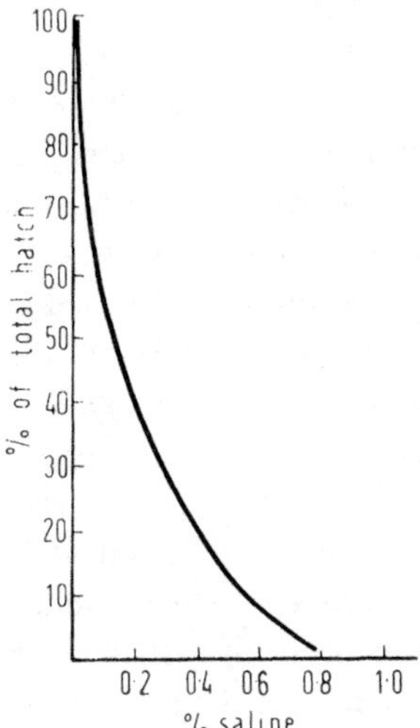

Fig. 13.13
Effect of osmotic pressure on hatching of eggs of *Schistosoma mansoni*. (After Standen, 1951.)

effect plays a major role in the prevention of premature hatching. Thus hatching of the eggs of *Schistosoma mansoni* is almost completely inhibited by 0.6 per cent NaCl (Fig. 13.13) and extensive hatching does not occur until a dilution of 0.1 per cent is reached. Hence eggs in blood, gut contents or urine will hatch only on reaching water. The mechanism of this inhibition is not known.

Gas phase. Although oxygen appears to be generally necessary for embryonation, its presence is inhibitory to hatching, at least in *Dicrocoelium*; the p_{CO_2} may also be important but does not appear to stimulate hatching (Ractliffe, 1968).

○ 13.3.2 Egg hatching, within snails

Some eggs, such as those of *Dicrocoelium dendriticum* (p. 214) hatch only on ingestion by the snail intermediate host. This phenomenon has been surprisingly little investigated, probably on account of the technical difficulties involved. The process takes place fairly rapidly, requiring only 20 min in *D. dicrocoelium*. In other species it takes longer: 1 h for the eggs of *Haplometra intestinalis* in the snail *Physa* spp. (Schell, 1961), and 2 h for the eggs of *Haematoloechus breviplexus* in *Gyraulus similaris* (Schell, 1965). The mechanism of hatching is not understood, but it appears to be an active process stimulated largely by the action of the snail's digestive enzymes. This has been demonstrated by the fact that the eggs of *Plagitura salamandra* have been hatched *in vitro* in intestinal extracts of the snail *Pseudosuccinea columella* (Russell, 1954).

○ 13.3.3 Larval forms

Miracidium. A typical miracidium (Fig. 13.14) is essentially a swimming, sac-like larva. Each carries a number of germinal cells from which will arise subsequent generations of organisms. In some species (e.g. *Parorchis*) the next generation is formed before the molluscan host has been penetrated, but this is unusual.

Miracidia contain a number of glands whose role is uncertain. Chief of these is a large 'apical' gland (formerly referred to as a 'gut'; Fig. 13.14) which empties rapidly during penetration and is thought to release proteolytic enzymes which could aid in this process. Penetration is probably assisted by a boring action of the larva. After penetration, a miracidium normally sheds its ciliated plates and elongates to become a motile, vermiform sac-like larva, the *sporocyst*. The factors triggering differentiation of the miracidium into a

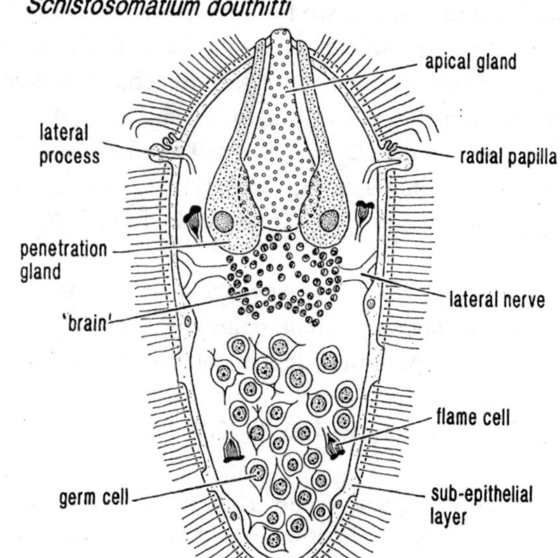

Fig. 13.14
Miracidium of the bird schistosome *Schistosomatium douthitti*. (After Price, 1931.)

sporocyst are not known, but it is thought that release of the apical gland material may serve as a trigger. A further pair of glands, the so-called *penetration glands* (= the *adhesive glands* of some authors; the nomenclature is confusing!) secrete a mucoid material which appears to assist in attachment to the snail tissue (Wajdi, 1966).

The ultrastructure of the tegument of the miracidium of *F. hepatica* reveals that the ciliated epithelial cells overlie a discontinuous layer of cytoplasm, which itself is an extension of vesiculated cells lying below layers of circular and longitudinal muscles.

There is some evidence that a miracidium is attracted to its molluscan host by chemotaxis (Smyth & Halton, 1983); but if this does take place it probably only operates over a very short range (Smyth, 1966; Cheng, 1968). Various substances present in 'snail-conditioned water' have been thought to be responsible for acting as miracidium attractants. These include mucus, serotonin (secreted from the snail's foot), amino or other organic acids and a substance referred to as 'miraxone'; this last substance may prove to be an inorganic ion such as Mg^{2+}. Early work on miracidial 'attraction' has been reviewed by Ulmer (1971) and more recent work by Smyth & Halton (1983). Chernin (1970) suggests that the response of miracidia is 'more nearly characterised as a *chemokinesis* rather than a chemotaxis'. It is likely that miracidia exhibit tropisms related to light, gravity, temperature, salinity and pH which could bring the miracidia into range of the molluscan host; thereafter chemotaxis could operate. The host specificity of larval trematodes for certain molluscan species is not, however, governed by chemotaxis alone and complex factors, such as immunity, are also involved; this problem has been reviewed by Cheng (1968).

The fact that some miracidia show marked positive and negative geotaxis permits these responses to be used to separate miracidia from eggs. A 'side-arm' flask has long been used to separate schistosome cercariae; a recent, and more efficient modification of this is shown in Fig. 16.4.

Sporocysts and rediae. Sporocysts are essentially germinal sacs containing germinal cells which have descended from the original ovum from which the miracidium developed. Within the sporocyst, the germinal cells multiply and form new germinal masses. These may either (*a*) produce daughter sporocysts like the parent sporocyst or (*b*) produce *rediae*, which are organisms with a rhabdocoele-like intestine, bearing a pharynx and a birth pore. Both of these generations produce embryos which develop into the final generation of organisms termed *cercariae*. If sporocysts give rise to daughter sporocysts, the latter give rise directly to cercariae and rediae are not formed. If sporocysts give rise to rediae before producing cercariae (Fig. 13.16), these may produce a second or even a third generation of rediae before producing cercariae. The reproductive power of these generations is enormous, and a single egg, by the multiplicative processes of the sporocysts and rediae, may ultimately give rise to up to a million cercariae. This is a reproductive phenomenon unsurpassed among the Metazoa.

The ultrastructure and cytochemistry of sporocysts and rediae during their transformation from miracidia have been particularly studied in schistosomes (Basch & DiConza, 1974; Meuleman, 1972; Meuleman *et al.*, 1978; Køie & Frandsen, 1976; Hockley, 1973) and in *Fasciola hepatica* (Køie *et al.*, 1976). Early work has been reviewed by Erasmus (1972) and later work by Smyth & Halton (1983).

Parasitic castration. The presence of large numbers of sporocysts and redial generations in the tissues of a snail can be expected to have a major mechanical and physiological effect on its biology and this has been found to be the case. The snail reproductive system is particularly affected: a wide range of effects from inhibition of gametogenesis to parasitic castration have been reported. These various effects have been reviewed by Erasmus (1972), Stadnichenko (1969) and Cheng (1968). A well-known example of parasitic castration is that of *Lymnaea stagnalis* parasitised by the bird schistosome, *Trichobilharzia ocellata* (p. 251). It has recently been shown that an agent, *schistosomin*, present in the haemolymph of the snail, inhibits the activation of the snail gonadotropic body hormone (DBH) – which normally stimulates oocyte growth and sexual maturity of the snail – thus preventing development of the reproductive system and resulting in parasitic castration (De Jong-Brink & Bergamin-Sassen, 1989).

Cercariae. These are essentially young flukes which develop parthenogenetically (Bednarz, 1973) in rediae and sporocysts. During their development, propagatory cells, derived from the original germ cell, give rise to the anlagen of the reproductive system of the adult fluke. Most cercariae bear many of the features of the adult fluke. Most are furnished with a tail, mouth, gut, suckers, flame cells and well-developed histolytic and cystogenous glands. The former glands secrete proteolytic enzymes for assisting penetration into the second intermediate host. The nature of these enzymes has not been fully investigated except in a few forms. In *Schistosoma mansoni*, hyaluronidase and a collagenase have definitely been detected. Cystogenous glands secrete a cyst wall in those forms which have an encysted stage in their life cycle. Other glands, sometimes known as 'escape' glands, assist in the escape of the cercaria from the snail but how they act is unknown. The histochemistry of the glands in cercariae has been studied in several species, e.g. *S. mansoni* (p. 242), *F. hepatica* (p. 209).

Progenetic cercariae. Most cercariae contain only the traces of the anlagen of the future genitalia which normally develop to various degrees in the metacercariae after entry into the intermediate host. A remarkable exception to this is the cercaria of *Transversotrema patialensis* (Fig. 13.15), which becomes almost fully sexually mature even to the extent of spermatozoa being present in the seminal vesicle, although the vitellaria are relatively undeveloped and eggs are not yet formed. On being released from its snail intermediate host, the cercariae attach themselves to a fish by means of adhesive pads, shed their tails and settle down under a scale. The vitellaria then rapidly mature and eggs are produced without the organism passing through a metacercarial stage (Cruz, 1956; Cruz *et al.*, 1964; Rao & Ganapati, 1967; Whitfield *et al.*, 1975; Whitfield & Wells, 1973). This organism is readily maintained in laboratory tanks using tropical fish such as the Pearl Danio (*Brachydanio alblineatus*), the Molly (*Molliensia* spp.) or the Flame Tetra (*Hyphessobrycon flammeus*) as definitive hosts. A suitable snail host is *Melanoides tuberculatus* (Whitfield & Wells, 1973). It would appear to be an

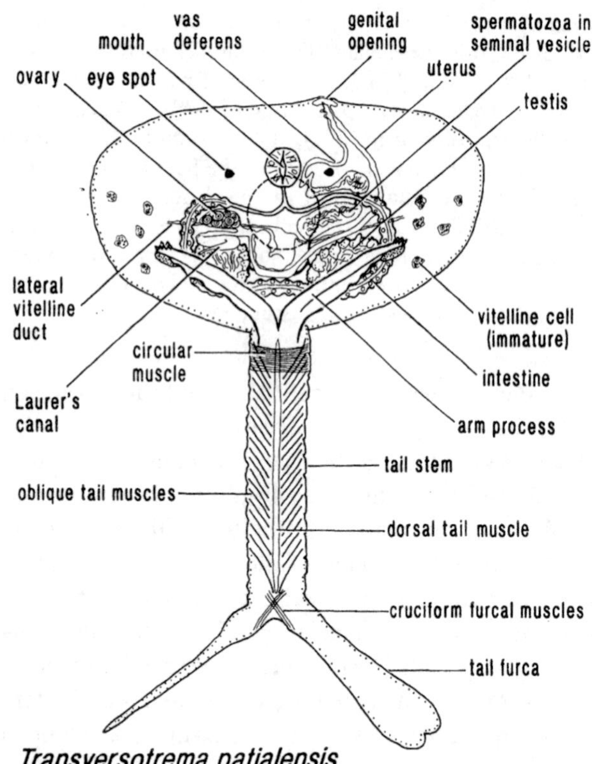

Fig. 13.15
Progenetic cercaria of *Transversotrema patialensis*. The male genitalia are fully developed and active spermatozoa are present in the seminal vesicle but the vitelline cells are still not fully matured. On release from the snail intermediate host, the cercaria attaches to the surface of a fish, sheds its tail and settles under a scale where the female genitalia complete their maturation and the organism (essentially an 'adult') produces eggs without passing through a metacercarial stage. (After Whitfield *et al.*, 1975.)

excellent experimental model for studies in trematode developmental biology, behaviour and ecology.

When fully developed, cercariae leave the molluscan host by several different mechanisms such as *active* escape, either through a fixed opening or through the intact tegument (e.g. *S. mansoni*), or *passive* extrusion of masses of cercariae (e.g. *F. hepatica*). In some cases (e.g. *Dicrocoelioides petiolatum*) *active* escape of sporocysts containing cercariae occurs. These sporocysts, which possess a thick internal wall which offers some protection, are then eaten by the second intermediate host (an isopod). The most important factors influencing the release of cercariae appear to be temperature and light, and these effects can be used to control the release for experimental purposes. Thus the cercariae of *Schistosomatium douthitti* normally emerge from their snail hosts in their natural habitats in the evening or in the laboratory after being placed in darkness after exposure to light. The emergence cycle may be reversed by reversing the light cycle (Fig. 16.11). It is possible that the emergence may be correlated in some way with the habits of the definitive hosts, but this question has never been fully investigated. Released cercariae usually show negative geotropism, positive phototropism and thermotropism but a complex behavioural pattern exists in different species (Smyth & Halton, 1983; Erasmus, 1972).

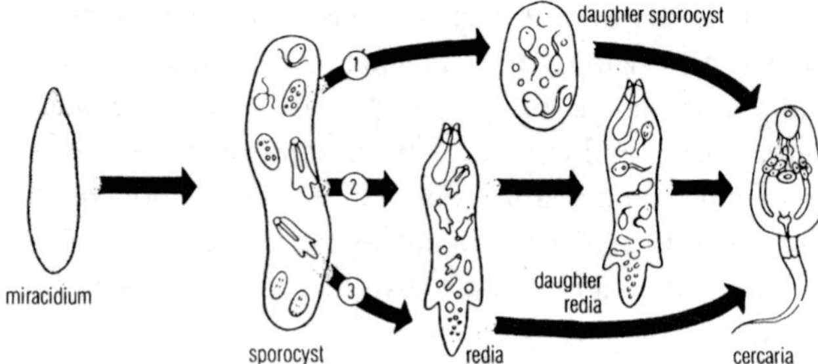

Fig. 13.16
Patterns of larval development in trematodes. (Original.)

Classification of cercariae. The structure of cercariae may provide valuable clues to relationships, often unsuspected, between adults apparently morphologically dissimilar. The number of flame cells, the type of tail (whether single or forked), and the presence of a stylet near the oral sucker, are all important diagnostic features. For a useful review of the value of cercariae in taxonomy, see Cable (1965). Various terms are used to describe cercariae, and these are based especially on the form of the tail. In addition, the adult characters of certain groups (gasterostomes, monostomes and amphistomes) are sufficiently apparent in their cercariae to make their structure unmistakable, so that these terms are applied to the cercariae also. Thus the cercaria of an amphistome fluke, which has a large posterior sucker, is easily recognised. The following classification is based on that of Lühe (1909); it is descriptive rather than systematic.

Classification of cercariae (mollusc host in parentheses). See Fig. 13.17.

1. **Gasterostome (= gasterstome) cercariae.** Two symmetrical tail furcae arising from posterior end of body. Mouth central on ventral surface. Intestine sac-like. Develop in sporocysts; e.g. cercaria of *Bucephaloides gracilescens* (*Abra alba*).

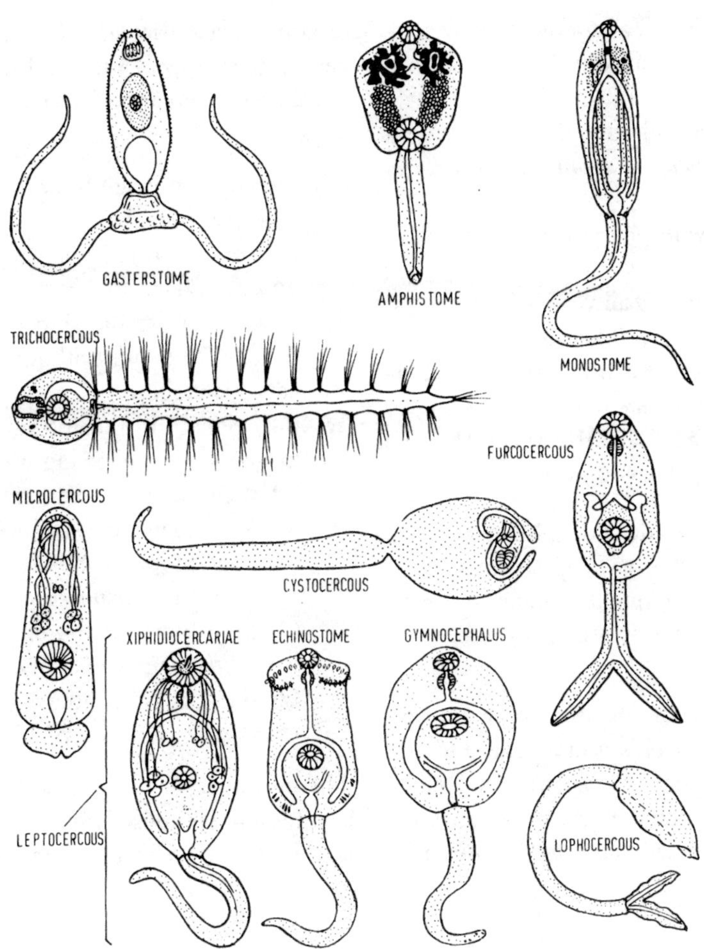

Fig. 13.17
Various types of cercaria. (Mainly after Dawes, 1946.)

2. **Monostome cercariae.** No ventral sucker. Pigmented, two or three eyes. Two excretory canals in body uniting near eyes, one in simple tail. No pharynx. Dense cystogenous glands. Develop in rediae, encyst in open; e.g. *Cercaria monostomi* (*Lymnaea peregra*).
3. **Amphistome cercariae.** Ventral sucker at root of slender tail. Develop in rediae, encyst in open; e.g. cercaria of *Diplodiscus subclavatus* (*Planorbis umbilicatus*).
4. **Distome cercariae.** The commonest form of cercaria with the ventral sucker some distance from the posterior end, lying in approximately the anterior third of the body.
5. **Rhopalocercous cercariae.** Tail when contracted is wider than body; e.g. *Cercaria isopori* (*Sphaerium corneum*).
6. **Cystocercous cercariae.** Body can be withdrawn into pocket at base of well-developed tail. Tail may be forked or not. Remainder of anatomy very variable. Unnatural group; e.g. *Cercaria macrocerca* (*Sphaerium corneum*).
7. **Gymnocephalous cercariae.** Two almost equal suckers, no stylet, well-developed pharynx, oesophagus and intestine, tail simple. A heterogeneous collection; e.g. cercaria of *Fasciola hepatica* (*Lymnaea truncatula*).
8. **Xiphidiocercariae.** Boring stylet with single point. Stylet glands, tail simple. A natural group with the same type of excretory system. Develop in sporocysts; e.g. *Cercaria ornata* (*Planorbis corneus*).
9. **Echinostome cercariae.** Collar of spines around anterior end, tail simple. A natural group; e.g. cercaria of *Echinostoma secundum* (*Littorina littorea*).
10. **Trichocercous cercariae.** Tail with rings of fine bristles; e.g. *Cercaria pectinata* (*Donax vittatus*).

11. **Furcocercous cercariae.** Tail forked distally containing branches of excretory duct. Flame cells in tail stem.
    (*a*) Blood fluke cercariae without pharynx; e.g. cercaria of *Schistosoma mansoni* (*Biomphalaria glabrata*).
    (*b*) Strigeid cercariae with pharynx, e.g. cercaria of *Diplostomum phoxini* (*Lymnaea peregra*).
12. **Microcercous cercariae.** Tail vestigial, unnatural group; e.g. *Cercaria micrura* (*Bithynia tentaculata*).
13. **Cercariaea.** Tail absent, unnatural group; e.g. cercaria of *Leucochloridium paradoxum* (*Succinea* sp.).
14. **'Rattenkönig' cercariae.** Cercariae tangled by tails to form colony. Marine, little known.

Gymnocephalous, echinostome and xiphidiocercariae are sometimes grouped together as 'Leptocercous' cercariae, i.e. cercariae with tail straight, slender and narrower than the body. Other terms are also used.

Metacercariae. Before becoming infective, most cercariae (except those of the Schistosomatidae and the Azygiidae: see below) must undergo a further developmental phase during which time they are known as *metacercariae*. The term *mesocercaria* is also used to describe prolonged cercarial stages which occur (rarely) in some genera (e.g. *Alaria*, Fig. 13.19).

Released cercariae behave in one of the following ways:

(i) they become ingested *directly* by the definitive host (e.g. Azygiidae).
(ii) they encyst directly on vegetation (e.g. *Fasciola hepatica*, p. 207).
(iii) they penetrate the skin of the definitive host and develop to adults without passing through a metacercaria stage.
(iv) they penetrate the intermediate host, and behave in one of the following ways:
    (*a*) they undergo some growth without encystment (e.g. *Diplostomum phoxini*, p. 256).
    (*b*) they encyst at the beginning of a growth phase (e.g. *Dicrocoelium dendriticum*, p. 214).
    (*c*) they encyst at the end of a growth phase (e.g. *Posthodiplostomum minimum*).
    (*d*) they encyst without a growth phase (e.g. *Echinostoma revolutum*, p. 228).

The structure of the cyst wall varies considerably, but is generally complex often involving lipid, polysaccharide and tanned protein layers (see p. 209 and Fig. 14.9).

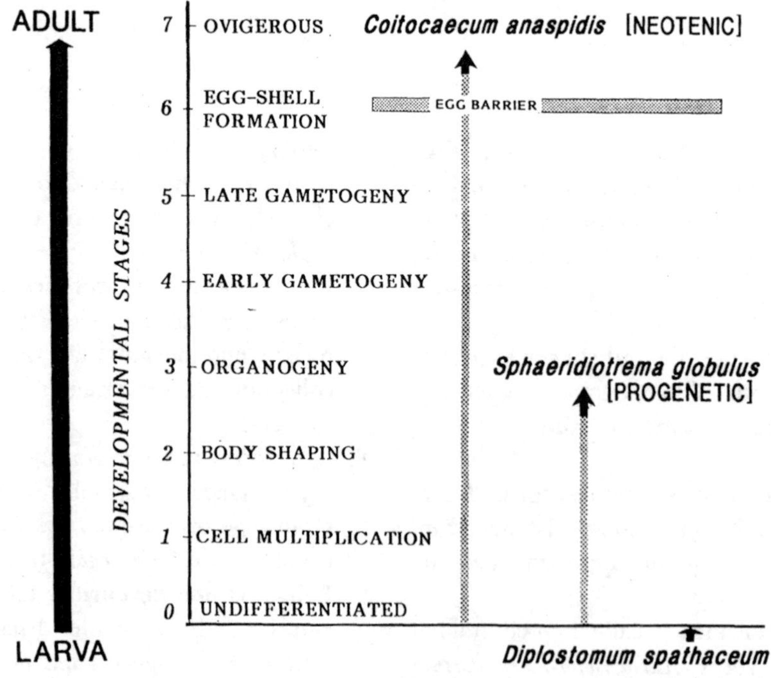

Fig. 13.18
Examples of the stages of development reached by different species of metacercaria in the intermediate host. The metacercariae of *D. spathaceum* are entirely undifferentiated; those of *S. globulus* are *progenetic* and possess well-formed genital anlagen; those of *C. anaspidis* are *neotenic* and reach full sexual maturity in the 'intermediate' crustacean host. (After Smyth, 1990.)

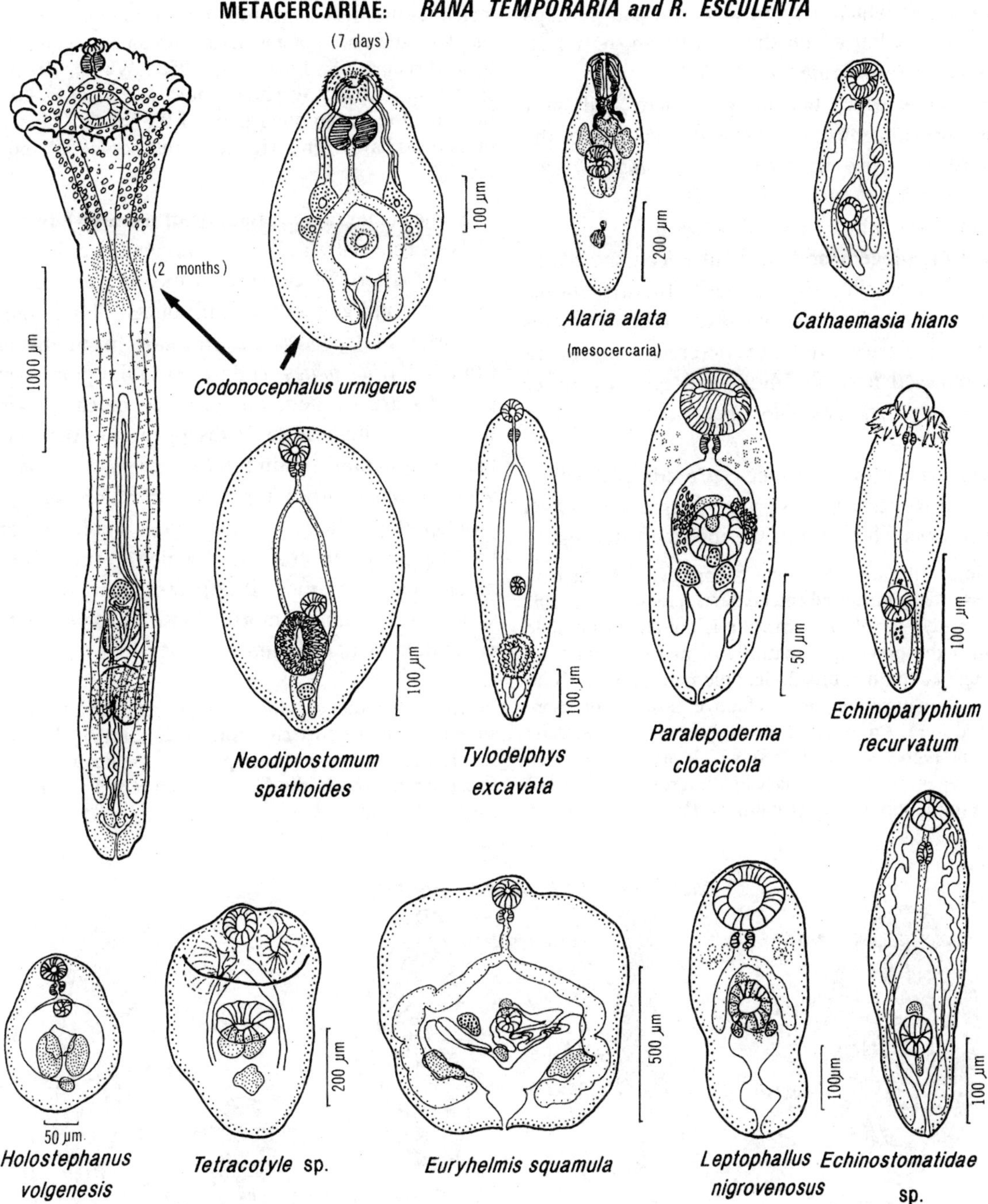

Fig. 13.19
Some species of metacercaria found in the European frogs *Rana temporaria* and *R. esculenta*. (After Smyth & Smyth, 1980.)

'Maturation' of metacercariae. The time required for metacercariae to become infective after encystment varies from one hour to several months (Dönges, 1969). The degree of morphological development, as compared with that of a cercaria, varies from no development to complete sexual maturation including egg production (i.e. progenesis: see below). The following series may be recognised:

(*a*) metacercariae which show little morphological difference from cercariae and are infective immediately after encystment (e.g. *Fasciola hepatica*);

(*b*) metacercariae which may grow in size and perhaps alter in shape, but which show no organogeny (e.g. *Diplostomum spathaceum*, *D. phoxini*);

(*c*) metacercariae which become progenetic and show marked organogeny and possess the anlagen of the major reproductive organs (e.g. *Cryptocotyle lingua*, *Sphaeridiotrema globulus*; Fig. 13.18);

(*d*) metacercariae which have developed beyond the stage of organogeny and show advanced spermatogenesis and possibly vitellogenesis. In such forms, mature spermatozoa may be present in the testes and sometimes even in the vas deferens, but ova are never released from the ovary and eggs are never formed (e.g. *Bucephaloides gracilescens*, see Fig. 14.1);

(*e*) metacercariae (often called 'pre-adults') which exhibit progenesis to a varying degree but which produce eggs while still in the intermediate host.

The meaning of the term *progenesis* is considered further on p. 310: it essentially means advanced development of genitalia in an (apparently!) larval form. Since it is not always possible to distinguish the precocious maturity of metacercaria from that of 'true' adult, the broad definition of progenesis to denote sexual maturity in non-definitive hosts (Jamieson, 1966) is used here. An example of a progenetic metacercaria is *Coitocaecum anaspidis* (Fig. 13.18), which produces a small number of eggs in the crustacean (intermediate?) host. Another example (in Great Britain) is *Proctoeces subtenuis*, which reaches sexual maturity in the lamellibranch *Scrobicularia plana* and does not appear to have an alternative definitive host (Freeman & Llewellyn, 1958; White, 1972). Other species of *Proctoeces* reaching maturity in other molluscs have been reported from other parts of the world. Other cases of progenesis have been reviewed by Buttner (1955) and Babero (1972).

Some remarkable variations of the above patterns are known, especially in the Gymnophallidae, in which the normal pattern of cercaria to metacercaria is reversed and a second generation is produced. Thus, the metacercariae of *Gymnophallus australis* (from the lamellibranch *Mytilus platensis*) develop to the point that adult genitalia are formed, but produce cercariae after the testes and vitellaria have disappeared! An even more bizarre situation is found in *Parvatrema homoeotecnum* in which furcocercariae (or tailed 'germinal sacs') in the gastropod produce a second generation of cercariae (= 'daughter germinal sacs') which in turn give rise to metacercariae (James, 1964); moreover, the 'germinal sacs' have a somatic morphology similar to that of the adult fluke which occurs in birds.

Source of metacercariae. Frogs are a particularly valuable source of metacercariae (Smyth & Smyth, 1980). Some species found in the European frogs *Rana esculenta* and *R. temporaria* and in the N. American *R. pipiens* are shown in Figs 13.19 and 13.20.

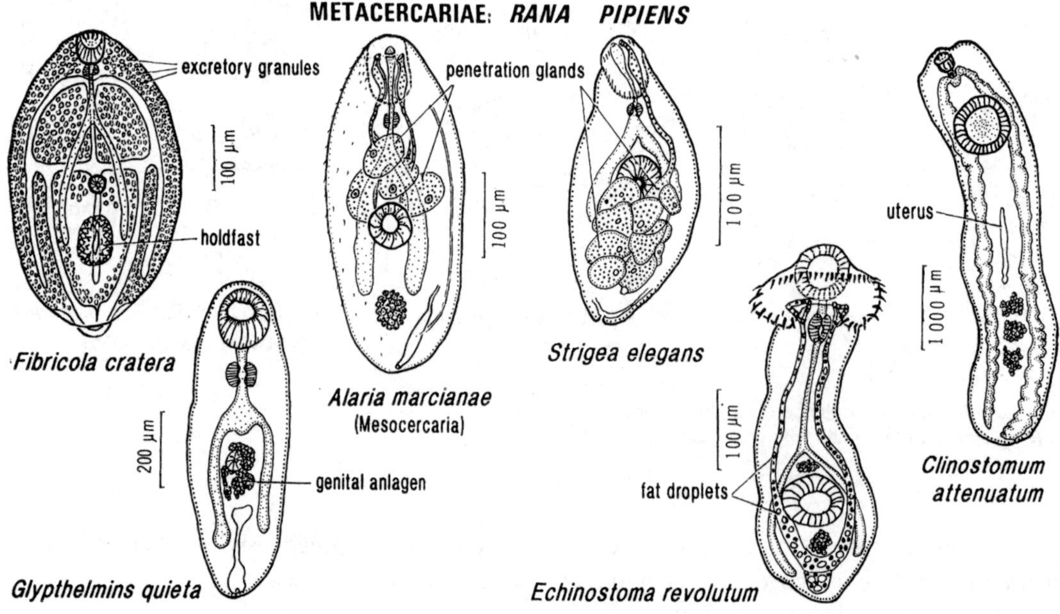

Fig. 13.20
Some species of metacercaria found in the North American frog *Rana pipiens*. (After Smyth & Smyth, 1980.)

The germinal cell cycle in trematodes. The form of larval reproduction in trematodes is a distinctive one and its interpretation has given rise to many controversies, now largely of historic interest. The main point of issue has been the origin and development of the germ cells in the various larval stages (reviewed by Bednarz, 1973). All the morphological and cytological evidence now available suggests that the reproductive cells in the sporocysts and rediae can be traced back to the fertilised ovum. These cells, termed *germinal* cells, rather than *germ* cells, remain separate from the somatic cells during the development of the germinal sacs, and undergo no reduction divisions. Therefore the multiplication of such cells is really a *polyembryony* of the original zygote. The actual tissues of the germinal sacs of miracidia, rediae and sporocysts are split off the cells of the germinal line. The only cells which undergo meiosis are the spermatozoa and ova in the adult reproductive cells.

○○○ 13.4 Classification

The classification of trematodes is still in a very uncertain state. The number and arrangement of suckers or other external features have given rise to terms which, although having no precise classificatory significance, are generally widely used and serve as convenient operative terms. The main trematode types are shown in Fig. 13.21.

Gasterostome. Intestine simple and sac-like (e.g. *Bucephaloides*, see p. 203). Mouth not terminal.

Monostome. Generally lacking one sucker, usually the ventral but may be the oral; alternatively one may be greatly reduced. Both suckers may be absent (e.g. *Notocotylus*).

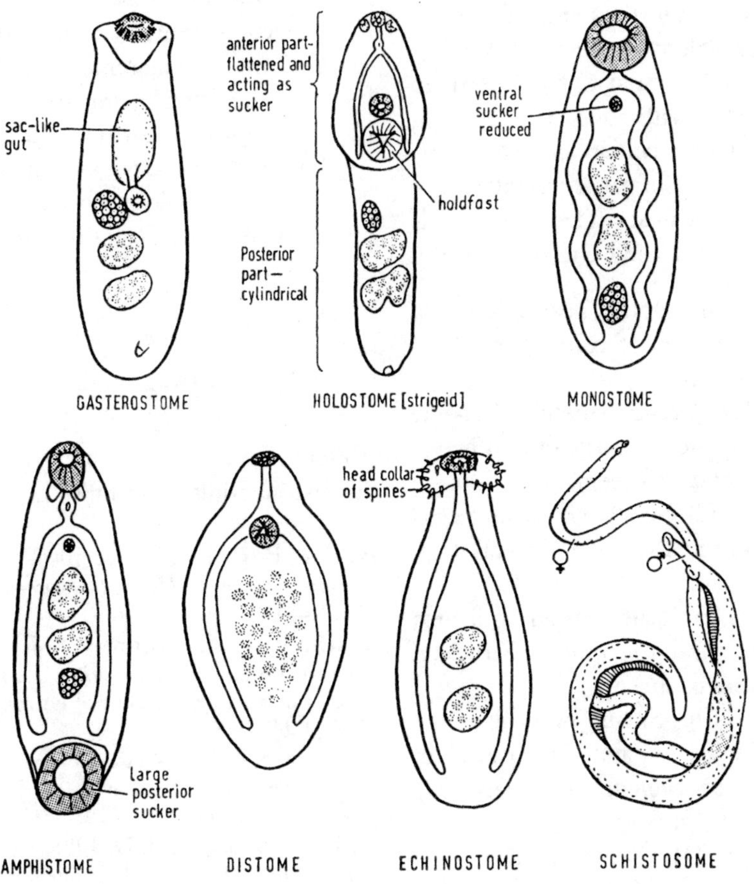

Fig. 13.21
Diagrammatic representation of the main types of trematode. (Partly after Cable, 1949.)

Amphistome. Body thick and fleshy with well-developed posterior sucker (e.g. *Paramphistomum*, see p. 260).

Distome. The mouth is surrounded by an oral sucker and the remaining sucker may be anywhere on the ventral surface except at the posterior end of the body (e.g. *Fasciola*, see p. 206).

Echinostome. The oral sucker is surrounded by a very characteristic collar of spines (e.g. *Echinostoma*).

Holostome (= strigeid). Body divided into anterior and posterior regions, the former containing an additional adhesive organ, the holdfast or tribocytic organ (e.g. *Diplostomum*, see p. 197).

Schistosome. Elongate dioecious, worm-like forms; parasitic in the blood stream (e.g. *Schistosoma*, see pp. 197, 236).

A precise classification should take into account the characteristics of the miracidiuim and cercaria as well as the adult. The development of the excretory system is of fundamental importance in determining relationships, and forms the basis of most systems of classification (see p. 182). The problem has been reviewed by La Rue (1957), Erasmus (1972) and Cable (1965). In the absence of an agreed classification, the group is considered here merely under certain families. Only those which are common, or are of particular interest or importance, are discussed. The order of families given is one of convenience and does not indicate relationship.

1. **Bucephalidae** (p. 203). Mouth near middle of body; sac-like gut; parasites of fishes (e.g. *Bucephaloides gracilescens* in fish).
2. **Fasciolidae** (p. 205). Large flattened leaf-like forms; much branched intestine; parasites of mammals (e.g. *Fasciola hepatica* in sheep).
3. **Opisthorchiidae** (p. 212). Small to medium transparent forms; parasitic in liver of mammals (e.g. *Clonorchis sinensis* in man).
4. **Dicrocoeliidae** (p. 214). Small transparent forms; vitellaria restricted to behind ventral suckers (e.g. *Dicrocoelium dendriticum* in sheep).
5. **Plagiorchiidae** (p. 219). Small, oval, flattened forms; simple intestine; mainly parasites of amphibia and birds, occasionally mammals (e.g. *Haematoloechus variegatus* in frog).
6. **Echinostomatidae** (p. 221). Echinostomes. Elongate forms with a reniform (kidney-shaped) collar surrounding the oral sucker and bearing spines (e.g. *Parorchis acanthus* in birds; but see p. 226).
7. **Heterophyidae** (p. 228). Small forms usually not more than 2 mm long and wider posteriorly than anteriorly. Ventral sucker usually central and may be absent (e.g. *Cryptocotyle lingua* in birds).
8. **Troglotrematidae** (p. 230). Large or medium-sized trematodes with suckers frequently poorly developed and the ventral sucker sometimes absent; cirrus sac usually absent; parasites of birds and carnivorous mammals, usually in pairs in cysts in various parts of the body (e.g. *Paragonimus westermani* in man).
9. **Strigeidae** (p. 254). Strigeids or holostomes. Body divided characteristically into an anterior flattened or cup-shaped portion, bearing an additional adhesive organ provided with histolytic glands (e.g. *Alaria alata* in dog, fox).
10. **Diplostomatidae** (p. 254). Similar to the Strigeidae, but anterior region of body more flattened and having two lateral tentacle-like processes bearing gland openings (e.g. *Diplostomum phoxini* in birds).
11. **Schistosomatidae** (p. 236). Schistosomes. Dioecious trematodes parasitic in the blood of various birds and mammals (e.g. *Schistosoma mansoni* in man and other mammals).
12. **Paramphistomatidae** (p. 260). Amphistomes. Large trematodes, usually thick and circular in transverse section; ventral sucker large and situated at or near the posterior end (e.g. *Paramphistomum cervi* in cattle).

○○○ References

(References of general interest have been included in this list, in addition to those quoted in the text.)

Babero, B. B. 1972. A record of progenesis in Trematoda. *Proceedings of the Helminthological Society of Washington*, **39:** 128–31.

Basch, P. F. & DiConza, J. J. 1974. The miracidium-sporocyst transition in *Schistosoma mansoni*: surface changes *in vitro* with ultrastructural correlation. *Journal of Parasitology*, **60:** 935–41.

Bednarz, S. 1973. The developmental cycle of the germ cells in several representatives of Trematoda (= Digenea). *Zoologica Poloniae*, **23:** 279–326.

Bell, E. J. & Smyth, J. D. 1958. Cytological and histochemical criteria for evaluating development of trematodes and pseudophyllidean cestodes *in vitro* and *in vivo*. *Parasitology*, **48:** 131–48.

Bobek, L. A., Loverde, P. T. & Rekosh, D. M. 1989.

Schistosoma haematobium: analysis of eggshell protein genes and their expression. *Experimental Parasitology*, **68**: 17–30.

Brunet, P. J. C. 1967. Sclerotins. *Endeavour*, **26**: 68–74.

Burton, P. R. 1972. Fine structure of the reproductive system of a frog lung fluke. III. The spermatozoa and its differentiation. *Journal of Parasitology*, **58**: 68–83.

Burton, P. R. 1973. Some structural and cytochemical observations on the axial filament complex of lung-fluke spermatozoa. *Journal of Morphology*, **140**: 185–96.

Burton, P. R. & Silveira, M. 1971. Electron microscopic and optical diffraction studies of negatively stained axial units of certain platyhelminth sperm. *Journal of Ultrastructure Research*, **36**: 757–66.

Buttner, A. 1955. Les distomes progénétiques sont-ils des pré-adultes ou des adultes véritables? Valeur évolutive de la progénèse chez les Digenea. *Comptes Rendus des Séances de la Société de Biologie*, **149**: 267–72.

Cable, R. M. 1965. 'Thereby hangs a tail'. *Journal of Parasitology*, **51**: 3–12.

Cheng, T. C. 1968. The compatibility and incompatibility concept as related to trematodes and molluscs. *Pacific Science*, **22**: 141–60.

Chernin, E. 1970. Behavioral responses of miracidia of *Schistosoma mansoni* and other trematodes to substances emitted by snails. *Journal of Parasitology*, **56**: 287–96.

Clegg, J. A. 1965. Secretion of lipoprotein by Mehlis' gland in *Fasciola hepatica*. *Annals of the New York Academy of Sciences*, **118**: 969–86.

Clegg, J. A. & Smyth, J. D. 1968. Growth, development and culture methods: parasitic platyhelminths. In: *Chemical Zoology* (ed. M. Florkin & B. T. Scheer), Vol. II, pp. 395–466. Academic Press, New York.

Coil, W. H. 1966. Egg shell formation in the notocotylid trematode, *Ogmocotyle indica* (Bhalerao, 1942) Ruiz, 1946. *Zeitschrift für Parasitenkunde*, **27**: 205–9.

Coil, W. H. 1972. Observations on the histochemistry of egg shell formation in *Lintonium vibex* (Trematode: Fellodistomidae). *Zeitschrift für Parasitenkunde*, **39**: 195–9.

Coil, W. H. & Kuntz, R. E. 1963. Observations on the histochemistry of *Syncoelium spathulatum* n.sp. *Proceedings of the Helminthological Society of Washington*, **30**: 60–5.

Cordingley, J. S. 1987. Trematode eggshells: novel protein biopolymers. *Parasitology Today*, **3**: 341–4.

Cruz, H. 1956. The progenetic trematode *Cercaria patialensis* Soparker in Ceylon. *Journal of Parasitology*, **42**: 245.

Cruz, H., Ratnayake, W. E. & Sathananthan, A. H. 1964. Observations on the structure and life cycle of the digenetic trematode *Transversotrema patialense* (Soparkar). *Ceylon Journal of Science (Biological Science)*, **5**: 8–17.

Dawes, B. 1946. *The Trematoda*. Cambridge University Press.

De Jong-Brink, M. & Bergamin-Sassen, M. J. M. 1989. *Trichobilharzia ocellata*: influence of infection on the interaction between the dorsal body hormone, a female gonadotropic hormone, and the follicle cells in the gonad of the intermediate snail host *Lymnaea stagnalis*. *Experimental Parasitology*, **68**: 93–8.

Dönges, J. 1969. Entwicklungs- und Lebensdauer von Metacercarien. *Zeitschrift für Parasitenkunde*, **31**: 340–66.

Ebrahimzadeh, A. 1966. Histologische Untersuchungen über den Feinbau des Oogenotop bei digenen Trematoden. *Zeitschrift für Parasitenkunde*, **27**: 127–68.

Eduardo, S. L. 1976. Egg shell formation in *Carmyerius synethes* (Fischoeder, 1901) and *Gastrothylax crumenifer* (Creplin, 1847) Digenea: Gastrothylacidae. *Philippine Journal of Veterinary Medicine*, **15**: 117–22.

Erasmus, D. A. 1969*a*. Studies on the host–parasite interface of strigeoid trematodes. IV. The ultrastructure of the lappets of *Apatemon gracilis minor* Yamaguti, 1933. *Parasitology*, **59**: 193–201.

Erasmus, D. A. 1969*b*. Studies on the host–parasite interface of strigeoid trematodes. V. Regional differentiation of the adhesive organ of *Apatemon gracilis minor* Yamaguti, 1933. *Parasitology*, **59**: 245–56.

Erasmus, D. A. 1972. *The Biology of Trematodes*. Edward Arnold, London.

Fairweather, I. & Halton, D. W. 1991. Neuropeptides in platyhelminths. *Parasitology*, **102**: S77–92.

Freeman, R. F. H. & Llewellyn, J. 1958. An adult digenetic trematode from an invertebrate host: *Proctoeces subtenuis* (Linton) from the lamellibranch *Scrobicularia plana* (Da Costa). *Journal of the Marine Biological Association of the United Kingdom*, **37**: 435–57.

Fried, B. & Stromberg, B. E. 1971. Egg-shell precursors in trematodes. *Proceedings of the Helminthological Society of Washington*, **38**: 262–4.

Fujino, T., Ishii, Y. & Mori, T. 1977. Ultrastructural studies on the spermatozoa and spermatogenesis in *Paragonimus* and *Eurytrema* (Trematoda: Digenea). *Japanese Journal of Parasitology*, **26**: 240–55.

Gerzeli, G. 1968. Aspetti istochimici della formazione degli involucri ovulari in *Aspidogaster conchicola* von Baer (Trematoda). *Instituto Lombardo Scienze e Lettere* (Rend. Sc.), **B102**: 263–76.

Gönnert, R. 1962. Histologische Untersuchungen über den Feinbau der Eibildungstätte (Oogenotop) von *Fasciola hepatica*. *Zeitschrift für Parasitenkunde*, **21**: 475–92.

Grant, W. C., Harkema, R. & Muse, K. E. 1976. Ultrastructure of *Pharyngostomoides procyonis* Harkema 1942 (Diplostomatidae). 1. Observations on the male reproductive system. *Journal of Parasitology*, **62**: 39–49.

Guildford, H. G. 1961. Gametogenesis, egg-capsule formation and early miracidial development in the digenetic trematode *Halipegus eccentricus* Thomas. *Journal of Parasitology*, **47**: 757–64.

Halton, D. W., Fairweather, I., Shaw, C. & Johnston, C. F. 1990. Regulatory peptides in parasitic platyhelminths. *Parasitology Today*, **6**: 284–90.

Halton, D. W., Shaw, C., Maule, A. G., Johnston, C. F. & Fairweather, I. 1992. Peptidergic messengers: a new perspective of the nervous system of parasitic platyhelminths. *Journal of Parasitology*, **78**: 179–93.

Harris, K. H. & Cheng, T. C. 1972. Presumptive neurosecretion in *Leucochloridiomorpha constantiae* (Trematoda) and its possible role in governing maturation. *International Journal for Parasitology*, **2**: 361–7.

Hathaway, R. P. & Herlevich, J. C. 1976. A histochemical study of egg shell formation in the monogenetic trematode *Octomacrum lanceatum* Mueller, 1932. *Proceedings of the Helminthological Society of Washington*, **43**: 203–6.

Ho, Y. H. & Yang, H. C. 1973. Histochemical localization of phenols and phenolase in *Schistosoma japonica. Acta Zoologica Sinica*, **19**: 1–10.

Hockley, D. J. 1973. Ultrastructure of the tegument of *Schistosoma. Advances in Parasitology*, **11**: 233–305.

Hockley, D. J. & McLaren, D. J. 1973. *Schistosoma mansoni*: changes in the outer membrane of the tegument during development from cercaria to adult worm. *International Journal for Parasitology*, **3**: 13–25.

James, B. L. 1964. The life cycle of *Parvatrema homoeotecnum* sp. nov. (Trematoda: Digenea) and a review of the family Gymnophallidae Morozov, 1955. *Parasitology*, **54**: 1–41.

Jamieson, B. G. M. 1966. *Parahemiurus bennettae* n.sp. (Digenea), a hemiurid trematode progenetic in *Salinator fragiles* Lamark (Gastropoda, Amphibolidae). *Proceedings of the Royal Society of Queensland*, **77**: 73–80.

Jennings, J. B. 1968. Nutrition and Digestion. Chapter 2. In: *Chemical Zoology* (ed. M. Florkin & B. T. Scheer), Vol. 2, pp. 305–26. Academic Press, New York.

Johri, L. N. & Smyth, J. D. 1956. A histochemical approach to the study of helminth morphology. *Parasitology*, **46**: 107–16.

Kandhaswami, D. W. W. 1980. Studies on *Ganeo tigrum* Mehra & Negi, 1928 from the amphibian *Rana hexadactyla* with reference to 'egg' formation. Ph.D. Thesis, University of Madras.

Kanwar, U. & Agrawal, M. 1977. Cytochemistry of the vitelline glands of the trematode *Diplodiscus amphichrus* (Tubangi, 1933) (Diplodiscidae). *Folia Parasitologica*, **24**: 123–7.

Kitajima, E. W., Paraense, W. L. & Correa, L. R. 1976. The fine structure of *Schistosoma mansoni* sperm (Trematoda: Digenea). *Journal of Parasitology*, **62**: 215–21.

Köhlmann, F. W. 1961. Untersuchungen zur Biologie, Anatomie, und Histologie von *Polystoma integerrimum* Fröhlich. *Zeitschrift für Parasitenkunde*, **20**: 495–524.

Køie, M. Christensen, N. Ø. & Nansen, P. 1976. Stereoscan studies of eggs, free-swimming and penetrating miracidia and early sporocysts of *Fasciola hepatica. Zeitschrift für Parasitenkunde*, **51**: 79–90.

Køie, M. & Frandsen, F. 1976. Stereoscan observations of the miracidium and early sporocyst of *Schistosoma mansoni. Zeitschrift für Parasitenkunde*, **50**: 335–44.

Köster, B., Dargatz, H., Schröder, J., Hirzmann, J., Haarmann, C., Symmons, P. & Kunz, W. 1988. Identification and localisation of the products of a putative eggshell precursor gene in the vitellarium of *Schistosoma mansoni. Molecular and Biochemical Parasitology*, **31**: 183–98.

La Rue, G. R. 1957. The classification of digenetic Trematoda: a review and a new system. *Experimental Parasitology*, **6**: 306–49.

Lal, M. B. & Johri, G. N. 1967. The function of the vitellocalycal glands in eggshell formation in *Fasciola indica* Verma, 1953. *Journal of Parasitology*, **53**: 989–93.

Llewellyn, J. 1965. The evolution of parasitic platyhelminths. *Symposia of the British Society for Parasitology*, **3**: 47–78.

Lühe, M. 1909. *Die Süsswasserfauna Deutschlands*, vol. 17, *Parasitische Plattwürmer*, Part I. Trematoden. Verlag Gustav Ascher, Jena.

Lumsden, R. D. 1975. Surface ultrastructure and cytochemistry of parasitic helminths. *Experimental Parasitology*, **37**: 267–339.

McKay, D. M., Halton, D. W., Allen, J. M. & Fairweather, I. 1990. The effects of cholinergic and serotoninergic drugs on motility *in vitro* of *Haplometra cylindracea* (Trematoda: Digenea). *Parasitology*, **99**: 241–52.

McKay, D. M., Halton, D. W., Johnston, C. F., Fairweather, I. & Shaw, C. 1990. The occurrence and distribution of putative neurotransmitters in the frog-lung parasite *Haplometra cylindracea* (Trematoda: Digenea). *Parasitology Research*, **76**: 509–17.

McLaren, D. J. & Hockley, D. J. 1976. *Schistosoma mansoni*: the occurrence of microvilli on the surface of the tegument during transformation from cercaria to schistosomulum. *Parasitology*, **73**: 169–87.

Ma, L. 1963. Trace elements and polyphenol oxidase in *Clonorchis sinensis. Journal of Parasitology*, **49**: 197–203.

Madhavi, R. 1966. Egg-shell in Paramphistomatidae (Trematoda: Digenea). *Experientia*, **22**: 93–4.

Madhavi, R. 1968. *Diplodiscus mehrai*: chemical nature of the egg-shell. *Experimental Parasitology*, **23**: 392–7.

Madhavi, R. 1971. Keratin in the egg shell of amphistomes: histochemical differentiation from quinone tanned protein in other Trematoda. *Stain Technology*, **46**: 105–9.

Madhavi, R. & Rao, K. H. 1971. *Orchispirium heterovitellatum*: chemical nature of the eggshell. *Experimental Parasitology*, **30**: 345–8.

Meuleman, E. A. 1972. Host–parasite interrelationships between the freshwater pulmonate *Biomphalaria pfeifferi* and the trematode *Schistosoma mansoni. Netherlands Journal of Zoology*, **22**: 355–427.

Meuleman, E. A., Lyaruu, D. M., Khan, M. A., Holzmann P. J. & Sminia, T. 1978. Ultrastructural changes in the body wall of *Schistosoma mansoni* during the transformation of a miracidium into a mother sporocyst in the snail host *Biomphalaria pfeifferi. Zeitschrift für Parasitenkunde*, **56**: 227–42.

Nollen, P. M. 1971. Digenetic trematodes: quinone tanning in egg-shells. *Experimental Parasitology*, **30**: 64–72.

Piva, N. & De Carneri, I. 1961. Studio istochimico sui vitellogeni di *Schistosoma mansoni. Parasitologia*, **3**: 235–8.

Price, E. W. 1931. Life history of *Schistosomatium douthitti* (Cort). *American Journal of Hygiene*, **13**: 685–727.

Ractliffe, L. H. 1968. Hatching of *Dicrocoelium lanceolatum* eggs. *Experimental Parasitology*, **23**: 67–78.

Ramalingam, K. 1970*a*. Relative role of vitelline cells and Mehlis' gland in the formation of egg-shell in trematodes. *Anales del Instituto de Biologia, Universidad Nacional Autónoma de Mexico*, **41**. (*Series Zoologíca* 1): 145–54.

Ramalingam, K. 1970*b*. Prophenolase and the role of Mehlis' gland in helminths. *Experientia*, **26**: 828.

Ramalingam, K. 1973. The chemical nature of the egg-shell of helminths. I. Absence of quinone tanning in the egg-shell of the liver fluke, *Fasciola hepatica. International Journal for Parasitology*, **3**: 67–75.

Rao, K. H. & Ganapati, P. N. 1967. Observations on *Transversotrema patialensis* (Soparkar, 1924) (Trematoda) from Waltair, Andhra Pradesh (India). *Parasitology*, **57**: 661–4.

Rees, W. J. 1936. The effect of parasitism by larval trematodes on the tissues of *Littorina littorea* (Linne). *Proceedings of the Zoological Society of London*, pp. 357–68.

Rennison, B. D. 1953. A morphological and histochemical study of egg-shell formation in *Diclidophora merlangi*. M.Sc. Thesis, Trinity College, University of Dublin.

Rigby, D. W. & Marx, R. A. 1962. A comparative histochemical study of the monogenetic trematode *Rajonchocotyle batis* Cerfontaine with the trypanorhynch cestode *Gilquinia squalia* (Fabricus). *Walla Walla College, Publications (Washington), Department of Biological Sciences*, **31**: 1–11.

Robinson, G. & Threadgold, L. T. 1975. Electron microscope studies of *Fasciola hepatica*. XII. The fine structure of the gastrodermis. *Experimental Parasitology*, **37**: 20–36.

Rohde, K. 1990. Phylogeny of platyhelminthes, with special reference to parasitic groups. *International Journal for Parasitology*, **20**: 979–1007.

Rothman, A. H. 1959. The physiology of tapeworms, correlated to structures seen with the electron microscope. *Journal of Parasitology*, **45** (Suppl.): 28.

Rowan, W. B. 1956. The mode of hatching of the egg of *Fasciola hepatica. Experimental Parasitology*, **5**: 118–37.

Rowan, W. B. 1957. The mode of hatching of the egg of *Fasciola hepatica*. II. Colloidal nature of the viscous cushion. *Experimental Parasitology*, **6**: 131–41.

Russell, C. M. 1954. The effects of various environmental factors on the hatching of eggs of *Plagitura salamandra* Hall (Trematoda: Plagiorchiidae). *Journal of Parasitology*, **40**: 461–4.

Schell, S. C. 1961. Development of mother and daughter sporocysts of *Haplometra intestinalis* Lucker, a plagiorchioid trematode of frogs. *Journal of Parasitology*, **47**: 493–500.

Schell, S. C. 1965. The life history of *Haematoloechus breviplexus* Stafford, 1902 (Trematoda: Haplometridae McMullen, 1937) with emphasis on the development of the sporocysts. *Journal of Parasitology*, **51**: 587–93.

Smyth, J. D. 1954. A technique for the histochemical demonstration of polyphenol oxidase and its application to egg-shell formation in helminths and byssus formation in *Mytilus. Quarterly Journal of Microscopical Science*, **95**: 139–52.

Smyth, J. D. 1966. *The Physiology of Trematodes*. 1st Edition. W. H. Freeman, San Francisco.

Smyth, J. D. (ed.) 1990. *In Vitro Cultivation of Parasitic Helminths*. CRC Press, Boca Raton, Florida.

Smyth, J. D. & Clegg, J. A. 1959. Egg-shell formation in trematodes and cestodes. *Experimental Parasitology*, **8**: 286–323.

Smyth, J. D. & Halton, D. W. 1983. *The Physiology of Trematodes*. 2nd Edition. Cambridge University Press.

Smyth, J. D. & Smyth, M. M. 1980. *Frogs as Host Parasite Systems I. An Introduction to Parasitology through the parasites of Rana temporaria, R. esculenta and R. pipiens*. The Macmillan Press, London.

Srivastava, M. & Gupta, S. P. 1975. Phosphatase activity in *Isoparorchis hypselobargi* (Trematoda). *Zoologischer Anzeiger*, **195**: 35–42.

Srivastava, M. & Gupta, S. P. 1978. Polyphenol oxidase and scleroprotein/melanin pigments of *Isoparorchis hypselobagri. Zeitschrift für Parasitenkunde*, **55**: 55–8.

Stadnichenko, A. P. 1969. (Effect of larval forms of trematodes on the reproductive organs of the intermediate gastropod hosts.) (In Russian.) (British Library Translation No: RTS 11230.) *Parasitologiya*, **3**: 53–7.

Standen, O. D. 1951. The effects of temperature, light, and salinity upon the hatching of the ova of *Schistosoma mansoni. Transactions of the Royal Society of Tropical Medicine and Hygiene*, **45**: 225–41.

Stunkard, H. W. 1946. Interrelationships and taxonomy of the digenetic trematodes. *Biological Reviews*, **21**: 148–58.

Threadgold, L. T. 1963. The tegument and associated structures of *Fasciola hepatica. Quarterly Journal of Microscopical Science*, **104**: 505–12.

Threadgold, L. T. 1967. Electron-microscope studies of *Fasciola hepatica*. III. Further observations on the tegument and associated structures. *Parasitology*, **57**: 633–7.

Threadgold, L. T. 1976. *Fasciola hepatica*: ultrastructure and histochemistry of the glycocalyx of the tegument. *Experimental Parasitology*, **39**: 119–34.

Thurston, J. P. 1964. The morphology and life-cycle of *Protopolystoma xenopi* (Price) Bychovsky in Uganda. *Parasitology*, **54**: 441–50.

Ulmer, M. J. 1971. Site-finding behaviour in helminths in intermediate and definitive hosts. In: *Ecology and Physiology of Parasites* (ed. A. M. Fallis), pp. 123–60. Adam Hilger, London.

Waite, J. H. 1990. The phylogeny and diversity of quinone-tanned glues and varnishes. *Comparative Biochemistry and Physiology*, **97B**: 19–29.

Waite, J. H. & Rice-Ficht, A. C. 1987. Presclerotized egg-shell protein from the liver fluke *Fasciola hepatica. Biochemistry*, **26**: 7819–25.

Waite, J. H. & Rice-Ficht, A. C. 1989. A histidine-rich protein from the vitellaria of the liver fluke *Fasciola hepatica. Biochemistry*, **28**: 6104–10.

Waite, J. H. & Rice-Ficht, A. C. 1992. Eggshell precursor proteins of *Fasciola hepatica*: II. Microheterogeneity in vitelline protein B. *Molecular and Biochemical Parasitology*, **54**: 143–51.

Wajdi, N. 1966. Penetration by the miracidia of *S. mansoni* into the snail host. *Journal of Helminthology*, **15**: 235–44.

Wells, K. E. & Cordingley, J. S. 1992. *Schistosoma mansoni*: Eggshell formation is regulated by pH and calcium. *Experimental Parasitology*, **73**: 295–310.

White, I. C. 1972. On the ecology of an adult digenetic trematode *Proctoeces subtenuis* from a lamellibranch host *Scrobicul-*

aria plana. Journal of the Marine Biological Association of the United Kingdom, **52**: 457–67.

Whitfield, P. J., Anderson, R. M. & Moloney, N. A. 1975. The attachment of cercariae of an ectoparasitic digenean, *Transversotrema patialensis*, to the fish host: behavioural and ultrastructural aspects. *Parasitology*, **70**: 311–29.

Whitfield, P. J. & Wells, J. 1973. Observations in the ectoparasitic digenean, *Transversotrema patialensis*. *Parasitology*, **67** (2): xxvii.

Wilson, R. A. 1967. The structure and permeability of the shell and vitelline membrane of the egg of *Fasciola hepatica*. *Parasitology*, **67**: 47–58.

Wilson, R. A. 1968. The hatching mechanism of the egg of *Fasciola hepatica*. *Journal of Parasitology*, **55**: 124–33.

Zurita, M., Bieber, D. & Mansour, T. E. 1989. Identification, expression and *in situ* hybridization of an eggshell protein gene from *Fasciola hepatica*. *Molecular and Biochemical Parasitology*, **37**: 11–18.

Zurita, M., Bieber, D., Ringold, G. & Mansour, T. E. 1987. Cloning and characterization of female genital complex cDNA from the liver fluke *Fasciola hepatica*. *Proceedings of the National Academy of Sciences, USA*, **84**: 2340–4.

14

Digenea: Bucephalidae, Fasciolidae, Opisthorchiidae, Dicrocoeliidae

○○○ 14.1 Family Bucephalidae

The bucephalids or *gasterostomes* characteristically have the mouth in the centre of the ventral surface. The adults occur in the gut of marine and fresh-water fish. The metacercariae encyst in smaller fish, sometimes in the nervous system.

Type example: *Bucephaloides gracilescens*

definitive host:	*Lophius piscatorius* (angler fish) and others?
location:	pyloric caeca, intestine
miracidium:	free-swimming
(presumed) molluscan hosts:	*Abra alba*
second intermediate host:	various gadoid fish (especially haddock)

Morphology. The morphology of the adult, which occurs in the pyloric caeca and duodenum of *Lophius*, is shown in Fig. 14.1. The unusual features presented by bucephalids in general are as follows.

The mouth is ventrally located (not terminal as in other trematodes) and the gut is sac-like, resembling the gut of rhabdocoele turbellarians. There is a large excretory bladder often containing pigment. The common genital pore is posterior. The reproductive organs characteristically (Fig. 14.1) have an elongate cirrus with glandular walls.

Life cycle. The adults produce large numbers of eggs with an unusually thick brown shell. Unlike most trematodes, this shell, although giving phenolic reactions (Table 13.1), does not contain phenolase and is not apparently tanned and hardened by the usual quinone-tanning system; its nature has not been determined.

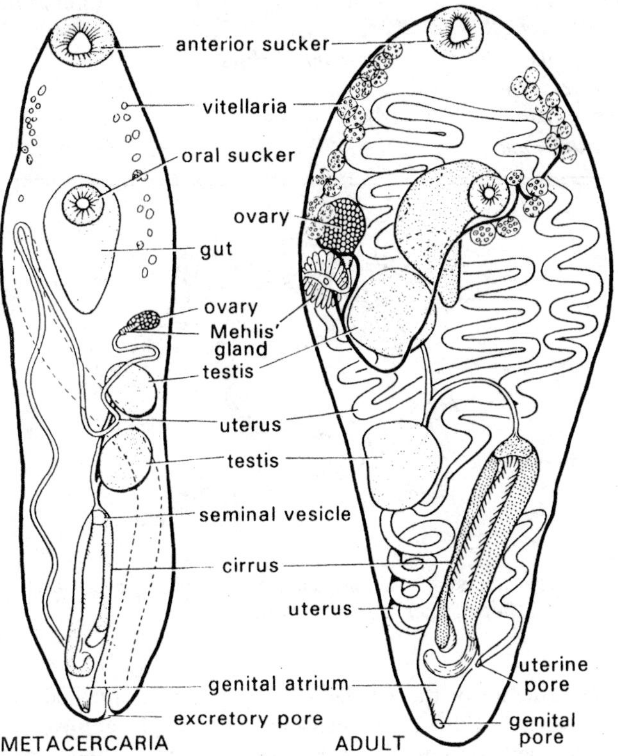

Fig. 14.1
Bucephaloides gracilescens. Progenetic metacercaria with mature spermatozoa in the testes; from brain of *Merlangius merlangus*. Adult from the gut of angler fish *Lophius piscatorius*. (From an original drawing by Dr D. I. Howie.)

Embryonation takes place in sea water. The miracidium which hatches is a grotesque organism with the cilia restricted to comb plates and protruding bars. Miracidia are presumably drawn into the molluscan intermediate host via the inhalant siphon and develop as in other species (Fig. 14.2) into branching sporocysts. The life cycle of *B. gracilescens*, which has often been confused with that of *Bucephalus haimeanus*, has been described by Matthews (1973, 1974).

Table 14.1. *Some representative species of bucephalids*

Species	Mollusc host	2nd Intermediate host	Definitive host	Location	Reference
Bucephaloides gracilescens	*Abra alba*	Ganoid fish	*Lophius piscatorius*	Europe	Matthews (1974)
Bucephalus haimeanus	*Cardium edule* (cockle) *Ostrea edulis* (oyster)	*Pomatoschistus microps*	*Morone labrax*	Europe	Matthews (1973)
Bucephalus cuculus (?)	*Crassostrea virginica*	?	?	USA	Cheng & Burton (1965)
Bucephalus polymorphus	*Dreissena polymorpha*	*Phoxinus phoxinus*	*Lucioperca lucioperca*	Europe	Wallet *et al.* (1985)
Bucephalus longicornutus	*Ostrea lutaria* (oyster)	?	?	New Zealand	Howell (1967)
Prosorhychus crucibulum	*Mytilus edulis*	?	*Conger conger*	Europe	Matthews (1973)

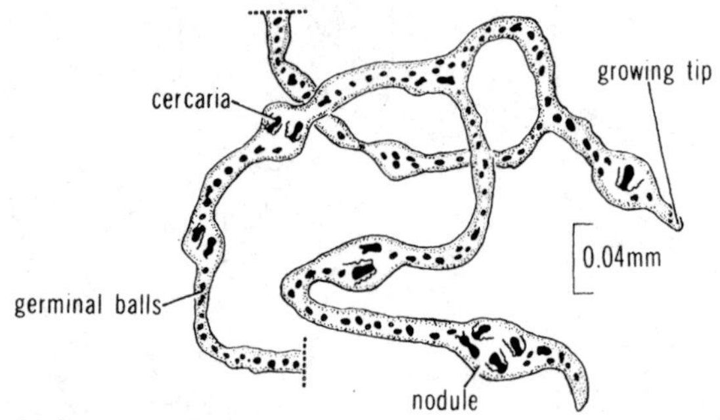

Fig. 14.2
Branching sporocyst of a bucephalid trematode (probably *Bucephalus haimeanus*). (After James & Bowers, 1967.)

The structure of the tubular sporocyst and the method of reproduction within it has been a matter of much controversy and has been variously described as (*a*) haploid parthenogenesis with germinal cells arising from somatic cells; (*b*) sexual reproduction; (*c*) germinal lineage with polyembryony; and (*d*) apomictic parthenogenesis (i.e. without meiosis). For consideration of these various possibilities see the review by James & Bowers (1967). Cytological studies have been made on various species of bucephalids by James & Bowers (1967), James *et al.* (1966), Cheng (1965) and Howell (1966).

The morphology of the cercaria (Fig. 14.3) has been described in detail by Matthews (1974) but the factors stimulating release of cercariae from the mollusc host have not been investigated. In a related species, *Bucephalus polymorphus* (Table 14.1), it has been shown experimentally that cercarial emergence exhibits a circadian rhythm of shedding with a peak in the dark period (Fig. 14.4) of a light:dark 12:12 h photoperiod (Wallet *et al.*, 1985).

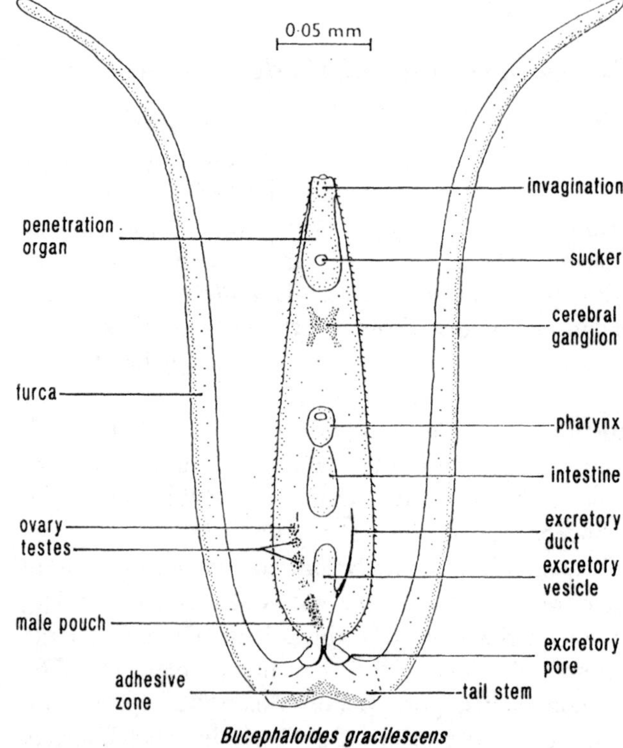

Fig. 14.3
Bucephaloides gracilescens, cercaria with furca contracted; ventral view. Note presence of anlagen of ovary and testes. (After Matthews, 1974.)

Metacercariae. The cercariae penetrate ganoid fish. Metacercariae have been reported from nine fish species (Matthews, 1974), showing a marked predilection to encyst in or on the nervous system. In Europe, Johnston

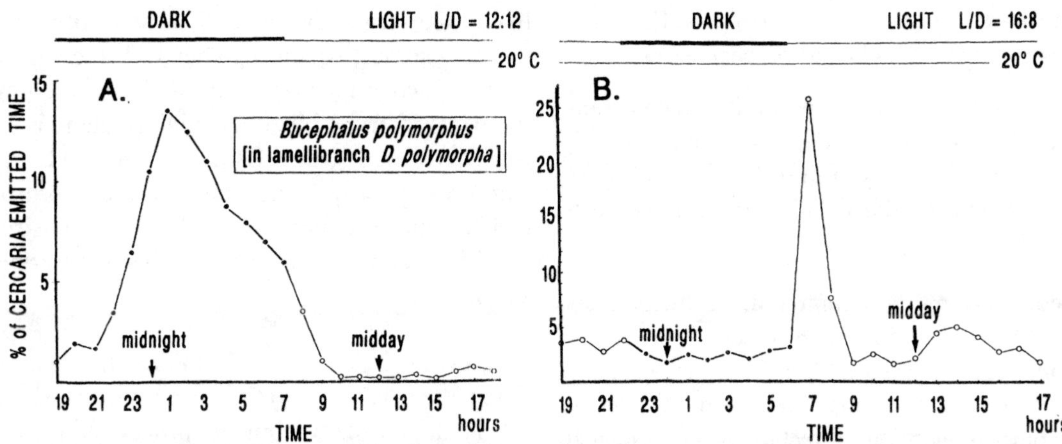

Fig. 14.4
Bucephalus polymorphus: emission of cercaria from the lamellibranch, *Dreissena polymorpha* during two experimental photoperiods: light : dark 12 : 12 h, with a peak during the dark period; and light : dark 16 : 8 h, with a peak at the beginning of the light period. (Modified from Wallet *et al.*, 1985.)

& Halton (1981*a*) reported prevalences in whiting (*Merlangius merlangus*), hake (*Merluccius merluccius*) and haddock (*Melanogrammus aeglefinus*) of 95, 100 and 100 per cent, respectively. Most cysts were found in the cranial fluid surrounding the brain, in the spinal nerves or in the auditory capsules of whiting and haddock, but in haddock they were also found in the spinal nerves posterior to the anus. The nutriment available in the brain cavity is probably low in protein but is sufficient for an advanced stage of morphological development to be achieved, and for partial progenesis (p. 310) to occur. In the most 'advanced' metacercariae, the major genitalia are differentiated (Fig. 14.1), *spermatozoa are present in the testes and seminal vesicles*, and the vitellaria contain shell precursors which give reactions for phenols. The nutritional level is thus sufficient to satisfy the demands of the gametogenesis level of organisation, but not sufficient for egg-production. On ingestion by *Lophius*, the metacercariae are carried to the stomach, duodenum or pyloric caeca, where the nutritive level is higher than that in the brain, and sufficient to overcome the nutritive barrier apparently separating it from maturation. *In vitro*, at 18 °C, maturation to the egg-producing stage required about twenty-one days. In later experiments, maturation at 14 days was achieved (Halton & Johnston, 1983).

Metacercarial cysts have an inner cyst wall of parasite material and an outer fibrous capsule of host origin (Halton & Johnston, 1982). Digestion in 0.5 per cent pepsin at pH 2.0 for 1 h removes the outer capsule and a second active phase is triggered by a pH rise to 7.2 in the presence of bile salts, excystment taking place within 5 min (Johnston & Halton, 1981*b*).

Other species. Many other species of bucephalids are known, but the taxonomy is in a somewhat confused state. In a major review of speciation, Stunkard (1974) concluded that 'lack of information on successive stages of particular species has led to confusion in systematic determinations'. Bucephalid metacercarial cysts can be found in freshwater and marine fish in most countries and often provide valuable material for teaching and research. Some representative species readily available are shown in Table 14.1. The species in oysters (*B. cuculus*?) in the USA can cause parasitic castration and degeneration of the ova in gravid females and changes in the carbohydrate and lipid composition of their hosts (Cheng & Burton, 1965). Oysters in other countries are also affected; for example, *B. longicornutus* causes parasitic castration in the mud-oyster, *Ostrea lutaria*, in New Zealand (Howell, 1967).

○○○ 14.2 Family Fasciolidae

○ 14.2.1 Genus *Fasciola*

Large distomes with broad, flat, muscular bodies. Most organs complicated by branching. Includes several species of economic importance.

Type example: ***Fasciola hepatica***

natural definitive hosts:	sheep, cattle and others
suitable laboratory host:	rabbit
habitat:	bile ducts
miracidium:	free-swimming

intermediate host: *Lymnaea truncatula* (Europe)
metacercariae: encysted on vegetation

The biology of this organism, which has been extensively studied, has been reviewed by Pantelouris (1965), Kendall (1965, 1970), Boray (1969), Dawes & Hughes (1962, 1970), Malone (1986), Smith (1981) and Wescott & Foreyt (1986).

Occurrence. Occurs most commonly in sheep, goats and cattle, although a wide range of other hosts has been reported; rabbits are readily infected experimentally. Man is not usually considered to be a host of *F. hepatica* but, in fact, this infection is not unusual in humans and infections have been reported in many countries including Europe and the USA (Facey & Marsden, 1960). Exceptionally, one human infection is reported to have lasted for 6 years! The eating of watercress appears to be a common source of human infection (Marsden & Warren, 1984; Jones *et al.*, 1977). The usual site of infection is the liver but in aberrant hosts (such as man) other sites, such as the lung, may be involved.

Morphology. The morphology of the species (Fig. 14.5) is too well known to warrant detailed description here. The tegument is well armed with backwardly directed spines, which together with the suckers serve as an effective mechanism to maintain the position of the parasite in the bile duct. Mehlis' gland is especially well developed, as is Laurer's canal. The egg-shell appears to be a quinone-tanned protein formed by the mechanism described on p. 184. The fine structure of *Fasciola* tegument has already been described (Fig. 13.2).

Life cycle. The life cycle follows the typical trematode pattern (Fig. 14.6). The eggs embryonate in about 14–17 days at 22 °C, but the process is much affected by moisture and other environmental factors (Rowcliffe & Ollerenshaw, 1960). Hatching only occurs on exposure to light, probably due to release of a 'hatching enzyme' (p. 207) which attacks the opercular cement. The life of a hatched miracidium is only about twenty-four hours. The most commonly infected snail in Europe is *Lymnaea truncatula* but in countries where *Fasciola* has been introduced in infected stock, such as North America, Canada and Australia and some Pacific Islands, a new parasitic relationship appears to have been established with indigenous snail populations (Boray, 1969). 'Physiological' strains of *F. hepatica*

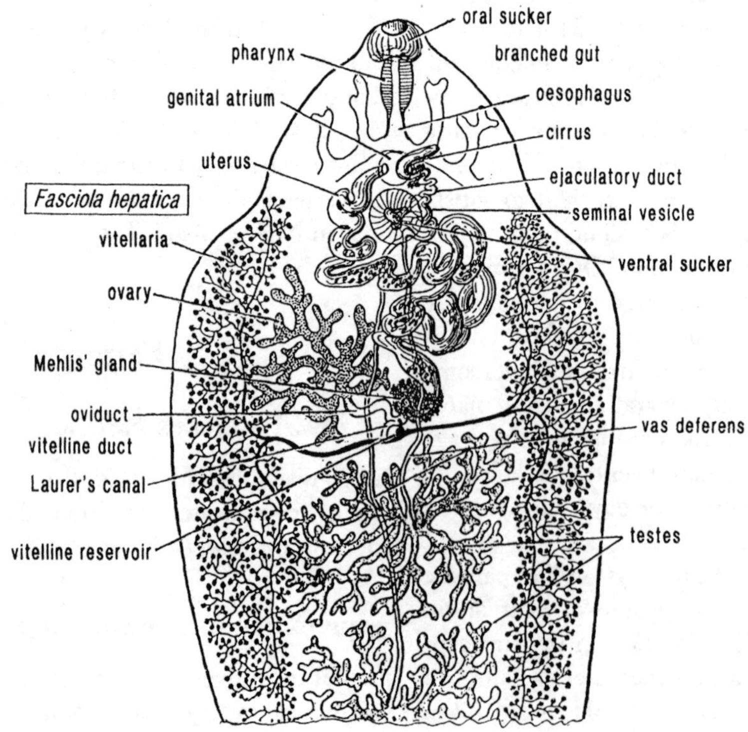

Fig. 14.5
Fasciola hepatica: morphology of adult fluke.

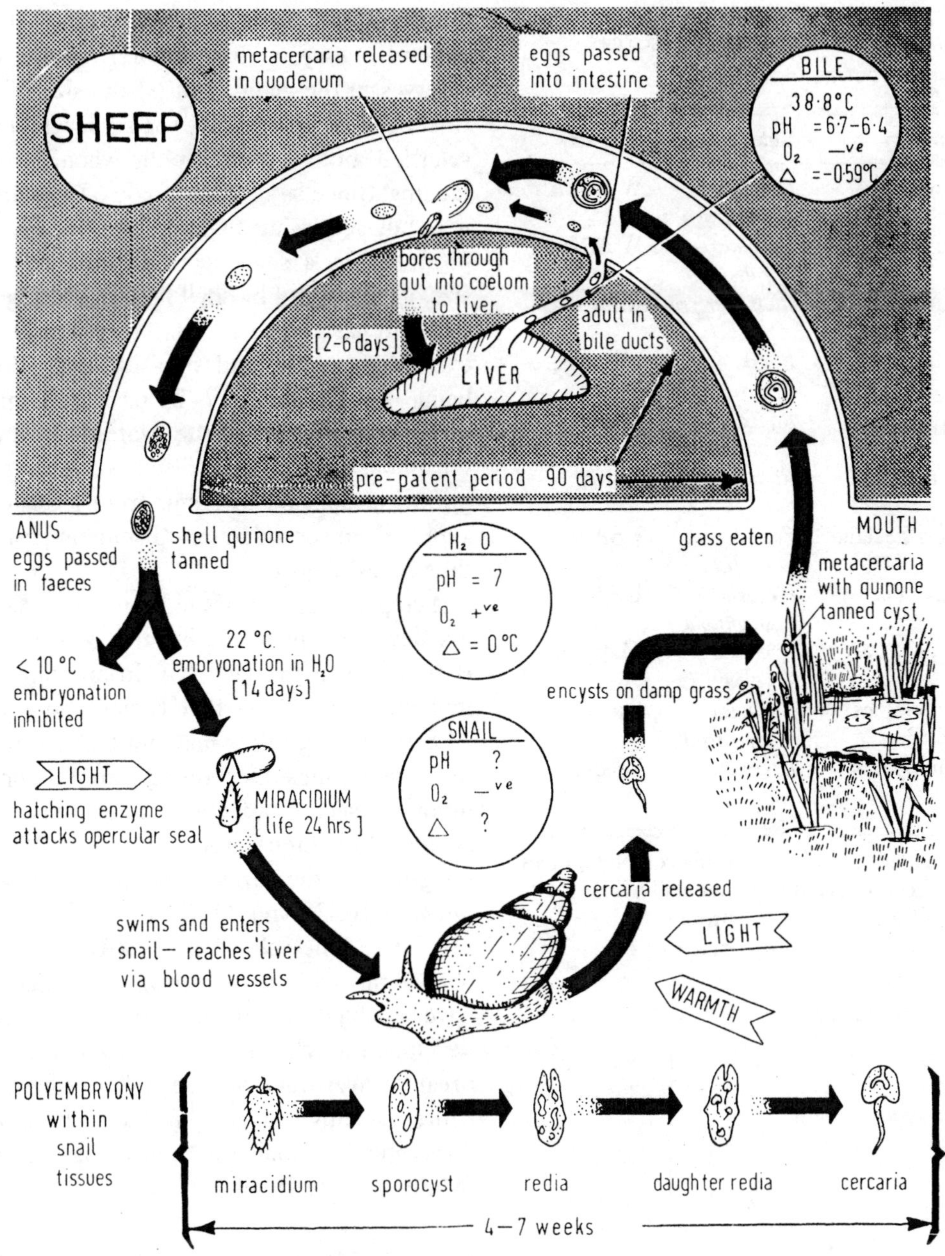

Fig. 14.6
Fasciola hepatica: life cycle and some of the physiological factors concerned in it. (Original.)

adapted to new molluscan hosts almost certainly exist (Kendall, 1965, 1970). Many species of *Lymnaea* are susceptible to some degree but only when young (Kendall, 1949; Boray, 1966). In some countries (e.g. China), species of *Radix* serve as intermediate hosts (Li *et al.*, 1988). Some vector molluscan species in different countries are listed in Table 14.2.

L. truncatula is a non-operculate pulmonate snail. It is essentially amphibious in habit but spends more time out of water than in; it is an inhabitant of temporary ponds and muddy places which may become dry for part of the year. It is thus admirably suited to transmit a parasite to grazing cattle. It has been shown experimentally that all stages – eggs, young and adults – can withstand desiccation for considerable periods, the adults for periods of a year or more, provided they are favourably covered with mud.

Development within the molluscan host. The delicate miracidium penetrates a snail assisted by the

Table 14.2. *Some intermediate hosts* of* Fasciola hepatica *in different countries*

Snail species	Country	Snail species	Country
Lymnaea truncatula	Europe	*Lymnaea natalensis*	W & E Africa
Lymnaea stagnalis	Europe	*Lymnaea rufescens*	W Africa
Lymnaea glabra	Europe	*Radix peregra*	Rumania
Lymnaea palustris	Europe	*Radix auricularia*	Rumania
Lymnaea viator	Argentina	*Radix cucunorica*	China
Lymnaea viatric	Peru	*Radix lagotis*	China
Lymnaea columella	New Zealand	*Fossaria modicella*	USA
Lymnaea tomentosa	Australia	*Fossaria stagnicola*	USA
Lymnaea cubensis	Venezuela	*Lymnaea rubiginosa*	Malaysia
Lymnaea philippensis	Philippines	*Lymnaea japonicum*	Japan
Lymnaea swinhoe	Philippines	*Lymnaea pervia*	Japan

*This list is representative but by no means complete; many other species have been infected experimentally. The taxonomy of some of the species is in dispute.

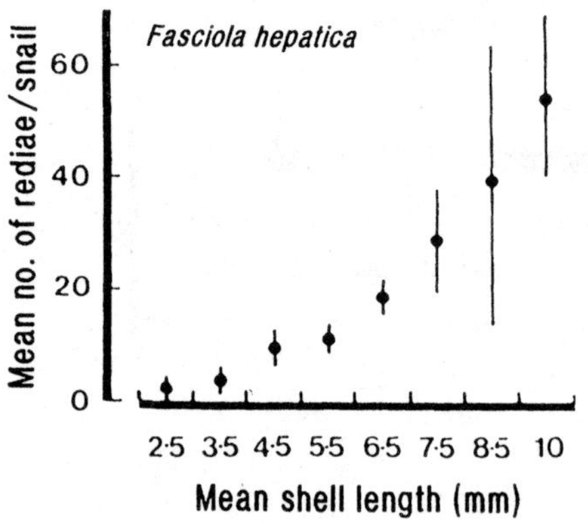

Fig. 14.7
Fasciola hepatica: relationship between size of naturally infected snails (*Lymnaea truncatula*) and intensity of redial infection. (After Smith, 1984.)

secretion of its histolytic glands and reaches the digestive gland via the lymph channels. The gland secreting the (putative) histolytic enzyme appears to be the apical gland (Fig. 13.14) and the associated lateral glands secrete a PAS-positive substance whose function is unknown (Buzzell, 1983). Sporocyst and redia generations develop. The rate and extent of development depend mainly on two factors, (*a*) the available food reserves in the gland, and (*b*) the size of the infection. In starved or hibernating snails, development is greatly retarded, but recovers rapidly when the snail resumes feeding. Since large snails possess large digestive glands it is not surprising to find that the number of rediae produced by a snail bears a direct relationship to its size, as measured by shell length; see Fig. 14.7 (Smith, 1984). It has also been shown that heavily infected snails exhibit gigantism and parasitic castration, the latter being brought about 17–21 days after infection by the direct consumption of the ovotestis by a proportion of the redial population (Wilson & Denison, 1980). The gigantism appears to result from a switch in nutrient supply from reproduction to somatic tissue growth and thus parasite growth.

Temperature appears to be the main factor influencing development of the larval stages in snails and this is reflected in the fact that, in any one year, the first infections in snails in the UK start about mid-May (Fig. 14.8B), although they may not be detected in the field until mid-June. Depending on environmental conditions, a second infection can also take place in mid-August. Laboratory studies (Nice & Wilson, 1974) on the growth of the intramolluscan stages in *Lymnaea truncatula* at 16, 20 and 24 °C have produced a computer-simulated timetable (Fig. 14.8A) very similar to the natural timetable (Fig. 14.8B). These ecological studies have enabled a valuable forecasting system to be developed which warns farmers when to expect an outbreak of liver fluke disease (Ollerenshaw, 1959, 1971). More recently, an accurate computerised system for forecasting annual prevalence has been developed (McIlroy *et al.*, 1990).

Cercaria production. A single miracidium can ultimately give rise to about 600 cercariae. Emergence commences some five to six weeks after the initial infection at average temperatures. The factors which stimulate cercaria production in *Fasciola* are not understood. Temperature has a threshold effect only. Below 9 °C emergence is entirely inhibited, and as the snails cannot survive for long at 26 °C, this is an upper limiting temperature for cercarial emergence also. Between these limits emergence is apparently unaffected by temperature. Factors such as light or darkness, pH and oxygen tension, have been shown experimentally to have no effect, and yet in the laboratory the act of transferring infected snails into fresh water invariably stimulates

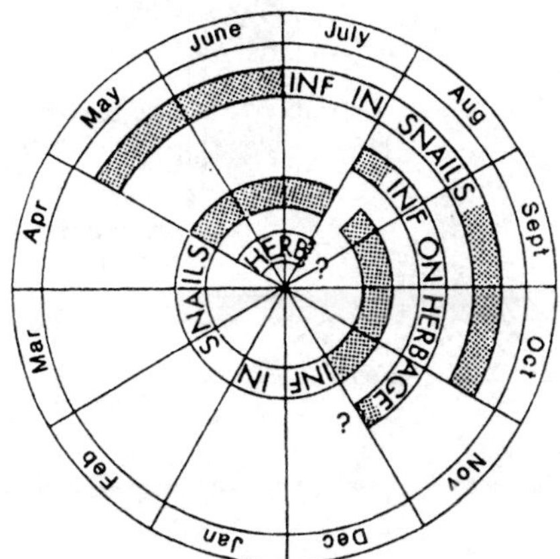

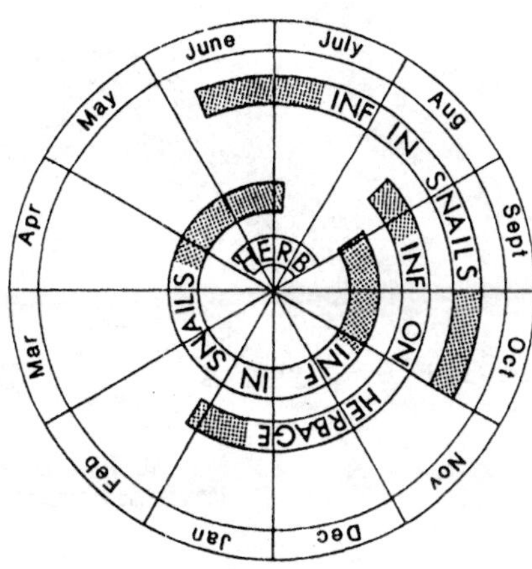

Fig. 14.8
Fasciola hepatica: seasonal time-tables showing the predicted occurrence of infections of larval stages (sporocyst/redia/cercaria) in the snail and of metacercarial cysts on the herbage, as occurring naturally (**B**: Ollerenshaw, 1959) or as simulated by computer from experimental data (**A**: Nice & Wilson, 1974). Although the first infections in snails occur first in May (in Europe, as the weather becomes warmer) they are rarely detected until mid-June. A second infection may also occur in mid-August/September depending on climatic conditions. The stippled areas show the presence of the parasite. The question marks (?) indicate that these periods may extend further depending on the survival of the metacercarial cysts on the herbage. (Modified from Nice & Wilson, 1974.)

Table 14.3. *The effect of various conditions on the emergence of cercariae of* F. hepatica

Fifty snails were used in each group.

Conditions	Number from which cercariae emerged
Snails undisturbed	1
Snail returned to same water after mechanical disturbance	2
Snails placed in water previously occupied by other snails	6
Snails placed in fresh pond water	27

Source: Data from Kendall & McCullough (1951).

cercaria production (Table 14.3). This stimulation is associated in some way with the activity of the host, but its basis is at present unknown. Rainfall has a similar stimulating effect in the field. In undisturbed experimental conditions, over a period of seven days, examination of cercarial shedding has revealed an infradian-type rhythm with a peak at 7 days. A circadial rhythm also occurs with a maximum production between midnight and 1 am (Audousset *et al.*, 1989).

The cercariae are gymnocephalous in type (p. 193) and have well-developed cystogenous glands. On emergence from the snail, a cercaria anchors itself by means of its oral sucker to a suitable substrate, such as grass, loses its tail and transforms into a *metacercaria*, which becomes enclosed in a metacercarial cyst secreted by its cystogenic glands.

Light and electron microscope studies have revealed that the cyst wall has a complex structure (Dixon, 1965; Dixon & Mercer, 1967; Mercer & Dixon, 1967). It consists (Fig. 14.9) essentially of four layers: Layer I, an external quinone-tanned protein, two layers (II and III) giving reactions of proteins and polysaccharides and an internal finely laminated layer (IV) of keratinised protein. The precursors of each of these layers are synthesised in a characteristic type of cystogenic gland in the cercaria.

Encysted metacercariae have considerable powers of resistance to low temperatures, being able to withstand the usual winter conditions in Europe and Australia, especially if infected pastures are covered by snow (Boray, 1969). Cysts are, however, very susceptible to desiccation.

Excystment. On ingestion by the definitive host, metacercariae excyst in the small intestine. Excystment in *Fasciola* is an *active* not a passive process, occurring in two stages: *activation* and *emergence* (Dixon, 1966; Smith & Clegg, 1981). Activation appears to be initiated by high concentrations of carbon dioxide and reducing conditions (i.e. a low E_h) and a temperature of about

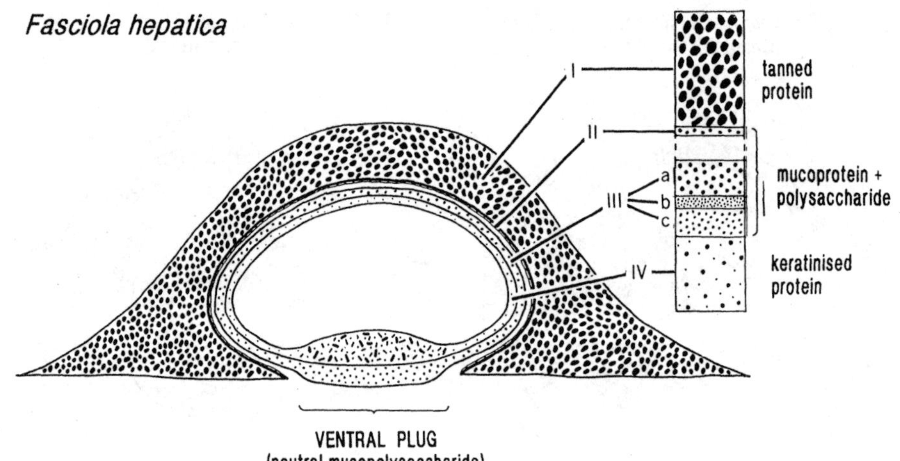

Fig. 14.9
Fasciola hepatica: structure of metacercarial cyst showing composition of the various layers. (After Dixon, 1965.)

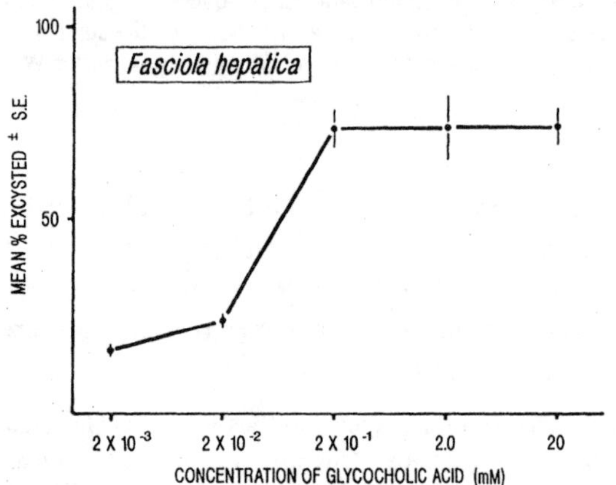

Fig. 14.10
Fasciola hepatica: excystment of metacercaria occurring with increasing (log) concentrations of glycocholic acid. A sharp threshold is seen to occur between 2×10^{-1} and 2×10^{-2} mM. (After Sukhedo & Mettrick, 1986.)

39 °C; the emergence phase is triggered by bile. The metacercaria excysts through a small circular hole in the ventral side of the cyst and it is possible that the organism releases a proteolytic enzyme, which digests the wall at this point. Although the release of enzymes has not been confirmed for *Fasciola*, evidence for the involvement of enzymes in the excystment process has been obtained for another species, the microphallid *Maritrema arenaria* (Irwin, 1983).

Bile probably plays a role of non-specific stimulation of muscular activity (Lackie, 1975). There is evidence (Sukhdeo & Mettrick, 1986) that glycine conjugates of cholic acid are more effective than the taurine conjugates, and glycocholic acid alone (Fig. 14.10) gives a 71.4 per cent emergence of metacercariae, very close to that of whole bovine bile (72.2 per cent; Table 14.4), which, from a herbivore, is known to contain largely glycine-conjugated bile salts (Smyth & Haslewood, 1963).

Table 14.4. *The effect of various bile salts on the emergence of the pre-activated metacercariae of* Fasciola hepatica in vitro

All concentrations were 10^{-3} M, except bile, which was 10 per cent.

Treatment	% Emergence (mean ± SE)
Control	0.0 ± 0.0[c]*
Deoxycholic acid	0.0 ± 0.0[c]
Taurochenodeoxycholic acid	0.0 ± 0.0[c]
Lithocholic acid	0.9 ± 0.7[c]
Taurodeoxycholic acid	1.9 ± 0.9[c]
Dehydrocholic acid	9.2 ± 1.5[c]
Glycochenodeoxycholic acid	45.2 ± 1.7[b]
Cholic acid	52.6 ± 6.2[ab]
Taurocholic acid	64.1 ± 10.1[ab]
Glycocholic acid	71.4 ± 3.23[a]
Bovine bile	72.2 ± 6.7[a]

*Values with different superscripts are significantly different ($p < 0.05$).
Source: Data from Sukhdeo & Mettrick (1986).

Experimental excystment of metacercarial cysts. For excystment of small numbers of metacercarial cysts, the following simple technique of Sewell & Purvis (1969) is useful: (*a*) place cysts in 0.05 M HCl at 39 °C in a screw-top vial; (*b*)

add an equal volume of 1% $NaHCO_3$, 0.8% NaCl and 20% bile (sheep, dog or pig); 1.2% crude (commercial) tauroglycocholate can be used instead of bile. The glycine-conjugated salts listed in Table 14.4 should also be effective. Metacercariae will excyst within 4–5 hours. A large-scale procedure for hatching metacercarial cysts (for biochemical work) is also available (Tielens *et al.*, 1981).

Development within definitive host. Development in the experimental mouse host has been studied by Dawes (1961*a,b*) and Bennett (1975). An excysted metacercaria burrows through the intestinal wall and reaches the abdominal cavity within 24 hours. After random wanderings, the liver is reached and penetrated, burrows being formed. The penetrating flukes eventually reach the bile ducts where they become permanently established. In sheep, the flukes reach the bile ducts in approximately six weeks; the histopathology in this host has been described by Dow *et al.* (1968).

It is of special interest to note that during its passage through the liver some liver enzymes such as glutamate-pyruvate-transaminase (GPT) or glutamic oxalo-acetic transaminase (GOT) are released into the bloodstream. Thus by monitoring the concentrations of enzyme in serum (Fig. 14.11) the time when an organism invades the liver and eventually reaches the bile duct (after which the tissue damage commences to recover) can be followed (Bundesen & Janssens, 1971; Campbell & Barry, 1970). The use of biochemical tracers in infections of this kind is likely to have considerable value in *in vivo* studies of this nature.

Nutrition. The nature of the food of *Fasciola* has long been in dispute, the main point of disagreement being whether or not the organism feeds on blood. Although there is evidence (Dawes, 1961*a,b*) that young flukes feed on tissue, it is difficult to believe that some blood is not inevitably ingested. This question has been reviewed by Jennings (1968), and is discussed further on pp. 265, 269.

F. gigantica

A trematode resembling *F. hepatica* but larger in size (2.5–7.5 cm × 1.2 cm). It is a common liver fluke of domestic and wild ruminants in many parts of Africa and the orient. The life cycle resembles that of *F. hepatica*. The intermediate snail hosts belong to the *Lymnaea auricularia* complex; they are true aquatics and hence are generally different from the snail hosts of *F. hepatica* (Kendall, 1965). *Lymnaea natalensis* is a convenient laboratory host. The morphology of the larval stages and the epidemiology of the parasite in Sierra Leone have been discussed by Asanji (1988); Fabiyi (1987) has reviewed the control and production losses caused by this parasite in the tropics.

○ 14.2.2 Genus *Fascioloides*

Fascioloides magna

Occurs in the liver (and occasionally lung) of deer in N. America, but can also infect horses, sheep, elk, yak, goat, buffalo, caribou, moose and cattle. In Canada, prevalence in caribou can reach 58 per cent (Lankester & Luttich, 1988).

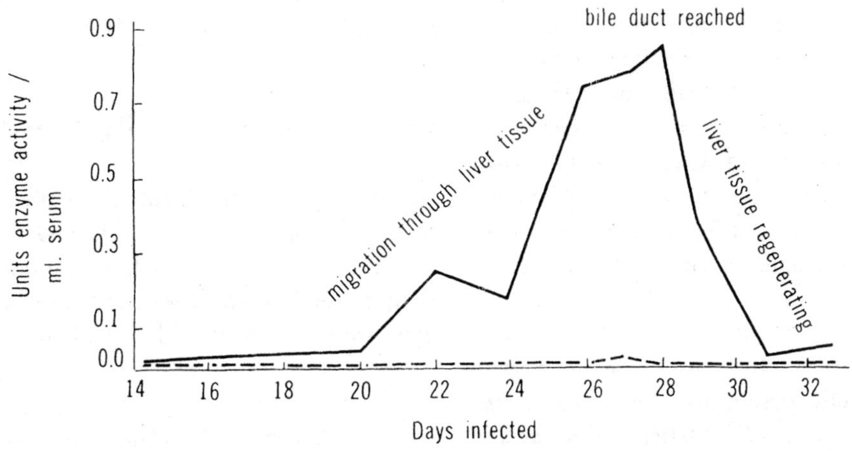

Fig. 14.11
Use of host enzyme activity in tracing tissue migration of *Fasciola hepatica*. Damaged liver tissue releases the enzyme GPT (glutamate-pyruvate-transaminase) and the activity of this increases in the host serum as the juvenile flukes burrow through the liver 'en route' for the bile ducts. Once the bile ducts are reached and entered (about 28 days) the liver tissue damage begins to heal and GPT rapidly falls. (After Bundesen & Janssens, 1971.)

Table 14.5. *Prevalence of the giant intestinal fluke,* Fasciolopsis buski, *in the human and pig population of six Indian villages*

Village	Human		Pigs	
	No. examined	Positive (%)	No. examined	Positive (%)
Shivsingh Khera	171	17	44	18.2
Bhabani Khera	45	15.5	0	0
Bhabuti Khera	48	16.6	0	0
Umed Khera	148	41.8	92	43.0
Saraiyan	302	22.1	55	25.4
Khusal Khera	94	8.5	40	15.0
Total	808	22.4	231	29.4

Source: Data from Chandra (1984).

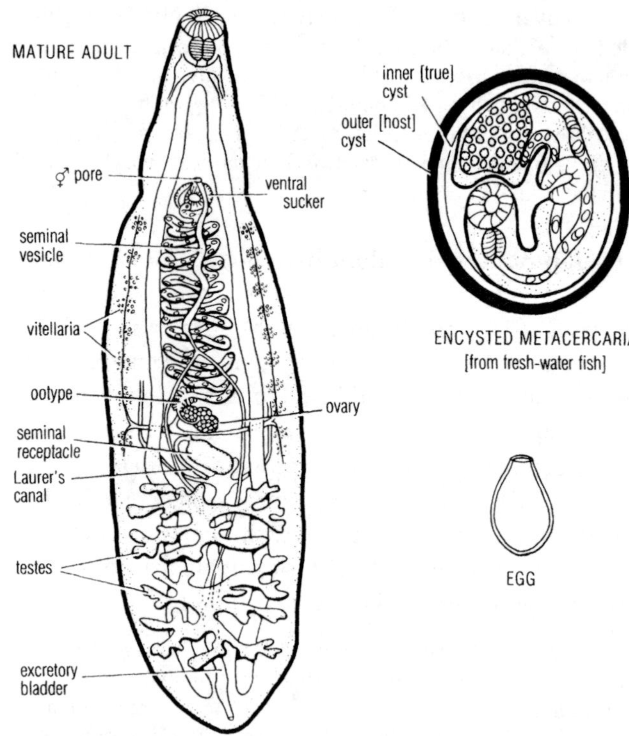

Fig. 14.12
Clonorchis sinensis: morphology of adult, metacercaria and egg. (After Brown, 1950; Komiya & Tajimi, 1940.)

In cattle, the flukes become entirely encapsulated before reaching maturity so that eggs are never released. The pathogenicity of these hosts is low. In deer, the capsule is developed later and eggs may escape. In sheep, the liver may become severely affected by migration and the presence of two or three of these large flukes may prove fatal. The parasites contain a characteristic black pigment related to melanin, presumably derived from haemoglobin, but whose composition is unknown. The life cycle is similar to that of *F. hepatica.* Some intermediate hosts are: *Fossaria parva*, *F. modicella*, *F. modicella rustica*, *Lymnaea bulimoides*, *L. columella* and *Stagnicola palustris nuttalliana.*

Fasciolopsis buski

A duodenal parasite occurring in man and pigs and referred to as the *giant intestinal fluke.* Its general morphology is similar to that of *Fasciola* but it lacks the thickened anterior cone and its intestinal caeca are unbranched. It is widespread in Asia, occurring mostly in China, but it is also endemic in Taiwan, Thailand, Vietnam, Laos, Bangladesh and India (Chandra, 1984). Its general biology and diagnosis have been reviewed by Hillyer (1988).

Life cycle. Resembles that of *F. hepatica* with members of the family Planorbidae, e.g. *Segmentina* (*Polypylis*) *hemisphaerula*, *S.* (*Trochorbis*) *trochoideus* and *Hippeutis contori*, as intermediate hosts. These snails feed on plants such as the water caltrop, *Trapa natans* and *T. bicornis*, the water chestnut *Eleocharis tuberosa* and other species, much cultivated for food and fertilised with human night soil. The nuts or bulbs of these plants are eaten raw, the nuts of the caltrops being peeled by the teeth so that heavy, and sometimes lethal, infections may occur. Pigs act as reservoir hosts, as is clearly shown in a recent survey (Table 14.5) of villages in India where the prevalence was 22.4 per cent in man and 29.4 per cent in pigs (Chandra, 1984).

○○○ 14.3 Family Opisthorchiidae

All members of this family are parasites of fish-eating mammals, particularly in Asia and Europe. They are unusually transparent in whole-mount preparations and their general anatomy is readily seen (Fig. 14.12). The eggs are embryonated when laid and contain miracidia.

Clonorchis sinensis

An important parasite of man and fish-eating mammals of the Far East, mainly Japan, Korea, China, Taiwan and Vietnam. The main non-human hosts are cats, pigs, rats, camels and dogs. Cats, rabbits and guinea pigs

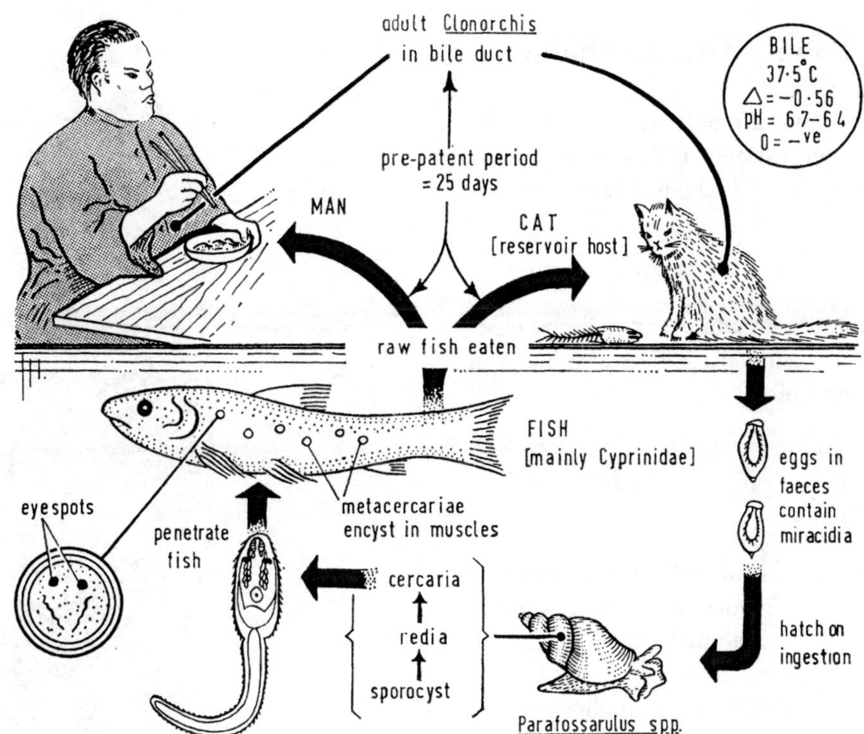

Fig. 14.13
Clonorchis sinensis: life cycle and some physiological factors relating to it. (Original.)

are useful experimental hosts. A comprehensive general account of this organism and the infection it produces, *clonorchiasis*, has been given by Komiya (1966) and Rim (1988). Kim & Kuntz (1964) have reviewed the epidemiology of this parasite in Taiwan.

Life cycle (Fig. 14.13). The adults live in the bile ducts, often occurring in vast numbers. The food consists of epithelial cells, leucocytes and other disintegrating cells. The flask-shaped, operculate eggs (like those of *Dicrocoelium*, p. 214) only hatch on ingestion by snails of which the following act as intermediate host: *Parafossarulus manchouricus*, *Bulimus fuchsianus* (= *Bithynia fuchsiana*) and *Alocima longicornis* (= *B. longicornis*). Sporocyst and redia generations occur. The cercariae are exceptional in possessing eye spots; penetration and cystogenous glands are also well developed. The cercariae penetrate the skin of fishes of the family Cyprinidae, in whose muscles they eventually settle, lose their tails and encyst. Infection is by eating uncooked, commonly smoked, fish and the excysted metacercariae make their way to the liver, not via the coelom as in *Fasciola*, but by direct migration up the bile duct.

Opisthorchis spp.

Opisthorchis felinus and *O. viverrini* are two other species which resemble *Clonochis sinensis* and have a similar life cycle, but differ from it in the shape of the testes and other features. The clinical picture of biliary tract infection which results from *Clonorchis* is similar to that with *Opistorchis*, and it is impossible to differentiate clinically between the diseases caused by these three species (Bunnag & Harinasuta, 1984). The snail intermediate hosts for *O. felinus* are *Bithynia leechi*, and for *O. viverrini*, *B. goniomphales*, *B. funiculata* and *B. leavis* (*B. siamensis*). The life cycles and morphological differencs between *O. felinus* and *O. viverrini* have been reviewed by Wykoff *et al.* (1965).

In highly endemic areas, the prevalence of *O. viverrini* in humans may reach more than 90 per cent (Brockelman *et al.*, 1987) and the fish intermediate host prevalence up to 97 per cent (Tesana *et al.*, 1985). In contrast, the prevalence in snails is remarkably low, only 0.11 per cent, suggesting that the host finding and penetration mechanisms of the cercaria, which have been investigated experimentally by Haas *et al.* (1990), are very efficient.

○○○ 14.4 Family Dicrocoeliidae

The members of this family are delicate, elongate, flattened, rather translucent distomes with suckers not far apart in the anterior region of the body. Flame-cell formula of adult:

$$2[(2 + 2 + 2) + (2 + 2 + 2)] = 24$$

The cercariae are xiphidiocercariae. Although the best-known species occur in mammals, species also occur in amphibians, reptiles and birds.

Type example: ***Dicrocoelium dendriticum*** (= the lancet fluke)

definitive hosts:	sheep, cattle, deer, woodchuck, rabbit and other mammals
location:	bile ducts
miracidium:	freed only in molluscan hosts
first intermediate host:	*Cionella lubrica* (USA), *Planorbis marginatus*, *P. complanatus*, *Arion* and *Limax* sp. (Europe)
second intermediate host:	the ant *Formica fusca* (USA)

Occurrence. The lancet fluke is restricted in its distribution largely to Europe and Asia and only sparsely distributed in Africa, the New World and Australasia.

Technique for removal. The organisms may be obtained by liver dissection, but a much more satisfactory method is to force liquid into the liver under pressure and thus force out the flukes.

Morphology. The internal organs are widely spaced, and this, together with the transparency, makes the morphology especially easy to study in whole mounts. The anatomy, as shown in Fig. 14.14, is simple and will not be described here. The eggs are operculate and the egg shell probably a quinone-tanned protein.

Life cycle (Fig. 14.14). Some major problems in the life cycle and mode of transmission remained unsolved until the meticulous work of Krull & Mapes (1951–56) led to their elucidation.

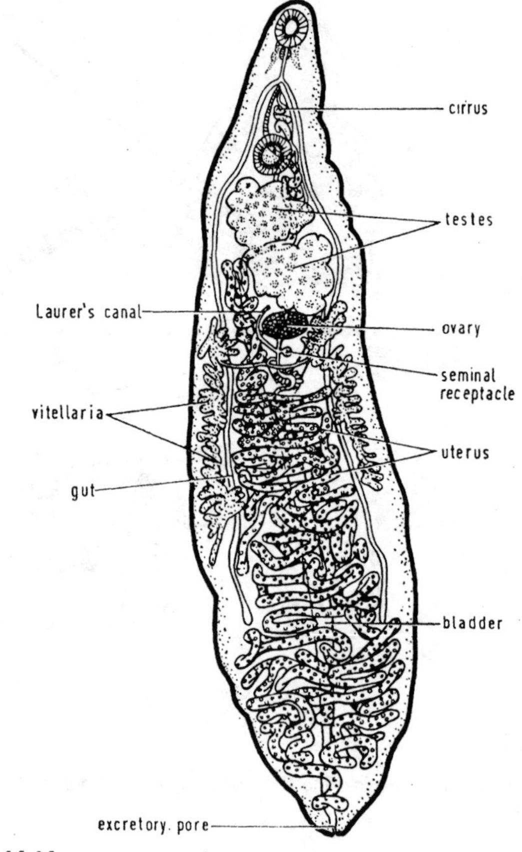

Fig. 14.14
Dicrocoelium dendriticum: morphology of adult. (After Neuhaus, 1938.)

It was formerly thought that mammals became infected by ingesting 'slimeballs', peculiar slimy secretions produced by terrestrial snails and known to contain cercariae. It was suspected by some that a second intermediate host was necessary to complete the life cycle. In 1952 Krull & Mapes showed that a second intermediate host, an ant, is involved, and infection of the definitive host is brought about only by ingestion of infected ants, and cannot take place by direct ingestion of slimeballs.

The operculate eggs passed in the faces are embryonated when laid, but do not hatch on exposure to light, as do the majority of trematode eggs. Hatching only takes place when the eggs are ingested by the appropriate molluscan host. This process has been extensively studied by Ractliffe (1968).

Intra-molluscan development. The intermediate host in the United States is the snail *Cionella lubrica*. In addition to those mollusc hosts mentioned above, other snail hosts involved are: *Helicella candidula*, *H. itala*,

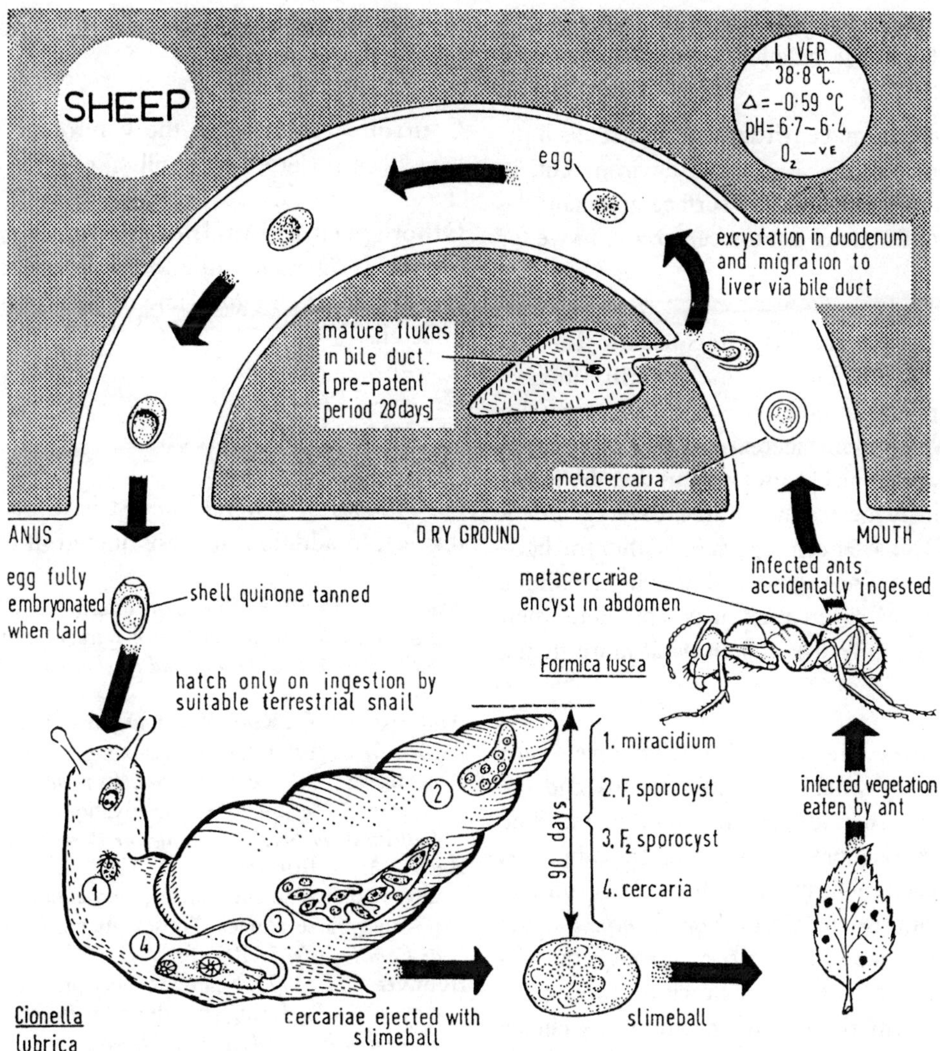

Fig. 14.15
Dicrocoelium dendriticum: life cycle and some physiological factors relating to it. (Based on the work of Krull & Mapes (original).)

H. ericetorum, *Zebrina detrita*, *Abida rumentum*, *Ena obscura*, *Euomphalia strigella* and *Theba carthusiana*. In France, *Cochlicella acuta*, *C. ventricosa* and *Conchicopa lubrica* act as hosts (Badie & Rondelaud, 1987) with a prevalence of 0.5–12 per cent. A survey of snails in Spain revealed a prevalence of 5.68 per cent (Manga-Gonzalez, 1987). Hatching takes place in the snail's gut. The miracidia make their way to a site near the albumen gland, between the liver follicles or along the intestine. Mother sporocysts and daughter sporocysts are produced, the latter giving rise to xiphidiocercariae, long known as *Cercaria vitrina*. Development from miracidium to cercaria takes approximately four to five months at average temperatures, but the rate of development may be influenced by other factors, such as season or age of the snail. Cercariae escape from the snail, and are collected in slime masses known as *slimeballs*, the formation of which is not fully understood.

The cercariae contain massive glands, formerly considered to be penetration glands, but now believed to produce the slime responsible for the formation of these balls. It is likely that the slime produced by the snail itself may contribute, at least partly, to the formation of a slimeball. Slimeballs can be released in at least two ways, either by expulsion of a fully formed slimeball from the respiratory pore, or by a method involving the use of the surface slime of the snail in which the cercariae, resulting from the sudden exodus from the snail (perhaps induced by a drop in temperature) are incorporated. The balls are sticky in consistency and adhere to vegetation and other debris. Usually only one slimeball is produced by a snail, but up to five may be produced; each ball may contain 100–400 cercariae.

In an excessively humid environment, slimeballs liquefy, and the cercariae die. In an arid environment, in which the snails normally live, the surface dries and the balls shrink to some extent, but the cercariae survive for a period of time.

Development in the second intermediate host. On ingestion by the ant (*Formica fusca* in North America, and other ant species in other countries) the cercariae become transformed to metacercariae in the body cavity of the abdomen, presumably after penetrating the intestine of the ant; thirty metacercariae are an average infection, but up to 218 have been reported. Within the body cavity the metacercariae grow, although remaining semi-transparent. Metacercariae may complete their growth and become infective in somewhat more than a month.

Natural definitive hosts. The definitive host becomes infected by swallowing infected ants and the metacercariae are encysted in the duodenum. The released metacercariae make their way up the bile ducts and the flukes reach maturity in the liver. This route to the liver is in contrast to that of *Fasciola hepatica*, the excysted metacercariae of which bore through the alimentary canal and reach the coelom within two hours.

Dicrocoelium is remarkably non-specific in its choice of hosts and infections from some fifty different hosts have been reported. In the United States, at least, deer, woodchuck and the cottontail rabbit serve as definitive reservoirs for *Dicrocoelium*.

Laboratory hosts. In the laboratory, *Dicrocoelium dendriticum* is most readily maintained in the golden hamster or the rabbit, but the development can also occur in the guinea pig and albino mouse, although the last two are poor hosts. The albino rat is refractive to infection. For experimental purposes, the hamster is the host of choice for not only does the parasite reach a fair size, but in this host its distribution is confined to main bile ducts at the surface of the liver, and in the gall bladder.

Physiology. Little is known regarding the nutrition and general physiology of *D. dendriticum*. Haemoglobin has been detected in its tissue, so that it can be assumed that, as in *Fasciola*, blood actually forms part of the ingested food material.

Control. Problems of the control of *D. dendriticum* have been reviewed by Guilhon (1987).

Other species. A related species, *Dicrocoelium hospes*, occurs in Nigeria in cattle with a prevalence of 20.48 per cent (Ogbe, 1984); goats and sheep are also occasionally infected.

○○○ References

(References of general interest have been included in this list, in addition to those quoted in the text.)

Asanji, M. F. 1988. The snail intermediate host of *Fasciola gigantica* and the behaviour of miracidia in host selection. *Bulletin of Animal Health and Production in Africa*, **36**: 245–50.

Audousset, J. C., Rondelaud, D., Dreyfuss, G. & Vareille-Morel, C. 1989. Les emissions cercariennes de *Fasciola hepatica* L. chez le mollusque *Lymnaea truncatula* Müller. A propos de quelques observations chronobiologiques. *Bulletin de la Société Française de Parasitologie*, **7**: 217–28.

Badie, A. & Rondelaud, D. 1987. Les mollusques hôtes intermédiares de *Dicrocoelium lanceolatum* Rudolphi. A propos de quinze années d'observations. *Bulletin de la Société de Parasitologie*, **5**: 105–8.

Bennett, C. E. 1975. Scanning electron microscopy of *Fasciola hepatica* L. during growth and maturation in the mouse. *Journal of Parasitology*, **61**: 892–8.

Boray, J. C. 1966. Studies on the relative susceptibility of some lymmnaeids to infection with *Fasciola hepatica* spp. *Annals of Tropical Medicine and Parasitology*, **60**: 114–24.

Boray, J. C. 1969. Experimental fascioliasis in Australia. *Advances in Parasitology*, **7**: 95–210.

Brockelman, W. Y., Upatham, E. S., Viyanant, V. & Hirunkraks, A. 1987. Measurement of incidence of the human liver fluke, *Opisthorchis viverrini* in northeast Thailand. *Transactions of the Royal Society of Tropical Medicine and Hygiene*, **81**: 327–35.

Bundesen, P. G. & Janssens, P. A. 1971. Biochemical tracing of parasitic infections. I. *Fasciola hepatica* L. in mice – a qualitative study. *International Journal for Parasitology*, **1**: 7–14.

Bunnag, D. & Harinasuta, T. 1984. Opisthorchiasis, Clonorchiasis, and Paragonimiasis. In: *Tropical and Geographical Medicine* (ed. K. S. Warren & A. A. F. M. Mahmoud), pp. 461–70. McGraw-Hill, New York.

Buzzell, G. R. 1983. Composition, secretion and fate of the glands in the miracidia and sporocysts of *Fasciola hepatica*. *Journal of Helminthology*, **57**: 79–84.

Campbell, W. C. & Barry, T. A. 1970. A biochemical

method for the detection of anthelminthic activity against liver fluke (*Fasciola hepatica*). *Journal of Parasitology*, **56**: 325–31.

Chandra, S. S. 1984. Epidemiology of *Fasciolopsis buski* in Uttar Pradesh. *Indian Journal of Medical Research*, **79**: 55–9.

Cheng, T. C. 1965. On the structure and alteration of cellular polarity of the sporocysts of *Bucephalus* sp. as a function of age. *Proceedings of the Pennsylvania Academy of Science*, **39**: 180–6.

Cheng, T. C. & Burton, R. S. 1965. Relationships between *Bucephalus* sp. and *Crassostrea virginica*: histopathology and sites of infection. *Chesapeake Science*, **6**: 3–16.

Dawes, B. 1961*a*. The juvenile stages of *Fasciola hepatica* during the early penetration into the liver of the mouse. *Transactions of the Royal Society of Tropical Medicine and Hygiene*, **55**: 310–11.

Dawes, B. 1961*b*. On the early stages of *Fasciola hepatica* penetrating into the liver of an experimental host, the mouse: a histological picture. *Journal of Helminthology, R. T. Leiper Supplement*, 41–52.

Dawes, B. & Hughes, D. L. 1962, 1970. Fascioliasis: the invasive stages of *Fasciola hepatica* in mammalian hosts. *Advances in Parasitology*, **2**: 97–168; **8**: 259–74.

Dixon, K. E. 1965. The structure and histochemistry of the cyst wall of the metacercariae of *Fasciola hepatica* L. *Parasitology*, **55**: 215–26.

Dixon, K. E. 1966. The physiology of excystment of the metacercariae of *Fasciola hepatica* L. *Parasitology*, **56**: 431–56.

Dixon, K. E. & Mercer, E. H. 1967. The formation of the cyst wall of the metacercaria of *Fasciola hepatica* L. *Zeitschrift für Zellforschung*, **77**: 345–60.

Dow, C., Ross, J. G. & Todd, J. R. 1968. The histopathology of *Fasciola hepatica. Parasitology*, **58**: 129–35.

Fabiyi, J. P. 1987. Production losses and control of helminths in ruminants of tropical regions. *International Journal for Parasitology*, **17**: 435–42.

Facey, R. V. & Marsden, P. D. 1960. Fascioliasis in man: an outbreak in Hampshire. *British Medical Journal*, **2**: 619–25.

Guilhon, J. 1987. Histoire de la lutte contre la petite douve (1936–1986). *Bulletin de l'Académie Vétérinaire de France*, **60**: 187–97.

Haas, W., Granzer, M. & Brockelman, C. P. 1990. *Opisthorchis viverrini*: finding and recognition of the fish host by the cercariae. *Experimental Parasitology*, **71**: 422–31.

Halton, D. W. & Johnston, B. R. 1982. Functional morphology of the metacercarial cyst of *Bucephaloides gracilescens* (Trematoda: Bucephalidae). *Parasitology*, **85**: 45–52.

Halton, D. W. & Johnston, B. R. 1983. Development *in vitro* of the metacercaria of *Bucephaloides gracilescens* (Trematoda: Bucephalidae). *International Journal for Parasitology*, **13**: 157–64.

Hillyer, G. V. 1988. Fascioliasis and fasciolopsiasis. In: *Laboratory Diagnosis of Infectious Diseases*, Vol. 1 (ed. A. Balows *et al.*), pp. 856–62. Springer-Verlag, New York.

Howell, M. 1966. A contribution to the life history of *Bucephalus longicornatus* (Manter, 1954). *Zoology Publications from Victoria University of Wellington*, **40**: 1–42.

Howell, M. 1967. The trematode, *Bucephalus longicornatus* (Manter, 1954) in the New Zealand Mud-Oyster, *Ostrea lutaria. Transactions of the Royal Society of New Zealand*, **8**: 221–37.

Irwin, S. W. B. 1983. *In vitro* excystment of the metacercariae of *Maritrema arenaria* (Digenea: Microphallidae). *International Journal for Parasitology*, **13**: 191–6.

James, B. L. & Bowers, E. A. 1967. Reproduction in the daughter sporocyst of *Cercaria bucephalopsis haimeana dichotoma* (Lacaze-Duthiers, 1854) (Bucephalidae) and *Cercaria dichotoma* Lebour, 1911 (Von Müller) (Gymnophallidae). *Parasitology*, **57**: 607–25.

James, B. L., Bowers, E. A. & Richards, J. G. 1966. The ultrastructure of the daughter sporocyst of *Cercaria bucephalopsis haimeana* Lacaze-Duthiers, 1854 (Digenea: Bucephalidae) from the edible cockle, *Cardium edule* L. *Parasitology*, **56**: 753–62.

Jennings, J. B. 1968. Nutrition and digestion. In: *Chemical Zoology* (ed. M. Florkin & B. T. Scheer), Vol. II, Chapter 2, pp. 303–26. Academic Press, New York.

Johnston, B. R. & Halton, W. 1981*a*. Occurrence of *Bucephaloides gracilescens* metacercariae in three species of ganoid fish. *Journal of Fish Biology*, **18**: 685–91.

Johnston, B. R. & Halton, D. W. 1981*b*. Excystation *in vitro* of *Bucephaloides gracilescens* metacercaria (Trematoda: Bucephalidae). *Zeitschrift für Parasitologie*, **65**: 71–8.

Jones, A. E., Kay, J. M., Milligan, H. P. & Owens, D. 1977. Massive infection with *Fasciola hepatica* in man. *American Journal of Medicine*, **63**: 836–42.

Kendall, S. B. 1949. Nutritional factors affecting the rate of development of *Fasciola* in *Limnaea truncatula. Journal of Helminthology*, **23**: 179–90.

Kendall, S. B. 1965, 1970. Relationship between the species of *Fasciola* and their molluscan hosts. *Advances in Parasitology*, **3**: 59–98; **8**: 251–8.

Kendall, S. B. & McCullough, F. S. 1951. The emergence of the cercariae of *Fasciola hepatica* from the snail *Limnaea truncatula. Journal of Helminthology*, **24**: 77–92.

Kim, D. C. & Kuntz, R. E. 1964. Epidemiology of helminth diseases: *Clonorchis sinensis* (Cobbald, 1875) Looss, 1907 on Taiwan (Formosa). *Chinese Medical Journal*, Peking, **11**: 29–47.

Komiya, Y. 1966. *Clonorchis* and clonorchiasis. *Advances in Parasitology*, **4**: 53–106.

Krull, W. H. & Mapes, C. R. 1951–56. Studies on the biology of *Dicrocoelium dendriticum* (Rudolphi, 1819, Looss, 1899) Trematoda: including its relation to its intermediate host, *Cionella lubrica* (Müller). I–X. *The Cornell Veterinarian*, Vols **41–45**.

Lackie, A. M. 1975. The activation of infective stages of endoparasites of vertebrates. *Biological Reviews*, **50**: 285–323.

Lankester, M. W. & Luttich, S. 1988. *Fascioloides magna* (Trematoda) in woodland caribou (*Rangifer tarandus caribou*) of the George River Herd, Labrador. *Canadian Journal of Zoology*, **66**: 475–9.

Li, G. Q., Jin, J. S., Cai, X. P. & Duan, Z. Q. 1988. (A survey of the species of intermediate hosts of *Fasciola hepatica* in the rangeland in South Ganse.) (In Chinese.) *Chinese Journal of Veterinary Science & Technology*, **9**: 21–5. (*Helminthology Abstracts*, **58**: No. 2458.)

McIlroy, S. G., Goodall, E. A., Stewart, D. A., Taylor, S. M. & McCracken, M. 1990. A computerised system for the accurate forecasting of the annual prevalence of fasciolosis. *Preventative Veterinary Medicine*, **9**: 27–35.

Malone, J. B. 1986. Fascioliasis and cestodiasis in cattle. *Veterinary Clinics of North America, Food Animal Practice*, **2**: 261–75.

Manga-Gonzalez, M. Y. 1987. Some aspects of the biology and helminthofauna of *Helicella* (*Helicella*) *italia* (Linnaeus, 1758) (Mollusca). Natural infection by Dicrocoeliidae (Trematoda). *Revista Ibérica de Parasitologie*, Vol. Extraordinario, 131–48.

Marsden, P. D. & Warren, K. S. 1984. Fascioliasis. In: *Tropical and Geographical Medicine* (ed. K. S. Warren & A. A. F. Mahmoud), pp. 458–60. McGraw-Hill, New York.

Matthews, R. A. 1973. The life-cycle of *Bucephalus haimeanus* Lacaze-Duthiers, 1854 from *Cardium edule* L. *Parasitology*, **67**: 341–50.

Matthews, R. A. 1974. The life-cycle of *Bucephaloides gracilescens* (Rudolphi, 1819) Hopkins, 1954 (Digenea: Gasterostomata). *Parasitology*, **69**: 1–12.

Mercer, E. H. & Dixon, K. E. 1967. The fine structure of the cystogenic cells of the cercaria of *Fasciola hepatica* L. *Zeitschrift für Zellforschung*, **77**: 331–44.

Nice, N. G. & Wilson, R. A. 1974. A study of the effect of temperature on the growth of *Fasciola hepatica* in *Lymnaea truncatula*. *Parasitology*, **68**: 47–56.

Ogbe, M. G. 1984. *Dicrocoelium hospes* Looss, 1907 (Digenea: Dicrocoeliidae) infection in cattle from abattoirs in Lagos, Nigeria. *Biologia Africana*, **1**: 35–7.

Ollerenshaw, C. B. 1959. The ecology of the liver fluke (*Fasciola hepatica*). *Veterinary Record*, **71**: 957–63.

Ollerenshaw, C. B. 1971. Forecasting liver fluke disease in England and Wales 1958–1968 with a comment on the influence of climate on the incidence of disease in some other countries. *Veterinary Medical Review*, **2/3**: 289–312.

Pantelouris, E. M. 1965. *The Common Liver Fluke, Fasciola hepatica L.* Pergamon Press, Oxford.

Ractliffe, L. H. 1968. Hatching of *Dicrocoelium lanceolatum* eggs. *Experimental Parasitology*, **23**: 67–78.

Rim, H. J. 1988. Clonorchiasis. In: *Laboratory Diagnosis of Infectious Diseases* (ed. A. Barlow *et al.*), Vol. 1, pp. 801–10. Springer-Verlag, New York.

Rowcliffe, S. A. & Ollerenshaw, C. B. 1960. Observations of the egg of *Fasciola hepatica*. *Annals of Tropical Medicine and Parasitology*, **54**: 172–81.

Sewell, M. M. H. & Purvis, G. M. 1969. *Fasciola hepatica*: the stimulation of excystation. *Parasitology*, **59**: 4P–5P.

Smith, G. 1981. A three years study of *Lymnaea truncatula* habits, disease foci of fascioliasis. *British Veterinary Journal*, **137**: 398–410.

Smith, G. 1984. The relationship between the size of *Lymnaea truncatula* naturally infected with *Fasciola hepatica* and the intensity and maturity of redial infection. *Journal of Helminthology*, **58**: 123–7.

Smith, M. A. & Clegg, J. A. 1981. Improved culture of *Fasciola hepatica in vitro. Zeitschrift für Parasitenkunde*, **665**: 9–15.

Smyth, J. D. & Haslewood, G. A. D. 1963. The biochemistry of bile as a factor in determining host specificity in intestinal parasites, with particular reference to *Echinococcus granulosus*. *Annals of the New York Academy of Sciences*, **113**: 234–60.

Stunkard, H. W. 1974. The trematode family Bucephalidae – problems of morphology, development, and systematics: description of *Rudolphinus gen. nov. Transactions of the New York Academy of Sciences*, Series II, **36**: 143–70.

Sukhdeo, M. V. K. & Mettrick, D. F. 1986. The behaviour of juvenile *Fasciola hepatica*. *Journal of Parasitology*, **72**: 492–7.

Tesana, S., Kaewkes, S., Sriswangwonk, T. & Phinlaor, S. 1985. Distribution and density of *Opisthorchis viverrini* metacercariae in cyprinoid fish from Khon Kaen province. *Journal of the Parasitology and Tropical Medicine Association of Thailand*, **8**: 36–9.

Tielens, A. G. M., Van der Meer, P. & Van den Berg, S. G. 1981. *Fasciola hepatica*: simple, large-scale, *in vitro*, excystment of metacercariae and subsequent isolation of juvenile liver flukes. *Experimental Parasitology*, **51**: 8–12.

Wallet, M., Theron, A. & Lambert, A. 1985. Rythme d'émission des cercariae de *Bucephalus polymorphus* Baer, 1827 (Trematoda, Bucephalidae). *Annals de Parasitologie Humaine et Comparée*, **60**: 675–84.

Wescott, R. B. & Foreyt, W. J. 1986. Epidemiology and control of trematodes in small ruminants. *Veterinary Clinics of North America, Food Animal Practice*, **2**: 373–81.

Wilson, R. A. & Denison, J. 1980. The parasitic castration and gigantism of *Lymnaea truncatula* infected with the larval stages of *Fasciola hepatica*. *Zeitschrift für Parasitenkunde*, **61**: 109–19.

Wykoff, D. E., Harinasuta, C., Juttijudata, P. & Winn, M. M. 1965. *Opisthorchis viverrini* in Thailand – the life cycle and comparison with *O. felinus*. *Journal of Parasitology*, **51**: 207–14.

15

Digenea: Plagiorchiidae, Echinostomatidae, Heterophyidae, Troglotrematidae

15.1 Family Plagiorchiidae

This family contains several species of trematodes readily available for laboratory study. Many are commonly occurring parasites in the amphibian or reptile types normally dissected in zoology courses. A few species infect homoiothermic animals. The best known genera are those found in frogs, such as *Rana temporaria* and *R. esculenta* in Europe and *Rana pipiens* in North America; other amphibian species can also act as hosts (Schell, 1985; Smyth & Smyth, 1980). All species have stylet cercariae (xiphidiocercariae) which generally encyst in arthropods (usually insects) but occasionally in vertebrates.

15.1.1 Type example: *Haplometra cylindracea*

definitive hosts: *Rana* spp., *Bufo* spp., *Bombina* spp.
location: lungs
molluscan hosts: *Lymnea ovata, L. palustris, L. stagnalis, L. truncatula, Radix auricula*

Morphology. This is a common species in European frogs, being reported from *Rana temporaria* (Table 15.1), *R. esculenta, R. ridibunda, R. dalmaltina, R. arvalis, Bufo* spp. and from other amphibia in other parts of the world. It is unusual in being almost cylindrical in transverse section. Its general anatomy is shown in Fig. 15.1 but the occurrence of minor morphological variations may indicate the existence of intraspecific variants (Grabda-Kazubska & Combes, 1981).

Life history (Fig. 15.2). The lung provides an environment with a high oxygen tension; it is also rich in mucus and has an abundant blood supply. The presence of the fluke in the lung apparently does little damage; the epithelium, which is pavement in the uninfected lungs, becomes columnar. *Haplometra* feeds exclusively upon blood drawn from the lung capillaries; large amounts of esterase are produced during feeding and some histolysis of host tissue occurs (Halton, 1967). Haemoglobin is broken down into soluble, colourless compounds and not to haematin as in some trematodes (e.g. *Schistosoma*).

The life cycle has been studied in considerable detail by Combes (1968), Grabda-Kazubska (1970, 1974) and Dobrovolskij & Raĭkhel (1973). The eggs of *Haplometra*, which are embryonated and contain a miracidium when laid, rapidly turn brown on exposure to air and give reactions for tanned protein (Table 13.1). The eggs hatch only on ingestion by a suitable snail host; the morphology of the miracidium has been studied in detail (Dobrovolskij, 1965). The sporocysts develop in the hepatopancreas and produce xiphidiocercariae; there is no redial stage. It was formerly thought that aquatic insects served as a second intermediate host. It is now known that cercariae penetrate tadpoles directly, and encyst especially in the antero-ventral region and in the buccal cavity. After 3–4 days, the metacercariae burst out of their cysts and make their way to the lungs from which they can be recovered within 48 hours. However, abbreviated life cycles in which the metacercarial phase is either reduced or eliminated (Fig. 15.2, I, II, III) are believed to occur.

15.1.2 Other plagiorchids in amphibia

Haematoloechus spp. (Fig. 15.3)

Various species of *Haematoloechus* are very common in the lungs of *Rana, Bufo,* and *Triturus* in Europe and other genera in other countries. A typical life cycle is

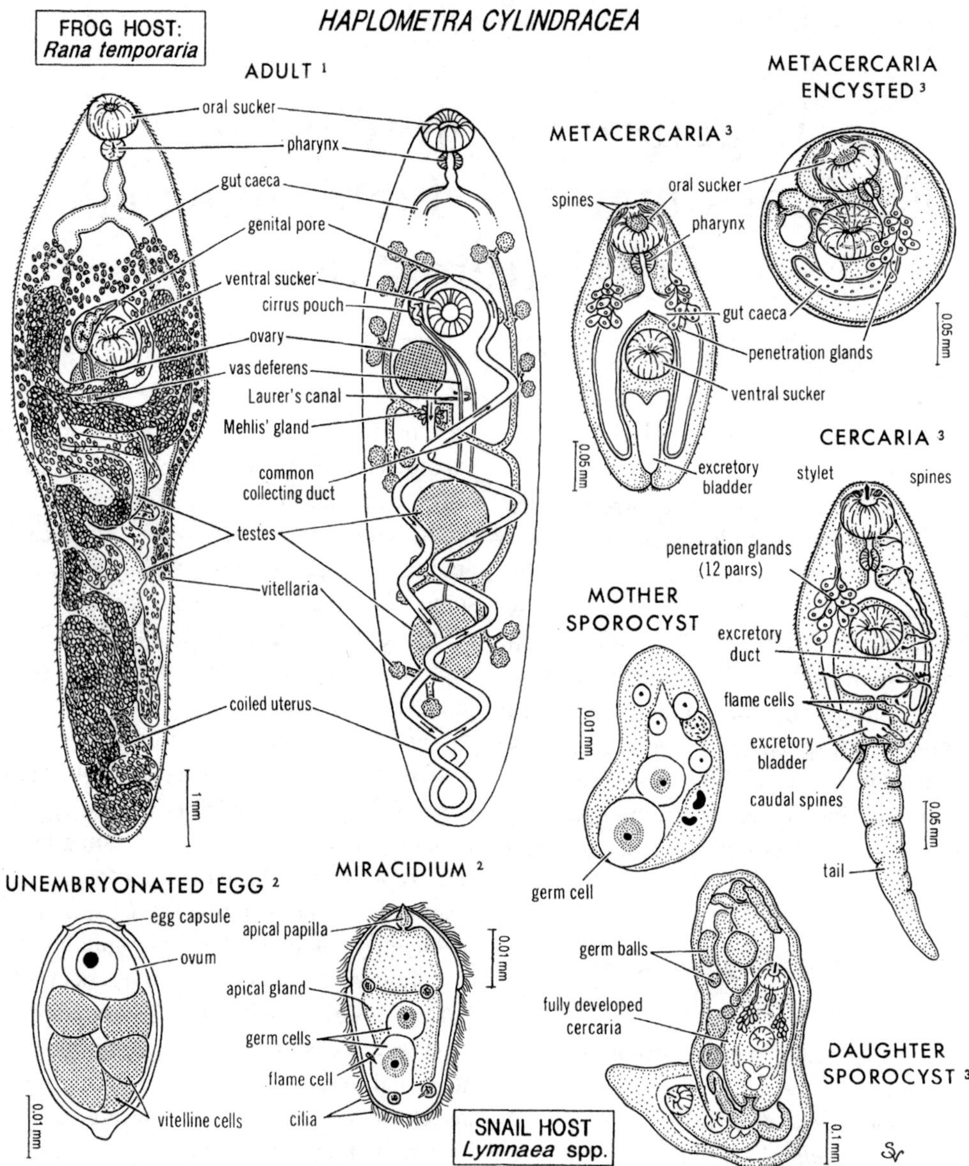

Fig. 15.1
Haplometra cylindracea, from the lungs of *Rana temporaria* and *Rana* spp. Details based on [1]Combes (1968), [2]Dobrovolskij (1965), [3]Grabda-Kazubska (1970, 1976). (From Smyth & Smyth, 1980.)

shown in Fig. 15.4; aquatic larvae of several insect orders (especially dragonflies) serve as the second intermediate host. Species of *Planorbis* generally act as molluscan hosts, exceptionally species of *Lymnea* or other genera. The life cycle has been reviewed by Combes (1968) and details of the laboratory maintenance have been given by Bailey & McDaniel (1965). The best known species in amphibia in North America are probably *Haematoloechus medioplexus* (Krull, 1930, 1931) (Fig. 15.3), *H. breviplexus* (Schell, 1965) and *H. coloradensis* (Dronen, 1975). Other species are listed in Table 15.2; the genera *Ostiolum* and *Haematoloechus* are considered to be synonyms by some workers (Prudhoe & Bray, 1982), but separate genera by others, e.g. Odening (1964).

Opisthioglyphe ranae

This is a common intestinal species in Europe (Frandsen, 1974), occurring in *Rana temporaria*, *R. esculenta*, *R. ridibunda*, *R. terrestris* and the grass snake, *Natrix natrix*. This species shows an unusual degree of variation in specimens taken from different hosts; these

Table 15.1. *Some adult trematodes* found in* Rana temporaria *and* R. esculenta

Species	Site	*R. esculenta*	*R. temporaria*
Brandesia turgida	small intestine	fairly common	—
Cephalogonimus retusus	intestine	common	rare
Diplodiscus subclavatus	rectum and bladder	common	rare
Dolichosaccus rastellus	small intestine	common	common
Gorgodera cygnoides	bladder	common	fairly common
Gorgodera microovata	bladder	rare	rare
Gorgodera pagenstecheri	bladder	fairly common	fairly common
Gorgoderina vitelliloba	bladder	common	common
Haematoloechus variegatus	lung	common	rare
Haematoloechus similis	lung	common	rare
Halipegus ovocaudatus	stomach, oesophagus	common	rare
Haplometra cylindracea	lung	common	fairly common
Opisthioglyphe ranae	small intestine	common	common
Opisthodiscus nigrivasis	rectum	fairly common	—
Pleurogenes claviger	small intestine	common	rare
Pleurogenes hepaticola	liver, gall bladder	rare	—
Pleurogenes loossi	intestine	rare	rare
Pleurogenoides medians	small intestine	fairly common	rare
Pleurogenoides stromi	intestine	rare	—
Prosotocus confusus	intestine	fairly common	rare
Prosotocus fuelleborni	intestine	fairly common	—
Polystoma integerrimum (Monogenea)	bladder	rare	fairly common

*Only general indications of the occurrence of the trematodes can be attempted here, as infection depends on ecological factors in the area where the frogs are collected.
Source: Data from Smyth & Smyth (1980).

were previously named *O. ranae*, *O. endoloba*, and *O. natricis* but are now considered to represent a single species, *O. ranae* (Grabda-Kazubska, 1967).

The morphology is clear from Fig. 15.5. *Lymnaea stagnalis* and *L. palustris* are the common intermediate hosts but numerous other molluscan species can also serve as hosts. In the 'normal' life cycle, cercariae encyst in tadpoles, snails or insect larvae, which are eaten by frogs. An 'abbreviated' cycle can also occur in which metacercariae initially encyst in the buccal cavity of tadpoles; after a few days the cysts become detached and pass down into the intestine where they excyst. Tadpoles can thus act as both intermediate and definite hosts (Grabda-Kazubska, 1969).

Dolichosacchus spp.

Species of *Dolichosacchus* are intestinal parasites of amphibia; *D. rastellus* is the common species in Europe and *D. schmidti* is found in the USA (Fischtal & Kuntz, 1975). Some authors include the species in the genus *Opisthioglyphe*. The life cycle is almost identical with that of *O. ranae* (Grabda-Kazubska, 1969; Combes, 1968; Smyth & Smyth, 1980).

○○○ 15.2 Family Echinostomatidae

Possession of a head collar surrounding the oral sucker separates the echinostomes morphologically from the remainder of the trematodes. There is no other outstandingly unusual feature. All members of this family are parasites of birds or mammals. Their habitats range along the entire length of the alimentary canal from the duodenum to the caecum and rectum; the bile duct is invaded by some species. The life cycle differs from that of *Fasciola* in but one minor point: the cercariae encyst either within the tissues of the same molluscan host in which sporocysts and rediae develop, or penetrate and encyst in other animals such as planarians, molluscs, amphibians or fish. Like the adults, cercariae bear collars of spines.

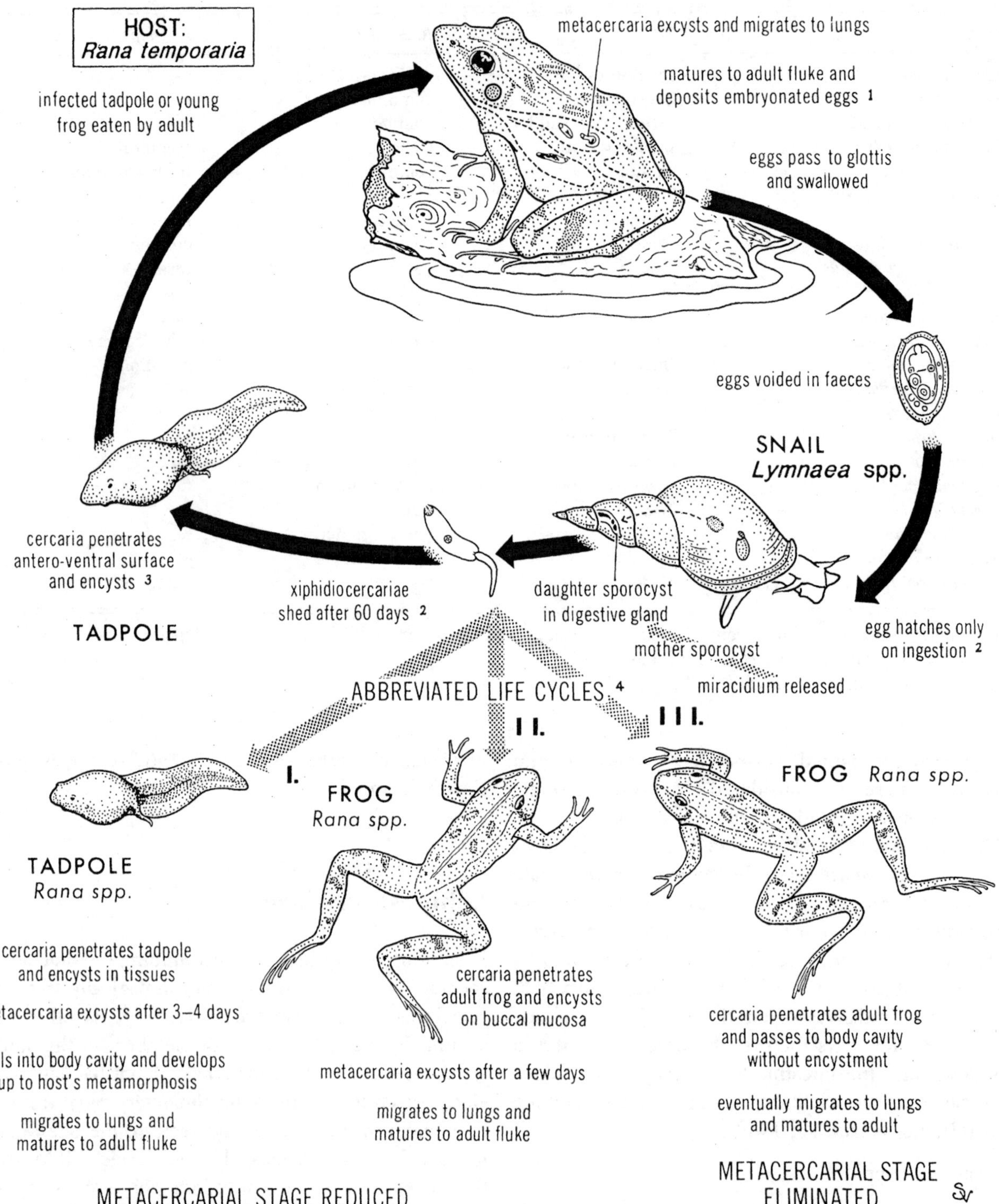

Fig. 15.2
Haplometra cylindracea: life cycle. Details based on [1]Dobrovolskij (1965), [2]Dobrovolskij & Răkhel (1973), [3,4]Grabda-Kazubska (1970, 1976). (From Smyth & Smyth, 1980.)

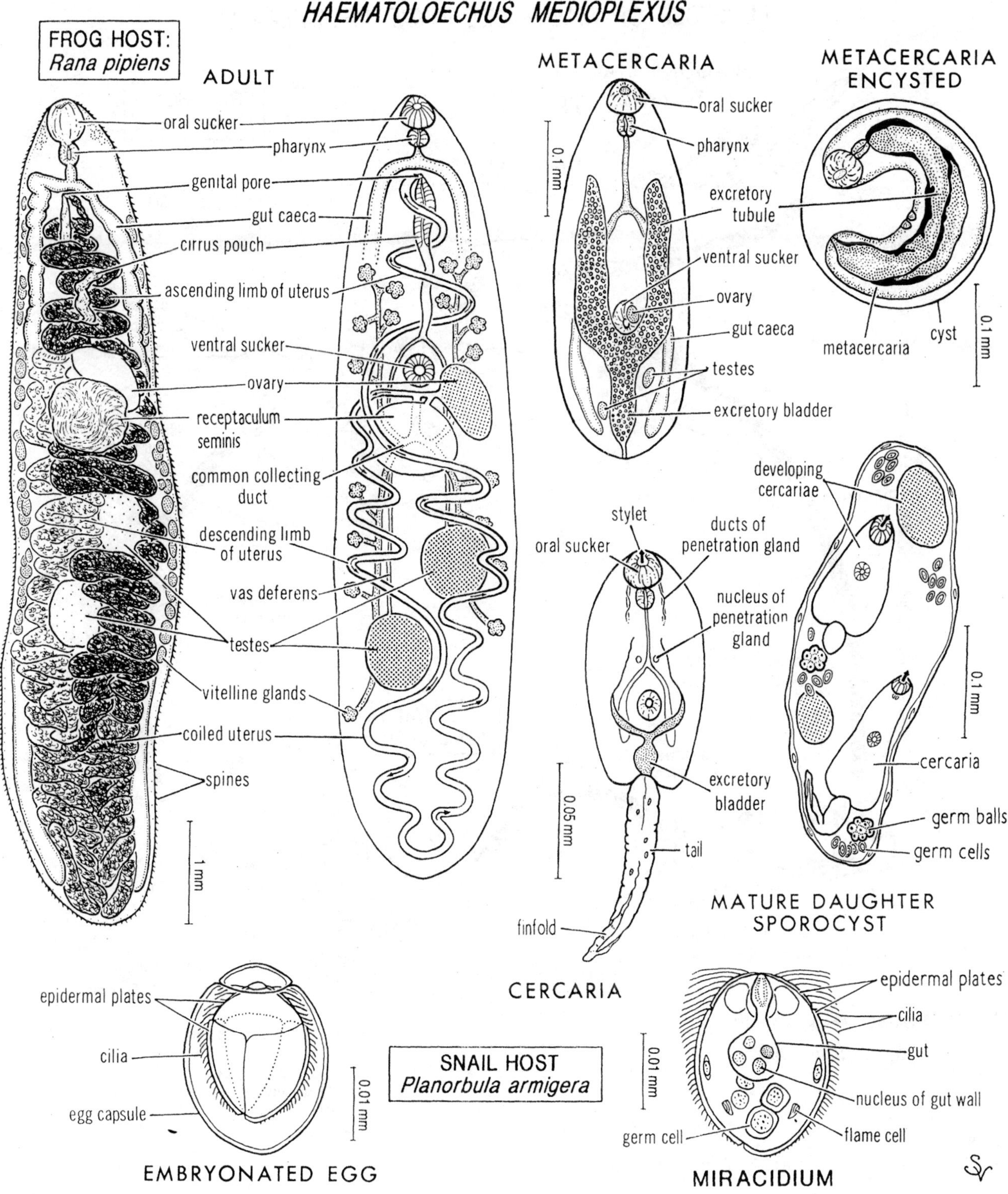

Fig. 15.3
Haematoloechus medioplexus, from the lungs of *Rana pipiens*. Details based on [1]Krull (1930, 1931), [2]Underwood & Dronen (1977). (From Smyth & Smyth, 1980.)

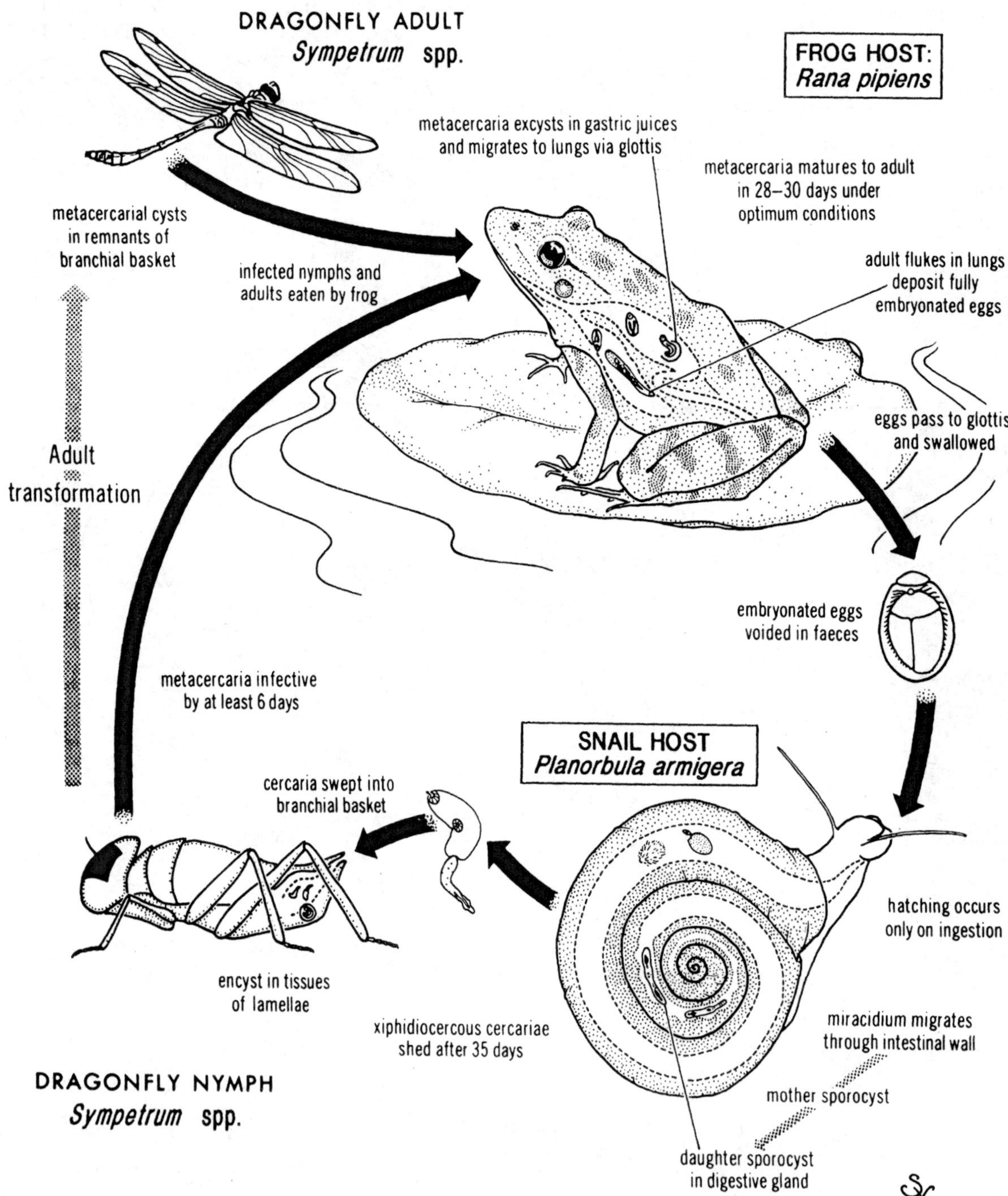

Fig. 15.4
Haematoloechus medioplexus: life cycle. (From Smyth & Smyth, 1980.)

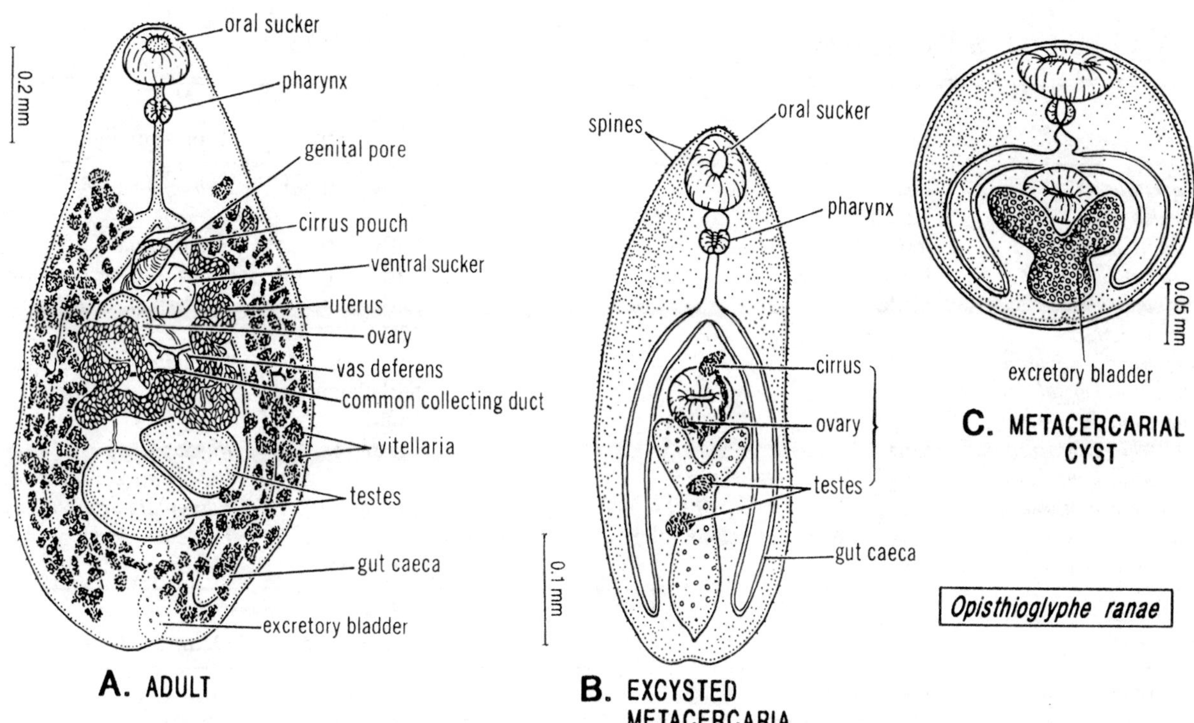

Fig. 15.5
Opisthioglyphe ranae, from the gut of *Rana temporaria*. A. Adult. B. Excysted metacercaria showing the presence of genital anlagen. C. Metacercarial cyst. (From Smyth & Smyth, 1980.)

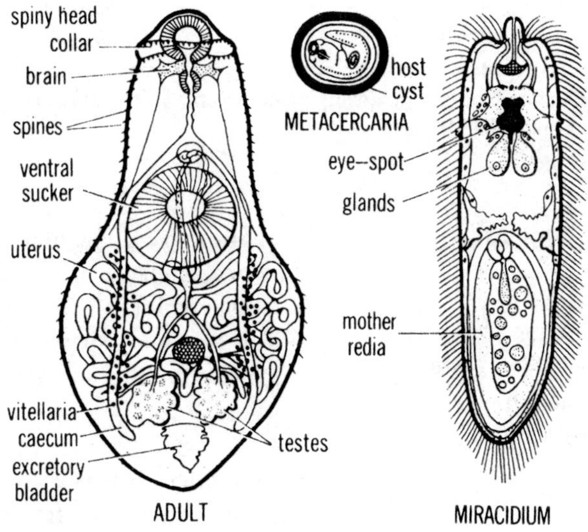

Fig. 15.6
Parorchis acanthus from rectum or bursa Fabricii of gull. Morphology of adult, miracidium and metacercaria. (After Rees, 1939, 1940.)

Table 15.2. **Trematodes in* Rana pipiens

Adults Species	Site	*Occurrence	Typical molluscan host
Cephalogonimus sp.	intestine	fairly common	*Helisoma* spp.
Glypthelmins facioi	intestine	no data	?
Glypthelmins quieta	intestine	fairly common	*Physa gyrina*
Gorgodera amplicava	bladder	fairly common	*Musculium partumeium*
Gorgoderina attenuata	bladder	fairly common	*Sphaerium occidentale*
Gorgoderina simplex	bladder	rare	?
Haematoloechus breviplexus	lung	common	*Gyraulus similaris*
Haematoloechus longiplexus	lung	fairly common	*Gyraulus similaris*
Haematoloechus varioplexus	lung	common	?
Haematoloechus colordensis	lung	common	*Physa virgata*
Haematoloechus complexus	lung	rare?	*Pseudosuccinea columella*
Haematoloechus medioplexus	lung	common	*Planorbula armigera*
Halipegus sp.	mouth cavity, eustachian tube	rare	*Physa* sp., *Helisoma* sp.
Megalodiscus temperatus	rectum	common	*Helisoma* spp.
Metaceracariae Species	**Site**	**Occurrence**	**Definitive host**
Alaria sp. (mesocercariae)	body cavity	common	carnivore
Apharyngostrigea pipientis	clusters on liver/mesenteries/bladder/intestine	common	bittern
Cephalogonimus sp.	skin, viscera	no data	frog
Clinostomum marginatum	peritoneum, lymph spaces, mesenteries	no data	bird
Clinostomum attenuatum	sub-cutaneous, lung surface, mesenteries	rare	bittern
Echinostoma revolutum	kidney	no data	bird
Fibricola cratera	pelvic muscles, coelom	fairly common	mouse, raccoon
Glypthelmins quieta	skin	no data	frog
Megalodiscus temperatus	skin	no data	frog
Strigea elegans (mesocercariae)	body cavity, muscles	no data	bird

*Only the most general indications of occurrence can be given, as the habitat of '*R. pipiens* complex' can be so varied.
Source: Data modified from Smyth & Smyth (1980).

○ 15.2.1 Type example: *Parorchis acanthus**

definitive hosts: herring gull, common gull, tern, flamingo, long-billed curlew
location: rectum, bursa Fabricii
molluscan hosts: *Nucella* (*Purpura*) *lapillus*, *N. lamellosa*
transport hosts: shelled molluscs
distribution: cosmopolitan

*Some authorities (e.g. Yamaguti, 1958) consider this an aberrant echinostome and place it in a separate family, the Philophthalamidae.

Morphology. The morphology is shown in Fig. 15.6. The head collar bears about sixty spines and the cuticle is heavily spined in the region anterior to the ventral sucker becoming sparser behind. Except that the cirrus is spined, the reproductive system presents no unusual features.

Occurrence. This is an important species biologically since it is one of the few trematodes in which gametogenesis and larval development have been investigated cytologically in great detail (Rees, 1939, 1940). Infections of up to about twelve flukes can occur in a single host. In Europe the larval stages occur in the marine snail *Nucella lapillus* and in N. America in *N. lamellosa* (Ching, 1978), a high proportion of which are commonly infected.

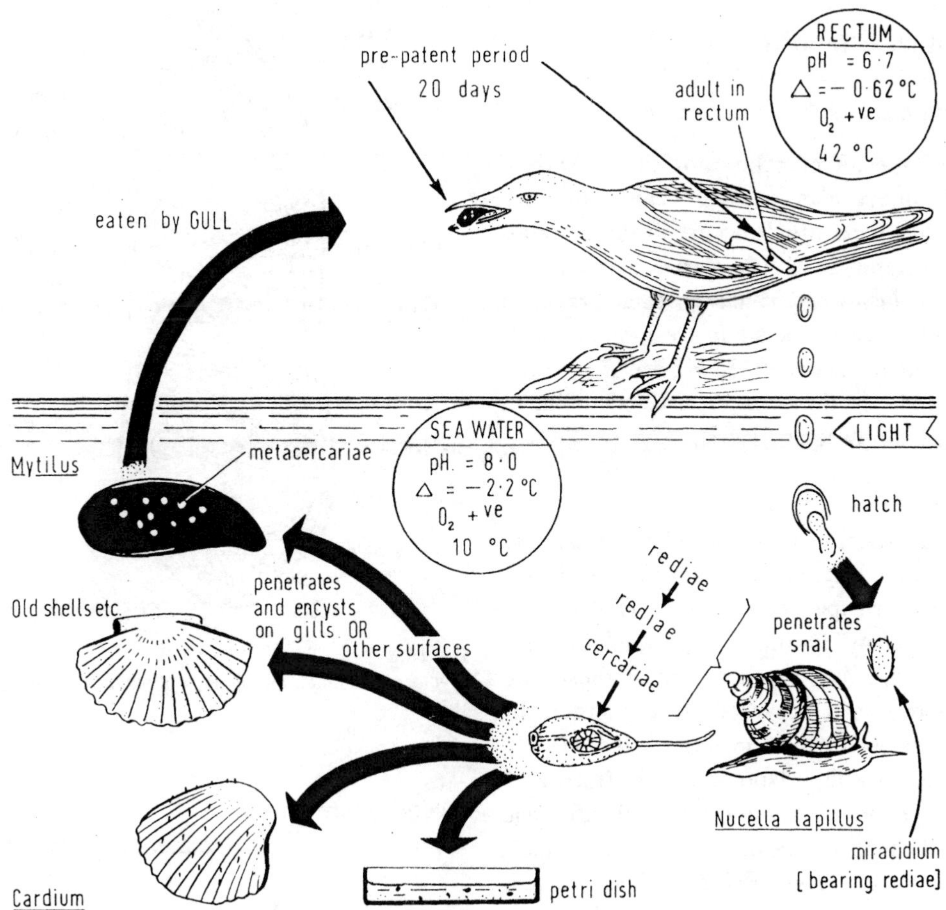

Fig. 15.7
Parorchis acanthus: life cycle and some physiological factors relating to it. (Original.)

Life cycle (Fig. 15.7). This has been described by Stunkard & Cable (1932). Whatever mechanism retards embryonation in most trematode eggs while within the host, it is lacking here; as the eggs pass up the uterus they increase slightly in size, and the whole process of larval development becomes enormously accelerated. Mature miracidia develop in eggs still *in utero*, and hatching may even occur there. Hatching normally, however, takes place in sea water shortly after the eggs are laid. A miracidium, in addition to containing the usual larval features such as eye spots, penetration glands, and a vestigial gut, is almost unique in containing a mother redia with a well-developed pharynx, an intestine and germinal balls (Fig. 15.6). The miracidium penetrates *Nucella* and the mother redia is liberated. The germinal balls contained in the latter form daughter rediae which ultimately reach the digestive gland, finally producing typical echinostome cercariae (Fig. 15.7). The structure and ultrastructure of the rediae have been examined by Rees (1966).

Released cercariae will encyst on almost any surface, such as snail shells, petri dishes or the walls of the aquarium. Rees (1967) has described the structure of the cercaria and the metacercarial cyst wall.

The cyst wall consists of an outer wall with three layers and an inner wall with two layers; these contain acid and neutral mucopolysaccharide, protein, lipoprotein and glycoprotein.

Excystment takes place in the bird's duodenum and the released metacercariae make their way to the rectum or into the bursa Fabricii. A wide range of shore birds is infected, but experiments have failed to grow worms in guinea pigs, rats or mice.

○ 15.2.2 Other echinostomes

Species infecting man

Numerous other species of echinostomes are known, some of which infect man causing echinostomiasis, e.g. *Echinostoma ilocanum* which occurs in the Ilocanos people of the Philippines. Metacercariae occur in the large snail *Piola luzionica*, which is eaten raw. In Celebes, the metacercariae occur in fresh-water mussels, which are eaten uncooked. In Korea, *E. hortense* infects humans and rats, the metacercariae occurring in fresh-water fish (Ahn & Ryan, 1987).

Other species

The most studied species in birds and mammals are probably *E. trivolis*, *E. caproni* and *E. revolutum*, which are useful models for experimental work. The biology, life histories, infectivity, immunology, pathology, epidemiology, physiology and biochemistry of these and other species have been surveyed in a wonderfully comprehensive review by Huffman & Fried (1990). *E. revolutum* develops well in laboratory host (rats, rabbits and guinea pigs); it can also infect man. Its life cycle is very non-specific in its choice of hosts, infecting snails of the genera *Lymnaea*, *Physa*, *Paludina*, *Helisoma* and *Segmentina*. Metacercariae encyst in fish, tadpoles, molluscs and planaria.

○○○ 15.3 Family Heterophyidae

These are very small (sometimes less than 0.5 mm long) egg-shaped trematodes, usually parasitic in fish-eating animals. They show a striking morphological peculiarity in the possession of a *gonotyl* (Fig. 15.8) or genital sucker, a retractile sucker-like structure which assists in copulation and which may either be incorporated in the ventral sucker or lie on one side of it. The life cycle closely resembles that of the Opisthorchiidae, the eggs when laid containing miracidia. Hatching does not take place until the eggs are ingested by the molluscan host.

○ 15.3.1 Type example: *Cryptocotyle lingua*

definitive hosts:	fish-eating birds and mammals
situation:	intestine
molluscan host:	*Littorina littorea*, *L. scutulata*
second intermediate host:	numerous marine fish

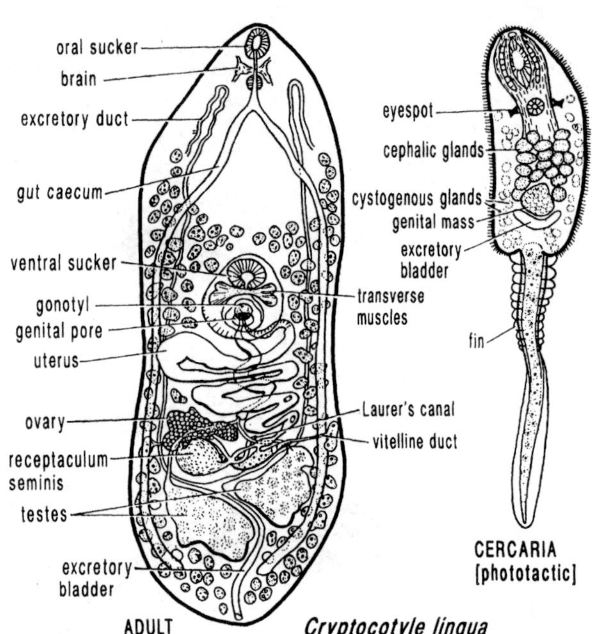

Fig. 15.8
Cryptocotyle lingua: morphology of adult and cercaria. (After Brown, 1950.)

Morphology. Apart from the possession of the gonotyl already referred to, this fluke shows no special morphological peculiarity. Its anatomy is shown in Fig. 15.8.

Occurrence. The natural hosts of this widely distributed parasite are fish-eating birds, especially those inhabiting sea coasts (e.g. greater black-backed gull, herring gull, lesser black-backed gull, slavonic grebe, night heron, common tern, razorbill, and kittiwake; maturation is not achieved in the duck). One of the characteristics of the fluke is its unspecificity regarding definitive hosts; many laboratory mammals such as rats, cats and guinea pigs may be infected, but the degree of resistance varies enormously. Wild rats, foxes and cats probably serve as important reservoir hosts in nature.

Life history (Fig. 15.9). The adult flukes live deep between the villi of the small intestine especially in the anterior region; some variation in the distribution of the parasites between adult birds and chicks has been described (Threlfall, 1967). The flukes are capable of producing vast numbers of eggs. These embryonate in

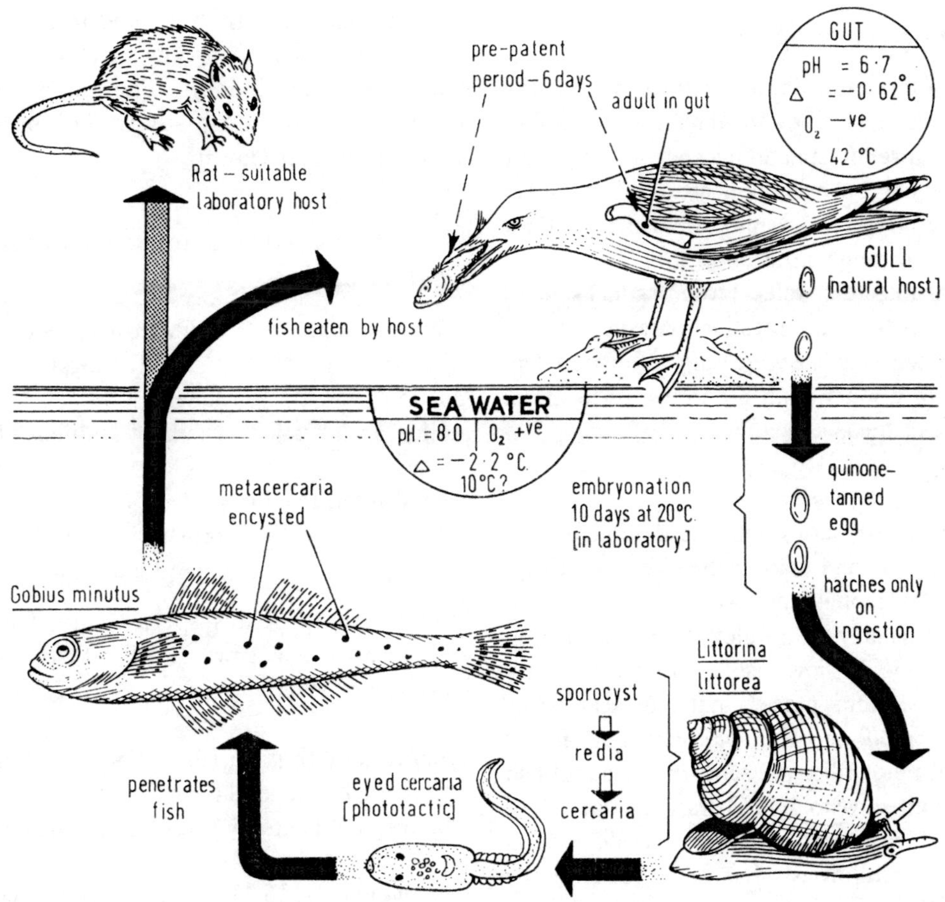

Fig. 15.9
Cryptocotyle lingua: life cycle and some physiological factors relating to it. (Original.)

sea water at 20 °C in about ten days. Hatching has not been observed, and probably only takes place on ingestion by the molluscan host. In Europe, miracidia penetrate into the common shore periwinkle *Littorina littorea* and pass through the usual sporocyst and redia stages; *L. scutulata* is infected in Canada (Ching, 1978). The released cercariae have prominent eye spots, are markedly phototropic and well supplied with both cystogenous and histolytic glands. Infected snails can release enormous numbers of cercariae: 3000 a day have been reported. There is a well-authenticated case of a single periwinkle releasing a total of several million cercariae over a period of five years. The cercariae will readily penetrate into a variety of shore fish (e.g. cunner, gudgeon, etc.). A prevalence of 85 per cent in charr has been reported (Kristoffersen, 1988). Although the tissues can be penetrated at any site, the process requires some two hours to accomplish; cercariae show a predilection for the cartilage fin rays and frequently appear there in enormous numbers.

The cyst secreted by a cercaria is thin and flexible and later becomes surrounded by connective tissue as a result of a tissue reaction by the host. As with many trematode cysts in fish, an orange-red or black pigment develops in the cyst site. Encysted metacercariae may grow to some two to three times their initial size and many remain viable for several years. The immune response in the mullet, *Chelon labrosus*, has been investigated by Wood & Matthews (1987).

The factors involved in excystment of the metacercariae appear to follow those required for *Fasciola* (p. 210), i.e. pre-treatment in pepsin followed by trypsin + bile; carbon dioxide and a reducing condition enhance the process (McDaniel, 1966).

○ 15.3.2 Other Heterophyidae

Many species of Heterophyidae are characterised by the wide range of hosts in which they can mature. A number of species can mature in man and may occur in a variety of unusual tissue sites such as the heart or other viscera where they deposit eggs and may do damage. They reach these sites through infiltration into the general circulation via the intestinal walls. Two intestinal species may be regarded as 'normal' intestinal parasites of man since, although occurring in a wide range of mammalian hosts, they do not wander from the intestinal location. The commonest of these are:

Heterophyes heterophyes

A tiny fluke whose size varies with the definitive host, common in cats, dogs and man in the Far East, Egypt and Palestine. The molluscan host is *Pirenella conica*, and the fish host usually the mullet, *Mugil cephalus*, in which metacercariae of low tissue organisation are found; the genital sucker is present in the metacercariae. The related *H. aequalis* and *H. dispar*, which are found in dogs, cats and rats, utilise the same mollusc and fish hosts but are not infective to man (Taraschewski, 1985).

The biology of this species and the prevalence and transmission of heterophyiasis has been reviewed by Taraschewski (1984, 1985).

Metagonimus yokogawai

A common parasite of dogs and cats in the Far East, the Northern Provinces of Siberia and the Balkan States. In human infections, large numbers of flukes sometimes occur, the location being the duodenum. The snail intermediate hosts are species of Thiaridae (formerly Maleniidae). A wide range of fish hosts is infected, beneath the scales of which the metacercariae encyst. The hamster is the most suitable laboratory host, but the worms can also become established in rats, in particular the cotton rat (Yokogawa & Sano, 1968).

○○○ 15.4 Family Troglotrematidae

These are medium or fairly large 'fleshy' trematodes with poorly developed suckers. The absence of strongly developed suckers clearly limits the distribution of these flukes to environments where currents such as occur in the intestine, or other mechanical stresses, are lacking. They are almost entirely limited to cyst-like spaces in various mammalian organs such as lungs, frontal sinuses, skin and kidneys. There is no species readily available for laboratory study; the species described below is the best known.

○ 15.4.1 Type example: *Paragonimus westermani*

definitive hosts:	man, carnivores (especially cat family, dogs)
location	lungs
molluscan hosts:	snails of families Thiaridae, Pleuroceridae, Hydrobiilidae
second intermediate host:	crustacea, especially crabs and crayfish

Useful reviews of this species and of paragonimiasis are those of Yokogawa (1965, 1969) and Bunnag & Harinasuta (1984).

Morphology (Fig. 15.10). The adult flukes are rather egg-shaped in general form, thick and reddish in colour. The internal anatomy presents no unusual feature.

Life cycle (Fig. 15.11). Although the majority of adult specimens occur in the lungs, their digestive system can cope with a variety of food materials, because they occasionally occur in tissues such as brain, liver, spleen,

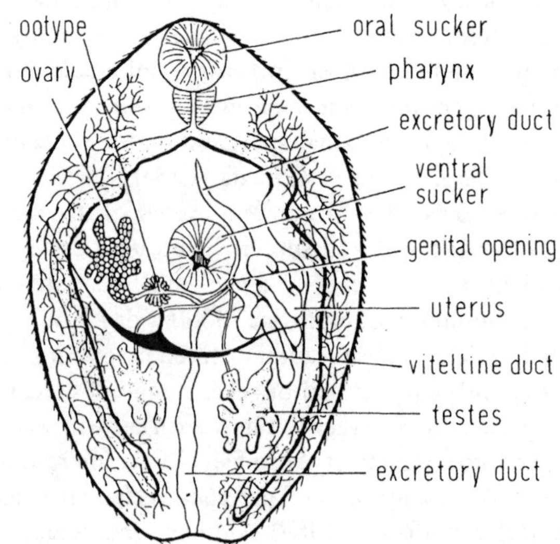

Fig. 15.10
Paragonimus westermani: morphology. (After Chandler, 1955.)

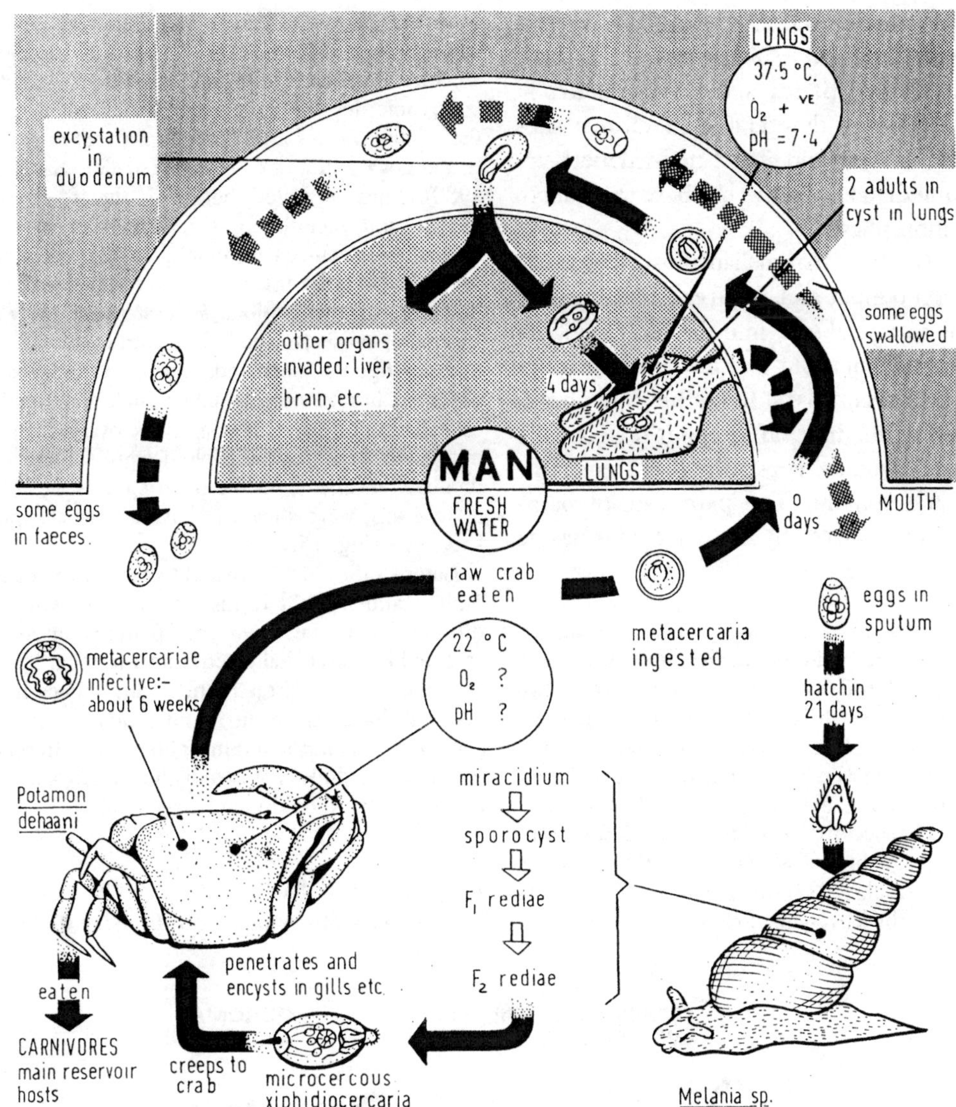

Fig. 15.11
Paragonimus westermani: life cycle and some physiological factors relating to it. (Original.)

intestinal wall, eye, muscles or even kidneys. The occurrence in this wide range of habitats suggests that man may not be the natural host. The lung cysts in which the worms most commonly occur usually contain 1–3 flukes. The eggs are freed into the bronchial tubes and pass out with sputum, but they may also appear in the faeces in large numbers, as a result of being swallowed. The number of eggs produced varies greatly with species (Yokogawa, 1965).

The most important endemic areas occur in Korea, Japan, Taiwan, central China and the Philippines. Eggs may most easily be embryonated using the Harada–Mori filter paper culture method (Beaver *et al.*, 1964). At 27 °C eggs embryonate within two weeks and hatch in about three weeks. A change in temperature and light generally induces hatching but the process does not appear to take place easily (Yokogawa, 1965). The molluscan hosts vary with locality. The most important host is probably *Semisulcospira libertina* (*S. bensoni*); others implicated are *Melania amurensis* (*Hua amurensis*), *Melania obliquegranosa* (*Tarebia granifera*) and *Brotia asperata*.

Sporocyst and two redia generations occur, giving rise to creeping microcercous cercariae. These penetrate a number of fresh-water crustaceans, especially species of *Potamon*, *Paratelphusa* and *Eriocheir* in the Far East, *Cambarus* in the United States and *Pseudotelphusa* in South America, in which they encyst in various sites such as gills, muscles, heart and liver. Encysted metacercariae are not immediately infective but have to undergo a further maturation phase.

On ingestion by the mammalian host, cysts are digested in the duodenum and the freed metacerariae bore through the intestinal wall into the body cavity to reach the pleural cavity in about four days and the lungs in about fourteen to twenty days. In the lungs, a fibrous capsule is formed by the host; after about 6 weeks the worms mature and produce eggs, which characteristically appear in the sputum. The pathology of paragoniamiasis has been reviewed by Bunnag & Harinasuta (1984).

Speciation. Although *P. westermani* is the recognised specific name for this parasite in the medical literature, there is now indisputable evidence that '*P. westermani*' exists in two biological forms, one of which is diploid ($2n = 22$) and can be regarded as the 'true' '*P. westermani*', and one which is triploid ($3n = 33$) (Fig. 15.12). Some consider that the latter should be regarded as an independent species, *P. pulmonalis* (Miyazaki, 1987). The occurrence of the triploid form was first observed by Sakaguchi & Tada (1975, 1976). This was later confirmed by Terasaki (1977) who in later studies (Terasaki, 1980) found that the karyotype of *P. westermani* collected from the Akita Prefecture in Japan was diploid ($2n = 22$), whereas that collected from Korea was triploid ($3n = 33$).

More detailed cytological studies of gametogenesis and embryonic development were carried out later on these and the related species *P. iloktsuenensis* (from rats) and *P. sadoensis* (from dogs and rats) (Terasaki, 1980; Sakaguchi & Tada, 1980). These revealed that in the diploid *P. westermani* both mitotic and meiotic figures were seen in the metaphase, normal spermatozoa developed in the testes and numerous spermatozoa were present in the seminal vesicle. In contrast, however, in the triploid *P. westermani* (= *P. pulmonalis*?) spermatogenesis proceeded only up to the 32-cell stage, only mitotic figures occurred, and spermatozoa were never observed although the seminal vesicle contained some spherical cells but no spermatozoa. In the ovary, the oocytes divided mitotically resulting in triploid eggs, which developed parthenogenetically. Both the other species, *P. iloktsuenensis* and *P. sadoensis*, were shown to have a normal diploid ($2n = 22$) karyotype (Fig. 15.12).

Later studies have revealed further differences between the diploid and triploid forms, the most striking being marked differences in the isoenzyme patterns of five enzymes, as revealed by starch-gel electrophoresis (Fig. 15.13) (Agatsuma & Habe, 1985). Ultrastructural differences in metacercarial cyst wall have also been found (Higo & Ishii, 1987). These genetic, biochemical and morphological differences seem sufficient to conclude that two distinct species, *P. westermani* and *P. pulmonalis*, should now be recognised.

Experimental hosts. *P. westermani* readily develops in many laboratory hosts but the size of worms and eggs

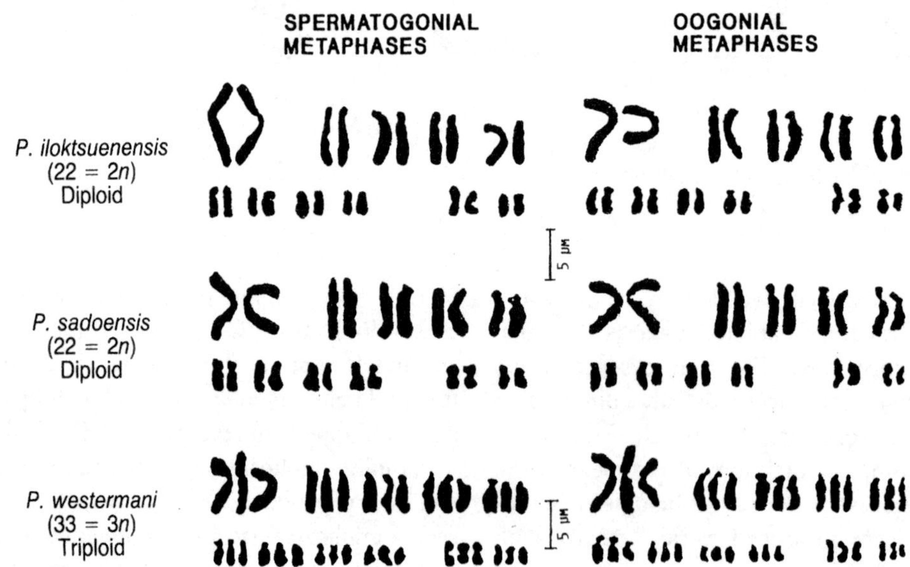

Fig. 15.12
Karyotype analysis of spermatogonial and oogonial metaphases in three species of *Paragonimus*: *P. iloktsuenensis* (from rats); *P. sadoensis* (from dogs and rats); *P. westermani* (from man). Some authors conclude that the triploid variety of *P. westermani* should be designated a separate species, *P. pulmonalis*. (Adapted from Sakaguchi & Tuda, 1980.)

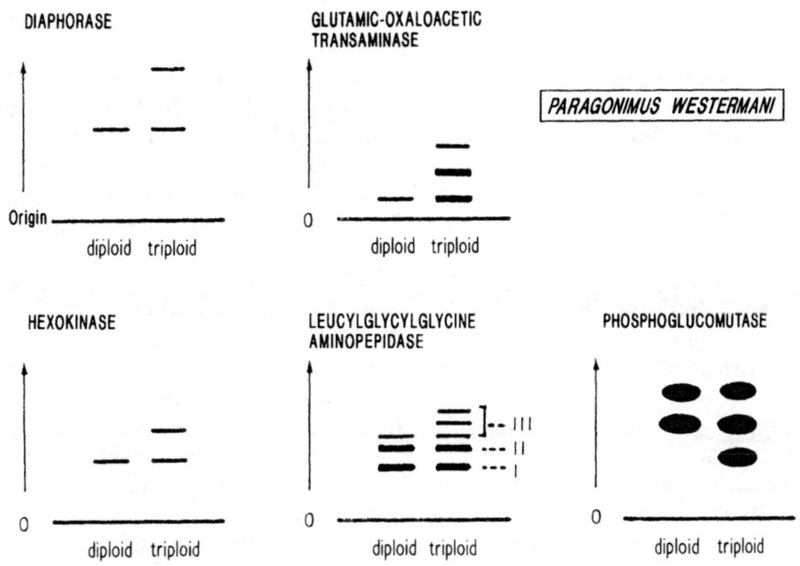

Fig. 15.13
Paragonimus westermani: diagram of the electrophoretic patterns of five enzymes in the diploid ($2n$ = 22) and triploid ($3n$ = 33) varieties, which reveal clear differences between them. Some workers conclude that the diploid and triploid varieties should be designated *P. westermani* and *P. pulmonalis*, respectively. (Adapted from Agatsuma & Habe, 1985.)

varies with the host species. Pan (1966) has described the development in cats, rats, dogs and monkeys.

Nanophyetus salmincola

An interesting trematode whose metacercariae transmit the rickettsia-like organism *Neorickettsia helminthoeca*, causing 'salmon poisoning' in dogs. The adult flukes are unusually non-specific; about thirty mammalian and three avian natural and experimental hosts have been identified (Schlegel *et al.*, 1968; Millemann & Knapp, 1970; Foreyt *et al.*, 1987).

The life cycle resembles that of *Paragonimus*. The penetration and migration of cercariae cause some pathological effects in the intermediate fish hosts (Baldwin *et al.*, 1967). The biology of *N. salmincola* has been comprehensively reviewed by Milleman & Knapp (1970).

○○○ References

(References of general interest have been included in this list, in addition to those quoted in the text.)

Agatsuma, T. & Habe, S. 1985. Electrophoretic studies in enzymes of diploid and triploid *Paragonimus westermani*. *Parasitology*, **91**: 489–97.

Ahn, Y. K. & Ryan, G. S. 1987. (Experimental and epidemiological studies on the life cycle of *Echinostoma hortense* Asada, 1926 (Trematoda: Echinostomatidae).) (In Korean.) *Korean Journal of Parasitology*, **24**: 121–36. (Helminth. Abstracts **57**, No. 243.)

Bailey, H. H. & McDaniel, J. S. 1965. Living trematodes in the biological laboratory. *Biologist*, **47**: 55–61.

Baldwin, N. L., Millemann, R. E. & Knapp, S. E. 1967. 'Salmon poisoning' disease. III. Effect of experimental *Nanophyetus salmincola* infection on the fish host. *Journal of Parasitology*, **53**: 556–64.

Beaver, P. C., Malek, E. A. & Little, M. D. 1964. Development of *Spirometra* and *Paragonimus* eggs in Harada-Mori cultures. *Journal of Parasitology*, **50**: 664–6.

Bunnag, D. & Harinasuta, T. 1984. Opisthorchiasis, Clonorchiasis, and Paragonimiasis. In: *Tropical and Geographical Medicine* (ed. K. S. Warren & A. A. F. M. Mahmoud), pp. 461–70. McGraw-Hill, New York.

Ching, H. L. 1978. New marine hosts for *Parorchis acanthus*, *Cryptocotyle lingua*, *Maritrema megametrios* and *Maritrema gratiosum*, trematodes of birds from British Columbia, Canada. *Canadian Journal of Zoology*, **86**: 1877–9.

Combes, C. 1968. Biologie, écologie des cycles et biogéographie de digènes et monogènes d'amphibiens dans les Pyrénées. *Mémoires du Muséum National d'Histoire Naturelle*, **51**: 1–191.

Dobrovolskij, A. A. 1965. Über die Einheitlichkeit des Bauplanes von Mirazidien der Überfamilie *Plagiorchioidea*. *Angewandte Parasitologie*, **6**: 157–65.

Dobrovolskij, A. A. & Raĭkhel, A. S. 1973. (The life cycle of *Haplometra cylindracea* Zeder 1800 (Trematoda, Plagiorchiidae).) (In Russian: translation No: RTS 10612, British

Library, London.) *Vestnik Leningradskii Unversitet Seriia Biologii*, **28**: 5–13.

Dronen, N. O. 1975. The life cycle of *Haematoloechus coloradensis* Cort 1915 (Digenea: Plagiorchiidae), with emphasis on host susceptibility to infection. *Journal of Parasitology*, **61**: 657–60.

Fischtal, J. H. & Kuntz, R. E. 1975. Some trematodes of amphibians and reptiles from Taiwan. *Proceedings of the Helminthological Society of Washington*, **42**: 1–13.

Foreyt, W. J., Gorham, J. R., Green, J. S., Leathers, C. W. & LeaMaster, B. R. 1987. Salmon poisoning disease in juvenile coyotes: clinical evaluation and infectivity of metacercariae and rickettsiae. *Journal of Wildlife Diseases*, **23**: 412–17.

Frandsen, F. 1974. A study of Danish amphibians' parasitic fauna. *Acta Parasitologica Polonica*, **22**: 49–66.

Grabda-Kazubska, B. 1967. Morphological variability of adult *Opisthioglyphe ranae* (Frölich, 1791) (Trematoda, Plagiorchiidae). *Acta Parasitologica Polonica*, **15**: 15–33.

Grabda-Kazubska, B. 1969. Studies on the abbreviation of the life cycle of *Opisthioglyphe ranae* (Frolich, 1791) and *O. rastellus* (Olsson, 1876) (Trematoda: Plagiorchiidae). *Acta Parasitologica Polonica*, **16**: 249–69.

Grabda-Kazubska, B. 1970. Studies on the life cycle of *Haplometra cylindracea* (Zeder, 1800) (Trematoda: Plagiorchiidae). *Acta Parasitologica Polonica*, **18**: 497–512.

Grabda-Kazubska, B. 1974. Observations on penetration of cercariae and further development of *Haplometra cylindracea* (Zeder, 1800) (Trematoda, Plagiorchiidae) in final host. *Acta Parasitologica Polonica*, **22**: 393–400.

Grabda-Kazubska, B. 1976. Abbreviations of the life cycles of plagiorchid trematodes. General remarks. *Acta Parasitologica Polonica*, **24**: 125–41.

Grabda-Kazubska, B. & Combes, C. 1981. Morphological variability of *Haplometra cylindracea* (Zeder, 1800) (Trematoda, Plagiorchiidae). *Acta Parasitologica Polonica*, **28**: 39–63.

Halton, D. W. 1967. Observations on the nutrition of digenetic trematodes. *Parasitology*, **57**: 639–60.

Higo, H. & Ishii, Y. 1987. Comparative studies on the surface ultrastructure of newly encysted metacercariae of Japanese lung flukes. *Parasitology Research*, **73**: 541–9.

Huffman, J. E. & Fried, B. 1990. *Echinostoma* and echinostomiasis. *Advances in Parasitology*, **29**: 215–69.

Kristoffersen, R. 1988. A new species of parasite in Arctic charr, *Salvelinus alpinus* (L.) in seawater cages in Scandinavia. *Aquaculture*, **71**: 187–91.

Krull, W. H. 1930. The life history of two North American frog lung flukes. *Journal of Parasitology*, **16**: 207–12.

Krull, W. H. 1931. Life history of two frog lung flukes, *Pneumonoeces medioplexus* and *Pneumobites parviplexus*. *Transactions of the American Microscopical Society*, **50**: 215–77.

McDaniel, J. S. 1966. Excystment of *Cryptocotyle lingua* metacercaria. *Biological Bulletin*, **130**: 369–77.

Milleman, R. E. & Knapp, S. E. 1970. Biology of *Nanophyetus salmincola* and 'salmon poisoning' disease. *Advances in Parasitology*, **8**: 1–41.

Miyazaki, I. 1978. Two types of lung flukes which have been called *Paragonimus westermani* (Kerbert, 1971). *Medical Bulletin of Fukuoka University*, **5**: 251–63.

Odening, K. 1964. Zur Taxonomie der Trematodenunterordnung Plagiorchiata. *Monatsberichte der Deutschen Akademie der Wissenschaften zu Berlin*, **6**: 191–8.

Pan, P. C. 1966. Comparative studies on some problems of *Paragonimus westermani* in experimental animals. *Chinese Medical Journal*, **13**: 370–86.

Prudhoe, S. & Bray, R. A. 1982. *Platyhelminth Parasites of the Amphibia*. Oxford University Press.

Raĭkhel, A. S. 1971. (Development of parthenogenetic generations of *Haplometra cylindracea* Zeder, 1800.) *Parazitologiya*, **5**: 441–5. (In Russian.)

Rees, F. G. 1939. Studies on the germ-cell cycle of the digenetic trematode *Parorchis acanthus* Nicoll. Part I. Anatomy of the genitalia and gametogenesis of the adult. *Parasitology*, **31**: 417–33.

Rees, F. G. 1940. Studies on the germ-cell cycle of the digenetic trematode *Parorchis acanthus* Nicoll. Part II. Structure of the miracidium and germinal development in the larval stages. *Parasitology*, **32**: 372–91.

Rees, F. G. 1966. Light and electron microscope studies of the redia of *Parorchis acanthus* Nicoll. *Parasitology*, **56**: 589–602.

Rees, F. G. 1967. The histochemistry of the cystogenous gland cells of *Parorchis acanthus* Nicoll, and some details of the morphology and fine structure of the cercaria. *Parasitology*, **57**: 87–110.

Sakaguchi, Y. & Tada, I. 1975. Studies on the chromosomes of helminths (1), A comparative study of the karyotypes of *Paragonimus ohirai* and *Paragonimus miyazakii*. *Japanese Journal of Parasitology*, **24**, (2, suppl.): 44–5.

Sakaguchi, Y. & Tada, I. 1976. Chromosomes of lung fluke, *Paragonimus westermani*. *Chromosome Information Service*, **20**: 23–4.

Sakaguchi, Y. & Tada, I. 1980. Karyotypic studies of lung flukes, *Paragonimus iloktsuenensis*, *P. sadoensis* and *P. westermani* with special reference to gametogenesis in *P. westermani*. *Japanese Journal of Parasitology*, **29**: 251–6.

Schell, S. C. 1965. The life history of *Haematoloechus breviplexus* Stafford, 1902 (Trematoda: Haplometridae McMullen, 1937), with emphasis on the development of the sporocysts. *Journal of Parasitology*, **51**: 587–93.

Schell, S. C. 1985. *Handbook of Trematodes of North America north of Mexico*. University Press of Idaho, Idaho, USA.

Schlegel, M. W., Knapp, S. E. & Milleman, R. E. 1968. 'Salmon poisoning disease'. V. Definitive hosts of the trematode vector, *Nanophyetus salmincola*. *Journal of Parasitology*, **54**: 770–4.

Smyth, J. D. & Smyth, M. M. 1980. *Frogs as Host–Parasite Systems I*. The Macmillan Press, London.

Stunkard, H. W. & Cable, R. M. 1932. The life history of *Parorchis avitus*, a trematode from the cloaca of the gull. *Biological Bulletin*, **62**: 328–38.

Taraschewski, H. 1984. Heterophyiasis, an intestinal fluke infection of man and vertebrates transmitted by euryhaline gastropods and fish. *Helgoländer Meeresuntersuchungen*, **37**: 463–78.

Taraschewski, H. 1985. Transmission experiments on the host specificity of *Heterophyes* species. *Zeitschrift für Parasitenkunde*, **71**: 505–18.

Terasaki, K. 1977. Studies on chromosomes of lung fluke in Japan. *Japanese Journal of Parasitology*, **26**: 222–9.

Terasaki, K. 1980. Comparative studies on the development of germ cells between *Paragonimus westermani* (Kerbert, 1878) and *P. pulmonalis* (Baez, 1880). *Japanese Journal of Parasitology*, **29**: 127–36.

Threlfall, W. 1967. Studies on the helminth parasites of the herring gull, *Larus argentatus* Pontopp., in Northern Caernarvonshire and Anglesey. *Parasitology*, **57**: 431–53.

Underwood, H. T. & Dronen, N. O. 1977. The molluscan intermediate hosts for species of *Haematoloechus* Looss 1899 (Digenea: Plagiorchiidae) from Raniid Frogs of Texas. *Journal of Parasitology*, **63**: 122.

Whitelaw, A. & Fawcett, A. R. 1982. Biological control of the liver fluke. *Veterinary Record*, **110**: 500–1.

Wood, B. P. & Matthews, R. A. 1987. The immune response of the thick-lipped grey mullet, *Chelon labrosus* (Risso, 1826), to metacercarial infections of *Cryptocotyle lingua* (Creplin, 1825). *Journal of Fish Biology*, **31** (Suppl. A): 175–83.

Yamaguti, S. 1958. *Systema Helminthum*. Vol. I, Pts I & II. *The Digenetic Trematodes*. Interscience Publishers, New York.

Yokogawa, M. 1965, 1969. *Paragonimus* and paragonimiasis. *Advances in Parasitology*, **3**: 99–158; **7**: 375–87.

Yokogawa, M. & Sano, M. 1968. (Studies on intestinal flukes. IV. On the development of the worm in the experimentally infected animals with metacercariae of *Metagonimus yokogawai*.) (In Japanese; English summary.) *Japanese Journal of Parasitology*, **17**: 540–5.

16

Digenea: Schistosomatidae

16.1 General account

Members of this family show morphological and physiological peculiarities which set them apart from all other trematodes. Firstly, they are dioecious, the male bearing the female in a ventral canal, the *gynaecophoric* canal, and secondly, they live in the blood stream of warm-blooded hosts, being the only trematodes to do so.

Only one other family shows sexual dimorphism, the Didymozoidae, cyst-dwelling trematodes of fish, and only two other trematode families, the Spirorchidae and the Sanguicolidae, occur in the blood stream; these are parasites of cold-blooded hosts. Although the life cycles are known, in general, these organisms have not received much attention; Smith (1972) gives a valuable review of their biology.

In mammals, the peripheral blood stream as an environment is relatively poor in carbohydrates and protein break-down products of low molecular mass. On the other hand, the portal system, which carries intestinal break-down products from the duodenum, is rich in glucose and amino acids, so that together with the protein available in the plasma and blood cells, it would represent an environment capable of satisfying the metabolic demands of an egg-producing trematode. That this is so is evidenced by the efficiency with which a number of species of schistosomes grow and reproduce there.

16.1.1 Species infecting man

Five species of *Schistosoma* are pathogenic parasites of man. Of these the chief ones are *S. mansoni*, *S. haematobium* and *S. japonicum*, with *S. mekongi* and *S. intercalatum* having a more limited distribution. Other species, *S. matheei* and *S. bovis*, are occasionally parasites of man and *S. incognitum* may also prove to be infective to humans. The disease caused by schistosomes, *schistosomiasis* (= *bilharziasis*), is the most important disease of helminth origin and causes untold misery in some 75 countries (WHO, 1985). Some 200 million people are probably infected and 500–600 million more exposed to infection (Webbe, 1981).

A vast literature on schistosomiasis exists, with many thousands of new publications appearing annually. Recent valuable books or reviews on schistosomiasis or on the biology of schistosomes are those of Abdel-Wahab (1982), Bruce & Sornmani (1980), Bundy (1984), Chen & Mott (1989), Jordan (1985), Jordan & Webbe (1982), McLaren (1980), Rollinson & Simpson (1987), Smyth & Halton (1983), Webbe (1981, 1987) and WHO (1985).

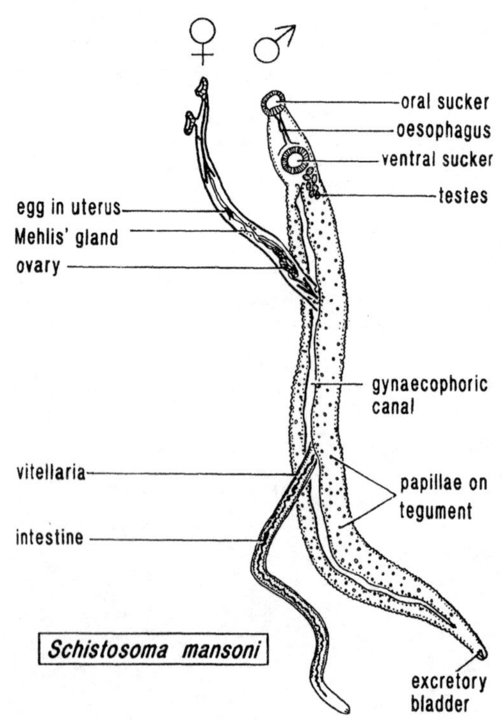

Fig. 16.1
Schistosoma mansoni: male and female worms (from mesenteric veins) *in copula*. (Adapted from various authors.)

Table 16.1. *The relative susceptibilities of common laboratory mammals to infections by* Schistosoma mansoni

Host	Maturing per cent	Fatality per cent	With eggs in faeces
Hamster	31.8	71	++
Mouse	22.1	49	+
Cotton rat	17.2	12	+
Guinea pig	6.4	9	0
Cat	10–45	0	0
White rat	2.6	1	0
Rabbit	0–18	0	0
Dog	0	0	0

Source: Data from Stirewalt *et al.* (1951).

Suitable laboratory species. One of the species attacking man, *Schistosoma mansoni* (Table 16.1), is readily maintained in the laboratory in mice, hamsters and cotton rats and because of its wide usage has been selected as the type example here. On account of its non-pathogenicity to man, the rodent parasite *Schistosomatium douthitti*, which has been less studied, is, however, a safer and often more convenient form for routine laboratory investigations. Its life cycle will be discussed later (p. 246).

○○○ 16.2 Type example: *Schistosoma mansoni*

definitive hosts: man, mice, hamsters (Table 16.1)
location: mesenteric veins
molluscan hosts: species of the genus *Biomphalaria*

○ 16.2.1 Morphology

General. The adult worms are beautifully adapted to life within blood vessels. The male is actually flat but has the sides of the body rolled ventrally to form a *gynaecophoric canal* bearing the long narrow body of the female. The cuticle of the male is covered with minute papillae. In the female these only occur at the anterior and posterior ends, the middle region being mainly held in the gynaecophoric canal of the male. Oral and ventral suckers are present, the latter being larger and more muscular in the male than the female and serve to maintain the position of the worms within the blood vessels against the circulatory current.

Intra-specific variation. It is well recognised that there is considerable intra-specific variation in the morphology, physiology, biochemistry and immunology of schistosomes found in different parts of the world. This has created the need for a special terminology. Rollinson & Southgate (1987) propose the following: *sample* = a specimen collected on a unique occasion at a specific location; *isolate* = 'viable parasites present in experimental definitive or intermediate hosts after the introduction of a sample'; *strain* = 'isolates of one species sharing particular characters and usually derived from known location'; *line* = 'laboratory parasite derived from an isolate and passaged in different hosts or selected under different conditions'. With some qualifications, these terms can be applied to many protozoan and helminth species showing intra-specific variations. This question is considered further on p. 243.

The alimentary canal is unusual in that the paired intestinal caeca rejoin about the middle of the body and continues as a single winding tube ending blindly posteriorly. The nervous, muscular and excretory systems present no unusual features.

Reproductive system. Unusually for trematodes, the sexes are separate but, as mentioned the female lives in permanent *copula* with the male, the latter curving its body into a *gynaecophoric canal* in which the female is grasped (Fig. 16.1).

Male. There are 6–9 testes with efferent ducts leading to a vas deferens which swells to form a seminal vesicle opening by a non-muscular cirrus tube into the genital pore situated just posterior to the ventral sucker. The diploid number of chromosomes is $2n = 16$.

Female. The vitellaria occupy the posterior part of the body and alternate on either side of a single median vitelline duct. The latter leads into a well-developed ootype with which the uterus and oviduct bear the usual relationship (Fig. 16.2). The proximal part of the oviduct acts as a receptaculum seminis. A Laurer's canal is lacking. The shape of the egg, which varies diagnostically in the three human species, appears to be determined by the shape of the ootype (Fig. 16.2); this varies strikingly in the different species (Gönnert, 1955*b*).

The egg of *S. mansoni* is characterised by a large, backward-pointing spine which is revealed under EM and scanning microscopy to be made up of minute spines 0.28 μm × 0.5 μm.

Female maturation: male/female interactions. This topic has been reviewed by Erasmus (1987). It has

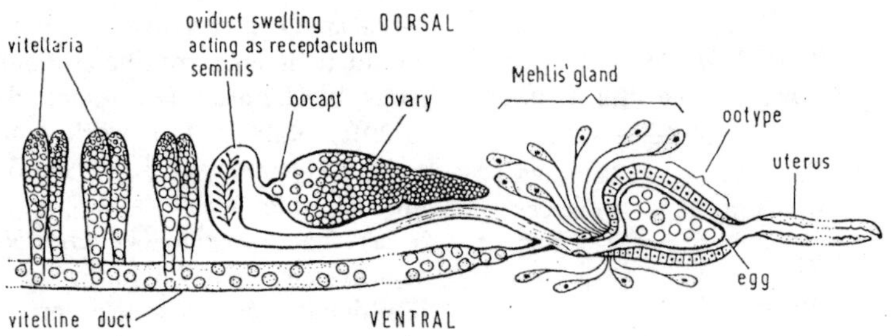

Fig. 16.2
Morphology of female genitalia of *Schistosoma mansoni*. (After Gönnert, 1955*b*.)

now been unequivocally demonstrated that maturation of the female is dependent on the process of pairing with the male worm. There is evidence of some form of chemical stimulus, passing between male and female, possibly involving a hormone or some nutrient material (Popiel & Basch, 1984; Torpier *et al.*, 1982). Extracts of male worms have also been shown to stimulate the maturation of the vitellaria in the female.

In an interesting approach in an attempt to isolate a possible stimulating substance, Basch & Nicolas (1989) experimented with artificial 'surrogate worms' made of alginate fibres which were found to be grasped and held by male worms *in vitro*. Examination of these fibres by polyacrylamide gel electrophoresis revealed the presence of two substances at 40 and 46 kDa which appeared to represent materials released by the male during pairing. The nature of such substances has yet to be determined.

Pairing between worms is not confined to male + female, and '*homosexual*' and *multiple* pairing between males/males and males/females can occur. *Inter-specific* mating can also occur and a number of crosses between different species have been obtained (Table 16.6). In unisexual infections involving a female only, the vitellaria develop but do not 'mature' (i.e. do not produce egg-shell material). In male single-sex infections, the worms become sexually mature. Some may develop female characteristics and are known as 'hermaphrodite males'; the latter may also occur in 'normal' mixed male/female infections.

Curiously enough, insemination is not needed for maturation of females and sperm need not be present; the only requirement appears to be prolonged residence in the gynaecophoric canal of a mature male (Basch & Basch, 1984).

Ultrastructure. Various aspects of the ultrastructure have been reviewed by Hockley (1973) and McLaren (1980). In general, the fine structure is similar to that of most other trematodes (Chapter 13) with the exception of the tegument, which has been extensively studied (McLaren, 1980). The main point of interest has been the changes in the structure of the tegument during the cercaria–schistosomulum transformation, which occurs after penetration of the host skin by a cercaria. This process, which involves the thick glycocalyx and subsurface layer in the schistosomulum being transformed into a remarkable heptalaminate membrane, has already been described (p. 181; Fig. 13.3).

○ 16.2.2 Life cycle (Fig. 16.3)

Almost every stage of the life cycle of schistosomes presents problems of enthralling biological interest and the life cycle of *S. mansoni* has been especially studied.

Laboratory maintenance. Technical details for the large-scale maintenance of *S. mansoni* in the laboratory are given in considerable detail by Lewis *et al.* (1988) who emphasise that different strains may require different culture conditions. Simplified maintenance procedures for this species and for the rodent *Schistosomatium douthitti* are also summarised in Table 16.2.

Adult worms. The adult worms normally reside in pairs in the mesenteric veins; but the females need to be carried there by the males. In unisexual infections (p. 250) the male produces viable sperm in the absence of the female. As mentioned earlier, females do not achieve sexual maturation in the absence of the male, since the vitellaria and ovary do not become functional. The females lay only one egg at a time, and they retreat slightly along the venule after each egg is deposited. The worm is so orientated when oviposition is taking place that an egg is laid with the spine pointing back-

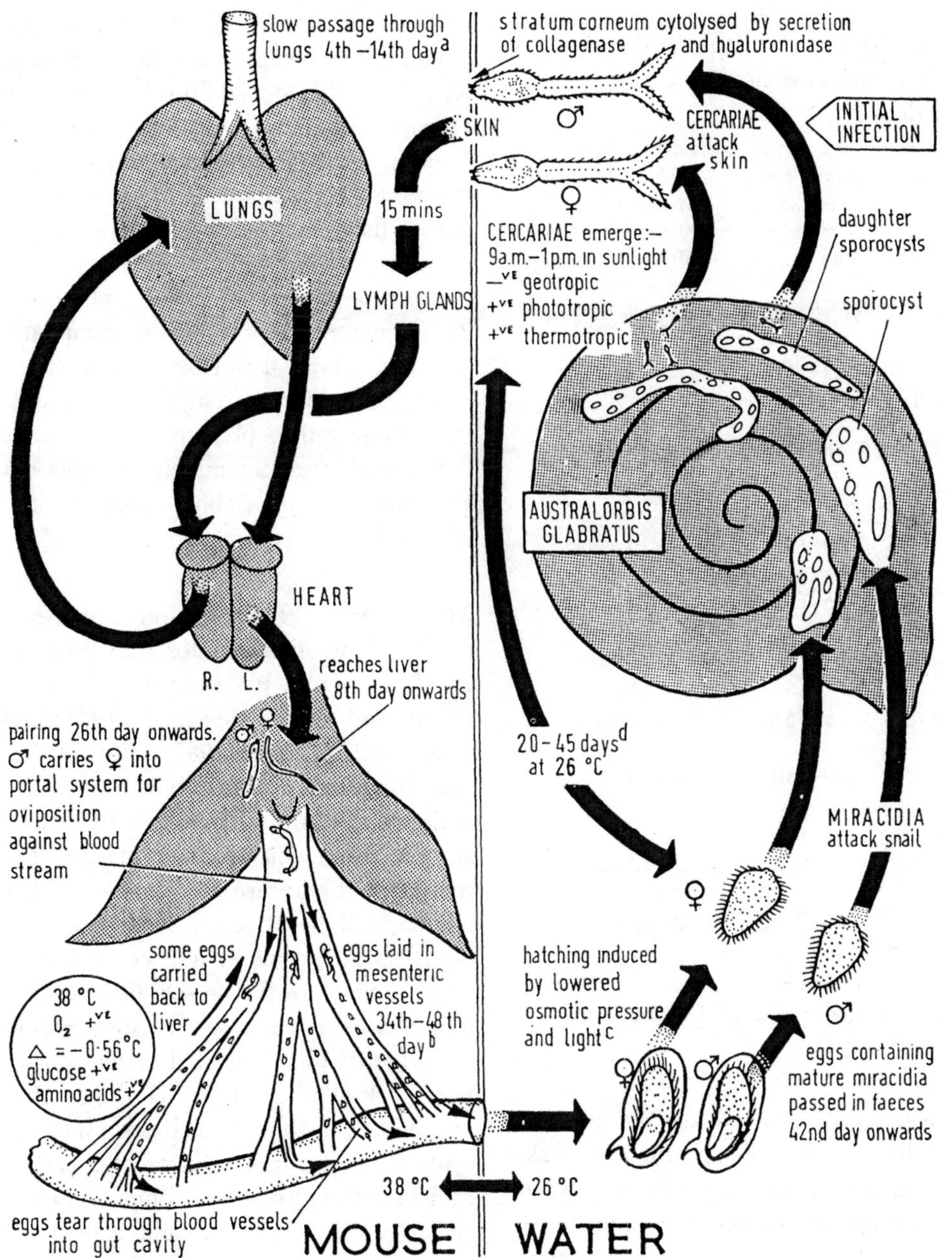

Fig. 16.3
Life cycle of *Schistosoma mansoni* in the albino mouse. Details of developmental times taken from the following: (*a*) Olivier (1952); (*b*) Standen (1953*a*); (*c*) Standen (1951); (*d*) Standen (1949). (Original.)

wards; this spine tends to catch on the intima of the blood vessels and prevents the egg being swept away by the blood current. When laden with eggs, the vessels eventually rupture so that eggs find their way into the mucosa and submucosa, and hence via the intestinal lumen to the faces. Some eggs are enclosed in scar tissue and destroyed, and others are carried by the venous current to the liver and spleen, and even to other organs such as the lungs, brain and spinal cord. A pair of worms produce about 100–300 eggs daily. The pathology has been reviewed at length by Chen & Mott (1989). Qualitative techniques for the detection of eggs in faeces in clinical practice are reviewed by Jordan & Webbe (1982).

Table 16.2. *Summary of practical details for the laboratory maintenance of* Schistosoma mansoni *and* Schistosomatium douthitti *in mice*

	Schistosoma mansoni	*Schistosomatium douthitti*
Definitive laboratory host	mouse	mouse
Molluscan laboratory host	*B. glabrata*	*L. palustris*
Maintenance temperature (snail breeding)	24–25 °C	22–24 °C
Maintenance temperature (snail infection)	26–28 °C	22–24 °C
Age of snails for infecting	10–14 weeks	3–10 weeks
Exposure time of snails to miracidia	5–6 hours	20–24 hours
Number of miracidia per snail	15	3–5
Time for cercariae to emerge	15–75 days	37–52 days
Number of cercariae per mouse for infection	130–150	50–75
Time of attainment of sexual maturity	28 days	10–12 days
Time for eggs to appear in faeces	5–7 weeks	26 days

Source: Figures based on Kagan *et al.* (1954) and Standen (1949).

Fate of eggs. Eggs which fail to reach the intestinal lumen are eventually phagocytosed or, more rarely, calcified, in the various tissue sites in which they occur. Eggs when laid, are unembryonated and require 6–7 days to embryonate in the tissues. After an infection has been patent for some time eggs in the liver contain miracidia; only 18–34 per cent of such eggs embryonate in man and experimental animals. The total life of a miracidium within an egg is about 20 days, after which time the miracidium degenerates and the egg-shell bursts and becomes phagocytosed by giant cells.

Hatching. Early work (reviewed by Smyth, 1976) suggested that light, temperature and osmotic pressure were major factors in the hatching of *S. mansoni* eggs. More critical recent work (Kassim & Gilbertson, 1976) has demonstrated that no significant difference could be observed between the hatching of eggs placed in the light and those kept in the dark, the experiments being terminated by the addition of 10 per cent formalin. However, other work has confirmed early experiments (see Fig. 13.13) that osmotic pressure is the major factor inducing hatching (Kusel, 1970), even dead eggs (showing no flame cell activity) hatching in solutions of low osmotic pressure.

Schistosoma eggs hatch at an optimum temperature of 28 °C; the process is almost completely inhibited at 4 °C and at body temperature 37 °C (Fig. 13.11). There is no operculum, so that if enzymes are concerned in the hatching they presumably attack and weaken the entire shell and not merely an operculum-cementing substance as probably happens in some other trematodes.

Miracidium. The miracidium of *S. mansoni* is similar to that of the other schistosomes (Figs 13.14, 16.6) and its ultrastructure has been described in detail by Pan (1980) and Eklu-Natey *et al.* (1985) and reviewed by Jourdane & Therón (1987). Most of the details of the fine structure are clear from Fig. 16.5. The anterior end is drawn out into a *terebratorium* (the *anterior papilla* of some authors) which contains at least 12 ciliated sensory receptors. There are well developed longitudinal and circular muscle systems, beneath the external ciliated layer. The nervous system consists of a neural mass which innervates the muscle fibres and numerous peripheral sensory papillae. The surface of the miracidium is covered with 21 ciliated plates which are separated from each other by epidermal ridges which connect the cell bodies (cytons), which are syncytial, with narrow cytoplasmic bridges. During the early stages of the transformation from miracidium to sporocyst, the cytons show greatly increased cellular activity with the appearance of membrane-bound bodies. These vesicles appear to store the membrane which is utilised in the formation of the tegument of the mother sporocyst. For further details of the fine structure, see Pan (1980). Miracidia are markedly phototropic and move to the bright side of the container in which hatching occurs. This behaviour may be used to obtain concentrations of miracidia.

A darkened flask with an illuminated 'side arm' has long been used to separate hatched miracidia of schistosomes (and other trematodes) from faecal debris containing eggs. A more effective modification of this, designed by Rau *et al.* (1972), is shown in Fig. 16.4.

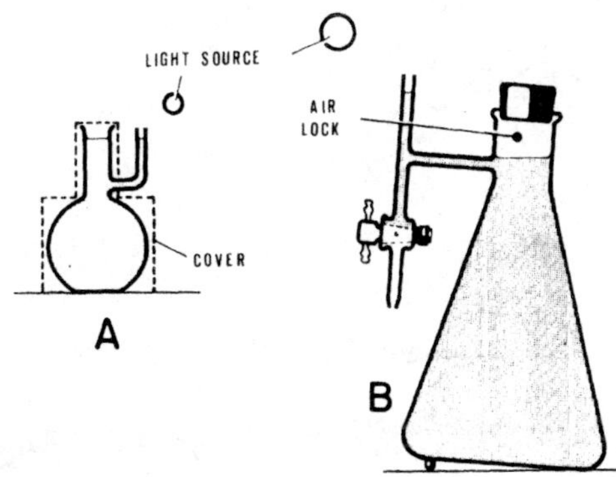

Fig. 16.4
Side-arm flasks for the recovery of newly hatched miracidia; these exploit the negative geo- and positive phototaxes of miracidia, which move upwards towards the light source. A. Original flask of McMullen & Beaver (1945). B. Modified flask of Rau *et al.* (1972); this utilises a 2 l flask (uncovered), slightly tilted. Miracidia move to the surface, then along the side arm towards the light and into the vertical arm; they may be then withdrawn from the latter by the stopcock. (After Rau *et al.*, 1972.)

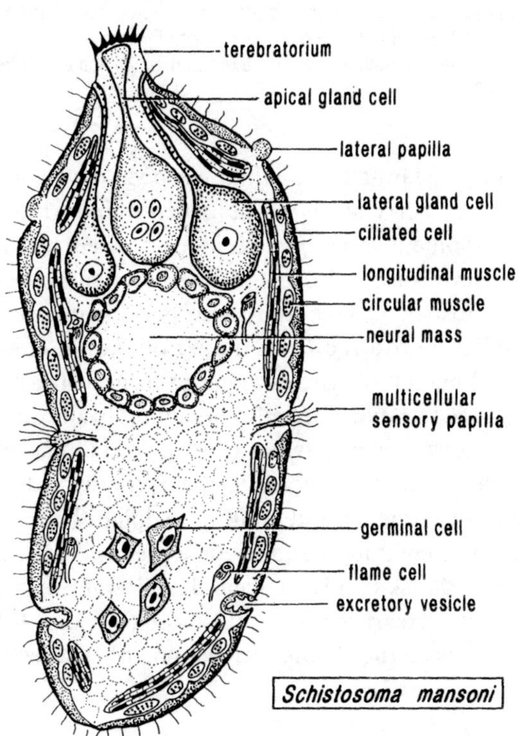

Fig. 16.5
A schematic cellular architecture of the miracidium of *Schistosoma mansoni*. (Adapted from Jourdane & Therón, 1987, based on Pan, 1980.)

Molluscan host. The molluscan hosts are species of the genus *Biomphalaria*, which is widely distributed in Africa and occurs in Egypt, Israel and some parts of the Arab world, except Iraq. It is also found in the Caribbean Islands, the southern United States and in S. America. The taxonomy of the genus is complex with much intra-specific variation. Although many species have been infected experimentally, only three species in the Western hemisphere (*B. glabrata*, *B. straminea* and *B. tenagophila*) are found naturally infected. *B. glabrata* is widely used in the laboratory for experimental infections.

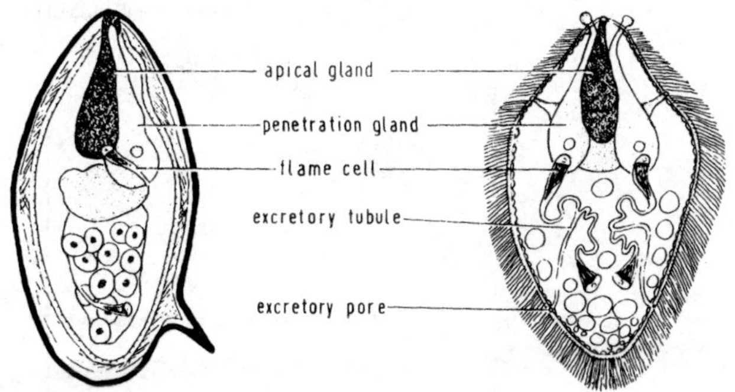

Fig. 16.6
Schistosoma mansoni: mature egg and hatched miracidiuim. (After Faust, 1939.)

Intra-molluscan development. Factors involved in the penetration of miracidia of *S. mansoni* into the snail host have been discussed by Wajdi (1966) and Pan (1980). Only a small proportion of miracidia which penetrate develop to mature mother sporocysts; these are usually located in the foot. It was formerly thought that mother sporocysts simply gave rise to daughter sporocysts which, after migrating to the digestive gland, gave rise to furcocercariae. More recent studies have revealed that the pattern of intra-molluscan development is apparently much more complex (Jourdane & Therón, 1987). Thus, under certain conditions, daughter sporocysts can produce further generations of daughter sporocysts (sporocystogenesis) and/or undergo cercaria production (cercariogenesis). The results of these studies are illustrated in Fig. 16.7.

In *B. glabrata*, these intra-molluscan stages develop at an optimum temperature range of 26–28 °C (Table 16.2). At 26 °C, the time required for development from miracidium to emergence of cercariae varies from 15 to 75 days in *B. glabrata* and 15 to 100 days in *Biomphalaria boissyi*.

Emergence of cercariae is periodic, and in nature tends to occur in direct sunlight between 09.00 and 14.00 hours, but this process is inhibited or partly inhibited by temperatures lower than about 21 °C.

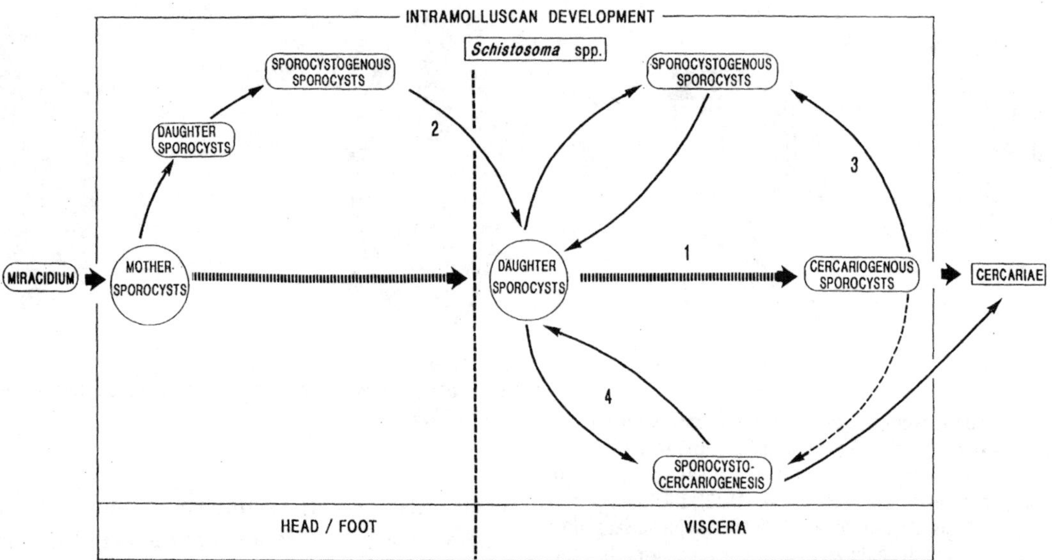

Fig. 16.7
A proposed new interpretation of the intracellular larval development of schistosomes for which several pathways are possible:
1. *Cercariogenous* sporocysts producing only cercariae;
2. *Sporocystogenous* sporocysts producing a new generation of daughter sporocysts by direct *sporocystogenesis*, or by
3. *Sporocystogenesis* taking place after *cercariogenesis*; or by
4. Simultaneous *sporocysto-cercariogenesis*. (Adapted from Jourdane & Therón, 1987; reprinted with permission from *Biology of Schistosomes* (ed. R. Rollinson & J. C. Simpson), p. 87. Academic Press, London, 1987.)

Cercariae. It is claimed that minor differences exist between the cercariae of the three schistosomes of man, although some workers believe that no reliable criteria for their differentiation exist. In general, the gross morphology follows the normal trematode pattern (Chapter 13). The fine structure of the cercaria has been described in detail by McLaren (1980) and the structural changes which take place during the cercaria–schistosomulum transformation have already been described (Chapter 13, Fig. 13.3). As might be predicted, when a cercaria passes from its aqueous environment into a tissue environment and undergoes transformation to a schistosomulum, substantial physiological and biochemical adaptations become necessary for survival. For example, a cercaria (which is water-tolerant) on penetrating the skin and transforming to a schistosomulum, immediately becomes water-intolerant. The numerous other adaptations which occur in the cercaria of *S. mansoni* are summarised in Table 16.3. Those occurring in *S. japonicum* have also been studied (Irie & Yasuraoka, 1981).

Invasion and maturation. Human infection is brought about by bathing or wading in infected waters. Laboratory animals may be infected by exposing them to water containing cercariae or simply by subcutaneous injection. A quantitative method has been developed by Smithers & Terry (1965). The penetration of the skin and development and migration in mammalian hosts has been reviewed by Wilson (1987). Cercariae appear to locate a host by chance and there is no evidence that chemotaxis is involved. Penetration of skin is effected by a combination of lytic outpouring from 5–6 pairs of cephalic (penetration?) glands containing hyaluronidase and collegenase-like enzymes (Smyth & Halton, 1983). Cercariae attach themselves to the epidermis, destroy the stratum corneum and stratum malpighi by enzyme action, and penetrate first at right angles and then later parallel to the skin surface (Standen, 1953*b*). In mice, only a few cercariae die in the skin (Wilson, 1987). Cercariae reach the venous circulation either directly or via the lymph vessels.

The stages of development are best known in the mouse; the following details are those of Clegg (1965) and Wilson (1987). In the venous system, schistosomula are carried to the lungs within 3–4 days and reach their highest concentration at 4–6 days (Miller & Wilson, 1980). From eight days onwards, schistosomula can be found in the portal vessels of the liver. Coupling of worms occurs for the first time in the mesenteric and portal veins at 28 days but the majority of worms leave the liver on about the 30th day. Egg-shell protein can first be detected in vitellaria on the 30th day (i.e. they become Fast Red $+^{ve}$; see Table 13.1). The first eggs are produced on the

Table 16.3. *The cercaria–schistosomulum transformation: differential characteristics of cercariae and schistosomula of* Schistosoma mansoni

	Cercariae	Schistosomula
1	Tail present[a]	Tail separated from body[b]
2	Water-adapted	Water-intolerant[b]
3	Serum complement sensitive	Serum complement insensitive[b]
4	Silhouette precise	Silhouette not rigidly defined; body flaccid and worm-like[b]
5	Glycocalyx intact	Glycocalyx lost, interrupted or modified[b]
6	CHR-positive in antiserum	CHR-negative in antiserum[b]
7	RBC binding loose; oral	RBC binding tight; overall
8	Surface membrane trilaminate; not pitted; no surface microvilli	Surface membrane heptalaminate; pitted; transient surface microvilli[b]
9	Surface PAS and alcian blue-positive	Surface PAS and alcian blue-negative[b]
10	Membranous inclusions in tegument few	Membranous inclusions in tegument numerous[b]
11	Surface unstable in selected chemicals	Surface stable in selected chemicals
12	Amino acid uptake low	Amino acid uptake high
13	Fluoride-insensitive	Fluoride-sensitive
14	Not cultured except in Rose chambers	Cultured without dialysis membranes[b]
15	Oral sucker alternately protruded and retracted	Oral sucker permanently protruded[b]
16	Pre- and post-acetabular glands full	Pre- and post-acetabular glands evacuated or nearly so[b]
17	Enzyme activity in extract high	Enzyme activity in extract absent or low
18	No ingestion	Ingestion
19	Metabolism for energy	Metabolism for synthesis
20	Locomotion by alternate attachment of suckers	Locomotion restricted, worm-like; constant elongation and contraction[b]

CHR = Cercarienhüllenreaktion; RBC = red blood cells; PAS = periodic acid–Schiff.
[a]Characteristics 1–13 are related to the surface change as cercariae transform to schistosomula.
[b]Characteristics *in vivo* of post-penetration larvae, which are schistosomula by definition.
Source: Data from Stirewalt (1974).

34–35th day. In normal mixed infections in mice, eggs appear in the faeces with 5–7 weeks.

Choice of laboratory host. The relative susceptibilities of some laboratory mammals to infections of *S. mansoni* are listed in Table 16.1. Some 54 species of host have been infected. Although mice, hamsters and cotton rats are satisfactory hosts, each has its advantages or limitations (Stirewalt *et al.*, 1951). For particular lines of investigation, therefore, the most suitable host must be selected. Thus, although the hamster has the highest percentage of maturing schistosomes and the largest number of available eggs in the faeces, its high death rate is a disadvantage. Where large well-developed adult worms are required, the cotton rat is the most suitable choice. However, mice are the most widely used laboratory host.

Intra-specific and inter-specific genetic variation. All species of schistosome exhibit much variation between different species, and also intra-specific variation in isolates collected from different field situations. Important characteristics which show variation are: egg size and shape, infectivity to snails, patterns of cercarial emergence, response to chemotherapy, development in experimental hosts, growth rate, egg productivity, pre-patent period, immunogenicity and pathology (Rollinson & Simpson, 1987). Genetic variation between species and/or strains has been especially examined by isoenzyme analysis and, more recently, by DNA restriction enzyme analysis. Some relevant studies on isoenzyme analyses of isolates of different *Schistoma* species are those of Agatsuma & Suzuki (1983), Fletcher *et al.* (1981), Richards (1975) and Viyanant & Upatham (1985). An example of isoenzyme analysis showing inter- and intra-specific differences between some species is shown in Fig. 16.8.

○○○ 16.3 Other species attacking man

○ 16.3.1 *Schistosoma haematobium*

Geographical distribution. This 'species' is prevalent in Africa, especially in the Nile Valley where up to 95 per cent of the local rural population may be infected. As with *S. mansoni*, its distribution runs parallel to irrigation projects and in general wherever local conditions

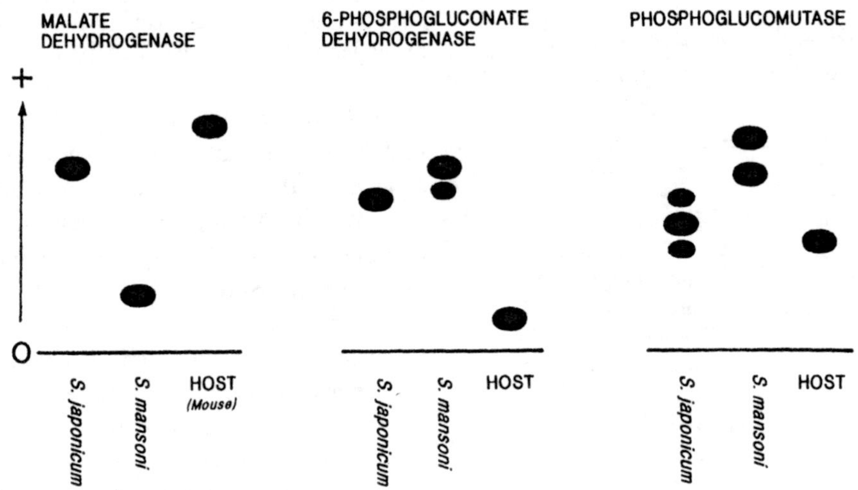

Fig. 16.8
Diagrammatic representation of electrophoretic pattern of three isoenzymes in *Schistosoma mansoni*, *S. japonicum* and the mammalian host. (Adapted from Agatsuma & Suzuki, 1983.)

are favourable for the molluscan vectors. It also occurs in the islands of Mauritius and Malagasy and has been reported from India.

Morphology. In general, it resembles *S. mansoni*, but differs from it in a number of minor points:

(*a*) both male and female worms are longer;
(*b*) there are only four large testes (6–9 in *mansoni*);
(*c*) the ovary is in the posterior third of the body;
(*d*) the uterus contains large numbers of eggs, 10–100 at one time (*mansoni*: 1–4 eggs only);
(*e*) the eggs possess a terminal spine and are passed in urine, rarely in faeces;
(*f*) it is much more host-specific.

Habitat. Mature worms live *in copula* mainly in the tributaries of the inferior mesenteric veins and the females deposit their eggs in the walls of the urinary bladder, ureters or urethra, through which they slowly infiltrate into the urine *in which they are passed*; a few eggs may reach the rectum and appear in the faeces. As in *S. mansoni* some eggs are carried back to the liver and sometimes to other viscera.

Life cycle. Very similar to that of *S. mansoni*. Eggs hatch rapidly on dilution of urine and on exposure to light. Times for the various developmental stages to appear are incompletely known. Sexual maturity is reached in about 4–5 weeks, but eggs may not appear in the urine until 10–12 weeks, or even considerably later.

S. haematobium is almost exclusively a parasite of man, and infections which have been reported in other hosts are probably of little importance in relation to transmission (contrast *S. japonicum*, see below). It is transmitted by snails of the *Bulinus africanus* group and the *B. forskalii* group. A number of geographical strains exist and it is common practice to speak of the *S. haematobium* group (Rollinson & Simpson, 1987). *S. capense* used to be considered to be a separate species but is now considered to be a synonym of *S. haematobium*. The taxonomy is also complicated by the fact that *S. haematobium* is known to hybridise with *S. intercalatum*, a species infecting man which is endemic in NE Zaire, Gabon and Cameroun and transmitted by snails of the *Bulinus africanus* group.

There is some evidence to suggest that *S. haematobium* may be involved in the induction of cancer of the bladder in man, but only in some areas and not in others. The pathology has been reviewed in detail by Jordan & Webbe (1982) and Chen & Mott (1989).

Laboratory hosts. Albino mice, hamsters, monkeys, baboons are good hosts for experimental studies. Albino rats, cotton rats, guinea pigs are poor hosts; the numbers of parasites are characteristically reduced in the early stages. Rabbits, dogs and cats are almost completely refractory to infection.

Of these hosts, the hamster is the best all-round for general studies but egg-production varies with the strain of *S. haematobium* (Webbe & James, 1971). In it, the schistosomes reach the veins in the vicinity of the urinary bladder as well as those of the lower levels of the large intestine.

○ 16.3.2 *Schistosoma japonicum*

Schistosomiasis due to *S. japonicum* differs from that due to *S. mansoni* and *S. haematobium* in that it is a *zoonosis* in which a large number of mammalian species serve as reservoir hosts – some 31 wild species being involved – domestic cattle, dogs and pigs playing a major role (Kumar, 1987; Rollinson & Southgate, 1987). It is confined in its distribution to the Far East, the infected areas being China, Japan, the Philippines, Celebes, Thailand and Laos. In man and animals, it causes a grave, chronic form of intestinal schistosomiasis. China is the largest area of endemicity, but the number of infected cases has fallen within recent years, owing to effective control measures (Kumar, 1987).

Morphology. Differs from *S. mansoni* in a number of minor points:

(*a*) male longer and narrower;
(*b*) integument free from tubercles (except for minute spines on suckers and gynecophoric canal);
(*c*) testes 6–7, characteristically compressed into a single column;
(*d*) the ovary is in the middle of the body;
(*e*) uterus has 50 or more eggs at any one time; eggs are laid in clumps;
(*f*) eggs with only short lateral spines and passed in faeces;
(*g*) has a wider host spectrum (also wider than *S. haematobium*).

Habitat. The adult worms live *in copula* chiefly in the superior mesenteric veins and deposit some 1500–3500 eggs daily in the vessels of the intestinal walls. Eggs infiltrate through the tissues and are passed *in the faeces*. As with other species, eggs are carried to the liver and may reach the other viscera.

Life cycle. This is not significantly different from that of *S. mansoni*. The molluscan vectors are various subspecies of *Oncomelania hupensis*. Numerous strains of *S. japonicum* have been isolated and characterised by isoenzyme analysis (Viyanant & Upatham, 1985); see Fig. 16.8. Sexual maturity is reached in about four weeks and eggs may appear in faeces as early as five weeks.

Pathology. The chief lesions due to *S. japonicum* occur in the intestine and liver. In the intestine, the eggs sequestered in the mucosa or submucosa initiate granulatomatous reactions resulting in the formation of pseudotubercles (Kumar, 1987). Similar lesions may occur in the liver.

Schistosoma mekongi. This 'species' is a *S. japonicum*-like schistosome causing an infection in Thailand, Laos, Kampuchea and Malaysia. The eggs are smaller and more subspherical than those of *S. japonicum* and the pre-patent period is longer. Dogs serve as reservoir hosts and *Tricula aperta* is the intermediate snail host. All aspects of schistosomiasis due to this species have been reviewed in detail by Bruce & Sornmani (1980).

○○○ 16.4 Schistosomiasis as a world problem

Schistosomiasis is undoubtedly the most important disease of helminth origin in the world today. Detailed consideration of its epidemiology is beyond the scope of this book, but as pointed out (p. 236) the extent of the problem can be seen from the fact that in some 75 countries or islands in which it is endemic 200 million of 354 million individuals are infected. The disease is well known from early human history and schistosome antigens have been found in Egyptian mummies from 1198–1150 BC (Deeler *et al.*, 1990).

Control of schistosomiasis depends basically on the control of the snail vectors, a problem which presents an extremely complex ecological situation, and one which has been reviewed in a number of major publications (Jordan & Webbe, 1982; Nozais, 1990; Rollinson & Simpson, 1987; WHO, 1985). Major control programmes have been established in a number of countries, the best documented probably being that carried out on the island of St Lucia in the Caribbean, which was subjected to an extensive well-supported project. The results have been published in a book (Jordan, 1985) which provides a valuable guide to the problems

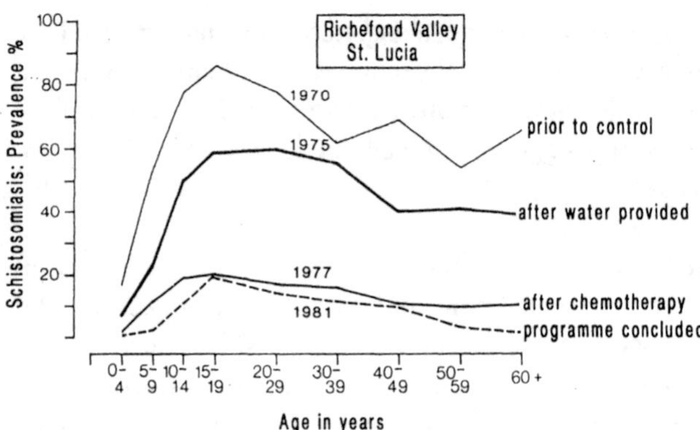

Fig. 16.9
Effects of control programme in reducing the prevalence of schistosomiasis in the Richefond Valley on St Lucia. (Modified from Jordan, 1985.)

Table 16.4. *Development of the rodent schistosome,* Schistosomatium douthitti, *in different hosts*

Host	Development
Musk rat	normal
Lynx	normal
Deer mouse	normal
Field mouse	normal
Hamster	normal
Albino mouse	normal
Rat	abnormal
Cat	abnormal
Rabbit	abnormal
Monkey	Immunity developed in three weeks

involved and the procedures adopted and serves as a valuable model for other projected control programmes. Its success is well illustrated in the fall in prevalence over the years 1970–81, as shown in Fig. 16.9.

Because of the difficulties of controlling the snail vector, enormous research efforts in many laboratories throughout the world are directed towards developing a vaccine against the adult worms. Although substantial progress has been made using experimental animals, there is little evidence that a vaccine against the human disease is likely to be obtained in the near future (Butterworth *et al.*, 1987). This question is discussed further in Chapter 32.

Other species. *S. intercalatum* occurs in man in a restricted area – the Katanga region of the Congo. Other species of economic importance are *S. bovis*, *S. spindale*, *S. nasalis* and *S. mattheei*, which are primarily parasites of ungulates; *S. mattheei* is prevalent as a human infection in southern Africa. *S. rodhaini* is found in rodents in eastern Africa; it grows well in mice, hamsters and guinea pigs (Fripp, 1967). *S. margrebowiei* occurs in a variety of mammals in Mali, Zaire, Zambia, Botswana, Chad and SW Africa (Southgate & Knowles, 1975). It can be grown in hamsters, mice and gerbils and is a valuable laboratory model (Ogbe, 1983, 1985*a*,*b*).

○○○ 16.5 Type example: *Schistosomatium douthitti*

natural definitive hosts: numerous rodents (e.g. field mice, deer, mice, albino mice, musk rats, nutria, red-backed mouse)
suitable laboratory hosts: mice, hamsters (Tables 16.2, 16.4)
location: hepatic portal system
molluscan hosts: *Lymnaea* spp., *Physa* spp., *Stagnicola* spp., *Pseudosuccinea columella* (Table 16.5)

The biology and epidemiology of this species has been reviewed by Malek (1977).

Although the human blood flukes are readily maintained in the laboratory in experimental hosts, the danger of an accidental infection is always present. Moreover, certain governments lay down strict quarantine laws regarding the passage and importation of pathological material, so that the establishment of laboratory cultures may be difficult. For these reasons, the maintenance of the rodent species *Schistosomatium douthitti*, which is non-pathogenic to man, offers many advantages. This schistosome has essentially the same life cycle as those species parasitising humans, and it is now extensively used in many laboratories for experimental work on schistosomes. An excellent account of its laboratory maintenance is given by Kagan *et al.* (1954); see Table 16.2.

○ 16.5.1 Morphology (Fig. 16.10)

Male: length, 1.9–6.3 mm. The body is divided into two distinct parts, a *prebody* which is flattened and occupies two-fifths of the organism, and a *hind-body* occupying the remaining three-fifths and forming a gynaecophoric canal. There are 14–16 testes situated between the intestinal caeca of the anterior end of the hind body. Each testis follicle opens via a short vas efferens into a single median vas deferens leading to a

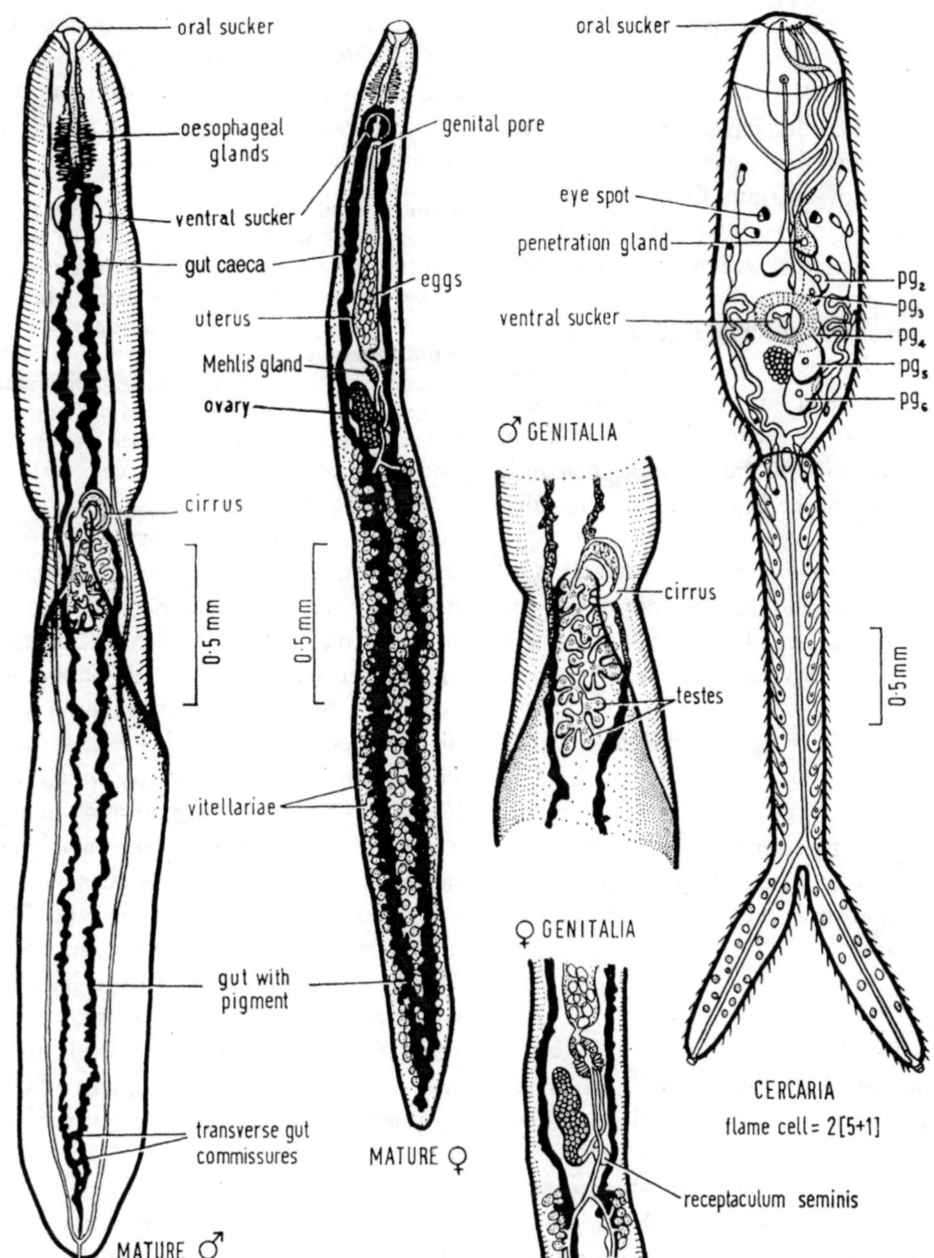

Fig. 16.10
Morphology of the rodent schistosome *Schistosomatium douthitti*. (After Price, 1931.)

seminal vesicle. The latter, together with the cirrus, is enclosed in a cirrus pouch. Spines cover most of the body.

Female: length, 1.1–5.4 mm. The essential female organs are present: ovary, oviduct, vitellaria, vitelline duct, seminal receptacle, ootype, Mehlis' gland and uterus. A Laurer's canal has not been described. The arrangement of the genitalia is shown in Fig. 16.10. The ootype is oval in shape and lined with large refractive cells. As in other species of schistosomes, the spines on female worms are limited in distribution, covering the lateral edges of the dorsal and ventral surface, and extending from the anterior part of the body posteriorly to the beginning of the ovary. The oral sucker also has spines.

When *in copula*, a female worm is held within the gynaecophoric canal, its dorsal surface against the ventral surface of the male with only its anterior end protruding.

Digestive system. The oesophagus is characterised by the presence of conspicuous oesophageal glands, presumably secreting digestive enzymes for the break-down of the blood. The oesophagus bifurcates into two intestinal caeca which have numerous lateral diverticula; these caeca unite near the posterior end of the body. Various anomalies of the gut can occur. The digestive system is better developed in the female than in the male, the intestinal caeca being wider and holding more blood. This dimorphism is clearly related to the nutritional requirements of the female and the enormous demands made on its metabolism by the egg-producing genitalia.

Egg. The egg of *S. douthitti*, which measures 42–80 μm × 50–58 μm, is smaller than that of the schistosomes of man (e.g. *S. mansoni*, 150 μm × 60 μm) and unlike them is free from a lateral spine.

Habitat. The adults live *in copula* within the mesenteric veins.

Karyotype. The karyotype consists of seven chromosome pairs ($2n = 14$) with a heteromorphic pair (ZW) in the female and a homomorphic pair (ZZ) in the male (Puente & Short, 1985).

○ 16.5.2 Life cycle

This has been described by Price (1931), El-Gindy (1950) and Malek (1977). Some practical details for laboratory maintenance are summarised in Table 16.2.

In spite of lacking a spine, the eggs are able to penetrate the intestinal wall and pass out with the faeces. The majority of eggs, however, pass to the liver where they collect and from which they may be obtained in the laboratory by maceration. Eggs hatch when the medium (faeces, urine or saline) is diluted, provided light is present; the side-arm flask technique is effective (Fig. 16.4). The morphology of the miracidium does not differ significantly from that of *S. mansoni*.

Molluscan stages. The natural snail hosts from which cercariae of this species have been obtained in the United States are listed in Table 16.5. The prevalence of natural infection is small, within the range of 3–7 per cent (Kagan *et al.*, 1954; El-Gindy, 1950). On account of easier availability, *Lymnaea stagnalis appressa* and *L. palustris* are the snails of choice for laboratory study; both are easily maintained in an aquarium at 22–24 °C. Experimental infections usually result in snail infections of 25–80 per cent.

Although snails of all ages can be infected, young snails (3–10 weeks old) are more readily infected than older ones. As with the schistosomes of man, there is no redia stage in the developmental cycle within the snail, miracidia giving rise to mother sporocysts which produce daughter sporocysts, which in turn release cercariae. The incubation period is dependent on temperature; at a maintenance temperature of 22–24 °C cercariae emerge within 37–52 days after the initial infection. As in other schistosomes, cercariae from a single miracidium produce worms of one sex (but see p. 251).

Cercariae. The morphology (Fig. 16.10) of the cercaria differs from that of the *S. mansoni* cercaria only in minor respects. Thus, there are two well-developed eye spots in *S. douthitti* (absent in human schistosomes) and the cercariae are markedly phototropic. There are six pairs of flame cells, five in the body and one in the tail. The formula is thus $2(5 + 1) = 12$, in contrast to the schistosomes of man in which the formula is $2(3 + 1)$. The behaviour pattern of emergence is, however, different. In *S. mansoni*, possibly correlated with the habits of their hosts, shedding of cercaria takes place in direct sunlight between 09.00 and 14.00 hours, but in *Schistosomatium douthitti* shedding only occurs in the evening or at night (Fig. 16.11). The reason for this pattern of periodic behaviour is unknown (Oliver, 1951). Although laboratory-infected snails tend to produce smaller numbers (500–1500), naturally infected snails have been known to produce up to 5000 cercariae in one evening.

Development within the definitive host. The natural definitive hosts in North America are the musk rat (*Ondatra zibethica*), the deer mouse (*Peromyscus maniculatus*), the meadow mouse (*Microtus pennsylvanicus*) and (more recently) the red-black mouse, *Clethrionomys gapperi* (Malek, 1977). In the laboratory, hamsters, albino mice and nutria are the most suitable hosts. In rhesus monkeys, although sexual maturity is reached, immunity develops within three weeks after which time the worms are killed. In rats, cats and rabbits development is poor or abnormal.

As far as is known, the mechanism of skin penetration resembles that of the schistosomes of man, and the five pairs of penetration glands probably produce

Table 16.5. *Molluscan intermediate hosts of the rodent schistosome* Schistosomatium douthitti *in the USA and Canada*

Snail species	Locality	Per cent infected	References[a]
Stagnicola reflexa	Near Chicago, Ill.	6.8	Cort (1914)
S. palustris	Near Ann Arbor, Mich.		Price (1931)
S. palustris	Near Boston, Mass.		Tanabe (1923)
S. palustris	Near Ann Arbor, Mich.	3	Kagan *et al.* (1954)
S. palustris	Near Ann Arbor, Mich.	14.2	Malek (1977)
S. exilis	Near Ann Arbor, Mich.		Cort (1936)
S. emarginata angulata	Douglas Lake area, Mich.		Cort (1936)
S. emarginata canadensis	Northern Mich.		van der Schalie, personal communication
Lymnaea stagnalis appressa	Douglas Lake area, Mich.		Cort (1918); Miller (1926)
L. stagnalis appressa	Gull Lake, Man.	26.5	Farley (1962*a*)
L. stagnalis appressa	Near Peterborough, Ont.	8.6	Bourns (1963)
L. stagnalis appressa	Lake Itasca area, Minn.	2.9	Malek (1977)
L. stagnalis lillianae	Wisconsin		Brackett (1940)
L. stagnalis jugularis	Wisconsin		Brackett (1940)
L. stagnalis sanctaemariae	Wisconsin		Brackett (1940)
Physa gyrina eliptica	Near Ann Arbor, Mich.		Price (1931)
Physa parkeri	Douglas Lake area, Mich.		Cort (1918)

[a]See Malek (1977).
Source: Data from Malek (1977).

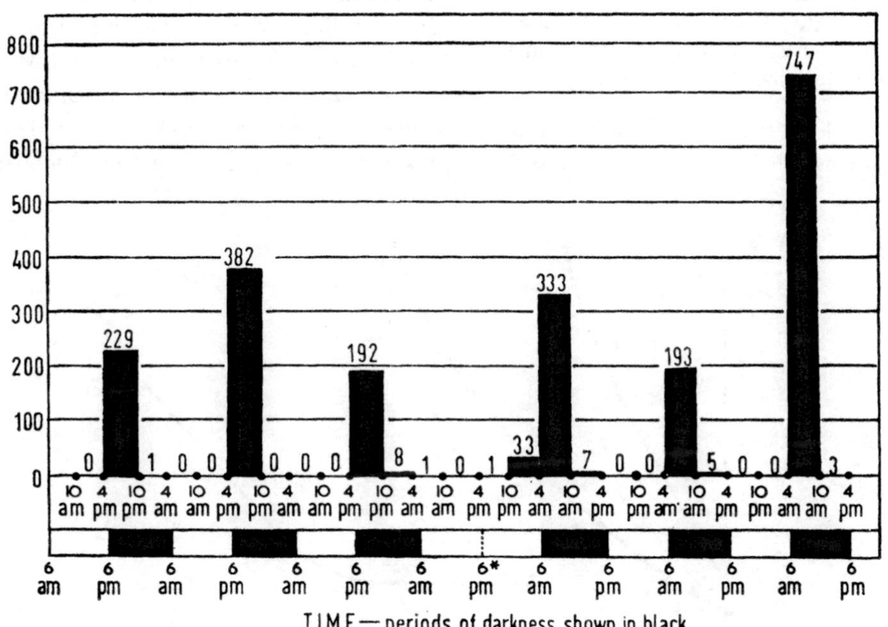

Fig. 16.11
Effect of light on the emergence of cercariae of *Schistosomatium douthitti*. Note how the pattern changes when, at the point *, the snails are kept in the dark between 4 and 10 a.m. (After Olivier, 1951.)

hyaluronidase and collagenase. The route of migration from the skin to the portal system is uncertain; cercariae may reach the blood system directly by piercing capillaries, or entry may be via the lymph system. According to one view the liver is reached through the pleural cavity. The schistosomulae develop within the vessels of the liver from which they migrate into the mesenteric veins about 10–11 days after infection; by about the 13th day practically all worms are in the extra-hepatic portal vessels, and are either sexually mature, or nearly so. All worms are usually mature by 20 days.

This rapid maturation contrasts strangely with that of *S. mansoni* where sexual maturity requires more than twice as long (28 days). This marked discrepancy may have at least partly a nutritional basis. In *S. mansoni* the passage through the lungs is slow (4–14 days) and the majority of schistosomulae do not reach the liver before about the eighth day after infection, which means that they do not come into contact with the veins of the liver carrying their rich supply of amino acids and carbohydrates until that time. In *S. douthitti*, on the other hand (and in *S. japonicum*), passage through the lungs is rapid and the liver is reached about the fourth day, so that the rich nutriment of the liver is available earlier than in *S. mansoni*. There is also evidence that *S. douthitti*, which causes severe lung lesions, is feeding on particulate food in the lungs, as worms removed from the lungs contain brown granular material in their gut. In specimens of *S. mansoni* from the lungs, the gut is usually colourless. Hence in the lungs, *S. douthitti* may be obtaining more food than *S. mansoni*.

The growth rates of *S. mansoni*, *S. douthitti* and *S. japonicum* as found by Olivier (1952) are compared in Fig. 16.12. Note that in *S. mansoni* the growth rate after about 16 days rises suddenly and thereafter compares favourably with that of *S. japonicum* and *S. douthitti*.

○○○ 16.6 Unisexual infections

○ 16.6.1 General account

Since schistosomes are dioecious it is possible to produce experimental infections involving worms of one sex only.

There is considerable variation in development between species in unisexual infections. Thus, in mice infected with female cercariae of *S. mansoni* or *S. japonicum* only, maturation of adult females is normally greatly inhibited; the ovary develops but the vitellaria do not mature (i.e. do not produce shell material as evidenced by failure to stain for phenols with Fast Red Salt B (see p. 162)) and virtually no migration into the mesenteric veins occurs. The addition of males has a striking effect and pairing occurs as soon as the males are mature; the paired worms eventually move out into the portal veins and egg production occurs. In a few cases, malformed eggs may be formed by *S. mansoni* females in single-sex infections in mice.

In contrast, both sexes of *S. douthitti* reach maturity in the normal time (10–12 days) independently of the presence of the other. Eggs from such single-sex female

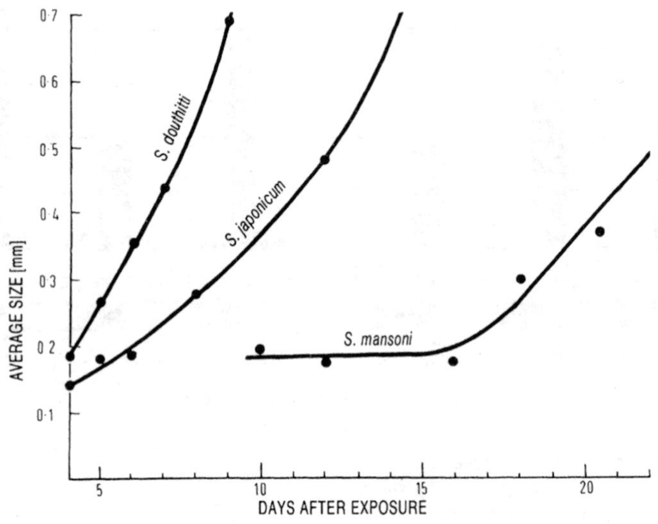

Fig. 16.12
Comparative growth of *Schistosomatium douthitti*, *Schistosoma mansoni* and *Schistosoma japonicum* in mice. Maturation of *S. douthitti* is achieved in about 11 days. (After Olivier, 1952.)

infections of *S. douthitti* develop parthenogenetically.

The cattle and sheep schistosome, *S. mattheei*, reaches maturity in single-sex infections, but the females are not as well developed as normal paired females and produce only non-viable eggs, a situation referred to as *incomplete parthenogenesis* (Taylor *et al.*, 1969).

○○○ 16.7 Hybridisation in schistosomes

When a host is penetrated by cercariae of different species, inter-specific pairing followed by hybridisation may result, often with the production of parthenogenetic eggs.

Natural hybrids. Numerous natural hybrids have been reported, not only between the schistosomes of man, but also between these and animal schistosomes. Natural hybrids between *S. haematobium* × *S. matheei* (cattle), *S. haematobium* × *S. intercalatum* and *S. curassoni* (sheep) × *S. bovis* (cattle) are typical examples (Rollinson & Simpson, 1987). The occurrence of such natural hybrids can clearly cause difficult taxonomic problems, often only resolved by techniques such as isoenzyme analysis.

Experimental hybrids. Numerous experiments with cross-mating of different species have been carried out, some of which are listed in Table 16.6. Thus when mice were infected with female cercariae of *S. mansoni* and male cercariae of the rodent *Schistosomatium douthitti*, many mixed worm pairs developed. The male *S. douthitti* developed well-formed testes but no sperm were found in the female *S. mansoni* partners, but lateral-spined eggs, typical of *S. mansoni*, developed. The haploid miracidia which hatched gave rise to haploid sporocysts, which produced cercariae infective to mice (Basch & Basch, 1984).

○○○ 16.8 Cercarial dermatitis: 'Swimmers' itch'

Many bird schistosomes are responsible for a condition known as 'swimmers' itch' in which cercariae penetrate the skin of man (an unsuitable host), fail to develop but produce a severe rash. The chief species producing this condition are species of the genera *Dendritobilharzia*, *Heterobilharzia*, *Microbilharzia*, *Ornithobilharzia* and *Trichobilharzia*. Probably the most common species causing the problem is *T. ocellata*, which is a cosmopolitan trematode of water birds such as teal and duck with larval stages in *Lymnaea* spp. Many lakes in holiday resorts are spoilt by the occurrence of this parasite and its control is difficult as the use of molluscides is limited on account of their potential effect on fish in the same waters. Some relevant reviews are those of Cort (1950), Kirschenbaum (1979) and Rind (1980). It should be noted that a cercarial dermatitis due to the strigeid *Diplostomum spathaceum* in birds (Fig. 17.6) has also been reported (Sevcova *et al.*, 1987).

Table 16.6. *Experimental cross-specific pairings of schistosomes*

Female		Male	Reference[a]
S. mansoni	×	*S. haematobium*	Vogel (1941)
S. mansoni	×	*S. japonicum*	Vogel (1941, 1942)
S. mansoni	×	*S. rodhaini*	Le Roux (1954)
S. mansoni	×	*Schistosomatium douthitti*	Armstrong (1965)
S. mansoni	×	*Heterobilharzia americana*	Armstrong (1965)
S. haematobium	×	*S. mansoni*	Vogel (1941)
S. haematobium	×	*S. japonicum*	Vogel (1941)
S. japonicum	×	*S. mansoni*	Vogel (1941)
S. rodhaini	×	*S. mansoni*	Le Roux (1954)
S. douthitti	×	*S. mansoni*	Short (1948)
S. mattheei	×	*S. mansoni*	Taylor (1970)

[a]References earlier than 1965 are given in Taylor *et al.* (1969).
Source: Data from Taylor (1970).

○○○ References

(References of general interest have been included in this list, in addition to those quoted in the text.)

Abdel-Wahab, M. F. 1982. *Schistosomiasis in Egypt*. CRC Press, Boca Raton, Florida.

Agatsuma, T. & Suzuki, N. 1983. Isozyme patterns of *Schistosoma japonicum* and *S. mansoni*. *Journal of Helminthology*, **57**: 27–30.

Basch, P. F. & Basch, N. 1984. Intergeneric reproductive stimulation and parthenogenesis in *Schistosoma mansoni*. *Parasitology*, **89**: 369–76.

Basch, P. F. & Nicolas, C. 1989. *Schistosoma mansoni*: pairing of male worms with artificial females. *Experimental Parasitology*, **68**: 202–7.

Brown, D. S. 1980. *Freshwater Snails of Africa and their Medical Importance*. Taylor & Francis, London.

Bruce, J. I. & Sornmani, S. (eds) 1980. *The Mekong Schistosome. Malacological Review* (Suppl. 2), pp. 1–282.

Bundy, D. A. P. 1984. Caribbean schistosomiasis. *Parasitology*, **89**: 377–406.

Butterworth, A. E., Wilkins, H. A., Capron, A. & Sher, A. 1987. The control of schistosomiasis – is there need for a vaccine? *Parasitology Today*, **3**: 1–2.

Chen, M. G. & Mott, K. E. 1989. Progress in assessment of morbidity due to schistosomiasis; reviews of the recent literature. *Tropical Diseases Bulletin*, **86**: R1–56.

Clegg, J. A. 1965. *In vitro* cultivation of *Schistoma mansoni*. *Experimental Parasitology*, **16**: 133–47.

Cort, W. W. 1950. Studies on schistosome dermatitis. XI. Status of knowledge after more than twenty years. *American Journal of Hygiene*, **52**: 251–307.

Deeler, A. M., Miller, R. L., De Jong, N. & Krijger, F. W. 1990. Detection of schistosome antigens in mummies. *Lancet*, **335** (Mar. 24), 724–5.

Eklu-Natey, D. T., Wuest, J., Swiderski, Z., Striezel, H. P. & Huggel, H. 1985. Comparative scanning electron microscope (SEM) study of miracidia of four human schistosome species. *International Journal for Parasitology*, **15**: 33–42.

El-Gindy, M. S. 1950. Biology of *Schistosomatium douthitti* (Cort; 1914) Price, 1931 (Trematoda), in its host. Unpublished thesis, University of Michigan. (Available on microfilm, University Microfilms, Ann Arbor, Michigan, No. 2004.)

Erasmus, D. A. 1987. The adult schistosome: structure and reproductive biology. In: *The Biology of Schistosomes, from Genes to Latrines* (ed. D. Rollinson & A. J. G. Simpson), pp. 51–82. Academic Press, London.

Fletcher, M., LoVerde, P. T. & Woodruff, D. S. 1981. Genetic variation in *Schistosoma mansoni*: enzyme polymorphisms in populations from, Africa, South West Asia, South America and the Caribbean. *The American Journal of Tropical Medicine and Hygiene*, **30**: 406–21.

Fripp, P. 1967. On the morphology of *Schistosoma rodhaini* (Trematoda, Digena, Schistosomatidae). *Journal of Zoology, London*, **151**: 322–52.

Gönnert, R. von. 1955*a*. Schistosomiasis studies, I. Beiträge zur Anatomie und Histologie von *Schistosoma mansoni*. *Zeitschrift für Tropenmedizin und Parasitologie*, **6**: 18–33.

Gönnert, R. von. 1955*b*. Schistosomiasis Studies, II. Über die Eibildung bei *S. mansoni* und das Schicksal der Eier im Wirtsorganismus. *Zeitschrift für Tropenmedizin und Parasitologie*, **6**: 33–52.

Gupta, B. C. & Basch, P. F. 1988. Surface maturation in female *Schistosoma mansoni*, *S. matheei* and *Schistosomatium douthitti*. *International Journal for Parasitology*, **18**: 275–80.

Hockley, D. J. 1973. Ultrastructure of the tegument of *Schistosoma*. *Advances in Parasitology*, **11**: 233–305.

Irie, Y. & Yasuraoka, K. 1981. *Schistosoma japonicum*: ultrastructural changes in the tegument during the cercaria-schistosomulum transformation. *Japanese Journal of Experimental Medicine*, **51**: 53–63.

Jordan, P. 1985. *Schistosomiasis. The St Lucia Project*. Cambridge University Press.

Jordan, P. & Webbe, G. 1982. *Schistosomiasis: Epidemiology, Treatment and Control*. William Heinemann Medical Books, London.

Jourdane, J. & Therón, A. 1987. Larval development: eggs to cercariae. In: *The Biology of Schistosomes, from Genes to Latrines* (ed. D. Rollinson & A. J. G. Simpson), pp. 93–113. Academic Press, London.

Kagan, I. G., Short, R. B. & Nez, M. M. 1954. Maintenance of *Schistosomatium douthitti* (Cort, 1914) in the laboratory. *Journal of Parasitology*, **40**: 1–16.

Kassim, O. & Gilbertson, D. E. 1976. Hatching of *Schistosoma* eggs and observations on motility of miracidia. *Journal of Parasitology*, **62**: 715–20.

Kirschenbaum, M. B. 1979. Swimmer's itch. A review and case report. *Cutis*, **23**: 212–16.

Kumar, V. 1987. Zoonotic trematodiasis in south-east and far-east Asian countries. In: *Helminth Zoonoses* (ed. S. Geerts, V. Kumar & J. Brandt), pp. 106–18. Martin Nijhoff Publishers; Kluwer Academic Publishers Group, Dordrecht, The Netherlands.

Kusel, J. R. 1970. Studies on the structure and hatching of the eggs of *Schistosoma mansoni*. *Parasitology*, **60**: 79–88.

Lewis, F. A., Stirewalt, M. A., Souza, C. P. & Gazzinelli, G. 1988. Large-scale laboratory maintenance of *Schistosoma mansoni* with observations on three schistosome/snail host combinations. *Journal of Parasitology*, **72**: 813–29.

McLaren, D. J. 1980. *Schistosoma mansoni: the parasite surface in relation to host immunity*. Research Studies Press (J. Wiley & Sons), New York.

Malek, E. A. 1977. Geographic distribution, hosts, and biology of *Schistosoma douthitti* (Cort, 1914) Price, 1931. *Canadian Journal of Zoology*, **55**: 661–71.

Miller, P. & Wilson, R. A. 1980. Migration of the schistosomula of *Schistosoma mansoni* from the lungs to the hepatic portal system. *Parasitology*, **80**: 267–88.

Nozais, J. P. 1990. Les bilharzioses humaine dans le monde méditerranéen, au Proche-Orient. Historique et répartitions actuelles. *Bulletin de la Société de Pathologie Exotique*, **83**: 72–81.

Ogbe, M. G. 1983. *In vivo* and *in vitro* development of *Schistosoma margrebowiei* Le Roux, 1933. *Journal of Helminthology*, **57**: 231–6.

Ogbe, M. G. 1985*a*. Aspects of the life cycle of *Schistosoma margrebowiei* infection in laboratory animals. *International Journal for Parasitology*, **15**: 141–5.

Ogbe, M. G. 1985*b*. Histopathology of experimental *Schistosoma margrebowiei* infection in TFI mice. *Nigerian Journal of Science and Technology*, **3**: 91–7.

Olivier, L. 1951. Influence of light on the emergence of *Schistosomatium douthitti* cercariae from their snail host. *Journal of Parasitology*, **37**: 201–4.

Olivier, L. 1952. A comparison of infections in mice with three species of schistosomes, *Schistosoma mansoni*, *Schistosoma japonicum*, and *Schistosomatium douthitti*. *American Journal of Tropical Medicine and Hygiene*, **55**: 22–35.

Pan, S. Chia-Tung 1980. The fine structure of the miracidium of *Schistosoma mansoni*. *Journal of Invertebrate Pathology*, **36**: 307–72.

Popiel, I. & Basch, P. F. 1984. Putative polypeptide transfer from male to female *Schistosoma mansoni*. *Molecular and Biochemical Parasitology*, **11**: 179–88.

Price, H. E. 1931. Life history of *Schistosomatium douthitti* (Cort). *American Journal of Hygiene*, **13**: 685–727.

Puente, H. S. & Short, R. B. 1985. Redescription of chromosomes of *Schistosomatium douthitii* (Trematoda: Schistosomatidae). *Journal of Parasitology*, **71**: 345–8.

Rau, M. E., Bourns, T. K. R. & Ellis, J. C. 1972. An improved method for collecting miracadia. *International Journal for Parasitology*, **2**: 279–80.

Richards, C. S. 1975. Genetic studies on variation in infectivity of *Schistosoma mansoni*. *Journal of Parasitology*, **61**: 233–6.

Rind, S. 1980. The biology of 'Wanaka Itch' – a report of progress. In: *The Resources of Lake Wanaka* (ed. B. T. Roberton & I. D. Blair), pp. 40–4. Tussock Grasslands and Mountain Lands Institute, Lincoln College, South Island, New Zealand.

Rollinson, D. & Simpson, A. J. G. (eds) 1987. *The Biology of Schistosomes, from Genes to Latrines*. Academic Press, London.

Rollinson, D. & Southgate, V. R. 1987. The genus *Schistosoma*: a taxonomic appraisal. In: *The Biology of Schistosomes, from Genes to Latrines* (ed. D. Rollinson & A. J. G. Simpson), pp. 1–49. Academic Press, London.

Sevcova, M., Kolarova, L. & Gottwaldov, A. V. 1987. (Cercarial dermatitis.) (In Czech.) *Ceskosloven ská Dermatologie*, **62**: 369–74. (*Helm. Abstr.* **58**: 949.)

Sluiters, J. F. 1981. Development of *Trichobilharzia ocellata* in *Lymnaea stagnalis* and the effects of infection on the reproductive system of the host. *Zeitschrift für Parasitenkunde*, **64**: 303–19.

Smith, J. W. 1972. The blood flukes (Digenea: Sanguinicolidae and Spirorchidae) of cold-blooded vertebrates and some comparison with schistosomes. *Helminthological Abstracts*, **41**: 161–204.

Smithers, S. R. & Terry, R. J. 1965. The infection of laboratory hosts with cercariae of *Schistosoma mansoni* and the recovery of the adult worms. *Parasitology*, **55**: 695–700.

Smyth, J. D. 1976. *An Introduction to Animal Parasitology*. 2nd Edition. Edward Arnold, London.

Smyth, J. D. & Halton, D. W. 1983. *The Physiology of Trematodes*. 2nd Edition. Cambridge University Press.

Southgate, V. R. & Knowles, R. J. 1975. On *Schistosoma margrebowiei* Le Roux, 1933: the morphology of the egg, miracidium and cercaria, the compatibility with species of *Bulinus*, and the development in *Mesocricetus auratus*. *Zeitschrift für Parasitenkunde*, **54**: 233–50.

Standen, O. D. 1949. Experimental schistosomiasis. II. Maintenance of *Schistosoma mansoni* in the laboratory, with some notes on experimental infections with *S. haematobium*. *Annals of Tropical Medicine and Parasitology*, **43**: 268–83.

Standen, O. D. 1951. The effects of temperature, light and salinity upon the hatching of the ova of *Schistosoma mansoni*. *Transactions of the Royal Society of Tropical Medicine and Hygiene*, **45**: 225–41.

Standen, O. D. 1953*a*. The relationship of sex in *Schistosoma* to migration with in the hepatic portal system of experimentally infected mice. *Transactions of the Royal Society of Tropical Medicine and Hygiene*, **47**: 139–45.

Standen, O. D. 1953*b*. The penetration of the cercariae of *Schistosoma mansoni* into the skin and lymphatics of the mouse. *Transactions of the Royal Society of Tropical Medicine and Hygiene*, **47**: 292–8.

Stirewalt, M. A. 1974. *Schistosoma mansoni*: cercaria to schistosomule. *Advances in Parasitology*, **12**: 115–82.

Stirewalt, M. A., Kuntz, R. E. & Evans, A. S. 1951. The relative susceptibilities of the commonly used laboratory mammals to infection by *Schistosoma mansoni*. *American Journal of Tropical Medicine*, **31**: 57–82.

Taylor, M. G. 1970. Hybridization experiments with five species of African schistosomes. *Journal of Helminthology*, **44**: 253–314.

Taylor, M. G., Amin, M. B. A. & Nelson, G. S. 1969. 'Parthenogenesis' in *Schistosoma mattheei*. *Journal of Helminthology*, **44**: 197–206.

Torpier, G., Hirn, M., Nirde, P., de Reggi, M. & Capron, A. 1982. Detection of ecdysteroids in the human trematode, *Schistosoma mansoni*. *Parasitology*, **84**: 123–30.

Viyanant, V. & Upatham, E. S. 1985. Isoenzyme analyses of Malaysian *Schistosoma*, *S. mekongi* and *S. japonicum* by isoelectric focusing in polyacrylamide gel. *Southeast Asian Journal of Tropical Medicine and Public Health*, **16**: 539–45.

Wajdi, N. 1966. Penetration by the miracidia of *S. mansoni* into the snail host. *Journal of Helminthology*, **40**: 235–44.

Webbe, G. 1981. Schistosomiasis: some advances. *British Medical Journal*, **283**: 1–8.

Webbe, G. 1987. Treatment of schistosomiasis. *European Journal of Clinical Pharmacology*, **32**: 433–6.

Webbe, G. & James, C. 1971. A comparison of two geographical strains of *Schistosoma haematobium*. *Journal of Helminthology*, **45**: 271–84.

WHO, 1985. The control of schistosomiasis. *Technical Report Series* **728**, 113 pp.

Wilson, R. A. 1987. Cercariae to liver worms: development and migration in the mammalian host. In: *The Biology of Schistosomes, from Genes to Latrines* (ed. D. Rollinson & A. J. G. Simpson), pp. 115–46. Academic Press, London.

17

Digenea: Strigeidae, Diplostomatidae, Paramphistomatidae

○○○ 17.1 Families Strigeidae and Diplostomatidae

These families are conveniently considered together as they constitute a group of trematodes loosely referred to as 'strigeids' or 'holostomes'. In general, they are distomes in which the body (Fig. 17.1) is clearly divided by a constriction into a flattened or cup-shaped portion, which acts essentially as an adhesive organ, and a posterior cylindrical portion containing the genitalia. A large 'holdfast' or tribocytic organ provided with histolytic glands is usually present. The life cycles in general resemble those of schistosomes to which, on the basis of the morphology of the cercaria (fork-tailed) and the flame-cell pattern (1 + 1) in the miracidium, they appear to be closely related. The morphological difference between the two families is slight, the anterior region being more flattened in the Diplostomatidae than in the Strigeidae. Both groups produce peculiar metacercariae, known as '*diplostomulum*' and '*tetracotyle*' respectively, which occur in definite tissue sites and undergo morphological development slightly in advance of that found in a cercaria. Species of *Diplostomulum* occur in the eye, brain and spinal cord of fishes and amphibia. They lack cystogenous glands, and do not form a cyst wall, but can move actively in the host tissues. On the other hand, species of *Tetracotyle* occur in both invertebrates and vertebrates and form definite cyst walls. In many cases only the larval forms are known and may be frequently given characteristic names.

The nomenclature has been confusing in the past. The name *Diplostomulum* is now used for larval forms and the name *Diplostomum* for the adult. Thus the specific name given to a larva of the *Diplostomulum* group will disappear once its adult stage is known. *Diplostomulum spathaceum*, for example, is the larval stage of *Diplostomum spathaceum*.

○ 17.1.1 Type example: *Diplostomum phoxini*

definitive hosts:	*Mergus merganser merganser*, *Anas platyrhyncha*, *Cairina moschata*
location:	duodenum
molluscan hosts:	*Lymnaea peregra ovata*, *L. auricularia*
second intermediate host:	*Phoxinus phoxinus* (European minnow), *P. laevis*; larvae in brain

Occurrence. The larvae of *Diplostomum phoxini* are among the commonest of trematode larval stages available for laboratory study in the British Isles. The brains of minnows are 65–100 per cent infected (Erasmus, 1962); Rees (1955, 1957) has given detailed morphological descriptions of adult and larvae.

Morphology. The body shape is typically 'strigeid' with a flattened oval, anterior region with leaf-like edges, the whole serving as an adhesive organ. The anterior extremity (Fig. 17.1) is trilobed; the central region is occupied by the oral sucker and between it and the two lateral lobes are two *pseudosuckers* or *lappets*. Associated with these structures are two types of holocrine gland, which probably secrete (*a*) an esterase and (*b*) a carbohydrate complex; the cytology and fine structure of the glands have been studied in detail (Lee, 1962; Erasmus, 1969). These secretions probably play a part in extra-cellular digestion and the pseudosuckers may also assist in maintaining attachment to the gut, an action which leaves the mouth free to browse on the semisolid contents of the intestine. The tegument contains minute spines which do not appear to penetrate the surface. There is a well-developed *adhesive organ* or *holdfast* provided with glands, the secretions of which

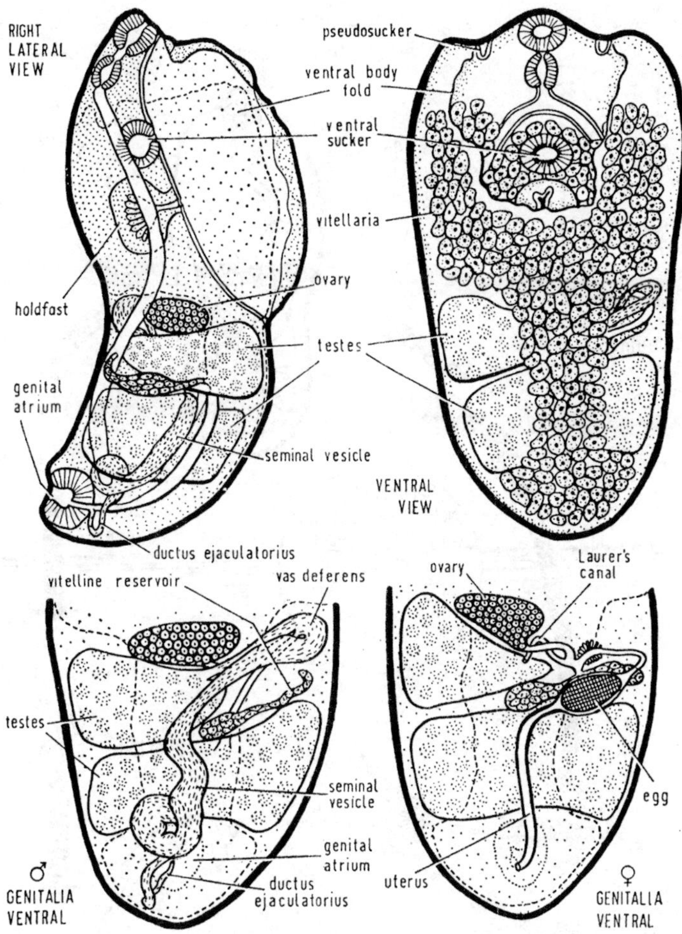

Fig. 17.1
Diplostomum phoxini: general morphology of adult from the intestine of gulls. (After Rees, 1955.)

are believed to be histolytic in nature; the fine structure has been examined by Erasmus (1970). The alimentary canal has a pre-pharynx, a pharynx and a well-developed pair of intestinal caeca. The arrangement of all genitalia, which, except for the vitellaria, all occur in the non-flattened posterior region, is clear from Fig. 17.1. A few specific points of interest may be noted:

Male

The testes are large and occupy the anterior two-thirds of the posterior region. Sexual amphitypy is common, the anterior testis lying on the right or left side. There is no cirrus, cirrus sac or prostate glands, the ejaculatory duct opening into the posterior genital atrium. The latter is a deep cavity surrounded by a special arrangement of muscles, the whole being superficially sucker-like but not so clearly defined.

Female

The female genitalia present no unusual features. The vitelline cells are extensive and form three lobes, the anterior two extending nearly to the base of the pharynx. The eggs are relatively large for so small a worm; only 1–5 are found in the uterus at a time.

Life cycle (Fig. 17.2). The eggs are unembryonated when laid. They embryonate in seven days at 20 °C (Dönges, 1969) and hatch when stimulated by light.

The miracidium presents no unusual features. The intermediate hosts are *Lymnaea peregra ovata* (Dönges, 1969) and *L. auricularia*. The larval development in the snail does not appear to have been described.

Cercaria. It is important to be able to distinguish this cercaria from others (such as that of *D. spathaceum*, also

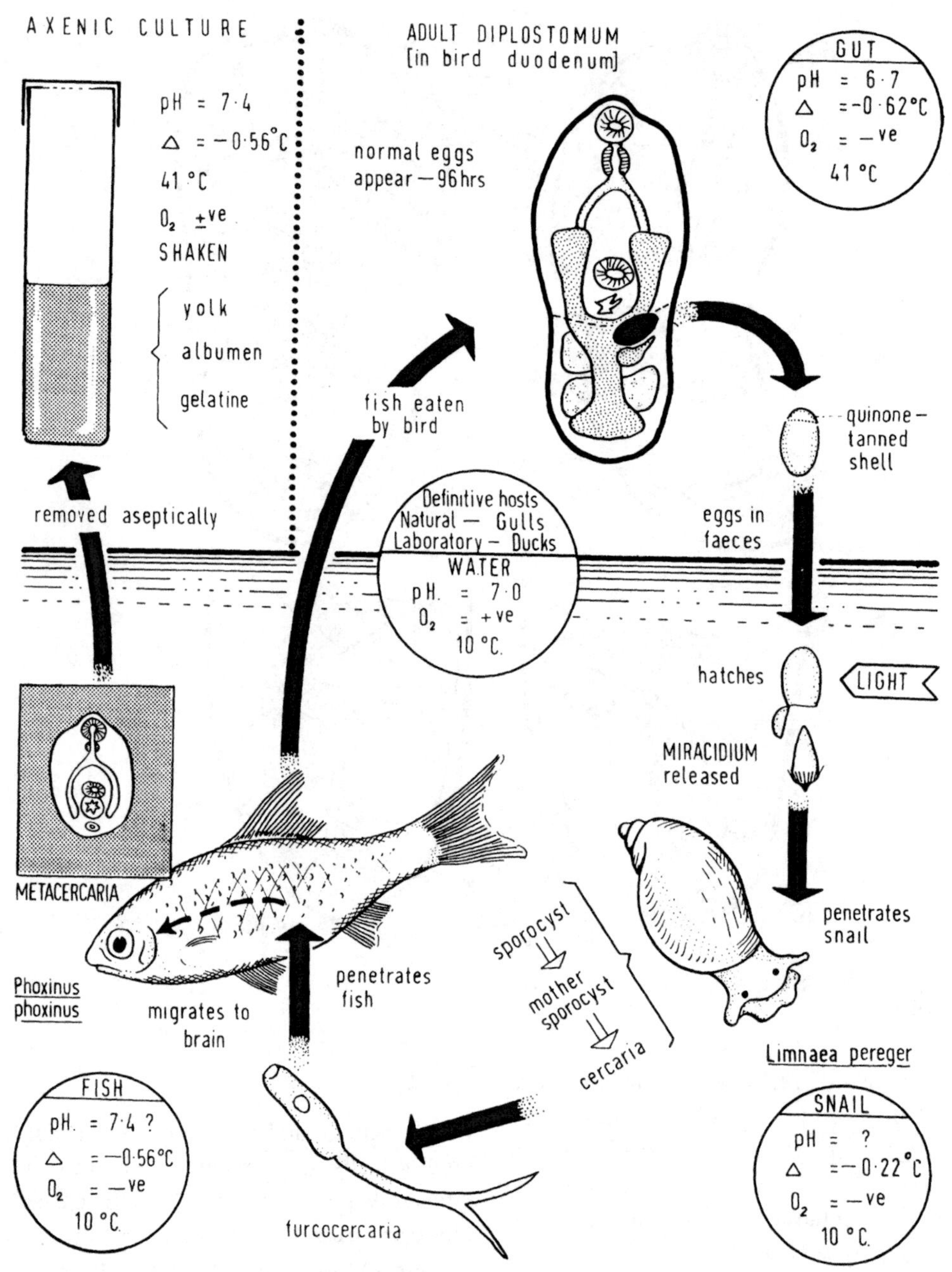

Fig. 17.2
Life cycle of the strigeid *Diplostomum phoxini* and some physiological factors relating to it. For details of *in vitro* culture, see p. 504. (Original.)

common in *L. peregra*; see p. 258). Excellent keys for cercariae in Great Britain are available (Blair, 1977; Nasir & Erasmus, 1964). The morphology has been described in detail by Rees (1957) and Dönges (1969).

The nature of the penetration-gland secretions has not been determined but, as in other species (p. 191), is probably hyaluronidase and a collagenase.

The flame-cell formula (Fig. 17.3) is:

$$2[(1a + 1b + 2) + (3 + 4a + 4b^{1} + (4b^{11*} + 4b^{11**}))] = 16$$

There are six pairs of caudal bodies in the tail. The function of these is uncertain, but it is believed they may give buoyancy to the cercaria. The tail also bears

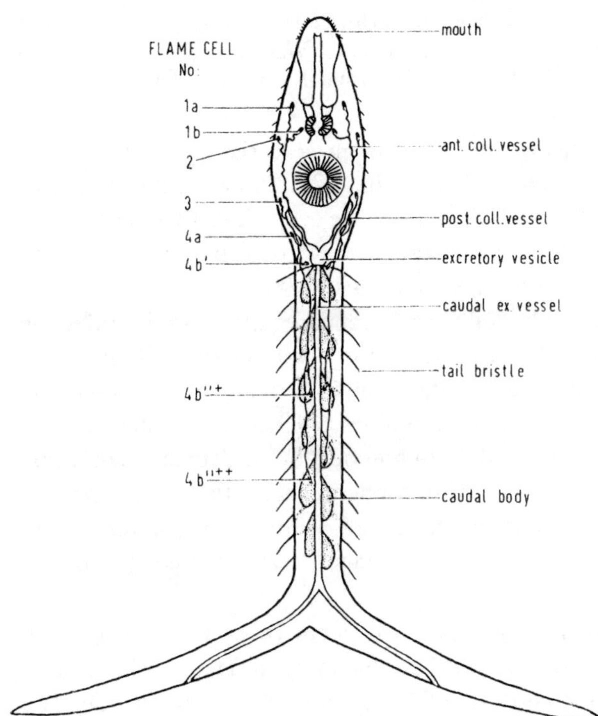

Fig. 17.3
Cercaria of *Diplostomum phoxini*, showing application of flame-cell formulae. (After Rees, 1957.)

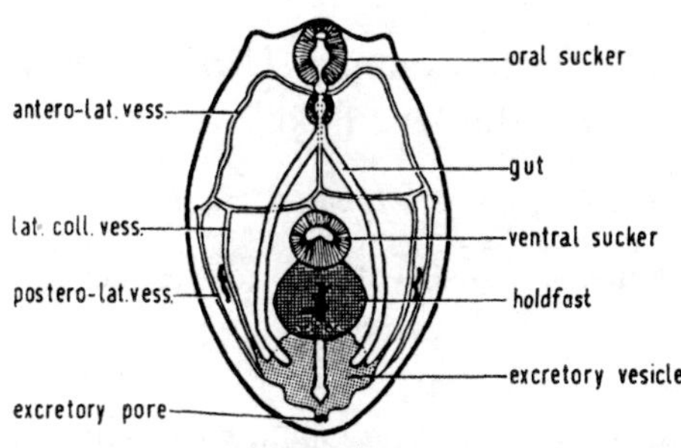

Fig. 17.4
Metacercaria of *Diplostomum phoxini* from brain of minnow, *Phoxinus phoxinus*. (Modified from Rees, 1955.)

twelve bristles. The cercaria is an intermittent swimmer and is activated by a 'shadowing' effect but also shows spontaneous activity with a one-second active phase (Dönges, 1969).

Penetration of minnow. Cercariae easily penetrate the thin skin of minnows, lose their tails and rapidly make their way to the brain, probably via the blood stream, where slight haemorrhages are observable after a recent attack. They accumulate particularly in the fourth ventricle, aquiductus sylvii, optic lobes, third ventricle, lobi inferiores, sometimes in the corpora striata and under the epithelium over certain areas of the brain and spinal cord.

Development of metacercaria. Within the brain, the tailless cercariae become metacercariae (*Diplostomulum phoxini*) which require a further period of some 28 days (at 10–15 °C) before they reach maximum development.

A fully developed metacercaria (Fig. 17.4) superficially bears the body shape of an adult. Internally it shows a number of advances over the cercaria, namely (*a*) the gut caeca are well developed, (*b*) the flame cells increase in number, and (*c*) the secondary excretory system, often referred to as the '*reserve bladder system*', develops. The sixteen flame cells in the cercaria give rise, by repeated division, to 104 flame cells. The cell formula of the metacercaria is:

$$2[(4 + 4 + 4) + (4 + 4 + 4 + 4 + 4) + (4 + 4 + 4 + 4 + 4)] = 104$$

The flame-cell system is often difficult to make out in the living condition on account of the reserve bladder system, the terminations of which are associated with calcareous corpuscles. The system is similar in the related species *D. spathaceum*.

An unusual behaviour pattern has been observed in minnows infected with *D. phoxini* (Dönges, 1969). On the 22nd day after being exposed to cercariae, fish tend to lift their heads and show an increased breathing frequency with the jaws gasping. In stormy weather conditions, fish may even jump right out of the aquarium!

Since the larvae become infective after 28 days in the fish, it is tempting to speculate that this abnormal behaviour attracts the attention of the bird host, thus increasing the probability of a fish being caught and eaten with subsequent maintenance of the life cycle of the trematode.

Maturation of metacercariae. When infected fish are eaten by birds, the metacercariae become firmly attached to the duodenal mucosa and rapidly reach maturation. Domestic ducks serve as suitable experimental hosts, although other anserines such as *Mergus merganser merganser* serve as natural definitive hosts.

D. phoxini may be cultured axenically, almost to maturity *in vitro* (Chapter 34).

Ultrastructure. The ultrastructure of the epidermis and associated structures in the metacercaria, cercaria and sporocyst has been described by Bibby & Rees (1971).

○ 17.1.2 Type example: *Diplostomum spathaceum* (the 'eye fluke')

(Synonyms: *D. volvens*, *D. horonense*, *D. flexicaudum*)

definitive hosts:	fish-eating birds, especially *Larus* spp.
location:	duodenum
molluscan hosts:	*Lymnaea* spp., especially *L. stagnalis*, *L. palustris*, *L. peregra*
second intermediate host:	numerous salmonoid fish; metacercariae in lens and aqueous humour of eye
distribution:	cosmopolitan, especially N. America and Europe

Occurrence. Metacercariae of *D. spathaceum* have been reported from 105 species of freshwater fish in Europe and N. America and 23 species in the UK. Natural infections have also been found in wild amphibia, reptiles and mammals, including (but rarely) man (Palmieri *et al.*, 1977). The South African clawed toad, *Xenopus laevis*, has also been experimentally infected.

It should be noted that the metacercariae of several other species found in the eye may be confused with those of *D. spathaceum*; notably, in the USA, *D. adamsi* in the retina (Lester & Freeman, 1976), and in Europe, *D. gasterostei* in the retina of the stickleback (Williams, 1966) and *Tydelphys clavata* in the eye of the trout (Kennedy, 1984).

Metacercaria. Although cercariae penetrate all over the body of a fish, within 12 h 80 per cent of the larvae are found in the lens of the eye, suggesting the existence of a powerful chemotaxis. Most fish have 1–20 larvae in a single lens but up to 100 may occur; these are non-progenetic and remain unencysted. Infections (diplostomiasis) results in some degree of blindness in the fish and stunted growth and may be a serious problem in fish farms (Stables & Chappell, 1986*a*,*b*). There are reports of small numbers of metacercariae occasionally becoming progenetic in snails; these are capable of developing directly in gulls without first passing through the fish intermediate stage (Lester & Lee, 1976).

Morphology and life cycle (Figs 17.5, 17.6). The morphology of the adult and larval stages – which closely resemble those of *D. phoxini* – are shown in Fig. 17.5. The life cycle is also similar to that of *D. phoxini* except that the metacercariae accumulate in the eye instead of the liver. The cercariae of *D. spathaceum* can readily be identified as they hang in the water in a very characteristic position, with their furca spread out at an angle of about 120° with the tail stem: see Fig. 17.6 (Smyth, 1990).

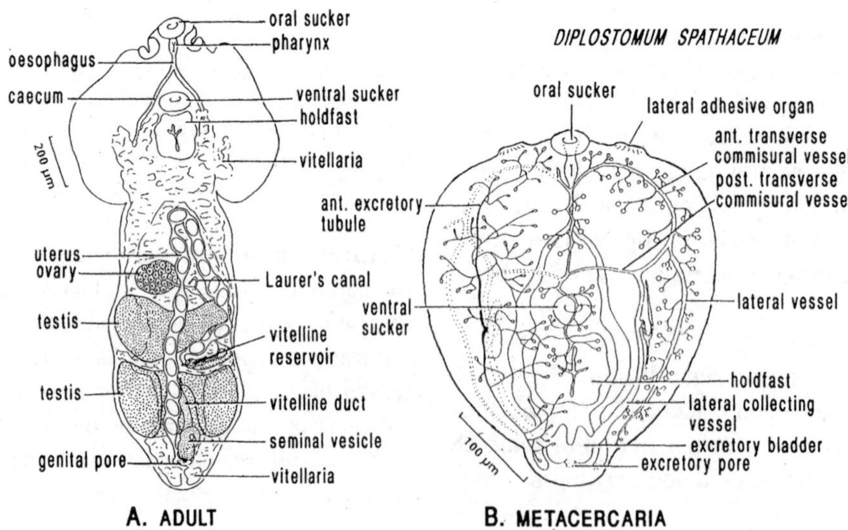

Fig. 17.5
Diplostomum spathaceum: A. Adult from intestine of gull; B. metacercaria from lens of three-spined stickleback, *Gasterosteus aculeatus*. (From Smyth, 1990; after Heatley, 1958.)

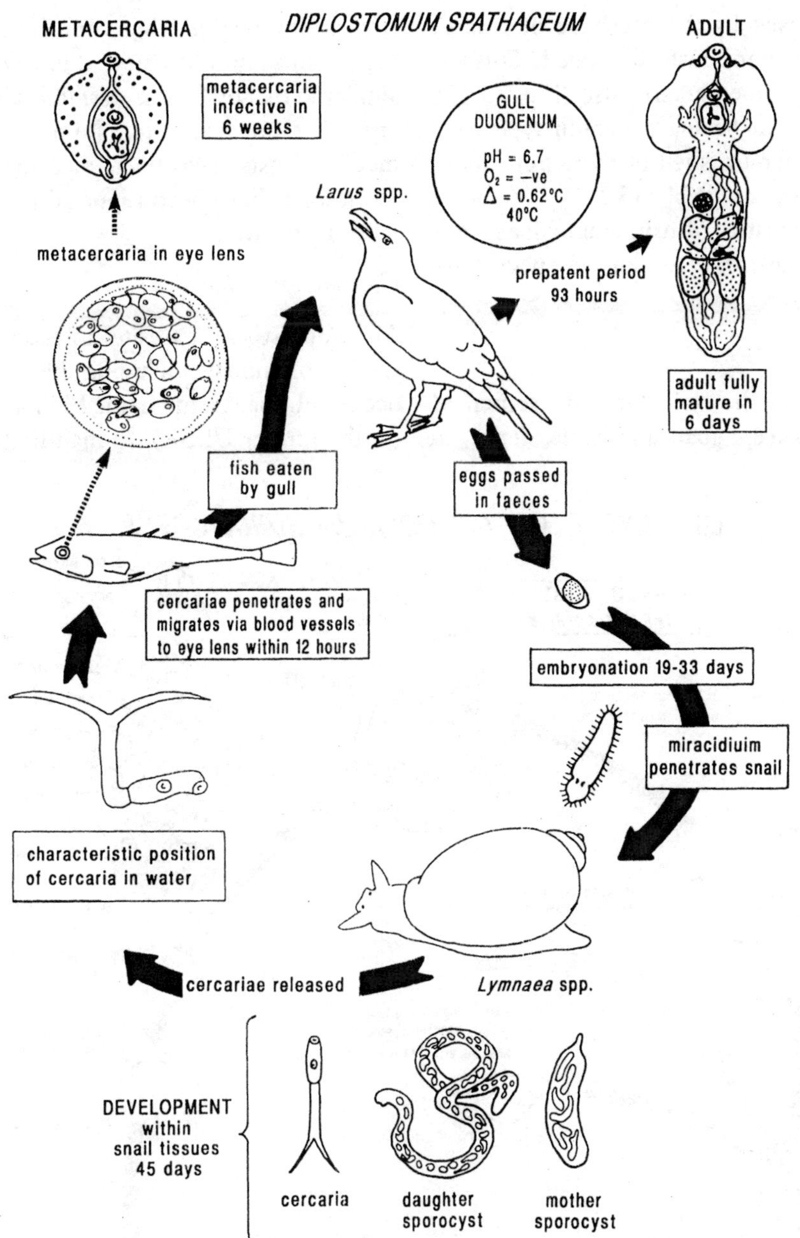

Fig. 17.6
Diplostomum spathaceum: life cycle. (From Smyth, 1990; after Heatley, 1958.)

***D. spathaceum* as an experimental model.** This species is a most valuable experimental model for teaching and research purposes, for it can be grown to sexual maturity *in vitro* (Kannangara & Smyth, 1974; Smyth, 1990) (see Chapter 34) and *in ovo*, on the chick chorioallantois (Leno & Holloway, 1986). Moreover, the cercariae can be transformed to metacercariae by a relatively simple *in vitro* technique (Whyte *et al.*, 1988).

Some useful studies on the biology, epidemiology and immunology of *D. spathaceum* are those of Brassard *et al.* (1982), Chubb (1979), Dubois (1970), Gaten (1987), Stables (1989) and Stables & Chappell (1986*a*,*b*).

○○○ 17.2 Paramphistomatidae

Members of this family, commonly referred to as *amphistomes*, are long thick forms almost circular in transverse section and characterised by the presence of a large posterior sucker, often enormously developed. This latter feature is present in the cercaria also (Fig. 13.17). The eggs are unusual in having a keratin type egg-shell instead of the sclerotin shell of most trematodes (Madhavi, 1966, 1968); see Table 13.1.

Amphistomes are parasites of fish, amphibians, reptiles, birds and mammals; the best-known species are from domestic animals.

Paramphistomum cervi

A pathogenic species in the reticulum and rumen of ruminants, especially sheep, goats and cattle, mainly in Africa, but also in other countries. The general shape is pear-like and it is pale red in colour. The life cycle is similar to that of *Fasciola hepatica*, with the metacercariae encysting on vegetation. Numerous molluscan intermediate hosts have been incriminated. *P. scotiae* and *P. hiberniae* have been reported from Scotland and Ireland, respectively.

Gastrodiscoides hominis

An amphistome parasite infecting man and several species of animals including the pig. Human infections occur in India, Vietnam, the Philippines and Kazakhstan (in the former USSR). In man up to 41 per cent of the

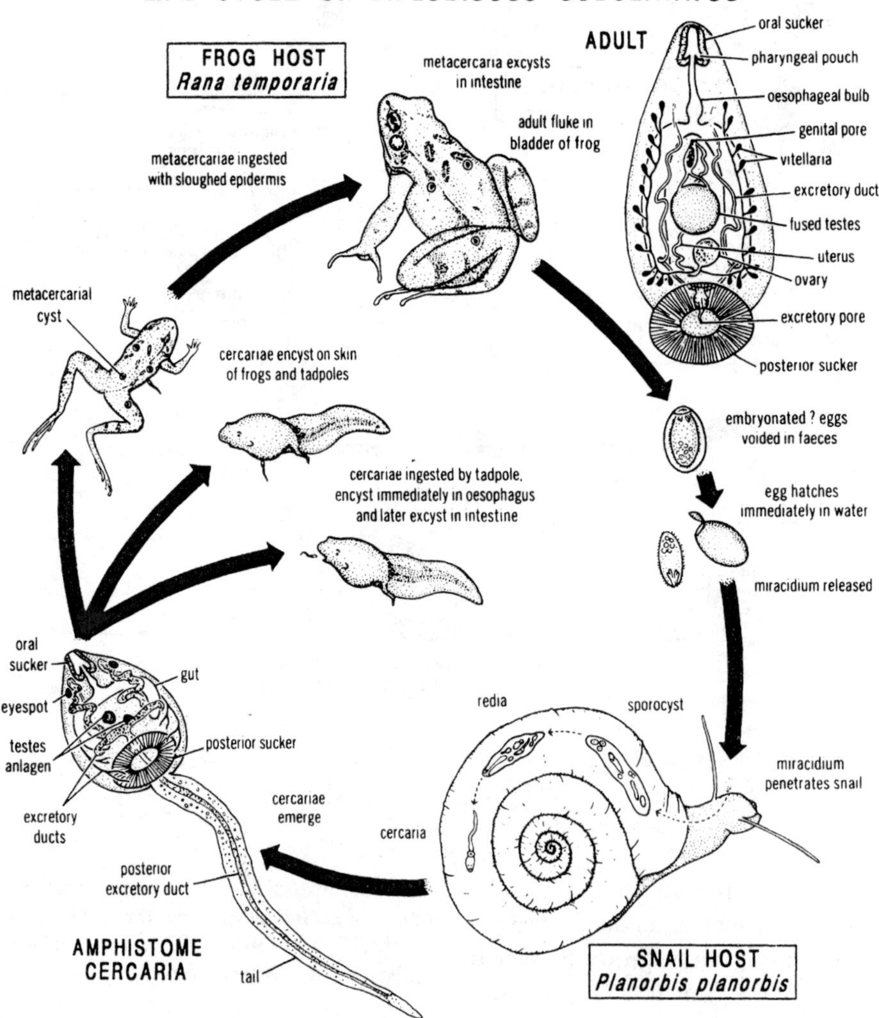

Fig. 17.7
Diplodiscus subclavatus: life cycle in Europe. The life cycle of *Megalodiscus temperatus* in North America is similar. (From Smyth & Smyth, 1980.)

inhabitants of some areas have been reported as infected; the parasite is often found in association with *Fasciolopsis buski*, with which it probably shares snail hosts. The life cycle is imperfectly known but *Helicorbis coenosus* acts as the intermediate host in India (Dutt & Srivastava, 1966). The distribution, pathology and systematics have been reviewed by Buckley (1964).

Paramphistomes in amphibia. Two species are common parasites of the rectum and urinary bladder of frogs, *Diplodiscus subclavatus* (in Europe) and *Megalodiscus temperatus* (in the USA). The morphology and life cycle of *D. subclavatus* is shown in Fig. 17.7; that of *M. temperatus* is similar. The relevant literature on these species is reviewed by Bourgat & Kulo (1977), Brooks (1976) and Smyth & Smyth (1980).

○○○ References

(References of general interest have been included in this list, in addition to those quoted in the text.)

Bell, E. J. & Hopkins, C. A. 1956. The development of *Diplostomum phoxini* (Strigeida, Trematoda). *Annals of Tropical Medicine and Parasitology*, **50**: 275–82.

Bell, E. J. & Smyth, J. D. 1958. Cytological and histochemical criteria for evaluating development of trematodes and pseudophyllidean cestodes *in vivo* and *in vitro*. *Parasitology*, **48**: 131–48.

Berrie, A. D. 1960. The influence of various definitive hosts on the development of *Diplostomum phoxini* (Strigeida, Trematoda). *Journal of Helminthology*, **34**: 205–10.

Bibby, M. C. & Rees, G. 1971. The ultrastructure of the epidermis and associated structures in the metacercaria, cercaria and sporocysts of *Diplostomum phoxini* (Faust, 1918). *Zeitschrift für Parasitenkunde*, **37**: 169–86.

Blair, D. 1977. A key to cercariae of British strigeoids (Digenea) for which the life cycles are known, and notes on the characters used. *Journal of Helminthology*, **51**: 155–66.

Bourgat, R. & Kulo, S. D. 1977. Recherches sur le cycle biologique d'un Paramphistomidae (Trematoda) d'amphibians en Afrique. *Annals de Parasitologie Humaine et Comparée*, **52**: 7–12.

Brassard, P., Rau, M. E. & Curtis, M. A. 1982. Infection dynamics of *Diplostomum spathaceum* cercariae and parasite-induced mortality of fish hosts. *Parasitology*, **85**: 489–93.

Brooks, D. R. 1976. Parasites of amphibians of the Great Plains. Part 2. Platyhelminths of amphibians in Nebraska U.S.A. *Bulletin of the University of Nebraska State Museum*, **10**: 1–92.

Buckley, J. J. 1964. The problem of *Gastrodiscoides hominis* (Lewis & McConnell, 1876) Leiper, 1913. *Journal of Helminthology*, **38**: 1–6.

Chubb, J. C. 1979. Seasonal occurrence of helminths in freshwater fish. II. Trematodes. *Advances in Parasitology*, **17**: 141–313.

Dönges, J. 1969. *Diplostomum phoxini* (Faust, 1918) (Trematoda). Morphologie des Miracidiums sowie Beobachtungen an weiteren Entwicklungsstadien. *Zeitschrift für Parasitenkunde*, **30**: 120–7.

Dubois, G. 1970. Synopsis des Strigeidae des Distomatidae (Trematoda). *Mémoires de la Société neuchâteloise des Sciences Naturelles*, **10**: 259–727.

Dutt, S. C. & Srivastava, H. D. 1966. The intermediate host and the cercaria of *Gastrodiscoides hominis* (Trematoda: Gastrodiscidae). Preliminary Report. *Journal of Helminthology*, **40**: 45–52.

Erasmus, D. A. 1962. Distribution of certain strigeid trematodes in Great Britain. *Nature, London*, **195**: 828–9.

Erasmus, D. A. 1969. Studies on the host-parasite interface of strigeoid trematodes. VI. Ultrastructural observations on the lappets of *Diplostomum phoxini* Faust, 1918. *Zeitschrift für Parasitenkunde*, **32**: 48–58.

Erasmus, D. A. 1970. The host–parasite interface of strigeoid trematodes. VII. Ultrastructural observations on the adhesive organ of *Diplostomum phoxini* Faust, 1918. *Zeitschrift für Parasitenkunde*, **33**: 211–24.

Gaten, E. 1987. Aggregation of the eye fluke *Diplostomum spatheceum* (Digenea: Diplostomatidae) in the lens of various species of fish. *Journal of Fish Diseases*, **10**: 69–74.

Heatley, W. 1958. Studies on the development of helminths. Observations on the development of *Diplostomum spathaceum* (Rud.) *in vivo* and *in vitro*. M.Sc. Thesis, Trinity College, University of Dublin, Ireland.

Kannangara, D. W. W. & Smyth, J. D. 1974. *In vitro* cultivation of *Diplostomum spathaceum* and *Diplostomum phoxini* metacercariae. *International Journal for Parasitology*, **4**: 667–73.

Kennedy, C. R. 1984. The use of frequency distributions in an attempt to detect host mortality induced by infections of diplostomatid metacercariae. *Parasitology*, **89**: 209–20.

Lee, D. L. 1962. Studies on the function of the pseudosuckers and holdfast organ of *Diplostomum phoxini* Faust (Strigeida, Trematoda). *Parasitology*, **55**: 103–12.

Leno, G. H. & Holloway, H. L. 1986. The culture of *Diplostomum spathaceum* metacercariae on the chick chorioallantois. *Journal of Parasitology*, **72**: 555–8.

Lester, R. J. G. & Freeman, R. S. 1976. Survival of two trematode parasites (*Diplostomum* spp.) in mammalian eyes and associated pathology. *Canadian Journal of Ophthalmology*, **11**: 229–34.

Lester, R. J. G. & Lee, T. D. G. 1976. Infectivity of progenetic metacercariae of *Diplostomum spathaceum*. *Journal of Parasitology*, **62**: 832–3.

Madhavi, R. 1966. Egg-shell in Paramphistomatidae (Trematoda: Digenea). *Experientia*, **22**: 93–4.

Madhavi, R. 1968. *Diplodiscus mehrai*: chemical nature of eggshell. *Experimental Parasitology*, **23**: 392–7.

Nasir, P. & Erasmus, D. A. 1964. A key to the cercariae from British freshwater molluscs. *Journal of Helminthology*, **38**: 245–68.

Öhman, C. 1965. The structure and function of the adhesive organ in strigeid trematodes. Part II. *Diplostomum spathaceum* Braun, 1893. *Parasitology*, **55**: 481–502.

Palmieri, J. R., Heckman, R. A. & Evans, R. S. 1977. Life history and habitat analysis of the eye fluke *Diplostomum spathaceum* (Trematoda: Diplostomatidae) in Utah. *Journal of Parasitology*, **63**: 427–9.

Rees, F. G. 1955. The adult and diplostomulum stage (*Diplostomulum phoxini* (Faust) of *Diplostomum pelmatoides* Dubois, and an experimental demonstration of part of the life cycle. *Parasitology*, **45**: 295–312.

Rees, F. G. 1957. Cercaria *Diplostomum phoxini* (Faust) a furco-cercaria which develops into *Diplostomulum phoxini* in the brain of the minnow. *Parasitology*, **47**: 126–38.

Smyth, J. D. 1990. *In Vitro Cultivation of Parasitic Helminths*. CRC Press, Boca Raton, Florida.

Smyth, J. D. & Smyth, M. M. 1980. *Frogs as Host–Parasite Systems. I.* The Macmillan Press, London.

Stables, J. N. 1989. Studies on the speciation, epidemiology and immunology of *Diplostomum spathaceum* in fishes. *Dissertation Abstracts International, B* (*Sciences & Engineering*), **49**: 5081. (Order No: BRDX4519.)

Stables, J. N. & Chappell, L. H. 1986*a*. *Diplostomum spathaceum* (Rud. 1818): effects of physical factors on the infection of rainbow trout (*Salmo gairdneri*) by cercariae. *Parasitology*, **93**: 71–9.

Stables, J. N. & Chappell, L. H. 1986*b*. The epidemiology of diplostomiasis in farmed rainbow trout from north-east Scotland. *Parasitology*, **92**: 699–710.

Whyte, S. K., Chappell, L. H. & Secombes, C. J. 1988. *In vitro* transformation of *Diplostomum spathaceum* (Digenea) cercariae and short-term maintenance of post-penetration larvae *in vitro*. *Journal of Helminthology*, **62**: 293–302.

Williams, M. O. 1966. Studies on the morphology and life-cycle of *Diplostomum* (*Diplostomum*) *gasterostei* (Strigeida: Trematoda). *Parasitology*, **56**: 693–706.

18

Physiology of trematodes

Most studies on the physiology and biochemistry of trematodes have concentrated on those genera which are of medical or veterinary importance, especially *Schistosoma* and *Fasciola*. It should be emphasised, however, that many other species also present interesting and fundamental problems and research on these should not be overlooked; moreover, many of these species are more easily maintained in the laboratory and hence more readily available for experimental studies. This chapter deals largely with selected areas of the physiology and biochemistry of the Digenea with occasional examples from the Monogenea. Areas not dealt with are covered in the reviews listed.

Earlier literature (pre-1980) has been reviewed by Coles (1975) and Smyth & Halton (1983). More recent books or reviews are those of Barrett (1981), Bennet *et al.* (1989), Bryant & Behm (1989), Chappell (1980), Coles (1975, 1984), Coles *et al.* (1980), Dunne (1990), Earle (1984), Fairweather & Halton (1991), Frayha & Smyth (1983), Köhler (1985), Halton *et al.* (1992), Lloyd (1986), Smyth & Halton (1983) and Ward (1982).

18.1 Chemical composition

The chemical analysis of parasites seldom provides significant information unless it is related to the nutritional state of the host and the stage of maturity of the parasite. Chemical analysis data are usually expressed as a percentage of the dry mass; some available data for Digenea are shown in Table 18.1. However, such data must be viewed with caution as the dry : fresh mass ratio may alter substantially during the life cycle of a worm, owing to the loss of major constituents such as glycogen. In metabolic studies, this problem can be overcome by basing calculations on the absolute content of a particular substance (e.g. glycogen) per worm or number of worms, as in Fig. 18.3, rather than the dry mass.

18.2 Nutrition

Feeding mechanisms. Trematodes are essentially suctorial feeders showing little variation in their feeding mechanisms (Halton, 1967; Smyth & Halton, 1983), a situation which reflects the relative uniformity of their diet of fluids, semi-fluids and soft host tissues (Table 18.2). Most trematodes suck in food through the efficient action of the oral sucker and pharynx, both of which are muscular structures. In some apharyngate genera (e.g. *Diplodiscus*, *Gorgoderina*) the oesophageal musculature is thickened and simulates a pharynx.

Tegumental feeding. The ultrastructure of the trematode tegument (Fig. 13.2) has much in common with that of cestodes (Fig. 20.2) and it is therefore not surprising to find that, as in cestodes, it has considerable absorption potential. Thus, in addition to ingesting and digesting food in its alimentary tract, those species which lie in a medium containing soluble nutrients also absorb nutrient through the tegument.

In many species – especially *Schistosoma* and *Fasciola* – the uptake of a wide range of substances through the tegument has been demonstrated experimentally. These include: carbohydrates (glucose, fructose, mannose, glucosamine, ribose); amino acids (methionine, glutamate, glycine, arginine, alanine, trypthophan); and pyrimidines (cytosine, thymidine and uracil) (Pappas & Read, 1975; Smyth & Halton, 1983). Different substances appear to be taken up by different transport mechanisms; uptake by diffusion, active transport, mediated transport, facilated diffusion or – in some cases – a combination of these, has been reported. Although the absorption of these low-molecular mass substances has been demonstrated, the importance of the process, relative to intestinal absorption, is more difficult to estimate. In *Schistosoma mansoni*, for example, there is evidence

Table 18.1. *Chemical composition of trematodes*

Species	Host	Location	Dry matter as % fresh mass	Composition (per cent dry mass) Inorganic ash	Carbohydrate (glycogen)	Protein	Lipid	References[a]
Azygia lucii	Pike	Stomach	—	—	0.57[d]	—	11.0	Vystotskaya *et al.* (1972)
	Burbot	Stomach	—	—	—	—	21.2	Vystotskaya *et al.* (1973)
Clinostomum complanatum	Fish	Body cavity	41.6[b]	—	18.3	36.2	35.9	Siddiqi & Nizami (1981)
Cotylophoron cotylophorum	Buffalo	Rumen	31.4	—	—	—	27.3	Yusufi & Siddiqi (1987)
Dicrocoelium dendriticum	Sheep	Bile duct	20.2	6.0	28.5	43	6.5	Eckert & Lenner (1971)
Diplodiscus subclavatus	Frog	Rectum	16.5	4.0	15.5	62	—	Halton (1964, 1967*d*)
Echinostoma malayanum	Pig	Small intestine	24.7		—	—	39.0	Yusufi & Siddiqi (1976)
Echinostoma revolutum	Bird	Rectum	—	—	—	—	15.0	Fried & Boddorff (1988)
Eurytrema pancreaticum	Cattle	Pancreas	—	—	—	—	15.7	Vykhrestyuk & Yarygina (1975)
Fasciola gigantica	Buffalo	Bile duct	22.4[b]	—	25.4	66	12.8	Goil (1961), Smyth (1966)
Fasciola hepatica	Sheep	Bile duct	19.5	5.5	16.0	59	12–13	Halton (1964, 1967*d*), Hrzenjak & Ehrlich (1975), Von Brand (1966)
Fasciolopsis buski	Pig	Small intestine	17.6	—	—	—	50.4	Yusufi & Siddiqi (1976)
Gastrodiscoides hominis	Pig	Caecum	—	—	—	—	14.5	Yusufi & Siddiqi (1977)
Gastrothylax crumenifer	Buffalo	Reticulum	25.7[b]	—	25.1	49	10.5	Yusufi & Siddiqi (1976)
Gigantocotyle explanatum	Buffalo	Bile duct	20.6	—	—	—	34.4	Yusufi & Siddiqi (1976)
Gorgodera cygnoides	Frog	Bladder	—	—	9.0	—	—	Halton (1967*d*)
Gorgoderina vitelliloba	Frog	Bladder	19.3	4.5	8.7	—	—	Halton (1964, 1967*d*)
Gynaecotyla adunca	Bird	Bile duct	20.2	6.0	28.5	43	6.5	Vernberg & Hunter (1956)
Haematoloechus medioplexus	Frog	Lungs	17.5	6.1	9.0	66	12.8	Cain & French (1975), Halton (1964, 1967*d*)
Haplometra cylindracea	Frog	Lungs	16.7	5.3	9.5	63	—	Halton (1964, 1967*d*)
Isoparorchis hypselobagri	Cat fish	Swim bladder	82.4	—	—	—	29.5	Yusufi & Siddiqi (1976)
Mehraorchis ranarum	Frog	Liver	25.7[b]	—	17.6	—	—	Karyakarte *et al.* (1976)
Paramphistomum cervi	Cattle	Reticulum	—	—	—	—	4.8	Patil *et al.* (1977), Vykhrestyuk & Vadgina (1975)
Paramphistomum explanatum	Buffalo	Bile duct	23.9[b]	—	30.3	30.3	4.5	Goil (1961), Smyth (1966)
Paramphistomum microbothrium	Buffalo	Bile duct	—	—	—	—	2.5[c]	Hrzenjak & Ehrlich (1975)
Schistosoma bovis	Mouse	Venous system	—	—	3.93[c]	—	—	Magzoub (1973)
Schistosoma haematobium	Gerbil	Venous system	—	—	2.00[c]	—	—	Magzoub (1973)
Schistosoma mansoni (♂)	Mouse	Venous system	—	—	14–29	50	(33.9 paired)	Smith & Brooks (1969), Von Brand (1966)
Schistosoma mansoni (♀)	Mouse	Venous system	—	—	3–5	65	—	Von Brand (1966)

[a]References given in Smyth & Halton (1983) and Von Brand (1973).
[b]Calculated from author's data.
[c]Fresh mass.
[d]Glucose.
Source: Data mainly from Smyth & Halton (1983).

that the tegument serves as a major route for the uptake of glucose and amino acids from the blood (Smyth & Halton, 1983; Asch & Read, 1975; Chappell, 1974; Rogers & Bueding, 1975). Some of these results have been obtained by ligaturing the pharynx of a worm so that it is unable to feed via the digestive system. However, other workers (Nollen & Nadakavukaren, 1974; Halton & Arme, 1971) have demonstrated, by ultra-

Table 18.2. *Nature of food and hydrolytic enzymes present in the gut of some trematodes*

Species	Host	Location	Food (caecal contents)	Enzymes present									
				Protease	Aminopeptidase	Lipase	Esterase	Alkaline phosphatase	Acid phosphatase	β-glucuronidase	*N*-acetyl-β-D-glucosaminidase	ATPase	Glucose-6-phosphatase
Alaria marcianae	Dog	Intestine	Mucus, tissue	+	+	−	+	0	+	+	−	−	−
Alaria mustelae	Mink	Intestine	Mucus, tissue	−	−	−	+	−	−	−	−	−	−
Apatemon gracilis minor	Duck	Caecum	Mucus, tissue	−	0	−	+	0	+	−	−	−	−
Cotylophoron cotylophorum	Sheep	Rumen	Rumenal contents	−	−	+	+	−	+	−	−	+	+
Cyathocotyle bushiensis	Duck	Rectum	Tissue	−	+	0	+	0	+	−	−	−	−
Derogenes varicus	Fish	Stomach	Mucus, tissue	−	0	−	−	0	+	−	−	−	−
Dicrocoelium lanceatum	Cattle	Bile duct	—	−	0	+	+	−	−	−	−	−	−
Diplodiscus subclavatus	Frog	Rectum	Tissue, blood	−	0	0	0	0	+	−	−	−	−
Diplostomum spathaceum	Gull	Intestine	Mucus, tissue	−	0	−	0	0	+	−	−	−	−
Echinostoma revolutum	Chicken	Intestine	Tissue, blood	−	−	−	−	−	−	−	−		
Fasciola gigantica	Buffalo	Bile duct	—	−	−	−	−	0	+	−	−	−	−
Fasciola hepatica	Sheep	Bile duct	Blood, tissue	+*	+	+	+	0	+	0	−	+	+
Gorgodera cygnoides	Frog	Bladder	Tissue, blood	−	0	0	0	0	+	−	−	−	−
Gorgoderina attenuata	Frog	Bladder	—	−	−	−	−	−	+	−	−	−	−
Gorgoderina vitelliloba	Frog	Bladder	Tissue, blood	−	0	0	0	0	+	−	−	−	−
Haematoloechus medioplexus	Frog	Lungs	Blood	−	0	0	+	0	+	−	−	−	−
Haplometra cylindracea	Frog	Lungs	Blood	+*	0	+	+	0	+	−	−	−	−
Hemiurus communis	Fish	Stomach	—	−	0	−	+	0	+	−	−	−	−
Holostephanus lühei	Duck	Rectum	Mucus, tissue	−	+	−	+	0	+	−	−	−	−
Isoparorchis hypselobagri	Fish	Air bladder	—	−	−	−	−	0	+	−	−	−	−
Leucochloridiomorpha constantiae	Chicken	Bursa Fabricius	Tissue	+	−	−	−	−	−	−	−	−	−
Megalodiscus temperatus	Frog	Rectum	—	−	−	−	−	+	+	−	+	−	−
Opisthioglyphe ranae	Frog	Intestine	Tissue, blood	−	0	0	0	0	+	−	−	−	−
Paragonimus kellicotti	Cat	Lungs	—	−	−	−	−	−	+	−	−	−	−
Philophthalmus burrili	Chicken	Eye	Lacrimal secretion	+	0	0	+	0	+	0	−	−	−
Schistosoma haematobium	Hamster	Venous system	Blood	−	−	−	−	−	−	0	−	−	−
Schistosoma mansoni	Mouse	Venous system	Blood	+*	0	0	+	0	+	0	+	−	−
Schistosoma rodhaini	Mouse	Venous system	Blood	−	0	−	−	−	−	0	−	−	−
Schistosomatium douthitti	Mouse	Venous system	Blood	−	−	−	−	−	0	−	0	−	−

+ = present; 0 = absent; − = not tested for; *enzyme determined biochemically.
Source: Data from Smyth & Halton (1983).

structure studies, that such ligatures can readily damage the worm tissue. Hence results obtained by ligaturing must be viewed with caution and some may prove to be unreliable.

Intestinal digestion. The process of digestion in digenetic trematodes is predominantly an extracellular process taking place in the lumen of the caeca (Halton, 1967), the soluble breakdown products of digestion being absorbed and the process completed in the gastrodermis. Digestive enzymes so far identified are listed in Table 18.2. In *Schistosoma*, the protease identified as an optimal activity at pH 3.9–4.5, and is nearly five times more active in the female than the male. In *Fasciola*, the protease in the caeca, which attacks globin more rapidly than albumin or globulin (producing peptides), has a maximum activity at pH 3.9–4.0 (Simpkin *et al.*, 1980).

The major acid phosphatase activity detected in both *Schistosoma* and *Fasciola* is localised at the lumenal surface of the gastrodermis and is associated with the plasma membrane of the lamellar folds (Fig. 13.4) (Smyth & Halton, 1983). Despite the widespread occurrence of acid phosphatases in the animal kingdom, their precise physiological function has yet to be determined.

'Placental' digestion. One additional factor operates in trematode digestion, namely the fact that certain adhesive organs serve not only as organs of attachment, but as extracorporeal organs of digestion. In strigeids, the holdfast has been shown to be both digestive and 'placental', a phenomenon which has been most studied in *Apatemon*, *Cyathocotyle*, *Alaria*, *Holostephanus* and *Diplostomum*. The gland cells of the strigeid holdfast are rich in protein and RNA and secrete, at the site of attachment, material which contains hydrolytic enzymes. Well-developed glands are also associated with the lappets in some species, e.g. *Diplostomum phoxini*. There is now abundant evidence that both the lappets and holdfast in strigeids digest host tissue and produce finely fragmented material which is then absorbed (Bhatti & Johnson, 1972; Erasmus & Ohman, 1963; Lee, 1962; Ohman, 1966; Smyth & Halton, 1983).

○○○ 18.3 Energy metabolism

Most of the detailed studies on the energy metabolism of trematodes have been confined to the three genera *Fasciola*, *Schistosoma* and *Dicrocoelium*. As with all helminths studied, adult trematodes utilise carbohydrate as a primary source of energy and the metabolism is predominantly anaerobic (i.e. fermentative). Glucose is the primary respiratory substrate and this is absorbed and converted into glycogen or metabolised directly via the glycolytic series of reactions.

○ 18.3.1 *Fasciola* spp.

Fasciola hepatica and *F. gigantica* contain large amounts of glycogen stored chiefly in the parenchyma and muscle cells. This glycogen is rapidly depleted during starvation but readily synthesised if glucose becomes available. It has long been established that adult *F. hepatica* ferments glucose via glycolysis and a fumarate reductose pathway to propionate and acetate (Prichard, 1989). A small amount of PEP is converted to pyruvate by the action of pyruvate kinase and then reduced to lactate (Fig. 18.1). Under these anaerobic conditions, propionate and acetate are produced in a 2 : 1 ratio, achieving redox balance. Under aerobic conditions, however, the proportion of propionate decreases and more of the reducing equivalents from the production of acetate are oxidised by oxygen.

It is particularly interesting to note that – as in many other parasites (e.g. *Trypanosoma brucei*, Fig. 5.12) – the metabolic pathway changes substantially during the life cycle and the early liver stages have been shown to have a predominantly aerobic metabolism. This change in metabolism was investigated by incubation, in D-[6-^{14}C]glucose, of the various developmental stages, from a newly encysted juvenile to a mature adult fluke from the bile duct (Tielens *et al.*, 1984). It was shown that the major part of ATP production during aerobic incubation is contributed by three different pathways of glucose breakdown (Fig. 18.2). The Krebs cycle (Table 24.8) is the main energy-yielding pathway in the juvenile fluke (up to about 24 days), and this is gradually replaced by aerobic acetate formation and finally by anaerobic dismutation reactions of the adult fluke. Moreover, it was shown that the Krebs cycle activity of the developing fluke was directly proportional to the surface area of the fluke, indicating that this activity is limited by the diffusion of oxygen. This result suggests that in the small, juvenile stages, the oxygen could penetrate to most of the cells of the organism, and a Krebs cycle could operate, whereas in a large adult fluke, oxygen could probably only reach the outer layers of the worm (Tielens *et al.*, 1984). Hence, by about 114 days (Fig. 18.2) acetate production ceases, the Krebs cycle plays only a minor role and the major energy-production pathway is anaerobic.

The anaerobic catabolism of *F. gigantica* follows closely that of *F. hepatica*, all the major enzymes having been identified (Umezurike & Anya, 1980).

○ 18.3.2 *Schistosoma* spp.

The metabolism of *S. mansoni*, in particular, has been the subject of numerous studies over some 40 years. In spite of intensive experimentation, how this species obtains the bulk of its energy is still a matter of controversy (see below). Less attention has been paid to the metabolism of other *Schistosoma* species, probably on account of their general unavailability, and the difficulty of maintaining them in the laboratory; this is an area which clearly calls for further research.

Some relevant recent studies or reviews are: *S. mansoni*, Barrett (1981), Bueding & Fisher (1982), Coles (1975, 1984), Cornford (1987), Cornford *et al.* (1983, 1988), Köhler (1985), Lane *et al.* (1987), McManus (1989), Smyth & Halton (1983), Tielens *et al.* (1984); *S. japonicum*, Bueding & Fisher (1982), Huang (1980); *S. haematobium*, Cornford *et al.* (1983); *S. margrebowei*, Earle (1984); *Schistosomatium douthitti*, Rogers (1976). Earlier work is reviewed by Smyth & Halton (1983).

Glucose metabolism. In *S. mansoni*, at 7 weeks post infection, the glycogen is approximately 14–17 per cent of the dry mass of males and 4–5 per cent of that of females. It has also been shown that in this species, and

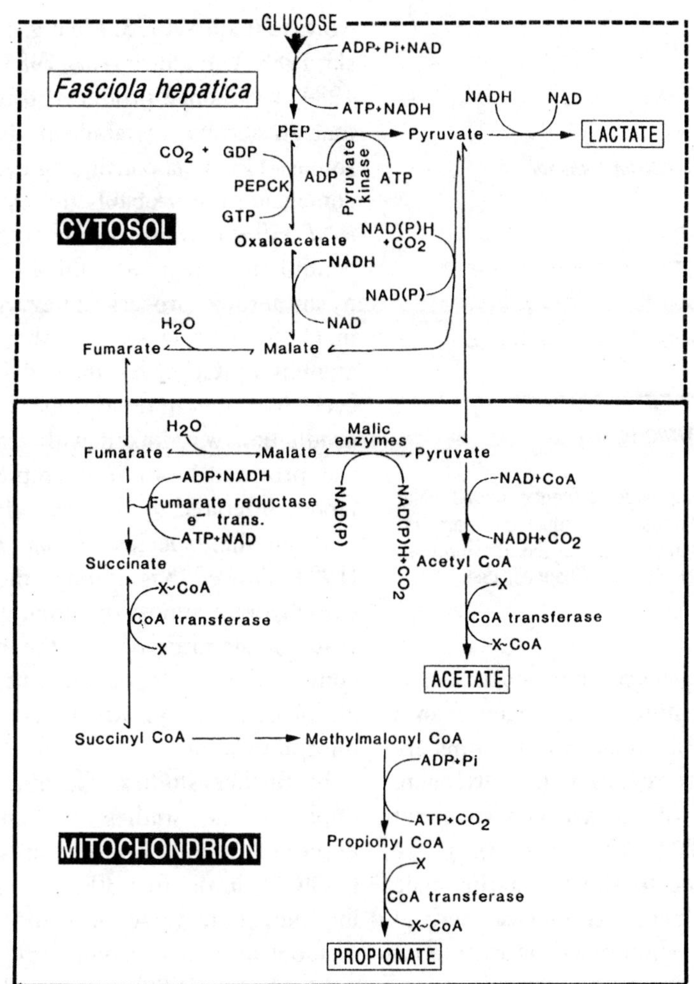

Fig. 18.1
Anaerobic energy metabolism in the adult liver fluke, *Fasciola hepatica*. Note that during the early phase of development, the metabolism is predominantly aerobic and a Krebs cycle operates; see Fig. 18.2. (Modified from Prichard, 1989.)

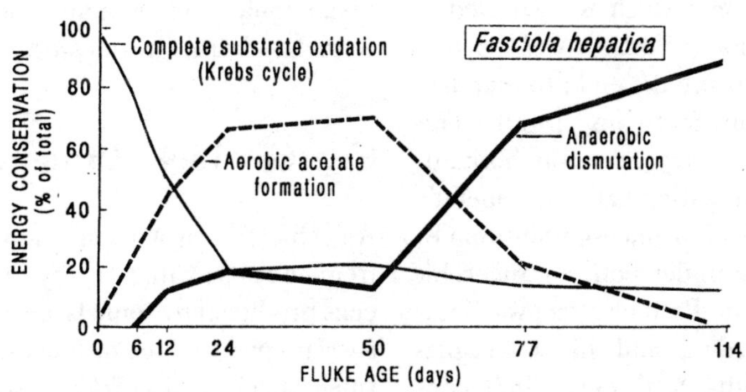

Fig. 18.2
Quantitative relationship between the various energy conserving pathways operating during the development of the liver fluke, *Fasciola hepatica*, in experimental rats: the energy conservation of the theoretical yields of ATP in each of the three pathways of glucose breakdown. The early phases of growth are aerobic but the later (adult) stage is anaerobic, as in Fig. 18.1. (After Köhler, 1985; modified from Tielens *et al.*, 1984.)

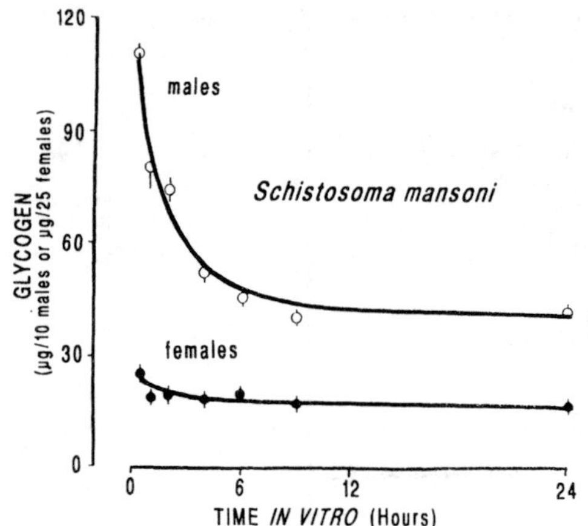

Fig. 18.3
The effect of *in vitro* maintenance on glycogen content of paired male and female *Schistosoma mansoni*. Worms were incubated as pairs for periods of up to 24 hours and then separated and analysed as groups of 10 male or 25 female worms. (After Mercer & Chappell, 1986.)

in *S. japonicum* and *S. haematobium*, the unmated male always contains greater quantities of glycogen than a mated male (Cornford, 1987). When paired worms are cultured *in vitro*, females lose relatively little glycogen, but the male utilises most of its glycogen in about 9 hours (Mercer & Chappell, 1986) (Fig. 18.3). There is also considerable evidence to show that the male schistosome can transfer glucose or glucose-derived metabolites to the female, the latter receiving an amount equal to 75 per cent of her free glucose requirements per minute (Cornford, 1987).

Metabolic pathways. The early work of Bueding (1950) proposed that this species was a 'homolactic acid fermenter', i.e. it exclusively converted glucose into lactic acid by glycolysis, a process which is restricted to the cytosol. Energy generation by this process is thus wholly independent of oxygen and the end product, lactic acid, is secreted (Bryant & Behm, 1989). This implies that the parasite is solely dependent on this pathway and no other energy-generating pathways operate. Bueding (1950) found that *in vitro* glucose utilisation by adult *S. mansoni* was similar under both an anaerobic and aerobic metabolism, i.e. no Pasteur effect was indicated. Later work by Bueding and his colleagues (Schiller *et al.*, 1975; Bueding & Fisher, 1982) confirmed their conclusion that schistosomes are homolactic fermenters and obtain no energy from oxidative pathways.

However, in contrast to these results, a number of workers (Coles, 1975; Huang, 1980; Moczon & Swiderski, 1983; Van Oordt *et al.*, 1985, 1989; Thompson *et al.*, 1984) have since produced direct or indirect evidence that an aerobic metabolism also functions in schistosomes. The most convincing evidence in support of this conclusion are probably the experiments of Van Oordt *et al.* (1985) whose results suggest that under aerobic conditions at least one-third of the energy is obtained by an aerobic process. They incubated worms *in vitro* in D-[6-^{14}C]glucose, and showed that a considerable amount of CO_2 was formed indicating that a Krebs cycle was operating. Further evidence that the $^{14}CO_2$ production was linked with oxidative phosphorylation was provided by the observation that the reaction was largely inhibited by the drug oligomycin.

In another species, *S. margrebowiei* (p. 246), Earle (1984) showed that although the male worms converted carbohydrate almost quantitatively to lactate, with glucose : lactate ratios of 1 : 2, the female and paired worms converted a substantial amount of the glucose via non-glycolytic pathways, which may have included oxidative phosphorylation.

In further studies, Cornford (1987) attempted to duplicate the studies of Van Oordt *et al.* (1985) described above, but was unable to find significant [^{14}C]CO_2 in the first 30 minutes after incubation, leaving the controversy over the occurrence of aerobic metabolism in schistsomes still unresolved.

McManus (1989) sums up the situation by pointing out that *in vitro* studies are notoriously variable and that variations in the media constituents, gas phase, culture vessel, incubation times, and other factors, can greatly influence results. He concludes that the matter is unlikely to be resolved until the problem – at present insuperable – of obtaining bulk preparations of intact mitochondria from schistosomes, is resolved.

○○○ 18.4 Protein metabolism

An efficient protein metabolism clearly operates in trematodes as witnessed by the astonishing numbers of eggs produced by some large species, such as *Fasciolopsis buski* (in pigs), which produces 25 000 eggs per day; even small species, like *Schistosoma japonica*, are estimated to produce 1800 eggs per day (Kawanaka, 1991). The potential protein synthesis of larval trematodes is even greater: a single miracidium of *S. mansoni* may give rise to more than 200 000 cercariae.

There are relatively few data available on the precise amino acid requirements of trematodes and whether or not any are 'essential', i.e. cannot be synthesised by the parasite. The limited information on the free pool of amino acids has been reviewed by Barrett (1981), Gupta & Gupta (1977) and Von Brand (1973).

In *Schistosoma mansoni*, *Fasciola hepatica* and *Fascioloides magna*, amino acid absorption appears to take place primarily via the tegument. In *S. mansoni*, amino acids are taken in by a mixture of mediated transport and diffusion, whereas in *Fasciola hepatica* and *Fascioloides magna* only diffusion appears to be involved (Barrett, 1981). The amino acid uptake sites have been poorly investigated. In male *S. mansoni*, there is evidence for at least five transport systems, including a highly specific transport locus for proline and one for acid amino acids (Asch & Read, 1975). The existence of a specific transport site for proline was indicated by the fact that unlabelled proline almost completely inhibited labelled protein.

For most of the neutral amino acids, there are probably at least two transport systems, each of broad and overlapping specificity. Basic amino acids also appear to be taken up by complex transport systems. Trematodes appear to be relatively poor in transaminases; in *S. mansoni* only 5 out of 16 amino acids were converted to other amino acids and incorporated (Chappell & Walker, 1982; Bruce *et al.*, 1972). Amino acids may play an important part as a source of energy in blood-feeding parasites such as *Fasciola* and *Schistosoma*. Thus, the catabolic pathway of arginine in *Fasciola* has been shown to result in the regeneration of NAD (Smyth & Halton, 1983), large quantities of proline being produced by this reaction, and *Schistosoma* likewise produces large amount of proline. This result is supported by the finding that *Fasciola* and *Schistosoma* possess a very active arginine catabolic pathway. In the cytosol, arginine is converted to ornithine and urea; in the mitochondria, arginine is not metabolised, but converted into ornithine and into proline, via ornithine–alanine transaminase and Δ'-pyrroline-5'-carboxylate reductase (Smyth & Halton, 1983).

Much attention has been paid to the protein which forms the basis of the egg-shell of trematodes (Table 13.1; Fig. 13.9) which in most species is a quinone-tanned protein, the phenolic components of which are converted to quinone via phenol oxidase (Cordingley, 1987). It is not surprising, therefore, to find that the phenolic amino acid tyrosine is avidly taken up by the vitelline cells of female *S. mansoni* and localised in the endoplasmic reticulum and in the 'vitelline' droplets, which – as shown in Figs 13.8 and 13.9 – fuse to form the egg-shell. In *S. mansoni*, it has been shown that [^{14}C]tyrosine is selectively incorporated into the vitelline cells *in vitro* and that two days after exposure the eggs in the uterus become radioactive (Kawanaka, 1991). It was also shown that a 34 kDa female protein is a precursor of the egg-shell and forms a much larger protein molecule as a result of the cross-linking during the process of tanning. Other recent studies on the molecular biology and biochemistry of the egg-shell protein are reviewed by Cordingley (1987) and Wells & Cordingley (1991). Relatively little information is available on the nature of other proteins of *Schistosoma* spp. although some differences between species have been found. In adults, electrophoresis has revealed the existence of 74 proteins, only 5 of which are male-specific and 4 female-specific (Atkinson & Atkinson, 1982). One particular protein in the female (M_r 66 000) is of especial interest as it constitutes some 10 per cent of the female protein and yet is not synthesised by it but by the male worm, indicating that it must be transferred from male to female (Atkinson & Atkinson, 1980). From this it can be concluded that continuous copulation is essential for the survival metabolism of the female.

Other aspects of trematode protein metabolism have been reviewed by Barrett (1981), Coles (1984) and Smyth & Halton (1983).

○○○ 18.5 Lipid metabolism

The lipid metabolism of adult and larval trematodes has been surveyed in considerable detail by Frayha & Smyth (1983); more specific reviews, dealing especially with the lipids and lipid metabolism of *Fasciola* and *Schistosoma*, are those of Barrett (1981), Coles (1984) and Furlong (1991).

Quantitative data on the total lipid content of adult trematodes are given in Table 18.1. Some classes of lipids which play an important role in the biology and physiology of *S. mansoni* are listed in Table 18.3.

Trematodes, like cestodes (Chapter 24), have lost their oxygen-dependent pathways for the synthesis of those sterols and unsaturated fatty acids which occur in free-living invertebrates. They have also lost the ability to synthesise long-chain fatty acids *de novo* (Meyer *et al.*, 1970) but are, however, able to modify fatty acid chain length and synthesise fatty acid metabolites (Furlong, 1991).

Table 18.3. Schistosoma mansoni*: important roles played by some lipids during the life cycle*

Lipid class	Function
Fatty acids and metabolites	Skin penetration by cercariae Secreted by schistosomula and adults
Phospholipids and sterols	Membrane components (both cellular and tegumental membranes)
Glycolipids	Membrane components and antigens
Fatty acids and phospholipids	Lipid anchors for protein antigens
Phospholipids, sterols and lipoproteins	Defence from host immunity
Ecdysones	Sexual development; marker of infection
Dolichols	Egg production

Source: Data from Furlong (1991).

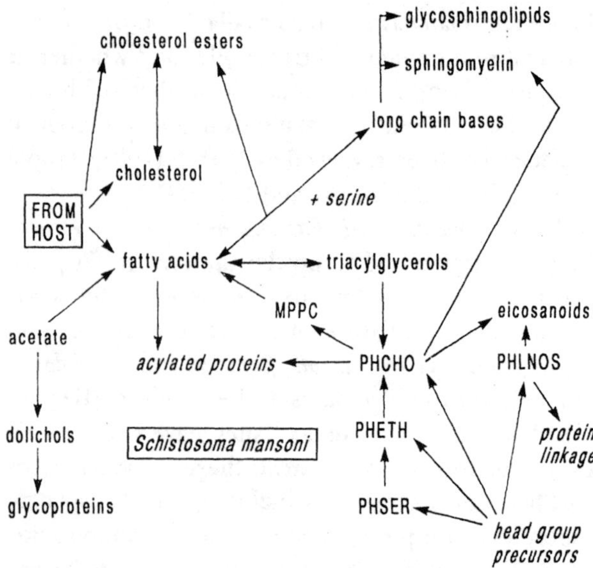

Fig. 18.4
The major biosynthetic pathways of lipids in *Schistosoma mansoni*. Like all trematodes, the organism is unable to synthesise sterols or fatty acids *de novo* and is dependent on these being obtained from the fluids or tissues of the host. Many of the details of these synthetic pathways remain to be clarified. PHCHO, phosphatidylcholine; MPPC, monopalmitoylphosphatidylcholine; PHETH, phosphatidylethanolamine; PHSER, phosphatidylserine; PHLNOS, phosphatidylinositol. (After Furlong, 1991.)

Since trematodes are unable to synthesise most essential lipids, they have to acquire exogenous lipids from the fluids, tissues or cell membranes of the host. The biosynthetic pathways have been best investigated in *S. mansoni* and the probable interrelationships between these are shown in Fig. 18.4.

○○○ 18.6 Neurobiology

Until recently, relatively few studies had been carried out on the physiology and biochemistry of the nervous system of trematodes and even the basic morphology and nerve distribution is poorly documented. Two cytochemical techniques have revolutionised this situation: (*a*) the indoxyl acetate technique for esterases which, when applied to whole worms, clearly and specifically displays the morphology of the nervous system; (*b*) the immunocytochemical technique, developed for the demonstration of regulatory peptides in vertebrate tissues, which has been applied to trematodes and cestodes with considerable success.

The indoxyl acetate technique, which shows up cholinesterase, even in whole worms, was first applied, some years ago, to the monogenean *Diplozoon paradoxum* and clearly displayed the distribution of the nerve cords and nerve tissues (Halton & Jennings, 1964). Since then, the technique has been applied with much success to numerous trematode and cestode species.

Neurosecretion. Early workers used the positive staining of nerve cells with paraldehyde fuchsin to identify putative neurosecretory cells. Since then, neurosecretory cells have been described in Monogenea, Digenea and Aspidogastrea. Staining in paraldehyde fuchsin by itself, without other supporting evidence, is generally regarded as unreliable, and results using this technique must be interpreted with caution. Data on the species investigated have been reviewed by Fairweather & Halton (1991) and Smyth & Halton (1983).

Regulatory peptides: immunocytochemistry. In vertebrates, our understanding of the functional complexity of the nervous system was significantly enhanced by the discovery of a large number of peptidergic molecules which appear to have a neuroregulatory function and coordinate the activities of the nervous system and the endocrine system (Fairweather & Halton, 1991; Halton *et al.*, 1990, 1992). This conclusion arose from the finding that many of the messenger molecules produced by endocrine cells in the form of hormones, such

Table 18.4. *Peptide immunoreactivities demonstrated in parasitic platyhelminths*

	Monogenea		Digenea					Cestoda			
Peptide antisera	*Diclidophora merlangi*	*Gyrodactylus salaris*	*Schistosoma* spp.	*Fasciola hepatica*	*Echinostoma* spp.	*Haplometra cylindracea*	*Gorgoderina vitelliloba*	*Triloculаria acanthiaevulgaris*	*Diphyllobothrium dendriticum*	*Schistocephalus solidus*	*Hymenolepis* spp.
Pancreatic polypeptide (PP)	+	−	+	+	·	+	+	+	+	·	+
Peptide YY (PYY)	+	·	+	+	·	+	+	+	−	·	+
Neuropeptide Y (NPY)	+	·	+	−	·	−	+	−	−	·	·
Glucagon	−	·	·	−	·	+	−	·	−	·	−
Vasoactive intestinal polypeptide (VIP)	+	·	·	−	+	·	−	+	−	−	−
Peptide histidine isoleucine (PHI)	−	·	·	+	·	+	−	+	+	·	+
Gastrin	−	·	+	−	−	−	−	−	+	·	−
Cholecystokinin (CCK)	+	·	+	−	−	−	·	−	−	·	·
Gastrin-releasing peptide (GRP)	−	·	·	+	·	+	·	−	+	·	+
Substance P (SP)	+	−	+	+	+	+	+	+	−	·	−
Neurokinin A (NKA)	+	·	·	−	·	+	·	·	·	·	·
Eledoisin	+	·	·	·	·	·	·	·	·	·	·
Leu-enkephalin	·	+	+	−	−	·	·	·	+	+	·
Met-enkephalin	·	·	·	−	·	·	·	−	+	+	·
β-Endorphin	·	−	+	·	·	·	·	·	−	·	·
Adrenocorticotropin hormone (ACTH)	−	·	+	·	·	·	·	·	−	·	−
α-Melanocyte-stimulating hormone (α-MSH)	·	·	+	·	·	·	·	·	·	·	·
Neurotensin (NT)	−	·	−	−	−	−	−	−	+	·	−
Somatostatin (SRIF)	−	·	·	−	−	−	·	+	−	·	−
Urotensin I	·	+	−	·	·	·	·	·	·	·	·
Oxytocin	·	·	−	−	−	·	·	·	+	·	·
Vasotocin	·	·	−	·	·	·	·	·	+	·	·
Prolactin (PRL)	·	·	−	−	−	·	·	·	·	·	+
Growth hormone-releasing factor (GRF)	·	+	+	·	·	·	·	·	+	·	·
Luteinising hormone-releasing hormone (LHRH)	·	·	+	−	−	·	·	·	·	·	·
Human chorionic gonadotropin (hCG)	·	·	+	+	−	·	·	·	·	·	·
FMRFamide	+	+	+	+	−	+	+	+	+	+	·
RFamide	·	+	·	·	·	·	·	·	+	·	·
Small cardiac peptide B (SCPb)	·	·	·	·	·	·	·	·	+	·	·

+, Present; −, absent; ·, not tested.
Source: Data from Fairweather & Halton (1991); see also references in Halton *et al.* (1992).

as cholecystokinin (CCK) from the gastrointestinal tract, are also produced by neurones and are now considered to be potential neurotransmitters and neuromodulators. Other peptides (e.g. enkephalins and endorphins), present in the mammalian brain, have been found to be common in the adrenal medulla and – like adrenaline – play the role of an endocrine secretion. Adrenaline has been shown to act as either a hormone or a neurotransmitter.

In addition to the eight well-known transmitters – acetylcholine, adrenaline, noradrenaline, dopamine, 5-hydroxytryptamine (5-HT), glutamate, octopamine and histamine – some 29 other transmitter substances have now been identified (Table 18.4). These biologically active messengers are now usually referred to as '*regulatory peptides*', which are small molecules (2–80 amino acid residues). To date, some 50 or so are currently recognised. Further details of the synthesis, secretion and degradation of regulatory peptides are reviewed by Halton *et al.* (1990). The cytochemical technique involved in the demonstration of the regulatory peptides involves the application of antisera against specific peptides; this results in the 'staining' of the sites of individual peptides.

Specific peptidergic molecules were first detected in platyhelminths by Gustafsson *et al.* (1985) in the plerocercoid of the bird cestode *Diphyllobothrium dendriticum* (Fig. 22.3) using antisera developed against mammalian peptides. Since then, the homologues of some 30 vertebrate and 3 invertebrate peptides have been identified in trematodes and cestodes. In trematodes, immunocytological studies on the serotoninergic, cholinergic and peptidergic components in a number of species have been carried out, of which the following are representative (see also Table 18.4):

MONOGENEA: *Diclidophora merlangi* (Maule *et al.*, 1990*a*,*b*; Halton *et al.*, 1991). DIGENEA: *Echinostoma paraensei* (Basch & Gupta, 1988); *Fasciola hepatica* (Basch & Gupta, 1988; Magee *et al.*, 1989, 1991); *Gorgoderina vitelliloba* (McKay *et al.*, 1991); *Haplometra cylindracea* (McKay *et al.*, 1990); *Schistosoma mansoni* (Gustafsson, 1987; Basch & Gupta, 1988); *Schistosoma japonicum* (Basch & Gupta, 1988); *Schistosomatium douthitti* (Basch & Gupta, 1988). Cestode species are shown in Table 18.4.

Peptide distribution and function. The distribution of peptide immunoreactivity in trematode tissues has been succinctly surveyed by Fairweather & Halton (1991) and Halton *et al.* (1992). The reactions occur throughout the central nervous system with some regional differences in the immunostaining of some peptides. Since these immunocytological reactions can be visualised by indirect fluorescent microscopy, the distribution of the reacting cells is readily visualised and easy to record photographically, as seen in the superb micrographs published in the papers cited above. The resolution of these reactions has been further clarified by the use of confocal scanning laser microscopy (Johnson *et al.*, 1990) which allows whole-mount preparations to be scanned and computerised images from different levels to be produced. A diagrammatic interpretation of the use of this technique with the monogenean *Diclidophora merlangi* (p. 169) is shown in Fig. 18.5.

All the studies show that the peripheral nervous system, the somatic musculature and the cells associated with the reproductive system show particular reactivity to the peptide antisera. Although the peptides have been described most commonly in the nervous tissues, they have also been identified in non-nervous tissue, and are consistently found in the cells and muscles of

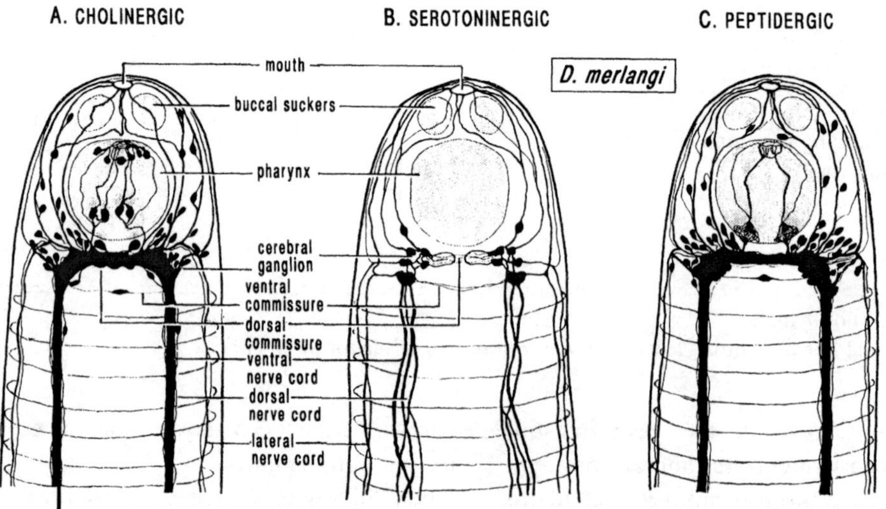

Fig. 18.5
Diclidophora merlangi: diagrammatic interpretations of the cholinergic, serotoninergic and peptidergic nervous systems, based largely on immunocytological reactions (see Table 18.4) in conjunction with confocal laser microscopy, the cholinergic system being demonstrated by cholinesterase staining. The cholinergic system closely parallels the peptidergic system but the serotoninergic system has a distinct construction. (Modified from Maule *et al.*, 1990*b*.)

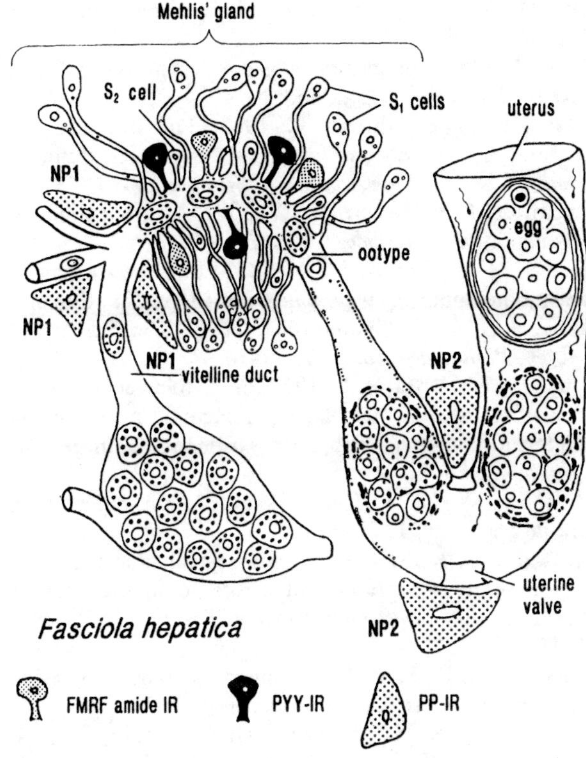

Fig. 18.6
Diagram showing the distribution of PP-, PYY-, and FRMFamide-immunoreactive cells in the ootype of the Mehlis' gland complex of *Fasciola hepatica* (see Fig. 13.8). **NP1**, cell body in nerve plexus 1; **NP2**, cell body in nerve plexus 2. S_1, S_1-type of Mehlis' gland cell; S_2, S_2-type Mehlis' gland cell. (After Magee *et al.*, 1989.)

the egg-forming system, e.g. as in *Fasciola* (Fig. 18.6) (Magee *et al.*, 1989).

The precise functions played by the peptides identified in trematode tissues is still a matter for speculation, although it can generally be concluded that they play a part in complex system of internal signalling and regulatory systems. Their possible roles have been discussed by Fairweather & Halton (1991).

Molecular biology of regulatory peptides. Substantial progress has been made in our understanding of the molecular mechanisms underlying the development of regulatory peptides and a possible mechanism for their biosynthesis is shown in Fig. 18.7. The various problems involved are reviewed by Halton *et al.* (1992).

○○○ References

(References of general interest have been included in this list, in addition to those quoted in the text.)

Asch, H. L. & Read, C. P. 1975. Membrane transport in *Schistosoma mansoni*: transport of amino acids by adult males. *Experimental Parasitology*, **38**: 123–35.

Atkinson, K. H. & Atkinson, B. G. 1980. Biochemical basis for continuous copulation of female *Schistosoma mansoni*. *Nature, London*, **283**: 478–9.

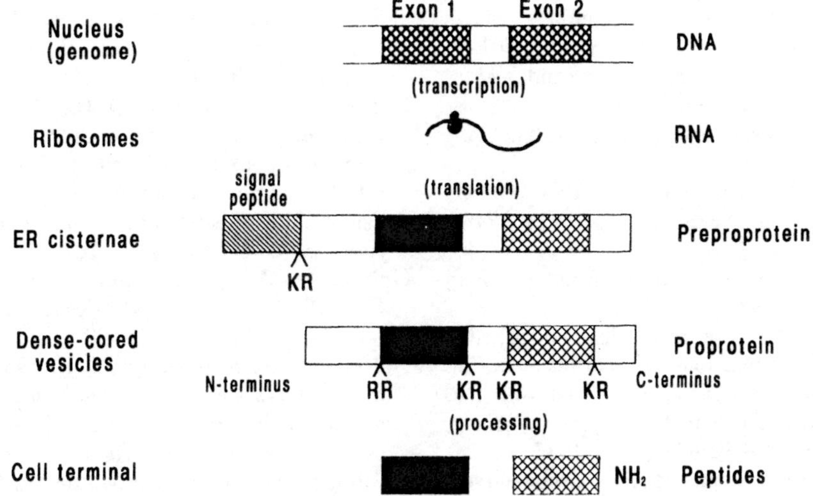

Fig. 18.7
Biosynthesis of regulatory neuropeptides. This involves gene action and DNA transcription in the nucleus and RNA at the ribosomes producing a large polypeptide precursor, known as a preproprotein which may contain several copies of individual peptides from the encoded exons. A hydrophobic signal in the N-terminus of the precursor protein routes the chain of peptides from the ribosomes to the ER cisternae. The signal peptide is removed enzymatically and the released proprotein becomes incorporated into dense cored vesicles for transport to the cell terminal. The protease cleavage sites are marked by pairs of basic amino acids: K, lysine; R, arginine. The two peptide fragments may be modified further by amidation (NH_2) of the C-terminus, before being released from the cell terminal. (After Halton *et al.*, 1992.)

Atkinson, B. G. & Atkinson, K. H. 1982. *Schistosoma mansoni*: one- and two-dimensional electrophoresis of proteins synthesised by males, females and juveniles. *Experimental Parasitology*, **53**: 26–38.

Barrett, J. 1981. *Biochemistry of Parasitic Helminths*. Macmillan Publishers, London. ISBN 0 333 25669 7.

Basch, P. F. & Gupta, B. C. 1988. Immunocytochemical localization of regulatory peptides in six species of trematode parasites. *Comparative Biochemistry and Physiology*, **91**C: 565–70.

Bennet, E.-M., Behm, C. & Bryant, C. (eds) 1989. *Comparative Biochemistry of Parasitic Helminths*. Chapman & Hall, London. ISBN 0 412 32730 9.

Bhatti, I. & Johnson, A. D. 1972. Enzyme histochemistry of the holdfast organ and forebody gland cells of *Alaria marcianae* (La Rue, 1917) (Trematoda: Diplostomatidae). *Proceedings of the Helminthological Society of Washington*, **39**: 78–87.

Bruce, J. I., Ruff, M. D., Belusko, R. J. & Werner, J. K. 1972. *Schistosoma mansoni* and *Schistosoma japonicum*: utilization of amino acids. *International Journal for Parasitology*, **2**: 425–30.

Bryant, C. & Behm, C. A. 1989. *Biochemical Adaptation in Parasites*. Chapman & Hall, London. ISBN 0 412 32530 6.

Bueding, E. 1950. Carbohydrate metabolism of *Schistosoma mansoni*. *Journal of General Physiology*, **33**: 475–95.

Bueding, E. & Fisher, J. 1982. Metabolic requirements of schistosomes. *Journal of Parasitology*, **68**: 208–12.

Chappell, L. H. 1974. Methionine uptake by larval and adult *Schistosoma mansoni*. *International Journal for Parasitology*, **4**: 361–9.

Chappell, L. H. 1980. *Physiology of Parasites*. Blackie, London.

Chappell, L. H. & Walker, E. 1982. *Schistosoma mansoni*: incorporation and metabolism of protein amino-acids *in vitro*. *Comparative Biochemistry and Physiology*, **73**B: 701–7.

Coles, G. C. 1975. Fluke biochemistry – *Fasciola* and *Schistosoma*. *Helminthological Abstracts*, **44**: 147–62.

Coles, G. C. 1984. Recent advances in schistosome biochemistry. *Parasitology*, **89**: 603–37.

Coles, G. C., Simpkin, K. G. & Barrett, J. 1980. *Fasciola hepatica*: energy sources and metabolism. *Experimental Parasitology*, **49**: 122–7.

Cordingley, J. S. 1987. Trematode eggshells: novel protein biopolymers. *Parasitology Today*, **3**: 341–4.

Cornford, E. M. 1987. Glucose metabolism in parasitic trematodes: unique specializations of schistosomes. In: *Molecular Paradigms for Eradicating Helminthic Parasites* (ed. A. J. MacInnis), pp. 457–76. Alan R. Liss, New York.

Cornford, E. M., Diep, C. P. & Rowley, G. A. 1983. *Schistosoma mansoni*, *S. japonicum*, *S. haematobium*: glycogen content and glucose uptake in parasites from fasted and control hosts. *Experimental Parasitology*, **56**: 397–408.

Cornford, E. M., Fitzpatrick, A. M., Quirk, T. L., Diep, C. P. & Landaw, E. M. 1988. Tegumental glucose permeability in male and female *Schistosoma mansoni*. *Journal of Parasitology*, **74**: 116–28.

Dunne, D. W. 1990. Schistosome carbohydrates. *Parasitology Today*, **6**: 45–8.

Earle, C. A. 1984. Energy metabolism in schistosomes. Ph.D. Thesis, University of London.

Erasmus, D. A. & Ohman, C. 1963. The structure and function of the adhesive organ in strigeid trematodes. *Annals of the New York Academy of Sciences*, **113**: 7–35.

Fairweather, I. & Halton, D. W. 1991. Neuropeptides in platyhelminths. *Parasitology*, **102**: S77–92.

Frayha, G. J. & Smyth, J. D. 1983. Lipid metabolism in parasitic helminths. *Advances in Parasitology*, **22**: 309–87.

Furlong, S. T. 1991. Unique roles for lipids in *Schistosoma mansoni*. *Parasitology Today*, **7**: 59–62.

Gupta, S. P. & Gupta, R. C. 1977. Free amino acid composition of *Opisthorchis pedicellata* (Trematoda) from the gall bladder of fresh-water fish, *Rita rita* (Ham.). *Zeitschrift für Parasitologie*, **54**: 83–7.

Gustafsson, M. K. S. 1985. Cestode neurotransmitters. *Parasitology Today*, **1**: 72–5.

Gustafsson, M. K. S. 1987. Immunocytochemical demonstration of neuropeptides and serotonin in the nervous system of adult *Schistosoma mansoni*. *Parasitology Research*, **74**: 168–74.

Gustafsson, M. K. S. 1992. The neuroanatomy of parasitic flatworms. *Advances in Neuroimmunology*, **2**: 267–86.

Gustafsson, M. K. S., Wikgren, M. C., Karhi, T. J. & Schot, L. P. C. 1985. Immunocytochemical demonstration of neuropeptides and serotonin in the tapeworm *Diphyllobothrium dendriticum*. *Cell and Tissue Research*, **240**: 255–60.

Halton, D. W. 1967. Observations on the nutrition of digenetic trematodes. *Parasitology*, **57**: 639–60.

Halton, D. W. & Arme, C. 1971. *In vitro* technique for detecting tegument damage in *Diclidophora merlangi*: possible screening method for selection of undamaged tissues or organisms prior to physiological investigation. *Experimental Parasitology*, **30**: 54–7.

Halton, D. W., Brennan, G. P., Maule, A. G., Shaw, C., Johnston, C. F. & Fairweather, I. 1991. The ultrastructure and immunogold-labelling of the pancreatic polypeptide-like immunoreactive cells associated with the egg-forming apparatus of a monogenean parasite, *Diclidophora merlangi*. *Parasitology*, **102**: 429–36.

Halton, D. W., Fairweather, I., Shaw, C. & Johnston, C. F. 1990. Regulatory peptides in parasitic platyhelminths. *Parasitology Today*, **6**: 284–90.

Halton, D. W. & Jennings, J. B. 1964. Demonstration of the nervous system in the monogenetic trematode *Diplozoon paradoxum* Nordman by the indoxyl acetate method for esterases. *Nature*, **202**: 510–11.

Halton, D. W., Shaw, C., Maule, A. G., Johnston, C. F. & Fairweather, I. 1992. Peptidergic messengers: a new perspective of the nervous system of parasitic platyhelminths. *Journal of Parasitology*, **78**: 179–93.

Huang, T.-Y. 1980. The energy metabolism of *Schistosoma japonicum*. *International Journal of Biochemistry*, **12**: 457–64.

Johnston, C. F., Shaw, C., Halton, D. W. & Fairweather, I. 1990. Confocal scanning laser miscroscopy and helminth neuroanatomy. *Parasitology Today*, *6:* 305–8.

Kawanaka, M. 1991. Identification of a putative eggshell precursor protein of the female *Schistosoma japonicum. International Journal for Parasitology*, **21:** 225–31.

Köhler, P. 1985. The strategies of energy conservation in helminths. *Molecular and Biochemical Parasitology*, **17:** 1–18.

Kurlec, B. 1975. Catabolic path of arginine and NAD regeneration in the parasite *Fasciola hepatica. Comparative Biochemistry and Physiology*, **51B:** 151–6.

Lane, C., Foster, L., Pax, R. & Bennett, J. 1987. Evidence that aerobic generation of energy in *S. mansoni* is not physiologically relevant. *Journal of Cellular Biochemistry, Supplement*, **11A:** 154.

Lee, D. L. 1962. Studies on the functions of the pseudosuckers and holdfast organ of *Diplostomum phoxini* Faust (Strigeida, Trematoda). *Parasitology*, **52:** 103–12.

Lennox, R. W. & Schiller, E. L. 1972. Changes in the dry weight and glycogen content as criteria for measuring the postcercarial growth and development of *Schistosoma mansoni. Journal of Parasitology*, **58:** 489–94.

Lloyd, G. M. 1986. Energy metabolism and its regulation in the adult liver fluke *Fasciola hepatica. Parasitology*, **93:** 217–48.

McKay, D. M., Halton, D. W., Johnston, C. F., Fairweather, I. & Shaw, C. 1990. Occurrence and distribution of putative neurotransmitters in the frog-lung parasite *Haplometra cylindracea* (Trematoda: Digenea). *Parasitology Research*, **76:** 509–17.

McKay, D. M., Halton, D. W., Johnston, C. F., Fairweather, I. & Shaw, C. 1991. Cytochemical demonstration of cholinergic, sertoninergic and peptidergic nerve elements in *Gorgoderina vitelliloba* (Trematoda: Digenea). *International Journal for Parasitology*, **21:** 71–80.

McManus, D. P. 1989. Carbohydrate and energy metabolism in adult schistosomes: a reappraisal. In: *Comparative Biochemistry of Parasitic Helminths* (ed. E.-M. Bennet, C. Behm & C. Bryant), pp. 79–94. Chapman & Hall, London.

Magee, R. M., Fairweather, I., Johnston, C. F., Halton, D. W., & Shaw, C. 1989. Immunocytochemical demonstration of neuropeptides in the nervous system of the liver fluke, *Fasciola hepatica* (Trematoda, Digenea). *Parasitology*, **98:** 227–38.

Magee, R. M., Fairweather, I., Shaw, C., McKillop, J. M., Montgomery, W. I., Johnston, C. F. & Halton, D. W. 1991. Quantification and partial characterisation of regulatory peptides in the liver fluke, *Fasciola hepatica*, from different laboratory hosts. *Comparative Biochemistry and Physiology*, **99C:** 201–7.

Maule, A. G., Halton, D. W., Johnston, C. F., Shaw, C. & Fairweather, I. 1990*a*. A cytochemical study of the serotoninergic, cholinergic and peptidergic components of the reproductive system in the monogenean parasite, *Diclidophora merlangi. Parasitology Research*, **76:** 409–19.

Maule, A. G., Halton, D. W., Johnston, C. F., Shaw, C. & Fairweather, I. 1990*b*. The serotoninergic, cholinergic and peptidergic components of the nervous system in the monogenean parasite, *Diclidophora merlangi*: a cytological study. *Parasitology*, **100:** 255–73.

Mercer, J. G. & Chappell, L. H. 1985. *Schistosoma mansoni*: effect of maintenance *in vitro* on the physiology and biochemistry of adult worms. *Parasitology*, **90:** 339–49.

Mercer, J. G. & Chappell, L. H. 1986. The effect of maintenance *in vitro* of glucose uptake and the incorporation of glucose by adult *Schistosoma mansoni. International Journal for Parasitology*, **16:** 253–61.

Meyer, F., Meyer, H. & Beuding, E. 1970. Lipid metabolism in the parasitic and free-living flatworms *Schistosoma mansoni* and *Dugesia dorotocephala. Biochimica et Biophysica Acta*, **210:** 257–66.

Moczon, T. & Swiderski, Z. 1983. *Schistosoma haematobium* oxidoreductase histochemistry and ultrastructure of niridazole-treated females. *International Journal for Parasitology*, **13:** 225–32.

Nahih, I. & El-Ansary, A. 1982. Metabolic end-products in parasitic helminths and their intermediate hosts. *Comparative Biochemistry and Physiology*, **101B:** 499–508.

Nollen, P. M. & Nadakavukaren, M. J. 1974. Observations on ligated adults of *Philophthalmus megalurus*, *Gorgoderina attenuata*, and *Megalodiscus temperatus* by scanning electron microscopy and autoradiography. *Journal of Parasitology*, **60:** 921–4.

Ohman, C. 1966. The structure and function of the adhesive organ in strigeid Trematodes. IV. *Holostephanus luhei* Szidat, 1936. *Parasitology*, **56:** 481–91.

Pappas, P. W. & Read, C. P. 1975. Membrane transport in helminth parasites: a review. *Experimental Parasitology*, **37:** 469–530.

Pena, S. D., Macedo, A. M., Braga, V. M. M., Rumjanek, F. D. & Simpson, A. J. G. 1990. F10, the gene for the glycine-rich major eggshell protein of *Schistosoma mansoni* recognises a family of hyperviable minisatellites in the human genome. *Nucleic Acids Research*, **18:** 7466.

Prichard, R. K. 1989. How do parasitic helminths use and survive oxygen and oxygen metabolites? In: *Comparative Biochemistry of Parasitic Helminths* (ed. E.-M. Bennett, C. Behm & C. Bryant), pp. 67–78. Chapman & Hall, London.

Rogers, S. H. 1976. *Schistosomatium douthitti*: carbohydrate metabolism of adults. *Experimental Parasitology*, **40:** 397–405.

Rogers, S. H. & Bueding, E. 1975. Anatomical localization of glucose uptake by *Schistosoma mansoni* adults. *International Journal for Parasitology*, **5:** 369–71.

Schiller, E. L., Bueding, E., Turner, V. M. & Fisher, J. 1975. Aerobic and anaerobic carbohydrate metabolism and egg production of *Schistosoma mansoni in vitro. Journal of Parasitology*, **61:** 385–9.

Simpkin, K. G., Chapman, C. R., Coles, G. C. 1980. *Fasciola hepatica*: a proteolytic digestive enzyme. *Experimental Parasitology*, **49:** 281–7.

Skuce, P. J., Johnston, C. F., Fairweather, D. W., Halton, D. W., Shaw, C. & Buchanan, K. D. 1990. Immunoreactivity to the pancreatic polypeptide family in the nervous system of the adult human blood fluke, *Schistosoma mansoni. Cell & Tissue Research*, **261**: 573–81.

Smyth, J. D. & Halton, D. W. 1983. *The Physiology of Trematodes*. 2nd Edition. Cambridge University Press. ISBN 0 521 29434 7.

Thompson, D. P., Morrison, D. D., Pax, R. A. & Bennett, J. L. 1984. Changes in glucose metabolism and cyanide sensitivity in *Schistosoma mansoni* during development. *Molecular and Biochemical Parasitology*, **13**: 39–51.

Tielens, A. G. M., Van den Heuvel, J. M. & Van den Bergh, S. G. 1984. The energy metabolism of *Fasciola hepatica* during its development in the final host. *Molecular and Biochemical Parasitology*, **13**: 301–7.

Umezurike, G. M. & Anya, A. O. 1980. Carbohydrate energy metabolism in *Fasciola gigantica. International Journal for Parasitology*, **10**: 175–80.

Van Oordt, B. E. P., Tielens, A. G. M. & Van de Bergh, S. G. 1988. The energy metabolism of *Schistosoma mansoni* during its development in the hamster. *Parasitology Research*, **75**: 31–5.

Van Oordt, B. E. P., Tielens, A. G. M. & Van de Bergh, S. G. 1989. Aerobic to anaerobic transition in the carbohydrate metabolism of *Schistosoma mansoni* cercariae during transformation *in vitro. Parasitology*, **98**: 409–15.

Van Oordt, B. E. P., Van den Heuvel, J. M., Tielens, A. G. M. & Van den Bergh, S. G. 1985. The energy production of adult *Schistosoma mansoni* is for a large part aerobic. *Molecular and Biochemical Parasitology*, **16**: 117–26.

Von Brand, Th. 1973. *Biochemistry of Parasites*. Academic Press, New York.

Ward, P. F. V. 1982. Aspects of helminth metabolism. *Parasitology*, **84**: 177–94.

Wells, K. E. & Cordingley, J. S. 1991. *Schistosoma mansoni*: eggshell formation is regulated by pH and calcium. *Experimental Parasitology*, **73**: 295–310.

19

Cestoda: Cestodaria

19.1 General account

The Cestoda form a group of worms which, with a few exceptions, exhibit two striking morphological features; they possess an elongated tape-like body and they lack an alimentary canal. The elongated shape precludes them from habitats whose spatial arrangement provides no elongated axis. Thus, with the exception of the subclass Cestodaria and a few neotonic larval forms in oligochaetes (e.g. *Archigetes*) adult cestodes occur only in tubular habitats, usually the alimentary canal, but occasionally in the bile or pancreatic ducts (e.g. *Stilesia*, *Atriotaenia*). These are habitats of high nutritional levels, a fact associated with cestodes' high rate of growth.

The larval habitat, on the other hand, shows a wide range of variation, and larvae can be found in almost any organ of both vertebrate and invertebrate hosts, although most larvae show a predilection for a particular site.

The lack of an alimentary canal separates cestodes markedly from trematodes and nematodes, and this feature, which is shared with the Acanthocephala, dominates the physiology of the group. Thus, the body covering or tegument (Fig. 20.1) must serve not only as a protective coating but also as a metabolically active layer through which nutritive material can be absorbed, and secretions and waste materials transported. The ultrastructure of the tegument bears a striking resemblance to that in trematodes (Fig. 13.2) with which it is clearly homologous.

Except in certain 'primitive' species, each tapeworm is a string of individuals having a complete set of reproductive organs in progressive degrees of sexual maturity and budding off from a body attached to the host tissue by a head or scolex. In one subclass, the Cestodaria, and in the family Caryophyllaeidae there is only one set of reproductive organs, and budding does not occur.

How this budding takes place, and indeed the whole question of tissue growth and differentiation in cestodes, is very imperfectly understood. In *Echinococcus* spp. remarkable non-strobilated, 'monozoic' forms have been grown *in vitro* (see Chapter 34). The group thus provides interesting material for the experimental embryologist.

With the exception of *Hymenolepis nana* (p. 324), which can develop directly in the same host, all cestodes require one or sometimes two intermediate hosts. These follow no general pattern, and can be either vertebrate or invertebrate, warm-blooded or cold-blooded.

Classification

The group may be divided into two subclasses as follows:

Cestodaria: Cestoda which are not divided into segments and which contain only one set of reproductive organs; scolex lacking; 10-hooked larva (decacanth).

Eucestoda: Cestoda which become distinctly divided into segments or proglottids (except the Caryophyllaeidae), each containing a set of male and female reproductive organs; scolex usually present; 6-hooked larva (hexacanth). Major reference works on the Cestoda are those of Arme & Pappas (1983*a*,*b*), Joyeux & Baer (1961), Schmidt (1970), Smyth & McManus (1989), Wardle & McLeod (1952) and Yamaguti (1959).

19.2 Subclass Cestodaria

This is a group of worms of uncertain affinities, and although today they are restricted to a few species, they may have been more abundant in ancient times. The group is divided into two orders:

Order 1. **Amphilinidea**
Order 2. **Gyrocotylidea**

Yamaguti (1959) includes a third order, the Caryophyllidea, in this subclass. Joyeux & Baer (1961) include them in the Eucestoda (Order Pseudophyllidea) and the latter classification has been followed here (but see Chapter 20).

○ 19.2.1 Amphilinidea

Amphilinids are unusual among the Cestoda in being parasitic in the body cavity. The growth and metabolic rates of these worms must be sufficiently low for their nutritional requirements to be satisfied by the constituents of the coelomic fluid.

Type example: *Amphilina foliacea*

definitive host: *Acipenser* sp. (sturgeon)
intermediate host: *Gammarus* spp. or *Dikerogammarus* spp.
location: body cavity

Morphology. The body is leaf-shaped and creamy-white in colour. The arrangement of the genitalia is shown in Fig. 19.1. The male opening is in the middle of the posterior end. The cirrus is well developed and armed with ten hooks. The ultrastructure of the reproductive system has been studied by Xylander (1988).

Life history. The thin-shelled eggs are elongated and bear a tiny stalk at one pole. When laid, they contain a curious ciliated larva termed a *lycophora* (Fig. 19.2). A mucous substance is secreted by the egg (probably from the well-developed glands of the lycophora) and swells when the egg comes into contact with water. This mechanism may enable the eggs to keep afloat and be more readily available for the intermediate hosts, which are fresh-water amphipods of the genera *Gammarus* and *Dikerogammarus*. The eggs, on ingestion, are ruptured by the crushing action of the mandible. The lycophora bores its way through the intestine into the haemocoele, and develops into a *procercoid* and later a *plerocercoid* larva, resembling the adult. Fish become infected by ingesting infected amphipods (Janicki, 1930). The prepatent period in the definitive hosts is not known, but in view of the low nutritional level of the coelomic cavity is likely to be comparatively long. Eggs, when released, escape from the body cavity of the fish via the abdominal pores. Amphilinids can also penetrate the body wall of the host and protrude slightly through the pore formed, releasing their eggs into the water.

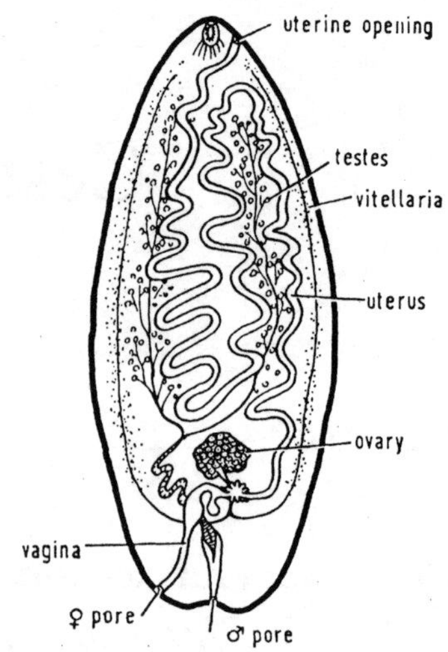

Fig. 19.1
Amphilina foliacea from the body cavity of the sturgeon. (After Bercham, 1901.)

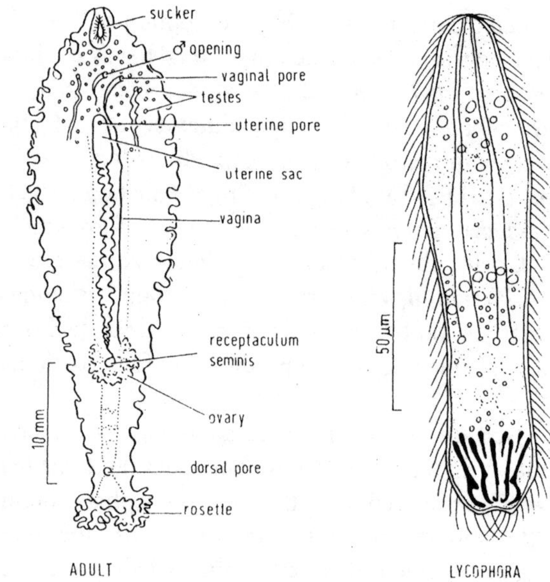

Fig. 19.2
Gyrocotyle urna: adult and larva. (After Lynch, 1945.)

Other species. The life cycle, morphology and ultrastructure of the Australian species *Austramphilina elongata* has been studied in some detail by Rohde & Georgi (1983) & Rohde & Watson (1986). Adults occur in the turtle *Chelodina longicollis*, with larval stages in shrimps and crayfish, the infective stages being reached in the latter intermediate host.

○ 19.2.2 Gyrocotylidea

Adult gyrocotylids are grotesque-looking parasites of chimaerids (deep-sea fish). The contracted body of a gyrocotylid is thrown into folds and bears a curious 'rosette' organ at its posterior end (Fig. 19.2). This secretes a PAS-positive material and is clearly adhesive (Lyons, 1969). Although morphologically gyrocotylids do not resemble amphilinids, they are clearly related to them as witnessed by their lycophora larva; some workers, however, group them with the Monogenea. Some species (e.g. *G. major*) are relatively enormous and reach the length of 300 mm (Land & Templeman, 1968).

The group has been relatively little studied, but species on which some detailed morphological, ecological and genetical information is available are *Gyrocotyle urna* (Halvorsen & Williams, 1967–8; Lynch, 1945), *G. fimbriata* and *G. parvispinosa* (Simmons & Laurie, 1972; Simmons *et al.*, 1972).

The best-known species is probably *Gyrocotyle urna* (synonyms: *G. confusa*, *G. nybelini*) (Colin *et al.*, 1986), which is a common parasite of *Chimaera monstrosa*; its biology has been studied in detail by Halvorsen & Williams (1967–8). Although the incidence in small fish (130–200 mm) was only 11 per cent, in fish longer than 300 mm it was 96 per cent. The parasites occur in the spiral valve and commonly only two *Gyrocotyle* are found. It has been speculated that the infection develops from a mass larval infection from which only two organisms survive. Since pairs have been found consisting of (*a*) a large functional 'female' with eggs in the uterus and no distinct testes and (*b*) a smaller 'male' with distinct testes and no eggs in the uterus, it has been suggested (Halvorsen & Williams, 1967–8) that protandry coupled with cross-fertilisation occurs; the position, however, is not clear. Knowledge of the life cycle is fragmentary; the eggs of most species require embryonation and hatch in sea water. The failure of hatched larvae to penetrate possible intermediate hosts, such as molluscs or crustacea, has led to the speculation that the life cycle is direct (Manter, 1951).

Various aspects of the ultrastructure have been studied by Allison (1980), Lyons (1969) and Xylander (1989).

○○○ References

(References of general interest have been included in this list, in addition to those quoted in the text.)

Allison, F. R. 1980. Sensory receptors of the rosette organ of *Gyrocotyle rugosa*. *International Journal for Parasitology*, **10:** 342–53.

Arme, C. & Pappas, P. W. (eds) 1983*a*,*b*. *Biology of the Eucestoda*. Vols 1, 2. Academic Press, London.

Colin, J. A., Williams, H. H. & Halvorsen, O. 1986. One or three gyrocotylideans (Platyhelminthes) in *Chimera monstrosa* (Holocephali)? *Journal of Parasitology*, **72:** 10–21.

Halvorsen, O. & Williams, H. H. 1967/8. Studies on the helminth fauna of Norway. IX. *Gyrocotyle* (Platyhelminthes) in *Chimaera monstrosa* from Oslo Fjord, with emphasis on its mode of attachment and a regulation in the degree of infection. *Nytt Magasin for Zoologi*, **15:** 130–42.

Janicki, C. von. 1930. Ueber die jüngsten Zustände von *Amphilina foliacea* in der Fischleibeshöhle, sowie Generelles zur Auffassung des genus *Amphilina* Wagener. *Zoologischer Anzeiger*, **90:** 190–205.

Joyeux, Ch. & Baer, J. G. 1961. Classe des Cestodaires. In: *Traité de Zoologie*, T.IV, Fasc. 1, pp. 327–560.

Land, J. van der & Templeman, W. 1968. Two new species of *Gyrocotyle* (Monogenea) from *Hydrolagus affinis* (Brito Capello) (Holocephali). *Journal of the Fisheries Research Board of Canada*, **25:** 2365–85.

Lynch, J. E. 1945. Redescription of the species of *Gyrocotyle* from the ratfish *Hydrolagus coliei* (Lay and Bennett) with notes on the morphology and taxonomy of the genus. *Journal of Parasitology*, **31:** 418–46.

Lyons, K. M. 1969. The fine structure of the body wall of *Gyrocotyle urna*. *Zeitschrift für Parasitenkunde*, **33:** 95–109.

Manter, H. W. 1951. Studies on *Gyrocotyle rugosa* Diesing, 1850, a cestodarian parasite of the elephant fish, *Callorhynchus milii*. *Zoology Publications from Victoria University College*, No. **17:** 1–11.

Rohde, K. & Georgi, M. 1983. Structure and development of *Austramphilina elongata* Johnston, 1931 (Cestodaria: Amphilinidea). *International Journal for Parasitology*, **13:** 273–87.

Rohde, K. & Watson, N. 1986. Ultrastructure of the sperm ducts of *Austramphilina elongata* (Platyhelminthes, Amphilinidae. *Zoologischer Anzeiger*, **217:** 23–30.

Schmidt, G. D. 1970. *How to know the Tapeworms*. Wm. C. Brown, Publishers, Dubuque, Iowa, USA.

Riser, N. W. 1948. *Amphilina bipunctata* n.sp., a North American cestodarian. *Journal of Parasitology*, **34:** 479–85.

Simmons, J. E., Buteau, G. H., MacInnis, A. J. & Kilejian, A. 1972. Characterization and hybridization of DNA's of gyrocotylidean parasites of chimaeroid fishes. *International Journal for Parasitology*, **2:** 273–8.

Simmons, J. E. & Laurie, J. S. 1972. A study of *Gyrocotyle* in the San Juan Archipelago, Puget Sound, U.S.A., with observations on the host, *Hydrolagus colliei* (Lay & Bennett). *International Journal for Parasitology*, **2**: 59–77.

Smyth, J. D. & McManus, D. P. 1989. *Physiology and Biochemistry of Cestodes*. Cambridge University Press.

Wardle, R. A. & McLeod, J. A. 1952. *The Zoology of Tapeworms*. Minnesota Press.

Xylander, W. E. R. 1988. Ultrastructural studies on the reproductive system of Gyrocotylidea and Amphilinidea (Cestoda). *Parasitology Research*, **74**: 363–70.

Xylander, W. E. R. 1989. Ultrastructural studies on the reproductive system of Gyrocotylidea and Amphilinidea (Cestoda): spermatogenesis, spermatozoa, testes, and vas deferens of *Gyrocotyle*. *International Journal for Parasitology*, **19**: 897–905.

Yamaguti, S. 1959. *Systema Helminthum*. Vol. II. *The Cestodes of Vertebrates*. Interscience Publishers, New York and London.

20

Eucestoda: general account

20.1 Classification

Only four orders of Eucestoda are universally recognised: the Tetraphyllidea, Trypanorhyncha, Pseudophyllidea and Cyclophyllidea. Authors are not generally in agreement as to what other orders should be set up, and those marked with an asterisk * are not considered here. The classification given below follows that of Yamaguti (1959).

Order 1 **Tetraphyllidea.** Scolex with four bothridia (lappet-like outgrowths); vitellaria located in lateral margins of proglottides; genital pores lateral. Parasites of elasmobranchs (e.g. *Anthobothrium* spp.).

Order 2 **Lecanicephalidea.*** Reproductive system similar to Tetraphyllidea, but scolex divided into an upper disc-like or globular part and a lower collar-like part bearing four suckers. Mainly parasites of elasmobranchs (e.g. *Lecanicephalum* sp.).

Order 3 **Proteocephalidea.** Scolex with four cyclophyllidean-like suckers and sometimes a fifth terminal one: vitellaria located in lateral margins; genital pores lateral. Mainly parasites of cold-blooded vertebrates (e.g. *Proteocephalus* sp.).

Order 4 **Diphyllidea.*** Two bothridia, each sometimes divided into two by a median longitudinal ridge. Large rostellum armed with a dorsal and ventral group of large hooks. Cephalic peduncle bears longitudinal rows of T-shaped hooks. Genital pore median. In elasmobranchs. Only a single genus: *Echinobothrium*.

Order 5 **Trypanorhyncha.** Scolex with four spiny eversible proboscoides and two or four bothridia: vitellaria in continuous sleeve-like distribution. Parasites of elasmobranchs (e.g. *Grillotia erinaceus*).

Order 6 **Pseudophyllidea.** Scolex with two elongated shallow bothria; one dorsal and one ventral, segmented or unsegmented (caryophyllaeid† cestodes). Genital pore lateral or median. Vitellaria lateral or extending across proglottid encircling other organs. Parasites of teleosts and land vertebrates (e.g. *Diphyllobothrium latum*).

Order 7 **Nippotaeniidae.*** Scolex bears a single apical acetabulum. Parasites of freshwater fish. Only a single genus: *Nippotaenia*.

Order 8 **Cyclophyllidea.** Scolex with four acetabula; uterine pores lacking; a single compact vitellarium posterior to the ovary. Mainly parasites of birds and mammals (e.g. *Hymenolepis diminuta*).

Order 9 **Aporidea.*** Rare forms lacking sex ducts or genital openings; ovary probably germovitellaria. Parasites of swans (e.g. *Nematoparataenia* sp.).

Order 10 **Spathebothriidea.*** Scolex without true bothria or suckers; strobila with internal segmentation but no external segmentation. Adult may be a neotenic procercoid, sexually mature in fish. Parasites of marine teleosts (e.g. *Cyathocephalus truncatus*).

All aspects of the biology of the Eucestoda have been reviewed in detail (in two volumes) by Arme & Pappas (1983), and their physiology and biochemistry by Smyth & McManus (1989). For tapeworm identification, see Schmidt (1986).

†Some workers include these in a separate Order Caryophyllidea; see p. 310.

○○○ 20.2 General charactersitics

○ 20.2.1 General morphology

External characters. Typically a cestode is divided into a *scolex*, bearing attachment organs, followed by a short unsegmented region, the *neck*, succeeded by a chain of *proglottides* termed the *strobila*. The presence of a uterine pore usually defines the ventral surface, but external differentiation of surfaces may be difficult. Using internal features, the surface in proximity to the female system is defined as ventral.

The organs of attachment which occur on the scolex are of three types (Fig. 20.1):

Bothria (typical of the Pseudophyllidea) are long, narrow grooves of weak muscularity. In life, a bothrium may become extremely flattened to form an efficient sucking organ.

Bothridia (phyllidea) (typical of the Tetraphyllidea) are broad, leaf-like structures with thin, flexible margins. They may be extremely variable, very mobile, stalked or sessile.

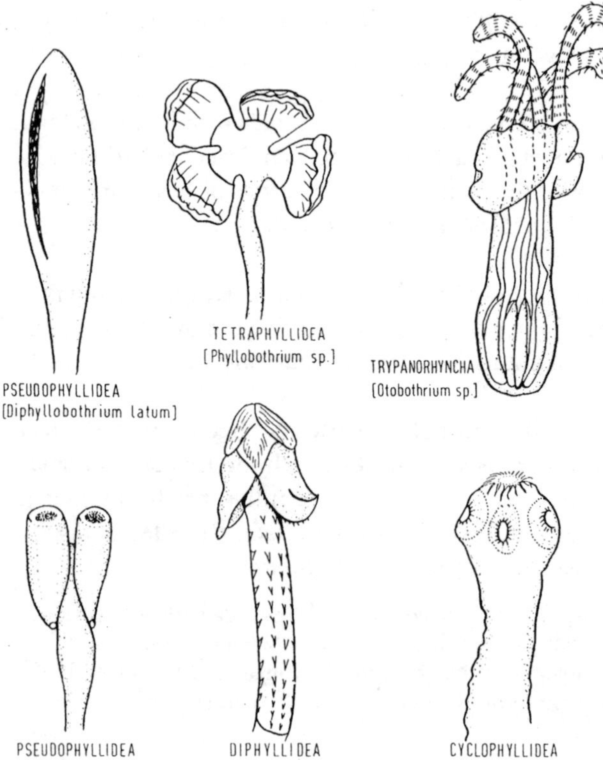

Fig. 20.1
Types of cestode scoleces. (After various authors.)

Acetabula (suckers) (typical of the Cyclophyllidea) are true sucking organs, similar in structure to the suckers of the digenetic trematodes.

The scolex may be additionally armed with *hooks*. In the taenioid scolex, a mobile cone or *rostellum*, usually armed and retractable, is present.

Proglottides.† The strobila of most adult cestodes is divided into *segments* or *proglottides* (singular *proglottis* or *proglottid*) which appear to arise by a series of external transverse constrictions. It was formerly thought that these external constrictions were continued internally, thus separating each proglottis by a membranous partition. TEM studies by Mehlhorn *et al.* (1981) have shown that this is not, in fact, the case: the external constrictions are not continued internally and no such internal partitions are formed. As new proglottides form, the strobila elongates so that in some forms enormous lengths are achieved. In many groups (typically in the Cyclophyllidea) the most posterior proglottides become ripe first and when fully mature consist mainly of a branched uterus packed with eggs. Such a segment is said to be *gravid*. Gravid segments are often shed into the intestine and pass out with the faeces. Cestodes which shed ripe proglottides in this manner are said to be *apolytic* and those which retain them throughout life are termed *anapolytic*. Other terms used are: *euapolytic* (segments detach when nearly gravid); *hyperapolytic* (segments detach much earlier and have a free existence in gut of host); *pseudoapolytic* (eggs liberated through uterine pore, segments then detach in groups and degenerate).

A unique form of reproduction is found in the carnivore parasite *Mesocestoides corti*, the adult of which has the ability to undergo *asexual* development in the gut, either by the scolex splitting longitudinally and separating to form a new strobila or by lateral buds arising from a proglottis (Fig. 23.28). This was first demonstrated by Eckert *et al.* (1969) who fed 1000 larvae (tetrathyridia, p. 342) to a dog and on autopsy, 11 weeks later, recovered 40 000 worms from the gut! Further details of this unusual species are given on p. 342.

Body wall (tegument) (Fig. 20.2). The ultrastructure of the tegument has been extensively studied; for details of the species investigated see the reviews of

†The term *proglottids* (plural) instead of *proglottides* is widely (although incorrectly) used in the literature, including in the 2nd Edition of this text! The singular terms *proglottis* or *proglottid* both appear to be acceptable.

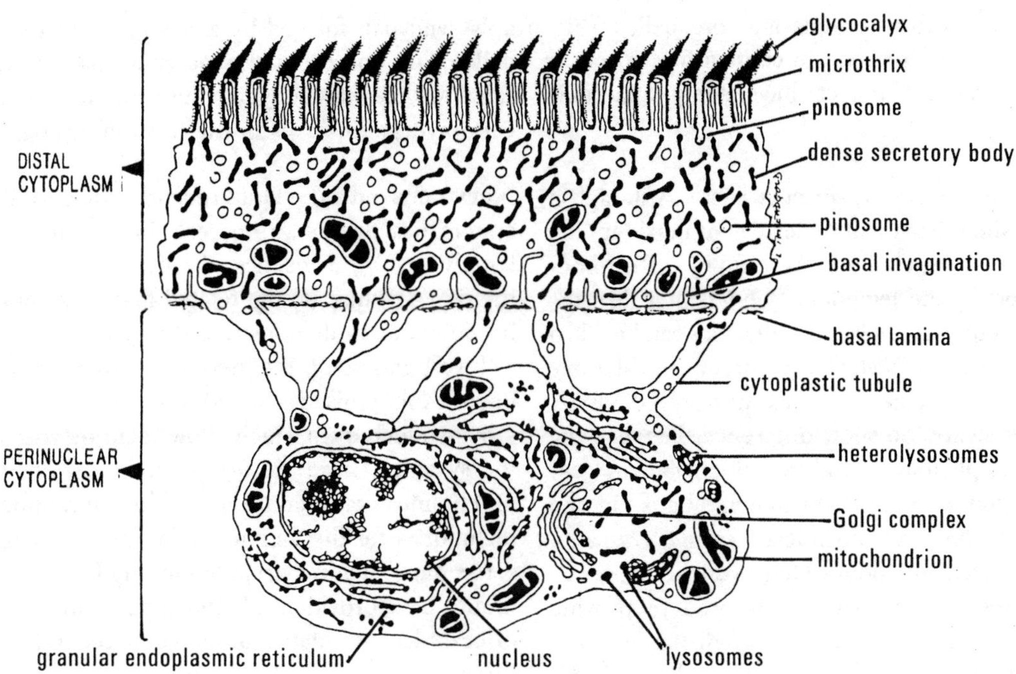

Fig. 20.2
Diagrammatic representation of the tegument of a typical adult cestode, based on several species. (After Threadgold, 1984.)

Smyth & McManus (1989) and Threadgold (1984). Early literature is reviewed in the 2nd Edition of this text. In general, the ultrastructure follows that of the Digenea (Fig. 13.2) and the Monogenea, i.e. it consists of a syncytial outer layer formed from cytoplasmic extensions which come from an inner nucleated layer containing *tegumental cells*. Based on the terminology used for trematodes these regions have been termed *distal cytoplasm* and *perinuclear cytoplasm*, respectively. The terminology used by different authors is confusing; other terms used for the inner (nucleated) zone (in addition to *perinuclear cytoplasm*, used here) are: *perikarya*, *cytoplasm*, *tegumental cells*, *cytons*. The term *tegumental cytons* is used here for the basic cells which make up the syncytial epithelium.

Microtriches. A major morphological adaptation for absorptive biological membranes is the amplification of the surface areas by folds or extensions. In the mammalian intestine, this is achieved by a 'brush border' of microvilli which amplifies the surface by as much as 26–30 times (Béguin, 1966; Lumsden & Murphy, 1980). In cestodes, this amplification is achieved by the presence of *microtriches* (singular: *microthrix*, Fig. 20.2) each consisting of a shaft borne on an electron-dense base. The spine-like nature of the microtriches may also assist in maintaining the position of the worm in the gut. The ultrastructure has been examined in many species; see reviews by Threadgold (1984), Smyth & McManus (1989) and Threadgold & Robinson (1984).

Glycocalyx. Like most absorptive cells, the surface of the microtriches is covered by a 'fuzzy' layer or *glycocalyx* consisting of mucopolysaccharides and glycoproteins. The glycocalyx appears to be synthesised in the Golgi apparatus in the perinuclear cytoplasm and transferred to the surface. In *Hymenolepis diminuta* the glycocalyx has a turnover rate of about 6 h (Lumsden, 1975*b*) and also has the potential to bind *host* enzymes (e.g. amylase) to its surface, which may allow 'contact digestion' (= 'membrane' digestion) to take place (Smyth & McManus, 1989).

The distal and perinuclear cytoplasm. The distal cytoplasm contains numerous 'vesicles', 'vacuoles' and 'rhabdiform organelles' whose functions are largely unknown. Some of these bodies may represent secretory products (originating in the Golgi apparatus) *en route* to the surface and/or some may represent absorbed materials. The cytoplasm contains mitochondria, Golgi bodies and the normal cytoplasmic organelles (Fig. 20.2). The mitochondria tend to contain few cristae, a condition which may reflect the largely anaerobic metabolism.

Uptake by the tegument; pinocytosis? Although it is well established that small molecules of nutrient (e.g. glucose, amino acids, etc.) are taken up by diffusion or active transport by the tegument (see Chapter 24), the question of whether or not larger molecules can be taken up by pinocytosis (endocytosis) is one of considerable dispute, which, to date, remains unresolved. Early experiments appeared to show differences between the uptake by cyclophyllidean and pseudophyllidean cestodes in relation to this phenomenon. Thus, in the cysticercus of the cyclophyllidean *Taenia crassiceps*, smooth and coated micropinocytotic vesicles were reported as being present, which appeared to provide indirect evidence of pinocytosis although this was not confirmed experimentally (Dunn & Threadgold, 1984; Threadgold & Dunn, 1983, 1984). Attempts to demonstrate pinocytosis in *adult* cyclophyllideans proved negative until recently, when the uptake of ruthenium red marker was described to occur in *Hymenolepis nana* (= *fraterna*) but only, apparently, in immature proglottides (Poljakova-Krusteva *et al.*, 1984). These authors speculated that there may be fundamental differences between different regions of the strobila and that, in the posterior regions, nutrition may be carried out by membrane digestion (p. 349). In contrast, in the pseudophyllidean cestodes *Schistocephalus solidus* (adult and plerocercoid) and *Ligula intestinalis* (plerocercoid), pinocytosis 'on a huge scale', using ruthenium red, lanthanum nitrate and horseradish peroxidase, was reported (Threadgold & Hopkins, 1981; Hopkins *et al.*, 1978). More recently, however, these results have been challenged by Conradt & Peters (1989) who – using the same markers – were unable to identify any pinocytotic activity in *S. solidus*. Moreover, they pointed out that ruthenium red is unsuitable for the demonstration of pinocytosis as it 'leads to the production of electron-dense artifacts in the tegument that are not related to endocytotic uptake of the tracer'. In addition, it was claimed that toxic effects were produced in the tegument by ruthenium red. The whole topic of endocytosis in cestodes thus remains controversial and clearly needs further experimental investigation.

Parenchyma. This fills all the spaces between the internal organs. As in trematodes it is a meshwork, probably syncytial, formed by anastomosis of mesenchymal cells. The spaces in the parenchymal meshwork are filled with a parenchymal fluid, often containing rich carbohydrate reserves in the form of glycogen.

Nervous system. Until recently, the cestode nervous system has proved to be very difficult to study owing largely to (*a*) the absence of a delimiting membrane on the nerve trunks. (*b*) the related failure of routine staining methods to demonstrate the system.

Investigation of the nervous system, however, has been revolutionised – in addition to electron microscopy – by two techiques which allow accurate visualisation of the system. These are (*a*) cytochemical demonstration of cholinesterase and (*b*) specific immunocytological techniques for the demonstration of neuropeptides and serotonin (5-hydroxytryptamine, 5HT). All these techniques have provided valuable data on the neurobiology of cestodes and platyhelminthes in general.

Useful basic reviews on the neurobiology of cestodes are those of Gustafsson *et al.* (1985), Falkmer *et al.* (1985) and Smyth & McManus (1989). More recent studies are those of Fairweather *et al.* (1988, 1990*a*,*b*), Halton *et al.* (1990) and Fairweather & Halton (1991).

However, the gross anatomy of the nervous system has been worked out for only a few species, e.g. *Mesocestoides* sp. and *Hymenolepis nana* (Fig. 20.3). The physiology of the nervous system is discussed further in Chapter 24.

Muscular system. Cestodes are well supplied with longitudinal, circular and transverse muscle systems. A cestode muscle cell consists of (*a*) a non-contractile cytoplasmic portion or *myocyton*, containing a nucleus, and (*b*) a myofibrila portion containing myofilaments. Further details on the ultrastructure are given by Lumsden & Hildreth (1983) and Lumsden & Specian (1980). The neuromuscular system has been particularly studied in the Trypanorhyncha (see Chapter 21).

Excretory system. Essentially this consists of a nephridial system similar to that of Turbellaria with flame cells (bulbs) as excretory units. Typically, there are dorsal and ventral vessels on each side, but additional vessels may be present. The dorsal vessel is continuous with the ventral vessel of its own side forming a loop in the scolex. These two loops may be joined by a ring or a system of branches. In some cases (e.g. Pseudophyllidea) the excretory vessels may have a much more complicated arrangement, there being six longitudinal vessels plus a more superficial network.

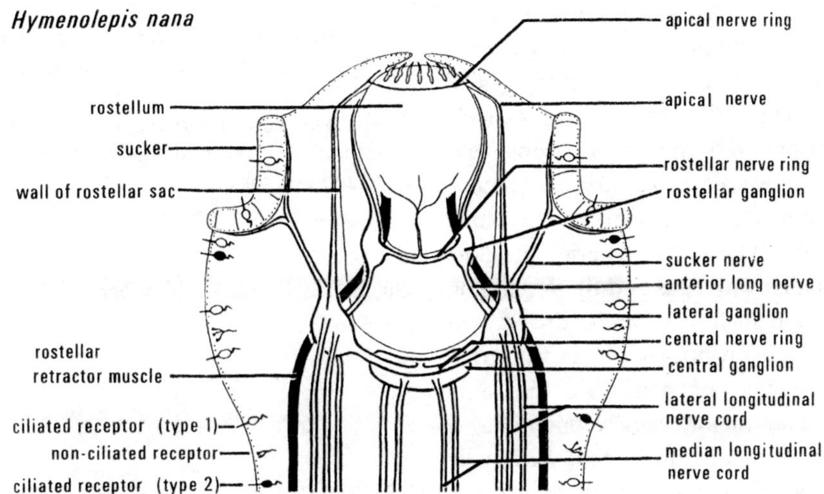

Fig. 20.3
Nervous system of *Hymenolepis nana*. (After Fairweather & Threadgold, 1983.)

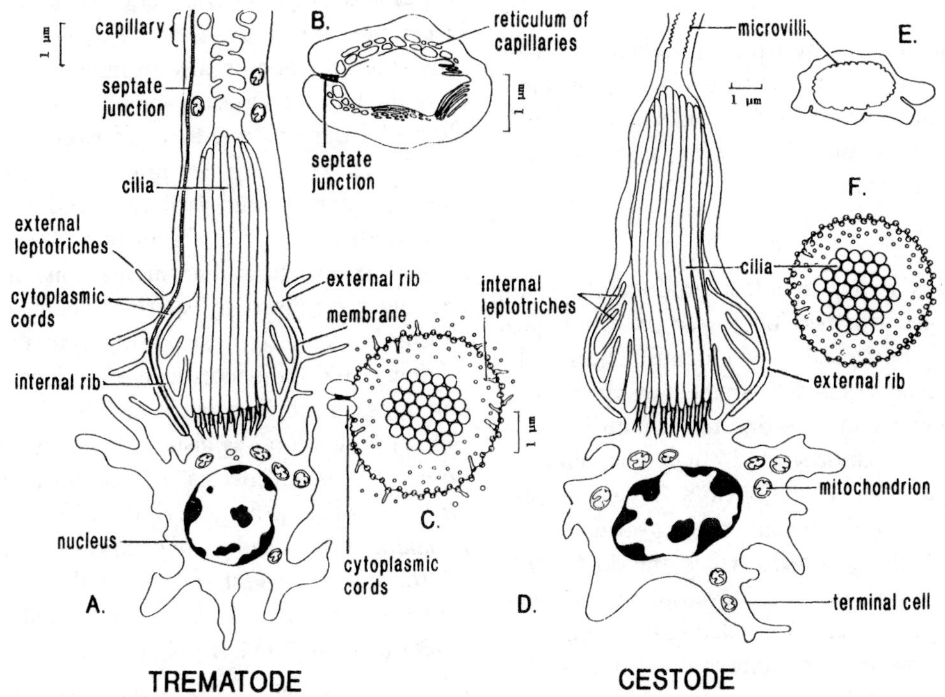

Fig. 20.4
Comparison of the flame bulbs (cells) and protonephridial capillaries of the Trematoda and Cestoda. Note that the weir in the trematodes has two longitudinal cytoplasmic cords connected by a septate junction, whereas these are lacking in cestodes. A, D. Longitudinal section through the flame bulb. B, E. Cross-section through the capillary. C, F. Cross-section through the weir. (Modified from Rohde, 1990.)

The flame cells (bulbs) of cestodes (Fig. 20.4) show some differences at the ultrastructural level from those of trematodes (Rohde, 1990). One of the main differences is the presence, in trematodes, of a septate junction along the protonephridial capillaries and lamellae and/or a reticulum, whereas in cestodes there is a continuous wall of capillaries and short microvilli occur.

○ 20.2.2 Reproductive system

General. With the exception of the cyclophyllidean genus *Dioecocestus*, which is dioecious and dimorphic, cestodes are monoecious. The reproductive system follows the platyhelminth pattern. The female system in general resembles that of the digenetic trematodes, but in the Cyclophyllidea, the vitellaria are much reduced. The vagina of cestodes is homologous with Laurer's canal of the Digenea and with the vagina of Monogenea.

The reproductive system differentiates progressively from anterior to posetrior end of the strobila. Most cestodes show a tendency towards protandry, the receptaculum seminis becoming filled with spermatozoa while the ovaries are maturing. The most posterior proglottides usually become laden with eggs and the remaining genitalia almost disappear; such proglottides are termed 'gravid'. In the Pseudophyllidea, however, large numbers of proglottides become mature at the same time. The cirrus sac and vagina usually open into a common genital atrium which in turn opens externally by a single gonopore which may be lateral or central. In some species, the reproductive system is duplicated in each proglottid.

Male

Spermatogenesis follows the typical platyhelminth pattern. Testes may be in small dispersed groups or large well-formed bodies. Cirrus and cirrus sac are usually highly developed. Peculiarities arising are dealt with under the separate orders.

Female

Except in species with a double reproductive system, the ovary is single with two lobes, and its histological structure follows the typical platyhelminth pattern. The uterus takes on its characteristic shape only after sexual maturity is reached and it begins to fill with eggs. In tetraphyllids and trypanorhynchids, it is characteristically a tube capable of distension to a sac which does not open by a pore to the outside when gravid.

In pseudophyllids, it usually takes the form of a much-coiled tube opening to the exterior by a median pore. In cyclophyllids, an external uterine opening is lacking and embryos are freed only by the disintegration of the shed proglottid. In these forms, the uterus develops extensive lateral diverticula, often of diagnostic value. In some Davaineidae and a few Anoplocephalidae, a *paruterine organ* is present; this is a fibrous capsule surrounding the uterus or isolated pieces of uterus containing eggs.

Insemination. Most cestodes are hermaphrodite, but surprisingly little is known regarding the exact mode of sperm transfer. This question has been reviewed in detail by Williams & McVicar (1968), Smyth (1982) and Smyth & McManus (1989). Although self-insemination has been well established (see below), cross-insemination is also common, especially in the Tetraphyllidea. Cross-insemination has also been elegantly demonstrated in *Hymenolepis diminuta* by Schiller (1974) using rats infected both with irradiated worms, which showed variations, and normal (i.e. non-irradiated) worms. He found that the offspring of the normal worms showed an increase in variants, i.e. they had been fertilised by spermatozoa from the variant worms. Nollen (1975), also using irradiated *H. diminuta*, demonstrated the occurrence of both self-insemination and cross-insemination in this species (Table 20.1). Self-insemination has also been demonstrated in some other well-known species of Cyclophyllidea, e.g. *Echinococcus granulosus* (Smyth & Smyth, 1969) (see Fig. 20.5), *E. multilocularis*, and *E. oligarthus* (Kumaratilake *et al.*, 1986), and in several species of Pseudophyllidea, e.g. *Ligula intestinalis* (Flockart, 1979) and *Schistocephalus solidus* (Smyth, 1954, 1982).

Table 20.1. *Demonstration of self-insemination and cross-fertilisation in* Hymenolepis diminuta

Adult *Hymenolepis diminuta* were exposed for 3 h to tritiated thymidine and returned singly with or without unlabelled worms to rats for 3 days.

Unlabelled worms per experiment	Number of experiments	Insemination by labelled worms	
		Self	Cross
0	8	8	—
1	5	5	4
2	4	4	6
3	4	4	12

Source: Data from Nollen (1975).

In the Pseudophyllidea, the strobila has a median (i.e. not a lateral) genital pore and compression against the gut mucosa appears to be essential to allow the cirrus to bend backwards into the vagina for insemination to take place. This has clearly been demonstrated experimentally (Fig. 34.6) in *Schistocephalus solidus* (p. 312) by the fact that *in vitro* cultured worms compressed within cellulose tubing during maturation produced fertile eggs, whereas those matured free in the medium did not (Smyth, 1954); see also Chapter 34.

The occurrence of self-fertilisation in hermaphrodite organisms, such as cestodes, raises most interesting questions in relation to the genetics of cestodes, for

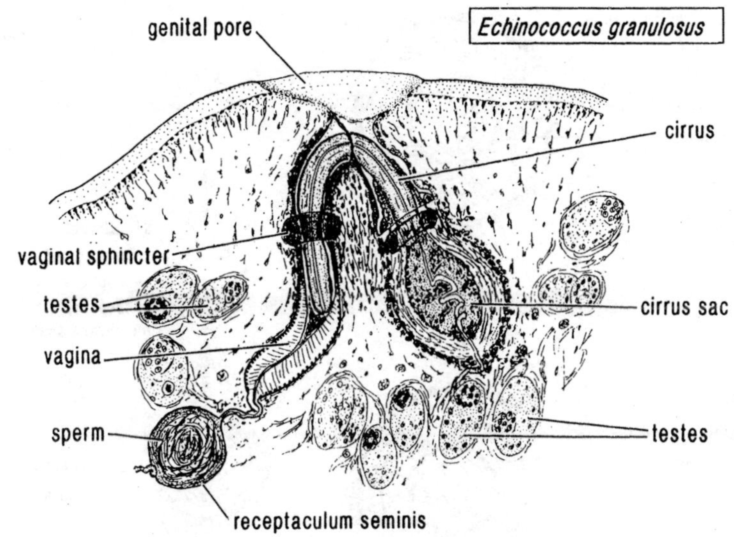

Fig. 20.5
Self-insemination in *Echinococcus granulosus*. The cirrus is inserted into the vagina and probably held in position by the sphincter muscle. The latter is lacking in *E. multilocularis* in which self-insemination appears less common; see text. (Based on Smyth & Smyth, 1969.)

it implies that cestodes must therefore necessarily be homozygous. Since the homozygous condition is generally considered to be 'evolutionarily detrimental', other mechanisms may operate to assure genetic diversity. The process also has considerable implications in the production of new 'strains' for the coming together of two mutant genes (being hermaphrodite) could theoretically produce an 'instant' new strain especially in forms (e.g. *Echinococcus*) with an asexually multiplying larval stage (= hydatid cyst) (Smyth & Smyth, 1964); but see Thompson & Lymbery (1988).

○ 20.2.3 Eggs and egg-envelopes (membranes)

The pre-larval development of only a few cestodes has been worked out in detail; the segmented egg gives rise to an oval embryo, the *oncosphere*, or hexacanth larva, so-called because it possesses three pairs of hooks at the posterior pole; it also possesses a pair of flame cells and some muscle fibres. In the taenioids, the embryo develops to the oncosphere stage in the uterus of mature proglottides, but in the pseudophyllids embryonation only takes place when the eggs reach water. The larva hatching from the embryonated pseudophyllid egg is a *coracidium* (Fig. 20.8), that is an oncosphere surrounded by a ciliated embryophore. Unlike the miracidium of digenetic trematodes, the coracidium does not enter the intermediate host actively but must be eaten by it.

A number of 'envelopes' (membranes) are formed during the development of the embryonated egg and a complex (and confusing!) nomenclature has been used to describe them. Rybicka (1966) has given a comprehensive review of this problem and pre-larval development in general. Her terminology is followed below. There are only *four primary* envelopes which may, however, give rise to a number of other *secondary* envelopes (hence the confusion in terminology). These are as follows (Fig. 20.6):

(*a*) **The capsule** (= 'egg-shell' of most authors): well developed in the Pseudophyllidea, Tetraphyllidea, Trypanorhyncha, and moderately well developed in the Proteocephaloidea. It forms a waterproof coating (generally of *sclerotin* (Fig. 13.9) and is poorly developed or absent in the Cyclophyllidea.

(*b*) **The 'outer envelope'**: formed from 2–8 macromeres. A complex layer fills the space between the capsule and the inner envelope (Fig. 20.6).

(*c*) **The 'inner envelope'**: part of this forms the embryophore (a secondary envelope, Fig. 20.6).

(*d*) **The oncospheral membrane**: a thin membrane immediately surrounding the oncosphere.

Cestode eggs have been divided (Löser, 1965*a*) into the following four types, which may be considered as falling into two groups (Smyth & McManus, 1989).

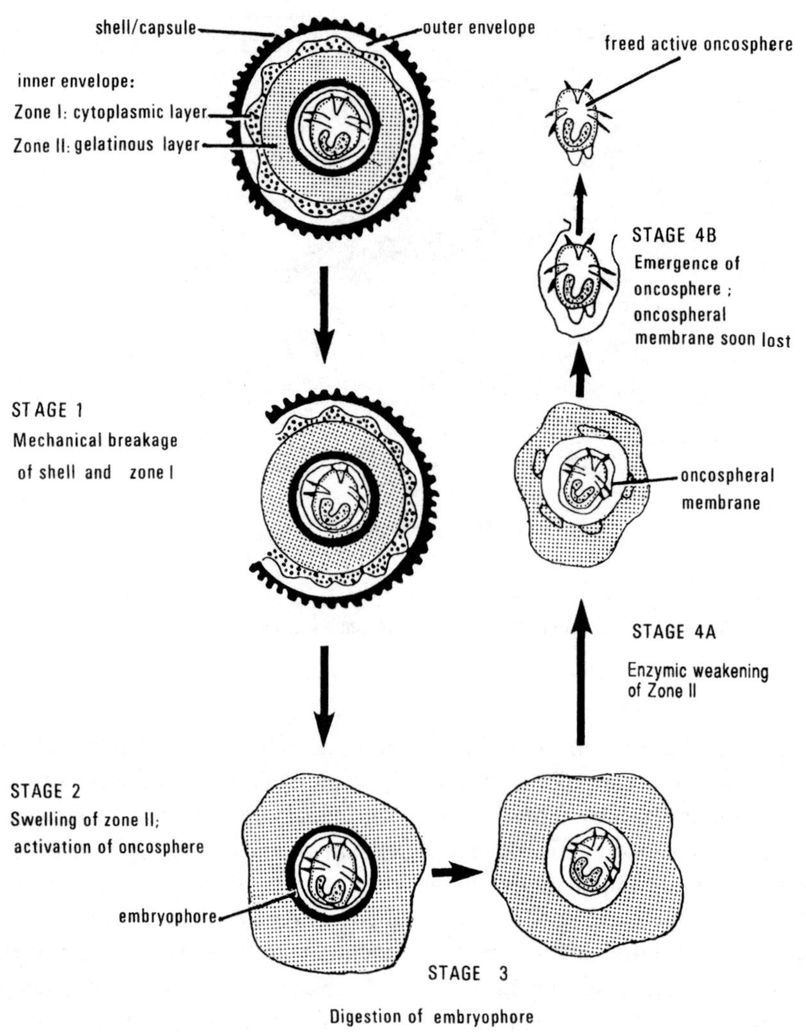

Fig. 20.6
Hymenolepis diminuta: process of hatching of the egg and the fate of the egg envelopes (membranes). (Modified from Holmes & Fairweather, 1982.)

(i) *Pseudophyllidea*-type egg	Group I (generally with aquatic stages)
(ii) *Dipylidium*-type egg (iii) *Taenia*-type egg (iv) *Stilesia*-type egg	Group II (aquatic stage lacking)

Consideration of the various egg types is beyond the scope of this book; these are reviewed in detail by Smyth & McManus (1989) and Ubelaker (1983).

A brief account of the general difference between the two Groups is given below.

Group I. Cestodes, excepting Cyclophyllidea; with extensive vitellaria (Fig. 20.8). All the cestodes in this group lay their eggs in water and the first larval form passes into an aquatic intermediate host. Only the pseudophyllidean egg is well known and in some cases has a thick operculate capsule (= shell) like that of digenetic trematodes (Fig. 13.10).

In most species examined in this group (e.g. *Diphyllobothrium* spp.) the capsule is a quinone-tanned protein, sclerotin, formed by a process almost identical to that utilised by trematodes (Figs 13.8, 13.9). As in the trematodes, the egg-shell (= capsule) material (presclerotin) is formed in the cells of the vitellaria and released in the ootype. The vitelline cells thus give positive reactions for sclerotin precursors: protein, phenol and phenolase (Table 13.1). Löser (1965*a,b*) and Smyth (1956) have studied the process in a number of species.

In the Tetraphyllidea and the Proteocephaloidea, the

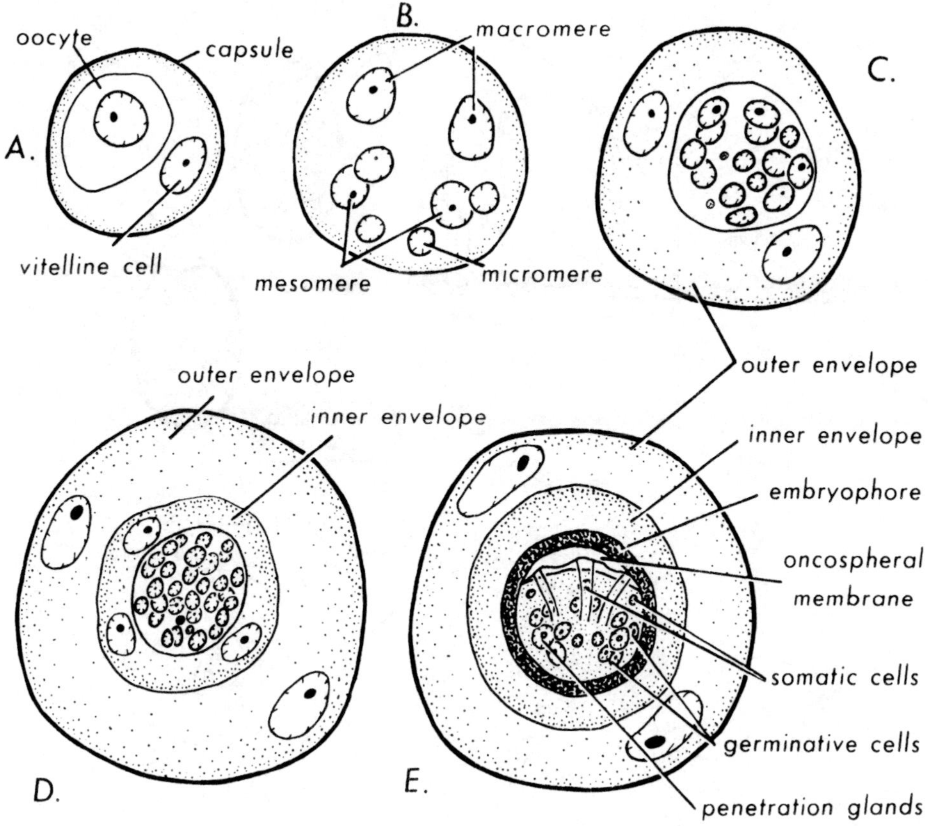

Fig. 20.7
Embryonic development in the Cyclophyllidea; diagrammatic.
A. fertilized oocyte; B. cleaving embryo; C. early preoncosphere; D. late preoncosphere; E. oncosphere. (After Rybicka, 1964.)

egg-capsule is thin and the eggs mature while still within the uterus and are ready to hatch or to be eaten whole on reaching water. It is possible that the thickness of the capsule may be closely related to the site where the egg matures, either in the water or in the uterus. According to this view, it is considered that an egg which matures in the uterus does not need the protection of the thick capsule necessary for eggs which embryonate in water. This generalisation is based only on limited examples and may not hold when further species are investigated.

Group II. The eggs in this group are usually embryonated when laid and only rarely have an aquatic stage in the life cycle. The most studied species are in the Cyclophyllidea. A capsule is present in some cases (e.g. *Hymenolepis*, Fig. 20.6) but does not appear to be sclerotin. In the Anoplocephalidae the embryophore has a pair of horns forming a 'pyriform' apparatus. In the Mesocestoididae, the embryophore is a thin cellular membrane separated from the capsule by a granular layer.

In the Taeniidae, the capsule is represented by a very thin membrane normally not seen in faecal eggs. The embryophore consists of a thick structure made up of keratin blocks (Morseth, 1966) and is often referred to as a 'shell' (the very thin outer capsule being lost).

The structure, ultrastructure and formation of the hymenolepid egg has been most studied and has been reviewed in detail by Ubelaker (1980). These aspects are best known for *Hymenolepis diminuta* (Holmes & Fairweather, 1982; Ubelaker, 1980, 1983) and *H. nana* (Fairweather & Threadgold, 1981).

The hatching of the egg of *H. diminuta* is shown in detail in Fig. 20.6. Hatching in *H. nana* is described as taking place more readily, a process which may be related to its ability to hatch in the rodent host intestine as well as in its arthropod intermediate host.

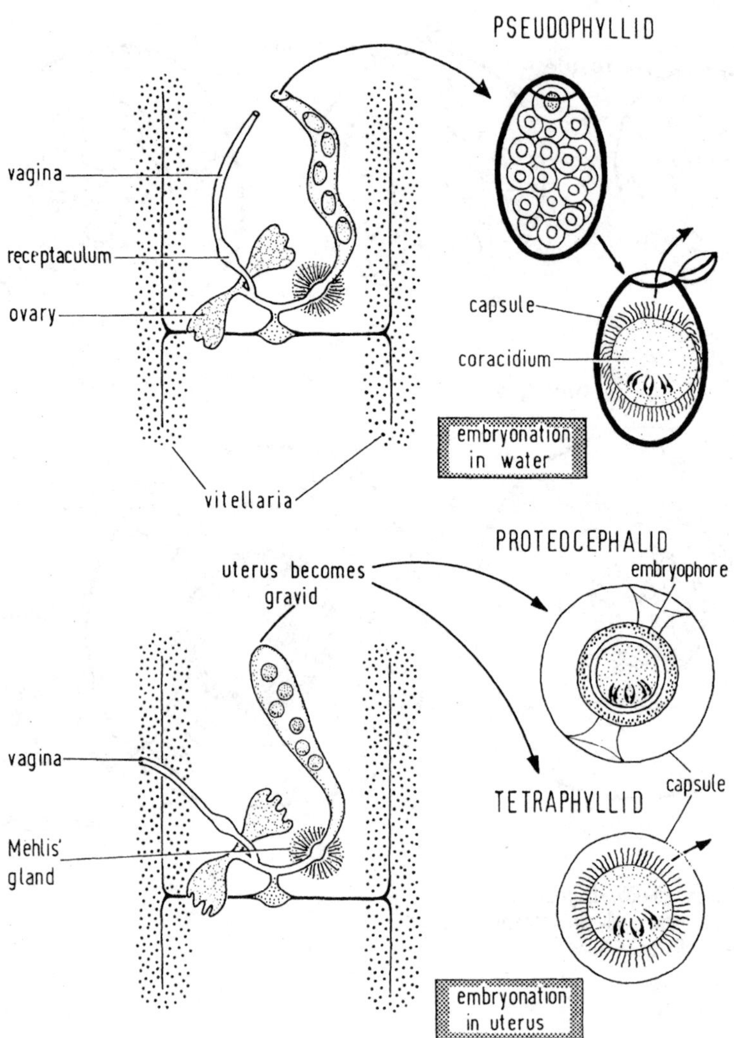

Fig. 20.8
Types of egg formed in cestodes (Group I; see text) with extensive vitellaria. (After Smyth & Clegg, 1959.)

○ 20.2.4 Egg hatching

Eggs of Group I (above) and Group II in general hatch under different environmental conditions; the problems of egg-hatching have been reviewed by Smyth & McManus (1989).

Group I eggs (above) are generally unembryonated when laid, but embryonate in the presence of oxygen and water; the embryonation time depends on temperature. As in trematodes, most embryonated eggs in this group require light for hatching. The precise mechanism of hatching is not known, but it can be speculated that, as in the trematode egg, light may release an enzyme which attacks the opercular seal and permits the escape of the coracidium through the operculum. It is particularly interesting to note that in *Diphyllobothrium latum* ultraviolet (300 nm) and yellow (600 nm) light give peak hatching, a result which may have biochemical implications (Smyth & McManus, 1989).

Group II eggs. In general, eggs in this group must undergo (*a*) *hatching* (i.e. removal of capsule and embryophore) and (*b*) *activation* (i.e. stimulation of the embryo to motility). In cestodes with an insect intermediate host (e.g. *Hymenolepis diminuta*) hatching appears to be brought about partly by the mechanical action of the mouth parts, which break the capsule,

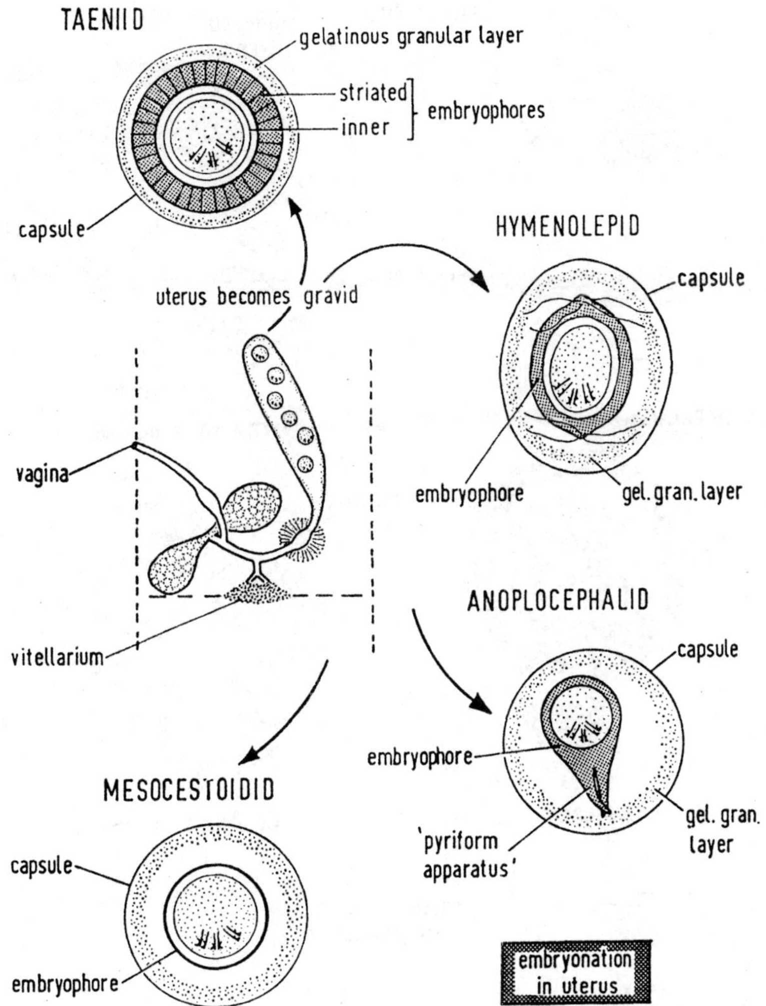

Fig. 20.9
Types of eggs formed in cyclophyllidean cestodes. (After Smyth & Clegg, 1959.)

followed by digestion of the remaining egg layers in the midgut lumen; proteolytic enzymes appear to be largely involved, although these have not been identified (Lethbridge, 1971, 1972). Physico-chemical factors appear to be largely involved in the activation process.

In eggs with a greatly thickened embryophore (e.g. those of the Taeniidae) hatching is brought about largely by the action of digestive enzymes such as pepsin and pancreatin (Smyth & McManus, 1989). The requirements vary somewhat between species, pancreatin alone being sufficient in some species, e.g. *T. pisiformis* (Silverman, 1954*b*); other species such as *T. saginata* require only pepsin (Silverman, 1954*a*). With such eggs, maximum hatching is probably achieved by both pepsin and pancreatin. Activation, after hatching, seems largely dependent on the presence of bile and suitable physiological conditions. Although the latter have not been examined in detail, factors such as pH, E_h and pCO_2 are likely to be important. Details of protocols for the experimental hatching of taeniid-type eggs, using enzyme mixtures, are given in Chapter 34. Hatching can also be achieved by treatment with sodium hypochlorite.

○○○ 20.3 Life cycles

With the exception of *Hymenolepis nana*, whose eggs can develop directly within the rat intestine, cestodes require

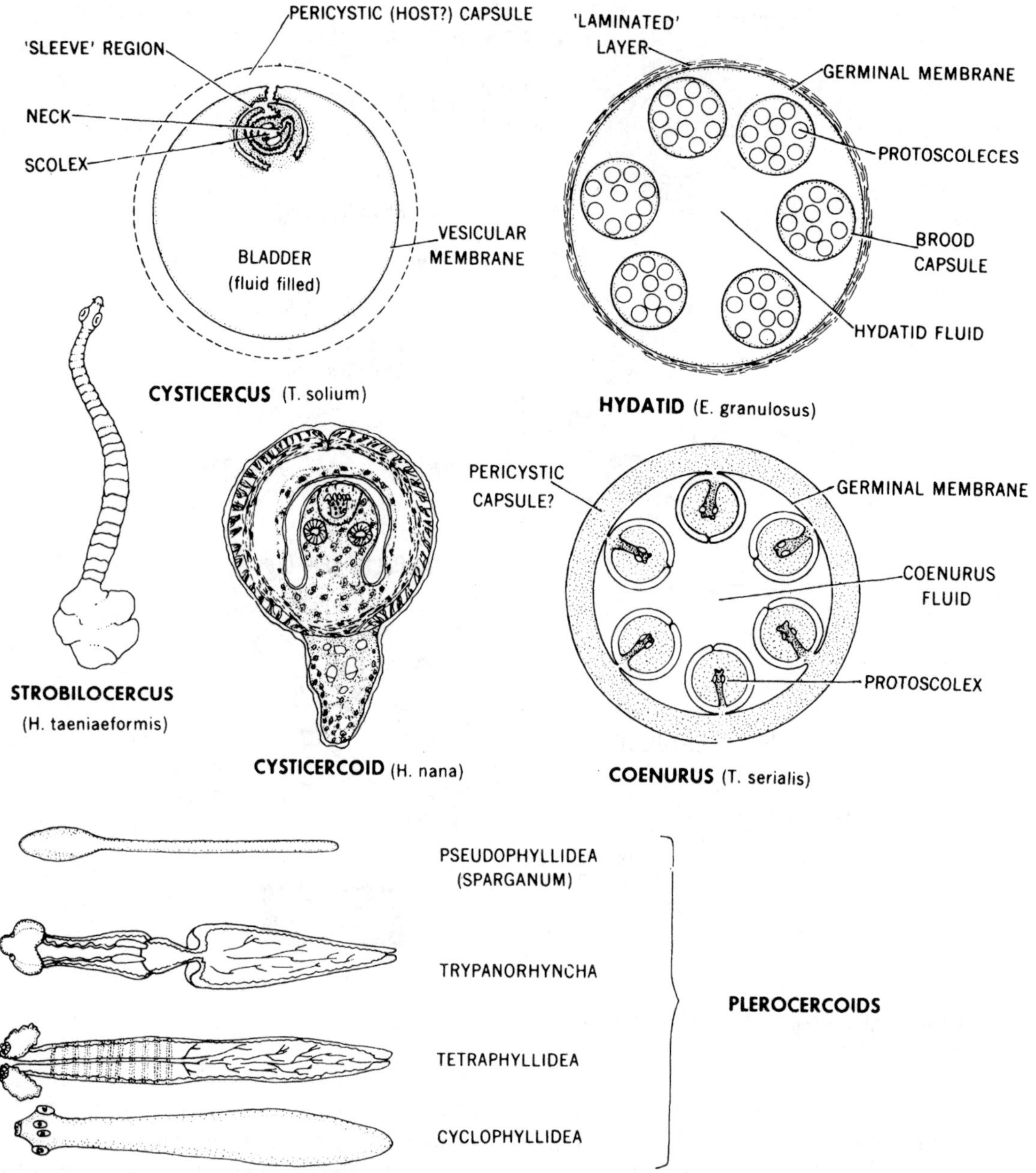

Fig. 20.10
The most common types of cestode larva. (After Smyth & Heath, 1970.)

an intermediate host for completion of the life cycle. A wide range of homoiothermic and poikilothermic vertebrate and invertebrate intermediate hosts is utilised, the cyclophyllidean cestodes using mainly terrestrial species and the remaining orders mainly aquatic species.

In all cases, infection of the intermediate host takes place orally, the ingested oncopshere penetrating the gut by means of its hooks and, perhaps, the secretions of its 'penetration' glands. The following types of larvae are recognised (Fig. 20.10).

Larval types

The terminology used for describing the larval forms of cestodes is very controversial and no general consensus on these has been reached. The term *metacestode* has been widely used as a general term to include all the pre-adult stages of cestodes (except the oncosphere) and was previously used in the 2nd Edition of this text. Its use is no longer recommended for the following reasons:

The term 'metacestode' can said to have been 'invented' by Wardle & McCleod (1952) in their *Zoology of Tapeworms*, who introduced it without attempting to justify its etymology. Its use was further supported and recommended by Freeman (1970, 1973) and Arme (1984). Although the term is now deeply entrenched in the literature, it is a misleading one since the prefix 'meta-' means 'after' (Greek), suggests an intermediate stage between a tapeworm and some other organism and has no basic meaning (Pojmanska, 1986). The correct use of the prefix is seen in the Trematoda in the term 'metacercaria', which accurately describes a stage after the cercaria, and also in the word 'metanauplius', which accurately denotes the stage after a nauplius. The original term 'larva' as defined in most dictionaries – 'An immature stage intervening between the egg and the adult' (Chambers, 1961) – appears to be quite adequate for use in cestode oncology. As stressed by several authors (Arme, 1984; Thompson, 1985) there is, however, a need for international standardisation on several aspects of cestode terminology.

Neotenic larvae occur in some species. A number of types of larvae are found and these may be grouped according to whether or not they show budding (Fig. 20.10). Terminology below generally follows Yamaguti (1959):

(*a*) those which do not show budding:
- (i) **procercoid** (Fig. 22.15): a small spindle-shaped larva with a solid body with posterior hooks; the first larval stage of pseudophyllideans or trypanorhynchids (e.g. larva of *Schistocephalus solidus* in copepods).
- (ii) **plerocercoid:** the second larval stage of Tetraphyllidea, Pseudophyllidea and Trypanorhyncha; also (more rarely) in taenioids (e.g. *Paruterina* sp.). A solid larva possessing an adult scolex but lacking the embryonic hooks of the procercoid from which it was developed (e.g. larva of *Diphyllobothrium* sp.).
- (iii) **cysticercoid:** a larva consisting of an anterior vesicle containing the scolex, which is not invaginated, and a tail-like posterior region containing the larval hooks (which may persist for some time) (e.g. larva of *Hymenolepis diminuta* in the grain beetle *Tenebrio molitor*).
- (iv) **cryptocystis:** a cysticercoid in which the tail is present early in development only (e.g. larva of *Dipylidium caninum* in the flea *Ctenocephalus canis*).
- (v) **cysticercus:** a bladder enclosing a single scolex retracted and invaginated within itself. Usually in vertebrates (e.g. larva of *Taenia pisiformis* in rabbits).
- (vi) **strobilocercus:** scolex usually not invaginated and connected by a long, solid, segmented strobila to a small bladder (e.g. larva of *Hydatigera taeniaeformis* in rodents).

(*b*) those which do show budding:
- (i) external budding:
  urocystis (= staphylocystis): budding cysticercoid.
  urocystidium: budding strobilocercus.
- (ii) internal budding:
  monocercus, polycercus: of cysticercoid type; scolex or scoleces are budded from wall and become free in the bladder (e.g. larvae of *Paricterotaenia paradoxa* in earthworm).
  coenurus: of cysticercus type; groups of scoleces are budded from wall of bladder and remain connected to wall (e.g. larva of *Taenia multiceps*).
  hydatid: of cysticercus type; scoleces do not develop in walls of bladder but within vesicles termed 'brood capsules' (e.g. larva of *Echinococcus granulosus*).

Valuable reviews on the developmental biology and ontology of cestodes are those of Freeman (1973) and Ubelaker (1983).

○○○ References

(References of general interest have been included in this list, in addition to those quoted in the text.)

Arme, C. 1984. The terminology of parasitology: the need for uniformity. *International Journal for Parasitology*, **14:** 539–40.

Arme, C. & Pappas, P. W. (eds) 1983. *Biology of the Eucestoda*. 2 vols. Academic Press, London.

Béguin, F. 1966. Étude au microscope électronique de la cuticle et de ses structures associées chez quelques cestodes. Essai d'histologie comparée. *Zeitschrift für Zellforschung und Mikroskopische Anatomie*, **72:** 30–46.

Chambers, 1961. *Chambers's Technical Dictionary*. W. R. Chambers, London.

Conradt, U. & Peters, W. 1989. Investigations on the occurrence of pinocytosis in the tegument of *Schistocephalus solidus*. *Parasitology Research*, **75:** 630–5.

Dunn, J. & Threadgold, L. T. 1984. *Taenia crassiceps*: temperature, polycations, and cysticercal endocytosis. *Experimental Parasitology*, **58:** 110–24.

Eckert, J., Brand, T. von & Voge, M. 1969. Asexual multiplication of *Mesocestoides corti* (Cestoda) in the intestine of dogs. *Journal of Parasitology*, **55:** 241–9.

Fairweather, I. & Halton, D.W. 1991. Neuropeptides in platyhelminthes. *Parasitology*, **102:** S77–92.

Fairweather, I., Macartney, G. A., Johnston, C. F., Halton, D. W. & Buchanan, K. D. 1988. Immunocytochemical demonstration of 5-hydroxytryptamine (serotonin) and vertebrate neuropeptides in the nervous system of excysted larvae of the rat tapeworm. *Hymenolepis diminuta* (Cestoda, Cyclophyllidea). *Parasitology Research*, **74**: 371–9.

Fairweather, I., Mahendrasingam, S., Johnston, C. F., Halton, D. W., McCullough, J. S. & Shaw, C. 1990*a*. An ontogenetic study of the cholinergic and serotinergic nervous systems in *Trilocularia acanthiaevulgaris* (Cestoda, Tetraphyllidea). *Parasitology Research*, **76**: 487–96.

Fairweather, I., Mahendrasingam, S., Johnston, C. F., Halton, D. W. & Shaw, C. 1990*b*. Peptidergic nerve elements in three developmental stages of the tetraphyllidean tapeworm *Trilocularia acanthiaevulgaris*. An immunocytological study. *Parasitology Research*, **76**: 497–508.

Fairweather, I. & Threadgold, L. T. 1981. *Hymenolepis nana*: the fine structure of the embryonic envelopes. *Parasitology*, **82**: 429–43.

Fairweather, I. & Threadgold, L. T. 1983. *Hymenolepis nana*: the fine structure of the adult nervous system. *Parasitology*, **86**: 89–103.

Falkmer, S., Gustafsson, M. K. S. & Sundler, F. 1985. Phylogenetic aspects of the neuroendocrine system. *Nordisk Psykiatrsik Tidsskrift*, **39** (suppl. 11): 21–30.

Flockart, H. A. 1979. *Ligula intestinalis* (L.1758) – *in vitro* and *in vivo* studies. Ph.D. Thesis, University of London.

Freeman, R. S. 1970. Terminology of cestode development. *Journal of Parasitology*, **56** (II): 106–7.

Freeman, R. S. 1973. Ontogeny of cestodes and its bearing on their phylogeny and systematics. *Advances in Parasitology*, **11**: 481–557.

Gustafsson, M. K. S., Wikgren, M. C., Karhi, T. J. & Schot, L. P. C. 1985. Immunocytochemical demonstration of neuropeptides and serotonin in the tapeworm *Diphyllobothrium dendriticum*. *Cell and Tissue Research*, **240**: 255–60.

Halton, D. W., Fairweather, I., Shaw, C. & Johnston, C. F. 1990. Regulatory peptides in parasitic platyhelminthes. *Parasitology Today*, **6**: 284–90.

Holmes, S. D. & Fairweather, I. 1982. *Hymenolepis diminuta*: the mechanism of egg hatching. *Parasitology*, **85**: 237–50.

Hopkins, C. A., Law, L. M. & Threadgold, L. T. 1978. *Schistocephalus solidus*: pinocytosis by the plerocercoid tegument. *Experimental Parasitology*, **44**: 161–72.

Kumaratilake, L. M., Thompson, R. C. A., Eckert, J. & Alessandro, A. D. 1986. Sperm transfer in *Echinococcus* (Cestoda: Taeniidae). *Zeitschrift für Parasitenkunde*, **72**: 265–9.

Lethbridge, R. C. 1971. The hatching of *Hymenolepis diminuta* eggs and penetration of the hexacanths in *Tenebrio molitor* beetles. *Parasitology*, **62**: 445–56.

Lethbridge, R. C. 1972. The *in vitro* hatching of *Hymenolepis diminuta* eggs in *Tenebrio molitor* extracts and in defined enzyme preparations. *Parasitology*, **64**: 389–400.

Löser, E. 1965*a*. Der Feinbau des Oogenotop bei Cestoden. *Zeitschrift für Parasitenkunde*, **25**: 413–58.

Löser, E. 1965*b*. Die Eibildung bei Cestoden. *Zeitschrift für Parasitenkunde*, **25**: 556–80.

Lumsden, R. D. 1975*a*. The tapeworm tegument: a model system for studies of membrane structure and function in host–parasite relationships. *Transactions of the American Microscopical Society*, **94**: 501–7.

Lumsden, R. D. 1975*b*. Surface ultrastructure and cytochemistry of parasitic helminths. *Experimental Parasitology*, **37**: 267–339.

Lumsden, R. D. & Hildreth, M. B. 1983. The fine structure of adult tapeworms. In: *The Biology of the Eucestoda* (ed. C. Arme & P. W. Pappas), Vol. 1, pp. 177–233. Academic Press, London.

Lumsden, R. D. & Murphy, W. A. 1980. Morphological and functional aspects of the cestode surface. In: *Cellular Interactions in Symbiosis and Parasitism* (ed. C. B. Cook, P. W. Pappas & E. D. Rudolph), pp. 95–130. Ohio State University Press, Ohio.

Lumsden, R. D. & Specian, R. D. 1980. The morphology, histology, and fine structure of the adult stage of the cyclophyllidean tapeworm *Hymenolepis diminuta*. In: *The Biology of Hymenolepis diminuta* (ed. H. P. Arai), pp. 157–280. Academic Press, New York.

Mehlhorn, H., Becker, B., Andrews, P. & Thomas, H. 1981. On the nature of the proglottids of cestodes: light and electron microscope study of *Taenia*, *Hymenolepis* and *Echinococcus*. *Zeitschrift für Parasitenkunde*, **65**: 243–59.

Morseth, D. J. 1966. Chemical composition of the embryonic blocks of *Taenia hydatigena*, *Taenia ovis* and *Taenia pisiformis* eggs. *Experimental Parasitology*, **18**: 347–54.

Nollen, P. M. 1975. Studies on the reproductive system of *Hymenolepis diminuta* using autoradiography and transplantation. *Journal of Parasitology*, **61**: 100–4.

Poljakova-Krusteva, O., Stoitsova, S. & Mizinska-Boevska, Ya. 1984. Pinocytosis in the tegument of *Hymenmolepis fraterna*. *Khelminthologiya*, **17**: 52–7.

Pojmanska, T. 1986. Terminology in parasitology. *International Journal for Parasitology*, **16**: 433–4.

Rohde, K. 1990. Phylogeny of platyhelminthes, with special reference to parasitic groups. *International Journal for Parasitology*, **20**: 979–1007.

Rybicka, K. 1964. Gametogenesis and embryonic development in *Dipylidium caninum*. *Experimental Parasitology*, **15**: 293–313.

Rybicka, K. 1966. Embryogenesis in cestodes. *Advances in Parasitology*, **4**: 107–86.

Schiller, E. L. 1974. The inheritance of X-irradiation-induced effects in the rat tapeworm, *Hymenolepis diminuta*. *Journal of Parasitology*, **60**: 35–46.

Schmidt, G. D. 1986. *Handbook of Tapeworm Identification*. CRC Press, Boca Raton, Florida.

Silverman, P. H. 1954*a*. Studies on the biology of some tapeworms of the genus *Taenia*. I. Factors affecting hatching and activation of taeniid ova, and some criteria of their viability. *Annals of Tropical Medicine and Parasitology*, **48**: 207–15.

Silverman, P. H. 1954*b*. Studies on the biology of some tapeworms of the genus *Taenia*. II. The morphology and development of the taeniid hexacanth embryo and its enclosing membranes, with some notes on the state of development and propagation of gravid segments. *Annals of Tropical Medicine and Parasitology*, **48**: 355–66.

Smyth, J. D. 1954. Studies on tapeworm physiology. VII. Fertilization of *Schistocephalus solidus in vitro. Experimental Parasitology*, **3**: 64–71.

Smyth, J. D. 1956. Studies on tapeworm physiology. IX. A histochemical study of egg-shell formation in *Schistocephalus solidus* (Pseudophyllidea). *Experimental Parasitology*, **5**: 519–40.

Smyth, J. D. 1982. The insemination-fertilization problem in cestodes cultured *in vitro*. In: *Aspects of Parasitology* (ed. E. Meerovitch), pp. 393–406. McGill University, Montreal.

Smyth, J. D. & Clegg, J. A. 1959. Egg-shell formation in trematodes and pseudophyllidean cestodes. *Experimental Parasitology*, **8**: 286–323.

Smyth, J. D. & Heath, D. D. 1970. Pathogenesis of larval cestodes of mammals. *Helminthological Abstracts*, **39**: 1–23.

Smyth, J. D. & McManus, D. P. 1989. *The Physiology and Biochemistry of Cestodes*. Cambridge University Press.

Smyth, J. D. & Smyth, M. M. 1964. Natural and experimental hosts of *Echinococcus granulosus* and *E. multilocularis*, with comments on the genetics of speciation in the genus *Echinococcus. Parasitology*, **54**: 493–514.

Smyth, J. D. & Smyth, M. M. 1969. Self-insemination in *Echinococcus granulosus in vivo. Journal of Helminthology*, **43**: 383–8.

Thompson, R. C. A. 1985. Terminology in parasitology. *International Journal for Parasitology*, **15**: 477.

Thompson, R. C. A. & Lymbery, A. J. 1988. The nature, extent and significance of variation within the genus *Echinococcus. Advances in Parasitology*, **27**: 209–58.

Threadgold, L. T. 1984. Parasitic platyhelminths. In: *Biology of the Integument* (ed. J. Bereiter-Hahn, A. G. Maltoltsy & K. S. Richards), pp. 132–91. Springer-Verlag, Berlin.

Threadgold, L. T. & Dunn, J. 1983. *Taenia crassiceps*: regional variations in ultrastructure and evidence of endocytosis in the cysticercus tegument. *Experimental Parasitology*, **55**: 121–31.

Threadgold, L. T. & Dunn, J. 1984. *Taenia crassiceps*: basic mechanisms of endocytosis in the cysticercus. *Experimental Parasitology*, **58**: 263–9.

Threadgold, L. T. & Hopkins, C. A. 1981. *Schistocephalus solidus* and *Ligula intestinalis*: pinocytosis by the tegument. *Experimental Parasitology*, **51**: 444–56.

Threadgold, L. T. & Robinson, A. 1984. Amplification of the cestode surface: a sterological analysis. *Parasitology*, **89**: 523–35.

Ubelaker, J. E. 1980. Structure and ultrastructure of the larvae and metacestodes of *Hymenolepis diminuta*. In: *Biology of the Tapeworm Hymenolepis diminuta* (ed. H. P. Arai), pp. 59–156. Academic Press, New York.

Ubelaker, J. E. 1983. The morphology, development and evolution of tapeworm larvae. In: *Biology of the Eucestoda*, Vol. 1 (ed. C. Arme & P. W. Pappas), pp. 235–96. Academic Press, London.

Wardle, R. A. & McCleod, J. A. 1952. *The Zoology of Tapeworms*. The University of Minnesota Press, Minneapolis.

Williams, H. H. & McVicar, A. 1968. Sperm transfer in the *Tetraphyllidea* (Platyhelminthes: Cestoda). *Nytt Magasin for Zoologi*, **16**: 61–71.

Yamaguti, S. 1959. *Systema Helminthum*. Vol. II. *The Cestodes of Vertebrates*. Interscience Publishers, New York.

21

Eucestoda: minor orders

Order 1 **Tetraphyllidea.** The tetraphyllids are mostly small cestodes, exclusively parasitic in elasmobranchs. Their life cycles are not completely known.

The scolex is characterised by the possession of four bothridia (Fig. 20.1) which may vary in size and shape. The proglottides possess a single reproductive system with a lateral genital opening. The cirrus is usually armed with hooks, hairs or spines. The vitellaria are lateral.

The taxonomy, ecology and host-specificity of the family Phyllobothriidae have been particularly studied (Williams, 1966, 1968).

Hyperapolysis (Fig. 21.1). This remarkable phenomenon is a characteristic of the Order. It involves the release of individual proglottides in an immature condition which then develop independently of the strobila and from each other. The process has been particularly studied in *Trilocularia acanthiaevulgaris*, a common parasite of the spiny dogfish, *Squalus acanthias* (McCullough & Fairweather, 1983, 1984*a*,*b*; McCullough *et al.*, 1986). Three forms (Fig. 21.1) are involved in the developmental sequence within the fish gut: (*a*) a *plerocercoid-like* juvenile, which occurs in the stomach, but which migrates to the spiral valve in the early summer (May–July) and develops into (*b*) an immature *near-adult* form which in (June–July) releases (*c*) *free* (but immature) *proglottides*, which reach sexual maturity in a few weeks as individual organisms in the intestine. A remarkable adaptation is the development of numerous large, backwardly projecting spines (McCullough & Fairweather, 1983) on the anterior end of the free proglottides; these must greatly assist the latter to maintain their position in the intestine. When mature, the gravid proglottides become detached from the mucosa and pass out of the rectum, releasing their eggs on contact with sea water.

The seasonal strobilisation of the adult spiral valve worms and development of the released free proglottides may be related to the water temperature and the levels of reproductive hormone activity associated with mating of the fish (McCullough *et al.*, 1986).

Order 2 **Lecanicephalidea.** A small order, exclusively parasites of elasmobranchs. The scolex is divided into two portions: an upper disc-like or globular part, sometimes bearing tentacles, and a lower collar-like part, bearing suckers like those of the cyclophyllideans. The reproductive system is practically identical with that of tetraphyllids. The life cycle is largely unknown (e.g. *Lecanicephalum* sp. from spiral valve of elasmobranchs).

Order 3 **Proteocephalidea.** The proteocephalids combine a tetraphyllid type of reproductive system with a somewhat cyclophyllidean type of scolex.

The principle genus is *Proteocephalus* (Fig. 21.2) with numerous species in fresh-water fish, amphibians and reptiles. A species commonly available in Great Britain is *Proteocephalus filicollis* (syn. *Ichthyotaenia filicollis*) in the sticklebacks *Gasterosteus aculeatus* and *Pungitius pungitius* (Rødland, 1983); in *G. aculeatus*, the adult worms are distributed evenly between the anterior and posterior halves of the intestine. Some 'spatial' competition of *P. filicollis* appears to occur with the acanthocephalean *Neoechinorhynchus rutili*, which occurs in the posterior intestine and rectum, because, in concurrent infections with these two parasites (Table 21.1), a significantly greater proportion of worms occurred in the anterior half of the intestine (Chappell, 1969). Competitive inhibition between *P. ambloplitis* and the acanthocephalan *Neoechinorhynchus* sp. has also been reported (Durborow *et al.*, 1988).

Life cycle. Until recently, the life cycles of proteocephalids have been somewhat of a mystery because – at least in some species – there appeared to be one stage missing in the life cycle. Studies by Wood (1965) on *Ophiotaenia filaroides* and Fischer & Freeman (1969) on

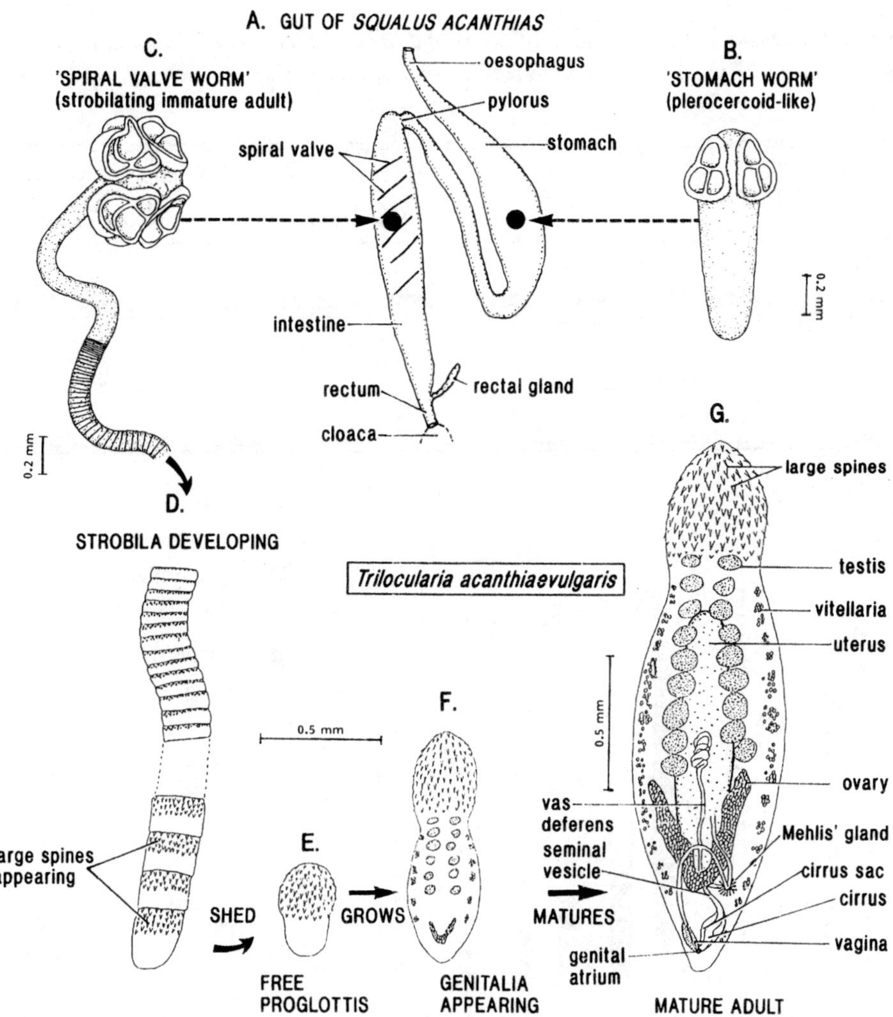

Fig. 21.1
Trilocularia acanthiaevulgaris: stages of development in the gut of the spiny dogfish, *Squalus acanthias*. The plerocercoid-like larva (B) found in the stomach migrates, in the early summer (May–July), the spiral valve where it strobilates (C) but does not become mature. As the sea temperature rises and sexual activity develops in the fish host (June–July), the strobila releases free – but not mature – proglottides (E). These attach to the intestinal mucosa by means of anterior spines and develop genitalia (F) eventually becoming sexually mature (G). The gravid proglottides pass out of the rectum and, on reaching sea water, split and release their eggs. (Based on McCullough & Fairweather, 1984*a* and Fairweather, personal communication.)

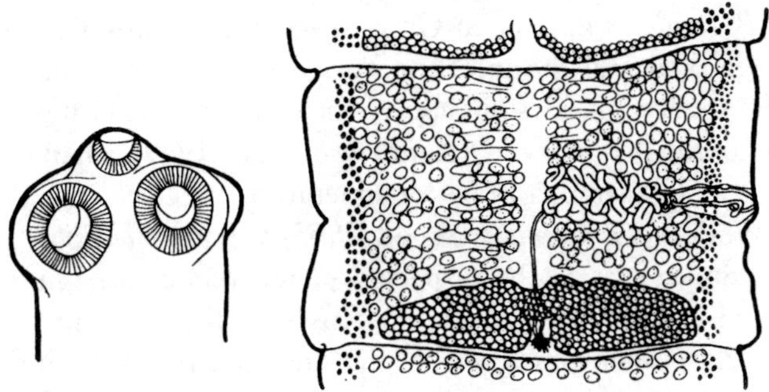

Fig. 21.2
Proteocephalus osculatus (order Proteocephalidea): morphology of scolex and proglottid. (After Nybelin, 1942.)

Proteocephalus ambloplitis have shown that in these species (and possibly in all) the definitive fish host *also* acts as an intermediate host!

The eggs of proteocephalids, which are probably quinone-tanned (Table 13.1), are embryonated when laid and hatch on ingestion by the appropriate copepod host. In *P. fluviatilis*, hatched eggs, freed of all membranes, have been found in the gut within 15 minutes and in the haemocoele within one hour (Fischer, 1968). Within the haemocoele oncospheres differentiate into procercoids (sometimes known as plerocercoid I!) as in the pseudophyllidean cestodes (Chapter 22).

Table 21.1. *Numbers of the cestode* Proteocephalus filicollis *and the acanthocephalan* Neoechinorhynchus rutili *attached in the regions of the gut of the 3-spined stickleback,* Gasterosteus aculeatus, *in single-species and concurrent infections*

Species	Type of infection	Number of worms		
		Anterior intestine	Posterior intestine	Rectum
Proteocephalus filicollis (adults)	single species	91	77	0
	concurrent	20	4	0
Proteocephalus filicollis (plerocercoids)	single species	74	138	0
	concurrent	18	10	0
Neoechinorhynchus rutili	single species	0	23	19
	concurrent	0	11	26

Source: Data from Chappell (1969).

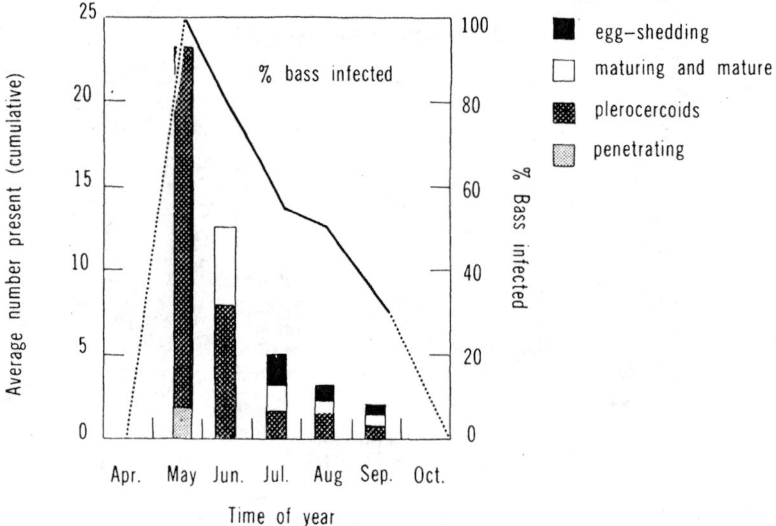

Fig. 21.3
Occurrence of *Proteocephalus ambloplitis* in the gut of smallmouth bass (7.6 cm or greater) from Lake Opeongo, Ontario. (After Fischer & Freeman, 1969.)

It is the next stage in the life cycle about which confusion has arisen. Because plerocercoids are frequently found in the *gut* of the definitive fish hosts it has been generally assumed that in the life cycle *either* (*a*) *no* second intermediate host was required and development from procercoid (ingested in a copepod) to plerocercoid to adult could take place in the definitive fish host *or* (*b*) a second intermediate host containing (encysted) pleroceroids was involved. Moreover, the occurrence of adults in the definitive host gut (as opposed to plerocercoids) has been shown to be seasonal (Fig. 21.3), egg production reaching a peak in late spring or early summer and a new infection either overlapping the old generation, or taking place shortly after its disappearance. In *P. torulosus* in dace in Great Britain, the maturation is seasonal and the infection period is very short; the new generation does not appear for six months after the loss of the previous one (Kennedy & Hine, 1969). Many authors believe this to be a temperature effect.

Work on *O. filaroides* (Wood, 1965) and *P. ambloplitis* has shown that, in these species, plerocercoids *encyst* in the viscera or on the outer (serosal) wall of the gut. During certain times of the year and as a response to temperature, these visceral plerocercoids *migrate through the gut wall* and become parenteral plerocercoids, and ultimately become sexually mature. Fischer & Freeman

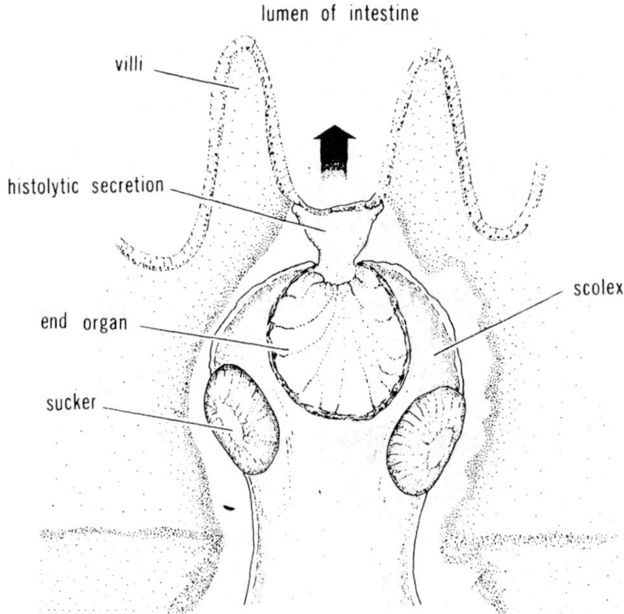

Fig. 21.4
Diagrammatic representation of the penetration of the fish gut by the plerocercoid of *Proteocephalus ambloplitis* migrating from the viscera. The end organ releases a histolytic exocrine secretion which aids in penetration. (Based on photographs by Fischer & Freeman, 1969.)

(1969) in a beautiful experimental demonstration, showed that when fish (bass) were raised from 5.5 to 7 °C massive migration of plerocercoids to the gut occurred! This rapid penetration is apparently aided by action of a large *end organ* (Fig. 21.4), which is essentially an exocrine gland secreting an aminopeptidase and possibly other histolytic secretions (Wood, 1965).

P. ambloplitis is a well-recognised pathogen of fish, especially the breeding stocks of the smallmouth bass, *Micropterus dolomieui*. Severe pathogenic changes occur, largely due to the migrating plerocercoids, the liver being particularly affected (Esch & Huffines, 1973; Joy & Madan, 1989), but other tissues, even the ovary, may also be invaded (McCormick & Stokes, 1982).

Order 4 **Diphyllidea.** A little-studied order containing only the single genus *Echinobothrium* (Fig. 20.1); main morphological features are given on p. 281. Adults in elasmobranchs; larval stages in crustacea and molluscs.

Order 5 **Trypanorhyncha.** Easily distinguished from other orders by the presence of four spiny *proboscides* which lie in sheaths and can be everted from the scolex. Rees (1944, 1950) has made detailed anatomical studies of several species.

Type example: ***Grillotia erinaceus***

definitive host:	*Raia* spp.
location:	intestine
first intermediate host:	*Acartia longiremis*, *Pseudocalanus elongatus*, *Paracalanus parvis*, *Temora longicornis*
second intermediate host:	*Gadus* spp., *Trigla* spp., *Lophius piscatorius* and other marine teleosts
distribution:	widespread

This parasite is most widely known through its larva which occurs encysted in many marine fish. It is among the commoner helminth parasites of fish from the Irish Sea. It is rare for the whiting or the gurnard to be free from infection.

Morphology. The morphology of the adult from the ray is shown in Fig. 21.6. Specimens rarely exceed 60 mm in length. The posterior proglottides are continually becoming detached and are often found free; they show no exceptional morphological features. Each *proboscis* is a tube lined internally with a variety of different-sized hooks so that when everted the spiny

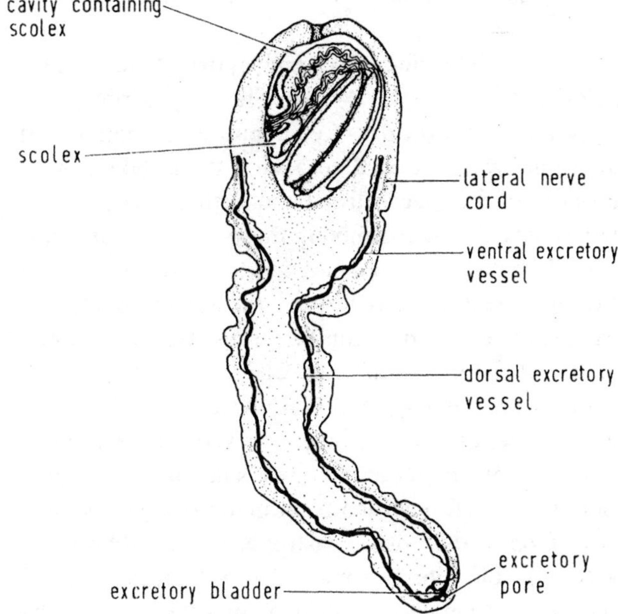

Fig. 21.5
Plerocercoid larva of *Grillotia heptanchi* from the hake (*Merluccius merluccius*); note that the scolex is invaginated in the anterior part of the body of the larva. (After Rees, 1950.)

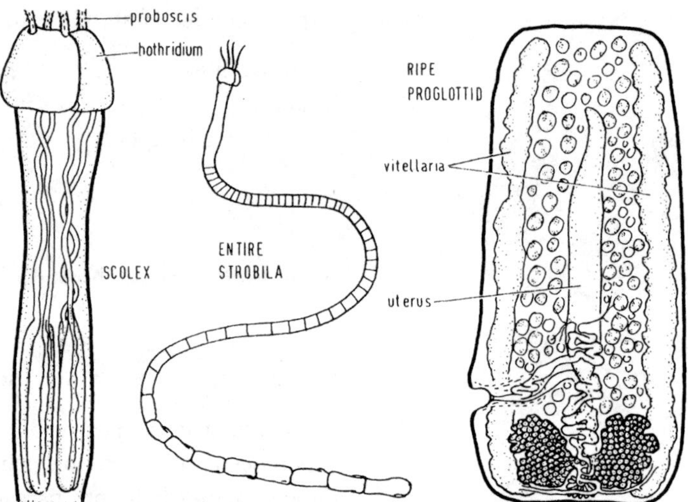

Fig. 21.6
Grillotia erinaceus (order Trypanorhyncha): plerocercoid removed from cyst (from body cavity of *Gadus* sp.) and adult strobila and proglottid (from intestine of rat). (After Johnstone, 1911.)

side is outwards. It is contained in a tube or *proboscis sheath* which terminates in an oval muscular mass, the *proboscis bulb*. The whole apparatus is filled with a fluid and the proboscoides are everted by contraction of the muscular bulbs. A retractor muscle effects the withdrawal of the proboscis into the sheath. The ultrastructure and physiology of the muscular system in this species has been studied by Ward *et al.* (1986*a*,*b*).

Life cycle. The life cycle resembles that of the Pseudophyllidea (Ruszkowski, 1934). The eggs embryonate in sea water in eight days (temperature not stated) and hatch out typical ciliated coracidia. When taken in by copepods of the species listed above, these develop into procercoids closely resembling those of pseudophyllids. When infected copepods are ingested by teleosts the larvae are freed and bore through the wall of the alimentary canal to encyst beneath the serous layer or the peritoneum of the body wall (Fig. 21.5).

Numerous other species occur in sharks and rays, but their life cycles have not been worked out. Advanced larvae, mainly plerocercoids but sometimes cysticercoids, have been recorded from numerous marine animals (copepods, fish, gastropods, lamellibranchs, medusae). The life cycles are poorly known. In some species a procercoid is formed, in others, coracidia appear to be able to develop directly to a plerocercoid stage (Mudry & Dailey, 1971). Rees (1941, 1950) has given a detailed description of the anatomy of a plerocercoid.

Order 7 **Nippotaeniidea***. This order contains only the single genus *Nippotaenia*, small cestodes from Japanese fish. They have no proper scolex but only a single acetabulum.

Order 9 **Aporidea***. This order contains rare monozoic forms, first described in swans. The ovary appears to be a germovitellaria and the testes form a mass running most of the body length. There appear to be no sex ducts.

○○○ References

(References of general interest have been included in this list, in addition to those quoted in the text.)

Chappell, L. H. 1969. Competitive exclusion between two intestinal parasites of the three-spined stickleback, *Gasterosteus aculeatus* L. *Journal of Parasitology*, **55**: 775–8.

Duborow, R. M., Rogers, W. A. & Klesius, P. H. 1988. Interaction of bass tapeworm, *Proteocephalus ambloplitis*, and *Neoechinorhynchus* sp. (Acanthocephala) in largemouth bass, *Micropterus salmoides*. *Journal of Parasitology*, **74**: 1056–9.

Esch, G. W. & Huffines, W. J. 1973. Histopathology associated with endoparasitic helminths in bass. *Journal of Parasitology*, **59**: 306–13.

Fischer, H. 1968. The life cycle of *Proteocephalus fluviatilis* Bangham (Cestoda) from smallmouth bass, *Micropterus dolomieui* Lacépède. *Canadian Journal of Zoology*, **46**: 569–79.

Fischer, H. & Freeman, R. S. 1969. Penetration of parenteral plerocercoids of *Proteocephalus ambloplitis*

*For convenience these orders are included here.

(Leidy) into the gut of smallmouth bass. *Journal of Parasitology*, **55**: 766–74.

Johnstone, J. 1911. *Tetrarhynchus erinaceus* van Beneden. I. The structure of larva and adult worm. *Parasitology*, **4**: 364–415.

Joy, J. E. & Madan, E. 1989. Pathology of black bass hepatic tissue infected with larvae of the tapeworm *Proteocephalus ambloplitis*. *Journal of Fish Biology*, **35**: 111–18.

Kennedy, C. R. & Hine, P. M. 1969. Population biology of the cestode *Proteocephalus torulosus* (Batsch) in dace *Leuciscus leuciscus* (L.) of the River Avon. *Journal of Fish Biology*, **1**: 209–19.

McCormick, J. H. & Stokes, G. N. 1982. Intraovarian invasion of smallmouth bass oocytes by *Proteocephalus ambloplitis* (Cestoda). *Journal of Parasitology*, **68**: 973–5.

McCullough, J. S. & Fairweather, I. 1983. A SEM study of the cestodes *Trilocularia acanthiaevulgaris*, *Phyllobothrium squali* and *Gilquina squali* from the spiny dogfish. *Parasitology*, **69**: 655–65.

McCullough, J. S. & Fairweather, I. 1984*a*. A comparative study of *Trilocularia acanthiaevulgaris* Olsson 1867 (Cestoda, Tetraphyllidea) from the stomach and spiral valve of the spiny dogfish. *Zeitschrift für Parasitenkunde*, **70**: 797–807.

McCullough, J. S. & Fairweather, I. 1984*b*. Growth and development of free proglottides of *Trilocularia acanthiaevulgaris*. *Parasitology*, **89**: LXVII.

McCullough, J. S., Fairweather, I. & Montgomery, W. I. 1986. The seasonal occurrence of *Tricularia acanthiaevulgaris* (Cestoda: Tetraphyllidea) from spiny dogfish in the Irish Sea. *Parasitology*, **93**: 153–62.

Mudry, D. R. & Dailey, M. D. 1971. Postembryonic development of certain tetraphyllidean and trypanorhynchan cestodes with a possible alternative life cycle for the order Trypanorhyncha. *Canadian Journal of Zoology*, **49**: 1249–53.

Rees, G. 1941. The musculature and nervous system of the plerocercoid larva *Dibothriorhynchus grossum*. *Parasitology*, **33**: 373–89.

Rees, G. 1944. A new cestode of the genus *Grillotia* from a shark. *Parasitology*, **35**: 180–5.

Rees, G. 1950. The plerocercoid larva of *Grillotia heptanchi* (Vaullegeard). *Parasitology*, **40**: 265–72.

Rødland, J. T. 1983. A redescription of the cestodes *Proteocephalus filicollis* (Rudolphi) from *Gasterosteus aculeatus* Pl., and *P. ambiguus* from *Pungitius pungitius* (L.). *Zoologica Scripta*, **12**: 19–23.

Ruszkowski, J. S. 1934. Études sur le cycle évolutif et sur la structure des cestodes de mer. III. Le cycle évolutif du Tétrarhynque *Grillotia erinaceus*. *Mémoires de l'Académie Polonaise des Sciences*, B. **6**: 1–9.

Ward, S. M., Allen, J. M. & McKerr, G. 1986*a*. Neuromuscular physiology of *Grillotia erinaceus* metacestodes (Cestoda: Trypanorhyncha) *in vitro*. *Parasitology*, **93**: 121–32.

Ward, S. M., McKerr, G. & Allen, J. M. 1986*b*. Structure and ultrastructure of muscle systems within *Grillotia erinaceus* metacestodes (Cestoda: Trypanorhyncha). *Parasitology*, **93**: 587–97.

Williams, H. H. 1966. The ecology, functional morphology and taxonomy of *Echeneibothrium* Beneden, 1849 (Cestoda: Tetraphyllidea), a revision of the genus and comments on *Discobothrium* Beneden, 1870, *Pseudanthobothrium* Baer, 1956, and *Phormobothrium* Alexander, 1963. *Parasitology*, **56**: 227–85.

Williams, H. H. 1968. The taxonomy, ecology and host-specificity of some Phyllobothriidae (Cestoda: Tetraphyllidea), a critical revision of *Phyllobothrium* Beneden, 1849 and comments on some allied genera. *Philosophical Transactions of the Royal Society of London*, B**253**: 231–307.

Wood, D. E. 1965. Nature of the end organ in *Ophiotaenia filaroides* (La Rue). *Journal of Parasitology*, **51**: 541–4.

22 Eucestoda: Pseudophyllidea

Members of this order, commonly referred to as pseudophyllids, are chiefly parasites of fish-eating mammals, birds and fish other than elasmobranchs. In typical forms, the scolex is characterised by two shallow elongated bothria situated one dorsally and one ventrally (Fig. 22.4), and proglottides are flattened dorsoventrally. The median genital pores are dorsal or ventral in position. This order contains a number of species of value in experimental work, or of medical or economic importance. Most of these belong to the family Diphyllobothriidae.

The plerocercoid larvae of three species of the genus *Diphyllobothrium*, *Diphyllobothrium dendriticum*, *D. ditremum* and *D. latum*, are common parasites of lacustrine (lake-dwelling) teleost fish in Europe, the USA and Canada. The adults of the first two species are chiefly parasites of fish-eating birds and occasionally mammals, but *D. latum* is an important human parasite (p. 307). However, the development of industrial salmonid aquaculture has meant that *D. dendriticum* and *D. ditremum*, whose plerocercoids can seriously damage fish, can also now be regarded as important parasites of economic importance (Soutar, 1990).

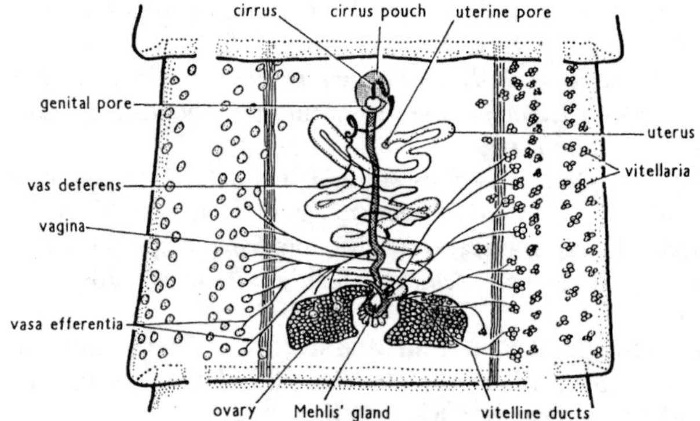

Fig. 22.1
Morphology of proglottid to illustrate the general features of diphyllid organisation. (After Brown, 1950.)

22.1 Genus *Diphyllobothrium*

Although *D. dendriticum* is taken here as a type example, much of the data relating to it are also relevant to the closely related species *D. ditremum*. The latter also occurs in the same ecological situation, and it is often difficult to distinguish specimens of one species from the other (see p. 307). *D. dendriticum* makes a valuable experimental cestode model, all the more so because full details of its laboratory maintenance are available (Sharp *et al.*, 1990).

22.1.1 Type example: *Diphyllobothrium dendriticum*

definitive hosts:	natural: various fish-eating birds, especially gulls; occasionally mammals (including, rarely, man) experimental: golden hamsters, rats
location:	small intestine
first intermediate host:	freshwater copepods, *Cyclops* spp., *Diaptomus* spp.
second intermediate host:	various freshwater teleosts, especially the 3-spined stickleback (*Gasterosteus aculeatus*) and various trout species; paratenic hosts common

distribution: widespread, especially Europe and N America

This species has a structure and life cycle closely paralleling that of the 'fish' tapeworm of man, *Diphyllobothrium latum* (p. 307), but is more readily available and has the advantage of developing in the rat, the golden hamster and the common gull, all of which can be maintained in the laboratory (Archer & Hopkins, 1958; Halvorsen & Andersen, 1974; Yamane *et al.*, 1988). Sharp *et al.* (1990) have published a valuable, detailed protocol for the routine laboratory maintenance of this species.

Morphology. The morphology of a diphyllobothriid proglottid (Fig. 22.1), is too well known to require detailed description. Both on morphological and on histochemical grounds, the genitalia bear a striking resemblance to those of the Trematoda (Fig. 22.2) with which they clearly have affinities.

There are three genital openings: the vagina, the uterus and the cirrus. The cirrus and vaginal openings are in a common atrium in some species. The vagina runs posteriorly from its opening and joins the oviduct before entering an ootype surrounded by the cells of Mehlis' gland. The vitellaria are scattered throughout lateral fields and their ducts join to form a short median duct, which swells to form a vitelline reservoir before entering the ootype. The egg, which is quinone-tanned, is formed by a mechanism, almost identical with that operating in trematodes such as *Fasciola* (Figs 13.8, 22.2), the vitellaria giving positive reactions for proteins, phenols and phenolase (Table 13.1).

There is a bilobed ovary leading by an oviduct into the ootype. The uterus consists of a series of coiled tubes increasing in size posteriorly. The male system follows a typical pattern with testes capsules widespread and the cirrus and cirrus sac well developed.

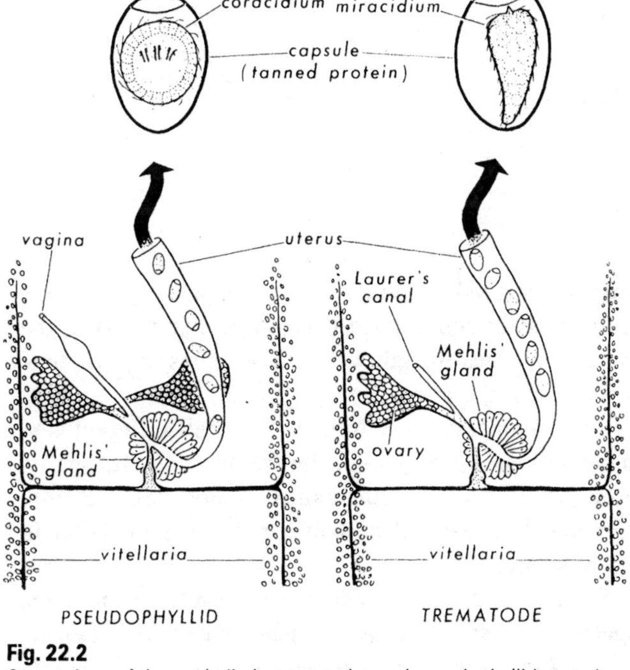

Fig. 22.2
Comparison of the genitalia in trematodes and pseudophyllid cestodes. (Original.)

Life cycle (Fig. 22.3). A large proportion of proglottides, which are pseudoapolytic (p. 282), mature at the same time, and quantities of eggs appear in the faeces. For obvious ecological reasons, faeces from fish-eating birds have a good chance of reaching water. At 15 °C, eggs embryonate in thirty days; at 25 °C, in eight days.

Embryonated eggs hatch only on exposure to light. Presumably the mechanism here is the same as in the eggs of *Fasciola*, in which there is some evidence that a light-released enzyme may attack the opercular seal (p. 207).

The wavelength which stimulates hatching in *D. dendriticum* is not known, but in *D. latum* there are two peak stimulating wavelengths, 300 nm (UV) and 600 nm (yellow).

Details of embryonation of eggs and the copepod infection procedures are given by Sharp *et al.* (1990). Eggs may be embryonated by placing about 10 000 in 5 cm Petri dishes in aerated, distilled water at pH 7.0 with no added antibiotics. They are then incubated in the dark, in black plastic bags at 25 °C and examined daily after exposure to light for 10 min, until hatching occurs.

The retardation of hatching by darkness is useful in laboratory procedures as release of coracidia from embryonated eggs can be withheld until suitable cultures of *Cyclops* are available. The hatched larva is a *coracidium* (Fig. 22.2), essentially a hexacanth embryo enclosed in a ciliated embryophore. It swims actively with a ciliate-like motion and exhibits a tendency towards negative geotropism. It contains few food reserves and dies if not eaten by a copepod within about twelve hours.

Infection of copepod. Sharp *et al.* (1990), who give details of the laboratory maintenance of copepods, used *Cyclops abyssorum* (in the UK) as the first intermediate host, but several other species of *Cyclops*, e.g. *C. strenuus*,

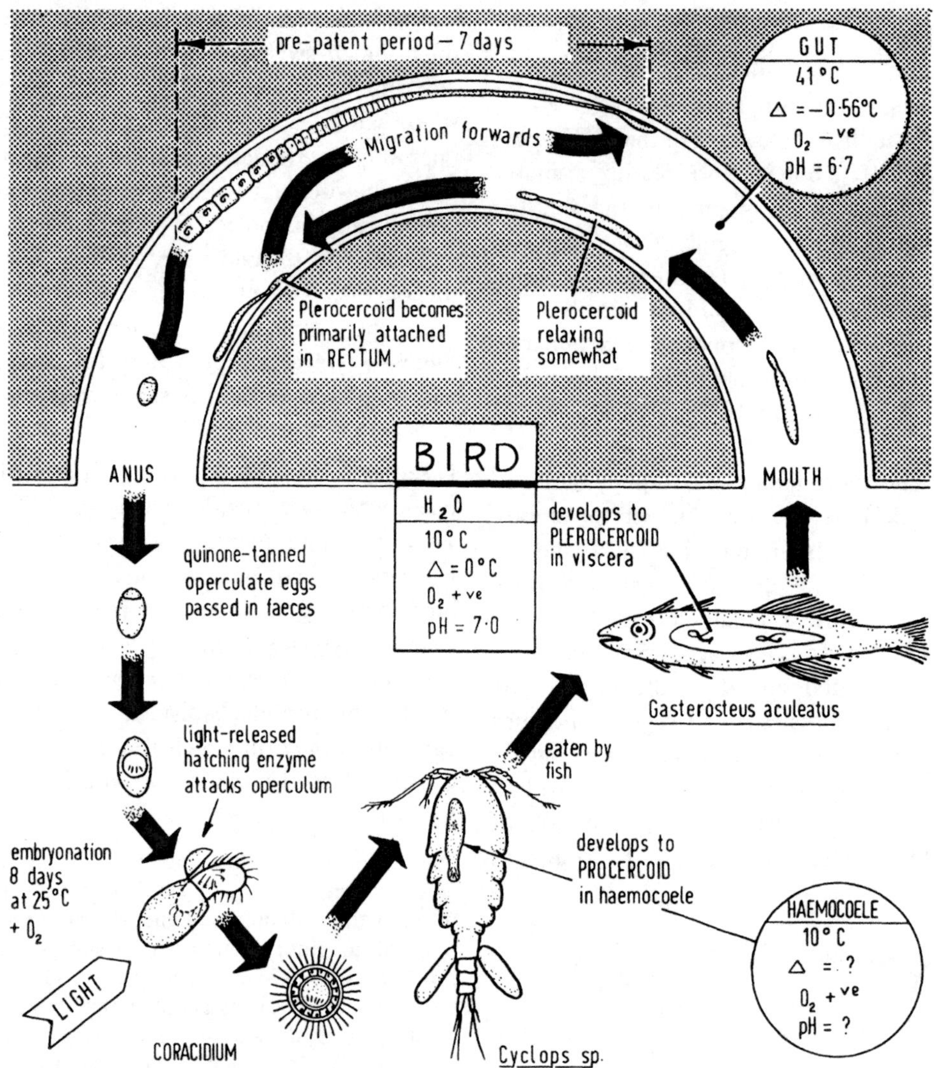

Fig. 22.3
Diphyllobothrium dendriticum: life cycle and some physiological factors relating to it. (Original.)

are also suitable. Species of *Diaptomus* are especially good hosts (Kühlow, 1953) but are difficult to maintain in the laboratory. The copepodid stages are more easily infected, the mature copepods often being refractory to infection.

When ingested by a copepod, the embryophore is shed and the contained hexacanth bores its way rapidly through the intestine into the haemocoele, to become a *procercoid* larva, the scolex of which contains glands (Kuperman & Davydov, 1982).

A procercoid reaches an infective stage when the hooks become separated into a constricted posterior region termed a *cercomer* (Fig. 22.4).

Infection of fish. When a *Cyclops* containing a procercoid in the infective stage is eaten by a stickleback, the larva penetrates the gut wall and develops into the final larval stage, the *plerocercoid* (Fig. 22.4). This is a white, opaque, elongate structure, with a well-differentiated scolex. The latter is usually contracted and partly invaginated, and its true nature is not revealed until stimulated to expand and evaginate by immersion in warm saline.

Plerocercoids may become encapsulated in some species (e.g. sticklebacks) or may remain free. During hot weather, the latter may become active and migrate through the tissues, inducing pathogenic effects or even

Table 22.1. Diphyllobothrium dendriticum *and* D. ditremum*: distribution on and/or in different organs in 535 char* (Salvelinus alpinus) *caught during 1971–74*

Organ	*D. dendriticum* (%) n = 4890	*D. ditremum* (%) n = 23 486
Stomach	43.0	65.5
Pyloric caeca	16.4	13.7
Intestine	1.7	0.4
Liver	10.5	2.1
Kidney	3.5	0.0
Gonads	6.5	1.7
Swimbladder	4.6	0.3
Mesenteries	1.4	0.1
Musculature	5.3	0.0
Free in coelom	7.0	16.2
Other	0.1	0.0

Source: Data from Hendricson (1977).

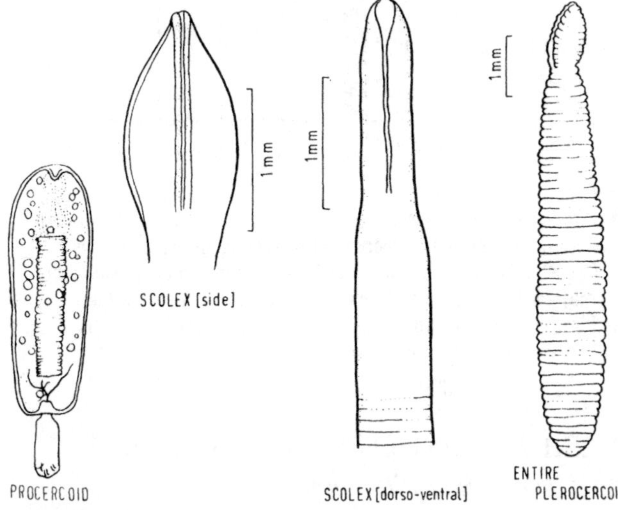

Fig. 22.4
Diphyllobothrium dendriticum: anatomy of plerocercoid, procercoid and adult scolex. (After Kuhlow, 1953.)

killing the fish. Although the alimentary canal and liver are the most favoured sites, plerocercoids may be found in almost any organ or tissue (Table 22.1).

It has been shown in some species that if infected fish are eaten by larger fish, the pleroceroids penetrate the gut and become re-encapsulated in its vicinity. Hundreds of plerocercoids may thus accumulate in this second fish, which is known as a *paratenic host*. If the second fish is a large one, such as a pike, not normally eaten by a bird, the plerocercoids it accumulates will have no chance of developing to maturity and are thus withdrawn from the life cycle.

Experimental infections have shown that only about 5–6 per cent of *D. dendriticum* plerocercoids can re-establish themselves in a new host on ingestion of the original fish host, and in *D. ditremum* only one larva was found to do so; in contrast, 50 per cent of *D. latum* plerocercoids readily established themselves when fed to new fish (Halvorsen & Wissler, 1973).

In relation to the penetration of the fish gut, it should be noted that TEM studies have shown that two types of glands occur in the scolex, one of which may possibly secrete an enzyme which could assist in penetration (Kuperman & Davydov, 1982; Gustafsson & Vaihela, 1981). In the plerocercoid scolex of a related species, *Spirometra erinacei* (p. 310), the presence of such a proteolytic enzyme has been confirmed (Kwa, 1972*b*).

Adult worms: growth in the definitive host. The common definitive hosts of *D. dendriticum* are gulls, especially the common gull, *Larus canus*, and the herring gull, *L. argentatus*. A wide range of other natural hosts have been reported, including fish-eating birds, such as pelicans, ravens and herons, and mammals, such as the Arctic fox, the black bear, the rat and man (Andersen *et al.*, 1987). In addition to gulls, experimental infections include the golden hamster and the laboratory rat (Archer & Hopkins, 1958; Halvorsen & Andersen, 1974; Sharp *et al.*, 1990; Yamane *et al.*, 1988).

Development *in vivo*. Eggs appear in both natural and experimental hosts in 6–8 days (Table 22.3). In the rat, worms become first established in the large intestine, but by 3–7 days undergo a forward migration to become established in the duodenum (Archer & Hopkins, 1958). A similar migration, it should be noted, occurs in *Hymenolepis diminuta* (p. 322). A comparable situation occurs in the hamster (Fig. 22.5) where, after 2–8 h, most worms become established in the second or third third-part of the small intestine, but undergo a forward migration, so that by 66 h all worms are found in the anterior third (Halvorsen & Andersen, 1974).

In this latter experiment, the pre-patent period was not influenced by the tapeworm population size, but much larger worms (839–1936 mg) were obtained from 1-plerocercoid infections than from 13-plerocercoid infections, which gave rise to worms weighing only 142–482 mg.

The developmental phases up to maturation are

Table 22.2. *Developmental phases of* Diphyllobothrium dentriticum *during maturation in the intestine of the rat*

Body temperature was 38.8 °C.

Phase	Incubation time (days)	Characteristics
(1) Cell multiplication	0–1	numerous mitoses
(2) Segmentation	1–2	appearance of proglottides
(3) Organogeny	2–3	appearance of uterus and testes primordia
(4) Early gametogeny	4–5	appearance of early stages in spermatogenesis
(5) Late gametogeny	5–6	appearance of mature spermatozoa
(6) Egg-shell formation	6–7	appearance of egg-shell precursors in 'vitelline' cells
(7) Oviposition	7–8	appearance of fully formed egg

Source: From Bell & Smyth (1958).

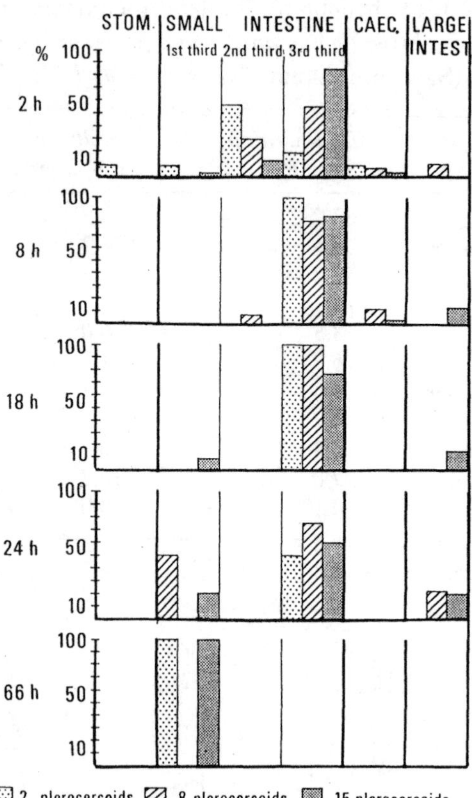

Fig. 22.5
Diphyllobothrium dendriticum: establishment in the gut of the golden hamster after experimental feeding of plerocercoids. Most worms become established in the 2nd or 3rd 'third' of the small intestine, but later migrate forwards until, by 66 h, all worms are found in the anterior 'third'. STOM, stomach; CAEC, caecum; INTEST, intestine. (After Halvorsen & Andersen, 1974.)

shown in Table 22.2. A characteristic of diphyllobothriid plerocercoids is that on ingestion by the definitive host, the 'larval body' (i.e. the abothrial region posterior to the scolex) is shed. This process is most marked in the plerocercoid of *D. latum* (Table 22.4), where shedding can be detected at 2 h post infection and by 48–60 h is almost completed (Andersen, 1978; Yamane *et al.*, 1988). This process was originally thought not to take place in *D. dendriticum* or *D. ditremum* (Andersen, 1978) but later work found that some degree of shedding takes place in both these species (Yamane *et al.*, 1988). Thus, *D. dendriticum* exhibited a low (8.7 per cent) shedding rate in hamsters, but a high (34.9 per cent) rate in the rat. In *D. ditremum*, in hamsters, a shedding rate of 42.9 per cent was recorded.

A remarkable characteristic of the mitotic activity during the development of *D. dendriticum* is that it exhibits a circadian rhythm (Wikgren *et al.*, 1970), presumably related to the metabolic activities of the host, the latter, in turn, being associated with the availability of nutritive materials.

○ 22.1.2 Other Diphyllobothriidae in fish: *Diphyllobothrium ditremum*

The adult of *D. ditremum* – in contrast to *D. dendriticum* – has a limited host range, which is restricted to fish-eating birds: loons, mergansers, cormorants and grebes. Various features of its morphology are compared with those of *D. dendriticum* and *D. latum* in Table 22.3.

Identification of the plerocercoids of *Diphyllobothrium* species from freshwater fish has always been recognised as a difficult problem but recent papers providing useful criteria and/or valuable keys are those of Andersen (1972), Andersen & Gibson (1989), Andersen *et al.* (1987), Sergeeva *et al.* (1987) and Chubb *et al.* (1987). The speciation situation is further complicated by the fact that, like many other cestode species (e.g. *Echinococcus granulosus*, p. 336), isoenzyme analysis has shown

Table 22.3. *Comparison of morphological characters from* Diphyllobothrium dendriticum, D. ditremum *and* D. latum *reared in the hamster*

The morphometric measurements are from single infections or two worm populations.

	D. dendriticum	*D. ditremum*	*D. latum*
Eggs appear in hamster faeces	Days 6–8	Days 8–10	Days 16–18
Shape of scolex	Lanceolate to spathulate	Cordate to lanceolate, often with curved bothrial edges	Spathulate
Scolex length (mm)	1.63 (1.25–2.00)	1.75 (1.58–2.08)	2.03 (1.59–2.50)
Scolex dorsoventral height (mm)	0.90 (0.75–1.00)	1.15 (0.83–1.33)	1.08 (0.92–1.25)
Neck	Absent	Absent	Present
Total length of strobila (cm)	42 (21–63)	23 (10–43)	91 (61–138)
Maximum width of segments (mm)	11.2 (7.1–18.3)	4.5 (2.7–6.4)	9.3 (6.7–14.1)
Length of segments in the widest part of the strobila (mm)	1.7 (0.90–2.80)	1.5 (0.7–2.5)	3.0 (2.10–5.00)
Max. no. of segments found in a strobila	388	167	650
Shape of ovary	Very variable, dumbbell shape	Variable, usually with anterior and posterior horns	Variable, rounded or elongate lobes
Position of cirrus sac in sagittal sections	Oblique	Oblique	Horizontal
Position of seminal vesicle in relation to cirrus sac	Dorsal; not seen from the ventral side	Dorsal and caudal; seen also from the ventral side	Dorsal and caudal: seen also from the ventral side
Type of boundary between neighbouring segments	No distinct boundary; vitellaria and testes confluent	No distinct boundary; vitellaria and testes confluent	Constriction between the segments and an area without testes and vitellaria

Source: Data pooled from Andersen (1972) and Andersen *et al.* (1987).

that marked intra-specific variation also occurs, even in a local population (de Vos *et al.*, 1990).

The general biology of the family Diphyllobothriidae has been reviewed by Delyamure *et al.* (1985).

○ 22.1.3 *Diphyllobothrium latum*: the human 'broad' tapeworm

General account. The largest tapeworm found in man. It has a length of 3–25 m and a width of 10–20 mm. It occurs especially in countries whose populace is addicted to eating fish in an uncooked or partly cooked (i.e. smoked) condition. The term 'broad' relates to the fact that the proglottides are usually wider than they are long.

D. latum also develops in many other mammals, especially the dog, the cat and the pig. It has never been recorded from birds; experimental feeding of birds has also proved unsuccessful (Dick & Poole, 1985). In countries, such as Alaska, Siberia and Finland, where fish is a main foodstuff and the offal is accessible to domestic animals, high levels of infection have been reported. Levels of 100 per cent have been reported in pigs and 34–47 per cent in dogs (Witenberg, 1964); dogs, however, are probably the most important source of dissemination. The biology, clinical manifestations, pathogenesis and control of this species have been reviewed in a valuable monograph by Von Bonsdorff (1977).

Distribution. *D. latum* has a worldwide distribution but it occurs especially in countries bordering the Baltic Sea (Finland, Sweden etc.); it also occurs in Russia, Switzerland and North America. Occasional records from other countries have been reported.

Morphology. The general pattern of the morphology differs from that of *D. dendriticum* only in a number of

Table 22.4. *Development of* Diphyllobothrium latum *in hamsters*

	Worm age (hours post infection)					Total
	24	36	48	60	84	
Number of plerocercoids given	8	8	8	8	8	80
Number of hamsters autopsied	2	2	2	2	2	10
Number of worms recovered	6	11	12	5	5	39
Recovery rate (per cent)	37.5	68.8	75.0	31.3	31.3	48.8
Growth pattern of recovered worms						
Non-shedding	3	2	2	0	0	7
Shedding	3	9	10	5	5	32
Shedding rate (per cent)	50.0	81.8	83.3	100.0	100.0	82.1 (mean)

Source: Data from Yamane *et al.* (1988).

minor details, summarised in Table 22.3. The neck of the scolex is longer, the total number of segments higher and the total length much greater; there are also differences in the shape of the ovary and the position of the cirrus sac and seminal vesicle (Andersen, 1972; Andersen *et al.*, 1987; Andersen & Gibson, 1989).

Life cycle. The life cycle differs from *D. dendriticum* only in detail, the main point of difference being that the plerocercoids do not encyst but lie usually in the muscles, gut walls or other viscera. The distribution between viscera and muscles, however, appears to vary not only between different fish hosts but between hosts collected from different areas (Wikgren, 1963); 'strain' differences with slightly different behaviour patterns may account for such results. There are also marked differences in size in different fish hosts. Many species of fish have been incriminated and in some countries (e.g. parts of Russia) practically all food fish are infected. The larger carnivorous fish are paratenic hosts, becoming infected by eating smaller plankton-eating fish. The larger fish involved are different in different countries: in Europe, trout, perch and turbot; in the USA and Canada, pike and walleyes; in the Far East and Chile, trout; in Africa, the barbel.

***Diphyllobothriasis*: pathology.** The chief physical symptoms of the infection are fatigue and weakness, but the major, and most serious, manifestation is the onset of pernicious anaemia. This is caused by the extensive absorption of vitamin B_{12} by the adult worm and the absorption of cobalamins from the host intestine (Pawlowski, 1984; von Bonsdorff, 1977).

Growth rate. This is known from experimental infections of dogs and hamsters, but this species fails to develop in the rat. In the dog, the pre-patent period is reported to be 21 days (Wardle & Green, 1941), but only 16–18 days in the hamster (Table 22.3) (Andersen, 1972; Dick & Poole, 1985). In the intestine, before growth commences, the plerocercoid sheds its total abothrial extremities – piece by piece – usually during the first 40–60 h of infection, but later in some cases (Table 22.4) (Yamane *et al.*, 1988). The adult strobila thus develops only from the scolex and neck region of the larva (Andersen, 1978; Yamane *et al.*, 1988). This appears to be an active process rather than the result of digestion by the host; the discarded fragments are either digested or expelled with the faeces. This shedding process also occurs in *Spirometra* (see below) but only to a limited extent, and in many species not at all, e.g. in *D. dendriticum* (see earlier).

Other *Diphyllobothrium* species in man. Although *D. latum* is recognised as the common species infecting man, other species have also been reported from man, e.g. *D. pacificum* in Peru and Chile (Pawlowski, 1984) and *D. nikonkaiense* in Japan. These, and other reported species, may prove to be 'strains' of *D. latum*, although the immunoelectrophoregrams of *D. nikonkaiense* have been shown to be significantly different from those of *D. latum* (Fukumoto *et al.*, 1987). It should also be pointed out that there is now evidence to suggest that some human infections may be due to *D. dendriticum*, especially if the infections were diagnosed from faecal samples only (Andersen *et al.*, 1987).

○ 22.1.4 *Spirometra* spp.: 'sparganum' infections

In the viscera of man and numerous animal species (e.g. frogs, snakes, hogs, hedgehogs etc.), especially in Asia, long, white, ribbon-like larvae are sometimes found. These larvae, originally known as *spargana*, are now known to be plerocercoids, probably of several species of pseudophyllid belonging to the genera *Diphyllobothrium* and *Spirometra*. The pathological condition induced by spargana is known as *sparganosis*; reviews of human sparganosis have been given by Huang & Kirk (1962), Smyth & Heath (1970) and Witenberg (1964). Spargana are ribbon-like and ivory white in colour; the scolex is poorly developed (if at all) and the strobila is unsegmented. Infection takes place in three ways: (*a*) by direct ingestion of infected copepods; (*b*) by ingestion of amphibia, reptilia, birds or mammals (which in these species serve

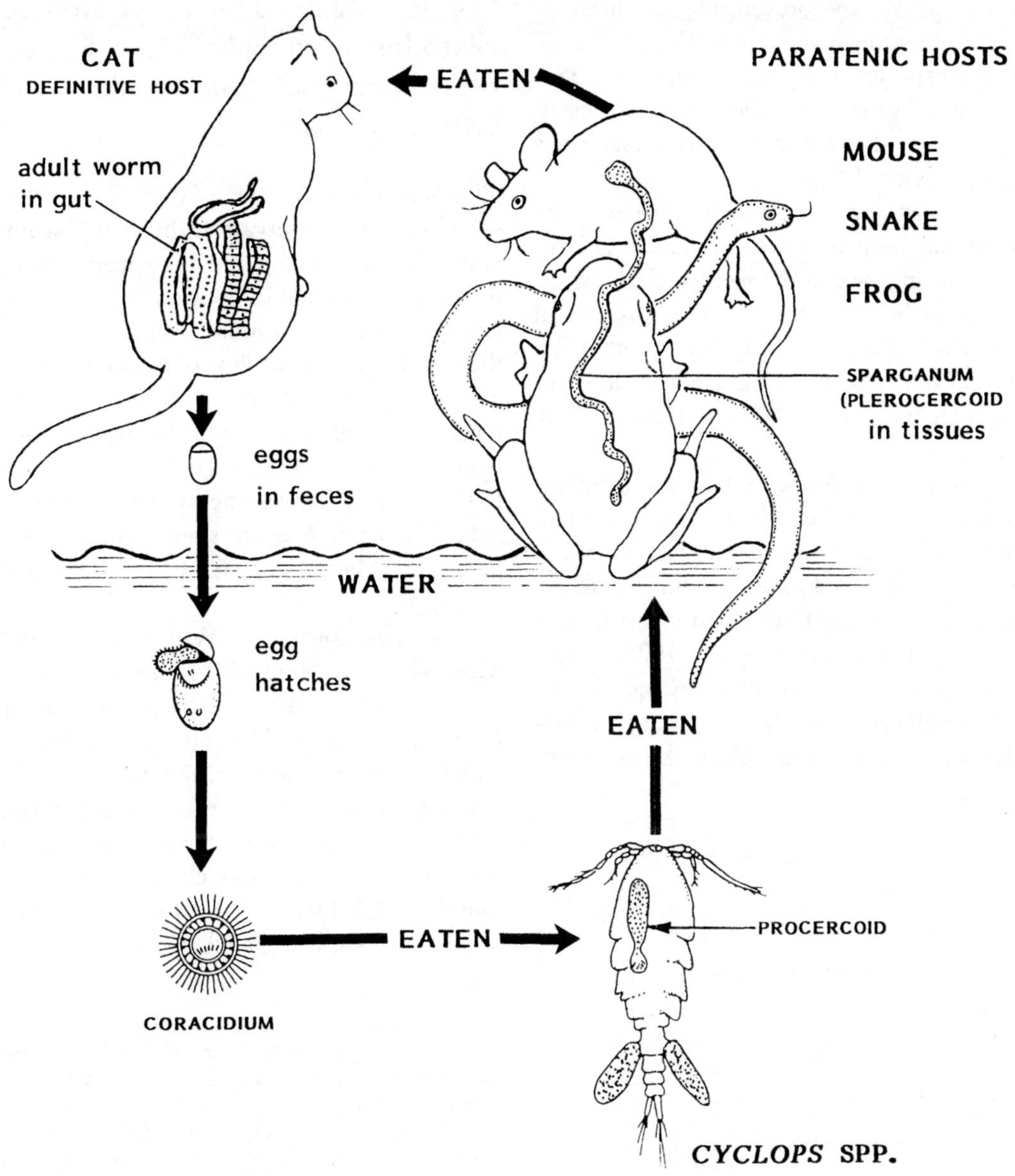

Fig. 22.6
Spirometra mansonoides: life cycle. The water snake (*Natrix*) is probably the natural intermediate (and paratenic) host, but spargana are also found in fish, amphibia and mammals (including man). (Modified from Mueller, 1974.)

as intermediate host instead of fish) and the subsequent intestinal penetration of the released plerocercoids; and (*c*) by local application of flesh (e.g. split frogs) as poultices to wounds or (especially) to sore eyes; the larvae migrate into the human tissue from the poultice.

Although the taxonomy of plerocercoids in sparganosis is in a confused state, many cases of human infection are probably due to *Spirometra mansoni*. The plerocercoid stage of this species occurs in snakes and the adult stage in dogs. Plerocercoids can readily pass from one intermediate host to another. If plerocercoids are fed to mice, they penetrate the gut within 40 minutes; glands which are presumably histolytic in nature have been identified in the scolex (Kwa, 1972*a,b*). A further Oriental species is *S. erinacei*, but the distinction between this species and *S. mansoni* is not clear. The ultrastructure of the tegument in *S. erinacei* has also been studied (Kwa, 1972*c*).

A useful experimental species is *S. mansonoides*, the biology of which has been reviewed by Mueller (1974). The adult develops in cats and the plerocercoids (spargana) in numerous hosts, especially mice, frogs and snakes, all of which can act as paratenic hosts (Fig. 22.6). Man can also be a host and human infections are marked by a high level of eosinophila (Campbell & Beals, 1977).

When mice are infected (by subcutaneous injection) a remarkable effect occurs: the mice gain mass at a rate which cannot be accounted for by the mass of the growing parasites or the associated tissue reaction (Mueller, 1974). A similar (but not identical) effect has also been described for *S. erinacei* (Hirai *et al.*, 1978). Most workers concluded that this phenomenon was due to the release by the parasite of a 'growth factor', which was at first thought to be an insulin-like hormone. More recent studies point to its being a substance resembling a mammalian growth hormone (Phares, 1987; Shiwaku *et al.*, 1983). The substantial literature on this unusual effect has been reviewed by Smyth (1969), Smyth & McManus (1989) and Odening (1979).

A further unusual physiological feature of *S. mansonoides* is that – like *Diphyllobothrium latum* – it has a remarkable capacity for absorbing vitamin B_{12} (Tkachuck *et al.*, 1976; Marchiondo *et al.*, 1989); see Fig. 22.7.

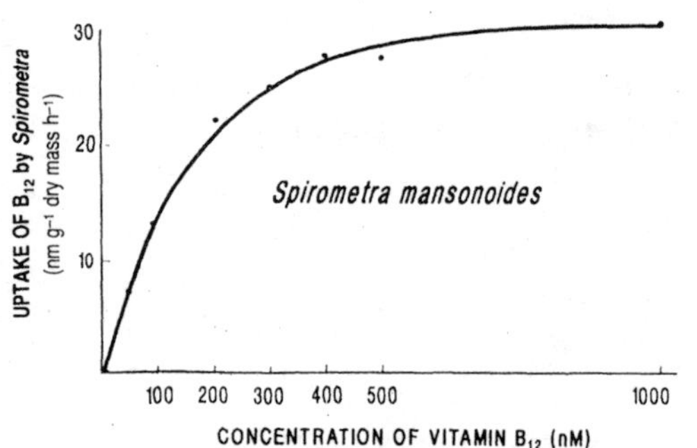

Fig. 22.7
Spirometra mansonoides: uptake, by spargana, of ^{57}Co-vitamin B_{12} with respect to concentration (Tkachuck *et al.* 1976.)

○○○ 22.2 The caryophyllaeid cestodes: pseudophyllidean-like cestodes with progenetic* procercoids? (= Order Caryophyllidea?)

This is an unusual group of cestodes which has a single set of reproductive organs within a non-segmented body and which utilises aquatic oligochaetes as intermediate hosts. The taxonomic position of this group has long been a matter of dispute, some workers considering them to be pseudophyllidean cestodes with an abbreviated life cycle, and others placing them in a separate Order Caryophyllidea. The biology of the group has been the subject of an excellent review by Mackiewicz (1972) who discusses the systematic position in considerable detail. Keys to some families have been prepared by Mackiewicz & Blair (1978) and Williams (1978).

The best-known genus is undoubtedly *Archigetes* (Fig. 22.8) whose 'procercoid' reaches maturity in the body cavity of tubificid oligochaetes of the genus *Limnodrilus* (Kennedy, 1965*b*). The adult stage may also (apparently!) reach maturity in the intestine of freshwater teleosts, especially those of the family Cyprinidae. To date, however, it has not been found possible to infect fish experimentally, although naturally infected fish are found. The failure to infect fish with *Archigetes* species may be due to the existence of physiological strains from

*The terms *progenesis* and *neoteny* are almost synonymous. Maturation of the gonads in a larval animal is known as *neoteny*. Advanced development of genitalia (which may or may not reach actual maturation) is known as *progenesis*. An advanced progenetic condition clearly becomes neoteny. *Archigetes* is strictly speaking a *neotenic* larva, but the terms *neotenic* and *progenetic* are rather loosely used and have been interpreted slightly differently by different authors. For a valuable discussion on this terminology, see Mackiewicz (1972).

Table 22.5. *The occurrence of* Archigetes limnodrili *in species of tubificid*

Only tubificid species occurring in the same locality as *A. limnodrili* are considered. C = Shropshire Union Canal; T = River Thames; E = elsewhere; + = infection present in nature or established experimentally; O = infection absent or failed to be established; A = tubificid absent from this locality; N = infection not attempted.

Tubificid	Natural infections			Experimental infections
	C	T	E	
Limnodrilus hoffmeisteri	+	+	+	+
L. udekemianus	A	O	O	O
L. claparedeanus	O	+	+	+
L. cervix	O	+	+	+
Euilyodrilus bavaricus	A	O	O	N
E. hammoniensis	O	O	O	O
E. moldaviensis	O	O	O	O
Tubifex tubifex	A	O	O	O
T. ignotus	A	O	O	N
T. templetoni	A	A	O	N
Psammoryctes barbatus	A	O	O	O
P. albicolus	A	O	O	N
Aulodrilus pluriseta	O	O	O	N
Branchiura sowerbyi	A	O	A	N
Rhyacodrilus coccineus	A	O	O	N

Source: Data from Kennedy (1965*a*).

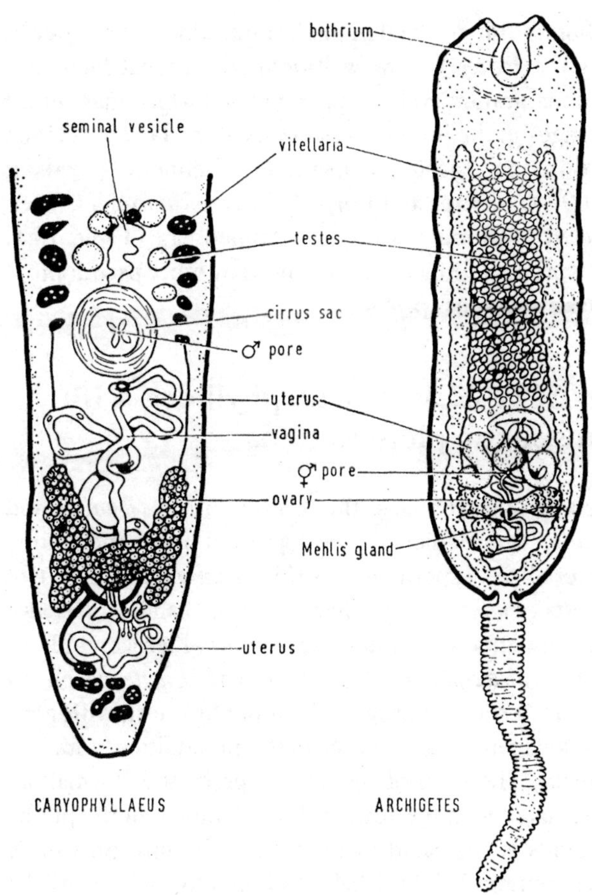

Fig. 22.8
'Archigetes' sieboldi from the body cavity of freshwater oligochaetes. *Caryophyllaeus* sp. from the gut of fresh-water teleosts. (After Wisniewski, 1930; Hunter, 1927.)

oligochaete cycles which are incapable of infecting fish. Mackiewicz (1972) concludes that 'The relationship of the progenetic, coelom-dwelling stages of *Archigetes* in oligochaetes to the intestinal stages in fish has yet to be clarified . . .'.

When adult *Archigetes* was first discovered in a fish it was assigned to the genus *Biacetabulum*. Later, however, it was found that the larva of *Biacetabulum* (*a*) is unable to undergo progenetic development and (*b*) infects only the oligochaete *Tubifex* but (unlike *Archigetes*) is incapable of infecting the other oligochaete genus *Limnodrilus* (Calentine, 1965).

○ 22.2.1 *Archigetes* spp.

The best-described species are probably *A. sieboldi* (Fig. 22.8) (Wisniewski, 1930), *A. iowensis* and *A. limnodrili* (Kennedy, 1965*a*).

Life cycle. The eggs embryonate in water and hatch only on ingestion by the tubificid worm. The hatched oncospheres penetrate the body cavity and develop into 'procercoid' larvae; in *A. limnodrili*, gravid adults develop in about 140 days. Eggs are only released by the death of the worm. Mature *Archigetes* also occur in fish, but, as indicated above, the mode of infection has not been unequivocally determined; infection presumably takes place by ingestion of infected tubificids.

In England, the degree of infection of species of *Limnodrilus* is generally low and rarely reaches 12 per cent. The occurrence in different species is shown in Table 22.5.

○ 22.2.2 *Caryophyllaeus* spp. (Fig. 22.8)

As indicated above (p. 310), larvae of species of this genus develop genitalia while within tubificids but maturation is only reached in the fish intestine. Some species (e.g. *C. laticeps*) show well-defined seasonal incidence and maturation cycles, a phenomenon which may reflect changes in the immune responses of the host in relation to different seasonal temperatures (Kennedy & Walker, 1969). *C. laticeps* and *Caryophyllaeides fennica* have been shown to produce a quinone-tanned egg (Table 13.1) by a process similar to that utilised by other pseudophyllidean cestodes (p. 288).

○○○ 22.3 Pseudophyllidea with progenetic plerocercoids

This group contains the genera *Schistocephalus* and *Ligula*, whose plerocercoids present most interesting material for experimental work *in vitro*. The effect of *Schistocephalus* on the behaviour of the fish host has also received much attention: see below. Plerocercoids of both *Schistocephalus* (Fig. 22.9) and *Ligula* reach an advanced stage of progenesis in the fish host, with genital anlagen clearly observable; in addition, *Schistocephalus* shows well-developed proglottid formation. Associated with the advanced development of the plerocercoids is the rapid maturation in the bird host or *in vitro* (Smyth, 1990*a*, 1954): 36 hours for *Schistocephalus* (Hopkins & Smyth, 1951; Orr & Hopkins, 1969*a*) and 67–72 hours for *Ligula* (Smyth, 1947, 1949). These organisms are discussed in further detail below.

○ 22.3.1 *Schistocephalus solidus* (Ligulidae)

Occurrence. The adults occur in a wide range of bird hosts. The worm is non-specific in its choice of definitive hosts and a wide range of natural bird hosts has been reported; it has also been reported from the otter and other fish-eating mammals.

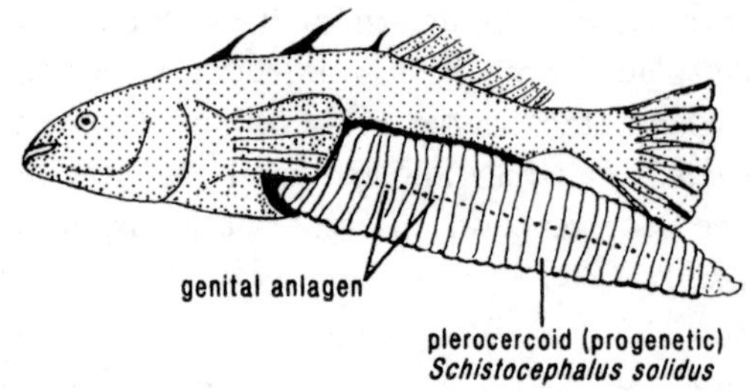

Fig. 22.9
Three-spined stickleback (*Gasterosteus aculeatus*) with body cavity cut open to release the (progenetic) plerocercoid of *Schistocephalus solidus*. (Original.)

This species deserves a special place in the history of parasitology, as it was used by the Danish worker Peter Abildgaard in 1790, to carry out the first demonstration of a parasite life cycle. By feeding sticklebacks infected with the plerocercoids to ducks, and, on autopsy several days later, obtaining the adult worms, he demonstrated the transmission of a parasite from one host to another (Smyth, 1990*a*,*b*).

Both bird and mammal hosts (rat and hamsters) have been successfully infected, but the position taken up in the gut, as well as the longevity, varies from host to host (Fig. 22.10) (McCaig & Hopkins, 1963). One- to four-week-old ducks (over 50 per cent established) and 2–5-week-old chicks (nearly 40 per cent established) appear to be the best laboratory hosts. For practical details of the laboratory maintenance, see Orr & Hopkins (1969*a*).

The wide host spectrum is presumably related to the progenetic nature of the plerocercoid (Fig. 22.9), which has sufficient food reserves (Table 24.1) to enable it to mature without the assistance of exogenous food material; a temperature of 35–40 °C and suitable physico-chemical conditions suffice to assure maturation *in vitro* (Smyth, 1954; p. 509).

Morphology. The adult (Figs 22.11, 22.15) is a lanceolate-shaped worm with a size range of 50–80 mm by 10 mm. Bothria on the scolex are represented by a short median groove which appears to possess no adhesive powers. The adult strobila must thus find it difficult to hold on to the intestinal wall. The lack of bothria, then, may be correlated with the rapid rate of maturation (36 hours), although worms have been found to remain in the host intestine for as long as 18 days (hamsters). That it can remain so long in the gut is presumably due to an ability to brace itself by muscular action against the peristalsis and the intestinal current. In relation to this, it may be significant that this species has an additional band of circular muscles. The morphology of the mature female genitalia is shown in Fig. 22.12.

Plerocercoid. The plerocercoid (Fig. 22.9) has the main features of the adult, i.e. (*a*) division into proglottides (62–92 in number) and (*b*) the presence of genital anlagen in the organogeny state of development; testes are present, but spermatozoa are not formed. Unlike *Diphyllobothrium*, the plerocercoid is extremely specific in its host, developing *only* in the body cavity of the marine and fresh-water forms of the three-spined

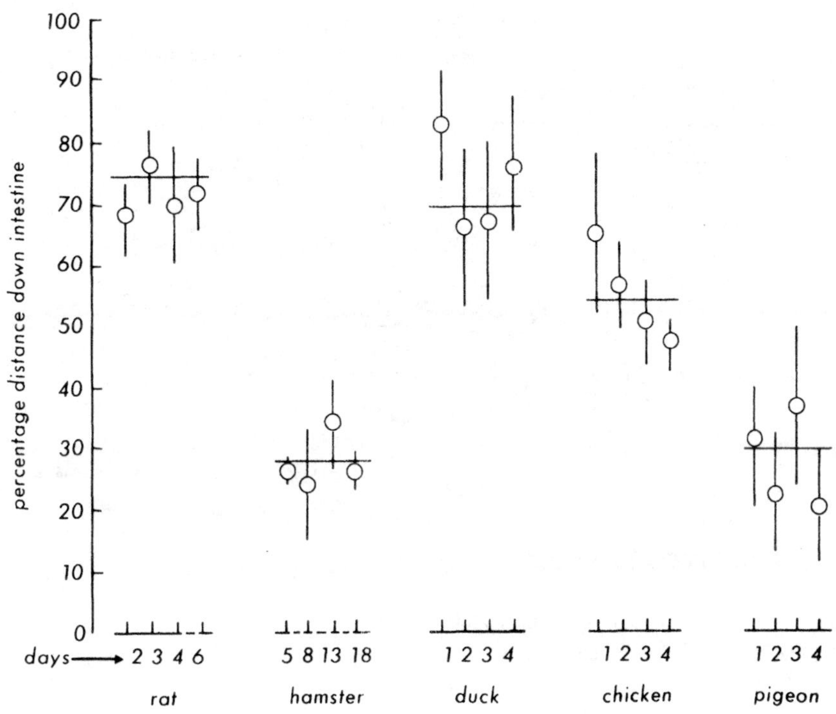

Fig. 22.10
Position of *Schistocephalus solidus* when established in different hosts. (After McCaig & Hopkins, 1963.)

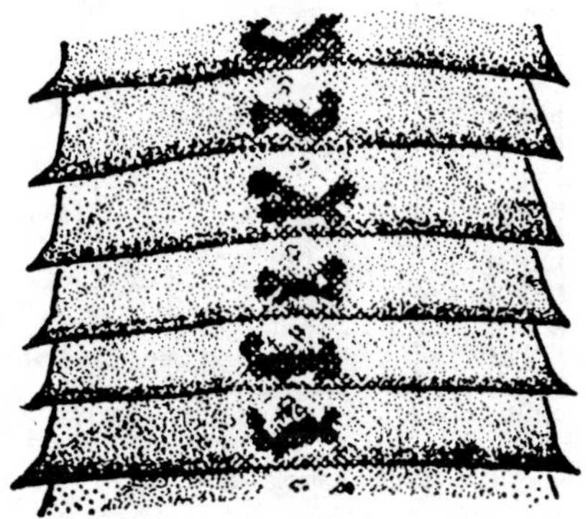

Fig. 22.11
Schistocephalus solidus. Morphology of adult, from specimens matured *in vitro* (see p. 509). (Original.)

stickleback, *Gasterosteus aculeatus*. In Britain, the nine-spined stickleback *Pungitius pungitius* has also been infected experimentally (Orr *et al.*, 1969) but growth was inhibited and the plerocercoids died within 10–14 days. In Canada and Russia, in contrast, *P. pungitius* acts as a suitable host for *Schistocephalus* (*solidus*?) (Dubinina, 1966; Curtis, 1981) and it may be that host and/or parasite strain differences are involved.

The majority of infected *G. aculeatus* contain only one to four large-sized plerocercoids (Figs 22.9, 22.13), but much higher numbers may occasionally occur, the total even exceeding the mass of the fish. Plerocercoids become infective in about 2 months (Orr & Hopkins, 1969*a*). The plerocercoids lie free in the body cavity from which they can, if required, be removed in a sterile condition and matured by *in vitro* culture, thus removing the necessity of having to use experimental animal hosts (Smyth, 1947, 1954); see p. 509.

Frequency distribution. In common with many other host–parasite systems (see Chapter 1 of Crofton, 1971), the distribution of *Schistocephalus* plerocercoids in sticklebacks was found to be overdispersed. In this case, the distribution did not closely fit the negative binomial, but was better described by a log normal (Fig. 22.13) (Pennycuick, 1971).

Pathogenic effects on fish. Although plerocercoid infections may be heavy, unlike *Ligula* (see below), parasitic castration of sticklebacks host does not occur, nor is the activity of the pituitary gland affected (Kerr, 1948). Nevertheless, some suppression of the growth and maturation of the gonads occurs, as the ovaries develop

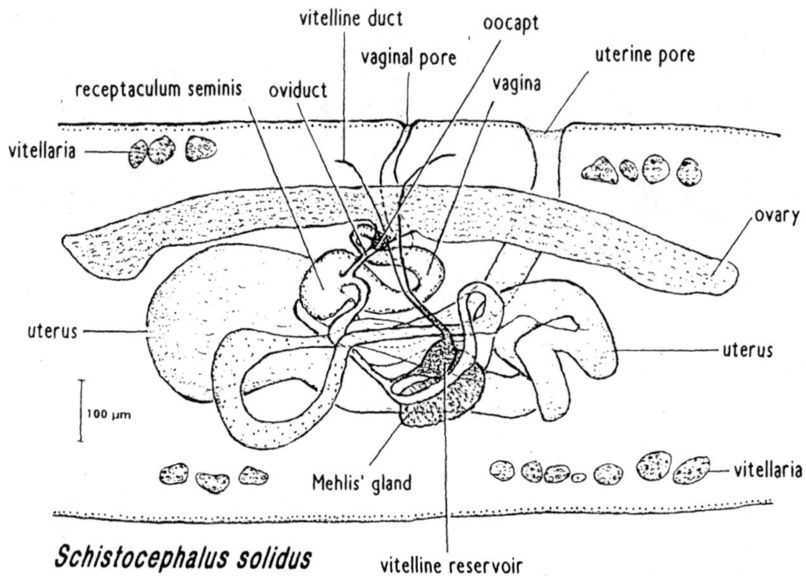

Fig. 22.12
Schistocephalus solidus: anatomy of female genitalia. From adult worm cultured to maturity *in vitro*. (After Hopkins & Smyth, 1951.)

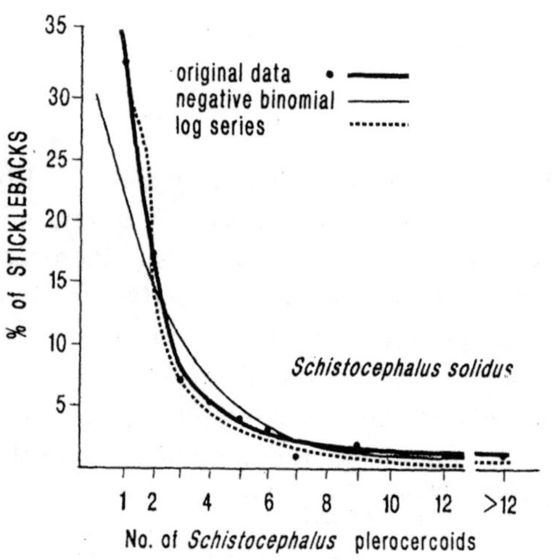

Fig. 22.13
Schistocephalus solidus: frequency distribution of plerocercoids in a sample of 3-spined sticklebacks (*Gasterosteus aculeatus*) collected on November 12, 1966, from Priddy Pond, Somerset, UK. In this sample, although overdispersed, the distribution can best be described by a log normal, rather than a negative binomial. The overdispersion was reduced because fish with many pleroceroids died, which resulted in a truncated distribution in some samples. (Adapted from Pennycuick, 1971.)

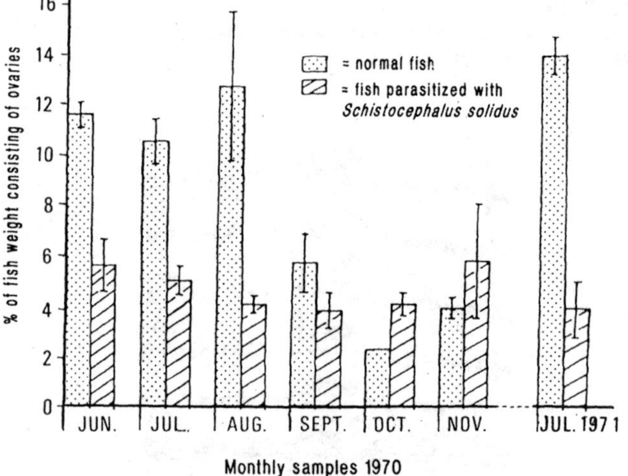

Fig. 22.14
Schistocephalus solidus: effect of plerocercoids on the seasonal development of the ovaries of the 3-spined stickleback (*Gasterosteus aculeatus*). Although the size of the ovaries are reduced and oocyte maturation is delayed, normal reproduction can take place and parasitic castration does not occur (contrast *Ligula*, Fig. 22.17). (After Meakins, 1974.)

less well in parasitised fish than in normal fish (Fig. 22.14) (Meakins, 1974).

Life cycle. The life cycle (Fig. 22.15) follows the typical cycle already described for *D. dendriticum* with corresponding differences in times for maturation in the definitive hosts as already stressed (Clarke, 1954; Hopkins & Smyth, 1951). The procercoid can develop in many copepod species but *Eucyclops* (= *Cyclops*?) *agilis* appears to be most suited for laboratory conditions (Orr & Hopkins, 1969*a*); with some species the copepodid stages are more readily infected than the adult copepod.

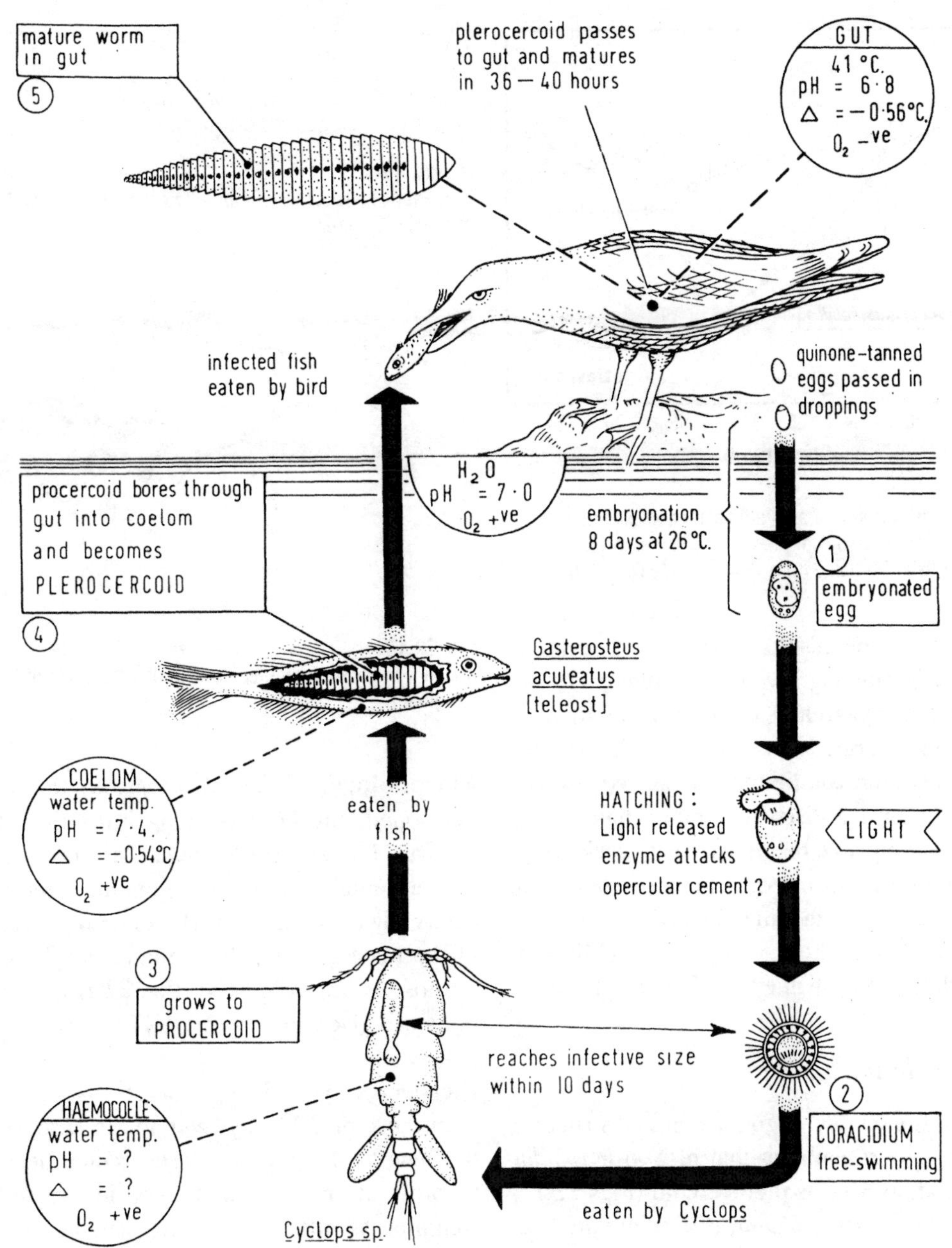

Fig. 22.15
Life cycle of *Schistocephalus solidus* and some physiological factors relating to it. (From Smyth, 1959.)

At 23–25 °C, cercomer formation takes place after 4–5 days, calcareous corpuscles appear on the seventh day and by the tenth day procercoids (which now show 'penetration glands') are infective to sticklebacks (Orr & Hopkins, 1969*a*).

Fish become naturally infected by ingesting infected copepods. Experimentally, they may be infected by introducing copepods via a stomach tube or by inoculation of procercoids into the body cavity.

When infected fish are eaten by birds, plerocercoids mature rapidly (within 36–48 hours) and release eggs. Maturation is also readily obtained *in vitro* (Chapter 34); the stimulus inducing maturation of genitalia appears to be only a rise in temperature, providing other suitable physiological conditions are maintained (Smyth, 1947, 1954, 1959).

Effect of *Schistocephalus* on fish behaviour. The stickleback–*Schistocephalus* system has been widely used as a model to study the effect of parasitism on fish

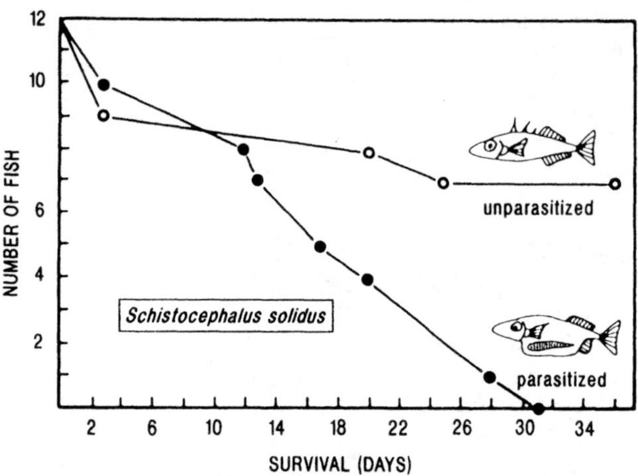

Fig. 22.16
Schistocephalus solidus: survival of starved 3-spined sticklebacks (*Gasterosteus aculeatus*) infected with plerocercoids compared with that of unparasitised fish. (Modified from Walkey & Meakins, 1970.)

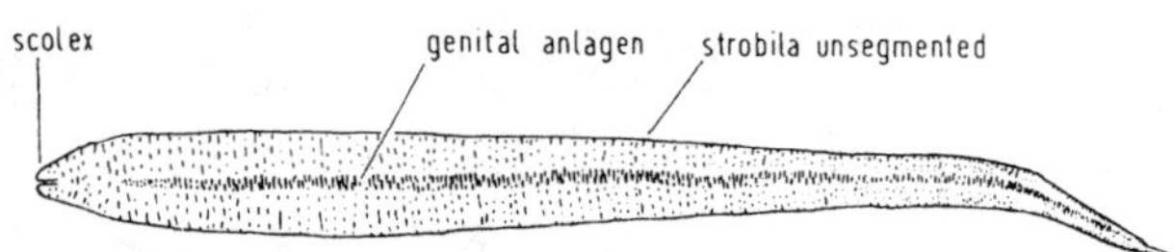

Fig. 22.17
Progenetic pleroceroid of *Ligula intestinalis* removed from body cavity of roach. (Original.)

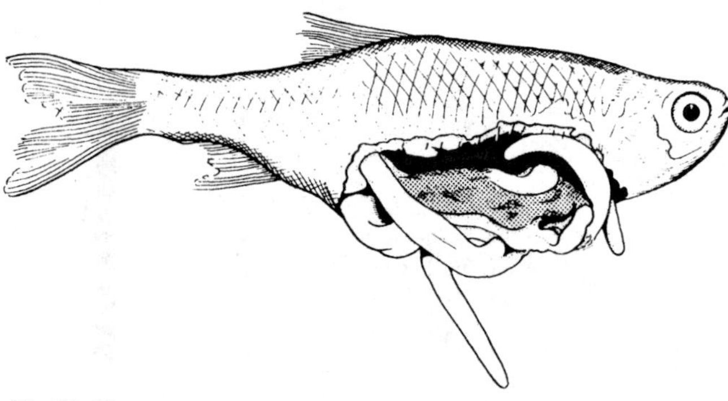

Fig. 22.18
Roach (*Rutilus rutilus*) infected with plerocercoids of *Ligula intestinalis*. (Original.)

behaviour; many of the results have been succinctly summarised in a fascinating review by Milinski (1990). The plerocercoid is apparently more efficient than its host in converting energy. However, as might be predicted, under starvation conditions, parasitised fish die more rapidly than unparasitised fish, more than 50 per cent of the latter being able to survive for 5 weeks (Fig. 22.16). Various other aspects of fish behaviour have been studied, such as feeding behaviour, oxygen depletion (hypoxia) and host competition (Milinski, 1984, 1990; Giles, 1987; Walkey & Meakins, 1970).

○ 22.3.2 *Ligula intestinalis*

This species of the family Ligulidae has a structure and life cycle closely resembling that of *Schistocephalus solidus*. It is best known for its plerocercoid (Figs 22.17, 22.18) which, unlike *Schistocephalus*, is unsegmented (as is the adult strobila) and which often reaches the length of one metre!

Occurrence. The presence of the plerocercoid in fish is best documented in Europe, Russia, Canada and the USA but it is also found in other countries, including Australia (Pollard, 1974) and New Zealand (Weekes & Penlington, 1986). Some 70 fish species have been reported as intermediate hosts, including cyprinids, catastomids and percids (Szalai *et al.*, 1989). The adult worm has been reported from numerous species of fish-eating birds, including herons, ducks, pelicans, terns and gulls.

Morphology. Like *Schistocephalus*, the scolex is poorly developed, the bothria being little more than shallow grooves. Unlike *Schistocephalus*, however, the adult strobila, although containing segmented genitalia, remains externally unsegmented when mature. The morphology of the male and female genitalia closely resembles that of those of *Schistocephalus* (Fig. 22.12); detailed descriptions are given in Smyth (1947, 1990*c*).

Plerocercoid. The plerocercoids are much larger than those of *Schistocephalus* and in large fish may reach the length of a metre. Plerocercoids become infective to birds after about 6 months in the fish and larvae weighing as little as 0.5 g are capable of maturation *in vitro* (Wyatt & Kennedy, 1988). The intensity of infection varies greatly in different ecological situations depending on the season, the populations of birds, copepods and fish. Fish typically have only small numbers of plerocercoids. For example, in Dauphin Lake, Canada, in which the major fish host is the spottail shiner (*Notropis hudsonius*), 88 per cent of infected fish had only a single larva, 11 per cent had two and 1 per cent had four (Szalai *et al.*, 1989). In older fish, however, infections as high as 50 have been reported (Arme & Owen, 1968).

In N America, a wide range of fish hosts have been reported. In Europe, the roach (*Rutilus rutilus*) is most commonly infected, but *L. intestinalis* has also been reported from gudgeon, brown trout and minnow. In the laboratory, tropical cyprinid fish such as giant danios (*Danio malabaricus*), Schuberti barbs (*Barbus sachsi*) and White Cloud Mountain minnows (*Tanichthys albonubes*) are suitable experimental hosts; plerocercoids develop genital anlagen after 60 days at 25 °C (Orr & Hopkins, 1969*b*).

Life cycle. As mentioned, the life cycle of *Ligula* parallels that of *Schistocephalus*, with adults in birds, procercoid stages in copepods and plerocercoids in fish. Procedures for the laboratory maintenance have been summarised by Orr & Hopkins (1969*b*). Ducklings serve as suitable definitive hosts, but ducklings older than 2–7 days or those which have been previously infected are not suitable (Flockart, 1978). *Diaptomus fragilis* and *Mesocyclops leuckarti* serve as suitable copepod hosts. Eggs can also be obtained by cutting large plerocercoids into small pieces (as for *in vitro* culture, see Fig. 34.7) and implanting them in the body cavity of mice; some 11–21 per cent produce fertile eggs but eventually become encapsulated (Flockart, 1978). Isoenzyme analysis suggests that cross-fertilisation as well as self-fertilisation may occur in *Ligula* (McManus, 1985).

Frequency distribution. The ecological situation in which *Ligula* infections in fish are found are so variable that the frequency distribution must generally be regarded as a random one. In most cases, the prevalence varies from year to year, and no statistically significant seasonal patterns can be recorded. Recent studies on distribution are those of Szalai *et al.* (1989) in N America and Wyatt & Kennedy (1988, 1989) and Bean & Winfield (1989) in Europe.

Pathogenic effect on fish. The plerocercoids of *Ligula* have a marked pathogenic effect on the fish host and suppress the growth of the gonads, resulting in *parasitic castration*. The ovary is reduced to the level found in a 'spent' fish with only oogonia and early oocytes present, whereas only germ cells and some spermatogonia are present in the testis. The effects are associated with the reduction in size and granulation of basophil cells in the middle glandular region of the pituitary gland (Arme, 1975; Kerr, 1948; Smyth & McManus, 1989), but the mechanism of this phenomenon is not understood. Pathogenic effects on the spleen and pronephros in roach have also been observed (Taylor & Hoole, 1989).

○○○ References

(References of general interest have been included in this list, in addition to those quoted in the text.)

Andersen, K. 1972. Studies of the helminth fauna of Norway XXIV: The morphology of *Diphyllobothrium ditremum* (Creplin, 1825) from the golden hamster (*Mesocrisetus auratus*) (Waterhouse, 1839) and a comparison with *D. dendriticum* (Nitzsch, 1824) and *D. latum* (L. 1758) from the same final host. *Norwegian Journal of Zoology*, **20**: 255–64.

Andersen, K. 1978. The development of the tapeworm *Diphyllobothrium latum* (L. 1756) (Cestoda: Pseudophyllidea) in its definitive hosts, with special reference to the growth patterns of *D. dendriticum* (Nitzsch, 1824) and *D. ditremum* (Creplin, 1827). *Parasitology*, **77**: 111–20.

Andersen, K., Ching, H. L. & Vik, R. 1987. A review of freshwater species of *Diphyllobothrium* with redescriptions and the distribution of *D. dendriticum* (Nitsch, 1824) and *D. ditremum* (Creplin, 1825) from North America. *Canadian Journal of Zoology*, **65**: 2216–28.

Andersen, K. & Gibson, D. I. 1989. A key to the three species of larval *Diphyllobothrium* Cobbald, 1858 (Cestoda: Pseudophyllidea) occurring in European and North American freshwater fish. *Systematic Parasitology*, **13**: 3–9.

Archer, D. M. & Hopkins, C. A. 1958. Studies on cestode metabolism. III. Growth pattern of *Diphyllobothrium* sp. in a definitive host. *Experimental Parasitology*, **7**: 125–44.

Arme, C. 1975. Tapeworm–interactions. *Symposia of the Society for Experimental Biology*, **29**: 505–32.

Arme, C. & Owen, R. W. 1968. Occurrence and pathology of *Ligula intestinalis* infections in British fishes. *Journal of Parasitology*, **54**: 272–80.

Bean, C. W. & Winfield, I. J. 1989. Biological and ecological effects of a *Ligula intestinalis* (L.) infestation of the gudgeon (*Gobio gobio*) (L.) in Lough Neagh, Northern Ireland. *Journal of Fish Biology*, **34**: 135–47.

Bell, E. J. & Smyth, J. D. 1958. Cytological and histochemical criteria for evaluating development of trematodes and pseudophyllidean cestodes *in vivo* and *in vitro*. *Parasitology*, **48**: 131–48.

Calentine, R. L. 1965. The biology and taxonomy of *Biacetabulum* (Cestoda: Caryophyllaeidae). *Journal of Parasitology*, **51**: 243–8.

Campbell, E. W. & Beals, C. 1977. Striking eosinophila in sparganosis. *Postgraduate Medicine*, **62**: 138–40.

Chubb, J. C., Pool, D. W. & Veltkamp, C. J. 1987. A key to the species of cestodes (tapeworms) parasitic in British and Irish freshwater fishes. *Journal of Fish Biology*, **31**: 517–43.

Clarke, A. S. 1954. Studies on the life cycle of the pseudophyllidean cestode *Schistocephalus solidus. Proceedings of the Zoological Society of London*, **124**: 257–302.

Crofton, H. D. 1971. A quantitative approach to parasitism. *Parasitology*, **62**: 179–93.

Curtis, M. A. 1981. Observations on the occurrence of *Diplostomum spathaceum* and *Schistocephalus* sp. in ninespine sticklebacks (*Pungitius pungitius*) from the Belcher Islands, Northwest Territories, Canada. *Journal of Wildlife Diseases*, **17**: 241–6.

Delyamure, S. L., Skryabin, A. & Serdyukova, A. M. 1985. (*Principles of Cestodology*. Vol. II: *Diphyllobothriids – Tapeworms of Man, Mammals and Birds*.) (In Russian.) Nauka, Moscow.

Dick, T. A. & Poole, B. C. 1985. Identification of *Diphyllobothrium dendriticum* and *Diphyllobothrium latum* from some freshwater fishes of central Canada. *Canadian Journal of Zoology*, **63**: 196–201.

Dubinina, M. N. 1966. *Cestoda: Ligulidae*. Publishing House of Science, Moscow and Leningrad.

Flockart, H. 1978. *Ligula intestinalis* (L. 1758) – *in vivo* and *in vitro* studies. Ph.D. Thesis, University of London. (Ref No: D.28674/79.)

Fukumoto, S., Yazaki, S., Kamo, H., Yamane, Y. & Tsuji, M. 1988. Distinction between *Diphyllobothrium nikonkaiense* and *Diphyllobothrium latum* by immunoelectrophoresis. *Japanese Journal of Parasitology*, **37**: 91–5.

Fukumoto, S., Yazaki, S., Nagai, D., Takechi, M., Kamo, H. & Yamane, Y. 1987. Comparative studies on soluble protein profiles and isoenzyme patterns in 3 related species of the genus *Diphyllobothrium. Japanese Journal of Parasitology*, **36**: 222–30.

Giles, N. 1987. A comparison of the behavioural responses of parasitized and non-parasitized three-spined sticklebacks, *Gasterosteus aculeatus* L., to progressive hypoxia. *Journal of Fish Biology*, **30**: 631–8.

Gustafsson, M. K. S. & Vaihela, B. 1981. Two types of frontal glands in *Diphyllobothrium dendriticum* (Cestoda, Pseudophyllidea) and their fate during the maturation of the worm. *Zeitschrift für Parasitenkunde*, **66**: 145–54.

Halvorsen, O. & Andersen, K. I. 1974. Some effects of population density in infections of *Diphyllobothrium dendriticum* (Nitzsch) in golden hamster (*Mesocricetus auratus* Waterhouse) and common gull (*Larus larus* L.). *Parasitology*, **69**: 149–60.

Halvorsen, O. & Wissler, K. 1973. Studies of the helminth fauna of Norway XXVIII: An experimental study of the ability of *Diphyllobothrium latum* (L.), *D. dendriticum* (Nitsch) and *D. ditremum* (Creplin) (Cestoda, Pseudophyllidea) to infect paratenic hosts. *Norwegian Journal of Zoology*, **21**: 201–10.

Hendricson, J. 1977. The abundance and distribution of *D. dendriticum* (Nitsch) and *D. ditremum* (Creplin) in the char *Salvelinus alpinus* (L.) in Sweden. *Journal of Fish Biology*, **11**: 231–48.

Hirai, K., Nishida, H., Shiwaku, K. & Okuda, H. 1978. Studies of the plerocercoid growth factor of *Spirometra erinacei* (Rudolphi, 1819) with special reference to the effect of lipid mobilization *in vitro. Japanese Journal of Parasitology*, **27**: 527–33.

Hopkins, C. A. & Smyth, J. D. 1951. Notes on the morphology and life history of *Schistocephalus solidus* (Cestoda: Diphyllobothriidae). *Parasitology*, **41**: 283–91.

Huang, C. T. & Kirk, R. 1962. Human sparganosis in Hong Kong. *Journal of Tropical Medicine and Hygiene*, **65**: 133–8.

Kennedy, C. R. 1965*a*. The life history of *Archigetes limnodrili* (Yamaguti) (Cestoda: Caryophyllaeidae) and its development in the invertebrate host. *Parasitology*, **55**: 427–37.

Kennedy, C. R. 1965*b*. Taxonomic studies on *Archigetes* Leuckart, 1878 (Cestoda: Caryophyllaeidae). *Parasitology*, **55**: 439–51.

Kennedy, C. R. & Walker, P. J. 1969. Evidence for an immune response by dace, *Leuciscus leuciscus*, to infections by the cestode *Caryophyllaeus laticeps. Journal of Parasitology*, **55**: 579–82.

Kerr, T. 1948. The pituitary in normal and parasitized roach (*Leuciscus rutilus* Flem). *Quarterly Journal of Microscopical Science*, **89**: 129–37.

Kühlow, F. 1953. Über die Entwicklung und Anatomie von *Diphyllobothrium dendriticum* Nitzsch 1824. *Zeitschrift für Parasitenkunde*, **16**: 1–35.

Kuperman, B. I. & Davydov, V. G. 1982. The fine structure of glands in oncospheres, procercoids and plerocercoids of Pseudophyllidea (Cestoidea). *International Journal for Parasitology*, **12**: 135–44.

Kwa, B. H. 1972*a*. Studies on the sparganum of *Spirometra erinacei* I. The histology and histochemistry of the scolex. *International Journal for Parasitology*, **2**: 23–8.

Kwa, B. H. 1972*b*. Studies on the sparganum of *Spirometra erinacei* II. Proteolytic enzyme(s) in the scolex. *International Journal for Parasitology*, **2**: 29–34.

Kwa, B. H. 1972*c*. Studies on the sparganum of *Spirometra erinacei*. III. The fine structure of the tegument of the scolex. *International Journal for Parasitology*, **2**: 35–43.

McCaig, M. L. O. & Hopkins, C. A. 1963. Studies on *Schistocephalus solidus*. II. Establishment and longevity in the definitive host. *Experimental Parasitology*, **13**: 273–83.

Mackiewicz, J. S. 1972. Caryophyllidea (Cestoidea): a review. *Experimental Parasitology*, **31**: 417–512.

Mackiewicz, J. S. & Blair, D. 1978. Balanotaeniidae fam. n. and *Balanotaenia newquinensis* sp.n. (Cestoidea; Caryophyllidea) from Tandus (Siluriformes; Plotosidae) in New Guinea. *Journal of Helminthology*, **52**: 199–203.

McManus, D. P. 1985. Enzyme analyses of natural populations of *Schistocephalus solidus* and *Ligula intestinalis. Journal of Helminthology*, **59**: 323–32.

Marchiondo, A. A., Weinstein, P. P. & Mueller, J. F. 1989. Significance of the distribution of ^{57}Co-vitamin B_{12} in *Spirometra mansonoides* (Cestoidea) during growth and differentiation in mammalian intermediate and definitive hosts. *International Journal for Parasitology*, **19**: 119–24.

Meakins, R. H. 1974. A quantitative approach to the effects of the plerocercoid of *Schistocephalus solidus* Muller 1776 on the ovarian maturation of the three-spined stickleback *Gasterosteus aculeatus. Zeitschrift für Parasitenkunde*, **44**: 73–9.

Milinski, M. 1984. Parasites determine a predator's optimal feeding strategy. *Behavioural Ecology and Sociobiology*, **15**: 35–7.

Milinski, M. 1990. Parasites and host decision-making. In: *Parasitism and Host Behaviour* (ed. C. J. Barnard & J. M. Behnke), pp. 95–116. Taylor & Francis, London.

Mueller, J. F. 1974. The biology of *Spirometra*. *Journal of Parasitology*, **60**: 3–14.

Odening, K. 1979. Zum Erforschungsstand des 'sparganum growth factor' von *Spirometra*. *Angewandte Parasitologie*, **20**: 185–92.

Orr, T. S. C. & Hopkins, C. A. 1969*a*. Maintenance of *Schistocephalus solidus* in the laboratory with observations on the rate of growth of, and proglottid formation in, the plerocercoid. *Journal of the Fisheries Research Board of Canada*, **26**: 741–52.

Orr, T. S. C. & Hopkins, C. A. 1969*b*. Maintenance of the life cycle of *Ligula intestinalis* in the laboratory. *Journal of the Fisheries Research Board of Canada*, **26**: 2250–1.

Orr, T. S. C., Hopkins, C. A. & Charles, G. H. 1969. Host specificity and rejection of *Schistocephalus solidus*. *Parasitology*, **59**: 683–90.

Pawlowski, Z. S. 1984. Cestodiases: taeniasis, diphyllobothriasis, hymenolepiasis, and others. In: *Tropical and Geographical Medicine* (ed. R. S. Warren & A. A. F. Mahmoud), pp. 471–86. McGraw-Hill, New York.

Pennycuick, L. 1971. Frequency distribution of parasites in a population of three-spined sticklebacks, *Gasterosteus aculeatus* L., with particular reference to the negative binomial distribution. *Parasitology*, **63**: 389–406.

Phares, C. K. 1987. Plerocercoid growth factor: a homologue of human growth hormone. *Parasitology Today*, **3**: 346–9.

Pollard, D. A. 1974. The biology of a landlocked form of the normally catadromous salmoniform fish *Galaxias maculatus* (Jenyns) VI. Effects of cestodes and nematode parasites. *Australian Journal of Marine and Freshwater Research*, **25**: 105–20.

Sergeeva, E. G., Savina, N. A., Shevchenko, S. F., Freze, V. I. & Dalin, M. V. 1987. On methods of obtaining diphyllobothriid surface antigens. *Angewandte Parasitologie*, **28**: 73–9.

Sharp, G. J. E., Secombes, C. J. & Pike, A. W. 1990. The laboratory maintenance of *Diphyllobothrium dendriticum*. *Parasitology*, **101**: 153–61.

Shiwaku, K., Hirai, K., Torii, M. & Tsuboi, T. 1983. Effects of *Spirometra erinacei* plerocercoids on the growth of Snell dwarf mice. *Parasitology*, **87**: 447–53.

Smyth, J. D. 1947. Studies on tapeworm physiology. II. Cultivation and development of *Ligula intestinalis in vitro*. *Parasitology*, **38**: 173–81.

Smyth, J. D. 1949. Studies on tapeworm physiology. IV. Further observations on the development of *Ligula intestinalis in vitro*. *Journal of Experimental Biology*, **24**: 374–86.

Smyth, J. D. 1954. Studies on tapeworm physiology. VII. Fertilization of *Schistocephalus solidus in vitro*. *Experimental Parasitology*, **3**: 64–71.

Smyth, J. D. 1959. Maturation of larval pseudophyllidean cestodes and strigeid trematodes under axenic conditions; the significance of nutritional levels in platyhelminth development. *Annals of the New York Academy of Sciences*, **77**: 102–25.

Smyth, J. D. 1969. *The Physiology of Cestodes*. 1st Edition. W. H. Freeman, San Francisco.

Smyth, J. D. 1990*a*. Parasitological serendipity: from *Schistocephalus* to *Echinococcus*. *International Journal for Parasitology*, **20**: 411–23.

Smyth, J. D. 1990*b*. Peter Abildgaard – forgotten pioneer of parasitology. *Parasitology Today*, **6**: 337–9.

Smyth, J. D. 1990*c*. *In vitro Cultivation of Parasitic Helminths*. CRC Press, Boca Raton, Florida.

Smyth, J. D. & Heath, D. D. 1970. Pathology of larval cestodes in mammals. *Helminthological Abstracts*, **39**: 1–23.

Smyth, J. D. & McManus, D. P. 1989. *The Physiology and Biochemistry of Cestodes*. Cambridge University Press.

Soutar, R. 1990. Veterinary involvement in smolt-producing farms. *Irish Veterinary News*, **12**: 32–6.

Szalai, A. J., Yang, X. & Dick, T. A. 1989. Changes in numbers and growth of *Ligula intestinalis* in the spotted shiner (*Notropis hudsonius*), and their roles in transmission. *Journal of Parasitology*, **75**: 571–6.

Taylor, M. & Hoole, D. 1989. *Ligula intestinalis* (L.) (Cestoda: Pseudophyllidea): plerocercoid-induced changes in the spleen and pronephros of roach, *Rutilus rutilus* (L.) and *Gobio gobio* (L.). *Journal of Fish Biology*, **34**: 583–96.

Tkachuck, R. D., Weinstein, P. P. & Mueller, J. F. 1976. Comparison of the uptake of vitamin B_{12} by *Spirometra mansonoides* and *Hymenolepis diminuta* and the functional groups of B_{12} analogues affecting uptake. *Journal of Parasitology*, **62**: 94–101.

Von Bonsdorff, B. 1977. *Diphyllobothriasis in Man*. Academic Press, London.

Vos, T. de, Szalai, A. J. & Dick, T. A. 1990. Genetic and morphological variability in a population of *Diphyllobothrium dendriticum* (Nitsch, 1824). *Systematic Parasitology*, **16**: 99–105.

Walkey, M. & Meakins, R. H. 1970. An attempt to balance the energy budget of a host–parasite system. *Journal of Fish Biology*, **3**: 361–72.

Wardle, R. A. & Green, N. K. 1941. The rate of growth of the tapeworm *Diphyllobothrium latum*. *Canadian Journal of Research*, D **19**: 245–57.

Weekes, P. J. & Penlington, B. 1986. First records of *Ligula intestinalis* (Cestoda) in rainbow trout, *Salmo gairdneri*, and common bully, *Gobiomorphus cotidianus*, in New Zealand. *Journal of Fish Biology*, **28**: 183–90.

Wikgren, B.-J. P. 1963. A contribution to the occurrence of plerocercoids of *Diphyllobothrium latum* in Finland. *Societas Scientiarum Fennica Commentationes Biologicae*, **26**: 1–11.

Wikgren, B.-J. P., Knuts, G. M. & Gustafsson, M. K. S. 1970. Circadian rhythm of mitotic activity in the adult gull-tapeworm *Diphyllobothrium dendriticum* (Cestoda). *Zeitschrift für Parasitenkunde*, **34**: 242–50.

Williams, D. D. 1978. A key to the caryophyllidean cestodes of Iowa fishes. *Iowa State Journal of Research*, **52**: 401–9.

Wisniewski, L. W. 1930. Das Genus *Archigetes* R. Leuck.

Eine Studie zur Anatomie, Histogenase, Systematik und Biologie. *Mémoires Académie Polonaise des Sciences et des Lettres Classe des Sciences Mathematiques et Naturelles*, Série B, **2**: 1–168.

Witenberg, G. G. 1964. Cestodiases. In: *Zooparasitic Diseases* (ed. J. van der Haeden), pp. 648–707. Elsevier, Amsterdam.

Wyatt, R. J. & Kennedy, C. R. 1988. The effects of a change in the growth rate of roach, *Rutilus rutilus* (L.), on the biology of the fish tapeworm *Ligula intestinalis* (L.). *Journal of Fish Biology*, **33**: 45–57.

Wyatt, R. J. & Kennedy, C. R. 1989. Host-constrained epidemiology of the fish tapeworm *Ligula intestinalis* (L.). *Journal of Fish Biology*, *35:* 215–27.

Yamane, Y., Bylund, G., Abe, K., Fukumoto, S. & Yazaki, S. 1988. Early development of four *Diphyllobothrium* species in the final host. *Parasitological Research*, **74**: 463–8.

○○○○○○○○○○○○○○○○○○○ 23 ○○○○○○○○○○○○○○○○○○○

Eucestoda: Cyclophyllidea

The Cyclophyllidea are mainly parasites of birds and mammals, rarely occurring in reptiles and amphibians. They are characterised by possessing four well-formed hemispherical acetabula symmetrically placed around the small, rounded scolex, which is typically armed with hooks.

They differ from the pseudophyllids in that (*a*) the vitellarium is median, single and not follicular, (*b*) only one vitelline cell takes part in the formation of an egg, (*c*) intermediate hosts are usually terrestrial animals and rarely aquatic, (*d*) only the most posterior proglottides are mature and (*e*) they have an apolytic (p. 282) strobila.

This order contains many species of medical, veterinary or biological importance. Probably the most worked on and most easily available cestodes are the rodent species *Hymenolepis diminuta*, *H. microstoma* and *H. nana*, which are considered below.

○○○ 23.1 Type example: *Hymenolepis diminuta*

definitive hosts: rats, other rodentia. Occasionally found in other mammals including man
location: intestine (but see below)
intermediate host: some thirty species of insects or their larvae. Commonly *Ceratophyllus fasciatus*, *Ctenocephalides canis*, *Tenebrio molitor* (larva), *Tribolium confusum* (larva)
distribution: cosmopolitan

Literature. A vast literature exists on this species, much of which has been comprehensively reviewed in a valuable monograph by Arai (1980), dealing with the life cycle, morphology, ultrastructure, migratory activity, developmental biology, biochemistry and *in vitro* culture. The biochemistry and physiology have also been reviewed by Smyth & McManus (1989), the early embryology by Coil (1986) and the larval structure and ultrastructure by Ubelaker (1980).

Speciation. Speciation in *H. diminuta* is a complex issue and it is now recognised that workers in different countries may be working with different 'strains' or 'varieties'. This is not surprising, for many years of passage through different strains or species of rats and intermediate hosts is likely to lead to different selection pressures and intra-specific variations could result. There is now unequivocal evidence that fundamental differences in metabolism exist between different isolates and that there are at least 3–4 'sorts' of *H. diminuta* (Bryant & Flockart, 1987; Kohlhagen *et al.*, 1985). Differences in morphology, infectivity, growth and fertility have also been recorded (Kino & Kennedy, 1987; Pappas & Leiby, 1986).

Many of the specimens used in laboratories throughout the world came from an original isolate (the TEXAS strain) obtained from *Rattus norvegicus* in Houston, Texas, and maintained in Sprague–Dawley rats and the flour beetle, *Tenebrio molitor*, ever since. Other, more recent, isolates have been obtained from wild rats in Australia (ANU strain) (Andrews *et al.*, 1989) and Japan (JAPAN strain) (Kino & Kennedy, 1987); other 'recognised' strains are those from Canada (TOR strain) and N America (OHIO strain). Using alloyzme electrophoresis, Andrews *et al.* (1989) concluded that, after examining the ANU, TOR and OHIO strains, there was no evidence to suggest that they represented more than one species, although all these strains were distinguishable on electrophoretic profiles.

Adult morphology (Fig. 23.1). The scolex bears four deep acetabula and an *unarmed* (contrast *H. nana*) pyriform rostellum which can be withdrawn into a rostellar

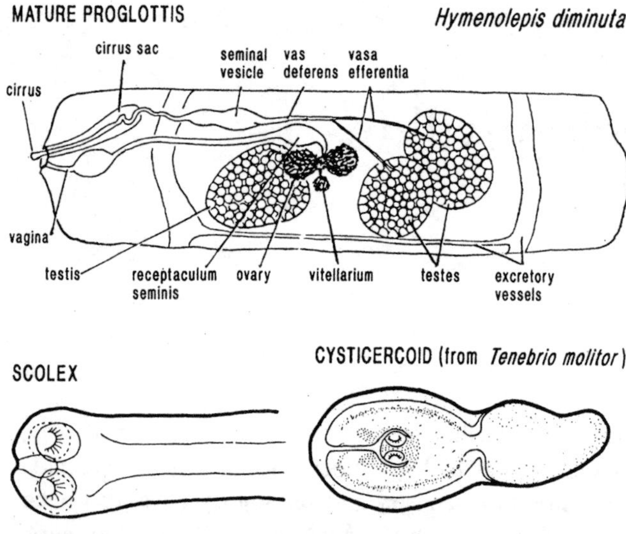

Fig. 23.1
Hymenolepis diminuta: morphology of proglottis, scolex and cysticercoid. (After various authors.)

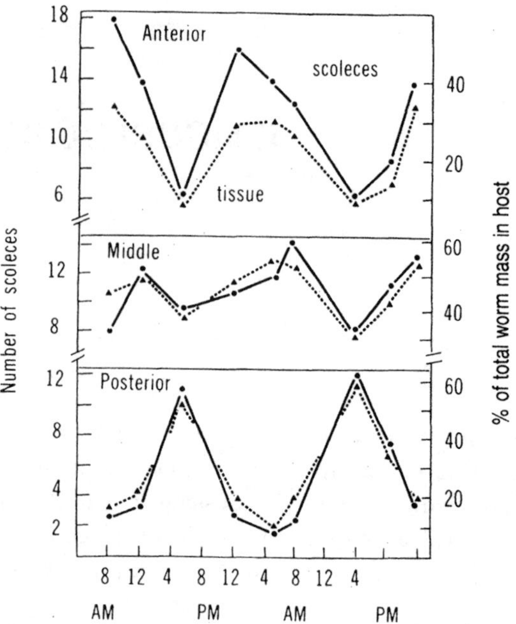

Fig. 23.2
The distribution of scoleces and of *Hymenolepis diminuta* tissue in the rat intestine at various times of the day. 'Anterior' refers to the first ten inches (25.4 cm), 'Middle' to the second ten inches and 'Posterior' to the remainder of the small intestine. Each point is a mean of determinations from four host animals, representing 110–120 worms. (After Read & Kilejian, 1969.)

sac. The strobila measures 20–60 μm with about 1000 proglottides, each of which is wider than long. The most striking feature, which is characteristic of the genus, is the possession of three ovoid testes, the remaining genitalia possessing no unusual features.

Migratory behaviour. Adult *H. diminuta* exhibit a complex pattern of migratory behaviour within the intestine (Arai, 1980). Descriptions of this have been somewhat conflicting, a situation which may be related to the fact that different groups of workers have used different worm burdens within the range 1 to 100 per rat. Thus, using single-worm infections Braten & Hopkins (1969) found that worms attached initially 30–40 per cent of the distance down the small intestine, but from day 7 to 14 migrated forwards into the 10–20 per cent region. Using 10-worm infections, Cannon & Mettrick (1970) found that 5-day worms attached in the 39 per cent region, 7-day worms attached in the 15 per cent region but that after 7 days there was a gradual spreading of scolex attachment sites, which could occur anywhere in the anterior 75 per cent region. Over the same period the mean mass distribution (i.e. the worm biomass) moved from a point 44 to 23 per cent of the intestinal length. The situation, however, is further complicated by the fact that a marked daily (circadian) migratory pattern has also been shown to occur, a result which may make some earlier work open to re-examination (Read & Kilejian, 1969). Thus, the maximum number of scoleces and worm tissue were found to occur in the anterior 250 mm of the intestine during the period 24.00 to 04.00 hours and the minimum number at 16.00–17.00 hours (Fig. 23.2). Associated inverse worm distribution changes occurred in the posterior part of the small intestine.

That this diurnal migratory behaviour was related to feeding behaviour can readily be demonstrated by the fact that when the feeding time of the rats was changed from day to night the migratory pattern was reversed (Fig. 23.3) (Read & Kilejian, 1969). The stimulus for migration is not known but it has been suggested that cestodes appear to be able to detect a specific region and recognise when they are not in this region (Braten & Hopkins, 1969). It has also been posulated that the anterior migration enables a worm to lie with its body in the region of the intestine most favourable for absorption (Crompton & Whitfield, 1969).

Life cycle. The eggs, which retain the outer capsule (p. 291), are fairly resistant to both chemicals and desiccation and remain viable in the faeces for some six months. Although a large number of insects can act as

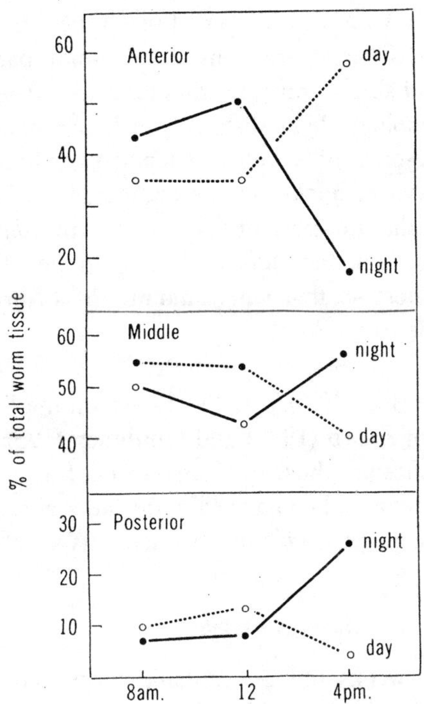

Fig. 23.3
The effects of day versus night feeding on the distribution of worm tissue in the small intestine. Each point is the mean from four rats. (After Read & Kilejian, 1969.)

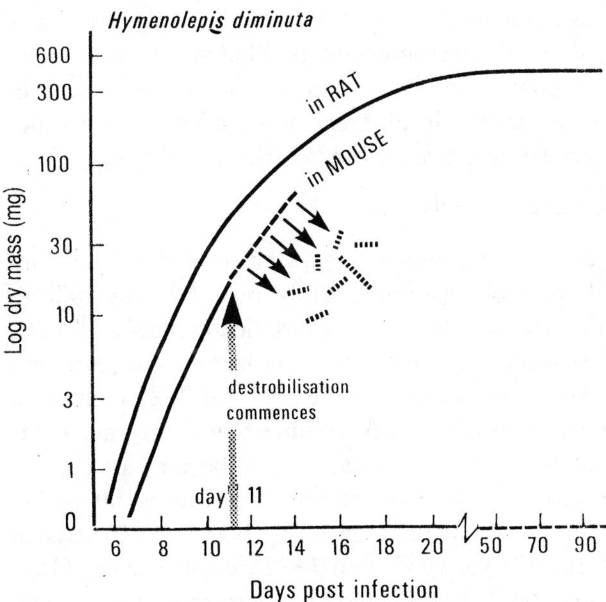

Fig. 23.4
Hymenolepis diminuta: comparative development in Sprague–Dawley rats (a natural definitive host) and SPF-CFLP mice (a non-permissive host). As in most strains of mice, growth ceases on day 10–11 and destrobilation occurs, leaving only a scolex and a very short neck. (Modified from Hopkins *et al.*, 1972.)

intermediate hosts, the flour beetle *Tribolium confusum* serves as a convenient laboratory host. Starved beetles rapidly ingest eggs and the hatched oncosphere penetrates the gut and develops into a *cysticercoid* in the haemocoele. A detailed morphological description of the cysticercoid (Fig. 23.1) has been given by Voge & Heyneman (1957); the ultrastructure has been reviewed by Ubelaker (1980). The optimum temperature for development in *Tenebrio molitor* is 30 °C, an infective cysticercoid developing in about 180 hours.

Development in definitive host. *H. diminuta* 'normally' develops in albino rats; it will also develop in certain strains of mice and hamsters, hooded rats and man and has been reported from a 3-year-old child in the USA (Levi *et al.*, 1987). In most varieties of laboratory mice, although worms are able to establish themselves initially, they are rejected on or about day 11 post infection, when destrobilisation commences (Fig. 23.4), and do not reach maturity. This phenomenon, which appears to be largely the result of an immune reaction, has been much investigated (Andreassen, 1981; Wakelin, 1984) (see Chapter 32); increase in histamine and mucus may be involved (McKay *et al.*, 1990). The prepatent period in these hosts varies between 13 and 21 days. Development is markedly affected by the size of the infection, the diet of the host – especially the quantity of carbohydrate present – and the presence or absence of other parasites, e.g. acanthocephala (Smyth & McManus, 1989).

Growth of *H. diminuta* is most rapid during the first seven days and thereafter the effect of population density (i.e. the 'crowding effect') becomes particularly apparent. It is especially interesting to note that the onset of maturation does not appear to be affected by population size although the size and mass of worms are markedly affected (Roberts, 1961).

Development in intermediate host. Eggs of *H. diminuta* can hatch, on ingestion, in a number of insect intermediate hosts in which they develop into *cysticercoids* (Fig. 23.1). The most commonly used laboratory hosts are the flour beetles *Tribolium confusum* and *Tenebrio molitor*, but numerous insect species have been reported as hosts (Arai, 1980). Eggs kept at room temperature retain their infectivity for up to 60 days (Keymer & Anderson, 1979) and eggs may be separated from the faeces by various gradient methods such as those of Lethbridge (1971) or Pappas & Leiby (1986).

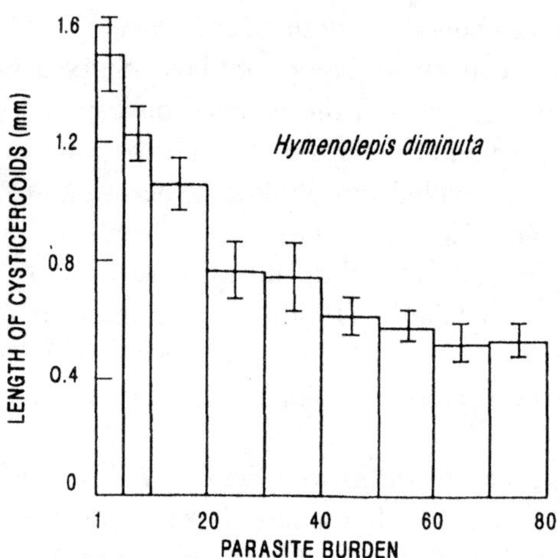

Fig. 23.5
Hymenolepis diminuta: growth of cysticercoid in the confused flour beetle *Tribolium confusum*. Relationship between parasite burden and cysticercoid length. The histogram bars are observed mean values, and the vertical bars represent the 95 per cent confidence limits of the means. (After Keymer, 1981.)

Beetles used for infection should be about 20–24 days old, starved for 6 days prior to infection and maintained at about 30 °C with 70 per cent humidity. At this temperature, larvae develop to infectivity, in *Tribolium confusum*, in about 8–9 days. Starved beetles rapidly ingest the eggs, hatching oncospheres rapidly penetrating the gut wall and developing in the haemocoele. The size of the cysticercoids is markedly density-dependent, large cysticercoids being obtained at low infections (Fig. 23.5); rather surprisingly, no density-dependent constraints were found on larval parasite establishment with parasite burdens of up to 60 larvae per beetle (Keymer, 1981).

○○○ 23.2 Family Hymenolepididae

The diagnostic features of this large family are: scolex armed with one circlet of five hooks; 1–3 large testes and sacciform uterus. Its members occur primarily in birds, but also in mammals. The genus *Hymenolepis* with three testes has over 400 species (Yamaguti, 1959). In addition to *H. diminuta* described above, three other species, *H. nana*, *H. microstoma* and *H. citelli*, have been much used for physiological and ultrastructural studies on cestodes.

○ 23.2.1 *Hymenolepis* (= *Vampirolepis*) *nana*

Some workers believe, for historical reasons, that this species should be renamed *Vampirolepis nana* but this designation is not in widespread use. Often referred to as the 'dwarf' tapeworm, this is a common parasite of rodents and also, of man, in whom it causes *hymenolepiasis*. Morphologically, it differs from *H. diminuta* chiefly by the presence of a well-developed rostellum armed with a crown of hooks. Its life cycle is also of unusual interest in that the larva of this species can *either* utilise an insect intermediate host *or* develop in the villi of the definitive host so that what amounts to a 'direct' life cycle results (Fig. 23.6).

Life cycle (Fig. 23.6). Early studies on the life cycle are those of Shorb (1933) and Hunninen (1935); more recent studies are those of Kumazawa & Suzuki (1983) and Henderson & Hanna (1987), the latter authors providing valuable protocols for the maintenance of the life cycle.

Adult development from cysticercoids

Mice 5–6 weeks old are suitable hosts and can be infected with cysticercoids (after light anaesthesia) by means of a stomach tube attached to a syringe.

The cestodes establish themselves initially in the anterior small intestine but subsequently migrate to the lower ileum, where they remain for the duration of the infection (Henderson & Hanna, 1987). Length measurements (Fig. 23.7) show exponential growth during days 1–7 post infection (p.i.) but slower, more constant growth, in days 8–15 p.i. Some differences in growth patterns in different hosts (mice, hamsters and squirrels) have been recorded (Schiller, 1959).

Adult development from eggs

In a direct egg infection, eggs hatch in the duodenum and oncospheres immediately penetrate the villi in which they develop into fully formed cysticercoids (Fig. 23.8) within 93–96 hours. The highest concentration of cysticercoids occurs in the region 100–200 mm from the pyloric sphincter. At about 102 hours cysticercoids break out of the villi, evaginate and become attached in the posterior third of the gut. The pre-patent period varies somewhat between hosts, being 11–16 days in the rat (Shorb, 1933) and 14–25 days in mice (Hunninen, 1935). In rats, maximum egg production occurs between 14 and 16 days.

The life span is short, 12–24 days in rats and 2–56 days in mice. From the experimental point of view it is important to note that the 'take' of eggs (i.e. the percentage becoming established) is low, 1–13 per cent (mean 4 per cent)

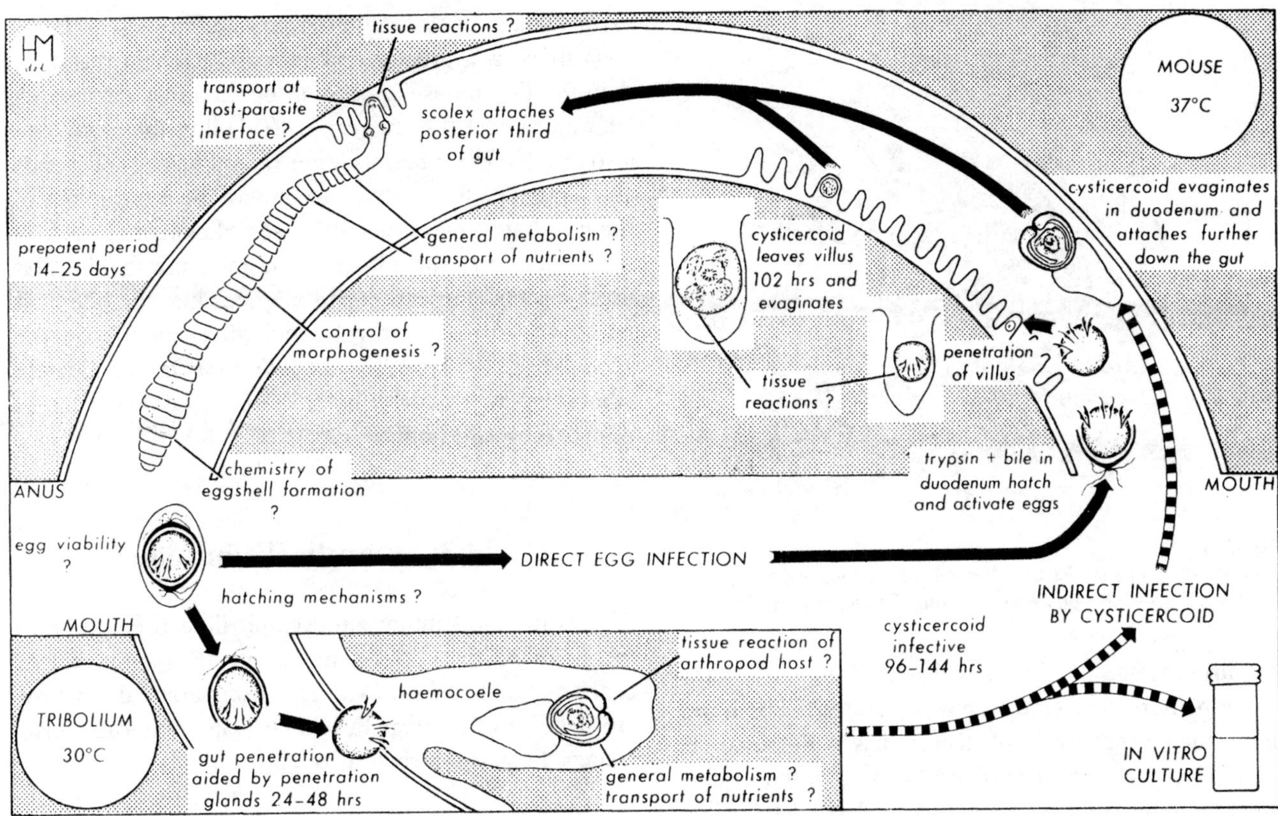

Fig. 23.6
Life cycle of *Hymenolepis nana* in the mouse. (Based on Smyth, 1969*b*.)

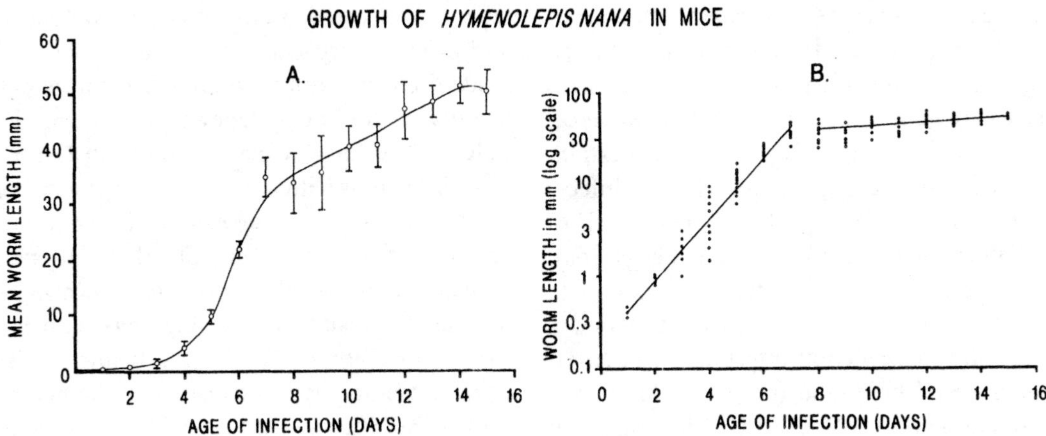

Fig. 23.7
A. Growth of *Hyemnolepis nana* in mice during the first 15 days of infection. **B.** Log_{10} of length of *H. nana* recovered at daily intervals over the same period. (After Henderson & Hanna, 1987.)

developing into cysticercoids and 1–8 per cent (mean 3 per cent) developing into adults (Hunninen, 1935).

For maintenance of the 'indirect' life cycle, *Tribolium confusum* is a suitable laboratory host. Development can also occur in a number of other insects, e.g. *Tenebrio molitor*, *Pulex irritans*, *Xenopsylla cheopis* and *Ctenocephalides canis*.

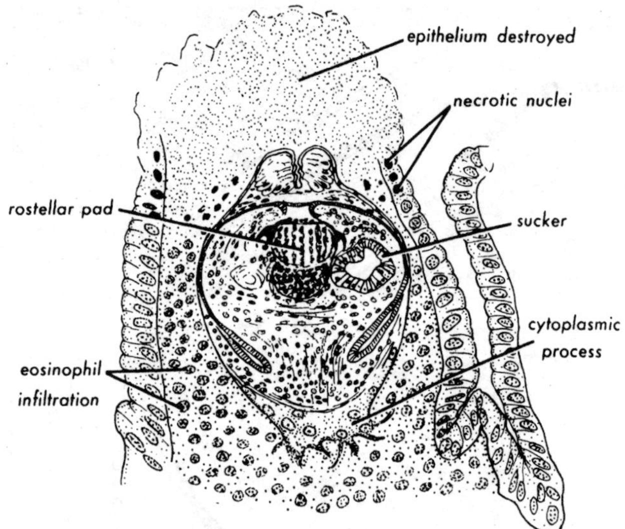

Fig. 23.8
Vertical section of a cysticercoid of *Hymenolepis nana* in the intestinal villus of a mouse; 96 h post-infection. (After Smyth, 1969*b*.)

***H. nana* in man.** *H. nana* is also a cosmopolitan parasite of man, being found chiefly in children. The infection is the most common of all cestodiases. For example, up to 16 per cent *prevalence* has been recorded in children in some rural communities in Australia (Meloni *et al.*, 1988), but much higher figures have been reported elsewhere. The clinical symptoms are restlessness, irritability and diarrhoea. For a review of human hymenolepiasis see Pawlowski (1984). Perhaps the most interesting problem in the epidemiology of *H. nana* is the role of animals in the spread of human infection. Two subspecies are generally recognised, *H. nana nana* in man and *H. nana fraterna* in rodents, but the question is very controversial and feeding eggs from human infections to mice has provided conflicting results. For example, Al-Baldawi *et al.* (1989), in Iraq, fed eggs from man to mice but failed to infect them. In contrast, Ferretti *et al.* (1981), in Sardinia, fed eggs from a woman patient to mice and succeeded in infecting them. After five passages in mice, the *q* ratio (no. of parasites : no. of eggs used) was low (0.14) but after 10 passages it was practically the same as in two murine strains (0.40, 0.48). They concluded that *H. nana nana* and *H. nana fraterna* are synonomous. Local strain differences in both parasites and experimental animals may account for these contrasting results.

○ 23.2.2 *Hymenolepis microstoma*

This is an unusual species in that it occupies an extra-intestinal site – namely the common bile duct and the extra-hepatic ducts – although in some hosts (e.g. hamster) there is a greater tendency to attach in the duodenum. The mouse appears to be the most satisfactory laboratory host (Dvorak *et al.*, 1961), but the worm can also develop in numerous other rodent hosts such as rats and hamsters. For a useful account of the morphology of the adult and the cysticercoid, see Hickman (1964).

In mice, the worm lives in the intestine for the first three days post infection and migrates into the bile duct on the fourth day; gravid proglottides appear on the 16th day (de Rycke, 1966). *Tribolium confusum* is a useful laboratory intermediate host; the cysticercoid becomes infective by 180 hours.

○○○ 23.3 Family Taeniidae

This is the most important cyclophyllidean family from the point of view of containing species of medical or economic importance. With rare exceptions, the worms are large and the larvae develop in homoiothermic hosts (especially carnivores and rodents).

○ 23.3.1 Genus *Taenia*: taeniasis and cysticercosis in man

Man acts as the definitive host for two species of this family, *Taenia saginata* – the 'beef' tapeworm – which is acquired by eating undercooked 'measly' beef, and *T. solium* – the 'pork' tapeworm – acquired by eating undercooked 'measly' ham or pork. In Europe and the USA, infections with adult *T. solium* have become virtually extinct, while *T. saginata* is the commonest adult tapeworm found in man. The larva (cysticercus) of *T. solium* (but not of *T. saginata*) can also develop in man (as the result of ingesting eggs) and is a serious parasite in many countries (see below). It is now the major cause of a parasitic disease of the nervous system in the USA (Earnest *et al.*, 1987). An infection due to an adult *Taenia*, in man or animals, is referred to as *taeniasis* and that due to the larval stage as *cysticercosis*. In man, when the brain is invaded the resulting disease is referred to as *neurocysticercosis* (see p. 331).

Literature. FAO/UNDP/WHO have produced a comprehensive handbook of 'Guidelines' which deals with all aspects of the biology, diagnosis, epidemiology, chemotherapy, treatment and control of taeniasis and cysticercosis (WHO, 1983). Other valuable reviews are those of Pawlowski (1990), Pawlowski & Schultz (1972), Hird & Pullen (1979), Flisser (1985,

Table 23.1. Taenia saginata*: survival of eggs in sewage and sludge*

Medium	Conditions	Survival (days)	Infectivity remaining
Sewage	Laboratory, 18 °C	16	Yes
Sewage	Exp. plant, trickling filter	42	No
Sewage	Septic tank	40	Not tested
Sewage	Plant, raw waste water	Yes	Yes
Sludge	Lab, anaerobic digestion, 25–29 °C	200	Not tested
Sludge	Lab, anaerobic digestion, 35 °C	<5	No
Sludge	Plant, anaerobic digestion, 26–28 °C	56	Not tested
Sludge	Plant, anaerobic digestion, cold	38	Not tested
Sludge	Exp. plant, activated sludge	42	Not tested
Effluent	Plant, activated sludge	Yes	Yes
Effluent	Plant, trickling filter + lagoon	No?	No?
Effluent	Plant, lagoons	No	No

Source: Data from Bürger (1984).

1988), Flisser *et al.* (1982, 1985), Cook (1988) and Earnest *et al.* (1987).

Taenia saginata

T. saginata has a cosmopolitan distribution, but is more common in developing countries, where hygiene may be poor and where the inhabitants traditionally eat raw or insufficiently cooked meat. However, it is also a problem in developed countries where many 'rare' (i.e. undercooked) beef steaks are consumed. This situation has been summarised by saying (Pawlowski & Schultz, 1972) that *T. saginata* is a problem both for the poor countries *because they are poor*, i.e. with comparatively lower standards of hygiene, but it is also a problem for the rich countries *because they are rich* (and greedy!), i.e. their sewage treatment facilities are overtaxed. It is significant to note that eggs have been shown to survive almost all stages of sewage treatment (Table 23.1) (Bürger, 1984). It is significant, too, that even the high standard of meat inspection in abattoirs – which should identify 'measly' beef carcases – has not succeeded in eliminating this parasite from highly developed countries such as the USA and this remains a major health and veterinary problem.

A high prevalence of *T. saginata* occurs in Africa where cattle, rather than pigs, are commonly grazed. The parasite appears to be specific to cattle, while wild animals appear to play no part as intermediate hosts there (Harrison & Sewell, 1991). In Ethiopia, *T. saginata* is so common that infected individuals do not even consult physicians but treat themselves (Tesfa-Yohannes, 1990). Praziquantel appears to be the effective drug of choice.

T. saginata is also common in parts of Europe, especially Poland and Russia, and in South America.

Distinction between *T. saginata* and *T. solium*. The scolex of *T. saginata* is without a rostellum or hooks and has four suckers. The morphology of the genitalia in a proglottis (Fig. 23.9) follows closely that of the typical taeniid type (Fig. 23.10). It is important – especially in clinical situations – to be able to distinguish specimens (passed by patients) of *T. saginata* from those of *T. solium*, since the eggs of *T. solium* are themselves infective to man causing cysticercosis (see below) and an infected person is potentially capable of infecting himself and immediate contacts as well as pigs. If the scolex is recovered from the patient, the absence of hooks on *T. saginata* distinguishes it immediately from the hooked scolex of *T. solium*, but, often, the scolex remains in the gut (and may regenerate) or is otherwise lost. A much-used criterion is the number of lateral branches of the uterus in gravid proglottides (Fig. 23.11), observed by compressing a specimen between two glass slides. In *T. saginata* these number 14–32 and in *T. solium* 7–11. However, since these ranges overlap, they cannot be relied on absolutely, but a proglottis with 10 branches or fewer can be considered likely to be *T. solium* and one with more than 16 branches to be *T. saginata* (WHO, 1983). The vagina in *T. saginata* has a sphincter muscle, but this is lacking in *T. solium* (Fig. 23.12). Other morphological or biological differences are listed in Table 23.2.

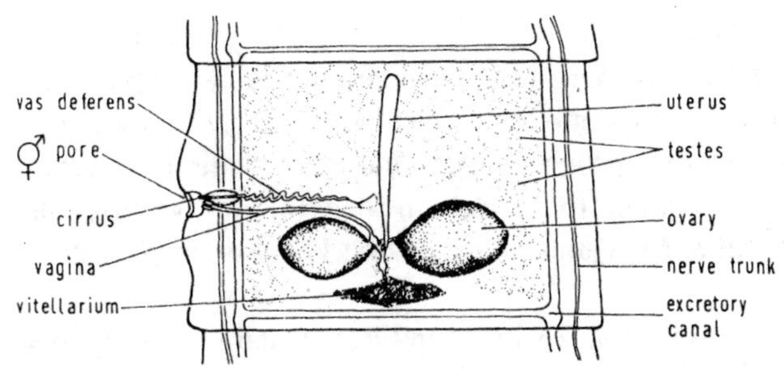

Fig. 23.9
Taenia saginata: morphology of a proglottis. The vaginal sphincter muscle (see Fig. 23.12) is not shown. (After Faust, 1949.)

Table 23.2. *Some characteristics for differentiating between* Taenia saginata *and* Taenia solium

The speciation of the *Taenia* sp. found in Taiwan is in doubt, as it shows some characteristics of both species.

Characteristic	*Taenia saginata*	*T. solium*	Taiwan *Taenia*
Intermediate hosts	cattle, reindeer	pig, wild boar	pig, cattle, goat, wild boar
Site of development	muscle, viscera	brain, skin, muscle	liver (exclusively)
Scolex: adult worm	no hooks	hooks	no hooks
Scolex: cysticercus	no rostellum	rostellum + hooks	rostellum + hooks
Proglottis: uterine branches	23 (14–32)*	8 (7–11)*	20 (11–32)
Passing of proglottides	single, spontaneously	in groups, passively	single, spontaneously
Ovary	2 lobes	3 lobes	2 lobes
Vagina: sphincter muscle	present	absent	present

*There is no universal agreement on the numbers of uterine branches in these two species. WHO (1983) gives a figure of 7–16 for *T. solium* which means that the numbers for the two species overlap and specimens with 10–16 branches cannot be accurately diagnosed. As a rough guide, specimens with over 16 branches are likely to be those of *T. saginata* and those with fewer than 10 branches those of *T. solium.*

Source: Data from WHO (1983) and Fan (1988).

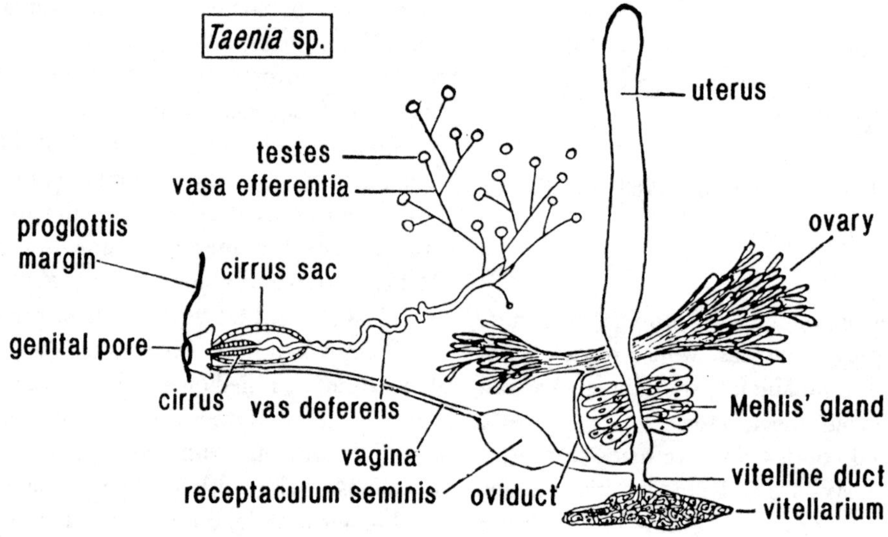

Fig. 23.10
Taenia sp. Generalised diagram to show the arrangements of the genitalia. Some species (e.g. *T. saginata*) have a sphincter muscle on the vagina (see Fig. 23.12).

Recently, more precise methods have emerged with the development of molecular biological techniques, and using DNA probes an accurate identification of even a tiny fragment (preserved in alcohol) is now possible (Rishi & McManus, 1987, 1988; Harrison *et al.*, 1990).

Life cycle (Fig. 23.13). In Great Britain and the USA cattle are the normal intermediate hosts, but in the tropics several other ruminants (e.g. goat, sheep, llama, giraffe, etc.), may serve. Human infections with cysticerci are exceedingly rare.

On ingestion by a ruminant, the thick embryophore of the ova remains unaffected in its passage through the first three compartments of the stomach. On reaching the abomasum, it is exposed to the action of pepsin, which attacks the cementing substance. On reaching the duodenum, it is attacked further by the pancreatic secretion, and disintegrates, releasing the oncosphere

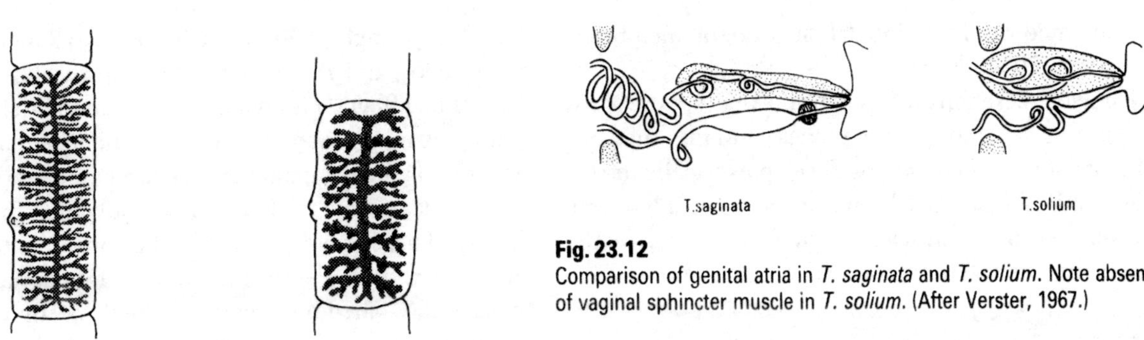

Fig. 23.12
Comparison of genital atria in *T. saginata* and *T. solium*. Note absence of vaginal sphincter muscle in *T. solium*. (After Verster, 1967.)

Fig. 23.11
Comparison of the gravid proglottides of *Taenia saginata* and *T. solium*; comparative details given in Table 23.2. (After several authors.)

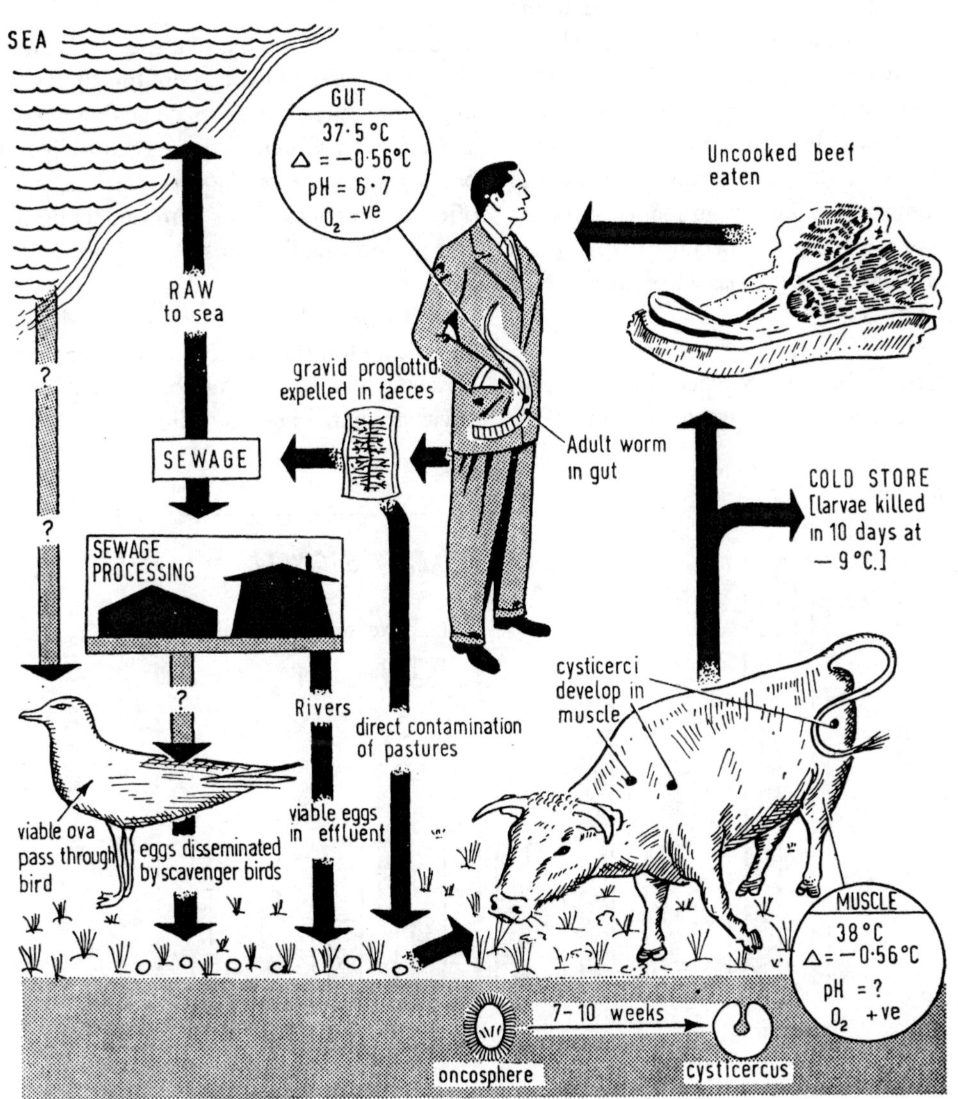

Fig. 23.13
Life cycle of the 'beef' tapeworm, *Taenia saginata*, based on the work of Silverman & Griffiths (1955). (Original.)

still contained within its lipoidal oncospheral membrane (Silverman, 1954*a,b,c*).

Histolytic secretions released by the oncosphere assist it to bore its way through the mucosa and into the general circulation. The embryos develop especially in voluntary and cardiac muscle, but also in fat. They can, however, occur in muscles all over the body, in the diaphragm, shoulder, tongue and in other parts.

Infection of cattle. The mode of infection of the bovine host is imperfectly understood, as the majority of human infections occur in urban districts where the carriers have no opportunity of directly infecting pastures with egg-laden faeces. As pointed out (Table 23.1), cestode (and other) eggs are able to negotiate the various sewage processes and may be disseminated over a wide area; in highly developed countries this is probably the chief means of infection of cattle. Disposal of raw sewage into the sea, release of treated effluent into rivers, and use of sewage sludge as fertiliser all seem to assist in the dissemination. This is probably aided by scavenger birds such as seagulls, because it has been shown that *T. saginata* eggs can pass through the bird gut and retain viability.

Human infection. The size reached by the adult worm is related to the number of worms present (Fig. 23.14). In a single-worm infection a worm 3 metres long can develop and this may produce up to 600 proglottides. The effect on human health is generally slight and symptoms may be vague or absent. The most noticeable symptom is the spontaneous discharge of one or several proglottides, which often show individual muscular activity. These may creep out of the anus onto the perianal skin and may even migrate over the clothes or on the ground, shedding eggs as they go. In contrast, *T. solium* releases its proglottides in groups without muscular activity. Other symptoms recorded are abdominal pains, headache and increased appetite (Tesfa-Yohannes, 1990).

Taenia solium

The adult of *T. solium*, like that of *T. saginata*, is exclusively a parasite of man where the adult strobila develops in the intestine. The larval stage (known by veterinarians as *Cysticercus cellulosae*) occurs naturally in pigs and wild boars but man can also be infected, with serious and often fatal results. Thus man can act both as a definitive and an intermediate host.

Life cycle. The life cycle is similar to that of *T. saginata* but the pig acts as intermediate host. A cysticercus, which has a white opalescent colour, may lodge anywhere in the body but shows predilection for muscles

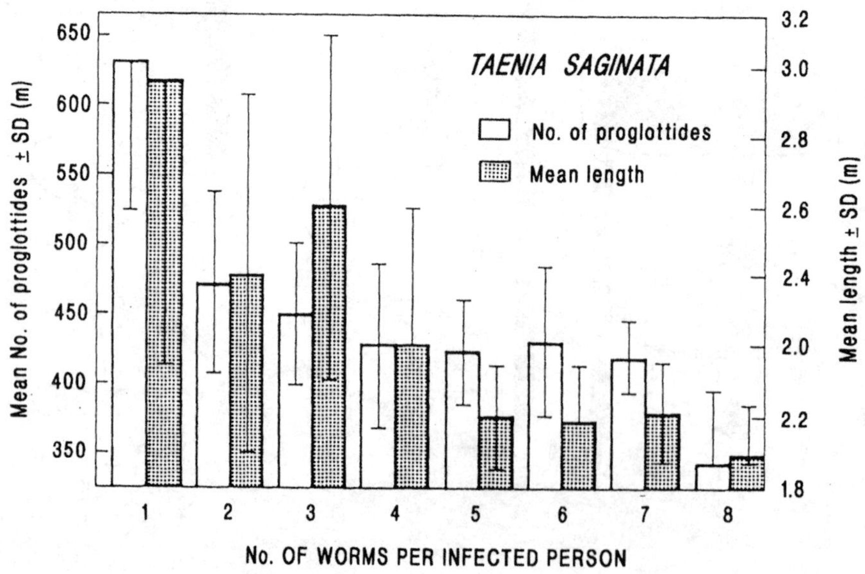

Fig. 23.14
Taenia saginata: the mean length and number of proglottides in adult worms from patients with single and multiple infections. (After Tesfa-Yohannes, 1990.)

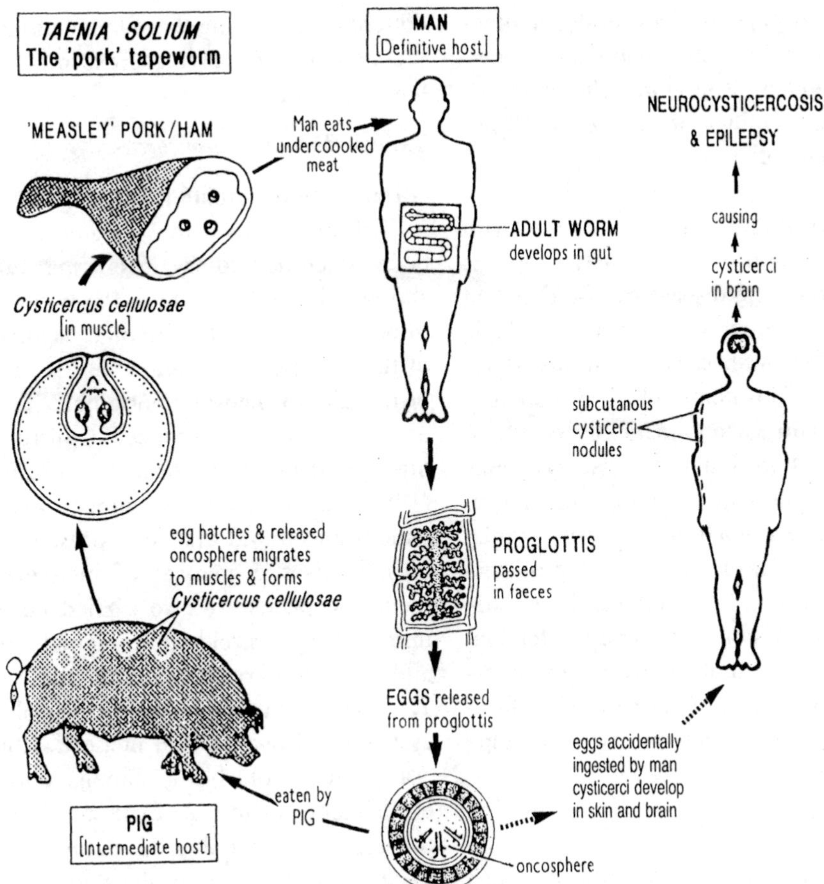

Fig. 23.15
Life cycle of the 'pork' tapeworm, *Taenia solium*. Note that, unlike the 'beef' tapeworm, *Taenia saginata*, the larval stage (cysticercus) can also develop in man causing the dangerous and often fatal disease neurocysticercosis. (Original.)

of the jaws, heart, neck, tongue, elbow, shoulders and the thighs. When 'measly pork' is eaten by man, the scolex (activated by bile) evaginates and becomes attached to the duodenum, the 'bladder' is digested, and the strobila grows. Its rate of growth has not been studied.

Literature. Two major books dealing with all aspects of *T. solium*, biology, epidemiology, diagnosis, pathology, immunology and control, are those of Flisser *et al.* (1982) and WHO (1983). Other valuable reviews are those of Cook (1988), Earnest *et al.* (1987), Flisser (1985, 1988), Flisser *et al.* (1985, 1990), Hird & Pullen (1979), Pawlowski & Schultz (1972).

Occurrence. *T. solium* occurs in most pork-eating countries where sanitary conditions are primitive and scavenging pigs have access to human faecal material. The UK has been free of the infection for many decades; a survey of 13 000 pigs in 1972 revealed not a single infection. It has also virtually disappeared from East and Central Europe.

In many countries, *neurocysticercosis* due to the larval stage in the brain is a major health hazard and has been reported from Africa, China, Japan, Indonesia, Poland, Italy, Spain, USA and throughout Latin America, especially Mexico and Central and South America. Although not a common disease, it is now the most prevalent disease of the central nervous system in the USA, although most of the patients appear to have their origin from Mexico or Latin America (Earnest *et al.*, 1987).

The disease has received special attention in Mexico, from Flisser and her colleagues (1982–90), where autopsies show a clinical incidence of cysticercosis of 7 per cent (Flisser, 1988). In Africa, the infection is probably more widespread than is apparent, as it is poorly documented. In West Cameroon, for example, postmortem examination of pigs showed a prevalence of 24 per cent with 2 per cent of the human population showing *Taenia solium* antibodies (Zoli *et al.*, 1987). In the Ekari population of New Guinea, it may reach 18 per cent prevalence (see below).

Human infection; pathology. Human infection with cysticerci occurs by ingestion of contaminated food or drink or by autoinfection. The disease can be detected by X-ray, computerised tomography and immunologic methods, but it should be emphasised that up to 40 per cent of patients may show no immunological response. Some patients may remain asymptomatic, especially in the early days of an infection. Later, because cysts may occur anywhere in the nervous system, the clinical features are varied. These include 'seizure, cranial nerve palsy, stroke, paralysis, or a combination of syndromes' (Earnest *et al.*, 1987). Although cerebral cysticercosis is the most commonly occurring infection, cysticerci can occur anywhere in the body, the eye commonly being infected. Cysticerci under the skin form visible subcutaneous nodules, which are largely asymptomatic.

Cysticercosis epidemic in New Guinea

A remarkable outbreak of *T. solium* cysticercosis – representing a rare example of a man-made disease – occurred in the Ekari people of West New Guinea in recent years (Gajdusek, 1977, 1978). In 1974, two Indonesian doctors were astonished to find an epidemic of severe burns in patients attending the Enarotali Hospital on Lake Paniai. These burns had occurred at night when patients apparently fell into the fire around which they slept in their huts, as the nights are cold. This was remarkable, for healthy sleepers would naturally wake up and withdraw from the fire and avoid being burnt. Those patients who were burnt were incapable of doing so due to apparent 'epilepsy'. Clinical examination of these patients revealed the presence of nodules in the skin and, in those who died, cysticerci of *T. solium* in the brain. Since there has been no previous reports of *T. solium* in the area, the origin of the infection was investigated. It was then found that the parasite had been brought in in pigs which had been presented by the Indonesian Government as a gift; these pigs were also found to contain the larvae of the nematode *Trichinella spiralis* (see p. 390). In West New Guinea, pigs are highly prized and a pig feast (at which undercooked pork is eaten) is held every year. The infection has rapidly disseminated among the human and pig population and the situation now presents exceptionally difficult problems of control.

Taiwan Taenia: a new species?

An anomalous situation has been recognised to occur in Taiwan where both *T. saginata* and *T. solium* have been documented as occurring among the native aboriginal population (Fan, 1988; Fan *et al.*, 1990). Most infections appeared to be of *T. saginata*, but the source of the infection was confusing because beef is not part of the diet of Taiwan aboriginals. These people prefer to eat raw or undercooked squirrel, serow, monkey, muntjac, deer or wild boar. In a major experiment, Fan (1988) exposed a range of animals to eggs from a *Taenia* patient and found cysticerci present in the livers of 30 pigs, 3 calves, 5 goats and 3 monkeys. Further experiments on pigs were also carried out and, by 28 days, immature cysticerci had developed a rostellum, suckers and rudimentary hooks, i.e. the characteristics of *T. solium*. A human volunteer was also infected with a cysticercus from the liver of a calf. Comparing the general features of the specimens obtained from these experiments (Table 23.2) it was found that, although the general morphological features resembled those of *T. saginata*, they did not develop well in bovids and the scolex showed the characteristics of *T. solium* (a rostellum with hooks). Moreover, in contrast to *T. saginata* and *T. solium*, which develop in muscle or viscera, the cysticerci of Taiwan developed exclusively in the liver. This evidence strongly suggests that, in this isolated situation, a new species may be evolving by selection pressure during many years of passage through diverse intermediate hosts and man. A similar complex speciation situation may occur in the Philippines (Arambulo *et al.*, 1976) and Indonesia (Fan *et al.*, 1989).

Taenia multiceps (syn. *Multiceps multiceps*)

A tapeworm of carnivores with a cosmopolitan distribution. The intermediate stages develop in the brain and spinal cord of ungulates, especially sheep, and cause a disease called 'staggers' or 'gid' which affect the balancing powers of animals. The dog is the common definitive host in the USA and the UK. A major outbreak has recently been recorded in Ireland (Doherty *et al.*, 1989).

The cysticercus (*Cysticercus cerebralis*) possesses unusual powers of asexual multiplication, forming a bladder or *coenurus* which may give rise to many hundreds of daughter protoscoleces directly from its inner wall (Fig. 20.10); in this way it differs from a hydatid cyst (Fig. 23.19). The coenurus may similarly develop in man causing severe damage and even death, but only a dozen or so human cases have been reported.

A related 'species', *T. serialis*, in which the protoscoleces in the larva (*C. serialis*) form in rows or series (hence the name) is well known in various rodents, especially rabbits; some authors, however, consider this to be a 'strain' of *T. multiceps* (Smyth & Heath, 1970).

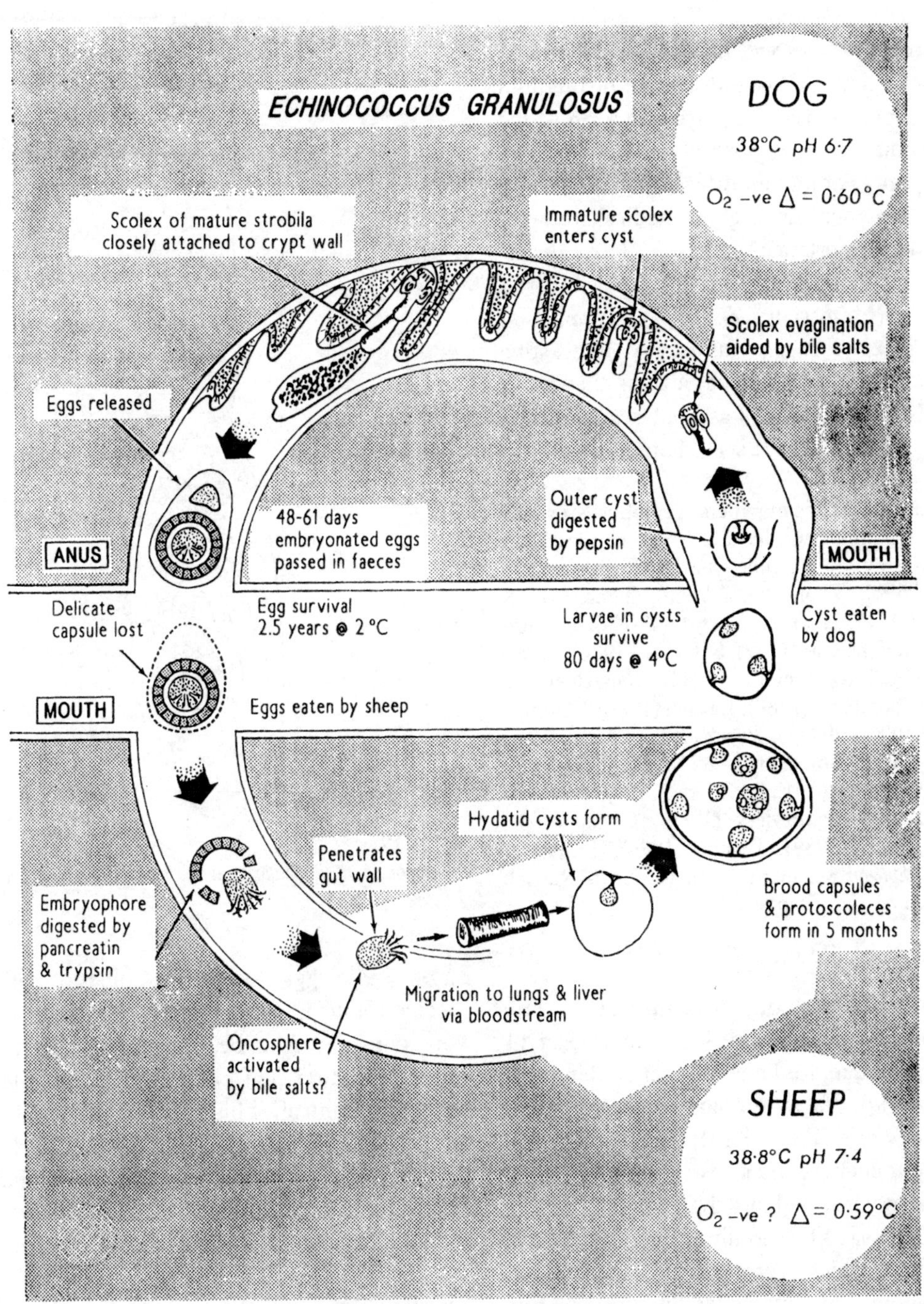

Fig. 23.16
Echinococcus granulosus, the hydatid organism. Life cycle and some of the physiological factors relating to it. (After Smyth, 1964*a*.)

○ 23.3.2 Genus *Echinococcus*: hydatid disease

Species of the genus *Echinococcus* are small (2–8 mm) cestodes of carnivores (especially dogs). The larval stage, or *hydatid cyst*, occurs in animals and man causing *hydatid disease*, also known as *hydatidosis* or *echinococcosis*. Although there are four species generally recognised (*E. granulosus*, *E. multilocularis*, *E. oligarthrus* and *E. vogeli*), only the first two will be dealt with here as they are the chief agents of hydatid disease throughout the world. *E. vogeli* gives rise to a polycystic form of the disease in Columbia and *E. oligarthrus* has not yet been identified as the cause of human disease (WHO, 1981). Hydatid disease is a zoonosis of great medical, veterinary and economic importance. It has a global distribution and is insidiously spreading into countries previously free of it (Matossian *et al.*, 1977). *E. granulosus* and *E. multilocularis* are also unusually interesting species from the biological point of view, as they serve as superb models for the study of asexual/sexual differentiation *in vitro* (Smyth, 1987; see Chapter 34.

Literature. Many thousands of papers are published yearly on all aspects of this important disease. Much of these data have been reviewed under the following headings. *General*: Smyth (1964*a*, 1969*a*), Thompson (1986*a*), Thompson & Allsop (1988), WHO (1981). *Biology*: Smyth (1964*a*), Thompson (1986*b*). *Epidemiology*: Dar (1987), Schwabe (1986), Gemmell (1990), Gemmell & Lawson (1985). *Speciation and 'strains'*: Eckert & Thompson (1988), Rausch (1986), Thompson & Lymbery (1988, 1990*a*,*b*). *Physiology and biochemistry*: McManus & Bryant (1986), Smyth & McManus (1989). *Immunobiology*: Heath (1986), Lightowlers (1990). *Immunodiagnosis*: Rickard and Lightowlers (1986). *Chemotherapy*: Eckert (1986), Horton (1989), Webbe (1986), Smyth & Barrett (1980). *Pathology*: Schantz (1984), Wilson & Rausch (1980). *In vitro culture*: see Chapter 34.

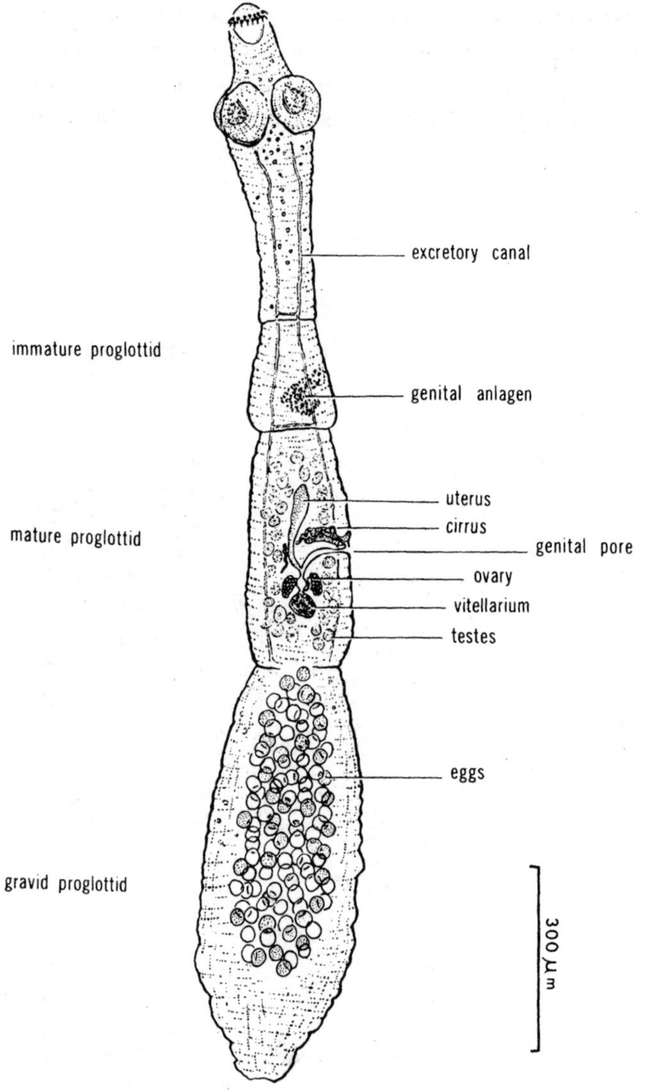

Fig. 23.17
Echinococcus multilocularis; entire worm from dog. (Original.)

Echinococcus granulosus: life cycle (Fig. 23.16)

Adult worm: morphology and development. The adult worm commonly develops in dogs, but at least 11 other species of Canidae are known to act as definitive hosts (Smyth, 1964*a*). Its morphology closely follows that of *E. multilocularis* (Fig. 23.17); see Table 23.3. The scolex has four suckers and a rostellum with hooks, the latter being tightly inserted into the crypts of Lieberkühn in the gut mucosa. The rostellum contains a *rostellar gland* (Fig. 23.18) which secretes a substance, which appears to be a lipoprotein, but whose function remains a mystery (Smyth, 1964*b*). The mature strobila has only 3–4 proglottides; when gravid, these or shed eggs are expelled in the faeces.

Eggs. The eggs closely resemble those of other *Taenia* species found in dogs from which, until recently, they were considered to be morphologically indistinguishable, a situation that created major problems in diagnosis and control. This problem has now been solved by the development by Craig *et al.* (1986) of a specific *Echinococcus* anti-oncospheral monoclonal antibody, which specifically identifies *Echinococcus* eggs, thus resolving a long-standing problem in parasitology.

The larval (hydatid) cyst. Man and animals become infected by eating faecal eggs contaminating water or foodstuffs or from handling infected dogs (see epidemiology and control below). The eggs hatch in the intestine

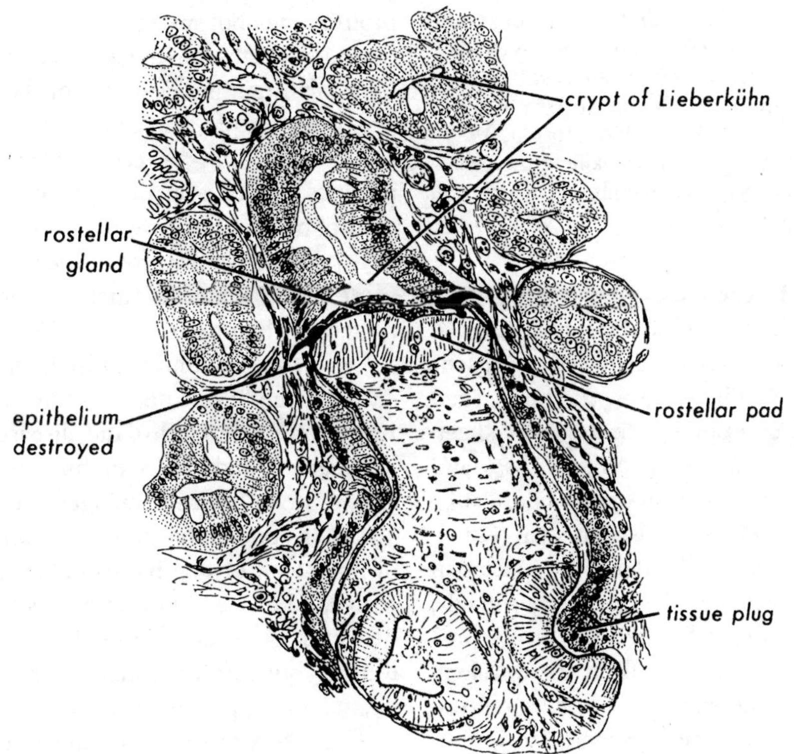

Fig. 23.18
Scolex of *Echinococcus granulosus* attached to dog gut; the rostellum extends into a crypt of Lieberkühn, the suckers grip the base of the villi. (After Smyth, 1969*b*.)

and the released oncosphere penetrate the mucosa. On entering a mesenteric venule or a capillary, oncospheres are transported to and trapped largely in the liver, but may reach and develop in any of the viscera. The larva develops into a large cystic form or *hydatid cyst*, referred to as a *unilocular cyst*, which gives rise to *unilocular hydatid disease* in man. This is characterised as having only one bladder or many completely isolated bladders, each enclosed in its own well-developed envelope. The latter consists (Fig. 23.19) of several layers, the most prominent being the *laminated layer*, a thick laminated structure rich in polysaccharides and hence PAS-positive in stained sections. Within this again is the *germinal membrane* from which the *brood capsules* arise, inside which develop thousands of larvae or *protoscoleces*, the whole being suspended in a *hydatid fluid*. The ultrastructure of a hydatid cyst has been studied in detail by numerous authors, e.g. Morseth (1967), Richards *et al.* (1983) and Lascano *et al.* (1975).

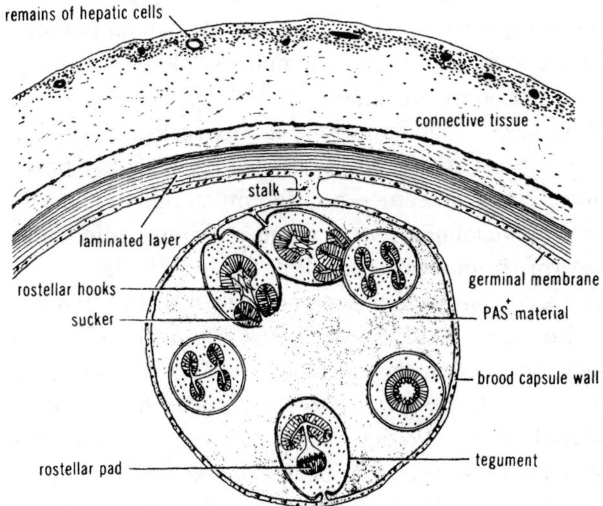

Fig. 23.19
T.S. of hydatid cyst of *Echinococcus granulosus*; diagrammatic. (After Morseth, 1967.)

Pathology and chemotherapy. Detailed consideration of these are beyond the scope of this book and reference should be made to Wilson & Rausch (1980), Schantz (1984), Schwabe (1986) or Pawlowski (1984). The cysts in man often grow to an enormous size (up to 50 cm in diameter) and may contain many litres of hydatid fluid and up to a million protoscoleces. A unilocular cyst can be removed surgically but a major difficulty is to prevent spillage of the contained protoscoleces, each of which can regenerate into a new cyst and cause *secondary hydatidosis* (Fig. 34.8). Within recent years, the newly introduced benzimidazole compounds *mebendazole* and *albendazole* have been used with some success, but conflicting results have been given by different workers, possibly related to different protocols (Eckert, 1986; Horton, 1989). One difficulty is that these drugs, although killing the protoscoleces, may not kill the germinal membrane which can regenerate when the drug is withdrawn.

Speciation: 'strain' differences. It has long been recognised that slight morphological differences existed between adult *E. granulosus* from dogs fed protoscoleces of different host origin; some workers regarded these as subspecies. For example, those from sheep origin were referred to as *E. granulosus granulosus* and those of horse origin as *E. granulosus equus*. These subspecies are not now generally recognised as such (Kumaratilake *et al.*, 1986*b*), as it is realised that the speciation problem is much more complex than formerly thought and that numerous intra-specific varieties or 'strains' probably exist. This situation arose from the development of a technique for the successful *in vitro* culture of adult worms from protoscoleces (Smyth, 1967–90). The early experiments were carried out with hydatids of sheep origin and adult worms obtained. When, however, the identical technique was used with protoscoleces of horse origin, the organisms grew only slightly, failed to strobilate and did not reach sexual maturity. This unexpected result, at first, was blamed on possible faulty technique or media components. When, however, over a period of two years, some 200 cultures of horse protoscoleces failed to grow, it was realised that this result represented a new phenomenon and that horse *Echinococcus* represented a different 'strain' with nutritional or metabolic requirements different from those of worms from sheep (Smyth & Davies, 1974*a*).

This conclusion was soon confirmed by biochemical and isoenzyme studies (McManus & Smyth, 1978–86) which revealed major metabolic differences between sheep and horse protoscoleces. Since then hydatid material from a variety of hosts in many countries throughout the world has been examined by numerous workers, especially in Australia and Europe, using isoenzymes and DNA probes (McManus & Rishi, 1989). The numerous data collected have been analysed in a valuable, very detailed, review by Thompson & Lymbery (1990*b*). Evidence is accumulating that a number of other strains, in addition to the sheep and horse strains occur, e.g. camel, pig, etc.

In addition to differences in morphology, isoenzyme profiles, metabolism, etc., unexpected differences in the developmental biology have been identified which may have important implications for the control of this parasite. For example, it is generally accepted that protoscoleces of sheep origin reach sexual maturity in dogs and produce shelled eggs in approximately 40 days p.i., and control programmes are based on this figure. However, in Switzerland, it was found that protoscoleces of Swiss cattle origin matured much earlier, producing eggs in 35 days (Thompson *et al.*, 1984).

Other, as yet unrecognised, differences may exist, such as whether or not all strains are infective to man. For example, horse hydatid disease is now widespread in England and packs of hunting hounds have been shown to be heavily infected (Thompson & Smyth, 1975). In spite of this potential hazard there is no record of huntsmen or dog handlers having a high prevalence of hydatid disease. This (admittedly circumstantial) evidence may indicate that the horse strain is not infective to man, but further epidemiological evidence is clearly needed in support of this hypothesis.

The nature of strain variation in *Echinococcus* is a complex issue and much more information is needed on the genetics of the genus. Since the adult *E. granulosus* is a self-fertilising hermaphrodite (Smyth & Smyth, 1964; Kumaratilake *et al.*, 1986*a*) it is apparent that a recessive gene could appear in both ova and spermatozoa at the same time so that at fertilisation a double recessive could readily be formed (Fig. 23.20). Furthermore, since a hydatid cyst is developed by asexual budding (polyembryony) from a single egg, many thousands of genetically identical individuals (protoscoleces) can be formed. Thus, if established in a new host, (theoretically) the conditions exist for a new 'strain' to appear (Smyth & Smyth, 1964). Although it is likely that some – perhaps most – new strains are developed in this way, this is almost certainly a simplification of a complex genetic situation involving many other factors. For instance, cross-fertilisation may occasionally take place, but has not as yet been reported. In contrast, self-fertilisation appears to be rare in *E. multilocularis* (p. 339) which may explain why only a few strains have been found in this species, although this conclusion may be premature, as very few isolates of this species have been examined (Kumaratilake *et al.*, 1986*a*).

Control and epidemiology. Human and animal infections with hydatid cysts are caused by intimacy with infected dogs, especially on farms, resulting in the inges-

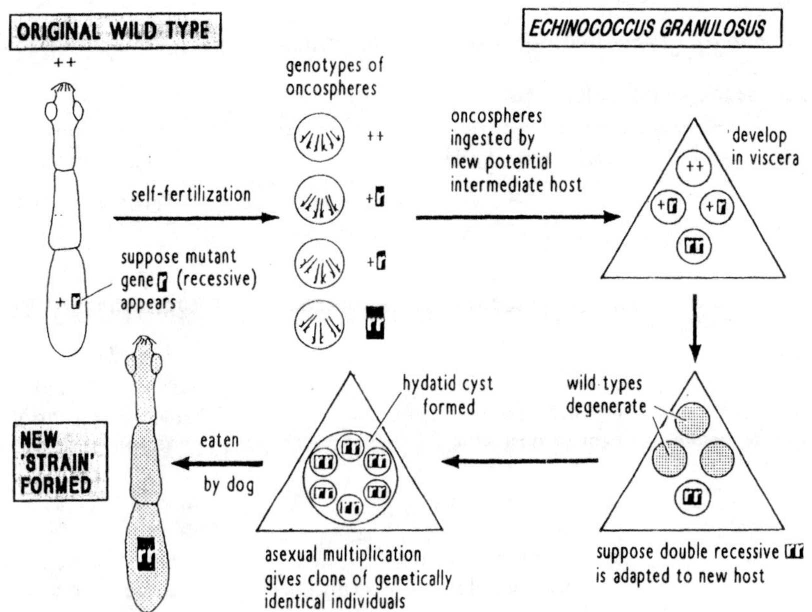

Fig. 23.20
Genetics of speciation in *Echinococcus granulosus*. Because it is a hermaphrodite, a mutant gene (normally recessive) can theoretically appear in both sperm and ova at the same time with the result that when self-fertilisation occurs, a double recessive could immediately be formed. If such an oncosphere, bearing a double recessive, invades and establishes itself in a new host species, asexual reproduction within the hydatid cyst would produce a clone of genetically identical individuals, i.e. a new 'strain' would be formed. (After Smyth & Smyth, 1964.)

tion of eggs. Dogs become infected with the strobilar stage by being fed sheep viscera containing hydatid cysts – a common practice in farming communities – or by having access to abattoirs or by scavenging dead sheep in remote areas. Control methods are (theoretically) simple but, in practice, difficult to enforce on account of farming habits or social traditions: careful washing after contact with dogs, boiling of offal before using it to feed dogs, and the regular, periodic de-worming of dogs. Application of these and other strict measures have greatly reduced or eliminated the disease from some countries, notably Iceland, where it was once endemic (Beard, 1973), but more recently from New Zealand, Tasmania and Cyprus. In the latter country, for example, the deliberate destruction of stray or unwanted dogs has rapidly reduced the prevalence of the disease in animals and man to zero (Polydorou, 1987). Many of the programmes now in use were based on experience in New Zealand (Fig. 23.21) and Tasmania where the main components of the control campaigns have been education, dog surveillance and regular dosing programmes. The problems and techniques of those control measures have been reviewed in detail by Gemmell & Lawson (1985), Nelson (1989), Polydorou (1987) and WHO (1981).

The highest prevalence of hydatid disease in the world occurs in Kenya where unusual conditions prevail, in that the Turkana tribe traditionally have a very close association with dogs. Most households keep a dog which they utilise to 'lick'clean' their babies after defecation, an understandable situation as water is so scarce. A further factor is that human dead bodies (which may contain hydatid cysts) are often buried in shallow graves in the bush where they may be scavenged by dogs or wild carnivores. The problems of control in this region thus are especially demanding but an active control programme, involving immunodiagnosis, ultrascan surveys, dog purging and chemotherapy, is now in operation and is achieving some success (Macpherson & Craig, 1991; Macpherson *et al.*, 1986).

Echinococcus multilocularis: alveolar (multivesicular) hydatid disease

The larva of this very dangerous species is the cause of alveolar (multilocular) hydatid disease in man and animals. The adult worm mainly parasitises the intestine of the red fox (*Vulpes vulpes*) and the arctic fox (*Alopex lagopus*) but other species of foxes and wolves can act as definitive hosts in addition to domestic dogs and cats, when they have access to infected rodents. The general

Table 23.3. *Comparative characteristics of* Echinococcus granulosus *and* Echinococcus multilocularis

No allowances have been made for 'strain' differences.

Characteristic	*Echinococcus granulosus*	*Echinococcus multilocularis*
Length of strobila (mm)	2–7	1.2–3.7
Rostellar hooks, length (μm)		
Large hooks (mean)	31–49 (37–42)	28–34 (31)
Small hooks (mean)	22–39 (29–34)	28–31 (27)
Number of proglottides (range)	3 (4–6)	4–5 (2–6)
Number of testes (range)	25–80 (32–68)	16–35 (18–26)
Distribution of testes relative to genital pore	equally anterior/posterior	majority posterior
Position of genital pore relative to middle of proglottis	near or posterior	anterior
Form of uterus	lateral sacculations	sac-like
Vaginal sphincter muscle	present	absent
Definitive hosts	dogs and other canids	mainly foxes, but also other wild canids
Intermediate hosts	primarily ungulates (especially sheep); also marsupials, primates and man	primarily arvicolid rodents (especially voles) and man

Source: Data from Thompson (1986*b*) and WHO (1981).

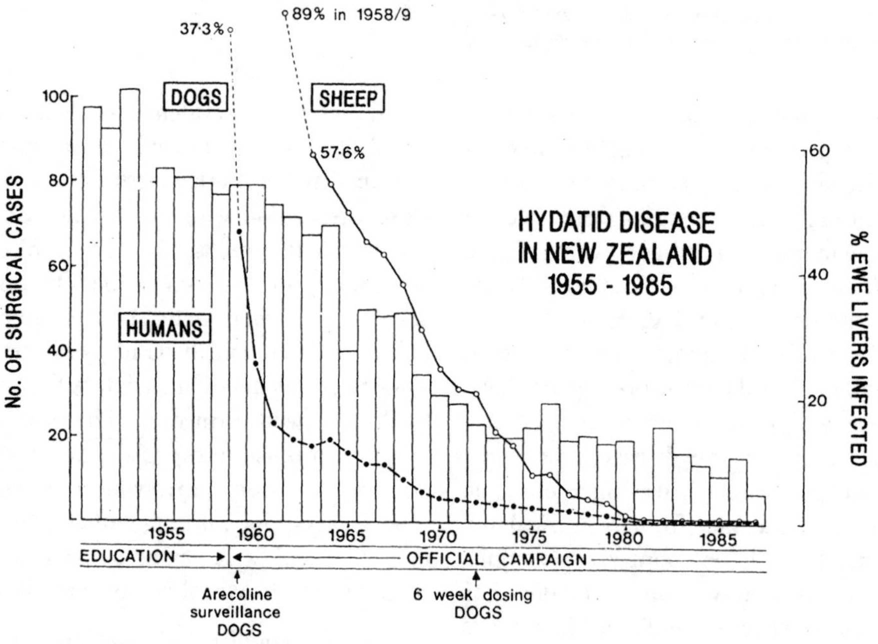

Fig. 23.21
Control of hydatid disease in man and sheep in New Zealand based on educational and dog-dosing programmes. (After Gemmell, 1990.)

morphology (Fig. 23.22) is similar to that of *E. granulosus*, but the adult is much smaller; the characters of both species are compared in Table 23.3.

Life cycle. The life cycle is essentially a sylvatic one, the commonest intermediate hosts being arvicolid rodents, especially the voles *Microtus* spp. and *Clethrionomys* spp. The larval cysts have also been reported from a wide range of other wild species: mice, rats, hamsters, squirrels, jerboas and picas (WHO, 1981). In the laboratory, the larval stage is very easily maintained in cotton rats or gerbils by intra-peritoneal injection of protosco-

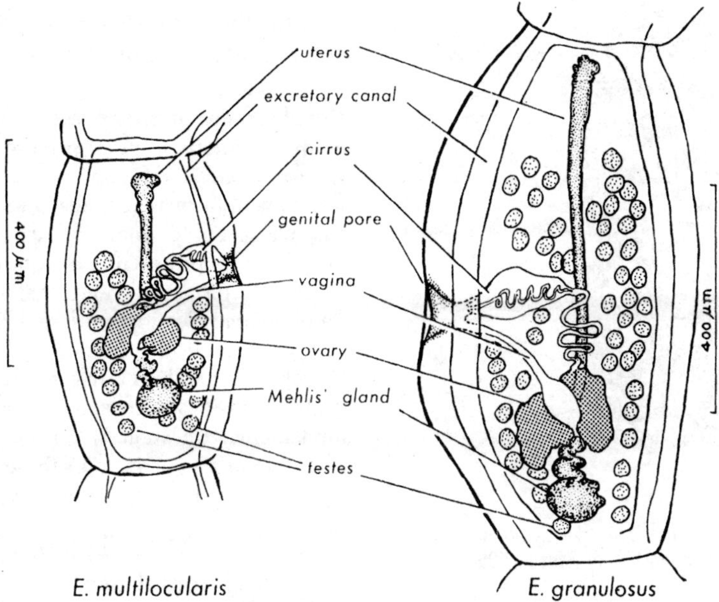

Fig. 23.22
Comparison of mature proglottides of *Echinococcus granulosus* and *E. multilocularis* (based on photographs by Vogel, 1957). The differences may not always be as great as they appear to be here, as the size and morphology of the adult worms may vary considerably with the origin of the cystic stages (i.e. whether from sheep, horse, cotton rats, etc.) and the species of definitive host, whether dog, cat, fox or other carnivore; see text.

leces (Lubinsky, 1960); masses of cysts are produced in about 3 months.

Owing to the extremely dangerous nature of the infective eggs of the adult, the latter has rarely been maintained in dogs in the laboratory and this has been a great barrier to research on alveolar disease. In an early pioneer experiment, the Russian worker Kovalenko (1976) showed that both *E. multilocularis* and *E. granulosus* could establish themselves in mice, cotton rats and hamsters and develop to early genital pore formation. Unfortunately, this important work was overlooked until recently, when Kamiya & Sato (1990*a*) achieved a major advance in the laboratory maintenance of *E. multilocularis* by establishing the adult worm in immunosuppressed golden hamsters and obtaining adult worms and infective eggs. This important result should greatly stimulate research in this field. Since the larval stage can also be maintained (by intraperitoneal injection) in gerbils, the entire life cycle can now be maintained in the laboratory in one host species (Kamiya & Sato, 1990*b*).

The technique used by Kamiya & Sato (1990*a*) involved the treatment of 6-week-old golden hamsters with an immunosuppressant, prednisolone tertiary butylacetate (PTBA), followed by the oral administration of 20 000 protoscoleces of *E. multilocularis*. The development of the adult worms to sexual maturity compared favourably with that in dogs, as did the egg production. Development up to 25 days p.i. also took place in non-treated hamsters, but not to the same extent as in the treated animals.

Alveolar (multilocular) hydatid disease in man. The larval cyst, referred to as an *alveolar* or *multilocular* hydatid cyst, which develops in man and animals, does not form a uniform, well-defined cyst (as in *E. granulosus*) but a multicystic structure made up of proliferating vesicles embedded in a dense fibrous stroma (Fig. 23.23), the whole resembling – and often mistaken for – a hepatic sarcoma. In older cysts the hydatid fluid is replaced by a jelly-like mass; a laminated membrane is either lacking or poorly developed.

Pathology. Detailed consideration of this is beyond the scope of this book. The cyst is normally inoperable in man and most of the non-resectable alveolar cyst cases die within 10 years, making it one of the most lethal parasitic diseases known. Drugs, such as albendazole, have greatly helped to stabilise the pathology of the disease although they do not kill the parasite tissues but merely inhibit their growth. The use of chemotherapy has been reviewed by Eckert (1986), Horton (1989) and Webbe (1986) and the clinical features of the disease by Wilson & Rausch (1980).

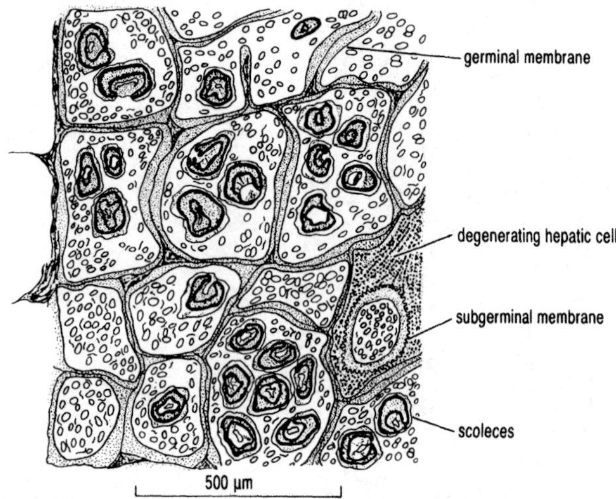

Fig. 23.23
Larval mass of *Echinococcus multilocularis* in voles five months after infection: infective scoleces are abundant. (After Rausch, 1954.)

Distribution. The distribution of *E. multilocularis* is much more limited than that of *E. granulosus*, and is almost entirely confined to the Northern Hemisphere (WHO, 1981; Rausch, 1986). The endemic regions in Europe include France, Switzerland, S Germany, W Austria, Bulgaria, Turkey, the former Soviet Union and the Japanese Islands (especially Redun Island).

Literature. For further references on the biology, epidemiology, physiology and pathology, see the reviews listed on p. 334.

Biology of cystic and strobilar differentiation in *Echinococcus*. One of the most fascinating problems concerning *Echinococcus* is related to the unusual property of a protoscolex being capable of differentiating in one of two directions (Fig. 34.8). Thus, if protoscoleces 'leak' out of a cyst in a host they can become established in another organ and 'dedifferentiate' into a further hydatid cyst; this produces a condition known as *secondary hydatiasis*. On the other hand, if cysts are eaten by a dog, the scoleces evaginate in the gut, become attached to the mucosa and grow into adult strobila.

In vitro studies have shown (Chapter 34) that the stimulus 'triggering' development of an evaginated scolex of *E. granulosus* in a strobilar direction appears to be contact with a suitable substrate, i.e. the gut cell surface, although other factors may also play an important role. In the absence of evagination and subsequent contact, development only in a cystic direction results (Fig. 34.10). *E. multilocularis*, however, will develop to sexually mature forms *in vitro* in a monophasic medium, and a 'contact' stimulus does not appear to be necessary.

Development of 'monozoic' worms *in vitro*. Under optimum conditions *in vitro*, protoscoleces of *E. granulosus* and *E. multilocularis* develop into segmented, sexually mature worms with normal genitalia, although insemination does not take place. Under some culture conditions – at present not understood – protoscoleces occasionally develop into bizarre 'monozoic' forms which are characterised by containing a full set of genitalia without undergoing segmentation (Fig. 23.24) (Smyth, 1971, 1987, 1990; Smyth & Davies, 1975; Smyth & Barrett, 1979). This clearly indicates that somatic growth and genital growth can proceed independently of each other. This fundamental question is further discussed in Chapter 34, when the *in vitro* culture of these species is considered.

○ 23.3.3 Miscellaneous Taeniidae suitable for experimental use

There are a number of species of *Taenia* which may be conveniently maintained in the laboratory and whose developmental stages have been worked out in detail. The following two species have been widely utilised for experimental work.

T. pisiformis (= *T. serrata*)

This is one of the commonest tapeworms of dogs and several other wild carnivores, but it rarely occurs in the cat. In the general anatomy of its scolex and proglottids it resembles *T. solium*. Its larval stages, known as *Cysticercus pisiformis*, are well known in rabbits where they develop in the liver and mesenteries. After penetrating the intestinal wall and reaching a blood vessel, the oncospheres are carried to the liver where they develop into actively moving larvae which – within 3 days – produce characteristic haemorrhagic tunnels in the liver tissue (Smyth & Heath, 1970). The developing cysticerci continue to migrate through the liver tissue for 14–18 days, eventually leaving the liver after penetrating the capsule. After reaching the coelom, the larvae become attached to the mesentery and become enclosed in a host adventitious cyst. The development of the adult has been studied experimentally in dogs, foxes and cats (Beveridge & Rickard, 1985). In dogs (Fig. 23.25), it grew well and had a pre-patent period of 35 days. In foxes, although a high percentage of worms became established, only a few gravid proglottides were obtained after 71–75 days p.i. In cats, both establishment and development was poor and worms never became gravid.

Both larval and adult stages have been cultured *in vitro* (see Chapter 34).

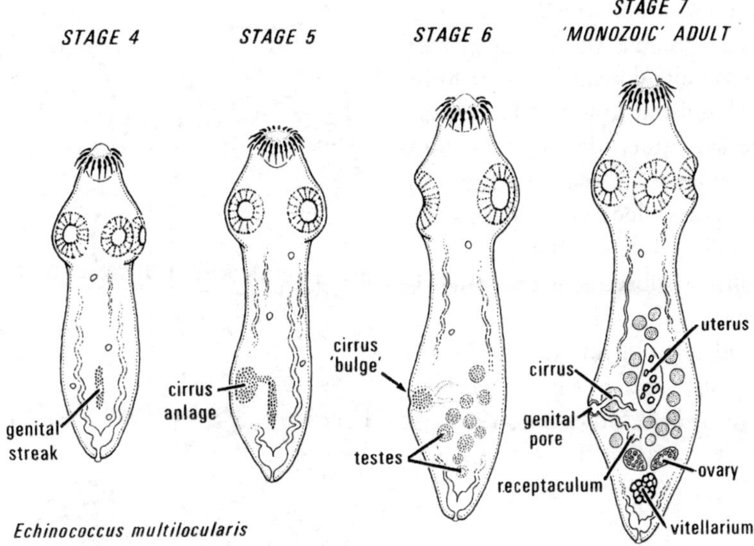

Fig. 23.24
Echinococcus multilocularis: development of 'monozoic' forms *in vitro* (earlier stages 1–3 not shown). These forms may appear under abnormal culture conditions; they develop a full set of genitalia but fail to strobilate. Similar 'monozoic' forms have been found in cultures of *E. granulosus*. (From Smyth & Barrett, 1979.)

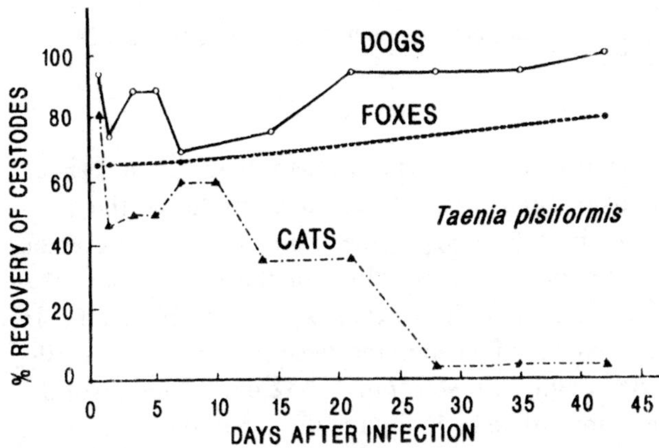

Fig. 23.25
Taenia pisiformis: percentage recovery of adult worms from experimentally infected dogs, foxes and cats. (After Beveridge & Rickard, 1975.)

Taenia crassiceps

The adult of this species is a common parasite of the red fox in Europe and North America. Its life cycle and morphology have been described by Freeman (1962), Miyaji *et al.* (1990) and Rietschel (1981). Its cysticercus stage is unusual in that it reproduces by exogenous budding in the body cavity of rodents, and hence is readily maintained by intra-peritoneal injection; mice are the usual laboratory hosts. The cystic stage has been widely used as a model for the study of cestode physiology, biochemistry and morphogenesis. It appears to undergo mutation fairly readily and several strains are now recognised. The best known of these are the ORF and KBF strains with diploid chromosome numbers of 14 and 16, respectively (Smith *et al.*, 1972). The pre-patent period of a German isolate in foxes was found to be 31–32 days, and the most successful intermediate host was the vole *Microtus arvalis* in which the larva developed in the brain (Rietschel, 1981). A Japanese isolate from the vole *Microtus montebelli* grew well in dogs and with a shorter pre-patent period of 27–31 days (Miyaji *et al.*, 1990). The usefulness of this species as a laboratory teaching and research model has been greatly enhanced by the fact that it has now been grown to sexual maturity in golden hamsters which had been immunosuppressed. The treatment involved subcutaneous injections of prednisolone tertiary butylacetate at 2–4 day intervals (Sato & Kamiya, 1989). Thus the entire life cycle of this cestode can now be maintained in the laboratory using only small rodents.

Taenia taeniaeformis (= *Hydatigera taeniaeformis*)

The synonym (*H. taeniaeformis*) is often used, but detailed analysis of the characters by Esch & Self (1965) suggest that *Taenia* is the appropriate generic name. This tapeworm occurs in the small intestine of the cat and other carnivores, such as the stoat, lynx and fox. The scolex is characterised by possessing prominent suckers and no distinct neck region. The chief interest of this species lies in its larva, *Cysticerus fasciolaris*, which does not form a cysticercus but a *strobilocercus* (Fig. 20.10). This may induce sarcoma in the host liver (Smyth, 1969*b*). The intermediate hosts are said to be rats, mice, rabbit, squirrel and muskrat, but Hutchison (1959) found that, in the laboratory, development took place most readily in the mouse. On reaching the liver in the intermediate host, the strobilocercus develops and grows rapidly, becoming infective after 30 days. A high degree of immunity against a subsequent challenge is induced by the presence of a single larva in the liver. This species has been widely used for immunological studies and its ultrastructure and biochemistry have been much investigated (Smyth & McManus, 1989). A cDNA library has recently been constructed (Cougle *et al.*, 1991).

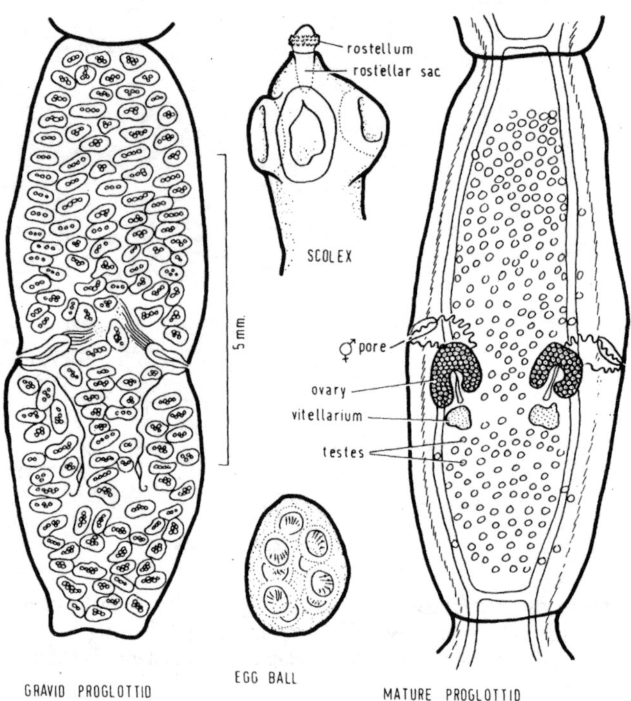

Fig. 23.26
Dipylidium caninum: morphology of scolex and proglottides. (After Chandler, 1955.)

○○○ 23.4 Other Cyclophyllidea

Diphylidium caninum

A tapeworm of dogs, cats, foxes, and occasionally children, it is found all over the world. The characteristics of the scolex and proglottides are shown in Fig. 23.26. The genitalia are duplicated in each proglottid. The gravid proglottides when passed in faeces may show some activity. *D. caninum* occurs in the ileum but as the worm burden rises, the posterior segment of the jejunum is occupied. The dog flea, *Ctenocephalides canis*, the cat flea, *C. felis*, the human flea, *Pulex irritans*, and the dog louse, *Trichodectes canis*, can act as intermediate hosts. The factors affecting larval development have been examined in detail by Pugh & Moorhouse (1985). Temperature and the saturation deficit (at 3 mmHg) were important factors, but the parasite development was found to be independent of the stage of host development. No perceptible growth occurred at 20 °C and maximum growth occurred at 35 °C. The life cycle, intermediate host development, experimental infection, diagnosis, vector control and human dipylidiasis have been reviewed in detail by Boreham & Boreham (1990).

More than 20 cases of dipylidiasis have been recorded in children in the USA.

Genus *Moniezia*

Two species of this genus, *M. expansa* and *M. benedeni*, occur in sheep and other ruminants in many parts of the world. They are long worms (up to 0.6 m) with duplicate sets of genitalia in mature proglottides. The occurrence of the worm is seasonal in Great Britain. Lambs or sheep acquire the infection during the summer months (pre-September), and the eggs are eaten by mites on the grass. Mites of the genera *Galumna*, *Scheloribates* and *Zygoribatula* are the chief vectors. In young lambs, *M. expansa* reaches sexual maturity in 30–48 days, and *M. benedeni* in 30–52 days. The life of the worm appears to be limited to three months, an unusual characteristic for a cestode, and one probably related to the physiology of the host.

Mesocestoides corti

The genus *Mesocestoides* is characterised by the presence of a *paruterine organ*: a thickened sac in which the delicate oncospheres are protected and in which they develop (Fig. 23.27). The adult occurs mainly in carnivores such as the fox, dog and skunk (Voge, 1955). *M. corti* is one of the most useful organisms for teaching and research purposes as its larva or *tetrathyridium* (Fig. 23.28) can easily be maintained in mice by intraperitoneal injection (Specht & Voge, 1965). Moreover, the tetrathyridia can be cultured *in vitro* to develop either sexually or asexually (Smyth, 1990). Its life cycle is also unique in that the adult worm in the gut can (remark-

Table 23.4. *Relation between length, mass and sexual maturity and worm load of* Dipylidium caninum *in 40 dogs*

	Growth data			Stage of development (per cent)			
Worm load	Mean length (cm)	Mean fresh mass per worm (g)	Mean fresh mass of whole population (g)	Scolex only	With genital anlagen	With sexual segments	With gravid segments
10	37	0.47	2.5	0	0	18	82
50	13	0.20	5.5	0	13	33	54
200	7.2	0.07	11.1	15	25	36	34
500	3	0.03	15.3	24	27	45	4

Source: Data from Zebrowska (1974).

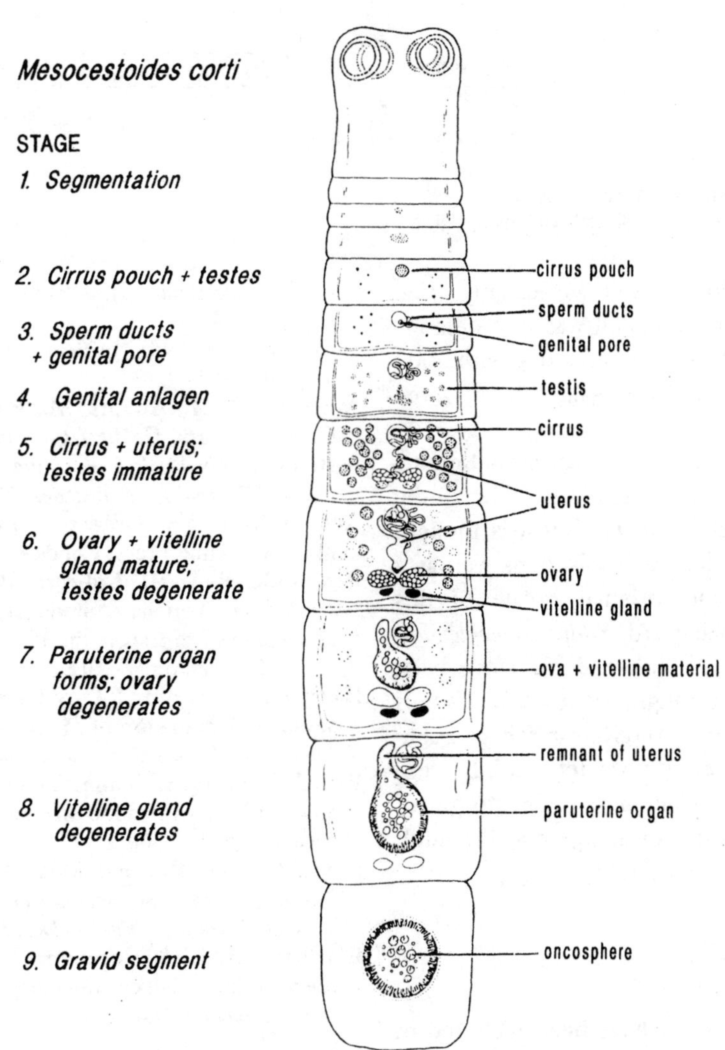

Fig. 23.27
Mesocestoides corti: diagrammatic representation of the stages of sexual differentiation of an adult worm grown from a tetrathyridium *in vitro*. (After Barrett *et al.*, 1982.)

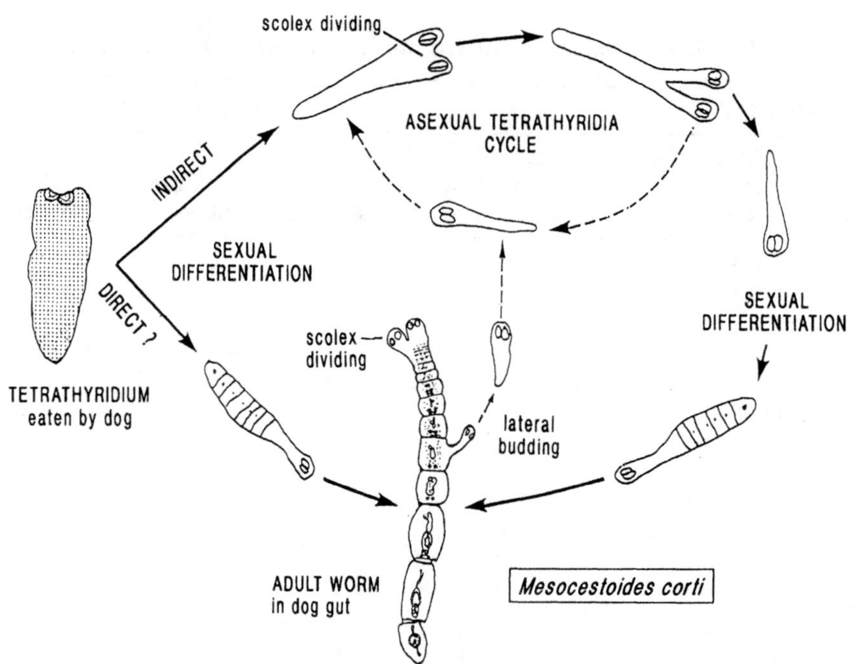

Fig. 23.28
Mesocestoides corti: pattern of growth and asexual–sexual differentiation in the intestine of a dog. (Modified from Smyth, 1987.)

ably!) reproduce asexually as well as sexually (Fig. 23.28). This surprising phenomenon was first demonstrated by Eckert *et al.* (1969) who fed a dog with 1000 tetrathyridia and on autopsy, 11 weeks later, found some 40 000 worms in the gut.

The life cycle and the mode of asexual/sexual multiplication is shown in Fig. 23.28. The remainder of the life cycle is still incompletely known, but it is thought that another intermediate host is involved, as well as a rodent (Smyth, 1990). The original isolate of this species came from the fence lizard, *Scalporus occidentalis biseriatus* (Specht & Voge, 1965), and has since been distributed to laboratories throughout the world. Details of *in vitro* culture techniques are given in Chapter 34.

A related species, *M. lineatus*, does not multiply asexually in the intermediate rodent host or in the gut of the definitive host. It has the advantage that the adult will develop in hamsters (Conn, 1990).

○○○ References

(References of general interest have been included in this list, in addition to those quoted in the text.)

Al-Baldawi, F. A., Mahdi, N. K. & Abdul-Hafidh, B. A. 1989. Resistance of mice to infection with the human strain of *Hymenolepis nana*. *Annals of Tropical Medicine and Parasitology*, **83**: 275–7.

Andreassen, J. 1981. Immunity to adult cestodes. *Parasitology*, **82**: 153–9.

Andrews, R. H., Adams, M., Baverstock, P. R., Behm, C. A. & Bryant, C. 1989. Genetic characterization of three strains of *Hymenolepis diminuta* at 39 enzyme loci. *International Journal for Parasitology*, **19**: 515–18.

Arai, H. (ed.) 1980. *Biology of the Tapeworm Hymenolepis diminuta*. Academic Press, London.

Arambulo, P. V. III, Cabrera, B. D. & Tongson, M. S. 1976. Studies on the zoonotic cycle of *Taenia saginata* taeniasis and cysticercosis in the Philippines. *International Journal of Zoonoses*, **3**: 77–104.

Barrett, N. J., Smyth, J. D. & Ong, S.-J. 1982. Spontaneous sexual differentiation of *Mesocestoides* tetrathyridia *in vitro*. *International Journal for Parasitology*, **12**: 315–22.

Beard, T. C. 1973. Elimination of echinococcosis from Iceland with consideration of special circumstances. *WHO Bulletin*, **48**: 653–60.

Beveridge, I. & Rickard, M. D. 1975. The development of *Taenia pisiformis* in various definitive host species. *International Journal for Parasitology*, **5**: 633–9.

Boreham, R. E. & Boreham, P. F. L. 1990. *Dipylidium caninum*: life cycle, epizootiology, and control. *Compendium of Continuing Education for the Practicing Veterinarian*, **12**: 667–71.

Braten, T. & Hopkins, C. A. 1969. The migration of *Hymenolepis diminuta* in the rat's intestine during normal development and following surgical transplantation. *Parasitology*, **59**: 891–905.

Bryant, C. & Flockhart, H. A. 1987. Biochemical strain variation in parasitic helminths. *Advances in Parasitology*, **25**: 275–319.

Bürger, A.-J. 1984. Survival of *Taenia saginata* eggs in sewage and on pasture. In: C.E.C. (Agriculture), *Report of Parasitological Symposium* (Lyons, 1983), pp. 155–65. Official Publication of the European Communities, Luxemburg.

Cannon, C. E. & Mettrick, D. F. 1970. Changes in the distribution of *Hymenolepis diminuta* (Cestoda: Cyclophyllidea) within the rat intestine during the prepatent development. *Canadian Journal of Zoology*, **48**: 761–9.

Coil, W. H. 1986. The early embryology of *Hymenolepis diminuta. Proceedings of the Helminthological Society of Washington*, **53**: 38–47.

Conn, D. B. 1990. The rarity of asexual reproduction among *Mesocestoides* tetrathyridia (Cestoda). *Journal of Parasitology*, **76**: 453–5.

Cook, G. C. 1988. Neurocysticercosis: parasitology, clinical presentation, diagnosis, and recent advances in management. *Quarterly Journal of Medicine*, **68**: 575–83.

Cougle, W. G., Lightowlers, M. W., Bogh, H. O., Rickard, M. D. & Johnson, K. S. 1991. Molecular cloning of *Taenia taeniaeformis* oncosphere antigen genes. *Molecular and Biochemical Parasitology*, **45**: 137–46.

Craig, P. S., Macpherson, C. N. L., Nelson, G. S. 1986. The identification of eggs of *Echinococcus* by immunofluorescence using a specific anti-oncospheral monoclonal antibody. *American Journal of Tropical Medicine and Hygiene*, **35**: 152–8.

Crompton, D. W. T. & Whitfield, P. J. 1969. A hypothesis to account for the anterior migrations of adult *Hymenolepis diminuta* (Cestoda) and *Moniliformis dubius* (Acanthocephala) in the intestine of rats. *Parasitology*, **58**: 227–9.

Dar, F. K. 1987. Human hydatid disease in the Arab World. In: *Paediatric Infections in Arab Countries* (ed. A. Y. Elzouki), pp. 191–7. John Wiley & Sons, New York.

Doherty, M. L., Bassett, H. F., Breathnach, R., Monaghan, M. L. & McErlean, B. A. 1989. Outbreak of acute coenuriasis in adult sheep in Ireland. *Veterinary Record*, **125**: 185.

Dvorak, J. A., Jones, A. W. & Kuhlman, H. H. 1961. Studies on the biology of *Hymenolepis microstoma* (Dujardin, 1845). *Journal of Parasitology*, **47**: 833–8.

Earnest, M. P., Reller, L. B., Filley, C. H. & Grek, A. J. 1987. Neurocysticercosis in the United States: 35 cases and a review. *Reviews of Infectious Diseases*, **9**: 961–79.

Eckert, J. 1986. Prospects for treatment of the metacestode stage of *Echinococcus*. In: *The Biology of Echinococcus and Hydatid Disease* (ed. R. C. A. Thompson), pp. 250–84. George Allen & Unwin, London.

Eckert, J., Brand, T. von & Voge, M. 1969. Asexual multiplication of *Mesocestoides corti* (Cestoda) in the intestine of dogs. *Journal of Parasitology*, **55**: 241–9.

Eckert, J. & Thompson, R. C. A. 1988. *Echinococcus* strains in Europe. *TropenMedizin und Parasitologie*, **39**: 1–8.

Esch, G. W. & Self, J. T. 1965. A critical study of the taxonomy of *Taenia pisiformis* Block, 1780; *Multiceps multiceps* (Leske, 1780); and *Hydatigera taeniaeformis* (Batsch, 1786). *Journal of Parasitology*, **51**: 932–7.

Fan, P. C. 1988. Taiwan *Taenia* and taeniasis. *Parasitology Today*, **4**: 86–8.

Fan, P. C., Chung, W. C., Lin, C. Y. & Wu, C. C. 1990. The pig as an intermediate host for Taiwan *Taenia* infection. *Journal of Helminthology*, **64**: 223–31.

Fan, P. C., Lin, C. Y. & Kosin, E. 1989. Experimental infection of Indonesia *Taenia* (Samosir strain) in domestic animals. *International Journal for Parasitology*, **19**: 809–12.

Ferretti, G., Gabriele, F. & Palmas, C. 1981. Development of human and mouse strain of *Hymenolepis nana* in mice. *International Journal for Parasitology*, **11**: 424–30.

Flisser, A. 1985. Cysticercosis: a major threat to human health and livestock production. *Food Technology*, **39**: 61–4.

Flisser, A. 1988. Neurocysticercosis in Mexico. *Parasitology Today*, **4**: 131–7.

Flisser, A. *et al.* 1982. *Cysticercosis: Present State of Knowledge and Perspectives*. Academic Press, New York. ISBN 0 12 260740 6.

Flisser, A. *et al.* 1985. The immunology of human and animal cysticercosis: a review. *WHO Bulletin*, **57**: 839–56.

Flisser, A. *et al.* 1990. New approaches in the diagnosis of *Taenia solium* cysticercosis and taeniasis. *Annales de Parasitologie Humaine et Comparée*, **65** (Suppl. **1**): 95–8.

Freeman, R. S. 1962. Studies on the biology of *Taenia crassiceps* (Zeder, 1800) Rudolphi, 1810 (Cestoda). *Canadian Journal of Zoology*, **40**: 969–90.

Gajdusek, D. C. 1977. Urgent opportunistic observations: the study of changing, transient and disappearing medical interest in disrupted primitive human communities. In: *Health and Disease in Tribal Societies* (Ciba Foundation Symposium No. 49), pp. 69–102. Elsevier, Amsterdam.

Gajdusek, D. C. 1978. Introduction of *Taenia solium* into West New Guinea with a note on an epidemic of burns from cysticercosis epilepsy in the Ekari people of the Wissel Lakes area. *Papua and New Guinea Medical Journal*, **21**: 329–42.

Gemmell, M. A. 1990. Australian contributions to an understanding of the epidemiology and control of hydatid disease caused by *Echinococcus granulosus* – past, present and future. *International Journal for Parasitology*, **20**: 431–56.

Gemmell, M. A. & Lawson, J. D. 1985. Epidemiology and control of hydatid disease. In: *The Biology of Echinococcus and Hydatid Disease* (ed. R. C. A. Thompson), pp. 189–216. George Allen & Unwin, London.

Harrison, L. J. S. & Sewell, M. M. H. 1991. The zoonotic *Taenia* of Africa. In: *Parasitic Helminths and Zoonoses in Africa* (ed. C. N. L. Macpherson & P. S. Craig), pp. 54–82. Unwin Hyman, London.

Harrison, L. J. S., Delgardo, J. & Parkhouse, R. M. E. 1990. Differential diagnosis of *Taenia saginata* and *Taenia soliium* with DNA probes. *Parasitology*, **100**: 459–61.

Heath, D. D. 1986. Immunobiology of *Echinococcus* infections. In: *The Biology of Echinococcus and Hydatid Disease* (ed. R. C. A. Thompson), pp. 164–88. George Allen & Unwin, London.

Henderson, D. J. & Hanna, R. E. B. 1987. *Hymenolepis nana* (Cestoda: Cyclophyllidea): migration, growth and development in the laboratory mouse. *International Journal for Parasitology*, **17**: 1249–56.

Hickman, J. L. 1964. The biology of *Hymenolepis microstoma* (Dujardin). *Papers and Proceedings of the Royal Society of Tasmania*, **98**: 73–7.

Hird, D. W. & Pullen, M. M. 1979. Tapeworms, meat and man: a brief review and update of cysticercosis caused by *Taenia saginata* and *Taenia solium*. *Journal of Food Production*, **42**: 58–64.

Hopkins, C. A., Subramian, G. & Stallard, H. 1972. The development of *Hymenolepis diminuta* in primary and secondary infections in mice. *Parasitology*, **64**: 401–12.

Horton, R. J. 1989. Chemotherapy of *Echinococcus* infection in man with albendazole. *Transactions of the Royal Society of Tropical Medicine and Hygiene*, **83**: 97–102.

Hundley, D. F. & Berntzen, A. K. 1969. Collection, sterilization and storage of *Hymenolepis diminuta* eggs. *Journal of Parasitology*, **55**: 1095–6.

Hunninen, A. V. 1935. Studies on the life history and host –parasite relations of *Hymenolepis fraterna* (*H. nana* var. *fraterna Stiles*) in white mice. *American Journal of Hygiene*, **22**: 414–43.

Hutchison, W. M. 1959. Studies on *Hydatigera* (*Taenia*) *taeniaeformis*. Growth of the adult phase. *Experimental Parasitology*, **8**: 557–67.

Kamiya, M. & Sato, H. 1990*a*. Survival, strobilization and sexual maturation of *Echinococcus multilocularis* in the small intestine of golden hamsters. *Parasitology*, **100**: 125–30.

Kamiya, M. & Sato, H. 1990*b*. Complete life-cycle of the canid tapeworm *Echinococcus multilocularis*, in laboratory rodents alone. *FASEB Journal*, **4**: 3334–9.

Keymer, A. 1981. Population dynamics of *Hymenolepis diminuta* in the intermediate host. *Journal of Animal Ecology*, **50**: 941–50.

Keymer, A. & Anderson, R. M. 1979. The dynamics of infection of *Tribolium confusum* by *Hymenolepis diminuta*: the influence of infective stage density and spatial distribution. *Parasitology*, **79**: 195–207.

Kino, H. & Kennedy, C. R. 1987. Differences in biological characteristics of two strains of the cestode *Hymenolepis diminuta*. *Parasitology*, **95**: 97–110.

Kohlhagen, S., Behm, C. A. & Bryant, C. 1985. Strain variation in *Hymenolepis diminuta*: enzyme profiles. *International Journal for Parasitology*, **15**: 479–83.

Kovalenko, F. P. 1976. Survival of *Echinococcus multilocularis* and *E. granulosus* in the intestine of laboratory animals. *Meditsinskaya Parazitologiya i Parazitarnye Bolezni*, **45**: 350–2.

Kumaratilake, L. M., Thompson, R. C. A., Eckert, J. & D'Alessandro, A. D. 1986*a*. Sperm transfer in *Echinococcus* (Cestoda: Taeniidae). *Zeitschrift für Parasitenkunde*, **72**: 265–9.

Kumaratilake, L. M., Thompson, R. C. A. & Eckert, J. 1986*b*. *Echinococcus granulosus* of equine origin from different countries possess uniform morphological characteristics. *International Journal for Parasitology*, **16**: 529–40.

Kumazawa, H. & Suzuki, N. 1983. Kinetics of proglottid formation, maturation and shedding during development of *Hymenolepis nana*. *Parasitology*, **86**: 275–89.

Lascano, E. F., Coltorti, E. A. & Varela-Diaz, V. M. 1975. Fine structure of the germinal membrane of *Echinococcus granulosus cysts*. *Journal of Parasitology*, **61**: 853–60.

Lethbridge, R. C. 1971. The hatching of *Hymenolepis diminuta* eggs and penetration of the hexacanths in *Tenebrio molitor* beetles. *Parasitology*, **62**: 445–56.

Levi, M. H., Raucher, B. G., Teicher, E., Sheehan, D. J. & McKitrick, J. C. 1987. *Hymenolepis diminuta*: one of three enteric pathogens isolated from a child. *Diagnostic Microbiology and Infectious Disease*, **7**: 255–9.

Lightowlers, M. W. 1990. Immunology and molecular biology of *Echinococcus* infections. *International Journal for Parasitology*, **20**: 471–8.

Lubinsky, G. 1960. The maintenance of *Echinococcus multilocularis* without the definitive host. *Canadian Journal of Zoology*, **38**: 149–51.

McKay, D. M., Halton, D. W., McCaig, M. D., Johnston, C. F., Fairweather, I. & Shaw, C. 1990. *Hymenolepis diminuta*: intestinal goblet cell response to infection of male C57 mice. *Experimental Parasitology*, **71**: 9–20.

McManus, D. P. & Bryant, C. 1986. Biochemistry and Physiology of *Echinococcus*. In: *The Biology of Echinococcus and Hydatid Disease* (ed. R. C. A. Thompson), pp. 114–42. George Allen & Unwin, London.

McManus, D. P. & Rishi, A. K. 1989. Genetic heterogeneity within *Echinococcus granulosus*: isolates from different hosts and geographical areas characterized with DNA probes. *Parasitology*, **99**: 17–29.

McManus, D. P. & Smyth, J. D. 1978. Differences in the chemical composition and carbohydrate metabolism of *Echinococcus granulosus* (horse and sheep strains) and *E. multilocularis*. *Parasitology*, **77**: 103–9.

McManus, D. P. & Smyth, J. D. 1979. Isoelectric focusing of some enzymes from *Echinococcus granulosus* (horse and sheep strain) and *E. multilocularis*. *Transactions of the Royal Society of Tropical Medicine and Hygiene*, **73**: 259–65.

McManus, D. P. & Smyth, J. D. 1986. Hydatidosis: changing concepts in epidemiology and speciation. *Parasitology Today*, **2**: 163–8.

Macpherson, C. N. L. & Craig, P. S. 1991. *Parasitic Helminths and Zoonoses in Africa*. Unwin Hyman, London.

Macpherson, C. N. L., Wachira, T. M., Zeyhle, E., Romig, T. & Macpherson, C. 1986. Hydatid disease: research and control in Turkana, IV. The pilot control programme. *Transactions of the Royal Society of Tropical Medicine and Hygiene*, **80**: 196–200.

Matossian, R. M., Rickard, M. D. & Smyth, J. D. 1977. Hydatidosis: a global problem of increasing importance. *Bulletin of the World Health Organization*, **55**: 499–507.

Meloni, B. P., Lymbery, A. J., Thompson, R. C. A. & Gracey, M. 1988. 1990. High prevalence of *Giardia lamblia* in children from a West Australian Aboriginal community. *Medical Journal of Australia*, **149**: 715.

Miyaji, S., Oku, Y., Kamiya, M., Okamoto, M., Ohbayashi, M., Uchida, A. & Rausch, R. L. 1990. Growth of a Japanese isolate of *Taenia crassiceps* in intermediate and definitive hosts. *Parasitology Research*, **76**: 351–4.

Morseth, D. J. 1967. The fine structure of the hydatid cyst and the protoscolex of *Echinococcus granulosus. Journal of Parasitology*, **53**: 312–25.

Nelson, G. S. 1989. Microepidemiology, the key to the control of parasitic infections. *Transactions of the Royal Society of Tropical Medicine and Hygiene*, **84**: 3–13.

Pappas, P. W. & Leiby, D. A. 1986. Variation in the size of eggs and oncospheres and the numbers and distribution of testes in the tapeworm, *Hymenolepis diminuta. Journal of Parasitology*, **72**: 383–91.

Pawlowski, P. W. 1984. Cestodiases: Taeniasis, Diphyllobothriasis, Hymenolepiasis, and others. In: *Tropical and Geographical Medicine* (ed. K. S. Warren & A. F. Mahmoud), pp. 471–86. McGraw-Hill, New York. ISBN 0 07 068327 1.

Pawlowski, Z. S. 1990. Perspectus and control of *Taenia saginata. Parasitology Today*, **6**: 371–2.

Pawlowski, Z. S. & Schultz, M. G. 1972. Taeniasis and cysticercosis (*Taenia saginata*). *Advances in Parasitology*, **10**: 269–343.

Polydorou, K. 1987. Control of echinococcosis in Cyprus. In: *Helminth Zoonoses* (ed. S. Geerts, V. Kumar & J. Brandt), pp. 60–7. Martinus Nijhoff, Dordrecht, Netherlands. ISBN 0 89838 896 1.

Pugh, R. E. & Moorhouse, D. E. 1985. Factors affecting the development of *Dipylidium caninum* in *Ctenocephalides felis felis* (Bouché, 1835). *Zeitschrift für Parasitenkunde*, **71**: 765–75.

Rausch, R. L. 1986. Life-cycle patterns and geographic distribution of *Echinococcus* species. In: *The Biology of Echinococcus and Hydatid Disease* (ed. R. C. A. Thompson), pp. 44–80. George Allen & Unwin, London.

Read, C. P. & Kilejian, A. Z. 1969. Circadian migratory behaviour of a cestode symbiote in the rat host. *Journal of Parasitology*, **55**: 574–8.

Richards, K. S., Arme, C. & Bridges, J. F. 1983. *Echinococcus granulosus granulosus*: an ultrastructural study of the laminated layer, including changes on incubating cysts in various media. *Parasitology*, **86**: 399–405.

Rickard, M. D. & Lightowlers, M. W. 1986. Immunodiagnosis of hydatid disease. In: *The Biology of Echinococcus and Hydatid Disease* (ed. R. C. A. Thompson), pp. 217–49. George Allen & Unwin, London.

Rietschel, G. 1981. Beitrag zur Kenntnis von *Taenia crassiceps* (Zeder, 1800) Rudolphi, 1810 (Cestoda, Taeniidae). *Zeitschrift für Parasitenkunde*, **65**: 309–15.

Rishi, A. K. & McManus, D. P. 1987. DNA probes which unambiguously distinguish *Taenia solium* from *Taenia saginata. The Lancet*, **ii** (No. 8570): 1275–6.

Rishi, A. K. & McManus, D. P. 1988. Molecular cloning of *Taenia solium* genomic DNA and characterization of taeniid cestodes by DNA analysis. *Parasitology*, **97**: 161–76.

Roberts, L. S. 1961. The influence of population density on patterns and physiology of growth in *Hymenolepis diminuta* (Cestoda: Cyclophyllidea) in the definitive host. *Experimental Parasitology*, **11**: 332–71.

Rycke, P. H. de 1966. Development of the cestode *Hymenolepis microstoma* in *Mus musculus. Zeitschrift für Parasitenkunde*, **27**: 350–4.

Sato, H. & Kamiya, M. 1989. Viable egg production of *Taenia crassiceps* developed in the intestine of prednisolone-treated golden hamsters. *Japanese Journal of Parasitology*, **38**: 46–53.

Schantz, P. M. 1984. Echinococcosis (Hydatidosis). In: *Tropical and Geographical Medicine* (ed. K. S. Warren & A. A. F. Mahmoud), pp. 487–97. McGraw-Hill, New York.

Schiller, E. L. 1959. Experimental studies on morphological variation in the cestode genus *Hymenolepis*. II. Growth, development and reproduction of the strobilate phase of *H. nana* in different mammalian host species. *Experimental Parasitology*, **8**: 215–35.

Schwabe, C. W. 1986. Current status of hydatid disease: a zoonosis of increasing importance. In: *The Biology of Echinococcus and Hydatid Disease* (ed. R. C. A. Thompson), pp. 81–113. George Allen & Unwin, London.

Shorb, D. A. 1933. Host–parasite relations of *Hymenolepis fraterna* in the rat and the mouse. *American Journal of Hygiene*, **18**: 74–113.

Silverman, P. H. 1954*a*. Studies on the biology of some tapeworms of the genus *Taenia*. I. Factors affecting hatching and activation of taeniid ova, and some criteria of their viability. *Annals of Tropical Medicine and Parasitology*, **48**: 207–15.

Silverman, P. H. 1954*b*. The hatching stimuli of taeniid ova and a description of secreting glands in the activated hexacanth embryo. *Transactions of the Royal Society of Tropical Medicine and Hygiene*, **48**: 287.

Silverman, P. H. 1954*c*. Studies on the biology of some tapeworms of the genus *Taenia*. II. The morphology and development of the taeniid hexacanth embryo and its enclosing membranes, with some notes of the state of development and propagation of gravid segments. *Annals of Tropical Medicine and Parasitology*, **48**: 356–66.

Smith, J. K., Esch, G. W. & Kuhn, R. E. 1972. Growth and development of larval *Taenia crassiceps* (Cestoda). I. Aneuploidy in the anomalous ORF strain. *International Journal for Parasitology*, **2**: 261–3.

Smyth, J. D. 1964*a*, 1969*a*. The biology of the hydatid organisms. *Advances in Parasitology*, **2**: 169–219; **7**: 327–47.

Smyth J. D. 1964*b*. Observations on the scolex of *Echinococcus granulosus*, with special reference to the occurrence of secretory cells in the rostellum. *Parasitology*, **54**: 515–26.

Smyth, J. D. 1967. Studies on tapeworm physiology. XI. *In vitro* cultivation of *Echinococcus granulosus* from the protoscolex to the strobilate stage. *Parasitology*, **57**: 111–33.

Smyth, J. D. 1969*b*. *The Physiology of Cestodes*. W. H. Freeman, San Francisco.

Smyth, J. D. 1971. Development of monozoic forms of *Echinococcus granulosus* during *in vitro* culture. *International Journal for Parasitology*, **1**: 121–4.

Smyth, J. D. 1982. The insemination–fertilization problem in cestodes cultured *in vitro*. In: *Aspects of Parasitology* (ed. E. Meerovitch), pp. 393–406. McGill University, Montreal.

Smyth, J. D. 1987. Asexual and sexual differentiation in cestodes: especially *Mesocestoides* and *Echinococcus*. In: *Molecular Paradigms for Eradicating Parasites* (ed. A. MacInnis). *UCLA Symposium on Molecular and Cellular Biology*, New Series, Vol. 60, pp. 19–34. Alan R. Liss, New York.

Smyth, J. D. 1990. *In Vitro Cultivation of Parasitic Helminths*. CRC Press, Boca Raton, Florida, USA.

Smyth, J. D. & Barrett, N. J. 1979. *Echinococcus multilocularis*: further observations on strobilar differentiation *in vitro*. *Revista Ibérica de Parasitologia*, **39**: 39–53.

Smyth, J. D. & Barrett, N. J. 1980. Procedures for testing the viability of human hydatid cysts following surgical removal: especially after chemotherapy. *Transactions of the Royal Society of Tropical Medicine and Hygiene*, **74**: 649–52.

Smyth, J. D. & Davies, Z. 1974*a*. Occurrence of physiological strains of *Echinococcus granulosus* demonstrated by *in vitro* culture of protoscoleces from sheep and horse hydatid cysts. *International Journal for Parasitology*, **4**: 443–5.

Smyth, J. D. & Davies, Z. 1974*b*. *In vitro* culture of the strobilar stage of *Echinococcus granulosus* (Sheep strain): a review of basic problems and results. *International Journal for Parasitology*, **4**: 631–44.

Smyth, J. D. & Davies, Z. 1975. *In vitro* suppression of segmentation in *Echinococcus multilocularis* with morphological transformation of protoscoleces into monozoic adults. *Parasitology*, **71**: 125–35.

Smyth, J. D. & Heath, D. D. 1970. Pathogenesis of larval cestodes in mammals. *Helminthological Abstracts*, **39**: 1–23.

Smyth, J. D. & McManus, D. P. 1989. *The Physiology and Biochemistry of Cestodes*. Cambridge University Press. ISBN 0 521 35557 5.

Smyth, J. D. & Smyth, M. M. 1964. Natural and experimental hosts of *Echinococcus granulosus* and *E. multilocularis*, with comments on the genetics of speciation in the genus *Echinococcus*. *Parasitology*, **54**: 493–514.

Smyth, J. D. & Smyth, M. M. 1969. Self-insemination in *Echinococcus granulosus in vivo*. *Journal of Helminthology*, **43**: 383–8.

Specht, D. & Voge, M. 1965. Asexual multiplication of *Mesocestoides* tetrathyridia in laboratory animals. *Journal of Parasitology*, **51**: 268–72.

Tesfa-Yohannes, T.-M. 1990. Effectiveness of praziquantel against *Taenia saginata* infections in Ethiopia. *Annals of Tropical Medicine and Parasitology*, **84**: 581–5.

Thompson, R. C. A. (ed.) 1986*a*. *The Biology of Echinococcus and Hydatid Disease*. George Allen & Unwin, London. ISBN 0 04 591020 0.

Thompson, R. C. A. 1986*b*. Biology and systematics of *Echinococcus*. In: *The Biology of Echinococcus and Hydatid Disease* (ed. R. C. A. Thompson), pp. 5–43. George Allen & Unwin, London.

Thompson, R. C. A. & Allsopp, C. E. 1988. *Hydatidosis: Veterinary Aspects and Annotated Bibliography*. C.A.B. International, Wallingford, UK. ISBN 0 85198 610 2.

Thompson, R. C. A., Kumaratilake, L. M. & Eckert, J. 1984. Observations on *Echinococcus granulosus* of cattle origin in Switzerland. *International Journal for Parasitology*, **14**: 283–91.

Thompson, R. C. A. & Lymbery, A. J. 1988. The nature, extent and significance of variation within the genus *Echinococcus*. *Advances in Parasitology*, **27**: 209–58.

Thompson, R. C. A. & Lymbery, A. J. 1990*a*. Intraspecific variation in parasites – what is a strain? *Parasitology Today*, **6**: 345–8.

Thompson, R. C. A. & Lymbery, A. J. 1990*b*. *Echinococcus*: biology and strain variation. *International Journal for Parasitology*, **20**: 457–70.

Thompson, R. C. A. & Smyth, J. D. 1975. Equine hydatidosis: a review of the current status in Great Britain and the results of an epidemiological survey. *Veterinary Parasitology*, **1**: 107–27.

Ubelaker, J. E. 1980. Structure and ultrastructure of the larvae and metacestodes of *Hymenolepis diminuta*. In: *Biology of the Tapeworm Hymenolepis diminuta* (ed. H. P. Arai), pp. 59–156. Academic Press, New York.

Ubelaker, J. E. 1983. The morphology, development and evolution of tapeworm larvae. In: *Biology of the Eucestoda*, Vol. 1 (ed. C. Arme & P. W. Pappas), pp. 235–96. Academic Press, London. ISBN 0 12 062101 0.

Voge, M. 1955. North American cestodes of the genus *Mesocestoides*. *University of California Publications in Zoology*, **59**: 125–56.

Voge, M. & Heyneman, D. 1957. Development of *Hymenolepis nana* and *Hymenolepis diminuta* (Cestoda: Hymenolepididae) in the intermediate host *Tribolium confusum*. *University of California Publications in Zoology*, **59**: 549–80.

Wakelin, D. 1984. *Immunity to Parasites. How Animals Control Parasitic Infections*. Edward Arnold, London.

Webbe, G. 1986. Cestode infections of man. In: *Chemotherapy of Parasitic Diseases* (ed. W. C. Campbell & R. S. Rew), pp. 457–77. Plenum Press, New York.

***WHO** (1981). Guidelines for surveillance, prevention and control of Echinococcosis/Hydatidosis. Ref: VPH/81.28.

***WHO** (1983). Guidelines for surveillance, prevention and control of Taeniasis/ Cysticercosis. Ref: VPH/83.49.

Wilson, J. F. & Rausch, R. L. 1980. Alveolar hydatid disease. A review of clinical features of 33 indigenous cases of *Echinococcus multilocularis* infection in Alaskan Eskimos. *American Journal of Tropical Medicine and Hygiene*, **29**: 1340–55.

Yamaguti, S. 1959. *Systema Helminthum*. Vol. II. *Cestodes of Vertebrates*. Interscience Publishers, New York.

Zebrowska, D. 1964. The intra-population aspects of *Dipylidium caninum* (Linné, 1758) in the alimentary tract of dogs. *Wiadomości Parazytologiczne*, **10**: 553–8.

Zoli, A., Geerts, S. & Vervoort, T. 1987. An important focus of porcine and human cysticercosis. In: *Helminth Zoonoses* (ed. S. Geerts, V. Kumar & V. Brandt), pp. 85–91. Martinus Nijhoff, Dordrecht, Netherlands. ISBN 0 89838 896 1.

*Copies obtainable from WHO, quoting title and reference number. An Arabic edition of the echinococcosis/hydatidosis volume is available.

24

Physiology of cestodes

24.1 General account

Cestodes may be considered to be biological models of exceptional physiological interest on account of (*a*) the unusual habitat in which they live, (*b*) their basic morphology, which is characterised by the lack of an alimentary canal, and (*c*) their pattern of strobilar growth.

Expanding these points further, it can be seen that:

(*a*) Cestodes live in an environment whose physico-chemical properties and nutritional level may vary in relation to the feeding pattern and nutritional condition of the host; some species (e.g. *Hymenolepis diminuta*) undergo diurnal migration patterns in relation to the host feeding pattern (p. 322).

(*b*) Cestodes lack a digestive tract and a defined circulatory system; their outer body covering (the tegument) is essentially a 'naked' protoplasmic surface (Fig. 20.2) through which all substances must both enter and leave the body; *active* rather than passive transport appears to be generally involved in these processes (p. 355). The occurrence of this naked surface implies that cestodes would be strongly antigenic (Chapter 32).

(*c*) It has generally been assumed that, for nutritional purposes, cestodes utilise small molecules derived from the digested food sources of the host. The situation appears to be more complex, however, for it is now known that cestodes are capable of fixing carbon dioxide (p. 356) and that this process serves as a valuable source of carbon atoms. Furthermore, there is some evidence that cestodes may be capable of utilising larger molecules, such as protein, either by direct uptake or by a process known as membrane (= contact) digestion (Dubovskaya, 1970; Smyth, 1972); this hypothesis is, however, by no means proved. In the latter process (p. 349; Fig. 24.1) membrane-bound enzymes are believed to be capable of digesting large molecules with which they come in contact.

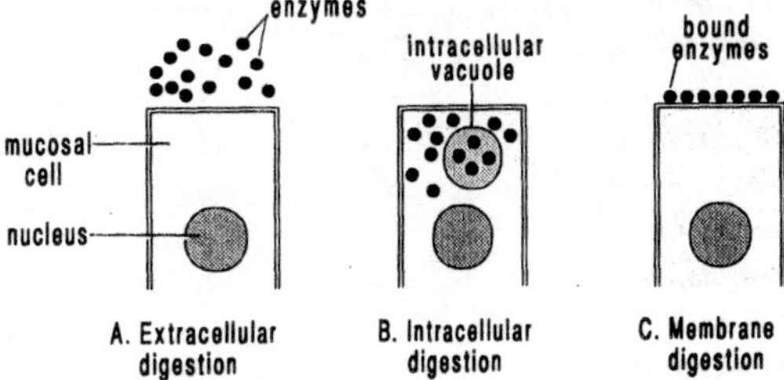

Fig. 24.1
Comparison of membrane digestion with extracellular and intracellular digestion. (After Ugolev, 1968.)

As discussed on p. 284, whether pinocytosis occurs in cestodes is a much debated question; to date, unequivocal evidence in support of this process is still lacking.

(*d*) Possibly in relation to the process of membrane digestion, mentioned above, it is likely that in some cestodes (e.g. *Echinococcus granulosus*) the scolex acts as an organ of nutrition (i.e. has a placenta-like function) as well as of attachment.

(*e*) The growth of cestodes is unusual in that it results in the continuous production of *embryonic* tissue (i.e. the strobila) which differentiates into a string of sexual individuals which eventually produce fertile eggs. Certain species (*E. granulosus*, *Taenia serialis*, and probably *T. crassiceps*) are capable of *de*differentiation and, under certain conditions, cystic forms can give rise to either strobilar or cystic forms depending on which stimulus has been applied (Figs 34.8–34.10).

The biochemistry and physiology of cestodes has been comprehensively reviewed by Smyth & McManus (1989) and specific aspects have been reviewed by Arai

Table 24.1. *Analyses of cestodes, expressed as a percentage of dry mass*

For figures to be of value, the nutritional condition of the host and the analytical method used should be known. Figures in italics have been calculated from authors' data.

Species	Dry mass as % of fresh mass	Glycogen	Lipid	Protein	Inorganic substances	Stage	References*
Anoplocephala magna (Taenia plicata)	27.5	6	33.1	—	1.22	adult	von Brand (1952)
Calliobothrium verticillatum	27.6	—	—	—	—	adult	Read *et al* (1959)
Cittotaenia perplexa	27.1	—	—	20.60	—	adult	Campbell (1960*b*)
Diphyllobothrium latum	—	17.9	—	—	—	plerocercoid	Markov (1939)
Diphyllobothrium latum	9.0	20	*16.6*	60	4.8	adult	von Brand (1952)
Diphyllobothrium sp.	29.9	31.5	—	41	—	plerocercoid	Archer & Hopkins (1958*b*)
Diphyllobothrium sp.	30.8	36.2	—	48	—	adult	Archer & Hopkins (1958*b*)
Diphyllobothrium dendriticum	27.0	36.2	—	—	—	plerocercoid	Reuter (1967)
Dipylidium caninum	20.4	—	—	—	—	adult	von Brand (1952)
Echinococcus granulosus	14.8	19.8	13.6	62.5	*13.5*	protoscoleces	Agosin *et al.* (1957)
Eubothrium rugosum	—	22.8	—	—	—	adult	von Brand (1952)
Hydatigera (Taenia) taeniaeformis	22.3	*19.7*	*4.2*	*27.1*	*29.0*	strobilocercus (mice)	von Brand & Bowman (1961)
Hydatigera (Taenia) taeniaeformis	28	43.3	5.3	*26.3*	18.1	strobilocercus (mice)	Hopkins (1960); Hopkins & Hutchison (1960)
Hydatigera (Taenia) taeniaeformis	29.5	*24.9*	*3.1*	*28.9*	*28.4*	strobilocercus (rats)	von Brand & Bowman (1961)
Hydatigera (Taenia) taeniaeformis	20	—	6.9	*40.6*	—	adult	Hopkins (1960); Hopkins & Hutchison (1958)
Hydatigera (Taenia) taeniaeformis	26.7	*23.2*	*6.3*	*45.0*	*22.0*	adult	von Brand & Bowman (1961)
Hymenolepis diminuta	*22.4*	*45.7*	*20.1*	*31.0*	—	adult	Fairbairn *et al.* (1961)
Hymenolepis citelli	—	—	16.1	—	—	adult	Harrington (1965)
Ligula intestinalis	29.0	38–52	—	35–45	—	plerocercoid	Markov (1939)
Moniezia expansa	9.2–11.0	24–32	30.1	21.8	10.5	adult	von Brand (1952, 1966); Campbell (1960*b*)
Multiceps multiceps (Coenurus cerebralis)	25.3	—	—	—	27.4	scolex	von Brand (1952)
Multiceps multiceps	12.4	—	—	—	4.1	membranes	von Brand (1952)
Raillietina cesticillus	20.5	31.8	15.5	*36.4*	11.5	adult	Reid (1942)
Schistocephalus solidus	31.8	50.9	—	35.8	5.8	plerocercoid	Hopkins (1950)
Schistocephalus solidus	38	28.0	—	—	—	adult	Hopkins (1960)
Spirometra mansonoides	—	—	16	—	—	plerocercoid	Meyer *et al.* (1966)
Spirometra mansonoides	—	—	24	—	—	adult	Meyer *et al.* (1966)
Taenia crassiceps	20.0	27.5	—	—	—	larva	Taylor *et al.* (1966)
Taenia hydatigena (marginata)	16–26	28	4.9	—	—	adult	Featherston (1969); von Brand (1952)
Taenia saginata	12.2	48.8	11.2	32.0	*5.3*	adult	von Brand (1952); Machnicka-Roguska (1961)
Taenia solium	8.7	*25.4*	*16.2*	46	6.4	adult	von Brand (1952)
Thysanosoma actinioides	16.3	—	—	29.0	—	adult	Campbell (1960*b*)
Triaenophorus nodulosus	—	13.8	—	—	—	adult	von Brand (1952)

*Most references given in Smyth (1969) and Smyth & McManus (1989).

Table 24.2. *The biochemical composition (in μg/mg⁻¹ dry mass ± s.e.) of protoscoleces and adults of* Echinococcus granulosus *from Kenya and the UK*

Host origin	Protein	Polysaccharides	Lipids	RNA	DNA
Horse[a]	550 ± 10	177 ± 2	109 ± 4	60 ± 4	5 ± 0.1
Sheep[a]	625 ± 12	169 ± 1	88 ± 3	89 ± 4	5 ± 0.1
Sheep	592 ± 16	166 ± 4	76 ± 4	74 ± 6	7 ± 1
Cattle	547 ± 14	146 ± 3	118 ± 6	52 ± 4	6 ± 1
Goat	569 ± 16	166 ± 4	95 ± 3	54 ± 4	6 ± 1
Camel	550 ± 17	215 ± 6	113 ± 4	53 ± 4	5 ± 1
Human	668 ± 16	164 ± 4	122 ± 4	57 ± 5	4 ± 1
Dog[b]	712 ± 18	139 ± 5	143 ± 4	96 ± 6	6 ± 1
Dog (natural)	584 ± 16	213 ± 8	122 ± 4	203 ± 7	4 ± 1

[a]From the UK.
[b]Experimental infection with protoscoleces of human origin.
Source: Data from McManus & Smyth (1978) and McManus (1981).

(1980), Arme & Pappas, (1983*a*,*b*), Barrett (1981), McManus (1987) and McManus & Bryant (1986).

○○○ 24.2 Chemical composition

○ 24.2.1 General comment

Some available information is summarised in Table 24.1. As emphasised elsewhere, data on the chemical analysis of parasitic worms are of little value unless qualifying information is provided on the nutritional state of the host at the time of autopsy. The carbohydrate content especially is liable to fluctuate, and as this affects the dry mass, changes in other constituents, such as protein, when expressed as a percentage, may be more apparent than real. The lipid content is also directly related to that of the host. Moreover, the chemical composition may vary with the 'strain' of both parasite and host. It can also vary in specimens from different regions of the intestine. This has been clearly demonstrated in *Echinococcus granulosus* from a wide range of hosts, in which substantial differences in the protein, polysaccharide, lipid, RNA and DNA contents have been recorded (Table 24.2).

The data must also be accepted with caution on technical grounds, because some of the older analytical methods have been shown, by more modern workers, to be unreliable.

Major chemical constituents. The proportions of the main tissue constituents – protein, lipid and carbohydrate – show a somewhat different pattern from most other invertebrates in that the carbohydrate content (largely in the form of glycogen) tends to be high and the protein relatively low. Larval cestodes are particularly rich in glycogen, which may reach astounding levels (over 50 per cent in the plerocercoids of *Ligula*) (Table 24.1).

Calcareous corpuscles. Cestodes (and trematodes) often contain enormous numbers of curious bodies termed *calcareous corpuscles*, made up of an organic base together with inorganic material. They vary much in size; in some species they are very large, 16–32 μm (*Spirometra mansonoides*, *Echinococcus granulosus*), in most species they may be as small as 12 μm. The organic material contains RNA, DNA, proteins, polysaccharides and alkaline phosphatase (Smyth, 1969). The inorganic materials are mainly calcium, magnesium, phosphorus and carbon dioxide with traces of other metals, but these constituents, especially phosphate, can vary considerably in relation to metabolic conditions and to origin. Ultrastructural studies have shown that in *T. taeniaeformis*, the corpuscles are formed intra-cellularly and a single corpuscle is formed in one cell, which is apparently destroyed in the process (Nieland & von Brand, 1969).

Species in which the calcareous corpuscles have been especially studied are: *Mesocestoides corti* (Baldwin *et al.*, 1978; Kegley *et al.*, 1969, 1970); *Hymenolepis microstoma* (Chowdhury & De Rycke, 1977); *Diplogonoporus grandis* (Ishii, 1984); *Taenia taeniaeformis* (Von Brand & Weinbach, 1975).

The role of calcareous corpuscles is not clear but it

has been speculated that they act as major reserves for (*a*) phosphates and (*b*) other organic ions together with carbon dioxide. These materials may be called on suddenly – phosphates for phosphorylation, the ions to act as enzyme catalysts and the carbon dioxide for CO_2 fixation – such as when a larva enters a host intestine and immediately requires a large amount of energy for its establishment process; such processes frequently involve muscular attachment. Another view is that the calcareous corpuscles buffer anaerobically produced acids.

○ 24.2.2 Carbohydrates

As with trematodes (Chapter 18) and nematodes (Chapter 30) the main carbohydrate reserve in cestodes is glycogen, which is a typical energy reserve of helminths inhabiting biotopes with a low oxygen tension. The properties of cestode glycogens resemble those of vertebrate glycogens; for example, that of *Moniezia expansa* is a highly branched structure consisting of α-1,4- and α-1,6-linked glucopyranose units (Orpin *et al.*, 1976).

The data presented in Table 24.1 show that the total glycogen content of cestodes may vary in the range 6–48 per cent of the dry mass. Larval cestodes generally show a higher and more constant glycogen content than the corresponding adults; this may reflect the more stable intermediate host environment, usually the coelomic cavity or tissues. The glycogen content can show marked variation within any species in relation to the nutrition of the host and the degree of maturity of the strobila (or part of it).

Mucopolysaccharides form the most common structural carbohydrates in cestodes. These are often complexed with proteins to form mucoproteins or glycoproteins, which form the major components of the cestode glycocalyx (Fig. 20.2). Another much-studied structure is the laminated layer of the hydatid cyst of *Echinococcus granulosus* (Fig. 23.19), which is PAS-positive and consists of complex polysaccharide-protein with the carbohydrate component consisting of glucose, galactose, galactosamine and glucosamine (Rishi & McManus, 1987). Complex polysaccharides are also produced by the eggs and adults of *Hymenolepis diminuta*. These stages are refractory to proteolysis and it has been suggested (Robertson & Cain, 1984) that these carbohydrates, in conjunction with specially resistant proteins, serve to protect the worm against the proteolytic action of the enzymes in the host gut.

○ 24.2.3 Protein

Figures for protein content are given in Table 24.1, but like the glycogen content, the figures are meaningful only against the age, degree of maturation and previous metabolic history of the worm. The chief structural proteins in cestodes are (*a*) the sulphur-rich keratins (Fig. 13.7) of which the hooks and embryophores of the taeniid cestodes are composed and (*b*) sclerotin (a quinone-tanned protein, Fig. 13.7) which is the principal component of the egg-shell of pseudophyllidean cestodes (Fig. 22.2) and trematodes (Figs 13.8, 13.9) and also, possibly, in the egg-shells of other groups.

Collagen also occurs in cestodes and has been characterised in the cysticercus of *Taenia solium* (*Cysticercus cellulosae*) (Torre-Blanco & Toledo, 1981). In general, its composition resembles that of vertebrate collagen, with the difference that hydroxyproline is lacking. A variety of soluble proteins have been isolated from cestodes, many of which are antigens. The most highly characterised of the latter is the lipoprotein Antigen 5, which is specific to *Echinococcus granulosus* and much used in the diagnosis of hydatid disease (Shepherd & McManus, 1987). An array of membrane-bound proteins are also found embedded in the lipid bilayer of the cestode tegument (Pappas, 1983).

○ 24.2.4 Lipid

The lipid content of cestodes is given in Table 24.1 and clearly varies considerably from species to species. Moreover, the lipid content may vary considerably even in the same species grown in the same host species fed on different diets. This is related to the fact that host intestinal fatty acids and sterols are *directly* (i.e. without further digestion) absorbed by cestodes and the qualitative composition of cestode lipids generally follows that of the host. The lipid content can also vary with the age of the proglottid.

Early data on cestodes must be treated with some caution as recently developed techniques, such as thin-layer and gas chromatography, and infra-red and ultra-violet spectroscopy, have completely revolutionised the field. Useful reviews are those of Barrett (1983), Frayha & Smyth (1983), Smirnov (1982), Smirnov & Bogdan (1982) and Smyth & McManus (1989). In general, all the major neutral lipids and phospholipid fractions have been demonstrated in cestodes and the lipid composition of this group thus appears to be similar to that of

Table 24.3. *Echinococcus granulosus: intra-specific variation in end products of carbohydrate metabolism excreted during aerobic incubation in vitro*

	Per cent of total excreted end products			
	Acetic acid	Lactic acid	Succinic acid	Ethanol
Protoscoleces: Origin				
Sheep, Kenya[a]	38	23	38	—
Sheep, UK[a]	55	17	28	—
Sheep, UK[b]	58	22	14	7
Sheep, Tasmania	37	21	11	32 (n = 1)
Cattle, Kenya[a]	35	10	55	—
Goat, Kenya[a]	48	24	28	—
Horse, UK[b]	25	59	8	8
Camel, Kenya[a]	13	38	48	—
Man, Kenya[a]	25	45	30	—
Adults: Origin				
Natural/dog, Kenya[a]	18	31	51	—
Man/dog, Kenya (experimental)[a]	33	16	51	—
Sheep/dog, NSW (isolates from different locations)[c]	23	21	36	20 (n = 1)
	26	17	42	15 (n = 1)
	21	27	21	31 (n = 4)
Sheep/dog, Tasmania[c]	31	26	9	34 (n = 2)
Wallaby/dingo, Queensland[c]	21	36	0.4	43 (n = 1)

—, Not determined.
[a]Calculated from McManus (1981).
[b]Calculated from McManus & Smyth (1978).
[c]Behm, Bryant & Thompson (unpublished data).
Source: After Bryant & Behm (1989).

other organisms. The lipid metabolism is discussed later (section 24.6) and summarised in Fig. 24.7.

○○○ 24.3 Carbohydrate metabolism

From a series of studies carried out on both cyclophyllidean and pseudophyllidean cestodes it is possible to build up a general picture of their carbohydrate metabolism. These experiments fall into two groups: (*a*) *in vivo* experiments: carried out by feeding the host on experimental diets and observing the effect on the chemical composition of a worm; (*b*) *in vitro* experiments: concerned largely with the type and quantity of carbohydrate utilised and the end products of carbohydrate metabolism.

The carbohydrate metabolism of cestodes dominates all other aspects of metabolism; this area of cestode physiology has been extensively reviewed by Barrett (1981), McManus & Bryant (1986), Smyth & McManus (1989) and Roberts (1983). Consideration of all aspects of metabolism is complicated by the fact that certain species occur as complexes of intra-specific variants or strains and substantial qualitative and quantitative differences in metabolism, especially the carbohydrate metabolism, have been recorded (see below).

Intra-specific metabolic differences. Intra-specific differences have been especially investigated in *Echinococcus granulosus* and *Hymenolepis diminuta*. In *E. granulosus*, marked differences in the chemical composition and metabolism (Tables 24.2, 24.3) of protoscoleces from hydatid cysts from different hosts (especially from sheep and horse) have been found (McManus & Smyth, 1978). This difference is reflected in the fact that *in vitro* the protoscoleces of the sheep strain will differentiate sexually whereas those from horse will not, suggesting that major nutritional differences exist between the two strains (p. 515).

The metabolism of *H. diminuta* maintained in different laboratories in different countries has also been extensively investigated (Kohlhagen *et al.*, 1985; Mettrick & Rahman, 1984),

Table 24.4. Hymenolepis diminuta*: acidic end products produced by the adults of two strains, UT (Canada) and ANU (Australia), under aerobic and anaerobic conditions; intermediate host* Tribolium confusum

Results are expressed as μmol g^{-1} fresh mass h^{-1}, mean ± s.d. Figures in parentheses are percentages.

Condition	Acid	UT	ANU
Aerobic (*n* = 3)	Succinic	16.4 ± 0.32 (39.2)	9.5 ± 0.41 (22.5)
	Acetic	24.1 ± 0.22 (57.7)	8.2 ± 0.19 (9.4)
	Lactic	1.3 ± 0.04 (3.1)	24.5 ± 0.67 (58.1)
	Total	*41.8*	*42.2*
Anaerobic (*n* = 3)	Succinic	25.9 ± 0.73 (55.3)	41.5 ± 1.2 (41.6)
	Acetic	15.5 ± 0.17 (33.1)	34.7 ± 0.84 (34.8)
	Lactic	5.4 ± 0.3 (11.6)	23.6 ± 0.47 (23.6)
	Total	*51.8*	*99.8*
Aerobic[a] (*n* = 3)	Succinic	21.4 (28.4)	18.3 (23.3)
	Acetic	23.2 (30.8)	15.3 (19.5)
	Lactic	30.8 (40.8)	45.0 (57.2)
	Total	*75.4*	*78.6*

[a]Mettick & Rahman (1984).
Source: Data from Kohlhagen *et al.* (1985).

Table 24.5. Hymenolepis diminuta*: acidic end products produced aerobically by the adults of two strains, UT (Canada) and ANU (Australia), the larvae of which were maintained in two different intermediate hosts, the beetles* Tribolium confusum *and* Tenebrio molitor

Strain	Intermediate host	Acidic end products (per cent of total)			References
		Succinic	Acetic	Lactic	
UT	*Tribolium confusum*	28	31	41	Mettrick & Rahman (1984)
ANU	*Tribolium confusum*	23	20	57	Mettrick & Rahman (1984)
UT	*Tenebrio molitor*	35	38	27	Mettrick & Rahman (1984)
ANU	*Tenebrio molitor*	27	31	42	Mettrick & Rahman (1984)
UT	*Tribolium confusum*	39	58	3	Kohlhagen *et al.* (1985)
ANU	*Tribolium confusum*	23	9	58	Kohlhagen *et al.* (1985)

Source: Data from Bryant & Behm (1989).

especially one strain of *H. diminuta* maintained in Canberra, Australia (the ANU strain) and one in Toronto, Canada (the UT strain). It has been shown that the respiratory end products (succinic, acetic and lactic acids) from each differed substantially (Table 24.4). The ANU strain produced more lactic acid and less succinic acid under aerobic and anaerobic conditions. The ANU strain possessed higher activities of hexokinase, pyruvate kinase, lactate dehydrogenase, malic enzyme and glycerophosphate dehydrogenase, and the UT strain showed high activities of fumarase, succinate dehydrogenase and fumarate reductase. Differences in the activities of alkaline phosphatase and phosphodiesterase have also been recorded (Pappas & Leiby, 1986).

The strain problem in *H. diminuta* is also accentuated by the fact that adults, grown from cysticercoids raised in different beetles (*Tribolium confusum* and *Tenebrio molitor*), have been shown to produce different ratios of the acidic end products (Table 24.5). The biochemical strain variation in the above species, and others, has been reviewed by Bryant & Flockart (1986).

The level of infection also has a direct effect on the carbohydrate metabolism. This effect is related to the 'crowding effect' which has been especially examined in *Hymenolepis* spp. and is characterised by an inverse correlation between worm size and increasing number of worms per host (Fig. 24.2).

Most species studied have been found to utilise glucose: these include *H. diminuta*, *H. microstoma*, *Callibothrium verticillatum*, *Taenia crassiceps* and *T. taeniformis* (Pappas, 1983). The uptake mechanisms have been

Table 24.6. *End products of carbohydrate breakdown in cestodes, besides* CO_2

Species	Aerobic	Anaerobic
Echinococcus granulosus (Adult)	Succinate, lactate, acetate	Similar to aerobic
(Protoscolex)	Lactate, pyruvate, acetate, succinate, ethanol	Similar to aerobic; larva with more succinate, no pyruvate
Hymenolepis diminuta (Adult)	Lactate, acetate, succinate	Similar to aerobic with more succinate
Hymenolepis microstoma (Adult)	Lactate, succinate, acetate, propionate	Similar to aerobic with more succinate
Moniezia expansa (Adult)	Lactate, succinate	Similar to aerobic with more succinate
Cotugnia digonopora (Adult)	Lactate, pyruvate, succinate, malate	—
Spirometra mansonoides (Adult)	—	Acetate, propionate, traces of lactate, succinate
(Larva)	—	Acetate, lactate, traces of propionate, succinate
Schistocephalus solidus (Plerocercoid)	Acetate, propionate	Propionate, acetate
Diphyllobothrium dendriticum (Adult and plerocercoid)	Succinate, lactate	Similar to aerobic
Ligula intestinalis (Adult and plerocercoid)	Lactate, succinate, acetate, propionate, pyruvate, malate	Similar to aerobic
Mesocestoides corti (Tetrathyridium)	Lactate, succinate, acetate	Similar to aerobic with more succinate

Source: Data from Barrett (1981), Rahman & Mettrick (1982), Pampori *et al.* (1984*a,b*) and McManus & Sterry (1982).

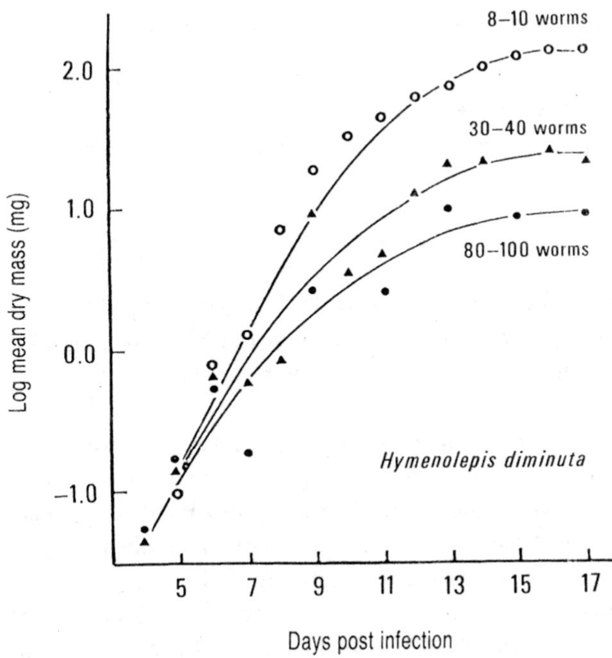

Fig. 24.2
The 'crowding effect' in *Hymenolepis diminuta* in rats. The growth is measured as dry mass at different worm loads. (After Roberts, 1961.)

much studied and glucose has been shown to be taken up by active transport; this system is characterised by Michaelis–Menten kinetics and operates against a concentration gradient. The theoretical and practical aspects of uptake mechanisms generally have been reviewed by Barrett (1981) and Smyth & McManus (1989). Galactose can also be utilised by cestodes; like glucose, it is taken up by active transport. Although fructose can also enter some species by passive diffusion (but not, apparently, *H. diminuta*), it is not, in fact, metabolised.

Useful studies on intermediary metabolism have been made on *Monieza expansa*, *Hymenolepis* spp., *Echinococcus* spp., *Mesocestoides corti*, *Cotugnia digonopora*, *Spirometra mansonoides*, *Diphyllobothrium dendriticum*, *Schistocephalus solidus* and *Ligula intestinalis* (Smyth & McManus, 1989).

A feature of carbohydrate breakdown in cestodes is the range of complex end products, mainly organic acids, which are produced under aerobic and anaerobic conditions (Table 24.6). This situation contrasts sharply with predominately aerobic organisms (such as free-living metazoa) where the pyruvic acid produced during glycolysis is converted into lactic acid. The lactic acid produced by muscular activity under anaerobiosis is converted under aerobic conditions to CO_2 and water in the Krebs cycle. If oxygen is not available, an 'oxygen debt' is built up.

Numerous studies in cestodes have confirmed that the classical Embden–Meyerhof pathway of glycolysis (Table 24.7) operates in cestodes. Most studies have been carried out on species readily maintained in the

Table 24.7. *Embden–Meyerhof pathway of anaerobic glycolysis*[a]

1. Glycogen + $n\text{H}_3\text{PO}_4 \xrightleftharpoons{\alpha\text{-glucan phosphorylase}}$ *n* glucose 1-phosphate
2. Glucose 1-phosphate + glucose 1, 6-diphosphate $\xrightleftharpoons{\text{Phosphoglucomutase}}$ glucose 1, 6-diphosphate + glucose 6-phosphate

Alternatively, starting with glucose, 1 and 2 are replaced by 3.

3. Glucose + ATP $\xrightleftharpoons{\text{Hexokinase}}$ glucose 6-phosphate + ADP
4. Glucose 6-phosphate $\xrightleftharpoons{\text{Glucosephosphate isomerase}}$ fructose 6-phosphate
5. Fructose 6-phosphate + ATP $\xrightleftharpoons{\text{Phosphofructokinase}}$ fructose 1, 6-diphosphate + ADP
6. Fructose 1, 6-diphosphate $\xrightleftharpoons{\text{Aldolase}}$ dihydroxyacetone phosphate + D-glyceraldehyde 3-phosphate
7. Dihydroxyacetone phosphate $\xrightleftharpoons{\text{Triosephosphate isomerase}}$ D-glyceraldehyde 3-phosphate
8. D-glyceraldehyde 3-phosphate + dehydrogenase + NAD $\xrightleftharpoons[\text{dehydrogenase}]{\text{Glyceraldehydephosphate}}$ 3-phospho-D-glyceric acid-enzyme complex + NADH_2
9. 3-phospho-D-glyceric acid-enzyme complex + $\text{H}_3\text{PO}_4 \xrightleftharpoons[\text{dehydrogenase}]{\text{Glyceraldehydephosphate}}$ 1, 3-diphospho-D-glycerate + dehydrogenase
10. 1, 3-diphospho-D-glycerate + ADP $\xrightleftharpoons{\text{Phosphoglycerate kinase}}$ 3-phospho-D-glycerate + ATP
11. 3-phospho-D-glycerate + 2, 3-diphospho-D-glycerate $\xrightleftharpoons{\text{Phosphoglyceromutase}}$ 2, 3-diphospho-D-glycerate + 2-phospho-D-glycerate
12. 2-phospho-D-glycerate $\xrightleftharpoons{\text{Phosphopyruvate hydratase}}$ phosphoenolpyruvate + H_2O
13. Phosphoenolpyruvate + ADP $\xrightleftharpoons{\text{Pyruvate kinase}}$ pyruvate + ATP

If O_2 is absent, the NADH_2 from Reaction 8 is oxidised in muscle by Reaction 14 and the end product is lactate:

14. Pyruvate $\xrightleftharpoons{\text{Lactate dehydrogenase}}$ L-lactate

Sum of reactions 3–14: glucose + 2 ADP + $2\text{H}_3\text{PO}_4$ = 2 lactic acid + 2 ATP + $2\text{H}_2\text{O}$

[a]Enzyme terminology as recommended by the Enzyme Commission.

laboratory or available from abattoirs or fisheries, especially *Moniezia expansa* (Behm & Bryant, 1975*a,b,c*), *Schistocephalus solidus* (Beis & Barrett, 1979), *Echinococcus granulosus* (McManus & Smyth, 1982), *Ligula intestinalis* (McManus & Sterry, 1982) and *Hymenolepis microstoma* (Rahman & Mettrick, 1982).

Many of the enzymes of the glycolytic pathway in cestodes have been purified and/or characterised, especially hexokinase, glycogen phosphorylase, phosphofructokinase, pyruvate kinase, lactate dehydrogenase (reviewed by Smyth & McManus, 1989). No significant differences between these and the corresponding vertebrate enzymes have so far been detected, but differencs between these in cestode species have been identified. Thus, the characteristics of LDH (lactate dehydrogenase) and MDH (malate dehydrogenase) in *H. microstoma* and *H. diminuta* show some differences; both exist in 2–3 isoenzymes (Pappas & Schroeder, 1979). Both enzymes show peak activities at pH 6.4 and still show 70 per cent activity at pH 8.4. Freezing and thawing had little effect on their activities.

Respiratory end products. As a result of their respiratory metabolism, cestodes produce a range of end products (Table 24.6). According to Bryant & Flockart (1986) three patterns of respiratory metabolism can be identified in parasitic helminths. The cestodes fit broadly into the first two categories of biochemical classification illustrated in Fig. 24.3; Type 3 occurs in *Ascaris*. Although one strain of *H. diminuta* apparently fits into Type 1, most cestodes conform to Type 2, which is characterised by CO_2 fixation. Carbohydrate is degraded as far as PEP in classical (vertebrate) glycolysis; at this point it is exposed to the competing enzymes pyruvate kinase and phosphoenolpyruvate carboxykinase (PEPCK) and a branch point occurs (Fig.

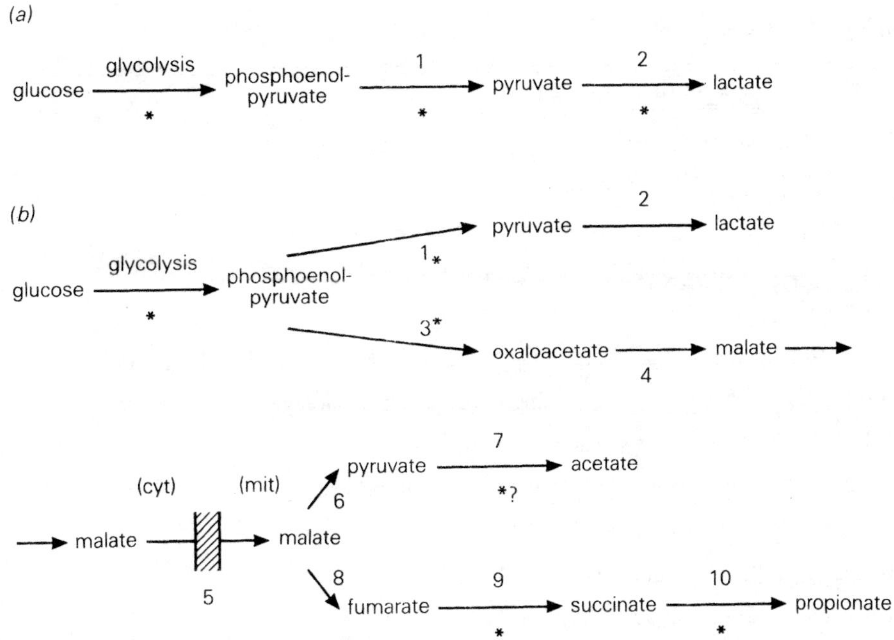

Fig. 24.3
Two types of energy metabolism in cestodes. (*a*) *Type 1*: homolactate fermentation. (*b*) *Type 2*: Malate dismutation. Reaction 3 involves a carboxylation step; decarboxylation occurs at 6, 7 and 10. Reducing equivalents are generated at reactions 6 and 7; one reducing equivalent is used at reaction 9. Thus, when the mitochondrial compartment is in redox balance and malate is the sole substrate, twice as much propionate as acetate is produced. Key: 1, pyruvate kinase; 2, lactate dehydrogenase; 3, phosphoenolpyruvate carboxykinase; 4, malate dehydrogenase; 5, mitochondrial membrane; 6, malic enzyme; 7, pyruvate dehydrogenase complex; 8, fumarase; 9, fumarate reductase; 10, succinate decarboxylase complex. *Indicate reactions at which ATP is synthesised from ADP; cyt, cytosol; mit, mitochondrion. (After Bryant & Flockart, 1986.)

24.3). It is the relative activities of these two enzymes which determine the fate of PEP and the nature of the end products (Table 24.6).

The branchpoint has been most studied in *Moniezia expansa* and *H. diminuta*. In *M. expansa*, the major end products of carbohydrate metabolism are lactate and succinate (Table 24.6), the relative amount of each being related to the $p\mathrm{O}_2$ and the presence or absence of glucose or fumarate. Anaerobiosis favours the production of lactate which is accompanied by a fall in the intra-cellular concentration of malate, whereas under aerobic conditions malate concentrations increase, pyruvate kinase activity is inhibited and lactate production is decreased. A model (Fig. 24.4) to explain these events has been proposed by Bryant & Behm (1976). These authors point out that TEM studies show that sections of mitochondria are very heterogeneous, and they postulate the existence of 'aerobic' and 'anaerobic' mitochondria. It is suggested that succinate (or NADH) is oxidised in 'aerobic' mitochondria, which contain a functional electron transport system linked to oxygen, which is translocated to the cytoplasm. Simultaneously, in 'anaerobic' mitochondria (which lack cytochrome oxidase), malate is metabolised to succinate via the fumarate reductase system. This unusual hypothesis clearly requires confirmation from further work.

In *Hymenolepis diminuta*, as mentioned (p. 353) the data are difficult to interpret owing to the remarkably different results obtained in different laboratories utilising different intra-specific variants or strains (Tables 24.4, 24.5). Some experiments showed that the ambient $p\mathrm{CO}_2$ apparently determined the amounts of lactate and succinate secreted (Podesta *et al.*, 1976). However, other workers have found that (as in *Moniezia*, above) other environmental factors, as well as strain differences, can substantially influence the concentrations of organic acids secreted and the $p\mathrm{CO}_2$ alone does not cause a major shift in the ratio of lactic to succinic acid produced (Ovington & Bryant, 1981).

Krebs (TCA) cycle (Table 24.8). This cycle is essentially the metabolic centre at which carbohydrate, fat and protein metabolisms make contact and are able to exchange intermediate compounds. Although the

Table 24.8. *The citric acid or Krebs cycle*[a]

1. Acetyl-CoA + oxaloacetate $\xrightleftharpoons{\text{Citrate synthase}}$ citrate + CoA
2. Citrate $\xrightleftharpoons{\text{Aconitate hydratase}}$ *cis*-aconitate + H_2O
3. *cis*-aconitate + H_2O $\xrightleftharpoons{\text{Aconitate hydratase}}$ L_s-isocitrate
4. L_s-isocitrate + NADP $\xrightleftharpoons{\text{Isocitrate dehydrogenase}}$ oxalosuccinate + $NADPH_2$
5. $NADPH_2$ + ½O_2 $\xrightleftharpoons[\text{and cytochrome oxidase}]{\text{NADPH}_2\text{ cytochrome } c \text{ reductase}}$ NADP + H_2O
6. Oxalosuccinate $\xrightleftharpoons{\text{Isocitrate dehydrogenase}}$ 2-oxoglutarate + CO_2
7. 2-oxoglutarate + oxidized lipoate $\xrightleftharpoons[\text{pyrophosphate}]{\text{Oxoglutarate dehydrogenase with thiamine}}$ 6-S-succinyl-hydrolipoate
8. 6-S-succinyl-hydrolipoate $\xrightleftharpoons{\text{Lipoate acetyltransferase}}$ succinyl-CoA + dihydrolipoate
9. Succinyl-CoA + H_2O $\xrightleftharpoons{\text{Succinyl-CoA hydrolase}}$ succinate + CoA
10. Succinate + ½O_2 $\xrightleftharpoons{\text{Succinate dehydrogenase and cytochrome oxidase}}$ fumarate + H_2O
11. Fumarate + H_2O $\xrightleftharpoons{\text{Fumarate hydratase}}$ L-malate
12. L-malate + NAD $\xrightleftharpoons{\text{Malate dehydrogenase}}$ oxaloacetate + $NADH_2$

Sum of reactions: Acetyl-CoA + $2O_2$ = $2CO_2$ + H_2O + CoA

[a]Enzyme terminology as recommended by the Enzyme Commission.

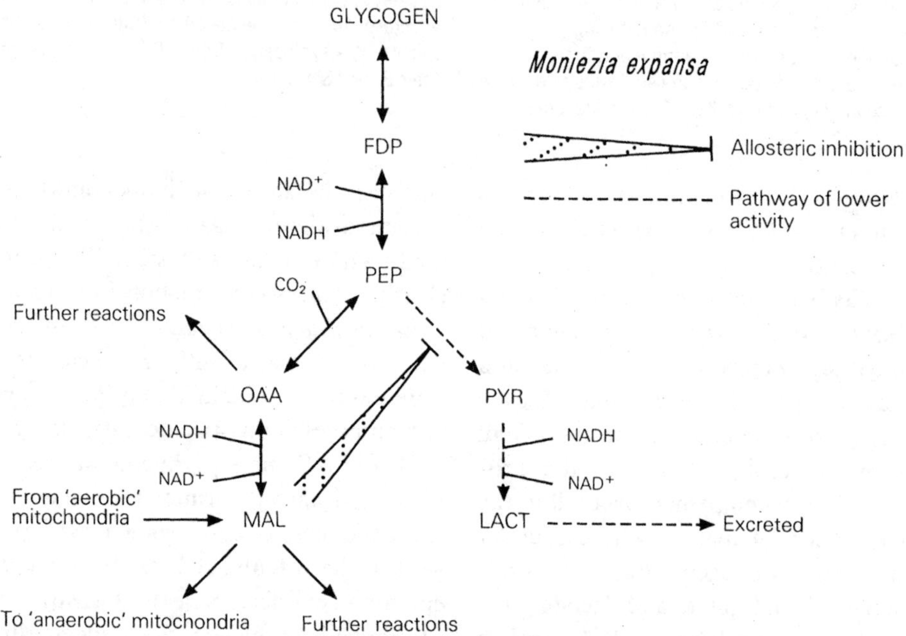

Fig. 24.4
Proposed pathways of metabolism in *Moniezia expansa* scoleces under aerobic conditions: cytosolic reactions. OAA, oxaloacetate; PYR, pyruvate; MAL, malate; LACT, lactate; FDP, fructose-1,6 diphosphate. (After Bryant & Behm, 1976.)

importance of the Krebs cycle in the energy budget of cestodes in general is difficult to estimate, there is evidence that species such as *Schistocephalus solidus* (Körting & Barrett, 1977) and *Echinococcus* spp. (McManus & Smyth, 1978) are capable of catabolising substantial amounts of carbohydrate via a functional Krebs cycle. In addition to these species, the specific activities of many of the Krebs cycle enzymes (Table 24.8) have been measured in *H. diminuta* (Ward & Fairbairn, 1970), *Bothriocephalus gowkonensis* (Körting, 1976), *Khawia sinensis* (Körting, 1976), *Triaenophorus crassus* (Körting, 1976), *Ligula intestinalis* (McManus, 1975) and *Mesocestoides corti* (Köhler & Hanselmann, 1974).

○○○ 24.4 Electron transport

Electron transport in cestodes has been comprehensively reviewed by Cheah (1983) and earlier by Bryant (1970), Smyth (1969) and Smyth & McManus (1989). The main species investigated have been *Moniezia expansa*, *Taenia hydatigena*, *T. taeniaeformis* and *H. diminuta*. The mitochondria in these species generally resemble those in mammalian cells with some variation being noted in different regions of the strobila (Lumsden, 1965).

The system has been most studied in *M. expansa* in which it has been demonstrated that this differs from the mammalian system in being branched and possessing multiple terminal oxidases (Fig. 24.5). A minor pathway resembles the classical mammalian chain with cytochrome a_3 as the terminal oxidase. The major, alternative, pathway branches at the level of rhodoquinone or vitamin K and involves cytochrome *o*, which occurs in microorganisms, parasitic protozoa and plants. Cytochrome *c* does not bond cyanide as strongly as does cytochrome a_3. The evidence indicates that the minor pathway is capable of oxidative phosphorylation, i.e. the aerobic production of ATP.

The pentose phosphate pathway. Evidence for the existence of this pathway is fragmentary in most species, but a complete sequence of enzymes has been demonstrated in larval *Echinococcus granulosus* (Smyth & McManus, 1989).

○○○ 24.5 Protein metabolism

○ 24.5.1 General account

It is evident from the remarkable growth rate of some species that rapid protein synthesis takes place. It must be borne in mind, however, that cestodes are essentially made up of a string of embryos and hence different parts of a strobila may not only be synthesising at a different rate, but may also have different nutritional demands depending on the degree of maturation of the proglottides in question.

Our concept of the way in which protein or its breakdown products are taken up by an absorptive surface, such as the intestinal mucosa, has undergone drastic revision within recent years (p. 13). It is now known that not only are small molecules, such as amino acids, absorbed, but also larger ones such as dipeptides or even whole protein molecules may be ingested. The

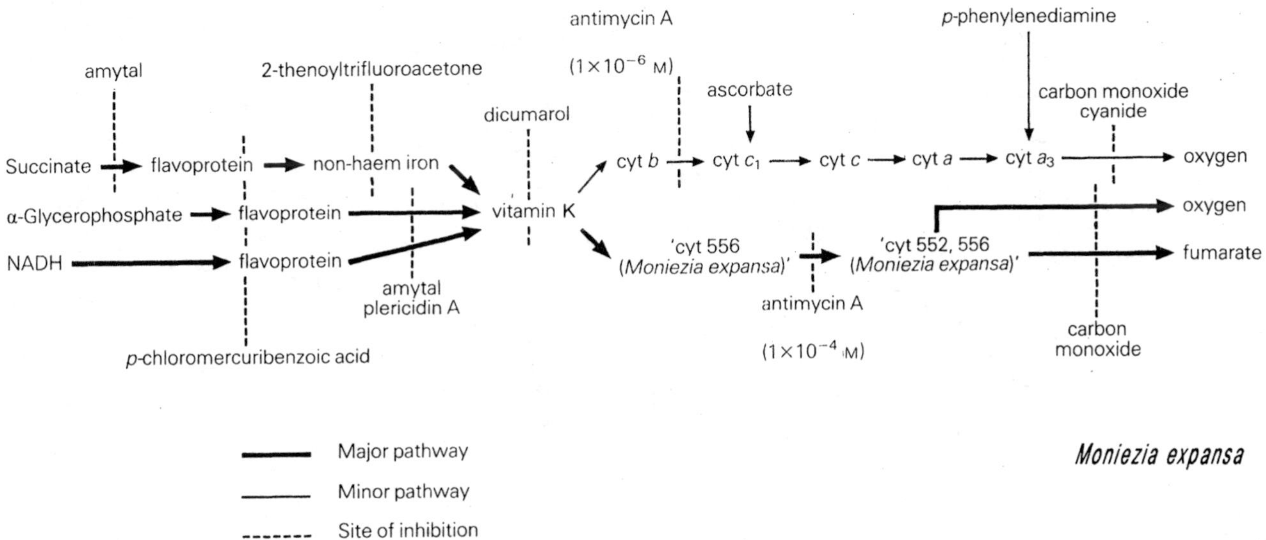

Fig. 24.5
Moniezia expansa: proposed electron transport system. (After Cheah, 1983.)

same processes may take place in the cestode tegument, although there is, as yet, no unequivocal evidence that pinocytosis – which would be necessary for the ingestion of large molecules – actually occurs in cestodes (p. 284).

○ 24.5.2 Amino acid uptake

The uptake of amino acids by cestodes has been extensively studied and it is now known to take place largely by active transport, although a diffusion component may also be involved (Fig. 24.6B). This may generally be recognised experimentally by the fact that the reciprocal of the velocity of uptake ($1/V$) against the reciprocal of the concentration ($1/S$) (a Lineweaver–Burk plot) is a straight line (Fig. 24.6A). The evidence for this is based on the facts that:

(*a*) cestodes can accumulate amino acids against a gradient;
(*b*) the uptake of one amino acid can inhibit the uptake of another (i.e. there is 'competition' for the carrier locus in the amino acid transport system).

○ 24.5.3 Amino acid metabolism

Little is known of the precise amino acid requirements of cestodes, i.e. which amino acids are 'essential' and cannot be synthesised by the host. *In vitro* studies (Chapter 34) should ultimately provide the answer to this question.

A valuable review of early work is that of Pappas & Read (1975); more recent work has been summarised by Smyth & McManus (1989). Some recent studies on amino acid uptake are those on *H. diminuta* (Jeffs & Arme, 1985*a,b*; Lussier *et al.*, 1982; Webb, 1986) and *E. granulosus* (Jeffs & Arme, 1986).

In general, uptake of amino acids shows most of the well-known features of mediated transport which obey Michaelis–Menten kinetics, i.e. they are stereospecific (for the amino-group), they can be inhibited both competitively and non-competitively, and they have a large temperature coefficient and show a distinct pH optimum.

Amino acid interconversions

Cestodes appear to have little ability to catabolise amino acids. For example, out of 10 amino acids tested, *H. diminuta* was shown to generate $^{14}CO_2$ only from [^{14}C]asparate and [^{14}C]alanine *in vitro* (Wack *et al.*, 1983). Two main pathways are involved in amino acid metabolism in cestodes: transamination and oxidative deamination.

Transamination. Compared with vertebrates, cestodes have a very limited capacity for performing transaminations, a situation which probably reflects the fact that the intestinal environment in which they live is rich in amino acids. However, transaminases have been demonstrated in the following species: *Anoplocephala magna*, *Hymenolepis* spp., *Raillietina cesticillus*, *Taenia taeniae-*

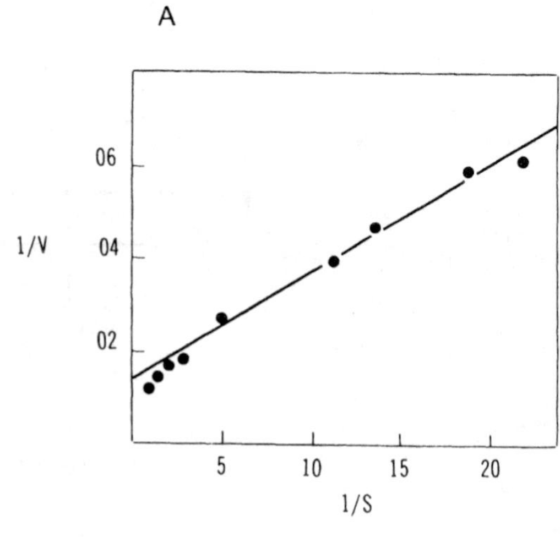

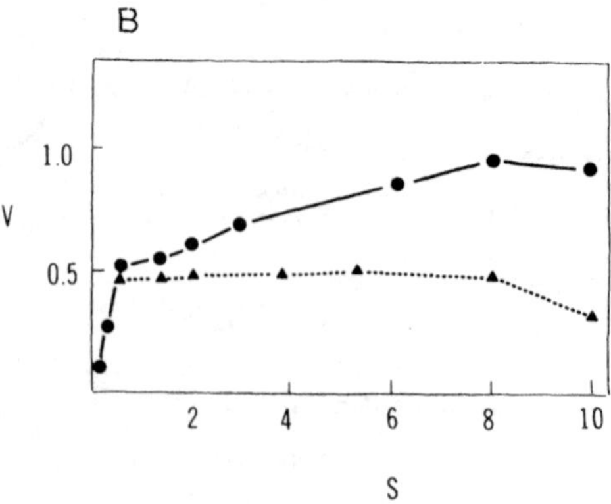

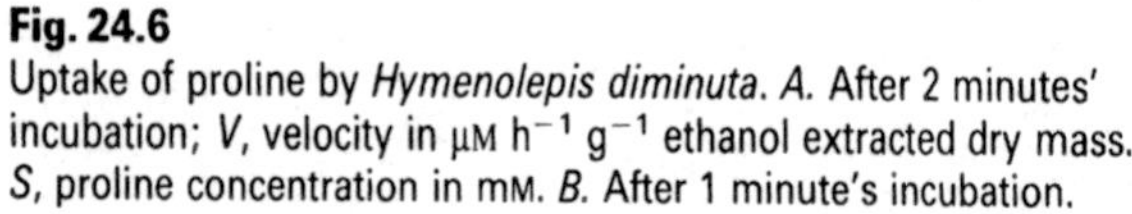

Fig. 24.6
Uptake of proline by *Hymenolepis diminuta*. *A*. After 2 minutes' incubation; *V*, velocity in μM h^{-1} g^{-1} ethanol extracted dry mass. *S*, proline concentration in mM. *B*. After 1 minute's incubation. *V*, velocity in μM min^{-1} g^{-1} worm water. *S*, proline concentration in mM. Solid circles, observed values; triangles, values corrected for 'diffusion'. (After Kilejian, 1966.)

formis, *Moniezia expansa*, and *Lytocestus indicus* (Barrett, 1981; Rasheed, 1981; Smyth, 1969).

Oxidative deamination. Glutamate dehydrogenase, which catalyses the oxidative deamination of glutamate:

$$\text{glutamate} + NAD^+(NADP^+) + H_2O \rightarrow$$
$$\text{2-oxoglutarate} + NADH(NADPH) + NH_4$$

has been characterised from *H. diminuta* (Mustafa *et al.*, 1978). In this reaction, the ammonia which is produced is excreted.

Protein synthesis

Detailed consideration of protein synthesis and molecular biology is beyond the scope of this text and is only briefly discussed below. These areas have been reviewed, in depth, by Harris (1983) and Smyth & McManus (1989).

The chief methods utilised to investigate protein synthesis have been (*a*) incorporation of radioactively labelled precursors into the proteins of intact worms and (*b*) the use of cell-free protein-synthesising systems. The first approach has been used to study protein synthesis in larval *H. diminuta* (Jeffs & Arme, 1984). The incorporation of four amino acids into stage V cysticercoids, in the order leucine > phenylalanine > alanine > proline, was demonstrated.

In cell-free systems, it has been shown that protein synthesis is essentially similar to that in mammals, useful studies having being carried out on larval *Echinococcus granulosus* (Agosin & Repetto, 1967) and larval *T. crassiceps* (Náquira *et al.*, 1977) and *H. diminuta* (Parker & MacInnis, 1977). Studies have also been carried out using rabbit reticulocytes to translate cestode messenger RNA (mRNA) *in vitro*. Translation of the total RNA from *E. granulosus*, *E. multilocularis* and *T. crassiceps* showed significant incorporation of [^{35}S]methionine into synthesised proteins (McManus *et al.*, 1985).

Nucleic acids. As with many other parasites (e.g. *Entamoeba histolytica*, Chapter 3), cestodes in general appear to lack the capacity to synthesise purines although there is indirect evidence that this may occur in *Mesocestoides corti* (Heath & Hart, 1970). In contrast, it would appear that some species (e.g. *H. diminuta*) have the capacity to synthesise pyrimidines (Hill *et al.*, 1981). However, very little is known on the possible salvage pathways for purine and pyrimidines.

The isolation of cestode nucleic acids, the construction of genetic libraries and the use of DNA probes have been reviewed by Smyth & McManus (1989).

○ 24.5.4 End products of nitrogen metabolism

The intermediary metabolism of cestodes has been little studied. Although the nitrogenous compounds excreted are known for some species, the metabolic pathways whereby these products are produced are poorly known. The end products most commonly reported are urea, uric acid and ammonia, but numerous other compounds have been found.

The production of urea points to the occurrence of an ornithine cycle and two important enzymes of this cycle, ornithine transcarbamylase and arginase, have been detected (Smyth & McManus, 1989). Despite the presence of these enzymes, a *functional* ornithine cycle could not be detected in *Moniezia expansa*, *Dipylidiuim caninum*, *T. pisiformis* and *E. granulosus* (Janssens & Bryant, 1969) and it is doubtful if a complete cycle operates in cestodes.

○○○ 24.6 Lipid metabolism

The limited data available on the lipid metabolism of cestodes have been reviewed by Barrett (1981, 1983), Frayha & Smyth (1983) and Smyth & McManus (1989). In general, this can be briefly summarised as follows.

The ability to synthesise fatty acids *de novo* has been completely lost, with the result that cestodes are entirely dependent on their hosts for these substances. Hence, fatty acid synthesis is restricted to the chain lengthening of host-derived fatty acids by sequential addition of acetyl-CoA. It has been shown that *Hymenolepis diminuta* can convert palmate and stearate into saturated fatty acid chains as long as 26 carbons (Jacobsen & Fairbairn, 1967); elongation of C_{16}, C_{18}, $C_{18.1}$, $C_{18.2}$, $C_{18.3}$ to C_{20} and C_{22} acids has also been demonstrated in *Spirometra mansonoides* (Meyer *et al.*, 1966).

In mammals, the action of lipases yields fatty acids, glycerol and other components. The fatty acids are then broken down by beta-oxidation to give NADH, reduced flavoprotein and acetyl-CoA, which then enters the Krebs cycle. Although lipases have been identified in several species (Matskási & Hajdú, 1983; Singh *et al.*, 1977) there is no evidence that an active beta-oxidation

Table 24.9. *Specific activities of the tricarboxylic acid cycle in a variety of cestodes*

Specific activities are expressed as nmol min^{-1} mg^{-1} protein at 30 °C[a–e] or nmol min^{-1} mg^{-1} fresh mass at 25 °C[f].

Enzyme	*Hymenolepis diminuta* (adults)[a]	*Bothriocephalus gowkonensis* (adults)[b]	*Khawia sinensis* (adults)[b]	*Triaenophorus crassus* (adults)[b]	*Ligula intestinalis* (plerocercoids)[c]	*Schistocephalus solidus* (plerocercoids)[d]	*Echinococcus multilocularis* (protoscoleces)[e]	*Echinococcus granulosus* (protoscoleces, ovine strain)[e]	*Mesocestoides corti* (tetrathyridia)[f]
Citrate synthase	4	8	10	0	17†	59	23	36	1512†
Aconitase	0	2	0	0	8†	3	43	62	56†
Isocitrate dehydrogenase (NAD)	—	6†	0†	0†	0†	1	16	29	0†
Isocitrate dehydrogenase (NADP)	0	3	2	4	21†	5	193	73	682†
Oxoglutarate dehydrogenase	—	—	—	—	420†	22	46	20	30†
Succinate dehydrogenase	65	100	20	42	111†	119	11	7	1535†
[Fumarate reductase]	[45]	[7]†	[0]†	[0]†	[25]†	[5]	[1]	[2]	[363]†
Fumarase	45	157	112	188	101	59	39	39	468†
Malate dehydrogenase	2803	4360	9861	6313	1596	2316	7023	6809	114050

†, Mitochondrial fraction.
Source: Data from [a]Ward & Fairbairn (1970); [b]Körting (1976); [c]McManus (1975); [d]Körting & Barrett (1977); [e]McManus & Smyth (1982); [f]Köhler & Hanselmann (1974).

sequence operates although beta-oxidation enzymes have been detected in *H. diminuta*, *Schistocephalus solidus* and *Ligula intestinalis* (Barrett, 1983).

The chief sterol in cestodes is cholesterol but, as with fatty acids, cestodes are unable to synthesise this or any other sterol *de novo*. Cholesterol is taken up by the hydatid cysts of *Echinococcus granulosus* by simple diffusion (Bahr *et al.*, 1979) and by adult *H. diminuta*, in part, by a mediated system specific for sterols (Pappas, 1983).

Many of the pathways and enzyme systems involved in lipid metabolism have been little studied; a tentative diagram of the predominant metabolic pathways is shown in Fig. 24.7.

○○○ 24.7 Neurobiology

The neurobiology of platyhelminth parasites in general has already been surveyed in Chapter 18, to which reference should be made. Useful reviews of the field are those of Fairweather & Halton (1991), Halton *et al.* (1990, 1992) and Gustafsson (1985, 1992).

The identification of serotoninergic, cholinergic and peptidergic molecules, at the cellular level, has been revolutionised, in both vertebrates and invertebrates, by the development and application of immunocytochemical techniques using antisera developed against these substances in vertebrates. Specific peptidergic substances were first detected in platyhelminths in the bird cestode *Diphyllobothrium dendriticum* by Gustafsson *et al.* (1985); they have since been detected in other cestodes, e.g. the pseudophyllidean *Schistocephalus solidus* (Wikgren *et al.*, 1986), the cyclophyllideans *Hymenolepis nana* (Kumazawa & Moriki, 1986) and *H. diminuta* (Fairweather *et al.*, 1988) and the tetraphyllidean *Trilocularia acanthiaevulgaris* (Fairweather *et al.*, 1990). The peptide immunoreactivities of these species are listed in Table 18.4 together with those found in trematodes. The distribution of neuropeptides has been especially studied in *D. dendriticum* (Gustafsson, 1985;

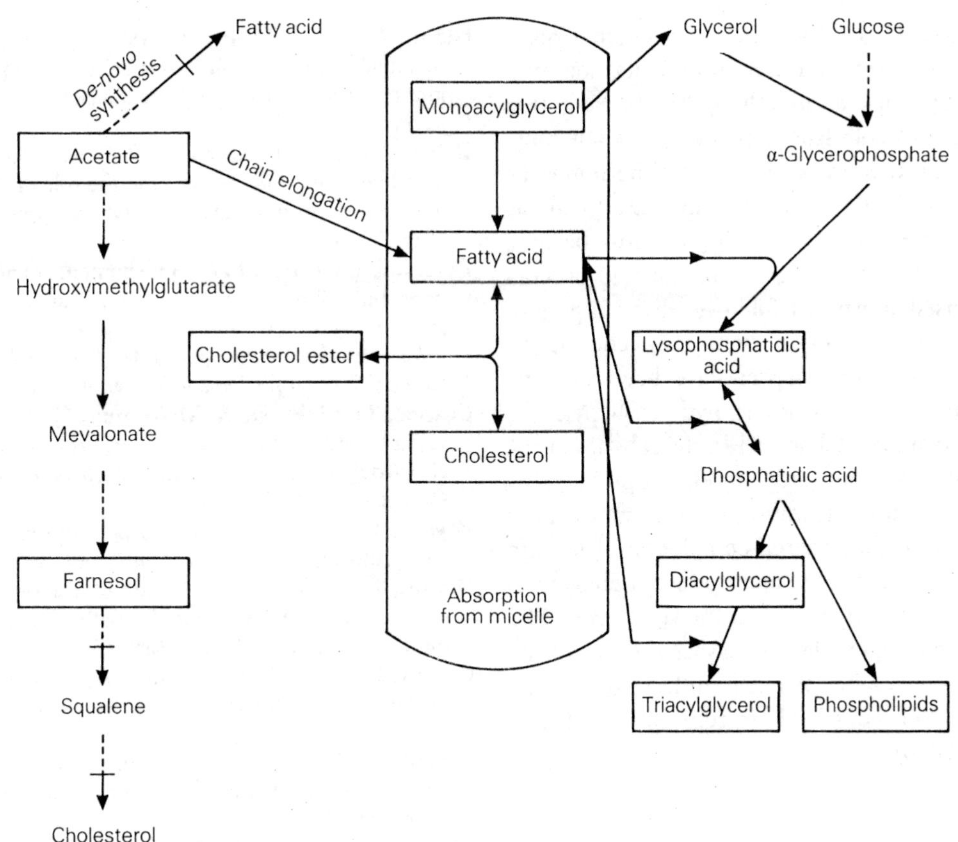

Fig. 24.7
Predominant metabolic pathways of the major classes of lipid in cestodes. (After Frayha & Smyth, 1983.)

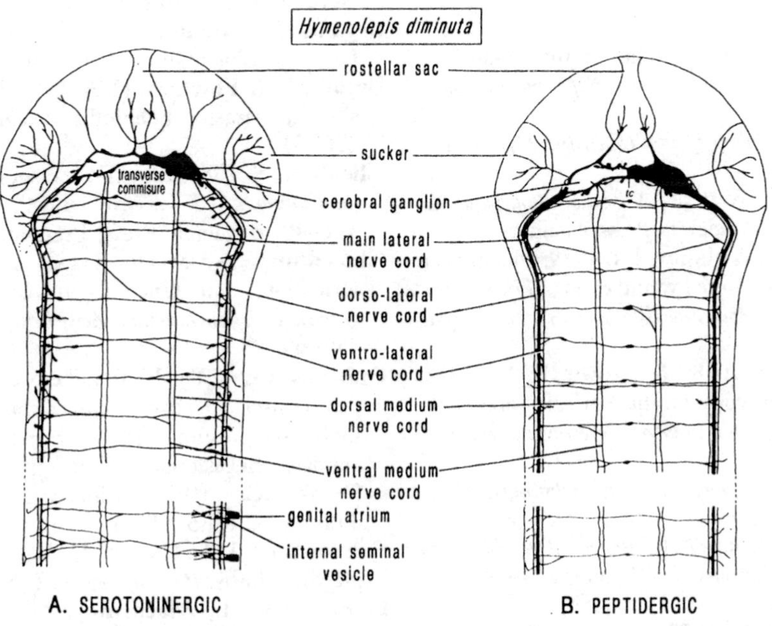

Fig. 24.8
Hymenolepis diminuta: diagrammatic representation (not to scale) of the distribution of serotoninergic and peptidergic nerve elements in the scolex and strobila, based on immunocytological reactions together with confocal laser scanning technology and epifluorescence microscopy. (Modified from McKay *et al.*, 1991.)

Gustafsson *et al.*, 1985). That the cells reacting positively were, in fact, true neurosecretory cells, was first demonstrated by Gustafsson (1985) by showing that when plerocercoid larvae from a fish (i.e. at a low temperature) were cultured *in vitro* at 38 °C (the temperature of the warm-blooded bird host) increased cellular activity began after 5 min of cultivation and became marked by 1 h. Moreover, the number of neurosecretory cells increased at least 10-fold over that 1 h period.

In addition to peptidergic neurones, cholinergic neurones and aminergic synapses have been demonstrated in a number of species (Smyth & McManus, 1989). As in trematodes (Chapter 18), the role of neuropeptides in cestodes is not understood, but presumably they are potential neurotransmitters or neuromodulators which play a part in the ordered completion of the life cycle and assure that the appropriate behavioural responses occur in the right host at the right time. The occurrence of neuropeptides in association with the reproductive ducts in other platyhelminths suggests that they may play a 'hormonal' role in reproduction (Fairweather *et al.*, 1990).

○○○ References

(References of general interest have been included in this list, in addition to those quoted in the text.)

Agosin, M. & Repetto, Y. 1967. Studies on the metabolism of *Echinococcus granulosus* protoscoleces. *Experimental Parasitology*, **21**: 195–208.

Arai, H. P. (ed.) 1980. *Biology of the Tapeworm Hymenolepis diminuta.* Academic Press, New York.

Arme, C. & Pappas, P. W. (eds) 1983*a*,*b*. *Biology of the Eucestoda.* Vols 1 and 2. Academic Press, London.

Bahr, J. M., Frayha, G. J. & Hajjar, J. J. 1979. Mechanism of cholesterol absorption by the hydatid cysts of *Echinococcus granulosus* (Cestoda). *Comparative Biochemistry and Physiology*, **62**A: 485–9.

Baldwin, J. L., Berntzen, A. K. & Brown, B. W. 1978. *Megocestoides corti*: cation concentration in calcareous corpuscles of tetrathyridia grown *in vitro*. *Experimental Parasitology*, **44**: 190–6.

Barrett, J. 1981. *Biochemistry of Parasitic Helminths.* Macmillan, London.

Barrett, J. 1983. Lipid metabolism. In: *Biology of the Eucestoda* (ed. C. Arme & P. W. Pappas), Vol. 2, pp. 391–419. Academic Press, London.

Behm, C. A. & Bryant, C. 1975*a*. Studies of regulatory metabolism in *Moniezia expansa*: general conditions. *International Journal for Parasitology*, **5**: 209–17.

Behm, C. A. & Bryant, C. 1975*b*. Studies of regulatory metabolism in *Moniezia expansa*: the role of phosphofructokinase (with a note on pyruvate kinase). *International Journal for Parasitology*, **5**: 339–46.

Behm, C. A. & Bryant, C. 1975*c*. Studies on regulatory metabolism in *Moniezia expansa*: the role of phosphophoenolpyruvate carboxykinase. *International Journal for Parasitology*, **5**: 347–54.

Beis, I. & Barrett, J. 1979. The contents of adenine nucleotides and glycolytic and tricarboxylic acid cycle intermediates in activated and non-activated plerocercoids of *Schistocephalus solidus* (Cestoda: Pseudophyllidea). *International Journal for Parasitology*, **9**: 465–8.

Bowles, J., Blair, D. & McManus, D. P. 1992. Genetic variants within the genus *Echinococcus* identified by mitochondrial DNA sequencing. *Molecular and Biochemical Parasitology*, **54**: 165–73.

Bryant, C. 1970. Electron transport in parasitic helminths and protozoa. *Advances in Parasitology*, **8**: 139–72.

Bryant, C. 1975. Carbon dioxide utilisation, and the regulation of respiratory metabolic pathways in parasitic helminths. *Advances in Parasitology*, **13**: 35–69.

Bryant, C. 1978. The regulation of respiratory metabolism in parasitic helminths. *Advances in Parasitology*, **16**: 311–31.

Bryant, C. 1983. Intraspecific variations of energy metabolism in parasitic helminths. *International Journal for Parasitology*, **13**: 327–32.

Bryant, C. & Behm, C. A. 1976. Regulation of respiratory metabolism in *Moniezia expansa* under aerobic and anaerobic conditions. In: *Biochemistry of Parasites and Host–parasite Relationships* (ed. H. Van den Bossche), pp. 89–94. North-Holland, Amsterdam.

Bryant, C. & Behm, C. A. 1989. *Biochemical Adaptation in Parasites.* Chapman & Hall, London.

Bryant, C. & Flockart, H. A. 1986. Biochemical strain variation in parasitic helminths. *Advances in Parasitology*, **25**: 276–319.

Cheah, K. S. 1983. Electron transport systems. In: *Biology of the Eucestoda* (ed. C. Arme and P. W. Pappas), Vol. 2, pp. 421–40. Academic Press, London.

Chowdhury, N. & De Rycke, P. H. 1977. Structure, formation, and functions of calcareous corpuscles in *Hymenolepis microstoma. Zeitschrift für Parasitenkunde*, **53**: 159–69.

Dubovskaya, A. Ya. 1970. (The utilization of protein by the fish cestode *Bothriocephalus scorpii*.) (In Russian.) (All-Union symposium (1st) on parasites and diseases of marine animals, Sevastopol, 1970), pp. 21–4. Izdat Naukova Dumka, Kiev. (British Library Lending Division. English Translation RTS 8231.)

Fairweather, I. & Halton, D. W. 1991. Neuropeptides in platyhelminths. *Parasitology*, **102**: S77–92.

Fairweather, I., Macartney, G. A., Johnston, C. F., Halton, D. W. & Buchanan, K. D. 1988. Immunocytochemical demonstration of 5-hydroxytryptamine (serotonin) and vertebrate neuropeptides in the nervous system of excysted cysticercoid larvae of the rat tapeworm, *Hymenolepis diminuta* (Cestoda, Cyclophyllidea). *Parasitology Research*, **74**: 371–9.

Fairweather, I., Mahendrasingam, S., Johnston, C. F., Halton, D. W. & Shaw, C. 1990. Peptidergic nerve elements in three developmental stages of the tetraphyllidean tapeworm *Trilocularia acanthiaevulgaris*. An immunocytological study. *Parasitology Research*, **76**: 497–508.

Frayha, G. J. & Smyth, J. D. 1983. Lipid metabolism in parasitic helminths. *Advances in Parasitology*, **22**: 309–87.

Gustafsson, M. K. S. 1985. Cestode neurotransmitters. *Parasitology Today*, **1**: 72–5.

Gustafsson, M. K. S. 1992. The neuroanatomy of parasitic flatworms. *Advances in Neuroimmunology*, **2**: 267–86.

Gustafsson, M. K. S., Wikgren, M. C., Karhi, T. J. & Schot, L. P. C. 1985. Immunocytochemical demonstration of neuropeptides and serotonin in the tapeworm *Diphyllobothrium dendriticum*. *Cell and Tissue Research*, **240**: 255–60.

Halton, D. W., Fairweather, I., Shaw, C. & Johnston, C. F. 1990. Regulatory peptides in parasitic platyhelminths. *Parasitology Today*, **6**: 284–90.

Halton, D. W., Shaw, C., Maule, A. G., Johnston, C. F. & Fairweather, I. 1992. Peptidergic messengers: a new perspective of the nervous system of parasitic platyhelminths. *Journal of Parasitology*, **78**: 179–93.

Harris, B. G. 1983. Protein metabolism. In: *Biology of the Eucestoda* (ed. C. Arme & P. W. Pappas), Vol. 2, pp. 335–41. Academic Press, London.

Heath, D. D. & Hart, J. L. 1970. Biosynthesis *de novo* of purines and pyrimidines in *Mesocestoides* (Cestoda). II. *Journal of Parasitology*, **56**: 340–5.

Hill, B., Kilsby, J., Rogerson, G. W., McKintosh, R. T. & Ginger, C. D. 1981. The enzymes of pyrimidine biosynthesis in a range of parasitic protozoa and helminths. *Molecular and Biochemical Parasitology*, **2**: 123–34.

Hope, M., Bowles, J. & McManus, D. P. 1991. A reconsideration of the *Echinococcus granulosus* strain situation in Australia following RFLP analysis of cystic material. *International Journal for Parasitology*, **21**: 471–5.

Ishii, A. I. 1984. Fe-rich corpuscles in *Diplogonoporus grandis* detected using x-ray microanalysis. *Zeitschrift für Parasitenkunde*, **70**: 199–202.

Jacobsen, N. S. & Fairbairn, D. 1967. Lipid metabolism in helminth parasites. III. Biosynthesis and interconversion of fatty acids by *Hymenolepis diminuta* (Cestoda). *Journal of Parasitology*, **53**: 355–61.

Janssens, P. A. & Bryant, C. 1969. The ornithine–urea cycle in some parasitic helminths. *Comparative Biochemistry and Physiology*, **30**: 261–72.

Jeffs, S. A. & Arme, C. 1984. *Hymenolepis diminuta*: protein synthesis in cysticercoids. *Parasitology*, **88**: 351–7.

Jeffs, S. A. & Arme, C. 1985*a*. *Hymenolepis diminuta*: characterization of the neutral amino acid transport loci of the metacestode. *Comparative Biochemistry and Physiology*, **81**A: 387–90.

Jeffs, S. A. & Arme, C. 1985*b*. *Hymenolepis diminuta* (Cestoda): uptake of cycloleucine by metacestodes. *Comparative Biochemistry and Physiology*, **81**A: 495–9.

Jeffs, S. A. & Arme, C. 1986. *Echinococcus granulosus*; absorption of cycloleucine and α-aminoisobutyric acid by protoscoleces. *Parasitology*, **92**: 153–63.

Kegley, L. M., Baldwin, J., Brown, B. W. & Bernstzen, A. K. 1970. *Mesocestoides corti*: environmental cation concentration in calcareous corpuscles. *Experimental Parasitology*, **27**: 88–94.

Kegley, L. M., Brown, B. W., & Bernstzen, A. K. 1969. *Mesocestoides corti*: inorganic components in calcareous corpuscles. *Experimental Parasitology*, **25**: 85–92.

Kilejian, A. 1966. Permeation of L-proline in the cestode *Hymenolepis diminuta*. *Journal of Parasitology*, **52**: 1108–15.

Köhler, P. 1985. The strategies of energy conservation in helminths. *Molecular and Biochemical Parasitology*, **17**: 1–18.

Köhler, P. & Hanselmann, K. 1974. Anaerobic and aerobic metabolism in the larvae (tetrathyridia) of *Mesocestoides corti*. *Experimental Parasitology*, **36**: 178–88.

Kohlhagen, S., Behm, C. A. & Bryant, C. 1985. Strain variation in *Hymenolepis diminuta*: enzyme profiles. *International Journal for Parasitology*, **15**: 479–83.

Körting, W. 1976. Metabolism in parasitic helminths of freshwater fish. In: *Biochemistry of Parasites and Host–Parasite Relationships* (ed. H. Van den Bossche), pp. 95–100. North-Holland, Amsterdam.

Körting, W. & Barrett, J. 1977. Carbohydrate catabolism in the plerocercoids of *Schistocephalus solidus* (Cestoda: Pseudophyllidea). *International Journal for Parasitology*, **7**: 411–17.

Kumazawa, H. & Moriki, T. 1986. Immunoenzymatic demonstration of a presumptive prolactin-like substance in *Hymenolepis nana*. *Zeitschrift für Parasitenkunde*, **72**: 137–9.

Lumsden, R. D. 1965. Macromolecular structure of glycogen in some cyclophyllidean and trypanorhynch cestodes. *Journal of Parasitology*, **51**: 501–15.

Lussier, P. E., Podesta, R. B. & Mettrick, D. F. 1982. *Hymenolepis diminuta*: the non-saturable component of methionine uptake. *International Journal for Parasitology*, **12**: 265–70.

McKay, D. M., Fairweather, I., Johnston, C. F., Shaw, C. & Halton, D. W. 1991. Immunocytochemical and radioimmunometrical demonstration of serotonin- and neuropeptide-immunoreactivities in the adult rat tapeworm. *Hymenolepis diminuta* (Cestoda, Cyclophyllidea). *Parasitology*, **103**: 275–89.

McManus, D. P. 1975. Tricarboxylic acid cycle enzymes in the plerocercoid of *Ligula intestinalis* (Cestoda: Pseudophyllidea). *Zeitschrift für Parasitenkunde*, **45**: 319–22.

McManus, D. P. 1981. A biochemical study of adult and cystic stages of *Echinococcus granulosus* of human and animal origin from Kenya. *Journal of Helminthology*, **55**: 21–7.

McManus, D. P. 1987. Intermediary metabolism in parasitic helminths. *International Journal for Parasitology*, **17**: 79–95.

McManus, D. P. & Bryant, C. 1986. Biochemistry and physiology of *Echinococcus*. In: *The Biology of Echinococcus and*

Hydatid Disease (ed. R. C. A. Thompson), pp. 114–42. George Allen & Unwin, London.

McManus, D. P., Knight, M. & Simpson, A. J. G. 1985. Isolation and characterisation of nucleic acids from the hydatid organisms, *Echinoccus* spp. (Cestoda). *Molecular and Biochemical Parasitology*, **16**: 251–66.

McManus, D. P. & Smyth, J. D. 1978. Differences in the chemical composition and carbohydrate metabolism of *Echinococcus granulosus* (horse and sheep strains) and *E. multilocularis*. *Parasitology*, **77**: 103–9.

McManus, D. P. & Smyth, J. D. 1982. Intermediary carbohydrate metabolism in protoscoleces of *Echinococcus granulosus* (horse and sheep strains) and *E. multilocularis*. *Parasitology*, **84**: 351–66.

McManus, D. P. & Sterry, P. R. 1982. *Ligula intestinalis*: intermediary carbohydrate metabolism in plerocercoids and adults. *Zeitschrift für Parasitenkunde*, **67**: 73–85.

Matskási, I. & Hajdú, É. 1983. Studies on the lipase activity of parasitic platyhelminths. *Parasitologia Hungarica*, **16**: 53–7.

Mettrick, D. F. & Rahman, M. S. 1984. Effects of parasite strain and intermediate host species on carbohydrate intermediary metabolism in the rat tapeworm, *Hymenolepis diminuta*. *Canadian Journal of Zoology*, **62**: 355–61.

Meyer, F., Kimura, S. & Mueller, J. L. 1966. Lipid metabolism in the larval and adult forms of the tapeworm, *Spirometra mansonoides*. *Journal of Biological Chemistry*, **241**: 4224–32.

Mustafa, T., Komuniecki, R. & Mettrick, D. F. 1978. Cytosolic glutamate dehydrogenase in adult *Hymenolepis diminuta* (Cestoda). *Comparative Biochemistry and Physiology*, B, **61**: 219–22.

Náquira, C., Paulin, J. & Agosin, M. 1977. *Taenia crassiceps*: protein synthesis in larvae. *Experimental Parasitology*, **42**: 359–69.

Nieland, M. L. & von Brand, Th. 1969. Electron microscopy of cestode calcareous corpuscle formation. *Experimental Parasitology*, **24**: 279—89.

Orpin, C. G., Huskisson, N. S. & Ward, P. F. V. 1976. Molecular structure and morphology of glycogen isolated from the cestode, *Moniezia expansa*. *Parasitology*, **73**: 83–95.

Ovington, K. S. & Bryant, C. 1981. The role of carbon dioxide in the formation of end-products of *Hymenolepis diminuta*. *International Journal for Parasitology*, **11**: 221–8.

Pampori, N. A., Singh, G. & Srivastava, V. M. L. 1984*a*. *Cotugnia digonopora*: carbohydrate metabolism and effect of anthelmintics on immature worms. *Journal of Helminthology*, **58**: 39–47.

Pampori, N. A., Singh, G. & Srivastava, V. M. L. 1984*b*. Energy metabolism in *Cotugnia digonopora* and the effect of anthelmintics. *Molecular and Biochemical Parasitology*, **11**: 205–13.

Pappas, P. W. 1983. Host–parasite interface. In: *Biology of the Eucestoda* (ed. C. Arme & P. W. Pappas), Vol. 2, pp. 297–334. Academic Press, London.

Pappas, P. W. & Leiby, D. A. 1986. Variation in the sizes of eggs and oncospheres and the numbers and distributions of testes in the tapeworm *Hymenolepis diminuta*. *Journal of Parasitology*, **72**: 383–91.

Pappas, P. W. & Read, C. P. 1975. Membrane transport in helminth parasites: a review. *Experimental Parasitology*, **37**: 469–530.

Pappas, P. W. & Schroeder, L. L. 1979. *Hymenolepis microstoma*: lactate and malate dehydrogenases of the adult worm. *Experimental Parasitology*, **47**: 134–9.

Parker, R. D. & MacInnis, A. J. 1977. *Hymenolepis diminuta*: isolation, purification, and reconstruction *in vitro* of a cell-free system for protein synthesis. *Experimental Parasitology*, **41**: 2–16.

Podesta, R. B., Mustafa, T., Moon, T. W., Hulbert, W. C. & Mettrick, D. F. 1976. Anaerobes in an aerobic environment: role of CO_2 in energy metabolism of *Hymenolepis diminuta*. In: *Biochemistry of Parasites and Host–Parasite Relationships* (ed. H. Van den Bossche), pp. 81–8. North-Holland, Amsterdam.

Rahman, M. S. & Mettrick, D. F. 1982. Carbohydrate intermediary metabolism in *Hymenolepis microstoma* (Cestoda). *International Journal for Parasitology*, **12**: 155–62.

Rahman, M. S., Mettrick, D. F. & Podesta, R. B. 1982. 5-Hydroxytryptamine, glucose uptake, glycogen utilization and carbon dioxide fixation in *Hymenolepis microstoma* (Cestoda). *Comparative Biochemistry and Physiology*, B, **73**: 901–6.

Rasheed, U. 1981. Transaminase activity in *Lytocestus indicus* and its host. *Proceedings of the Indian Academy of Parasitology*, **2**: 115–16.

Rishi, A. K. & McManus, D. P. 1987. Genetic cloning of human *Echinococcus granulosus DNA*: isolation of recombinant plasmids and their use as genetic markers in strain characterization. *Parasitology*, **94**: 369–83.

Roberts, L. S. 1961. The influence of population density on patterns and physiology of growth in *Hymenolepis diminuta* (Cestoda: Cyclophyllidea) in the definitive host. *Experimental Parasitology*, **11**: 332–71.

Roberts, L. S. 1983. Carbohydrate metabolism. In: *Biology of the Eucestoda* (ed. C. Arme & P. W. Pappas), Vol. 2, pp. 343–90. Academic Press, London.

Robertson, N. P. & Cain, G. D. 1984. Glycosaminoglycans of tegumental fractions of *Hymenolepis diminuta*. *Molecular and Biochemical Parasitology*, **12**: 173–83.

Shepherd, J. C. & McManus, D. P. 1987. Specific and cross-reactive antigens of *Echinococcus granulosus* hydatid cyst fluid. *Molecular and Biochemical Parasitology*, **25**: 143–54.

Singh, B. B., Singh, K. S. & Dwarkanath, P. K. 1977. Lipase activity in *Thysaniezia giardi*. *Indian Journal of Parasitology*, **1**: 69–70.

Smirnov, L. P. 1982. (Lipids of helminths.) (In Russian.) *Helminthological Abstracts*, **52**: 2044.

Smirnov, L. P. & Bogdan, V. V. 1982. (Comparative study of the lipid composition of some cestodes and of their hosts.) (In Russian.) *Helminthological Abstracts*, **52**: 2043.

Smyth, J. D. 1969. *The Physiology of Cestodes*. W. H. Freeman, San Francisco.

Smyth, J. D. 1972. Changes in the digestive–absorptive surface of cestodes during larval/adult differentiation. *Symposia of the British Society for Parasitology*, **10**: 41–70.

Smyth, J. D. & McManus, D. P. 1989. *The Physiology and Biochemistry of Cestodes*. 2nd Edition. Cambridge University Press.

Torre-Blanco, A. & Toledo, I. 1981. The isolation, purification and characterization of the collagen of *Cysticercus cellulosae*. *Journal of Biological Chemistry*, **256**: 5926–30.

Ugolev, A. M. 1968. *Physiology and Pathology of Membrane Digestion*. Plenum Press, New York.

Von Brand, Th. & Weinbach, E. G. 1975. Incorporation of calcium into the soft tissues and calcareous corpuscles of larval *Taenia taeniaeformis*. *Zeitschrift für Parasitenkunde*, **48**: 53–63.

Wack, M., Komuniecki, R. & Roberts, L. S. 1983. Amino acid metabolism in the rat tapeworm, *Hymenolepis diminuta*. *Comparative Biochemistry and Physiology*, B, **74**: 399–402.

Ward, C. W. & Fairbairn, D. 1970. Enzymes of beta-oxidation and the tricarboxylic acid cycle in adult *Hymenolepis diminuta* (Cestoda) and *Ascaris lumbricoides* (Nematoda). *Journal of Parasitology*, **56**: 1009–12.

Webb, R. A. 1986. The uptake and metabolism of L-glutamate by tissue slices of the cestode *Hymenolepis diminuta*. *Comparative Biochemistry and Physiology*, C, **85**: 151–62.

Wikgren, M., Reuter, M. & Gustafsson, M. 1986. Neuropeptides in free-living and parasitic flatworms (Platyhelminths). An immunocytochemical study. *Hydrobiologica*, **132**: 93–9.

25 Nematoda: general account

The nematodes, or 'round worms', make up a large assemblage of relatively simple structure with a widespread distribution, their cylindrical non-segmented bodies distinguishing them easily from other helminths. They occur in fresh water, in the sea and in soil, and are among the most successful parasites of plants and animals. Most of the free-living nematodes are microscopic, as are many of the parasitic species invading the body fluids such as the blood or lymph channels of their hosts. Those species which live in the intestine are generally larger, while some in tissue habitats (e.g. the kidney) grow to relatively enormous lengths.

Nematodes exhibit a wide range of feeding habits. Many feed entirely on the microorganisms present in decaying vegetable matter, others live on the outsides of plants and suck their juices, while others venture within the plant and wander destructively through its tissues. In vertebrates, they may parasitise the eye, mouth, tongue, alimentary canal, liver, lungs or body cavity, often causing destructive and revolting diseases and producing untold hardship. A much quoted figure, given by Stoll (1947) in an entertaining survey of helminth parasites, estimated that in a world of 2200 million inhabitants, there were some 2000 million human nematode infections, a tribute, indeed, to the efficiency of nematode life cycles.

The life cycles range from the very simple to the extremely complicated. Most nematodes are dioecious, producing eggs with tough resistant coverings. Some species are monoecious while in some dioecious species the males are rare or as yet unknown. The monoecious species may be either parthenogenetic or self-fertilising hermaphrodites. The majority of nematodes are oviparous, but some are ovoviviparous (see p. 380). The successful development of nematode eggs outside the host is largely dependent on environmental conditions, particularly oxygen and temperature. All hatched nematode larvae, whether they hatch in water, soil or within an animal host, must undergo a series of four ecdyses (moults) before reaching maturity.

Literature. Various aspects of the general biology, systematics, behaviour, ecology and physiology of nematodes have been reviewed in the following books or reviews: Adamson (1987), Anderson (1988, 1992), Anderson *et al.* (1974–83), Bird (1971, 1984), Bird & Bird (1991), Croll (1970, 1976), Croll & Matthews (1977), Gibbons (1986), Kassai (1982), Lee & Atkinson (1977), Levine (1980), Maggenti (1981), Poinar (1983), Stone *et al.* (1983), Wharton (1980, 1986).

25.1 Classification

An agreed classification of nematodes has been difficult to devise and agreement is still not universal. The classification below is based on a set of keys produced by the Commonwealth Institute of Helminthology (CIH) and edited by Anderson *et al.* (1974–83). Some revision of this system has been proposed by Adamson (1987).

Nematodes have been traditionally divided broadly into the *Aphasmidea* and the *Phasmidea*, depending on whether caudal sense organs, the *phasmids* (p. 373), are present or not. Unfortunately, phasmids are often difficult to detect in living or preserved specimens, so that from a practical point of view, this division is rather unsatisfactory although the terms are still loosely used (and retained below). Only those groups of interest to the animal parasitologist are included below.

Sub-Class I. Adenophorea (= 'Aphasmidea' = aphasmid nematodes)

Phasmids absent or few in number. Excretory system without lateral canals and terminal duct not lined with cuticle. Pharynx (oesophagus) a long fine tube forming a *stichosome* (p. 388, Fig. 26.2) or *trophosome*. Eggs usually unsegmented with a plug at either pole (Fig. 26.1) or hatching *in utero*. First-stage larva often with stylet and usually infective to final host.

Order **Enoplida**

Superfamily 1. **Trichuroidea** ('capillarids'). Pharynx a stichosome (Fig. 26.2) or a trophosome. Females with only one ovary; males with one spicule or none (e.g. *Trichuris*, *Trichinella*, *Capillaria*).

Superfamily 2. **Dioctophymatoidea.** Large worms parasitic in the kidneys of vertebrates. Pharynx cylindrical and well developed: stichosome or trophosome absent. Female with one ovary; male with cup-shaped bursa-like expansion with rays (e.g. *Dioctophyma*).

Sub-Class II. Secernentea (= 'Phasmidea' = phasmid nematodes)

Phasmids present. Caudal papillae almost always numerous (basic number 21). Excretory system well developed with lateral canals and terminal canal lined with cuticle. Pharynx never in the form of stichosome. Eggs without polar plugs, rarely operculate at one pole. The beginning of the third larval stage infective to the final host. Saprophages in soil (rarely water) and parasitic in vertebrates and invertebrates.

Order 1 **Rhabditida** (Chapter 27). Small transparent meromyarian (see p. 370) worms. Pharynx usually with a posterior bulb and frequently a prebulbar swelling. Mouth with three to six lips. Females oviparous or viviparous and may be parthenogenetic or hermaphroditic. Majority free-living, some with both free-living and parasitic phases (e.g. *Rhabditis*, *Strongyloides*).

Order 2 **Ascaridida** (Chapter 27). Mouth with three lips (one dorsal and two sub-ventral); pharynx bulbed or cylindrical; vagina elongate; male usually with ventrally curled tail and two spicules; alae may be present (e.g. *Ascaris*, *Enterobius*).

Order 3 **Oxyurida** (Chapter 27). Males with reduced number of caudal papillae. Generally only one spicule. Parasites of colon or rectum. Body short and stout. Pharynx with posterior bulb. Female with large embryonated eggs often flattened on one side (e.g. *Oxyuris*).

Order 4 **Strongylida** (Chapter 28). Male with copulatory bursa; mouth simple without lips, frequently with a buccal capsule and teeth; usually meromyarin; pharynx muscular, club-shaped or cylindrical (e.g. *Ancylostoma*, *Nippostrongylus*, *Syngamus*).

Order 5 **Spirurida** (Chapter 29). Pharynx composed of two parts, both cylindroid, an anterior muscular portion and a posterior glandular portion; males usually with two spicules and well-developed alae on spirally coiled tail; mouth with no lips, rudimentary lips, or with two or four paired lips. Long, thin nematodes, parastic in vertebrates in adult stage. Mainly viviparous (e.g. *Wuchereria*, *Dracunculus*).

○○○ 25.2 Type example: *Rhabditis maupasi*

A number of the most readily obtainable nematodes belong to the genus *Rhabditis*. They occur in soil, water and decaying organic materials, and are facultative parasites in the larval stage. *R. maupasi* occurs commonly in earthworms. Other common species in the same host are *R. terrestris*, *R. pellio* and *R. anomala*, but it is difficult (even for those working on this group) to identify the species from the female or protandrous hermaphrodite. Males may be rare in some species. Young encysted forms (sometimes known as '*dauer*' larvae, although the term is used rather differently by various authors) occur in the ovoid brown bodies lying in the coelomic spaces of the posterior somites. Free forms may usually be found in the seminal vesicles or coelomic spaces and are especially common in the nephridia.

In the earthworm, rhabditids never develop beyond the juvenile stage, presumably owing to the lack of certain growth factors, but with the available nutriment, are just able to survive without further differentiation. When an earthworm dies, however, the decaying flesh, and the bacteria which thrive on it, provide additional food to raise the nutritional level to a degree which permits the juveniles to reach sexual maturity within a few days. Flourishing laboratory cultures are readily obtained by allowing a portion of the body wall of an earthworm, containing nephridia, to undergo degeneration on nutrient agar plates (Fig. 25.2).

External features. Both sexes are cylindrical and elongated with a very thin, almost hair-like caudal ter-

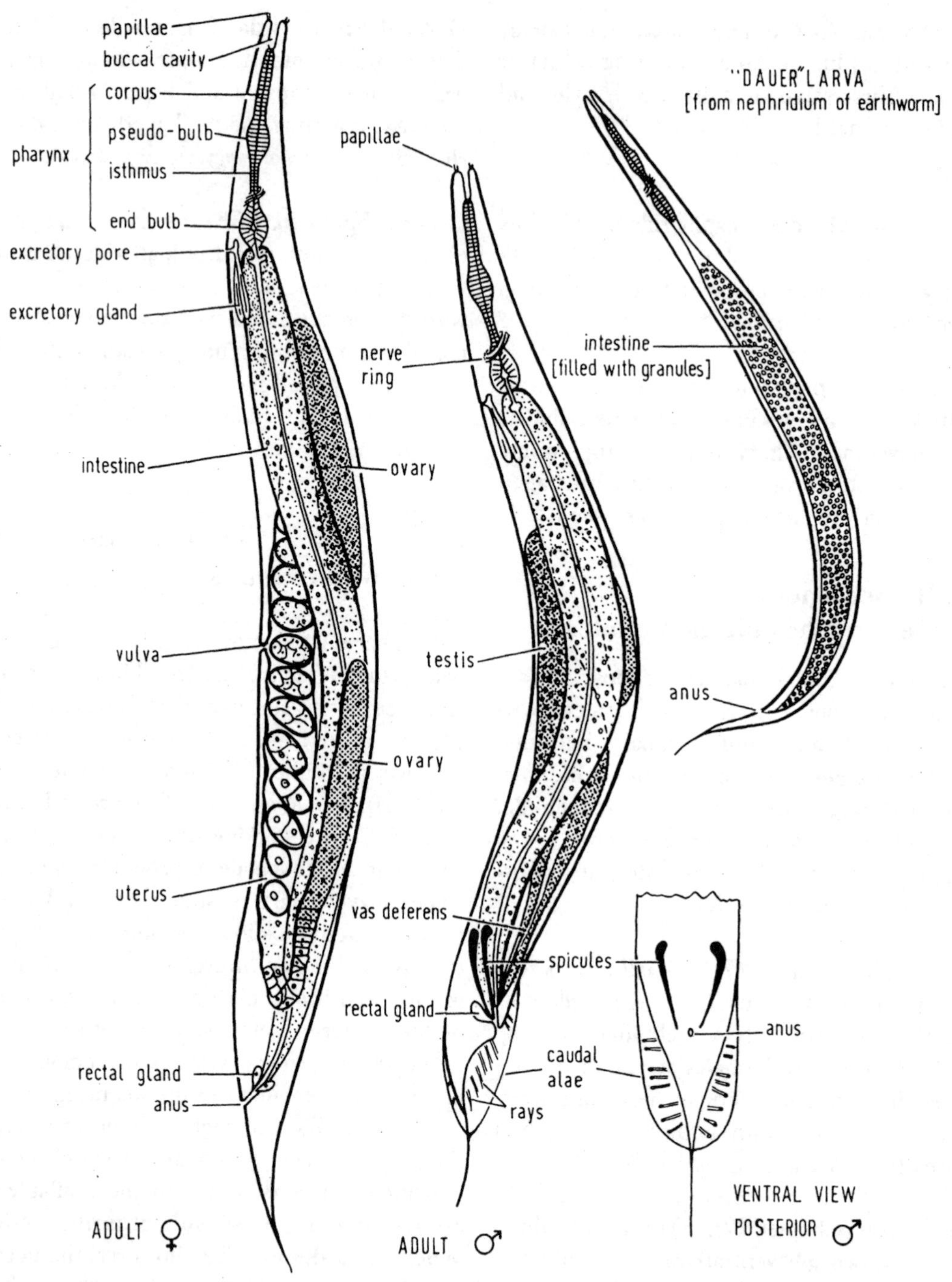

Fig. 25.1
Rhabditis maupasi: morphology of free-living males and females from laboratory cultures. (Original.)

mination or post-anal tail. The males (which in some 'races' are rare in laboratory cultures) are characterised by possessing a flared posterior end forming an accessory copulatory organ (Fig. 25.1).

Body wall. A thin transparent cuticle covers the body. This is clearly secreted by the hypodermis, a subjacent layer containing a small number of nuclei. Beneath the hypodermis is a muscular layer consisting of a single sheet of longitudinal cells. *Rhabditis* is a *meromyarian* nematode (Fig. 25.17), that is, it has few muscle cells in each sector. These muscle cells are unusual in that they have processes extending to the motor nerves instead of nerve processes extending to the muscles.

Digestive system. The intestine has a fine basement membrane supporting a single layer of cells. The pharynx and rectum have a cuticle, continuous with that on the exterior, which is shed at ecdysis. The mouth is enclosed in one dorsal and two ventral lips, each of which is further subdivided into two lobes, bearing a pair of sensory papillae. The narrow buccal cavity is heavily cuticularised and bears a ridge-like construction near its posterior end; the ridge bears fine cuticular teeth of taxonomic value. There is a muscular *pharynx* (= oesophagus) with a wide anterior region or *corpus* leading to a middle *pseudobulb*, connected by a narrow *isthmus* to a posterior *end bulb*, containing a tripartite valve; the bulb wall contains three gland cell nuclei. The arrangement of these cells, which ramify throughout the oesophagus, is of taxonomic importance.

Excretory system. The so-called 'excretory system' consists of a pair of lateral canals extending from the pharyngeal pro-corpus region nearly to the end of the worm, where they end blindly. These canals lead into an excretory sinus, which opens by a cuticular duct to the excretory pore. Also connected with the excretory sinus is a pair of sub-ventral cells. The excretory sinus bears a single nucleus; it is usually considered that the lateral canals are extensions of the sub-ventral cells (see p. 377).

Nervous system. The general features of the nervous system are very similar in all nematodes. Only those parts easily seen in living unstained specimens will be described here. A nerve ring, composed chiefly of nerve fibres, surrounds the isthmus portion of the pharynx. Nerve cells lie in contact with the nerve ring and dorsal and ventral cords join it.

Male reproductive system. There is a single tubular testis reflexed at its anterior end and filled with spermatozoa in all stages of development. Spermatozoa are spherical and tail-less, a typical nematode feature. A well-developed vas deferens leads from the testis to the cloaca, in association with which are cuticular spicules with characteristic knobbed ends. These spicules appear to act as holdfast organs during copulation and are moved by protractor and retractor muscles. The tail of the male is further differentiated by being drawn out into longitudinal ridges, the *caudal alae*. These are thin-walled extensions of the postero-lateral body margins and also assist in holding the female in apposition during copulation. They are supported by nine pairs of *genital papillae*, finger-like projections of the ventral surface. Caudal alae are sometimes referred to as 'bursae', but this term, strictly speaking, is reserved for the wide caudal alae of the Strongylida (p. 412).

Female reproductive system. This is built on the typical nematode plan with paired ovaries converging on the vulva. The ovaries contain developing oocytes packed tightly together. The ends are retroflexed and each becomes constricted to form an oviduct in which spermatozoa may be found. Fertilisation takes place in this region, and eggs accumulate in the paired uteri, which connect by a short ventrally directed vagina to the vulva.

In 'races' which are protandrous hermaphrodites, the worm has a genital system of the female form. Spermatozoa are first produced, then ova, which are fertilised by the stored spermatozoa.

Life cycle. This is shown diagrammatically in Fig. 25.2. When an earthworm dies, the contained third-stage 'dauer' larvae feed on the bacteria and fluid of the decaying organic material and develop into adults. These reproduce either bisexually or hermaphroditically, and larvae produced from hatched eggs migrate into the soil to penetrate another earthworm via the nephridiopores or other apertures. Although the coelomic forms encyst, those in the nephridia and other sites remain free but, presumably due to the absence of suitable nutrients, they do not grow beyond the 'dauer' larva stage.

○○○ 25.3 General morphology of nematodes

Nematodes are remarkable for exhibiting a high degree of uniformity in basic structural organisation. Their shape is essentially that of an elongated cylinder, inside which runs the body musculature consisting entirely of longitudinal fibres. Harris & Crofton (1957) have pointed out that a system of muscles of this kind can only operate against a suitable antagonistic force; since circular muscles are lacking, the return mechanism must be provided by alternative means such as internal hydrostatic pressure. They have shown that many of the morphological features of excretory, alimentary and reproductive systems can be correlated with this fact and the mechanical problems of organisation which result. When nematode morphology is viewed in this light the

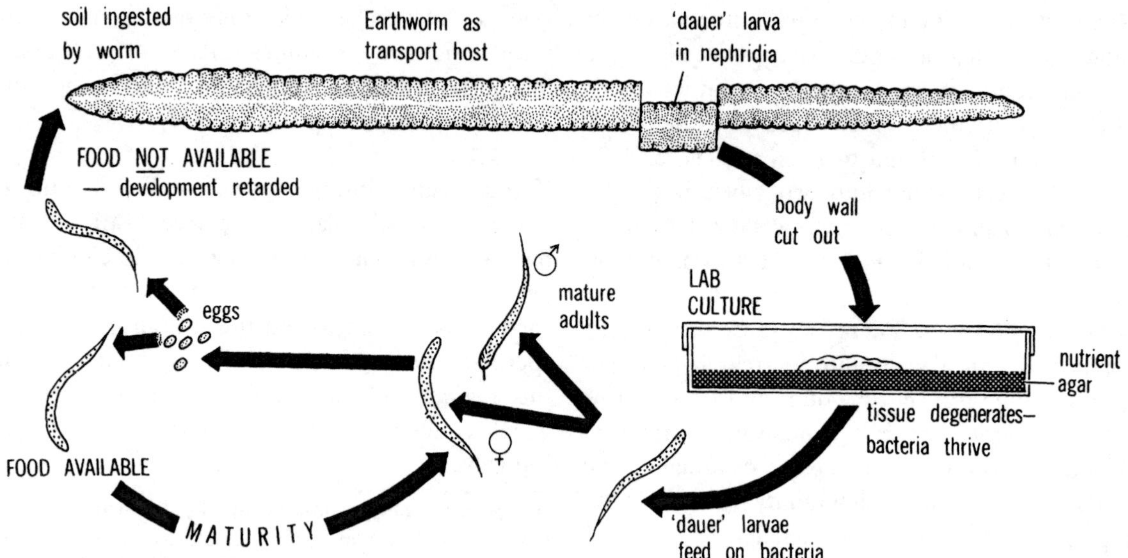

Fig. 25.2
Life cycle and laboratory culture of *Rhabditis maupasi*. (Original.)

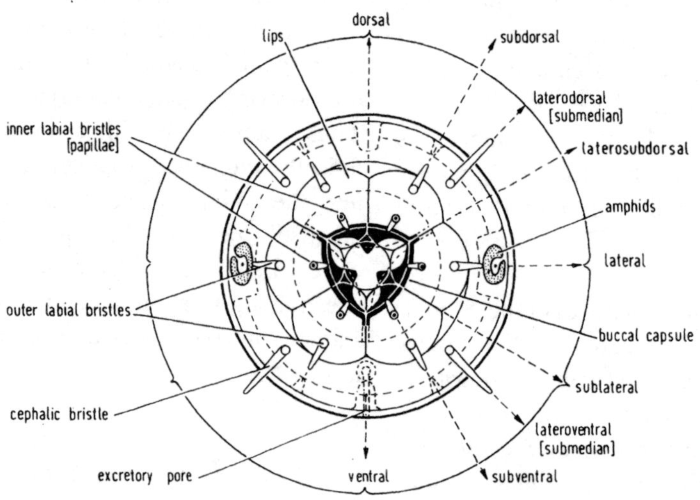

Fig. 25.3
Generalised diagram of end-on view of primitive nematode showing the hexamerous symmetry on the sensory bristles. (Modified from de Coninck, 1942.)

result is a most interesting integration of form and function. Although it is early yet to say whether these views will stand up to detailed examination when a number of species are examined, they may, if confirmed, revolutionise our ideas on nematode morphology.

○ 25.3.1 External features

Recent studies have revealed that a marked hexamerous symmetry is superimposed on the original bilateral symmetry of primitive nematodes. This hexamerous symmetry supports the view that nematodes may have evolved through a semi-sessile stage in which they were fastened posteriorly by caudal secretions to objects, and waved their bodies about in the water. Today, many Aphasmidia can attach themselves by the secretions of the caudal glands.

The cephalic structures, which are of value for taxonomic purposes, are best studied from *en face* sections (Fig. 25.3). Primitively, the mouth is surrounded by six lips or labia each bearing a papilla, which may be modified to a bristle, so that a circle of six *inner labial papillae* is formed. Outside of this is an *outer labial circle* of six papillae, and outside of this again, in the cephalic region, is a *cephalic circlet* of four papillae. The total number of anterior sense organs is thus considered to be sixteen, a condition found in some free-living marine nematodes. In parasitic (and free-living, terrestrial) nematodes the number of anterior sense organs is usually greatly reduced and the two outer circlets reduced to one, so that at the most only two of the three circlets are present. In some species (e.g. *Ascaris*), the inner circlet is vestigial. Various other cuticular specialisations may be present in the head region.

○ 25.3.2 Alimentary canal

Buccal cavity. This is variable in shape and size, and in degree of differentiation. In some forms, its cuticular lining may be moulded into rods or plates, often bearing teeth. The latter are particularly well developed in intestinal forms such as hookworms which rely on sucking blood from the mucous membrane (Fig. 28.5).

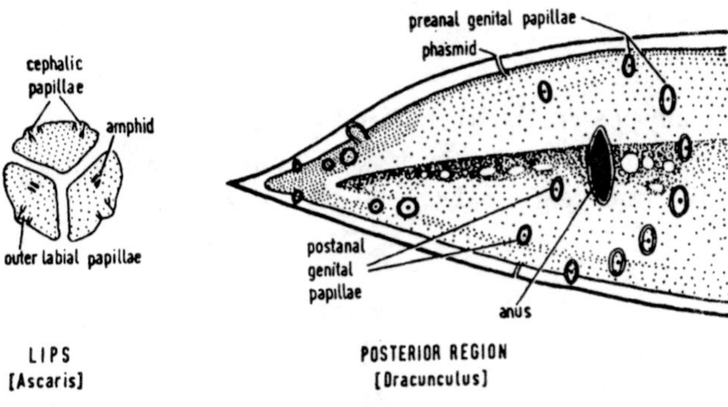

Fig. 25.4
Lips of *Ascaris lumbricoides* showing amphids, and tail of *Dracunculus medinensis* showing phasmids. (After Thornton, 1924; Moorthy, 1937.)

Pharynx (oesophagus). The pharynx is one of the most characteristic features of nematode morphology. Although the terms 'pharynx' and 'oesophagus' are used as being synonymous, *pharynx* is used in preference to *oesophagus* as it is used for homologous structures in related groups. Since nematodes generally appear to have a high internal pressure, and since the intestinal wall of nematodes is only one cell thick (and therefore liable to collapse under turgor pressure), the intestine can only be filled by a powerful pumping organ. The pharynx which serves as this organ was originally thought to be syncytial because cells walls could not be detected by early workers. TEM studies later showed that the pharynx is, in fact, a cellular structure with distinct muscle cells, supporting cells, nerve cells and gland cells (Bird, 1971) (Fig. 25.5). It is lined with cuticle continuous with that of the buccal cavity and is, therefore, part of the stomadaeum; the cuticular lining is shed at ecdysis. The median and posterior regions of the pharynx are generally swollen into one or two muscular bulbs, but these are lacking in some nematodes. The pharynx shows considerable variation in both its structure and its mode of operation, a diversity correlated largely with feeding habits, chief of which are microbivorous (i.e. on bacteria), saprophagous (i.e. on decaying organic material), phytophagous (i.e. on fungi,

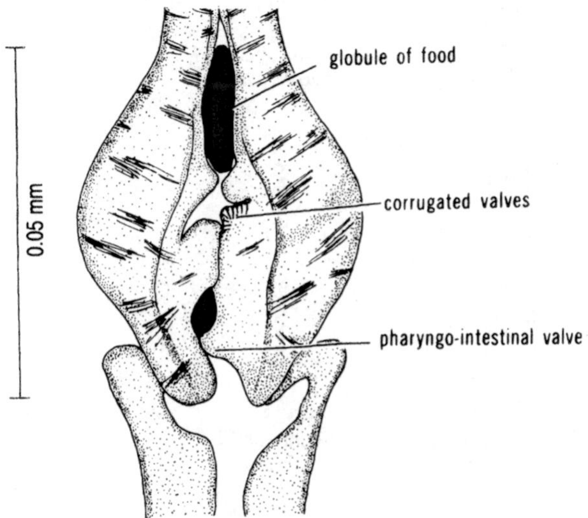

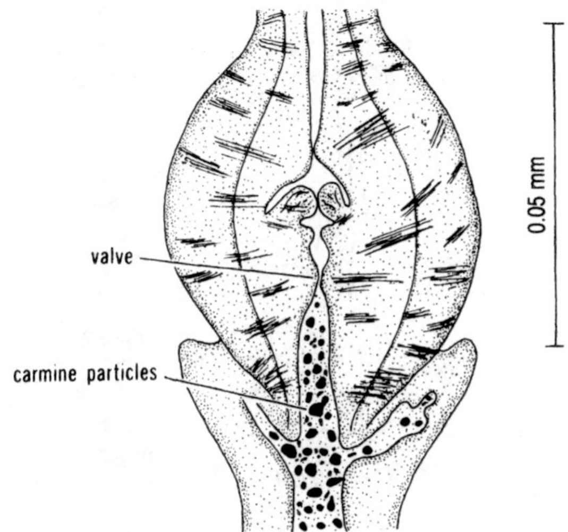

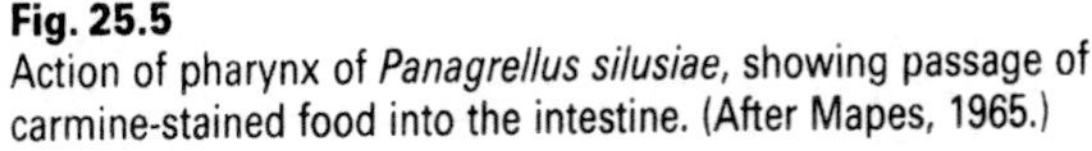
Fig. 25.5
Action of pharynx of *Panagrellus silusiae*, showing passage of carmine-stained food into the intestine. (After Mapes, 1965.)

algae and cells of higher plants) or carnivorous (e.g. on Protozoa). Although most of these terms refer to free-living nematodes, it must be pointed out that many nematode parasites (e.g. *Ancylostoma*, p. 418) have free-living stages.

Most nematode parasites of vertebrates feed on the semi-liquid contents of the alimentary canal, the tissues of the mucosa, lysed tissues of other organs, or blood or other body fluids, and their mouth structure and pharynx are appropriately adapted. For detailed consideration, see Bird (1971), Croll & Matthews (1977) or Wharton (1986).

In addition to muscular bulbs, many nematodes possess a well-developed system of valves which prevent possible regurgitation of food. The passage of food may often be conveniently traced by the use of carmine particles (Fig. 25.5).

A complex terminology has grown up in relation to the nematode pharynx. The following definitions are based on Hyman (1951):

bulb: a muscular swelling containing sclerotinized valves capable of partly closing the lumen
pseudobulb: a muscular swelling lacking a valvular arrangement
cylindrical: of the same diameter throughout
dorylaimoid: slender anteriorly and wider posteriorly
oxyuroid (bulboid): provided with an end bulb
rhabditoid: with a wide anterior region usually followed by a median pseudobulb, an isthmus and an end bulb
ventricular (glandular) region: a part of the pharynx devoid of muscle fibres

The form of the pharynx in the various orders is as follows.

Enoplida

Markedly adapted for sucking. The wall, although muscular, is so reduced that it is virtually a capillary tube. A column of cells, the *stichosome*, lies along the oesophagus; the latter is sometimes embedded in it (e.g. *Trichuris*, *Dioctophyma*) (Fig. 25.6).

Rhabditida

Usually possess the 'rhabditoid' type of pharynx, characteristically with an anterior pseudobulb and a posterior valved bulb (e.g. *Rhabditis*, Figs 25.1, 25.6).

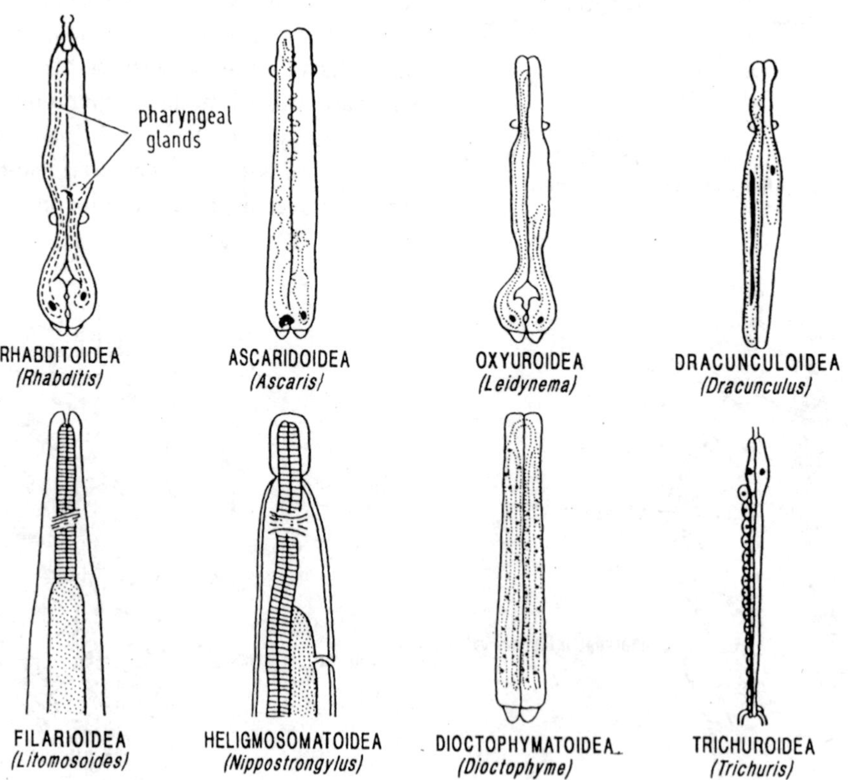

Fig. 25.6
Various forms of the pharynx in the major superfamilies of nematodes. (After several authors.)

Ascarida

Form of the oesophagus varies more than any other group. In the Oxyuroidea, it is basically of the rhabditoid type. In the Ascaridoidea, the pharynx is usually cylindrical and lacks a bulbar enlargement. Posterior pharyngeal diverticula, known as appendices or intestinal pouches, are not uncommon. Larval ascarids may show traces of a rhabditoid pharynx in their early larval stages (e.g. *Ascaris*, *Leidynema*, Fig. 25.6).

Strongylida

Oesophagus simple with little or no differentiation into regions (e.g. *Nippostrongylus*, Fig. 25.6).

Spirurida

Pharynx usually divisible into a narrow anterior muscular portion and a broader posterior glandular portion (e.g. *Dracunculus*, *Litomosoides*, Fig. 25.6).

Intestine. The morphology of the intestine does not show great variation and, in general, consists of a straight tube showing little differentiation into regions.

Layers

The wall of the intestine (Fig. 25.7), which lacks an intestinal cuticular lining, is made up of a single layer of epithelial cells which, on their intestinal surface, bear a border of fine hair-like structures often referred to as the *bacillary layer* (*bordeur en brousse*, *Stabchensaum*). The composition of this layer has long been a matter of interest to cytologists as it was thought that the structures might be cilia. Ultrastructure studies on a number of nematodes (e.g. *Nippostrongylus*, *Ascaris*) have demonstrated quite unequivocally that these are microvilli similar to those which characterise the lining of the gut of many vertebrates and invertebrates (Bird, 1971; Kassai, 1982).

In *Nippostrongylus*, the microvilli are about 1 μm long and are covered with a fibrous material, the glycocalyx. The usual cytoplasmic inclusions are present and large amounts of glycogen are commonly found. The basal (pseudocoele) surface of the intestine is covered by a thin basal lamella. Many elongate vesicles lie amongst the mitochondria and these may represent secretory (or ingested?) bodies. In many species, peculiar spherical crystals termed *sphaerocrystals* occur (Fig. 25.7). These are composed of a carbohydrate material ('rhabditin') and are of unknown function, although it has been suggested that these may be lysosomes (Bird, 1971).

Rectum

The hindgut is lined with cuticle continuous with the external cuticle. In the female it becomes a rectum, but in the male the reproductive system always opens into it so that it becomes a true cloaca. A number of unicellular rectal glands (three to six) open into the rectum in the majority of species (except in the Aphasmidia).

Nervous system. The nervous system, the morphology of which is remarkably constant in different species, is best known in *Ascaris lumbricoides*, where it was first described in the classical studies of Goldschmidt (1908). His original description (Fig. 25.8) has

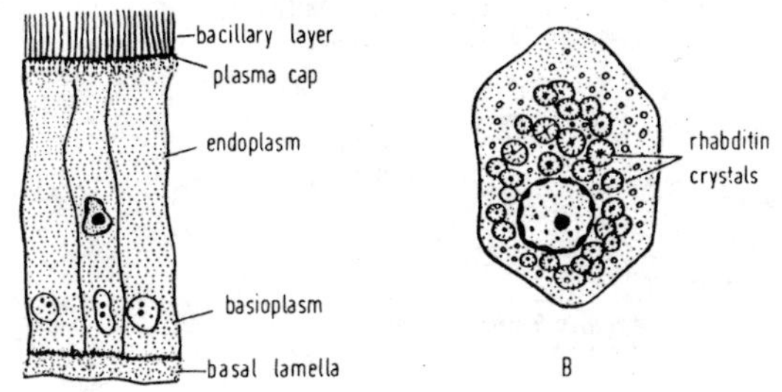

Fig. 25.7
A. Intestinal cells of *Ascaris* showing bacillary layer. B. Intestinal cell containing sphaerocrystals of rhabditin. (After Chitwood & Chitwood, 1950.)

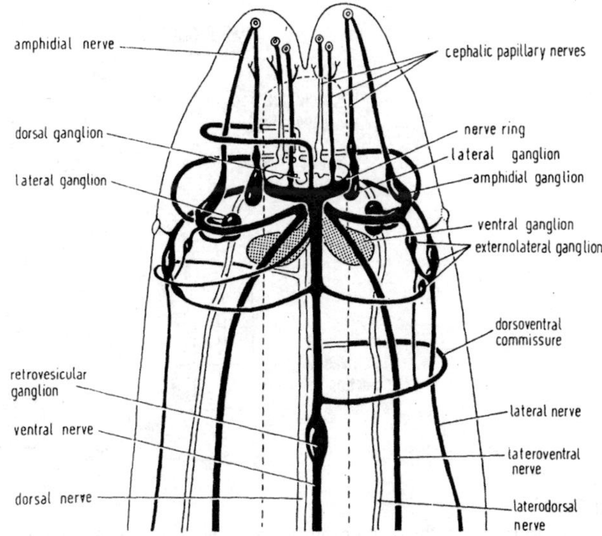

Fig. 25.8
Anterior part of nervous system of *Ascaris lumbricoides*. (After Goldschmidt, 1908.)

been largely confirmed by later work (Stretton *et al.*, 1978, 1985) which has also demonstrated that the nervous system is composed of about 90 neurones and interneurones. In *Ascaris* the system contains (*a*) a nerve ring and associated ganglia at the anterior end, (*b*) major dorsal and ventral nerve cords together with smaller subdorsal and subventral cords, and (*c*) another (smaller) set of ganglia in the posterior region. For a very detailed survey of the nervous system of nematodes in general, see Bird (1971).

The structure and physiology of the nervous system of *Ascaris* has excited the interest of neurophysiologists who have demonstrated that – even though nematodes are not segmented worms – the nervous system is made up of repeating units or segments, each containing eleven motoneurones which can be classified into seven different types (Stretton *et al.*, 1978). These consist of three types of synapse onto the ventral nerve cells and four onto muscle cells.

Neuromuscular system. The musculature of *Ascaris* is composed of large cells (Figs 25.9, 27.4) which show features rarely combined in other phyla. Chief of these is the finding that the nerves do not branch out to the muscles (as is usual); instead, branches of the muscles reach the nerves. Thus the somatic musculature consist of a single layer of spindle-shaped structured (contractile) fibres, each connected to a noncontractile muscle cell body which projects into the pseudocoele and contains the muscle cell nucleus (Fig. 25.9). This system has been termed a muscle syncytium or *sarcopile* (Del Castillo *et al.*, 1989). The ultrastructure of the syncytium and its electrical properties have been much investigated; for details, see the reviews of Del Castillo *et al.* (1989) and Stretton *et al.* (1985). The different types of muscle-cell organisation found in different species are shown in Fig. 25.17 and discussed further on p. 384.

Sense organs. Numerous 'sense' organs have been described in nematodes although there is no electrophysical evidence to support that these are, in fact, sensory receptors. Putative sense organs include the following:

Amphids

These are complex organs opening on each side of the head and are typically large pits filled with modified cilia, known collectively as 'sensilla'. They are believed to be chemoreceptors. Their ultratsructure and functional significance have been reviewed by Altner & Prillinger (1980).

Phasmids

These are generally similar to amphids, but are found at the posterior end. The presence or absence of these organs has given rise to the original division of the Nematoda into the Aphasmida and the Phasmida (p. 368).

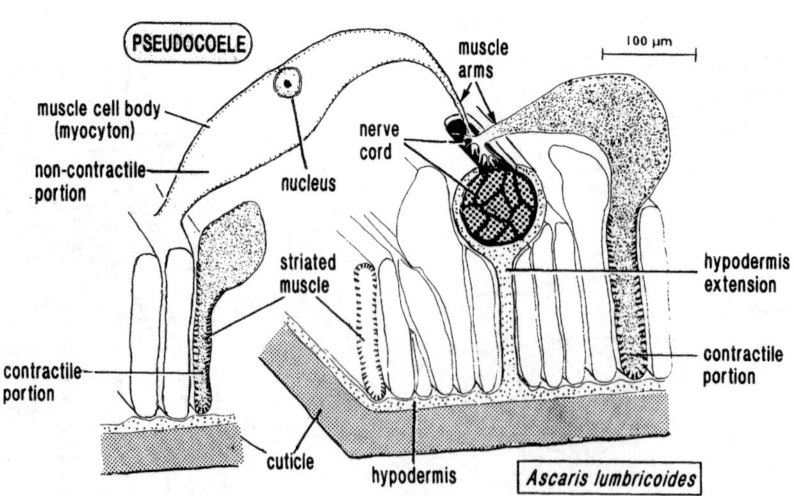

Fig. 25.9
Ascaris lumbricoides: the relationship between the muscle cells and the longitudinal nerve cord. The somatic musculature consists of longitudinal muscle fibres beneath the hypodermis (= epidermis, p. 384); the striations have only been shown in a few fibres. Each striated fibre is connected to a non-contractile muscle cell, which projects into the pseudocoele and contains the muscle cell nucleus. A muscle cell arm extends from the muscle cell to the longitudinal nerve cord. See also Fig. 25.17. (Modified from Debell, 1965.)

Cephalic and caudal papillae

A series of papillae occur around the mouth and cloaca of nematodes. The cephalic papillae are probably tactile and the caudal papillae appear to be involved in copulation.

Spicules

The males of some species, e.g. *Nippostrongylus brasiliensis*, have spicules through which runs nervous material. These spicules are used in holding open the vulva during copulation and appear to be sensory in function.

Ocelli

Some free-living nematodes possess ocelli (eye spots).

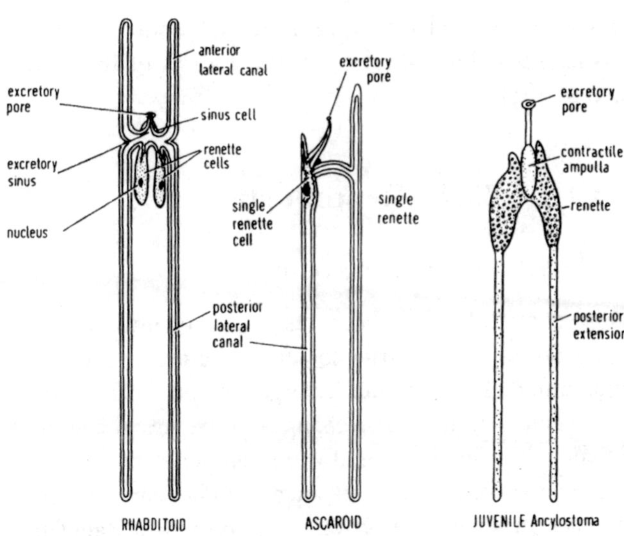

Fig. 25.10
Nematode excretory systems. (After Chitwood & Chitwood, 1950; Stekhoven, 1927.)

○ 25.3.3 Excretory system

The so-called 'excretory system' is perhaps the most varied anatomical feature of nematodes. It possesses no flame cells, protonephridia or other current-producing mechanisms. The structures to which an excretory function has been ascribed are of two sorts, glandular and canal.

Glandular system. This is best developed in marine nematodes, which are generally considered the most primitive of living nematodes. It consists essentially of one or two gland cells situated in the posterior ventral region of the pharynx and opening via a long neck to the excretory pore (Fig. 25.10); the whole is termed a *renette*.

Canal system. This is related to the glandular system and probably developed from it. The renette is often poorly developed in the adult stages although well formed in the larval stages. Thus, in young *Ancylostoma*, the renette cells are drawn out into tubular extensions and it is tempting to assume that they give rise to the adult tubular excretory system. The canal system is most frequently H-shaped, but the anterior limbs, or one side of the system, may atrophy, giving an asymmetrical arrangement (Fig. 25.10).

In larval nematodes, the canal system may pulsate rhythmically and in some species it has been shown to play a major part in the maintenance of water balance, thus probably serving the same function as the contractile vacuole in protozoans. The rate of pulsation varies

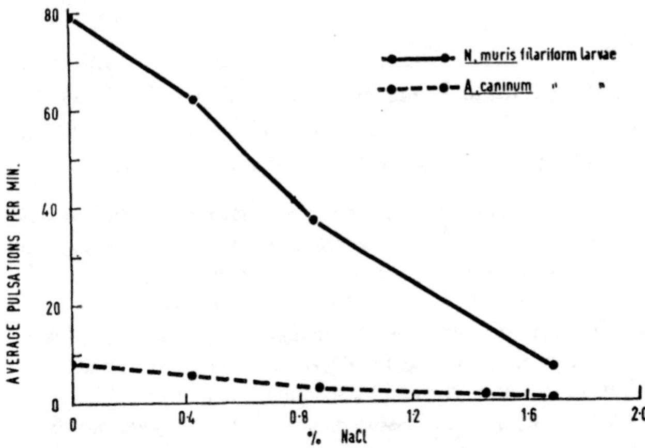

Fig. 25.11
Pulsation rate of the excretory vesicle in filariform larvae of *Nippostrongylus brasiliensis* and *Ancylostoma caninum* in solutions of different osmotic pressure. (Redrawn from Weinstein, 1952.)

inversely with the salt concentration in the medium (Fig. 25.11) in some species, e.g. *N. muris* and *A. caninum* (Weinstein, 1952), but in others, e.g. *Ancylostoma tubaeforme*, the relationship is not linear (Croll *et al.*, 1972).

There is little evidence to indicate that these systems are, in fact, 'excretory'. Studies on the pulsation cycle and ultrastructure of the excretory ampulla of the third-stage larva of *Haemonchus contortus* have provided data to support the above view that one of its functions is osmoregulatory (Wharton & Sommerville, 1984). However, ultrastructure studies on the excretory cells (renette, Fig. 25.10) suggest that electron-dense granules in these cells may also be the source of the 'exsheathing fluid' involved in the ecdysis of the larva (see p. 381).

○○○ 25.4 Reproduction

○ 25.4.1 General account

Most nematodes are dioecious, but protandrous hermaphrodites and parthenogenetic females are not uncommon. Experimental changes in the environment have been used to induce changes in the sexual characteristics of individuals, and even complete sex reversal has been achieved by this means. The chromosome changes in such cases have been studied in detail by Nigon (1949). Males are usually distinguishable from females by their smaller size, posterior curvature and presence of accessory copulatory structures such as spicules, bursae or genital papillae.

Male system. The male system was undoubtedly originally double, but is single in the majority of Phasmidia. The male system of *R. maupasi*, already described (p. 371), is fully representative of the nematodes.

With a few exceptions (e.g. *Trichinella spiralis*, p. 390), males possess *copulatory spicules* secreted by and lodged in cloacal evaginations termed *spicule pouches*. There are usually two such pouches, which unite before entering the cloaca. Spicules, which are hard sclerotinised cuticular structures, are typically two in number, but occasionally single. They show great variations in size and shape. The spicule pouch may be thickened dorsally to form an accessory piece of plate, the *gubernaculum*, which serves to guide the spicules (Fig. 25.12). A ventral thickening of the ventral and lateral cloacal wall, termed a *telamon*, may also occur; these terms are often confused in the literature.

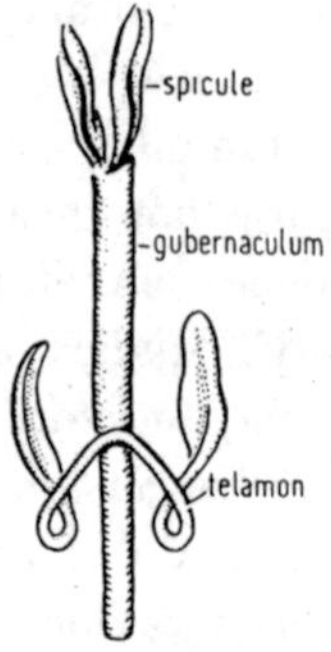

Fig. 25.12
Male armature of a trichostrongyloid, with telamon. (After Hall, 1921.)

In nematodes of the order *Strongylida* the cuticle in the posterior region is expanded into a structure, supported by ribs or rays, termed a *bursa* (Fig. 28.5), and species possessing this structure are said to be 'bursate'. Many authors restrict the term 'bursa' to the structure found in the male strongyloid, but the term is also loosely used to describe any similar posterior male expansion such as the cuticular alae in the rhabditoids and the bell-like muscular organ of the Dioctophymatida.

The spermatozoa of nematodes possess a number of unusual features. Many are non-flagellate but amoeboid. Even in those which appear to have a 'head' and 'tail', these structures do not carry out the functions usually attributed to these regions. Thus EM studies on the spermatozoon of *Aspicularis tetraptera* have shown that the 'head' contains no nucleus and the DNA occurs in the 'tail' region where it is associated with an electron-dense sheath and microtubules (Lee & Anya, 1967). The tail does not apparently move, but movement is achieved by means of pseudopodia (often up to 10–15 μm in length!). Much of the 'tail' is occupied by a single, elongate mitochondrion which extends from within the 'head' almost to the end of the 'tail'. Similar studies on other species such as *Nippostrongylus brasiliensis* (Jamuar, 1967) and *Nematospiroides dubius* (Sommerville & Weinstein, 1964) have also reported the DNA to be located in the 'tail' region.

Female system. There are usually two ovaries, rarely one or more than two. They are generally oriented in opposite directions and may be straight, reflexed or, when long, wound backwards and forwards. The ovary is made up of a continuous sac consisting of an epithelial layer and a *germinal cord*. The epithelial lining consists of a single layer of squamous epithelium which in the oviduct region becomes transformed into high columnar epithelium. The germinal cord contains the maturing oogonia, which often lie round a central strand or rachis whose functions are uncertain. The epithelium of the ovary continues distally as the oviduct with characteristic columnar epithelium. This, in turn, passes into the wider uterus, made up mainly of squamous epithelium. Sperms are stored in the lower part of the uterus where fertilisation occurs. The uteri lead into a short muscular vagina, which opens by the vulva, usually situated about one-third of the body distance from the anterior end. There is sometimes an undivided uterine part between the uterine tubes and the vagina. In many strongyloids and spiruroids, the posterior end of the female may become heavily muscularised to form an ovijector for the expulsion of eggs.

○ 25.4.2 The nematode egg-shell

The egg-shell is particularly well developed in most parasitic nematodes and serves as a highly protective stage in the life cycle. It shows a high degree of complexity and variability which can be regarded as adaptations to increase survival of the enclosed larva or embryo in different environments. It is, in fact, one of the most variable characteristics of nematode morphology. Wharton (1980) has given a comprehensive review of the structure and general biology of egg-shell in different groups. The egg-shell may consist of one to five layers, but typically has three basic layers: an (outer) vitelline layer, a middle chitinous layer and an inner lipid layer. The structure in different orders is summarised in Fig. 25.13, and the composition of the layers is briefly summarised below.

(*a*) **The (outer) vitelline layer.** A lipoprotein layer derived from the vitelline membrane of the fertilised oocyte.

(*b*) **The chitinous layer.** Usually the thickest layer of the egg-shell and the one that provides structural strength. X-ray, infra-red spectroscopy (Wharton & Jenkins, 1978) and histochemical studies have demonstrated the presence of chitin; protein is frequently, but not always, associated with this layer. There is some evidence that quinone-tanning (p. 184) may occur in this layer as evidenced by the browning of the egg-shell in faeces and by other reactions (Monné, 1962). Anya (1964), however, could not detect the quinone-tanning precursors (phenols, polyphenol oxidase) in the egg of *Aspicularis tetraptera.*

(*c*) **The lipid layer.** The presence of this layer is responsible for the remarkable impermeability of some nematode egg-shells. In *Ascaris*, the layer is a lipoprotein made up of 25 per cent protein and 75 per cent lipid. The lipid fraction contains a mixture of α-glycosides called *ascarosides*, which have been identified in the eggs of many other ascarids (e.g. *Toxocara canis*). This layer is probably responsible for the extreme resistance of ascarid eggs to adverse environmental conditions.

(*d*) **Uterine and rectal secretions.** 'Sticky coats' are produced by these structures around some nematode

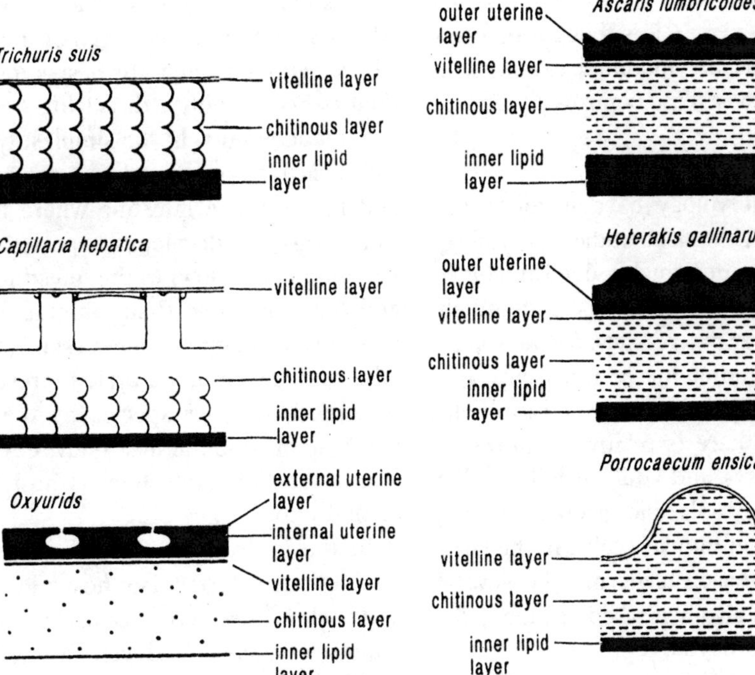

Fig. 25.13
The variety of structure of the nematode egg-shell. (Modified from Wharton, 1980.)

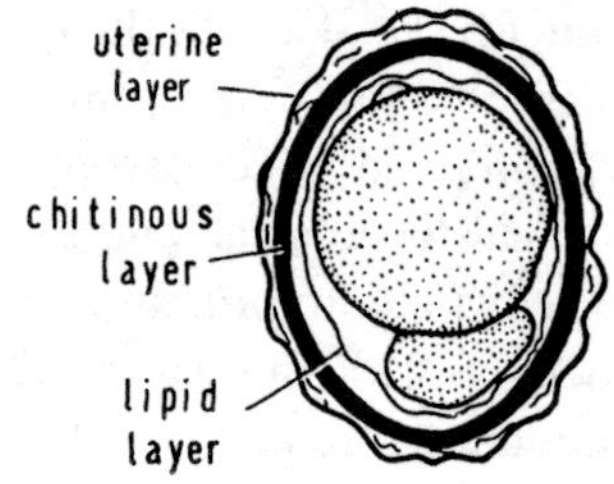

Fig. 25.14
Egg of *Ascaris lumbricoides* (see p. 380). (After Chitwood & Chitwood, 1950.)

eggs. In *Ascaris*, for example, there is an additional 'uterine' layer deposited around the egg; this is often referred to as the *protein* or *mammalated* layer and appears to consist of an acid–mucopolysaccharide complex (Bird, 1971).

Egg-laying habits. The majority of parasitic nematodes are oviparous, the eggs being discharged outside the female body. Some are *ovoviviparous*, in which case the eggs hatch while still within the uterus. This is a widespread habit in parasitic forms. Many of the so-called viviparous species are actually ovoviviparous, and it is questionable whether viviparity in the true sense of the word ever occurs.

The so-called microfilariae of the Filarioidea (p. 423) are enclosed in an elastic 'sheath' (Fig. 29.6) which takes on the contours of the larva. It is thought that this 'sheath', in fact, represents the modified egg-shell in these forms, although there is some controversy on this point.

Development. The embryology of nematodes has been the subject of several classical studies in cytology and cell lineage. Development is highly determinant in character and the fate of each cell can be definitely followed. It is remarkable for two reasons: (*a*) the separation of the germ cell line after the first cell cleavage, and (*b*) the absence (usually) of somatic cell multiplication after hatching. As there is relatively enormous size difference between larva and adult, it follows that an unusual amount of cytoplasmic growth occurs, resulting in the production of monster cells. In the large nematodes, for instance, muscle cells may be several millimetres long. There is also a marked tendency for nematode tissues to form syncytia.

The process of growth in size of nematodes, like that of arthropods, is accompanied by ecdysis. As far as is known, there are always four moults and the nematode formed as the fifth stage is the adult, although some growth in size may follow. The stages in the post-embryonic growth of nematodes are usually referred to as 'larvae', although according to some authors (notably L. Hyman) 'juveniles' is a more suitable term for such stages.

The terminology of larval development is straightforward. After the first moult, larvae are termed second-stage larvae; after the second, third-stage, and so on. The stage at which the larvae are capable of living in the host is termed the *infective* stage. In some species, two moults take place within the egg, so that two cuticles are present when the larvae hatch. In some cases, the cast sheath is retained by the larva as a loose-fitting coat and this probably has some protective function.

The degree of development of the egg when laid varies considerably with species. In some, the eggs are laid in an unsegmented condition (e.g. *Ascaris* and *Trichuris*), while in others, early segmentation has commenced (e.g. hookworms). Still others (e.g. *Syphacia*, Fig. 27.11) may reach a stage with a coiled larva visible, and in the ovoviviparous species and the so-called viviparous species, fully developed larvae are released by the females (e.g. *Dracunculus*).

Life cycles. The life cycles of different species show interesting variations on the theme egg-larvae-adult. Although almost every organ of the host has been used as a nematode habitat, the best-known species are intestinal parasites with a direct life cycle not involving an intermediate host. In the simplest type of life cycle (e.g. *Enterobius*), embryonated eggs are swallowed by the host and hatch in the intestine where they reach maturity. Some directly developing species (e.g. *Ascaris*), after hatching (see below) in the intestine, undergo a curious and unaccountable tissue migration through the body before returning to become established in the intestine. The simple type of life cycle represented by *Ascaris* may be varied in other species by the first-stage embryo hatching outside the host body, developing to an infective larva, and reinfecting its host by skin penetration or oral entry.

In the indirect type of life history, use is made of invertebrate intermediate hosts: insects (usually beetles and cockroaches), crustaceans (especially copepods) and molluscs (snails and slugs). In one group (the Mermithoidea) the larval stage is a parasite of an invertebrate and the adult free-living. Detailed accounts of individual life cycles will be dealt with later.

In yet another type of life history, there may be an alternation of a parasitic generation with a free-living one. The best known are those of various species of the genus *Strongyloides*, in which the free-living larvae may develop to free-living adults if the external nutritional conditions are suitable.

Some genera (e.g. *Neoaplectana*, p. 400) are facultative parasites. They may pass a number of generations in dead insects, but when food is exhausted, resting larval stages remain capable of invading living insects as parasites.

○ 25.4.3 Physiology of egg hatching and larval moulting

Egg hatching. Much work has been carried out on the mechanism of egg hatching in nematodes. It is clearly important that eggs hatch in response to appropriate hatching stimuli so that they will hatch in a suitable environment, either the *external* environment (into which the free-living stages will normally be released) or an *internal* one (into which the parasitic stages are released). Thus, the eggs of the sheep nematode *Nematodirus*, even if laid in the autumn, hatch only in spring: a season when uninfected lambs normally graze on pastures. Temperature, oxygen and moisture are particularly important factors in the hatching of those eggs of species which have free-living stages (e.g. *Ancylostoma*, p. 418, *Nippostrongylus*, p. 412). In species with parasitic larval stages, like many other parasites with intestinal-hatching stages (e.g. the cysts of protozoans, p. 23, or trematodes, p. 209), nematode eggs hatch only in a specific intestinal region in a specific host where the combination of factors essential to induce hatching occurs.

Of the various physiological characteristics of the intestine, the presence of carbon dioxide, under strongly reducing conditions, at a suitable pH (at 38 °C) appears to be one of the most important. The exact role played by carbon dioxide in this mechanism or the form in which it produces the maximum 'signal' is not yet determined although a considerable amount of work has been carried out on this problem. Carbon dioxide has the advantage that it rapidly penetrates many membranes and cells and hence the partial pressure in the environment could readily be monitored by an egg. Rogers (1966) gives a valuable review of early work on this

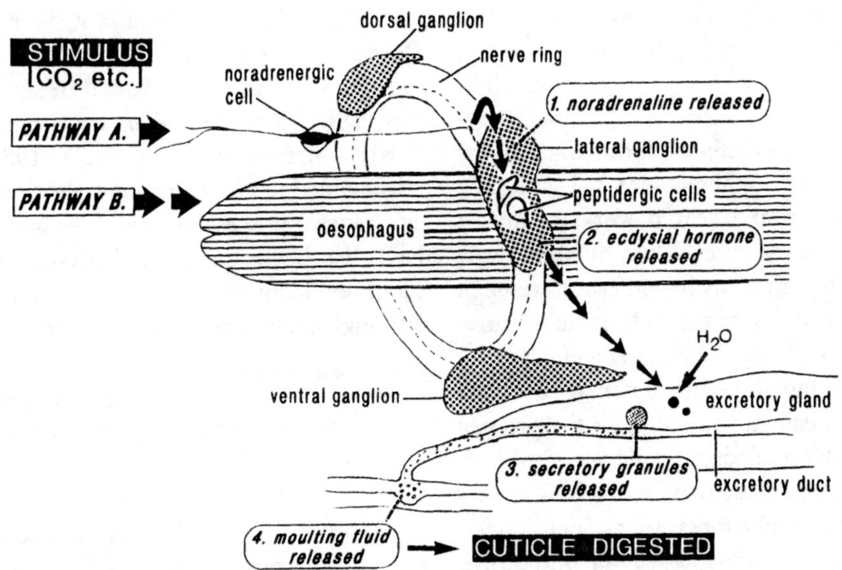

Fig. 25.15
Hormonal control of ecdysis (moulting) in nematodes: a working hypothesis by Davy (1988). A stimulus (such as CO_2) impinges on the sense organs served by the cephalic papillary ganglion. This activates *two* pathways. **Pathway A** involves the noradrenergic cells which results in the release of noradrenaline from terminals in the nerve ring. This in turn activates peptidergic cells acting as a neurotransmitter, a neuromodulator or a neurohormone; these cells are here arbitrarily shown to be located in the lateral ganglion. The peptidergic cells release the ecdysial hormone, which acts on the excretory duct to permit the entry of water. The increased water content of the excretory gland results in the activation of the contents of paracrystalline secretory granules which release their contents, via the excretory duct, into the space between the two cuticles. The moulting fluid digests the cuticle locally. **Pathway B**, as yet undetermined, leads ultimately to the oesophagus which loses fluid. It may be nervous in character; all that is certain is that it is a different pathway. (After Davey, 1988.)

question; more recent work has been summarised by Wharton (1986).

Most work on egg hatching has been carried out on *Ascaris lumbricoides*. Barrett (1976) demonstrated the presence of protease, chitinase, α-glucosidase and β-glucosidase activity in the hatching fluid of the egg of this species. Ward & Fairbairn (1972) showed that the chitinase needed for hatching was synthesised between 10 and 21 days of development and stored in the embryos; it was eventually released in response to the hatching stimulus. Hinck & Ivey (1976) showed that a protease was also involved and that its activity increased markedly 18 days after embryonation had begun.

After the hatching stimulus – CO_2 plus reducing conditions – has been applied, the process of hatching also appears to involve a change in the permeability of the 'lipid' (ascaroside) membrane (Fig. 25.14) (Barrett, 1976; Clarke & Perry, 1980) which allows the enzymes in the hatching fluid, released by the larvae, to pass through the liquid layer and weakens the outer layer and so permits hatching to take place.

It should be noted that an *oxidising* agent – such as iodine – can reversibly inhibit the hatching of the infective of eggs of *Ascaris*; the capacity to hatch can be restored by exposing to the *reducing* agent hydrogen sulphide (Hurley & Sommerville, 1982). The same effect has been observed with the exsheathment of infective larva.

Larval exsheathment (ecdysis = moulting). The process of exsheathment in nematodes has attracted considerable attention from research workers, but in spite of these efforts, very little is known definitively about the process, although many hypotheses have been put forward. Useful reviews of the field include those of Bird (1984), Davey (1976, 1988), Maggenti (1981) and Wharton (1986). One of the fundamental questions to which answers were sought was whether the process of moulting was under endocrinal control or not. A major difficulty has been that, until recently, the evidence for the existence of an endocrine system has been fragmentary and almost entirely based on descriptive methods (Davey, 1988). These methods were based on whether or not the occurrence or non-occurrence of the event under investigation could be correlated with the cytological changes in the cells suspected of producing a hormone.

Earlier identification of such (so-called) neurosecretory cells was based on positive staining with paraldehyde fuchsin or chrome-haemotoxylin. Although widely used, these stains were recognised as being unreliable and non-specific. However, the introduction of electron microscopy enabled such cells to be identified unequivocally, as they possessed specific characteristics: abundant granular endoplasmic reticulum, a permanent Golgi apparatus and a population of membrane-bound vesicles. Extension of these studies allowed aminergic cells (containing small vesicles) and peptidergic cells (containing large vesicles) to be identified, although the size criteria quoted cannot be regarded as specifically diagnostic. Catecholaminergic cells have also been identified in several species by using fluorescent microscopy (Davey, 1988). The very specific immunocytological techniques, which have been so successful with cestodes (Chapter 20), do not appear to have been applied yet to nematodes.

The most detailed studies of the occurrence of aminergic neurones in nematodes have been carried out on *Pseudoterranova decipiens* whose adult is in the gut of seals and the dormant infective stage in the muscles of the cod. The latter larva moults to the adult stage on reaching the seal digestive tract. What makes this species such a useful experimental model is that the moulting process can also be carried out *in vitro* and the parasite is large enough for the neurones to be dissected out (Davey, 1988; Townsley *et al.*, 1963). Moulting has also been investigated in several other species, especially *Haemonchus contortus* (Rogers & Petronijevic, 1982; Wharton & Sommerville, 1984) in which noradrenaline increases during exsheathment and in which substances possessing juvenile hormone-like activity have also been identified. Ecdysteroids have also been identified in many species of nematode (Spindler, 1988), although their exact role is unknown.

Ecdysteroids are a class of hormones whose main function is related to the regulation of ecdysis in arthropods. *Ecdysteroid* is a generic name for some 90 steroids present in plants and/or invertebrates, the best known being *ecdysone* involved in moulting in insects.

A number of hypotheses have been put forward over the years regarding the mechanisms of exsheathment and its possible hormonal control. A recent hypothesis by Rogers & Petronijevic (1982) argues that each larval stage would have a characteristic set of genes. These authors proposed that the infective larva is interrupted in its development because the action of the gene(s) controlling its development is suppressed. A CO_2 stimulus could reverse the inhibition and the process could then proceed. A generalised diagram (excluding the possible gene action) presenting a possible working hypothesis of the mechanism of hormonal control of ecdysis in nematodes, as visualised by Davey (1988), is shown in Fig. 25.15, the legend of which is self-explanatory.

○○○ 25.5 Body wall

The body wall of nematodes consists of (*a*) cuticle, (*b*) epidermis (= hypodermis) and (*c*) muscle layers. The cuticle also extends into the vagina, the excretory pore, stomatodaeum and proctodaeum.

○ 25.5.1 Cuticle

The structure of this has long been controversial, especially with regard to the question whether or not it should be regarded as a cell membrane. Some relevant reviews are those of Bird (1984), Maggenti (1981), Ouazana (1982), Wharton (1986) and Wright (1987).

The nematode cuticle shows considerable diversity in structure, not only between different genera but even within a single species. Thus, in the females of the genus *Bradynema*, which live in the haemocoele of the fly *Megaselia halterata*, the cuticle is almost non-existent and microvilli are present (Riding, 1970). In contrast, the cuticle of the horse gut nematode, *Strongylus equinus*, is 100 μm thick (Bird, 1984).

There is no general agreement on the terminology used for describing the various layers of the cuticle although there is general agreement that it consists of three layers covered by a triple-layered outer membrane. Based on ultrastructural and cytochemical studies, some authors (e.g. Bird, 1980) have concluded that the outer covering is a 'modified cell membrane' and refer to it as the *epicuticle.* This concept of the outside layer being a cell membrane has been disputed by Wright (1987) who points out that the surface structure is likely to vary considerably in different species as a result of different functional and anatomical requirements. Nevertheless, the term *epicuticle* now appears to be accepted for the outside layer.

Regarding the remaining layers of the cuticle, there now seems to be agreement that a generalised cuticle (Fig. 25.16) would consist of an epicuticle, a cortical zone, a medium zone, and a basal zone which overlies the epidermis (= hypodermis). The following brief descriptions are based on Bird (1984).

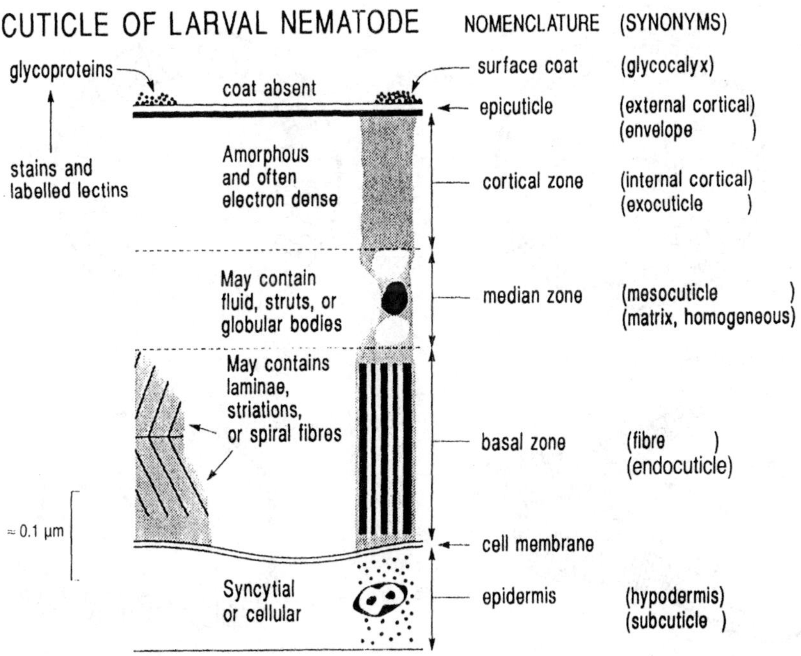

Fig. 25.16
Generalised diagram of the zones or layers of the cuticle of a typical larval nematode; the cuticle is *ca.* 0.5 μm thick. Slightly different terminology has been used by different workers. (After Bird & Bird, 1991.)

Epicuticle. This structure shows great morphological variability, with a range of thickness of 6–40 nm. It appears to consist of glycoproteins and may contain some lipid.

Cortical zone. The later layers are referred to as 'zones' because, although they may contain several layers, the distinction between them is not always clear. The cortical zone is generally amorphous and electron-dense and its structure varies greatly between species.

Median zone. This zone is not very clear in some species and may be absent in some. It often contains fluid in which there may be fibres, struts or globular material.

Basal zone. This may be made up of spiral fibres, laminae or striations; a variety of names have been used for these structures. The fibres consist of a collagen-like protein and in *Ascaris* runs at an angle of 70° to the long axis of the worm. Thus any two sets of fibres enclose a system of minute parallelograms (Fig. 25.18). On the 'lazy tongs' principle these parallelograms may be distorted by shearing and the worm can become longer and thinner or shorter and thicker (Harris & Crofton, 1957).

Epidermis (hypodermis). Classically, the cell layer beneath the cuticle has been referred to as the *hypodermis*, but the term *epidermis* now appears to be preferred (Wright, 1987). The epidermis bulges into the pseudocoele to form ridges of cords (Fig. 25.17). The lateral cords contain the excretory canal.

Musculature. The relationship between the muscles and the nervous system has been described earlier, where it was pointed out that each muscle fibre is one elongated cell which has striated (contractile) and non-striated (non-contractile) components (Fig. 25.9).

○ 25.5.2 Epidermis (Hypodermis)

This consists essentially of a cellular or syncytial layer beneath the cuticle; it bulges into the pseudocoele to form four ridges or cords (Fig. 25.17). The lateral cords contain the excretory canals. In the hypodermis, nuclei occur only in the cords.

The cytology of the nematode muscle cell is of special interest, for each muscle fibre is only one enormously elongated cell. Each cell has a *fibrillated* (contractile) and a *cytoplasmic* or *sarcoplasmic* (non-contractile) zone (Fig. 25.17). The fibrillated zone consists of longitudinal bands of myofibrils between supporting elements. The sarcoplasmic zone contains the

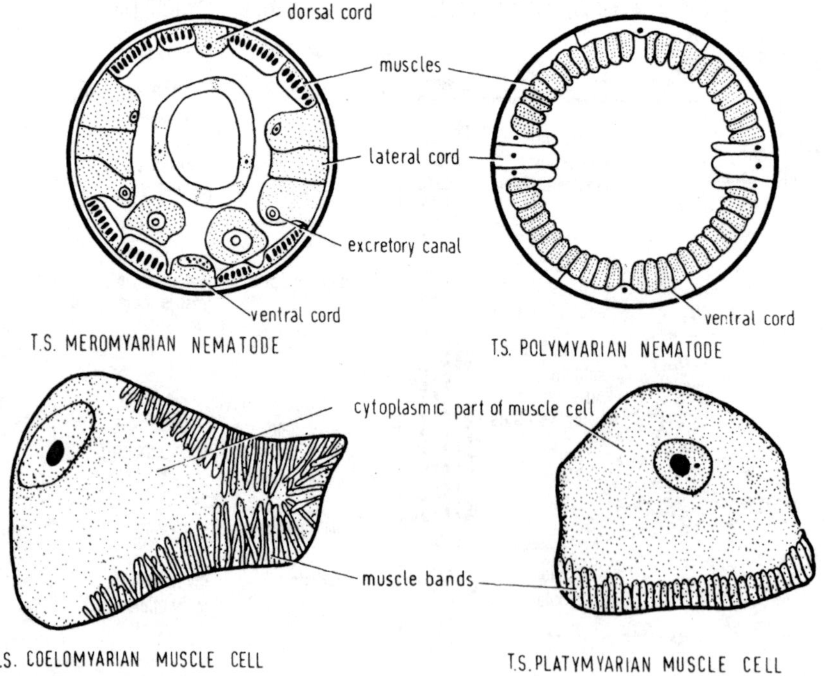

Fig. 25.17
Types of muscle-cell organisation in nematodes. (After various authors.)

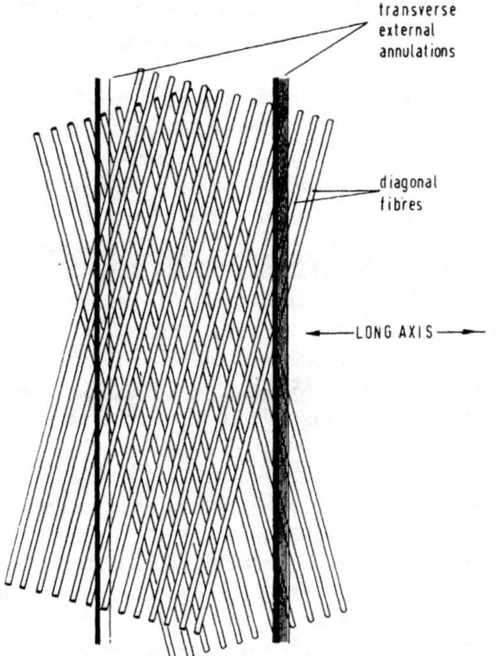

Fig. 25.18
A schematic diagram illustrating the orientation of the two fibril layers in the cuticle of *Ascaris lumbricoides* in relation to two of the transverse external annulations. (Adapted from Harris & Crofton, 1957.)

nucleus, mitochondria and reserves (fat and glycogen) (see Fig. 25.9).

In some species the muscle cells are flat with fibrils confined to the region parallel to the hypodermis, a condition known as *platymyarian* (Fig. 25.17). In others, the protoplasmic zone bulges into the pseudocoele with the fibrillated zone extending along the side; this condition is termed *coelomyarian*. This is usually clearly seen in sections of *Ascaris*.

It has been pointed out that circular muscles are lacking. The longitudinal muscles must operate against some antagonistic force which, according to Harris & Crofton (1957) is provided by the internal hydrostatic pressure of the pseudocoelomic fluid. These authors also point out that, as a consequence of the lateral cords dividing the muscles into dorsal and ventral regions, nematodes bend *dorso-ventrally* and not laterally as do fish and annelids.

Pseudocoele. The body cavity between the body wall and viscera contains a fluid, the coelomic fluid or haemolymph, which in *Ascaris lumbricoides* has a mean hydrostatic pressure of about 70 mmHg with a range of 16–225 mmHg (Harris & Crofton, 1957; Harpur, 1964). The pseudocoele also usually contains fibrous tissue and fixed cells (*pseudocoelomocytes* or *coelomocytes*), which vary in number from two to six, but usually occupy fixed positions in relation to the longitudinal cords. These cells are often highly branched or stellate and associated with fibrous strands or membranes which constitute a kind of mesenchyme. Their function is uncertain but may be related to the secretion of oxidative enzymes.

○○○ 25.6 General considerations of structure and function

It has been pointed out (p. 371) that, in the absence of circular muscles, the longitudinal muscles act against forces exerted by the internal fluid on the body wall. According to the views of Harris & Crofton (1957), many of the morphological features of nematodes can be interpreted as representing adaptations to a changing body length and a high internal pressure.

Thus, the alimentary canal is thin-walled, flexible and capable of extension. The presence of dilator muscles (the *depressor ani*) would be correlated with a high internal pressure; without such a mechanism the gut contents would be forced out. The triradiate pharynx could act as a valve (Fig. 25.5) anteriorly and also serve to pump food into the gut, against the internal pressure.

The excretory tubes embedded in lateral cords would be well adapted to resist collapse under pressure. In the reproductive system there is present a peristaltic musculature and true sphincter muscles. This allows for controlled emission of eggs under pressure by a combination of peristaltic movement (which pinches off the selected group of eggs) and a sphincter muscle which releases the eggs at the appropriate time.

These views are based on observations on *Ascaris lumbricoides* and it is difficult to know how far they can be extended to other nematodes. These principles probably operate in other large nematodes which similarly possess a system of fibres in the cuticle. Some smaller nematodes, however, lack these but possess toughened regions of the external cortical layer in the form of rings or annuli, an alternative arrangement which could provide diametrical resistance but which would permit longitudinal extension.

○○○ References

(References of general interest have been included in this list, in addition to those quoted in the text.)

Adamson, M. L. 1987. Phylogenetic analysis of the higher classification of the Nematoda. *Canadian Journal of Zoology*, **65**: 1478–82.

Altner, H. & Prillinger, L. 1980. Ultrastructure of invertebrate chemo-, thermo-, and hydroreceptors and its functional significance. *International Review of Cytology*, **67**: 69–139.

Anderson, R. C. 1988. Nematode transmission patterns. *Journal of Parasitology*, **74**: 30–45.

Anderson, R. C. 1992. *Nematode Parasites of Vertebrates: Their Development and Transmission*. CAB International, Oxford, UK. ISBN 0 851 98799 0.

Anderson, R. C., Chabaud, A. G. & Willmott, S. (eds) 1974–83. *C.I.H. Keys to the Nematode Parasites of Vertebrates*. Nos 1–10. Commonwealth Agriculture Bureaux, Farnham Royal, Bucks, England.

Anya, A. O. 1964. Studies on the structure of the female reproductive system and egg-shell formation in *Aspicularis tetraptera* Schulz, (Nematoda; Oxyuroidea). *Parasitology*, **54**: 699–719.

Barrett, J. 1976. Studies on the induction of permeability in *Ascaris lumbricoides* eggs. *Parasitology*, **73**: 109–21.

Bird, A. F. 1971. *The Structure of Nematodes*. Academic Press, London and New York.

Bird, A. F. 1980. The nematode cuticle and its surfaces. In: *Nematodes as Biological Models* (ed. B. M. Zuckerman), Vol. 2, pp. 213–36. Academic Press, New York.

Bird, A. F. 1984. Nematoda. In: *Biology of the Integument* (ed. J. Bereiter-Hahn, G. Maltoltsy & K. S. Richards), Vol. 1 (*Invertebrates*), pp. 212–33. Springer-Verlag, New York. ISBN 3 540 13062 4.

Bird, A. F. & Bird, J. 1991. *The Structure of Nematodes*. 2nd Ed. Academic Press, New York.

Chitwood, B. G. & Chitwood, M. B. 1950. *An Introduction to Nematology*. Monumental Printing Co., Baltimore.

Clarke, A. J. & Perry, R. N. 1980. Egg-shell permeability and hatching of *Ascaris suum*. *Parasitology*, **80**: 447–56.

Croll, N. A. 1970. *The Behaviour of Nematodes*. Edward Arnold, London. ISBN 0 713 12286 2.

Croll, N. A. (ed.) 1976. *The Organization of Nematodes*. Academic Press, New York. ISBN 0 121 96850 2.

Croll, N. A. & Matthews, B. E. 1977. *Biology of Nematodes*. Blackie, London. ISBN 0 216 90294 0.

Croll, N. A., Slater, L. & Smith, J. M. 1972. *Ancylostoma tubaeforme*: osmoregulatory ampulla of larvae. *Experimental Parasitology*, **31**: 356–60.

Davey, K. G. 1976. Hormones in nematodes. In: *The Organization of Nematodes* (ed. N. A. Croll), pp. 273–91. Academic Press, New York.

Davey, K. G. 1988. Endocrinology of nematodes. In: *Endocrinology of Selected Invertebrate Types*, Vol. 2 (*Invertebrate Endocrinology*) (ed. H. Laufer & R. G. H. Downer), pp. 63–86. Alan R. Liss, New York.

Debell, J. T. 1965. A long look at neuromuscular junctions in nematodes. *Quarterly Review of Biology*, **40**: 233–51.

Del Castillo, J., Rivera, A., Solórzano, S. & Serrato, J. 1989. Some aspects of the neuromuscular system of *Ascaris*. *Quarterly Journal of Experimental Physiology*, **74**: 1071–87.

Gibbons, L. M. 1986. *SEM Guide to Morphology of Nematode Parasites of Vertebrates*. CAB International, Oxford, UK. ISBN 0 851 98569 6.

Goldschmidt, R. 1908. Das nervensystem von *Ascaris lumbricoides* und *megalocephala*. Ein Versuch, in den Aufbau eines einfachen Nervensystems einzudringen, Erster Teil. *Zeitschrift für Wissenschaftliche Zoologie*, **90**: 73–136.

Harpur, R. P. 1964. Maintenance of *Ascaris lumbricoides in vitro*. III. Changes in the hydrostatic skeleton. *Comparative Biochemistry and Physiology*, **13**: 71–85.

Harris, J. E. & Crofton, H. D. 1957. Structure and function in the nematodes: internal pressure and cuticular structure in *Ascaris*. *Journal of Experimental Biology*, **34**: 116–30.

Hinck, L. W. & Ivey, M. H. 1976. Proteinase activity in *Ascaris suum* eggs, hatching fluid, and excretions-secretions. *Journal of Parasitology*, **62**: 771–4.

Hurley, L. C. & Sommerville, R. I. 1982. Reversible inhibition of hatching of infective eggs of *Ascaris suum* (Nematoda). *International Journal for Parasitology*, **12**: 463–5.

Hyman, L. 1951. *The Invertebrates*. Vol. III. *Acanthocephala, Aschelminthes, and Entoprocta. The Pseusocoelomate Bilateraia*. McGraw-Hill, London.

Jamuar, M. P. 1967. Studies of spermiogenesis in a nematode, *Nippostrongylus brasiliensis*. *Journal of Cell Biology*, **31**: 381–96.

Kassai, T. 1982. *Handbook of Nippostrongylus brasiliensis*. Commonwealth Agricultural Bureaux and Akadémiai Kiadó, Budapest.

Lee, D. L. 1965. *The Physiology of Nematodes*. 1st Edition. Oliver & Boyd, Edinburgh.

Lee, D. L. & Anya, A. O. 1967. The structure and development of the spermatozoon of *Aspicularis tetraptera* (Nematoda). *Journal of Cell Science*, **2**: 537–44.

Lee, D. L. & Atkinson, H. J. 1977. *The Physiology of Nematodes*. 2nd Edition. Columbia University Press, New York.

Levine, N. D. 1980. *Nematode Parasites of Domestic Animals and of Man*. 2nd Edition. Burgess Publishing Co., Minneapolis.

Maggenti, A. R. 1981. *General Nematology*. Springer-Verlag, New York and Heidelberg.

Monné, L. 1962. On the formation of the egg-shell of the Ascaroidea, particularly *Toxascaris leonina*. *Arkiv för Zoologi*, **15**: 277–84.

Nigon, V. 1949. Modalités de la reproduction et determinisme du sexe chez quelques nématodes libres. *Annales des Sciences Naturelles, Zoologie, Série* II, pp. 1–132.

Ouazana, R. 1982. Structure et composition chimique du tegument cuticulaire des nématodes. *Bulletin de la Société Zoologique de France*, **107**: 419–26.

Perry, R. N. & Clarke, A. J. 1981. Hatching mechanisms of nematodes. *Parasitology*, **83**: 435–49.

Poinar, G. O. Jr 1983. *The Natural History of Nematodes*. Prentice-Hall, New Jersey. ISBN 0 136 09925 4.

Riding, I. 1970. Microvilli on the outside of a nematode. *Nature*, **226**: 179–80.

Roberts, N. B., Steffens, W. L. & Bowen, J. M. 1988. The ultrastructure of neuromuscular and neural–neural relationships in *in vitro* maintained *Dirofilaria immitis. Micron and Microscopica Acta*, **19**: 67–72.

Rogers, W. P. 1966. Exsheathment and hatching mechanisms in helminths. In: *Biology of Parasites* (ed. E. J. L. Soulsby), pp. 33–40. Academic Press, New York and London.

Rogers, W. P. & Petronijevic, T. 1982. The infective stage and the development of nematodes. In: *Biology and Control of Endoparasites* (ed. L. E. A. Symons & A. D. Donald), pp. 3–28. Academic Press, New York.

Schulte, F. 1989. Life history of *Rhabditis* (*Pelodera*) *orbitalis. Proceedings of the Helminthological Society of Washington*, **56**: 1–7.

Sommerville, R. I. & Weinstein, P. P. 1964. Reproductive behaviour of *Nematospiroides dubius in vivo* and *in vitro. Journal of Parasitology*, **50**: 401–9.

Spindler, K.-D. 1988. Parasites and hormones. In: *Parasitology in Focus* (ed. H. Mehlhorn), pp. 463–76. Springer-Verlag, Berlin.

Stoll, N. R. 1947. This wormy world. *Journal of Parasitology*, **33**: 1–18.

Stone, A. R., Platt, H. M. & Khalil, L. F. 1983. *Concepts in Nematode Systematics*. Academic Press, London. ISBN 0 126 72680 9.

Stretton, A. O. W., Fishpool, R. M., Southgate, E., Donmoyer, J. E., Walrond, J. P., Moses, J. E. R. & Kass, I. S. 1978. Structure and physiological activity of the motorneurons of the nematode *Ascaris. Proceedings of the National Academy of Sciences of the USA*, **75**: 3493–7.

Stretton, A. O. W., Davis, R. E., Angstadt, J. D., Donmoyer, J. E. & Johnson, C. D. 1985. Neural control of behavior in *Ascaris. Trends in Neurosciences*, **8**: 294–300.

Townsley, P. M., Wight, H. G., Scott, M. A. & Hughes, M. L. 1963. The *in vitro* maturation of the parasitic nematode *Terranova decepiens* from cod muscle. *Journal of the Fisheries Research Board of Canada*, **20**: 743–7.

Ward, K. A. & Fairbairn, D. 1972. Chitinase in the developing eggs of *Ascaris suum* (Nematoda). *Journal of Parasitology*, **58**: 546–9.

Weinstein, P. P. 1952. Regulation of water balance as a function of the excretory system of the filariform larvae of *Nippostrongylus brasiliensis* and *Ancylostoma caninum. Experimental Parasitology*, **1**: 363–76.

Wharton, D. A. 1980. Nematode egg-shells. *Parasitology*, **81**: 447–63.

Wharton, D. A. 1986. *A Functional Biology of Nematodes*. Croom Helm, London.

Wharton, D. A. & Jenkins, T. 1978. Structure and chemistry of the egg-shell of a nematode (*Trichuris suis*). *Tissue and Cell*, **10**: 427–40.

Wharton, D. A. & Sommerville, R. I. 1984. The structure of the excretory system of the infective larva of *Haemonchus contortus. International Journal for Parasitology*, **14**: 591–600.

Willett, J. D. 1980. Control mechanisms in nematodes. In: *Nematodes as Biological Models* (ed. B. M. Zuckerman), Vol. 2, pp. 237–95. Academic Press, London and New York.

Wright, K. A. 1987. The nematode's cuticle – its surface and the epidermis: function, homology, analogy – a current consensus. *Journal of Parasitology*, **73**: 1077–83.

Zuckerman, B. M. (ed.) 1980. *Nematodes as Biological Models*. 2 vols. Academic Press, New York and London. ISBN 0 127 82401 4.

26

Aphasmid Nematoda

○○○ 26.1 Superfamily Trichuroidea

Worms of this superfamily are characterised by the possession of an enormously elongated capillary-like oesophagus or pharynx, whose internal cavity is so small that it is difficult to locate even in sections (Fig. 26.2). The oesophagus is embedded in a long column of single cells (*stichocytes*) which may serve as oesophageal glands, the whole structure being referred to as a *stichosome*. Ultrastructure studies show that the oesophagus is made up of a single tubular cell separated from the stichocytes by a membranous sheath (Sheffield, 1963). The cuticle is also unusual in possessing a 'bacillary band' (Fig. 26.3) consisting of bacillary cells which penetrate the cuticle over about one-third of its circumference. Ultrastructure studies suggest that these cells secrete antigenic substances (Campbell, 1983).

The anus is at the extreme posterior tip, and not just in front of the hind end, so that there is no tail. The male has one spicule or none at all; if present, it is enclosed in a protrusible membranous sheath.

The major families are diagnosed as follows.

Trichuridae. The 'whip worms'. Oviparous. Adults in intestine. Males with protrusible spiny sheath and spicule (e.g. *Trichuris trichiura*).

Trichinellidae. Viviparous. Male with no spicule or spicule sheath. Larvae in tissue; adult in intestine (only one species: *Trichinella spiralis*).

○ 26.1.1 Family Trichuridae

Genus: *Trichuris*

These are the so-called 'whip worms', a term derived from the whip-like form of the body.

The correct generic name for members of this group is *Trichuris*, a term which gives a misleading description, for it literally means 'hair tail'. The name was applied before it was realised that the thin thread-like part which looked like the tail was, in fact, the head. A more apt name, then, is *Trichocephalus*, which was later applied, but the earlier name *Trichuris* has zoological priority.

The best-known species are as follows:

Trichuris muris: rodents.

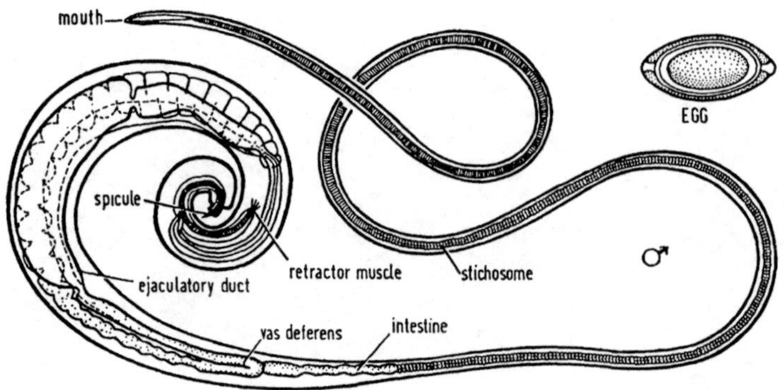

Fig. 26.1
Trichuris trichiura: morphology of male and egg. (After Brown, 1950.)

T. ovis: sheep, goats, cattle and other herbivores.
T. trichiura: man and monkeys.
T. suis: pigs.
T. vulpis: dogs, cats.

General morphology. This does not vary much. The thread-like oesophageal region occupies about two-thirds the length of the body. The mouth has no lips. The vulva is at the junction of the thread-like and thickened regions of the body. The eggs have a characteristic barrel shape with an opercular plug at each end (Fig. 26.1).

Life cycle. The life cycle is similar in all species. Eggs require a warm moist environment for embryonation (such as that provided by rain-soaked soil), but once embryonated are exceptionally resistant to environmental conditions. Embryonated eggs hatch near the caecum into which the larvae migrate. Possibly related to scarcity of food materials in this site, development is slow.

Trichuris muris. This is a useful laboratory model which can be maintained in laboratory mice (Wakelin, 1969). Eggs containing infective larvae hatch in the intestine within 30–60 min and the larvae pass to the caecum where, within 4 h, they enter the caecal mucosa. After 4–5 days, the larvae move to a more superficial site in the mucosa and eventually become free in the caecal lumen. Moulting larvae can be recovered on days 9–11 p.i.

***Trichuris trichiura:* human trichuriasis.** This species is a much more common human parasite than is generally appreciated and it is reported to infect up to 800 million people throughout the tropical and temperate areas (Cooper & Bundy, 1988). Its pathogenicity has long been a question of dispute and for that reason has been largely neglected. There is now evidence that children are especially prone to infection and 'While massive trichuriasis may present with distinctive features like rectal prolapse or digital clubbing, the majority of infections are chronic and mild, with vague, nonspecific symptoms like chronic diarrhoea, pallor, growth retardation, abdominal discomfort and anaemia' (Kan, 1986).

As the eggs of both *Trichuris* and *Ascaris* (p. 403) require the same conditions for embryonation – high humidity and access to oxygen – these two species frequently occur together in humans. This is reflected in the global occurrence of these two species as shown in Table 26.1. Useful reviews on the biology, pathology and epidemiology of human trichuriasis are those of Bundy & Cooper (1989), Bundy *et al.* (1987), Cooper & Bundy (1988) and Forrester *et al.* (1990).

○ 26.1.2 Family Trichinellidae

This family contains only the single genus *Trichinella* (syn. *Trichina*) and was originally thought to contain only one species, *Trichinella spiralis*, the cause of a serious and often fatal disease in man known as *trichinosis* (trichinellosis). The parasite has been known for centuries in Europe where it was recognised that the chief hosts were primarily rats, pigs, carnivores and man.

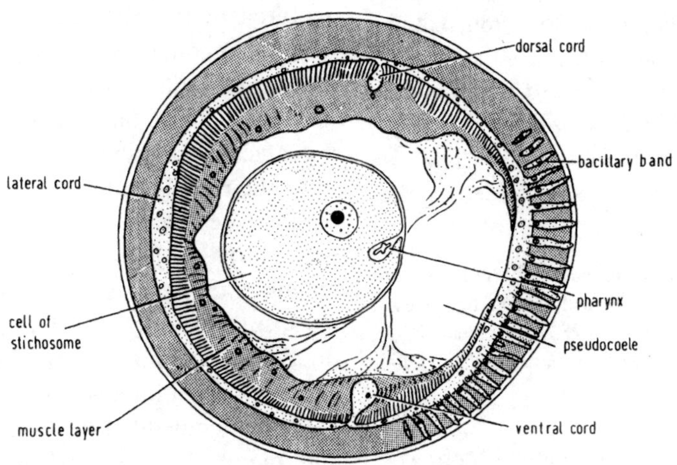

Fig. 26.2
Trichuris trichiura: section through pharyngeal region, showing the narrow pharynx embedded in the stichosome. (After Ranther, 1918.)

Table 26.1. *Estimates of prevalence of* Trichuris trichiura *and* Ascaris lumbricoides *in parts of Africa, America and Asia*

Locality	Sample size	Prevalence (per cent) Trichuris	Ascaris
America			
Brazil	*a*	32	55
Caribbean	15962	57	14
Mexico	*a*	21	26
USA	*a*	1	2
Africa			
Cameroon	1095	70	69
Nigeria	5595	76	74
Ethiopia	4469	4	6
Kenya	166415	2	17
Zimbabwe	1543	1	2
Asia			
India (N)	52113	1	4
India (S)	59489	12	15
Indonesia	13566	73	72
Japan	1663145	0.9	0.6
Korea	40581	74	58
Malaysia	25246	33	19
Thailand	6724	32	17
Philippines	18262	74	48

a National estimate based on a large series of surveys.
Source: Data from Cooper & Bundy (1988).

Table 26.2. *Trichinosis: source of infections reported in the USA during 1975–81*

Food	No. of cases
Pork products	
Domestic swine	740
Sausage	515
Other preparation	135
Unspecified	90
Wild (feral) swine	9
Subtotal	*749*
Non-pork products	
Bear meat	64
Other wild animal	68
Hamburger	66
Subtotal	*198*
Unknown	119
Total	*1066*

Source: Data from Schantz (1983).

It is now generally accepted that four 'varieties' of this parasite exist which many (perhaps most) workers now recognise as sibling species: *T. spiralis*, *T. nelsoni*, *T. nativa* and *T. pseudospiralis* (Flockart, 1986; Campbell, 1988, 1991) and which are recognised as such in this text. Other workers prefer to regard these as only sub-species designated *T. spiralis spiralis*, *T. s. nelsoni*, *T. s. nativa* and *T. s. pseudospiralis* (Lichtenfels *et al.*, 1983). The question of speciation is discussed further below.

An enormous literature on trichinosis exists, the more recent of which has been surveyed in the following books or reviews: Campbell, 1983, 1988, 1991; Dick, 1983; Flockart, 1986; Kan, 1986; Kim & Pawlowski, 1978; Schantz, 1983, 1991; Tanner *et al.*, 1989.

Trichinella spiralis. This is essentially a 'domestic' or 'synanthropic' parasite long recognised to cause a zoonosis transmitted to man by the ingestion of infected pork (see below) and it can be accepted as the basic type.

Morphology. Both sexes are small, the male (1.4–1.6 mm) being smaller and more slender than the female (3–4 mm). Spicules are absent in the male, but the cloaca is everted during copulation and its opening flanked by two papillae. The female genital system is single (Fig. 26.4). The ultrastructure has been described in Campbell (1983).

Life history (Fig. 26.5). Infection in animals and man acquired by the host eating raw or undercooked flesh (e.g. pork) containing encapsulated larvae. In the domestic environment, rats are probably the most highly infected 'natural' host and pigs become infected by eating infected pork scraps or occasionally rats which penetrate their stalls. Sausages are a dangerous source of the parasite (Table 26.2) as a small fragment of infected pork, after mincing, may become widely distributed among a number of sausages. When flesh containing encapsulated larvae is eaten, larvae are released in the duodenum and may be found there as early as one hour after the initial infection.

Deeply embedded in the mucosa, larvae feed, grow and rapidly become sexually mature. Female worms have been found inseminated at 36 hours which suggests that copulation first takes place at about 30–35

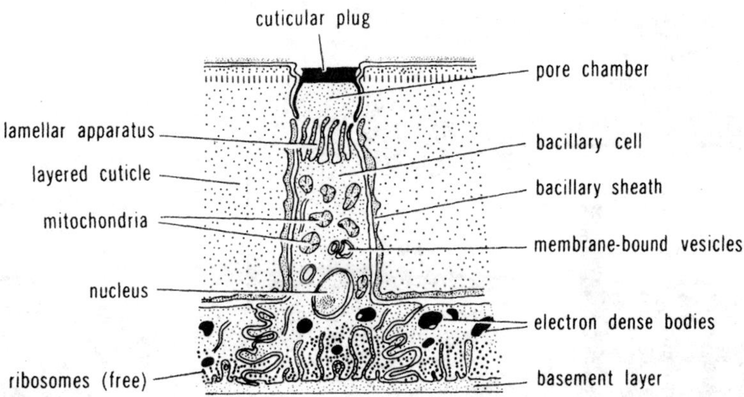

Fig. 26.3
Trichuris suis: ultrastructure of bacillary cell. (Based on Jenkins, 1969.)

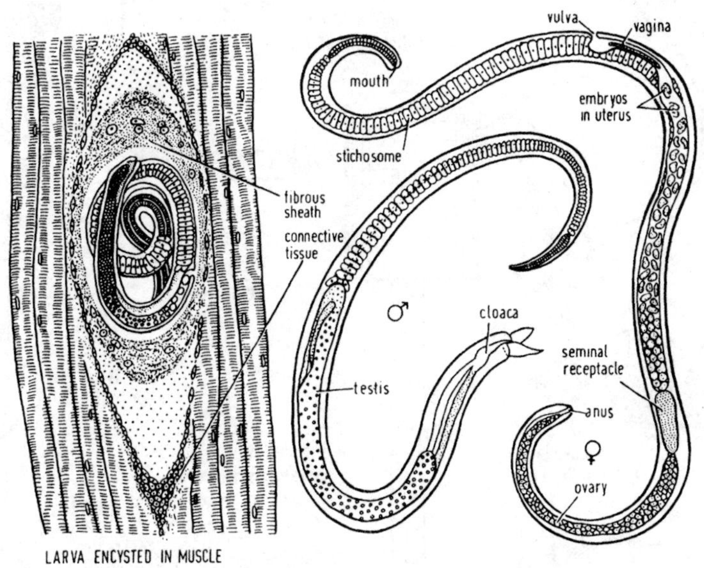

Fig. 26.4
Trichinella spiralis: morphology of male and female and encysted larvae in muscle fibres. (After Brown, 1950.)

hours. There is evidence from *in vitro* studies that a chemically mediated sexual attraction operates between male and female worms (Belosevic & Dick, 1980). The effectiveness of this attraction mechanism (*in vivo*) is remarkable and even mice inoculated orally with only *two Trichinella* larvae have been shown to develop patent infections (Campbell & Yakstis, 1969). The females are ovoviviparous and up to 1500 larvae may be produced by a single worm.

Released larvae penetrate the mucosa and are carried to the voluntary muscles, especially those of the diaphragm, jaws, tongue, larynx and eye. According to Shanta & Meerovitch (1967) the development phase in muscle is as follows: larvae commence penetrating the diaphragm muscle on the sixth day, piercing the sarcolemma of the fibres and initially lying extended. On the seventeenth day, they begin to coil and on the twenty-first day the formation of a cyst wall can be detected and encapsulation follows (Fig. 26.4), the larvae becoming resistant to pepsin digestion at 19 days. Calcification of cysts may occur as early as five months, but usually begins after 6–18 months.

Larvae may remain viable for several years within the cysts but do not develop further. Even at this stage, however, the genital primordia are well developed although differences between males and females cannot

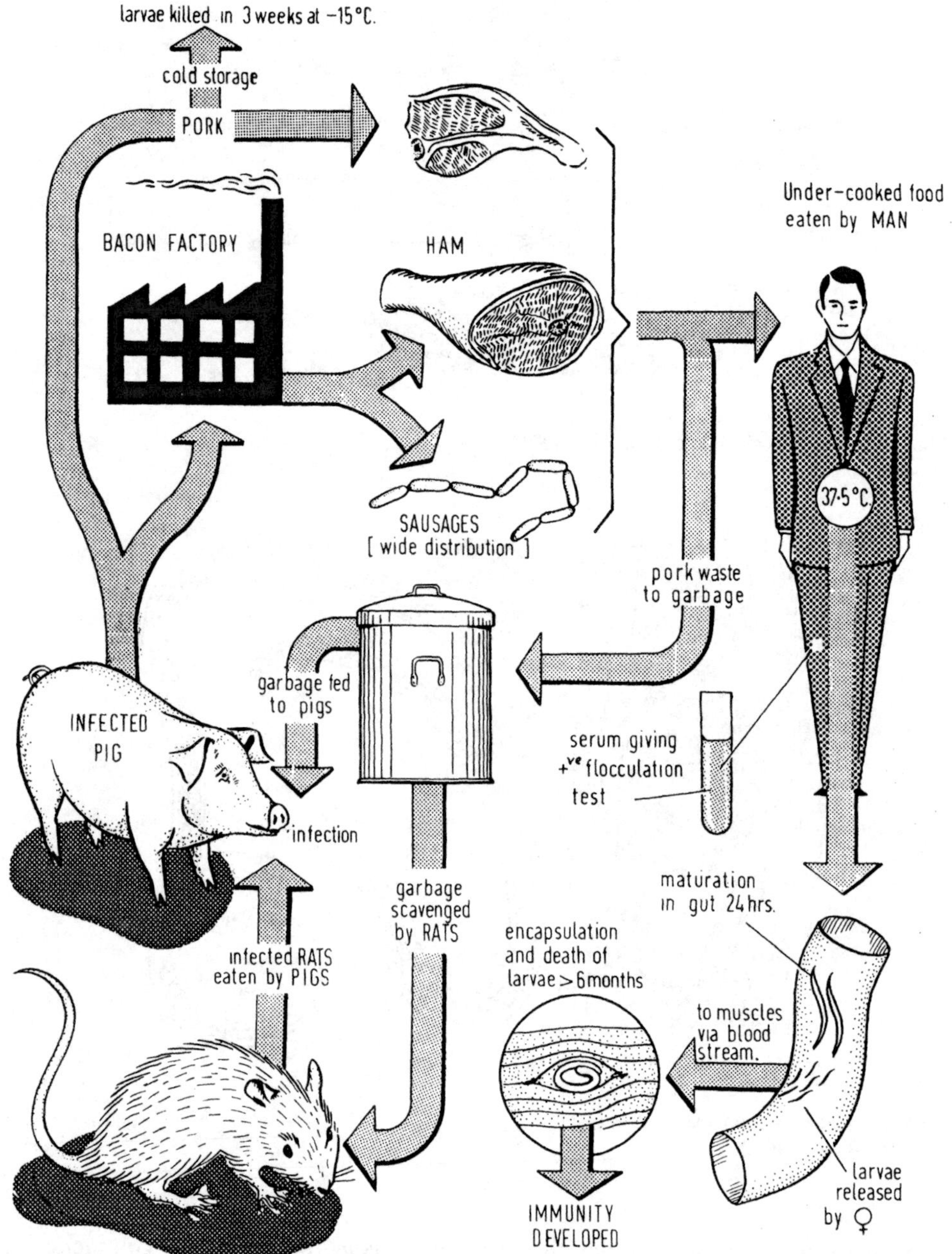

Fig. 26.5
Trichinella spiralis: life cycle and probable mode of transmission in European and American communities. (Original.)

be detected. When infected muscles are eaten by the definitive host, sexual maturation in the intestinal phase rapidly occurs, as explained earlier.

Speciation and epidemiology. The life cycle has been largely studied in Europe where it was thought that only one species, *Trichinella spiralis*, existed and that this was a domestic or synanthropic parasite (i.e. one associated with human dwellings). It is now recognised (Campbell, 1988) that cycles other than the accepted domestic rat–pig cycle occur in other parts of the world and speciation is more complex than previously thought.

Table 26.3. *Results of mating* Trichinella spiralis *and* T. pseudospiralis in vivo

Trials were conducted in C_{57}Br mice; results were recorded 35 days after infection.

Trichinella pairs in cross-breeding	No. of mice	No. infected	Mean number of larvae recovered
T. spiralis × *T. pseudospiralis*	20	0	0
T. pseudospiralis × *T. spiralis*	20	0	0
T. spiralis × *T. spiralis*	25	18 (72%)	481 ± 51
T. pseudospiralis × *T. pseudospiralis*	10	6 (60%)	354 ± 12.2

Source: Data from Bessonov *et al.* (1974, 1978).

The sylvatic cycle, tropic variant: T. nelsoni

The first new species to be recognised was in Africa where an outbreak of human trichinosis (Forrester *et al.*, 1961) led to the discovery of a parasite of wildlife origin, the patients in question having eaten the flesh of a bush pig (*Potamochoerus porcus*). This isolate was found to have a low infectivity for domestic pigs and laboratory rats and for this and other reasons was designated a new species, *T. nelsoni*. It was later found that wild carnivores and other scavengers served as chief hosts in the life cycle.

The sylvatic cycle, arctic variant: T. nativa

In contrast to the domestic *T. spiralis*, which is killed by freezing, a parasite in walruses and seals was found to have considerable resistance to freezing and appears to be a further 'variety' now designated *T. nativa*, although the species rank is not universally accepted (Campbell, 1988).

The sylvatic cycle, a 'cystless' species: T. pseudospiralis

This remarkable species, unlike any of the others, *does not induce capsule formation in host muscles*. Although infecting mammals, it appears to be primarily a parasite of birds. It was first isolated from the skeletal muscles of a raccoon (*Procyon lotor*) shot in Daghestan (Northern Caucasus, Russia) and was found to develop readily in most laboratory animals (Garkavi, 1972). In addition to the absence of a host capsule, other minor differences have been recorded; for example, the immunological response appears to be weaker than that of *T. spiralis* (Flockart, 1986).

A major criterion for the separation of two species, recognised by geneticists, is if they fail to interbreed. This has been clearly demonstrated in *T. pseudospiralis* and *T. spiralis* by Bessonov *et al.* (1974, 1978). They infected mice with one male and one female of each species (Table 26.3). If they had been capable of interbreeding, on becoming sexually mature in the intestine, larvae would have been produced which would have been found in the muscles on autopsy. However, although adults of *T. spiralis* and *T. pseudospiralis* show some attraction towards each other *in vitro* (Belosevic & Dick, 1980), no larvae were recovered, demonstrating the failure of interbreeding to take place. Dick & Chadee (1983) obtained similar results. This result was in striking contrast to the control mice, which were each fed a male and a female larva of the *same* species: in both cases, substantial numbers of larvae were recovered (Table 26.3).

'Species' identification. It is evident from the above survey that a complex speciation pattern exists in the genus *Trichinella*. The aberrant species, *T. pseudospiralis*, can be identified unequivocally, as it is smaller than the others and does not form cysts. However, the remaining species, *T. spiralis*, *T. nelsoni* and *T. nativa* are morphologically indistinguishable and thus fit the description of 'sibling species' (Flockart, 1986). Numerous attempts have been made to distinguish between these species by the use of techniques such as scanning electron microscopy (Lichtenfels *et al.*, 1983), isoenzyme electrophoresis (Flockart *et al.*, 1982), sexual attraction *in vitro* (Belosevic & Dick, 1980), DNA studies (Murrell *et al.*, 1987) and karyotype analysis (Mutafova *et al.*, 1982). Thus, isoenzyme analysis of four enzymes (AK, PGM, MP, GPI) showed that *T. spiralis* could clearly be differentiated from *T. nelsoni* and *T. nativa*, both of which had identical enzyme profiles (Flockart *et al.*, 1982). Comparison of the karyotypes has shown that all four species have a chromosome number of $2n = 6$ for female specimens and $2n = 5$ for male specimens (Mutafova

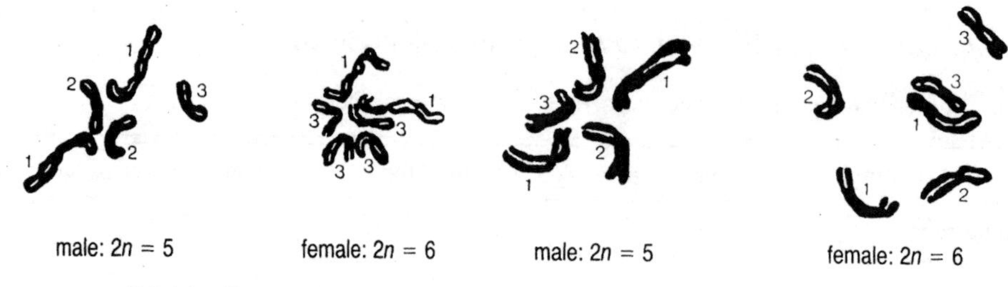

Fig. 26.6
Karyotopes of *Trichinella spiralis* and *T. pseudospiralis*. In the female *T. spiralis*, the second and third chromosome pairs are difficult to distinguish from each other and both have been labelled as 3. (After Mutafova *et al.*, 1982.)

et al., 1982) (Fig. 26.6). However, although there are minor differences in the relative lengths of the chromosomes of individual species, these differences are not significant and not of value for taxonomic purposes.

Human trichinosis. Trichinosis in man occurs in the Arctic, temperate climates and the tropics, occurring especially in Europe, N America, Latin America, Africa and Asia. Its epidemiology has been mostly studied in the USA where it is still recognised as an important health problem although its incidence is on the decline (Schantz, 1991). Sporadic outbreaks occur from time to time in most countries, the commonest source of infection probably being pork and sausages (Table 26.2). It is difficult to obtain accurate figures for the incidence of trichinosis as the disease often remains undetected during life. Diagnosis of trichinosis depends on clinical signs, such as myalgias, periorbital oedema, fever and eosinophilia in a patient with a history of eating undercooked pork or sausages. Serological tests (such as the bentonite flocculation assay) are available but may be negative if carried out within 3–4 weeks of the infection. Muscle biopsy can also be used, the muscle samples being (*a*) digested in pepsin, which frees the encapsulated larvae or (*b*) examined by a 'trichinoscope', a simple device which compresses the muscle sample between glass plates to make it semi-transparent. In cadavers examined in the USA in the period 1936–41, an astonishing incidence of 16 per cent was revealed by this method (Zimmerman *et al.*, 1968). However, infections have since declined substantially and only an average of 57 cases are now reported in the USA each year (Bailey & Schantz, 1990).

Control. It is beyond the scope of this book to discuss the detailed prophylaxis of trichinosis; much of this is clear from considerations of the life cycle. Avoidance of feeding infected pork scraps in pig food is an obvious precaution, as are measures preventing rats penetrating piggeries where they are often cannibalised by pigs. Human infections can be entirely avoided by assuring that pork or sausages are thoroughly cooked before eating. Deep freezing is also an effective control. Periods of freezing required to kill larvae in pig meat at different temperatures are: 30 min at −40 °C; 24 h at −25 °C; 3 d at −20 °C and 6 d at −15 °C. The US Bureau of Animal Industry makes use of this knowledge and requires uncooked pork to hang at −15 °C for twenty days before distribution.

The problems of effective control have recently been highlighted in a very detailed study of a recent outbreak on a farm in the USA (Schad *et al.*, 1987; Murrell *et al.*, 1987). This study confirmed rats as the main source of infection but also implicated wildlife, such as skunks, opossums and raccoons, in carrying the identical infection as determined by DNA studies. This important result is in contrast to studies elsewhere, which implied that sylvatic infections played little or no part in synanthropic infections. Recent work on the epidemiology and control of trichinosis in different countries is reviewed in detail in the *Proceedings of the 7th International Conference on Trinchinellosis* (Tanner *et al.*, 1989).

○ 26.1.3 Family Trichosomoididae

This is an interesting family about which little is known. *Trichosomoides crassicauda* is a parasite of the rat bladder, widely distributed throughout the world in wild rat populations and laboratory colonies (Cornish *et al.*, 1988). The adult female lives attached to the mucosa; the male is remarkable for living in the uterus of the female, 3–4 males occurring within a single female. The life cycle is not known in detail; after hatching, released larvae probably pass to the lungs and are carried by the blood stream to the bladder.

○○○ 26.2 Superfamily Dioctophymatoidea

Members of this order are large forms parasitic in the kidneys or occasionally the peritoneal cavity of mammals, including man.

Dioctophyme renale (= *renala*)

Morphology. This is one of the largest nematodes known, the female being up to 1030 mm long with a diameter of 5–12 mm; the male is considerably smaller (350 mm × 3–4 mm). The female has one ovary, an anterior vulva and a terminal anus; the male has a terminal bell-shaped bursa with a spicule 5–6 mm long. The worms are blood-red in colour.

Hosts. These include man, the dog, wolf, puma, glutton, raccoon, coati, mink, marten, skunk, weasel, otter, seal, cat, ox and horse. Hosts seldom have more than two or three worms and the left kidney is rarely infected. A case has been reported, however, involving 27 worms with 14 in the right kidney, 10 in the peritoneal cavity and 3 in the peritoneum (Sadighian & Amini, 1967).

D. renale is rare in Britain, but in some countries such as Canada, it may be of economic importance as a parasite of mink; it is also common in Russia.

Life cycle. The life cycle in mink has been studied by Woodhead (1950). Eggs, which contain the first-stage larva, are eaten by an oligochaete in which they hatch and undergo three moults to reach the fourth larval stage, which is directly infective to the final host. Larvae may pass to various fish hosts in which no further larval development occurs; centrarchid fish have been reported as paratenic hosts in Canada (Measures & Anderson, 1985). Dioctophymosis as a zoonosis has been reviewed by Battisti (1982).

○○○ References

(References of general interest have been included in this list, in addition to those quoted in the text.)

Bailey, T. M. & Schantz, P. M. 1990. Trends in the incidence and transmission patterns of trichinosis in humans in the United States: comparison of the periods 1975–1981 and 1982–1986. *Reviews of Infectious Diseases*, **12**: 1–11.

Battisti, G. 1982. (A little-known zoonosis: dioctophymosis.) (In Italian.) *Obiettivi e Documenti Veterinari*, **3**: 29–31.

Belosevic, M. & Dick, T. A. 1980. Chemical attraction in the genus *Trichinella*. *Journal of Parasitology*, **66**: 88–93.

Bessonov, A. S., Penkova, R. A. & Gumenshchikova, V. P. 1978. *Trichinella pseudospiralis* Garkavi, 1972: morphological and biological characteristics and host specificity. In *Trichinellosis* (ed. C. W. Kim & Z. S. Pawlowski), pp. 81–93. (Proceedings of the 4th International Conference on Trichinellosis, Poland, 1976.) University Press of New England, New Hampshire, USA. ISBN 0 874 51156 9.

Bessonov, A. S., Penkova, R. A., Uspenskii, A. V. & Belozyorov, S. N. 1974. On the self-dependent position of *Trichinella* species. *Proceedings of the 3rd International Congress of Parasitology, Munich*, **2**: 662–4. Facta Publication, Austria.

Bundy, D. A. P. & Cooper, E. S. 1989. *Trichuris* and trichuriasis in humans. *Advances in Parasitology*, **28**: 107–73.

Bundy, D. A. P., Cooper, E. S., Thompson, D. E., Didier, J. M., Anderson, R. M. & Simmons, I. 1987. Predisposition to *Trichuris trichiura* infections in humans. *Epidemiology and Infection*, **98**: 65–71.

Campbell, W. C. (ed.) 1983. *Trichinella and Trichinosis*. Plenum Press, New York. ISBN 0 306 41140 7.

Campbell, W. C. 1988. Trichinosis revisited – another look at modes of transmission. *Parasitology Today*, **4**: 83–6.

Campbell, W. C. 1991. *Trichinella* in Africa and the *nelsoni* affair. In: *Parasitic Helminths and Zoonoses in Africa* (ed. C. N. L. Macpherson & P. S. Craig), pp. 83–100. Unwin Hyman, London.

Campbell, W. C. & Yakstis, J. J. 1969. Mating success and fecundity of pairs of *Trichinella* larvae administered to mice. *Wiadomósci Parazytologiczne*, **15**: 526–32.

Cornish, J., Vanderwee, M. A., Findon, G. & Miller, T. E. 1988. Reliable diagnosis of *Trichosomoides crassicauda*, in the urinary bladder of the rat. *Laboratory Animals*, **22**: 162–5.

Cooper, E. S. & Bundy, D. A. P. 1988. *Trichuris* is not trivial. *Parasitology Today*, **4**: 301–6.

Dick, T. A. 1983. Species and intraspecific variation. In: *Trichinella and Trichinosis* (ed. W. C. Campbell), pp. 31–73. Plenum Press, New York.

Dick, T. A. & Chadee, K. 1983. Interbreeding and gene flow in the genus *Trichinella*. *Journal of Parasitology*, **69**: 176–80.

Flockart, H. A. 1986. *Trichinella* speciation. *Parasitology Today*, **2**: 1–3.

Flockart, H. A., Harrison, S. E., Dobinson, A. R. & James, E. R. 1982. Enzyme polymorphism in *Trichinella*. *Transactions of the Royal Society of Tropical Medicine and Hygiene*, **76**: 541–5.

Forrester, A. T. T., Nelson, G. S. & Sander, G. 1961. The first record of an outbreak of trichinosis in Africa south of the Sahara. *Transactions of the Royal Society of Tropical Medicine and Hygiene*, **55**: 503–13.

Forrester, J. E., Scott, M. E., Bundy, D. A. P. & Golden, M. H. N. 1990. Predisposition of individuals and families in Mexico to heavy infection with *Ascaris lumbricoides* and *Trichuris trichiura. Transactions of the Royal Society of Tropical Medicine and Hygiene*, **84**: 272–6.

Garkavi, B. L. 1972. Species of *Trichinella* isolated from wild carnivores. *Veterianariya*, **10**: 90–1. (In Russian.)

Jenkins, T. 1969. Electron microscope observations of the body wall of *Trichuris suis*, Schrank, 1788 (Nematoda: Trichuroidea). I. The cuticle and bacillary band. *Zeitschrift für Parasitenkunde*, **32**: 374–87.

Kan, S. P. 1986. Public health significance of Trichuriasis. WHO mimeographed document PDP/EC/WP/86, 12.

Kim, C. W. & Pawlowski, Z. S. (eds) 1978. *Trichinellosis*. University of New England Press, New Hampshire, USA. ISBN 0 874 51156 9.

Lichtenfels, J. R., Murrell, K. D. & Pilitt, P. A. 1983. Comparison of three sub-species of *Trichinella spiralis* by scanning electron microscopy. *Journal of Parasitology*, **69**: 1131–40.

Measures, L. N. & Anderson, R. C. 1985. Centrarchid fish as paratenic hosts of a giant kidney worm, *Dioctophyme renale* (Goeze, 1782), in Ontario, Canada. *Journal of Wildlife Diseases*, **21**: 11–19.

Murrell, K. D., Stringfellow, F., Dame, J. B., Leiby, D. A., Duffy, C. H. & Schad, G. A. 1987. *Trichinella spiralis* in an agricultural ecosystem. II. Evidence for natural transmission of *Trichinella spiralis* from domestic swine to wildlife. *Journal of Parasitology*, **73**: 103–9.

Mutafova, T., Dimitrova, Y. & Komandarev, S. 1982. The karyotype of four *Trichinella* species. *Zeitschrift für Parasitenkunde*, **67**: 115–20.

Sadighian, A. & Amini, F. 1967. *Dioctophyme renale* (Geoze, 1782) Stiles, 1901 in stray dogs and jackals in Shahsavar Area, Caspian Region, Iran. *Journal of Parasitology*, **53**: 961.

Schad, G. A., Duffy, C. H., Leiby, D. A., Murrell, K. D. & Zirkle, E. W. 1987. *Trichinella spiralis* in an agricultural ecosystem: transmission under natural and experimentally modified on-farm conditions. *Journal of Parasitology*, **73**: 95–102.

Schantz, P. M. 1983. Trichinosis in the United States – 1947–1981. *Food Technology*, March: 83–6.

Schantz, P. M. 1991. Parasitic zoonoses in perspective. *International Journal for Parasitology*, **21**: 161–70.

Shanta, C. S. & Meerovitch, E. 1967. The life cycle of *Trichinella spiralis* II. The muscle phase of development and its possible evolution. *Canadian Journal of Zoology*, **45**: 1261–7.

Sheffield, H. G. 1963. Electron microscopy of the bacillary band and stichosome of *Trichuris muris* and *T. vulpis. Journal of Parasitology*, **49**: 998–1009.

Tanner, C. E., Martinez-Fernandez, A. R. & Bolas-Fernandez, F. (eds) 1989. *Proceedings of the Seventh International Conference on Trichinellosis, Alicante, Spain, 1988*. Consejo Superior de Investigaciones Cientificas Press. ISBN 84 00 06985 4.

Wakelin, D. 1969. The development of the early stages of *Trichuris muris* in the albino laboratory mouse. *Journal of Helminthology*, **43**: 427–36.

Woodhead, A. E. 1950. Life cycle of the giant kidney worm, *Dioctophyme renale* (Nematoda), of man and many other mammals. *Transactions of the American Microscopical Society*, **69**: 21–46.

Zimmerman, W. J., Steele, J. H. & Kagan, I. G. 1968. The changing status of trichinosis in the U.S. population. *Public Health Reports*, **83**: 957–66.

27

Phasmid Nematoda: Rhabditida, Ascaridida and Oxyurida

The characteristics of the sub-class Phasmidea have already been defined (p. 369).

27.1 Order Rhabditida

From the point of view of the biologists, this order is of outstanding interest, for it contains both parasitic and non-parasitic species and also many which are on the borderline between a free-living and a parasitic existence. Many parasitic species have free-living stages in their life cycle, a fact which some workers consider represents a phylogenetic recapitulation. Three genera have been especially well studied, *Rhabditis*, *Strongyloides* and *Rhabdias*, and although these contain forms which are not all strictly parasitic, they are worth studying both on account of the ease with which they can be obtained for laboratory study and for the light which they throw on stages in the life cycles of other important parasitic forms. *Rhabditis* may readily be maintained through all its stages *in vitro* (Fig. 25.2) and offers superb experimental material, especially in the field of nutritional studies.

The members of the order are characterised by possessing a 'rhabditoid' pharynx with an anterior swelling followed by a posterior bulb (Fig. 25.1).

27.1.1 Family Rhabditidae

The morphology of a typical example of this family, *R. maupasi*, has already been studied (p. 369). Earthworms all over the world are parasitised by rhabditid larvae of this and related species in the 'dauer larva' stage and form a readily obtainable source of material. Invertebrates other than earthworms are also utilised by juvenile *Rhabditis*. Coprophagous insects are particularly favoured as hosts. Thus, *R. coarcta* is parasitic in or on dung beetles, and makes use of them as carriers to fresh dung in which final sexual development is achieved. *R. dubia* similarly makes use of dung-feeding psychodid flies. *R. strongyloides* may establish itself in cutaneous ulcers in dogs. Some species of *Rhabditis* can infect mammals, e.g. *R. orbitalis* whose third-stage larva is commonly found in the conjunctival sacs of mice and voles (Schulte, 1989). Larvae of the soil nematode *Pelodera strongyloides* (Rhabditidae) are found in the eyes and skin of the wood mouse, *Apodemus sylvaticus*, in Europe (Hominick & Aston, 1981).

A method for *in vitro* culture of *R. maupasi* has been given on p. 370; comparable methods are available for other species. Factors which affect the growth rate in monoxenic culture include temperature and associated bacterial fauna; the sex ratio is influenced in some species by different cultures of bacteria (Sohlenius, 1968). The systematics and phylogeny of the superfamily Rhabditoidea have been reviewed by Anderson & Bain (1982).

27.1.2 Family Strongyloididae

Many vertebrates, particularly mammals but also reptiles, birds and amphibians, are parasitised by members of this family. The most readily available source of laboratory material is probably the faeces of farm animals which contain eggs and larvae. The best-known species are probably *S. stercoralis*, in man, cat and dog; *S. fulleborni*, in primates other than man; *S. papillosus*, in ruminants; *S. ratti*, in rodents; *S. ransomi*, in pigs; and *S. westeri* in equines. In man, *S. stercoralis* causes strongyloidiasis, a major intestinal infection (Grove, 1989) which is widely distributed in the tropics and subtropics. The life cycle is similar in all species.

Morphology. Adult *Strongyloides* are largely localised in the duodeno-jejunal region. Worms burrow deeply

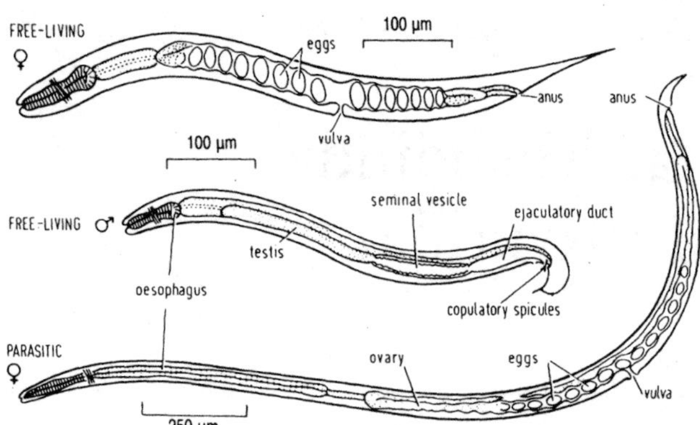

Fig. 27.1
Strongyloides stercoralis: morphology of parasitic and free-living generations. (After Faust, 1949.)

into the mucosa and may tunnel out galleries. In form they are slender and small and characterised by the possession of an unusually long oesophagus which in the female extends $^1/_3$ – $^2/_5$ of the body length (Fig. 27.1). The anal opening is ventral and the tail is pointed. In the parasitic generation, the adults are always female and usually, if not always, parthenogenetic; they appear to be triploid (Galliard, 1967). The males probably only appear in the free-living (see below) part of the cycle (Fig. 27.2).

Life cycle. The life cycle consists of two phases, the parasitic (or *endogenous*) and the free-living (or *exogenous*). The thin-shelled transparent eggs are either passed in the host's faeces (e.g. *S. westeri*) or hatch within the mucosa from which larvae reach the lumen and are passed in the faeces. Larvae are rhabditoid but without the typical median bulbar swelling found in the true rhabditiform type (e.g. *Rhabditis*).

Larval development (Fig. 27.2). The fate of larvae passed in the faeces is determined to a large extent by the nutriment available in the environment. Indirect (heterogonic) or direct (homogonic) development can take place. In indirect development, larvae rapidly show sexual differentiation, and within two to five days develop into free-living sexually mature male and female worms which resemble *Rhabditis* in their general morphology (Fig. 27.1). Rhabditoid larvae develop from the released eggs. After three to four days larvae moult and develop into the elongate infective filariform larvae. These infective larvae are clearly distinguishable from other rhabditoid stages by the very long cylindrical pharynx.

Under environmental conditions not defined, the rhabditoid larvae shed in the faeces may undergo 'direct' development into infective filariform larvae without undergoing sexual maturity. Both types of infective larvae become established in the host either by penetrating the skin or by oral ingestion. Those which penetrate the skin undergo a migration from the dermal tissues to the venous circulation and from there through the heart to the lungs, then up the bronchi and trachea where they are swallowed and pass down into the intestine.

The females on reaching the duodenal mucosa burrow deeply into the intestinal villi and even into the epithelium of the crypts of Lieberkühn. In the mucosa, they grow and produce eggs. In *Strongyloides stercoralis*, the parasitic males are apparently unable to penetrate the mucosa so that copulation, if it occurs at all, must occur in the lumen. In this species, also, the rhabditoid larvae of the parasitic female become transformed into the infective filariform type and penetrate the intestinal mucosa or perianal region without external development being required (Fig. 27.2). This is an example of 'auto-infection'.

Factors controlling heterogonic or homogonic development. The factors which determine whether the life cycle of *Strongyloides stercoralis*, and related species, will be heterogonic or homogonic are not properly understood. It was formerly thought that purely homogonic and heterogonic strains might exist but this has been disproved, as apparently heterogonic strains can give rise to homogonic strains and *vice versa*. The most recent work suggests that although the type of life cycle is determined by genetic factors in the eggs of parasitic females, external environmental factors play a considerable role in determining the final pattern of development. Parthenogenetic female parasites appear to lay two types of eggs: *haploid* which give rise to free males and *diploid* which develop either to the rhabditoid female stage or to the infective strongyloid stage. Thus the predominance of either cycle would be influenced by environmental factors such as temperature and available nutrients. Under 'unfavourable' environmental conditions there appears to be a tendency to develop into infective filariform larvae, whereas under 'favourable' conditions the tendency is towards free-living forms.

Pathology in man. The pathology of strongyloidiasis has been reviewed in detail by Galliard (1967), Georgi (1982) and Grove (1989). Its main clinical symptoms are abdominal pain, diarrhoea and urticaria.

○ 27.1.3 Genus *Rhabdias*

The best known of this genus is probably *R. bufonis*, a common parasite in the lungs of frogs, toads and other

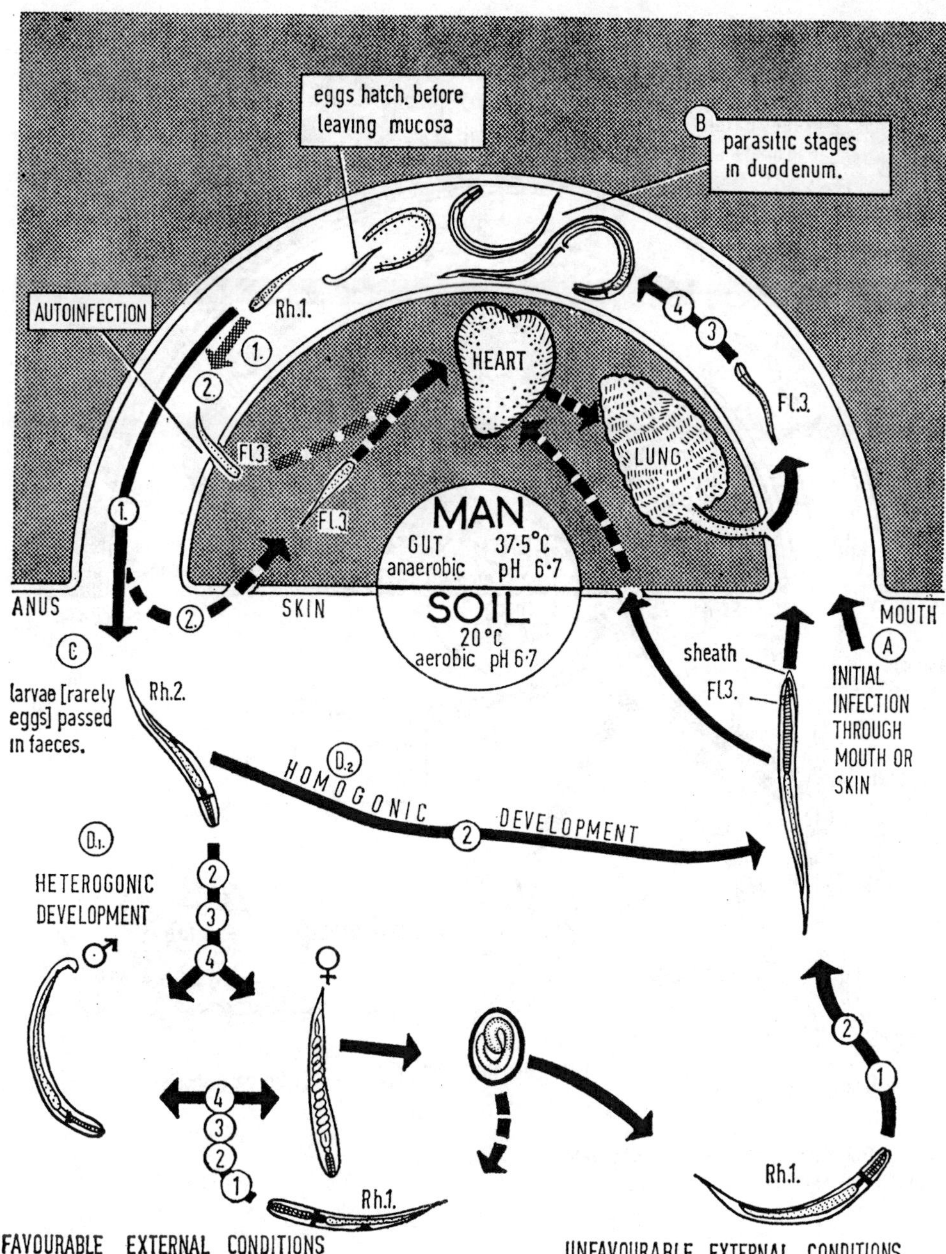

Fig. 27.2
Strongyloides stercoralis: life cycle in man. The numbers in the circles refer to the number of larval moults but the precise position of some of these is imperfectly known. (Original.)

amphibians; other common species are *R. americana* and *R. ranae* (Baker, 1979). It is a protandrous hermaphrodite with the structure of a female. During the early male phase of development, sperms are produced and stored in the receptaculum seminis. During the later phase of maturation, eggs are released, and pass from the lungs via the buccal cavity into the alimentary canal. Here they hatch into rhabditoid larvae with a typical rhabditoid pharynx. These larvae are passed with the faeces and develop in the soil into males and females. After insemination, the females viviparously produce filariform larvae which devour the mother worm and escape as infective forms. These penetrate the skin of the amphibians and reach sexual maturity in the lungs;

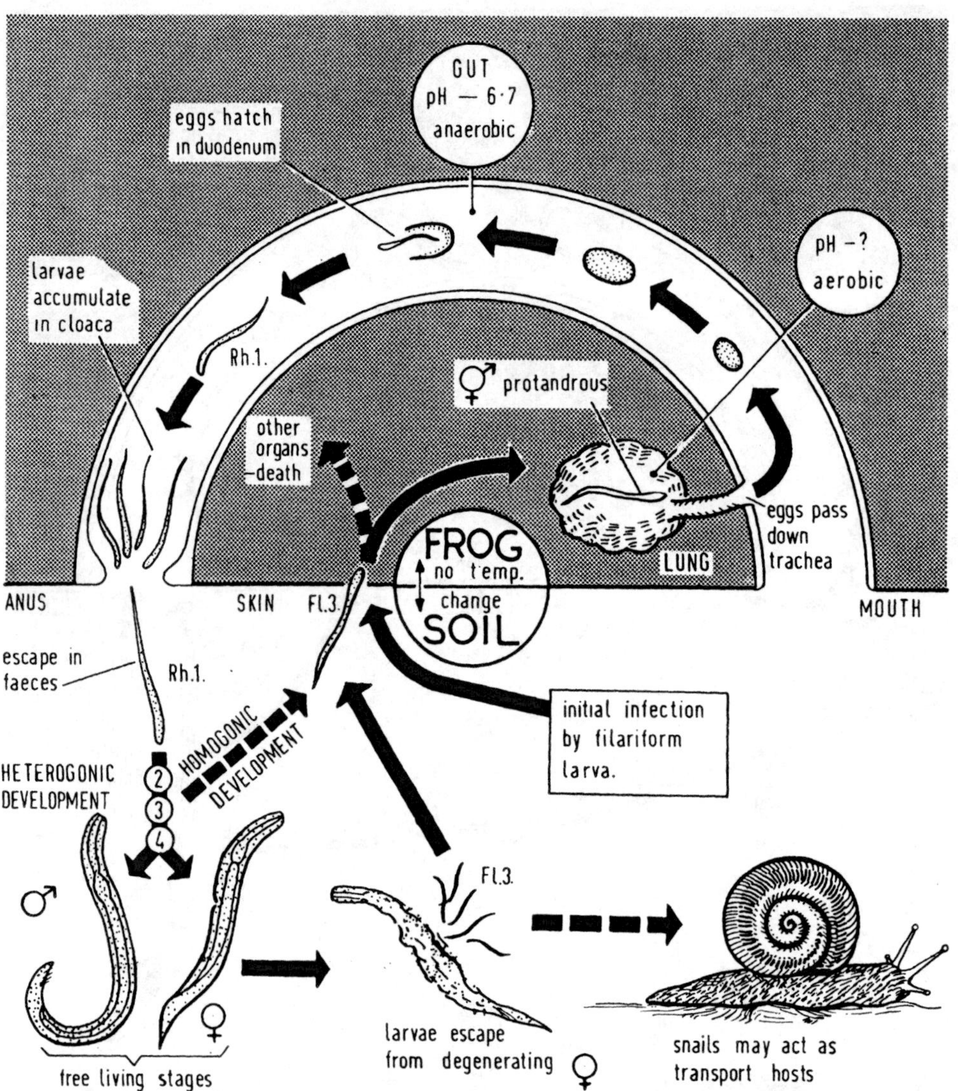

Fig. 27.3
Rhabdias bufonis: life cycle in frog. The numbers in the circles refer to the number of larval moults. (Original.)

snails may act as transport hosts. The genus *Rhabdias* contains about 47 species (Baker, 1987).

○ 27.1.4 Genus *Neoaplectana*

This genus contains forms parasitic in insects. Like many other rhabditoids, it is a facultative parasite which may repeatedly complete its life cycle in decaying insect bodies, but when the food supply has been exhausted it will produce an infective filariform larva which penetrates into living beetles, both larvae and adults. Within the beetle gut, larvae become sexually mature and reproduce ovoviviparously. An infected beetle generally dies and the young nematodes feed on its carcase. The best-known species is *Neoaplectana glaseri*, which has been cultured through all its life cycle *in vitro*.

○○○ 27.2 Order Ascaridida

○ 27.2.1 Superfamily Ascaridoidea

The members of this group, to which the term ascaridoids is loosely applied, are relatively large nematodes which are obligatory intestinal parasites of vertebrates. They are mostly large, stout polymyarian worms. The eggs differ from those of the nematodes already described, being thick-shelled with an uneven surface (Fig. 25.14). Unlike those of the Rhabditida, the eggs rarely hatch externally but require to embryonate in an aerobic environment before development to an infective stage can occur. Hatching takes place only on ingestion by another animal.

Formerly it was thought that the life cycle was direct, but it has now been shown that several species utilise intermediate hosts in which larvae become encysted. For example, *Amplicaecum robertsi*, which occurs in a python (*Morelia spilotes*) can utilise earthworms, snails, fish, tadpoles, reptiles, birds and mammals as intermediate hosts (Sprent, 1963*a*,*b*).

It is also important to note that the *larvae* of several ascaridoid species – notably *Toxocara canis* of cats and dogs – can survive in abnormal hosts (such as man) wandering through tissues as a *larva migrans* which often causes serious pathological effects; see p. 404.

○ 27.2.2 Genus *Ascaris*

Type examples: ***Ascaris lumbricoides, A. suum***

The best-known species of this genus is the so-called 'large roundworm', *Ascaris lumbricoides*, which is a parasite of man, some apes and pigs all over the world. Much debate has taken place concerning the taxonomic status of the human and pig *Ascaris*. Some authors have claimed to detect small morphological differences between worms from these different hosts; others have failed to confirm these differences. Scanning electron microscopy has, however, confirmed that differences in the shape of lips and denticles do occur. Some workers refer to them as separate species, *A. lumbricoides* (man) and *A. suum* (pig), or subspecies, *A. lumbricoides lumbricoides* and *A. lumbricoides suum*; the first mentioned nomenclature is followed here. Antigenic analysis has shown that *A. suum* shares eleven antigens with

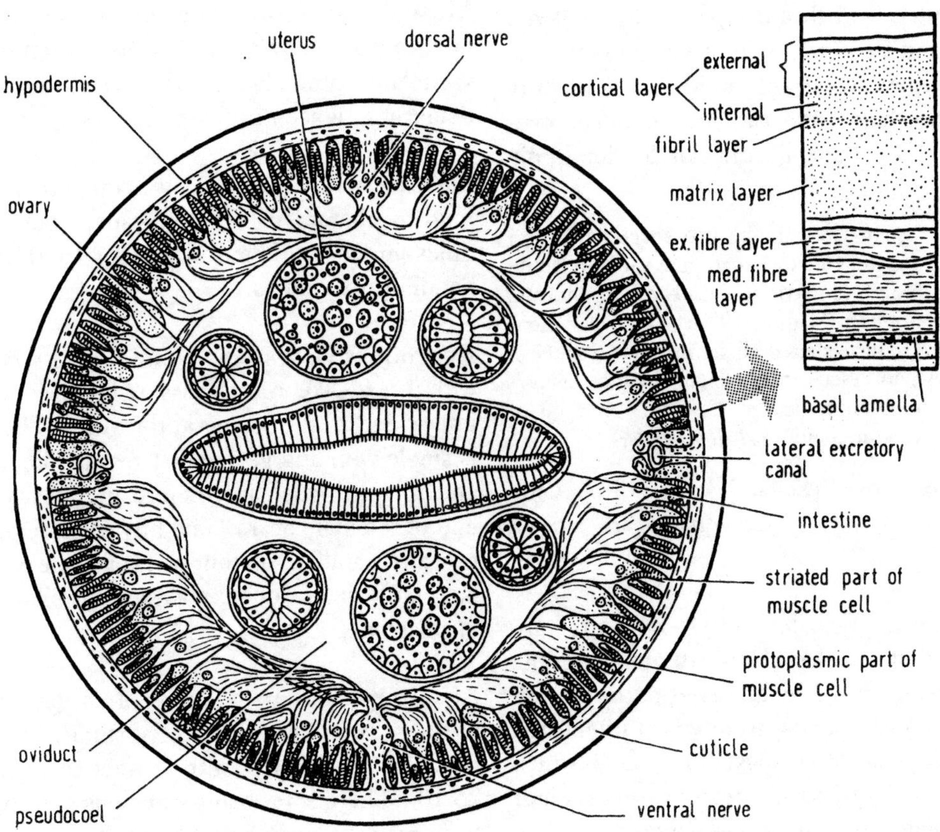

Fig. 27.4
Ascaris lumbricoides. Diagrammatic transverse section showing general morphology; the figure on the top right is an enlarged section of cuticle. (After Brown, 1950.)

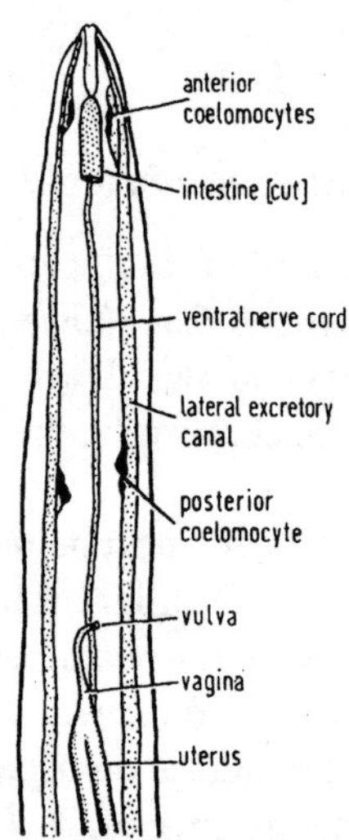

Fig. 27.5
Ascaris lumbricoides: anterior end dissected. (After Chitwood, 1950.)

A. lumbricoides, but in addition, has three specific antigens (Imperato *et al.*, 1968). Small biochemical differences have also been reported (Kurimoto, 1974). Although it is often said that the pig 'strain' will not develop in man and *vice versa*, there is now unequivocal evidence from experimental and accidental laboratory infections that the 'pig' form can indeed infect man (Lýsek, 1967) and the 'human' form can develop in pig (Galvin, 1968).

Examination of the isoenzymes in the two species with gel electrophoresis showed that 14 loci that were monomorphic in *A. suum* were found to have identical electrophoretic mobilities in *A. lumbricoides* (Nadler, 1987). Three polymorphic loci were found to have shared alleles in the two species. However, one enzymatic locus, SOD-1 (superoxide dismutase) – which was monomorphic for *A. suum* – may be of different electrophoretic mobility in *A. lumbricoides*.

Reviewing all the above results, Nadler (1987) concluded that *A. suum* and *A. lumbricoides* are sibling species.

External features. Females (200–350 mm) are much larger than the males (150–310 mm) and the latter are also distinguished by a curved posterior end bearing a slit-like anal opening from which often protrudes a pair of copulatory spicules. The cuticle, which bears transverse markings, is sometimes brown in colour owing to the presence of quinone-tanned proteins. The position of the excretory canals is marked by two broad brownish lateral lines, and the position of the main nerve cords is similarly marked by dorsal and ventral lines just visible as white lines through the cuticle. Some 2 mm from the anterior end is an excretory pore, ventral in position and visible only in exceptional specimens. In the female, a narrow constricted band may be seen about one-third of the distance from the anterior end, and this bears the slit-like opening of the vulva. In the male, there is a sub-terminal cloaca.

Digestive system. The mouth is surrounded by three cuticulate lips, which are best seen in *en face* sections (Fig. 25.4). The dorsal lip bears a pair of papillae which represent the two papillae of the external circle (p. 372). The paired subventral lips bear (*a*) amphids, and (*b*) fused ventro- and latero-ventral papillae. The inner edges of the lips are provided with dentigerous ridges whose structure is believed to differ in the human and pig varieties. The alimentary canal is the usual nematode type (p. 373). As in other species (p. 375) ultrastructure studies have shown the presence of well-developed microvilli.

Muscular system. This is built on the polymyarian plan with about 150 muscle cells in each quadrant. The vacuolated protoplasmic parts of the muscles are particularly well seen in transverse sections (Fig. 27.4).

Pseudocoele. This cavity contains a protein-rich fluid of characteristic unpleasant odour, which sometimes induces allergic symptoms in those frequently handling specimens.

Reproductive system. The female system is double with thread-like ovaries passing into oviducts and finally into wider uteri opening at the vulva. The male system is single with a large seminal vesicle joining the rectum and spicule pouch just before the cloaca. The caudal end of the male bears some fifty or more pairs of simple pre-anal papillae and four pairs of single post-anal papillae, one pair of double post-anal papillae and one pair of phasmids.

Excretory system. This is often described as being of the ∩ type, but more recent studies suggest that it is of the H type, that is, with distinct anterior canals (Fig. 25.10). There is probably a nucleus present in the short terminal duct leading to the excretory pore and another, the *giant sinus nucleus*, associated with the left lateral canal. Four coelomocytes occur in the body cavity situated anterior to the vulva, as shown in Fig. 27.5.

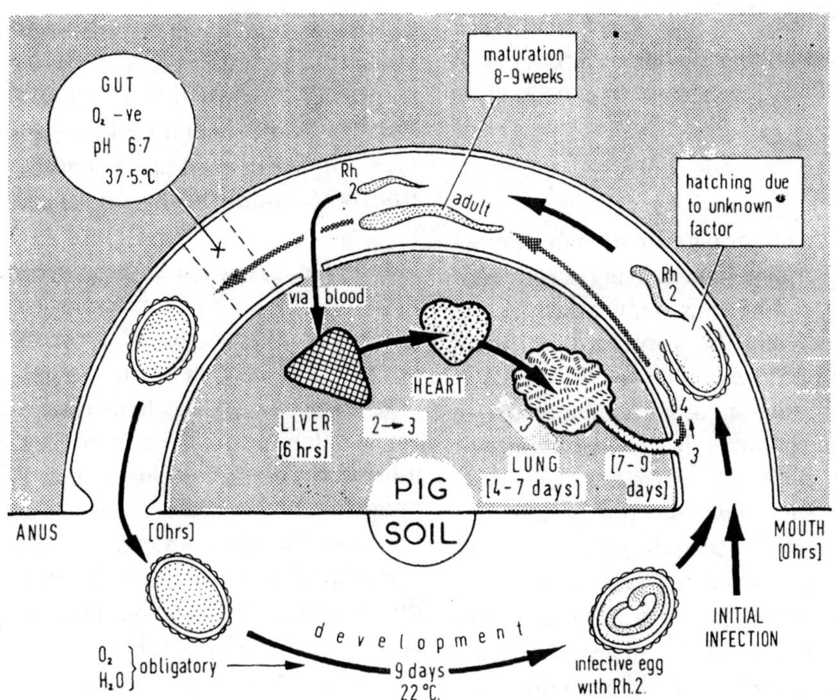

Fig. 27.6
Ascaris suum: life cycle in pig and some physiological factors relating to it. Hatching has been shown* (Rogers, 1959) to be due to enzymes released under the stimulus of p_{CO_2} of the gut. Numbers refer to the earliest larval stage appearing at any site. Times of development are based on Douvres *et al.* (1969). (Original.)

Life cycle (Fig. 27.6). Comprehensive reviews of the life cycle have been given by Crompton & Pawlowski (1985) and Crompton (1989). When eggs are passed in faeces, their further development is largely dependent on the oxygen tension, moisture content and temperature of their environment. In a cold dry atmosphere, it is said that eggs may remain viable for up to six years and in infected areas they may be widely distributed. At a temperature of 22–33 °C, the eggs give rise to coiled rhabditoid larvae within 9–13 days, but do not hatch (except accidentally) until taken into the definitive host. The shell layers of the egg (p. 380) provide a remarkably resistant structure which can withstand many chemical agents, and in the laboratory eggs are normally embryonated experimentally in 2 per cent formal solution.

The eggs can show considerable variation in shape and size (52–84 μm × 45–67 μm). Infection is brought about by ingestion of viable eggs, which are triggered to hatch under the influence of the intestinal conditions (especially p_{CO_2}) (Clarke & Perry, 1988); hatching largely occurs in the duodenum but some takes place in the stomach. The released larva is either a second-stage larva (i.e. it has moulted once within the egg) or a larva still within the sheath of its first moult. On release within the intestine, the larvae undergo a migration so remarkable that it is difficult to believe that it is not a phylogenetic reminiscence (see below), the parasite reliving its life cycle in a once intermediate but now definitive host. On hatching, larvae burrow into the mucosa, penetrate blood vessels and appear as second-stage larvae in the liver within six hours post-infection. Some larvae penetrate lymphatics but apparently become inhibited and it is doubtful if these develop further (Douvres *et al.*, 1969). They remain in the liver for a few days and develop to the early third-stage larva. Larvae then continue to the heart and are carried via the pulmonary arteries to the lungs, appearing there within four to seven days. Seven or eight days post-infection, larvae break out of the capillaries into the alveoli and finally work their way up the trachea to the pharynx and reach the small intestine on the eighth to tenth day post-infection. Within the intestine, larvae begin the third moult on the ninth day and are in the fourth stage by the tenth day. The pre-patent period of *A. suum* in pigs (40–53 days) is less than that of

A. lumbricoides (54–61 days) in pigs (Galvin, 1968). It should be noted, however, that slightly different developmental times from those given above have been given by different authors.

Ascariasis

A. lumbricoides is an important parasite of man; it often occurs in high levels in populations living under conditions of poor hygiene. The eggs require oxygen and moisture to embryonate and the worm is often found associated with *Trichuris trichiura* – see Table 26.1. Ascariasis is relatively rare in urban communities of most highly developed countries. It has been estimated that there are about 1000 million cases of ascariasis worldwide, with average prevalences in the range of 32–60 per cent (Crompton *et al.*, 1989). In some surveys of children between the ages of 6 and 12 years, the infection rate was as high as 90 per cent. The infection results in malnutrition and retardation of growth in children, but other symptoms associated with both the larval (tissue) and adult (intestinal) stages include pneumonitis, asthma, diarrhoea, nausea, abdominal pain and anorexia.

Control. The obvious basic control strategy is to interrupt transmission of infective eggs to water, food or human hands. This clearly implies good sanitation, a safe water supply, education, chemotherapy and socio-economic development, difficult targets to achieve in developing countries. All aspects of the control of ascariasis are discussed by Crompton *et al.* (1985, 1989). Drugs in general use for treatment include lavamisole, pyrantel, mebendazole and albendazole. More recently, ivermectin has been shown to be 100 per cent effective against ascariasis (Ottesen, 1990).

○ 27.2.3 Other ascaridoids

Toxocara canis: visceral larva migrans

It is well known that the larvae of a number of species of nematode develop in the tissues of hosts other than their 'normal' definitive host. The most important species infecting man in this way is *Toxocara canis*, primarily a parasite of dogs, whose larva causes *visceral larva migrans*, the infection sometimes being referred to as *toxocariasis* or *toxocaral disease*. The extensive literature on this zoonoses has been succinctly reviewed by Gillespie (1988) and Schantz (1989). *Toxocara canis* is a common infection in dogs throughout the world. Parks are favourite walking-out areas for dogs, so it is not surprising to learn that in 1970, 35 per cent of soil samples from London parks were positive for *T. canis* eggs. Similar or even higher figures have been reported from other countries, e.g. 66 per cent in rural Germany.

Life cycle. The life cycle is shown in Fig. 27.7. It is important to note that faecal eggs are almost entirely passed by puppies up to 5 weeks old and lactating bitches but, in general, not older animals. Although embryonated eggs hatch in older dogs, the L_2 larvae which hatch remain dormant in the tissues so that patent infections do not develop unless the animal becomes pregnant. In pregnant bitches, dormant L_2 larvae are reactivated by the hormonal changes during pregnancy, and migrating larvae infect the developing foetuses via the placenta. In lactating bitches, puppies can also become infected by larvae in the milk. Lactating bitches themselves develop patent infections by ingesting eggs from their puppies' faeces. Eggs require oxygen and a warm temperature to embryonate and become infective; this takes about 2–6 weeks in temperate climates.

Human infection. Humans become infected by ingesting infective eggs. The L_2 larvae, which hatch in the gut, migrate through the tissues causing *visceral larval migrans* (VLM) or, if they become trapped in the eye, *ocular larval migrans* (OLM). Children playing with puppies, or in parks or in other areas contaminated with dog faeces, are particularly at risk.

The symptoms due to the migrating larvae are often unrecognised and individuals may not develop overt clinical disease. VLM is the result of an inflammatory response and is characterised by persistent eosinophilia, leucocytosis, fever, and hepatosplenomegaly and bronchospasm. OVM is potentially more serious as the retina may be damaged and in extreme cases may result in loss of vision or severe ocular inflammation. Some visual defects may go unnoticed and there is increasing evidence that the effects of the infection are wider than is generally appreciated (Taylor *et al.*, 1988). In the UK alone, more than 300 cases of toxocariasis are diagnosed each year and in one US study, the infection accounted for 37 per cent of retinal disease in paediatric patients (Gillespie, 1988).

Animal models for toxocariasis. Experiments on mice, using ^{75}Se-labelled L_2 larvae introduced into the

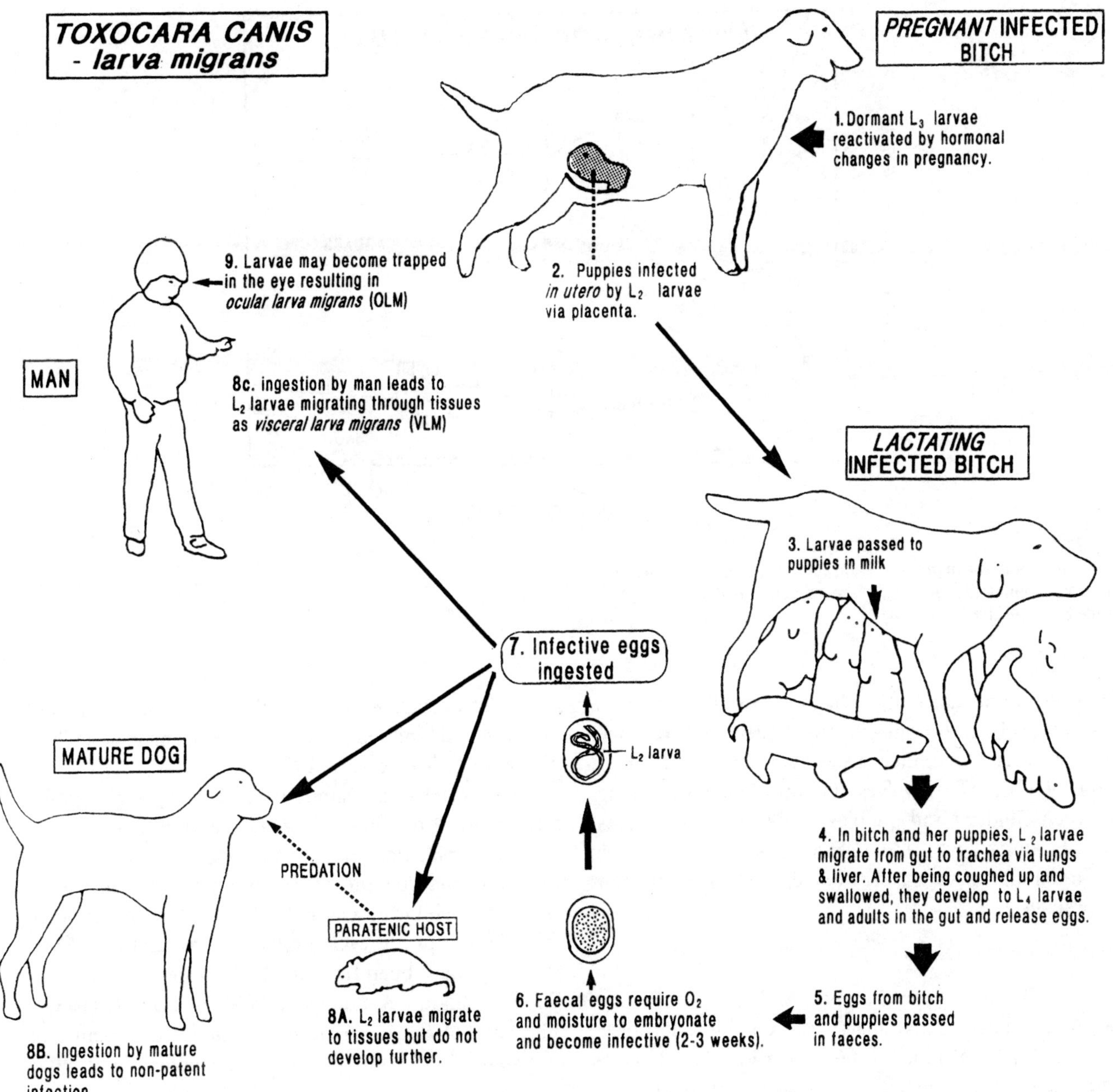

Fig. 27.7
Life cycle of the dog nematode, *Toxocara canis*, the causative organism of *visceral larva migrans* (*VLM*) and *ocular larva migrans* (*OLM*) in man. (Modified from Gillespie, 1988.)

stomach, have revealed a remarkable picture of how widespread larvae can become in the tissues; see Fig. 27.8 (Wade & Georgi, 1987). The carcase accumulated the greatest number of larvae, followed by the CNS, liver and lung, in that order. Similar experiments by Olson (1976) have shown that after 34 days' infection, 90 per cent of mice had larvae in their eyes, the retina being largely infected. If the distribution in mice reflects the distribution in human infections, it is clear that toxocariasis must be considered as a seriously invasive zoonosis.

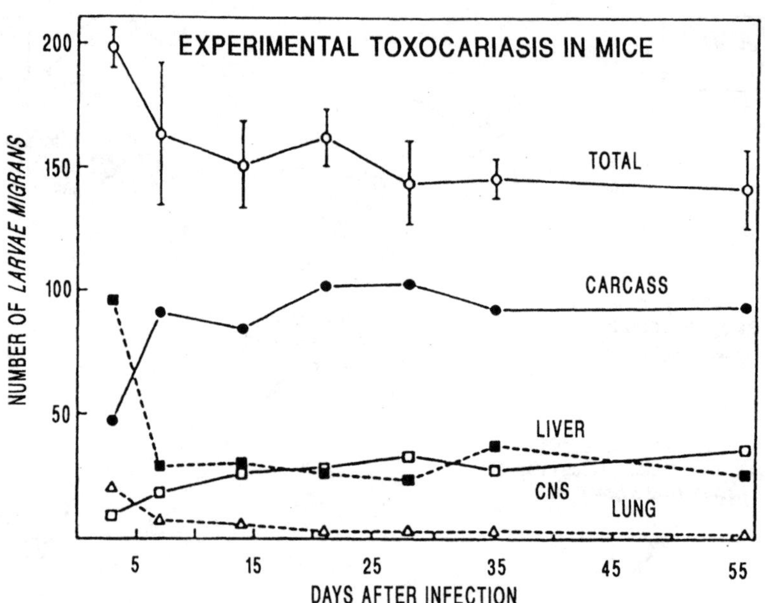

Fig. 27.8
Distribution of *larvae migrans* of *Toxocara canis* in male CD-1 mice after intragastric administration of ^{75}Se-labelled L_3 larvae. (Modified from Wade & Georgi, 1987.)

Diagnosis and control. Diagnosis used to present a difficult problem but an effective ELISA test to detect antibodies is now widely used (Savigny *et al.*, 1978). Control obviously involves the prevention of contact between children and puppies and sites of dog faeces. Dog-proofing parks and regular de-worming of dogs are other obvious measures but difficult to implement in practice.

Anisakidae: Anisakiasis

This family contains several species, the adults of which occur in various sea mammals, seals, dolphins, porpoises and whales (Van Thiel, 1966) but whose larval stages cause a serious zoonosis – *anisakiasis* – in man. Eggs passed in the definitive host faeces embryonate and hatch out in sea water where, if eaten by the tiny crustaceans *Euphausia* and *Thysanoessa*, they develop into L_2 and L_3 larvae. If these infected crustaceans are eaten by marine fishes or squid, the larvae encyst; over 100 species of fish can act as second intermediate hosts in this way. Humans acquire the infection by eating raw or undercooked fish, squid or sea-food, especially in dishes such as sashi, sashimi, caviche, and lomi-lomi. Smoked salmon has even been found to be infected (Gardiner, 1990). Although especially prevalent in Japan, where over 1000 cases are recorded each year, it has also been reported from the Netherlands, France, Belgium, the United Kingdom, USA, Canada and Chile (Sakanari & McKerrow, 1989).

Two species of nematode have been especially implicated worldwide, *Anisakis simplex* and *Pseudoterranova* (= *Phocanema*) *decipiens*; a third species, *Contracaecum osculatum*, may be the cause of human anisakiasis in Germany (Ishikura & Namiki, 1989). The morphology (Fig. 27.9) and life cycle of *Anisakis* sp. and *Pseudoterranova* sp. have been reviewed by Beverley-Burton *et al.* (1977), Ishikura & Namiki (1989) and Smith (1983). The larva of *Pseudoterranova* is distinguished from that of *Anisakis* by the possession of a caecum (Fig. 27.9).

Anisakiasis. In man, larvae are found commonly in the stomach or duodenum but can be found in any part of the alimentary canal or outside the gut in various viscera (Smith & Wootten, 1978). Diagnosis can only be definitely confirmed by endoscopy and the removal of the worm by biopsy forceps and microscopical identification.

Because the symptoms of anisakiasis are often vague, it is easily misdiagnosed. Clinical symptoms of gastric anisakiasis include sudden epigastric pain, vomiting, nausea, diarrhoea and urticaria, especially occurring after an infected meal, usually within 6 h.

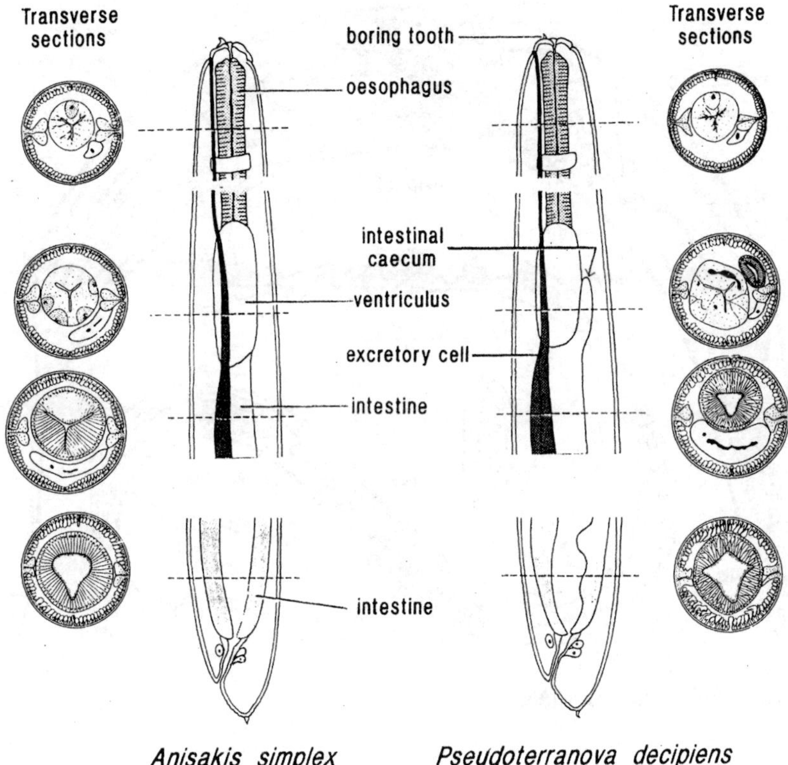

Fig. 27.9
Diagrammatic representation of the morphology of L_3 larvae of *Anisakis simplex* and *Pseudoterranova decipiens* (from fish), causative organisms of *anisakiasis* in man. (Modified from Ishikuri & Namiki, 1989.)

All aspects of the biological and medical aspects of gastric anisakiasis are covered in a (coloured) monograph by Ishikura & Namiki (1989). Other valuable general reviews are those of Oshima (1987), Sakanari & McKerrow (1989), Smith (1983) and Smith & Wootten (1978). Anisakiasis in the USA has been reviewed by Fontaine (1985) and Kwee & Sautter (1987) and in the UK by Lewis & Shore (1985).

Control. Avoiding the eating of raw or undercooked fish is an obvious measure that individuals can take to avoid infection. WHO (1988) recommended that fish should be cleaned by gutting as soon as possible after they are caught: the larvae migrate into the muscle tissue after the death of the fish. In the commercial fishing industry, freeze-blasting at at least −35 °C for 15 h is an extremely effective way of killing all larvae and moreover causes little change in the flavour of the fish. As this process is already in use in some countries for the preservation of fish, this is both an efficient and a cost-effective method of control (Deardorff & Throm, 1988).

Toxocara cati

This species closely resembles *T. canis*, but as cats bury their faeces, human infections with larva migrans are less common.

○○○ 27.3 Order Oxyurida

The oxyuroid nematodes (Table 27.1) are chiefly characterised by possessing an 'oxyuroid' (= bulboid) type of pharynx, that is one provided with an end bulb (Fig. 25.6). They are small, rather transparent, meromyarian worms, with pointed tails. The species parasitising vertebrates inhabit the colon, caecum, or posterior region of the intestine, all regions of poor nutriment. Those parasitising invertebrates inhabit the intestine or malpighian tubules of insects.

Type example: ***Leidynema appendiculata***

definitive host: *Periplaneta americana* (cockroach)
location: midgut

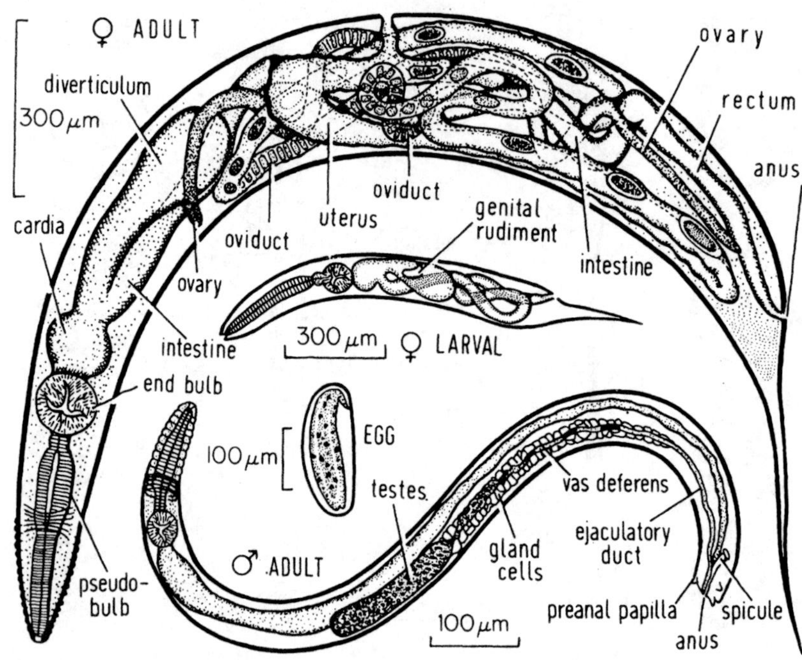

Fig. 27.10
Leidynema appendiculata, an oxyuroid of the cockroach. (After Dobrovolny & Ackert, 1934.)

Table 27.1. *Some common oxyuroid nematodes*

Species	Host
Enterobius vermicularis	man
Syphacia obvelata	rat and mouse
Aspiculuris tetraptera	mouse
Hammerschmidthiella diesingi	*Periplaneta americana*
Leidynema appendiculata	*Periplaneta americana*
Thelastoma pteroton	*Periplaneta americana*
Thelastoma icemi	*Periplaneta americana*
Blatticola blattae	*Blatella germanica*

Morphology. The general morphology of the female, which has been described by Dobrovolny & Ackert (1934), is shown in Fig. 27.10. Points of special interest are the closely annulated cuticle and the presence of a large intestinal diverticulum. The male likewise has an annulated cuticle but no intestinal diverticulum. The tail is characteristically curved. The testis is a compact structure near the middle of the body. There is a single spicule. Both have a cylindrical oesophageal pseudobulb.

Life cycle. The life history is direct as in other oxyuroids. The eggs when laid are unsegmented, but develop rapidly under the usual requirements of oxygen and moisture. The embryonic development has been studied by Adamson & Clease (1989).

Other species

The second commonest species of oxyuroid in the cockroach is *Hammerschmidthiella diesingi*, which may be distinguished from *L. appendiculata* by possessing an oval instead of a cylindrical pseudobulb. Its life cycle is similar to that of *Leidynema*. Both *H. diesingi* and *L. appendiculata* are cosmopolitan in distribution, but a number of other species in cockroaches are known (Chitwood, 1932).

Genus *Syphacia*: oxyuroids of rodents

Two oxyuroids, *S. obvelata* and *S. muris*, are common parasites of laboratory rats and mice. The worms lie in the caecum where they feed by sucking (Singhvi & Johnson, 1989), but are not attached to the mucosa (Panter, 1969; Gulden, 1967). Like the oxyuroid of man, *Enterobius vermicularis* (see below), worms show a marked diurnal rhythm in their egg production. Thus, *S. muris* leaves the gut about noon and passes out of the rectum in the afternoon to lay eggs in the perianal region (Gulden, 1967).

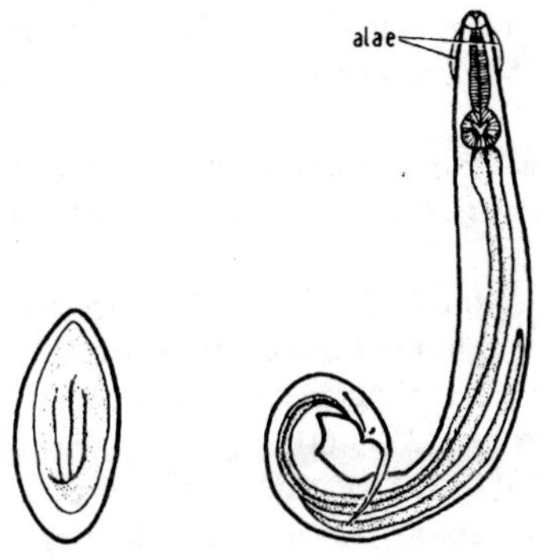

Fig. 27.11
Syphacia obvelata, male. An oxyuroid of rats and mice. (Adapted from Yorke & Maplestone, 1926.)

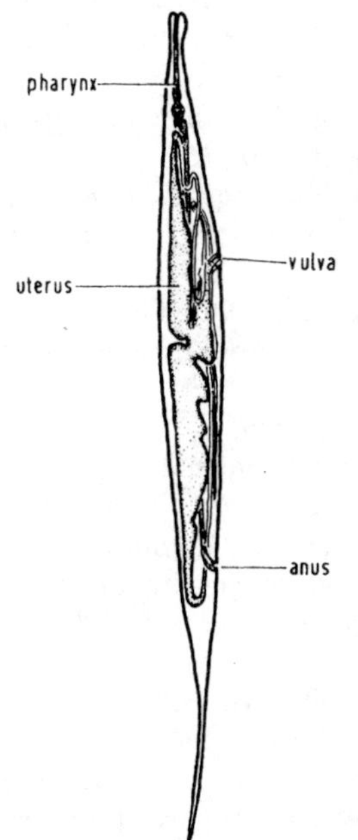

Fig. 27.12
Enterobius vermicularis, female. The 'pinworm' of man. (After Faust, 1939.)

Enterobius vermicularis: an oxyuroid of man (Fig. 27.12)

This is the human 'pinworm' or 'seatworm' widespread in temperate climates but, unlike most helminths, rare in the tropics. Probably every child in the temperate-zone areas has been infected not once, but many times in early childhood. Once it reaches a household, it is likely to infect every member of a family. Fortunately it is relatively harmless, although it may cause restlessness and irritability in young children.

Morphology. The worms, which have an opalescent white appearance, are small and active, and have a typical oxyuroid morphology. The females measure 8–13 mm and the males 2–5 mm. The adults live in the caecum.

Life cycle. By some mechanism, not understood, the females are capable of restricting their egg laying to certain periods. At night, possibly stimulated by the extra warmth of bedclothes, egg-laden females migrate down to the anus where, on contact with air, eggs are laid and the females retreat to the caecum once more. The newly laid eggs are sticky and adhere to the skin. Some worms fail to return to the anus, become dried on the skin and 'explode' to release quantities of eggs.

One of the features of infections with this worm is that eggs are rarely found in the faeces but may be found in skin scrapings near to the anal region. The standard medical practice is to press a piece of Scotch tape on to the skin and stick this to a microscope slide.

The eggs are like those of *Syphacia*, slightly flattened on one side and containing a rhabditoid larva. In a small number of cases, eggs may hatch in the anal region and migrate up the caecum to reach maturity. Such a mode of infection is termed 'retrofection'. In children, the intense itching caused by the migrating worms at night may cause them to scratch the anal region, so that the eggs may become lodged under the finger nails and so reach the mouth. This is, however, probably the least important means of infection. The eggs are exceptionally light and easily airborne, and rapidly distributed throughout a normal household or institute, often in spite of the most elaborate hygienic precautions. In one household examined, up to 90 per cent of samples of dust taken from different parts contained eggs of *Enterobius* (Nolan & Reardon, 1939). It is of interest to note that eggs of *E. vermicularis* have even been found in human coprolites (petrified faeces) and desiccated material from caves in Utah and Tennessee (Faulkner *et al.*, 1989; Fry & Moore, 1969).

A recent Anthelminthic Study Group on Enterobiasis (1984) reported the success rates of various drugs as follows: pyrantel, 94.1 per cent; mebendazole, 67.6 per cent; piperazine, 67 per cent. Ivermectin has been reported to be 86 per cent effective (Ottesen, 1990).

○○○ References

(References of general interest have been included in this list, in addition to those quoted in the text.)

Adamson, M. L. & Clease, D. F. 1989. Morphological changes during *in ovo* development in the Thelastomatoidea: description and functional considerations. *Journal of Parasitology*, **75**: 728–34.

Anderson, R. C. & Bain, O. 1982. Keys to the genera of the superfamilies Rhabditoidea, Dioctophymatoidea, Trichinelloidea and Muscpiceoidea. *C.I.H. Keys to the Nematode Parasites of Vertebrates*, No. 2 (ed. R. C. Anderson, A. G. Chabaud & S. Wilmott). Commonwealth Agricultural Bureaux, Farnham, England.

Anthelminthic Study Group on Enterobiasis. 1984. A comparative evaluation of mebendazole, piperazine and pyrantel. *Indian Pediatrics*, **21**: 623–8.

Baker, M. R. 1987. Synopsis of the nematodes parasitic in amphibians and reptiles. *Memorial University of Newfoundland, Occasional Papers in Biology*, No. 11.

Baker, M. R. 1979. The free-living and parasite development of *Rhabdias* sp. (Nematoda: Rhabdiasidae) in amphibians. *Canadian Journal of Zoology*, **57**: 179–83.

Beverley-Burton, M., Nyman, O. L. & Pippy, J. H. C. 1977. The morphology, and some observations on the population genetics of *Anasakis simplex* larvae (Nematoda: Ascaridata) from fishes of the North Atlantic. *Journal of the Fisheries Research Board of Canada*, **34**: 105–12.

Chitwood, B. G. 1932. Synopsis of nematodes parasitic in insects of the family Blattidae. *Zeitschrift für Parasitenkunde*, **5**: 14–50.

Clarke, A. J. & Perry, R. N. 1988. The induction of permeability in egg-shells of *Ascaris suum* prior to hatching. *International Journal for Parasitology*, **18**: 987–90.

Crompton, D. W. T. 1989. Biology of *Ascaris lumbricoides*. In: *Ascariasis and its Prevention and Control* (ed. D. W. T. Crompton, M. C. Nesheim & Z. S. Pawlowski), pp. 9–44. Taylor & Francis, London.

Crompton, D. W. T., Nesheim, M. C. & Pawlowski, Z. S. (eds) 1985. *Ascariasis and its Public Health Significance*. Taylor & Francis, London. ISBN 0 850 66317 2.

Crompton, D. W. T., Nesheim, M. C. & Pawlowski, Z. S. (eds) 1989. *Ascariasis and its Prevention and Control*. Taylor & Francis, London. ISBN 0 850 66424 1.

Crompton, D. W. T. & Pawlowski, Z. S. 1985. Life history and development of *Ascaris lumbricoides* and the persistence of human ascariasis. In: *Ascariasis and its Public Health Significance* (ed. D. W. T. Crompton, M. C. Nesheim & Z. S. Pawlowski), pp. 9–23. Taylor & Francis, London.

Deardorff, T. L. & Throm, R. 1988. Commercial blast-freezing of third-stage *Anisakis simplex* larvae encapsulated in salmon and rockfish. *Journal of Parasitology*, **74**: 600–3.

Dobrovolny, C. & Ackert, J. 1934. Life history of *Leidynema appendiculata* (Leidy), a nematode of cockroaches. *Parasitology*, **26**: 468–80.

Douvres, F. W., Tromba, F. G. & Malakatis, G. M. 1969. Morphogenesis and migration of *Ascaris suum* larvae developing to fourth stage in swine. *Journal of Parasitology*, **55**: 689–712.

Faulkner, C. T., Patton, S. & Johnson, S. S. 1989. Prehistoric parasitism in Tennessee: evidence from the analysis of desiccated fecal material collected from Big Bone Cave, Van Buren Country, Tennessee. *Journal of Parasitology*, **75**: 461–3.

Fontaine, R. E. 1985. Anisakiasis from the American perspective. *Journal of the American Medical Association*, **253**: 1024–5.

Fry, G. F. & Moore, J. G. 1969. *Enterobius vermicularis*: 10 000-year-old human infection. *Science*, **166**: 1620.

Galliard, H. 1967. Pathogenesis of *Strongyloides*. *Helminthological Abstracts*, **36**: 247–60.

Galvin, T. J. 1968. Development of human and pig *Ascaris* in the pig and rabbit. *Journal of Parasitology*, **54**: 1085–91.

Gardiner, M. A. 1990. Survival of *Anisakis* in cold smoked salmon. *Canadian Institute of Food Science and Technology Journal*, **23**: 143–4.

Georgi, J. R. 1982. Strongloidiasis. In: *CRC Handbook Series in Zoonoses. Section C: Parasitic Zoonoses*, Vol. II (ed. M. G. Schultz), pp. 257–67. CRC Press, Boca Raton, Florida, USA. ISBN 0 849 32917 5.

Gillespie, S. H. 1988. The epidemiology of *Toxocara canis*. *Parasitology Today*, **4**: 180–2.

Grove, D. I. (ed.) 1989. *Strongyloidiasis. A Major Roundworm Infection of Man*. Taylor & Francis, London. ISBN 0 850 66732 1.

Gulden, W. J. I. van der. 1967. Diurnal rhythm in egg production by *Syphacia muris*. *Experimental Parasitology*, **21**: 344–7.

Hominick, W. M. & Aston, A. J. 1981. Association between *Pelodera strongyloides* (Nematoda: Rhabditidae) and wood mice, *Apodemus sylvaticus*. *Parasitology*, **83**: 67–75.

Imperato, S., Foresi, C. & Martinetto, P. 1968. Comparative analysis of antigenic constitution of *Ascaris lumbricoides* var. *hominis* and var. *suum*. *Revista dell'Instituto Sieroterapico Italiano*, **43**: 235–40.

Ishikura, H. & Namiki, M. (eds) 1989. *Gastric Anisakiasis in Japan*. Springer-Verlag, Tokyo.

Kennedy, T. J., Bruer, D. J., Marchionondo, A. A. & Williams, J. A. 1988. Prevalence of some parasites in major hog producing areas of the United States. *Agri-Practice*, **9**: 25–32.

Knapen, F. van. 1989. Control systems of sylvatic and domestic animal trichinellosis. *Wiadomosci Parazytologiczne*, **35**: 475–81.

Kurimoto, H. 1974. Morphological, biochemical and immunological studies on the differences between *Ascaris lumbricoides* Linnaeus, 1758 and *Ascaris suum* Goeze, 1782. *Japanese Journal of Parasitology*, **23**: 251–67.

Kwee, H. G. & Sautter, R. L. 1987. Anisakiasis. *American Family Physician*, **36**: 137–40.

Leslie, J. F., Cain, G. D., Meffe, G. K. & Vrijenhoek, R. C. 1982. Enzyme polymorphism in *Ascaris suum* (Nematoda). *Journal of Parasitology*, **68**: 576–87.

Lewis, R. & Shore, J. H. 1985. Anisakiasis in the United Kingdom. *Lancet*, **ii**: 1019.

Lýsek, H. 1967. On the host specificity of ascarids of human and pig origin. *Helminthologia*, **8**: 309–12.

Nadler, S. A. 1987. Biochemical and immunological systematics of some ascaridoid nematodes: genetic divergence between congeners. *Journal of Parasitology*, **73**: 811–16.

Nolan, M. & Reardon, L. 1939. Distribution of the ova of *Enterobius vermicularis* in household dust. *Journal of Parasitology*, **25**: 173–7.

Olson, L. J. 1976. Ocular toxocariasis in mice: distribution of larvae and lesions. *International Journal for Parasitology*, **6**: 247–51.

Oshima, T. 1987. Anisakiasis – is the sushi bar guilty? *Parasitology Today*, **3**: 44–8.

Ottesen, E. A. 1990. Activity of ivermectin in human parasitic infections other than onchocerciasis. *Current Opinion in Infectious Diseases*, **3**: 834–7.

Panter, H. C. 1969. Studies on host–parasite relationships: *Syphacia obvelata* in the mouse. *Journal of Parasitology*, **55**: 74–8.

Perry, R. N. 1989. Dormancy and hatching of nematode eggs. *Parasitology Today*, **5**: 377–83.

Sakanari, J. A. & McKerrow, J. H. 1989. Anisakis. *Clinical Microbiology Reviews*, **2**: 278–84.

Savigny, D. H. de, Voller, A. & Woodruff, A. W. 1978. Toxocariasis: serological diagnosis by enzyme immunoassay. *Journal of Clinical Pathology*, **32**: 284–8.

Schantz, P. M. 1989. *Toxocara* larva migrans now. *American Journal of Tropical Medicine and Hygiene*, **41**: 21–34.

Schulte, F. 1989. Life history of *Rhabditis* (*Pelodera*) *orbitalis*. *Proceedings of the Helminthological Society of Washington*, **56**: 1–7.

Singhvi, A. & Johnson, S. 1989. Feeding apparatus and niche specialization in nematode parasites of the house rat. *Indian Journal of Helminthology*, **41**: 71–4.

Smith, J. W. 1983. *Anisakis simplex* (Rudolphi, 1809, det. Krabbe, 1878) (Nematoda: Ascaridoidea): morphology and morphometry of larvae from euphausiids and fish, and a review of the life history and ecology. *Journal of Helminthology*, **57**: 205–24.

Smith, J. W. & Wootten, R. 1978. *Anisakis* and anisakiasis. *Advances in Parasitology*, **16**: 93–163.

Sohlenius, B. 1968. Influence of micro-organisms and temperature upon some rhabditid nematodes. *Pedobiologia*, **8**: 137–45.

Sprent, J. F. A. 1963*a*,*b*. The life history and development of *Amplicaecum robertsi*, an ascaridoid nematode of the carpet python (*Morelia spilotes variegatus*). I, II. *Parasitology*, **53**: 7–38; 321–7.

Sprent, J. F. A. 1963*c*. Visceral larva migrans. *Australian Journal of Science*, **25**: 344–54.

Taylor, M. R., Keane, C. T., O'Conner, P., Mulvihill, E. & Holland, C. 1988. The expanded spectrum of toxocaral disease. *Lancet*, **i**: 692–5.

Van Thiel, P. H. 1966. The final hosts of the herringworm *Anisakis marina*. *Tropical and Geographical Medicine*, **18**: 310–28.

Wade, S. E. & Georgi, J. R. 1987. Radiolabelling and autoradiographic tracing of *Toxocara canis* in male mice. *Journal of Parasitology*, **73**: 116–20.

WHO. 1988. *Weekly Epidemiological Record*, **47**: 362–3.

28

Phasmid Nematoda: Strongylida

The most striking diagnostic feature of this group is the possession by the male of a copulatory 'bursa', a posterior umbrella-like cuticular expansion, which is absent from the males of all the other orders of nematodes (but see p. 371). Many of the strongylid nematodes are of considerable economic importance, attacking mammals, including man and domestic animals. The eggs hatch outside the body and the free-living larvae, after the usual moults, reinfect via the skin or mouth, or, more rarely, by means of an intermediate host. In particular, two species, *Nippostrongylus brasiliensis* in rats, and *Heligmosomoides polygyrus* (= *Nematospiroides dubius*) in mice, serve as useful experimental organisms. Both species are readily maintained in the laboratory and have been widely used as models for fundamental studies in immunology and for the testing of anthelminthics.

28.1 Type example: *Nippostrongylus brasiliensis**, the rat hookworm

This nematode, originally known as *Heligmosomum muris* (Yokagawa, 1920, 1922), occurs naturally as an intestinal parasite of the wild rat *Epimys rattus norvegicus*, but it may be transmitted experimentally to laboratory rats, cotton rats, rabbits and hamsters, although stunted or abnormal development may occur in some of these hosts. The adults occur in the anterior region of the intestine, either in contact with the mucosa or partly embedded in it. All aspects of the biology of *N. brasiliensis* and the extensive literature has been reviewed in a valuable monograph by Kassai (1982). Earlier work has been reviewed by Taylor (1969).

*Often written as *braziliensis*.

28.1.1 General morphology

The morphology of the worm is shown in Fig. 28.1. The males measure 3–4 mm and the females 4–6 mm in the wild rat. In aberrant hosts, such as the cotton rat, growth is stunted and the reported size ranges are: males 1.7–3.0 mm, females 1.7–3.4 mm. The head is small and bears a cephalic expansion of the cuticle. The cuticle has transverse striations and ten prominent longitudinal ridges. The mouth and buccal cavity are both small; the subcuticular part of the circumoral area has the form of two small teeth, one on each side. The armed buccal cavity enables the nematode to take a firm hold on the mucosa, from which it draws blood; recently fed worms thus appear blood-red in colour. The excretory system is well developed and hides the anterior region containing the gonads. It consists of two elongate sacs, which open externally by an excretory pore just in front of the base of the pharynx. The nerve ring lies just posterior to the excretory pore. The intestinal cells contain a melanin-like pigment.

Male. The characteristic external feature is the bursa, which forms an umbrella-like expansion surrounding the cloaca. It consists of two large lateral lobes and a small dorsal lobe. The lateral lobes, which are each supported by fleshy rays comparable to the ribs of an umbrella, are asymmetrical, the right being larger than the left. The left-lobe rays are more divergent than those of the right. The terminology used in naming the rays is shown in Fig. 28.1. The dorsal ray, supporting the small dorsal lobe, terminates in four digitations.

The testis occupies much of the anterior half of the body; its beginning is difficult to see, being hidden by the large excretory sac. It passes into the usual male organs, vas deferens, seminal vesicle and ejaculatory duct; the distal end of the latter contains two yellow-

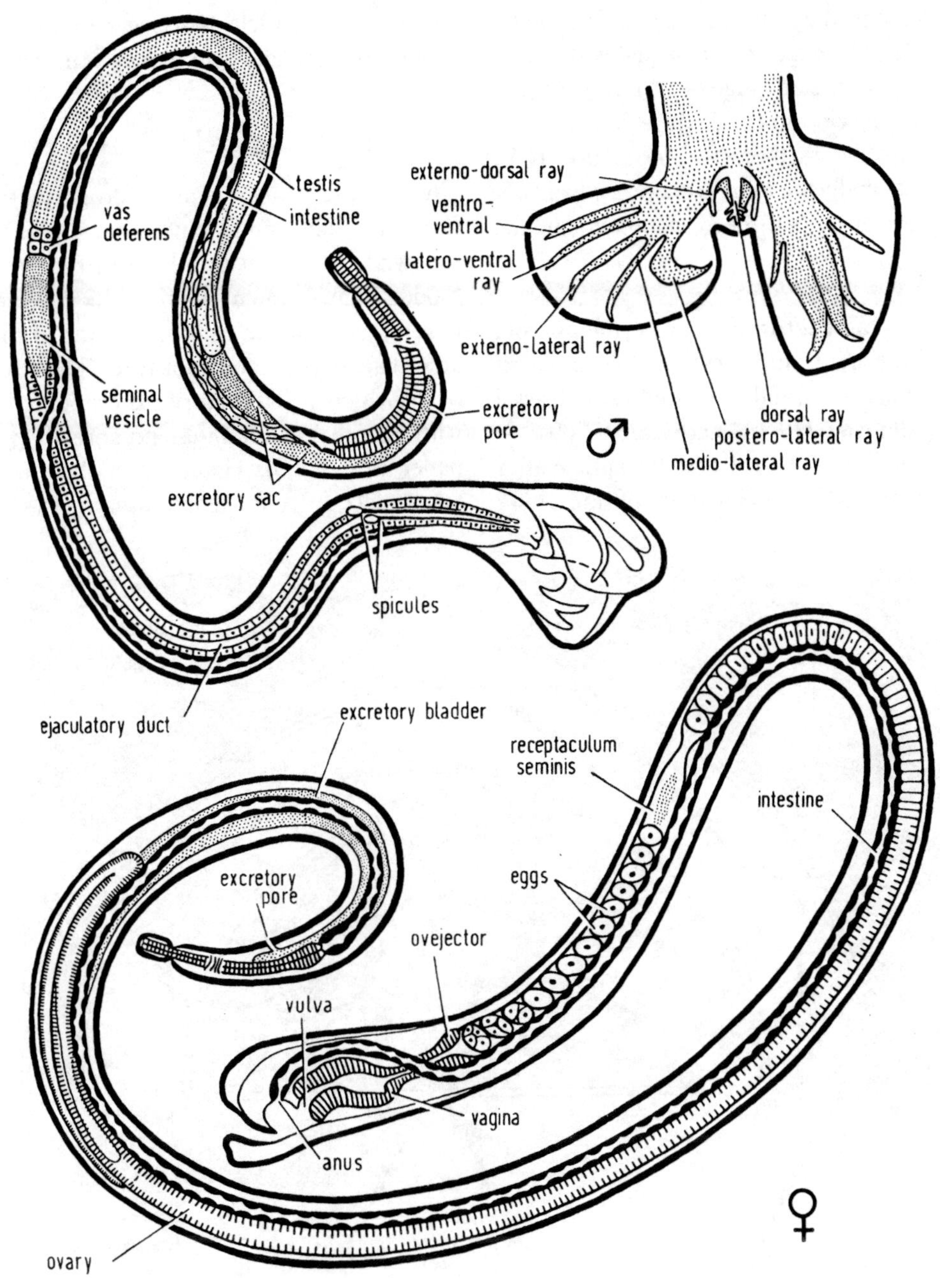

Fig. 28.1
Nippostrongylus brasiliensis: anatomy of male and female. (Adapted from Yokagawa, 1920.)

brown spicules. The distal end of the intestine joins the posterior end of the ejaculatory duct.

Female. The female is larger than the male. The vulva is posterior in position and opens on the ventral surface just in front of the anus. When contracted, the posterior body region is withdrawn to such an extent that the cuticle forms a sac surrounding the vulva and anus. The anterior end of the single ovary is bent and looped. The main part of the ovary is dorsal and is filled with a single row of developing oocytes gradually increasing in maturity as they approach the receptaculum seminis. The uterus, which occupies much of the posterior part of the body, is connected posteriorly with a muscular ovijector leading into the vagina. The latter is lined with cuticle. The eggs are ellipsoidal with a very thin shell, and an average size range of 58 μm × 33 μm. In the uterus, they develop to the 1–16 cell stage, and when passed in the faeces they may be at the 4–16 cells stage, or occasionally the morula stage. Schwartz & Alicata (1934) have described the development in the rat.

Cytology and ultrastructure. The cytology and ultrastructure of the adult and larva of *N. brasiliensis* have been the subject of a number of studies by Kassai (1982), Lee & Atkinson (1976), and Lee (1969). In particular, the formation of the collagenous cuticle after moulting has been studied. Formation of the outer layers of the cuticle is associated with an increase in rough endoplasmic reticulum in the hypodermis.

○ 28.1.2 Life cycle

The life cycle involves an external non-parasitic phase in an aerobic, humid environment at air temperatures followed by a parasitic phase in the anaerobic, warm-blooded environment provided by the intestine. Normal development of the eggs and larvae in the soil requires abundant oxygen and moisture. These conditions may be provided in laboratory experiments by mixing eggs with charcoal or alumina and spreading on moist filter paper (Fig. 28.2). Hatching of the rhabditiform larva takes place at room temperatures (18–22 °C) in about

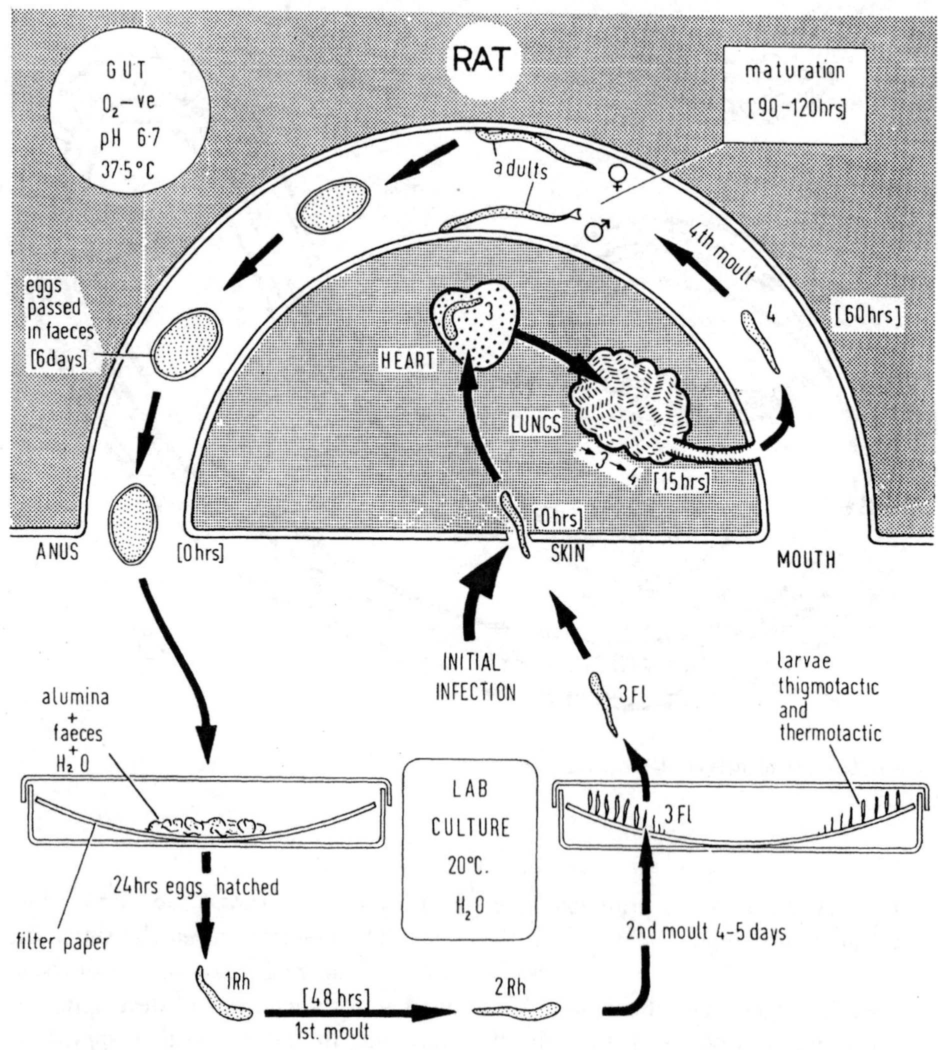

Fig. 28.2
Nippostrongylus brasiliensis: life cycle as carried out in the laboratory. (Original.)

18–24 hours. The first rhabditiform larva grows and moults to the second rhabditiform larva within about 48 hours. This in turn grows, moults within 4–5 days and gives rise to the third-stage filariform larva. This larva normally exsheaths while still on the surface of the charcoal culture, unlike many other strongylid larvae, which exsheath only when taken into the host gut. The cycle from egg to infective filariform larva normally requires about 4–5 days, but the time naturally varies with temperature.

The filariform larvae show tropisms which are characteristic of all strongylid larvae. They are markedly thermotactic and are rapidly stimulated into activity by the warmth of a nearby animal. They also show remarkable negative geotropism. This is clearly shown in laboratory faecal cultures. In such cultures, larvae migrate to the edge of the moist filter paper, where on reaching the highest points, they extend themselves into the air and wave back and forth slowly. In the natural state, they likewise climb to the top of soil particles and await a suitable host. Both these tropisms would tend to increase the chance of infecting a passing host.

Infection of the rat host is accomplished by placing larvae directly on the skin and allowing them to penetrate, or by hypodermic injection, or via the mouth, the last method being rather ineffective. Placing 1000–4000 larvae on the skin gives a heavy but not fatal infection. Hypodermically, a much smaller dose is effective. After entry into the blood stream, larvae are carried via the heart to the lungs within about 11–20 hours (usually about 15). Here the larvae feed on whole blood and undergo rapid growth and differentiation, culminating in the third moult to the fourth-stage larva. The latter are carried by ciliary action up the bronchi and trachea, finally passing down the pharynx to the intestine. The first larvae reach the intestine in about 41 hours and about 50 per cent arrive between 45 and 50 hours (Twohy, 1956). Here the fourth and final moult takes place resulting in fifth-stage male and female worms. Maturation is rapid and by the sixth day following infection, eggs appear in the faeces. Within the host gut, worms feed mainly on blood and tissue cells, but intestinal contents must also form part of the diet, since intestinal flagellates have been found within the gut lumen of the worms.

In the mouse, parasite recovery appears to be maximal on day 2 in the lungs and on day 4 in the intestine (Stadnyk *et al.*, 1990; Fig. 28.3). Development has been shown to be very similar in five strains of mice (Table 28.1). Rats infected with *N. brasiliensis* rapidly develop a substantial degree of immunity and towards the second week of infection exhibit a 'self-cure' reaction, in which most of the worms are expelled by day 14–15 (Fig. 28.4) and the rats resist a challenge infection (Ogilvie, 1969). The immunity is further discussed in Chapter 32.

○○○ 28.2 Order Strongylida

Important families:

Family **Trichostrongylidae** (trichostrongyles) (e.g. *Haemonchus contortus*)

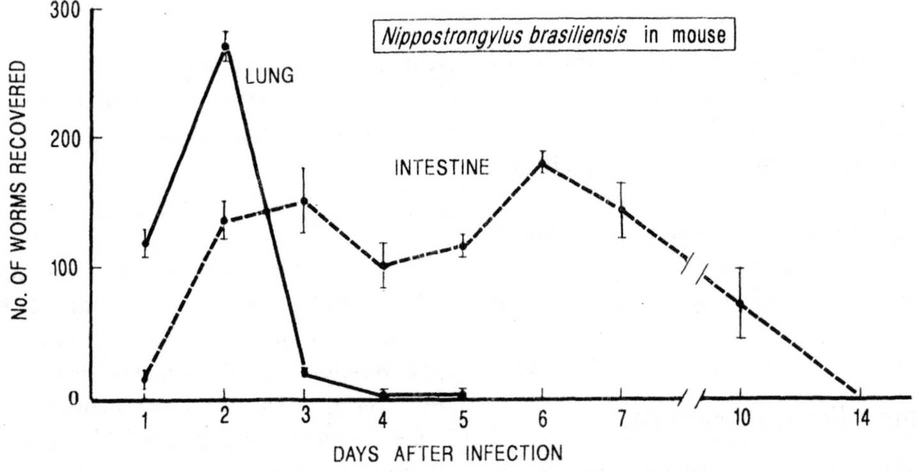

Fig. 28.3
Numbers of parasites recovered from the lungs and small intestine of CBA/J mice injected subcutaneously with 500 L_3 larvae of *Nippostrongylus brasiliensis*. Each point represents the mean of 5 mice. Bars indicate standard error of mean. (Modified from Stadnyk *et al.*, 1990.)

Table 28.1. *Recoveries of* Nippostrongylus brasiliensis *from different strains of mice after infection with 500* L_3 *larvae*[a]

	Percentage infective dose recovered					
		Intestine				
Mouse strain	Lungs day 2	day 4	day 6	day 8	day 10	day 15
CBA/J	38 ± 8[b]	30 ± 4	30 ± 6	37 ± 9	18 ± 9	0
C3H/HeJ	27 ± 7	28 ± 11	29 ± 5	17 ± 7	17 ± 6	1 ± 2
A/J	46 ± 1	23 ± 5	15 ± 2	29 ± 4	<1	<1
BALB/c	56 ± 4	21 ± 2	25 ± 6	43 ± 3	6 ± 6	1 ± 1
bg/bg	35 ± 6	22 ± 3	33 ± 5	19 ± 8	4 ± 6	2 ± 1

[a] n = 4–5 mice per day per strain.
[b] Mean ± s.e. of the mean.
Source: Stadnyk *et al.*, 1990.

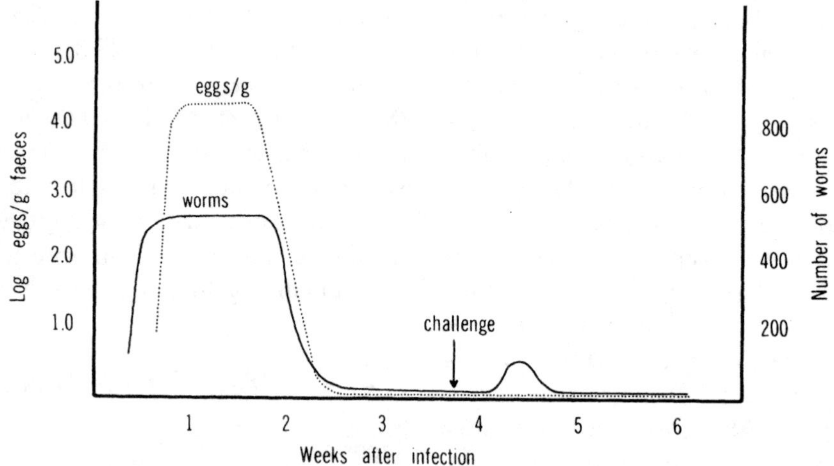

Fig. 28.4
Development of immunity to *Nippostrongylus brasiliensis* in rats. Eggs appear in the faeces on the sixth day and reach a peak of production on the tenth day. Thereafter, a 'self-cure' reaction sets in and most worms are expelled, but a low level of infection remains. A later challenge infection produces only a temporary low-level infection. See pp. 415 and 479. (Courtesy Dr R. F. Phillipson.)

Family **Heligmosomatidae** (e.g. *Nippostrongylus brasiliensis*)
Family **Strongylidae** (e.g. *Strongylus equinus*)
Family **Ancylostomatidae** (hookworms) (e.g. *Ancylostoma duodenale*)
Family **Metastrongylidae** (metastrongylids = 'lungworms') (e.g. *Metastrongylus* sp.)
Family **Syngamidae** (gapeworms) (e.g. *Syngamus trachea*).

○ 28.2.1 Family Trichostrongylidae

The trichostrongyles are small slender intestinal worms without leaf crowns (see Strongylidae, below) or cutting plates and with buccal capsules poorly developed or lacking. The family also contains a number of species of considerable veterinary importance, such as *Haemonchus contortus*; a limited number of these are considered below.

Haemonchus contortus

This species is a blood-sucking nematode which occurs in the fourth stomach (abomasum) of sheep and other ruminants, a highly acid environment with a pH of approximately 3.0. In general structure, it resembles *Nippostrongylus brasiliensis* and is often reddish or striped in colour owing to contained blood. Females are 18–30 mm long; males 10–20 mm. The ultrastructure has been examined by SEM and TEM by Gamble *et al.* (1989) and Lichtenfels *et al.* (1990).

Life cycle. Eggs are passed in the faeces and hatch in favourable climatic conditions. As in *Nippostrongylus*, three non-parasitic larval phases occur, but the third larva, after the second moult, retains its skin so that it is unable to feed, but survives at a low metabolic rate on its food reserves. The infective larva is unable to penetrate the skin and must be taken directly into the mouth; grazing animals ingest them accidentally with their food. On reaching the abomasum, it undergoes the third and fourth moults and reaches maturity. Many thousands of worms may occur in a single ruminant stomach, and it has been estimated that 4000 worms suck about 60 cm^3 of blood per day. Medium infections cause sheep and cattle to lose condition and heavy infections may cause death.

Haemonchus contortus will undergo limited development in rabbits, mice and guinea pigs and immunosuppressed jirds (Wagland *et al.*, 1989; Conder *et al.*, 1990). There is evidence that a poor diet greatly impairs host-protective immunity against haemonchosis (Roberts & Adams, 1990).

This species, like some 30 other species of gastrointestinal nematodes, exhibits a somewhat related phenomenon known as the 'spring rise' in which there is increased egg production in lactating ewes during early spring. This phenomenon, which appears to be related to the immune system of the host, is also discussed further in Chapter 32. Control of the gastroenteritis due to haemonchosis has been reviewed by Reinemeyer (1990).

Other Trichostrongyles

Farm animals are infected with a number of other Trichostrongylidae with essentially the same life cycle as *H. contortus*. They differ in their host preferences and the part of the alimentary canal parasitised. Their distribution, in a world-wide sense, is largely determined by the conditions under which the free-living larvae can develop and survive. The most studied species are probably *Trichostrongylus retortaeformis* (rabbits), *T. tenuis* (birds), *T. axei* (horses, sheep), *T. colubriformis* (goats, sheep), *Ostertagia ostertagi* (cattle). Fitzsimmons (1969) gives a valuable review of the pathogenesis of trichostrongyles.

○ 28.2.2 Family Heligmosomatidae

This family includes the rat 'hookworm', *Nippostrongylus brasiliensis* already described (p. 412) and the mouse 'hookworm' *Heligmosomoides polygyrus* (= *Nematospiroides dubius*?), both widely used as experimental organisms.

Heligmosomoides polygyrus (= *Nematospiroides dubius*?)

Taxonomy. The taxonomy of the genus is unusually complex and controversial and has been extensively reviewed (Behnke *et al.*, 1991; Durette-Desset, 1968; Durette-Desset *et al.*, 1972; Forrester, 1971; Forrester & Neilson, 1973; Quinnell *et al.*, 1991). So confusing has been the situation that it is likely that different laboratories may have been using different species or subspecies, referring to them as *Nematospiroides dubius* or *Heligmosomoides polygyrus*, after isolating them from or maintaining them in different hosts. It is now recognised that a species complex exists and that four sub-species can be distinguished morphologically and by their host range (Behnke *et al.*, 1991; Quinnell *et al.*, 1991) as follows: *H. polygyrus polygyrus* and *H. polygyrus corsicus* from Europe, and *H. p. bakeri* and *H. p. americanus* in North America. *H. p. polygyrus* is the normal parasite in the fieldmouse, *Apodemus sylvaticus*, in the UK, and *H. p. corsicus* is a similar worm from *Mus domesticus* in Corsica.

Experiments have shown that the wild strain of *H. p. polygyrus*, isolated from the fieldmouse, produces only low-level infections in laboratory mice. Conversely, established strains of *H. p. bakeri* in laboratory mice survive only for short periods in *A. sylvaticus* (Quinnell *et al.*, 1991). One of the most widely used strains of *H. polygyrus* is that isolated nearly 40 years ago from *Peromyscus maniculatus* in California by Ehrenford (1954) and now identified as *H. p. bakeri*. For the purpose of this general account, the terminology *Heligmosmoides polygyrus* is used here.

Life cycle. The population dynamics of *H. polygyrus* in the fieldmouse, *A. sylvaticus*, has been investigated and it has been shown that male mice had consistently higher burdens than female mice (Gregory *et al.*, 1990).

The adults live in the small intestine of mice and eggs appear in faeces nine to eleven days post-infection. Eggs may be hatched in charcoal or filter paper cultures using a technique similar to that for *N. brasiliensis* (Fig. 28.2), eggs hatching after five or six days at 25 °C (seven days at 22–24 °C). Infective larvae – which unlike *N. brasiliensis* cannot penetrate the skin – on oral ingestion migrate through the mucosa to the muscularis, where they encyst. After eight or nine days larvae return to the lumen, mature and mate. The life cycle thus takes about

13–16 days. Mice may be readily infected experimentally by allowing them to drink from a pipette containing about 400 infected larvae.

The rat is highly susceptible to *Heligmosomoides* during early post-natal life (one to three weeks) only, but rapidly becomes increasingly refractory and at five weeks is essentially resistant to infection. Both young and old hamsters are equally resistant. The immunology is discussed further in Chapter 32.

○ 28.2.3 Family Strongylidae

The typical features of this family are the well-developed buccal capsule, with a thickened ridge on its dorsal side, the *dorsal gutter*, which represents the opening of the dorsal oesophageal gland, and one or two rows of circum-oval extensions of the cuticle termed *leaf-crowns* or *corona radiata*. As in the Ancylostomidae, teeth are present. Many members of this family are pathogenic to domestic animals, and all have a life cycle in general similar to that of *Haemonchus contortus*, although the path followed by migrating larvae in some cases is still a matter of dispute.

One of the most interesting genera is *Strongylus*, which includes strongylids of horses sometimes termed 'red worms'. The early larval stages of some species develop in 'nodules' in the alimentary canal, from which the fourth-stage larvae escape to undergo complicated migrations; in some species, development occurs in the peritoneal cavity. The biology of *Strongylus vulgaris* in the horse has been reviewed in detail by Ogbourne & Duncan (1985).

○ 28.2.4 Family Ancylostomatidae

The hookworms which make up this family are chiefly distinguished from other strongylids by the possession of two ventro-lateral cutting plates often bearing teeth. The life cycle in general resembles that of *Nippostrongylus brasiliensis* (p. 412). There are a number of genera, the best known being *Ancylostoma* and *Necator*.

Hookworms of man

Hookworms must be classified as one of the most destructive of human helminth parasites, with estimates of some 900 million cases worldwide (Crompton, 1989). They insidiously undermine the health of their hosts, causing stunting of growth and general laziness often accompanied by acute mental distress. They occur in predictable areas where sanitary and environmental conditions favour the development of the eggs and larval infection. This is roughly throughout the tropical and sub-tropical world. The two major species attacking man are:

Ancylostoma duodenale: essentially a northern species occurring in the northern districts of China, Japan, India and SE Asia. Smaller foci occur in the USA, S America and the Caribbean Islands.

Necator americanus: this is the predominant species in the New World, but is also indigenous in Africa, India, SE Asia, China and some Pacific Islands. A less common species is *Ancylostoma ceylanicum*, normally found in Sri Lanka, SE Asia, the E Indies and the Philippines.

Literature. Valuable reviews on all aspects of hookworm disease are those of Anderson & Medley (1985), Crompton (1989), Gilles & Ball (1991), Goldsmid (1991), Miller (1979), Pritchard *et al.* (1990), Schad (1990, 1991), Schad & Warren (1990) and WHO (1987).

Morphology. The general morphology and details of the life cycle resemble those of *Nippostrongylus* but the worms are roughly twice the size (males 8–11 mm; females 10–13) and the vulva opening is almost one-third of the body length from the posterior end; the female has two ovaries. The conspicuous buccal cavity is armed with a pair of chitinous plates which either bear teeth (*Ancylostoma*) or have a sharp blade-like edge (*Necator*). The head is curved in both species, but is finer in necators than in the ancylostomes and more sharply bent. The bursae in the males of both species are well developed, that of *Necator* being distinguished from *Ancylostoma* by the split dorsal ray and close arrangement of the lateral rays (Fig. 28.5). In the field, however, morphological identification of species is difficult and attempts to develop immunological or biochemical criteria are now in progress (Pritchard *et al.*, 1990).

Life cycle. The life cycle closely follows that of *Nippostrongylus*. The adults live in the small intestine, firmly attached to the mucous membrane, and feed on blood and tissue. Large numbers of eggs are produced (up to 20 000 per day in the ancylostomes) and are passed out with the host faeces. In favourable conditions of moisture, temperature and oxygen they then hatch as rhabditiform larvae which feed on bacteria. After two moults, these larvae become infective and climb to the

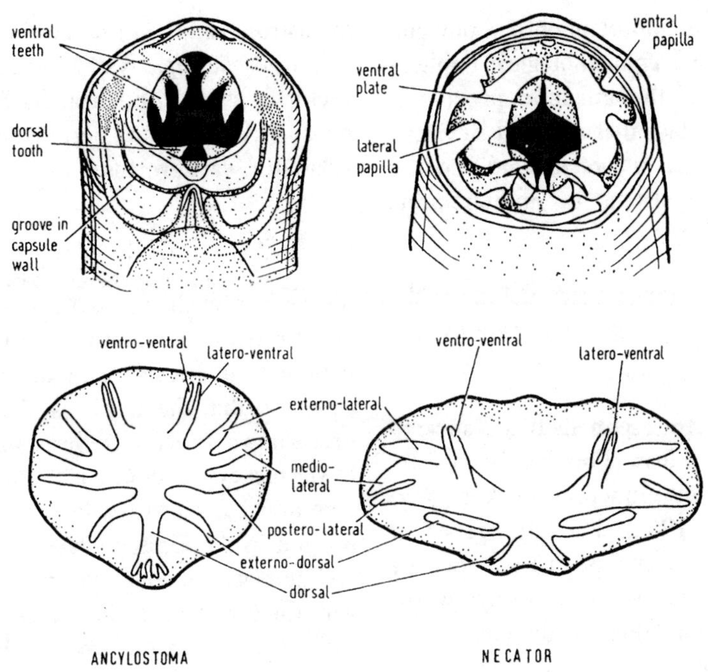

Fig. 28.5
Comparison of the buccal cavities and bursae of *Ancylostoma* and *Necator*. (Adapted from Brown, 1950; Chandler, 1955.)

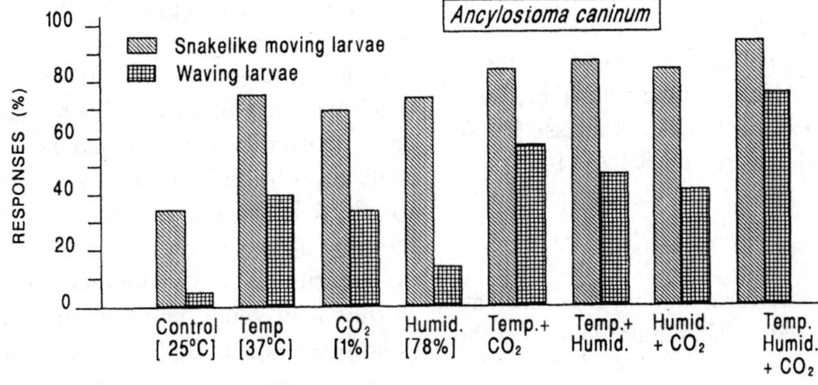

Fig. 28.6
Ancylostoma caninum: effect of temperature, CO_2 and humidity on the behavioural responses of the infective L_3 larvae. Control humidity 33 per cent; 100 replicates; 5–6 larvae each. (Modified from Granzer & Haas, 1991.)

highest part of the moist ground awaiting a barefooted victim. Like *Nippostrongylus*, they exhibit positive thermotaxis, moderate heat or touch stimulating them into activity.

These responses, which must greatly improve the chances of infecting hosts, have been examined experimentally in the dog hookworm *Ancylostoma caninum* (Granzer & Haas, 1991). Snake-like movements were stimulated by warmth and vibrations of the substratum; waving behaviour was stimulated by heat and CO_2 content, and creeping directions by heat gradients (Fig. 28.6).

On reaching the blood stream, a migration similar to that of *Nippostrongylus* occurs. Larvae swallowed may not show this migration.

Means of infection. It was originally thought that larvae normally infected humans by skin penetration, and whereas obligatory dependence on skin penetration holds for *N. americanus*, it is now realised that *A. duodenale* can successfully also infect man by oral, transmammary and (probably) transplacental routes

(Crompton, 1989). This is an important observation and could have major control and epidemiological implications. An understanding of the infection process is further complicated by the fact that arrested development (hypobiosis) can occur, in which condition the parasites become highly resistant to anthelminthic chemotherapy (Schad, 1990).

Chemotherapy. Anthelminthics used include mebendazole, thiabendazole and pyrantel embonate (Schad & Banwell, 1984).

Pathology. Hookworms are essentially blood suckers and can cause severe blood loss. This has been estimated by the use of ^{51}Cr by numerous workers (e.g. Neilson, 1969; Fraga *et al.*, 1965). The latter reported blood losses of 1.35 and 16.7 cm^3 per patient per 24 hours with 692 and 335 worms, respectively; worm burden and blood loss cannot thus be directly correlated. A small blood loss in a well-nourished host generally produces little or no effect, but can produce serious effects – including hookworm anaemia – in an undernourished host. Thus nutritional considerations must play a marked part in the evaluation of this organism as a pathogen. The depletion of blood must be made good by feeding iron in a suitable form and also assuring an adequate diet of protein. 'Hookworm anaemia' has been reviewed by Miller (1979) and Roche & Layrisse (1966) and the general clinical features by Schad (1991) and Gilles & Ball (1991).

Hookworms of dogs and cats

Several species of hookworms occur in dogs and cats and have been widely used as laboratory models for the study of ancylostomiasis.

Ancylostoma caninum is a common parasite of dogs but does not grow well in cats. The larvae of this and other animal species can penetrate human skin, giving rise to *cutaneous larva migrans* (Miller, 1979).

○ 28.2.5 Family Metastrongylidae

This family includes the so-called 'lungworms', slender worms which inhabit the respiratory tract of sheep, goats, cattle, horses and pigs. In contrast to the hookworms, the first larva, not the egg, is usually shed and after two moults becomes infective. Infection is by ingestion. The best-known genus is probably *Dictyocaulus*, which has a simple direct life cycle. Other metastrongylids require slugs or snails as carrier hosts. *Metastrongylus apri* is of biological interest as being the vector of swine influenza. Its life cycle is indirect, and it can survive for long periods (up to three years) encapsulated in earthworms.

○ 28.2.6 Family Syngamidae

These are bright red worms which live in the trachea and bronchi of mammals and birds, sometimes with pathological effects. The male and female remain *in copula* so that the whole is Y-like in appearance. The best-known species is *Syngamus trachea*, a parasite of fowl, turkeys and other domestic birds. It is also a common parasite of wild birds in European countries but not in the United States.

The eggs are coughed up, swallowed and passed out with the host faeces. The usual early larval stages occur while still within the egg, so that when hatching does occur, the infective third larva has developed. This may either be eaten directly by a bird host or by an invertebrate 'transport' host in which it encysts but undergoes no further development.

The earthworm is probably the major transport host, especially *Eisenia* spp. (Long *et al.*, 1987). Transport hosts not only protect larvae from climatic conditions to which they may be susceptible but also serve to distribute them over a wider area than the rather inactive larvae could cover by their own efforts. When ingested by suitable bird hosts, the infective larvae reach the lungs, probably by way of the circulatory system, and develop to maturity. The development and pathogenesis of *S. trachea* in young chickens and pheasants has been described in detail by Fernando *et al.* (1971).

The taxonomy of the family Syngamidae presents considerable difficulties on account of the large number of species and high intra-specific variability; it has been reviewed by Lengy (1969).

○○○ References

(References of general interest have been included in this list, in addition to those quoted in the text.)

Anderson, R. M. & Medley, G. F. 1985. Community control of helminth infections of man by mass and selective chemotherapy. *Parasitology*, **90:** 629–60.

Baker, N. F. 1962. The nature and etiology of the leukocytic response of Webster mice infected with *Nematospiroides dubius*. *Journal of Parasitology*, **48**: 438–41.

Baylis, H. A. 1926. On a trichostrongylid nematode from a wood-mouse (*Apodemus sylvaticus*). *Annals and Magazine of Natural History*, **18**: 455–64.

Behnke, J. M., Keymer, A. & Lewis, J. W. 1991. *Heligmosomoides polygyrus* or *Nematospiroides dubius*? *Parasitology Today*, **7**: 177–9.

Conder, G. A., Jen, L. W., Marbury, K. S., Johnson, S. S., Guimond, P. M., Thomas, E. M. & Lee, B. L. 1990. A novel anthelminthic model utilizing jirds, *Meriones unguiculatus*, infected with *Haemonchus contortus*. *Journal of Parasitology*, **76**: 168–70.

Crompton, D. W. T. 1989. Hookworm disease: current status and new directions. *Parasitology Today*, **5**: 1–2.

Durette-Desset, M.-C. 1968. Identification des strongyles des mulots et campagnols décrits par Dujardin. *Annales de Parasitologie Humaine et Comparée*, **43**: 387–404.

Durette-Desset, M.-C., Kinsella, J. M. & Forrester, D. J. 1972. Arguments en faveur de la double origine de nematodes néartiques de genre *Heligmosomoides* Hall 1916. *Annales de Parasitologie Humaine et Comparée*, **47**: 365–82.

Ehrenford, F. A. 1954. The life cycle of *Nematospiroides dubius* Baylis (Nematoda: Heligmosomoidae). *Journal of Parasitology*, **40**: 480–1.

Fernando, M. A., Stockdale, P. H. G. & Remmier, O. 1971. The route of migration, development, and pathogenesis of *Syngamus trachea* (Montagu, 1811) Chapin, 1925, in pheasants. *Journal of Parasitology*, **57**: 107–16.

Fitzsimmons, W. M. 1969. Pathogenesis of the trichostrongyles. *Helminthological Abstracts*, **38**: 139–90.

Forrester, D. J. 1971. *Heligmosomoides polygyrus* (= *Nematospiroides dubius*) from wild rodents of Northern California: natural infections, host specificity, and strain characteristics. *Journal of Parasitology*, **57**: 498–503.

Forrester, D. J. & Neilson, J. T. McL. 1973. Comparative infectivity of *Heligmosomoides polygyrus* (= *Nematospiroides dubius*) in three species of *Peromyscus*. *Journal of Parasitology*, **59**: 251–5.

Fraga de Azevedo, J. *et al.* 1965. A patogenicidade dos ancilostomideos estudada pelos radioisotopos. I. A perda de sangue peles fezes appreciada pelo ^{51}Cr. *Anais do Instituto de Medicina Tropical Lisbon*, **22**: 15–23.

Gamble, H. R., Lichtenfels, J. R. & Purchell, J. P. 1989. Light and scanning electron microscopy of the ecdysis of *Haemonchus contortus* infective larvae. *Journal of Parasitology*, **75**: 303–7.

Gilles, H. M. & Ball, P. A. J. (eds) 1991. *Hookworm Infections*. Elsevier Science Publishers, Amsterdam, The Netherlands.

Goldsmid, J. M. 1991. The African hookworm problem: an overview. In: *Parasitic Helminths and Zoonoses in Africa* (ed. C. N. L. MacPherson & P. S. Craig), pp. 101–37. Unwin Hyman, London. ISBN 0 044 45565 8.

Granzer, M. & Haas, W. 1991. Host-finding and host recognition of infective *Ancylostoma caninum* larvae. *International Journal for Parasitology*, **21**: 429–40.

Gregory, R. D., Keymer, A. E. & Clarke, J. R. 1990. Genetics, sex and exposure: the ecology of *Heligmosomoides polygyrus* (Nematoda) in the wood mouse. *Journal of Animal Ecology*, **59**: 363–78.

Kassai, T. 1982. *Handbook of Nippostrongylus brasiliensis (Nematoda)*. Commonwealth Agricultural Bureaux, Slough, UK and Akademémiai Kiadó, Budapest. ISBN 9 630 52976 9.

Lee, D. L. 1969. *Nippostrongylus brasiliensis*: some aspects of the fine structure and biology of the infective larva and the adult. *Symposia of the British Society for Parasitology*, **7**: 3–16.

Lee, D. L. & Atkinson, H. J. 1976. *Physiology of Nematodes*. 2nd Edition. Macmillan Press, London.

Lengy, J. 1969. Notes on the classification of Syngamidae (Nematoda) with new data on some of the species. *Israel Journal of Zoology*, **18**: 9–23.

Lichtenfels, J. R., Gamble, H. R. & Purchell, J. P. 1990. Scanning electron microscopy of the sheathed infective larva and parasitic third-stage larva of *Haemonchus contortus*. *Journal of Parasitology*, **76**: 248–53.

Long, P. L., Current, W. L. & Noblet, G. P. 1987. Parasites of the Christmas Turkey. *Parasitology Today*, **3**: 361–6.

Miller, T. A. 1979. Hookworm infection in man. *Advances in Parasitology*, **17**: 315–84.

Neilson, J. T. McL. 1969. Radioisotope studies on the blood loss associated with *Nippostrongylus brasiliensis* infection in rats. *Parasitology*, **59**: 123–7.

Ogbourne, C. P. & Duncan, J. L. 1985. *Strongylus vulgaris in the Horse: its Biology and Veterinary Importance*. Commonwealth Agricultural Bureaux, Slough, UK. ISBN 0 851 98547 5.

Ogilvie, B. M. 1969. Immunity to *Nippostrongylus brasiliensis*. *Symposia of the British Society for Parasitology*, **7**: 31–40.

Pritchard, D. I., McKean, P. G. & Schad, G. A. 1990. An immunological and biochemical comparison of hookworm species. *Parasitology Today*, **6**: 154–6.

Quinnell, R. J., Behnke, J. M. & Keymer, A. E. 1991. Host specificity of and cross-immunity between two strains of *Heligmosomoides polygyrus*. *Parasitology*, **102**: 419–27.

Reinemeyer, C. R. 1990. Prevention of parasitic gastroenteritis in dairy replacement heifers. *Compendium on Continuing Education for the Practicing Veterinarian*, **12**: 761–6.

Roberts, J. A. & Adams, D. B. 1990. The effect of level of nutrition on the development of resistance to *Haemonchus contortus* in sheep. *Australian Veterinary Journal*, **67**: 89–91.

Roche, M. & Layrisse, M. 1966. The nature and causes of 'hookworm anemia'. *American Journal of Tropical Medicine and Hygiene*, **15**: 1031–102.

Schad, G. A. 1990. Hypobiosis and related phenomena in hookworms. In: *Hookworm Disease: Current Status and New Directions* (ed. G. A. Schad & K. S. Warren), pp. 71–88. Taylor & Francis, London.

Schad, G. A. 1991. Presidential Address. Hooked on hookworms. *Journal of Parasitology*, **77**: 177–86.

Schad, G. A. & Banwell, J. G. 1984. Hookworms. In:

Tropical and Geographical Medicine (ed. K. S. Warren & A. A. F. Mahmoud), pp. 359–72. McGraw-Hill, New York.

Schad, G. A. & Warren, K. S. (eds) 1990. *Hookworm Disease: Current Status and New Directions*. Taylor & Francis, London.

Schwartz, B. & Alicata, J. 1934. The development of the Trichostrongyle, *Nippostrongylus muris* in rats following ingestion of larvae. *Journal of the Washington Academy of Science*, **24**: 334–8.

Stadnyk, A. W., McElroy, P. J., Gauldie, J. & Befus, A. D. 1990. Characterization of *Nippostrongylus brasiliensis* infection in different strains of mice. *Journal of Parasitology*, **76**: 377–82.

Taylor, A. E. R. (ed.) 1969. *Nippostrongylus* and *Toxoplasma. Symposia of the British Society for Parasitology*, vol. 7. Blackwell Scientific Publication, Oxford.

Twohy, D. W. 1956. The early migration and growth of *Nippostrongylus muris* in the rat. *American Journal of Hygiene*, **63**: 165–8.

Wagland, B. M., Abeydeera, L. R., Rothwell, T. L. W. & Ouwerk, D. 1989. Experimental *Haemonchus contortus* in guinea pigs. *International Journal for Parasitology*, **19**: 301–5.

Weinstein, P. P., Newton, W. L., Sawyer, T. K. & Sommerville, R. I. 1969. *Nematospiroides dubius*: development and passage in the germfree mouse, and a comparative study of the free-living stages in germfree feces and convential cultures. *Transactions of the American Microscopical Society*, **88**: 95–117.

Wescott, R. B. 1968. Experimental *Nematospiroides dubius* infection in germfree and convential mice. *Experimental Parasitology*, **22**: 245–9.

Wharton, D. A. & Sommerville, R. I. 1984. The structure of the excretory system of the infective larva of *Haemonchus contortus*. *International Journal for Parasitology*, **14**: 591–600.

WHO. 1987. *Technical Report Series*, No. 749.

Yokagawa, S. 1920. A new nematode from the rat. *Journal of Parasitology*, **7**: 29–33.

Yokagawa, S. 1922. Development of *Heligmosomum* (= *Nippostrongylus*) *muris* Yokagawa, a nematode from the intestine of the wild rat. *Parasitology*, **14**: 127–66.

○ ○ ○ ○ ○ ○ ○ ○ ○ ○ ○ ○ ○ ○ ○ ○ ○ ○ ○ 29 ○ ○ ○ ○ ○ ○ ○ ○ ○ ○ ○ ○ ○ ○ ○ ○ ○ ○

Phasmid Nematoda: Spirurida

Members of this order of nematodes require intermediate hosts for the completion of their life cycle. The pharynx is cylindroid with an anterior muscular portion and a posterior glandular portion (Fig. 25.6); the males have well-developed alae and spirally coiled tails. There are two suborders, Spirurina and Camallanina, of which only the major superfamilies are considered here. For a detailed classification, see Anderson *et al.* (1974).

○○○ 29.1 Suborder Spirurina: superfamily Filarioidea

The filariae represent a group which has successfully invaded the blood stream, connective tissue or serous cavities of vertebrates; they are long, thread-like forms. Many species are of medical or veterinary importance, attacking man and domestic animals to which they are transmitted by haematophagous arthropods, often mosquitoes. The sexually mature females release swarms of pre-larval stages, termed *microfilariae*, into the peripheral blood. Most species are ovoviviparous and some have 'sheathed' microfilaria; the sheath is probably the ruptured 'shell' still attached to the worm. Some filariae of birds are oviparous.

○ 29.1.1 Type example: *Litomosoides carinii*

definitive host:	the cotton rat, *Sigmodon hispidus*
intermediate host:	a mite, *Ornithonyssus bacoti*
location:	pleural cavity

The life cycle (Fig. 29.1) and biology of this worm have been much studied; the very extensive literature has been comprehensively reviewed by Bertram (1966). Early experimental and descriptive work includes that of McDonald & Scott (1953) and Scott *et al.* (1951).

Natural occurrence. This filariid occurs as a common natural parasite of various subspecies of the cotton rat, *Sigmodon hispidus*, a creature of open grasslands in the USA and S America. In Florida an incidence of 43 per cent has been reported from one area. Other natural hosts include *Sciurus* (squirrels) and *Neotoma* (rats) in Brazil, *Mus* in Venezuela and *Holochilus* in Argentina.

Laboratory hosts. Compared with cotton rats, other species are rather poor hosts. Laboratory rats, mice, hamsters, gerbils and field mice are all susceptible to some degree.

Site of infection. In natural infections, the adult worms occur massed together in the pleural cavity, but occasionally invade the peritoneal space, especially in laboratory infections established by subcutaneous injection.

Morphology. The worms are thin and long; females 60–120 mm; males 20–25 mm, but the size is influenced by many factors. The buccal cavity is narrow and the male tail lacks alae or papillae. In the female, there is a long ovijector, with a bulbous enlargement near the vulva. The histology and ultrastructure of the adult and the third-stage larva have been described in detail by Franz & Andrews (1986*a*,*b*) and Storey & Ogbogu (1991).

Life cycle (Fig. 29.1). The mature females ovoviviparously discharge slender *microfilariae* into the pleural cavity. These are essentially pre-larval stages, which will not undergo further development until taken into the haemocoele of the intermediate host. From the pleural cavity, the microfilariae (which are sheathed) migrate to the blood stream by a variety of routes, the most

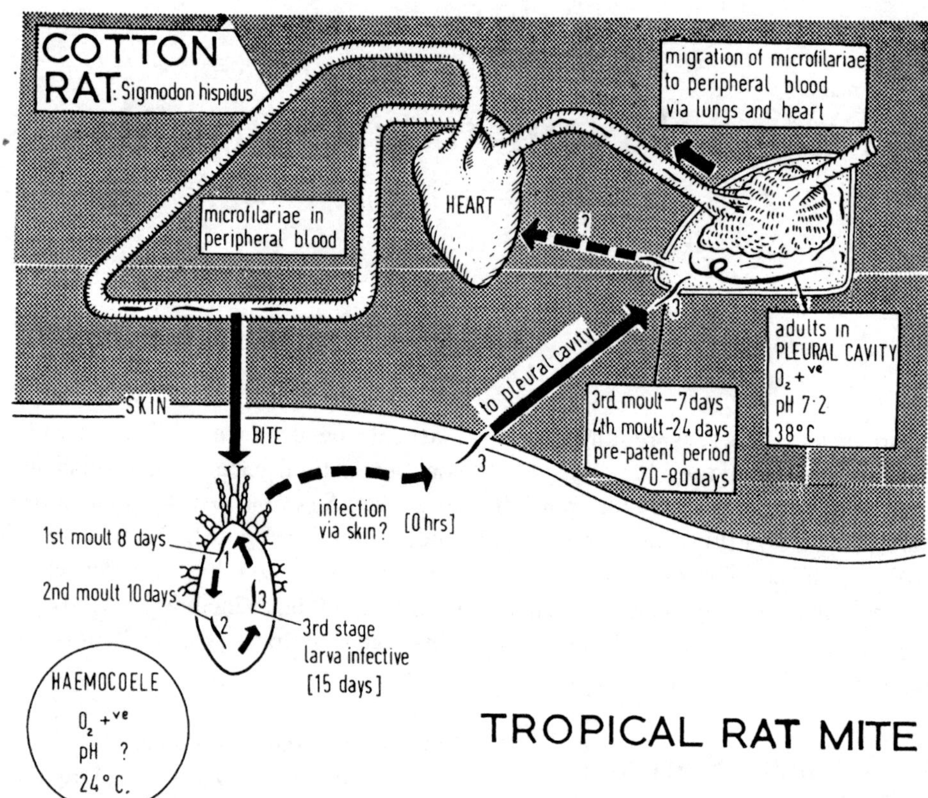

Fig. 29.1
Life cycle of *Litomosoides carinii*, a rodent filariid readily maintained in the laboratory; *Ornithonyssus bacoti* is the usual mite intermediate host. The times given for the various phases are minimal and considerable variations can occur. (Original.)

common being via the lungs. They can, however, burrow between and into muscle fibres and may be found in small numbers in the cardiac muscle fibres of the ventricles.

The intermediate host is the tropical rat mite, *Ornithonyssus bacoti*, in which microfilariae develop to infective third-stage larvae. The course of development in the mite has been described in detail by Renz & Wenk (1981) and Mössinger & Wenk (1986). When infected blood is taken up by the mite, the microfilariae pass to the gut and then migrate to the salivary glands, fat bodies and other tissues where they develop intracellularly (Fig. 29.2). At 27 °C, development from ingested microfilariae to the infective third-stage larva (L_3) is accomplished within 13 days, the two moults taking place at 6–7 and 10–12 days, respectively. At 24 °C, developmental times are slightly longer (Fig. 29.1).

The mode of infection of the rat is uncertain but it is probably through the skin rather than by ingestion of mites. Within the rat blood, an infective larva migrates to the pleural cavity and there continues its development for about one week before moulting to the fourth-stage larva. This larva undergoes considerable growth during the next 17 days, taking on the main characteristics of the male or female worm, with length ranges of approximately 6.4 ± 0.5 and 8.7 ± 0.2 mm, respectively. At the end of this period (i.e. about 23–24 days after the initial infection of the rat) the final moult takes place and growth to the adult stages commences. The worms become sexually mature within 70–80 days and microfilariae may then be detected in the peripheral blood.

Up to 15 000 microfilariae may be produced and an infection of 78 females may produce a microfilaria population of one million to several million (Bertram, 1966). The normal peak of microfilarial production is between the seventeenth and twentieth week after infection and the adults may live for approximately 60 weeks or more. *L. carinii*, unlike some of the other filarial worms, does not exhibit microfilarial periodicity.

L. carinii was originally used as a model for testing potential new filaricides, but because its metabolism appears to differ from some other filariids its use is now no longer recommended by WHO (Storey & Ogbogu, 1991).

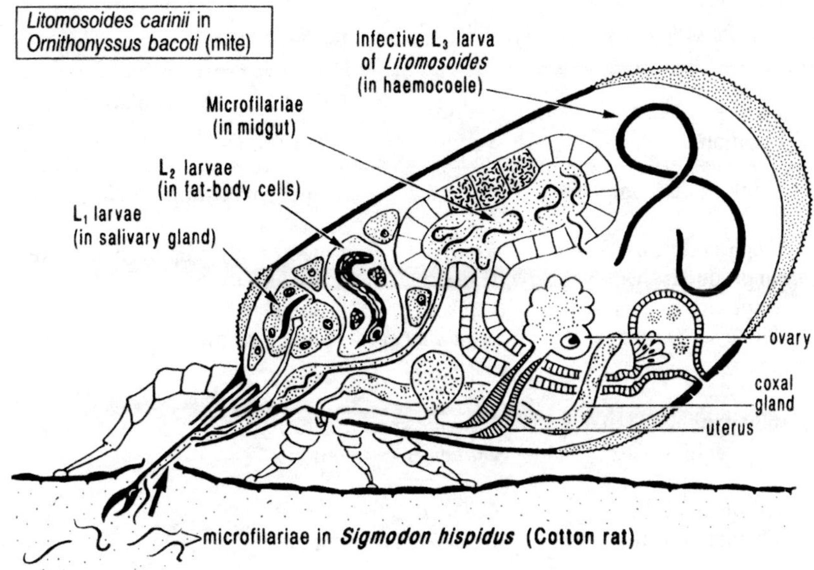

Fig. 29.2
Litomosoides carinii: ingestion and development of microfilariae in the mite *Ornithonyssus bacoti* (schematic). (Modified from Renz & Wenk, 1981.)

○ 29.1.2 Filariae in dogs: *Dirofilaria immitis*

This species, known as the 'heartworm', occurs in carnivores in the right ventricle and adjacent blood vessels including the pulmonary artery, right atrium and venae cavia and occasionally in ectopic sites. Although it is best known in dogs, where it causes 'heartworm disease' or *dirofilariasis*, it has a wide host range including cats, wolves, foxes and other carnivores. Its taxonomy, life cycle, epidemiology, biochemistry, pathology and clinical treatment have been comprehensively reviewed (Boreham & Atwell, 1988; Euzéby, 1990; Knight, 1987).

D. immitis was long considered to be a parasite limited to the tropics and subtropics, but its distribution is now recognised to be worldwide and is known to be enzooic in Canada and the USA where high prevalences in dogs, wolves and foxes have been reported from some areas (Boreham & Atwell, 1988). Some 60 species of mosquito have been recorded as vectors of this species, and – since many of these attack man – it is not surprising to find that man is occasionally infected, some 100 cases being reported in the literature, the heart and lung being mainly implicated (Boreham & Atwell, 1988).

Life cycle. Many species of mosquito can act as intermediate hosts. The general pattern of the life cycle resembles that of *Litomosoides*. The pre-patent period is about 32 weeks and when microfilariae become patent they show a partial periodicity. They reach a maximum and a minimum concentration at 18.00 and 06.00 hours, respectively (Fig. 29.9).

○ 29.1.3 Human filariasis

Eight species of filariae (Table 29.1) parasitise man. These occur in a wide range of tissue habitats: lymph glands, deep connective tissue, subcutaneous tissues or mesenteries. Habitats of this kind lend themselves particularly to inflammatory reactions, and this is a typical symptom in most human filarial infections. In some cases infections give rise to revolting fleshy deformities (Fig. 29.3) which are collectively known as *elephantiasis*. This is partly due to the inflammation of the walls of the lymphatics and the consequent hyperplasia, and partly due to mechanical blockage by the worms. Many species of filaria exhibit marked periodicity (Fig. 29.9).

It has been estimated that about one billion people in tropical and subtropical countries are exposed to infection risk and that at least 200 million are infected with filariasis (Sasa, 1976). Three of the species listed in Table 29.1 are primarily responsible for most of the cases of human filariasis: *Wuchereria bancrofti* and *Brugia malayi*, which are essentially lymphatic parasites, and *Onchocerca volvulus*, largely a subcutaneous parasite.

Table 29.1. *Characteristics of the common species of filariae of man and their microfilariae*

Species	Location	Vector	Microfilariae		Distribution
			Periodicity	Sheath	
Wuchereria bancrofti	lymph nodes, lymphatics	*Culicidae*	nocturnal, subperiodic	yes	Tropics (global)
Brugia malayi	lymph nodes, lymphatics	*Culicidae*	nocturnal, subperiodic	yes	SE Asia
Brugia timori	lymph nodes, lymphatics	*Culicidae*	nocturnal	yes	Indonesia
Brugia pahangi[a]	lymph nodes, lymphatics	*Culicidae*	nocturnal?	yes	Malaysia
Onchocerca volvulus	skin, eye, lymphatics	*Simulium* spp.	none or little	no	Africa, central and S America
Loa loa	subcutaneous tissue, eye	*Chrysops* spp.	diurnal	yes	Africa
Mansonella perstans	body cavities, tissues	*Culicoides* spp.	none	no	Africa, S America
Mansonella ozzardi	peritoneal cavity	*Culicoides* spp.	none	no	S America
Dipetalonema streptocerca	subcutaneous tissue	*Culicoides* spp.	none	no	Africa

[a]Normally a parasite of cats; man has been infected experimentally (Edeson *et al.*, 1960) and it may be a natural zoonotic species (Palmieri *et al.*, 1985).

The extensive literature on various aspects of filariasis has been the subject of numerous reviews of which the following are representative: Evered & Clark (1987), Greene (1990), Hillyer (1990), Mak (1983), Muller & Horsburgh (1987), Nelson (1990, 1991), Nanduri & Kazura (1989), Ottesen (1990), Southgate (1984), Sucharit & Supavej (1987), WHO (1987*a*,*b*, 1989*a*,*b*).

Wuchereria bancrofti

This species causes *bancroftian filariasis*, resulting in *elephantiasis* in man, causing revolting swellings often in the legs or genital system (Fig. 29.3). The adult worm occurs in tightly coiled nodular masses in the major lymphatic ducts.

The distribution is limited to the tropical and subtropical countries, chiefly in Asia, Africa, America and the Pacific, with a long intermediate-host season and high humidity. It also occurs in Australia and the Mediterranean area between 40°N and 30°S.

There is evidence that subspecies and strains of *Wuchereria* exist, but the taxonomy is difficult, especially when based on microfilariae; this problem is discussed further on p. 430.

Fig. 29.3
Advanced elephantiasis, due to *Wuchereria bancrofti*, in man on Society Islands. (Based on a photograph by Kessel, 1966.)

Life cycle. This is similar to that of the rat filariid *Litomosoides carinii* in most respects except that the arthropod hosts are various species of mosquito (see below) and that the appearance of the microfilariae in the peripheral blood exhibits nocturnal periodicity. Larvae reach their maximum concentration in the blood about midnight (Fig. 29.9) whereas during the daytime the embryos are concentrated in the lungs. Like the larvae of *Litomosoides*, the microfilariae (Fig. 29.6) are ensheathed in their egg membrane as are the majority of filariae of man showing periodicity. On ingestion by the mosquito host, the sheath is lost and the microfilariae migrate rapidly to the thoracic muscles where further growth and differentiation take place. In the muscles they become characteristically sausage-shaped. The infective stage is reached after the usual two moults, the process of maturation requiring some two weeks for completion. The infective larvae migrate into the proboscis and escape on to the skin and penetrate through the bite wound during the short period that a mosquito is biting. Maturation to a sexually mature worm requires about nine months. In the Polynesian Islands and the Philippines there exists a variety, sometimes known as the *pacifica* variety or *W. bancrofti* Pacific, with a different periodicity pattern (Fig. 29.9).

Man appears to be the only host for this species and – as there are no known reservoir hosts or laboratory hosts – research has been greatly restricted. Recently, however, transmission to the leaf monkey, *Presbytis metalophos*, of Malaysia has been achieved and this should considerably extend opportunities for laboratory research (Sucharit *et al.*, 1982).

Vectors. The most important vectors of *W. bancrofti* are members of the *Culex* (*C.*) *pipiens* 'complex' in urban areas and species of *Anopheles*, *Aedes* and (more rarely) *Mansonia*.

Control. The World Health Organisation (WHO, 1987*a*) has published a valuable practical manual on all aspects of filariasis control. Control depends largely on the elimination of mosquito breeding sites, which in bancroftian filariasis involves urban mosquitoes, a situation reviewed by Service (1990) and Southgate (1984). Mass chemotherapy has long involved the use of diethylcarbamazine (DEC) (Goodwin, 1984) which can be mixed with salt (Fan, 1990). Recent results suggests that ivermectin, which is very effective against the microfilariae, can replace DEC (Mak *et al.*, 1991; Roux *et al.*, 1989). Unfortunately, it does not kill the adult worms; this fact severely limits its value in the eradication of the disease (Campbell, 1981/82; Maren, 1990). Several mathematical models relating to the transmission dynamics have been produced (Rochet, 1990; Das *et al.*, 1990).

Pathology. This has been reviewed by Ottesen (1990) and Nanduri & Kazura (1989).

Brugia malayi (Table 29.1)

A species which resembles *Wuchereria bancrofti* and also parasitises the lymph nodes and lymphatics; the adults of the two species are indistinguishable. It causes Malayan filariasis and also infects monkeys, various wild cat species and the pangolin. It occurs in two forms (*a*) a *periodic* form in India, Sri Lanka, Thailand, Malaysia, Korea and Japan and (*b*) a *subperiodic* form in the east coast of Malaysia, Thailand and the Philippines. The major vectors belong to the subgenus *Mansonia* (*Mansonoides*), and a few species of *Anopheles*. Its biology, epidemiology and pathology have been reviewed by Ottesen (1990). Useful DNA probes have been developed to distinguish between *Brugia* species (Piessens *et al.*, 1987; Dissanayake *et al.*, 1991; Williams *et al.*, 1987) (see Fig. 29.4).

Brugia timori (Table 29.1)

A species in Indonesia sharing many features with *B. malayi* but with no known animal reservoirs.

Brugia pahangi (Table 29.1)

This species is a natural parasite of cats and dogs in Asia but has been experimentally transmitted to man and may prove to be a natural infection in man in Borneo (Palmieri *et al.*, 1985). Because it develops readily in jirds (Ash & Riley, 1970), golden hamsters (Malone *et al.*, 1974) and athymic nude mice (Suswillo *et al.*, 1980), it forms a valuable laboratory tool for research in filariasis (Denham & Fletcher, 1987). Its develop-

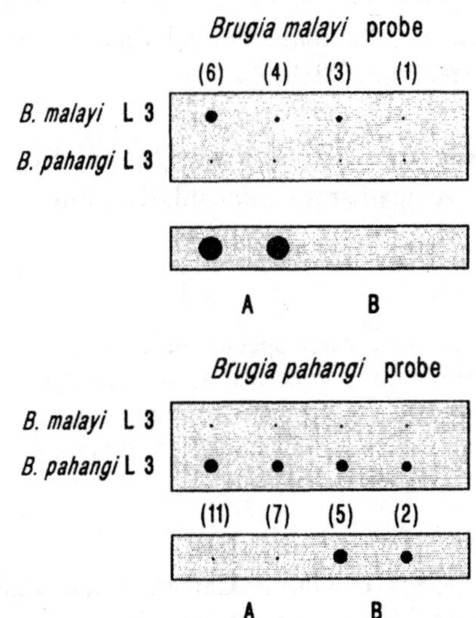

Fig. 29.4
Application of species-specific DNA probes to distinguish between third-stage larvae of *Brugia malayi* and *B. pahangi*. Graphic representation based on original radiographic filters of Williams *et al.* (1987). A and B represented unknown samples (provided by Dr Mak Joon Wah) and, as shown above, were found to be *B. malayi* and *B. pahangi*, respectively. (For technical details see the original paper.)

ment in cats, injected subcutaneously with microfilariae, has been described by Denham *et al.* (1972*a,b*); microfilariae appeared in the blood as early as 53 days and adult worms were recovered from the lymphatics.

Loa loa

The 'eye worm' of West and Central Africa, with estimates of human infection of 2–13 million (Pinder, 1988). Adults migrate in the subcutaneous tissues of man and monkeys, a habit which from time to time leads them to cross the eyeball under the conjunctiva. Various aspects of the biology, epidemiology, control and pathology have been reviewed by Noireau (1990), Ottesen (1990) and Pinder (1988).

Life cycle. The vectors are tabanid flies, especially *Chrysops silacea* and *C. diminiata.* Attempts to infect laboratory animals have not been very successful although adults (but not L_3 larvae) inoculated into jirds induced microfilaraemia which lasted a few months (Pinder, 1988). Some monkeys have also been infected. Like *W. bancrofti*, this species exhibits periodicity, but in this case it is diurnal (Fig. 29.9), the larvae almost disappearing from the blood at night. Associated with the infections are cutaneous swellings – 'Calabar swellings' – generally considered to be allergic reactions; they occur predominantly in the wrists and ankles. The swellings are especially developed in visitors to an endemic area, the lifetime inhabitants being largely asymptomatic (Nanduri & Kazura, 1989).

Chemotherapy. Mebendazole, DEC and ivermectin are all effective against the microfilariae, but as in other filariae, the adults are not affected.

Mansonella perstans

This is a mildly pathogenic species in man and apes which occurs in Africa and S America. It is located in the deep connective tissue and serous cavities and its unsheathed microfilariae do not exhibit periodicity. It is transmitted by midges of the genus *Culicoides*.

Mansonella streptocerca

This species occurs in Ghana, Cameroon and Zaïre. The adults are poorly known, and occur in the cutaneous tissue of man and chimpanzee. The microfilariae, like those of the previous species, do not exhibit periodicity. The vector is *Culicoides grahamii*, and probably other *Culicoides* spp.

Mansonella ozzardi

A non-pathogenic species occurring in the West Indies are northern parts of S America. The adults, which are thin, thread-like worms often 0.6 m long, occur in mesenteries and visceral fat; the male is almost unknown. Members of this genus are parasitic in man, horses, cattle and antelopes, the species in man and animals being morphologically indistinguishable. The microfilariae are unsheathed and occur only in the superficial layers of the skin which are penetrated by the midge vector, *Culicoides furens*. The parasites cause nodules in the skin of the vertebrate hosts.

Onchocerca volvulus: onchocerciasis, river blindness

This species occurs mainly in tropical and subtropical Africa, Asia and in the South Pacific. Lesser endemic areas are the Caribbean countries, such as Haiti and coastal Brazil. The disease it causes, *onchocerciasis*, is one of the world's most distressing diseases of helminth origin, often resulting in blindness, generally referred to as 'river blindness' since its main vector, blackflies of the genus *Simulium*, occupy rivierine habitats. Occasionally other species of *Onchocerca* from domestic animals have been recorded in man (Nelson, 1991). The adult worms are found in nodules or *onchodermata* in superficial sites, but may invade other tissues.

The extensive literature has been the subject of a number of reviews of which the following are representative: Aziz (1986), Bianco (1991), Duke (1990), Edungbola (1991), Evered & Clark (1987), Greene (1990), Nanduri & Kazura (1989), Nelson (1970, 1991).

Life cycle (Fig. 29.5). The life cycle is similar to that of *W. bancrofti*, except that the intermediate hosts are various species of blackflies and buffalo flies of the genus *Simulium*. The most important vector is *Simulium damnosum* which is recognised as a 'species-complex', for which keys are now available (Dang & Peterson, 1980). Development in the vector has been described by Bianco (1991) and Nelson (1991). After being ingested, on biting by a blackfly, microfilariae are carried to the midgut where they penetrate the epithelium and migrate, via the haemocoele, to the indirect flight muscles. Here they undergo two moults (L_1–L_3) and develop into infective L_3 larvae which move to the mouth parts. Development is completed in 6–9 days at 27 °C.

Transmission to another host occurs when infected flies take a blood meal and microfilariae are released from the mouth parts, but the mechanism for this and whether or not a trigger is involved has been little investigated (Bianco, 1991). Moreover, remarkably little is known about the subsequent development in man

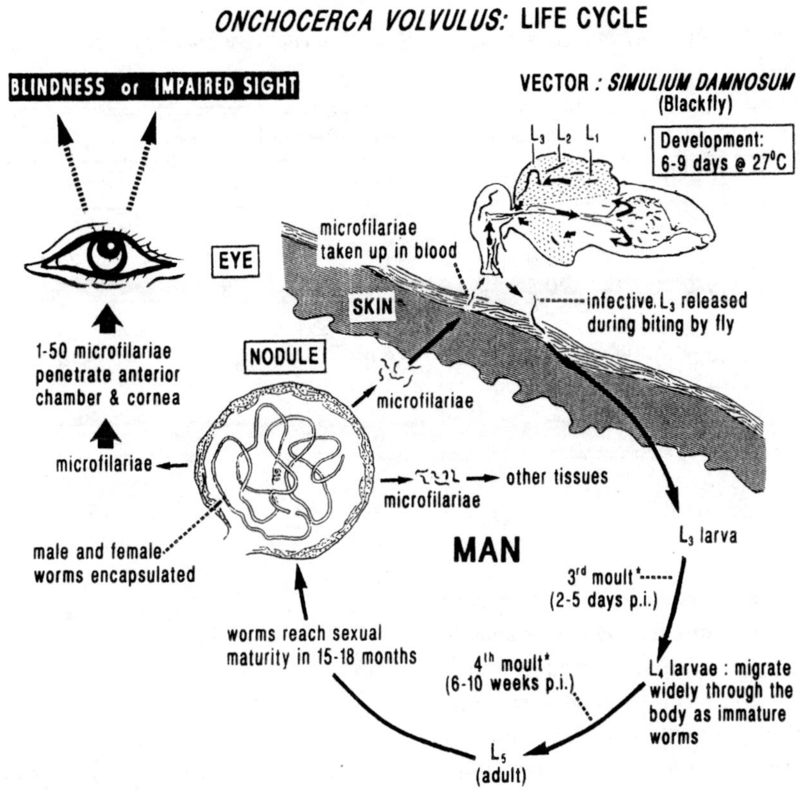

ONCHOCERCA VOLVULUS: LIFE CYCLE

Fig. 29.5
Life cycle of *Onchocerca volvulus* in the blackfly, *Simulium damnosum*, and in man. The moulting times marked* are based on data from Bianco (1991) for *O. lienalis* in cattle. (Modified from Nelson, 1991.)

(Duke, 1991) but the early stages of *O. lienalis* in cattle have been investigated and by analogy may reflect the development of *O. volvulus* in man (Bianco, 1991). In *O. lienalis*, moulting of L_3 to L_{4+} takes place within 2–5 days and the L_4 larvae migrate widely through the body under the skin and between muscles, ligaments and tendons. The final moult (L_4 to L_5) occurs at 1.5–2.5 months after entry. Males mature in about 4 months later. Female worms initiate the formation of a nodule and males and females may join the nodule later (Duke, 1991). The sexually mature females produce microfilariae which migrate out from nodule into the skin and other tissues and – most significantly – into the eye (Table 29.2).

Pathology. Detailed consideration of the pathology is beyond the scope of this book, but it is summarised briefly below; for reviews see Greene (1990), Bianco (1991) and Rodger (1977).

In contrast to *Wuchereria bancrofti* and *Brugia malayi*, it is the microfilariae of *O. volvulus*, rather than the adult worms, which are the principal cause of the pathology of onchocerciasis. Migration of these results in (*a*) onchocercal dermatitis, (*b*) onchocercomata (nodules) and (*c*) visual impairment or blindness. One of the most serious manifestations is lymphadenopathy in the inguinal and femoral regions, resulting in 'hanging groin'. The major complication of onchocerciasis, however, is the development of lesions in the eye which may result in blindness or other distressing ocular diseases.

Control. This problem has received much attention and has been the subject of numerous reviews (e.g. Edungbola, 1991; Duke, 1990). Ivermectin (Mectizan) is very effective against the microfilariae, a single dose being able 'to reduce skin microfilarial counts to undetectable levels within days' (Bradshaw, 1989); unfortunately, as in *W. bancrofti*, the adults are not killed.

Duke (1990) summarises the problems of onchocerciasis control as follows: (*a*) infections with *O. volvulus* are multiple and cumulative and may last for 10 years; (*b*) no immunity to reinfections takes place; (*c*) many infections are asymptomatic and require costly screening programmes to detect; (*d*) except where it causes

Table 29.2. *River blindness in Zaïre: percentage of total eyes affected in each age group by microfilariae in the cornea or the anterior chamber*

	Percentage of total eyes affected in each age group						
	18–20	20–30	30–40	40–50	50–60	60–70	70+
No. of microfilariae in cornea							
1–4	42.3	52.5	42.0	41.3	31.1	22.4	15.6
5–20	11.5	19.2	20.5	16.1	8.2	5.2	3.1
No. of microfilariae in anterior chamber							
1–19	26.9	44.0	54.9	50.0	39.2	28.3	—*
20–49	3.8	4.2	5.8	3.6	3.5	—	—
50+	3.8	—	—	—	—	—	—

*Dashes: no data.
Source: Data adapted from Rodger (1977).

much blindness, it is not rated as an important disease by governments in endemic countries; (*e*) drugs against the adults or vaccines have not been developed; (*f*) control of vectors is expensive and requires long-term commitments; (*g*) treatment of populations with ivermectin is never likely to be carried out on a sufficient scale to achieve eradication of infection.

Animal models for onchocerciasis reserach

Although there are some 30 known species of the genus *Onchocerca* only four have been generally used for laboratory research – *O. gutturosa, O. gibsoni, O. cervicalis* and *O. lienalis* – which are all parasites of cattle. Of these, only *O. lienalis* has been transmitted to laboratory animals (Bianco, 1991). *O. guttosa* and *O. gibsoni* are of value for chemotherapy research (Hutchinson, 1986).

○ 29.1.4 Identification of microfilariae

Microfilariae, as seen in blood films, show certain diagnostic characters. One of the most striking of these is the presence or absence of a *sheath*: a delicate, closely fitting membrane, probably the egg-capsule, which is often only detectable when it projects beyond the head or tail of the larva (Fig. 29.6). The positions of certain fixed points, shown in Fig. 29.7, have traditionally been used for identification but this method is open to certain technical difficulties. Histochemical methods for acid phosphatase have also proved of value (Yen & Mak, 1978). Thus the microfilariae of *Brugia malayi* show two prominent staining sites (with red azo dye) at the excretory pore and the anal pore; in contrast, microfilariae of *B. pahangi* stain diffusely over the entire body length (Redington *et al.*, 1975). However, with microfilariae of *Onchocerca volvulus* the method exhibits a large degree of variation and is of questionable reliability (Flockart *et al.*, 1986).

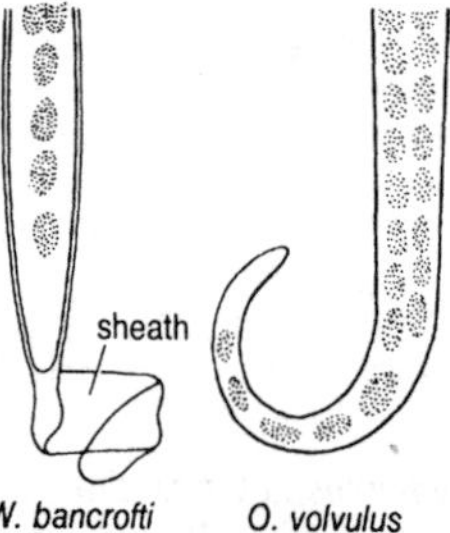

Fig. 29.6
Posterior end of a sheathed and an unsheathed microfilaria. (After Blacklock & Southwell, 1954.)

With some species, isoenzyme analysis is of value and the various stages of *Brugia pahangi, B. malayi* and *B. patei* can be distinguished by this technique (Flockart & Denham, 1984). The chromosomes of various species have also been examined and those of *Onchocerca volvulus* and *O. gutturosa* can readily be distinguished; see Fig. 29.8 (Procunier & Hirai, 1986). DNA probes are also proving of considerable value and probes have now been developed for species of *Brugia, Wuchereria, Onchocerca* and *Loa* (Dissanayake *et al.*, 1991; Dissanayake & Piessens, 1992; Harnett *et al.*, 1987; McReynolds *et al.*, 1991; WHO, 1989*b*; Williams *et al.*, 1987); see Fig. 29.4.

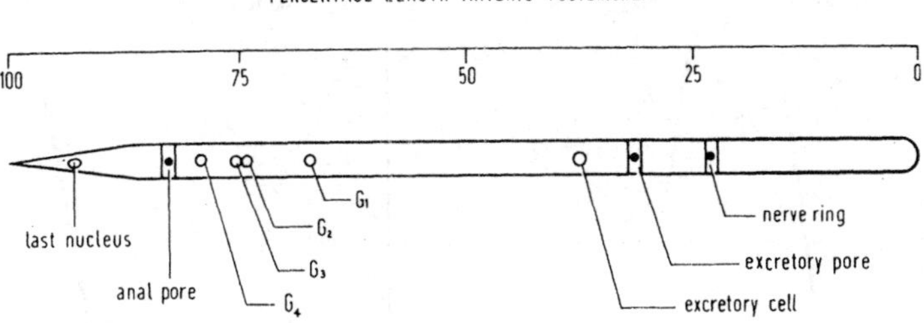

Fig. 29.7
Diagram of microfilaria, showing the chief points used in identification; G_1–G_4 represent genital rudiments. (Based on Newton & Wright, 1956.)

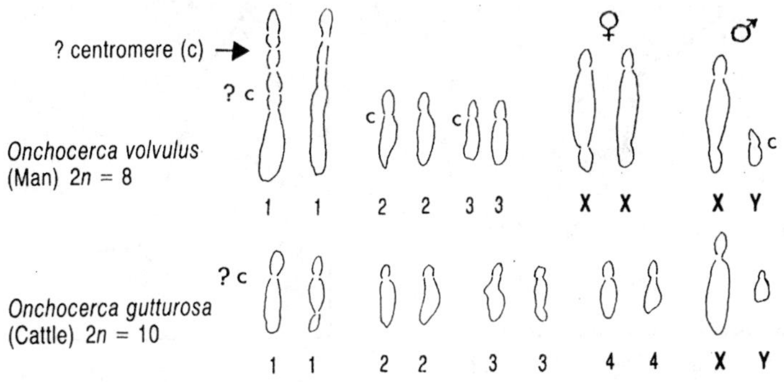

Fig. 29.8
Chromosomes of *Onchocerca volvulus* (from man) and *O. gutturosa* (from cattle); schematic interpretation. The position of the centromere (c) is not always obvious from the morphology of the mitotic chromosomes. The arrow indicates diagnostic chromomeric section of chromosome 1, which appears to vary between individuals. (Modified from Procunier & Hirai, 1986.)

○ 29.1.5 Periodicity of microfilariae

As already indicated, the microfilariae of various species exhibit periodicity to a varying extent and this phenomenon has been investigated extensively. In the human strains of *W. bancrofti* and *Loa loa* and various other species this phenomenon is well marked. The various theories put forward to explain periodicity have been comprehensively reviewed by Hawking (1967) and Ansari (1970). Hawking recognises four main periodicity patterns (Fig. 29.9):

(*a*) *W. bancrofti* with microfilariae reaching a peak about midnight (Fig. 29.9); transmitted by *night-biting* mosquitoes.
(*b*) *Dirofilaria immitis* in which microfilariae are present all the 24 hours (Fig. 29.9), but the maximum number (18.00 to 21.00 hours) is only two to five times as great as the minimum (09.00 hours); transmitted by *evening-biting* mosquitoes.
(*c*) *Loa loa* (Fig. 29.9) in which microfilariae are numerous at midday but almost absent at night; transmitted by a day-biting fly (*Chrysops*).
(*d*) *W. bancrofti* (so-called *pacific* type), a type morphologically indistinguishable from type (*a*) above but in which the microfilariae are always present, being more numerous in the afternoon than at night (Fig. 29.9); transmitted by day-biting mosquities.

Hawking (1967) points out that the behaviour of microfilariae appears to be adapted 'to promote transmission by arranging the maximum number of microfilariae in the peripheral blood at times when the arthropod vector is likely to bite'.

Filariae in animals other than man. Several species in mammals have already been dealt with but about 230 species infecting mammals are known. A number of species have also been reported from amphibia, reptiles and birds; in general the life cycles of these are poorly known. Mak (1983) gives useful host lists for Filaria in animals other than man.

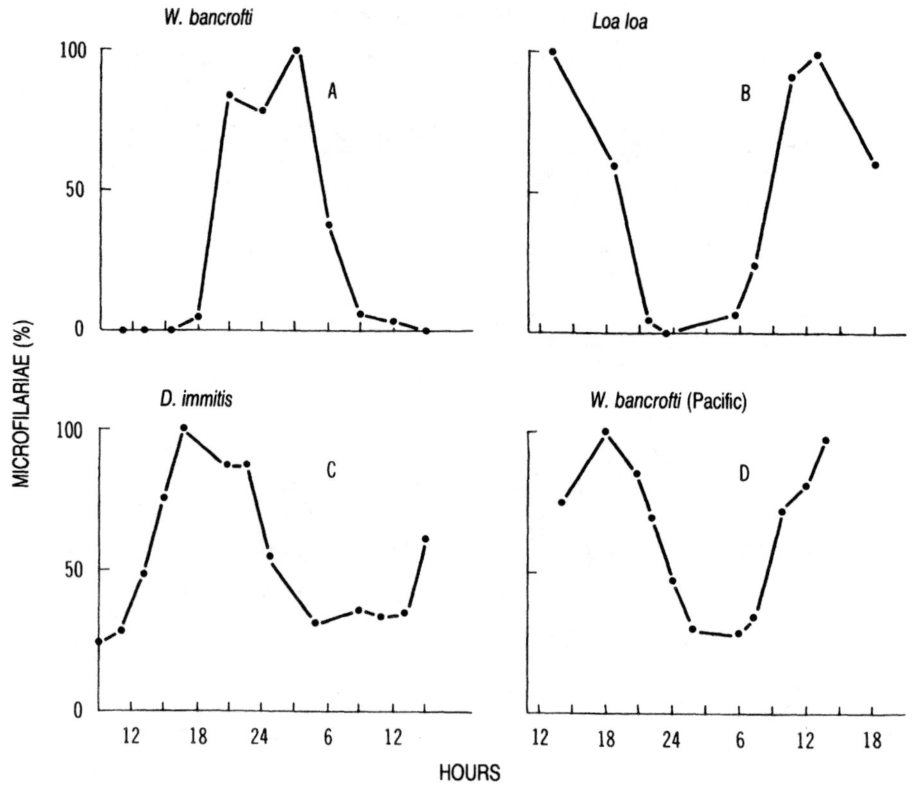

Fig. 29.9
Periodicity of microfilariae in four species of filaria. The number of microfilariae are expressed as percentages of the maximum number observed. (Based on Hawking, 1967.)

○○○ 29.2 Suborder Camallanina

These are usually referred to as the 'dracunculoid' worms, after the best-known genus *Dracunculus*. Formerly they were included with the Filariids. They are long filariform worms in which the mouth is simple or with lateral jaws. They are parasitic in the connective tissue or coelom of vertebrates. The life cycle usually involves a copepod host.

Dracunculus medinensis

The female worm is known as the 'guinea worm' and is probably the 'fiery serpent' referred to by Moses (Numbers XXI.6); it has been known since early recorded history. The structure, biology, epidemiology, pathology and bibliography have been reviewed by Hopkins (1983, 1987, 1990) and Muller (1971, 1985).

Geographical distribution. Widely distributed in tropical Africa, India, Pakistan, Saudi Arabia and Yemen, with probably 140 million persons at risk (Hopkins, 1990; WHO, 1987*c*, 1989*c*). The annual incidence in Africa is estimated at 3.32 million (Watts, 1987).

Morphology. The anatomy of the female is best known, the male remaining virtually unknown until 1936. The head bears a chitinous shield on which are six papillae (Fig. 29.10). The cuticle is smooth and milky white. The alimentary canal atrophies in gravid females and the body cavity becomes filled with uterus packed with larvae. The females measure up to one metre in length. The males range from 12 to 29 mm with a conical tail and ten pairs of genital papillae.

Habitat. The adults occur in the subcutaneous tissues, particularly those of the ankle and foot, arms and shoulders, but have been recorded from many other parts of the body (Muller, 1971).

Life cycle (Fig. 29.11). The site of copulation is unknown, but the males disappear rapidly after the process and the females, which require about 12 months

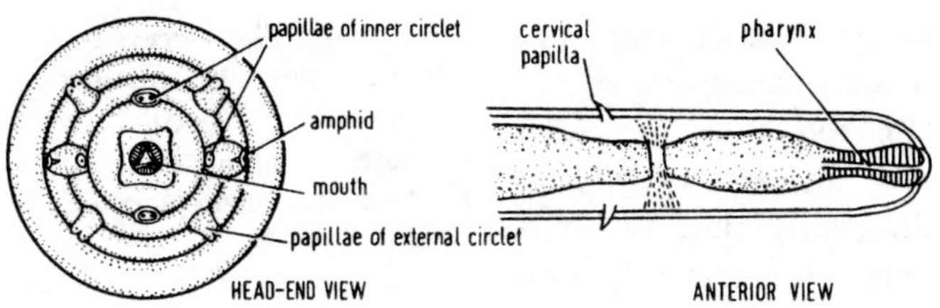

Fig. 29.10
Dracunculus medinensis: anatomy of anterior end of female. (Partly after Moorthy, 1937.)

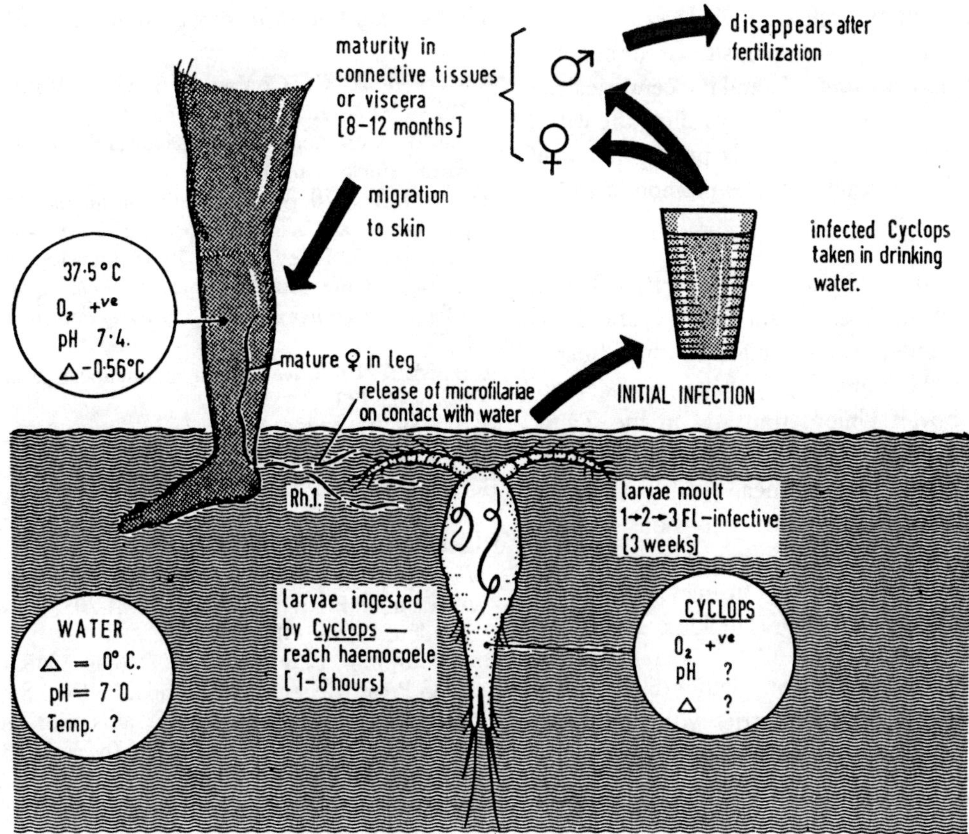

Fig. 29.11
Dracunculus medinensis: life cycle and some of the biological factors relating to it. (Original.)

to reach maturity, migrate to the superficial layers of the skin, especially to those regions liable to come in contact with water. Here the worm secretes a substance (?) which causes a blister on the skin to rise over its anterior end where it has pierced the lower layers. The blister bursts eventually and a small, shallow ulcer is formed. On contact with water, the uterus is projected out of the ulcer cavity, and a cloud of milky white secretion, containing hundreds of active larvae, is released. When the host leaves the water, the exposed end of the uterus dries and shrivels and so blocks the release of further larvae.

The rhesus monkey, *Macaca mulatta*, is a useful laboratory host for *D. medinensis* as the disease in these animals follows a similar pattern to that in humans (Muller, 1971). Dogs also serve as experimental hosts. The food reserves of the released rhabditoid larvae are small but sufficient to enable them to survive in water for several days. If ingested by an appropriate species of *Cyclops*, they break through the soft mid-intestine wall and come to lie in the body cavity. Here the usual two moults are completed, and larvae become infective in about three weeks. Human infection is brought about by accidentally taking in infected copepods in drinking water. Under the stimulus of the gastric juice, the larvae become active, penetrate the gut wall, migrate through the tissues, moult twice and come to lodge in the viscera, or subcutaneous tissues. Growth and development of the worm is slow and nearly a year is required before sexual maturity is reached and female worms are ready to migrate to the skin and release their larvae.

The disease caused by this parasite in humans is called *dracontiasis* or *dracunculiasis*, and the centuries-old remedy has been to remove it by gently rolling the worm daily around a small stick and slowly pulling it out of the skin. With a few precautions, this method is still in use.

Control. The major endemic countries have begun efforts to eliminate the disease, and anti-dracunculiasis programmes are under way for eight of the affected African countries (Hopkins, 1987). It has been eradicated from the Soviet Union, Iran and in the Tamil Nadu State of India and is in the process of being eliminated in Côte d'Ivoire and Guinea. Health education and *Cyclops*-free water supplies are obvious control measures. The use of filters to screen out the *Cyclops* (Duke, 1984) and chemical control (using Abate: 1 part per million) have proved of value (Hopkins, 1990).

Pathology. Although this is not a lethal disease, it is a serious problem in endemic countries with substantial adverse effects on health, education and agriculture (Hopkins, 1987). The pathology has been reviewed by Hopkins (1990).

○○○ 29.3 Suborder Spirurina

Genus *Spirocerca*

This genus contains a most interesting species, *Spirocerca lupi*, found in the oesophagus of dogs. *S. lupi* can induce the development of a sarcoma (Fox *et al.*, 1988), but the conditions under which this can develop are not understood. Research in this problem has been reviewed by Bailey (1972).

Genus *Thelazia*

These are filariid-like worms which live in the conjunctival sac, lachrymal ducts or nasal cavities of animals and sometimes man. The worms move across the eye intermittently and may injure the conjunctiva. The best-known species are *T. callipaeda*, *T. rhodesi*, *T. gulosa* and *T. lachrymalis*. The intermediate hosts are various species of flies of the genus *Musca*, the first larva being picked up from the eye-secretion of cattle by the flies.

○○○ References

(References of general interest have been included in this list, in addition to those quoted in the text.)

Anderson, R. C., Chabaud, A. G. & Wilmott, S. (eds) 1974. *C.I.H. Keys to the Nematode Parasites of Vertebrates*, No. 1. Commonwealth Agricultural Bureaux, Farnham Royal, Bucks, England.

Ansari, J. A. 1970. A review of the mechanism of microfilarial periodicity. *Zoologischer Anzeiger*, **185**: 387–92.

Ash, L. R. & Riley, J. M. 1970. Development of *Brugia pahangi* in the jird, *Meriones unguiculatus*, with notes on infections in other rodents. *Journal of Parasitology*, **56**: 962–8.

Aziz, M. A. 1986. Ivermectin vs. onchocerciasis. *Parasitology Today*, **2**: 233–5.

Bailey, W. S. 1972. *Spirocerca lupi*: a continuing inquiry. *Journal of Parasitology*, **58**: 3–22.

Bertram, D. S. 1966. Dynamics of parasitic equilibrium in cotton rat filariasis. *Advances in Parasitology*, **4**: 255–319.

Bianco, A. E. 1991. Onchocerciasis – river blindness. In: *Parasitic Helminths and Zoonoses in Africa* (ed. C. N. L. Macpherson & P. S. Craig), pp. 138–203. Unwin Hyman, London. ISBN 0 044 45565 8.

Boreham, P. F. L. & Atwell, R. B. (eds) 1988. *Dirofilariasis*. CRC Press, Boca Raton, Florida. ISBN 0 849 36488 4.

Bradshaw, H. 1989. Onchocerciasis and the mectizan donation programme. *Parasitology Today*, **5**: 63–4.

Campbell, W. C. 1981/82. Efficacy of the avermectins against filarial parasites: a short review. *Veterinary Research Communications*, **5**: 251–62.

Dang, P. T. & Peterson, B. V. 1980. Pictorial keys to the main species and species groups within the *Simulium damnosum* Theobald complex occurring in West Africa (Diptera: Simuliidae). *Tropenmedizin und Parasitologie*, **31**: 117–20.

Das, P. K., Manoharan, A., Srividya, A., Grenfell, B. T., Bundy, D. A. P. & Vanamail, P. 1990. Frequency distribution of *Wuchereria bancrofti* microfilariae in human populations and its relationship with age and sex. *Parasitology*, **101**: 429–34.

Denham, D. A. & Fletcher, C. 1987. The cat infected with *Brugia pahangi* as a model of human filariasis. In: *Filariasis* (ed. D. Evered & S. J. Clark), pp. 225–30. (Ciba Foundation Symposium, No. 127.) John Wiley & Sons, Chichester, UK.

Denham, D. A., Ponnudurai, T., Nelson, G. S., Guy, F. & Rogers, R. 1972*a*,*b*. Studies with *Brugia pahangi*. I. Parasitological observations on primary infections of cats (*Felis catus*). II. The effect of repeated infection on parasite levels in cats. *International Journal for Parasitology*, **2**: 239–47; 401–7.

Dissanayake, S., Min, X. & Piessens, W. F. 1991. Detection of amplified *Wuchereria bancrofti* DNA in mosquitoes with a nonradioactive probe. *Molecular and Biochemical Parasitology*, **45**: 49–56.

Dissanayake, S. & Piessens, W. F. 1992. Identification of filarial larvae in vectors by DNA hybridization. *Parasitology Today*, **8**: 67–9.

Duke, B. O. L. 1984. Filtering out the guinea worm. *World Health*, March: 29.

Duke, B. O. L. 1990. Onchocerciasis (river blindness) – can it be eliminated? *Parasitology Today*, **6**: 82–5.

Duke, B. O. L. 1991. Observations and reflections on the immature stages of *Onchocerca volvulus* in the human host. *Annals of Tropical Medicine and Parasitology*, **85**: 103–11.

Edeson, J. F. B., Wilson, T., Wharton, R. H. & Laing, A. B. G. 1960. Experimental transmission of *Brugia malayi* and *B. pahangi* to man. *Transactions of the Royal Society of Tropical Medicine and Hygiene*, **54**: 229–34.

Edungbola, L. D. 1991. Onchocerciasis control in Nigeria. *Parasitology Today*, **7**: 97–9.

Euzéby, J. 1990. *Dirofilaria immitis. Pratique Médicale et Chirurgicale de l'Animal de Compagnie*, **25** (3 suppl.): 283–91.

Evered, D. & Clark, S. (eds) 1987. *Filariasis.* (Ciba Foundation Symposium, 127.) John Wiley & Sons, Chichester, UK. ISBN 0 471 91093 7.

Fan, P. C. 1990. Eradication of bancroftian filariasis by diethylcarbamazine-medicated common salt in Little Kinmen (Liehyu District), Kinmen (Quemoy) Islands, Republic of China. *Annals of Tropical Medicine and Parasitology*, **84**: 25–33.

Flockhart, H. A., Cibulskis, R. E., Karam, M. & Albiez, E. J. 1986. *Onchocerca volvulus*: enzyme polymorphism in relation to the differentiation of forest and savannah strains of this parasite. *Transactions of the Royal Society of Tropical Medicine and Hygiene*, **80**: 285–92.

Flockart, H. A. & Denham, D. A. 1984. Differentiation of species and life cycle stages of *Brugia* spp. by isoenzyme analysis. *Journal of Parasitology*, **70**: 378–84.

Fox, S. M., Burns, J. & Hawkins, J. 1988. Spirocercosis in dogs. *Compendium on Continuing Education for the Practicing Veterinarian*, **10**: 807–22.

Franz, M. & Andrews, P. 1986*a*. Histology of adult *Litomosoides carinii* (Nematoda: Filarioidea). *Zeitschrift für Parasitenkunde*, **72**: 387–95.

Franz, M. & Andrews, P. 1986*b*. Fine structure of adult *Litomosoides carinii* (Nematoda: Filarioidea). *Zeitschrift für Parasitenkunde*, **72**: 537–47.

Garate, T., Harnett, W. & Parkhouse, R. M. E. 1990. Cloning a species-specific DNA probe from *Onchocerca gibsoni*. *International Journal for Parasitology*, **20**: 31–5.

Goodwin, L. G. 1984. Chemotherapy (of filariasis). *Transactions of the Royal Society of Tropical Medicine and Hygiene*, **78** (suppl.): 1–8.

Greene, B. M. 1990. Onchocerciasis. In: *Tropical and Geographical Medicine* (K. S. Warren & A. A. F. Mahmoud), 2nd Ed., pp. 429–39. McGraw-Hill, New York. ISBN 0 070 68327 1.

Harnett, W., Chambers, A. E. & Parkhouse, R. M. E. 1987. Cloning of an *Onchocerca volvulus* specific DNA sequence for use in speciation of L3's in blackflies. In: *Molecular Paradigms for Eradicating Helminthic Parasites* (ed. A. J. MacInnis), pp. 281–8. Alan R. Liss, New York. ISBN 0 845 12659 8.

Hawking, F. 1967. The 24-hour periodicity of microfilariae: biological mechanisms responsible for its production. *Proceedings of the Royal Society of London*, **B169**: 59–76.

Hillyer, G. V. 1990. Advances in schistosomiasis and filariasis. *Current Opinion in Infectious Diseases*, **3**: 427–33.

Hopkins, D. R. 1983. Dracunculiasis: an eradicable scourge. *Epidemiologic Reviews*, **5**: 208–19.

Hopkins, D. R. 1987. Dracunculiasis eradication: a mid-decade status report. *American Journal of Tropical Medicine and Hygiene*, **37**: 115–18.

Hopkins, D. R. 1990. Dracunculiasis. In: *Tropical and Geographical Medicine* (ed. K. S. Warren & A. A. F. Mahmoud), 2nd Ed., pp. 439–42. McGraw-Hill, New York.

Hutchinson, G. W. 1986. Onchocerciasis research in North Queensland. *Parasitology Today*, **2**: S14–15.

Knight, D. H. 1987. Heartworm infection. *Veterinary Clinics of North America, Small Animal Practice*, **17**: 1463–518.

McDonald, E. M. & Scott, J. A. 1953. Experimental filarial infections in cotton rats. *Experimental Parasitology*, **2**: 129–40.

McReynolds, L. A., Poole, C. B. & Williams, S. A. 1991. Filarial probes in Jakarta. (Report of WHO meeting on DNA probes.) *Parasitology Today*, **7**: 65–7.

Mak, J. W. (ed.) 1983. *Filariasis.* Institute for Medical Research, Kuala Lumpur, Malaysia.

Mak, J. W., Navartnam, V. & Ramachandran, C. P. 1991. Experimental chemotherapy of lymphatic filariasis. A review. *Annals of Tropical Medicine and Parasitology*, **85**: 131–71.

Malone, J. B., Leninger, J. R. & Thompson, P. E. 1974. *Brugia pahangi* in golden hamsters. *Transactions of the Royal Society of Tropical Medicine and Hygiene*, **68**: 170–1.

Maren, T. H. 1990. Ivermectin in lymphatic filariasis. *New England Journal of Medicine*, **323**: 917.

Meredith, S. E. O., Lando, G., Gbakima, A. A., Zimmerman, P. A. & Unnasch, T. R. 1991. *Onchocerca volvulus*: application of the polymerase chain reaction to identification and strain differentiation of the parasite. *Experimental Parasitology*, **73**: 335–44.

Mössinger, J. & Wenk, P. 1986. Fecundity of *Litomosoides carinii* (Nematoda, Filarioidea) *in vivo* and *in vitro*. *Zeitschrift für Parasitenkunde*, **72**: 121–31.

Muller, R. 1971. *Dracunculus* and dracunculiasis. *Advances in Parasitology*, **9**: 73–151.

Muller, R. 1985. Bibliography of *Dracunculus*. (1500 refs.) National Academy Press, Washington, D.C.

Muller, R. & Horsburgh, R. C. R. 1987. Bibliography of onchocerciasis (1941–1985). CAB International, Wallingford, Oxon, UK. ISBN 0 851 98604 8.

Nanduri, J. & Kazura, J. W. 1989. Clinical and laboratory aspects of filariasis. *Clinical Microbiology Reviews*, **2**: 39–50.

Nelson, G. S. 1970. Onchocerciasis. *Advances in Parasitology*, **8**: 173–224.

Nelson, G. S. 1990. Microepidemiology, the key to the control of parasitic infections. *Transactions of the Royal Society of Tropical Medicine and Hygiene*, **84**: 3–13.

Nelson, G. S. 1991. Human onchocerciasis: notes on the history, the parasite and the life cycle. *Annals of Tropical Medicine and Parasitology*, **83**: 85–95.

Nilsberg, A. & Wenk, P. 1988. Entrinsische Entwicklung und Infektiosität der metazyklischen Larven von *Litomosoides carinii* (Nematoda: Filarioidea) in *Sigmodon hispidus*. *Zeitschschrift für Angewandte Zoologie*, **75**: 429–39.

Noireau, F. 1990. Possibilités actuelles de lutte contre la filariose à *Loa loa*. *Annales de la Société Belge de Medécine Tropicale*, **70**: 167–72.

Ottesen, E. A. 1990. The filariases and tropical eosinophilia. In: *Tropical and Geographical Medicine* (ed. K. S. Warren & A. A. F. Mahmoud), 2nd Ed., pp. 407–29. McGraw-Hill, New York.

Palmieri, J. R., Ratiwayanto, S., Masbar, S., Tirtokusumo, S., Rusch, J. & Marwoto, H. A. 1985. Evidence of possible natural infections of man with *Brugia pahangi*, in South Kalimantan (Borneo), Indonesia. *Tropical and Geographical Medicine*, **37**: 239–44.

Piessens, W. F., McReynolds, L. A. & Williams, S. A. 1987. Highly repeated DNA sequences as species-specific probes for *Brugia*. *Parasitology Today*, **3**: 378–9.

Pinder, M. 1988. *Loa loa – a neglected filaria*. *Parasitology Today*, **4**: 279–84.

Procunier, W. S. & Hirai, H. 1986. The chromosomes of *Onchocerca volvulus*. *Parasitology Today*, **2**: 307–9.

Rajan, T. V. 1990. Molecular biology of human lymphatic filariasis. *Experimental Parasitology*, **70**: 500–3.

Redington, B. C., Montgomery, C. A., Jervis, H. R. & Hockmeyer, W. T. 1975. Histochemical differentiation of microfilariae of *Brugia pahangi* and subperiodic *Brugia malayi*. *Annals of Tropical Medicine and Parasitology*, **69**: 489–92.

Renz, A. & Wenk, P. 1981. Intracellular development of the cotton-rat filaria *Litomosoides carinii* in the vector mite *Ornithonyssus bacoti*. *Transactions of the Royal Society of Tropical Medicine and Hygiene*, **75**: 166–8.

Rochet, M. J. 1990. A simple deterministic model for bancroftian filariasis transmission dynamics. *Tropical Medicine and Parasitology*, **42**: 225–33.

Rodger, F. C. 1977. *Onchocerciasis in Zaire. A New Approach to the Problem of River Blindness*. Pergamon Press, Oxford. ISBN 0 080 20619 0.

Roux, J., Perolat, P., Cartel, J. L., Boutin, J. P., Sechan, Y., Lariviere, M. & Aziz, M. A. 1989. Etude de l'ivermectine pour le traitment de la filariose lymphatique due à *Wuchereria bancrofti* var. *pacifica* en Polynésia francaise. *Bulletin de la Société de Pathologie Exotique et de ses Filiales*, **82**: 72–81.

Sasa, M. 1976. *Human Filariasis. A Global Survey of Epidemiology and Control*. University Park Press, Baltimore.

Scott, J. A., McDonald, E. M. & Terman, B. 1951. A description of the stages in the life cycle of the filarial worm *Litomosoides carinii*. *Journal of Parasitology*, **37**: 425–32.

Senghor, J. E. & Samba, E. M. 1988. Onchocerciasis control programme – the human perspective. *Parasitology Today*, **4**: 332–3.

Service, M. W. 1990. Control of urban mosquitoes. *Pesticide Outlook*, **1**: 17–20.

Southgate, B. A. 1984. Recent advances in the epidemiology and control of filarial infections including entomological aspects of transmission. *Transactions of the Royal Society of Tropical Medicine and Hygiene*, **78**: 19–29.

Storey, D. M. & Ogbogu, V. C. 1991. Observations on third-stage larvae and adults of *Litomosoides carinii* (Nematoda: Filarioidea) by scanning and transmission electron microscopy. *Annals of Tropical Medicine and Parasitology*, **85**: 111–21.

Sucharit, S., Harinasuta, C. & Choochote, W. 1982. Experimental transmission of subperiodic *Wuchereria bancrofti* to the leaf monkey (*Presbytis melalophos*), and its periodicity. *American Journal of Tropical Medicine and Hygiene*, **31**: 599–601.

Sucharit, S. & Supavej, S. (eds) 1987. *Practical Entomology, Malaria and Filariasis*. Bangkok, Thailand, Museum and Reference Centre, Faculty of Tropical Medicine, Mahidol University, Bangkok, Thailand. ISBN 9 475 86315 7.

Suswillo, R. R., Owen, D. G. & Denham, D. A. 1980. Infections of *Brugia pahangi* in conventional and nude (athymic) mice. *Acta Tropica*, **37**: 327–35.

Tropis, M. *et al.* 1990. Effect of mass treatment of a human population with ivermectin on transmission of *Onchocerca volvulus* in Liberia, West Africa. *American Journal of Tropical Medicine and Hygiene*, **42**: 148–56.

Watts, S. J. 1987. Dracunculiasis in Africa in 1986: its geographical extent, incidence, and at-risk population. *American Journal of Tropical Medicine and Hygiene*, **37**: 119–225.

Wenk, P. 1991. The vector host link in filariasis. *Annals of Tropical Medicine and Parasitology*, **85**: 109–47.

WHO. 1987*a*. *Control of Lymphatic Filariasis. A Manual for Health Personnel*. WHO, Geneva. ISBN 9 241 54217 9.

WHO. 1987*b*. WHO expert Committee on onchocerciasis, 3rd Report. Technical Report Series, No. 752. WHO, Geneva.

WHO. 1987*c*. Dracunculiasis: global surveillance summary – 1987. *Weekly Epidemiological Record*, **63**: 375–9.

WHO. 1989*a*. *Tropical Diseases Progress in International Research, 1987–88*. (9th Programme Report of the UNDP/World Bank/WHO Special Programme for Research and Training in Tropical Diseases.) WHO, Geneva.

WHO. 1989*b*. Workshop on DNA diagnostics and filariasis and a symposium of filariasis and onchocerciasis. *WHO/TDR Mimeographed Series*, TDR/FIL/DNA/89.1.

WHO. 1989*c*. Dracunculiasis. *Weekly Epidemiological Record*, **64**: 207–10.

Williams, S. A., DeSimone, S. M., Poole, C. B. &

McReynolds, L. A. 1987. Development of DNA probes to identify and speciate filarial parasites. In: *Molecular Paradigms for Eradicating Helminthic Parasites* (ed. A. J. McKinnis), pp. 205–14. Alan R. Liss, New York. ISBN 0 845 12659 8.

Yen, P. F. K. & Mak, J. W. 1978. Histochemical differentiation of *Brugia*, *Wuchereria*, *Dirofilaria*, and *Breinlia* microfilariae. *Annals of Tropical Medicine and Parasitology*, **72**: 157–62.

30 Physiology of nematodes

Much of the work on nematode physiology and biochemistry has centred around those species readily available from local abattoirs, such as *Ascaris suum*, or those easily maintained in laboratory animals, such as *Nippostrongylus brasiliensis*, *Trichinella spiralis* and *Litomosoides carinii*.

Various aspects of nematode physiology and biochemistry have been reviewed by: Barrett (1981–88), Bryant & Behm (1989), Cox (1992), Croll & Matthews (1977), Köhler (1985, 1991), Lee & Atkinson (1976), Precious & Barrett (1989), Saz (1981*a*,*b*, 1989), Stewart (1983) and Ward (1982).

30.1 Chemical composition

The chemical composition of representative species of nematodes is given in Table 30.1.

Perienteric fluid. The body cavity, or *pseudocoelom*, of nematodes is filled with a fluid, the *perienteric* or *pseudocoelomic fluid*, sometimes referred to as *haemolymph*. Its composition and properties are largely known from studies on *Ascaris* and *Parascaris*. In *Ascaris*, the fluid has a pH of 6.2–6.4 and an osmotic pressure of 0.192M NaCl and contains – in addition to proteins, carbohydrates and fat – small quantities of haemoglobin, various enzymes and several organic acids. Of the carbohydrates present, rather surprisingly, glucose is present only in trace quantities but a disaccharide, trehalose (see below), is present (0.77 per cent) as is glycogen (0.4 per cent) (Fairbairn, 1960).

Many species contain different haemoglobins in their perienteric fluid and body wall. These have been especially studied in *Ascaris* where the perienteric fluid haemoglobin has the highest oxygen affinity reported for any haemoglobin: $p_s = 0.01$ mmHg at 37 °C. The role of nematode haemoglobin is not clear. Unlike mammalian pigments, the haemoglobin in *Ascaris* is not deoxygenated at low pressures; this fact makes it unlikely that the pigment functions in oxygen transport or as an oxygen reserve. However, the body wall haemoglobin in *Ascaris* is slowly deoxygenated under anaerobic conditions (Barrett, 1981).

Carbohydrates. As in trematodes and cestodes, the main polysaccharide in nematodes is a glycogen closely resembling mammalian glycogen. In adults, the greatest part of the glycogen occurs in the body-wall tissues, especially the muscles.

In some larval forms, there are very high levels of glycogen, as in the larvae of *Eustrongyloides ignotus* (in fish) in which levels as high as 28 per cent of dry mass have been reported. Another disaccharide, trehalose, occurs in a number of species, representing 4.0 per cent of the solids of *Ascaris*, where it forms 70–80 per cent of the total carbohydrate in the testes and seminal vesicle (Barrett, 1981).

Proteins. X-ray and chemical analysis have indicated that the chief structural proteins in nematodes are keratin, sclerotin (tanned protein) and collagen. Of all the structural proteins, collagen has been most thoroughly studied. Two types of collagen protein have been characterised in nematodes: basement membrane collagens and cuticular collagens. *Ascaris* adult cuticle collagen has been especially examined as it is easily solubilised by reducing agents. The collagenous components of adult *Ascaris* have been characterised by Winkfein *et al.* (1985) and substantial data on the molecular biology of nematode collagens have been accumulated (Table 30.2). Detailed consideration of this is beyond the scope of this text, but the field has been reviewed in depth by Cox (1992).

Table 30.1. *Chemical composition of nematodes*

	Glycogen		Lipid		Protein	
	% fresh mass	% dry mass	% fresh mass	% dry mass	% fresh mass	% dry mass
Ancylostoma caninum	1.6	—	—	—	—	—
Ascaridia galli ♂	4.2[a]	17.9[a]	1.2	—	—	—
Ascaridia galli ♀	4.8[a]	19.8	1.3	—	—	—
Ascaris lumbricoides ♂	3.3–8.7	14–24	1.3	—	—	48–57
Ascaris lumbricoides ♀	—	—	1.6	8.9	—	—
Dipetalonema gracilis	0.2	—	—	—	—	—
Dirofilaria immitis	1.9	10	—	—	—	—
Dirofilaria uniformis	1.7	—	—	—	—	—
Eustrongylides ignotus (larvae)	6.9	28	1.1	4.4	—	—
Heterakis gallinae	2.7	11.8	—	—	—	—
Litomosoides carinii	0.8	5	—	—	—	—
Parascaris equorum	2.1–3.8	10–23	—	—	—	—
Porrocaecum decipiens (larvae)	—	55	—	3.5	—	—
Strongylus vulgaris	3.5	—	—	—	—	—
Trichuris vulpis	2.3	—	—	—	—	—
Nippostrongylus brasiliensis (larva)	—	5–6[b]	—	15–20[b]	—	73–76[b]
Nippostrongylus brasiliensis (adult)	—	—	—	11.9	—	—
Trichinella spiralis (muscle larvae)	—	16.1[c]	—	5.5	—	44.1[c]
Trichinella spiralis (7-day adult)	—	7.2[c]	—	—	—	30.3[c]

Source: Data from von Brand (1973) and Fairbairn (1969) except where indicated; [a]Srivastava *et al.* (1970); [b]Wilson (1965); [c]Kilgore *et al.* (1986).

Table 30.2. *Characteristics of nematode basement membrane and cuticle collagens*

	Collagen type	
Characteristic	Basement membrane	Cuticle
Number of genes	2	20–150[a]
Introns per gene	11[b]	1–2
Genomic organization	Unlinked	Dispersed
RNA size	5.5 kb	1–1.5 kb
Polypeptide size[c]	*ca.* 180 kDa	*ca.* 30 kDa
Triple-helix length	469 nm	47 nm
Cysteine cross-links	Yes	Yes
Tyrosine cross-links	No	Yes
Tissues expressed	Intestine, muscle?	Epidermis

[a]Estimated.
[b]For the *clb-2* gene.
[c]Primary translation product.
Source: Data from Cox (1992).

Lipids. The detailed composition of lipids in nematodes has been reviewed by Fairbairn (1969) and Barrett (1981, 1983); some relevant data are given in Table 30.1. All the major classes of neutral and phospholipids have been identified. The fatty acid profiles are known for several species. For example, in the adult filarial worms *Dirofilaria immitis*, *Acanthocheilonema* (= *Dipetalonema*) *viteae* and *Litomosoides carinii*, a range of saturated and unsaturated fatty acids, from C_{12} to C_{26}, including branched chain and odd-numbered fatty acids, have been identified (Barrett, 1981). Unlike cestodes, which are largely dependent on the host for their lipids, the fatty acid profile of adult *D. immitis* showed little similarity with the fatty acid profiles of plasma from infected dogs (Hutchison *et al.*, 1976). Utilisation and biosynthesis of lipids has been especially examined in *Haemonchus contortus* (Kapur & Sood, 1984, 1987) and in *Trichuris globulosa* (Sarwal *et al.*, 1989) and is discussed later (p. 446).

A notable feature of nematodes is the occurrence of *ascarosides*: a series of unique α-glycosides which have been isolated from the eggs and female reproductive tract of ascarid and oxyurid nematodes, such as *Ascaris* spp., *Toxocara cati* and *Passalurus* sp. (Barrett, 1981).

○○○ 30.2 Respiration

As the life cycles of the majority of nematodes involve complex environmental changes, it is to be expected that marked differences between the respiratory metabolism of the different stages will exist. Eggs of a species such as *Ascaris suum*, for example, may embryonate and hatch in an aerobic environment (moist ground), pass through later larval stages similarly in an aerobic environment (e.g. the lungs) and finally produce eggs in the anaerobic environment prevailing in the alimentary canal. The differences in metabolism which occur between different species can be said to largely reflect responses to the differences in their respective environments, particularly with respect to the level of oxygen available.

Briefly, nematodes can be divided into three arbitrary groups based on their metabolic oxygen requirements (Saz, 1981*a*) as follows:

(*a*) Species such as *Ascaris suum*, the adults of which (but not the eggs or early larval stages) require no oxygen in their energy-yielding pathways and appear to survive *in vitro* equally well under anaerobic or aerobic conditions.

(*b*) Species which are obligate anaerobes, such as the rat hookworm, *Nippostrongylus brasiliensis*, and, as mentioned above, the larval (L_1–L_3) stages of *Ascaris*. With these species, removal of oxygen results in loss of motility and (usually) subsequent death.

(*c*) Larval stages which require an aerobic metabolism for motility but not for survival, e.g. the microfilariae of *Litomosoides carinii*, *Brugia pahangi* and *Acanthocheilonema* (= *Dipetalonema*) *viteae* and possibly the larvae of *Trichinella spiralis* (Saz, 1981*a*).

○○○ 30.3 Carbohydrate metabolism

Most of the work on the metabolism of nematodes has been carried out on ascarids (especially *Ascaris suum*) and filariids and it is the metabolism of these groups which is chiefly considered below. Valuable reviews of these areas are those of Saz (1981*a*) and Barrett (1981, 1983).

○ 30.3.1 Metabolism of *Ascaris*

Adult. *Ascaris* has an anaerobic pathway (Fig. 30.1) – often referred to as the fumarate reductase pathway – which is closely similar to that of some cestodes (e.g. *Hymenolepis diminuta*). This pathway involves CO_2 fixation and results in the accumulation of either succinate, which is the major fermentation product, or products derived from it. Glycogen is the main energy reserve in most adult nematodes, being about 20 per cent of the

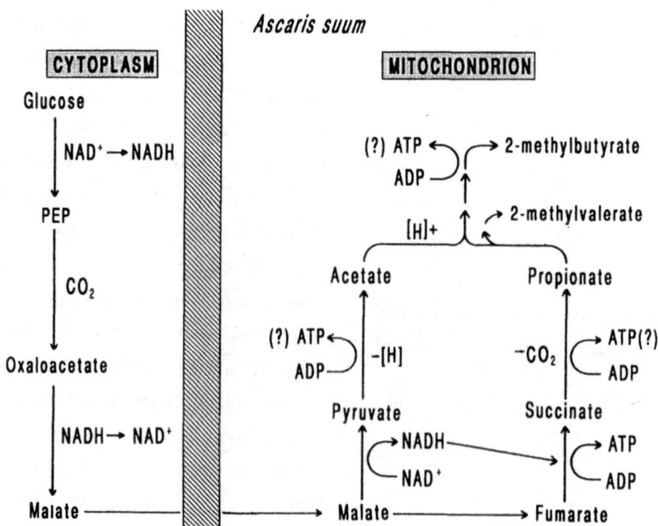

Fig. 30.1
Pathway of energy metabolism in *Ascaris suum* (= *lumbricoides*) involving the anaerobic dissimilation of carbohydrate. (Modified from Saz, 1981*a*.)

dry mass of the whole *Ascaris* and about 70 per cent of the dry mass of muscle (Lee & Atkinson, 1976). Under starvation conditions, *Ascaris* consumes 1.3 g of glycogen per 100 g body mass in 24 h under anaerobic conditions and 0.8 g under aerobic conditions.

The energy-yielding pathways in *Ascaris* are shown in Fig. 30.1. *Ascaris* muscle catabolises glucose as far as phosphoenolpyruvate (PEP) via the glycolytic pathway. Since the enzyme pyruvate kinase is only present in trace quantities, the muscle is incapable of forming significant quantities of pyruvate in the cytoplasm. Instead, CO_2 is fixed into PEP to form oxaloacetate (OAA), being catalysed by the enzyme PEP carboxykinase. OAA is then reduced by the active enzyme malate dehydrogenase (MDH), to malate. The *Ascaris* PEP carboxykinase appears to act in a reverse direction to that of the mammalian enzyme and it is interesting to note that a PEP carboxylase purified from the cestode *H. diminuta* has been shown to have a stronger affinity for PEP than the corresponding mammalian enzyme (Saz, 1981*a*).

The malate formed in the cytoplasm now penetrates the mitochondrial membrane and becomes the mitochondrial substrate where a dismutation takes place. A mole of malate is oxidised to pyruvate and CO_2, catalysed by the NAD^+-linked 'malic' enzyme, thereby generating NADH which in turn serves to reduce another mole of malate via fumarate to succinate. The pyruvate and succinate formed in the mitochondria serve as precursors for acetate and propionate which then serve as precursors for 2-methylbutyrate, tigate, and 2-methylvalerate (Fig. 30.1).

It should also be pointed out that *Ascaris* mitochondria differ significantly from mammalian mitochondria and are often referred to as 'anaerobic mitochondria'. They function primarily anaerobically and contain only very low concentrations of cytochromes *b*, *c*, and cytochrome oxidase; it is questionable whether these cytochromes have any physiological significance. The fact that *Ascaris* can survive as well under anaerobic as aerobic conditions, as well as other evidence, suggests that the cytochrome system plays little or no part in the muscle energy metabolism. Although the anaerobic pathway appears to be dominant in *Ascaris*, some oxygen may be required for synthetic purposes such as cuticle formation where the enzyme proline hydroxylase is involved (Cain & Fairbairn, 1971).

Eggs. In contrast to the adult worm, oxygen is required for the initiation and maintenance of the development of *Ascaris* eggs, processes which generally take place on pasture or in soil, normally highly aerobic environments. *Ascaris* eggs, which contain relatively more lipid than carbohydrate, have a complete glycolytic and Krebs cycle. Both trehalose and glycogen serve as energy sources during the early phases of development, but in later development, lipid metabolism becomes more important. Cytochrome activity has been shown to be absent or low during initial development, but on contact with air large amounts of cytochromes are formed and continue to increase during development (Barrett, 1976; Saz, 1981*a*).

Larvae. The lung-stage larvae (L_3) of *Ascaris* are obligate aerobes (Saz *et al.*, 1968), as evidenced by (*a*) the fact that both anaerobiosis and cyanide inhibit motility reversibly, (*b*) the presence of cytochrome oxidase, and (*c*) the fact that a definite Pasteur effect, (i.e. a more rapid utilisation of substrate anaerobically than aerobically) has been demonstrated. The activities of the enzymes pyruvate kinase, PEP carboxykinase, malate dehydrogenase and lactate dehydrogenase are essentially similar to those found in the adult muscle (Saz *et al.*, 1968).

Embryonated eggs form infective L_2 larvae at about 22 days, when eaten by an appropriate host, undergo a tissue migration (Fig. 27.6), moult to L_3 larvae and reach the lungs. L_2 larvae metabolise aerobically and have an active Krebs cycle and a cyanide-sensitive oxygen uptake (Komuniecki & Komuniecki, 1989). The L_3 larvae apparently make the transition from aerobic to anaerobic metabolism during the third moult (L_3 to L_4) as evidenced by the loss of cytochrome oxidase activity (Sylk *et al.*, 1974).

That L_3 larvae make the transition from an aerobic to an anaerobic metabolism has been further demonstrated by the *in vitro* experiments of Komuniecki & Vanover (1987). They showed that the L_3 larvae possess substantial malic enzyme activity and produce small quantities of reduced organic acids (propionate, 2-methylbutyrate and 2-methylvalerate) characteristic of the fermentation pathways of the adult (Fig. 30.2B). Moreover, this motility is cyanide-sensitive in the operation of a predominantly aerobic metabolism. In contrast, after ecdysis the L_4 larvae begin to utilise glucose at a greater rate (Fig. 30.2A) and the proportion of reduced organic acids substantially increases (Fig. 30.2B), indicating a transition to an anaerobic metabolism.

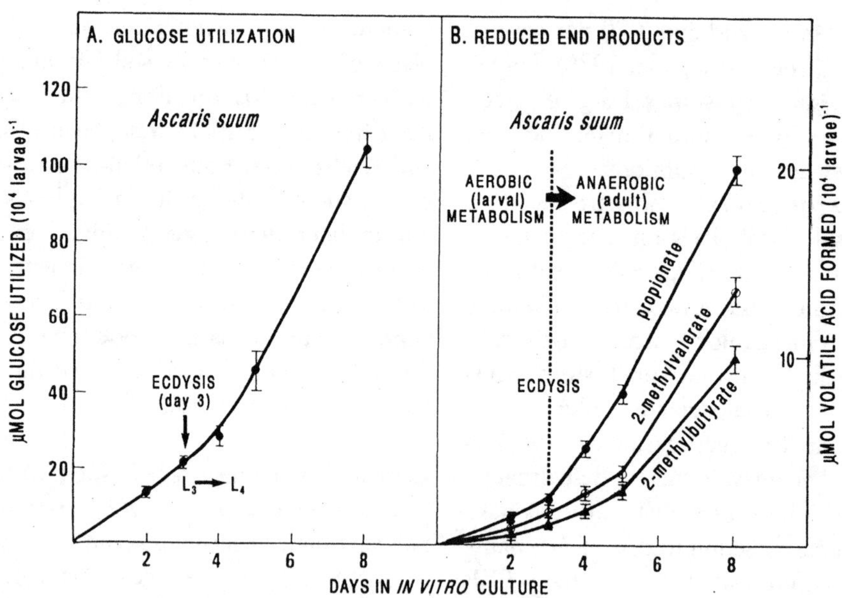

Fig. 30.2
Ascaris suum (= *lumbricoides*): (A) glucose utilisation and (B) volatile fatty acids produced by developing L_3 and L_4 larvae *in vitro* under a gas phase of 85 per cent N_2/10 per cent CO_2/5 per cent O_2. The third ecdysis of the L_3 larva to L_4 takes place on day 3, after which time a metabolic switch from an aerobic to an anaerobic metabolism, as in the adult worm, occurs. (Modified from Komuniecki & Vanover, 1987.)

○ 30.3.2 Other (non-filariid) species

The fumarate reductase metabolic pathway, described above for adult *Ascaris*, also occurs in several other species, such as *Heterakis gallinarum*, *Trichuris vulpis*, *Syphacia muris*, *Dictyocaulus viviparus* and juveniles of *Trichinella spiralis* and L_1 larvae of *Strongyloides ratti* (Bryant, 1975).

○ 30.3.3 Metabolism of filarial worms

Various aspects of the metabolism of filariae have been reviewed by Barrett (1983), Köhler (1991) and Saz (1981*b*).

Adults. Compared with non-filarial species, filariae have very low concentrations of glycogen, about 1–2 per cent of fresh mass (Barrett, 1983), and this may be lacking in the microfilariae of some species. However, varying concentrations of free glucose, galactose and trehalose have been recorded in some species.

Filariae appear to be able to synthesise glycogen from hexose precursors. In *Litomosoides carinii*, this takes place only under aerobic conditions. The main energy pathway in most species has generally been accepted as being by glycolysis – a view now modified (see below) – and a more or less complete sequence of glycolytic enzymes has been demonstrated in *Chandlerella hawkingi*, *Dirofilaria immitis*, *Litomosoides carinii* and *Setaria cervi*. Some isolated enzymes have been shown to be present in *Brugia pahangi*, *Onchocerca volvulus* and *Acanthocheilonema* (= *Dipetalonema*) *viteae* (Table 30.3). The adult worms possess a classical glycolytic pathway but species differ from one to another in the fate of pyruvate. For example, under both anaerobic and aerobic conditions, *A. viteae* and *B. pahangi* appear to be homolactic fermenters (Wang & Saz, 1974). However, *D. immitis* converts 55 per cent of glucose into lactate after 1 h, increasing to 93 per cent after 24 h incubation, but only traces of acetate and no propionate were detected (Barrett, 1983).

Litomosoides carinii has a markedly different metabolism from most filariae. It is an obligate aerobe and requires oxygen for motility and survival (Barrett, 1983). Under *aerobic* conditions, 30–40 per cent of the carbohydrate catabolised is converted into lactate and 25–35 per cent to acetate. Under *anaerobic* conditions, 80 per cent of the carbohydrate is catabolised to lactate and the remainder to acetate (Bueding, 1949). The adult worms show a rapid uptake of glucose during *in vitro* culture (Fig. 30.3) which is linear with time, but significantly higher under aerobic than anaerobic conditions (Ramp & Köhler, 1984). Under anaerobic conditions

Table 30.3. *Glycolytic enzymes identified in adult filarial worms*

+, Present; ·, no data.

Enzymes	Species					
	L. carinii	*B. pahangi*	*D. immitis*	*O. volvulus*	*A. viteae*	*S. cervi*
1. Phosphorylase	+	·	·	·	·	+
2. Hexokinase	+	·	+	·	·	+
3. Phosphoglucosemutase	+	·	+	·	·	+
4. Glucosephosphate isomerase	+	·	+	·	·	+
5. Phosphofructosekinase	+	+	+	·	+	+
6. Aldolase	+	+	+	·	+	+
7. Triophosphate isomerase	·	·	+	·	·	·
8. Glyceraldehyde-3-phosphate dehydrogenase	+	·	+	+	·	+
9. Phosphoglycerate kinase	+	·	+	·	·	·
10. Phosphoglyceromutase	+	·	+	·	·	·
11. Phosphopyruvate hydatase	+	·	+	·	·	+
12. Pyruvate kinase	+	·	+	+	·	+
13. Lactate dehydrogenase	+	·	+	+	·	+
14. Cytoplasmic glycerol-3-phosphate dehydrogenase	·	·	+	·	·	·

Source: Data from Barrett (1983).

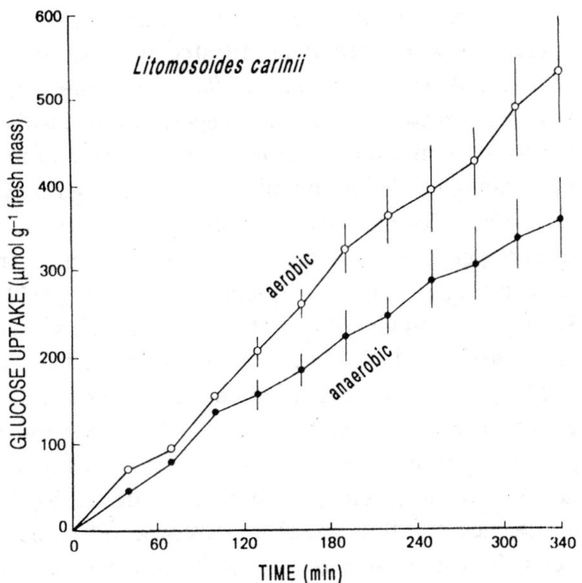

Fig. 30.3
Glucose uptake by the adult filariid *Litomosoides carinii* under aerobic and anaerobic conditions. All values represented means ±S.D. of four experiments. (Modified from Ramp & Köhler, 1984.)

(N_2), glucose was converted almost quantitatively to lactate, whereas under aerobic conditions appreciable quantities of CO_2, acetate and acetoin were formed in addition to lactate. The complete sequence of glycolytic enzymes has been detected in the cytosolic extracts. In addition, the entire set of enzymes of the Krebs cycle were identified.

Although it has been generally accepted that glucose metabolism appeared to be the chief energy source for filariids, there is now unequivocal evidence that the amino acid glutamine may act as the main energy source in *L. carinii* and some other species. The evidence for this is reviewed by Köhler (1991) and is based on the fact that when incubated with glutamine, *L. carinii* metabolises it primarily to CO_2 (Fig. 30.4). Moreover, the glutamine-dependent viability of the worm was affected by mitochondrial chain inhibitors, indicating the involvement of electron transport-associated phosphorylation. The production of CO_2 from glucose was found to be greatly stimulated by the presence of glutamine in the medium. Radiolabelling and related studies suggest that, in this species, glutamine provides carbon for the Krebs cycle, which in turn could explain the observed increase in the oxidation of glucose to CO_2 in the presence of glutamine. The significance of glutamine as an energy source has been confirmed by related studies on *A. viteae* and *Onchocerca* spp. (Comley & Rees, 1989; MacKenzie *et al.*, 1989; Van de Waa *et al.*, 1988).

Larval stages. Very little is known regarding the carbohydrate catabolism of larval stages. Under anaerobic conditions, the microfilariae *A. viteae*, *L. carinii* and

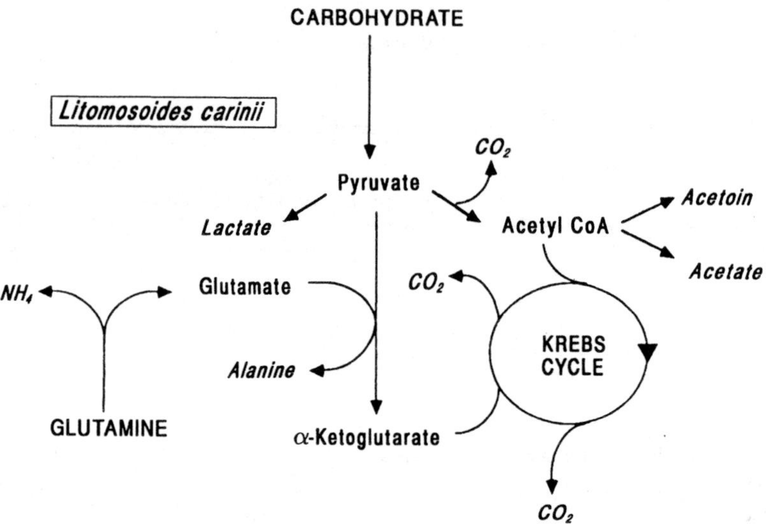

Fig. 30.4
Litomosoides carinii: proposed interactions between carbohydrate and glutamine metabolism. Substrates are shown in capitals and the end products in italics. (Modified from Köhler, 1991.)

B. pahangi ferment glucose entirely to lactate and the larvae lose their motility rapidly; this is resumed after exposure to air. Under aerobic conditions, only 80 per cent lactate is produced by *B. pahangi*; the rest is acetate.

○○○ 30.4 Krebs (TCA) cycle

Whether or not a fully functional Krebs cycle operates in nematodes has been the subject of much controversy. Since this pathway is the terminal oxidative pathway for end products of protein, carbohydrate and lipid metabolisms, as well as playing a major role in synthetic reactions, it is clearly important to determine the extent to which it operates. Earlier work generally failed to establish unequivocally a functional role for the cycle in nematodes (Barrett, 1981) but subsequent work has established that, in some species, under partly aerobic conditions, it plays an important metabolic role. Most or all of the enzymes of the cycle have now been identified in the eggs, larvae and adults of a number of species. Thus, a complete sequence of the enzymes has been detected in the eggs of *Ascaris*, which require oxygen for their embryonation and development (Barrett, 1981), although only a partial reversed cycle operates in the anaerobic adult. The enzymes have also been recorded in the infective larvae of *Ancylostoma tubaeforme* and *Haemonchus contortus*, although in the pre-infective L_1 and L_2 stages the enzymes of the early stages of the cycle, aconitate hydratase and the NAD-linked isocitrate dehydrogenase, are missing (Onwuliri, 1985).

A complete set of Krebs cycle (and glycolytic) enzymes have also been demonstrated in *Ancylostoma ceylanicum* and *Nippostrongylus brasiliensis* together with a glyoxylate bypass. All these enzymes showed activities similar to those in the rat liver homogenate, most of the enzymes being located in the mitochondria (Singh *et al.*, 1992*b*). The infective larvae of both species, however, were equipped only with the Krebs cycle enzymes. In the Filarioidea, a complete sequence of enzymes (Table 30.4) has also been identified in the cattle filariids *Onchocerca gutturosa* and *O. lienalis* (Dunn *et al.*, 1988). It should be pointed out that although the metabolisms of *O. gutturosa* and *O. lienalis* appear to be very similar they do show differences, especially in the production of lactate and the respiratory coefficients of the Krebs cycle enzymes. It is thus possible that similar differences between the cattle *Onchocerca* and the human parasite *Onchocerca volvulus* also exist so that the former species may not prove to be suitable models for the investigation of vaccines or biochemical targets for drugs against the human species; much work remains to be carried out on the biochemistry of the latter.

Table 30.4. *The activities of the Krebs (TCA) cycle enzymes in* Onchocerca gutturosa *and* Onchocerca lienalis

Enzyme	Activity: mean ± s.e., $n = 5$ (nmol min^{-1} mg^{-1} protein at 37 °C)	
	O. gutturosa	*O. lienalis*
Pyruvate dehydrogenase	0.84 ± 0.3[a]	2.2 ± 0.9[a]
Citrate synthase	3.98 ± 0.3	14.7 ± 1.3
Aconitase (citrate)	0.08 ± 0.01	0.13 ± 0.08
Aconitase (isocitrate)	0.43 ± 0.03	0.3 ± 0.02
Isocitrate dehydrogenase (NAD)	0.07 ± 0.01	0.31 ± 0.04
Isocitrate dehydrogenase (NADP)	1.7 ± 0.05	3.0 ± 0.36
2-Oxoglutarate dehydrogenase	1.5 ± 0.3	1.4 ± 0.05
Succinate dehydrogenase	2.7 ± 0.1	0.49 ± 0.1
Fumarate reductase	0.17 ± 0.05	0.45 ± 0.1
Fumarase	45.4 ± 5.9	15.9 ± 0.6
Malate dehydrogenase	928 ± 30	1306 ± 20

[a]nmol min^{-1} mg^{-1} mitochondrial protein.
Source: Data from Dunn *et al.* (1988).

Table 30.5. *Protease activity in nematodes, trematodes and cestodes, estimated by hydrolysis of haemoglobin at pH 3.7 and the inhibitory effect of pepstatin on the activity*

Source of enzyme	Specific activity in units mg^{-1} protein (n = no. of experiments)	Inhibitory effect of pepstatin (4.5 × 10^{-9} M) (per cent)
Angiostrongylus cantonensis	67.8 ± 22.4 (n = 14)	96
Dirofilaria immitis	10.4 ± 2.9 (n = 12)	94
Hymenolepis nana	3.6 ± 1.4 (n = 8)	89
Ascaris suum	2.4 (n = 2)	92[a]
Paramphistomum sp.	1.7 ± 0.04 (n = 6)	71
Diphyllobothrium erinacei	1.6 ± 0.5 (n = 10)	93
Trichuris muris	0.7 ± 0.2 (n = 6)	85

[a]Data from the intestinal homogenate.
Source: Maki *et al.* (1982).

○○○ 30.5 Protein metabolism

General reviews on the protein metabolism of nematodes are those of Barrett (1981, 1983) and Lee & Atkinson (1976).

Nematodes produce relatively enormous quantities of eggs – up to 200 000 a day in *Ascaris* (Lee & Atkinson, 1976) – and, clearly, an extensive metabolism of amino acids and proteins must operate. Proteins deposited in the oocyte are not used as sources of energy but as precursors for the proteins of the egg-shell. Protein metabolism has been a somewhat neglected area of nematode research, but the search for targets for the development of possible vaccines – especially against filariae – has recently served as a stimulus to work in this field.

Particular targets have been the proteinases, released by many species at various stages of their life cycle (for purposes such as skin penetration or food digestion), the cuticle of adult and larval worms, and the direct or indirect involvement of amino acids for synthetic or energy purposes.

○ 30.5.1 Proteolytic enzymes (Proteases)

Healer *et al.* (1991) divide proteolytic enzymes into four groups on the basis of active-site structure, mechanism of hydrolytic action and sensitivity to enzyme inhibitors, as follows: (*a*) metalloproteases, such as those secreted by the dog hookworm *Ancylostoma caninum* and by *Strongyloides stercoralis*; (*b*) serine proteases, such as the secreted elastinolytic enzyme from the microfilariae of *Onchocerca lienalis* and larvae of *Toxocara canis*; (*c*) proteases of the thiol (or cysteinyl) enzyme, which include a cysteine enzyme from *Haemonchus contortus*; (*d*) the acid protease of *Angiostrongylus cantonensis* and *Ascaris suum* (Maki *et al.*, 1982). The distribution of this enzyme and its activity on haemoglobin in the nematodes is shown in Table 30.5. In general, it can be stated that tissue-invasive helminth larvae secrete either serine or metalloproteases (McKerrow, 1989). Recent work on the characterisation of nematode proteolytic enzymes include studies on larval and adult *Nippostrongylus brasiliensis* (Healer *et al.*, 1991), *Ascaris suum* (Knox & Kennedy, 1988) and the larval and adult stages of the ovine gastro-intestinal nematodes *Nematodirus battus*, *Ostertagia circumcincta*, *Trichostrogylus colubriformis*, *T. vitrinus* and *Haemonchus contortus* (Knox & Jones, 1990). In general, enzyme activity was found to be higher in the L_3

larva than in the adult, and differences were observed between species and between stages.

○ 30.5.2 Amino acid metabolism

A variety of amino acids have been identified in hydrolysates of adult and larval nematodes, but no unusual patterns appear to occur. For example, in *Ascaris suum* and *A. lumbricoides* no differences between relative percentages of 17 amino acids were recorded although differences in the percentage of arginine and histamine between *Ascaris* spp. and *Toxocara canis* were observed (Lou *et al.*, 1987). Although it has generally been accepted that amino acids do not apparently act as an important source of energy in parasitic helminths, the recent finding that amino acids such as glutamine play a major role in the metabolism of the filariid *L. carinii* (Fig. 30.4), as described earlier (p. 443), and in other species, may cause this generalisation to be reviewed (see below).

The potential importance of amino acids in the energy metabolism is the observation that aerobically, but not anaerobically, amino acids can sustain motility as well as glycogen and ATP levels in *Nippostrongylus brasiliensis* as well as glucose can. The most effective amino acid in this case was found to be proline, which in combination with lysine, cysteine and phenylalanine, can completely replace glucose *in vitro* (Singh *et al.*, 1992*a*).

The first step in amino acid catabolism is the removal of the α-amino nitrogen, which can be achieved either by transamination or by oxidative deamination. The carbon skeletons remaining are converted to glycolytic or Krebs cycle or synthetic intermediates. Thus, *Ascaridia galli*, using 2-oxglutarate as an acceptor, transaminates asparate and alanine at significantly high rates, and valine, phenylalanine, leucine, isoleucine, arginine, tyrosine and methionine at moderate rates (Singh & Srivastava, 1983).

Heligmosomoides polygyrus has been shown to be capable of catabolising the branched-chain fatty acids leucine, isoleucine and valine, by pathways similar to those found in the mammalian liver (Grantham & Barrett, 1986*a,b*).

A number of 2-oxoglutarate transaminases have been reported in *Ascaris* spp. and other species, and the oxidative deamination of glutamate via glutamate dehydrogenase in *Haemonchus contortus* and *Ascaris* (Barrett, 1981). As mentioned above, glutamine plays a major role in the energy metabolism of *L. carinii*, a result deducted from the fact that the *in vitro* survival of adult *L. carinii* from the pleural cavity of the multimammate rat (*Mastomys natalensis*) was greatly enhanced by the presence of glutamine or alanine (Davies & Köhler, 1990). Moreover, the glutamine carbon could be completely recovered in the end products – CO_2, lactate and acetate – and the nitrogen in the additional excretory products, alanine and ammonia. Alanine, however, was only metabolised to a limited extent, the carbon balance being only 52 per cent. No other amino acids were found to support the worm's motility. Glutamine has also been shown to play a role in the energy metabolism of *Acanthocheilonema* (= *Dipetalonema*) *viteae*, *Onchocerca gibsoni*, *O. volvulus* and *Brugia pahangi* (Comley & Rees, 1989; MacKenzie *et al.*, 1989). Regarding the synthesis of proteins, the incorporation of amino acids into proteins has been demonstrated in both adult filariae and microfilariae, including *Setaria cervi*, *L. carinii* and *Dirofilaria immitis*, but little is known regarding the pathways involved (Barrett, 1983). Synthesis of collagen in the cuticle of nematodes, especially that of *Ascaris*, has received considerable attention; the relevant literature has been reviewed by Cox (1992) and Preston-Meek & Pritchard (1991). The structure and biosynthesis of cuticular proteins of lymphoid filarial parasites have been reviewed by Selkirk (1991).

The nitrogen metabolism of nematodes is essentially ammonitelic; ammonia may account for 27–71 per cent of the nitrogen excreted in different species. Appreciable quantities of urea (4–15 per cent), amino acids and peptides are also excreted by most species; small quantities of amines are also sometimes produced.

○○○ 30.6 Lipid metabolism

Studies of lipid metabolism in nematodes are important because lipids not only serve for storing and releasing energy in some species, but also form important constituents of cell membranes; in addition, large amounts of triacylglycerols and other lipids are incorporated into the oocysts and eggs. Nematodes which live under anaerobic conditions are not normally able to utilise lipids although a slight increase in total lipids in *Ascaris* during starvation has been reported (Beames *et al.*, 1968). Although carbohydrates form the major energy reserves in nematodes, under aerobic conditions lipids may be utilised for energy production, for example in the eggs of *Ascaris* and in the adult *Haemonchus contortus*, which has a fully functional Krebs cycle (Kapur & Sood, 1987; Ward & Huskisson, 1978).

Until recently, only limited studies on the lipid metabolism of nematodes had been carried out; early data are available for *Dirofilaria immitis*, *Ascaris lumbricoides*, *Setaria cervi* and *Litomosoides carinii* (Barrett, 1981). Most of these studies provided little information on the dynamics of lipid metabolism in nematodes. Recently, more detailed studies on *Haemonchus contortus* (Kapur & Sood, 1984, 1987) and *Trichuris globulosa* (an intestinal parasite of the goat) (Sarwal *et al.*, 1989) have provided more extensive data.

The adult males and females of *Trichuris globulosa* have been studied with regard to lipid composition, ability to incorpoate (Na)-1-[^{14}C]acetate into different lipid classes, and the activity of certain key enzymes of lipid metabolism (Sarwal *et al.*, 1989). A wide variety of lipids, including complex lipids, were identified. The females were found to contain more lipids than the males, particularly phospholipids and acylglycerols, probably to meet the requirements and synthetic structural needs for the daily production of large quantities of eggs. The presence of major enzymes of lipid biosynthesis were also established.

The metabolism of *H. contortus* has also been investigated in detail with reference to its ability to utilise lipids in relation to the total lipids, sterols, free fatty acids, acylglycerols and phospholipids produced during incubation *in vitro* (Kapur & Sood, 1987). All these components exhibited marked fluctuation (Fig. 30.5), sometimes increasing, sometimes decreasing, thus indicating that both lipid biosynthesis and utilisation were taking place. For example, after 4 h and 16 h, it was clear that a significant increase in total lipids occurred and this was followed by a decrease. Increase in lipid content may be accounted for by biosynthesis and the decrease, recorded during the initial incubation, by the expulsion of eggs and the turnover of membranes or lipids excreted/secreted. Changes in the fatty acid composition were also examined, and it was shown that *Haemonchus* was capable of using most of the fatty acids 12:1, 14:1, 16:0, 17:0, 18:0, 18:1, 18:2, 24:3 and 20, 21 and 22, at the same time synthesising 15:0, 16:0, 18, 20 and 21 carbon acids as evidenced by an increase in these. This species has also been shown to be capable of synthesising fatty acids from [^{14}C]acetate and glucose (Kapur & Sood, 1984).

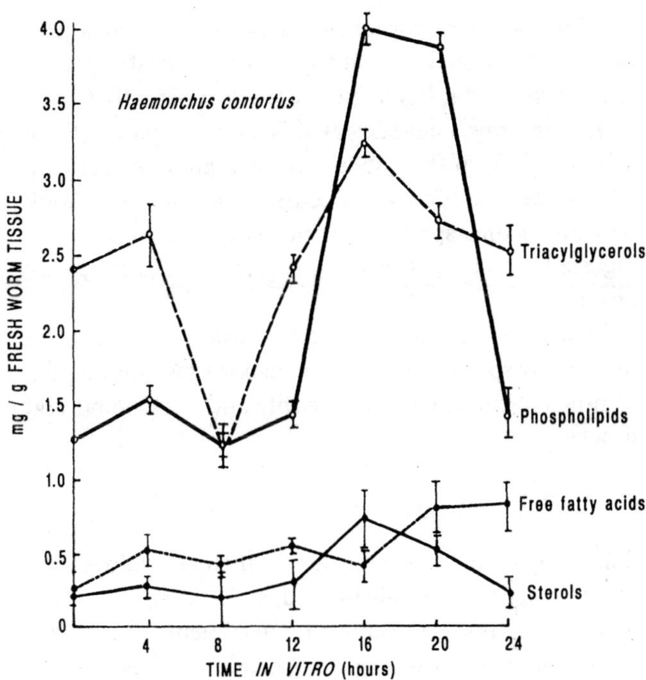

Fig. 30.5
Haemonchus contortus: quantitative changes in major fractions of lipid during incubation *in vitro*. (Modified from Kapur & Sood, 1987.)

○○○ 30.7 Nutrition

○ 30.7.1 Adults

The food of nematodes has already been discussed in Chapter 25. Very little detailed information, however, is available on the precise nature of the food requirements in chemical terms. The nature of such demands shows considerable variation during the life cycle. For example, an adult worm producing enormous quantities of eggs in the alimentary canal of a homoiothermic host will have greater nutritional demands than its free-living larva merely undergoing gradual growth and differentiation at air temperatures.

Carbohydrates and proteins form the main nutritional requirements of adult worms, with fats playing a relatively unimportant role (contrast eggs, p. 441). A high-carbohydrate diet is thus beneficial to the growth of most nematodes. Hormones, iron and certain unidentified growth factors may also be essential in most species.

A number of digestive enzymes, or enzymes related to feeding processes, have been detected in nematodes (Lee & Atkinson, 1976; McKerrow, 1989). These include amylase, invertase, pectinase, glycosidases, chitinase, cellulase, proteases (Table 30.5), esterases and lipases. Extra-corporeal digestion occurs as a preliminary to ingestion in many (perhaps most) species. In *A. lumbricoides*, holocrine secretion apparently occurs in the anterior intestine; in this process, parts of the cell detach from the gut and disintegrate in the lumen. Merocrine secretion also occurs in this species. Adult nematodes thus appear to be well equipped to deal with ingested food materials and contain batteries of enzymes.

In the strongylids, blood and tissue form the main diet. Many species, particularly hookworms, draw a plug of mucosa into their buccal cavity and suck blood and tissues.

○ 30.7.2 Larvae

The nematode egg contains sufficient endogenous reserves to enable a fully formed, active larva to develop without the absorption of further nutriment. Thereafter, food is required to supply energy and materials for synthesis of new tissue. Nematodes are peculiar in one respect, since, with a few exceptions, there is *no* somatic cell multiplication after hatching in spite of very considerable increase in size. This phenomenon of cell constancy is known as *eutely*. The major growth requirements of nematodes – prior to maturation – are thus for cytoplasmic rather than nuclear synthesis, and nucleic acids occur in nematodes in small quantities relative to the total tissue mass. In ovaries, the RNA : DNA ratio is about 4 : 1.

Free-living stages of nematodes (e.g. hookworms) must feed largely on bacteria; such stages are readily cultured in faecal–charcoal cultures, in which abundant bacteria are normally available.

○ 30.7.3 Filariae

The nutrition of filarial worms has been reviewed by Barrett (1983). Adult filariae have a functional gut. Erythrocytes have been observed in the digestive tract of *D. immitis* and the action of proteases determined (Table 30.5) (Maki *et al.*, 1982). In addition, adult worms appear to be able to take up nutrients via the cuticle. The transcuticular uptake of a range of substances has been demonstrated in various species. For example, it has been shown that *Brugia pahangi* can take up D-glucose, L-leucine, glycine, cycloglycine, adenosine, uracil, adenine, hypoxanthine and guanosine via the cuticle (Chen & Howells, 1981*a*,*b*; Barrett, 1983). In contrast to the adults, microfilariae have a non-functional gut and uptake must only take place via the cuticle. The uptake of glycine, uracil, adenine, hypoxanthine and guanine by *B. pahangi* microfilariae has been shown to take place (Chen & Howells, 1981*a*,*b*).

○○○ 30.8 Neurobiology

There is much evidence that acetylcholine and γ-aminobutyric acid can serve as neurotransmitters in nematodes; bigenic amines and peptide molecules probably also do so (reviews: Willett, 1980; Gration *et al.*, 1986). The immunocytochemical techniques which have provided such spectacular images of the nervous system in trematodes (Fig. 18.5) and cestodes (Fig. 24.8) have only recently been applied to nematodes (Davenport *et al.*, 1988). Thus an FMRFamide-like reaction has been detected in *Ascaris suum* using a peroxidase–anti-peroxidase technique. Positive reactions were recorded in the CNS and the peripheral nervous system. Immunoreactivity was also observed along the length of the main nerve cords and to lesser extent in the pharyngeal nerve cord. The strongest reactions were obtained in the anterior nerve ring, the cephalic papilla ganglion, the lateral ganglia and the dorsal-rectal ganglion. It has been speculated that the FMRFamide-like peptide could act as a neurotransmitter or neuromodulator of muscle contraction of the body wall, and possibly play an important role in nematode locomotion (Davenport *et al.*, 1988).

○○○ References

(References of general interest have been included in this list, in addition to those quoted in the text.)

Arevalo, J. I. & Saz, H. J. 1992. Effects of cholinergic agents on the metabolism of choline in muscle from *Ascaris suum*. *Journal of Parasitology*, **78**: 387–92.

Ash, L. R. 1989. Application of biotechnology to the study of nematodes. *Yonsei Reports on Tropical Medicine*, **20**: 68–72.

Barrett, J. 1976. Intermediate metabolism in *Ascaris* eggs. In: *Biochemistry of Parasites and Host–Parasite Relationships* (ed. H. van den Bossche), pp. 117–23. Elsevier/North Holland, Amsterdam.

Barrett, J. 1981. *Biochemistry of Parasitic Helminths*. Macmillan Publishers, London.

Barrett, J. 1983. Biochemistry of filarial worms. *Helminthological Abstracts*, **52**: 1–18.

Barrett, J. 1984. The anaerobic end-products of helminths. *Parasitology*, **88**: 179–98.

Barrett, J. 1988. The application of control analysis to helminth pathways. *Parasitology*, **97**: 355–62.

Barrett, J., Mendis, A. H., & Butterworth, P. E. 1986. Carbohydrate metabolism in *Brugia pahangi* (Nematoda: Filarioidea). *International Journal for Parasitology*, **16**: 465–9.

Beames, C. G. Jr., Jacobsen, N. S. & Harrington, G. W. 1968. Studies on the lipid metabolism of *Ascaris* during starvation. *Proceedings of the Oklahoma Academy of Science*, **47**: 40–4.

Brand, Th. von. 1973. *Biochemistry of Parasites*. Academic Press, New York.

Bryant, C. 1975. Carbon dioxide utilisation, and the regulation of respiratory metabolic pathways in parasitic helminths. *Advances in Parasitology*, **13**: 35–69.

Bryant, C. & Behm, C. A. 1989. *Biochemical Adaptation in Parasites*. Chapman & Hall, London.

Bueding, E. 1949. Studies on the metabolism of the filarial worm, *Litomosoides carinii*. *Journal of Experimental Medicine*, **89**: 107–30.

Cain, G. D. & Fairbairn, D. 1971. Procollagen proline hydroxylase and collagen synthesis in developing eggs of *Ascaris lumbricoides*. *Comparative Biochemistry and Physiology*, **40B**: 165–79.

Chen, S. N. & Howells, R. E. 1981*a*. The uptake of monosaccharides, disaccharide and nucleic acid precursors by adult *Dirofilaria immitis*. *Annals of Tropical Medicine and Parasitology*, **73**: 473–86.

Chen, S. N. & Howells, R. E. 1981*b*. *Brugia pahangi*: uptake and incorporation of nucleic acid precursors of microfilariae and macrofilariae *in vitro*. *Experimental Parasitology*, **51**: 296–306.

Comley, J. C. W. & Rees, M. J. 1989. Radiorespirometric detection of macrofilaricidal activity *in vitro*. *Parasitology*, **98**: 259–64.

Cox, G. N. 1992. Molecular and biochemical aspects of nematode collagens. *Journal of Parasitology*, **78**: 1–15.

Croll, N. A. & Matthews, B. E. 1977. *Biology of Nematodes*. Blackie, London.

Davenport, T. R. B., Lee, D. L. & Isaac, R. E. 1988. Immunocytochemical demonstration of a neuropeptide in *Ascaris suum* (Nematoda) using an antiserum to FMRFamide. *Parasitology*, **97**: 81–8.

Davies, K. P. & Köhler, P. 1990. The role of amino acids in the energy generating pathways of *Litomosoides carinii*. *Molecular and Biochemical Parasitology*, **41**: 115–24.

Dunn, T. S., Raines, P. S., Barrett, J. & Butterworth, P. E. 1988. Carbohydrate metabolism in *Onchocerca gutturosa* and *Onchocerca lienalis* (Nematoda: Filarioidea). *International Journal for Parasitology*, **18**: 21–6.

Fairbairn, D. 1960. The physiology and biochemistry of nematodes. In: *Nematology* (ed. J. N. Sasser & W. R. Jenkins), Chapter 30, pp. 267–96. University of North Carolina Press, Chapel Hill.

Fairbairn, D. 1969. Lipid components and metabolism of Acanthocephala and Nematoda. *Chemical Zoology*, **3**: 361–78.

Grantham, B. & Barrett, J. 1986*a*. Amino acid catabolism in the nematodes *Heligmosomoides polygyrus* and *Panagrellus redivivus* 1. Removal of the amino group. *Parasitology*, **93**: 481–93.

Grantham, B. & Barrett, J. 1986*b*. Amino acid catabolism in the nematodes *Heligomosomoides polygyrus* and *Panagrellus redivivus* 2. Metabolism of the carbon skeleton. *Parasitology*, **93**: 495–504.

Gration, K. A. F., Harrow, I. D. & Martin, R. J. 1986. GABA receptors in parasites of veterinary importance. In: *Neuropharmacology and Pesticide Action* (ed. M. G. Ford, G. C. Lunt, R. C. Reay & P. N. R. Usherwood), pp. 414–22. Ellis Horwood, Chichester, UK.

Healer, J., Ashall, F. & Maizels, R. M. 1991. Characterization of proteolytic enzymes from larval and adult *Nippostrongylus brasiliensis*. *Parasitology*, **103**: 305–14.

Hutchison, W. F., Turner, A. C., Grayson, D. P. & White, H. B. 1976. Lipid analysis of the adult dog heartworm, *Dirofilaria immitis*. *Comparative Biochemistry and Physiology*, **53B**: 495–7.

Kapur, J. & Sood, M. L. 1984. *Haemonchus contortus*: lipid biosynthesis from C^{14}-labelled acetate and glucose. *Zentralblatt für Veterinarmedizin, Reihe B*, **31**: 225–30.

Kapur, J. & Sood, M. L. 1987. Changes in lipids and free fatty acid fractions in adult *Haemonchus contortus* during incubation *in vitro*. *Veterinary Parasitology*, **23**: 95–103.

Kilgore, M. W., Stewart, G. L. & Lou, M. 1986. Chemical composition of newborn larvae, muscle larvae and adult *Trichinella spiralis*. *International Journal for Parasitology*, **16**: 455–60.

Knox, D. P. & Jones, D. G. 1990. Studies on the presence and release of proteolytic enzymes (proteinases) in gastrointestinal nematodes in ruminants. *International Journal for Parasitology*, **20**: 243–9.

Knox, D. P. & Kennedy, M. W. 1988. Proteinases released by the parasitic larval stages of *Ascaris suum*, and their inhibition by antibody. *Molecular and Biochemical Parasitology*, **28**: 207–16.

Köhler, P. 1985. The strategies of energy conservation in helminths. *Molecular and Biochemical Parasitology*, **17**: 1–18.

Köhler, P. 1991. The pathways of energy generation in filarial parasites. *Parasitology Today*, **7**: 21–5.

Komuniecki, R. & Komuniecki, P. R. 1989. *Ascaris suum*: a useful model for anaerobic mitochondrial metabolism and the aerobic–anaerobic transition in developing parasitic helminths. In: *Comparative Biochemistry of Parasitic Helminths* (ed. E.-M. Bennet, C. Behm & C. Bryant), pp. 1–12. Chapman & Hall, London.

Komuniecki, P. R. & Vanover, L. 1987. Biochemical changes during aerobic–anaerobic transition in *Ascaris suum*. *Molecular and Biochemical Parasitology*, **22**: 241–8.

Lee, D. L. & Atkinson, H. J. 1976. *Physiology of Nematodes*. 2nd Edition. The Macmillan Press, London.

Lou, Z. J. *et al.* 1987. Comparative study of amino acid contents and LDH isoenzymes by electrophoresis and slab-PAGE in adult worms of *Ascaris suum* and *Toxocara canis*. *Chinese Medical Journal*, **100**: 740–4.

MacKenzie, N. E., Waa, E. A. Van de, Gooley, P. R., Williams, J. F., Bennett, J. L., Bjorge, S. M., Baille,

T. A. & Geary, T. G. 1989. Comparison of glycolysis and glutaminolysis in *Onchocerca volvulus* and *Brugia pahangi* by ^{13}C nuclear magnetic resonance spectroscopy. *Parasitology*, **99**: 427–35.

McKerrow, J. H. 1989. Parasite proteases. *Experimental Parasitology*, **68**: 111–15.

Maki, J., Furuhashi, A. & Yanagisawa, T. 1982. The activity of acid proteases hydrolysing haemoglobin in parasitic helminths with special reference to interspecific and intraspecific distribution. *Parasitology*, **84**: 137–47.

Onwuliri, C. O. E. 1985. Energy metabolism in the developing larval stages of *Ancyclostoma tubaeforme* and *Haemonchus contortus*: glycolytic and tricarboxylic acid cycle enzymes. *Parasitology*, **90**: 169–77.

Oordt, B. E. P. Van, Tielens, A. G. M. & Berg, S. G. 1989. Aerobic to anaerobic transition in the carbohydrate metabolism of *Schistosoma mansoni* cercariae during transformation *in vitro*. *Parasitology*, **98**: 409–15.

Pasternak, J. J. 1988. Molecular biology of nematodes: some recent studies on *Panagrellus* and *Ascaris*. *Canadian Journal of Zoology*, **66**: 2591–9.

Precious, W. Y. & Barrett, J. 1989. Xenobiotic metabolism in helminths. *Parasitology Today*, **5**: 156–60.

Preston-Meek, C. M. & Pritchard, D. I. 1991. Synthesis and replacement of nematode cuticle components. In: *Parasitic Nematodes – Antigens, Membranes and Genes* (ed. M. W. J. Kennedy), pp. 84–94. Taylor & Francis, London.

Ramp, Th. & Köhler, P. 1984. Glucose and pyruvate catabolism in *Litomosoides carinii*. *Parasitology*, **89**: 229–44.

Samanta, S., Bose, M., Roy, M., Maity, C. R. & Majumdar, G. 1988. Some biochemical observations on the cuticle of *Ascaridia galli* (Nematoda: Heterakidae). *Indian Journal of Helminthology*, **5**: 47–9.

Sarwal, R., Sanyal, S. N. & Khera, S. 1989. Lipid metabolism in *Trichuris globulosa* (Nematoda). *Journal of Helminthology*, **63**: 287–97.

Saz, H. J. 1981*a*. Energy metabolisms of parasitic helminths: adaptations to parasitism. *Annual Review of Physiology*, **43**: 323–41.

Saz, H. J. 1981*b*. Biochemical aspects of filarial parasites. *Trends in Biochemical Science*, **6**: 117–19.

Saz, H. J. 1989. Helminths: primary models for comparative biochemistry. *Parasitology Today*, **6**: 92–3.

Saz, H. J., Lescure, O. L. & Beuding, E. 1968. Biochemical observations of *Ascaris suum* lung-stage larvae. *Journal of Parasitology*, **54**: 457–61.

Selkirk, M. E. 1991. Structure and biosynthesis of cuticular proteins of lymphatic filarial parasites. In: *Parasitic Nematodes – Antigens, Membranes and Genes* (ed. M. W. J. Kennedy), pp. 27–45. Taylor & Francis, London.

Singh, S. P., Gupta, S., Katiyar, J. C. & Srivastava, V. M. L. 1989. Amino acid decarboxylases of *Ancylostoma ceylanicum* and *Nippostrongylus brasiliensis*. *Current Science*, **58**: 1353–6.

Singh, S. P., Gupta, S., Katiyar, J. C. & Srivastava, V. M. L. 1992*a*. On the potential of amino acids to support survival and energy status of *Nippostrongylus brasiliensis in vitro*. *International Journal for Parasitology*, **22**: 131–3.

Singh, S. P., Katiyar, J. C. & Srivastava, V. M. L. 1992*b*. Enzymes of the tricarboxylic acid cycle in *Ancylostoma ceylanicum* and *Nippostrongylus brasiliensis*. *Journal of Parasitology*, **78**: 24–9.

Singh, G., Pampori, N. A. & Srivastava, V. M. L. 1983. Metabolism of amino acids in *Ascaridia galli* decarboxylation reactions. *International Journal for Parasitology*, **13**: 305–7.

Singh, G. & Srivastava, V. M. L. 1983. Metabolism of amino acids in *Ascaridia galli*: transamination. *Zeitschrift für Parasitenkunde*, **69**: 783–8.

Srivastava, V. M. L., Ghatak, S. & Murti, C. R. K. 1970. *Ascaridia galli*: lactic acid production, glycogen content, glycolytic enzymes and properties of purified aldolase, enolase and glucose-6-phosphate dehydrogenase. *Parasitology*, **60**: 157–80.

Stewart, G. L. 1983. Biochemistry. In: *Trichinella and Trichinosis* (ed. W. C. Campbell), pp. 153–72. Plenum Press, New York.

Sylk, S. R., Stromberg, B. E. & Soulsby, E. J. L. 1974. Development of *Ascaris suum* larvae from the third to the fourth stage *in vitro*. *International Journal for Parasitology*, **4**: 261–5.

Van de Waa, E. A., McKenzie, N. E., Geary, T. G., Bennett, J. & Williams, J. F. 1988. *Tropical and Medical Parasitology*, **39**: 85. (Abstract.) (Quoted by McKenzie *et al.*, 1989).

Vickery, A. C. 1989. Application of biotechnology to study of filarial parasites. *Yonsei Reports on Tropical Medicine*, **20**: 73–4.

Wang, E. J. & Saz, H. J. 1974. Comparative biochemical studies on *Litomosoides carinii*, *Dipetalonema vitae*, and *Brugia pahangi* adults. *Journal of Parasitology*, **60**: 316–21.

Ward, P. F. V. 1982. Aspects of helminth metabolism. *Parasitology*, **84**: 177–94.

Ward, P. F. V. & Huskisson, N. S. 1978. The energy metabolism of adult *Haemonchus contortus in vitro*. *Parasitology*, **77**: 255–71.

Willett, J. D. 1980. Control mechanism in nematodes. In: *Nematodes as Biological Models*, Vol. 1 (ed. B. H. Zuckermann), pp. 197–225. Academic Press, New York.

Wilson, P. A. G. 1965. Changes in lipid and nitrogen content of *Nippostrongylus brasiliensis* infective larvae at constant temperature. *Experimental Parasitology*, **16**: 190–4.

Winkfein, R. J., Pasternak, J., Mudry, T. & Martin, L. H. 1985. *Ascaris lumbricoides*: characterization of the collagenous components of the adult cuticle. *Experimental Parasitology*, **59**: 197–203.

Wisnewski, N., Saz, H. J., Mössinger, J., De Bruyn, B. S. & Weinstein, P. P. 1990. Biochemical analyses of secretory and excretory products of adult *Dipetalonema viteae* in culture. *Journal of Parasitology*, **76**: 302–6.

Wood, W. B. (ed.) 1988. *The nematode Caenorhabditis elegans*. Cold Spring Harbor Laboratory, Cold Spring Harbor, New York.

Zuckermann, B. B. 1980. *Nematodes as Biological Models*. 2 vols. Academic Press, New York.

31 Acanthocephala

The Acanthocephala represent a group of worms parasitic in all classes of vertebrates but especially in fish and birds. They show similarities with both the Platyhelminthes and the Nematoda, but no agreement has been reached on their phylogenetic position. They resemble Cestoda in lacking an alimentary canal. The chief diagnostic feature of the group is the presence of an invaginable proboscis armed with hooks from which the common name 'spiny-headed' worm is derived. They range in size from 1.5 mm to over 0.5 m. The majority of species are small, the largest being *Macracanthorhynchus hirudinaceus* from pigs. The females are nearly always larger than the males. The intermediate hosts are usually arthropods, and paratenic hosts frequently occur. In their general anatomy, life history and habits they show a remarkable degree of uniformity.

Literature. Early general reviews are those of Crompton (1970), Nicholas (1967, 1973) and Yamaguti (1963); more recent reviews are those of Amin (1985, 1987), Bauer & Skryabina (1987), Crompton (1989), Crompton & Nickol (1985), Khatoon & Bilqees (1991), Khokhlova (1986), and Parshad & Crompton (1981).

31.1 Occurrence

31.1.1 Host lists

Adult Acanthocephala are readily available, particularly from fish; some species commonly found in the USA, Canada and Europe are listed in Table 31.1. Other detailed host lists for these and other countries are those of Bauer & Skryabina (1987), Bilqees & Khan (1987), Chappell & Owen (1969), Chubb (1982) and Khokhlova (1986). Valuable lists of intermediate and paratenic hosts are those of Buron & Golvan (1986) and Schmidt (1985).

31.1.2 Useful experimental species

The most useful laboratory species are those in rodents and birds, such as:

Moniliformis moniliformis (= *dubius*)

This species has been widely used as an experimental model for research on the Acanthocephala, as its definitive host (the rat) and its intermediate host (the cockroach) are readily maintained in the laboratory. The adult worm in the rat gut releases *shelled acanthors* – usually referred to as '*eggs*' (Fig. 31.1) – which, when eaten by a cockroach, hatch in the lower midgut region, penetrate the tissues and develop in the haemocoele. The fully developed larva or *cystacanth* (Fig. 31.1) if eaten by a rat attaches and develops to an adult worm in about 5 weeks; other developmental stages, as described by Moore (1946) and King & Robinson (1967) are shown in Fig. 31.1. The laboratory maintenance, biology, ultrastructure, physiology and transmission dynamics have been described by numerous authors, e.g. Budziakowski *et al.* (1984), Crompton (1970), Crompton *et al.* (1984), Lackie (1972), Moore (1946), Monks & Nickol (1989) and Stoddart *et al.* (1991). The reproductive system is illustrated in detail in Asaolu (1980, 1981). Other species often readily available are:

Polymorphus spp.

Several species of *Polymorphus* which develop in birds also make suitable laboratory hosts. Some species have cystacanths which are brightly coloured orange-red, owing to the presence of a carotenoid, with the result that infected specimens of the intermediate hosts (*Gammarus* spp.) can be readily identified and collected. *Polymorphus minutus* is a common species in ducks in Europe but is less common in N America. Its life cycle

Table 31.1. *Some common species of Acanthocephala in fish in Europe and North America*

Species	Fish host[a]	Common intermediate host[b]	Distribution
Pomphorhynchus bulbocolli	Rainbow darter	*Gammarus* spp.	North America
Pomphorhynchus laevis	Roach, dace and others	*Gammarus* spp.	Europe
Echinorhynchus langeniformis	Starry flounder	*Corophium spinicorne*	North America
Echinorhynchus salmonis	Trout, eel	*Hyalella azteca*	Europe
Echinorhynchus truttae	Trout, roach and others	*Gammarus* spp.	Europe
Neoechinorhynchus cylindratus	Large-mouth bass and others	*Cypria globula*	North America
Neoechinorhynchus saginatus	Creek chub, fallfish	*Cypridopsis vidua*	North America
Neoechinorhynchus rutili	Trout, grayling, salmon	*Cypria turneri*	Europe
Acanthocephalus clavula	Grayling, chub and others	*Asellus meridianus*	Europe
Acanthocephalus dirus	Rainbow darter	*Asellus* spp.	North America
Acanthocephalus anguillae	Perch, trout and others	*Asellus aquaticus*	Europe
Acanthocephalus lucii	Perch, roach and others	*Asellus aquaticus*	Europe
Leptorhynchoides thecatus	Large-mouth bass and others	*Hyalella azteca*	North America

[a]Detailed lists of intermediate and definitive hosts are given in Brown *et al.* (1986), Buron & Golvan (1986), Chappell & Owen (1969), Chubb (1982), Golvan & Buron (1988), Schmidt (1985).

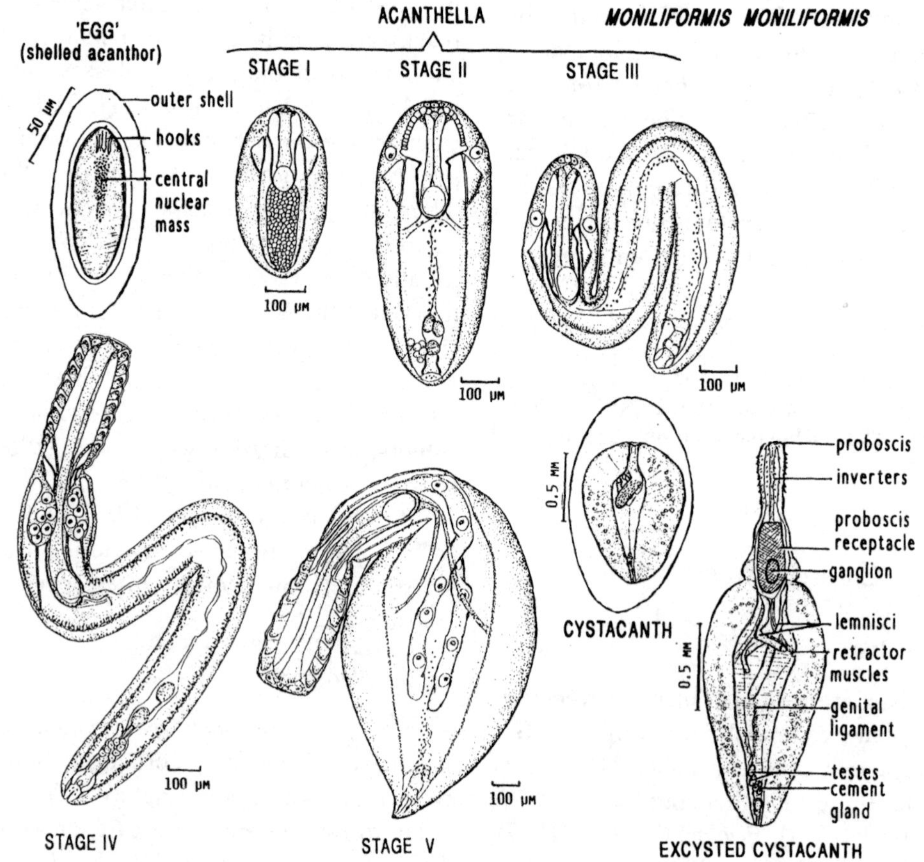

Fig. 31.1
Moniliformis moniliformis (= *dubius*). Stages of development of the cystacanth in the cockroach haemocoel; excystation takes place in the rat intestine. (Modified from Moore, 1946; and King & Robinson, 1967.)

and biology are well documented (Crompton, 1970; Crompton & Nickol, 1985; Hynes & Nicholas, 1963). In N America, the cystacanths of *P. contortus*, *P. marilis* and *P. paradoxus* in *Gammarus lacustris* are also bright orange; their life cycles have been described by Podesta & Holmes (1970).

Acanthocephalus ranae

A common species in *Rana temporaria*, *R. esculenta* and *Bufo bufo* (Crompton, 1970; Smyth & Smyth, 1980) whose morphology and life cycle have been described by Gassman (1972) and Grabda-Kazubska (1962). Its intermediate hosts are *Asellus aquaticus* or *Gammarus apulex* (Cox, 1971).

Macracanthorhynchus hirudinaceus

This is a cosmopolitan parasite of wild and domestic pigs with larval stages in scarabaeid beetles. Its large size has made it a valuable model for work on the nervous system and other fundamental studies (Crompton & Nickol, 1985). It occasionally infects man where it may cause intestinal perforation (Radomyos *et al.*, 1989).

Species in fish (Table 31.1). Many of the species in fish can be readily maintained in a suitably equipped laboratory. A list of those which have been maintained experimentally is given in Golvan & Buron (1988). For example, methods for *Leptorhynchoides thecatus* and *Neoechinorhynchus cylindricus*, common in N American fish, are given by Olsen (1967) and for *Acanthocephalus lucii* in Europe by Brattey (1986, 1988). Many species in fish cause intestinal pathology, e.g. *Neoechinorhynchus* sp. Szalai & Dick (1987) have reviewed the relevant literature.

○○○ 31.2 General account

○ 31.2.1 General morphology

The functional morphology has been comprehensively reviewed by Miller & Dunagan (1985); earlier reviews are those of Crompton (1970) and Nicholas (1967, 1973).

Body. The body consists of an anterior *praesoma* with a characteristic spiny proboscis and a neck free from spines, and a posterior *trunk* or *metasoma*. The spines on the proboscis are symmetrically arranged and of diagnostic value. The proboscis (and sometimes the neck) is usually retracted into a proboscis sac by being completely inverted, and the whole praesoma is also retractile. Retraction is carried out by means of retractor muscles (Fig. 31.2). Two lateral protrusions from the body wall at the base of the neck are known as the *lemnisci* (Fig. 31.2). These are concerned with fluid movement in relation to proboscis activity.

The body wall. This consists of a thick tegument which contains a system of *lacunar canals* or *channels* which seem to distribute absorbed food material (Fig. 31.3). It consists essentially of five layers, for which various authors have used different terminology. The account below is based on Crompton (1970). The outermost or *epicuticle* appears to contain largely mucopolysaccharide (Crompton, 1970), presumably equivalent to the glycocalyx of cestodes. Beneath the epicuticle lies a tough *cuticle* perforated by numerous pores which in turn lead into canals and ducts which make up the *striped layer*. The latter merges into a fibrous *felt layer* beneath which is a *radial layer* lacking cell walls but containing nuclei, mitochondria, ribosomes, folded plasma membranes, lipid and glycogen. This layer is probably the most metabolically active part of the body wall. Beneath

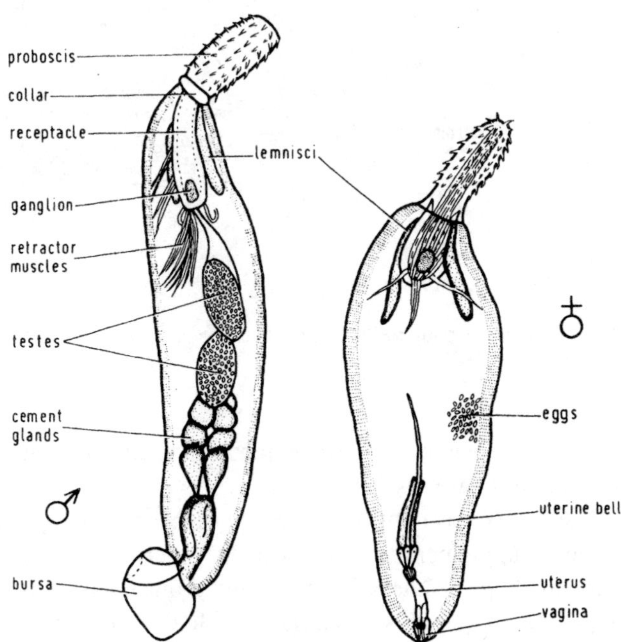

Fig. 31.2
Acanthocephalus sp.: morphology of adult male and female. (After Yamaguti, 1935; Van Cleave, 1915.)

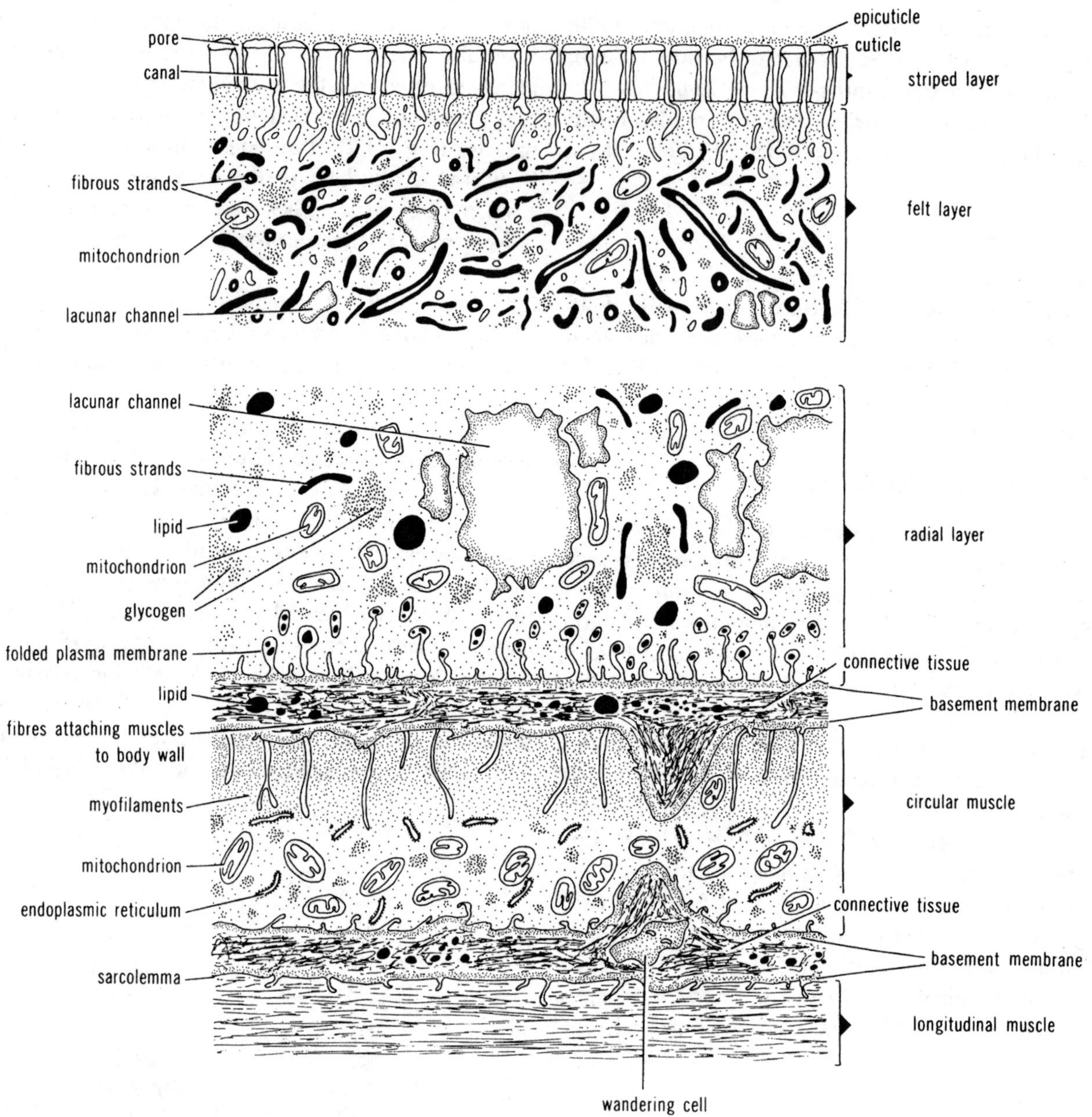

Fig. 31.3
Fine structure of the body wall of *Polymorphus minutus*. (Adapted from Crompton & Lee, 1965.)

these layers are a thick basement membrane and circular and longitudinal muscles.

The above account is based on light microscope studies supplemented by electron microscopy. A number of species have been examined but (as in cestodes) the body wall shows a remarkable degree of uniformity.

Representative ultrastructural studies are those on: *Polymorphus minutus* (Crompton & Lee, 1965), *Moniliformis dubius* (Nicholas & Mercer, 1965; Wright & Lumsden, 1970).

Digestive system. This is lacking, as in cestodes. Absorption of materials takes place directly through the body wall.

Excretory system. This has been reviewed by Miller & Dunagan (1985). Two types of flame cells may occur, the *dendritic* type and the *sac* type. The dendritic type has a central canal from which smaller canals lead to blind pouches containing cilia; the sac type consists of a bladder-like receptacle into which the flame cells open directly.

Nervous system. This is reduced to a cerebral ganglion in the wall of the proboscis and two nerve trunks which run posteriorly from the ganglion to the body wall. In the male there is a pair of genital ganglia in the penis base. The literature on the nervous system has been reviewed by Miller & Dunagan (1985) who point out that only some 14 studies have been carried out on this topic. They have shown (Dunagan & Miller, 1981) that in *Oligoacanthorhynchus tortuosa* (from a possum) the cerebral ganglion consists of 38 cells on each side with 7 cells centrally located and provide a map of the cell distribution. The fine structure of the nervous system of *M. moniliformis* has been described in detail by Budziakowski *et al.* (1984).

Nuclear constancy, chromosomes, sex determination. The group exhibits the interesting phenomenon of *eutely* or *nuclear constancy*, the number of nuclei in some species remaining constant throughout life (Miller & Dunagan, 1985). In the Neoechinorhynchidae, for example, there are only 6 nuclei in the tegument. In some species the nuclei fragment during later development. The number of chromosomes range from 6 to 12 and both the XY–XX and XX–XX sex determination systems are found (Crompton, 1985). Remarkably, polyploidy has been reported in the nuclei of all stages of *Moniliformis moniliformis* (Marshall *et al.*, 1973) with DNA contents of up to 343*n* being recorded (Table 31.2).

Table 31.2. *Polyploidy in Acanthocephala*

Ploidy values for the nuclei of larval stages (see Fig. 31.1) of *Moniliformis moniliformis (= dubius)* from the cockroach, as measured by Feulgen microspectrophotometry.

Type of nucleus and developmental stage	Ploidy (n = haploid DNA)
Muscle nuclei	
Acanthella stage III	4–5n
Cystacanth	5–8n
Cortical nuclei	
Acanthella stage I	39n
Acanthella stage II	147n
Acanthella stage III	112–220n
Acanthella stage IV	136n
Acanthella stage V	207n
Cystacanth	136–294n
Lemniscal nuclei	
Acanthella stage III	115–225n
Cystacanth	157–343n

Source: Data from Marshall *et al.* (1973).

○ 31.2.2 Reproductive system

This system has been reviewed by Crompton (1985, 1989) and Parshad & Crompton (1981).

Ligament sac and ligament strand. The ligament sacs are hollow tubes of connective tissue peculiar to the Acanthocephala; these enclose the gonads. In front, they are attached to the proboscis sac or the adjacent body wall, and posteriorly they connect to some part of the reproductive system. The number and arrangement of the ligament sacs varies in the different orders. Thus there are two (a dorsal and a ventral) in the females of the Archi- and Eoacanthocephala, but in the males of the same orders only the dorsal sac is present. In the Palaeacanthocephala, there is a single ligament sac in both sexes.

The *ligament strand* is a nucleated strand to which the gonads in both sexes are attached. This strand is believed to represent the endoderm and the body space between this and the body wall may be considered a pseudocoele.

Male system (Fig. 31.4). There are usually two testes attached to the ligament strand and enclosed in the ligament sac; monorchis males (with one testis) occur in some species. In the Archiacanthocephala and the Palaeacanthocephala, there occur *cement glands*, six to eight in number, with ducts leading into the common sperm duct. In the Eoacanthocephala, however, the cement glands form a syncytial mass from which a duct enters a *cement reservoir* from which, in turn, ducts enter the sperm duct (see below). The secretion of the cement glands seals the everted male bursa around the female gonopore.

Continuous with the ligament sac is a muscular tube, the genital sheath, which encloses the posterior ends of the sperm ducts, the cement glands and the pro-

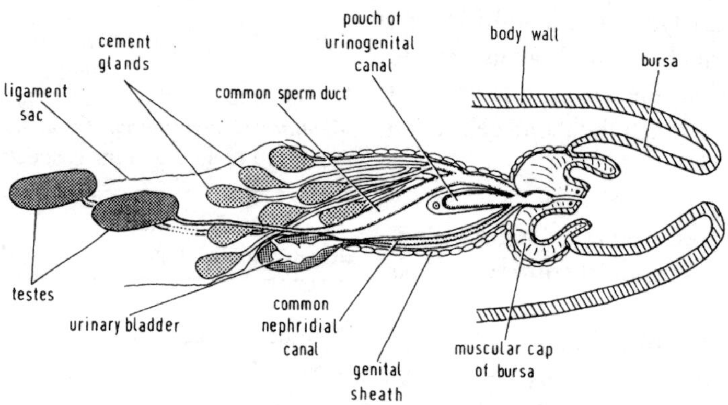

Fig. 31.4
Male reproductive system of *Hamanniella.* (Adapted from Kilian, 1932.)

tonephridial canals when present. There is a short penis which projects into a hemispherical or elongated bursa, a cavity made up of the in-turned body wall with a proximal muscular thickening, the muscular cap. The bursa can be everted to the exterior and is used in holding the female during copulation.

Female system (Fig. 31.2). This shows a number of peculiarities. The ovary or ovaries (two) which are visibly intact only early in life, break up into *ovarian balls*, each consisting of a syncytium from which ovogonia are formed. These soon come to be free; in the Palaeacanthocephala the sac ruptures and they fill the body cavity. The ligament strands lead to a muscular organ termed the *uterine bell*. Posteriorly this opens into a *uterine tube*, bearing two *bell pouches*, which connects to a muscular uterus and a non-muscular vagina opening to the exterior. The vaginal aperture is guarded by a sphincter.

The uterine bell serves as a selective apparatus. Ovic larvae are taken in by the uterine bell and passed through the uterus to the vagina and so to the exterior. Immature eggs are rejected and returned via an aperture in the ventral bell to the dorsal pseucocoele. The selective mechanism in *P. minutus* appears to operate on egg length, most eggs less than 100 μm being immature (Whitfield, 1970). The uterine bell is continuous with the dorsal ligament sac in the Archi- and Eoacanthocephala. In the Palaeacanthocephala, the single dorsal strand connects to the inner side of the uterine bell.

Copulation and fertilisation. The male bursa is used for grasping the female and the penis is inserted into the vagina and sperm discharged. The secretion from the cement glands is then poured over the whole posterior tip of the female and the gonopore securely sealed. The secretion persists for a time as a hard brown 'copulation cap' (Nicholas, 1967).

Structure of the egg-shell (Fig. 31.1). The basic structure of the egg-shell in Acanthocephala has been reviewed by Bilqees & Khatoon (1988), Crompton (1970) and Marchand (1984*a*) although comparisons are difficult as different authors have used different terminology. The ultrastructure has been reviewed by Marchand (1984*b*). Four membranes are usually present inside the shell: a thin outer membrane, a fibrillar coat, a fertilisation membrane and an inner membrane; the fertilisation membrane and the outer membrane are variously modified in different species.

○ 31.2.3 Life cycles

The life cycles of many species have been worked out in detail (Nickol, 1985). For those parasitic in land animals, the intermediate hosts are usually insects (larvae, beetles, cockroaches). For fish parasites, the intermediate hosts are amphipods (especially *Gammarus*) or isopods (especially *Asellus*); see Table 31.1. Some developmental stages of *Polymorphus minutus* are shown in Fig. 31.5. The time for hatching of the eggs (reviewed by Crompton, 1970) varies with species, two hours being the earliest reported time for *M. dubius* in the cockroach; eggs of *Neoechinorhynchus saginatus* hatch in ostracods within one hour.

Hatching is markedly affected by pH, a pH greater than 7.5 being necessary (Fig. 31.6) for optimal results; a molarity of over 0.2 is also essential. In the presence of carbon dioxide, however, hatching takes place at a

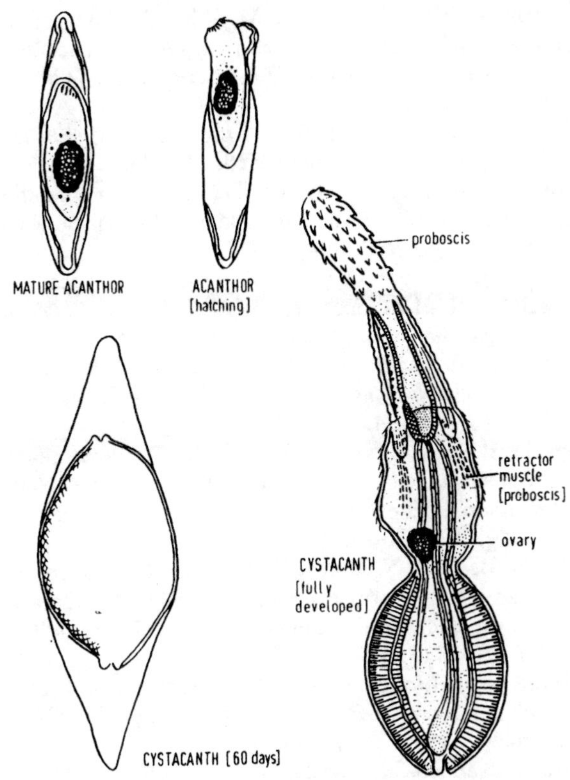

Fig. 31.5
Polymorphus minutus: morphology of developmental stages (after Hynes & Nicholas, 1957). Mature acanthor, from body cavity of female worm; acanthor, hatching in *Gammarus* gut; 60-day cystacanth, from body cavity of *Gammarus*; fully developed cystacanth, 24 h after being fed to a duck, hind end still retracted.

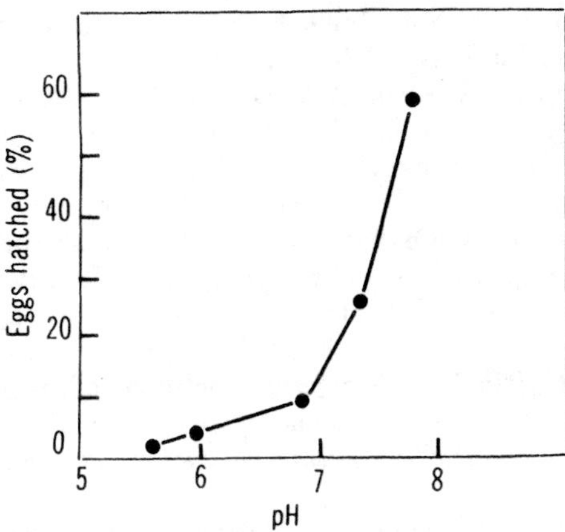

Fig. 31.6
Effect of pH on the hatching of the eggs of *Moniliformis dubius*. (After Edmonds, 1966.)

lower pH; a chitinase may be involved in the hatching process (Edmonds, 1966). In *P. minutus* bile salts appear to be the activating agents (Lackie, 1974).

○ 31.2.4 Classification

Recent reviews of the classification are those of Amin (1985, 1987) and Khatoon & Bilqees (1991). Three classes are generally recognised namely the Archiacanthocephala, Palaecanthocephala and the Eoacanthocephala. Amin (1987) has recommended that a new class – Polyacanthocephala should be created to accommodate the family Polyacanthorhynchidae. A valuable detailed classification down to species level is given in Amin (1985).

Class 1 **Archiacanthocephala.** Hooks on proboscis either in long rows or in a few circles; spines lacking on trunk; main lacunar vessels dorsal and ventral; both dorsal and ventral ligament sacs present in female; cement glands separate, with eight present in males; in terrestrial hosts, e.g. *Moniliformis dubius* (p. 451).

Class 2 **Palaeacanthocephala.** Hooks on proboscis usually in long rows; spines on trunk; nuclei in hypodermis usually fragmented; main lacunar vessels lateral; ligament sac in female single; cement glands separate; mostly aquatic hosts, e.g. *Leptorhynchoides thecatus* (in fish; Table 31.1).

Class 3 **Eoacanthocephala.** Hooks on proboscis usually in a few small circles; few nuclei (large) in hypodermis; main lacunar vessels dorsal and ventral; dorsal and ventral ligament sacs in female; cement glands syncytial e.g. *Neoechinorhynchus emydis* (in turtles).

○○○ References

(References of general interest have been included in this list, in addition to those quoted in the text.)

Amin, O. M. 1985. Classification. In: *Biology of the Acanthocephala* (ed. D. W. T. Crompton & B. B. Nickol), pp. 27–72. Cambridge University Press.

Amin, O. M. 1987. Key to the families and subfamilies of Acanthocephala, with the erection of a new class (Polyacanthocephala) and a new order (Polyacanthorhynchida). *Journal of Parasitology*, **73**: 1216–19.

Asaolu, S. O. 1980. Morphology of the reproductive system of female *Moniliformis dubius* (Acanthocephala). *Parasitology*, **81**: 433–46.

Asaolu, S. O. 1981. Morphology of the reproductive system of the male *Moniliformis dubius*. *Parasitology*, **82**: 297–309.

Bauer, O. N. & Skryabina, E. S. 1987. (Class Acanthocephala.) (In Russian.) Nauka, Leningrad.

Bilqees, F. M. & Khan, A. 1987. Acanthocephala of fishes in Pakistan. Biological Society of Pakistan, Monograph No. 13. Parasitological Section, Department of Zoology, University of Karachi, Karachi-32, Pakistan.

Bilqees, F. M. & Khatoon, N. 1988. Egg shell structure in Acanthocephala – a review. *Proceedings of Parasitology*, **5**: 59–73. Parasitology Section, Department of Zoology, University of Karachi, Karachi-32, Pakistan.

Brattey, J. 1986. Life history and population biology of larval *Acanthocephalus lucii* (Acanthocephala: Echinorhynchidae) in the isopod *Asellus aquaticus*. *Journal of Parasitology*, **72**: 633–45.

Brattey, J. 1988. Life history and population biology of adult *Acanthocephalus lucii* (Acanthocephala: Echinorhynchidae). *Journal of Parasitology*, **74**: 72–80.

Brown, A. F., Chubb, J. C. & Veltkamp, C. J. 1986. A key to the species of Acanthocephala parasitic in British freshwater fish. *Journal of Fish Biology*, **28**: 327–34.

Budziakowski, M. E., Mettrick, D. F. & Webb, R. A. 1984. Ultrastructural morphology of the nerve cells in the cerebral ganglion of the acanthocephalan *Moniliformis moniliformis*. *Journal of Parasitology*, **70**: 719–34.

Buron, I. De. & Golvan, Y. J. 1986. Les hôtes des acanthocéphales. I – Les hôtes intermediaires. *Annales de Parasitologie Humaine et Comparée*, **61**: 581–92.

Chappell, L. H. & Owen, R. H. 1969. A reference list of parasite species recorded in freshwater fish from Great Britain and Ireland. *Journal of Natural History*, **3**: 197–216.

Chubb, J. C. 1982. Seasonal occurrence of helminths in freshwater fishes. Part IV. Adult Cestoda, Nematoda and Acanthocephala. *Advances in Parasitology*, **20**: 1–292.

Cox, F. E. G. 1971. Parasites of British amphibians. *Journal of Biological Education*, **5**: 35–51.

Crompton, D. W. T. 1970. *An Ecological Approach to Acanthocephalan Physiology*. Cambridge University Press.

Crompton, D. W. T. 1985. Reproduction. In: *Biology of the Acanthocephala* (ed. D. W. T. Crompton & B. B. Nickol), pp. 213–71. Cambridge University Press.

Crompton, D. W. T. 1989. Acanthocephala. In: *Reproductive Biology of Invertebrates*, Vol. IV, Part A (Fertilization, development, and parental care) (ed. K. G. Adiyodi & R. G. Adiyodi), pp. 251–8. John Wiley & Sons, Chichester. ISBN 0 471 92269 2.

Crompton, W. T., Keymer, A. E. & Arnold, S. E. 1984. Investigating over-dispersion: *Moniliformis* (Acanthocephala) and rats. *Parasitology*, **88**: 317–31.

Crompton, D. W. T. & Lee, D. L. 1965. The fine structure of the body wall of *Polymorphus minutus* (Goeze, 1782) (Acanthocephala) from a wildfowl reserve in Kent. *Parasitology*, **55**: 345–55.

Crompton, D. W. T. & Nickol, B. B. (eds) 1985. *Biology of the Acanthocephala*. Cambridge University Press. ISBN 0 521 24674 1.

Denny, M. 1969. Life-cycles of helminth parasites using *Gammarus lacustris* as an intermediate host in a Canadian lake. *Parasitology*, **59**: 795–827.

Dunagan, T. T. & Miller, D. M. 1981. Anatomy of the cerebral ganglion in *Oligacanthorhynchus tortuosa* (Acanthocephala) from the opossum (*Delphis virginiana*). *Journal of Parasitology*, **67**: 881–5.

Edmonds, S. J. 1966. Hatching of the eggs of *Moniliformis dubius*. *Experimental Parasitology*, **18**: 216–26.

Gassmann, M. 1972. Étude des Trématodes et Acanthocéphales d'Amphibiens du Jura (Note préliminaire). *Revue Suisse de Zoologie*, **79**: 980–8.

Golvan, Y. J. & Buron, I. De. 1988. Les hôtes des acanthocéphalides. II – Les hôtes définitifs 1. Poissons. *Annales de Parasitologie Humaine et Comparée*, **63**: 349–75.

Grabda-Kazubska, B. 1962. On the validity of the species *Acanthocephalus falcatus* (Frölich, 1791). *Acta Parasitologica Polonica*, **10**: 377–94.

Hynes, H. B. N. & Nicholas, W. L. 1957. The development of *Polymorphous minutus* (Goeze, 1782) (Acanthocephala) in the intermediate host. *Annals of Tropical Medicine and Parasitology*, **51**: 380–91.

Hynes, H. B. N. & Nicholas, W. L. 1963. The importance of the acanthocephalan *Polymorphus minutus* as a parasite of domestic ducks in the United Kingdom. *Journal of Helminthology*, **37**: 185–98.

Khatoon, N. & Bilqees, F. M. 1991. Classification of Acanthocephala – a review. *Proceedings of Parasitology*, **11**: 22–70. Parasitology Section, Department of Zoology, University of Karachi, Karachi-32, Pakistan.

Khokhlova, I. G. 1986. (Acanthocephala of terrestrial vertebrates of the USSR.) (In Russian.) Nauka, Moscow.

King, D. & Robinson, E. S. 1967. Aspects of the development of *Moniliformis dubius*. *Journal of Parasitology*, **53**: 142–9.

Lackie, A. M. 1974. The activation of cystacanths of *Polymorphus minutus* (Acanthocephala) *in vitro*. *Parasitology*, **68**: 135–46.

Lackie, J. M. 1972. The course of infection and growth of *Moniliformis dubius* (Acanthocephala) in the intermediate host *Periplaneta americana*. *Parasitology*, **64**: 95–106.

Marchand, B. 1984*a*. The elaboration of the acanthor shell of *Acanthosentis acanthuri* (Acanthocephala). *Journal of Parasitology*, **70**: 712–18.

Marchand, B. 1984*b*. A comparative ultrastructural study of the shell surrounding mature acanthor larvae of 13 acanthocephalan species. *Journal of Parasitology*, **70**: 886–901.

Marshall, J., Call, R. N. & Nicholas, W. L. 1973. A microspectrophotometric study of the DNA of the embryonic and larval nuclei of *Moniliformis dubius* (Acanthocephala). *Journal of Parasitology*, **59**: 130–5.

Miller, D. M. & Dunagan, T. T. 1985. Functional morphology. In: *Biology of the Acanthocephala* (ed. D. W. T. Crompton & B. B. Nickol), pp. 73–123. Cambridge University Press.

Monks, S. & Nickol, B. B. 1989. Effect of *Moniliformis moniliformis* density on distribution within the definitive host population (*Rattus norvegicus*). *International Journal for Parasitology*, **19**: 865–74.

Moore, D. V. 1946. Studies on the life history and development of *Moniliformis dubius*, Meyer, 1933. *Journal of Parasitology*, **32**: 257–71.

Nicholas, W. L. 1967, 1973. The biology of the Acanthocephala. *Advances in Parasitology*, **5**: 205–46; **11**: 671–706.

Nicholas, W. L. & Hynes, H. B. N. 1958. Studies on *Polymorphus minutus* (Goeze, 1782) (Acanthocephala) as a parasite of the domestic duck. *Annals of Tropical Medicine and Parasitology*, **52**: 36–47.

Nicholas, W. L. & Mercer, E. H. 1965. The ultrastructure of the tegument of *Moniliformis dubius* (Acanthocephala). *Quarterly Journal of Microscopical Science*, **106**: 137–46.

Nickol, B. B. 1985. Epizootology. In: *Biology of the Acanthocephala* (ed. D. W. T. Crompton & B. B. Nickol), pp. 307–46. Cambridge University Press.

Olsen, R. E. 1967. *Animal Parasites: Their Biology and Life Cycles*. 2nd Edition. Burgess Publishing Co., Minneapolis. Library of Congress Cat. N: 62-17053.

Parshad, V. R. & Crompton, D. W. T. 1981. Aspects of acanthocephalan reproduction. *Advances in Parasitology*, **19**: 73–138.

Podesta, R. B. & Holmes, J. C. 1970. The life cycles of three polymorphids (Acanthocephala) occurring as juveniles in *Hyalella azteca* (Amphipoda) at Cooking Lake, Alberta. *Journal of Parasitology*, **56**: 1118–23.

Radomyos, P., Chobchuanchom, A. & Tungtrongchitr, A. 1989. Intestinal perforation due to *Macracanthorhynchus hirudinaceus* infection in Thailand. *Tropical Medicine and Parasitology*, **40**: 476–7.

Schmidt, G. D. 1985. Development and life cycles. In: *Biology of the Acanthocephala* (ed. D. W. T. Crompton & B. B. Nickol), pp. 273–305. Cambridge University Press.

Smyth, J. D. & Smyth, M. M. 1980. *Frogs as Host–Parasite Systems*. I. The Macmillan Press, London. ISBN 0 333 23565 7.

Starling, J. A. 1985. Feeding, nutrition and metabolism. In: *Biology of the Acanthocephala* (ed. D. W. T. Crompton & B. B. Nickol), pp. 125–212. Cambridge University Press.

Stoddart, R. C., Crompton, D. W. T. & Walters, D. 1991. Influence of host strain and helminth isolate on the first phase of the relationship between rats and *Moniliformis moniliformis* (Acanthocephala). *Journal of Parasitology*, **77**: 372–7.

Szalai, A. J. & Dick, T. A. 1987. Intestinal pathology and site specificity of the acanthocephalan *Neoechinorhynchus carpiodi* Dechtiar, 1968, in quillback, *Carpiodes cyprinus* (Lesueur). *Journal of Parasitology*, **73**: 467–75.

Whitfield, P. J. 1970. The egg sorting function of the uterine bell of *Polymorphus minutus* (Acanthocephala). *Parasitology*, **61**: 111–26.

Wright, R. D. & Lumsden, R. D. 1970. The acanthor tegument of *Moniliformis dubius*. *Journal of Parasitology*, **56**: 727–35.

Yamaguti, S. 1963. *Systema Helminthum*. V. *Acanthocephala*. John Wiley & Sons, New York.

32

Immunoparasitology

D. WAKELIN

The immunobiology of parasites is now a major field of study in parasitology. Not only has it contributed a great deal to our understanding of host–parasite interactions, it has also contributed in no small measure to immunology itself. This chapter provides a brief introduction to the basic concepts of immunology and reviews the immunoparasitology of selected protozoa and helminths.

32.1 Immunity and the immune response

All animals are exposed to the continuous threat of invasion by a wide variety of infectious organisms, ranging from viruses to parasitic worms. Defence mechanisms are therefore essential if hosts are to survive and reproduce. Immune responses are the most effective form of defence once pathogens have successfully gained entry, but they are not the only obstacles that have to be overcome. Many fixed structural, biochemical and physiological characteristics of the host provide formidable barriers to the initial processes of infection, and contribute to the *innate immunity* (natural resistance) that protects against many of the parasites with which hosts come into contact. For a parasite to develop at all it must necessarily be adapted to the characteristics of the environment of its natural host; such hosts are therefore fully *susceptible* to infection by their natural parasites. When failure occurs at the initial stages of infection it is usually because parasites are attempting to invade and establish in abnormal hosts to which they are not adapted; these hosts are *insusceptible* because of their innate immunity to these species.

Even when invasion and initial establishment have been achieved, the parasite is subjected to further innate defences (Albright & Albright, 1984; Mims, 1982). All animals, invertebrates and vertebrate, can mount cellular responses to foreign organisms entering their bodies. This common ability rests largely on the capacity of phagocytic and inflammatory cells to bind via membrane receptors to molecules (*ligands*) on the surfaces of the invader. This binding is not immunologically specific, and the cells can distinguish only between *self* and *not-self* (= foreign), not between different kinds of not-self. In invertebrates, recognition by phagocytic cells may result in phagocytosis and digestion, in cytotoxicity or in encapsulation, depending on the size and nature of the foreign target. In vertebrates, these cells can phagocytose or act as cytotoxic cells, as well as contribute to the complex inflammatory response.

Both invertebrates and vertebrates possess humoral recognition factors in addition to their cellular defences. In the former, these molecules include *agglutinins* and *lectins* that bind to glycoproteins on the surface of pathogens, causing agglutination and facilitating phagocytosis (Ratcliffe, 1985). Vertebrates, however, use a fundamentally different system of humoral recognition, involving immunoglobulin molecules or antibodies. These form part of a defence system that is restricted to vertebrates, the *adaptive immune response*.

(There are many excellent textbooks dealing with immunology. All will provide additional details of those aspects of the immune response which are relevant to immunoparasitology. Among those suitable for undergraduate and postgraduate reading are: Benjamin & Leskowitz, 1991; Klein, 1990; Male *et al.*, 1987; Roitt, 1991; Roitt *et al.*, 1989; Sell, 1987. General texts dealing with immunoparasitology are listed in section 32.2.)

32.1.1 The adaptive immune response

Adaptive immunity rests on the ability of a unique cell population, the lymphocytes, to recognise and respond to specific foreign molecules (*antigens*) associated with invading organisms. These immune responses are generated after experiencing infection (or vaccination) and provide the animal with an *acquired immunity*. In vertebrates, and particularly in birds and mammals, adaptive immune responses have become intimately linked with aspects of innate immunity, especially those involving the inflammatory response. It is therefore very difficult to separate the response to infection into distinct innate and acquired components.

Lymphocytes and antigen recognition. Lymphocytes develop from pluripotential *stem* cells that originate in the bone marrow. Two routes of development, involving the thymus and the bone marrow (or avian Bursa of Fabricius) result in two distinct populations of mature lymphocytes, known respectively as T and B cells. Both populations function through the recognition of antigens by membrane receptors, and both are regulated by soluble factors known as *cytokines*, which are released from lymphocytes themselves, as well as from a variety of other cells. T lymphocytes possess T cell receptors, *dimeric* glycoproteins that function in conjunction with a complex of related molecules (Fig. 32.1A). The role of these receptors is the recognition of antigenic determinants or *epitopes*, small fragments of antigens presented on the surfaces of cells after binding to a specialised set of molecules. These are coded for by a group of genes, the *major histocompatibility complex* (MHC), which is located on a specific chromosome in each species, for example chromosome 6 in humans and 17 in the mouse. B cells carry a different form of receptor, the immunoglobulin molecule. Again this is a glycoprotein, consisting of paired light and heavy chains (Fig. 32.1B). Like T cell receptors, immunoglobulins are expressed at the cell surface and bind the epitopes of antigens. Antigen recognition is the property of the N-terminal ends of the molecule (the Fab portion); the biological properties of the molecule are determined by the C-terminal ends (the Fc portion), which define the *isotype* (class or subclass) of antibody. Whereas T cells can recognise epitopes only after the antigen molecule has been appropriately processed by an antigen-presenting cell, B cell receptors can recognise epitopes while they are still part of the intact, native antigen

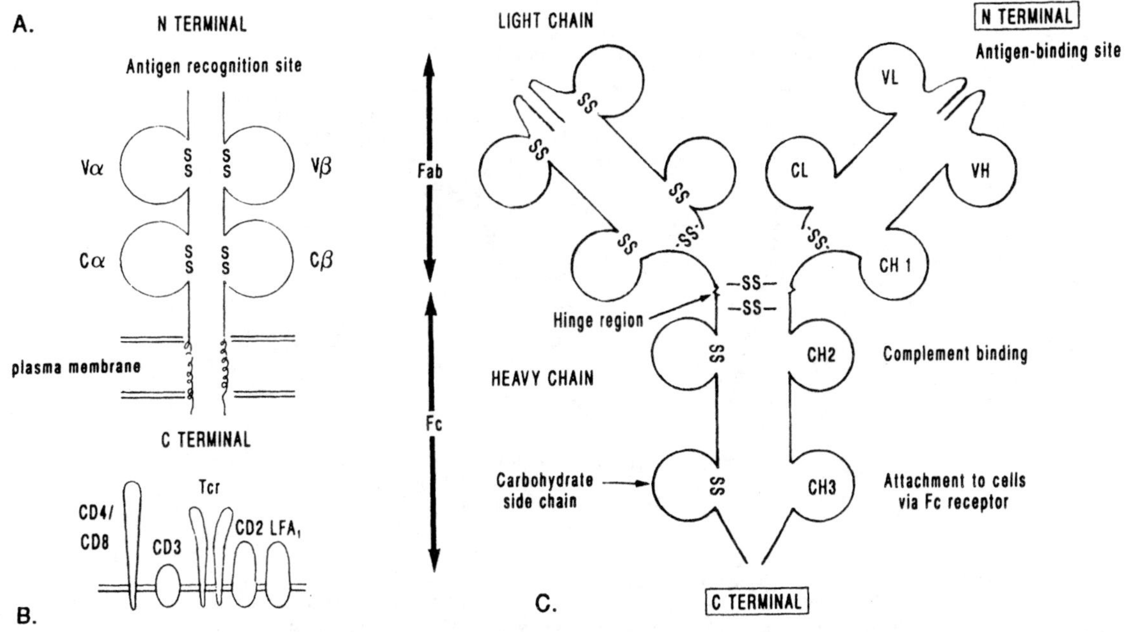

Fig. 32.1
Schematic drawing of (A) the molecular structure of the T cell receptor and (B) the cluster of surface molecules involved in antigen recognition by T cells. The receptor is formed by α and β chains, each being folded into domains in which the amino acid sequence is variable (V) or constant (C). The variable domains form the antigen-recognition site. The associated molecules are involved in binding to the MHC molecule on the antigen-presenting cell and in signal transduction. (C) Schematic drawing of the molecular structure of an immunoglobulin molecule (based on IgG). Each chain is folded into domains in which the amino acid sequence is variable (V) or constant (C). The variable domains of the light and heavy chains (VL, VH) form the antigen-binding site; the constant domains give the biological properties of the molecule.

molecule. Once stimulated, and after a period of division and differentiation, B cells mature into plasma cells. These synthesise large amounts of immunoglobulin, which is released as *antibody*.

T cell receptors and immunoglobulin are essentially monospecific, each receptor recognising only one epitope. Each cell carries receptors of one specificity only. The astounding recognition capacity of the immune system thus requires the generation of an enormous number of diverse receptor molecules. An important part of the immune response that follows antigen recognition is the clonal expansion of the recognising cells, resulting in greatly enlarged populations capable of responding to the epitope concerned.

During their development, T cell populations differentiate into two major subsets, known as *helper* and *cytotoxic* cells. These carry out largely separate functions and have quite distinct patterns of antigen recognition. T helper cells recognise antigen presented on class II MHC molecules and release cytokines capable of regulating the other components of the immune response. These cells characteristically carry the CD4 surface marker, which combines with class II molecules during antigen recognition. T cytotoxic cells, on the other hand, recognise antigen presented on class I MHC molecules and function directly as effector cells. These cells carry the CD8 marker. Only certain cells of the body express class II molecules, and only these (e.g. macrophages, dendritic cells, follicular cells, Langerhans cells, B cells) can process and present antigen to T helper cells. In contrast, class I molecules occur on all cells.

Antigen presentation and recognition. The initial binding of epitopes to MHC molecules takes place within the cells that will ultimately express them on their surface. The way in which this occurs, and the origin of antigens presented, differs between the two classes of MHC molecule (Fig. 32.2). Epitopes of antigens arising within the cell itself (*endogenous* antigens, e.g. from intracellular parasites) bind to class 1 molecules in the endoplasmic reticulum. Epitopes from antigens taken up from outside the cell (*exogenous* antigens, e.g. from extracellular parasites) bind to class II molecules in the endosome. Antigen recognition occurs when a T cell bearing the appropriate receptor comes into close contact with cells carrying the MHC–epitope complex. The process

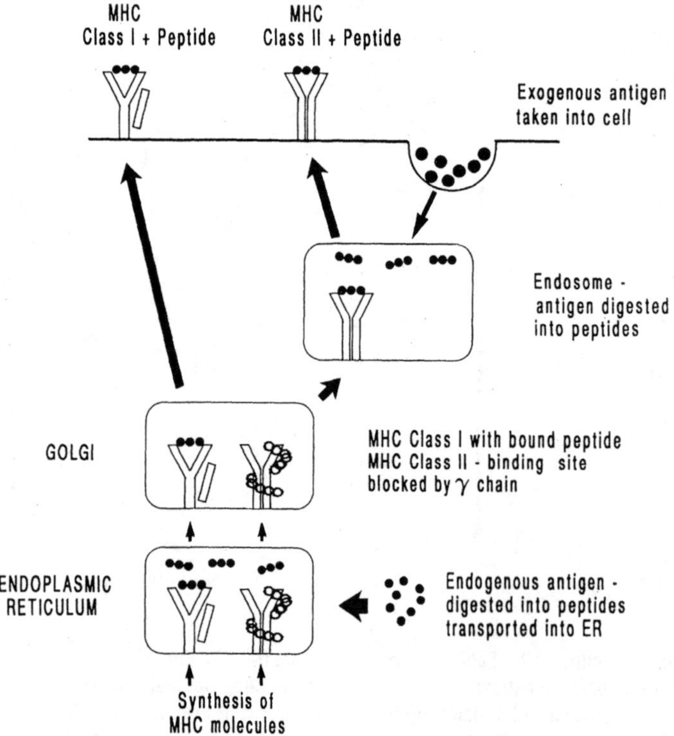

Fig. 32.2
Pathways by which exogenous and endogenous antigens are processed in antigen-presenting cells. Binding of exogenous peptides to Class II MHC molecules occurs in the endosomes; endogenous peptides bind to Class I MHC molecules in the endoplasmic reticulum.

Table 32.1. *Major cytokines released by T helper cells*

IL = interleukin; IFN = interferon; TNF = tumour necrosis factor; CSF = colony stimulating factor.

IL-2	Stimulates T cells, B cells and macrophages
IL-3	Stimulates B cells; controls mast cell development; multipotential CSF, stimulates formation of many cell types in bone marrow
IL-4	Stimulates T cells and B cells; induces IgE; controls mast cell development; enhances MHC expression; down regulates Th1 subset of T helper cells
IL-5	Stimulates B cells; induces IgA; controls eosinophil development
IL-6	Produced by many cell types; stimulates T cells, B cells and granulocytes; induces acute phase proteins
IL-7	Stimulates T cells and B cells
IL-8	Stimulates granulocytes
IL-9	Controls mast cell development
IL-10	Down regulates Th1 subset of T helper cells
TNF-β	Cytotoxic
IFN-γ	Activates macrophages; enhances MHC expression; stimulates B cells; induces acute phase proteins and fever; down-regulates Th2 subset of T helper cells
GM-CSF	Stimulates formation of granulocyte and macrophage colonies in bone marrow

requires not only successful interaction of receptor and epitope, but also binding of the CD4 or CD8 molecules to invariant regions of the MHC molecule involved.

Initiation, regulation and expression of immune responses. Almost all biologically relevant adaptive responses are initiated through T helper cells (although there are important exceptions). After antigen recognition has occurred, T helper cells (of which there are two subsets, Th 1 and Th 2) release cytokines that then regulate many of the subsequent events in the immune response (Table 32.1).

B cell responses are largely T cell-controlled, but certain antigens, notably polymeric carbohydrates, can stimulate antibody production that is T cell-independent. The antibody response to T-dependent antigens is tightly controlled by cytokines, which also determine the isotype of the immunoglobulin produced. Because of their differing biological properties, the different isotypes of antibodies can play a variety of roles in the immune response (Table 32.2), but their central function is combination with antigen, effectively labelling the antigen, or the organisms bearing those antigens, as foreign. The result of antigen–antibody binding may itself be sufficient to inactivate the antigen (e.g. if it is a functional component of a parasite such as an enzyme); more often antibody recognition serves to focus other components of the adaptive response onto the target. The two most important consequences of antibody recognition are the activation of *complement* and facilitation of recognition by phagocytes and inflammatory cells. Complement is the name given to a complex series of serum proteins that act in an enzyme-cascade fashion once the initial components have been activated. Intermediate products of the cascade can elicit inflammatory responses and interact with other components of the adaptive immune response. The final products of complement activation form lytic complexes which insert into and irreversibly damage cell membranes. Phagocytic and inflammatory cells carry membrane receptors for the Fc portion of the antibody molecule as well as for components (C3b) formed by complement activation. Binding of cells to antibody- or complement-labelled parasite surfaces can result in an *antibody-dependent cellular cytotoxicity* (ADCC).

The initiation and expression of T cell-mediated cytotoxic responses require T helper cell function, as do inflammatory responses, in each case help being given through the release of cytokines. The precursors of cytotoxic T cells require help before they can mature into effector cells capable of killing infected cells carrying foreign antigens. Killing is achieved in several ways, including the production of molecules (e.g. *perforin*) which become inserted into target-cell membranes and disrupt cellular integrity. The production, development and activities of phagocytes, granulocytes and amine-containing cells are all under cytokine control. These inflammatory cells (Table 32.3) can act directly in anti-parasitic ADCC; they also release a wide variety of mediators capable of damaging parasites in the immediate vicinity or of drastically changing the physicochemical environment in which the parasite lives. Inflammatory cells themselves can release cytokines, contributing to the complex network of regulatory factors through which acquired immunity is controlled.

○ 32.1.2 Parasite antigens

In a strict sense antigens are molecules that can be recognised by antibodies. Conventionally, the term is

Table 32.2. *Characteristics of immunoglobulins*

Isotype	Structure	Molec. mass (kDa)	% Total serum Ig	Characteristics
IgM	Pentamer	900	6	First isotype in response; good agglutinator, fixes complement
IgG	Monomer	150	80	Major isotype in body fluids; fixes complement, binds to macrophages and polymorphs, facilitates ADCC, crosses placenta, several subclasses
IgA	Monomer Dimer (+ secretory piece)	160 400	13	Major isotype at mucosal surfaces, crosses epithelial cells, secreted in milk
IgE	Monomer	200	0.002	Binds to mast cells and basophils, involved in immediate hypersensitivity, binds to eosinophils, facilitates ADCC

Table 32.3. *Inflammatory cells*

Neutrophils	In blood and tissues. Rapid response to infection. Carry receptors for antibody (IgG) and complement, used in ADCC. Have cytoplasmic granules that contain a variety of enzymes (e.g. myeloperoxidase, collagenase, lysozyme). Release inflammatory mediators. (e.g. leukotrienes, platelet-activating factor [PAF]).
Eosinophils	In blood and tissues. Carry receptors for antibody (IgM, IgG, IgE, IgA) and complement, used in ADCC. Have cytoplasmic granules that contain a variety of enzymes (e.g. myeloperoxidase, phospholipase) and a number of basic proteins (e.g. major basic protein). Produce cytokines. Release inflammatory mediators (e.g. leukotrienes, PAF).
Macrophages	In tissues, derived from blood monocytes. Carry receptors for antibody (IgM, IgG, IgE) and complement, used in ADCC. Major role in phagocytosis, produce variety of enzymes, oxygen radicals and nitric oxide. Release inflammatory mediators (e.g. IL-1, TNF-α, leukotrienes, prostaglandins, PAF).
Mast cells/ Basophils	Mast cells occur only in tissues; basophils occur in blood and tissues. Carry receptors for antibody (IgE) and complement. Major role in immediate hypersensitivity, through release of biologically active mediators (e.g. histamine, heparin, prostaglandins, leukotrienes, PAF). Also release a variety of enzymes.

also applied to molecules that initiate immune responses, although the term *immunogen* is often also used. Recognition by antibodies or by T cell receptors is a steric phenomenon, the 'shape' of part of the antigen (the *epitope*) fitting into the 'shape' of the receptor. Epitopes themselves are small molecular structures, 8–20 amino acids, but antigens are normally quite large with a molecular mass of over 8000 kDa; consequently, single antigen molecules can carry several epitopes, each of which may be separately recognised. Proteins, carbohydrates and glycoconjugates can all act as antigens, their antigenicity reflecting their structural complexity. These macromolecules form the major structural and functional components of all organisms and it is obvious, therefore, that parasites will present their hosts with a very large array of potential antigens, to which a variety of immune responses may be made. Many of these responses will play no part in protective immunity, because they will be directed against targets that are irrelevant to the continued survival of the parasite. Only if immune recognition inactivates functionally important molecules, results in direct cytotoxicity, or focuses harmful effectors is it likely to be host-protective. Antigens that act as crucial targets for such responses include molecules expressed at parasite surfaces, and molecules that have a significant functional role after release (e.g. in feeding, tissue penetration or communication). An important exception concerns antigens that trigger inflammatory responses, because in this case it is the indirect consequences of recognition, rather than the recognition itself, that are detrimental to parasite survival.

○ 32.1.3 Anti-parasite responses

To a large degree, the immune responses that can provide protection against a given parasite are determined by the type of parasite concerned and its location in the body of the host. Extra-cellular protozoa, for example, are vulnerable to phagocytosis, cytotoxicity, direct effects of antibody, and complement-dependent lysis. Intra-cellular species, on the other hand, are protected by their position from these defences. However, they can be attacked indirectly, by cytotoxic destruction of the infected host cell as a whole, and directly, through the action of cytokines or of immune-induced intracellular defence mechanisms (e.g. lysosomal enzymes, oxygen metabolites, nitric oxide). Worms provide different problems for immune defences because of their size and activity. Platyhelminths have plasma-membrane-bound outer surfaces, and these can be attacked by complement, by ADCC and by inflammatory mediators. Antibodies can also interfere with functionally important antigens. Nematodes, particularly larger species, gain some protection from their cuticle, although this is certainly antigenic and can be damaged by immune effectors. Tissue-dwelling worms are in intimate contact with immune components such as antibody, complement and cytotoxic cells, and these can therefore act directly against the parasite. However, this is not the case for intestinal species, and other mechanisms of defence, particularly inflammatory reactions, are thought to play more important protective roles in this environment.

○ 32.1.4 Evasion of immunity

Despite the fact that parasites present many immunogens to their hosts, and despite the fact that hosts mount a broad spectrum of immune and inflammatory responses, it is nevertheless the case that many parasites survive within their hosts for prolonged periods. The species that are of the greatest importance, because of their medical or veterinary significance, are precisely those where host immunity appears to be ineffective, or slow to develop, and these parasites often establish long-lasting, chronic infections. The reasons why levels of protective immunity are poor are many and varied, and reflect both characteristics of the host and properties of the parasite (Table 32.4).

Hosts exposed to parasitic infection may fail to respond effectively because their immune competence is reduced by environmental or physiological factors, such as concurrent disease, malnutrition, stress, immaturity or ageing. Genetic variation in immune capacity also influences the ability of particular individuals, races or breeds to respond protectively. It has been well-established, from many epidemiological studies, that the distribution of parasites (particularly helminths) within populations is almost always aggregated, some individuals showing much heavier parasite loads than others. Experimental evidence suggests that genetic influences on immune responsiveness play a contributory role in generating such aggregation.

Host factors are one source of deficiencies in protective immunity; equally important are the abilities of parasites to evade immunity and actively to suppress host responses. Evasion is achieved by many species as a result of changing or disguising target antigens, so that responses have little effect. Parasites also evade responses by occupying tissues in which immune responses are weak, or by inactivating components of the immune effector system. Many species induce a profound immune suppression in the host, through release of immunomodulatory factors, through interference with the structure and function of lymphoid organs, by diverting responses into irrelevant pathways, or by blocking immune effectors. These abilities of parasites to manipulate the immune system pose major problems for the host, and form significant obstacles to the development of immunological means of parasite control.

Table 32.4. *Factors underlying poor protective immunity to parasitic infections*

Host influences	Parasite influences
Age-related (very young or old)	Anticomplementary activity
Concurrent infection	Antigenic disguise
Genetic (low responsiveness)	Antigenic mimicry
Hormonal (pregnancy, lactation)	Antigenic polymorphism
Immunological (tolerance, blocking)	Immunomodulation of host
Nutritional	Inaccessibility to effectors

○○○ 32.2 Immunity in specific host–parasite systems

A major impetus for immunoparasitological studies has been the global importance of the major parasitic infections of humans, in particular those designated as target diseases in the UNDP/World Bank/WHO Special Programme for Research and Training in Tropical Diseases. These diseases, malaria, trypanosomiasis, leishmaniasis, schistosomiasis and filariasis, together with those caused by gastro-intestinal infections such as amoebiasis, giardiasis and the gastro-intestinal nematodes, afflict a substantial portion of the world's population and remain difficult to control by conventional means. Research has also been stimulated by the economic and social importance of parasitic infections in domestic animals, collectively responsible for enormous financial losses worldwide and a major constraint upon agricultural development in many countries. The focus of immunological research concerned with these human and animal diseases has been twofold: (*a*) the production of effective vaccines to control infection and to limit immunopathological reactions; and (*b*) the development of efficient diagnostic procedures. Progress towards these objectives has depended heavily upon the use of experimental model systems, in which the parasite of interest, or a closely related species, can be studied in a readily available and well-defined laboratory host. Such work has contributed a great deal to our fundamental understanding of the ways in which the immune system responds to and deals with infectious organisms, and has defined approaches to tackling the much more difficult problems involved in studying immune responses to parasitic infections in humans and in domestic animals.

There is now a very large body of literature on immunity to parasites, and much of this has been extensively reviewed in recent years.

Useful general texts include: Ash & Gallagher, 1991; Behnke, 1990; *Immunological Reviews*, 1989, 1992; Kierszenbaum, 1991; McAdam, 1989; Marchalonis, 1984; Soulsby, 1987; Wakelin, 1989; Warren, 1993; Wyler, 1990. Reviews of immunity to specific groups of parasites are included later.

It is impossible in this chapter to deal comprehensively with all the work that has been done, and so it is necessary to concentrate upon a restricted field. This section will select, from the many host–parasite systems available, those which provide information relevant to the parasitic diseases of greatest concern in medicine and veterinary science and those which have contributed significantly at a more fundamental level.

○ 32.2.1 Protozoa

As targets for the immune response, protozoan parasites are characterised by small size, which makes it possible for them to be taken up by phagocytic cells. They also possess an external plasma membrane, which is potentially vulnerable to attack by antibody and complement. However, many protozoans live intra-cellularly, within phagocytic and other cells, and once within this niche are beyond the reach of these particular effector mechanisms. The strategies of the immune response for dealing with extra-cellular and intra-cellular protozoans are therefore very different.

○ 32.2.2 Extra-cellular protozoa

Haemoflagellates. Trypanosomes provide good examples of extracellular protozoans against which antibody-, complement- and phagocyte-mediated responses can provide effective immunity. Such immunity is seen in infections with the stercorarian species *Trypanosoma lewisi* in rats and *T. musculi* in mice. The salivarian trypanosomes which infect humans (*T. brucei gambiense*, *T. b. rhodesiense*) and those which infect domestic animals (*T. b. brucei*, *T. congolense*, *T. vivax*) have evolved an efficient way of evading these immune mechanisms by changing the antigens that act as targets for the immune response (Roelants & Pinder, 1984; Mansfield, 1990).

Trypanosoma lewisi and *T. musculi*

In both of these species the course of infection follows a characteristic and predictable pattern (Albright & Albright, 1991). The trypanosomes multiply in the peripheral blood for a period of several days and parasitaemia rises rapidly. There is then a plateau, or declining plateau phase, before parasitaemia declines rapidly (Fig. 5.10). Termination of the exponential growth phase in *T. lewisi* and the initial decline in numbers at the first crisis are due to the production of antiparasite antibodies. These include *ablastin* (first identified by Taliaferro in 1924), an IgG antibody directed against surface antigens which inhibits reproduction (D'Alesandro, 1975), and a trypanocidal antibody which eliminates reproductive and dividing forms. At the second crisis,

IgM antibodies eliminate the non-dividing forms and the patent parasitaemia disappears.

Antibody equivalent to ablastin is not produced against *T. musculi*. Here the plateau phase reflects a dynamic balance between parasite destruction in the blood and renewal from dividing parasites in the body cavity. Clearance of *T. musculi* is known to involve IgG2a antibodies directed against surface antigens. These facilitate phagocytosis in the liver and spleen; antibodies of other isotypes may also contribute to this process (Vincendeau *et al.*, 1986). Animals from which the spleen has been removed are unable to control parasitaemia.

With both *T. lewisi* and *T. musculi*, experience of a primary infection results in complete immunity to re-infection. However, despite the efficiency with which parasitaemia is controlled, *T. musculi* is not wholly eliminated during the primary infection, parasites surviving for considerable periods in the capillaries of the kidney, presumably because they are within an immunologically privileged area in which protective responses operate with reduced efficiency.

Salivarian trypanosomes

The course of parasitaemia in human trypanosomiasis caused by *T. b. gambiense* or *T. b. rhodesiense*, or cattle trypanosomiasis caused by species such as *T. congolense*, is very different from that shown in Fig. 5.10 for *T. lewisi*. Characteristically, parasitaemia is chronic and fluctuating (Fig. 32.3), the numbers of trypanosomes rising and falling over prolonged periods (Barry & Turner, 1992). The explanation for this pattern of infection, and for the failure of the immune response successfully to control the parasitaemia, lies in the ability of parasite populations to express a sequence of antigenically different surface glycoprotein molecules, the variant specific glycoproteins (or VSGs; Fig. 5.5) and thus expose the host to a sequence of variant antigen types (or VATs). Each subpopulation expresses one such VSG against which the host mounts a protective antibody response. This response results in the destruction of the VAT concerned, but this merely allows another subpopulation of a different VAT to develop (Donelson & Turner, 1985).

It is known from work with *T. b. brucei* in mice that immune clearance of parasitaemia is possible, but only under unnatural conditions, in which trypanosomes of a defined VAT are exposed to non-limiting quantities of VSG-specific antibody. IgM antibodies directed against VSG epitopes allow the uptake and destruction of antibody-coated trypanosomes by Küpffer cells in the liver (Mansfield, 1990). Under normal conditions, the host is continually exposed to novel VATs and is never able to achieve parasite elimination in this way. As infection continues, the host becomes progressively immunosuppressed, and the ability to deal with the infection declines even further.

Giardia. The immune effectors potentially able to control trypanosome infections have direct access to the parasites. Antibodies can interact with the organisms in the bloodstream, and antibody-exposed trypanosomes circulate regularly past the phagocytic cells of the spleen and liver. In contrast, antibodies other than IgA and IgM have only limited access to extra-cellular parasites living in the intestine, such as *Giardia*, and these are effectively removed from the possibility of phagocytic activity. Given these circumstances, it becomes clear that different mechanisms of immunity are likely to operate.

Giardia lamblia is a widespread parasite of humans, against which effective immune responses can be

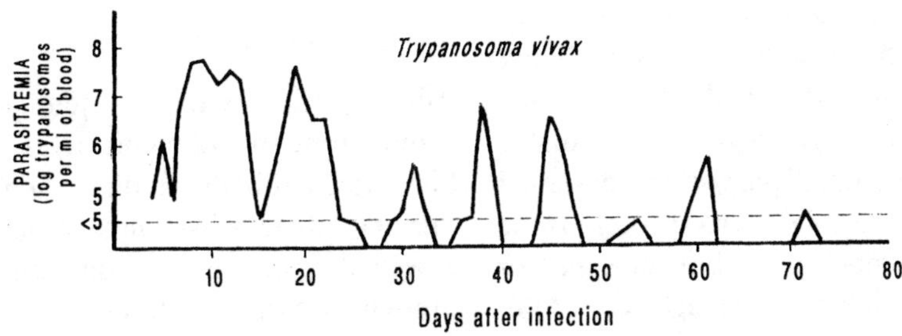

Fig. 32.3
Course of parasitaemia in a cow infected with *Trypanosoma vivax*, showing repeated peaks as antigenically different populations of trypanosomes appear, grow and are eliminated by immunity. (Redrawn from Barry, 1986.)

expressed. The majority of individuals exposed to infection suffer only mild consequences, control the parasite population in the small intestine, and are resistant to reinfection (Farthing, 1989). More severe consequences (chronic diarrhoea) and prolonged infection occur in immunocompromised individuals and specifically in patients suffering from deficiencies in production of the mucosal antibody IgA. This immediately points to a protective role for this isotype, and this conclusion is amply supported by experimental evidence from laboratory studies in mice (den Hollander *et al.*, 1988). A likely role for antibody is inhibition of adherence to the epithelial cells of the intestinal mucosa. Adherence is enhanced by receptor–ligand binding, involving lectins on the parasite membrane and sugars on the cell. Inhibition of adherence implies that protective antibodies recognise and combine with the parasite molecules concerned, and by doing so, interfere with their ability to bind to the host cell membrane. In addition to antibody-mediated immunity, cellular mechanisms are also thought to contribute to protection, but these are less well understood. *G. lamblia* has a limited but well-defined ability to undergo variation in its major surface antigen, although it is not clear to what extent this may contribute to chronicity of infection (Aggarwal & Nash, 1988).

○ 32.2.3 Intra-cellular protozoa

The range of cells parasitised by protozoans is considerable. A number of species live in red blood cells (*Plasmodium*, *Babesia*); others live in macrophages (*Leishmania*, *Toxoplasma*, *Trypanosoma cruzi*), enterocytes (*Cryptosporidium*, *Eimeria*), hepatocytes (*Plasmodium*) and muscles (*T. cruzi*). Although all of these niches provide protection against the defence mechanisms that can deal with extra-cellular organisms, it should be remembered that intra-cellular parasites all have extra-cellular stages. They are therefore vulnerable to extra-cellular defences at some point in their life cycle. Their ability to find and enter host cells can also be disrupted by the host's immune response. Some host cells, notably the macrophage, are 'professional' phagocytes, designed to kill ingested organisms; parasite survival within these cells therefore depends upon the ability to avoid or suppress intra-cellular defences (Hughes, 1988). Almost all nucleated host cells, however, can mount these defences if appropriately stimulated.

Malaria. Malaria is one of the world's commonest infectious diseases. In endemic areas, individuals are infected when very young and are repeatedly reinfected throughout much of their life, suggesting that immunity cannot be very effective, takes a long time to develop and gives poor protection. Interpretation of immunity in humans is in fact more complex than this. Children are the most heavily parasitised members of the community and are the age group most likely to suffer the effects of severe pathology. Adults show both a lower prevalence of infection and less severe disease, implying that they are protected. One complicating factor is that there is considerable strain variation within the species of *Plasmodium*, and immunity is known to be strain-specific (Jeffrey, 1966; Butcher, 1989). Individuals may therefore be immune to some strains but quite susceptible to others. Another complication is the fact that malaria, like many parasitic diseases, can be immunosuppressive, so that an individual's capacity to resist the parasite is substantially reduced by the infection itself. Within each species there is strong stage specificity, responses made to one stage of infection providing little resistance to other stages in the cycle.

Acquired immunity

Our present understanding of immune mechanisms comes from observations made in infected people, from *in vitro* studies, and from experimental work carried out in laboratory systems (see *Progress in Allergy*, 1988; *Immunology Letters*, 1990). None of the species which cause malaria in humans can be used in rodents, but there are several other *Plasmodium* species, such as *P. berghei*, which can be used in these hosts. Some primates have been used to study responses to the human plasmodia as well as to closely related species. The intensive search for an effective malaria vaccine (Mitchell, 1989) has also yielded a large body of data about immune responses to these parasites.

Sporozoites and liver stages

After injection by the mosquito, sporozoites spend only a short time in the blood before invading hepatocytes. However, antibody-mediated immunity can act against this stage before liver uptake occurs. Sporozoites have a well-defined surface antigen, the circumsporozoite protein (CSP), which elicits T cell-dependent antibodies (Zvala *et al.*, 1983). Under experimental conditions, i.e. injection of CSP-specific antibodies before injection of sporozoites, infection can be blocked very successfully. The antibody is known to bind to the sporozoite surface, and may therefore effectively prevent recognition and binding to target hepatocytes. How important such immunity is under conditions of natural exposure to infection is uncertain. Individuals in endemic areas have circulating anti-sporozoite antibodies, but are still subject to reinfection. CSP is one of the candidate antigens for malaria vaccines (Nussenzweig & Nussenzweig, 1989). The molecule itself and the coding gene have been fully sequenced (Fig. 32.4) and recombinant antigen produced. Peptide sequences from CSP have also been synthesised directly. CSP possesses both T- and B-cell epitopes, antibody being directed against an epitope defined by the repeated amino acid sequence NANP (asparagine–alanine–asparagine–proline). Both laboratory and field studies have shown that there is extensive polymorphism in T cell epitopes between different isolates of a species. Antibody response to CSP is entirely T-dependent, and cross-immunity at this stage is limited.

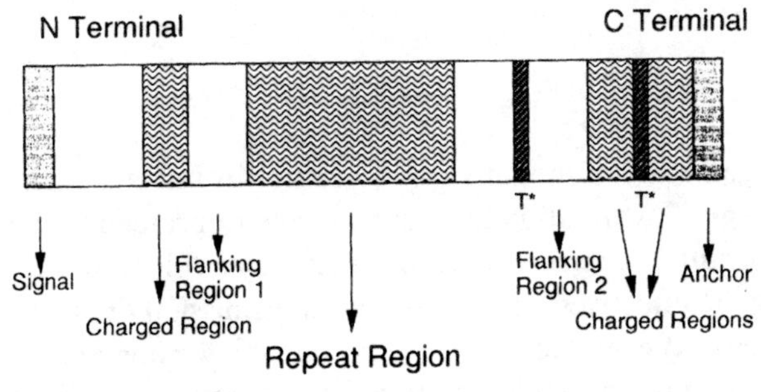

Fig. 32.4
Diagram of the circumsporozoite (CS) protein of *Plasmodium falciparum*, showing the major structural and functional regions and the amino acid sequences of the central repeat for *P. falciparum* and *P. vivax*. T* = T-cell epitope. (Modified from Mitchell, 1989.)

Once inside the hepatocyte the parasite undergoes schizogony, ultimately liberating the merozoites that will invade the red blood cells. Parasite antigen is expressed at the hepatocyte surface and infected cells can be the targets of immunologically based defence mechanisms (Fig. 32.5). Both T cell-mediated cytotoxicity and short-range mediator destruction involving nitric oxide and TNF, as well as other cytokines, have been implicated in this response (Schofield *et al.*, 1987; Nussler *et al.*, 1991).

Erythrocytic stages

Merozoites, like sporozoites, have surface antigens which are targets for protective antibody responses (Holder, 1988). In their brief extra-cellular phase, before red cells are invaded, combination of antibody with these antigens can prevent red cell recognition by interfering with ligand-receptor binding; antibodies may also cause merozoites to agglutinate. During the process of entry, peptides released from organelles such as the rhoptries and micronemes become associated with the red cell membrane (Fig. 32.6). Once inside the red cell, the merozoite grows and undergoes schizogony.

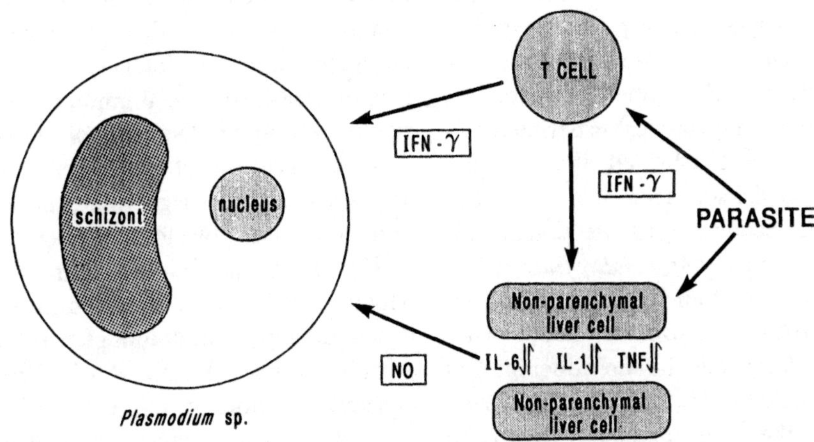

Fig. 32.5
Mechanisms of immune attack directed at the exoerythrocytic stages of the malaria parasite in hepatocytes. Immunity involves T cells and non-parenchymal liver cells, which communicate through a number of cytokines. Anti-parasite activity can be mediated indirectly via the action of interferon-gamma (IFN-γ) on the host cell, as well as directly by nitric oxide and other mediators.

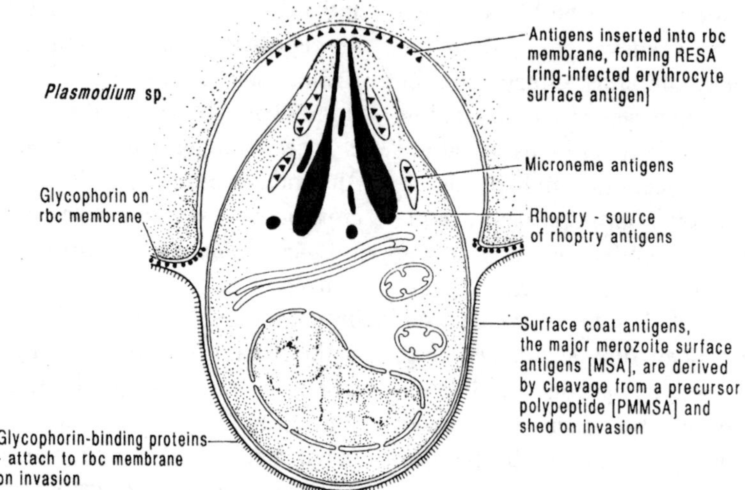

Fig. 32.6
Antigens of the malaria merozoite and their association with the host red cell during invasion. (Adapted from Wyler, 1990.)

Additional parasite antigens then become expressed at the cell surface (Perlmann *et al.*, 1984) and are also released into the plasma (Wilson *et al.*, 1975). A number of these molecules have been defined and sequenced, and some are key vaccine candidate antigens. Combination of antibody with these membrane-expressed antigens can result in agglutination and opsonisation of infected cells, allowing enhanced uptake by phagocytic cells, for example in the spleen. Evidence also suggests that intra-cellular stages may be killed by short-range mediators released from cells such as macrophages as infected erythrocytes pass through the spleen (Clark *et al.*, 1981). *In vitro* it is possible to show that antibodies against the erythrocyte stages can very effectively prevent the growth and replication of parasites cultured in red cells. How effectively such antibodies operate *in vivo* is less clear, although the classical experiments of Cohen (Cohen *et al.*, 1961) showed that transfer of gamma-globulin from the sera from immune adults conferred significant protection against infection in children, parasitaemia falling rapidly. In *P. falciparum*, infection results in the formation of knob-like structures on the surface of the parasitised erythrocyte. These structures are thought to play a role in the adherence of infected cells to endothelial cells (Berendt *et al.*, 1990) and thus contribute to the severe pathology associated with this species.

Antigens released from gametocytes can appear on the surface of infected red cells in exactly the same way as antigens from developing schizonts, and antibodies to gametocyte antigens can be found in infected individuals. Such antibodies have no role in protecting the individual against malaria, but may play a role in reducing the effectiveness of mosquito transmission (Targett, 1990). After uptake of red cells containing gametocytes, fertilisation takes place within the mosquito stomach as the gametes are released. Antibodies present in the plasma can interfere with this process, and effectively prevent infection of the mosquito. Gametocyte antigens are therefore a third group of vaccine candidates.

Immunity against infection and immunity against disease. The long-term nature of infections with malaria parasites is certainly explicable in terms of relatively ineffective immune defences developing slowly in the face of parasite-induced immunosuppression and parasite-programmed diversion of the immune response. The repeat regions of surface antigens are thought to act as a smokescreen, triggering production of redundant, non-protective polyclonal B cell responses (Anders, 1986). They also contain immunodominant epitopes which divert responses away from more critical targets in the antigen. In addition, plasmodia are not only genetically diverse (Kemp *et al.*, 1990), they have the capacity to undergo antigenic variation. Variation has been described in most detail in a variety of animal models with species such as *P. berghei*, *P. chabaudi* and *P. knowlesi*, but it also occurs in *P. falciparum* (Brown, 1974; Howard, 1987).

The pathological consequences of malaria are much

more severe in children than they are in adults who have experienced infection over many years. In part this reflects the operation of protective immunity in reducing levels of parasitaemia; in part it may also be the case that there is an immunity against molecules of the parasite that induce pathological changes. Evidence has accumulated (Clark *et al.*, 1991) that many aspects of malaria-related pathology can be explained by the activity of a variety of cytokines, of which TNF (tumour necrosis factor) is one of the most important. It seems that, like bacterial lipopolysaccharide, some parasite molecules selectively promote TNF release; antibody-mediated inactivation of these molecules might therefore effectively reduce the degree of pathological response (Playfair *et al.*, 1990).

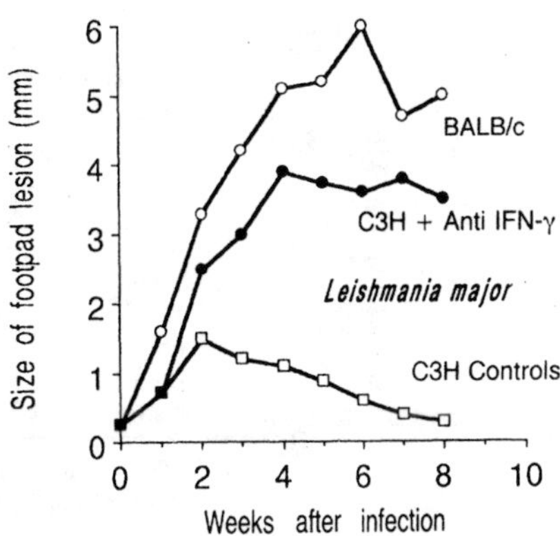

Fig. 32.7
Development of *Leishmania major* infection in a susceptible (BALB/c) and resistant (C3H) strain of mouse. The ability of C3H mice to control infection is dependent upon production of interferon-gamma (IFN-γ), as shown by the effect of depleting this cytokine by injection of an anti-IFN-γ monoclonal antibody). (Redrawn from Belosevic *et al.*, 1989.)

Leishmaniasis. *Leishmania* parasites live as amastigotes within macrophages (Fig. 5.18), where they grow and reproduce for prolonged periods. Ironically, entry of the promastigote stage into these cells can be facilitated by complement activation. Binding of C3b molecules to the parasite surface membrane allows interaction with C3b receptors on the macrophage, which, together with other receptor–ligand interactions, results in phagocytic uptake (Russell & Talamas-Rohana, 1989; Alexander & Russell, 1992). Macrophages are a first line of defence against many pathogens, possessing a variety of powerful intra-cellular defensive mechanisms. Long-term survival of any pathogen within these cells implies an ability to avoid or to turn off these defences. *Leishmania* lives within the phagolysosomes of the host cell, apparently obtaining protection against enzyme- and radical-mediated damage from the gp 63 and lipophosphoglycan molecules on its cell membrane. Infected cells also fail to produce the respiratory burst normally associated with phagocytic uptake, by which toxic oxygen metabolites and other effectors are produced. When macrophages are activated, by release of cytokines from immune T cells, intra-cellular killing does occur. Nitric oxide has recently been shown to be a potent killing factor (Green *et al.*, 1990; Liew & Cox, 1991).

Experimental analyses in mouse models, using *L. donovani* and *L. major*, have shown that there are strong genetically determined influences on the outcome of infection. In susceptible strains of mice, such as BALB/c, growth and proliferation of *L. major* amastigotes in permissive macrophages results in a failure to express immunity, so that severe lesions develop; resistant strains such as C3H control infection rapidly and lesions are controlled (Fig. 32.7). In BALB/c mice, infection appears to influence antigen presentation so that the immune response is biased towards the 'wrong' subset of T helper cells, with a consequent failure to produce the correct cytokine (interferon-gamma) for macrophage activation (Locksley & Scott, 1991; Chatelain *et al.*, 1992). Without this cytokine, infected macrophages cannot generate their intra-cellular defences, and parasite proliferation is unchecked (Fig. 32.7). The down-regulation of interferon-gamma production by T helper 1 cells seen in susceptible mice is linked with production of the cytokine IL-4 by T helper 2 cells; preventing IL-4 activity by giving anti-IL-4 antibody enables these mice to heal an otherwise uncontrollable lesion.

***Eimeria* and *Cryptosporidium*.** Species of the genus *Eimeria* are a major group of intra-cellular parasites, responsible for the diseases collectively called coccidiosis. The majority live within the epithelial cells of the intestinal mucosa, a niche which, until recently, has been little explored in terms of immune effector mechanisms. Coccidiosis is the major parasitic disease of poultry, infections causing mortality in a high proportion of untreated cases. Birds that survive infection acquire a very strong immunity to reinfection, but this immunity is species-specific (Wakelin & Rose, 1990). It is generally considered that immunity plays little part in controlling the course of initial infections.

The life cycle of the parasite (Fig. 6.8) suggests only

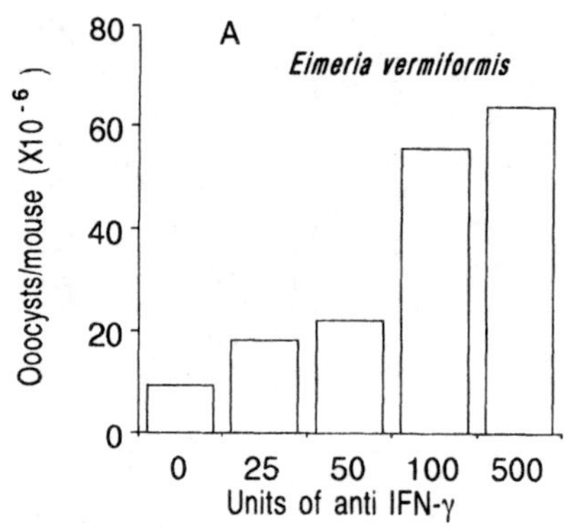

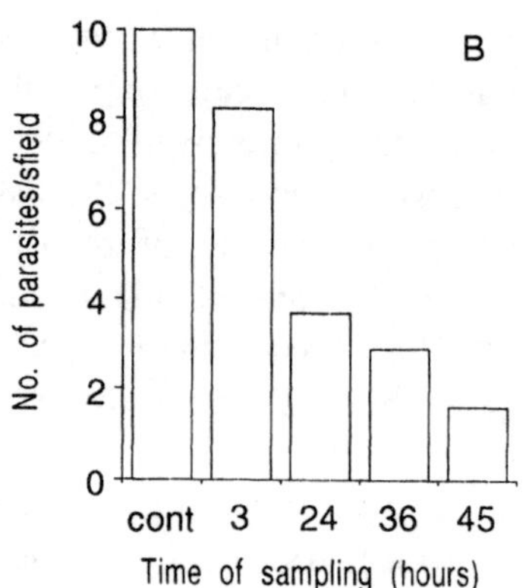

Fig. 32.8
Evidence that the cytokine interferon-gamma (IFN-γ) is involved in immunity of mice to *Eimeria vermiformis*. A. Output of *E. vermiformis* oocysts from infected mice treated by injection of increasing amounts of an anti-IFN-γ monoclonal antibody. (Data from Rose *et al.*, 1989.) B. Numbers of *E. vermiformis* developing *in vitro* in a fibroblast monolayer which was exposed to recombinant IFN-γ prior to inoculation of sporozoites at time 0 and sampled at intervals afterwards; cont, untreated (control) cells sampled at 45 hours. (Data from Rose *et al.*, 1991.)

three potential targets for effector mechanisms: the extracellular sporozoites, the merozoites, and the intracellular stages. It has been very difficult to establish a clear role for antibody-mediated mechanisms, although there is some indication that luminal antibody (IgA) may play a part in delaying movement of sporozoites into the epithelial cells after they are released from the infecting oocyst. Similarly, there is little evidence of an effective immunity against merozoites. Experimental studies, using *Eimeria* infections in mice, have shown on the other hand that immunity can directly affect the intracellular developmental stages in both initial and subsequent infections. This immunity relies on the production of the cytokine interferon-gamma, presumably activating an intracellular defence mechanism that is capable of preventing sporozoite and merozoite development (Rose *et al.*, 1989). Data obtained from observations made both *in vivo* (histological and parasitological) and *in vitro* (parasite growth in cell culture) support this hypothesis (Fig. 32.8).

Cryptosporidium belongs to the same group of Sporozoa as *Eimeria*. Like the coccidians, it invades the cells of the small intestine and undergoes schizogony and oocyst formation (Fig. 6.11). The intra-cellular stages live within a vacuole that is limited to the microvillal regions of the host enterocyte, so that infection is much more superficial than that of *Eimeria*. Infection produces a diarrhoeal condition that is usually self-limiting in immunologically normal individuals, implying immune control; in immunodeficient people, diarrhoea becomes severe and life-threatening (Curren & Owen, 1989). Cryptosporidiosis has become prominent as one of the many opportunistic infections that complicate AIDS. Little is known of the immunological mechanisms by which the infection is controlled, although recent experimental work in mice suggests that, as in coccidiosis, interferon-gamma may be involved (Ungar *et al.*, 1991). If this is correct, the T helper cell deficiency associated with AIDS would clearly contribute to the failure to clear infection from the body.

Toxoplasma gondii. Although, like the majority of *Eimeria* spp., *T. gondii* is an intra-cellular intestinal protozoan in its natural host, the cat (Fig. 6.14), its importance as a pathogen in humans reflects its ability to develop within deeper tissues. Infections are initiated by ingesting infective oocysts or by eating meat containing viable tissue cysts. In either case the invasive stages eventually become intra-cellular within the phagosomal vacuoles of macrophages. Like *Leishmania*, survival in this niche depends upon avoidance of the host cell's powerful defences. *T. gondii* achieves this by preventing lysosomal fusion with the phagosome. Immunity can control the intra-cellular stages only if the macrophage becomes successfully activated, which requires production of cytokines (interferon gamma) by T cells (Beaman *et al.*, 1992). Under these conditions the host cell produces oxygen metabolites and other short-range effectors which destroy the parasites present. The chronic nature of toxoplasmosis as a disease, and the long-term survival of parasites within cells, point to the success with which immunity is evaded. Equally, however, the severity of toxoplasmosis in the foetus and in immunocompromised individuals is proof that immunity does normally limit the pathological consequences of infection.

○ 32.2.4 Platyhelminthes: Trematoda (flukes)

Like all helminths, flukes are comparatively large organisms when adult and can present difficult targets for immune effector mechanisms. However, all invade the host as small larval stages, which are in general much more amenable to immune control. Both larval and adult worms have a plasma-membrane-bound external tegument. A variety of immunogenic molecules with which antibodies can interact are expressed at this surface. Such interactions can then initiate a variety of anti-parasite effects. For example, if target molecules play a critical structural or functional role this may be disrupted by antibody binding. Localisation of antibody at the parasite surface results in activation of complement, which may damage the tegument through the formation of lytic complexes. Bound antibody and complement can interact with the receptors of inflammatory cells, facilitating their adherence. Mediators released from these cells can also bring about direct tegumental damage. Much of our current knowledge of the operation of immune effectors against flukes has come from *in vitro* studies, and our understanding of *in vivo* anti-parasite responses is still limited.

○ 32.2.5 Schistosomes

Schistosomiasis is one of the major parasitic diseases of humans in tropical and subtropical countries, where it can also be an important infection of cattle. In endemic areas, the prevalence of infection is high and people remain infected for many years, implying that immunity is slow to develop and operates relatively inefficiently. This contrasts with data from many experimental systems where it is possible to demonstrate strong protective immune responses (Capron & Capron, 1986). It is only recently, in fact, that convincing evidence has been obtained for the existence of protective immunity against human schistosomiasis, and explanations proposed for its failure to control infection more effectively (Butterworth & Hagan, 1987).

Concomitant immunity. A series of now classical experiments carried out by Smithers and Terry in the 1960s (Smithers & Terry, 1969; Smithers *et al.*, 1969) using infections of *S. mansoni* in rhesus monkeys, showed that these animals were able to express an effective immunity against the parasite. However, although the immunity developed in response to an initial infection protected against subsequent infections, it did not influence the survival of the mature adult worms that had become established in the mesenteric veins: a situation described as *concomitant immunity*, by analogy with the situation in which tumours can survive in immune individuals. Immunity could be stimulated not only by infection, but also by exposure to irradiated larvae and by direct transplantation of adult worms. These results showed clearly that relevant antigens must be shared between the stages, but that these antigens were the targets of protective responses only in schistosomula, the adults in some way evading immune effector mechanisms.

One of the most important experiments which threw light on this evasion involved the transplantation of adult worms from monkey or mouse donors into normal monkeys or into monkeys that had been immunised against mouse antigens. In normal recipients 'monkey' worms survived quite happily and continued a normal pattern of egg production, whereas with 'mouse' worms, egg laying was interrupted for some time, suggesting that a period of adaptation was necessary in the new host species. In immunised monkeys, 'monkey' worms survived normally but 'mouse' worms were killed very rapidly, by mechanisms involving damage to the tegumental surface.

The conclusions drawn by Smithers and Terry from these and many other experiments were that during infection, as a result of exposure to the larval stages, the host develops antibodies against antigens expressed on the tegument. The fact that adult worms were normally not killed suggested that they must acquire onto their tegument a variety of host molecules, which prevent recognition by these antibodies, allowing them to evade immunity. Thus 'mouse' worms acquire mouse molecules and 'monkey' worms acquire monkey molecules, which protect them in each host; on transfer into normal monkeys, 'mouse' worms readapt by acquiring monkey molecules, but are killed if the host already has antibodies directed against their surface disguise.

Subsequent work showed that a wide variety of molecules can be acquired directly from the host, including glycolipid blood group antigens, immunoglobulins, histocompatibility antigens and skin inter-cellular substance-antigens. There is a continuing debate, however,

Table 32.5. *Surface antigens of schistosomes*

Molecular mass (kDa)	Epitope	Schistosomulum 3 hours	Schistosomulum 5 days	Adult worms	Eggs
>200	CHO[a]	+	–	–	+
38	CHO	+	–	–	+
38	Pep[a]	+	–	–	–
32	Pep	+	+	+	–
25	Pep	–	+	+	–
20	Pep	+	+	+	–
18	Pep	+	+	+	+
17	CHO	+	–	–	+
15	Pep	+	+	+	–

[a]CHO = carbohydrate; Pep = peptide.
Source: Simpson (1990).

as to whether, in addition, worms have the capacity independently to synthesise host molecules, i.e. whether they share gene sequences with their hosts, either from convergent evolution or by direct transfer. Whatever the mechanisms involved, this *molecular mimicry* plays a vital part in the host–parasite relationship (Damian, 1989).

Schistosome antigens. These early experiments focused attention on the schistosome surface as a primary source of parasite antigens and as a major site of immune attack. We now know that early schistosomula are the most vulnerable stage of the parasite, and can be killed in the skin and the lungs by immune effector mechanisms directed against their surface antigens (McLaren, 1989; Wilson & Coulson, 1989). As schistosomula mature, and move from the skin through the lungs to the liver, they become progressively less susceptible to immune attack, in part because of acquisition of host molecules, in part because of intrinsic tegumental changes and in part because of developmentally regulated antigen expression.

Several schistosome surface antigens have now been identified (Table 32.5) and a number have been cloned; many are glycoproteins, both the peptide and carbohydrate components acting as important epitopes. Some of these epitopes are shared between different life cycle stages, providing a molecular basis for phenomena such as concomitant immunity and generation of blocking antibodies (see below). A variety of other antigens of internal origin are released as worms feed and metabolise.

Protective responses. Antibodies made in response to tegumental antigens allow the host to mount ADCC reactions against the worms. A variety of antibody isotypes and of cell types are known to be involved in these responses (Fig. 32.9). An important finding by Capron and his co-workers has been that anti-worm IgE antibodies can bind to eosinophils and macrophages through their membrane FcE receptors, and thus allow specific adherence of these cells to the parasite surface (Capron *et al.*, 1989). Although production of IgE and circulating eosinophilia are characteristic responses to schistosome infection (as they are to many helminthiases), a positive, protective role in the host–parasite relationship for these components was previously unknown. The original evidence for such a role was obtained from *in vitro* experimentation, in which schistosomula were co-cultured with cells bearing surface anti-parasite IgE. Very recently, data from cross-sectional studies of infected populations in endemic areas have shown a significant age-related and inverse correlation between levels of specific IgE and degree of infection, for the first time providing convincing evidence for a role in protective immunity *in vivo* under conditions of natural exposure to infection (Hagan *et al.*, 1991; Rihet *et al.*, 1991; Dunne *et al.*, 1992).

The expression of immunity through the focusing of ADCC mechanisms onto the parasite tegument requires the availability of antibodies directed against appropriate antigens and inflammatory cell populations. Experimental data have shown that the presence of antibodies of an inappropriate isotype, or antibodies directed

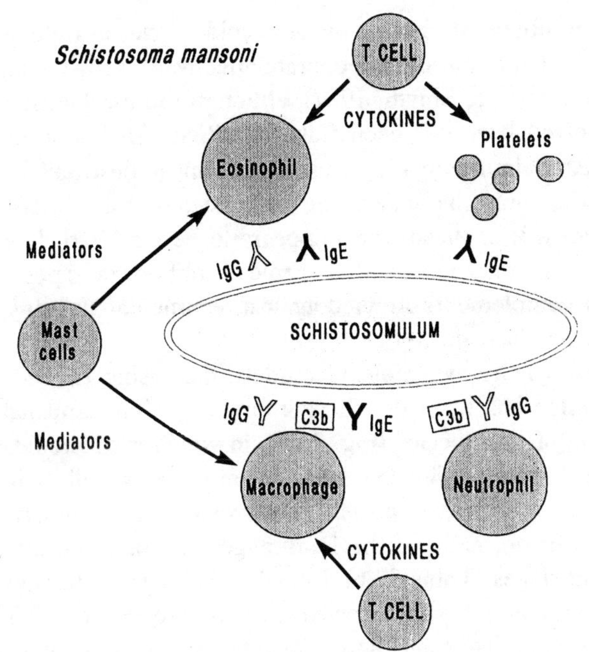

Fig. 32.9
Antibodies and cells involved in antibody-dependent cellular cytotoxicity (ADCC) reactions against the schistosomulum stage of schistosome parasites. IgA also mediates ADCC by eosinophils. (Adapted from Vignali *et al.*, 1989.)

against irrelevant epitopes of the target antigen, can effectively inhibit the cytotoxic activity of other antibody molecules. For example, in mice, IgM antibodies directed against carbohydrate epitopes on the surface antigens of schistosomula can block IgG antibody-mediated killing by eosinophils (Dunne *et al.*, 1987). Antigens from the eggs of schistosomes are rich in carbohydrates, and antibodies to these cross-react with the surface of the schistosomulum. Whereas hosts are exposed for relatively short periods to the larval stages, there is prolonged exposure to eggs and, consequently, maximum opportunity for production of blocking antibodies. These observations may help to explain the slow development of immunity in humans, anticarbohydrate blocking antibodies preventing the expression of immunity by other antibodies. Only when blocking activity declines can immunity be effective. The recent data on involvement of IgE in protective immunity have indeed shown that the age-related increase in immunity correlates not only with an increase in anti-worm IgE, but with a decrease in IgG4, an isotype with the potential to block IgE activity. It is now being understood that changes such as these in immune reactivity as schistosome infections progress are controlled by changes in the profile of T cell-derived cytokines (Pearce *et al.*, 1991).

Although antibody-mediated responses play a significant role, other effector mechanisms are almost certainly involved as well. Studies in mice immunised by exposure to irradiated larval stages have shown that macrophages activated by cytokines from T helper cells can function as important effector cells, being capable of damaging and killing schistosomula, by release of short-range mediators such as TNF and nitric oxide (James & Glaven, 1989; Smythies *et al.*, 1992). In immunised mice, such inflammatory reactions in the lungs contribute to the expression of immunity against challenge by interfering with normal migratory patterns. Paradoxically, the severe hepatic immunopathology associated with infection in mice also protects against challenge infections, because the development of arteriovenous shunts in the liver diverts developing worms away from their preferred site (Wilson *et al.*, 1983).

An interesting aspect of antibody-mediated recognition of schistosome antigens is the fact that the action of certain drugs (oxamniquine and praziquantel) is enhanced significantly when used in immune hosts, or in hosts given antibodies capable of recognising surface (or subsurface) antigens. It is thought that by damaging the tegument, the drug exposes target antigens on the parasite surface, allowing antibody binding to take place (Sabah *et al.*, 1985; Brindley & Sher, 1987).

Immunopathology. Infections with schistosomes provide a powerful immune stimulus to the host, the eggs being particularly immunogenic. Although the response to egg antigens is possibly beneficial in protecting against the histolytic enzymes released from the miracidium, it leads to the development of a damaging state of hypersensitivity. The cellular response to glycosylated antigens from eggs that become trapped in the body results in the formation of granulomata, which may then interfere with normal blood circulation in the organ involved. Inflammatory responses to eggs in the liver are the trigger for the sequence of events that lead to portal hypertension, hepato- and splenomegaly and the formation of varices, fragile, thin-walled swellings of blood vessels whose rupture can cause death through haemorrhage (Warren, 1987).

○ 32.2.6 Liver flukes

In addition to the schistosomes, there are several other species of fluke capable of causing disease, the liver flukes (*Clonorchis* in humans, *Dicrocoelium* and *Fasciola* in ruminants) being some of the most important. Each of these is known to elicit immune responses in its natural host, but in many cases there is little evidence of an effective immunity.

Detailed studies have been made of infections with *F. hepatica* in a variety of experimental hosts. In some of these, e.g. the rat, antibody-mediated cytotoxicity responses are capable of killing larval stages and young flukes as they migrate across the body cavity and through the liver (Hughes *et al.*, 1981). Adult flukes in the bile duct are unaffected by immunity, possibly because they have effective evasion strategies. In sheep there seems to be little immunity against the parasite at any stage. Certainly sheep respond immunologically, producing anti-fluke antibodies, but these appear to lack the capacity to mediate responses that can influence parasite survival (Rickard & Howell, 1982). The reasons for this lack of immunity are not understood, but could involve a failure to recognise critical tegumental antigens in the migrating flukes, a failure to make antibodies of the appropriate isotype, or a parasite-induced depression of immune responsiveness. Another possibility is that modern breeds of sheep have lost the genes necessary for anti-*Fasciola* responses, in the same way that genes for resistance to the nematode *Haemonchus contortus* have apparently been bred out of breeds such as the Merino.

○ 32.2.7 Platyhelminthes: cestodes

Separate consideration has to be given to immunity against adult tapeworms in the intestine and that against metacestode stages in the tissue of their intermediate hosts. Not only are the nature of immune responses different in the two sites, the parasites themselves also present rather different targets.

Adult tapeworms. The human *Taenia* tapeworms are notoriously long-lived parasites. It also seems to be the case that individuals rarely carry more than one or two worms. These two observations suggest that if immunity does operate, its role is to limit parasite numbers rather than to eliminate infection. Of course, other factors, such as 'crowding effects', might also explain limitation of numbers. Studies on hymenolepid species in rodents show that immunity can operate effectively against adult worms (Ito & Smyth, 1987), although the mechanisms involved have not been fully identified. An important effect of immunity is to cause the worms to destrobilate, leaving only the scolex and neck region (Fig. 32.10). Regrowth of these forms appears to be inhibited. It is tempting to assign an effector role to antibody (and possibly complement) targeted against tegumental antigens, but this has still to be proven.

Of all the hymenolepid models, that using *H. nana* in mice has been the most informative. The unusual cycle of this species (Fig. 23.6), in which cysticercoids can develop within the intestinal mucosa as well as in insect intermediate hosts, has allowed useful comparisons to be made of the immunogenicity of larval and adult stages (Table 32.6). Infection with eggs generates a rapid and highly effective immune response. This gives complete protection against further egg infections, because it prevents development of the mucosal cysticercoids, immunity being almost absolute within 2–3 days. Luminal stages are much less affected by this immunity, and cysticercoids given after an egg infection do establish. The luminal stages do generate protective responses, albeit more slowly than those initiated by egg infections, but the process of autoinfection complicates interpretation of this situation, as the host is exposed to both adult antigens and antigens from the tissue stages after egg release.

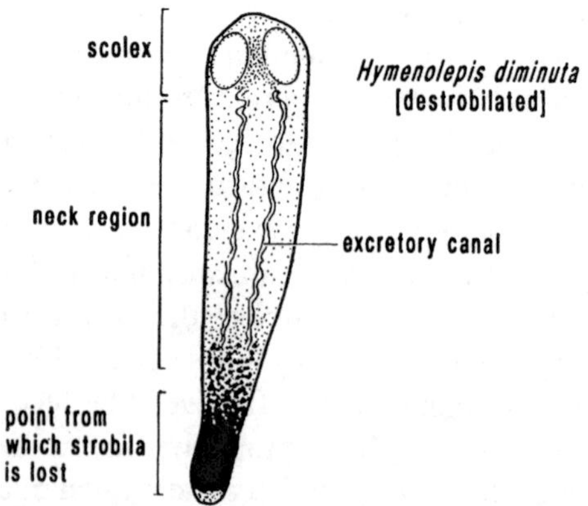

Fig. 32.10
A destrobilate hymenolepid cestode. Shedding of the strobila, leaving only the scolex, neck region and a short part of the body, is essentially a stress response. It occurs as the host develops immunity, and regrowth of the strobila is prevented unless the host is immune-suppressed. (Based on a micrograph by Befus & Threadgold, 1975.)

Table 32.6. *Immunity to the mucosal (cysticercoid) and lumenal (adult) stages of* Hymenolepis nana *in mice*

Primary (1°) infection	Challenge (2°) infection	Expression of immunity
Egg	None	Egg-induced immunity prevents autoinfection, i.e. reinfection by eggs from 1° adults
Cysticercoid	None	No immunity against autoinfection
Egg	Egg	Strong immunity, few 2° cysticercoids, no 2° adults
Egg	Cysticercoid	Moderate immunity, reduced numbers of 2° adults
Cysticercoid	Egg	Moderate immunity, reduced numbers of 2° cysticercoids and 2° adults
Cysticercoid	Cysticercoid	No immunity if 1° autoinfection is prevented

Source: From Heyneman (1963).

Larval tapeworms. Before larval tapeworms can establish within the body of the intermediate host, the hexacanth stage must hatch and penetrate the intestinal wall to enter a blood vessel or cross the peritoneal cavity. This life cycle stage is powerfully immunogenic, and its antigens stimulate strong protective responses (Rickard & Williams, 1982). Immune mechanisms involving IgA or IgG antibodies may operate at the intestinal level and effectively prevent further development (= early immunity). Once located in an organ, the hexacanth transforms into the cystic form, and passes through a stage during which the surface tegument is vulnerable to attack by antibody, activating complement or facilitating cytotoxicity (= late immunity). Subsequently, however, the parasite may express molecules which are anticomplementary in action, i.e. prevent the formation of complement components that can attach to the parasite to allow cell adherence or prevent the formation of lytic complexes that can damage the tegument directly (Leid, 1987). The parasite is then not susceptible to immune attack and its survival is assured (Fig. 32.11). The formation of host-derived layers around the developing larva provides additional protection against host immunity. It is interesting that, despite the formation of substantial cyst walls, there is still a considerable interchange of material between host and parasite. The fluid within cysts is known to contain host components, and parasite antigens are known to leak out into the host. The sensitisation that results from the gradual release of cyst fluid antigens can be very dangerous in situations (such as surgical removal of hydatid cysts) where the cyst wall is ruptured and large amounts of antigen are released into the body. Under these circumstances severe anaphylactic shock may result.

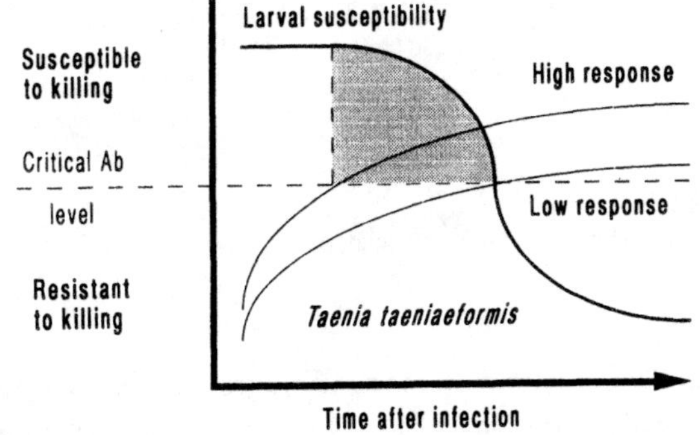

Fig. 32.11
Larval susceptibility to antibody-mediated immune attack and host antibody production determines the outcome of the relationship between *Taenia taeniaeformis* metacestodes and the mouse. Larvae lose their susceptibility as they acquire an anti-complementary activity, preventing the tegumental damage that normally follows antibody-mediated activation of the complement pathway. In high-responders, antibody levels rise to a point at which larvae can be killed before they become insusceptible (the shaded zone). In low responders, even though antibodies rise above the critical level, the worms are no longer susceptible. (Based on data from Mitchell *et al.*, 1980.)

○ 32.2.8 Nematodes

Some of the most important nematode parasites are very long-lived in their hosts. Chronic infections are characteristic of the tissue-dwelling filarial worms, where worms may live for more than ten years, as well as of many of the gastro-intestinal species (Behnke *et al.*, 1992). This ability to survive for prolonged periods may suggest that nematodes are intrinsically non-antigenic, but the converse is true, and nematodes are known to present the host with a wide range of immunologically active molecules (Fig. 32.12). Although the possession

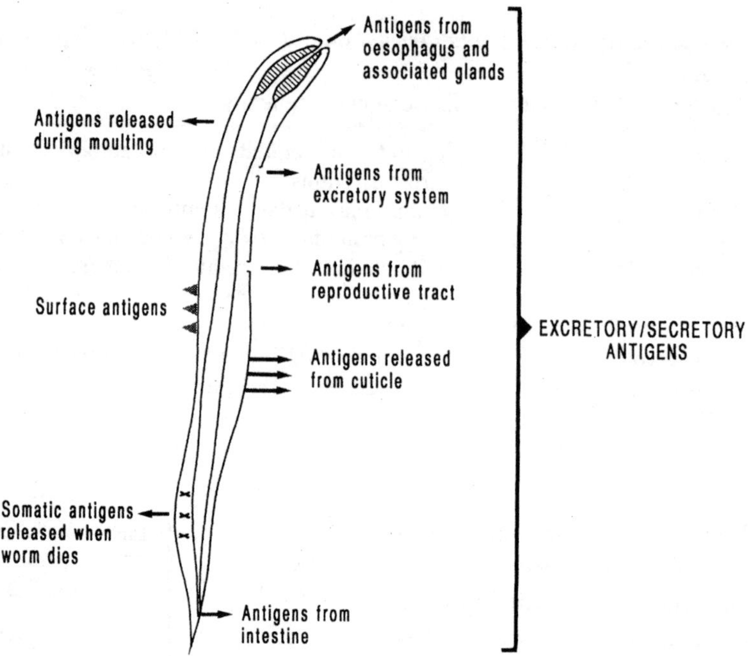

Fig. 32.12
Sources of antigens in parasitic nematodes.

of a collagenous cuticle certainly does confer a degree of protection, this layer is no longer seen as providing an immunologically inert interface with the host. In all species studied, the cuticle expresses and releases antigens against which immune responses can be made (Kennedy, 1991). Nematodes also release a wide variety of excretory/secretory (E/S) antigens that are powerfully immunogenic (Lightowlers & Rickard, 1988). It is therefore clear that the long-term survival of many species cannot be explained by an inability to stimulate immune responses.

Filarial nematodes. Although filarial infections with the species causing lymphatic disease (*Wuchereria* and *Brugia*) can be very common in areas of endemic transmission, it is well documented that, even where transmission rates are high, not all individuals are parasitologically positive, i.e. have circulating microfilaria. Those that are microfilaraemic may show a spectrum of pathological conditions. Three categories of individual in endemic areas are of particular interest:

(*a*) those who are amicrofilaraemic, have no pathology, but show immunological evidence of exposure to infection;
(*b*) those who are amicrofilaraemic but suffer pathological changes resulting from obstructive disease of the lymphatic system, such as elephantiasis;
(*c*) those who are microfilaraemic but show little pathology.

There is now good evidence that separation of individuals on the basis of these parasitological and pathological criteria reflects fundamental differences in their immune responses to infection (King & Nutman, 1991; Maizels & Lawrence, 1991). Those in the first category may be identified as immune, having made appropriate and functionally protective responses, which have eliminated infection and prevented reinfection. Those in category (*b*) have presumably made immune responses against adult worms, preventing microfilaraemia, but have become hyper-responsive, i.e. have made inappropriate responses which have led to immunopathological changes. Microfilaraemic individuals showing little or no pathology can be considered as hyporesponsive. Their infections have not been controlled, but neither have there been immunological responses.

Most of the available experimental evidence, from work with animal models and *in vitro* studies, suggests that ADCC mechanisms directed against cuticular antigens are an important component of immunity against filarial nematodes (Philipp *et al.*, 1984). A variety of antibody isotypes and inflammatory cells have been shown to kill the microfilarial and infective larval stages, and it is probable that adult worms are similarly vulnerable. In functional terms it is obviously best that the immunity should be directed against the infective larvae transmitted from mosquitoes. By killing this stage of the life cycle, the host avoids both the pathological and the immunopathological consequences associated with later stages.

With a few exceptions (notably the cat–*Brugia pahangi* system), animal models have thrown little light on the events associated with hyper-responsiveness and the development of obstructive disease in humans. T cells from such individuals proliferate very strongly in response to parasite antigens, and it seems probable that persistence of responsiveness, together with an altered pattern of cytokine production leading to hypersensitivity responses, may underlie the sequence of tissue changes leading to lymphatic pathology (Ottesen, 1992).

In contrast to hyper-responsive individuals, hyporesponsive patients show depressed T cell proliferation to parasite antigens (Nutman *et al.*, 1987). Often antibodies to microfilarial surface antigens cannot be detected (although they appear when the host becomes amicrofilaraemic) and the isotype profile of anti-worm antibodies differs markedly from that seen in individuals in other categories (Hussain *et al.*, 1987). A plausible explanation is that in hyporesponders, for genetic and parasitological reasons, normal T cell responses to the worm are modulated away from the pattern that would lead to protective immunity. For example, the high levels of IgG4 antibodies seen in microfilaraemic hosts presumably reflect selective T cell activation. IgG4 is known to block the activity of IgE, which may help to reduce immediate-type hypersensitivity, but is undesirable immunologically because, as in schistosomiasis, IgE may well mediate protective ADCC responses.

Gastro-intestinal nematodes. Far more is known about immunological aspects of infection with intestinal nematodes than about any other group. Although these worms are important and prevalent parasites of humans and domestic animals, the impetus for experimental studies has come largely from the convenience with which intestinal species can be studied in laboratory models. A variety of models is available, allowing analysis of acute and chronic infections. Two species in particular have provided detailed pictures of the events associated with the development and expression of protective immunity, *Nippostrongylus brasiliensis* and *Trichinella spiralis*; *Heligmosomoides polygyrus* and *Trichuris muris* have given useful data about the immunological aspects of more persistent intestinal infections (Wakelin *et al.*, 1993).

Nippostrongylus brasiliensis

Infections in the rat generate strong T cell-dependent immune responses that bring about expulsion of the worms from the intestine (see Fig. 28.4). After migration of larvae from the skin, through the lungs to the intestine, there is a plateau phase, during which the number of adult worms remains more or less stable. Egg production rises to a peak at about day 10, and then begins to decline. This decline is followed by a dramatic loss of worms, the process of 'self' or, more accurately, 'spontaneous' cure, which results in the loss of all but a small number of parasites (Jarrett *et al.*, 1968). The rat is then effectively resistant to subsequent infection, incoming larvae being trapped in skin or lungs before reaching the intestine, or, if they do migrate successfully, failing to grow and being eliminated quickly. This pattern of response is seen most clearly when rats are given single, pulse infections with relatively large numbers of larvae (up to 500). When small (50 larvae) infections are given, or when rats are exposed to repeated 'trickle' infections, a different picture is seen. With small infections, expulsion is slower; with trickle infections, expulsion may not occur (Jenkins & Phillipson, 1971). However, small infections, even as few as 10 worms, can stimulate a good immunity to reinfection.

Rats undergoing spontaneous cure have intestines that are highly inflamed. The gut is swollen, filled with fluid and mucus, the wall is heavily infiltrated with inflammatory cells, of which mast cells are a dominant component, and the normal architecture of the mucosa is disrupted (Miller, 1984; Rothwell, 1989). It seems self-evident that worm expulsion must in some way be causally correlated with inflammation, but it is still not clear exactly what the correlation is. Infection elicits high levels of IgE production, not only against antigens of the parasite, but also against other antigen specificities. The coincidence of IgE and mast cells suggests that an immediate hypersensitivity response might well underlie expulsion, either by changing the intestinal environment dramatically or by allowing other isotypes of anti-worm antibody (IgG for example) to enter the intestinal lumen (the 'leak–lesion' hypothesis; Murray, 1972) but it has been difficult to prove this directly.

Table 32.7. *Intestinal inflammatory changes during nematode infection*

1. Cellular infiltration into mucosa
2. Increase in intra-epithelial lymphocytes
3. Decrease in enterocyte transit time, appearance of immature cells on villus, loss of enterocyte brush-border enzymes
4. Increased epithelial permeability
5. Villus atrophy and crypt hyperplasia
6. Increased blood flow to mucosa, and increased vascular permeability
7. Increase in mucosal and luminal content of leucocyte enzymes and biologically active mediators
8. Change in mucosa–lumen fluid flux, resulting in net fluid secretion into gut
9. Increase in goblet cells and mucus secretion
10. Altered gut motility, increased peristalsis

It is possible to transfer immunity with immune sera, implying a role for anti-worm antibody, or with immune T cells, implying that non-antibody effects are also important (Nawa *et al.*, 1978; Miller, 1980). Under some conditions there is a good correlation between numbers of mast cells, and their functional activity as measured by release of granule-associated serine proteases, and the process of worm expulsion (Woodbury *et al.*, 1984), but this is not seen in all circumstances. In the mouse model, neither mast cells nor IgE appear to be essential for expulsion to occur.

It may well be the case that expulsion is multiphasic, and that multiple components of the host immune response, both specific and non-specific, may be able to contribute. Intestinal inflammation is a very complex process (Table 32.7), resulting not only in profound structural and functional changes, but also in the release of a wide variety of mediators, each potentially capable of damaging the worm. Degranulation of mast cells releases preformed vasoactive amines that can interact with the tissues of both host and parasite. At the same time other mediators, e.g. leukotrienes and prostaglandins, are generated. Mast cells are not the only cells infiltrating the mucosa during infection: lymphocytes and granulocytes, particularly eosinophils, are also present. These cells release cytokines, enzymes and basic proteins, all of which could affect the viability of the worm. Recently a correlation has been shown between worm expulsion and production of reactive oxygen radicals (Smith & Bryant, 1989). It therefore seems probable that the host has many ways of removing the worm, some of which may predominate under particular circumstances, but all of which have the potential to bring about worm expulsion. The expression of immunity in already immune hosts may be rather different. Not only may larvae be destroyed during their parenteral migration to the intestine, but if they reach the gut, they may become trapped in intestinal mucus, which has different properties in immune animals (Miller, 1987). This trapping results in their immune exclusion from the preferred mucosal niche, and the larvae are then removed by peristalsis.

Nippostrongylus is not entirely at the mercy of its host. It does survive as a natural infection in wild rats, and when given experimentally as a trickle infection, worms can live for prolonged periods, showing a degree of adaptation to host immunity. In common with many other species, *N. brasiliensis* releases large amounts of the enzyme acetylcholinesterase (Sanderson & Ogilvie, 1971), and it has been suggested that this may affect the host in ways that help to prolong worm survival. The enzyme may interfere with host neuromuscular transmission, preventing local peristaltic movements (act as a biological 'holdfast') or reduce the release of mucus. Acetylcholinesterase is certainly a target for the immune response, the host making antibodies against it, but worms appear to be able to modify the isoenzyme pattern as immunity develops (Edwards *et al.*, 1971).

Trichinella spiralis

Like *N. brasiliensis*, *T. spiralis* has a life cycle which has both intestinal and parenteral phases (Fig. 26.5), but the sequence is different. In *Trichinella* the parenteral phase occurs after the intestinal, and is the longest component of the cycle. Immunity operates almost exclusively against the intestinal and migratory stages, there being very little immune control of muscle larvae once they have become established.

Experimental infections can be established in a wide variety of hosts, but the majority of experimental data comes from work with mice and rats. In both of these species, primary infections follow a similar pattern. The infective larvae mature rapidly, adult worms mate and the females liberate newborn larvae, and there is then a strong T cell-dependent host response which results in a virtually complete expulsion of worms from the intestine (Fig. 32.13). As with *Nippostrongylus*, expulsion

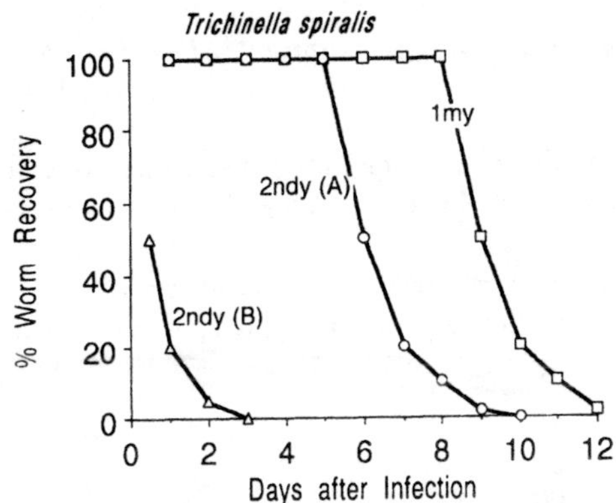

Fig. 32.13
Diagrammatic time course of primary (1my) and secondary (2ndy) infections of *Trichinella spiralis* in high-responder mice. When a secondary infection is given some weeks after the primary, worm loss is accelerated by a few days (2ndy (A)), but when given immediately after the primary there is rapid expulsion of the secondary infection (2ndy (B)).

is coincident with a profound inflammatory response, and many of the changes that occur are similar, but the evidence that inflammation and expulsion are causally connected is much stronger in *Trichinella* (Wakelin & Denham, 1983). For example, passive transfer of immunity with immune serum is ineffective, whereas adoptive transfer using $CD4^+$ T helper cells is highly effective, as long as the recipient is capable of mounting an inflammatory response. Significant correlations exist between worm expulsion and mast cell activity, as measured by release of serine proteases (Tuohy *et al.*, 1990), and the capacity to expel worms can be restored to T cell- or mast cell-deficient mice when the mast cell response is reconstituted (Reed, 1989). Antibodies probably play a secondary role, reducing the viability of the parasite through an interference with feeding and behaviour.

Once animals have experienced an initial infection they show a strong resistance to reinfection. In rats this is expressed as a rapid expulsion of incoming larvae within a few hours, and is thought to involve some form of trapping in intestinal mucus (Miller, 1987). Recently it has been shown that a similar process can be induced in neonatal rats after transfer of monoclonal antibodies. In mice, resistance to reinfection is more often shown as an accelerated form of the primary response, i.e. adult worms develop, but are stunted and reproduce only poorly before being expelled.

The newborn larvae released into the intestinal mucosa can act as targets for protective responses, and may well be killed by antibody-mediated cytotoxicity mechanisms (Lee, 1991). Once they have invaded the muscle cells, however, the parasites are effectively protected against immune attack.

Trichinella is one of the best-documented nematodes in terms of antigenic make-up. Like all members of the Trichuroidea, this species possesses an anterior stichosome (Fig. 26.4), a chain of large glandular cells, surrounding the oesophagus. These cells contain membrane-bound granules, the contents of which are known to be released from the worm, presumably to assist tissue penetration or feeding. The granule contents are potently immunogenic, and a number of components (notably the 43–45 and 50–55 kDa molecules) immunise animals very effectively. The worms also possess a number of stage-specific cuticular antigens (Almond *et al.*, 1986). These can act as targets for immune effectors *in vitro*, and are also known to elicit protective responses *in vivo*.

A number of studies have been carried out on immunity to *Trichinella* in pigs, and it is clear that there are some important differences from the patterns established using rodent hosts (Marti & Murrell, 1986). Primary infections are more prolonged, lasting several weeks, and immunity is most marked in its effect on reproduction. Worm antigens have been used successfully to vaccinate pigs against infection.

Heligmosomoides polygyrus and *Trichuris muris*

Although *Nippostrongylus* and *Trichinella* have provided a great deal of useful information about the ways in which intestinal immunity acts to control nematode infections, it can be argued that a more important question is to establish why such immunity does not operate more effectively against the species that, by virtue of their long-term survival, constitute the major medical and veterinary problems (Behnke *et al.*, 1992). Two model systems provide some information about this point. In many strains of mice *H. polygyrus* is a chronic infection, the worms surviving for many months; in others, infection is controlled by immunity. Analysis of the differences in response between susceptible and resistant mice has thrown light on the mechanisms that determine the level of immunity, and it seems certain that the balance of cytokine responses plays a decisive role (Finkelman *et al.*, 1991). One important finding has been that *H. polygyrus* is potently immunosuppressive,

Table 32.8. *Suppression by transplanted adult* Heligmosomoides polygyrus *of immunity stimulated by prior exposure to irradiated infective larvae*

One hundred adult worms were implanted directly into the small intestine on day −7, group 1 mice being sham-operated; 200 irradiated larvae were given orally on day 0. After anthelminthic on days 42 and 44, all mice, together with controls, were challenged on day 49 with 200 normal larvae. Worm counts were made 28 days later.

Group of mice	Adult worms transplanted	Irradiated L_3 on day 0	Challenge worm counts (as per cent of control)
1	−	+	2
2	+	−	88
3	+	+	80
Control	−	−	100

Source: Data taken from Behnke *et al.* (1983).

presumably by release of immunomodulatory molecules. Adult worms not only suppress the immunity generated by larval infections, but they also interfere with the response to immunisation with irradiated infective larvae, normally capable of generating a complete immunity to challenge (Table 32.8). This ability to suppress can also affect responses to heterologous antigens as well as to other pathogens.

Trichuris muris is a useful model for the whipworm *T. trichiura*, one of the most prevalent chronic worm infections in humans. As with *H. polygyrus*, it is possible to generate a spectrum of host–parasite relationships in different strains of mice, some responding rapidly and expelling the parasite, others failing to do so and harbouring long-term infections. Recent work has shown that high- and low-responder strains show contrasting cytokine response profiles, the former being associated with T helper cell (Th) 1 type responses, the latter with Th 2 responses (Else & Grencis, 1991). Significantly, low responders appear to switch from Th 1 to Th 2 during infection, presumably under the influence of worm-derived immunomodulatory molecules. If these data can be extrapolated to the situation in humans, they imply that the outcome of infection is not predetermined, but reflects the interaction of parasitological and immunological influences of both host and parasite origin.

Trichostrongyles in sheep

Domestic ruminants can be parasitised by a number of species of trichostrongyle. Infections cause significant economic losses, and are presently controlled by pasture management and chemotherapy. The intensive use of anthelminthics has resulted in the development of considerable drug resistance, and attention has turned again to the possibility of using immunological methods of control. Much of what is known about immunity to these nematodes has come from work with two of the species of greatest importance, *Haemonchus contortus* and *Trichostrongylus colubriformis*.

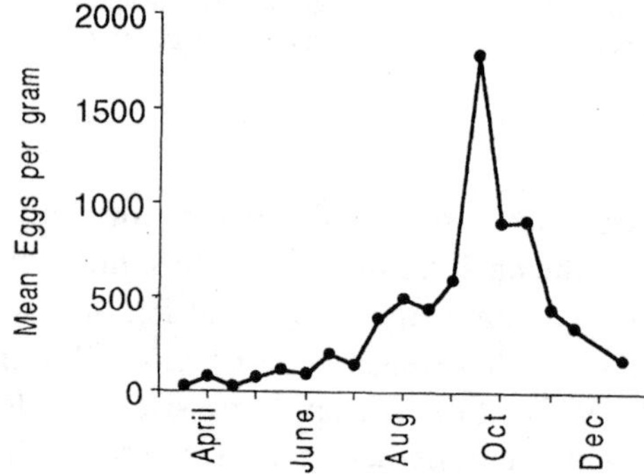

Fig. 32.14
Numbers of trichostrongyle eggs passed by lambs on infected pasture during one year. The steep fall during late September reflects the development of a self cure response, triggered by the increasing intake of infective larvae during the summer. (Redrawn from Crofton, 1955.)

The observations of 'self cure' in *H. contortus*-infected sheep, initially made by Stoll in 1929, led to some of the first systematic studies of immunity to a gastro-intestinal infection. Self cure is typically seen in sheep on infected pasture, exposed to a continuous intake of infective larvae. The animals develop a mature, egg-producing infection, but then, quite suddenly, undergo a series of intestinal changes which result in worm expulsion (Fig. 32.14). These changes have many of the characteristics of an immediate hypersensitivity reaction, and are presumed to reflect a response in primed animals to antigens released by the incoming larvae as they undergo moulting (Lloyd & Soulsby,1987). Following self cure the animal is relatively resistant to further infection, but this resistance then wanes.

Despite a great deal of experimental and field observation we still know little about the mechanisms involved in self cure. Only very recently, for example, has it been proven that immunity to *H. contortus* is functionally dependent upon T helper lymphocytes, deletion of this population preventing the development of resistance (Gill *et al.*, 1992). The intestinal changes accompanying self cure show suggestive similarities with those associated with spontaneous cure in rodent nematodes. Mucosal infiltration by mast cells and eosinophils, evidence of altered intestinal permeability, mucus trapping and immune exclusion of incoming larvae from the mucosa of immune hosts are some of the common characteristics (Miller, 1984; Rothwell, 1989).

Immunity to *T. colubriformis* appears overall to be similar to that against *H. contortus*, in that immune-mediated inflammatory mechanisms again seem to be important. Unlike *Haemonchus*, *T. colubriformis* can be studied experimentally in a laboratory host – the guinea pig – which has allowed more detailed studies to be made (Rothwell, 1989). Immunity in this host results in worm expulsion and again appears to involve inflammatory mechanisms. It is T cell-dependent, and can be transferred both adoptively with immune cells and passively with immune serum. Worm loss can be induced by administration of inflammatory mediators, and prevented by antagonists. As with *H. contortus*, a correlation has been established both in sheep and in guinea pigs between levels of genetically determined resistance to infection with *T. colubriformis* and the infiltration of the mucosa by inflammatory cells. All of these data point to the likelihood that protective mechanisms against infection with intestinal nematodes will have an underlying similarity in all hosts, this similarity reflecting the limited number of ways in which the gut can respond to immunological challenge.

○○○ 32.3 Vaccination and diagnosis

In recent years there has been considerable progress towards the development of effective anti-parasite vaccines and the achievement of improved diagnosis. This progress has been made possible by a greater understanding of the nature of immune responses to parasitic infections, but has depended heavily upon technical advances in the identification, isolation and production of antigens. The approaches of molecular biology have been particularly important in allowing determination of antigen structure and identification of coding genes, thus making it possible to produce synthetic and recombinant antigens. However, these techniques by themselves would have been less fruitful had they not gone hand-in-hand with improvements in the ability to pinpoint antigens and epitopes of major importance in immune responses.

○ 32.3.1 Vaccination

Relatively few vaccines are currently in commercial production, and of these the irradiated larval vaccine against the lungworm *Dictyocaulus viviparus* in cattle has the longest history of success in the field (Poynter *et al.*, 1960). Vaccination depends upon administration of irradiated infective third-stage larvae, which initiate, but do not complete, their development in the host. The fourth-stage larvae and juvenile worms provide the antigenic stimulus necessary for the development of protective immunity, but die before maturing into the adult stages responsible for the pathology of the disease. This is an excellent example of an empirical vaccine, in which neither the antigens nor the immune response necessary for immunity are known. Many attempts to produce similar vaccines, by using irradiated stages, have either failed to achieve significant immunity under field conditions or have suffered from drawbacks that have reduced their success. Of the latter, the commercial failure of the irradiated larval vaccine against canine hookworm is perhaps the best known (Miller, 1978). The vaccine protected well against the disease associated with infection, but did not give complete protection against infection itself, allowing low-level worm burdens to develop.

Vaccines against babesiosis and theileriosis have been developed successfully, but both depend upon the use of attenuated organisms (Morrison, 1989). Although provision of parasites is not necessarily an insuperable obstacle in all cases, as these vaccines show, the difficulties of obtaining sufficient parasite material, either for production of attenuated stages or as a source of parasite antigens, have imposed severe limitations on the development of vaccines against other infections. Equally, although it is acceptable to use live vaccines in animals, there are major ethical constraints on their use in humans. In general, vaccine research is currently focused almost exclusively on the use of recombinant or synthetic antigens. Brief summaries of progress made with four infections, malaria, schistosomiasis, larval cestodes and gastrointestinal nematodes, are given here.

There are many excellent reviews available which give more detailed information: Anders, 1988; Barbet, 1989; Bergquist, 1990; Bloom, 1989; Capron *et al.*, 1987; Gamble & Murrell, 1987; Good *et al.*, 1988; McLaren, 1989; Nussenzweig & Nussenzweig, 1989; Smith & Munn, 1990; Targett, 1991; Wakelin, 1994.

Malaria. Three stages in the malaria life cycle are seen as targets for vaccine-induced immunity: the sporozoites, the asexual erythrocytic stages, and the sexual stages. Immunity against the first two would protect against the development of infection and the onset of disease; immunity against the sexual stages would not protect the individual but would interfere with transmission and thus protect the community. The development of a sporozoite vaccine is supported by studies, both in humans and in animals, which have shown that injection of irradiated sporozoites can provide substantial protection against infection with this stage (although giving no immunity against subsequent stages of the infection). Many experimental studies and a number of human trials have now been carried out to determine whether effective vaccination can be achieved using synthetic and recombinant peptides based upon the immunodominant circumsporozoite (CS) antigen. These peptides have been given by injection as well as via live (*Salmonella*) vectors. Although positive results have been obtained, particularly in experiments with animals, there are several problems inherent in using such peptides. One is that antibodies directed against the immunodominant epitopes on this antigen may not be protective; another is that not all individual hosts are capable of seeing the same range of small peptides (because of MHC restriction); a third is the existence of considerable polymorphism in CS protein sequences. In practical terms anti-sporozoite immunity has to be 100 per cent effective if it is to prevent development of infection; the fact that individuals living in endemic areas do show anti-CS immune responsiveness, yet are frequently reinfected, suggests that under natural conditions, vaccine-induced anti-sporozoite immunity alone would not prove effective.

Vaccination against the erythrocytic stages of malaria should provide good protection against infection, but even if incompletely effective, would certainly help to reduce the level of parasitaemia and so reduce the severity of disease. Candidate antigens include those on the merozoite itself, especially those derived from the surface coat antigen precursor and the rhoptry antigens, as well as those appearing on the surface of the infected red cell. Of the latter, trials with the 155 kDa RESA have given promising results and there is much interest in the 'knob'-associated antigens expressed in cells infected with *P. falciparum* and which are responsible for cytoadherence. Some of the most dramatic vaccination trials have been carried out in Colombia, using mixtures of synthetic peptides derived from the blood stage forms (Pattarroyo *et al.*, 1988). Considerable success in controlling infection has been reported and more extensive trials are now being undertaken in other malaria endemic areas.

Schistosomiasis. Recognition of the fact that incoming larvae act as the primary targets of immunity has focused attention on the antigens of the early and late schistosomula as likely vaccine candidates. Vaccination with irradiation-attenuated larvae has been shown to be effective in rodent and primate models and has also been used in large scale field trials against infection with *S. bovis* in the Sudan (Taylor *et al.*, 1990). Both logistic and ethical considerations prevent the use of such attenuated vaccines in humans and attention is focused instead on the many recombinant antigens that are now available. The most promising of these are the gp 28 kDa antigens known to have glutathione S transferase activity, which have given very promising results in tests involving infections with *S. mansoni* in rodents and primates (Boulanger *et al.*, 1991). This protection has been seen both in terms of infection (vaccination reducing worm burdens as much as 70 per cent) and in terms of pathology, the reduction in egg output giving markedly less tissue damage.

Larval cestodes. Development of the larval stages of *Taenia ovis* and *T. hydatigera* in sheep is an important economic problem, as the infected carcasses lose much of their economic value. It has been known for many years that sheep develop strong protective immunity after initial infection, and experimental studies have pointed to antigens from the oncosphere as playing an important role in generating this immunity. Work on *T. taeniaeformis* in mice (see section 32.2.7) provided a sound basis for believing that a vaccine that raised high antibody levels against this stage of the ovine species should give substantial protection, and this prediction has been borne out by many experiments in sheep using attenuated parasites and antigens derived from parasite material. Recently, success has been achieved with a recombinant vaccine against *T. ovis* (Rickard, 1989). A 45 kDa antigen known to occur in oncospheres has been expressed in *E. coli*, using the pGEX-1 vector. When used with a number of adjuvants it has given high levels of protection under both experimental and field conditions. The use of the pGEX-1 vector gives expression of the target molecule as a fusion protein with *Schistosoma japonicum* glutathione S-transferase (GST) allowing easy purification of the product on a glutathione–agarose affinity column.

Gastro-intestinal nematodes. The early success of the irradiated vaccine used against lungworm raised hopes that similar vaccines would be effective against the trichostrongyle species responsible for gastro-intestinal infections in cattle and sheep. Although effective vaccination has been described under controlled conditions, attenuated vaccines have not been successful under field conditions. A number of programmes are now using defined and recombinant antigens as the basis of vaccines, principally targeted at the important parasites of sheep, such as *Haemonchus contortus* and *Trichostrongylus colubriformis*. Significant progress has been made using molecules expressed at the surface of the worm's intestinal cells as the vaccine antigen. These cryptic antigens raise very little immune response in natural infections, but sheep immunised against them show high levels of immunity against subsequent challenges. Vaccination has been achieved successfully using both parasite-derived and recombinant molecules, and has been found to be effective even in comparatively young lambs.

Successful vaccination against parasites faces many problems (Williams, 1986; Sher, 1988; Mitchell, 1989). One, especially relevant to use in humans, concerns the way in which vaccines are administered. Few of the conventional adjuvants are really suitable, and there is a need for novel, safe formulations that will promote precisely the type of immune response required. Alternatives to adjuvants are the use of microparticles, into which antigens can be incorporated and from which controlled release is possible, and of genetically engineered live vectors, such as attenuated *Salmonella* or vaccinia virus, which can be constructed to carry the genes for vaccine antigens.

○ 32.3.2 Diagnosis

Diagnosis of parasitic infections still depends heavily on parasitological findings, e.g. parasites in blood smears or biopsies and eggs or cysts in faecal samples. Although these tests, if successful, provide unequivocal evidence of infection they are limited, especially when applied to low-level infections or to infections where the parasite occurs in deeper tissues. Immunological diagnostic tests have a valuable role to play, not only for these more difficult situations, but also in large-scale examinations of populations in endemic areas. Many immunologically based tests have been used in the past, some with a great deal of success, but the provision of appropriate antigens, and achievement of sufficient sensitivity and specificity, have created considerable problems. Advances in knowledge of parasite antigens, and the ability to synthesise and clone suitable molecules, have considerably improved the situation. It is possible now to select antigens that are known to have high specificity, and to use these in a variety of sensitive tests, of which the enzyme-linked immunoabsorbent assay (ELISA) and its derivatives are among the most widely used. ELISA tests can be used to measure antibody responses to defined antigens, and the pattern and level of isotype response can be used to provide an accurate picture of the patient's experience of infection, not merely to show that there has been some exposure at some time in the past. ELISA can also be used to detect the presence of parasite antigens in body fluids or in faecal material. This provides not only a more certain indication that the patient is currently infected, but can be adapted to provide quantitative data on the level of infection present. The efficiency of ELISA and many other immunological diagnostic tests has been enhanced by the availability of monoclonal antibody reagents, which give a greater specificity and allow a greater degree of quality control, giving better reproducibility. ELISA testing is amenable to automation and also for rapid colorimetric applications which require minimal expertise, such as 'dip-stick' tests.

Specific diagnosis is improved by being able to show precisely which antigens are recognised by a patient's serum. Immunoblotting tests, in which parasite antigens are separated and transferred to nitrocellulose paper before reaction with sera, make it possible to see patterns of antigen recognition. Pre-prepared papers can be used under field conditions, either to obtain this kind of detailed information or simply to see whether there are any antibodies present against unseparated parasite antigens (the 'dot–blot' test).

○○○ References

(References of general interest have been included in this list, in addition to those quoted in the text.)

IMMUNOLOGY

Albright, J. F. & Albright, J. W. 1984. Natural resistance to animal parasites. *Contemporary Topics in Immunobiology*, **12**: 1–52.

Benjamin, E. & Leskowitz, S. 1991. *Immunology A Short Course*. 2nd Edition. John Wiley, New York.

Klein, J. 1990. *Immunology*. Blackwell Scientific Publications, Oxford.

Male, D., Champion, B. & Cooke, A. 1987. *Advanced Immunology*. Gower Medical Publishing, London.

Mims, C. A. 1982. Innate immunity to parasitic infections. In: *Immunology of Parasitic Infections* (ed. K. S. Warren & S. Cohen), pp. 3–27. Blackwell Scientific Publications, Oxford.

Ratcliffe, N. A. 1985. Review: Invertebrate immunity – a primer for the non-specialist. *Immunology Letters*, **10**: 253–70.

Roitt, I. 1991. *Essential Immunology*, 7th Edition. Blackwell Scientific Publications, Oxford.

Roitt, I., Brostoff, & Male, D. 1989. *Immunology*. 2nd Edition. Gower Medical Publications, London.

Sell, S. 1987. *Basic Immunology: Immune Mechanisms in Health and Disease*. Elsevier, New York and Amsterdam.

GENERAL IMMUNOPARASITOLOGY

Ash, C. & Gallagher, R. B. (eds) 1991. *Immunoparasitology Today*. Elsevier Trends Journals, Cambridge.

Behnke, J. M. (ed.) 1990. *Parasites: Immunity and Pathology. The Consequences of Parasitic Infection in Mammals*. Taylor & Francis, London.

Immunological Reviews, vol. **112**, 1989. Immunology of Parasites.

Immunological Reviews, vol. 127, 1992. Cytokines in Infectious Diseases.

Kierszenbaum, F. (ed.) 1991. *Immunology and Immunosuppression in Infectious Diseases*. Marcel Dekker, New York.

McAdam, K. W. P. (ed.) 1989. *New Strategies in Parasitology*. Churchill Livingstone, Edinburgh.

Marchalonis, J. J. (ed.) 1984. Immunobiology of parasites and parasitic infections. *Contemporary Topics in Immunobiology*, 12.

Soulsby, E. J. L. (ed.) 1987. *Immune Responses in Parasitic Infections: Immunology, Immunopathology and Immunoprophylaxis*. Vol. 1 *Nematodes*; Vol. 2 *Trematodes and Cestodes*; Vol. 3 *Protozoa*. CRC Press, Boca Raton, Florida.

Wakelin, D. 1989. *Immunity to Parasites*. Edward Arnold, London. (A revised edition of this volume is in preparation and will be published by Cambridge University Press in 1995).

Warren, K. (ed.) 1993. *Immunology and Molecular Biology of Parasitic Infections*. Blackwell Scientific Publications, Oxford.

Wyler, D. J. (ed.) 1990. *Modern Parasite Biology. Cellular, Immunological and Molecular Aspects*. W. H. Freeman, New York.

PROTOZOA: EXTRA-CELLULAR

Aggarwal, A. & Nash, T. E. 1988. Antigen variation of *Giardia lamblia in vivo. Infection and Immunity*, **45**: 700–7.

Albright, J. W. & Albright, J. F. 1991. Rodent trypanosomes: their conflict with the immune system of the host. *Parasitology Today*, **7**: 137–40.

Barry, J. D. 1986. Antigenic variation during *Trypanosoma vivax* infection of different host species. *Parasitology*, **92**: 51–65.

Barry, J. D. & Turner, C. M. R. 1992. The dynamics of antigenic variation and growth of African trypanosomes. *Parasitology Today*, **7**: 207–11.

D'Alesandro, P. A. 1975. Ablastin: the phenomenon. *Experimental Parasitology*, **38**: 303–8.

den Hollander, N., Riley, D. & Befus, A. D. 1988. Immunology of giardiasis. *Parasitology Today*, **4**: 124–31.

Donelson, J. E. & Turner, M. J. 1985. How the trypanosome changes its coat. *Scientific American*, **252**: 44–51.

Farthing, M. J. G. 1989. *Giardia lamblia*. In *Enteric Infection, Mechanisms, Manifestations and Manaegment* (ed. M. J. G. Farthing & G. T. Keusch), pp. 397–413. Chapman and Hall, London.

Mansfield, J. M. 1990. Immunology of African trypanosomiasis. In: *Modern Parasite Biology. Cellular, Immunological and Molecular Aspects* (ed. D. J. Wyler), pp. 222–46. W. H. Freeman, New York.

Roelants, G. E. & Pinder, M. 1984. Immunobiology of African trypanosomiasis. *Contemporary Topics in Immunobiology*, **12**: 225–74.

Taliaferro, W. H. 1924. A reaction product in infections with *Trypanosoma lewisi* which inhibits the reproduction of the trypanosomes. *Journal of Experimental Medicine*, **39**: 171–90.

Vincendeau, P., Daeron, M. & Daulouede, S. 1986. Identification of antibody classes and Fc receptors responsible for phagocytosis of *Trypanosoma musculi* by mouse macrophages. *Infection and Immunity*, **53**: 600–5.

PROTOZOA: INTRA-CELLULAR

Alexander, J. & Russell, D. G. 1992. The interaction of *Leishmania* species with macrophages. *Advances in Parasitology*, **31**: 175–254.

Anders, R. F. 1986. Multiple cross-reactivities amongst antigens of *Plasmodium falciparum* impair the development of protective immunity against malaria. *Parasite Immunology*, **8**: 529–39.

Beaman, M. H., Wong, S.-Y. & Remington, J. S. 1992. Cytokines, *Toxoplasma* and intracellular parasitism. *Immunological Reviews*, **127**: 97–117.

Belosevic, M., Finbloom, D. S., van der Meide, P. H., Slayter, M. V. & Nacy, C. A. 1989. Administration of monoclonal anti-IFN-γ antibodies *in vivo* abrogates natural resistance of C3H/HeN mice to infection with *Leishmania major*. *Journal of Immunology*, **143**: 266–74.

Berendt, A. R., Ferguson, D. J. P. & Newbold, C. I. 1990. Sequestration in *Plasmodium falciparum* malaria: sticky cells and sticky problems. *Parasitology Today*, **6**: 247–54.

Brown, K. N. 1973. Antibody induced variation in malaria parasites. *Nature*, **242**: 49.

Butcher, G. A. 1989. Mechanisms of immunity to malaria and the possibilities of a blood stage vaccine: a critical approach. *Parasitology*, **98**: 315–27.

Chatelain, R., Varkila, K. & Coffman, R. L. 1992. IL-4 induces a Th2 response in *Leishmania major*-infected mice. *Journal of Immunology*, **148**: 1182–7.

Clark, I. A., Rockett, K. A. & Cowden, W. B. 1991. Proposed link between cytokines, nitric oxide and human cerebral malaria. *Parasitology Today*, **7**: 205–7.

Clark, I. A., Virelizier, J.-L., Carswell, E. A. & Wood, P. R. 1981. The possible importance of macrophage derived mediators in acute malaria. *Infection and Immunity*, **32**: 1058–66.

Cohen, S., McGregor, I. A. & Carrington, S. C. 1961. Gamma globulin and acquired immunity to human malaria. *Nature*, **192**: 733–7.

Current, W. L. & Owen, R. L. 1989. Cryptosporidiosis and microsporidiosis. In: *Enteric Infection. Mechanisms, Manifestations and Management* (ed. M. J. G. Farthing & G. T. Keusch), pp. 223–49. Chapman and Hall, London.

Day, K. P., Koella, J. C., Nee, S., Gupta, S. & Read, A. F. 1992. Population genetics and dynamics of *Plasmodium falciparum*: an ecological view. *Parasitology*, **104**: S35–52.

Green, S. J., Meltzer, M. S., Hibbs, J. B. & Nacy, C. A. 1990. Activated macrophages destroy intracellular *Leishmania major* amastigotes by an L-arginine-dependent killing mechanism. *Journal of Immunology*, **144**: 278–83.

Holder, A. A. 1988. The precursor to major merozoite surface antigens: structure and role in immunity. *Progress in Allergy*, **41**: 72–97.

Howard, R. J. 1987. Vaccination against malaria: recent advances and the problems of antigenic diversity and other parasite evasion mechanisms. *International Journal for Parasitology*, **17**: 17–29.

Hughes, H. P. A. 1988. Oxidative killing of intracellular parasites mediated by macrophages. *Parasitology Today*, **4**: 340–7.

Immunology Letters, vol. **25**, 1990. Fogarty/WHO International Conference on Cellular Mechanisms in Malaria Immunity.

Jeffrey, G. M. 1966. Epidemiological significance of repeated infections with homologous and heterologous strains and species of *Plasmodium*. *Bulletin of the World Health Organization*, **35**: 873–82.

Kemp, D. J., Cowman, A. F. & Walliker, D. 1990. Genetic diversity in *Plasmodium falciparum*. *Advances in Parasitology*, **29**: 75–149.

Liew, F. Y. & Cox, F. E. G. 1991. Nonspecific defence mechanism: the role of nitric oxide. In: *Immunoparasitology Today* (ed. C. Ash & R. B. Gallagher), pp. A17–21. Elsevier Trends Journals, Cambridge.

Locksley, R. M. & Scott, P. 1991. Helper T-cell subsets in mouse leishmaniasis: induction, expansion and effector function. In: *Immunoparasitology Today* (ed. C. Ash & R. B. Gallagher), pp. A158–61. Elsevier Trends Journals, Cambridge.

Mitchell, G. H. 1989. An update on candidate malaria vaccines. *Parasitology*, **98**: S29–47.

Nussenzweig, V. & Nussenzweig, R. S. 1989. Rationale for the development of an engineered sporozoite malaria vaccine. *Advances in Immunology*, **45**: 283–334.

Nussler, A., Drapier, J.-C., Renia, L., Pied, S. Miltgen, F., Gentilini, M. & Mazier, D. 1991. L-arginine-dependent destruction of intrahepatic malaria parasites in response to tumour necrosis factor and/or interleukin 6 stimulation. *European Journal of Immunology*, **21**: 227–30.

Perlmann, H., Berzins, K., Wahlgren, M., Carlsson, J., Bjorkman, A., Patarroyo, M. E. & Perlmann, P. 1984. Antibodies in malarial sera to parasite antigens in the membrane of erythrocytes infected with early asexual stages of *Plasmodium falciparum*. *Journal of Experimental Medicine*, **159**: 1686–704.

Playfair, J. H. L., Taverne, J., Bate, C. A. W. & de Souza, J. B. 1990. The malaria vaccine: anti-parasite or anti-disease? *Immunology Today*, **11**: 25–7.

Progress in Allergy, vol. **41**, 1988. Malaria Immunology.

Rose, M. E., Smith, A. L. & Wakelin, D. 1991. Gamma interferon-mediated inhibition of *Eimeria vermiformis* growth in culture fibroblasts and epithelial cells. *Infection and Immunity*, **59**: 580–6.

Rose, M. E., Wakelin, D. & Hesketh, P. 1989. Gamma interferon controls *Eimeria vermiformis* primary infection in BALB/c mice. *Infection and Immunity*, **57**: 1599–603.

Russell, D. G. & Talamas-Rohana, P. 1989. *Leishmania* and the macrophage: a marriage of inconvenience. *Immunology Today*, **10**: 328–33.

Schofield, L., Villaquiran, J., Ferreira, A., Schellekens, H., Nussenzweig, R. S. & Nussenzweig, V. 1987. γ-interferon, $CD8^+$ T cells and antibodies required for immunity to malaria sporozoites. *Nature*, **330**: 664–6.

Targett, G. A. T. 1990. Immunity to sexual stages of human malaria parasites: immune modulation during natural infections, antigenic determinants, and the induction of transmission-blocking immunity. *Scandinavian Journal of Infectious Diseases* (Supplement), **76**: 79–88.

Ungar, B. L. P., Kao, T.-C., Burris, J. A. & Finkelman, F. D. 1991. *Cryptosporidium* infection in an adult mouse model. Independent roles for IFN-γ and CD4⁺ T lymphocytes in protective immunity. *Journal of Immunology*, **147**: 1014–22.

Wakelin, D. & Rose, M. E. 1990. Immunity to coccidiosis. In: *Coccidiosis* (ed. P. L. Long), pp. 281–306. CRC Press, Boca Raton, Florida.

Wilson, R. J. M., McGregor, I. A. & Williams, K. 1975. Occurrence of S-antigens in serum in *Plasmodium falciparum* infections in man. *Transactions of the Royal Society of Tropical Medicine and Hygiene*, **69**: 453–9.

Zvala, F., Cochrane, A. H., Nardin, E., Nussenzweig, R. S. & Nussenzweig, V. 1983. Circumsporozoite proteins of malaria parasites contain a single immunodominant region with two or more identical epitopes. *Journal of Experimental Medicine*, **157**: 1947–57.

PLATYHELMINTHES

Befus, A. D. & Threadgold, L. T. 1975. Possible immunological damage to the tegument of *Hymenolepis diminuta* in mice and rats. *Parasitology*, **71**: 525–34.

Brindley, P. J. & Sher, A. 1987. The chemotherapeutic effect of Praziquantel against *Schistosoma mansoni* is dependent upon host antibody response. *Journal of Immunology*, **139**: 215–20.

Butterworth, A. E. & Hagan, P. 1987. Immunity in human schistosomiasis. *Parasitology Today*, **3**: 11–16.

Capron, A., Capron, M., Grangette, C. & Dessaint, J.-P. 1989. IgE and inflammatory cells. *Ciba Foundation Symposium*, **147**: 153–70.

Capron, M. & Capron, A. 1986. Rats, mice and men: models for effector mechanisms against schistosomiasis. *Parasitology Today*, **2**: 69–75.

Damian, R. T. 1989. Molecular mimicry: parasite evasion and host defence. *Current Topics in Microbiology and Immunology*, **145**: 101–15.

Dunne, D. W., Bickle, Q. D., Butterworth, A. E. & Richardson, B. A. 1987. The blocking of human antibody-dependent eosinophil-mediated killing of *Schistosoma mansoni* by monoclonal antibodies which cross react with a polysaccharide-containing egg antigen. *Parasitology*, **94**: 269–80.

Dunne, D. W., Butterworth, A. E., Fulford, A. J. C., Kariuki, H. C., Langley, J. G., Ouma, J. H., Capron, A., Pierce, R. J. & Sturrock, R. F. 1992. Immunity after treatment of schistosomiasis mansoni: association between IgE antibodies to adult worm antigens and resistance to reinfection. *European Journal of Immunology*, **22**: 1483–94.

Hagan, P., Blumenthal, U. J., Dunn, D., Simpson, A. J. G. & Wilkins, H. A. 1991. Human IgE, IgG4 and resistance to reinfection with *Schistosoma haematobium*. *Nature*, **349**: 243–5.

Heyneman, D. 1963. Host–parasite resistance patterns – some implications from experimental studies with helminths. *Annals of the New York Academy of Sciences*, **113**: 114–29.

Hughes, D. L., Harness, E. & Doy, T. G. 1981. The different stages of *Fasciola hepatica* capable of inducing immunity and the susceptibility of various stages to immunological attack in the sensitized rat. *Research in Veterinary Science*, **30**: 93–8.

Ito, A. & Smyth, J. D. 1987. Adult cestodes. Immunology of the lumen-dwelling cestode infections. In: *Immune Responses in Parasitic Infections: Immunology, Immunopathology and Immunoprophylaxis*, Vol. 2 (*Trematodes and Cestodes*) (ed. E. J. L. Soulsby), pp. 115–63. CRC Press, Boca Raton, Florida.

James, S. L. & Glaven, J. 1989. Macrophage cytotoxicity against schistosomula of *Schistosoma mansoni* involves arginine-dependent production of reactive nitrogen intermediates. *Journal of Immunology*, **143**: 4208–12.

Leid, R. W. 1987. Parasite defense mechanisms for evasion of host attack: a review. *Veterinary Parasitology*, **25**: 147–62.

McLaren, D. J. 1989. Will the real target of immunity to schistosomiasis please stand up. *Parasitology Today*, **5**: 279–82.

Mitchell, G. F., Rajasekariah, G. R. & Rickard, M. D. 1980. A mechanism to account for strain variation in resistance to the larval cestode *Taenia taeniaeformis*. *Immunology*, **39**: 481–9.

Pearce, E. J., Caspar, P., Grzych, J. M., Lewis, F. A. & Sher, A. 1991. Down regulation of Th1 cytokine production accompanies induction of Th2 responses by a parasitic helminth *Schistosoma mansoni*. *Journal of Experimental Medicine*, **173**: 159–66.

Rickard, M. D. & Howell, M. J. 1982. Comparative aspects of immunity in fascioliasis and cysticercosis in domesticated animals. In: *Biology and Control of Endoparasites* (ed. L. E. A. Symons, A. D. Donald & J. K. Dineen), pp. 343–73. Academic Press, Australia.

Rickard, M. D. & Williams, J. F. 1982. Hydatidosis/cysticercosis: immune mechanisms and immunization against infection. *Advances in Parasitology*, **21**: 229–96.

Rihet, P., Demeure, C. E., Bourgois, A., Prata, A. & Dessein, A. J. 1991. Evidence for an association between resistance to *Schistosoma mansoni* and high anti-larval IgE levels. *European Journal of Immunology*, **21**: 2679–86.

Sabah, A. A., Fletcher, C., Webbe, G. & Doenhoff, M. J. 1985. *Schistosoma mansoni*: reduced efficiency of chemotherapy in infected T-cell deprived mice. *Experimental Parasitology*, **60**: 348–54.

Simpson, A. J. G. 1990. Schistosome surface antigens: developmental expression and immunological function. *Parasitology Today*, **6**: 40–5.

Smithers, S. R. & Terry, R. J. 1969. The immunology of schistosomiasis. *Advances in Parasitology*, **7**: 41–93.

Smithers, S. R., Terry, R. J. & Hockley, D. J. 1969. Host antigens in schistosomiasis. *Proceedings of the Royal Society of London*, **B171**: 483–94.

Smythies, L. E., Pemberton, R. M., Coulson, P. S., Mountford, A. P. & Wilson, R. A. 1992. T cell-derived cytokines associated with pulmonary immune mechanisms in mice vaccinated with irradiated cercariae of *Schistosoma mansoni*. *Journal of Immunology*, **148**: 1512–18.

Vignali, D. A. A., Bickle, Q. D. & Taylor, M. G. 1989. Immunity to *Schistosoma mansoni in vivo*: contradiction or clarification? *Immunology Today*, **10**: 410–16.

Warren, K. S. 1987. Determinants of disease in human schistosomiasis. In: *Balliere's Clinical Tropical Medicine and Communicable Diseases* (ed. A. A. F. Mahmoud), pp. 301–13. Balliere, London.

Wilson, R. A. & Coulson, P. S. 1989. Lung-phase immunity to schistosomes: a new perspective on an old problem. *Parasitology Today*, **5**: 274–8.

Wilson, R. A., Coulson, P. S. & McHugh, S. M. 1983. A significant part of the concomitant immunity of mice to *S. mansoni* is the consequence of a leaky hepatic portal system, not immune killing. *Parasite Immunology*, **5**: 595–601.

NEMATODES

Almond, N. M., McLaren, D. J. & Parkhouse, R. M. E. 1986. A comparison of the surface and secretions of *Trichinella pseudospiralis* and *T. spiralis*. *Parasitology*, **93**: 163–76.

Behnke, J. M., Barnard, C. J. & Wakelin, D. 1992. Understanding chronic nematode infections: evolutionary considerations, current hypotheses and the way forward. *International Journal for Parasitology*, **22**: 861–907.

Behnke, J. M., Hannah, J. & Pritchard, D. I. 1983. *Nematospiroides dubius* in the mouse: evidence that adult worms depress the expression of homologous immunity. *Parasite Immunology*, **5**: 397–408.

Crofton, H. D. 1955. Nematode parasite populations in sheep on lowland farms. II. Worm egg counts in lambs. *Parasitology*, **45**: 99–115.

Edwards, A. J., Burt, J. S. & Ogilvie, B. M. 1971. The effect of immunity upon some enzymes of the parasitic nematode *Nippostrongylus brasiliensis*. *Parasitology*, **62**: 339–47.

Else, K. J. & Grencis, R. K. 1991. Helper T-cell subsets in mouse trichuriasis. *Parasitology Today*, **7**: 313–16.

Finkelman, F. D., Pearce, E. J., Urban, J. F. & Sher, A. 1991. Regulation and biological function of helminth-induced responses. In: *Immunoparasitology Today* (ed. C. Ash & R. B. Gallagher), pp. A62–6. Elsevier Trends Journals, Cambridge.

Gill, H. S., Watson, D. L. & Brandon, M. R. 1992. Monoclonal antibody to CD 4^+ cells abrogates genetic resistance to *Haemonchus contortus* in sheep. *Immunology*, **78**: 43–9.

Hussain, R., Grogl, M. & Ottesen, E. A. 1987. IgG antibody subclasses in human filariasis: differential subclass recognition of parasite antigens correlates with different clinical manifestations of infection. *Journal of Immunology*, **139**: 2794–8.

Jarrett, E. E. E., Jarrett, W. F. H. & Urquart, G. M. 1968. Quantitative studies on the kinetics of establishment and expulsion of intestinal nematode populations in susceptible and immune hosts. *Nippostrongylus brasiliensis* in the rat. *Parasitology*, **58**: 625–39.

Jenkins, D. C. & Phillipson, R. F. 1971. The kinetics of repeated low-level infections of *Nippostrongylus brasiliensis* in the laboratory rat. *Parasitology*, **62**: 457–65.

Kennedy, M. W. (ed.) 1991. *Parasitic Nematodes – Antigens, Membranes and Genes*. Taylor & Francis, London.

King, C. L. & Nutman, T. B. 1991. Regulation of the immune response in lymphatic filariasis and onchocerciasis. In: *Immunoparasitology Today* (ed. C. Ash & R. B. Gallagher), pp. A54–8. Elsevier Trends Journals, Cambridge.

Lee, T. D. G. 1991. Helminthotoxic responses of intestinal eosinophils to *Trichinella spiralis* newborn larvae. *Infection and Immunity*, **59**: 4405–11.

Lightowlers, M. W. & Rickard, M. D. 1988. Excretory–secretory products of helminth parasites: effects on host immune responses. *Parasitology*, **96**: S123–66.

Lloyd, S. & Soulsby, E. J. L. 1987. Immunobiology of gastrointestinal nematodes of Ruminants. In: *Immune Responses in Parasitic Infections: Immunology, Immunopathology and Immunoprophylaxis*, Vol. 1 (*Nematodes*) (ed. E. J. L. Soulsby), pp. 1–41. CRC Press, Boca Raton, Florida.

Maizels, R. M. & Lawrence, R. A. 1991. Immunological tolerance: the key feature in human filariasis? *Parasitology Today*, **7**: 271–6.

Marti, H. P. & Murrell, K. D. 1986. *Trichinella spiralis*: antifecundity and antinewborn larvae immunity in swine. *Experimental Parasitology*, **62**: 370–5.

Miller, H. R. P. 1980. Expulsion of *Nippostrongylus brasiliensis* from rats protected with serum. I. The efficacy of sera from singly and multiply infected donors related to time of administration and volume of serum injected. *Immunology*, **40**: 325–34.

Miller, H. R. P. 1984. The protective mucosal response against gastrointestinal nematodes in ruminants and laboratory animals. *Veterinary Immunology and Immunopathology*, **6**: 167–259.

Miller, H. R. P. 1987. Gastrointestinal mucus, a medium for survival and for elimination of parasitic nematodes and protozoa. *Parasitology*, **94**: S77–100.

Murray, M. 1972. Immediate hypersensitivity effector mechanisms – II. *In vivo* reaction. In: *Immunity to Animal Parasites* (ed. E. J. L. Soulsby), pp. 155–90. Academic Press, New York.

Nawa, Y., Parish, C. R. & Miller, H. R. P. 1978. The protective capacities of fractionated immune thoracic duct lymphocytes against *Nippostrongylus brasiliensis*. *Cellular Immunology*, **37**: 41–50.

Nutman, T. B., Kumaraswami, V. & Ottesen, E. A. 1987. Parasite-specific anergy in human filariasis: insights after analysis of parasite-antigen driven lymphokine production. *Journal of Clinical Investigation*, **79**: 1516–23.

Ottesen, E. A. 1992. Infection and disease in lymphatic filariasis: an immunological perspective. *Parasitology*, **104**: S71–9.

Philipp, M., Worms, M. J., Maizels, R. M. & Ogilvie, B. M. 1984. Rodent models of filariasis. *Contemporary Topics of Immunobiology*, **12**: 275–32.

Reed, N. D. 1989. Function and regulation of mast cells in parasite infections. In: *Mast Cell and Basophil Differentiation and Function in Health and Disease* (ed. S. J. Galli & K. F. Austen), pp. 92–110. Raven Press, New York.

Rothwell, T. L. W. 1989. Immune expulsion of parasitic nematodes from the alimentary tract. *International Journal for Parasitology*, **19**: 139–68.

Sanderson, B. E. 1969. Acetylcholinesterase activity in *Nippostrongylus brasiliensis. Comparative Biochemistry and Physiology*, **29**: 1207–13.

Sanderson, B. E. & Ogilvie, B. M. 1971. A study of acetylcholinesterase throughout the life cycle of *Nippostrongylus brasiliensis. Parasitology*, **62**: 367–73.

Smith, N. C. & Bryant, C. 1989. Free radical generation during primary infection of *Nippostrongylus brasiliensis. Parasite Immunology*, **11**: 147–60.

Stoll, N. R. 1929. Studies on the strongylid nematode *Haemonchus contortus. American Journal of Hygiene*, **10**: 384–417.

Tuohy, M., Lammas, D. A., Wakelin, D., Huntley, J. F., Newlands, G. F. J. & Miller, H. R. P. 1990. Functional correlations between mucosal mast cell activity and immunity to *Trichinella spiralis* in high and low responder mice. *Parasite Immunology*, **12**: 675–85.

Wakelin, D. & Denham, D. A. 1983. The immune response. In: *Trichinella and Trichinellosis* (ed. W. C. Campbell), pp. 265–308. Plenum Press, New York.

Wakelin, D., Harnett, W. & Parkhouse, R. M. E. 1993. Nematodes. In: *Immunology and Molecular Biology of Parasitic Infections* (ed. K. Warren), pp. 496–526. Blackwell Scientific Publications, Oxford.

Woodbury, R. G., Miller, H. R. P., Huntley, J. F., Newlands, G. F. J., Palliser, A. C. & Wakelin, D. 1984. Mucosal mast cells are functionally active during spontaneous expulsion of intestinal nematode infections in rat. *Nature*, **312**: 450–2.

VACCINATION

Anders, R. F. 1988. Antigens of *Plasmodium falciparum* and their potential as components of a malaria vaccine. In: *The Biology of Parasitism* (ed. P. T. Englund & A. Sher), pp. 201–24. Alan R. Liss, New York.

Barbet, A. F. 1989. Vaccines for parasitic infections. *Advances in Veterinary Science and Comparative Medicine*, **33**: 345–75.

Bergquist, R. 1990. Prospects of vaccination against schistosomiasis. *Scandinavian Journal of Infectious Diseases* (Suppl.), **76**: 60.

Bloom, B. R. 1989. Vaccines for the third world. *Nature*, **342**: 15–120.

Boulanger, D., Reid, C. D. F., Sturrock, R. F., Wolowczuk, I., Balloul, J. M., Grezel, D., Pierce, R. J., Otieno, M. F., Guerret, J. A., Grimaoud, J. A., Butterworth, A. E. & Capron, A. 1991. Immunization of mice and baboons with the recombinant Sm28 GST affects both worm viability and fecundity after experimental infection with *Schistosoma mansoni. Parasite Immunology*, **13**: 473–90.

Capron, A., Dessaint, J. P., Capron, M., Ouma, J. H. & Butterworth, A. E. 1987. Immunity to schistosomes: progress towards a vaccine. *Science*, **238**: 1065–72.

Gamble, H. R. & Murrell, K. D. 1987. Progress in the development of vaccines against parasitic diseases. *Immunology Letters*, **16**: 329–36.

Good, M. F., Bersofsky, J. A. & Miller, L. H. 1988. The T cell response to the malaria circumsporozoite protein: an immunological approach to vaccine development. *Annual Reviews of Immunology*, **6**: 663–88.

McLaren, D. J. (ed.) 1989. *Vaccines and vaccination strategies.* (Symposia of the British Society for Parasitology, vol. 26.) *Parasitology*, **98**.

Miller, T. A. 1978. Industrial development and field uses of the canine hookworm vaccine. *Advances in Parasitology*, **16**: 333–42.

Mitchell, G. F. 1989. Problems specific to parasite vaccines. *Parasitology*, **98**: S19–28.

Morrison, W. I. 1989. Immunological control of ticks and tick-borne parasitic diseases of livestock. *Parasitology*, **98**: S69–85.

Pattarroyo, M. E., Amador, R., Clavijo, P. *et al.* 1988. A synthetic vaccine protects humans against challenge with asexual blood stages of *Plasmodium falciparum* malaria. *Nature*, **332**: 158–61.

Poynter, D., Jones, B. V., Peacock, A. M. R., Nelson, R., Robinson, J., Silverman, P. H. & Terry, R. J. 1960. Recent experiences with vaccination. *Veterinary Record*, **72**: 1078–90.

Rickard, M. D. 1989. A success in veterinary parasitology: cestode vaccines. In: *New Strategies in Parasitology* (ed. K. P. McAdam), pp. 3–14. Churchill Livingstone, Edinburgh.

Sher, A. 1988. Vaccination against parasites: special problems imposed by the adaptation of parasitic organisms to the host immune response. In: *The Biology of Parasitism* (ed. P. T. Englund & A. Sher), pp. 169–82. Alan R. Liss, New York.

Smith, T. S. & Munn, E. A. 1990. Strategies for vaccination against gastro-intestinal nematodes. *Scientific and Technical Review of the Office International des Epizooties*, **9**: 577–95.

Targett, G. A. T. (ed.) 1991. *Malaria: Waiting for the Vaccine.* John Wiley, Chichester, UK.

Taylor, M. G., Hussein, M. F. & Harrison, R. A. 1990. Baboons, bovines and Bilharzia vaccines. In: *Parasitic Worms, Zoonoses and Human Health in Africa* (ed. C. N. L. Macpherson & P. S. Craig), pp. 237–59. Unwin Hyman, London.

Wakelin, D. 1994. Vaccines against intestinal helminths. In: *Enteric Infection: Mechanisms, Manifestations and Management*, Vol. 2 (*Intestinal Helminths*) (ed. M. J. Farthing, G. T. Keusch & D. Wakelin). Chapman & Hall, London. (In press.)

Williams, J. F. 1986. Prospects for prophylaxis of parasitism. *International Journal for Parasitology*, **17**: 711–19.

33

In vitro cultivation of endoparasites: general principles; Protozoa

33.1 General principles

33.1.1 Advantages of *in vitro* culture

It is self-evident that once a parasite can be cultured *in vitro* – even for short periods – the horizons of research on the basic biology of the organism are greatly extended. Within recent years, the commercial availability of standard culture media, sterile plasticware and wide-spectrum antibiotics have greatly simplified the establishment of basic culture techniques and many major protozoan (e.g. *Plasmodium* spp.) and helminth (e.g. *Echinococcus* spp.) parasites can now be cultured through part or all of their life cycles *in vitro*.

The advantages of *in vitro* culture can be briefly summarised as follows.

(*a*) The use of experimental animals for the routine laboratory maintenance of parasites is dispensed with: a desirable end in its own right.
(*b*) It enables the physiology and biochemistry of parasites to be investigated independently of interference by host-related factors.
(c) It enables the collection of excretory/secretory products of parasites to be collected without contamination by host metabolic products. Many of these have been widely exploited as 'E/S antigens' for immunological studies especially as possible vaccines against parasites of human or veterinary importance.
(*d*) As many parasites serve as elegant models for the study of differentiation – especially asexual/sexual differentiation – it enables the possible factors controlling such fundamental processes to be controlled and investigated in a culture tube.
(*e*) It is widely used for the preliminary screening of anti-parasitic drugs.
(*f*) It is a valuable aid to the teaching of parasitology in the laboratory as it provides interesting and exciting opportunities for students to handle living material and study biological phenomena first-hand.

Table 33.1. *Terminology for growth of organisms, mostly under known (gnotobiotic) conditions*

Terms	Number of associated organisms
Gnotobiotic	None, or known species only
Axenic	None
Synxenic	One, or more, known species
Monoxenic	One known species
Dixenic	Two known species
Trixenic	Three known species
Polyxenic	Several, to many, known species
Agnotobiotic, or xenic	Unknown

Source: From Dougherty (1959).

Comprehensive texts dealing with the techniques of *in vitro* cultivation of parasites are those of Jensen (1985), Taylor & Baker (1968, 1978, 1987) and Smyth (1990).

33.1.2 Terminology

Cultivation of an organism in the absence of other organisms is termed 'axenic' (Greek *a* = free from; *xenos* = a stranger). A culture in which one other species of organism is present is called 'monoxenic'; and, if many are present, 'polyxenic'. The term 'axenic' is replacing older, less precise terms such as 'aseptic' or 'sterile'; nevertheless the time-honoured term '*in vitro*' (meaning *in glass*) is worth retaining (and is used here), as it has come to be associated with a special type of culture involving a liquid or solid medium in a tube or flask-type container. Culture terminology has been discussed by Dougherty (1959) and that now in general use is summarised in Table 33.1.

○ 33.1.3 Basic problems

The primary aim of culture *in vitro* is to provide artificial conditions under which the life cycle of the parasite in question can be entirely completed outside the host, under controlled conditions; moreover the rate of development, maturation and reproduction should at least equal that which takes place *in vivo*. Hence, mere *survival*, without growth or differentiation, cannot be termed 'culture' although clearly survival is a justifiable aim in preliminary attempts with difficult or new material. Ideally, too, the composition and physico-chemical conditions of the culture medium should be completely defined. Unfortunately, these criteria are seldom met and, like those used in organ or tissue culture, most media used have contained natural biological products such as serum, blood, etc. Also, in many culture systems utilised by various authors, the exact physico-chemical conditions were either not known or not clearly stated.

Briefly, in attempting to culture a parasitic organism *in vitro*, the following problems are faced:

Sterility. Many parasites (especially intestinal forms) live in habitats containing microflora and, when removed from the host, are contaminated with micro-organisms. This problem, which perhaps was previously the major stumbling block to the development of *in vitro* methods, is now generally easily overcome. Repeated washing with isotonic media containing broad-spectrum antibiotics (such as gentamicin or tetracycline) has now virtually eliminated this problem. Streptomycin and penicillin are also routinely added in most culture protocols. Care must, of course, be taken to check that the antibiotics concerned are not inhibiting to the parasite.

Alternatively, cultures can be commenced by using a stage in the life cycle where the parasite occupies a sterile environment (such as the blood stream, liver, muscles, etc.). Typical of such stages are the cysticerci of cestodes (in muscle or body cavity), the metacercariae of trematodes (in lens of eye or brain tissue), or the blood stages of protozoa.

Physico-chemical characteristics. Parasites live in biological habitats such as the intestine, blood stream, lymph channels, viscera, etc., whose physico-chemical characteristics are often imperfectly known and whose properties may be difficult to reproduce *in vitro*. It is beyond the scope of this text to discuss the properties of the environments in any detail but some consideration has been given to this question in Chapter 2. Parameters to which attention must be given are: size and shape of culture vessel, temperature, pH, gas phase (especially $p\mathrm{O}_2$, $p\mathrm{CO}_2$), oxidation–reduction potential (E_h), presence or absence of supporting matrix, and osmotic pressure. Although characteristics such as temperature are obviously fundamental in most culture systems, some parasites may have a wide range of tolerance with regard to requirements such as the $p\mathrm{CO}_2$, $p\mathrm{O}_2$ and E_h; in contrast, others may have very precise requirements.

Nutrition. *In vivo*, parasites feed on complete biological materials such as intestinal contents, bile, blood, mucus and tissue exudates. It is self-evident that the nutritional properties of these are difficult to replace as artificial components in a medium. On account of their convenience and commercial availability, tissue culture media such as NCTC 109 or 135, Parker 858 or 199, RPMI 1640 and Eagle's Medium have been extensively used as basic media in many culture systems. Depending on the parasite concerned these media usually require supplementation of additions such as yeast extract, vitamins, blood cells, etc., in addition to serum, which is almost a universal requirement of culture media. Furthermore, parasites with a well-developed gut (especially nematodes and trematodes) may require the presence of solid or semi-solid food on which they can feed. This food may need to be in a particular form, with relation to consistency, viscosity, particle size, etc., before it can be ingested.

Supporting matrix. Whereas many parasites may be satisfactorily cultured in liquid media, certain species (e.g. *Entamoeba*, p. 494, *Echinococcus*, p. 513) may have special requirements as regards the provision of a supporting matrix. Thus, the cystic stage of *E. granulosus* can be satisfactorily grown in liquid medium, whereas the protoscolex stage requires contact with a suitable (proteinaceous?) substrate in order to differentiate in a strobilar direction (Fig. 34.10). Similarly, the cestode *Schistocephalus solidus* can only produce fertile eggs when its surface is compressed (e.g. within a cellulose tube Fig. 34.6).

Elimination of toxic waste materials. *In vivo*, the metabolic waste products are generally readily removed from the immediate environment of the parasite as a result of the natural circulation of body fluids in the host. A successful *in vitro* method must similarly provide conditions which permit the rapid removal of toxic waste products. In many culture systems, it suffices to renew the media every two to four days or transfer the parasites to new media at intervals. In some cases, however, circulating or continuous flow culture systems (Fig. 33.7) have been found to be more satisfactory.

Trigger stimuli. Many parasites have complex life cycles, sometimes involving several (up to four in the case of trematodes!) hosts. Trigger stimuli, similar to those provided by the host environment (e.g. the $p\mathrm{CO_2}$, $p\mathrm{O_2}$, presence or absence of bile, temperature, etc.) may be required at specific points in the life cycle before the organism can transform from one stage to another.

○○○ 33.2 Intestinal protozoa

Many methods for the culture of protozoa are related to or adapted from those used for bacteria or other microorganisms, and like those, offer many variants or improvements of the original techniques. It is not intended here to review all the methods used to culture parasitic protozoa or to give practical details of the preparation of media. Details of many of these are given in Jensen (1985) and Taylor & Baker (1968, 1978, 1987). The main object of this section is to outline the principles behind the various culture methods and deal with a small number of selected examples.

○ 33.2.1 General account

Physico-chemical conditions in the vertebrate intestine are complex, and it might be expected that *in vitro* culture of intestinal forms would prove to be difficult. In practice, however, by using monoxenic or polyxenic cultures, it has been found possible to establish relatively stable conditions; these have been sufficiently close to those occurring in the gut to permit growth and multiplication of numerous species of intestinal protozoa using relatively simple methods. More elaborate axenic techniques are also available.

A variety of media have been developed. The early, simple method of Boeck & Drbohlav (1925) made use of a culture tube as shown in Fig. 33.1. The medium is diphasic, the solid phase being in the form of an ordinary bacteriological slant covered to a variable extent by the liquid phase. One of the most reliable is Dobell's 'HSre' medium. In 'HSre' medium, the solid phase is composed of heat-coagulated protein (i.e. horse serum, inspissated at 80 °C), and the liquid phase of egg-albumen diluted with isotonic saline. A loopful of sterile rice starch is added as a source of carbohydrate. Details of the preparation of media are given in Taylor & Baker (1968).

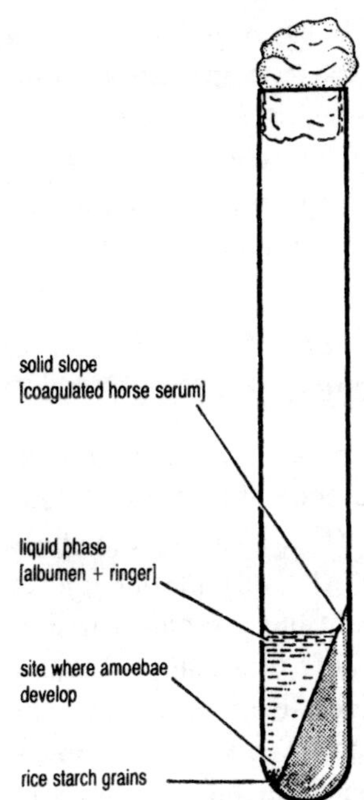

Fig. 33.1
Tube with Boeck–Drbohlav diphasic medium for *in vitro* culture of intestinal entamoebae, and certain other intestinal protozoa. (After Dobell, 1942.)

When a culture of an intestinal protozoon (e.g. *Entamoeba histolytica*) is initiated by inoculating a tube with a loopful of faeces containing cysts, or better still with a loopful from another culture, the organisms sink to the bottom and grow mainly in the interface region between the two phases where the starch has settled (Fig. 33.1). The cultures commence by being aerobic, but become rapidly anaerobic owing to the usage of available oxygen by the aerobic bacteria. A drop of methylene blue, for example, is rapidly bleached. In a flourishing culture, a complex microflora of aerobic and anaerobic bacteria occurs and more or less 'balanced' environmental conditions prevail for a time. Acid or alkali produced by concomitant bacteria is rapidly neutralised within wide limits by the system itself, so that other adjustments are unnecessary. The presence of bacteria is essential for the growth of many species of *Entamoeba*; their role in the nutrition of the amoebae is further considered below.

The Boeck–Drbohlav system, or modifications of it, have been most successful when used with protozoa of the caecum and large intestine, such as species of

Entamoeba, *Trichomonas* and *Balantidium*, whose main diet consists of bacteria and intestinal debris. Although this simple system is still used, a number of better defined, monophasic media have also been developed. *In vitro* techniques for some of the most commonly cultured species are considered below.

○ 33.2.2 *Entamoeba histolytica* and other intestinal amoebae

For many years *Entamoeba histolytica* could only be cultured in the presence of bacteria; the Boeck–Drbohlav diphasic system (Fig. 33.1), or modifications of it, was extensively used. This simple method is still useful in laboratories where more sophisticated systems are not available. A number of axenic and monoxenic complex media have since been developed in which most common species of *Entamoeba* can now be cultured.

The oxidation–reduction potential (E_h) is a major controlling factor in such systems (Chang, 1946; Harinasuta & Harinasuta, 1955): multiplication only takes place at low negative potentials of −300 to −400 mV (Fig. 33.2). In the intestine, this is achieved by the presence of bacteria and in culture media by the use of reducing agents such as L-cysteine and ascorbic acid (Fig. 33.3).

Axenic culture. This has been achieved using a number of media, a widely used one being TYI-S-33 (TYI = trypticase–yeast–iron–serum) (Diamond *et al.*, 1978) detailed in Table 33.2. In addition to *E. histolytica*, the medium is suitable for *E. invadens* (from snakes), *E. terrapine* (from turtles) and *E. moshkovskii* (from sewage, p. 31). It does not, however, support the growth of *E. gingivalis* (from the human mouth), but, when somewhat modified (Gannon & Linke, 1991) this has been successfully used for the axenic culture of this species, a major modification being the replacement of glucose by Dextran I (Table 33.3).

Monoxenic cultivation. Many *Entamoeba* spp. can be successively grown in monoxenic culture, the associated organisms being a bacterium (*Fusobacterium symbiosum*) or trypanosomatid (*Trypanosoma cruzi* or *Crithidia* sp.). Technical details are given in Diamond (1987).

Xenic cultivation. The liquid medium from TYI-S-33 (Table 33.2) has been used for the isolation and maintenance, under xenic conditions, of *E. histolytica*, *E. coli*, *Dientamoeba fragilis* and *E. hartmanni*, as well as the ciliate *Balantidium coli* and the flagellate *Pentatrichomonas hominis* (Taylor & Baker, 1987).

***Naegleria* sp.** The isolation and culture of the pathogenic strains of this species (Chapter 3) is described by Culbertson *et al.* (1968).

○ 33.2.3 Intestinal flagellates

Most media developed for the xenic culture of *Entamoeba*, even the simple Boeck–Drbohlav tube (Fig. 33.1),

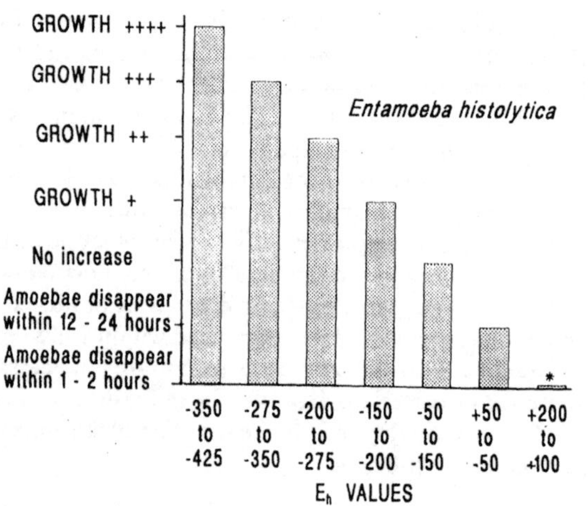

Fig. 33.2
Effect of oxidation–reduction potential (E_h) on the growth of *Entamoeba histolytica in vitro* for 24 h; *1–2 h culture. (After Chang, 1946.)

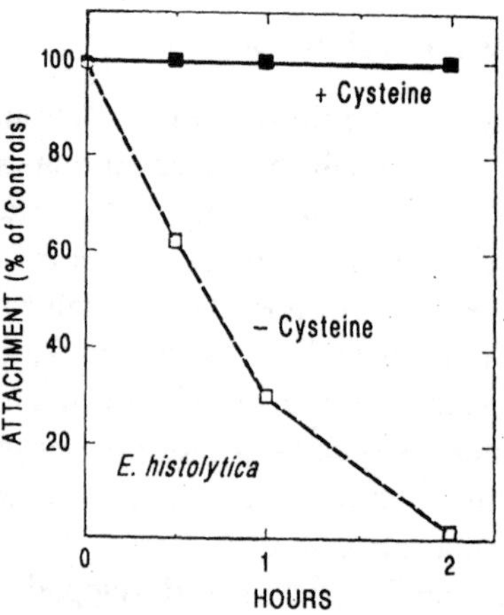

Fig. 33.3
Entamoeba histolytica: development *in vitro* following withdrawal of cysteine from the culture medium. Trophozoites (100 000 in 4 ml) were allowed to attach for 2 h in culture medium + 6 mM cysteine. The medium was then replaced by medium with or without cysteine. (After Gillin & Diamond, 1981.)

Table 33.2. *TYI-S-33 medium for axenic culture of* Entamoeba histolytica *and* Entamoeba *spp.*

(*a*) Nutrient Broth TY1	
K_2HPO_4	1.0 g
KH_2PO_4	0.6 g
NaCl	2.0 g
Casein digest peptone	20.0 g
Yeast extract	10.0 g
Glucose	10.0 g
L-cysteine-HCl	1.0 g
Ascorbic acid	0.2 g
Ferric ammonium citrate[a]	22.8 mg
Glass-distilled H_2O	600 ml

Bring final volume to 870 ml.
Adjust to pH 6.8 with 1N NaOH.
Filter through No. 1 paper.
Autoclave at 121 °C for 15 min.

(*b*) Vitamin mixture No. 13[b]

(*c*) Complete medium TYI-S-33	
TYI Broth (see *a*)	87 ml
Bovine serum (heat-inactivated)	10 ml
Vitamin mixture No. 13	2.0 ml
[pH 6.6; osmolality 370–400 mosmol kg^{-1}]	

(*d*) Culture preparation
Use 16 × 125 mm glass screw-top tubes, filling to 75–80 per cent of capacity. As the system is sensitive to light, culture in the dark, at 40 °C.

[a]Only the brown form (Mallinckrodt) can be used.
[b]For preparation details see Diamond *et al.* (1978) or Taylor & Baker (1987).
Source: Data from Diamond *et al.* (1978).

Table 33.3. *In vitro culture of* Entamoeba gingivalis*: the effect of various carbon sources on growth*

Major carbon source ($1\ gl^{-1}$ medium)	pH 24 h	pH 48 h	Growth of *E. gingivalis*[a]
Glucose	6.4	5.1	0
Dextran I (M_r 185 000)	6.7	6.3	++++
Dextran II (M_r 581 000)	6.6	6.1	+
Levan	6.8	6.2	+++
Maltose	6.2	4.8	0
Maltotriose	6.6	6.0	0
Melezitose	6.7	6.1	++
Rice starch	6.8	6.4	++++

[a]No. of trophozoites per field: + = 1–5, ++ = 6–10, +++ = 11–15, ++++ = 16–20.
Source: Data from Gannon & Linke (1991).

Table 33.4. *TYM*[a] *medium for the axenic culture of trichomonads*

Trypticase	2.0 g
Yeast extract	1.0 g
Maltose	0.5 g
L-cysteine hydrochloride	0.1 g
L-ascorbic acid	0.02 g
Distilled H_2O to make	90 ml

[a]TYM = trypticase–yeast–maltose. For preparation details see Diamond (1957) or Taylor & Baker (1968).
Source: Data from Diamond (1957).

can also be used for the cultivation of trichomonads in a mixed bacterial fauna. These methods have been summarised by Diamond (1985) and Taylor & Baker (1968–87). The axenic culture of trichomonads was first achieved by Diamond (1957) using Medium TYM (Table 33.4), which also proved suitable for *T. hominis*, *T. vaginalis* and *T. foetus* but not *T. tenax*. *T. muris* (p. 49) can also be cultured in this medium with the addition of a mouse caecal extract (Saeki *et al.*, 1983). A defined medium for *T. vaginalis* has been developed by Linstead (1981).

Giardia spp.

Species of *Giardia* remained refractory to axenic culture for many years until species from rabbit, chinchilla and cat were grown successfully in the presence of the yeast *Saccharomyces cerevisiae* from which the trophozoites were separated by a dialysis membrane (Meyer, 1970). *G. lamblia* (from man) has since been grown axenically in a number of media (reviewed Diamond, 1987; Meyer, 1990; Meyer & Radulescu, 1984), such as TYI-S-33 (Table 33.2). However, the latter required modifying

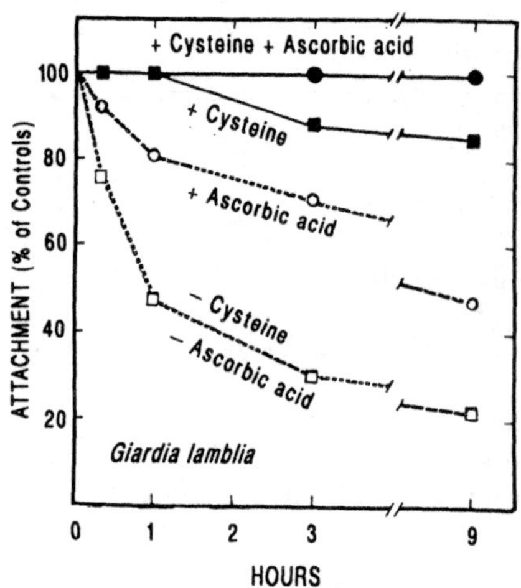

Fig. 33.4
Giardia lamblia: effects of removal of reducing agents on attached trophozoites *in vitro*. Trophozoites (400 000 in 4 ml) were allowed to attach in four sets of vials for 2 h in TP-S-1 medium containing both cysteine and ascorbic acid. The medium was then replaced by medium with or without cysteine or L-ascorbic acid (6 mM) as indicated. (After Gillin & Diamond, 1981.)

(Keister, 1983) by (*a*) doubling the concentration of cysteine, (*b*) adding 500 mg l^{-1} of dehydrated bovine bile, (*c*) adjusting the pH to 7.0–7.1, (*d*) sterilising the broth by filtration (rather than by autoclaving) and (*e*) eliminating the vitamin supplement. As in *Entamoeba*, growth and attachment of trophozoites also requires a low oxidation–reduction potential, which can be achieved by the addition of L-cysteine and ascorbic acid (Gillin & Diamond, 1981) (see Fig. 33.4).

It should be noted that one of the most commonly available species from laboratory animals, *Giardia muris* (Fig. 4.6), although isolated axenically from the mouse gut (Tillotson *et al.*, 1991), has so far defied all attempts to culture it *in vitro*. *G. agilis* (from tadpoles) (Fig. 4.6) has likewise not been cultured. Presumably, these species require special growth factors and/or conditions, as yet unidentified.

○ 33.2.4 Intestinal ciliates

Generally these have proved rather more refractory to *in vitro* culture than intestinal flagellates. In general, modifications of the Boeck–Drbohlav system have been used for blood-temperature ciliates such as *Balantidium*. Many species of rumen ciliates (in cattle, Fig. 10.11) have proved to be relatively easy to culture using basic components such as rumen fluid, dried grass, straw, etc. The methods available have been reviewed by Coleman (1987).

Ciliates from the rectum of a frog form useful class material and may be readily cultivated *in vitro*; suitable methods for *Opalina* (flagellate? p. 145) and *Nyctotherus* are given by Mofty & Smyth (1964) and Gaumont & Rémy (1967).

○○○ 33.3 Blood and tissue protozoa

○ 33.3.1 General account

Major advances have been made, within recent years, in the cultivation of blood stream and tissue stages of most genera of medical importance; several species of *Trypanosoma*, *Leishmania* and *Plasmodium* can now be cultured through all or part of their life cycles *in vitro*. Comprehensive general reviews on recent techniques are those of Jensen (1985) and Taylor & Baker (1987). Earlier work is reviewed in Taylor & Baker (1968, 1978).

○ 33.3.2 Haemoflagellates

Techniques for the *in vitro* culture of this group are reviewed in Baker (1987), Evans (1987) and Jensen (1985).

Salivarian trypanosomes (e.g. *T. brucei*). In early culture systems, organisms were found to transform only into the stages occurring in the insect vector and attempts to develop the blood stream forms (trypomastigotes) were unsuccessful. This problem was eventually solved by growing them in tissue culture, using a feeder layer of mammalian cells plus tissue culture medium. This was first achieved for *T. brucei* by Hirumi *et al.* (1977) and various modifications of this system have since been employed. More recently, cell-free culture systems for *T. brucei* have also been developed.

Many different mammalian cell lines can be used as 'feeder' layers, slow-growing cell lines being preferable. For *T. b. rhodesiense* and *T. b. gambiense* human embryonic fibroblasts are suitable as are several commercial cell lines. A suitable medium is RPMI 1640 + Eagle's minimum essential medium (EMEM) + serum. As is well recognised in culture work, batches of foetal bovine serum, which is commonly used, can vary substantially in their growth properties, some supporting growth and others failing to do so. Various kinds of plastic vessels are suitable for use such as T-flasks or multiwell plates.

The basic details of this technique and other suitable methods are reviewed in Brun & Jenni (1985, 1987).

Regarding non-human trypanosomes, fish frequently contains species which provide valuable material for culture experiments, e.g. *Trypanosoma catostomi* and *T. phaleri* from N American freshwater fish (Jones & Woo, 1991).

Stercorarian trypanosomes (e.g. *T. cruzi*). Cell culture systems have also been found to be suitable for the culture of *Trypanosoma cruzi* through the entire vertebrate phase of its life cycle. In its mammalian host, *T. cruzi* (Fig. 5.16) has two distinct developmental forms, the trypomastigote – which is non-dividing – and the amastigote, which proliferates in immunologically tolerant sites. Intracellular development of amastigotes followed by differentiation into trypomastigotes has been achieved by numerous authors (reviewed by Baker, 1987) using a number of cell lines such as HeLa cells, macrophages, monkey Vero cells, various muscle cells and kidney cells. A valuable system is that of Hudson *et al.* (1984) utilising a murine cell line (S2) and liquid phase comprising Dulbecco's modification of Eagle's minimum essential medium, with the addition of 10 per cent (inactivated) foetal bovine serum and antibiotics (penicillin, 200 iu ml^{-1}; streptomycin, 100 μg ml^{-1} buffered with $NaHCO_3$ (22 g l^{-1}) and 5 per cent CO_2 in air. In this system, a clear cycling of trypomastigote and amastigote forms occurred; by selecting the time of harvest, it was possible to obtain extracellular parasites more than 95 per cent of which were of one morphological type. Such infected S2 cultures have remained in continuous culture for up to 18 months and the amastigotes produced low-grade infections in normal mice and acute, fatal infections in irradiated mice. Full details of this technique are given by Hudson *et al.* (1984) and Baker (1987). Culture of the insect stages of *T. cruzi* is relatively simple as it grows readily in both liquid and diphasic media, a popular medium being LIT medium (Taylor & Baker, 1978).

Leishmania. The two morphological stages which are found in the two-host life cycle (Fig. 5.19) are the amastigote found in the macrophage and similar cells in the vertebrate host and the extracellular flagellated promastigote forms in the sandfly vector; both forms can now be cultured *in vitro*. A major difficulty is that different isolates have been found to have different growth requirements and it is impossible to predict which particular medium is suitable for any one isolate. A relatively simple medium, developed many years ago, is NNN medium (after Novy, MacNeal and Nicolle); this medium is still used.

Table 33.5. *MEM:FCS:EBLB medium for axenic culture of* Leishmania

Minimum Eagle's Medium (MEM) with modified salts (Gibco)	100 ml
$NaHCO_3$ solution (7.5% w/v)	3.0 ml
Evans blood lysate broth (EBLB)[a]	5.0 ml
Foetal bovine (calf) serum (heat-inactivated)	10 ml

[a]For preparation see Evans (1987).
Source: Data from Evans *et al.* (1984).

NNN medium consists of 14 g agar and 6 g NaCl dissolved in 900 ml water. This is dispersed in tubes and one-third volume of sterile defibrinated rabbit's blood added to the agar cooled to 45–50 °C. Slants are prepared without butts and the organisms are inoculated into the water of condensation formed on the surface. More recently, numerous liquid or diphasic media, most of which require the addition of foetal bovine serum, have been used. The various media and the appropriate techniques for isolating and cloning different isolates of *Leishmania* are reviewed by Evans (1987). A suitable rich medium for growing almost any isolate of *Leishmania* is Medium MEM:FCS:EBLB of Evans *et al.* (1984); its content and preparation is shown in Table 33.5.

○ 33.3.3 Malarial parasites: *Plasmodium* spp., erythrocytic stages

Plasmodium falciparum

The first successful method for the continuous culture of human malarial parasite, *Plasmodium falciparum*, was that of Trager & Jensen (1976) who established that the basic requirements for the successful culture involved use of (*a*) human erythrocytes, (*b*) human serum, (*c*) a low p_{O_2} and (*d*) frequent changes of culture medium, owing to the production of large amounts of lactic acid. These conditions were achieved by using the simple 'candle-jar' method (Fig. 33.5) in which infected erythrocytes, contained in a liquid medium in Petri dishes, were placed in a desiccator containing a lit candle, in an incubator (at 37 °C). When the flame went out, the stopcock was closed and the gas phase – now low in O_2 – was suitable for the development of the parasite.

Although this simple system has been replaced by

much more sophisticated systems (Fig. 33.7) it is still a valuable technique and is especially suitable for use in poor countries as desiccators are cheap and readily obtainable. The importance of a gas phase with a low pO_2 is shown in Fig. 33.6, which demonstrates that growth is poor in normal air but high in 5 per cent O_2, a system reflected in the candle-jar method (Mirovsky, 1989). The range of gas mixtures tolerated by *P. falciparum* appears to be 1–10 per cent O_2 in N_2. Ponndurai *et al.* (1986) used 3 per cent O_2, 4 per cent CO_2 in N_2 in the system shown in Fig. 33.7. *In vitro*, the rate of multiplication of the asexual blood stages is high up to a parasitema of about 15 per cent, and some isolates have been cultured for several years. Moreover, the organisms can be cryopreserved in liquid nitrogen and reconstituted as required. In the early culture systems, only small numbers of gametocytes were produced, but this problem was solved in later systems (see below).

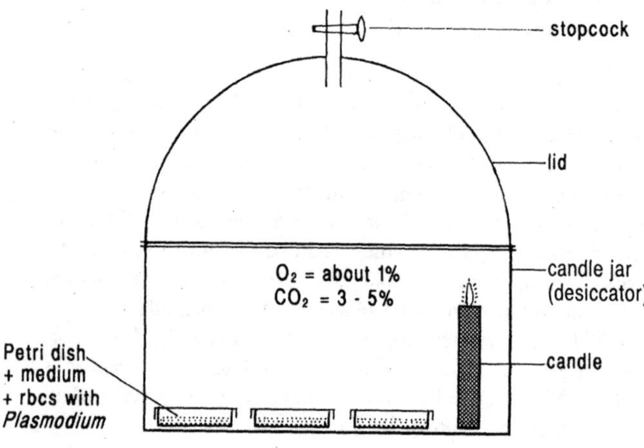

Fig. 33.5
Candle-jar system, such as used by Trager & Jensen (1976) for the first successful *in vitro* culture of *Plasmodium falciparum*. The candle is lit with the stop-cock open and the latter is closed when the flame goes out, providing a gas phase low in O_2 and CO_2. See also Fig. 33.6. (Original.)

Culture procedures: erythrocytic stages. A simplified account of the basic techniques utilised by Trager & Jensen (1976) is summarised in Taylor & Baker (1978). Its essential details are as follows.

(*a*) Centrifuge infected blood (1000 *g*, 10 min), remove the plasma and buffy coat and wash the cells twice in RMPI 1640 only.
(*b*) Resuspend the cells in RMPI + 25 mM HEPES + 10 per cent (v/v) type A/B human serum.
(*c*) Dilute the suspension with uninfected erythrocytes to give a final parasite level of 0.1–0.2 per cent.
(*d*) Dispense 1.5 ml aliquots in 35 mm plastic Petri dishes and place in a desiccator together with a candle (Fig. 33.5).
(*e*) Light the candle; only close the stopcock when the light goes out, then place the desiccator in an incubator at 37.5 °C.
(*f*) Change the medium (1.5 ml) every 24 h, keeping the disturbance of the settled cells to a minimum; replace in a candle-jar and treat as above.

Various modifications of this simple basic system have been developed using more complicated automated suspension systems (Butcher, 1981; Palmer *et al.*, 1982; Ponnudurai *et al.*, 1982*a*–86); others are reviewed in Ponnudurai (1987). As could be expected, parasites multiply more rapidly in suspension systems than in static (i.e. candle-jar type) cultures, since better contact of the parasitised cells with gas and medium is main-

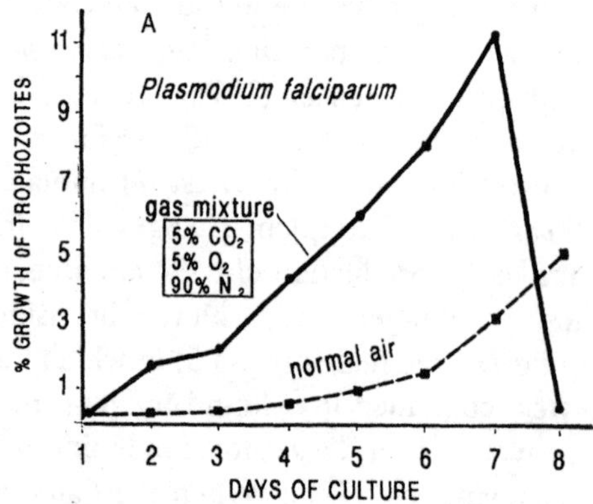

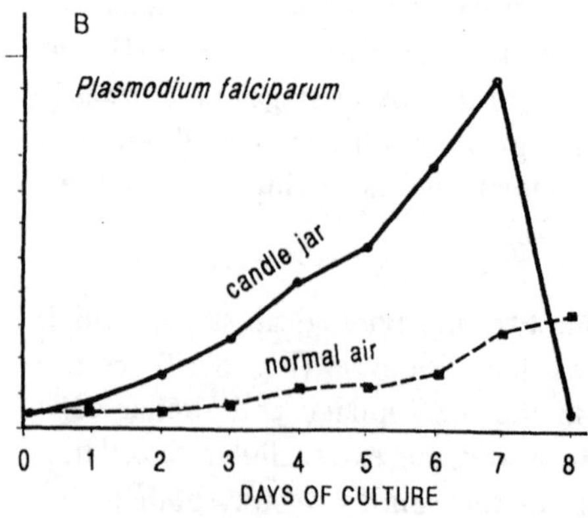

Fig. 33.6
Plasmodium falciparum: comparison of the average growth of trophozoites (A) in a stabilised gas mixture (5 per cent CO_2, 5 per cent O_2, 90 per cent N_2) and (B) in a candle-jar system (Fig. 33.5). The culture medium consisted of RMPI 1640 supplemented with 25 mmol l^{-1} HEPES buffer, 0.2 per cent $NaHCO_3$, 10 per cent serum and 50 μg ml^{-1} gentamicin. (Modified from Mirovsky, 1989.)

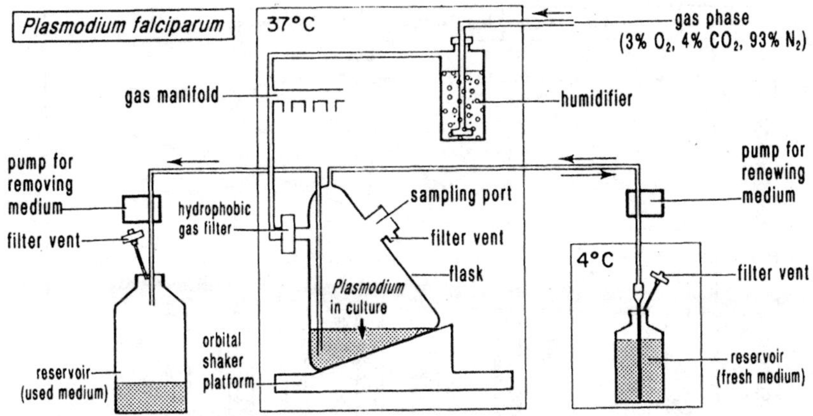

Fig. 33.7
Diagram of an automated suspension system for the continuous *in vitro* cultivation of *Plasmodium falciparum*. The orbital shaker rotates at 100 rev min^{-1} and stops twice daily for a 2 h period; this allows the red blood cells to sediment and the supernatant medium to be pumped off into the waste reservoir and replaced by 10 ml of fresh medium. (Modified from Ponnudurai *et al.*, 1986.)

tained. The results are further improved when the medium is replaced more frequently using automated systems such as that shown in Fig. 33.7.

Gametocyte stages. Although the early systems were reasonably successful in producing the asexual stages for long periods, very few gametocytes developed so that it was not possible to obtain stages for the routine infections of mosquitoes. However, the discovery that hypoxanthine was an essential requirement for the process of gametogenesis to be completed (Ifediba & Vanderberg, 1981) has revolutionsed the production of gametocytes. Thus, the medium used by Ponnudurai *et al.* (1986) in the automated suspension system shown in Fig. 33.7 contained 50 μg ml^{-1} of hypoxanthine and high densities of gametocytes were produced. As can be seen from Fig. 33.8, the development of gametocytes is accompanied by a slowing down in the production of asexual stages as the switch to gametogenesis takes place.

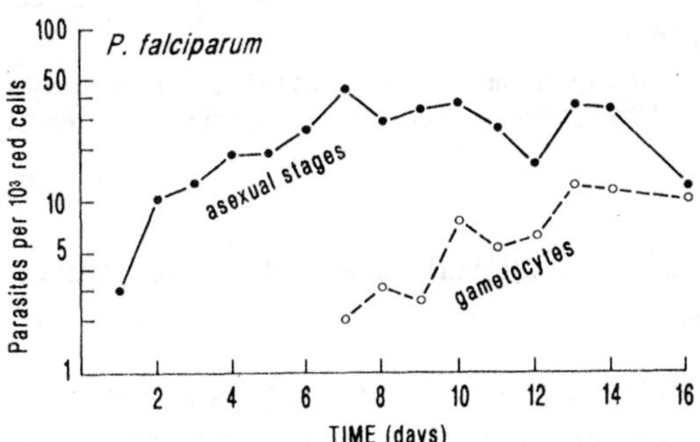

Fig. 33.8
In vitro cultivation of *Plasmodium falciparum* showing counts of asexual and gametocyte stages when parasite density is not reduced by the addition of red cells. Gametogenesis continued over a period of 8–10 days and exflagellation was demonstrated later. (After Ponnudurai *et al.*, 1982*b*.)

Other *Plasmodium* species

Attempts to culture the other species of human malaria, *P. vivax*, *P. malariae* and *P. ovale*, have been only partly successful; *P. falciparum* is the only species that can be maintained *in vitro* for long periods. However, useful advances have beem made in the cultivation of the rodent *P. berghei*, widely used as an experimental organism. Although difficult to culture at first, it was successfully cultured by Mons *et al.* (1983) using the simple suspension system shown in Fig. 33.9. Over 3–4 schizogonic cycles were achieved, resulting in 2–6-fold multiplication (Fig. 33.10); cultures were maintained for 6–9 days. The exo-erythrocytic (EE) stages have also been successfully cultured (see below).

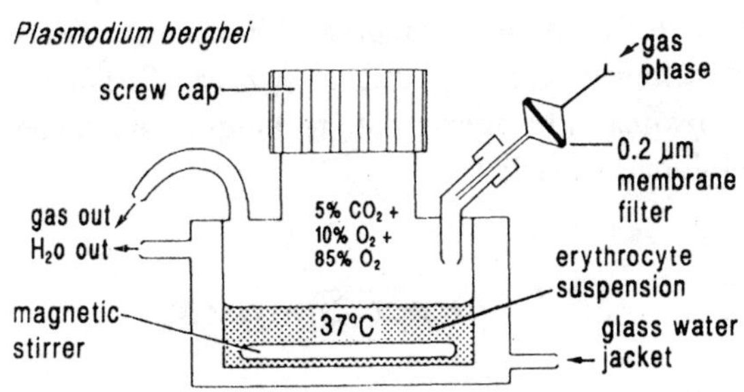

Fig. 33.9
Simple suspension system for the *in vitro* culture of the rodent malaria species *Plasmodium berghei*. (After Mons *et al.*, 1983.)

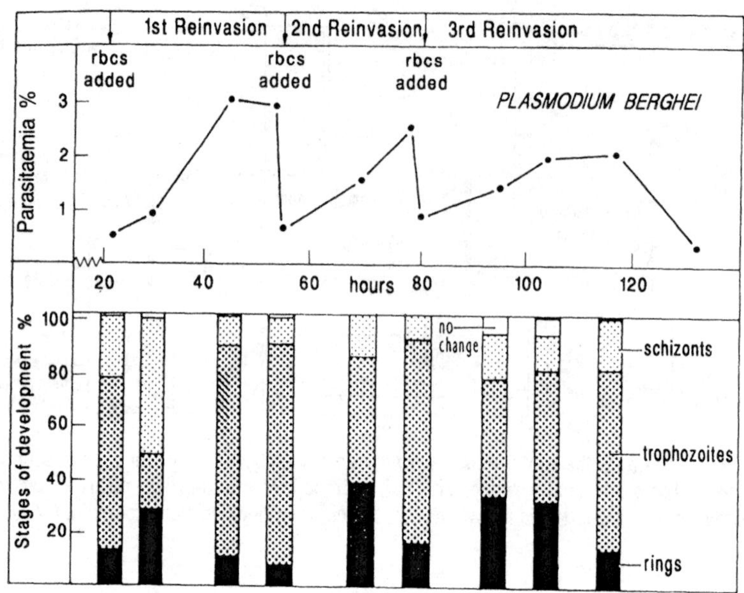

Fig. 33.10
Plasmodium berghei: the course of parasitaemia and the changes in the ratio of the different developmental asexual stages (per cent of total) in a typical *in vitro*, culture experiment using the suspension system shown in Fig. 33.9. (After Mons *et al.*, 1983.)

○ 33.3.4 Malarial parasites: exo-erythrocytic forms

Extensive progress has been made in the *in vitro* culture of the EE forms of both human and animal species. Although early success was achieved with the invasion by many species into various tissue culture cells, full maturation of parasites did not take place. This was first achieved in W138 cells (primary human embryonic cells) from which infectious EE merozoites were released (Hollingdale *et al.*, 1981). Primary hepatocytes were later used and complete EE cycles have now been achieved for *P. yoelii*, *P. berghei*, *P. vivax* and *P. falciparum*. The appropriate techniques are reviewed by Hollingdale (1987).

○○○ 33.4 Other Sporozoea

A reasonable amount of success has also been achieved with the *in vitro* culture of other Sporozoea, including species of *Eimeria*, *Babesia* and *Toxoplasma*. Methods for these are reviewed by Jensen (1985) and Taylor & Baker (1968–87).

○○○ References

(References of general interest have been included in this list, in addition to those quoted in the text.)

Baker, J. R. 1987. *Trypanosoma cruzi* and other stercorarian trypanosomes. In: *In Vitro Methods for Parasite Cultivation* (ed. A. E. R. Taylor & J. R. Baker), pp. 76–93. Academic Press, London.

Boeck, W. D. & Drbohlav, J. 1925. The cultivation of *Endamoeba histolytica. American Journal of Hygiene*, **5**: 371–407.

Beale, G. H., Thaithong, S. & Siripool, N. 1991. Isolation of clones of *Plasmodium falciparum* by micromanipulation. *Transactions of the Royal Society of Tropical Medicine and Hygiene*, **85**: 37.

Brun, G. H. & Jenni, L. 1985. Cultivation of African and South American trypanosomes of medical and veterinary importance. *British Medical Bulletin*, **41**: 122–9.

Brun, R. & Jenni, L. 1987. Salivarian trypanosomes: bloodstream forms (trypomastigotes). In: *In Vitro Methods for Parasite Cultivation* (ed. A. E. R. Taylor & J. R. Baker), pp. 94–117. Academic Press, London.

Butcher, G. A. 1981. A comparison of static thin layer and suspension cultures for the maintenance *in vitro* of *Plasmodium falciparum. Annals of Tropical Medicine and Parasitology*, **75**: 7–17.

Chang, S. L. 1946. Studies on *Entamoeba histolytica*. IV. The relation of oxidation-reduction potentials to the growth, encystation and excystation of *Entamoeba histolytica* in culture. *Parasitology*, **37**: 101–12.

Coleman, G. S. 1987. Rumen entodiniomorphid Protozoa.

In: *In Vitro Methods for Parasite Cultivation* (ed. A. E. R. Taylor & J. R. Baker), pp. 29–51. Academic Press, London.

Couatarmanach, A., Andre, P., Minous, D. Le., Martin, L., Robert, R. & Deunff, J. 1991. *In vitro* culture and cloning of *Toxoplasma gondii* in a newly established cell line derived from TG180. *International Journal for Parasitology*, **21**: 129–32.

Culbertson, C. G., Ensminger, P. W. & Overton, W. M. 1968. Pathogenetic *Naegleria* sp. – a study of a strain isolated from human cerebrospinal fluid. *Journal of Protozoology*, **15**: 353–63.

Diamond, L. S. 1957. The establishment of various trichomonads of animals and man in axenic conditions. *Journal of Parasitology*, **43**: 488–90.

Diamond, L. S. 1985. Lumen dwelling protozoa: *Entamoeba*, trichomonads and *Giardia*. In: *In Vitro Cultivation of Protozoan Parasites* (ed. J. B. Jensen), pp. 65–110. CRC Press, Boca Raton, Florida.

Diamond, L. S. 1987. *Entamoeba, Giardia* and *Trichomonas*. In: *In Vitro Methods for Parasite Cultivation* (ed. A. E. R. Taylor & J. R. Baker), pp. 1–28. Academic Press, London.

Diamond, L. S., Harlow, D. R. & Cunnick, C. 1978. A new medium for the axenic cultivation of *Entamoeba histolytica* and other *Entamoeba*. *Transactions of the Royal Society of Tropical Medicine and Parasitology*, **72**: 431–2.

Dougherty, E. C. (ed.) 1959. Axenic culture of invertebrate metazoa: a goal, *Annals of the New York Academy of Sciences*, **77**: 25–406.

Evans, D. A. 1987. *Leishmania*. In: *In Vitro Methods for Parasite Cultivation* (ed. A. E. R. Taylor & J. R. Baker), pp. 53–75. Academic Press, London.

Evans, D. A., Lanham, S. M., Baldwin, C. I. & Peters, W. 1984. The isolation and isoenzyme characterization of *Leishmania braziliensis* subsp. from patients with cutaneous leishmaniasis acquired in Belize. *Transactions of the Royal Society of Tropical Medicine and Hygiene*, **78**: 35–42.

Enders, B. 1988. *In vitro* cultivation of certain parasitic Protozoa. In: *Parasitology in Focus* (ed. H. Mehlhorn), pp. 702–18. Springer-Verlag, Berlin. ISBN 0 387 17838 4.

Gannon, J. T. & Linke, H. A. B. 1991. An antibiotic-free medium for the xenic cultivation of *Entamoeba gingivalis*. *International Journal for Parasitology*, **21**: 403–7.

Gaumont, R. & Rémy, M.-F. 1967. Essai de culture *in vitro* quelques ciliés du rectum de la grenouille. *Protistologica*, **3**: 67–71.

Gillin, F. D. & Diamond, L. S. 1981. *Entamoeba histolytica* and *Giardia lamblia*: effects of cysteine and oxygen tension on trophozoite attachment to glass and survival in culture medium. *Experimental Parasitology*, **52**: 9–17.

Harinasuta, C. & Harinasuta, T. 1955. Studies on the growth *in vitro* of strains of *Entamoeba histolytica*. *Annals of Tropical Medicine and Parasitology*, **49**: 331–50.

Hirumi, H., Doyle, J. J. & Hirumi, K. 1977. African trypanosomes: cultivation of animal-infective *Trypanosoma brucei in vitro*. *Science*, **196**: 992–4.

Hollingdale, M. R. 1987. Plasmodiidae: exoerythrocytic forms. In: *In Vitro Methods for Parasite Cultivation* (ed. A. E. R. Taylor & J. R. Baker), pp. 180–98. Academic Press, London.

Hollingdale, M. R., Leef, J. L., McCullough, M. & Beaudoin, R. L. 1981. *In vitro* cultivation of the exoerythrocytic stages of *Plasmodium berghei* from sporozoites. *Science*, **213**: 1021–2.

Hudson, L., Snary, D. & Morgan, S.-J. 1984. *Trypanosoma cruzi*: continuous cultivation with murine cell lines. *Parasitology*, **88**: 283–94.

Ifediba, T. & Vanderberg, J. P. 1981. Complete *in vitro* maturation of *Plasmodium falciparum* gametocytes. *Nature*, **294**: 354–96.

Janse, C. J., Mons, B., Croon, J. J. A. B. & Van Der Kaay, H. J. 1984. Long-term *in vitro* cultures of *Plasmodium berghei* and preliminary observations on gametogenesis. *International Journal for Parasitology*, **14**: 317–20.

Jensen, J. B. (ed.) 1985. *In Vitro Cultivation of Protozoan Parasites*. CRC Press, Boca Raton, Florida.

Jensen, J. B. & Trager, W. 1977. *Plasmodium falciparum*: use of outdated erythrocytes and description of the candle jar method. *Journal of Parasitology*, **63**: 883–6.

Jones, S. R. M. & Woo, P. T. K. 1991. Culture characteristics of *Trypanosoma catostomi* and *Trypanosoma phaleri* from North American freshwater fishes. *Parasitology*, **103**: 237–43.

Keister, D. B. 1983. Axenic culture of *Giardia lamblia* in TYI-S-33 medium supplemented with bile. *Transactions of the Royal Society of Tropical Medicine and Hygiene*, **77**: 487–8.

Linstead, D. 1981. New defined and semi-defined media for the cultivation of the flagellate *Trichomonas vaginalis*. *Parasitology*, **83**: 125–37.

Meyer, E. A. 1970. Isolation and axenic cultivation of *Giardia* trophozoites from the rabbit, chinchilla and cat. *Experimental Parasitology*, **27**: 179–83.

Meyer, E. A. 1990. *Giardiasis*. Elsevier Science Publishers, The Netherlands.

Meyer, E. A. & Radulescu, S. 1984. *In vitro* cultivation of *Giardia* trophozoites. In: *Giardia and Giardiasis, Biology, Pathogenesis and Epidemiology* (ed. S. L. Erlandsen & E. A. Meyer), Plenum Press, New York. ISBN 0 306 41539 9.

Mirovsky, P. 1989. Continuous culture of *Plasmodium falciparum* asexual stages in 'normal' air atmosphere. *Folia Parasitologica*, **36**: 107–12.

Mofty, M. M. El & Smyth, J. D. 1964. Endocrine control of encystation in *Opalina ranarum* parasitic in *Rana temporaria*. *Experimental Parasitology*, **15**: 185–99.

Mons, B., Croon, J. J. A. B., Van Der Star, W. & Van der Kaay, H. J. 1988. Erythrocytic schizogony and invasion of *Plasmodium vivax in vitro*. *International Journal for Parasitology*, **18**: 307–11.

Mons, B., Janse, C. J., Croon, J. J. A. B. & Van Der Kaay, H. J. 1983. *In vitro* culture of *Plasmodium berghei* using a new suspension system. *International Journal for Parasitology*, **13**: 213–17.

Palmer, K. L., Hui, G. S. N. & Siddiqui, W. A. 1982. A large scale *in vitro* production system for *Plasmodium falciparum*. *Journal of Parasitology*, **68**: 1180–3.

Perrin, L. 1987. *In vitro* culture of malaria parasites and its significance for vaccine production. *Zentralblatt für Bakteriologia, Mikrobiologie und Hygiene*, A, **267**: 277–8.

Ponnudurai, T. 1987. Plasmodiidae: erythrocytic stages. In: *In Vitro Methods for Parasite Cultivation* (ed. A. E. R. Taylor & J. R. Baker), pp. 153–79. Academic Press, London.

Ponnudurai, T., Lensen. A. H. W., Leeuwenberg, A. D. E. M. & Meuwissen, J. H. E. Th. 1982*a*. Cultivation of fertile *Plasmodium falciparum* gametocytes in semiautomated systems. 1. Static cultures. *Transactions of the Royal Society of Tropical Medicine and Hygiene*, **76**: 812–18.

Ponnudurai, T., Lensen, A. H. W., Meis, J. F. G. M. & Meuwissen, J. H. E. Th. 1986. Synchronization of *Plasmodium falciparum* gametocytes using an automated suspension culture system. *Parasitology*, **93**: 263–74.

Ponnudurai, T., Lensen, A. H. W. & Meuwissen, J. H. E. Th. 1983. An automated large-scale culture system of *Plasmodium falciparum* using tangential flow filtration for medium change. *Parasitology*, **87**: 439–45.

Ponnudurai, T., Meuwissen, J. H. E. Th., Leeuwenberg, A. D. E. M., Verhave, J. P. & Lensen, A. H. W. 1982*b*. The production of mature gametocytes of *Plasmodium falciparum* in continuous cultures of different isolates infective to mosquitoes. *Transactions of the Royal Society of Tropical Medicine and Hygiene*, **76**: 242–50.

Saeki, H., Togo, M., Imai, S. & Ishii, T. 1983. A new method for the serial cultivation of *Tritrichomonas muris*. *Japanese Journal of Veterinary Science*, **45**: 151–6.

Smyth, J. D. 1990. *In Vitro Cultivation of Parasitic Helminths*. CRC Press, Boca Raton, Florida.

Taylor, A. E. R. & Baker, J. R. 1968. *The Cultivation of Parasites in Vitro*. Blackwell Scientific Publications, Oxford.

Taylor, A. E. R. & Baker, J. R. (eds) 1978. *Methods of Cultivating Parasites in Vitro*. Academic Press, London.

Taylor, A. E. R. & Baker, J. R. (eds) 1987. *In Vitro Methods for Parasite Cultivation*. Academic Press, London. ISBN 0 126 83855 0.

Tillotson, K. D., Buret, A. & Olson, M. E. 1991. Axenic isolation of viable *Giardia muris* trophozoites. *Journal of Parasitology*, **77**: 505–8.

Trager, W. & Jensen, J. B. 1976. Human malaria parasites in continuous culture. *Science*, **193**: 674–5.

34

In vitro cultivation of endoparasites: helminths

34.1 General account

The cultivation of parasitic helminths presents a number of problems, many of which are common to those of parasite cultivation in general as discussed in the previous chapter (p. 492). The very size of the majority of helminths, however, as well as the complex nature of their life cycles, presents special problems which do not arise in the cultivation of protozoa. Moreover, these problems vary from group to group and from species to species. For example, cestodes possess no alimentary canal and all nutriment is taken in through the tegument; on the other hand, nematodes and trematodes have a well-developed gut and in some cases the food must be provided in particulate form to be ingested.

Again, many helminths require specific physical conditions (such as compression) before insemination can take place and fertile eggs produced (see especially *Schistocephalus*, Fig. 34.6). Furthermore, many helminths have very complex life cycles and the nutritional and physiological conditions required for development at each stage may vary enormously.

Within the past decade, techniques for *in vitro* cultivation have made remarkable advances and species from all major groups have been cultured through at least some stages in their life cycle.

The *in vitro* culture of parasitic helminths has been reviewed in detail by Smyth (1990) and earlier work by Taylor & Baker (1968–87).

In the section below no attempt has been made to survey the whole field, but merely to give selected examples from each group.

34.2 Trematodes

34.2.1 *Strigeid trematodes*

Strigeid metacercariae probably provide the most valuable and readily available material for the *in vitro* culture of trematodes. Many strigeid metacercariae occur in sterile situations in the intermediate host, especially in fish in which they may occur in great numbers. Moreover, many species (Table 34.1) exhibit advanced progenesis (i.e. have the genitalia partly differentiated) which means that they mature rapidly, often within 36–48 h.

Progenetic metacercariae. These are metacercariae whose genital anlagen are already in a somewhat advanced stage of development. The *in vitro* culture of many species is so simple that this can often be carried out even under classroom conditions, thus providing exciting experimental material for students.

Examples of species which have been cultured to egg production are *Posthodiplostomum minimum*, *Sphaeridiotrema globulus*, *Microphallus similis*, *Microphalloides japonicus*, *Parvatrema timondavidi* and *Amblosoma suwanese*. Details of hosts and references are given in Table 34.1. The detailed techniques for culturing the above species are reviewed in Smyth (1990). Basically they involve extracting the metacercariae from the intermediate host, excysting them (where necessary), washing the freed larvae in saline plus antibiotics (if not already in a sterile condition) and culturing them in screw-top cultures tubes in stationary or roller tube systems. Tissue culture media, e.g. Parker 199, NCTC 105, 135 or RMPI 1650, plus serum, are commonly used as culture media.

Species showing advanced progenesis, e.g. *Codonocephalus urniger*, will mature *in vitro* even in a non-nutrient medium (saline) at 40 °C, the rise in temperature being sufficient to stimulate the completion of spermatogenesis, vitellogenesis and oogenesis (Dollfus *et al.*, 1956).

Table 34.1. *Progenetic metacercariae that are easy to culture to maturity* in vitro

Species	Definitive host	Molluscan host	Intermediate host (for metacercaria)	Prepatent period	References (*in vitro* technique, life cycle)	Distribution
Posthodiplostomum minimum	birds, amphibia[a], mammals[a]	*Physa*	fish	34 h	Palmieri (1973); Smyth (1990)	USA, UK, Europe
Sphaeridiotrema globulus	birds, CAM[b], chickens[a], ducklings[a]	*Bithynia, Fluminicola, Gonobiosis*	encysted in mollusc host	68–72 h	Berntzen & Macy (1969); Smyth (1990)	USA, UK, Europe
Parvatrema timondavidi	birds, mice[a]	*Mytilus, Tapes*	encysted in mollusc host	36–48 h	Yasuraoka *et al.* (1974); Cable (1953); Smyth (1990)	USA, UK, Europe, Japan
Microphallus similis	gulls, mice[a], rats	*Littorina*	crabs	24–48 h	Davies & Smyth (1979); Stunkard (1957); Smyth (1990)	USA, UK, Europe
Microphalloides japonicus	rats, mice[a]	unknown	crabs	12–24 h	Fujino *et al.* (1977); Smyth (1990)	Japan
Amblosoma suwaense	birds? CAM[b]	unknown	in snails: *Sinotaia* (Japan), *Campeloma* (USA)	4–5 days	Schnier & Fried (1980); Fried *et al.* (1981); Smyth (1990); Shimazu (1974)	USA, Japan

[a]Experimental host.
[b]CAM = chorioallantois of chick (see Fig. 34.5).
Source: Data from Smyth (1990).

Most progenetic metacercariae, however, require longer periods of culture and a more nutrient medium. For example, the metacercariae of *Microphallus similis*, which is common in crabs in Europe (Davies & Smyth, 1979) and the USA (Stunkard, 1957), will just reach sexual maturity in saline alone, and produce a few eggs, but in saline augmented with serum or other nutrients many more eggs are produced. The eggs produced *in vitro* by this species were infertile but fertile eggs have been produced by some species, such as *Sphaeridiotrema globulus*, using a medium containing egg yolk (Berntzen & Macy, 1969) and *Parvatrema timondavidi* (Yasuraoka *et al.*, 1974), using NCTC 109 with 20 per cent bovine or chicken serum. A common problem in trematode culture is the failure of cultured worms to produce 'normal' fertilised eggs with properly formed and tanned egg-shells (p. 187), capable of embryonation. This situation is probably largely due to the failure to provide the necessary materials required for egg-shell synthesis, or the appropriate $p\mathrm{O}_2$, $p\mathrm{CO}_2$ or E_h.

Non-progenetic metacercariae. These are metacercariae in which the genitalia anlagen are not developed. Maturation of such a larva to an adult worm thus involves considerable tissue synthesis, i.e. in order to mature, the larva must develop from Stage 0 to Stage 7 (Fig. 34.1).

Diplostomum phoxini (p. 254; Figs 17.1, 17.4) and *D. spathaceum* (p. 259; Figs 17.5, 17.6)

These were the first non-progenetic metacercariae to be cultured to maturity *in vitro* (Bell & Smyth, 1958; Kannangara & Smyth, 1974). Both species failed to develop in a purely liquid medium but grew to maturity (Fig. 34.2) in a medium of 'cooked'-yolk : albumen : NTC 135 in a ratio of 1 : 1 : 1.5 although only abnormal and infertile eggs were produced.

Cotylurus spp.

Infertile eggs were also produced *in vitro* by *Cotylurus lutzi* in NCTC 135 + 50 per cent chick serum (Voge &

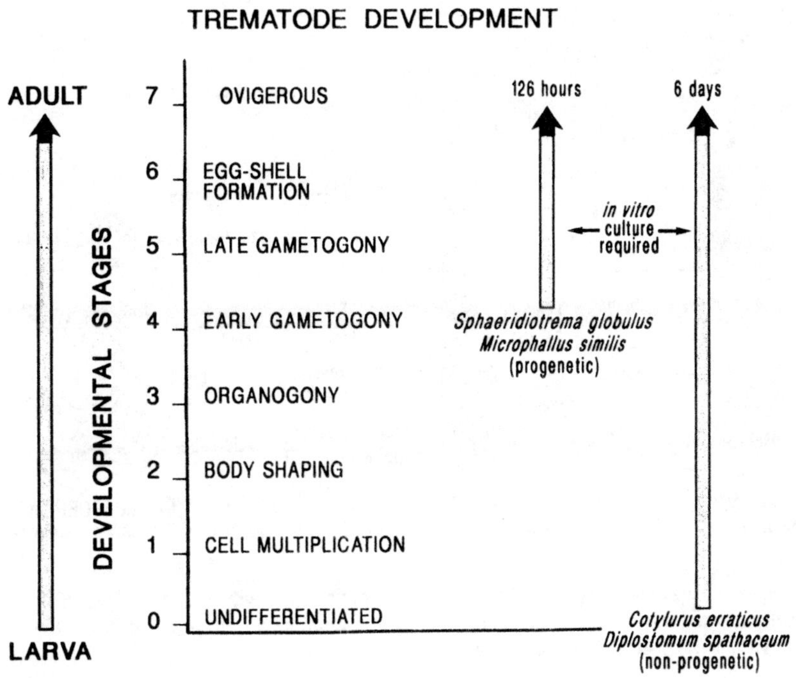

TREMATODE DEVELOPMENT

ADULT
7 OVIGEROUS
6 EGG-SHELL FORMATION
5 LATE GAMETOGONY
4 EARLY GAMETOGONY
3 ORGANOGONY
2 BODY SHAPING
1 CELL MULTIPLICATION
0 UNDIFFERENTIATED
LARVA
DEVELOPMENTAL STAGES
126 hours
6 days
in vitro culture required
Sphaeridiotrema globulus
Microphallus similis
(progenetic)
Cotylurus erraticus
Diplostomum spathaceum
(non-progenetic)

Fig. 34.1
Examples of species with progenetic and non-progenetic metacercariae, showing the stage of development in the larval stage and the time required to reach maturity *in vitro*.

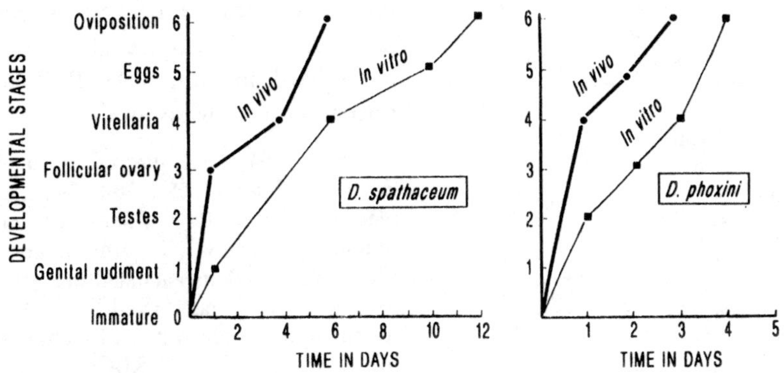

Fig. 34.2
In vitro development of the (non-progenetic) metacercaria of *Diplostomum spathaceum* and *D. phoxini* compared with that *in vivo*. (After Kannangara & Smyth, 1974.) For morphology and life cycles, see Figs 17.1–17.6.

Jeong, 1971) but fertile eggs, capable of embryonation, were later obtained using a modified medium which included chick intestine extract (Basch *et al.*, 1973). A related species, *Cotylurus erraticus*, the metacercariae of which are commonly found in fish in Europe and the USA (Fig. 34.3), has proved to be exceptionally easy to culture, the metacercaria growing to five times its size in 6 days (Mitchell *et al.*, 1978), a rate comparable with that *in vivo* in the gull (Table 34.2).

Fasciola hepatica

Attempts to culture an adult liver fluke from a metacercaria have been generally disappointing, only limited success being achieved by early workers (Davies & Smyth, 1978). More successful were the results of Smith & Clegg (1981) using RMPI 1640 + 50 per cent human serum + 2 per cent human red blood cells (RBCs). Although the parasites grew at an almost linear rate for 14 weeks, little differentiation occurred except in a few individuals which underwent early sexual devel-

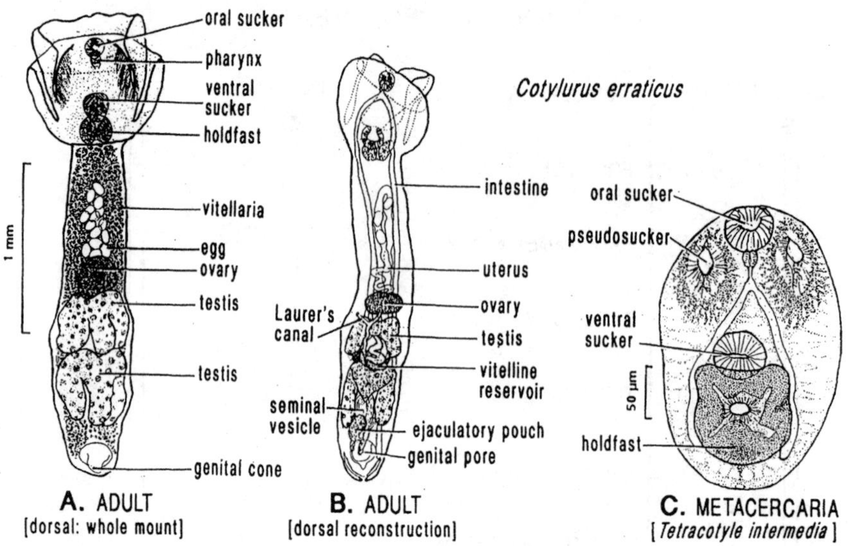

Fig. 34.3
Cotylurus erraticus: the (non-progenetic) metacercaria of this species, from fish, is relatively easy to culture to maturity *in vitro*. The adult occurs in gulls, the molluscan stages in *Valvata lewisi* and the metacercariae in the pericardial cavity of salmonid fish. (After Guberlet, 1922; Niewiadomska & Kozicka, 1970; Olson, 1970.)

Table 34.2. Cotylurus erraticus*: in vitro development in different media, compared with that* in vivo *(in the gull)*

Treatment	Period in culture/*in vivo* (days)	First egg production (days)
In vitro		
M199	6 (died)	—
NCTC135	6 (died)	—
M199 + 20% chicken serum	9	9
M199 + 40%	9	7
NCTC135 + 40%	9	8
M199 + 80%	9	6
NCTC135 + 80%	8	6
In vivo		
Grown in *L. ridibundus*	5	4–5

Source: Data from Mitchell *et al.* (1978).

opment. This involved the development of testes (with spermatozoa), cirrus, cirrus sac, vitellaria and uterus, but only a rudimentary ovary developed and no eggs were formed.

○ 34.2.2 Schistosomes

Schistosoma mansoni

Considerable success in culturing adult schistosomes from cercariae has been achieved by Basch and his colleagues (Basch, 1981*a,b*; Basch & Humbert, 1981; Basch & Rhine, 1983). Cercariae were transformed to schistosomula using a double syringe system by passing them backwards and forwards, which caused shearing of the tails; this process, after centrifugation and washing in MBE medium + HEPES, completed their transformation to schistosomula.

Schistosomula were cultured in small Petri dishes in 2.0–2.5 ml of Medium 169 (Table 34.3), eight small dishes being placed in larger Petri dishes and incubated at 36 °C in a humidified, flowing CO_2 atmosphere, a drop of RBCs (type O) being added to each culture after 24–48 h. When the worms reached stage 18 (Fig. 34.4), 5 male and 5 female worms were transferred to a Leighton tube in 1.5 ml medium, together with a drop of RBCs.

In this system, considerable growth and differentiation took place and pairing of male and female worms occurred after 7 weeks. Numerous adult worms developed, of which about 10 per cent produced eggs. Unfortunately – as in so many trematodes grown *in vitro* – the eggs were abnormal, although TEM studies showed that the structure of the vitelline cells was similar to that in worms grown *in vivo* (Irie *et al.*, 1983). It appears that *in vitro* development only became abnormal at, or near, the pairing stage, possibly owing to the lack of an appropriate stimulus from the male and/or deficiencies in the culture conditions.

Table 34.3. *Medium 169: medium developed by Basch for the successful* in vitro *culture of the schistosomula of* Schistosoma mansoni *to adult pairing worms*

	Component[a]	Stock conc.	Add stock	Working conc.
1 (i)	BME (E) liquid[b]		1 l	
	or			
1 (ii)	BME (E) powder[c]		1 packet	
	plus			
1 (iii)	Water 3 × distilled[c]		1 l	
2	Lactalbumin hydrolysate		1 g	$1\ g\ l^{-1}$
3	Glucose		1 g	$11.1\ mmol\ l^{-1}$
4	Hypoxanthine	$10^{-3}\ mol\ l^{-1}$[d]	0.5 ml	$5 \times 10^{-7}\ mol\ l^{-1}$
5	Serotonin	$10^{-3}\ mol\ l^{-1}$[d]	1 ml	$10^{-6}\ mol\ l^{-1}$
6 (i)	Insulin U-100[e]	$100\ U\ ml^{-1}$	2 ml	$0.2\ U\ ml^{-1}$
	or			
6 (ii)	Insulin, crystalline[f]	$8\ mg\ ml^{-1}$[d]	1 ml	$8\ \mu g\ ml^{-1}$
7	Hydrocortisone	10^{-3} M[d]	1 ml	$10^{-6}\ mol\ l^{-1}$
8	Triiodothyronine	2×10^{-4} M[d]	1 ml	$2 \times 10^{-7}\ mol\ l^{-1}$
9	MEM vitamins[g]	100 × [d]	5 ml	0.5 ×
10	Schneider's medium[h]	1 ×	50 ml	5%
11	HEPES		2.4 g	$10\ mmol\ l^{-1}$
12	Serum	1 × [d]	100 ml	10%
13	NaOH	$5\ mol\ l^{-1}$	*q.s.*	pH 7.4
14	$NaHCO_3$ (for 1 (ii) only)		2.2 g	$26\ mmol\ l^{-1}$
15	Water, 3 × distilled		*q.s.*	to $275\ mosmol\ l^{-1}$

[a]Biochemicals from Sigma, St. Louis, Missouri.
[b]Grand Island Biological (GIBCO) 320–1015.
[c]GIBCO 420–1100.
[d]Stocks frozen at −20°C. Working concentration is based upon a theoretical total volume of 1 l; dilution to $275\ mosmol\ l^{-1}$ and addition of other components increases the volume and reduces the nominal working concentrations.
[e]Lilly Iletin NDC 0002–1135–01.
[f]Bovine origin. Sigma 1–5500.
[g]GIBCO 320–1120.
[h]GIBCO 350–1720. (This medium was supplemented by human blood cells, Type O.)
Source: Data from Basch (1981*a*).

Other schistosome species

Some limited success has also been achieved with *S. haematobium* and *S. japonicum*, but, again, normal eggs were not obtained (Smyth, 1990; Wang & Zhou, 1987). The rodent species *Schistosomatium douthitti* (Fig. 16.10) has also been cultured to maturity (Basch & O'Toole, 1982) but only abnormal eggs were obtained.

○ 34.2.3 Other trematodes

Some limited success has also been obtained with other species: *Fasciolopsis buski*, *Echinostoma malayanum*, *Leucochloridiomorpha constantiae*, *Paragonimus westermani*, *P. miyazakii*, *P. ohira*, *Opisthorchis* (= *Clonorchis*) *sinensis*, *Isoparorchis hypselobagri* and *Orthocoelium scoliocoelium* (References in Smyth & Halton, 1983; Smyth, 1990).

○ 34.2.4 The CAM (chorioallantoic membrane) culture technique

Although not strictly speaking an *in vitro* technique, the culture of trematodes on the chorioallantoic membrane (CAM) of embryonating hen eggs (Fig. 34.5) provides a valuable additional method which can be used to replace animals for studies on the development of trematodes. The chick chorioallantoic membrane has been widely used for many years for the culture of viruses and protozoa but its application to trematode culture owes its origin to the pioneer work of Fried (1962) who first used it for the development of the bird trematode *Philophthalmus* sp. from a metacercaria to an adult fluke. Since that time, the growth of some 23 species from 14 families of the Digenea has been investigated (for review, see Fried & Stableford, 1991).

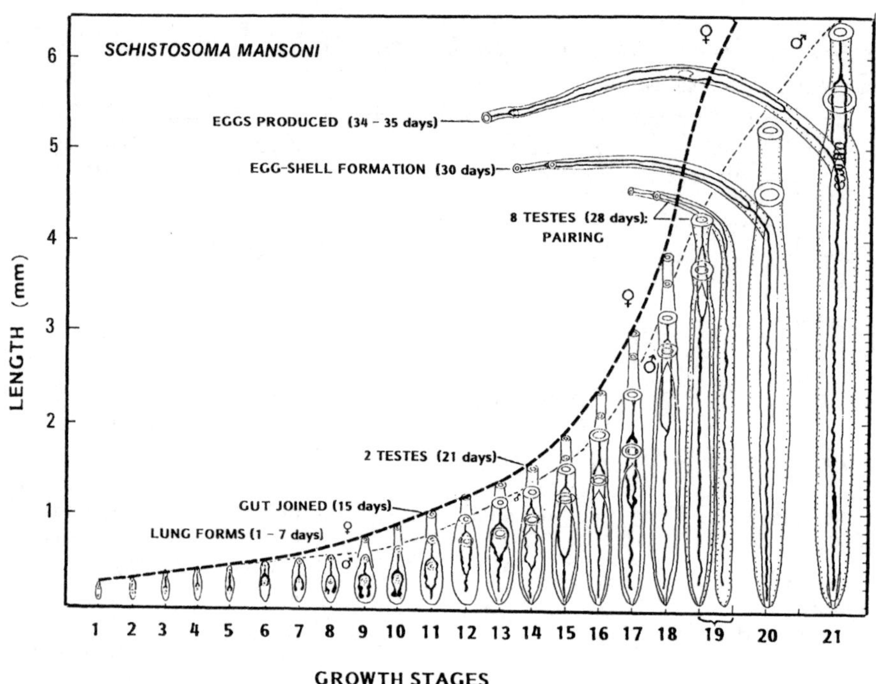

Fig. 34.4
Schistosoma mansoni: diagrammatic representation of growth stages, from schistosomulum to adult, in the mammalian host. (Adapted from Faust *et al.*, 1934; Basch, 1981*a*; Clegg, 1965.)

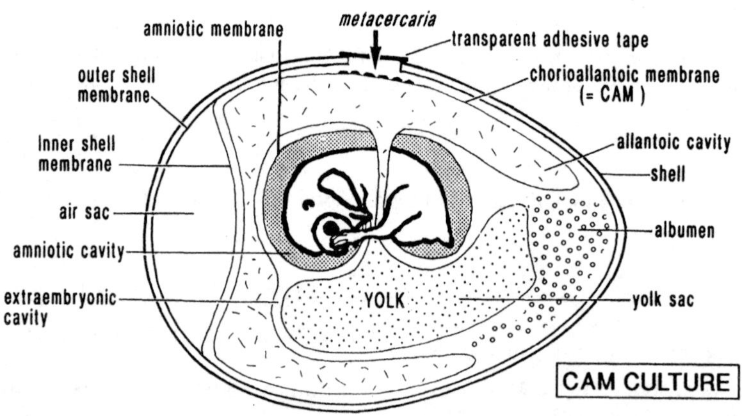

Fig. 34.5
Diagrammatic representation of a developing 9-day-old chick embryo, showing how metacercariae can be cultured on the chorioallantoic membrane (CAM). The metacercariae are inserted through a window cut in the shell, which is then sealed with transparent adhesive tape. (Partly after Saville & Irwin, 1991.)

Technique. The simplest technique to use is that of Zwilling (1959) as modified by Fried (1973) and redescribed in detail in Fried (1989) and Fried & Stableford (1991). In this, a window (1 cm × 1 cm) is cut in a 3-day-old fertile hen's egg, the shell and underlying shell membrane removed with a fine forceps, and about 2 ml of the albumen removed with a hypodermic syringe. This allows the embryo to move away from the shell membrane as the former develops. The shell membrane is moistened with a drop of saline and the window sealed with transparent adhesive tape and replaced in the humidified incubator at 38+ °C for another 3–5 days. On day 6–8, i.e. when the chorioallantois is developed, the tape is removed, the worms placed on the CAM in a drop of saline, and the window resealed with tape and the egg replaced in the incubator. As the CAM begins to dry out on about day 17–18, if a longer period of development is required, worms must be transferred to a new 8–10-day-old embryo.

A modification of the above technique is that of Irwin & Saville (1988*a,b*; Saville & Irwin, 1991), who examined the suballantois and allantois as habitats for the growth of microphallids and strigeids and also investigated the effect of adding a serum supplement.

Although only a limited number of species – especially avian digeneans with progenetic metacercaria, such as *Diplostomum spathaceum* (p. 259) and the species discussed earlier – have been grown to the ovigerous stage during CAM culture, much valuable basic biological information on various aspects of maturation, differentiation, fine structure, behaviour, pathology and interface studies has been gained from the use of this simple technique; it is also a valuable tool in chemotherapeutic studies. Details of all the species cultured are summarised in Fried & Stableford (1991).

○○○ 34.3 Cestodes

○ 34.3.1 General account

In vitro cultivation of cestodes presents special difficulties not encountered with other groups of helminths, owing mainly to the tapelike shape, the gigantic size of the strobila in some species, and the fact that they are almost exclusively intestinal parasites. In spite of these difficulties, substantial progress has been made with the *in vitro* cultivation of cestodes; a number of species can now be cultured through to maturity or near maturity, and some (e.g. *Hymenolepis diminuta*) through their whole life cycle. Recent work with this group has been reviewed by Arme (1987), Evans (1980), Smyth (1990) and Smyth & McManus (1989).

Cestodes differ from other helminths in that all nutrients must be taken in through the tegument and all waste materials excreted through it as they do not possess a gut. This places a limit on the size of molecules which can be absorbed by the tegument and severely limits the composition of the culture media which can be used.

A further complication is that since adult cestodes inhabit the vertebrate gut, which contains a rich microflora, it is difficult to obtain worms in a sterile condition although this is possible by repeated washing in antibiotics. The most successful cultures have been obtained by commencing with the larval stages, rather than the adult, as the former invariably occur in sterile tissue sites such as the body cavity of fish or the musculature of vertebrates from which they can be extracted using aseptic techniques. Such an approach has been successful with genera such as *Schistocephalus*, *Ligula* and *Diagramma* among the Pseudophyllidea and with *Hymenolepis*, *Echinococcus*, *Taenia* and *Mesocestoides* among the Cyclophyllidea.

○ 34.3.2 Pseudophyllidea

Progenetic plerocercoids. The larvae of Pseudophyllidea provided useful experimental models for early work in this field. This was largely related to the fact that several genera (e.g. *Schistocephalus*, *Ligula*) are progenetic and, moreover, the larvae contain sufficient carbohydrate and protein reserves to satisfy the energy and synthetic requirements of maturation without additional nutriment being provided.

Schistocephalus solidus

A detailed account of the technique for culturing this species is given in Smyth (1946, 1954, 1990).

Plerocercoid to adult. The plerocercoids of *S. solidus* (Fig. 22.9) are found in the coelomic cavity of the stickleback *Gasterosteus aculeatus*, from which they may be removed aseptically without difficulty and transferred to sterile culture medium.

Preliminary experiments established that plerocercoids rapidly develop into sexually mature adults if incubated in isotonic media such as saline or broth at 40 °C (temperature of definitive bird host). Development, when first obtained *in vitro*, was abnormal (Smyth, 1946), as evidenced by the fact that the eggs were infertile, spermatogenesis was abnormal as indicated by presence of necrotic cells, and insemination had not occurred as spermatozoa were lacking in the receptaculum.

It later became apparent that for normal maturation to take place *in vitro* more elaborate culture procedures, more nearly paralleling the physico-chemical conditions of the gut, were required. The procedures which were ultimately successful involve (*a*) the use of a highly buffered medium to counteract the effect of acidic metabolic products, (*b*) cultivation under semi-anaerobic (deep-tube) or anaerobic conditions, to prevent premature oxidation of the phenolase–phenol–protein complexes in the shell-producing vitellaria (p. 187), (*c*) compression within an artificial gut of seamless cellulose tubing, a necessary pre-requisite for insemination and fertilisation (Fig. 34.6) and (*d*) agitation of the culture

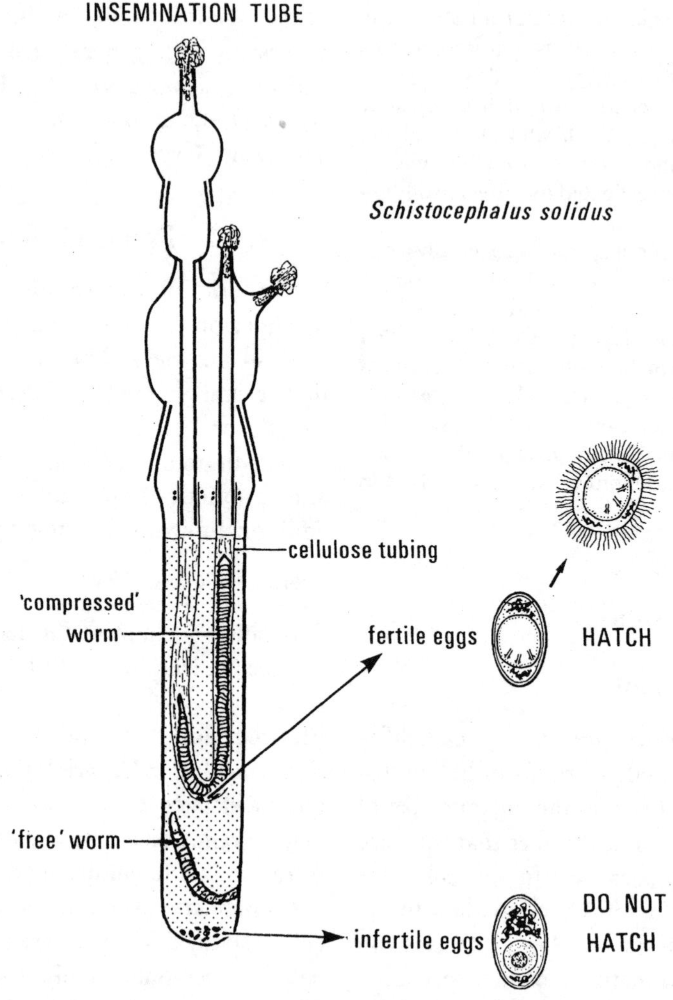

Fig. 34.6
Cellulose tube culture system which enables the cestode, *Schistocephalus solidus*, to undergo insemination during *in vitro* culture at 40°C. Fertile eggs are produced only by 'compressed' worms in which the cirrus is able to enter the vagina in each proglottis. Eggs from 'free' worms produce only infertile eggs. (After, Smyth, 1954; Smyth & McManus, 1989.)

medium by vigorous lateral shaking to assist in the diffusion of waste metabolites through the cellulose tubing.

Horse serum (which has strong buffering powers) was first used in the original culture (Smyth, 1946) but any modern tissue culture medium, such as Morgan 199, NCTC 109, NCTC 135 or RPMI 1640 – now readily available commercially – with HEPES added as a buffer and 10–20 per cent serum, is likely to be equally successful.

The form of the culture tube used is of importance. If the cellulose tubing is omitted and larvae cultured free within the medium, maturation proceeds normally since eggs and spermatozoa are formed, but *insemination does not take place* and sections of the mature strobila of cultured worms show that the receptaculum in each proglottid is empty. On the other hand, in histological sections of worms matured within cellulose tubing, the receptaculum always contains spermatozoa. For normal insemination to occur, gentle lateral compression is thus required. Presumably, the presence of a closely applied surface enables the cirrus to be bent round into the vaginal pore of the same or adjacent proglottides or into one of another worm in the same culture tube.

Eggs from larvae matured within cellulose tubing under optimal conditions show 50–88 per cent fertility and hatch out normal infective coracidia capable of continuing the *in vivo* cycle. Such eggs are unquestionably 'normal' so that the life cycle of this cestode can now be completed without the intervention of the natural definitive host (Smyth, 1959).

Ligula intestinalis

Like those of *Schistocephalus*, the plerocercoids of *Ligula intestinalis* (p. 316) are progenetic although they are not in quite such an advanced stage of development and require about 72 hours to reach maturity *in vivo* (p. 316). They can be readily matured *in vitro* by techniques very similar to those described above for *Schistocephalus* (Smyth, 1947, 1990). However, larger culture tubes than those used for *Schistocephalus* (Fig. 34.6) may be required and to assist the diffusion of the toxic, acidic waste products, the confining cellulose tube should be perforated by making a large number of small perforations with a sharp needle.

The plerocercoids are so large that they can be cut into short lengths (about 6–12 mm), each of which will mature and produce eggs. Such fragments (Fig. 34.7) are conveniently cultured in small Falcon tissue culture flasks; eggs appear on the third day of culture. This material is especially useful for class experiments.

Diagramma alterans

The progenetic plerocercoid of this species resembles *Ligula* except that it possesses a double set of genitalia. Using an *in vitro* technique similar to that used for *Ligula*, Awakura (1980) cultured it in horse serum and obtained eggs after 72–96 h.

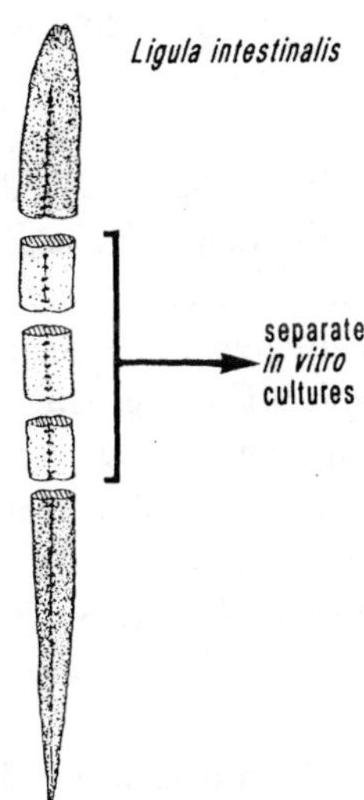

Fig. 34.7
Pieces of the plerocercoid of *Ligula intestinalis* (from fish) may be cut from the mid-region and cultured individually to sexual maturity *in vitro*. (After Smyth & McManus, 1989.)

Undifferentiated plerocercoids. Unlike those of *Schistocephalus*, the majority of pseudophyllidian plerocercoids are undifferentiated morphologically and poor in food reserves, with the result that they must carry out substantial tissue synthesis and differentiation before sexual maturity can be attained.

Spirometra mansonoides

Procercoid to plerocercoid. Mueller (1974) has developed a series of elegant techniques for infecting copepods and subsequently obtaining procercoids in a sterile condition and culturing them *in vitro* to infective plerocercoids. Procercoids obtained in this way were cultured in roller tubes in a liquid phase of Parker 199 plus ten per cent calf serum plus chick embryo extract. Plerocercoids obtained from such cultures were infective to cats.

Plerocercoid to adult. In a highly successful series of experiments, Berntzen & Mueller (1972) have grown plerocercoids of *Spirometra mansonoides* to egg-producing adults *in vitro*. This is the largest cestode species yet grown, the final strobila measuring nearly half a metre. Earlier attempts with this species were only partially successful; in later experiments, the medium used was essentially the same except that 0.1 cm^3 bile per litre was added to the basic medium (M 115).

○ 34.3.3 Cyclophyllidea

Recent work on the culture of cyclophyllidean cestodes has been reviewed by Arme (1987) and Smyth (1990). Three genera are dealt with in detail below – *Hymenolepis*, *Mesocestoides* and *Echinococcus* – with which most success has been achieved.

Hymenolepis diminuta

Cysticercoid to adult. The first successful cultivation of *H. diminuta* was carried out by Berntzen (1961), a result which represented a remarkable technical advance. His method involved an ingenious continuous flow apparatus and a complex culture medium; the gas phase was not stated but presumably carbon dioxide was available. There is some question as to the precise composition of the culture medium used, as other workers have been unable to prepare it according to the instructions published. With this technique cysticercoids were grown to sexually mature adults with gravid proglottides with oncospheres. Worms *in vitro* matured in 15 days, only two days longer than required *in vivo*.

Although these experiments were remarkably successful it is difficult to pinpoint the basic reasons why this was so. It may be that the success of the method with this species (and with *H. nana* and *S. mansonoides*) was due to the fact that conditions approached those pertaining in the rat intestine more nearly than in earlier experiments.

The requirements of *H. diminuta* may, however, be simpler than the results above suggest, for this species has also been grown in a much simpler system (Schiller, 1965). This consisted of a blood-agar base (NNN, such as that used for trypanosomes) in flasks overlaid with Hank's saline and a gas phase of three per cent carbon dioxide in nitrogen; the medium was shaken continually. Worms were transferred at frequent intervals to new flasks with fresh media; additional glucose was added from time to time. Worms developed in this system were smaller than those grown in rats and produced infertile proglottides. Nevertheless, the system is a useful and simple one, especially valuable for metabolic studies. The limits of the various physico-chemical factors such as pO_2, pCO_2, E_h or pH, etc., within which normal development of *H. diminuta* (and other species) can occur, is not known but with this species at least these may be wider than thought. Thus, in relation to the gas phase alone, a careful study by Roberts & Mong (1969) found that six-day-old *H. diminuta* (from rats) developed equally well in gas phases of zero, one, five or 20 per cent O_2 (all in nitrogen plus five per cent CO_2).

Oncosphere to cysticercoid. The oncosphere of *Hymenolepis diminuta* has also been cultured to a cysticercoid, thus enabling the entire life cycle of this species to be completed *in vitro*, a remarkable achievement (Arme, 1987; Smyth, 1990).

Other Hymenolepididae

Almost all stages of other common species of *Hymenolepis* (*H. nana*, *H. citelli* and *H. microstoma*) have been cultured *in vitro* with considerable success. One of the best results is that with *H. nana*, in which development *in vitro* only slightly lagged behind that *in vivo*. The relevant techniques are reviewed in Evans (1980), Taylor & Baker (1968–87) and Smyth (1990).

Mesocestoides corti

This species (Fig. 23.27) is unique in undergoing asexual division, not only in the larval stage in the intermediate host, but also in the adult phase in the carnivore gut (Fig. 23.28). It makes a particularly useful experimental model since the larval stage, the tetrathyridium (Fig. 23.28), is readily maintained by intra-peritoneal passage in mice. Moreover, it can be cultured to differentiate either asexually or sexually.

Table 34.4. *Medium S.10E.H: diphasic medium developed for the* in vitro *cultivation of the protoscoleces of* Echinococcus granulosus *and* E. multilocularis *to adult, sexually mature, worms*

Liquid phase	
Basic medium:	
CMRL 1066	260 ml
Fetal calf serum	100 ml
5% (w/v) yeast extract in CMRL 1066	36 ml
30% (w/v) glucose in distilled water	5.6 ml
5% (v/v) dog bile or 0.2% (w/v) Na taurocholate in Hanks' BSS	1.4 ml
Buffer:	
20 mM HEPES + 10 Mm $NaHCO_3$	
Antibiotics:	
100 μg Gentamicin and/or 100 i.u. penicillin + 100 μg streptomycin ml^{-1}	
Solid phase	
Bovine serum[a] coagulated at 76 °C for 30–60 min	

[a]Bovine serum may be sold commercially as 'newborn calf serum'. This should not be confused with 'foetal calf serum' (as used in the liquid phase), which does not coagulate on heating.

Source: Based on Smyth (1979, 1985).

***In vitro* asexual differentiation.** Although asexual multiplication of tetrathyridia will take place in a number of different media, it was later demonstrated that this process is greatly facilitated by culturing in a diphasic medium. The multiplication is greatly aided if holes – into which the larvae can burrow – are made in the coagulated serum base, suggesting that contact with a (protein?) substrate plays an important role in stimulating asexual differentiation (Ong & Smyth, 1986; Smyth, 1987).

***In vitro* sexual differentiation.** The successful development of sexual differentiation of tetrathyridia *in vitro* was reported simultaneously by Barrett *et al.* (1982) and Thompson *et al.* (1982) using a medium based on Medium S.10E.H (Table 34.4) as developed for the culture of *Echinococcus granulosus*. Although the optimum conditions for sexual differentiation have not been determined, experimental evidence suggests that anaerobic conditions and a gas phase of 5 per cent CO_2 are major controlling factors (Ong & Smyth, 1986). Although well-developed adult worms were developed in the above experiments, only low levels of self-insemination and fertilisation were achieved.

Mesocestoides lineatus

Unlike *M. corti*, the tetrathyridia of this species do not multiply asexually in the intermediate or definitive hosts. Kawamoto *et al.* (1986) showed that *in vitro* sexual differentiation in this species only occurred when the larvae were pretreated with trypsin or other proteolytic enzymes, a minimum of 24 h being necessary. After enzyme treatment, larvae were transferred to NCTC-135-Plus and cultured in plastic culture dishes, 10–15 larvae at a time in 8–15 ml of medium at 37 °C in a gas phase of 5 per cent CO_2 in air. In the best cultures, shelled eggs containing active oncospheres were obtained.

Echinococcus spp.

The cystic larval stage of the hydatid organisms (Figs 23.16, 23.19), *E. granulosus* and *E. multilocularis*, presents unusually interesting material for *in vitro* culture as it possesses the potential of being able to differentiate either sexually or asexually (Fig. 34.8): (*a*) *sexually* into a strobilated adult worm in the gut of a dog, or (*b*) *asexually* into a cystic form, which develops if a protoscolex leaks from the original hydatid cyst (perhaps during surgery) or is injected into the body cavity of an experimental host (such as a rodent). As hydatid cysts from sheep, horse, goat, pigs, camels, etc. – depending on the country – may often be obtained from local abattoirs, they form useful starting material for *in vitro* experiments.

Echinococcus granulosus. Protoscoleces cultured in liquid media eventually develop into miniature hydatid cysts, following a pattern which is now well established (Smyth, 1962–85; Rogan & Richards, 1986; Casado & Rodriguez-Caabeiro, 1989). Depending on the culture conditions, the types which appear during *in vitro* culture are as follows (Fig. 34.9):

(*a*) *Unevaginated protoscolex* (Fig. 34.9A)

In a few instances, organisms remain undifferentiated, undergoing no growth or evagination.

(*b*) *Vesicular type* (Fig. 34.9B)

A protoscolex evaginates and swells, taking on a 'cottage-loaf' profile; after some time, it secretes a sticky translucent layer which gradually comes to surround the organism as a laminated membrane, the whole developing into a miniature hydatid cyst (Fig. 34.9E).

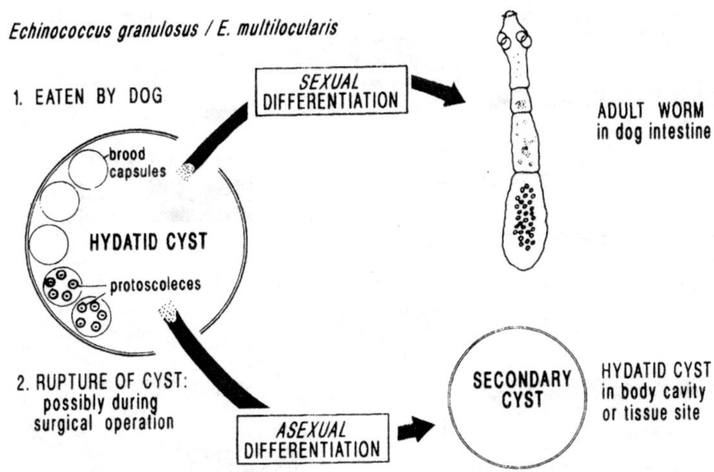

Fig. 34.8
Potential development of *Echinococcus granulosus* or *E. multilocularis* in different host habitats. 1. If a cyst is eaten by a dog, the scoleces evaginate, attach to the mucosa and differentiate into sexually mature worms. 2. If a cyst should accidentally burst – owing to injury or during the course of a surgical operation – each protoscolex has the potential to differentiate asexually into a (secondary) hydatid cyst. (After Smyth, 1987.)

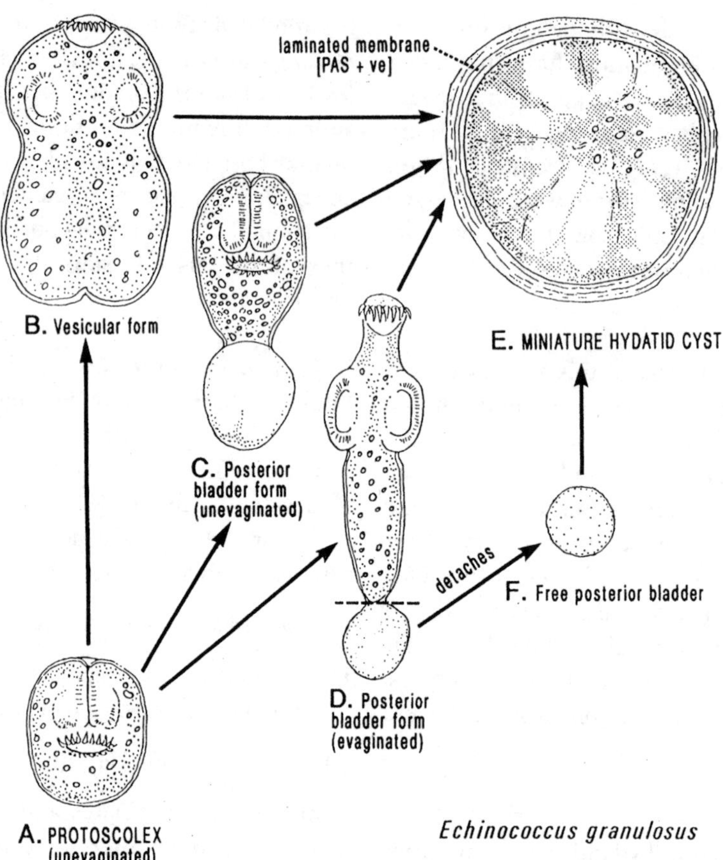

Fig. 34.9
The various forms which can differentiate from a protoscolex of *Echinococcus granulosus* when cultured *in vitro* in a liquid medium. (After Smyth & McManus, 1989.)

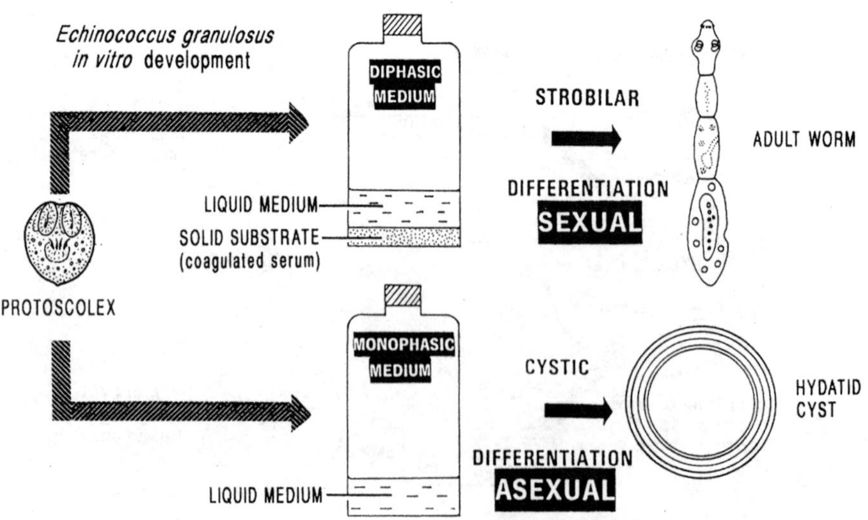

Fig. 34.10
In vitro culture of *Echinococcus granulosus*. In a diphasic medium protoscoleces differentiate into a sexually mature worm. In a monophasic, liquid medium (e.g. Medium S10E.H, Table 34.4) they may differentiate into the forms shown in Fig. 34.9, all of which eventually develop into hydatid cysts. (After Smyth, 1987.)

(*c*) *Posterior bladder type* (Fig. 34.9C)

A small bladder or vesicle develops in the posterior region of an unevaginated protoscolex, apparently arising from a few cells carried over from the germinal membrane to which it was originally attached (Fig. 23.19) in the brood capsule. The bladder develops a laminated membrane which gradually engulfs the whole organism, which becomes rounded and, as above, develops into a miniature hydatid cyst.

(*d*) *Evaginated protoscoleces with posterior bladders* (Fig. 34.9D). This form appears to arise from the previous type, becoming evaginated after the posterior bladder has formed; it also develops into a miniature hydatid cyst.

(*e*) *Free posterior bladder* (Fig. 34.9F)

In protoscoleces of horse origin (but not sheep) posterior bladders sometimes become separated, develop a laminated membrane and form a miniature hydatid cyst.

Sexual (strobilar) differentiation *in vitro*. In all early culture attempts (Smyth, 1962) organisms developed in a cystic direction, as described above (Fig. 34.9) and the determination of the factors required to induce strobilar (i.e. sexual) differentiation remained unsolved. This led to a further examination of the adult worm in the dog gut when it was found that the scolex contact with the mucosa was closer than previously realised (Fig. 23.18). This raised the possibility that contact with a surface in some way triggered strobilar development, possibly via a neurosecretory mechanism. This hypothesis proved to be valid (but contrast *E. multilocularis*, p. 516) as when protoscoleces were cultured in a diphasic medium, with coagulated serum providing a semi-solid surface, strobilar differentiation resulted and sexually mature adult worms developed (Fig. 34.10) (Smyth, 1967).

Since then, the original technique has been modified by varying the composition of the medium (Table 34.4) and increasing its buffering powers, and making other adjustments until reasonably consistent results were obtained. This technique is outlined in Fig. 34.11, but for detailed protocols, the original papers should be consulted (Smyth, 1979, 1985, 1990; Smyth & Davies, 1974*b*). Cultures need to be shaken or rotated in a roller tube system for consistent results; a more sophisticated system, using a simple 'lift' circulation system, is shown in Fig. 34.12.

This technique was successful in growing adult worms, but insemination (Fig. 20.5) failed to take place, with the result that fertile eggs were not produced. Although not intended, failure to produce fertile eggs – which are highly infective – means that such culture procedures can safely be carried out in the laboratory without risk of infection.

Culture difference in *E. granulosus* isolates All the above early culture experiments were carried out with *E. granulosus* protoscoleces of sheep origin but when applied to protoscoleces from horse hydatid cysts, the organisms, although evaginating and surviving for long periods, failed to grow significantly or strobilate. At first, this was attributed to faulty media or technique, but when after two years of experiments – involving some 200 cultures – the horse material failed to strobilate, it was realised that this could represent a new phenomenon, i.e. that *Echinococcus* of horse origin was an 'intra-specific variant' or 'strain' (see p. 336) with nutritional or other requirements different from those of strains of sheep origin (Smyth & Davies, 1974*a*). Biochemical studies confirmed this hypothesis as the chemical composition, isoenzymes and metabolism of the horse isolate were found to differ substantially from those of the sheep isolate (McManus & Smyth, 1978, 1979). This result led to a worldwide interest in the 'strain' problem in *E. granulosus* in other hosts, a question discussed in detail on p. 336.

The factors which are needed to induce the horse isolate to strobilate *in vitro* remain unidentified and remain a major challenge in hydatid research. The problem is especially intriguing, as isolates from other hosts – cattle, goats, buffalo, camel and man (Table 34.5) in addition to sheep – all readily develop sexually *in vitro* (Macpherson & Smyth, 1985).

Anomalous 'monozoic' development. Occasionally in *E. granulosus* cultures, probably owing to abnormal culture conditions, a few organisms developed sexually mature genitalia without becoming segmented into proglottides, i.e. they were *monozoic* in form (Smyth, 1971). Although sexually differentiated, such forms had undergone little growth in size, a result pointing to independent control of somatic growth and sexual differentiation (Fig. 34.13), either of which appears to be able to be blocked by unknown factors. In normal development the two processes are synchronised, but under certain conditions (not understood) either can be inhibited.

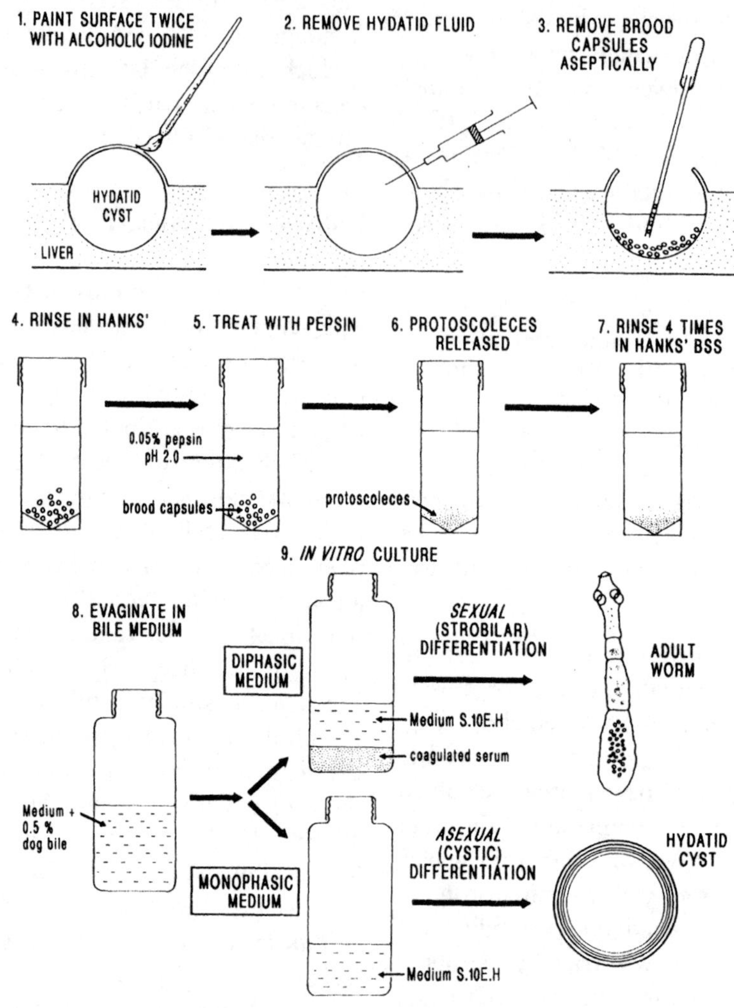

Fig. 34.11
Echinococcus granulosus: protocol for the preparation of *in vitro* cultures. All procedures are carried out under sterile conditions, ideally in a laminar flow cabinet. (After Smyth, 1979, 1990.)

Table 34.5. Echinococcus granulosus; in vitro *development of isolates from hydatid cysts from different hosts*

Host source	Sexual differentiation
Buffalo (India)	Yes
Camel (Kenya)	Yes
Cattle (Kenya)	Yes
Goat (Kenya)	Yes
Man (UK)	Yes
Sheep (UK)	Yes
Horse (UK)	No

Source: Data from Macpherson & Smyth (1985).

This unusual phenomenon occurs more readily in cultures of *E. multilocularis*, and is discussed further below.

Echinococcus multilocularis: sexual differentiation *in vitro*

Cystic material of this species is dense and jelly-like and the associated host tissue often heavily calcified. To release the protoscoleces requires a different treatment from that used for *E. granulosus*, involving more concentrated (0.5 per cent) pepsin and longer treatment (60–90 min). The tissue usually requires macerating in the pepsin solution in a Waring blender to break it up, before digestion; the pH must be adjusted repeatedly

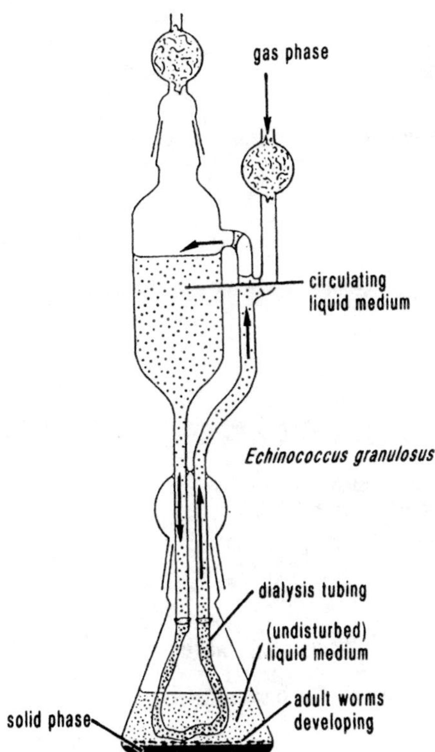

Fig. 34.12
A circulating 'lift' for the *in vitro* culture of adult *Echinococcus granulosus*, from protoscoleces. The circulating medium in the cellulose tubing is separated from the medium in the lower flask with the result that the interface between the worms and the solid substrate is undisturbed; nutrients and waste materials can be exchanged via the cellulose tubing. (After Smyth & McManus, 1989.)

to maintain a pH of 2.0 as the acid becomes neutralised by the released calcium. Details of the technique are described in detail by Smyth (1979, 1990).

In remarkable contrast to *E. granulosus*, protoscoleces differentiate readily to sexually mature adults in monophasic liquid medium (Medium S.10E.H; Table 34.4), i.e. without the presence of a solid protein (?) substrate, and may be readily cultured in a roller tube system. In early experiments, organisms showed a tendency to form remarkable 'monozoic' worms, as shown in Fig. 23.24, similar to those occasionally found in abnormal (?) *E. granulosus* cultures (Fig. 34.13). As culturing conditions improved, this tendency lessened and more normal worms with 2–5 proglottides developed, although the proglottides often showed a characteristic 'bulge' as the genitalia grew, indicating that some inhibition of somatic growth was still taking place. The fact that *E. multilocularis* can differentiate sexually in a monophasic medium, whereas *E. granulosus* requires a diphasic medium, raises fundamental questions for which no answers are at present forthcoming and presents an intriguing problem for further research.

Culture of cystic *E. granulosus* and *E. multilocularis*

The cystic stages of *E. multilocularis* have proved to be more amenable to culture than those of *E. granulosus*. Fragments of germinal membrane when grown in liquid medium vesiculated and formed protoscoleces within 29 days. These were infective to voles after 59 days (Rausch & Jentoft, 1957). 'Fertile' hydatid cysts – i.e. with brood capsules and protoscoleces – have never been grown from oncospheres or protoscoleces *in vitro*, although early hydatid cysts have been grown from oncospheres up to the appearance of the germinal membrane (Heath & Lawrence, 1976). However, in one experiment, starting with protoscoleces, after 8 months culture, the miniature cysts that formed (Fig. 34.9E) developed the anlagen of brood capsules before the culture was terminated (Smyth, 1985). Cysts developed from protoscoleces *in vitro* have been shown to be viable when injected into mice (Casado *et al.*, 1992).

Culture of various Taeniidae

With the exception of *Echinococcus* spp. described above, little progress has been made with the routine culture of taeniid cestodes and no species has yet been grown from cysticercus to gravid adult. This has nearly been achieved with *Taenia pisiformis*, which has been grown to sexual maturity from the cysticercus, but normal development or fertile eggs were not obtained (Osuna-Carrillo & Mascaro-Lazcanó, 1982). Early cysticerci of a number of Taeniidae have also been grown from oncospheres, the most successful results being obtained with *T. pisiformis*, but promising results have also been obtained with *T. hydatigena*, *T. ovis*, *T. serialis*, *T. taeniaeformis* and *T. saginata* (Heath, 1973; Heath & Smyth, 1970).

○○○ 34.4 Nematodes

○ 34.4.1 General account

Probably on account of their medical, veterinary and economic importance, extensive work has been carried out on the *in vitro* culture of nematodes. Recent work has been reviewed by Douvres & Urban (1987), Mössinger (1990) and Smyth (1990) and earlier work by Taylor & Baker (1968, 1978). Remarkable progress has

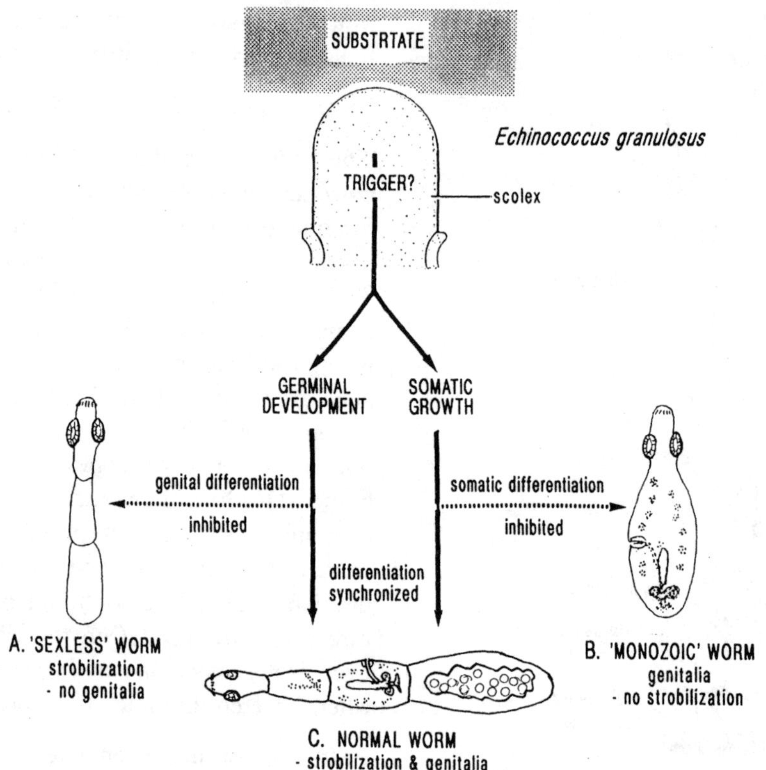

Fig. 34.13
Differentiation of *Echinococcus granulosus* and *E. multilocularis in vitro*. Normal segmented worms with proglottides containing genitalia develop in the dog gut, indicating that somatic growth and germinal development are closely coordinated. However, during *in vitro* culture, although in the best cultures 'normal' worms develop, *'sexless'* worms (i.e. worms that segment and form proglottides without genitalia) or *'monozoic'* worms (i.e. with one set of genitalia but not segmented) are occasionally formed. This indicates that somatic growth and germinal development can operate independently and either process can be inhibited, perhaps by a metabolic block present during abnormal culture conditions. (Original.)

been made in this field and several species can now be cultured through their entire cycle *in vitro*. A particularly significant advance has been the development of the filariid *Brugia malayi* (p. 427) from microfilariae to sexually mature adults by Riberu *et al.* (1990), thus achieving a long-standing goal in filariasis research. Representative examples of some of the best results achieved in nematode culture are listed in Table 34.6 and discussed briefly below.

○ 34.4.2 Techniques and media utilised

The presence of a thick cuticle (Fig. 25.16) on nematodes makes them more 'robust' than trematodes and cestodes, with the result that the adults of some species can often be 'maintained' for long periods in a reasonably good physiological condition. For example, adults of the human hookworm, *Ancylostoma duodenale*, have been maintained *in vitro* for over 200 days (Fig. 34.14), during which time egg production was maintained (Fenglin *et al.*, 1979). Similarly, large worms, such as *Ascaris suum*, can readily survive in basic saline solutions long enough for physiological or biochemical experiments to be carried out.

However, attempts to culture adult worms from eggs *in vitro* have proved to be much more demanding, particular problems being encountered with (*a*) transformation of L_3 larvae to L_4 and of L_4 to adult worms and (*b*) induction of copulation and fertilisation with the production of fertile eggs. Most successful culture procedures have proved to be complicated, some species requiring two- or even three-step treatments, involving the use of several different culture media and/or conditions, in order to satisfy the different requirements of the various free-living and parasitic stages.

Table 34.6. *Species of Nematoda (other than Filarioidea) which have been cultured* in vitro *with some success*

Species	Definitive host	In vitro culture achieved				References
		Egg to L_3	L_3 to L_4	L_4 to adult	Fertile eggs	
Necator americanus	man	yes	yes	yes	no	Weinstein & Jones (1959)
Ancylostoma caninum	dog	yes	yes	no	no	Banerjee (1972); Leland (1970)
Ascaris suum	man, pig	yes	yes	yes	no	Douvres & Urban (1983)
Trichinella spiralis	pig	ovoviviparous	yes	yes	ovoviviparous	Berntzen (1965); Sakamoto (1979)
Oesophagostomum radiatum	cattle	?	yes	partly	no	Douvres (1983)
Nippostrongylus brasiliensis (= muris)	rat	yes	yes	yes	no	Weinstein & Jones (1956–59)
Ostertagia ostertagia	cattle	yes	yes	yes	yes	Douvres & Malakatis (1977)
Haemonchus contortus	sheep, cattle	No?	yes	yes	no	Douvres (1983); Stringfellow (1986)
Cooperia punctata	cattle, sheep	yes	yes	yes	yes	Leland (1967); Zimmerman & Leland (1971)
Trichostrongylus colubriformis	ruminants	yes	yes	partly	no	Douvres (1980); Dorsman & Bijl (1985)

Source: Data from Douvres & Urban (1987) and Smyth (1990).

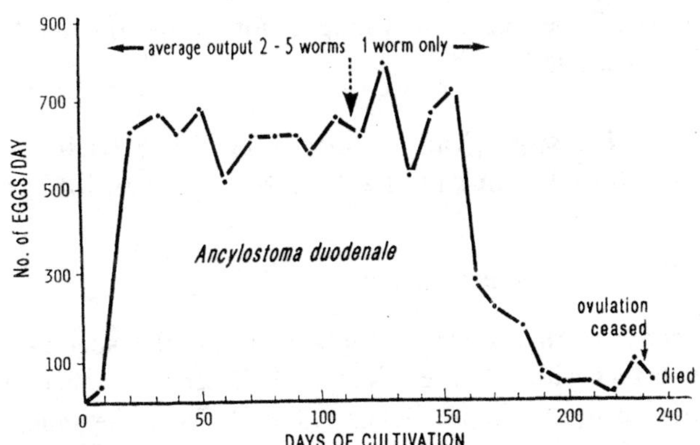

Fig. 34.14
In vitro maintenance of the adult of the human hookworm, *Ancylostoma duodenale*, showing the daily output of eggs over a period of 8 months. (After Fenglin *et al.*, 1979.)

○ 34.4.3 Trichostrongyloidea

This superfamily contains many genera of veterinary importance such as *Ostertagia*, *Haemonchus*, *Dictyocaulus*, *Cooperia* and *Trichostrongylus*. It also contains the rat hookworm, *Nippostrongylus brasiliensis* (Figs 28.1–28.3), which was the first parasitic nematode to be cultured to sexual maturity *in vitro* (Weinstein & Jones, 1956–59): a result which proved to be a great stimulus to others. This species is a valuable experimental model and the original techniques, although now more than 30 years old, are still valid and are briefly outlined below (Smyth, 1990).

Nippostrongylus brasiliensis. Eggs from faeces can be surface-sterilised in sodium hypochlorite (5 per cent w/v) and cultured in a medium containing chick embryo extract (CEE_{50}). For the preparation of the latter, which is much used in nematode culture, see Taylor & Baker (1968). Culture of L_3 larvae to adults was eventually achieved using CEE_{50} with 20 per cent rat or human serum in Earle's saline supplemented with vitamin or liver concentrate. Although sexually mature males and females were developed, copulation was not observed and only a few infertile eggs were produced.

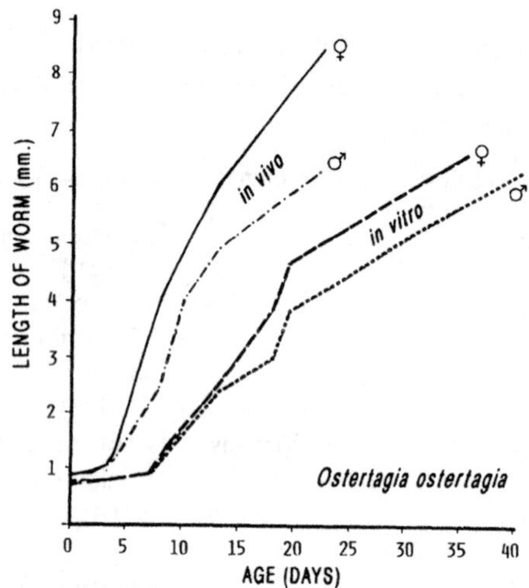

Fig. 34.15
Ostertagia ostertagia: comparison of growth and development from L_3 larvae *in vitro* and *in vivo*. (Modified from Douvres & Malakatis, 1977.)

Ostertagia ostertagia. This one of the few species for which L_3 larvae have been grown to adults with fertile egg production (Douvres & Malakatis, 1977), a step representing a major advance in nematode culture. A two-step culture system was necessary consisting of the use of two different complex media, Medium RFN and Medium API-1. Step 1 utilised Medium RFN in 95 per cent air, 5 per cent CO_2 at pH 7.3; step 2 utilised Medium API-1 + pepsin at pH 4.5 during day 3–9 and thereafter at pH 6.0. Although only 10–20 per cent of worms reached maturity and *in vitro* development was slower than that *in vivo* (Fig. 34.15), fertile eggs were produced, implying that copulation had taken place.

Haemonchus contortus. Using the basic medium API-1 (as above) plus various supplements, the L_3 stages of this species have been grown to egg-producing adults (Stringfellow, 1984, 1986). Forty thousand freshly exsheathed L_3 larvae were cultured in 40–42 ml medium in 1.8 l bottles rotated at 1 revolution per 1.5 min at 39 °C in 5 per cent O_2/10 per cent CO_2/85 per cent N_2 at pH 6.4 up to day 7 and pH 6.8 thereafter. In the best results (in API-1 + sheep gastric contents) adult males appeared in 28 days and ovipositing females in 36 days. A few eggs developed to the 2–8 cell stage.

Cooperia punctata. This species has been cultured through its entire life cycle *in vitro*. This was first achieved by Leland (1967) although only infertile eggs were produced. Using essentially the same techniques, Zimmerman & Leland (1971) cultured L_3 larvae to adults which produced large numbers of fertile eggs, only some of which developed to L_3. The basic medium was Medium 'Ae', which consisted of 50 per cent chick embryo extract (CEE_{50}), 15 per cent cysteine–casein solution, 5 per cent Eagle's vitamin mixture (100×), 14 per cent Earle's saline + antibiotics. Other variations of this were also tried. Cultures were incubated in screw-capped tubes on a roller system at 22–26 °C or 38.5 °C.

Trichostrongylus colubriformis. Hatched L_1 larvae of this species have been cultured to infective L_3 in several media, the best being a complex medium (Tc5) which contained CEE_{25}, NCTC 135, serum, yeast, extract, haemin, myoglobin and vitamin B6 (Dorsman & Bijl, 1985). This represented the first time the free-living rhabditiform larva of a parasitic nematode – which is food-dependent – has been so cultured. L_3 larvae have also been cultured but only to young adults (Douvres, 1980).

○ 34.4.4 Strongyloidea

Oesophagostomum radiatum. To date, culture of larvae to ovipositing adults has not been achieved. The best result was that in which L_3 were grown to young adults in medium API-1 plus glutathione or haemin in a one-step system; from an initial culture of 80 000 larvae, 25 per cent of young adults were obtained (Douvres, 1983).

***Strongylus* spp.** Various species of *Strongylus* have been cultured from L_3 to L_4 (Douvres & Urban, 1987).

○ 34.4.5 Ascaridoidea

Ascaris suum. Early attempts to culture this species, which met with a limited degree of success, have been reviewed by Taylor & Baker (1968, 1978). The most successful recent results have been those of Douvres & Urban (1983) who grew L_2 larvae, hatched from embryonated eggs, to mature adult worms. They used a 3-step system as follows:

Step 1: Medium KW-2 + 10 mM L-cysteine; gas phase for 4 days, 5 per cent CO_2 in N_2; thereafter 5 per cent O_2/10 per cent CO_2/85 per cent N_2.

Step 2: Medium API-18 from day 11 to day 18.

Step 3: Medium API-1 + 24 μg ml^{-1} haemin from days 18 onwards.

Only one male and two females reached maturity, one female producing over one million eggs from day 67 to day 125. Results were much better starting with L_3 larvae, as copulation occurred and fertilised eggs were obtained (Douvres & Urban, 1986).

○ 34.4.6 Trichinelloidea

***Trichinella spiralis*.** Early work on this species has been reviewed by Denham (1967), Sakamoto (1979) and Taylor & Baker (1968–78). Use of this organism has the advantages that (*a*) sterility is not a problem, as larvae occur encysted in sterile tissue sites; (*b*) they reach sexual maturity in three days. The most successful *in vitro* result was that of Berntzen (1965) who used a complex continuous flow apparatus. Encysted larvae were treated with pepsin followed by trypsin and pancreatin (to exsheath them) before culture in a complex medium (102B) with a gas phase of 10 per cent O_2/5 per cent CO_2/85 per cent N_2. Sexually mature worms were obtained with 120 h and embryos and juveniles after 8–10 days. However, there is some doubt as to the merits of this system, as other workers have been unable to reproduce his results (Denham, 1967).

A promising *in vitro* system for the culture of L_2 to L_3 larvae has been developed by Sakamoto (1979) using a complete sheet of tissue-cultured muscle cells on a collagen-covered surface. When L_2 larvae, isolated from adults using a 200-mesh sieve, were introduced, they penetrated the muscle cells and began coiling after 5 days.

○○○ 34.5 Filarioidea

○ 34.5.1 General comments

Recent progress in the *in vitro* culture of filarial worms has been comprehensively reviewed by Mössinger (1990) and earlier work by Taylor & Baker (1968–87). Attempts to culture filariids are complicated by the fact that the biphasic life cycle (Figs 29.1, 29.5) involves development in an arthropod intermediate host and a vertebrate definitive host. A microfilaria (L_1), when taken up by the arthropod host, is relatively undifferentiated morphologically, but develops intra-cellularly into a 'sausage stage' before moulting to L_2 and L_3 infective larvae (Fig. 34.16). In the definitive host, the L_3 moults to an L_4 and again to the adult stage with production of microfilariae.

Species utilised. Much of the basic experimental work on the culture of filariae has been carried out on species infecting animals, chiefly *Dirofilaria immitis* (dog, cat), *Brugia pahangi* (cats, monkeys, gerbils), *Acanthocheilonema vitae* (syn. *Dipetalonema vitae*) (gerbils), *Litomosoides carinii* (cotton rat) and *Onchocerca gutturosa* (cattle), but some species infecting man (*Wuchereria bancrofti*, *Brugia malayi*) have also been used.

Basic difficulties. In spite of the fact that microfilariae occur in the blood stream – and hence are readily available in sterile condition – attempts to culture filarial worms *in vitro* have, until recently, had only limited success. This is related to the fact that the factors that trigger or control the transformation of microfilariae (L_1) to L_2 larvae, L_2 to L_3, L_3 to L_4 and L_4 to adults are largely unknown. Extensive work has been carried out in an attempt to determine these factors.

General results. Results with different species have been so variable that they are not discussed in detail here, but a full account is given in Mössinger (1990). To date, no one has succeeded in transforming a microfilaria to an L_2 stage larva, although many workers have been able to grow it to a 'sausage' stage (Fig. 34.16). Results have been much better when cultures have been initiated using L_1 larvae that have already been taken up by the arthropod vector and become established in its cells. Limited success has also been achieved with several species (*L. carinii*, *D. immitis*, *W. bancrofti* and *O. volvulus*) in growing the infective L_3 larva to the L_{4+} stage. However, only the L_3 larva of *Brugia malayi* has been grown to a microfilaria-producing adult worm (Riberu *et al.*, 1990), a result which must be regarded as representing a milestone in filarial culture. This result, taken with the fact that the L_3 larvae of this species has also been cultured from L_3 (of arthropod muscle origin) (Nayar *et al.*, 1991*a*,*b*) means that almost the whole life cycle of this species can now be grown *in vitro*, as described below.

In vitro culture of *Brugia malayi*

L_1 larva to L_3. The first successful experiments with this species (and *B. pahangi*) commenced with 1-day-old intracellular-lodged L_1 larvae (Nayar *et al.*, 1991*a*) which were grown *in vitro* to the infective L_3 larva. The culture system involved excised thoraces of mosquito

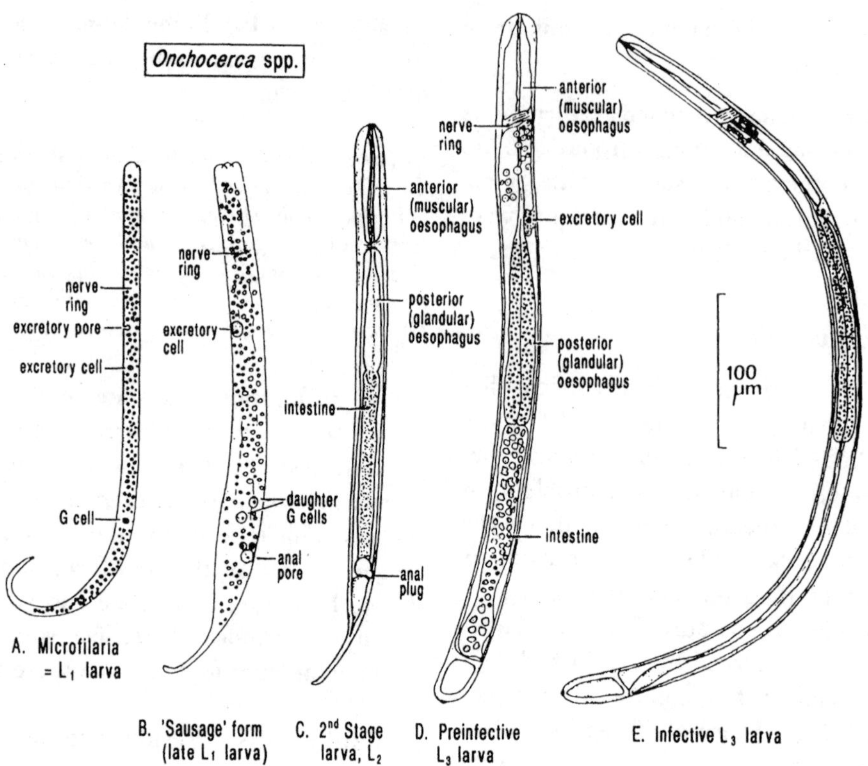

Fig. 34.16
Onchocerca spp.: development of stages in the vector from microfilaria (L_1 larva) to infective L_3 larva. (After Mössinger, 1990; courtesy of H. Schulz-Key.)

in a diphasic insect tissue culture medium. The latter consisted of nutrient agar base with a 1 : 1 mixture of Schneider's *Drosophila* medium and Grace's insect cell culture medium, supplemented with 20 per cent foetal bovine serum plus antibiotics. It was later shown that neither nutrient agar in the diphasic system nor the foetal bovine serum (Nayar *et al.*, 1991*b*) were essential in this system although development was better when these were present.

L_3 larvae to adult worms. Infective L_3 larvae, harvested from *Aedes togoi* 14 days after feeding on microfilaremic jirds, have been successfully cultured to sexually mature, microfilaria-producing adults (Riberu *et al.*, 1990). The culture technique was essentially that developed for *Dipetalonema viteae*, which enabled L_3 larvae to develop to the L_4 stage (Franke & Weinstein, 1984).

The culture medium consisted of a 1 : 1 mixture (v/v) of NCTC 135 and Iscove's modified Dulbecco's medium supplemented with 10 per cent human serum or plasma plus antibiotic and antimycotic agents. Cultures of 20–30 larvae in 2 ml medium were maintained at 37 °C in 5 per cent CO_2 in air.

This culture system was remarkably successful, some 85 per cent of worms reaching sexual maturity, each gravid female producing 200–1500 microfilariae after 75–100 days. How very close this *in vitro* system comes to reproducing *in vivo* growth conditions is reflected in the size and morphometric measurements of *in vitro* cultured worms when compared with those grown *in vivo* (Table 34.7). This remarkable result marks a new phase in the progress of *in vitro* culture of filariae and should prove a great stimulus to those attempting to culture other species.

In vitro maintenance of adult worms

The adult worms of many species have been maintained *in vitro* for up to several weeks or months, viability being generally assessed by measuring the numbers of microfilariae released. Such systems are useful for drug screening or the collection of E/S antigens. Examples include: *Dirofilaria immitis* (Tamashiro & Palumbo,

Table 34.7. Brugia malayi*: mean morphometric measurements of microfilariae taken from* in vitro *cultures (n = 10) compared with those taken from jirds in vivo (n = 10)*

	Measurements (μm)	
	in vitro	*in vivo*
Length	244	238
Width	7.5	7.3
Cephalic space	6.9	7.3
Caudal space	17	16
Nerve ring distance	50	49
Length of Innerkörper	38	37

Source: Data from Riberu *et al.* (1990).

1985), *D. uniformis* (Weinstein & Sawyer, 1961); *Acanthocheilonema viteae* (Mössinger *et al.*, 1987); *Litomosoides carinii* (Mössinger & Wenk, 1986; Mössinger *et al.*, 1987), *Onchocerca volvulus* (Engelbrecht & Schulz-Key, 1984; Schulz-Key & Karam, 1986); *O. gutturosa* (Townson *et al.*, 1986).

○○○ References

(References of general interest have been included in this list, in addition to those quoted in the text.)

Arme, C. 1987. Cestoda. In: *In Vitro Methods for Parasite Cultivation* (ed. A. E. R. Taylor & J. R. Baker), pp. 282–317. Academic Press, London.

Awakura, T. 1980. Cultivation of *Digramma alterans* (Rud.) Yamaguti, 1934 *in vitro*. *Scientific Reports of the Hokkaido Fish Hatchery*, **35**: 63–71. [HA/50/5486.]

Banerjee, D. 1972. *In vitro* cultivation of third stage larvae of *Ancylostoma caninum* (Ercolani, 1859). *Journal of Communicable Diseases*, **4**: 175–83.

Barrett, N. J., Smyth, J. D. & Ong, S. J. 1982. Spontaneous sexual differentiation of *Mesocestoides corti* tetrathyridia *in vitro*. *International Journal for Parasitology*, **12**: 315–22.

Basch, P. F. 1981*a*. Cultivation of *Schistosoma mansoni in vitro*. I. Establishment of cultures from cercariae and development until pairing. *Journal of Parasitology*, **67**: 179–85.

Basch, P. F. 1981*b*. Cultivation of *Schistosoma mansoni in vitro*. II. Production of infertile eggs by worm pairs cultured from cercariae. *Journal of Parasitology*, **67**: 186–90.

Basch, P. F., DiConza, J. J. & Johnson, B. E. 1973. Strigeid trematodes (*Cotylurus lutzi*) cultured *in vitro*: production of normal eggs with continuance of life cycle. *Journal of Parasitology*, **59**: 319–22.

Basch, P. F. & Humbert, R. 1981. Cultivation of *Schistosoma mansoni in vitro*. III. Implantation of cultured worms into mouse mesenteric veins. *Journal of Parasitology*, **67**: 191–5.

Basch, P. F. & O'Toole, M. L. 1982. Cultivation *in vitro* of *Schistosomatium douthitti* (Trematoda: Schistosomatidae). *International Journal for Parasitology*, **12**: 541–5.

Basch, P. F. & Rhine, W. D. 1983. *Schistosoma mansoni*: reproductive potential of male and female worms cultured *in vitro*. *Journal of Parasitology*, **69**: 567–9.

Bell, E. J. & Smyth, J. D. 1958. Cytological and histochemical criteria for evaluating development of trematodes and pseudophyllidean cestodes *in vivo* and *in vitro*. *Parasitology*, **48**: 131–48.

Berntzen, A. K. 1961. The *in vitro* cultivation of tapeworms. I. Growth of *Hymenolepis diminuta* (Cestoda: Cyclophyllidea). *Journal of Parasitology*, **47**: 351–5.

Berntzen, A. K. 1965. Comparative growth and development of *Trichinella spiralis in vitro* and *in vivo*, with redescription of the life cycle. *Experimental Parasitology*, **16**: 74–106.

Berntzen, A. K. & Macy, R. W. 1969. *In vitro* cultivation of the digenetic trematode *Sphaeridiotrema globulus* (Rudolphi) from the metacercarial stage to egg production. *Journal of Parasitology*, **55**: 136–9.

Berntzen, A. K. & Mueller, J. F. 1972. *In vitro* cultivation of *Spirometra* spp. (Cestoda) from the plerocercoid to the gravid adult. *Journal of Parasitology*, **58**: 750–2.

Cable, R. M. 1953. The life cycle of *Parvatrema boriquenae gen. et sp. nov.* (Trematoda: Digenea) and the systematic position of the subfamily Gymnophallinae. *Journal of Parasitology*, **39**: 408.

Casado, N., Criado, A., Jimenez, A., De Armas, C., Brasa, C., Perez-Serrano, J. & Rodriguez-Caabeiro, F. 1992. Viability of *Echinococcus granulosus* cysts in mice, following cultivation *in vitro*. *International Journal for Parasitology*, **23**: 335–9.

Casado, N. & Rodriguez-Caabeiro, F. 1989. Ultrastructural study of *in vitro* larval development of *Echinococcus granulosus* protoscoleces. *International Journal for Parasitology*, **19**: 21–8.

Clegg, J. A. 1965. *In vitro* cultivation of *Schistosoma mansoni*. *Experimental Parasitology*, **16**: 133–47.

Davies, C. & Smyth, J. D. 1978. *In vitro* cultivation of *Fasciola hepatica* metacercariae and of partly developed flukes recovered from mice. *International Journal for Parasitology*, **8**: 125–311.

Davies, C. & Smyth, J. D. 1979. The development of the metacercariae of *Microphallus similis in vitro* and in the mouse. *International Journal for Parasitology*, **9**: 261–7.

Denham, D. A. 1967. Applications of the *in vitro* culture of nematodes, especially *Trichinella spiralis*. *Symposium of the British Society for Parasitology*, **5**: 49–60.

Dollfus, R. Ph., Timon-David, J. & Rebecq, J. 1956. Maturité genitale provoquée experimentalement chez *Codonocephalus urniger* (Rudolphi) (Trematoda, Strigeidae). *Comptes Rendus Hebdomadaires des Séances de l'Académie des Sciences, Paris*, **242**: 2997–8.

Dorsman, W. & Bijl, A. C. 1985. Cultivation of free-living stages of *Trichostrongylus colubriformis* in media without bac-

LIBRARY, UNIVERSITY OF C

teria, animal tissue extract or serum. *Journal of Parasitology*, **71**: 200–3.

Douvres, F. W. 1980. *In vitro* development of *Trichostrongylus colubriformis* from infective larvae to young adult. *Journal of Parasitology*, **66**: 466–71.

Douvres, F. W. 1983. The *in vitro* cultivation of *Oesophagostomum radiatum*, the nodular worm of cattle. III. Effects of bovine heme on development in adults. *Journal of Parasitology*, **69**: 570–6.

Douvres, F. W. & Malakatis, G. M. 1977. *In vitro* cultivation of *Ostertagia ostertagia*, the medium stomach worms of cattle. I. Development from infective larvae to egg-laying adults. *Journal of Parasitology*, **63**: 520–7.

Douvres, F. W. & Urban, J. F. 1983. Factors contributing to the *in vitro* development of *Ascaris suum* from second-stage larvae to mature adults. *Journal of Parasitology*, **69**: 549–58.

Douvres, F. W. & Urban, J. F. 1986. Development of *Ascaris suum* from *in vivo* derived third-stage larvae to egg-laying adults *in vitro*. *Proceedings of the Helminthological Society of Washington*, **53**: 256–62.

Douvres, F. W. & Urban, J. F. 1987. Nematoda except parasites of insects. In: *In Vitro Methods for Parasite Cultivation* (ed. A. E. R. Taylor & J. R. Baker), pp. 318–78. Academic Press, London.

Engelbrecht, F. & Schulz-Key, H. 1984. Observations on adult *Onchocerca volvulus in vitro*. *Transactions of the Royal Society of Tropical Medicine and Hygiene*, **78**: 212–15.

Evans, W. S. 1980. The cultivation of *Hymenolepis in vitro*. In: *Biology of the Tapeworm Hymenolepis diminuta In Vitro* (ed. H. P. Arai), pp. 425–48. Academic Press, New York. ISBN 0 120 5898-0X.

Faust, E. C., Jones, C. A. & Hoffman, W. A. 1934. Studies on *schistosomiasis mansoni* in Puerto Rico. III. Biological studies; 2. The mammalian phase of the life cycle. *Puerto Rican Journal of Public Health and Tropical Medicine*, **10**: 133–96.

Fenglin, W., Rongjun, H., Xianxiang, H. & Xihui, Z. 1979. *In vitro* cultivation of *Ancylostoma duodenale* and *Ancylostoma caninum*. *Chinese Medical Journal (English Edition)*, **92**: 839–46.

Franke, E. D., Riberu, W. & Wiady, I. 1990. Evaluation of medium supplement for *in vitro* cultivation of *Wuchereria bancrofti*. *Journal of Parasitology*, **76**: 262–5.

Franke, E. D. & Weinstein, P. P. 1984. *In vitro* cultivation of *Dipetalonema vitae* third-stage larvae: effect of the gas phase. *Journal of Parasitology*, **70**: 493–8.

Fried, B. 1962. Growth of *Philophthalmus* sp. (Trematoda) on the chorioallantois of the chick. *Journal of Parasitology*, **48**: 545–50.

Fried, B. 1973. The use of 3-day-old chick embryos for the culture of *Leucochloridium variae* McIntosh, 1932 (Trematoda). *Journal of Parasitology*, **59**: 591–2.

Fried, B. 1989. Cultivation of trematodes in chick embryos. *Parasitology Today*, **5**: 3–5.

Fried, B., Heyer, B. L. & Pinski, A. K. 1981. Cultivation of *Amblosoma suwaense* (Trematoda: Brachylaimidae) in chick embryos. *Journal of Parasitology*, **67**: 50–2.

Fried, B. & Stableford, L. T. 1991. Cultivation of helminths in chick embryos. *Advances in Parasitology*, **30**: 108–65.

Fujino, T., Hamajima, F., Ishii, Y. & Mori, K. 1977. Development of *Microphalloides japonicus* (Osborn, 1919) metacercaria *in vitro* (Trematoda: Microphallidae). *Journal of Helminthology*, **51**: 125–9.

Guberlet, J. E. 1922. Three new species of Holostomidae. *Journal of Parasitology*, **9**: 6–17.

Heath, D. D. 1973. An improved technique for the *in vitro* culture of taeniid larvae. *International Journal for Parasitology*, **3**: 481–4.

Heath, D. D. & Lawrence, S. B. 1976. *Echinococcus granulosus*: development *in vitro* from oncosphere to immature hydatid cyst. *Parasitology*, **73**: 417–23.

Heath, D. D. & Smyth, J. D. 1970. *In vitro* cultivation of *Echinococcus granulosus*, *Taenia hydatigena*, *T. ovis*, *T. pisiformis* and *T. serialis* from oncosphere to cystic larva. *Parasitology*, **61**: 329–43.

Irie, Y., Basch, P. F. & Beach, N. 1983. Reproductive ultrastructure of adult *Schistosoma mansoni*, grown *in vitro*. *Journal of Parasitology*, **69**: 559–66.

Irwin, S. W. B. & Saville, D. H. 1988*a*. Cultivation and development of *Microphallus pygmaeus* (Trematoda: Microphallidae) in fertile chick eggs. *Parasitological Research*, **74**: 396–8.

Irwin, S. W. & Saville, D. H. 1988*b*. An alternative method for the culture of *Diplostomum spathaceum* (Trematoda) in chick embryos. *Journal of Parasitology*, **74**: 504–5.

Kannangara, D. W. W. & Smyth, J. D. 1974. *In vitro* cultivation of *Diplostomum spathaceum* and *Diplostomum phoxini* metacercariae. *International Journal for Parasitology*, **4**: 667–73.

Kawamoto, F., Fujioka, H. & Kumada, N. 1986. Studies on the post-larval development of cestodes of the genus *Mesocestoides*: trypsin-induced development of *M. lineatus in vitro*. *International Journal for Parasitology*, **16**: 333–40.

Leland, S. E. 1967. *In vitro* cultivation of *Cooperia punctata* from egg to egg. *Journal of Parasitology*, **53**: 1057–60.

Leland, S. E. 1970. *In vitro* cultivation of nematode parasites important to veterinary medicine. *Advances in Veterinary Science and Comparative Medicine*, **14**: 29–59.

McManus, D. P. & Smyth, J. D. 1978. Differences in the chemical composition and carbohydrate metabolism of *Echinococcus granulosus* (horse and sheep strains) and *E. multilocularis*. *Parasitology*, **77**: 103–9.

McManus, D. P. & Smyth, J. D. 1979. Isoelectric focusing of some enzymes frmo *Echinococcus granulosus* and *E. multilocularis* (horse and sheep strains). *Transactions of the Royal Society of Tropical Medicine and Hygiene*, **73**: 259–65.

Macpherson, C. N. L. & Smyth, J. D. 1985. *In vitro* culture of the strobilar stage of *Echinococcus granulosus* from protoscoleces of human, camel, sheep and goat origin from Kenya and buffalo origin from India. *International Journal for Parasitology*, **15**: 137–40.

Mitchell, J. S., Halton, D. W. & Smyth, J. D. 1978. Observations on the *in vitro* culture of *Cotylurus erraticus* (Trematoda: Strigeidae). *International Journal for Parasitology*, **8**: 389–97.

Mössinger, J. 1990. Nematoda: Filarioidea. In: *In Vitro Cultivation of Parasitic Helminths* (ed. J. D. Smyth), pp. 155–86. CRC Press, Boca Raton, Florida, USA.

Mössinger, J. & Wenk, P. 1986. Fecundity of *Litomosoides carinii* (Nematoda, Filarioidea) *in vivo* and *in vitro*. *Zeitschrift für Parasitenkunde*, **72**: 121–31.

Mössinger, J., Wenk, P. & Schulz-Key, H. 1987. *In vitro* maintenance of adult *Dipetalonema viteae* and *Litomosoides carinii* (Nematoda, Filarioidea): fecundity and survival in cell-free culture systems. *Zentralblatt für Bakteriologie, Parasitenkunde, Infectionskrankheiten und Hygiene 1, Abteilung 1, Originale A*, **267**: 303.

Mueller, J. F. 1974. The biology of *Spirometra*. *Journal of Parasitology*, **60**: 3–14.

Nayar, J. K., Crowder, C. G. & Knight, J. W. 1991*a*. *In vitro* development of *Brugia pahangi* and *Brugia malayi* in cultured mosquito thoraces. *Acta Tropica*, **48**: 173–84.

Nayar, J. K., Gunawardana, I. W. K. & Knight, J. W. 1991*b*. Effect of medium alterations on *in vitro* development of *Brugia malayi* larvae in cultured mosquito thoraces. *Journal of Parasitology*, **77**: 572–9.

Niewiadomska, K. & Kozicka, J. 1970. Remarks on the occurrence and biology of *Cotylurus erraticus* (Rudolphi, 1809) (Strigeidae) from the Mazuriaw Lakes. *Acta Parasitologica Polonica*, **18**: 42–50.

Olson, R. E. 1970. The life cycle of *Cotylurus erraticus* (Rudolphi, 1809) Szidat, 1928 (Trematoda: Strigeidae). *Journal of Parasitology*, **56**: 55–63.

Ong, S. J. & Smyth, J. D. 1986. Effects of some culture factors on sexual differentiation of *Mesocestoides corti* grown from tetrathyridia *in vitro*. *International Journal for Parasitology*, **16**: 361–8.

Osuna-Carrillo, A. & Mascaro-Lazcanó, M. C. 1982. The *in vitro* cultivation of *Taenia pisiformis* to sexually mature adults. *Zeitschrift für Parasitenkunde*, **67**: 67–71.

Palmieri, J. R. 1973. Additional natural and experimental hosts and interspecific variation in *Posthodiplostomum minimum* (Trematoda: Diplostomatidae). *Journal of Parasitology*, **59**: 744–5.

Rausch, R. L. & Jentoft, V. L. 1957. Studies on the helminth fauna of Alaska. XXXI. Observations on the propagation of the larval *Echinococcus multilocularis* Leuckart, 1863 *in vitro*. *Journal of Parasitology*, **43**: 1–8.

Riberu, W. A., Atmosoedjono, S., Purnomo, S., Tirtokusumo, S., Bangs, M. J. & Baird, J. K. 1990. Cultivation of sexually mature *Brugia malayi in vitro*. *American Journal of Tropical Medicine and Hygiene*, **43**: 3–5 (90–108).

Roberts, L. S. & Mong, F. N. 1969. Developmental biology of cestodes. IV. *In vitro* development of *Hymenolepis diminuta* in presence and absence of oxygen. *Experimental Parasitology*, **26**: 166–74.

Rogan, M. T. & Richards, K. S. 1986. *In vitro* development of hydatid cysts from posterior bladders and ruptured brood capsules of equine *Echinococcus granulosus*. *Parasitology*, **92**: 379–90.

Sakamoto, T. 1979. Development and behaviour of adult and larval *Trichinella spiralis* cultured *in vitro*. *Memorial Faculty of Agriculture of Kagoshima University*, **15**: 107–11.

Saville, D. H. & Irwin, S. W. B. 1991. *In ovo* cultivation of *Microphallus primas* (Trematoda: Microphallidae) metacercariae to ovigerous adults and the establishment of the life-cycle in the laboratory. *Parasitology*, **103**: 479–84.

Schiller, E. L. 1965. A simplified method for the *in vitro* cultivation of the rat tapeworm, *Hymenolepis diminuta*. *Journal of Parasitology*, **51**: 516–18.

Schnier, M. S. & Fried, B. 1980. *In vitro* cultivation of *Amblosoma suwaense* (Trematoda: Bracylaimidae) from the metacercaria to the ovigerous adult. *Internatinal Journal for Parasitology*, **10**: 391–5.

Schulz-Key, H. & Karam, M. 1986. Periodic reproduction of *Onchocerca volvulus*. *Parasitology Today*, **2**: 284–6.

Shimazu, T. 1974. A new digenetic trematode, *Amblosoma suwaense* sp. nov., the morphology of its adult and metacercaria (Trematoda: Brachylaimidae). *Japanese Journal of Parasitology*, **23**: 100–15.

Smith, M. A. & Clegg, J. A. 1981. Improved culture of *Fasciola hepatica in vitro*. *Zeitschrift für Parasitologie*, **66**: 9–15.

Smyth, J. D. 1946. Studies on tapeworm physiology. I. Cultivation of *Schistocephalus in vitro*. *Journal of Experimental Parasitology*, **23**: 47–70.

Smyth, J. D. 1947. Studies on tapeworm physiology. II. Cultivation and development of *Ligula intestinalis in vitro*. *Parasitology*, **38**: 173–81.

Smyth, J. D. 1954. Studies on tapeworm physiology. VII. Fertilization of *Schistocephalus solidus in vitro*. *Experimental Parasitology*, **3**: 64–71.

Smyth, J. D. 1959. Maturation of larval pseudophyllidean cestodes and strigeid trematodes under axenic conditions; the significance of levels of nutrition in platyhelminth development. *Annals of the New York Academy of Sciences*, **77**: 102–25.

Smyth, J. D. 1962. Studies on tapeworm physiology. X. Axenic cultivation of the hydatid organism, *Echinococcus granulosus*; establishment of a basic technique. *Parasitology*, **52**: 441–57.

Smyth, J. D. 1967. Studies on tapeworm physiology. XI. *In vitro* cultivation of *Echinococcus granulosus* from the protoscolex to the strobilate stage. *Parasitology*, **57**: 111–33.

Smyth, J. D. 1971. Development of monozoic forms of *Echinococcus granulosus* during *in vitro* culture. *International Journal for Parasitology*, **1**: 121–4.

Smyth, J. D. 1979. *Echinococcus granulosus* and *E. multilocularis*: *in vitro* culture of the strobilar stages from protoscoleces. *Angewandte Parasitologie*, **20**: 137–48.

Smyth, J. D. 1985. *In vitro* culture of *Echinococcus* spp. *Proceedings of the 13th International Congress of Hydatidology*, Madrid, 1985, pp. 84–9. Communidad de Madrid, Conserjía de Agricultura y Ganaderia, Madrid.

Smyth, J. D. 1987. Asexual and sexual differentiation in cestodes: especially *Mesocestoides* and *Echinococcus*. In: *Molecular Paradigms for Eradicating Parasites* (ed. A. MacInnis), pp. 19–33. Alan R. Liss, New York. ISBN 0 845 12659 8.

Smyth, J. D. 1990. *In Vitro Cultivation of Parasitic Helminths*.

CRC Press, Boca Raton, Florida, USA. ISBN 0 849 34586 3.

Smyth, J. D. & Davies, Z. 1974*a*. Occurrence of physiological strains of *Echinococcus granulosus* demonstrated by *in vitro* culture of protoscoleces from sheep and horse hydatid cysts. *International Journal for Parasitology*, **4**: 443–5.

Smyth, J. D. & Davies, Z. 1974*b*. *In vitro* culture of the strobilar stage of *Echinococcus granulosus* (sheep strain): a review of basic problems and results. *International Journal for Parasitology*, **4**: 631–44.

Smyth, J. D. & Halton, D. W. 1983. *The Physiology of Trematodes*. Cambridge University Press. ISBN 0 521 22283 3 & 0 521 29434 7.

Smyth, J. D. & McManus, D. P. 1989. *The Physiology and Biochemistry of Cestodes*. Cambridge University Press. ISBN 0 521 35557 5.

Stringfellow, F. 1984. Effects of bovine heme on development of *Haemonchus contortus in vitro*. *Journal of Parasitology*, **70**: 989–90.

Stringfellow, F. 1986. Cultivation of *Haemonchus contortus* (Nematoda: Trichostrongylidae) from infective larvae to adult male and egg-laying female. *Journal of Parasitology*, **72**: 339–45.

Stunkard, H. W. 1957. The morphology and life history of the digenetic trematode *Microphallus similis* (Jagerskiold, 1900) Baer, 1943. *Biological Bulletin*, **112**: 254–6.

Tamashiro, W. K. & Palumbo, N. E. 1985. Diethylcarbamazine: *in vitro* inhibitory effects of microfilarial production by *Dirofilaria immitis*. *Journal of Parasitology*, **71**: 381–3.

Taylor, A. E. R. & Baker, J. R. 1968. *The Cultivation of Parasites In Vitro*. Blackwell Scientific Publications, Oxford. ISBN 0 632 03980 9.

Taylor, A. E. R. & Baker, J. R. 1978. *Methods of Cultivating Parasites In Vitro*. Academic Press, London. ISBN 0 126 85550 1.

Taylor, A. E. R. & Baker, J. R. (eds) 1987. *In Vitro Methods for Parasite Cultivation*. Academic Press, London. ISBN 0 126 83855 0.

Thompson R. C. A., Jue Sue, L. P. & Buckley, S. P. 1982. *In vitro* development of the strobilar stage of *Mesocestoides corti*. *International Journal for Parasitology*, **12**: 303–14.

Townson, S., Connelly, C. & Muller, R. 1986. Optimization of culture conditions for the maintenance of *Onchocerca gutturosa* adult worms *in vitro*. *Journal of Helminthology*, **60**: 323–30.

Voge, M. & Jeong, K. 1971. Growth *in vitro* of *Cotylurus lutzi* Basch, 1969 (Trematoda: Strigeidae), from tetracotyle to patent adult. *International Journal for Parasitology*, **1**: 139–43.

Wang, W. & Zhou, S. L. 1987. (Studies on the cultivation *in vitro* of *Schistosoma japonicum* schistosomula transformed from cercariae by artificial methods.) (In Chinese.) *Acta Zoologica Sinica*, **33**: 144–50.

Weinstein, P. P. & Jones, M. F. 1956. The *in vitro* cultivation of *Nippostrongylus muris* to the adult stage. *Journal of Parasitology*, **42**: 215–36.

Weinstein, P. P. & Jones, M. F. 1957. The development of a study on the axenic growth *in vitro* of *Nippostrongylus muris* to the adult stage. *American Journal of Tropical Medicine and Hygiene*, **6**: 480–4.

Weinstein, P. P. & Jones, M. F. 1959. Development *in vitro* of some parasitic nematodes of vertebrates. *Annals of the New York Academy of Sciences*, **77**: 137–62.

Weinstein, P. P. & Sawyer, T. K. 1961. Survival of adults of *Dirofilaria uniformis in vitro* and their production of microfilariae. *Journal of Parasitology*, **47** (4, Sect. 2): 23.

Yasuraoka, K., Kaiho, M., Hata, H. & Endo, T. 1974. Growth *in vitro* of *Parvatrema timondavidi* Bartoli 1963 (Trematoda: Gymnophallidae) from the metacercarial stage to egg production. *Parasitology*, **68**: 293–302.

Zimmerman, G. L. & Leland, S. E. 1971. Completion of the life cycle of *Cooperia punctata in vitro*. *Journal of Parasitology*, **57**: 832–5.

Zwilling, E. 1959. A modified chorioallantoic grafting procedure. *Transplantation Bulletin*, **6**: 115–16.

Author index

Single and joint authors quoted in the text are listed in full. In the case of multiple-author publications, only the principal author is cited. Authors of general references are not listed here. Numbers in **bold** type indicate a full reference source.

Subject index

Note: Names in ***bold italics*** are parasites, those in *italics* only are hosts and vectors. A page number in bold type refers to an illustration, with or without accompanying text.

Lightning Source UK Ltd.
Milton Keynes UK
UKOW03f0705221214

243469UK00003BB/108/A